国外优秀物理著作
原版系列

运输过程的统一非局部理论——广义波尔兹曼物理动力学

Unified Non-Local Theory of Transport Processes—Generalized Boltzmann Physical Kinetics, 2e

（第2版）

[俄罗斯] Alexeev, B. V.（阿列克谢耶夫）著

HITP
哈爾濱工業大學出版社
HARBIN INSTITUTE OF TECHNOLOGY PRESS

黑版贸审字 08-2015-063 号

Unified Non-Local Theory of Transport Processes: Generalized Boltzmann Physical Kinetics, 2e
B. V. Alexeev
ISBN:978-0-444-63478-8

Elsevier (Singapore) Pte Ltd.
3 Killiney Road, #08-01 Winsland House I, Singapore 239519
Tel: (65)6349-0200 Fax: (65) 6733-1817
First Published 2016
2016 年初版

图书在版编目(CIP)数据

运输过程的统一非局部理论:广义波尔兹曼物理动力学:第 2 版 = Unified Non-Local Theory of Transport Processes: Generalized Boltzmann Physical Kinetics, 2e:英文/(俄罗斯)阿列克谢耶夫(Alexeev,B. V.)著. —哈尔滨:哈尔滨工业大学出版社,2016. 1
ISBN 978-7-5603-5757-7

Ⅰ. ①运… Ⅱ. ①阿… Ⅲ. ①波尔兹曼方程-动力学-研究-英文 Ⅳ. ①O175. 5

中国版本图书馆 CIP 数据核字(2015)第 291718 号

策划编辑 刘培杰
责任编辑 张永芹 杜莹雪
封面设计 孙茵艾
出版发行 哈尔滨工业大学出版社
社 址 哈尔滨市南岗区复华四道街 10 号 邮编 150006
传 真 0451-86414749
网 址 http://hitpress. hit. edu. cn
印 刷 哈尔滨市工大节能印刷厂
开 本 880mm×1230mm 1/16 印张 41. 25 字数 1329 千字
版 次 2016 年 1 月第 1 版 2016 年 1 月第 1 次印刷
书 号 ISBN 978-7-5603-5757-7
定 价 198. 00 元

(如因印装质量问题影响阅读,我社负责调换)

Contents

Preface

We are in front of the tremendous catastrophe in modern theoretical physics. Moreover, we have reached the revolutionary situation not only in physics but also in natural philosophy on the whole. Practically we are in front of the new challenge since Newton's *Mathematical Principles of Natural Philosophy* was first published in 1687. It is impossible to believe that in more than 300 years after Newton, we have the situation when 96% of matter and energy is of unknown origin. As it is shown in this monograph, the origin of difficulties consists in the total oversimplification inherent in local physics of the dissipative processes.

In the latter part of twentieth century, two very important results were obtained:

(1) The Irish physicist John Stewart Bell (1928–1990) was to show that all local statistical theories of dissipative processes are wrong in principal.

(2) The Russian physicist Boris V. Alexeev was to show that the derivation of kinetic equation with respect to one-particle distribution function from the BBGKY equations (prior to introducing any approximation destined to break the Bogolyubov chain) leads to additional terms of the non-local origin, generally of the same order of magnitude, appear in the Boltzmann equation. Then the passage to the Boltzmann equation means the neglect of non-local effects. These additional terms cannot be omitted even in the limit cases of kinetic theory, therefore Boltzmann equation is only a plausible equation.

Therefore, the case in point is of unprecedented situation in physics, when the fundamental physical equation is revised. During my stay in Marseille as invited professor, A.J.A. Favre reminds me Henri Poincaré's phrase after the death of a great Austrian physicist—"Boltzmann was wrong, but his mistake is equal to zero." It is a pity, but the situation in kinetic theory is much more serious.

The scientific community was convinced that the mentioned results could lead only to rather small corrections in the modern theoretical physics. So to speak—4% corrections to 96% of the known results, but not quite the reverse! Many scientists are aware that some way out will be achieved after creation of the unified theory of transport processes working from the structure of so-called elementary particles to the Universe evolution. This theory is in front of you.

This book reflects the scales of these alterations. One is safe to say that—as the main result of the non-local physical kinetics—this theory has showed it to be a highly effective tool for solving many physical problems in areas where the classical theory runs into difficulties.

Author is deeply indebted to V.L. Ginzburg and F. Uhlig for their interest in this work and in the subject in general. I am thankful to V. Mikhailov, I. Ovchinnikova, and A. Fedoseyev for cooperation.

This book is devoted to the memory of my mother.

January 2014

Historical Introduction and the Problem Formulation

"Alles Vergängliche
ist nur ein Gleichniss!"
Boltzmann's epigraph
for his "Vorlesungen über Gastheorie"

«Сотри случайные черты -
и ты увидишь — мир прекрасен»
(Obliterate the accidental features
And you will see: the world is splendid.)
Alexander Blok
"The Retribution"

In 1872, L Boltzmann, then a mere 28-year-old, published his famous kinetic equation for the one-particle distribution function $f(\mathbf{r}, \mathbf{v}, t)$ [1]. He expressed the equation in the form

$$\frac{\mathrm{D}f}{\mathrm{D}t} = J^{\mathrm{st}}(f), \tag{I.1}$$

where J^{st} is the collision (stoß) integral, and

$$\frac{\mathrm{D}}{\mathrm{D}t} = \frac{\partial}{\partial t} + \mathbf{v}\cdot\frac{\partial}{\partial \mathbf{r}} + \mathbf{F}\cdot\frac{\partial}{\partial \mathbf{v}} \tag{I.2}$$

is the substantial (particle) derivative, **v** and **r** being the velocity and radius-vector of the particle, respectively.

Equation (I.1) governs (in local approximation) the transport processes in a one-component gas which is sufficiently rarefied that only binary collisions between particles are of importance. While we are not concerned here with the explicit form of the collision integral (which determines the change of the distribution function f in binary collisions), note that it should satisfy conservation laws. For the simplest case of elastic collisions in a one-component gas we have

$$\int J^{\mathrm{st}}\psi_i \mathrm{d}\mathbf{v} = 0, (i = 1, 2, 3), \mathrm{d}\mathbf{v} = \mathrm{d}v_x \mathrm{d}v_y \mathrm{d}v_z, \tag{I.3}$$

where ψ_i are the collisional invariants ($\psi_1 = m$, $\psi_2 = m\mathbf{v}$, $\psi_3 = mv^2/_2$, m is the mass of the particle) related to the laws of conservation of mass, momentum, and energy.

Integrals of the distribution function (i.e. its moments) determine the macroscopic hydrodynamic characteristics of the system, in particular the number density of particles

$$n = \int f d\mathbf{v} \tag{I.4}$$

and the temperature T:

$$\frac{3}{2}k_{\mathrm{B}}nT = \frac{1}{2}m\int f(\mathbf{v} - \mathbf{v}_0)^2 d\mathbf{v}. \tag{I.5}$$

Here k_{B} is the Boltzmann constant and v_0 is the hydrodynamic flow velocity. It follows then that multiplying the Boltzmann integro-differential equation term by term by collisional invariants ψ_i, integrating over all particle velocities, and using the conservation laws (I.3) we arrive at the differential equations of fluid dynamics, whose general form is known as the hydrodynamic Enskog equations.

The Boltzmann equation (BE) is not of course as simple as its symbolic form above might suggest, and it is in only a few special cases that it is amenable to a solution. One example is that of a Maxwellian distribution in a locally, thermodynamically equilibrium gas in the event when no external forces are present. In this case, the equality

$$J^{st} = 0 \tag{I.6}$$

is met, giving the Maxwellian distribution function

$$f^{(0)} = n\left(\frac{m}{2\pi k_B T}\right)^{3/2} \exp\left(-\frac{mV^2}{2k_B T}\right), \tag{I.7}$$

where $\mathbf{V} = \mathbf{v} - \mathbf{v_0}$ is the thermal velocity.

It was much later, years after Boltzmann's death in 1906, that an analytic method for solving the Boltzmann equation was developed for the purpose of calculating transport coefficients. This method, developed in 1916-1917 by Chapman and Enskog [2–5], led to explicit expressions for the coefficients of viscosity, thermal conductivity, diffusion, and later thermal diffusion in a system with a small parameter (which for Chapman and Enskog's particular problem of a nonreacting gas was the Knudsen number, the ratio of the particle's mean free path to a characteristic hydrodynamic dimension).

However, even in Boltzmann's days there was a complete awareness that his equation acquires a fundamental importance for physics and that its range of validity stretches from transport processes and hydrodynamics all the way to cosmology—thus fully justifying the keen attention it attracted and debates it provoked.

Of the many results L Boltzmann derived from his kinetic equation, one of the most impressive is the molecular-kinetic interpretation of the second principle of thermodynamics and in particular of the statistical meaning of the concept of entropy. It turned out that it is possible to define the function

$$H = \int f \ln f d\mathbf{v}, \tag{I.8}$$

(H is the first letter in the English word *heat* and German word *Heizung*) which behaves monotonically in a closed system. If the relation between S, the entropy per unit volume of an ideal gas, and the H-function is written in the form

$$S = -k_B H + \text{const}, \tag{I.9}$$

then one can prove the inequality

$$\frac{\partial S}{\partial t} \geq 0. \tag{I.10}$$

The laconic formula

$$S = k \ln W, \tag{I.11}$$

connecting the entropy S and the thermodynamic probability W, is inscribed on Boltzmann's tombstone.

Ever since their creation, Boltzmann's physical kinetics and the Boltzmann equation have received severe criticism, much of which remains of interest even today. Let us elaborate on this.

To begin with, Boltzmann's contemporaries were very much in the dark regarding the relation between the Boltzmann equation and classical mechanics—in particular, with the Newton equation. The Boltzmann equation was obtained in a phenomenological manner based on convincing physical arguments and reflects the fact that the distribution function does not change along the particle's trajectory between collisions but rather changes as a result of an "instantaneous" interaction between colliding particles.

J Loschmidt noted in 1876 that the Boltzmann equation underlying the H-theorem includes only the first time derivative whereas the Newton equation contains the second (square of time) and hence the equations of motion are reversible in time. This means that if a system of hard-sphere particles starts a "backward" motion due to the particles reversing their direction of motion at some instant of time, it passes through all its preceding states up to the initial one, and this will increase the H-function whose variation is originally governed by reversible equations of motion. The essential point to be made here is that the observer cannot prefer one of the situations under study, the "forward" motion of the system in time, in favor of the second situation, its "backward" motion. In other words, the problem of the reversibility of time arises here.

Although somewhat differently formulated, essentially the same objection was made in 1896 by Planck's student E Zermelo, who noted that the H-theorem is inconsistent with Poincare's "recurrence" theorem proved in 1890 and stated that any physical system, even with irreversible thermodynamic processes operating in it, has a nonzero probability of returning to its original state. Boltzmann, himself fully aware of this possibility, wrote in the second part of his Lectures on the Theory of Gases (see Ref. [6], p. 251): "As a result of the motion of gas molecules, the H-function always decreases. The unidirectional nature of this process does not follow from the equations of motion, which the molecules obey. Indeed, these equations do not change if time changes sign."

There is a well-known example from probability theory, which Boltzmann employed as an argument in his discussions—sometimes very heated ones—with Zermelo, Planck and Ostwald. If a six-sided die is thrown 6000 times, one expects each side to turn up about 1000 times. The probability of, say, a six turning up 6000 times in a succession has a vanishing small value of $(1/6)^{6000}$. This example does not clear up the matter, however. Nor do the two papers which Boltzmann's student P Ehrenfest wrote in co-authorship with T Afanas'eva-Ehrenfest after the death of the great Austrian physicist.

Their first model, reported by Afanas'eva-Ehrenfest at the February 12, 1908 meeting of the Russian Physical-Chemical Society, involved the application of the H-theorem to the "plane" motion of a gas [7]. Suppose P-molecules, nontransparent to one another, start moving normally to axis y and travel with the same velocity in the direction of axis x. Suppose further that in doing so they undergo elastic collisions with Q-particles, squares with sides at an angle of 45° to axis y, which are nontransparent to the molecules and are all at rest.

It is readily shown that shortly after, all the molecules will divide themselves into four groups, and it is a simple matter to write down the change in the number of molecules P in each group in a certain time Δt and then to define a "planar-gas" H -function

$$H=\sum_{i=1}^{4} f_i \ln f_i, \tag{I.12}$$

where f_i is the number of molecules of the i-th kind, i.e., of those moving in one of the four possible directions. If all the velocities reverse their direction, the H-function starts to increase and reverts to the value it had when the P-molecules started their motion from the y axis. While this simple model confirms the Poincare-Zermelo theorem, it does not at all guarantee that the H-function will decrease when the far more complicated Boltzmann model is used.

P and T Ehrenfest's second model [8], known as the lottery's model, features two boxes, A and B, and N numbered balls to which there correspond "lottery tickets" placed in a certain box and which are all in box A initially. The balls are then taken one by one from A and transferred to B according to the number of a lottery ticket, drawn randomly. Importantly, the ticket is not eliminated after that but rather is returned to the box. In the event that the newly drawn ticket corresponds to a ball contained in B, the ball is returned to A. As a result, there will be approximately $N/2$ balls in either box.

Now suppose one of the boxes contains n balls—and the other accordingly $N-n$ balls—at a certain step s in the drawing process. We can then define Δ, a function, which determines the difference in the number of balls between the two boxes: $\Delta=n-(N-n)=2n-N$. In "statistical" equilibrium, $\Delta=0$ and $n=N/2$, the dependence $\Delta(s)$ will imitate the behavior of the H-function in a Boltzmann gas.

This example is also not convincing enough because this "lottery" game will necessarily lead to a fluctuation in the Δ function, whereas Boltzmann kinetic theory *excludes* completely fluctuations in the H-function. By the end of his life Boltzmann went over to fluctuation theory, in which the decrease of the H-function in time is only treated as the process the system is most likely to follow. This interpretation, however, is not substantiated by his kinetic theory since the origin of the primary fluctuation remains unclear (the galactic scale of such fluctuation included).

One of the first physicists to see that Boltzmann equation must be modified in order to remove the existing contradictions was J Maxwell. Maxwell thought highly of the results of Boltzmann, who in his turn did much to promote Maxwell electrodynamics and its experimental verification.

We may summarize Maxwell's ideas as follows. The equations of fluid dynamics are a consequence of the Boltzmann equation. From the energy equation, limiting ourselves to one dimension for the sake of simplicity and neglecting some energy transfer mechanisms (in particular, convective heat transfer), we obtain the well-known heat conduction equation

$$\frac{\partial T}{\partial t}=a^2\frac{\partial^2 T}{\partial x^2}. \tag{I.13}$$

The fundamental solution of Eqn (1.13) up to the dimensional constant is

$$T(x,t)=\frac{1}{2\sqrt{\pi a^2 t}}\exp\left(-\frac{x^2}{4a^2 t}\right) \tag{I.14}$$

and represents the temperature at point x at instant t provided at time $t=0$ an amount of heat $c\rho$, with ρ the density and a the thermal diffusivity of the medium, is evolved at the origin of coordinates. Defining an argument of a function T as $\theta=a^2t$ with the dimension of a coordinate squared we obtain

$$T=\frac{1}{2\sqrt{\pi\theta}}\exp\left(-\frac{x^2}{4\theta}\right). \tag{I.15}$$

The temperature distribution given by this equation is unsatisfactory physically. For small values of θ, the temperature at the heat evolution point $x=0$ is indefinitely large. On the other hand, at any arbitrarily distant point x the temperature produced by an instantaneous heat source will be different from zero for arbitrarily small times. While this difference may be small, it is a point of principal importance that it has a finite value.

As Landau and Lifshitz noted in their classical *Course of Theoretical Physics* ([9], p. 283), "The heat conduction process described by the equations obtained here has the property that any thermal perturbation becomes instantaneously felt over all space." This implies an infinitely fast propagation of heat, which is absurd from the point of view of molecular-kinetic theory. In the courses of mathematical physics this result is usually attributed to the fact that the heat conduction equation is derived phenomenologically, neglecting the molecular-kinetic mechanism of heat propagation. However, as has been already noted, the parabolic equation (I.13) follows from the Boltzmann equation. Some of Maxwell's ideas, phenomenological in nature and aimed at the generalization of the Boltzmann equation, are discussed in Woods' monograph [10].

Work on the hyperbolic equation of heat conduction was no longer related directly to the Boltzmann equation but rather was of a phenomenological nature. Without expanding the details of this approach, we only point out that the idea of the improvement of Eq. (I.13) was to introduce the second derivative with respect to time thus turning Eq. (I.13) into the hyperbolic form

$$\tau_r \frac{\partial^2 T}{\partial t^2} + \frac{\partial T}{\partial t} = a^2 \frac{\partial^2 T}{\partial x^2}, \tag{I.16}$$

where τ_r is treated as a certain relaxation kinetic parameter with the dimensions of time. For the first time in modern physics this idea was formulated by B Davydov [11] (see also interesting discussion about priority between C Cattaneo and P Vernotte [12–15]). The wave equation (I.16) leads to final propagation velocities for a thermal perturbation—although it should be remarked parenthetically that the quasi-linear parabolic equations may also produce wave solutions.

A breakthrough period in the history of kinetic theory occurred in the late 1930s and early 1940s, when it was shown through efforts of many scientists—of which Bogolyubov certainly tops the list—how, based on the Liouville equation for the multiparticle distribution function f_N of a system of N interacting particles, one can obtain a one-particle representation by introducing a small parameter $\varepsilon = n\upsilon_b$, where n is the number of particles per unit volume and υ_b is the interaction volume [16-20]. This hierarchy of equations is usually referred to as the Bogolyubov or BBGKY (Bogolyubov—Born—Green—Kirkwood—Yvon) chain.

We do not present the technical details in Introduction but refer the reader to the classical works cited above or, for example, to Ref. [21]. Some fundamental points of the problem are worth mentioning here, however.

(1) Integrating the Liouville equation

$$\frac{\partial f_N}{\partial t} + \sum_{i=1}^{N} \mathbf{v}_i \cdot \frac{\partial f_N}{\partial \mathbf{r}_i} + \sum_{i=1}^{N} \mathbf{F}_i \cdot \frac{\partial f_N}{\partial \mathbf{v}_i} = 0 \tag{I.17}$$

subsequently over phase volumes $d\Omega_{s+1}, \ldots, d\Omega_N$ ($d\Omega_j \equiv d\mathbf{r}_j d\mathbf{v}_j$), one obtains a kinetic equation for the s-particle distribution function, with the distribution function f_{s+1} in the integral part of the corresponding equation.
In other words, the set of integro-differential equations turns out to be a linked one, so that in the lowest-order approximation the distribution function f_1 depends on f_2. This means formally that, strictly speaking, the solution procedure for such a set should be as follows. First find the distribution function f_N and then solve the set of BBGKY equations subsequently for decreasingly lower-order distributions. But if we know the function f_N, there is no need at all to solve the equations for f_s and it actually suffices to employ the definition of the function

$$f_s = \int f_N(t, \Omega_1, \ldots, \Omega_N) d\Omega_{s+1} \ldots d\Omega_N. \tag{I.18}$$

We thus conclude that the rigorous solution to the set of BBGKY equations is again equivalent to solving Liouville equations. On the other hand, the seemingly illogical solution procedure involving a search for the distribution function f_1 is of great significance in kinetic theory and in non-equilibrium statistical mechanics. This approach involves breaking the BBGKY chain by introducing certain additional assumptions (which have a clear physical meaning, though). These assumptions are discussed in detail below.

(2) For a non-reacting gas, the Boltzmann equation is valid for two time scales of distribution functions: one of the order of the mean free time of the particles, and the other the hydrodynamic flow time. The Boltzmann equation is invalid for time lengths of the order of the collision times. Notice that a change from the time scale to the length scale can of course be made if desired.

(3) After the BBGKY chain is broken and f_2 represented as a product of one-particle distribution functions (which is quite reasonable for a rarefied gas), the Boltzmann equation cannot be written in a classical form with only one small parameter ε and it reduces instead to the Vlasov equation in a self-consistent field.

(4) Because the Boltzmann equation does not work at distances of the order of the particle interaction radius (or at the r_b scale), Boltzmann particles are pointlike and structureless, and it is one of the inconsistencies of the Boltzmann theory that the resulting collision cross sections of the particles enter the theory by the collision integral.

(5) Usually the one-particle distribution function is normalized to the number of particles per unit volume. For Boltzmann particles the distribution function is "automatically" normalized to an integer because a point-like particle may only be either inside or outside a trial contour in a gas — unlike finite-diameter particles which of course may overlap the boundary of the contour at some instant of time. Another noteworthy point is that the mean free path in Boltzmann kinetic theory is only meaningful for particles modeled by hard elastic spheres. Other models face difficulties related, though, to the level of one-particle description employed. The requirement for the transition to a one-particle model is that molecular chaos should exist prior to a particle collision.

The advent of the BBGKY chain led to the recognition that whatever generalization of Boltzmann kinetic theory is to be made, the logic to be followed should involve all the elements of the chain, i.e. the Liouville equation, the kinetic equations for s-particle distribution functions f_s, and the hydrodynamic equations. This logical construction was not generally adhered to.

In 1951, N Slezkin published two papers [22, 23] on the derivation of alternative equations for describing the motion of gas. The idea was to employ Meshcherskii's variable-mass point dynamics theory [24], well known for its jet propulsion applications.

The assumption of a variable-mass particle implies that at each point a liquid particle, close to this point and moving with a velocity **v**, adds or loses a certain mass, whose absolute velocity vector **U** differs, as Slezkin puts it, by a certain appreciable amount from the velocity vector **v** of the particle itself. Since there are different directions for this mass to come or go off, the associated mass flux density vector **Q** is introduced.

By applying the laws of conservation of mass, momentum, and energy in the usual way, Slezkin then proceeds to formulate a set of hydrodynamical equations, of which we will here rewrite the continuity equation for a one-component non-reacting gas:

$$\frac{\partial \rho}{\partial t} + \frac{\partial}{\partial \mathbf{r}} \cdot (\rho \mathbf{v} + \mathbf{Q}) = 0. \tag{I.19}$$

The mass flux density **Q** is written phenomenologically in terms of the density and temperature gradients.

Thus, the continuity equation is intuitively modified to incorporate a source term giving

$$\frac{\partial \rho}{\partial t} + \frac{\partial}{\partial \mathbf{r}} \cdot \rho \mathbf{v} = \frac{\partial}{\partial \mathbf{r}} \cdot \left(D \frac{\partial \rho}{\partial \mathbf{r}} + \beta \frac{\partial T}{\partial \mathbf{r}} \right), \tag{I.20}$$

where the coefficient D is that of self-diffusion, and β is related to thermal diffusion. Thus, we now have fluctuation terms on the right-hand side of Eq. (I.20), which are generally proportional to the mean free time τ_{mt} and, hence, after Eq. (I.20) is made dimensionless, to the Knudsen number which is small in the hydrodynamic limit.

At very nearly the time of the publication of Slezkin's first paper [23], Vallander [25] argued that the standard equations of motion are ill grounded physically and should therefore be replaced by other equations based on the introduction of additional mass Q_i and energy t_i fluxes $(i = 1,2,3)$ $Q_i = D_1 \frac{\partial \rho}{\partial r_i} + D_2 \frac{\partial T}{\partial r_i}$, $t_i = k_1 \frac{\partial \rho}{\partial r_i} + k_2 \frac{\partial T}{\partial r_i}$, where, to quote, "$D_1$ is the density self-diffusion coefficient, D_2 is the thermal self-diffusion coefficient, k_1 is the density heat conductivity, and k_2, the temperature heat conductivity."

Heuristic and inconsistent with Boltzmann's theory, the work of Siezkin and Vallander came under sufficiently severe criticism. Shaposhnikov [26] noted that in these papers, "which are almost identical in content ... the essential point is that instead of the conventional expression $\rho \mathbf{v}_0$, additional effects—'concentration self-diffusion' and 'thermal self-diffusion'—are introduced into the mass flux density which, in addition to the macroscopic mass transfer, cause a molecular mass transfer, much as the macroscopic energy and momentum transfer in a moving fluid goes in parallel with analogous molecular transport (heat conduction and viscosity)." Shaposhnikov then proceeds to derive the equation of continuity from the Boltzmann equation for a one-component gas and shows that the hydrodynamic equations of Siezkin and Vallander are in conflict with the Boltzmann kinetic theory.

Note that Siezkin and Vallander also modified the equations of motion and energy for a one-component gas in a similar way (by including self-diffusion effects). Possible consequences of additional mass transfer mechanisms for the Boltzmann kinetic theory were not analyzed by these authors.

Boltzmann's "fluctuation hypothesis" was repeatedly addressed by Ya Terletskii (see, for instance, Refs. [27, 28]) whose idea was to estimate fluctuations by using the expression the general theorems of Gibbs (see, for example, Ref. [29], pp. 85-88) yield for the mean-square deviation of an arbitrary generalized coordinate. To secure that fluctuations in statistical equilibrium be noticeable, Terletskii modifies the equation of perfect gas state by introducing a gravitational term, which immediately extends his analysis beyond the Boltzmann kinetic theory leaving the question about the irreversible change of the Boltzmann H -function unanswered.

Some comments concerning terminology should be done. In recent years, possible generalizations of the Boltzmann equation have been discussed widely in the scientific literature. Since the term "generalized Boltzmann equation" (GBE) has usually been given to any new modification published, we will only apply this term to the particular kinetic equation derived by me (for example) in Refs. [30–32] to avoid confusion. The corresponding equation is known also in the literature as Alexeev equation. Obviously it is not convenient for me to apply this term. Moreover in the following this kinetic equation will be transformed in the *basic equations of the unified theory* of transport processes (BEUT) valid in the tremendous diapason of scales—from the internal structures of so called "elementary particles" to the Universe expansion. Then GBE is only a particular case of BEUT which can be discussed in the subsequent chapters.

L Woods (see, e.g., Ref. [33]), following ideas dating back to Maxwell [34], introduces in his theory a phenomenological correction to the substantial first derivative on the left-hand side of the Boltzmann equation to take account of the further influence of pressure on transport processes. It is argued that the equation of motion of a liquid particle may be written as $\dot{\mathbf{v}} = \mathbf{F} + \mathbf{P}$, where $\mathbf{P}$ is a certain additional force, proportional to the pressure gradient: $\mathbf{P} = -\rho^{-1}\partial p/\partial \mathbf{r}$, with the result that the left-hand side of the Boltzmann equation becomes

$$\frac{\mathrm{D}f}{\mathrm{D}t} \equiv \frac{\partial f}{\partial t} + \mathbf{v}\cdot\frac{\partial f}{\partial \mathbf{r}} + \left(\mathbf{F} - \frac{1}{\rho}\frac{\partial p}{\partial \mathbf{r}}\right)\cdot\frac{\partial f}{\partial \mathbf{v}}, \tag{I.21}$$

whereas the collisional term remains unchanged. The phenomenological equation (I.21) has no solid foundation and does not fall into the hierarchy of Bogolyubov kinetic equations.

A weak point of the classical Boltzmann kinetic theory is the way it treats the dynamic properties of interacting particles. On the one hand, as the so-called "physical" derivation of the BE suggests [1, 6, 35, 36], Boltzmann particles are treated as material points; on the other hand, the collision integral in the BE brings into existence the cross sections for collisions between particles. A rigorous approach to the derivation of the kinetic equation for f_1 (KE_{f_1}) is based on the hierarchy of the Bogolyubov—Born—Green—Kirkwood—Yvon (BBGKY) equations. A KE_{f_1} obtained by the multi-scale method turns into the BE if one ignores the change of the distribution function (DF) over a time of the order of the collision time (or, equivalently, over a length of the order of the particle interaction radius). It is important to note [30–32] that accounting for the third of the scales mentioned above has the consequence that, prior to introducing any approximations destined to break the Bogolyubov chain, additional terms, generally of the same order of magnitude, appear in the BE. If the method of correlation functions is used to derive KE_{f_1} from the BBGKY equations, then a passage to the BE implies the neglect of non-local effects it time and space. Given the above difficulties of the Boltzmann kinetic theory (BKT), the following clearly interrelated questions arise.

First, what is a physically infinitesimal volume and how does its introduction (and, as a consequence, the unavoidable smoothing out of the DF) affect the kinetic equation [30]?

And second, how does a systematic account for the proper diameter of the particle in the derivation of the KE_{f_1} affect the Boltzmann equation? As it was mentioned before, we will refer to the corresponding KE_{f_1} as the generalized Boltzmann equation, or GBE.

Accordingly, our purpose in this introduction is first to explain the essence of the physical generalization of the BE and then to take a look at the specifics of the derivation of the GBE, when (as is the case in plasma physics) the self-consistent field of forces must of necessity be introduced. As the Boltzmann equation is the centerpiece of the theory of transport processes (TTP), the introduction of an alternative KE_{f_1} leads in fact to an overhaul of the entire theory, including its macroscopic (for example, hydrodynamic) aspects. Conversely, a change in the macroscopic description will inevitably affect the kinetic level of description. Because of the complexity of the problem, this interrelation is not always easy to trace when solving a particular TTP problem. The important point to emphasize is that at issue here is not how to modify the classical equations of physical kinetics and hydrodynamics to include additional transport mechanisms (in reacting media, for example); rather we face a situation in which, those involved believe, we must go beyond the classical picture if we wish the revised theory to describe experiment adequately. The alternative TTPs can be grouped conventionally into the following categories:

(1) theories that modify the macroscopic (hydrodynamic) description and neglect the possible changes of the kinetic description,
(2) those changing the kinetic description at the KE_{f_1} level without bothering much whether these changes are consistent with the structure of the entire BBGKY chain, and
(3) kinetic and hydrodynamic alternative theories consistent with the BBGKY hierarchy.

One of the pioneering efforts in the first line of research was a paper by Davydov [11], which stimulated a variety of studies (see, for instance, [37]) on the hyperbolic equation of thermal conductivity. Introducing the second derivative of temperature with respect to time permitted a passage from the parabolic to the hyperbolic heat conduction equation, thus allowing for a finite heat propagation velocity. However, already in his 1935 paper B I Davydov points out that his method "cannot be extended to the three-dimensional case"' and that "here the assumption that all the particles move at the same velocity would separate out a five-dimensional manifold from the six-dimensional phase space, suggesting that the problem cannot be limited to the coordinate space alone." We note, however, that also quasi-linear parabolic equations can produce wave solutions.

Therefore, to hyperbolize the heat conduction equation phenomenologically [13] is not valid unless a rigorous kinetic justification is given. The hyperbolic heat conduction equation appears when the BE is solved by the Grad method [38] retaining a term which involves a derivative of the heat flow with respect to time and to which, in the context of the Chapman-Enskog method, no particular order of approximation can be ascribed. Following its introduction, stable and high-precision computational schemes were developed for the hyperbolic equation of heat conduction [39], whose applications included, for example, two-temperature non-local heat conduction models and the study of the telegraph equation as a paradigm for possible generalized hydrodynamics [37, 40].

Major difficulties arose when the question of existence and uniqueness of solutions of the Navier–Stokes equations was addressed. O A Ladyzhenskaya has shown for three-dimensional flows that under smooth initial conditions a unique solution is only possible over a finite time interval. Ladyzhenskaya even introduced a "correction" into the Navier–Stokes equations in order that its unique solvability could be proved (see discussion in [41]). It turned out that in this case the viscosity coefficient should be dependent on transverse flow-velocity gradients—with the result that the very idea of introducing kinetic coefficients should be overhauled.

We shall now turn to approaches in which the KE_{f_1} can be changed in a way which is generally inconsistent with the BBGKY hierarchy. It has been repeatedly pointed out that using a wrong distribution function (DF) for charged particles may have a catastrophic effect on the macro-parameters of a weakly ionized gas.

Let us have a look at some examples of this. As is well known, the temperature dependence of the density of atoms ionized in plasma to various degrees was first studied by Saha [42] and Eggert [43]. For a system in thermodynamic equilibrium they obtained the equation

$$\frac{n_{j+1}n_e}{n_j}=\frac{s_{j+1}}{s_j}\frac{(2\pi m_e k_B T)^{3/2}}{h^3}\exp\left(-\frac{\varepsilon_j}{k_B T}\right), \tag{I.22}$$

where n_j is the number density of j-fold ionized atoms, n_e is the number density of free electrons, m_e is the electron mass, k_B is the Boltzmann constant, h is the Planck constant, s_j is the statistical weight for a j-fold ionized atom [44], and ε_j is the jth ionization potential. The Saha equation (I.22) is derived for the Maxwellian distribution and should necessarily be modified if another velocity distribution of particles exists in the plasma. This problem was studied in work [45], in which, for illustrative purposes, the values of $n_{j+1}n_e/n_j$ calculated with the Maxwell distribution function are compared with those obtained with the Druyvesteyn distribution function, the average energies for both distributions being assumed equal. Let $T=10^4 K$, $n_e=10^{14}\mathrm{cm}^{-3}$, $\varepsilon_j=10\,\mathrm{eV}$, the charge number $Z=1, s_{j+1}/s_j=1$. Then one arrives at [45]: $n_{j+1}n_e/n_j=6\times 10^2$ (calculation using the Druyvesteyn distribution), $n_{j+1}n_e/n_j=4.53\times 10^{16}$ (calculation using the Maxwellian distribution function by the Saha formula).

As E Dewan explained, "the discrepancy in fourteen orders of magnitude obtained above is clearly due to the fact that, unlike Maxwellian distribution, the Druyvesteyn distribution does not have a 'tail'."

In our second example, two quantities—the ionization rate constant and the ionization cross section—were calculated by Gryzinski et al. [46] using the two above-mentioned distributions. The ionization cross section σ_i is defined by the following interpolation formula known to match satisfactorily the experimental data

$$\sigma_i=\sigma_0 G_i(\xi,\varsigma)/\varepsilon_i^2, \tag{I.23}$$

where $\sigma_0=6.56\times 10^{-14}\mathrm{cm}^2(eV)^2$, ε_i is the ionization potential of the atom, and ξ is a dimensionless parameter characterizing the atomic electron shell:

$$\xi=W/\varepsilon_i, \tag{I.24}$$

where W is the average kinetic energy of the atomic electrons, given by the formula

$$W=\frac{1}{N_e}\sum_{j=1}^{N_e}\varepsilon_j, \tag{I.25}$$

in which N_e is the number of electrons in the atom, and ε_j are the ionization potentials for the atom successively stripped of its electrons. The parameter ς is defined by the expression

$$\varsigma=U_e/\varepsilon_i, \tag{I.26}$$

where U_e is the energy of the electrons bombarding the atom. The neutral particle velocities are assumed to be much lower than the average electron velocity, and the plasma is taken to be uniform. The average value of the ionization cross section is then given by

$$\overline{\sigma_i} = \int_0^\infty \sigma_i(v_e) f(v_e) dv_e, \tag{I.27}$$

and the ionization rates are evaluated by the formula

$$\overline{\sigma_i v_e} = \int_0^\infty \sigma_i(v_e) v_e f(v_e) dv_e, \tag{I.28}$$

provided the function $G_i(\xi, \varsigma)$ defined as [46]

$$G_i(\xi, \varsigma) = \frac{(\varsigma - 1)\left(1 + {}^2/_3\xi\right)}{(\varsigma + 1)(1 + \xi + \varsigma)}. \tag{I.29}$$

is known.

Table I.1 illustrates the calculated values of $\overline{\sigma_i}$ and $\overline{\sigma_i v_e}$ for $\xi = 1$ and various $\hat{T} = k_B T_e / \varepsilon_i$. It can be seen that the results obtained with different DFs can differ widely, indeed catastrophically so even for relatively small values of $\hat{T}$. Thus, the reliable computation of DFs remains a topic of intense current interest in plasma physics problems, the weak effect of the DF form on its moments being rather an exception than the rule. The use of collision cross sections which are "self-consistent" with kinetic equations is also suggested by the well-known Enskog theory of moderately dense gases [47]. Enskog's idea was to describe the properties of such gases by separating the *non-local part* out of the *essentially*

TABLE I.1 Comparison of ionization cross sections $\overline{\sigma_i}$ and ionization rates $\overline{\sigma_i v_e}$ calculated with the Maxwellian and Druyvesteyn DF

$\hat{T}$	Maxwellian DF		Druyvesteyn DF	
	$\overline{\sigma_i}$	$\overline{\sigma_i v_e}$	$\overline{\sigma_i}$	$\overline{\sigma_i v_e}$
0.1	4.206×10^{-6}	1.184×10^{-5}	1.278×10^{-27}	4.077×10^{-27}
0.2	8.262×10^{-4}	1.184×10^{-3}	4.382×10^{-9}	1.011×10^{-8}
0.3	5.029×10^{-3}	9.251×10^{-3}	2.128×10^{-5}	4.135×10^{-5}
0.4	1.259×10^{-2}	2.103×10^{-2}	5.403×10^{-4}	9.405×10^{-4}
0.5	2.194×10^{-2}	3.415×10^{-2}	2.773×10^{-3}	4.466×10^{-3}
0.6	3.180×10^{-2}	4.687×10^{-2}	7.305×10^{-3}	1.110×10^{-2}
0.7	4.143×10^{-2}	5.842×10^{-2}	1.376×10^{-2}	1.998×10^{-2}
0.8	5.047×10^{-2}	6.857×10^{-2}	2.145×10^{-2}	3.001×10^{-2}
0.9	5.875×10^{-2}	7.733×10^{-2}	2.973×10^{-2}	4.033×10^{-2}
1	6.624×10^{-2}	8.482×10^{-2}	3.813×10^{-2}	5.039×10^{-2}
2	1.079×10^{-1}	1.171×10^{-1}	9.918×10^{-2}	1.132×10^{-1}
3	1.195×10^{-1}	1.190×10^{-1}	1.233×10^{-1}	1.312×10^{-1}
4	1.209×10^{-1}	1.137×10^{-1}	1.311×10^{-1}	1.717×10^{-1}
5	1.185×10^{-1}	1.069×10^{-1}	1.320×10^{-1}	1.298×10^{-1}
6	1.146×10^{-1}	9.992×10^{-2}	1.299×10^{-1}	1.243×10^{-1}
7	1.102×10^{-1}	9.326×10^{-2}	1.263×10^{-1}	1.184×10^{-1}
8	1.056×10^{-1}	8.704×10^{-2}	1.222×10^{-1}	1.125×10^{-1}
9	1.010×10^{-1}	8.123×10^{-2}	1.179×10^{-1}	1.069×10^{-1}
10	9.662×10^{-2}	7.589×10^{-2}	1.137×10^{-1}	1.017×10^{-1}

local (!) Boltzmann collision integral. The transport coefficients obtained in this way for the hard-sphere model yielded an incorrect temperature dependence for the system's kinetic coefficients. To remedy this situation, the model of "soft" spheres was introduced to fit the experimental data (see, for instance, Ref. [48]).

In the theory of the so-called kinetically consistent difference schemes [49], the DF is expanded in a power series of time, which corresponds to using an incomplete second approximation in the "physical" derivation of the Boltzmann equation (see discussion in Refs. [41, 50]). The result is that the difference schemes obtained contain only an artificial ad hoc viscosity chosen specially for the problem at hand. Some workers followed the steps of Davydov by adding the term $\partial^2 f/\partial t^2$ to the kinetic equations for fast processes.

Bakai and Sigov [51] suggest using such a term in the equation for describing DF fluctuations in turbulent plasma. The so-called ordering parameter they introduce alters the very type of the equation. To describe spatial non-locality, Bakai and Sigov complement the kinetic equation by the $\partial^2 f/\partial x^2$ term and higher derivatives, including mixed time-coordinate partial derivatives—a modification which can possibly describe non-Gaussian random sources in the Langevin equations [52]. Vlasov [53] attempted to eliminate the inconsistencies of the Boltzmann theory through the inclusion of additional dynamical variables (derivatives of the velocity) in the one-particle distribution function $f(\mathbf{r}, \mathbf{v}, \dot{\mathbf{v}}, \ddot{\mathbf{v}}, \dddot{\mathbf{v}}, \ldots, t)$. However—due primarily to the reasonable complexity requirement which should be met for a theory to be useful in practice—this approach is, in our view, too early to try until all traditional resources for describing the DF are exhausted. The reader is referred to review [54] of some other theories of transport properties.

Clearly, approaches to the modification of the KE_{f_1} must be based on certain principles, and it is appropriate to outline these in brief here. Of the approaches we have mentioned above, the most consistent one is the third, which clearly reveals the relation between alternative KE_{f_1}'s and the BBGKY hierarchy. There are general requirements to which the generalized KE_{f_1} must satisfy.

(1) Since the artificial breaking of the BBGKY hierarchy is unavoidable in changing to a one-particle description, the generalized KE_{f_1} should be obtainable with the known methods of the theory of kinetic equations, such as the multi-scale approach, correlation function method, iterative methods, and so forth, or combinations of them. In each of these, some specific features of the particular alternative KE_{f_1} are highlighted.
(2) There must be an explicit link between the KE_{f_1} and the way we introduce the physically infinitesimal volume—and hence with the way the moments in the reference contour with transparent boundaries fluctuate due to the finite size of the particles.
(3) In the non-relativistic case, the KE_{f_1} must satisfy the Galileo transformation.
(4) The KE_{f_1} must ensure a connection with the classical H-theorem and its generalizations.
(5) The KE_{f_1} should not lead to unreasonable complexities in the theory.

Although the examples above are purely illustrative and the exhaustive list of difficulties faced by Boltzmann kinetic theory would of course be much longer, it should be recognized that after the intense debates of the early twentieth century, the search for an alternative kinetic equation for a one-particle distribution function has gradually leveled off or, perhaps more precisely, has become of marginal physical importance. Both sides of the dispute have exhausted their arguments. On the other hand, the Boltzmann equation has proven to be successful in solving a variety of problems, particularly in the calculation of kinetic coefficients. Thus, the development of Boltzmann kinetic theory has turned out to be typical for any revolutionary physical theory—from rejection to recognition and further to a kind of "canonization."

In the latter part of twentieth century two very important results were obtained:

(1) The Irish physicist John Stewart Bell (1928–1990) was to show (Bell's theorems [55]) that all local statistical theories of dissipative processes are wrong in principal.
(2) The Russian physicist Boris V Alexeev was to show that the derivation of KE_{f_1} from the BBGKY equations (*prior* to introducing any approximation destined to break the Bogolyubov chain) leads to additional terms of the non-local origin, generally of the same order of magnitude, appear in the BE. Then the passage to the BE means the neglect of non-local effects. These additional terms cannot be omitted even in the limit cases of kinetic theory, therefore BE is only a plausible equation.

It could be anticipated that we will obtain small corrections to the existing results. But in reality we are in front of the tremendous catastrophe in modern theoretical physics. Several extremely significant problems challenge modern fundamental physics, which can be titled as "Non-solved problems of the fundamental physics" or more precisely of *local physical kinetics* of dissipative processes, namely:

(1) Kinetic theory of entropy and the problem of the initial perturbation;
(2) Strict theory of turbulence;

(3) Quantum non-relativistic and relativistic hydrodynamics, theory of charges separation in the atom structure;
(4) Theory of ball lightning;
(5) Theory of dark matter;
(6) Theory of dark energy, Hubble expansion of the Universe;
(7) The destiny of anti-matter after the Big Bang;
(8) A unified theory of dissipative structures—from atom structure to cosmology.

In appearance these old and new problems (including the problems 5 and 6 in the list) mean that we have reached the revolutionary situation not only in physics but in natural philosophy on the whole. Practically we are in front of the new challenge since Newton's *Mathematical Principles of Natural Philosophy* was first published in 1687. It's impossible to believe that in more that three hundred years after Newton we have the following diagram obtained in astrophysics:

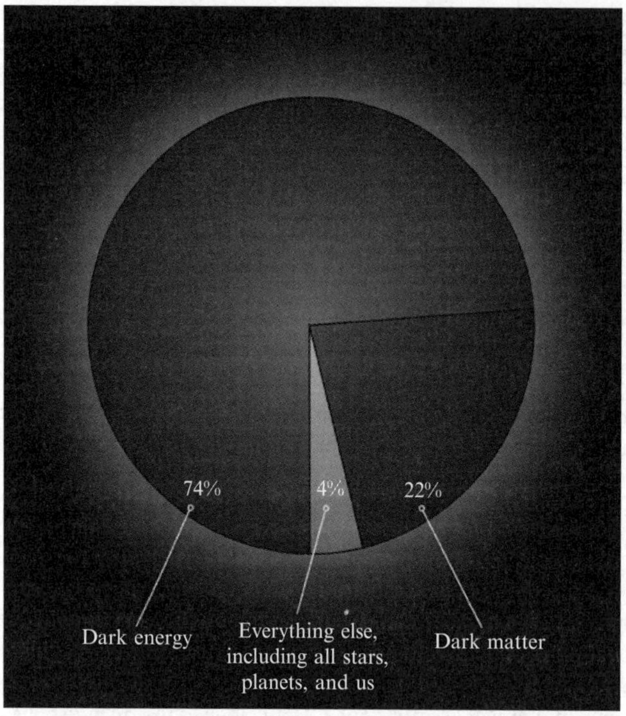

As result we have the Internet reaction like that: It is humbling, perhaps even humiliating, that we know almost nothing about 96% of what is "out there"!! But this situation is not so bad on its own. We have the worst case from position of local physics—dark matter and dark energy cannot be diagnosed by the known experimental methods. WHERE IS THE MAIN MISTAKE?

The origin of difficulties consists in the total oversimplification following from the principles of local physics and reflects the general shortcomings of the local kinetic transport theory.

Let us consider the fundamental methodic aspects of non-local physics from the qualitative standpoint, avoiding excessively cumbersome formulas.

Transport processes in open dissipative systems are considered in physical kinetics. Therefore, the kinetic description is inevitably related to the system diagnostics. Such an element of diagnostics in the case of theoretical description in physical kinetics is the concept of the physically infinitely small volume (PhSV). The correlation between theoretical description and system diagnostics is well known in physics. Suffice it to recall the part played by test charge in electrostatics or by test circuit in the physics of magnetic phenomena.

The traditional definition of PhSV contains the statement to the effect that the PhSV contains a sufficient number of particles for introducing a statistical description; however, at the same time, the PhSV is much smaller than the volume V of the physical system under consideration.

In a first approximation, this leads to local approach in investigating the transport processes. It is assumed in classical hydrodynamics that local thermodynamic equilibrium is first established within the PhSV, and only after that the transition occurs to global thermodynamic equilibrium if it is at all possible for the system under study.

Let us consider the hydrodynamic description in more detail from this point of view. Assume that we have two neighboring physically infinitely small volumes $PhSV_1$ and $PhSV_2$ in a non-equilibrium system. The one-particle distribution function (DF) $f_{sm,1}(\mathbf{r}_1, \mathbf{v}, t)$ corresponds to the volume $PhSV_1$, and the function $f_{sm,2}(\mathbf{r}_2, \mathbf{v}, t)$—to the volume $PhSV_2$. It is assumed in a first approximation that $f_{sm,1}(\mathbf{r}_1, \mathbf{v}, t)$ does not vary within $PhSV_1$, same as $f_{sm,2}(\mathbf{r}_2, \mathbf{v}, t)$ does not vary within the neighboring volume $PhSV_2$. It is this assumption of locality that is implicitly contained in the Boltzmann equation (BE). *However, the assumption is too crude.*

Indeed, a particle on the boundary between two volumes, which experienced the last collision in $PhSV_1$ and moves toward $PhSV_2$, introduces information about the $f_{sm,1}(\mathbf{r}_1, \mathbf{v}, t)$ into the neighboring volume $PhSV_2$. Similarly, a particle on the boundary between two volumes, which experienced the last collision in $PhSV_2$ and moves toward $PhSV_1$, introduces information about the DF $f_{sm,2}(\mathbf{r}_2, \mathbf{v}, t)$ into the neighboring volume $PhSV_1$. The relaxation over translational degrees of freedom of particles of like masses occurs during several collisions. As a result, "Knudsen layers" are formed on the boundary between neighboring physically infinitely small volumes, the characteristic dimension of which is of the order of path length.

Then a correction must be introduced into the DF in the PhSV, which is proportional to the mean time between collisions and to the substantive derivative of the DF being measured. Rigorous derivation is given for example in [21, 30–32, 56, 57].

Let a particle of finite radius be characterized as before by the position $\mathbf{r}$ at the instant of time t of its center of mass moving at velocity $\mathbf{v}$. Then, the situation is possible where, at some instant of time t, the particle is located on the interface between two volumes. In so doing, the lead effect is possible (say, for $PhSV_2$), when the center of mass of particle moving to the neighboring volume $PhSV_2$ is still in $PhSV_1$. However, the delay effect takes place as well, when the center of mass of particle moving to the neighboring volume (say, $PhSV_2$) is already located in $PhSV_2$ but a part of the particle still belongs to $PhSV_1$.

Moreover, even the point-like particles (starting after the last collision near the boundary between two mentioned volumes) can change the distribution functions in the neighboring volume. Adjusting of the particles dynamic characteristics for translational degrees of freedom takes several collisions. Therefore we experience a "Knudsen layer" effect between adjacent small volumes. This leads to fluctuations in mass and hence also in other hydrodynamic quantities. The existence of such "Knudsen layers" is not dependent on the choice of spatial nets and is fully defined by the reduced description for ensemble of particles of finite diameters in the conceptual framework of open physically small volumes, *i.e.*, it depends on the chosen method of measurement.

THE MAIN MISTAKE OF LOCAL PHYSICAL KINETICS CAN BE INDICATED AS FOLLOWS:

This entire complex of the mentioned effects defines non-local effects in space and time. The physically infinitely small volume (PhSV) is an *open* thermodynamic system *for any division of macroscopic system by a set of PhSVs.*

However, the Boltzmann equation (BE) fully ignores non-local effects and contains only the local collision integral J^B. The foregoing non-local effects are insignificant only in equilibrium systems, where the kinetic approach changes to methods of statistical mechanics.

This is what the difficulties of classical Boltzmann physical kinetics arise from. The rigorous approach to derivation of kinetic equation relative to one-particle DF f (KE_f) is based on employing the hierarchy of Bogoliubov equations. Low index "1" is omitted if it cannot lead to misunderstandings. Generally speaking, the structure of KE_f is as follows

$$\frac{Df}{Dt} = J^B + J^{nl}, \tag{I.30}$$

where J^{nl} is the non-local integral term.

An approximation for the second collision integral is suggested by me in *generalized* Boltzmann physical kinetics,

$$J^{nl} = \frac{D}{Dt}\left(\tau \frac{Df}{Dt}\right), \tag{I.31}$$

τ is non-local parameter (coinciding in a gas with the relaxation time τ_r proportional to the mean time τ_{mt} *between* collisions of particles); τ_{mt} is related in a hydrodynamic approximation with dynamical viscosity μ and pressure p,

$$\tau_{mt} p = \Pi \mu, \tag{I.32}$$

where the factor Π is defined by the model of collision of particles; for neutral hard-sphere gas, $\Pi = 0.8$, [35, 36]. Obviously in "the simplest version" τ_{mt} can be used in (I.31) instead of τ; it leads only to variety of Π-parameter in (I.32).

All of the known methods of deriving kinetic equation relative to one-particle DF f lead to the approximation (I.31), including the method of many scales, the method of correlation functions, and the iteration method.

We are faced in fact with the "price-quality" problem familiar from economics. That is, what price—in terms of the increased complexity of the kinetic equation—are we ready to pay for the improved quality of the theory? An answer to this question is possible only through experience with practical problems.

Extremely important:

1. Approximation $J^{\mathrm{nl}}=\frac{\mathrm{D}}{\mathrm{D}t}\left(\tau\frac{\mathrm{D}f}{\mathrm{D}t}\right)$ delivers local approximation of non-local collision integrals.
2. Approximation $J^{\mathrm{nl}}=\frac{\mathrm{D}}{\mathrm{D}t}\left(\tau\frac{\mathrm{D}f}{\mathrm{D}t}\right)$ return us to two level description (level of hydrodynamic processes + level of transport processes between collisions).
3. The generalized transport theory is not too complicated in applications.

One can draw an analogy with the Bhatnagar-Gross-Krook (BGK) approximation for local integral J^{B},

$$J^{\mathrm{B}}=\frac{f^{(0)}-f}{\tau_{\mathrm{r}}}, \tag{I.33}$$

(in the simplest case $\tau_{\mathrm{r}}\sim\tau$) the popularity of which in the case of Boltzmann collision integral is explained by the colossal simplification attained when using this approximation. The order of magnitude of the ratio between the second and first terms of the right-hand part of Eq. (I.30) is Kn^2, at high values of Knudsen number, these terms come to be of the same order. It would seem that, at low values of Knudsen number corresponding to hydrodynamic description, the contribution by the second term of the right-hand part of Eq. (I.30) could be ignored.

However, this is not the case. Upon transition to hydrodynamic approximation (following the multiplication of the kinetic equation by invariants collision and subsequent integration with respect to velocities), the Boltzmann integral part goes to zero, and the second term of the right-hand part of Eq. (I.30) *does not go to zero* after this integration and produces a contribution of the same order in the case of generalized Navier-Stokes description.

From the mathematical standpoint, disregarding the term containing a small parameter with higher derivative is impermissible. From the physical standpoint, the arising additional terms proportional to viscosity correspond to Kolmogorov small-scale turbulence; the fluctuations are tabulated. It turns out that the integral term J^{nl} is important from the standpoint of the theory of transport processes at both low and high values of Knudsen number.

Note the treatment of GBE from the standpoint of fluctuation theory,

$$\mathrm{D}f^a/\mathrm{D}t=J^{\mathrm{B}}, \tag{I.34}$$

$$f^a=f-\tau\mathrm{D}f/\mathrm{D}t. \tag{1.35}$$

Equations (I.34) and (I.35) have a correct free-molecule limit. Therefore, $\tau Df/Dt$ is a fluctuation of distribution function, and the notation (I.34) disregarding (I.35) renders the BE open. From the standpoint of fluctuation theory, Boltzmann employed the simplest closing procedure

$$f^a=f. \tag{I.36}$$

Fluctuation effects occur in any open thermodynamic system bounded by a control surface transparent to particles. Obviously the mentioned non-local effects can be discussed from viewpoint of breaking of the Bell's inequalities because in the non-local theory the measurement (realized in PhSV_1) has influence on the measurement realized in the adjoining space-time point in PhSV_2 and vice-versa.

The equation (GBE) reads

$$\frac{\mathrm{D}f}{\mathrm{D}t}=J^{\mathrm{B}}+\frac{\mathrm{D}}{\mathrm{D}t}\left(\tau\frac{\mathrm{D}f}{\mathrm{D}t}\right). \tag{I.37}$$

Here τ is non-local parameter, in the simplest case it is the mean time BETWEEN collisions (for plasma τ is mean time between close collisions), for plasma in $\mathrm{D}/\mathrm{D}t$ should be introduced the self consistent force F. It is interesting to note that the GBE also makes it possible to include higher derivatives of the DF (see the approximation (5.8) in Ref. [32]). For a multi species reacting gas, the generalized Boltzmann equation can be rewritten as

$$\frac{Df_\alpha}{Dt}-\frac{D}{Dt}\left(\tau_\alpha\frac{Df_\alpha}{Dt}\right)=J_\alpha^{B,el}+J_\alpha^{B,r}, \tag{I.38}$$

where f_α is distribution function for a particle of the α th kind, τ_α is non-local parameter for α species (in the simplest case τ_α is mean free time for a particle of the α th kind), and $J_\alpha^{B,el}, J_\alpha^{B,r}$ are the Boltzmann collision integrals for elastic and inelastic collisions, respectively. GBE (I.37) was derived in the theory of liquids, in this case τ is connected with the time of the particle residence in the Frenkel cell.

By the way derived by me GBE was presented in my lectures on physical kinetics given in Sophia University, Bulgaria, in the year 1987; the last monographs in Russian [58, 59].

Let us consider now some aspects of GBE application beginning with hydrodynamic aspects of the theory. Therefore, in the first approximation, fluctuations will be proportional to the mean free path λ (or, equivalently, to the mean time between the collisions). We can state that the number of particles in reference volume is proportional to cube of the character length L of volume, the number of particles in the surface layer is proportional to λL^2, and as result all effect of fluctuation can be estimated as ratio of two mentioned values or as $\lambda/L=\mathrm{Kn}$.

The important methodical question to be considered is how classical conservation laws fit into the GBE picture. Continuum mechanics conservation laws are derived on the macroscopic level by considering a certain reference volume within the medium, which is enclosed by an infinitesimally thin surface. Moving material points (gas particles) can be either within or outside the volume, and it is by writing down the corresponding balance equations for mass, momentum flux, and energy that the classical equations of continuity, motion, and energy are obtained.

Obviously the hydrodynamic equations will explicitly involve fluctuations proportional to τ. For example, the continuity equation changes its form and will contain terms proportional to viscosity. On the other hand, if the reference volume extends over the whole cavity with the hard walls, then the classical conservation laws should be obeyed.

However, we will here attempt to “guess” the structure of the generalized continuity equation using the arguments outlined above. Neglecting fluctuations, the continuity equation should have the classical form with

$$\rho^a=\rho-\rho^{fl}=\rho-\tau A,\ (\rho\mathbf{v}_0)^a=\rho\mathbf{v}_0-(\rho\mathbf{v}_0)^{fl}=\rho\mathbf{v}_0-\tau\mathbf{B}, \tag{I.39}$$

where ρ is density and $\mathbf{v}_0$ is hydrodynamic velocity. Strictly speaking, the factors A and $\mathbf{B}$ can be obtained from the generalized kinetic equation, in our case, from the GBE. Still, we can guess their form without appeal to the KE_f.

Indeed, let us write the generalized continuity equation

$$\frac{\partial}{\partial t}(\rho-\tau A)+\frac{\partial}{\partial \mathbf{r}}\cdot(\rho\mathbf{v}_0-\tau\mathbf{B})=0 \tag{I.40}$$

in the dimensionless form, using l, the distance from the reference contour to the hard wall (see Fig. I.1), as a length scale.

Then, instead of $\boldsymbol{\tau}$, the (already dimensionless) quantities A and $\mathbf{B}$ will have the Knudsen number $\mathrm{Kn}_l=\lambda/l$ as a coefficient. In the limit $l\to 0$, $\mathrm{Kn}_l\to\infty$ the contour embraces the entire cavity contained within hard walls, and there are no fluctuations on the walls. In other words, the classical equations of continuity and motion must be satisfied at the wall. Using hydrodynamic terminology, we note that the conditions $A=0$, $\mathbf{B}=0$ correspond to a laminar sub-layer in a turbulent flow.

Now if a local Maxwellian distribution is assumed, then the generalized equation of continuity in the Euler approximation is written as

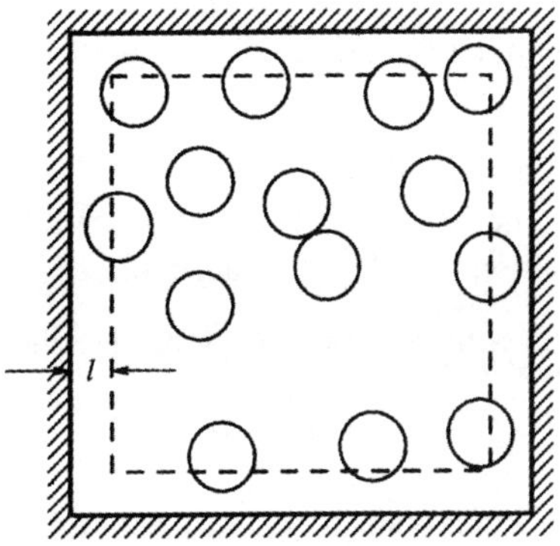

FIG. I.1 Closed cavity and the reference contour containing particles of a finite diameter.

$$\frac{\partial}{\partial t}\left\{\rho-\tau\left[\frac{\partial\rho}{\partial t}+\frac{\partial}{\partial\mathbf{r}}\cdot(\rho\mathbf{v}_0)\right]\right\}+\frac{\partial}{\partial\mathbf{r}}\cdot\left\{\rho\mathbf{v}_0-\tau\left[\frac{\partial}{\partial t}(\rho\mathbf{v}_0)+\frac{\partial}{\partial\mathbf{r}}\cdot\rho\mathbf{v}_0\mathbf{v}_0+\overset{\leftrightarrow}{\mathrm{I}}\cdot\frac{\partial p}{\partial\mathbf{r}}-\rho\mathbf{a}\right]\right\}=0, \tag{I.41}$$

where $\overset{\leftrightarrow}{\mathrm{I}}$ is the unit tensor, a is the acceleration due to an external field of forces.

GBE leads to generalized hydrodynamic equations (GHE), for example, to the continuity equation

$$\frac{\partial\rho^a}{\partial t}+\frac{\partial}{\partial\mathbf{r}}\cdot(\rho\mathbf{v}_0)^a=0, \tag{I.42}$$

where ρ^a, $\mathbf{v}_0{}^a$, $(\rho\mathbf{v}_0)^a$ are calculated in view of non-locality effect in terms of gas density ρ, hydrodynamic velocity of flow $\mathbf{v}_0$, and density of momentum flux $\rho\mathbf{v}_0$; for locally Maxwellian distribution; ρ^a a is the gas density, $\mathbf{v}_0{}^a$ is the hydrodynamic flow velocity, and $(\rho\mathbf{v}_0)^a$ is the momentum flux density obtained by neglecting fluctuations. These values ρ^a, $(\rho\mathbf{v}_0)^a$ are defined by the relations written in the relaxation form

$$(\rho-\rho^a)/\tau=\frac{\partial\rho}{\partial t}+\frac{\partial}{\partial\mathbf{r}}\cdot(\rho\mathbf{v}_0), \tag{I.43}$$

$$(\rho\mathbf{v}_0-(\rho\mathbf{v}_0)^a)/\tau=\frac{\partial}{\partial t}(\rho\mathbf{v}_0)+\frac{\partial}{\partial\mathbf{r}}\cdot\rho\mathbf{v}_0\mathbf{v}_0+\overset{\leftrightarrow}{\mathrm{I}}\cdot\frac{\partial p}{\partial\mathbf{r}}-\rho\mathbf{a}, \tag{I.44}$$

where $\overset{\leftrightarrow}{\mathrm{I}}$ is a unit tensor and **a** is the acceleration due to the effect of mass forces.

G Uhlenbeck, in his review of the fundamental problems of statistical mechanics [60], examines in particular the Kramers equation [61] derived as a consequence of the Fokker–Planck equation

$$\frac{\partial f}{\partial t}+\mathbf{v}\cdot\frac{\partial f}{\partial\mathbf{r}}+\mathbf{a}\cdot\frac{\partial f}{\partial\mathbf{v}}=\beta\left[\frac{\partial}{\partial\mathbf{v}}\cdot(\mathbf{v}f)+\frac{k_BT}{m}\frac{\partial}{\partial\mathbf{v}}\cdot\frac{\partial f}{\partial\mathbf{v}}\right], \tag{I.45}$$

where $f(\mathbf{r},\mathbf{v},t)$ is the distribution function of Brownian particles, **a** is the acceleration due to an external field of forces, and $m\beta$ is the coefficient of friction for the motion of a colloid particle in the medium. What intrigues Uhlenbeck is how Kramers goes over from the Fokker–Planck equation (I.45) to the Einstein–Smoluchowski equation

$$\frac{\partial\rho}{\partial t}+\frac{\partial}{\partial\mathbf{r}}\cdot\left(\frac{\mathbf{a}}{\beta}\rho-\frac{k_BT}{m\beta}\frac{\partial\rho}{\partial\mathbf{r}}\right)=0, \tag{I.46}$$

(ρ is the density) which has the character of the hydrodynamic continuity equation. In Uhlenbeck's words, "the proof of this change-over is very interesting, it is a typical Kramers-style proof. It is in fact very simple but at the same time some tricks and subtleties it involves make it very hard to discuss." The velocity distribution of colloid particles is assumed to be Maxwellian. The "trick," however, is that Kramers integrated along the line

$$\mathbf{r}+\frac{\mathbf{v}}{\beta}=\mathbf{r}_0, \tag{I.47}$$

and the number density of particles turned out to be given by the formula

$$n(\mathbf{r}_0,t)=\int f\left(\mathbf{r}_0-\frac{\mathbf{v}}{\beta},\mathbf{v},t\right)\mathrm{d}\mathbf{v}. \tag{I.48}$$

So what exactly did H Kramers do? Let us consider this change from the point of view of the generalized Boltzmann kinetic theory (GBKT) using, wherever possible, qualitative arguments to see things more clearly.

Now, having in mind the Kramers method, let us compare the generalized continuity equation (I.41) and the Einstein - Smoluchowski equation (I.46). Equation (I.41) reduces to Eq. (I.46) if

(a) the convective transfer corresponding to the hydro dynamical velocity $\mathbf{v_0}$ is neglected;
(b) the temperature gradient is less important than the gradient of the number density of particles, $n\frac{\partial T}{\partial\mathbf{r}}<<T\frac{\partial n}{\partial\mathbf{r}}$, and
(c) the temporal part of the density fluctuations is left out of account.

By integrating with respect to velocity **v** from $-\infty$ to $+\infty$ along the line

$$\mathbf{r}+\frac{\mathbf{v}}{\beta}=\mathbf{r}_0. \tag{1.49}$$

Kramers introduced non-local collisions without accounting for the time delay effect. In our theory, the coefficient of friction $\beta=\tau^{-1}$, which corresponds to the binary collision approximation. If the simultaneous interaction with many particles is important and must be accounted for, additional difficulties associated with the definition of the coefficient of friction β arise, and Einstein-Smoluchowski theory becomes semi-phenomenological. Overcoming these difficulties may require the use of the theory of non-Markov processes for describing Brownian motion [62].

Now several remarks of principal significance:

1. All fluctuations are found from the strict kinetic considerations and tabulated [21, 32, 57]. The appearing additional terms in GHE are due to viscosity and they correspond to the small-scale Kolmogorov turbulence. The neglect of formally small terms is equivalent, in particular, to dropping the (small-scale) Kolmogorov turbulence from consideration and is the origin of all principal difficulties in usual turbulent theory.
2. Fluctuations on the wall are equal to zero, from the physical point of view this fact corresponds to laminar sub-layer. Mathematically it leads to additional boundary conditions for GHE.
3. It would appear that in continuum mechanics the idea of discreteness can be abandoned altogether and the medium under study be considered as a continuum in the literal sense of the word. Such an approach is of course possible and indeed leads to Euler equations in hydrodynamics. But when the viscosity and thermal conductivity effects are to be included, a totally different situation arises. As is well known, the dynamical viscosity is proportional to the mean time τ between the particle collisions, and a continuum medium in the Euler model with $\tau=0$ implies that neither viscosity nor thermal conductivity is possible.
4. Many GHE applications were realized for calculation of turbulent flows with the good coincidence with the bench-mark experiments. GHE are working with good accuracy even in the theory of sound propagation in the rarefied gases where all moment equations based on the classical BE lead to unsatisfactory results.
5. The non-local kinetic effects listed above will always be relevant to a kinetic theory using one particle description—including, in particular, applications to liquids or plasmas, where self-consistent forces with appropriately cut-off radius of their action are introduced to expand the capability of GBE. The application of the above principles also leads to the modification of the system of the Maxwell electro-dynamic equations (ME).

While the traditional formulation of this system does not involve the continuity equation like (I.42) but for the charge density ρ^a and the current density $\mathbf{j}^a$, nevertheless the ME derivation employs continuity equation and leads to appearance of fluctuations (proportional to $\boldsymbol{\tau}$) of charge density and the current density. As a result, the system of Maxwell equations written in the standard notation, namely

$$\frac{\partial}{\partial \mathbf{r}}\cdot\mathbf{B}=0,\ \frac{\partial}{\partial \mathbf{r}}\cdot\mathbf{D}=\rho^a,\ \frac{\partial}{\partial \mathbf{r}}\times\mathbf{E}=-\frac{\partial \mathbf{B}}{\partial t},\ \frac{\partial}{\partial \mathbf{r}}\times\mathbf{H}=\mathbf{j}^a+\frac{\partial \mathbf{D}}{\partial t} \tag{I.50}$$

contains

$$\rho^a=\rho-\rho^{fl},\mathbf{j}^a=\mathbf{j}-\mathbf{j}^{fl}. \tag{I.51}$$

In rarefied media both effects lead to Johnson's flicker noise observed in 1925 for the first time by J.B. Johnson by the measurement of current fluctuations of thermo-electron emission. For plasma $\boldsymbol{\tau}$ is the mean time between "close" collisions of charged particles.

Finally we can state that introduction of control volume by the reduced description for ensemble of particles of finite diameters leads to fluctuations (proportional to Knudsen number) of velocity moments in the volume. This fact leads to the significant reconstruction of the theory of transport processes. The violation of Bell's inequalities [55] is found for local statistical theories, and the transition to non-local description is inevitable.

In the general case, the parameter $\boldsymbol{\tau}$ is the non-locality parameter; in quantum hydrodynamics, its magnitude is correlated with the "time-energy" uncertainty relation [63–66]. Now we can turn our attention to the quantum hydrodynamic description of individual particles. The abstract of the classical Madelung's paper [67] contains only one phrase: "It is shown that the Schrödinger equation for one-electron problems can be transformed into the form of hydrodynamic equations."

The following conclusion of principal significance can be done from the previous consideration [63–66]:

1. Madelung's quantum hydrodynamics is equivalent to the Schrödinger equation (SE) and leads to the description of the quantum particle evolution in the form of Euler equation and continuity equation. Quantum Euler equation contains additional potential of non-local origin which can be written for example in the Bohm form. SE is consequence of the Liouville equation as result of the *local* approximation of *non-local* equations.
2. Generalized Boltzmann physical kinetics leads to the strict approximation of non-local effects in space and time and *in the local limit* leads to parameter τ, which on the quantum level corresponds to the uncertainty principle "time-energy."

3. Generalized hydrodynamic equations (GHE) lead to SE as a deep particular case of the generalized Boltzmann physical kinetics and therefore of non-local hydrodynamics.

In principle GHE needn't in using of the "time-energy" uncertainty relation for estimation of the value of the non-locality parameter τ. Moreover the "time-energy" uncertainty relation does not lead to the exact relations and from position of non-local physics is only the simplest estimation of the non-local effects.

Really, let us consider two neighboring physically infinitely small volumes $PhSV_1$ and $PhSV_2$ in a non-equilibrium system. Obviously the time τ should tend to diminishing with increasing of the velocities u of particles invading in the nearest neighboring physically infinitely small volume ($PhSV_1$ or $PhSV_2$):

$$\tau = H/u^n. \tag{I.52}$$

But the value τ cannot depend on the velocity direction and naturally to tie τ with the particle kinetic energy, then

$$\tau = H/\left(mu^2\right), \tag{I.53}$$

where H is a coefficient of proportionality, which reflects the state of physical system. In the simplest case H is equal to Plank constant $\hbar$ and relation (I.53) becomes compatible with the Heisenberg relation. Possible approximations of τ-parameter in details are considered in this monograph and the general principle of quantization, (see Chapter 5).

As an additional explanation, we place the structure of the generalized transport theory.

The structure of the generalized transport theory

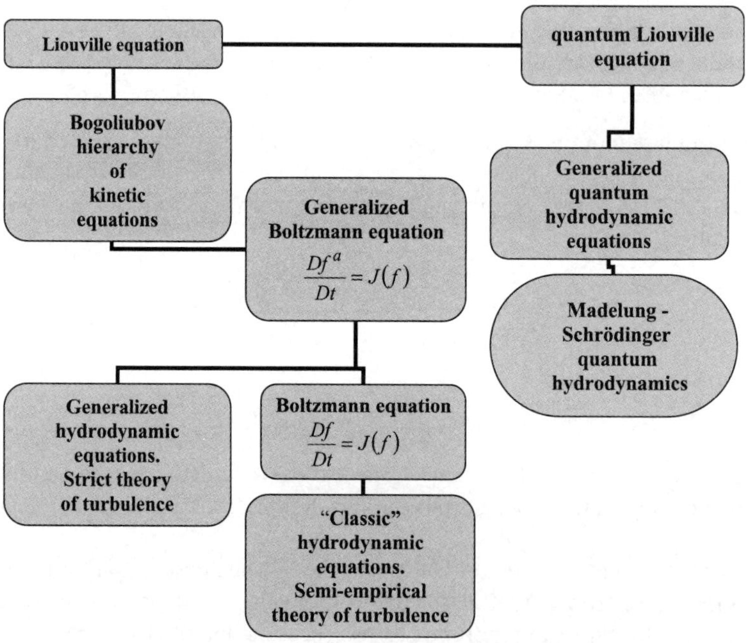

Chapter 1

Generalized Boltzmann Equation

ABSTRACT
In what follows, we intend to construct the generalized Boltzmann physical kinetics using the different methods of the kinetic equation derivations from the Bogolubov hierarchy.
Keywords: Generalized Boltzmann equation, Bogolubov hierarchy, Method of many scales

1.1 MATHEMATICAL INTRODUCTION—METHOD OF MANY SCALES

In the sequel asymptotic methods will be used, but first of all, we will look at the method of many scales. The method of many scales is so popular that Nayfeh in his book [68] written more than 40 years ago said that method of many scales (MMS) is discovered by different authors every half a year. As a result, there exist many different variants of MMS. As a minimum four variants of MMS are considered in [68]. We are interested only in the main ideas of MMS, which are used further in theory of kinetic equations.

From this standpoint we demonstrate MMS possibilities using typical example of solution linear differential equation which also has the exact solution for comparing the results [68]. But in contrast with usual consideration, which can be found in literature, we intend to bring this example up to table and a graph.

Therefore let us consider the linear differential equation

$$\delta\ddot{x}+\varepsilon\dot{x}+x=0. \tag{1.1.1}$$

We begin with the special case when $\delta=1$ and ε is a small parameter. Equation (1.1.1) has the exact solution

$$x=ae^{-\varepsilon t/2}\cos\left[t\sqrt{1-\frac{1}{4}\varepsilon^2}+\phi\right], \tag{1.1.2}$$

where a and ϕ are arbitrary constants of integrating. In typical case of small parameter δ in front of senior derivative—in this case it would be $\ddot{x}$—the effects of boundary layer can be observed. Using the derivatives

$$\dot{x}=-\frac{1}{2}\varepsilon x-ae^{-\varepsilon t/2}\sqrt{1-\frac{1}{4}\varepsilon^2}\sin\left[t\sqrt{1-\frac{1}{4}\varepsilon^2}+\phi\right],$$

$$\ddot{x}=-\frac{1}{2}\varepsilon\dot{x}-x\left(1-\frac{1}{4}\varepsilon^2\right)+\frac{1}{2}a\varepsilon e^{-\varepsilon t/2}\sqrt{1-\frac{1}{4}\varepsilon^2}\sin\left[t\sqrt{1-\frac{1}{4}\varepsilon^2}+\phi\right],$$

for substitution in Eq. (1.1.1) we find the identical satisfaction of Eq. (1.1.1).

We begin with a direct expansion in small ε, using series

$$x=x_0+\varepsilon x_1+\varepsilon^2x_2+\cdots, \tag{1.1.3}$$

and after differentiating

$$\dot{x}=\dot{x}_0+\varepsilon\dot{x}_1+\varepsilon^2\dot{x}_2+\cdots,$$

$$\ddot{x}=\ddot{x}_0+\varepsilon\ddot{x}_1+\varepsilon^2\ddot{x}_2+\cdots.$$

Substitute series (1.1.3) into (1.1.1) and equalize coefficients in front of equal powers of ε, having

$$\ddot{x}_0+x_0=0, \tag{1.1.4}$$

$$\ddot{x}_1+x_1=-\dot{x}_0, \tag{1.1.5}$$

Unified non-local theory of transport processes. http://dx.doi.org/10.1016/B978-0-444-63478-8.00001-4

$$\ddot{x}_2 + x_2 = -\dot{x}_1, \tag{1.1.6}$$
$$\ddot{x}_3 + x_3 = -\dot{x}_2, \tag{1.1.7}$$

and so on. The general solution of homogeneous Eq. (1.1.4) has the form

$$x_0 = a\cos(t+\phi). \tag{1.1.8}$$

Substitute (1.1.8) in (1.1.5):

$$\ddot{x}_1 + x_1 = a\sin(t+\phi). \tag{1.1.9}$$

General solution (1.1.1) should contain only two arbitrary constants. In this case, both constants a and ϕ are contained in the main term of expansion defined by relation (1.1.8). Then we need find only particular solution of Eq. (1.1.9); which can be found as follows

$$x_1 = -\frac{1}{2}at\cos(t+\phi). \tag{1.1.10}$$

Really,

$$\dot{x}_1 = -\frac{1}{2}a\cos(t+\phi) + \frac{1}{2}at\sin(t+\phi),$$
$$\ddot{x}_1 = \frac{1}{2}a\sin(t+\phi) - x_1 + \frac{1}{2}a\sin(t+\phi).$$

After substitution in (1.1.9), we find identity. Equation (1.1.6) can be rewritten as

$$\ddot{x}_2 + x_2 = \frac{1}{2}a\cos(t+\phi) - \frac{1}{2}at\sin(t+\phi), \tag{1.1.11}$$

and its solution

$$x_2 = \frac{1}{8}at^2\cos(t+\phi) + \frac{1}{8}at\sin(t+\phi). \tag{1.1.12}$$

Really,

$$\dot{x}_2 = \frac{1}{4}at\cos(t+\phi) - \frac{1}{8}at^2\sin(t+\phi) + \frac{1}{8}a\sin(t+\phi) + \frac{1}{8}at\cos(t+\phi),$$

$$\ddot{x}_2 = \frac{1}{2}a\cos(t+\phi) - \frac{5}{8}at\sin(t+\phi) - \frac{1}{8}at^2\cos(t+\phi).$$

Substitution into left-hand side of Eq. (1.1.11), lead to result

$$\begin{aligned}&\frac{1}{2}a\cos(t+\phi) - \frac{5}{8}at\sin(t+\phi) - \frac{1}{8}at^2\cos(t+\phi)\\&+\frac{1}{8}at^2\cos(t+\phi) + \frac{1}{8}at\sin(t+\phi)\\&=\frac{1}{2}a\cos(t+\phi) - \frac{1}{2}at\sin(t+\phi).\end{aligned}$$

Then we state the identical satisfaction of Eq. (1.1.11) by solution (1.1.12). In analogous way the solution of Eq. (1.1.7) is written as cubic polynomial in t. For the first three terms of Eq. (1.1.3) series the solution is

$$\begin{aligned}x &= a\cos(t+\phi) - \frac{1}{2}\varepsilon at\cos(t+\phi)\\&+\frac{1}{8}\varepsilon^2 a\left[t^2\cos(t+\phi) + t\sin(t+\phi)\right] + O(\varepsilon^3).\end{aligned} \tag{1.1.13}$$

At our desire the variable t can be considered as dimensionless time. Suppose, of course, that we wish to have a solution for arbitrary time moments. But it is not possible in the developed procedure, because the series construction regards the successive terms of the series to be smaller than the forgoing terms; in other case, it is impossible to speak about series convergence. But for fixed ε, the time moment can be found when successive term of expansion is no smaller than the forgoing term. Figure 1.1 contains comparison of the exact solution (1.1.2) for concrete parameters of calculations $a=1$, $\phi=0$, $\varepsilon=0.2$ with approximate solutions

$${}^{0}x = \cos t$$
$${}^{1}x = \cos t - 0.1t\cos t$$
$${}^{2}x = \cos t - 0.1t\cos t + 0.005[t^2\cos t + t\sin t]$$
$${}^{\mathrm{ex}}x = \mathrm{e}^{-0.1t}\cos\left(t\sqrt{0.99}\right).$$

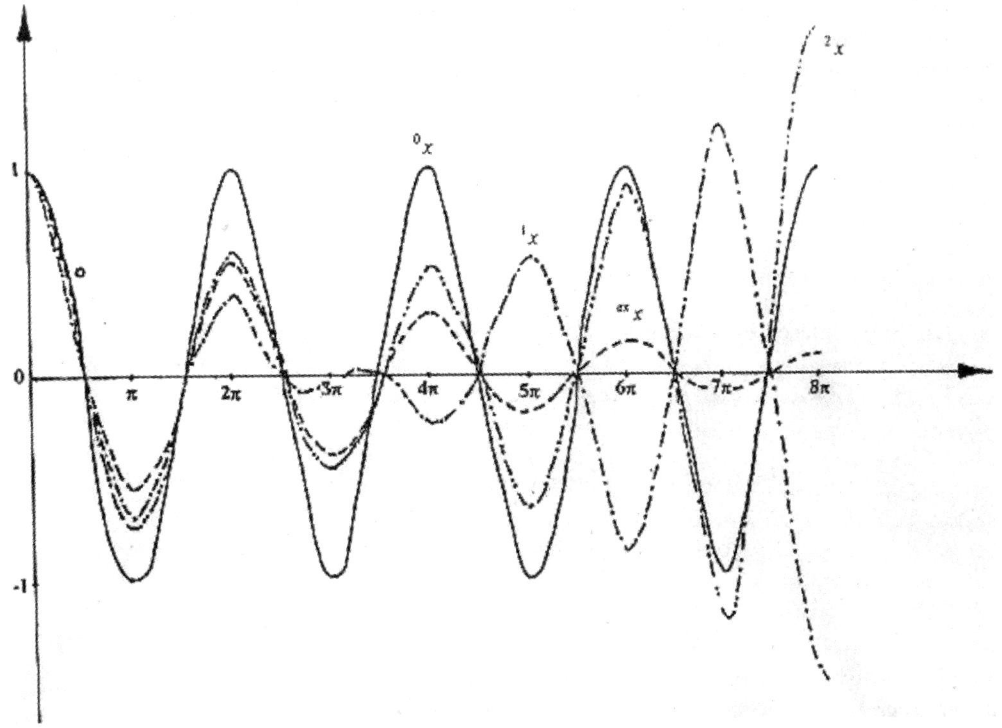

FIGURE 1.1 Comparison of solutions $^0x, ^1x, ^2x$, obtained by perturbation method with the many scales solution $^{ms}x_2$ and exact solution ^{ex}x for the case $a=1$, $\varphi=0$, $\varepsilon=0.2$.

As we could expect, the divergence of solution ^2x and exact solution ^{ex}x appears when t is of order 10; or, in the common case, if $t \sim \varepsilon^{-1}$. But as it follows from Fig. 1.1, the situation is much worse, because, for example, for $t=4\pi$ approximate solution ^1x gives wrong sign in comparison with the exact solution ^{ex}x. For the mathematical model of an oscillator with damping—which is reflecting by Eq. (1.1.1), it means that approximate solution ^1x forecasts a deviation in the opposite direction for the mentioned oscillator. By the way, solution ^1x is also worse in comparison with solution ^0x, in this sense the minor approximation is better than senior ones.

This poses the question how to improve the situation remaining in the frame of asymptotic methods. To answer this question, let us consider the exact solution (1.1.2). Exponential and cosine terms containing in this solution, can be expand in the following series for small ε and fixed t:

$$e^{-\varepsilon t/2} = 1 - \frac{1}{2}\varepsilon t + \frac{1}{8}(\varepsilon t)^2 - \frac{1}{64}(\varepsilon t)^3 + \cdots \tag{1.1.14}$$

$$\begin{aligned} \cos\left(t\sqrt{1-\frac{1}{4}\varepsilon^2}+\phi\right) &= \cos\left[t\left(1-\frac{1}{8}\varepsilon^2-\frac{1}{128}\varepsilon^4-\frac{1}{1024}\varepsilon^6-\cdots\right)+\phi\right] \\ &= \cos\left[t+\phi-\frac{1}{8}t\varepsilon^2-\frac{1}{128}t\varepsilon^4-\frac{1}{1024}t\varepsilon^6-\cdots\right] \\ &\cong \cos(t+\phi)+\left(\frac{1}{8}t\varepsilon^2+\frac{1}{128}t\varepsilon^4+\frac{1}{1024}t\varepsilon^6+\cdots\right)\sin(t+\phi) \\ &= \cos(t+\phi)+\frac{1}{8}t\varepsilon^2\sin(t+\phi)+\frac{1}{128}t\varepsilon^4\sin(t+\phi)+\cdots. \end{aligned} \tag{1.1.15}$$

Obviously, product of the first terms in expansions (1.1.14) and (1.1.15) gives ^0x, and retaining of terms of $O(\varepsilon^3)$ lead to result

$$\begin{aligned} x &\cong a\left(1-\frac{1}{2}\varepsilon t+\frac{1}{8}\varepsilon^2t^2\right)\left(\cos(t+\phi)+\frac{1}{8}t\varepsilon^2\sin(t+\phi)\right) \\ &\cong a\cos(t+\phi)-\frac{1}{2}a\varepsilon t\cos(t+\phi)+\frac{1}{8}at\varepsilon^2\sin(t+\phi)+\frac{1}{8}a\varepsilon^2t^2\cos(t+\phi). \end{aligned}$$

Then we state that the used construction of asymptotic solution is based in deed on the assumption that combination εt is small. If it is not so (for t having the order ε^{-1}), then expansions (1.1.14) and (1.1.15) are wrong or need to take into account all terms of expansions. But asymptotic expansion can be organized by another way, using additional variables:

$$T_1 = \varepsilon t \tag{1.1.16}$$

and

$$T_2 = \varepsilon^2 t. \tag{1.1.17}$$

In this case

$$\mathrm{e}^{-\varepsilon t/2} = \mathrm{e}^{-T_1/2}, \tag{1.1.18}$$

and expansion (1.1.15) is replaced by other ones:

$$\begin{aligned}
\cos\left(t\sqrt{1-\frac{1}{4}\varepsilon^2}+\phi\right) &= \cos\left[t+\phi-\frac{1}{8}t\varepsilon^2-\frac{1}{128}t\varepsilon^4-\frac{1}{1024}t\varepsilon^6-\cdots\right]\\
&= \cos\left[t+\phi-\frac{1}{8}T_2\right]+\left(\frac{1}{128}t\varepsilon^4+\frac{1}{1024}t\varepsilon^6+\cdots\right)\sin\left(t+\phi-\frac{1}{8}T_2\right)\\
&= \cos\left(t+\phi-\frac{1}{8}T_2\right)+\frac{1}{128}t\varepsilon^4\sin\left(t+\phi-\frac{1}{8}T_2\right)\\
&\quad+\frac{1}{1024}t\varepsilon^6\sin\left(t+\phi-\frac{1}{8}T_2\right)+\cdots
\end{aligned} \tag{1.1.19}$$

Expansion (1.1.19) "is working" with the accuracy $O(t\varepsilon^4)$ and therefore the movement in the direction of achieving of higher and higher accuracy leads to the appearance of new variables of the type

$$T_m = \varepsilon^m t, \quad m = 0,1,2\ldots, \tag{1.1.20}$$

or, which is the same, to new timescales $T_0, T_1, T_2, \ldots$. As result we obtain the asymptotic solution of the new type

$$x(t,\varepsilon) = x(T_0, T_1, T_2, \ldots, T_M; \varepsilon) = \sum_{m=0}^{M} \varepsilon^m x_m(T_0, T_1, \ldots, T_M) + O(\varepsilon T_M). \tag{1.1.21}$$

The application of expansion (1.1.21) inevitably leads to the system of equations in partial derivatives and the time derivative should be calculated, using the rule for differentiating of a composite function:

$$\frac{\mathrm{d}}{\mathrm{d}t} = \frac{\partial}{\partial T_0}+\varepsilon\frac{\partial}{\partial T_1}+\varepsilon^2\frac{\partial}{\partial T_2}+\cdots. \tag{1.1.22}$$

Then, with the accuracy $O(\varepsilon^2)$ we have:

$$\begin{aligned}
\frac{\mathrm{d}x}{\mathrm{d}t} &\cong \frac{\partial}{\partial T_0}(x_0+\varepsilon x_1+\varepsilon^2 x_2)+\varepsilon\frac{\partial}{\partial T_1}(x_0+\varepsilon x_1)+\varepsilon^2\frac{\partial x_0}{\partial T_2}\\
&\cong \frac{\partial x_0}{\partial T_0}+\varepsilon\left(\frac{\partial x_1}{\partial T_0}+\frac{\partial x_0}{\partial T_1}\right)+\varepsilon^2\left(\frac{\partial x_2}{\partial T_0}+\frac{\partial x_1}{\partial T_1}+\frac{\partial x_0}{\partial T_2}\right).
\end{aligned} \tag{1.1.23}$$

Using series (1.1.21) and the procedure of decomposition of the derivative (1.1.22) (see also [68]), we obtain the system of equations in partial derivatives defining x_0, x_1 and x_2 in series (1.1.21). Obviously

$$\begin{aligned}
\frac{\mathrm{d}^2x}{\mathrm{d}t^2} &\cong \left(\frac{\partial}{\partial T_0}+\varepsilon\frac{\partial}{\partial T_1}+\varepsilon^2\frac{\partial}{\partial T_2}\right)\left(\frac{\partial x}{\partial T_0}+\varepsilon\frac{\partial x}{\partial T_1}+\varepsilon^2\frac{\partial x}{\partial T_2}\right)\\
&\cong \frac{\partial^2 x}{\partial T_0^2}+\varepsilon\frac{\partial^2 x}{\partial T_0\partial T_1}+\varepsilon^2\frac{\partial^2 x}{\partial T_0\partial T_2}+\varepsilon\frac{\partial^2 x}{\partial T_0\partial T_1}+\varepsilon^2\frac{\partial^2 x}{\partial T_1^2}+\varepsilon^2\frac{\partial^2 x}{\partial T_0\partial T_2}\\
&= \frac{\partial^2 x}{\partial T_0^2}+2\varepsilon\frac{\partial^2 x}{\partial T_0\partial T_1}+\varepsilon^2\left(2\frac{\partial^2 x}{\partial T_0\partial T_2}+\frac{\partial^2 x}{\partial T_1^2}\right)\cong\frac{\partial^2 x_0}{\partial T_0^2}+\varepsilon\frac{\partial^2 x_1}{\partial T_0^2}\\
&\quad+\varepsilon^2\frac{\partial^2 x_2}{\partial T_0^2}+2\varepsilon\frac{\partial^2 x_0}{\partial T_0\partial T_1}+2\varepsilon^2\frac{\partial^2 x_1}{\partial T_0\partial T_1}+\varepsilon^2\left(2\frac{\partial^2 x_0}{\partial T_0\partial T_2}+\frac{\partial^2 x_0}{\partial T_1^2}\right)\\
&= \frac{\partial^2 x_0}{\partial T_0^2}+\varepsilon\left(\frac{\partial^2 x_1}{\partial T_0^2}+2\frac{\partial^2 x_0}{\partial T_0\partial T_1}\right)\\
&\quad+\varepsilon^2\left(\frac{\partial^2 x_2}{\partial T_0^2}+2\frac{\partial^2 x_1}{\partial T_0\partial T_1}+2\frac{\partial^2 x_0}{\partial T_0\partial T_2}+\frac{\partial^2 x_0}{\partial T_1^2}\right).
\end{aligned} \tag{1.1.24}$$

Equating now the coefficients for the same powers of ε in Eqs. (1.1.1) and (1.1.21), (1.1.24), we obtain a set of equations, written below, for terms containing $\varepsilon^0, \varepsilon^1, \varepsilon^2$.

$$\frac{\partial^2 x_0}{\partial T_0^2} + x_0 = 0, \tag{1.1.25}$$

$$\frac{\partial^2 x_1}{\partial T_0^2} + 2\frac{\partial^2 x_0}{\partial T_0 \partial T_1} + x_1 = -\frac{\partial x_0}{\partial T_0}, \tag{1.1.26}$$

$$\frac{\partial^2 x_2}{\partial T_0^2} + 2\frac{\partial^2 x_1}{\partial T_0 \partial T_1} + 2\frac{\partial^2 x_0}{\partial T_0 \partial T_2} + \frac{\partial^2 x_0}{\partial T_1^2} + x_2 = -\left(\frac{\partial x_1}{\partial T_0} + \frac{\partial x_0}{\partial T_1}\right). \tag{1.1.27}$$

The set of Eqs. (1.1.25)–(1.1.27) is not linked, it means, that equations served for obtaining the minor coefficients of expansion (1.1.21) do not contain the senior coefficients of the mentioned expansion. To underline this fact the next system of equations contains in the right-hand sides only the terms known from the previous approximations:

$$\frac{\partial^2 x_0}{\partial T_0^2} + x_0 = 0,$$

$$\frac{\partial^2 x_1}{\partial T_0^2} + x_1 = -\frac{\partial x_0}{\partial T_0} - 2\frac{\partial^2 x_0}{\partial T_0 \partial T_1}, \tag{1.1.26$'$}$$

$$\frac{\partial^2 x_2}{\partial T_0^2} + x_2 = -\frac{\partial x_0}{\partial T_1} - 2\frac{\partial^2 x_0}{\partial T_0 \partial T_2} - \frac{\partial^2 x_0}{\partial T_1^2} - \frac{\partial x_1}{\partial T_0} - 2\frac{\partial^2 x_1}{\partial T_0 \partial T_1}. \tag{1.1.27$'$}$$

The following step consists in a successive solution of Eqs. (1.1.25), (1.1.26'), (1.1.27'), and the corresponding next equations of this type. It is just this type of procedure, but for kinetic equations, will be realized in the Section 1.3.

For linear Eq. (1.1.1) chosen as an example, no reason to use this rather complicated procedure and it is better to avoid expansion (1.1.21). For this aim it is suffices to suppose that

$$x = x(T_0, T_1, T_2) \tag{1.1.28}$$

and calculate the first and second derivatives in Eq. (1.1.1) like Eqs. (1.1.23) and (1.1.24), i.e.

$$\frac{dx}{dt} = \frac{\partial x}{\partial T_0} + \varepsilon\frac{\partial x}{\partial T_1} + \varepsilon^2\frac{\partial x}{\partial T_2}, \tag{1.1.29}$$

$$\begin{aligned} \frac{d^2 x}{dt^2} &= \left(\frac{\partial}{\partial T_0} + \varepsilon\frac{\partial}{\partial T_1} + \varepsilon^2\frac{\partial}{\partial T_2}\right)\left(\frac{\partial x}{\partial T_0} + \varepsilon\frac{\partial x}{\partial T_1} + \varepsilon^2\frac{\partial x}{\partial T_2}\right) \\ &\cong \frac{\partial^2 x}{\partial T_0^2} + 2\varepsilon\frac{\partial^2 x}{\partial T_0 \partial T_1} + \varepsilon^2\left(\frac{\partial^2 x}{\partial T_1^2} + 2\frac{\partial^2 x}{\partial T_0 \partial T_2}\right). \end{aligned} \tag{1.1.30}$$

As result we have

$$\frac{\partial^2 x}{\partial T_0^2} + 2\varepsilon\frac{\partial^2 x}{\partial T_0 \partial T_1} + \varepsilon^2\left(\frac{\partial^2 x}{\partial T_1^2} + 2\frac{\partial^2 x}{\partial T_0 \partial T_2}\right) + x = -\varepsilon\left(\frac{\partial x}{\partial T_0} + \varepsilon\frac{\partial x}{\partial T_1}\right). \tag{1.1.31}$$

After the procedure of term equating described above, we reach the system of equations

$$\frac{\partial^2 x}{\partial T_0^2} + x = 0, \tag{1.1.32}$$

$$2\frac{\partial^2 x}{\partial T_0 \partial T_1} = -\frac{\partial x}{\partial T_0}, \tag{1.1.33}$$

$$2\frac{\partial^2 x}{\partial T_2 \partial T_0} = -\frac{\partial^2 x}{\partial T_1^2} - \frac{\partial x}{\partial T_1}. \tag{1.1.34}$$

A general solution of Eq. (1.1.32) has the form

$$x = A(T_1, T_2)e^{iT_0} + A^*(T_1, T_2)e^{-iT_0}. \tag{1.1.35}$$

Obviously

$$\frac{\partial x}{\partial T_0} = iA\mathrm{e}^{iT_0} - iA^*\mathrm{e}^{-iT_0}, \tag{1.1.36}$$

$$\frac{\partial^2 x}{\partial T_0 \partial T_1} = i\frac{\partial A}{\partial T_1}\mathrm{e}^{iT_0} - i\frac{\partial A^*}{\partial T_1}\mathrm{e}^{-iT_0}. \tag{1.1.37}$$

Substitute Eqs. (1.1.36) and (1.1.37) in Eq. (1.1.33):

$$\left(2\frac{\partial A}{\partial T_1} + A\right)\mathrm{e}^{iT_0} - \left(2\frac{\partial A^*}{\partial T_1} + A^*\right)\mathrm{e}^{-iT_0} = 0. \tag{1.1.38}$$

Equality (1.1.38) should be fulfilled for arbitrary values of independent variable T_0, but A and A^* are functions of another variables (T_1 and T_2). It can be realized if only the following relations are identities:

$$2\frac{\partial A}{\partial T_1} + A = 0, \tag{1.1.39}$$

$$2\frac{\partial A^*}{\partial T_1} + A^* = 0, \tag{1.1.40}$$

whence it follows that

$$A = b(T_2)\mathrm{e}^{-T_1/2}, \quad A^* = b^*(T_2)\mathrm{e}^{-T/2_1}. \tag{1.1.41}$$

We use Eq. (1.1.34) for definition of $b(T_2)$ and $b^*(T_2)$. Since now

$$x = b(T_2)\mathrm{e}^{-\frac{1}{2}T_1 + iT_0} + b^*(T_2)\mathrm{e}^{-\frac{1}{2}T_1 - iT_0}, \tag{1.1.42}$$

we have

$$\frac{\partial x}{\partial T_1} = -\frac{1}{2}x, \quad \frac{\partial^2 x}{\partial T_1^2} = \frac{1}{4}x, \tag{1.1.43}$$

$$\begin{aligned}\frac{\partial^2 x}{\partial T_0 \partial T_2} &= i\frac{\partial}{\partial T_2}\left[b(T_2)\mathrm{e}^{-\frac{1}{2}T_1 + iT_0} - b^*(T_2)\mathrm{e}^{-\frac{1}{2}T_1 - iT_0}\right] \\ &= i\left[\frac{\partial b}{\partial T_2}\mathrm{e}^{-\frac{1}{2}T_1 + iT_0} - \frac{\partial b^*}{\partial T_2}\mathrm{e}^{-\frac{1}{2}T_1 - iT_0}\right].\end{aligned} \tag{1.1.44}$$

and Eq. (1.1.34) after substitution of Eqs. (1.1.42)–(1.1.44) takes the form

$$\left(2i\frac{\partial b}{\partial T_2} - \frac{1}{4}b\right)\mathrm{e}^{-\frac{1}{2}T_1 + iT_0} - \left(2i\frac{\partial b^*}{\partial T_2} + \frac{1}{4}b^*\right)\mathrm{e}^{-\frac{1}{2}T_1 - iT_0} = 0. \tag{1.1.45}$$

Relation (1.1.45) should be fulfilled for all T_1 and T_0 (with b depending on another variable T_2), then terms in round brackets are equal to zero. Then

$$\begin{aligned} 2i\frac{\partial b}{\partial T_2} - \frac{1}{4}b = 0, \\ 2i\frac{\partial b^*}{\partial T_2} + \frac{1}{4}b^* = 0, \end{aligned} \tag{1.1.46}$$

and

$$\begin{aligned} b = b_0\mathrm{e}^{-i\frac{T_2}{8}}, \\ b^* = b_0^*\mathrm{e}^{+i\frac{T_2}{8}}. \end{aligned} \tag{1.1.47}$$

As result we have the solution

$$x = b_0\mathrm{e}^{-\frac{1}{2}T_1 + i\left(T_0 - \frac{T_2}{8}\right)} + b_0^*\mathrm{e}^{-\frac{1}{2}T_1 - i\left(T_0 - \frac{T_2}{8}\right)}. \tag{1.1.48}$$

For convenience b_0 will be written as

$$b_0 = \frac{1}{2} a e^{i\phi}. \tag{1.1.49}$$

Then

$$x = \frac{1}{2} a e^{-\frac{1}{2}T_1} e^{i\left(T_0 - \frac{T_2}{8} + \phi\right)} + \frac{1}{2} a e^{-\frac{1}{2}T_1} e^{-i\left(T_0 - \frac{T_2}{8} + \phi\right)}. \tag{1.1.50}$$

Another form of this solution is:

$$x = a e^{-\frac{1}{2}T_1} \cos\left(T_0 - \frac{T_2}{8} + \phi\right) \tag{1.1.51}$$

or, using variable t and parameter ε, we get:

$$x = a e^{-\varepsilon t/2} \cos\left(t - \frac{1}{8}\varepsilon^2 t + \phi\right). \tag{1.1.52}$$

In the previous variant MMS (method of many scales), which lead to the system (1.1.25), (1.1.26'), and (1.1.27'), the solution ${}^{\mathrm{ms}}x_2$ obtained with accuracy B $O(\varepsilon^2 t)$, can be written in the same form.

For the example mentioned above (ε=0.2; a=1; ϕ=0)

$${}^{\mathrm{ms}}x_2 = \mathrm{e}^{-0.1t} \cos(0.995t) \tag{1.1.53}$$

The comparison of exact solution ${}^{\mathrm{ex}}x$ with the approximate solution obtained by MMS ${}^{\mathrm{ms}}x_2$ in the region of "claimed" accuracy $O(\varepsilon^2 t)$ shows (see Table 1.1 and Fig. 1.1), that in considered range of t ($0 \div 8\pi$) MMS-solution is coincidental with the exact solution with the accuracy as a rule with three ciphers in spite of the fact that $\varepsilon = 0.2$ in not really very small parameter. This result significantly exceeds the accuracy of the traditional asymptotic method (Table 1.1 contains also solutions ${}^0x, {}^1x, {}^2x$). Since

$$\cos\left(t\sqrt{0.99}\right) - \cos\left(t\sqrt{0.995}\right) = 2\sin(t \times 0.994993719)\sin(t \times 0.000006281),$$

the divergence $O(1)$ could be appear only for $t \approx 10^5$, but, in this case, the difference of cosines is fully suppressed by the exponential term e^{-10^4}. Then, the growth of relative error $\frac{{}^{\mathrm{ex}}x - {}^{\mathrm{ms}}x_2}{{}^{\mathrm{ex}}x}$ is no of significance. The evolution of solution ${}^{\mathrm{ms}}x_2$ is not shown in Fig. 1.1 due to actual coincidence of the exact and many scales solutions.

We can recollect another analogy. In the kinetic theory of reacting gases, the calculation of kinetic coefficients (by explicit taking into account of inelastic collisions) is connected with the investigation of convergence Sonine polynomial expansions. In the region, where we can wait for a bad convergence, the perturbation effects are small, because they are suppressed by the exponential function $\mathrm{e}^{-\bar{E}}$ with large $\bar{E}$, where $\bar{E}$ is dimensionless activation energy. Only in the theory of reactions with high energy barriers it is not unreasonable to consider mentioned effects, but of course better on the basement of the generalized Boltzmann physical kinetics because, as it will be shown, the traditional Boltzmann physical kinetics leads to significant errors in calculations of high energetic "tails" of distribution functions.

Let us consider now the role of a parameter at the senior derivative in differential equations. The problem is significant, particularly, in the theory of generalized Boltzmann equation (GBE) solutions by the small Knudsen numbers. We use Eq. (1.1.1)

$$\delta \ddot{x} + \varepsilon \dot{x} + x = 0,$$

where δ, ε are some constant parameters. Non-trivial solution of Eq. (1.1.1) can be written as

$$x = u(t) \exp\left(-\frac{\varepsilon}{2\delta} t\right). \tag{1.1.54}$$

After the substitution of Eq. (1.1.54) for Eq. (1.1.1), we find

$$\ddot{u} - \frac{1}{\delta}\left(\frac{\varepsilon^2}{4\delta} - 1\right) u = 0. \tag{1.1.55}$$

Let us consider now Eq. (1.1.55) for the case $0 < \delta \ll 1, \varepsilon \approx 1$.

$$0 < \delta \ll 1, \varepsilon \sim 1. \tag{1.1.56}$$

TABLE 1.1 Comparison of solutions ^0x, ^1x, ^2x, obtained by perturbation method with the many scales solution $^{ms}x_2$ and exact solution ^{ex}x for the case $a=1$, $\phi=0$, $\varepsilon=0.2$

$^0x=\cos t$

$^1x=\cos t(1-0.1t)$

$^2x=\cos t(1-0.1t)+0.005t(t\cos t+\sin t)$

$^{ex}x=e^{-0.1t}\cos\left(t\sqrt{0.99}\right)=e^{-0.1t}\cos(t\times 0.99498744)$

$^{ms}x_2=e^{-0.1t}\cos(t\times 0.995)$

t	^0x	^1x	^2x	$^{ms}x_2$	^{ex}x
0	1	1	1	1	1
π/4	0.707	0.652	0.656	0.656	0.656
π/2	0	0	0.00785	0.00671	0.00673
3/4π	−0.707	−0.540	−0.552	−0.552	−0.552
π	−1	−0.686	−0.735	−0.730	−0.730
1.25π	−0.707	−0.429	−0.498	−0.487	−0.487
1.5π	0	0	−0.0236	−0.0147	−0.0147
1.75π	0.707	0.318	0.406	0.397	0.397
2π	1	0.372	0.569	0.533	0.533
2.25π	0.707	0.207	0.409	0.361	0.361
2.5π	0	0	0.0393	0.0179	0.0179
2.75π	−0.707	−0.0962	−0.329	−0.285	−0.285
3π	−1	−0.0575	−0.502	−0.389	−0.389
3.25π	−0.707	0.0149	−0.390	−0.267	−0.267
3.5π	0	0	−0.0550	−0.0183	−0.0183
3.75π	0.707	−0.126	0.323	0.204	0.204
4π	1	−0.257	0.533	0.284	0.284
4.25π	0.707	−0.237	0.440	0.198	0.198
4.5π	0	0	0.0707	0.0172	0.0172
4.75π	−0.707	0.348	−0.386	−0.147	−0.147
5π	−1	0.571	−0.663	−0.207	−0.207
5.25π	−0.707	0.459	−0.561	−0.147	−0.147
5.5π	0	0	−0.0864	−0.0153	−0.0154
5.75π	0.707	−0.570	0.520	0.105	0.105
6π	1	−0.885	0.892	0.151	0.151
6.25π	0.707	−0.681	0.751	0.108	0.108
6.5π	0	0	0.102	0.0133	0.0133
6.75π	−0.707	0.792	−0.722	−0.0754	−0.0753
7π	−1	1.199	−1.219	−0.110	−0.110
7.25π	−0.707	0.903	−1.011	−0.0803	−0.0803
7.5π	0	0	−0.118	−0.0111	−0.0112
7.75π	0.707	1.015	0.995	0.0540	0.0539
8π	1	−1.513	1.645	0.0804	0.0804

In this case Eq. (1.1.55) may be simplified:

$$\ddot{u}-\frac{\varepsilon^2}{4\delta^2}u=0. \tag{1.1.57}$$

Let there be $u(0)=1$, $u(\infty)=0$, the solution of Eq. (1.1.57) is

$$u(t)=\exp\left(-\frac{\varepsilon}{2\delta}t\right), \tag{1.1.58}$$

and taking into account Eq. (1.1.54)

$$x(t)=\exp\left(-\frac{\varepsilon}{\delta}t\right). \tag{1.1.59}$$

Let us omit now the first term in the left-hand side of Eq. (1.1.1), which is from the first glance is small as containing a small coefficient δ:

$$\varepsilon\dot{x}+x=0 \tag{1.1.60}$$

and

$$x=\exp\left(-\frac{1}{\varepsilon}t\right) \tag{1.1.61}$$

Solutions (1.1.59) and (1.1.61) have the same asymptotic $x\to 0$ if $t\to\infty$. But function x corresponding to Eq. (1.1.59) is decreasing very fast in comparison with function (1.1.61). It means that coefficient in front of senior derivative leads to formation of "boundary layer" with very fast function change.

Figures 1.2–1.6 contain the integral curves as results of the direct solution of Eq. (1.1.1) for the different parameters ε, δ, and the solution of Eq. (1.1.61).

The integral curves of Figs. 1.2–1.6 have the character features peculiar to the relaxation curves for BE and GBE. Namely:

1. Small parameter δ at the second derivative corresponds to the small Knudsen number in GBE theory. Integral curves on Figs. 1.2 and 1.3 have the different character—the small parameter provokes the "boundary layer" effect for the small t.
2. Effect of the boundary layer creation is enhanced by diminishing of δ, (see Fig. 1.4).

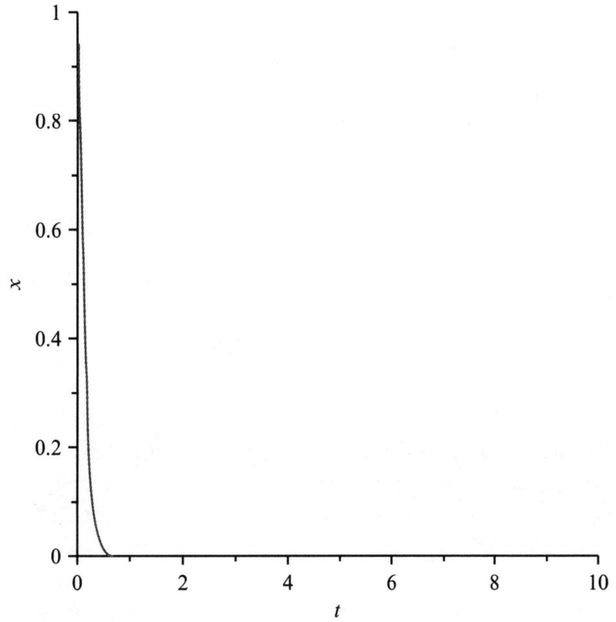

FIGURE 1.2 Solution of Eq. (1.1.1); $\delta=0.1, \varepsilon=1$.

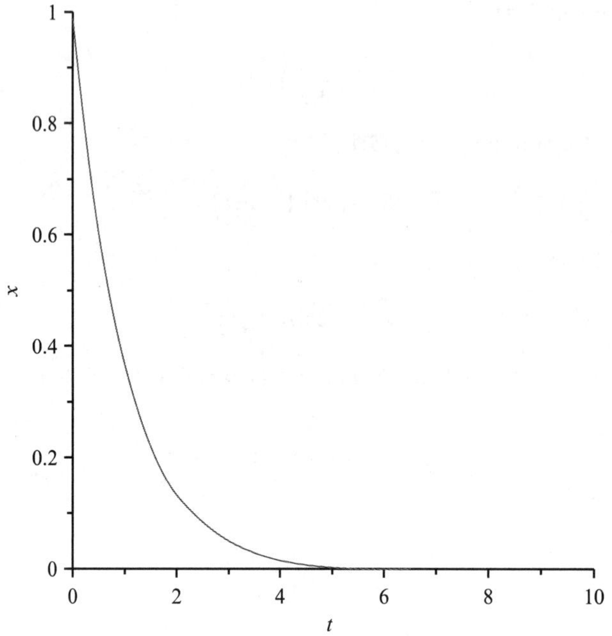

FIGURE 1.3 Solution of Eq. (1.1.61); $\varepsilon = 1$).

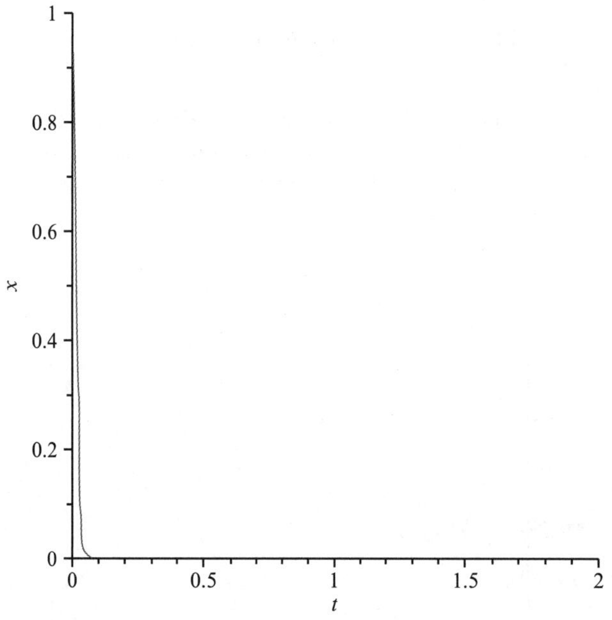

FIGURE 1.4 Solution of Eq. (1.1.1); $\delta = 0.01, \varepsilon = 1$.

3. For $\delta = 1$, $\varepsilon = 1$ the integral curve of the second-order differential equation demonstrates oscillations approaching to zero (Fig. 1.5). This case corresponds to the approximation along the time arrow in the theory of GBE.
4. For $\delta = -1$, $\varepsilon = 1$ (when the derivatives, contained in (1.1.1) have the different signs) the integral curve of Eq. (1.1.1) approach monotonically to zero (Fig. 1.6). This case in the GBE theory corresponds to the approximation against the time arrow. For the considered examples the integral curves, shown in Figs. 1.3 and 1.6 are rather close to each other. These peculiar features of the model equations have the corresponding reflection in the much more complicated equations of the kinetic theory. Then terms with small coefficients in front of senior derivatives cannot be omitted.

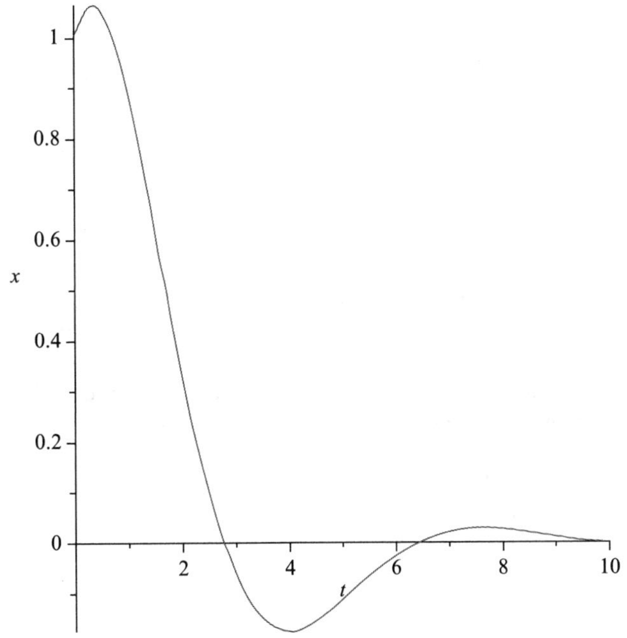

FIGURE 1.5 Solution of Eq. (1.1.1); $\delta=1, \varepsilon=1$.

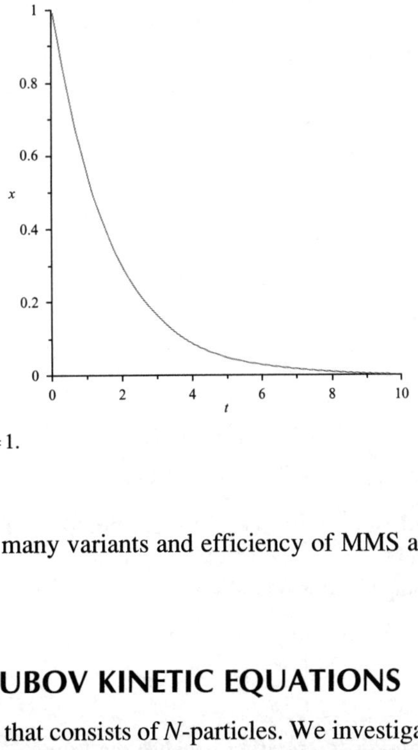

FIGURE 1.6 Solution of Eq. (1.1.1); $\delta=-1, \varepsilon=1$.

It remains only to notice that MMS has many variants and efficiency of MMS application related with specificity of the problem.

1.2 HIERARCHY OF BOGOLUBOV KINETIC EQUATIONS

Let us consider a closed physical system that consists of N-particles. We investigate non-relativistic gas motion adheres of laws of classical mechanics. Introduce in 6N-dimensional space the N-particles distribution function (DF) f_N, in such a way

$$\begin{aligned} dW &= f_N(t; \hat{\mathbf{r}}_1, \ldots, \hat{\mathbf{r}}_N; \hat{\mathbf{v}}_1, \ldots, \hat{\mathbf{v}}_N) d\hat{\mathbf{r}}_1 \ldots d\hat{\mathbf{r}}_N d\hat{\mathbf{v}}_1 \ldots d\hat{\mathbf{v}}_N \\ &= f_N d\Omega_1 \ldots d\Omega_N = f_N d\Omega, \end{aligned} \tag{1.2.1}$$

where

$$d\Omega_i = d\hat{\mathbf{r}}_i d\hat{\mathbf{v}}_i; \quad d\hat{\mathbf{r}}_i \equiv dr_{i1} dr_{i2} dr_{i3}; \quad d\hat{\mathbf{v}}_i \equiv dv_{i1} dv_{i2} dv_{i3}, \tag{1.2.2}$$

$$d\Omega = d\Omega_1 \ldots d\Omega_N, \tag{1.2.3}$$

as to dW is probability to find mentioned system in time moment t in element $d\Omega$ of phase space. Liouville equation is valid for the N-particles distribution function,

$$\frac{\partial f_N}{\partial t} + \sum_{i=1}^{N} \mathbf{v}_i \cdot \frac{\partial f_N}{\partial \mathbf{r}_i} + \sum_{i=1}^{N} \mathbf{F}_i \cdot \frac{\partial f_N}{\partial \mathbf{p}_i} = 0, \tag{1.2.4}$$

where $\mathbf{v}_i, \mathbf{p}_i$—velocity and momentum of particle numbered i, $\mathbf{F}_i$ is the force acting on i-th particle, which mass is equal to m_i.

Consider the physical sense of Eq. (1.2.4). Let at a time moment t there be a unit volume in the considered phase space defined by positions of vectors $\mathbf{r}_1, \ldots, \mathbf{r}_N; \mathbf{v}_1, \ldots, \mathbf{v}_N$. As it has written, the probability to find the system of this volume at the time moment t is f_N. Phase points—consisting at the time moment t the surface, that encloses the mentioned volume—change their positions at the time interval Δt; as a result, the position and form of the volume is also changing. At the time moment $t+\Delta t$, the volume will be found in the position in the vicinity of the $6N$-components vector with the coordinates

$$(\mathbf{r}_1 + \Delta\mathbf{r}_1, \ldots, \mathbf{r}_N + \Delta\mathbf{r}_N; \mathbf{v}_1 + \Delta\mathbf{v}_1, \ldots, \mathbf{v}_N + \Delta\mathbf{v}_N),$$

and the probability to find the system in this state is the same as at the time moment t, because the DF f_N does not change along the phase trajectory. Then

$$\begin{aligned} &f_N(t;\mathbf{r}_1, \ldots, \mathbf{r}_N;\mathbf{v}_1, \ldots \mathbf{v}_N) \\ &\quad = f_N(t+\Delta t;\mathbf{r}_1 + \Delta\mathbf{r}_1, \ldots, \mathbf{r}_N + \Delta\mathbf{r}_N;\mathbf{v}_1 + \Delta\mathbf{v}_1, \ldots, \mathbf{v}_N + \Delta\mathbf{v}_N), \end{aligned} \tag{1.2.5}$$

where the evolution of $\mathbf{r}_i$ and $\mathbf{v}_i$ $(i=1,\ldots,N)$ is defined by the equations of mechanical motion. Obviously Eq. (1.2.4) is also valid for multi-component mixture of gases when the system contains the particles of different masses. But the mixture should be a non-reacting one. Really, if chemical reactions take place, the number of particles can be changed in the system. This fact can lead to the appearance of a source term in Eq. (1.2.4). Another treatment is also valid when molecules are considered as bounded states of atoms which total number is the same during the system evolution. This approach is not simpler for the description of evolution on the kinetic level because, strictly speaking, the laws of classical mechanics are not sufficient for calculation of reacting gases. As result on the lowest level characterized by the one-particle DF, the problem arises in explicit form for the integral of inelastic collisions [21]. We begin with the chain of kinetic equations for non-reacting gases.

Introduce the s-particles distribution function in accordance with definition:

$$f_s = \int f_N d\Omega_{s+1} \ldots d\Omega_N = \int f_N(t, \Omega_1, \ldots, \Omega_N) d\Omega_{s+1} \ldots d\Omega_N. \tag{1.2.6}$$

Integrate Eq. (1.2.4) with respect to $\Omega_{s+1}, \ldots, \Omega_N$ with the aim of obtaining the equation for the s-particles distribution function (see [16–21])

$$\int \left(\frac{\partial f_N}{\partial t} + \sum_{i=1}^{N} \mathbf{v}_i \cdot \frac{\partial f_N}{\partial \mathbf{r}_i} + \sum_{i=1}^{N} \mathbf{F}_i \cdot \frac{\partial f_N}{\partial \mathbf{p}_i} \right) d\Omega_{s+1} \ldots d\Omega_N = 0 \tag{1.2.7}$$

Generally speaking on the right-hand side of Eq. (1.2.7) there could be a function $\psi(t, \Omega_1, \ldots, \Omega_s)$, but obviously $\psi(t, \Omega_1, \ldots, \Omega_s) = 0$. Suppose that gas is kept in a vessel which volume is known. This restriction is not of principal significance and can be removed at the final stage of consideration.

Consider

$$\int \frac{\partial f_N}{\partial t} d\Omega_{s+1} \ldots d\Omega_N. \tag{1.2.8}$$

We can change the order of integrating and differentiating because the limits of integration do not depend on time:

$$\int \frac{\partial f_N}{\partial t} d\Omega_{s+1} \ldots d\Omega_N = \frac{\partial}{\partial t} \int f_N d\Omega_{s+1} \ldots d\Omega_N = \frac{\partial f_s}{\partial t} \tag{1.2.9}$$

Precisely the same manner we change the order of operations in the second term for $i \le s$:

$$\int \sum_{i=1}^{s} \mathbf{v}_i \cdot \frac{\partial f_N}{\partial \mathbf{r}_i} d\Omega_{s+1} \ldots d\Omega_N = \sum_{i=1}^{s} \mathbf{v}_i \cdot \frac{\partial}{\partial \mathbf{r}_i} \int f_N d\Omega_{s+1} \ldots d\Omega_N, \tag{1.2.10}$$

and for $i \leq s$ one obtains

$$\int \sum_{i=1}^{s} \mathbf{v}_i \cdot \frac{\partial f_N}{\partial \mathbf{r}_i} \mathrm{d}\Omega_{s+1} \ldots \mathrm{d}\Omega_N = \sum_{i=1}^{s} \mathbf{v}_i \cdot \frac{\partial f_s}{\partial \mathbf{r}_i} \tag{1.2.11}$$

For $i \geq s+1$ we have

$$\begin{aligned} \int \mathbf{v}_i \cdot \frac{\partial f_N}{\partial \mathbf{r}_i} \mathrm{d}\Omega_{s+1} \ldots d\Omega_N &= \int \mathbf{v}_i \cdot \frac{\partial f_{s+1}(t, \Omega_1, \ldots, \Omega_s, \Omega_i)}{\partial \mathbf{r}_i} \mathrm{d}\Omega_i \\ &= \int \left[\int \left(\frac{\partial f_{s+1} v_{i1}}{\partial r_{i1}} + \frac{\partial f_{s+1} v_{i2}}{\partial r_{i2}} + \frac{\partial f_{s+1} v_{i3}}{\partial r_{i3}} \right) \mathrm{d}r_{i1} \mathrm{d}r_{i2} \mathrm{d}r_{i3} \right] \mathrm{d}\mathbf{v}_i \\ &= \int \left[\int f_{s+1} \mathbf{v}_i \cdot \mathbf{n} \mathrm{d}s \right] \mathrm{d}\mathbf{v}_i. \end{aligned} \tag{1.2.12}$$

In the last equality Gauss-Ostrogradskii theorem is used; the external—to the gas—direction is accepted as positive. Physical sense of the integral in the right side of Eq. (1.2.12) consists in the probability definition of the ith particle penetration through the wall of the vessel when s-particles are in the states $\Omega_1, \ldots, \Omega_s$ respectively. However, the walls of the vessel are accepted as impermeable for the gas molecules and, as a consequence, this integral is equal to zero. This integral is equal to zero also in the case when the gas is found in a restricted area of space, but the integration is realized over an infinitely distant surface.

Let us consider now the last term in Eq. (1.2.7). Let be $i \leq s+1$, then

$$\int \mathbf{F}_i \cdot \frac{\partial f_N}{\partial \mathbf{p}_i} \mathrm{d}\Omega_{s+1} \ldots \mathrm{d}\Omega_N = \mathbf{F}_i \cdot \frac{\partial}{\partial \mathbf{p}_i} \int f_N \mathrm{d}\Omega_{s+1} \ldots \mathrm{d}\Omega_N = \mathbf{F}_i \cdot \frac{\partial f_s}{\partial \mathbf{p}_i} \tag{1.2.13}$$

Suppose further that force $\mathbf{F}_i$ does not depend on velocity of the ith particle or depends in accordance with the Lorentz law. Then for $i \geq s+1$, the following transformation is true

$$\begin{aligned} \int \mathbf{F}_i \cdot \frac{\partial f_N}{\partial \mathbf{p}_i} \mathrm{d}\Omega_{s+1} \ldots \mathrm{d}\Omega_N &= \int \frac{\partial}{\partial \mathbf{p}_i} \cdot (\mathbf{F}_i f_N) \mathrm{d}\Omega_{s+1} \ldots \mathrm{d}\Omega_N \\ &= \int \left[\int \frac{\partial}{\partial \mathbf{p}_i} \cdot (\mathbf{F}_i f_{s+1}) \mathrm{d}\mathbf{v}_i \right] \mathrm{d}\mathbf{r}_i. \end{aligned} \tag{1.2.14}$$

The last integral is equal to zero. Really, the internal integral can be transformed from the volume form to the surface ones in the velocity space. Taking into account that the gas energy is finite, we conclude that probability density f_N should tend to zero, if $v_i \to \infty$. Then the surface integral over infinite distant surface in the velocity space, turns into zero.

Denote the external force acting on the i-particle as $\mathbf{F}_i^b$. We have

$$\mathbf{F}_i = \mathbf{F}_i^b + \sum_{j=1}^{N} \mathbf{F}_{ij}, \quad \mathbf{F}_{ii} = 0, \tag{1.2.15}$$

where $\mathbf{F}_{ij}(\mathbf{r}_i, \mathbf{r}_j)$ is the force acting on the ith particle from the particle j. Taking into account Eqs. (1.2.13)–(1.2.15) the third term of integrand in relation (1.2.7) is written as follows

$$\begin{aligned} \int \sum_{i=1}^{N} \mathbf{F}_i \cdot \frac{\partial f_N}{\partial \mathbf{p}_i} \mathrm{d}\Omega_{s+1} \ldots \mathrm{d}\Omega_N &= \int \sum_{i=1}^{s} \mathbf{F}_i \cdot \frac{\partial f_N}{\partial \mathbf{p}_i} \mathrm{d}\Omega_{s+1} \ldots \mathrm{d}\Omega_N \\ &= \sum_{i=1}^{s} \mathbf{F}_i^b \cdot \frac{\partial f_s}{\partial \mathbf{p}_i} + \sum_{i=1}^{s} \int \sum_{j=1}^{s} \mathbf{F}_{ij} \cdot \frac{\partial f_N}{\partial \mathbf{p}_i} \mathrm{d}\Omega_{s+1} \ldots \mathrm{d}\Omega_N \\ &\quad + \sum_{i=1}^{s} \int \sum_{j=s+1}^{N} \mathbf{F}_{ij} \cdot \frac{\partial f_N}{\partial \mathbf{p}_i} \mathrm{d}\Omega_{s+1} \ldots \mathrm{d}\Omega_N. \end{aligned} \tag{1.2.16}$$

Consider the terms in Eq. (1.2.16)

$$\sum_{i=1}^{s} \int \sum_{j=1}^{s} \mathbf{F}_{ij} \cdot \frac{\partial f_N}{\partial \mathbf{p}_i} \mathrm{d}\Omega_{s+1} \ldots \mathrm{d}\Omega_N = \sum_{i,j=1}^{s} \mathbf{F}_{ij} \cdot \frac{\partial f_s}{\partial \mathbf{p}_i}. \tag{1.2.17}$$

We transform the term

$$\sum_{i=1}^{s}\int\sum_{j=s+1}^{N}\mathbf{F}_{ij}\cdot\frac{\partial f_N}{\partial \mathbf{p}_i}\mathrm{d}\Omega_{s+1}\ldots\mathrm{d}\Omega_N, \tag{1.2.18}$$

at first for the particular case of one-species gas.

With this aim write down the integral

$$\int \mathbf{F}_{ij}f_{s+1}(t,\Omega_1,\ldots,\Omega_s,\Omega_j)\mathrm{d}\Omega_j,\ \ j\geq s+1,\ i=1,\ldots,s. \tag{1.2.19}$$

Physical sense of this integral relates to the averaged force acting on the i-particle from the particle j when s-particles are in the states $\Omega_1,\ldots,\Omega_s$. For one-species gas in the force of the particle identity, the value of this integral for $i\geq s+1$ does not depend on number j, and we obtain relation

$$\int \mathbf{F}_{ij}f_{s+1}(t,\Omega_1,\ldots,\Omega_s,\Omega_j)\mathrm{d}\Omega_j=\int \mathbf{F}_{i,s+1}f_{s+1}(t,\Omega_1,\ldots,\Omega_{s+1})\mathrm{d}\Omega_{s+1},$$
$$j\geq s+1,\ i=1,\ldots,s. \tag{1.2.20}$$

Now transform Eq. (1.2.18) using derived relation (1.2.20):

$$\sum_{i=1}^{s}\int\sum_{j=s+1}^{N}\mathbf{F}_{ij}\cdot\frac{\partial f_N}{\partial \mathbf{p}_i}\mathrm{d}\Omega_{s+1}\ldots\mathrm{d}\Omega_N=\sum_{i=1}^{s}\frac{\partial}{\partial \mathbf{p}_i}\cdot\int\sum_{j=s+1}^{N}\mathbf{F}_{ij}f_{s+1}\mathrm{d}\Omega_j$$
$$=\sum_{i=1}^{s}(N-s)\frac{\partial}{\partial \mathbf{p}_i}\cdot\int \mathbf{F}_{i,s+1}f_{s+1}\mathrm{d}\Omega_{s+1}. \tag{1.2.21}$$

Pay attention now to multi-component gas. In accordance with the chemical kinds of molecules, η groups of particles stand out. Within of each group all the particles are considered as identical. The transformation, analogous to Eq. (1.2.20), can be realized for condition $s\ll N$:

$$\sum_{j=s+1}^{N}\int \mathbf{F}_{ij}f_{s+1}(t,\Omega_1,\ldots,\Omega_s,\Omega_j)\mathrm{d}\Omega_j$$
$$=\sum_{\delta=1}^{\eta}N_\delta\int \mathbf{F}_{i,j\in N_\delta}f_{s+1}(t,\Omega_1,\ldots,\Omega_s,\Omega_{j\in N_\delta})\mathrm{d}\Omega_{j\in N_\delta},$$
$$i=1,\ldots,s. \tag{1.2.22}$$

Some comments to the relation (1.2.22). Le be, in mixture consisted of N-particles, N_δ particles belong to the same species of particles so that $\delta=1,\ldots,\eta$, and $N=\sum_{\delta=1}^{n}N_\delta$. In every group all particles are identical, therefore when integrating over entire physical space and the velocity space it does not matter what number of particle is used in every group. The integration over all $\Omega_{j\in N_\delta}$, means that integration is realized over all phase volume of one of the particles belonging to the group N_δ.

As result for one-species gas the following hierarchy of Bogolubov-Born-Green-Kirkwood-Yvon kinetic equations (BBGKY equations) takes place:

$$\frac{\partial f_s}{\partial t}+\sum_{i=1}^{s}\mathbf{v}_i\cdot\frac{\partial f_s}{\partial \mathbf{r}_i}+\sum_{i=1}^{s}\mathbf{F}_i^b\cdot\frac{\partial f_s}{\partial \mathbf{p}_i}+\sum_{i,j=1}^{s}\mathbf{F}_{i,j}\cdot\frac{\partial f_s}{\partial \mathbf{p}_i}$$
$$=-\sum_{i=1}^{s}(N-s)\frac{\partial}{\partial \mathbf{p}_i}\cdot\int f_{s+1}(t,\Omega_1,\ldots,\Omega_{s+1})\mathbf{F}_{i,s+1}\mathrm{d}\Omega_{s+1}. \tag{1.2.23}$$

The internal forces $\mathbf{F}_{ij}$, that do not depended on the velocity, can be factored out from the derivative by momentum $\mathbf{p}_i$.

Equation BBGKY-1 ($s=1$)

$$\frac{\partial f_1}{\partial t}+\frac{\mathbf{p}_1}{m}\cdot\frac{\partial f_1}{\partial \mathbf{r}_1}+\mathbf{F}_1^b\cdot\frac{\partial f_1}{\partial \mathbf{p}_1}=(1-N)\frac{\partial}{\partial \mathbf{p}_1}\cdot\int \mathbf{F}_{12}f_2\mathrm{d}\Omega_2. \tag{1.2.24}$$

Equation BBGKI-2 ($s=2$)

$$\frac{\partial f_2}{\partial t}+\left(\frac{\mathbf{p}_1}{m}\cdot\frac{\partial f_1}{\partial \mathbf{r}_1}+\frac{\mathbf{p}_2}{m}\cdot\frac{\partial f_2}{\partial \mathbf{r}_2}\right)+\mathbf{F}_1^b\cdot\frac{\partial f_2}{\partial \mathbf{p}_1}+\mathbf{F}_2^b\cdot\frac{\partial f_2}{\partial \mathbf{p}_2}+\mathbf{F}_{12}\cdot\frac{\partial f_2}{\partial \mathbf{p}_1}$$
$$+\mathbf{F}_{21}\cdot\frac{\partial f_2}{\partial \mathbf{p}_2}=(2-N)\left(\frac{\partial}{\partial \mathbf{p}_1}\cdot\int \mathbf{F}_{13}f_3\mathrm{d}\Omega_3+\frac{\partial}{\partial \mathbf{p}_2}\cdot\int \mathbf{F}_{23}f_3\mathrm{d}\Omega_3\right). \tag{1.2.25}$$

For multi-species gas, in the absence of chemical reactions, we have

$$\frac{\partial f_s}{\partial t}+\sum_{i=1}^{s}\mathbf{v}_i\cdot\frac{\partial f_s}{\partial \mathbf{r}_i}+\sum_{i=1}^{s}\mathbf{F}_i^b\cdot\frac{\partial f_s}{\partial \mathbf{p}_i}+\sum_{i,j=1}^{s}\mathbf{F}_{i,j}\cdot\frac{\partial f_s}{\partial \mathbf{p}_i}$$
$$=-\sum_{i=1}^{s}\sum_{j=s+1}^{N}\int \mathbf{F}_{ij}\cdot\frac{\partial}{\partial \mathbf{p}_i}f_{s+1}\left(t,\Omega_1,\ldots,\Omega_s,\Omega_j\right)\mathrm{d}\Omega_j, \tag{1.2.26}$$
$$s=1,\ldots,N.$$

If $s \ll N$ it is reasonable to write down Eq. (1.2.26) with an explicit extraction of species of particles in multi-species mixture of gases:

$$\frac{\partial f_s}{\partial t}+\sum_{i=1}^{s}\mathbf{v}_i\cdot\frac{\partial f_s}{\partial \mathbf{r}_i}+\sum_{i=1}^{s}\mathbf{F}_i^b\cdot\frac{\partial f_s}{\partial \mathbf{p}_i}+\sum_{i,j=1}^{s}\mathbf{F}_{i,j}\cdot\frac{\partial f_s}{\partial \mathbf{p}_i}$$
$$=-\sum_{i=1}^{s}\sum_{\delta=1}^{\eta}N_\delta\int \mathbf{F}_{i,j\in N_\delta}\cdot\frac{\partial}{\partial \mathbf{p}_i}f_{s+1}\left(t,\Omega_1,\ldots,\Omega_s,\Omega_{j\in N_\delta}\right)\mathrm{d}\Omega_{j\in N_\delta} \tag{1.2.27}$$

For reacting gases, Eq. (1.2.26) should be corrected taking into account an additional source integral term related to chemical reactions. Possible approximations for this term are discussed in [21], but generally speaking the form of this term should be based on the quantum theory of inelastic collisions.

The used assumption that, for convenience of consideration, the gas is taken as bounded in a vessel of volume V, is not significant. It is possible to use limit $V\to\infty, N\to\infty$ under the condition $n=N/V=\text{const}$. In this case hierarchy of equations has no changes.

The set of integro-differential equations turns out to be a linked one, so that in the lowest-order approximation the distribution function f_1 depends on f_2. This means formally that, strictly speaking, the solution procedure for such a set should be as follows. First find the distribution function f_N and then solve the set of BBGKY equations subsequently for decreasingly lower-order distributions. But if we know the function f_N, there is no need at all to solve the equations for f_s and it actually suffices to employ the definition (1.2.6) of the function f_s.

We thus conclude that the rigorous solution to the set of BBGKY equations is again equivalent to solving Liouville equations. On the other hand, the seemingly illogical solution procedure involving a search for the distribution function f_1 is of great significance in kinetic theory and in non-equilibrium statistical mechanics. This approach involves breaking the BBGKY chain by introducing certain additional assumptions (which have a clear physical meaning, though). These assumptions are discussed in detail below.

1.3 DERIVATION OF THE GENERALIZED BOLTZMANN EQUATION

We now proceed the derivation of the GBE [21, 30–32, 69] by applying Bogolyubov's procedure and writing down once more Eq. (1.2.26), introducing the forces acting on the unit of mass of the particles. We conserve the previous notations for this kind of forces that cannot lead to misunderstandings.

$$\frac{\partial f_s}{\partial t}+\sum_{i=1}^{s}\mathbf{v}_i\cdot\frac{\partial f_s}{\partial \mathbf{r}_i}+\sum_{i=1}^{s}\mathbf{F}_i^b\cdot\frac{\partial f_s}{\partial \mathbf{v}_i}+\sum_{i,j=1}^{s}\mathbf{F}_{ij}\cdot\frac{\partial f_s}{\partial \mathbf{v}_i}$$
$$=-\sum_{i=1}^{s}\sum_{j=s+1}^{N}\int \mathbf{F}_{ij}\cdot\frac{\partial}{\partial \mathbf{v}_i}f_{s+1}\mathrm{d}\Omega_j, \tag{1.3.1}$$

where $\mathrm{d}\Omega_j=\mathrm{d}\mathbf{r}_j\mathrm{d}\mathbf{v}_j$, f_N is normalized on a unit. It means that probability to find all N-particles of physical system at some points of physical space with some velocities is equal to a unit,

$$\int f_N \mathrm{d}\Omega_1 \mathrm{d}\Omega_2 \ldots \mathrm{d}\Omega_N = 1 \tag{1.3.2}$$

We now write down the dimensionless equation for the one-particle distribution function. In doing so, we follow the multi-scale method and introduce three groups of scales:

at the r_b level—the particle interaction radius r_b, $r_b = (V_b)^{1/3}$, where V_b is the volume of particles interaction, which could be non-spherical in the general case; the characteristic collision velocity v_{ob}, and the characteristic collision time r_b/v_{Ob};
at the λ level—the mean free path λ, the mean free-flight velocity $v_{0\lambda}$, and the characteristic time scale $\lambda/v_{0\lambda}$, and
at the L level—the characteristic hydrodynamic dimension L, the hydrodynamic velocity v_{0L}, and the hydrodynamic time L/v_{0L}.

Other notations: V—the character volume of physical system, F_0 is the scale of molecular interaction, $F_{o\lambda}$ is the scale of external forces.

A hat "∧"over a symbol means that the quantity labeled like this is made dimensionless. From the normalizing condition (1.3.2) follows, that the dimensionless s-particles distribution function $\hat{f}_s$ can be written as

$$\hat{f}_s = f_s v_{ob}^{3s} V^s. \tag{1.3.3}$$

For other values we have

$$\hat{F}_{ij} = \frac{F_{ij}}{F_0}, \hat{t}_b = \frac{t}{r_b v_{ob}^{-1}}, \hat{\mathbf{r}}_{ib} = \frac{\mathbf{r}_i}{r_b}, \hat{\mathbf{F}}_i = \frac{\mathbf{F}_i}{F_{0\lambda}}.$$

Equation (1.3.1) can be rewritten in the form:

$$\begin{aligned}&\frac{v_{ob}}{r_b}\frac{\partial \hat{f}_s}{\partial \hat{t}_b} + \frac{v_{ob}}{r_b}\sum_{i=1}^{s} \hat{\mathbf{v}}_{ib} \cdot \frac{\partial \hat{f}_s}{\partial \hat{\mathbf{r}}_b} + \frac{F_{0\lambda}}{v_{ob}}\sum_{i=1}^{s} \hat{\mathbf{F}}_i \cdot \frac{\partial \hat{f}_s}{\partial \hat{\mathbf{v}}_{ib}} + \frac{F_0}{v_{ob}}\sum_{ij=1}^{s} \hat{\mathbf{F}}_{ij} \cdot \frac{\partial \hat{f}_s}{\partial \hat{\mathbf{v}}_{ib}} \\ &= -\frac{F_0}{v_{ob}}\left(v_{ob}^{-3} V^{-1}\right) r_b^3 v_{ob}^3 \sum_{i=1}^{s}\sum_{j=s+1}^{N} \int \hat{\mathbf{F}}_{ij} \cdot \frac{\partial}{\partial \hat{\mathbf{v}}_{ib}} \hat{f}_{s+1} \mathrm{d}\hat{\Omega}_j\end{aligned} \tag{1.3.4}$$

and after dividing of both sides of equations by v_{ob}/r_b, we find

$$\begin{aligned}&\frac{\partial \hat{f}_s}{\partial \hat{t}_b} + \sum_{i=1}^{s} \hat{\mathbf{v}}_{ib} \cdot \frac{\partial \hat{f}_s}{\partial \hat{\mathbf{r}}_b} + \frac{F_{0\lambda}}{\left(v_{ob}^2/r_b\right)}\sum_{i=1}^{s} \hat{\mathbf{F}}_i \cdot \frac{\partial \hat{f}_s}{\partial \hat{\mathbf{v}}_{ib}} + \frac{F_0}{\left(v_{ob}^2/r_b\right)}\sum_{ij=1}^{s} \hat{\mathbf{F}}_{ij} \cdot \frac{\partial \hat{f}_s}{\partial \hat{\mathbf{v}}_{ib}} \\ &= -\frac{r_b^3}{V}\frac{F_0}{\left(v_{ob}^2/r_b\right)} \sum_{i=1}^{s}\sum_{j=s+1}^{N} \int \hat{\mathbf{F}}_{ij} \cdot \frac{\partial}{\partial \hat{\mathbf{v}}_{ib}} \hat{f}_{s+1} \mathrm{d}\hat{\Omega}_j.\end{aligned} \tag{1.3.5}$$

As a scale for the force of molecular interaction—recall that this force related to the unit of mass—choose the value v_{ob}^2/r_b, connected with scale of collision velocity; $F_0 = v_{ob}^2/r_b$ and then

$$\begin{aligned}&\frac{\partial \hat{f}_s}{\partial \hat{t}_b} + \sum_{i=1}^{s} \hat{\mathbf{v}}_{ib} \cdot \frac{\partial \hat{f}_s}{\partial \hat{\mathbf{r}}_{ib}} + \alpha\sum_{i=1}^{s} \hat{\mathbf{F}}_i \cdot \frac{\partial \hat{f}_s}{\partial \hat{\mathbf{v}}_{ib}} + \sum_{ij=1}^{s} \hat{\mathbf{F}}_{ij} \cdot \frac{\partial \hat{f}_s}{\partial \hat{\mathbf{v}}_{ib}} \\ &= -\beta \sum_{i=1}^{s}\sum_{j=s+1}^{N} \int \hat{\mathbf{F}}_{ij} \cdot \frac{\partial}{\partial \hat{\mathbf{v}}_{ib}} \hat{f}_{s+1} \mathrm{d}\hat{\Omega}_j,\end{aligned} \tag{1.3.6}$$

where introduced the dimensionless parameters

$$\alpha = F_{0\lambda}/F_0, \beta = r_b^3/V = V_b/V \tag{1.3.7}$$

For a one-particle DF $\hat{f}_1$ equation is valid

$$\frac{\partial \hat{f}_1}{\partial \hat{t}_b} + \hat{\mathbf{v}}_{1b} \cdot \frac{\partial \hat{f}_1}{\partial \hat{\mathbf{r}}_{ib}} + \alpha \hat{\mathbf{F}}_1 \cdot \frac{\partial \hat{f}_s}{\partial \mathbf{v}_{ib}} + \beta \sum_{j=2}^{N} \int \hat{\mathbf{F}}_{ij} \cdot \frac{\partial}{\partial \hat{\mathbf{v}}_1} \hat{f}_2(\Omega_1, \Omega_j, t) \mathrm{d}\hat{\Omega}_j. \tag{1.3.8}$$

If effects of the particle correlations could be completely omitted the simplest approximation for two-particles DF is valid

$$\hat{f}_2(\Omega_1,\Omega_j,t)=\hat{f}_1(\Omega_1,t)\hat{f}_{1j}(\Omega_j,t) \tag{1.3.9}$$

where j is number of particle, $j=2,\ldots,N$. From (1.3.8) it follows that

$$\begin{aligned}&\frac{\partial\hat{f}_1}{\partial\hat{t}_b}+\hat{\mathbf{v}}_{1b}\cdot\frac{\partial\hat{f}_1}{\partial\hat{\mathbf{r}}_{ib}}+\alpha\hat{\mathbf{F}}_1\cdot\frac{\partial\hat{f}_1}{\partial\hat{\mathbf{v}}_{ib}}\\&+\beta\frac{\partial\hat{f}_1}{\partial\hat{\mathbf{v}}_{ib}}\cdot\sum_{j=2}^{N}\int\hat{\mathbf{F}}_{1j}\hat{f}_{1,j}(\Omega_j,t)\,\mathrm{d}\hat{\Omega}_j=0.\end{aligned} \tag{1.3.10}$$

The integral term in Eq. (1.3.10) is connected with the averaged force $\hat{\mathbf{F}}_1^a$, acting on the first particle from the other particles of the system:

$$\hat{\mathbf{F}}_1^a=\sum_{j=2}^{N}\int\hat{\mathbf{F}}_{ij}\hat{f}_{1,j}(\Omega_j,t)\,\mathrm{d}\hat{\Omega}_j. \tag{1.3.11}$$

The relation for the averaged force can be rewritten for a multi-component system consisting of η species, as follows

$$\hat{\mathbf{F}}_1^a=\sum_{\delta=1}^{\eta}N_\delta\int\hat{\mathbf{F}}_{1,j\in N_\delta}\hat{f}_{1,j\in N_\delta}(\Omega_j,t)\,\mathrm{d}\hat{\Omega}_{j\in N_\delta}. \tag{1.3.12}$$

Using Eqs. (1.3.11) and (1.3.12), one obtains from Eq. (1.3.10)

$$\frac{\partial\hat{f}_1}{\partial\hat{t}_b}+\hat{\mathbf{v}}_{1b}\cdot\frac{\partial\hat{f}_1}{\partial\hat{\mathbf{r}}_{1b}}+\hat{\mathbf{F}}_1^{sc}\cdot\frac{\partial\hat{f}_1}{\partial\hat{\mathbf{v}}_{1b}}=0, \tag{1.3.13}$$

where the self-consistent force $\hat{\mathbf{F}}_1^{sc}$, acting on the particle numbered one, is introduced

$$\hat{\mathbf{F}}_1^{sc}=\alpha\hat{\mathbf{F}}_1+\beta\hat{\mathbf{F}}_1^a. \tag{1.3.14}$$

Kinetic equation (1.3.13) is a Vlasov equation widely used in plasma physics together with electrodynamics equations, which close the system of equations. In molecular dynamics of neutral gases the self-consistent forces significant only on the r_b-scale—usually the smallest ones from all possible kinetic scales—and equation (1.3.13) is the same as the free molecular limit of the Boltzmann equation.

In what follows, we intend, in particular, to construct the generalized Boltzmann physical kinetics outlining the statistical features of so-called, "rarefied" gases. Under the term "rarefied gases" we understand the physical systems for description of which one-particle distribution function f_1 is sufficient.

For multi-component mixture Eq. (1.3.8) can be written as

$$\begin{aligned}&\frac{\partial\hat{f}_1}{\partial\hat{t}_b}+\hat{\mathbf{v}}_{1b}\cdot\frac{\partial\hat{f}_1}{\partial\hat{\mathbf{r}}_{1b}}+\alpha\hat{\mathbf{F}}_1\cdot\frac{\partial\hat{f}_1}{\partial\hat{\mathbf{v}}_{1b}}\\&=-\frac{r_b^3N}{V}\sum_{\delta=1}^{\eta}\frac{N_\delta}{N}\int\hat{\mathbf{F}}_{ij\in N_\delta}\cdot\frac{\partial}{\partial\hat{\mathbf{v}}_b}\hat{f}_{2,j\in N_\delta}(\Omega_j,t)\,\mathrm{d}\hat{\Omega}_{j\in N_\delta}.\end{aligned} \tag{1.3.15}$$

For rarefied gas the value

$$\varepsilon=\frac{r_b^3N}{V}=nr_b^3=nV_b, \tag{1.3.16}$$

defining the particles number in the volume of interaction, is a small parameter and ratio N_δ/N the number density of species in the physical system is given.

In the kinetic theory another conditions of normalizations of DF f_s apart from Eq. (1.3.2), first of all the DF, are normalized to number density. Consider this question in details.

Let

$$\int\tilde{f}_N\mathrm{d}\Omega_1\ldots\mathrm{d}\Omega_N=N^N, \tag{1.3.2$'$}$$

where N is the total number of particles in the system which occupies the volume V, $\mathrm{d}\Omega_i=\mathrm{d}\mathbf{v}_i\mathrm{d}\mathbf{r}_i$.

For the s-particles distribution function, we have

$$\int \widetilde{f}_s \mathrm{d}\Omega_1 \ldots \mathrm{d}\Omega_s = N^s, \tag{1.3.17}$$

and finally for the one-particle DF we have

$$\int \widetilde{f}_1 \mathrm{d}\Omega_1 = N, \tag{1.3.18}$$

Since

$$\int n \mathrm{d}\mathbf{r} = N, \tag{1.3.19}$$

where n is the number density, we get

$$\int \left(\int \widetilde{f}_1 \mathrm{d}\mathbf{v}_1 \right) \mathrm{d}\mathbf{r} = N \tag{1.3.20}$$

and

$$\int \widetilde{f}_1 \mathrm{d}\mathbf{v}_1 = n. \tag{1.3.21}$$

With help of the DF $\widetilde{f}_s$ Eq. (1.3.1) is written as

$$\begin{aligned} &\frac{\partial \widetilde{f}_s}{\partial t_b} + \sum_{i=1}^{s} \mathbf{v}_i \cdot \frac{\partial \widetilde{f}_s}{\partial \mathbf{r}_i} + \sum_{i=1}^{s} \mathbf{F}_i \cdot \frac{\partial \widetilde{f}_s}{\partial \mathbf{v}_i} + \sum_{ij=1}^{s} \mathbf{F}_{ij} \cdot \frac{\partial \widetilde{f}_s}{\partial \mathbf{v}_i} \\ &= -\frac{1}{N} \sum_{i=1}^{s} \sum_{j=s+1}^{N} \int \mathbf{F}_{ij} \cdot \frac{\partial}{\partial \mathbf{v}_i} \widetilde{f}_{s+1} \mathrm{d}\Omega_{s+1}, \end{aligned} \tag{1.3.1$'$}$$

because $\widetilde{f}_s = N^s f_s$.

Write down Eq. (1.3.1′) in a dimensionless form; introduce the same scales, however recall that the DF $\widetilde{f}_s$ was normalized in another way (compare with Eq. (1.3.3)):

$$\hat{\widetilde{f}}_s = \widetilde{f}_s \frac{v_0^{3s}}{n^s}. \tag{1.3.3$'$}$$

Write down the analog of Eq. (1.3.4)

$$\begin{aligned} &\frac{n^s v_0}{v_0^{3s} r_b} \left\{ \frac{\partial \hat{\widetilde{f}}_s}{\partial \hat{t}_b} + \sum_{i=1}^{s} \hat{\mathbf{v}}_{ib} \cdot \frac{\partial \hat{\widetilde{f}}_s}{\partial \hat{\mathbf{r}}_{ib}} + \alpha \sum_{i=1}^{s} \hat{\mathbf{F}}_i \cdot \frac{\partial \hat{\widetilde{f}}_s}{\partial \hat{\mathbf{v}}_{ib}} + \sum_{ij=1}^{s} \hat{\mathbf{F}}_{ij} \cdot \frac{\partial \hat{\widetilde{f}}_s}{\partial \hat{\mathbf{v}}_{ib}} \right\} \\ &= -\frac{F_0}{v_0} \frac{n^{s+1}}{v_0^{3s+3}} \frac{1}{N} \sum_{i=1}^{s} \sum_{j=s+1}^{N} \int \hat{\mathbf{F}}_{ij} \cdot \frac{\partial}{\partial \hat{\mathbf{v}}_{ib}} \hat{\widetilde{f}}_{s+1} \mathrm{d}\hat{\Omega}_{s+1} v_0^3 r_b^3 \end{aligned} \tag{1.3.4$'$}$$

or, after canceling the factor before the curly brackets, we find

$$\begin{aligned} &\frac{\partial \hat{\widetilde{f}}_s}{\partial \hat{t}_b} + \sum_{i=1}^{s} \hat{\mathbf{v}}_{ib} \cdot \frac{\partial \hat{\widetilde{f}}_s}{\partial \hat{\mathbf{r}}_{ib}} + \alpha \sum_{i=1}^{s} \hat{\mathbf{F}}_i \cdot \frac{\partial \hat{\widetilde{f}}_s}{\partial \hat{\mathbf{v}}_{ib}} + \sum_{ij=1}^{s} \hat{\mathbf{F}}_{ij} \cdot \frac{\partial \hat{\widetilde{f}}_s}{\partial \hat{\mathbf{v}}_{ib}} \\ &= -n r_b^3 \frac{F_0}{\left(v_0^2/r_b\right)} \frac{1}{N} \sum_{i=1}^{s} \sum_{j=s+1}^{N} \int \hat{\mathbf{F}}_{ij} \cdot \frac{\partial}{\partial \hat{\mathbf{v}}_{ib}} \hat{\widetilde{f}}_{s+1} \mathrm{d}\hat{\Omega}_{s+1}. \end{aligned} \tag{1.3.22}$$

As before the value v_0^2/r_b is chosen as the scale for the force F_0 and for multi-component mixture we find the analog of Eq. (1.3.15)

$$\begin{aligned} &\frac{\partial \hat{\widetilde{f}}_1}{\partial \hat{t}_b} + \hat{\mathbf{v}}_{1b} \cdot \frac{\partial \hat{\widetilde{f}}_1}{\partial \hat{\mathbf{r}}_{1b}} + \alpha \hat{\mathbf{F}}_1 \cdot \frac{\partial \hat{\widetilde{f}}_1}{\partial \hat{\mathbf{v}}_{1b}} \\ &= -\varepsilon \sum_{\delta=1}^{\eta} \frac{N_\delta}{N} \int \hat{\mathbf{F}}_{1,j\in N_\delta} \cdot \frac{\partial}{\partial \hat{\mathbf{v}}_{1b}} \hat{\widetilde{f}}_{2,j\in N_\delta} \left(\Omega_1, \Omega_{j\in N_\delta}, t\right) d\hat{\Omega}_{j\in N_\delta}, \end{aligned} \tag{1.3.23}$$

In the following transformation of Eq. (1.3.23) for simplicity we omit the sign $\sim$ from DF returning to the details of normalization in the case of necessity.

Of the numerous scales involved in the gas kinetics problems, three major groups of scales pertaining to length, time, and velocity deserve a special consideration. In this case the particle interaction scale r_b presents only one of the scales (and the shortest) in the scale hierarchy in molecular systems, where the λ scale, related to the particle mean free path and the hydrodynamic L-scale—for example, the length or diameter of the flow channel, the characteristic size of the streamlined body, etc.—always exist.

In gas dynamics, the conditions

$$r_b \ll \lambda \ll L \tag{1.3.24}$$

are usually satisfied. If desired, inequalities (1.3.24) can be rewritten in terms of such parameters as the characteristic collision time, mean free time, and hydrodynamic flow time. Since the Boltzmann equation is valid only on the λ and L-scales, a fundamental problem arises here how to adequately describe kinetic processes at all the three scales of a system's evolution.

In Section 1.1 we were able to ascertain that standard perturbation methods can not present satisfactory results, if the system's evolution should be investigated on very different time scales. In this case, it is natural to apply the method of many scales.

We assume that the arguments of the s-particle function $\hat{f}_s$ are the above three groups of scaled variables and the mentioned small parameter $\varepsilon = nr_b^3$ (compare with Eq. (1.1.21)):

$$\hat{f}_s = \hat{f}_s(\hat{t}_b, \hat{\mathbf{r}}_{ib}, \hat{\mathbf{v}}_{ib}; \hat{t}_\lambda, \hat{\mathbf{r}}_{i\lambda}, \hat{\mathbf{v}}_{i\lambda}; \hat{t}_L, \hat{\mathbf{r}}_{iL}, \hat{\mathbf{v}}_{iL}; \varepsilon). \tag{1.3.25}$$

As it has been indicated, a hat over a symbol means that the quantity labeled like this is made dimensionless. We now write down an asymptotic series for the function $\hat{f}_s$:

$$\hat{f}_s = \sum_{\nu=0}^{\infty} \hat{f}_s^{\nu}(\hat{t}_b, \hat{\mathbf{r}}_{ib}, \hat{\mathbf{v}}_{ib}; \hat{t}_\lambda, \hat{\mathbf{r}}_{i\lambda}, \hat{\mathbf{v}}_{i\lambda}; \hat{t}_L, \hat{\mathbf{r}}_{iL}, \hat{\mathbf{v}}_{iL})\varepsilon^{\nu}, \tag{1.3.26}$$

which should be used for solution of equation

$$\begin{aligned} &\frac{\partial \hat{f}_s}{\partial \hat{t}_b} + \sum_{i=1}^{s} \hat{\mathbf{v}}_{ib} \cdot \frac{\partial \hat{f}_s}{\partial \hat{\mathbf{r}}_{ib}} + \sum_{ij=1}^{s} \hat{\mathbf{F}}_{ij} \cdot \frac{\partial \hat{f}_s}{\partial \hat{\mathbf{v}}_{ib}} + \alpha \sum_{i=1}^{s} \hat{\mathbf{F}}_{\mathbf{i}} \cdot \frac{\partial \hat{f}_s}{\partial \hat{\mathbf{v}}_{\mathbf{ib}}} \\ &= -\varepsilon \frac{1}{N} \sum_{i=1}^{s} \sum_{j=s+1}^{N} \int \hat{\mathbf{F}}_{i,j} \cdot \frac{\partial}{\partial \hat{\mathbf{v}}_{ib}} \hat{f}_{s+1}\left(\hat{t}, \hat{\Omega}_1, \ldots, \hat{\Omega}_s, \hat{\Omega}_j\right) d\hat{\Omega}_j. \end{aligned} \tag{1.3.27}$$

Let us take the derivatives on the left-hand side of the sth BBGKY equation according to the rules intended for taking the derivatives of the composite functions (compare with Eq. (1.1.22)):

$$\frac{d\hat{f}_s}{d\hat{t}_b} = \frac{\partial \hat{f}_s}{\partial \hat{t}_b} + \frac{\partial \hat{f}_s}{\partial \hat{t}_\lambda}\frac{\partial \hat{t}_\lambda}{\partial \hat{t}_b} + \frac{\partial \hat{f}_s}{\partial \hat{t}_L}\frac{\partial \hat{t}_L}{\partial \hat{t}_\lambda}\frac{\partial \hat{t}_\lambda}{\partial \hat{t}_b}, \tag{1.3.28}$$

$$\begin{aligned} \frac{d\hat{f}_s}{d\hat{\mathbf{r}}_{ib}} &= \frac{\partial \hat{f}_s}{\partial \hat{\mathbf{r}}_{ib}} + \sum_{k=1}^{3} \frac{\partial \hat{f}_s}{\partial \hat{r}_{i\lambda,k}} \frac{\partial \hat{r}_{i\lambda,k}}{\partial \hat{\mathbf{r}}_{ib}} \\ &+ \sum_{k,l=1}^{3} \frac{\partial \hat{f}_s}{\partial \hat{r}_{iL,k}} \frac{\partial \hat{r}_{iL,k}}{\partial \hat{r}_{i\lambda,l}} \frac{\partial \hat{r}_{i\lambda,l}}{\partial \hat{\mathbf{r}}_{ib}}, \end{aligned} \tag{1.3.29}$$

$$\begin{aligned} \frac{d\hat{f}_s}{d\hat{\mathbf{v}}_{ib}} &= \frac{\partial \hat{f}_s}{\partial \hat{\mathbf{v}}_{ib}} + \sum_{k=1}^{3} \frac{\partial \hat{f}_s}{\partial \hat{v}_{i\lambda,k}} \frac{\partial \hat{v}_{i\lambda,k}}{\partial \hat{\mathbf{v}}_{ib}} \\ &+ \sum_{k,l=1}^{3} \frac{\partial \hat{f}_s}{\partial \hat{v}_{iL,k}} \frac{\partial \hat{v}_{iL,k}}{\partial \hat{v}_{i\lambda,l}} \frac{\partial \hat{v}_{i\lambda,l}}{\partial \hat{\mathbf{v}}_{ib}}. \end{aligned} \tag{1.3.30}$$

Introduce the following parameters as ratio of the scale factors. Here no limitations are introduced for these parameters.

$$\varepsilon_1 = \frac{\lambda}{L} \text{ (Knudsen number)}, \tag{1.3.31}$$

$$\varepsilon_2 = \frac{v_{0\lambda}}{v_{0b}}, \varepsilon_3 = \frac{v_{0L}}{v_{0\lambda}}. \tag{1.3.32}$$

Then the approximated derivatives can be rewritten as follows:

$$\frac{d\hat{f}_s}{d\hat{t}_b} = \frac{\partial \hat{f}_s}{\partial \hat{t}_b} + \frac{\partial \hat{f}_s}{\partial \hat{t}_\lambda}\varepsilon\varepsilon_2 + \frac{\partial \hat{f}_s}{\partial \hat{t}_L}\varepsilon\varepsilon_1\varepsilon_2\varepsilon_3, \tag{1.3.33}$$

$$\frac{d\hat{f}_s}{d\hat{\mathbf{r}}_{ib}} = \frac{\partial \hat{f}_s}{\partial \hat{\mathbf{r}}_{ib}} + \frac{\partial \hat{f}_s}{\partial \hat{\mathbf{r}}_{i\lambda}}\varepsilon + \frac{\partial \hat{f}_s}{\partial \hat{\mathbf{r}}_{iL}}\varepsilon\varepsilon_1, \tag{1.3.34}$$

$$\begin{aligned} \frac{d\hat{f}_s}{d\hat{\mathbf{v}}_{ib}} &= \frac{\partial \hat{f}_s}{\partial \hat{\mathbf{v}}_{ib}} + \frac{\partial \hat{f}_s}{\partial \hat{\mathbf{v}}_{i\lambda}}\varepsilon\varepsilon_2\frac{F_0}{F_{0\lambda}} \\ &+ \frac{\partial \hat{f}_s}{\partial \hat{\mathbf{v}}_{iL}}\varepsilon\varepsilon_2\frac{F_0}{F_{0\lambda}}\varepsilon_3^{-1}, \end{aligned} \tag{1.3.35}$$

because, for example

$$\begin{aligned} &\frac{\partial \hat{t}_\lambda}{\partial \hat{t}_b} = \frac{M_{t_b}}{M_{t_\lambda}} = \frac{r_b}{v_{ob}}\frac{v_{o\lambda}}{\lambda} = \varepsilon_2\frac{r_b}{\lambda} = \varepsilon_2\frac{r_b}{\left(nr_b^2\right)^{-1}} = \varepsilon\varepsilon_2, \\ &\frac{\partial \hat{t}_L}{\partial \hat{t}_\lambda} = \frac{M_{t_\lambda}}{M_{t_L}} = \frac{v_{oL}\lambda}{v_{0\lambda}L} = \varepsilon_1\varepsilon_3, \\ &\frac{\partial \hat{v}_{i\lambda,k}}{\partial v_{ib,k}} = \frac{M_{v_{ob}}}{M_{v_{o\lambda}}} = \varepsilon\frac{M_{t_\lambda}}{M_{t_b}} = \varepsilon\frac{v_{o\lambda}}{F_{0\lambda}}\frac{F_0}{v_{ob}} = \varepsilon\varepsilon_2\frac{F_0}{F_{0\lambda}}, \end{aligned} \tag{1.3.36}$$

where M denotes the scales of values written as the index. The scales of free mean path and radius of interaction are associated by known relation

$$\lambda = \left(nr_b^2\right)^{-1}; \tag{1.3.37}$$

the following relations are written for scales on internal forces F_0 related to mass unit and the scale of velocity v_{ob}

$$F_0 = \frac{v_{ob}^2}{r_b} \tag{1.3.38}$$

and, analogously

$$F_{o\lambda} = \frac{v_{o\lambda}}{t_\lambda}. \tag{1.3.39}$$

Substitute now the series (1.3.26) in approximated derivations (1.3.33)–(1.3.35) and the obtained relations in Eq. (1.3.27). Furthermore we use the expression for $\hat{f}_{s+1}$ on the right side of Eq. (1.3.27) as the mentioned series. Equating the coefficients of ε^0 and ε^1, ε^2 now yieldsat ε^0:

$$\frac{\partial \hat{f}_s^0}{\partial \hat{t}_b} + \sum_{i=1}^{s}\hat{\mathbf{v}}_{ib}\cdot\frac{\partial \hat{f}_s^0}{\partial \hat{\mathbf{r}}_{ib}} + \sum_{ij=1}^{s}\hat{\mathbf{F}}_{ij}\cdot\frac{\partial \hat{f}_s^0}{\partial \hat{\mathbf{v}}_{ib}} + \alpha\sum_{i=1}^{s}\hat{\mathbf{F}}_i\cdot\frac{\partial \hat{f}_s^0}{\partial \hat{\mathbf{v}}_{ib}} = 0, \tag{1.3.40}$$

at ε^1:

$$\begin{aligned} &\frac{\partial \hat{f}_s^1}{\partial \hat{t}_b} + \sum_{i=1}^{s}\hat{\mathbf{v}}_{ib}\cdot\frac{\partial \hat{f}_s^1}{\partial \hat{\mathbf{r}}_{ib}} + \sum_{ij=1}^{s}\hat{\mathbf{F}}_{ij}\cdot\frac{\partial \hat{f}_s^1}{\partial \hat{\mathbf{v}}_{ib}} + \alpha\sum_{i=1}^{s}\hat{\mathbf{F}}_i\cdot\frac{\partial \hat{f}_s^1}{\partial \hat{\mathbf{v}}_{ib}} \\ &+ \varepsilon_2\frac{\partial \hat{f}_s^0}{\partial \hat{t}_\lambda} + \sum_{i=1}^{s}\hat{\mathbf{v}}_{ib}\cdot\frac{\partial \hat{f}_s^0}{\partial \hat{\mathbf{r}}_{i\lambda}} + \varepsilon_2\frac{F_0}{F_{0\lambda}}\sum_{ij=1}^{s}\hat{\mathbf{F}}_{ij}\cdot\frac{\partial \hat{f}_s^0}{\partial \hat{\mathbf{v}}_{i\lambda}} + \varepsilon_2\sum_{i=1}^{s}\hat{\mathbf{F}}_i\cdot\frac{\partial \hat{f}_s^0}{\partial \hat{\mathbf{v}}_{i\lambda}} \\ &+ \varepsilon_1\varepsilon_2\varepsilon_3\frac{\partial \hat{f}_s^0}{\partial \hat{t}_L} + \varepsilon_1\sum_{i=1}^{s}\hat{\mathbf{v}}_{ib}\cdot\frac{\partial \hat{f}_s^0}{\partial \hat{\mathbf{r}}_{iL}} + \frac{\varepsilon_2}{\varepsilon_3}\frac{F_0}{F_{0\lambda}}\sum_{ij=1}^{s}\hat{\mathbf{F}}_{ij}\cdot\frac{\partial \hat{f}_s^0}{\partial \hat{\mathbf{v}}_{iL}} + \frac{\varepsilon_2}{\varepsilon_3}\sum_{i=1}^{s}\hat{\mathbf{F}}_i\cdot\frac{\partial \hat{f}_s^0}{\partial \hat{\mathbf{v}}_{iL}} \\ &= -\frac{1}{N}\sum_{i=1}^{s}\sum_{j=s+1}^{N}\int\hat{\mathbf{F}}_{ij}\cdot\frac{\partial}{\partial \hat{\mathbf{v}}_{ib}}\hat{f}_{s+1}^0 d\hat{\Omega}_j, \end{aligned} \tag{1.3.41}$$

at ε^2:

$$\begin{aligned}
&\frac{\partial \hat f_s^2}{\partial \hat t_b}+\sum_{i=1}^{s}\hat{\mathbf v}_{ib}\cdot\frac{\partial \hat f_s^2}{\partial \hat{\mathbf r}_{ib}}+\sum_{ij=1}^{s}\hat{\mathbf F}_{ij}\cdot\frac{\partial \hat f_s^2}{\partial \hat{\mathbf v}_{ib}}+\alpha\sum_{i=1}^{s}\hat{\mathbf F}_{i}\cdot\frac{\partial \hat f_s^2}{\partial \hat{\mathbf v}_{ib}}\\
&+\varepsilon_2\frac{\partial \hat f_s^1}{\partial \hat t_\lambda}+\sum_{i=1}^{s}\hat{\mathbf v}_{ib}\cdot\frac{\partial \hat f_s^1}{\partial \hat{\mathbf r}_{i\lambda}}+\varepsilon_2\frac{F_0}{F_{0\lambda}}\sum_{ij=1}^{s}\hat{\mathbf F}_{ij}\cdot\frac{\partial \hat f_s^1}{\partial \hat{\mathbf v}_{i\lambda}}+\varepsilon_2\sum_{i=1}^{s}\hat{\mathbf F}_{i}\cdot\frac{\partial \hat f_s^1}{\partial \hat{\mathbf v}_{i\lambda}}\\
&+\varepsilon_1\varepsilon_2\varepsilon_3\frac{\partial \hat f_s^1}{\partial \hat t_L}+\varepsilon_1\sum_{i=1}^{s}\hat{\mathbf v}_{ib}\cdot\frac{\partial \hat f_s^1}{\partial \hat{\mathbf r}_{iL}}+\frac{\varepsilon_2}{\varepsilon_3}\frac{F_0}{F_{0\lambda}}\sum_{ij=1}^{s}\hat{\mathbf F}_{ij}\cdot\frac{\partial \hat f_s^1}{\partial \hat{\mathbf v}_{iL}}+\frac{\varepsilon_2}{\varepsilon_3}\sum_{i=1}^{s}\hat{\mathbf F}_{i}\cdot\frac{\partial \hat f_s^1}{\partial \hat{\mathbf v}_{iL}}\\
&=-\frac{1}{N}\sum_{i=1}^{s}\sum_{j=s+1}^{N}\int\hat{\mathbf F}_{ij}\cdot\left\{\frac{\partial \hat f_{s+1}^1}{\partial \hat{\mathbf v}_{ib}}+\frac{F_0}{F_{0\lambda}}\varepsilon_2\frac{\partial \hat f_{s+1}^0}{\partial \hat{\mathbf v}_{i\lambda}}+\frac{\varepsilon_2}{\varepsilon_3}\frac{F_0}{F_{0\lambda}}\frac{\partial \hat f_{s+1}^0}{\partial \hat{\mathbf v}_{iL}}\right\}\mathrm d\hat\Omega_j.
\end{aligned}\tag{1.3.42}$$

The next approximations are organized in an analogous way.

At $s=1$ one obtains ($F_{11}\equiv 0$)

$$\frac{\partial \hat f_1^0}{\partial \hat t_b}+\hat{\mathbf v}_{1b}\cdot\frac{\partial \hat f_1^0}{\partial \hat{\mathbf r}_{1b}}+\alpha\hat{\mathbf F}_1\cdot\frac{\partial \hat f_1^0}{\partial \hat{\mathbf v}_{1b}}=0,\tag{1.3.43}$$

$$\begin{aligned}
&\frac{\partial \hat f_1^1}{\partial \hat t_b}+\hat{\mathbf v}_{1b}\cdot\frac{\partial \hat f_1^1}{\partial \hat{\mathbf r}_{1b}}+\alpha\hat{\mathbf F}_1\cdot\frac{\partial \hat f_1^1}{\partial \hat{\mathbf v}_{1b}}+\varepsilon_2\frac{\partial \hat f_1^0}{\partial \hat t_\lambda}+\hat{\mathbf v}_{1b}\cdot\frac{\partial \hat f_1^0}{\partial \hat{\mathbf r}_{1\lambda}}+\varepsilon_2\hat{\mathbf F}_1\cdot\frac{\partial \hat f_1^0}{\partial \hat{\mathbf v}_{1\lambda}}\\
&+\varepsilon_1\varepsilon_2\varepsilon_3\frac{\partial \hat f_1^0}{\partial \hat t_L}+\varepsilon_1\hat{\mathbf v}_{1b}\cdot\frac{\partial \hat f_1^0}{\partial \hat{\mathbf r}_{1L}}+\frac{\varepsilon_2}{\varepsilon_3}\hat{\mathbf F}_1\cdot\frac{\partial \hat f_1^0}{\partial \hat{\mathbf v}_{1L}}\\
&=-\sum_{\delta=1}^{\eta}\frac{N_\delta}{N}\int\hat{\mathbf F}_{1,j\in N_\delta}\cdot\frac{\partial}{\partial \hat{\mathbf v}_{1b}}\hat f^0_{2,j\in N_\delta}\mathrm d\hat\Omega_{j\in N_\delta},
\end{aligned}\tag{1.3.44}$$

$$\begin{aligned}
&\frac{\partial \hat f_1^2}{\partial \hat t_b}+\hat{\mathbf v}_{1b}\cdot\frac{\partial \hat f_1^2}{\partial \hat{\mathbf r}_{1b}}+\alpha\hat{\mathbf F}_1\cdot\frac{\partial \hat f_1^2}{\partial \hat{\mathbf v}_{1b}}+\varepsilon_2\frac{\partial \hat f_1^1}{\partial \hat t_\lambda}+\hat{\mathbf v}_{1b}\cdot\frac{\partial \hat f_1^1}{\partial \hat{\mathbf r}_{1\lambda}}+\varepsilon_2\hat{\mathbf F}_1\cdot\frac{\partial \hat f_1^1}{\partial \hat{\mathbf v}_{1\lambda}}\\
&+\varepsilon_1\varepsilon_2\varepsilon_3\frac{\partial \hat f_1^1}{\partial \hat t_L}+\varepsilon_1\hat{\mathbf v}_{1b}\cdot\frac{\partial \hat f_1^1}{\partial \hat{\mathbf r}_{1L}}+\frac{\varepsilon_2}{\varepsilon_3}\hat{\mathbf F}_1\cdot\frac{\partial \hat f_1^1}{\partial \hat{\mathbf v}_{1L}}\\
&=-\sum_{\delta=1}^{\eta}\frac{N_\delta}{N}\int\hat{\mathbf F}_{1,j\in N_\delta}\cdot\left\{\frac{\partial \hat f_2^1}{\partial \hat{\mathbf v}_{1b}}+\frac{F_0}{F_{0\lambda}}\varepsilon_2\frac{\partial \hat f_2^0}{\partial \hat{\mathbf v}_{1\lambda}}+\frac{\varepsilon_2}{\varepsilon_3}\frac{F_0}{F_{0\lambda}}\frac{\partial \hat f_2^0}{\partial \hat{\mathbf v}_{1L}}\right\}\mathrm d\hat\Omega_{j\in N_\delta}.
\end{aligned}\tag{1.3.45}$$

Some conclusions can be done at this step of investigation.

It follows from Eq. (1.3.40) that on the r_b-scale the function $\hat f_s^0$ has no change along phase trajectory or, otherwise, after integrating on the r_b-scale

$$\hat f_s^0=\hat f_s^0(\hat t_\lambda,\hat{\mathbf v}_{i\lambda},\hat{\mathbf r}_{i\lambda};\hat t_L,\hat{\mathbf v}_{iL},\hat{\mathbf r}_{iL}).\tag{1.3.46}$$

If function (1.3.46) is known, the function $\hat f_s^1$ has to be found from Eq. (1.3.41). This is possible if certain additional assumptions are posed on the function $\hat f_{s+1}^0$ entering the integral right-hand side of expression (1.3.41). Equation (1.3.42) can be used for calculating $\hat f_s^2$ if not only functions $\hat f_s^0,\hat f_s^1$ are known from the preceding equations, but also $\hat f_{s+1}^1$. Thus, we see that the system of equations contains linked terms and if, in Eq. (1.3.41), we need to introduce an assumptions concerning the function $\hat f_{s+1}^0$, then in Eq. (1.3.42)—concerning $\hat f_{s+1}^1$.

In real life, the dependence (1.3.46) is unknown beforehand. Then Eq. (1.3.41) can serve to determine $\hat f_s^0$ on λ- and L-scales, but, in this case, it becomes doubly linked, with respect to both the lower index $(s+1)$ and the upper index (1). As a result, the problem of breaking the linked terms arises.

In the following we intend to deal with systems admitting the one-particle description by means of Eqs. (1.3.43)–(1.3.45). To this ends we transform the integral term in Eq. (1.3.44).

Let us write Eq. (1.3.40) at $s=2$ for two particles "1" and "j".

$$\begin{aligned}
&\frac{\partial \hat f_2^0}{\partial \hat t_b}+\hat{\mathbf v}_{1b}\cdot\frac{\partial \hat f_2^0}{\partial \hat{\mathbf r}_{1b}}+\hat{\mathbf v}_{j\in N_\delta,b}\cdot\frac{\partial \hat f_2^0}{\partial \hat{\mathbf r}_{j\in N_\delta,b}}+\hat{\mathbf F}_{1,j\in N_\delta}\cdot\frac{\partial \hat f_2^0}{\partial \hat{\mathbf v}_{1b}}\\
&+\hat{\mathbf F}_{j\in N_\delta,1}\cdot\frac{\partial \hat f_2^0}{\partial \hat{\mathbf v}_{j\in N_\delta}}+\alpha\hat{\mathbf F}_1\cdot\frac{\partial \hat f_2^0}{\partial \mathbf v_{1b}}+\alpha\hat{\mathbf F}_{j\in N_\delta}\cdot\frac{\partial \hat f_2^0}{\partial \hat{\mathbf v}_{j\in N_\delta,b}}=0.
\end{aligned}\tag{1.3.47}$$

Introducing a new variable $\hat{\mathbf{x}}_{1,j\in N_\delta} = \hat{\mathbf{r}}_{1b} - \hat{\mathbf{r}}_{j\in N_\delta,b}$, we find from Eq. (1.3.47)

$$\begin{aligned} -\hat{\mathbf{F}}_{1,j\in N_\delta} \cdot \frac{\partial}{\partial \hat{\mathbf{v}}_{1b}} \hat{f}^0_{2,j\in N_\delta} &= \frac{\partial \hat{f}^0_2}{\partial \hat{t}_b} + \hat{\mathbf{v}}_{1b} \cdot \frac{\partial \hat{f}^0_2}{\partial \hat{\mathbf{r}}_{1b}} + \left(\hat{\mathbf{v}}_{1b} - \hat{\mathbf{v}}_{j\in N_\delta,b}\right) \cdot \frac{\partial \hat{f}^0_2}{\partial \hat{\mathbf{x}}_{1,j\in N_\delta}} \\ &+ \hat{\mathbf{F}}_{j\in N_\delta,1} \cdot \frac{\partial \hat{f}^0_2}{\partial \hat{\mathbf{v}}_{j\in N_\delta}} + \alpha \hat{\mathbf{F}}_1 \cdot \frac{\partial \hat{f}^0_2}{\partial \hat{\mathbf{v}}_{1b}} + \alpha \hat{\mathbf{F}}_{j\in N_\delta} \cdot \frac{\partial \hat{f}^0_2}{\partial \hat{\mathbf{v}}_{j\in N_\delta,b}}. \end{aligned} \tag{1.3.48}$$

Using the last equation, we obtain the following representation for the integral in Eq. (1.3.44)

$$\begin{aligned} &-\int \hat{\mathbf{F}}_{1,j\in N_\delta} \cdot \frac{\partial}{\partial \hat{\mathbf{v}}_{1b}} \hat{f}^0_{2,j\in N_\delta} \mathrm{d}\hat{\Omega}_{j\in N_\delta} \\ &= \int \left(\hat{\mathbf{v}}_{1b} - \hat{\mathbf{v}}_{j\in N_\delta,b}\right) \cdot \frac{\partial \hat{f}^0_2}{\partial \hat{\mathbf{x}}_{1,j\in N_\delta}} \mathrm{d}\hat{\Omega}_{j\in N_\delta} + \int \left(\frac{\partial \hat{f}^0_2}{\partial \hat{t}_b} + \hat{\mathbf{v}}_{1b} \cdot \frac{\partial \hat{f}^0_2}{\partial \hat{\mathbf{r}}_{1b}} + \alpha \hat{\mathbf{F}}_1 \cdot \frac{\partial \hat{f}^0_2}{\partial \hat{\mathbf{v}}_{1b}} + \alpha \hat{\mathbf{F}}_{j\in N_\delta} \cdot \frac{\partial \hat{f}^0_2}{\partial \hat{\mathbf{v}}_{j\in N_\delta,b}} \right) \mathrm{d}\hat{\Omega}_{j\in N_\delta} \\ &+ \int \hat{\mathbf{F}}_{j\in N_\delta,1} \cdot \frac{\partial \hat{f}^0_2}{\partial \hat{\mathbf{v}}_{j\in N_\delta}} \mathrm{d}\hat{\Omega}_{j\in N_\delta} \end{aligned} \tag{1.3.49}$$

The last integral on the right-hand side of Eq. (1.3.49) can be written in the form

$$\begin{aligned} &\int \hat{\mathbf{F}}_{j\in N_\delta,1} \cdot \frac{\partial \hat{f}^0_2}{\partial \hat{\mathbf{v}}_{j\in N_\delta}} \mathrm{d}\hat{\Omega}_{j\in N_\delta} \\ &= \int \left[\int \frac{\partial}{\partial \hat{\mathbf{v}}_{j\in N_\delta}} \cdot \left(\hat{\mathbf{F}}_{j\in N_\delta,1} \hat{f}^0_2\right) \mathrm{d}\hat{\mathbf{v}}_{j\in N_\delta} \right] d\hat{\mathbf{r}}_{j\in N_\delta}. \end{aligned} \tag{1.3.50}$$

But the inner integral can be transformed by the Gauss theorem into an integral over an infinitely distant surface in the velocity space, which vanishes because $\hat{f}^0_2 \to 0$ as $\hat{v}_j \to \infty$.

Let us introduce an assumption for rarefied gas of neutral particles that in that λ- and L-scales the positions and velocities of particles 1 and j are not correlated, i.e.

$$\hat{f}^0_2\left(\hat{t}, \hat{\Omega}_1, \hat{\Omega}_j\right) = \hat{f}^0_1\left(\hat{t}, \hat{\Omega}_1\right) \hat{f}^0_{j\in N_\delta}\left(\hat{t}, \hat{\Omega}_{j\in N_\delta}\right). \tag{1.3.51}$$

We have

$$\begin{aligned} &\int \left(\frac{\partial \hat{f}^0_2}{\partial \hat{t}_b} + \hat{\mathbf{v}}_{1b} \cdot \frac{\partial \hat{f}^0_2}{\partial \hat{\mathbf{r}}_{1b}} + \alpha \hat{\mathbf{F}}_1 \cdot \frac{\partial \hat{f}^0_2}{\partial \mathbf{v}_{1b}} + \alpha \hat{\mathbf{F}}_{j\in N_\delta} \cdot \frac{\partial \hat{f}^0_2}{\partial \hat{\mathbf{v}}_{j\in N_\delta,b}} \right) \mathrm{d}\hat{\Omega}_{j\in N_\delta} \\ &= \int \left[\hat{f}^0_{j\in N_\delta} \left(\frac{\partial \hat{f}^0_1}{\partial \hat{t}_b} + \hat{\mathbf{v}}_{1b} \cdot \frac{\partial \hat{f}^0_1}{\partial \hat{\mathbf{r}}_{1b}} + \alpha \hat{\mathbf{F}}_1 \cdot \frac{\partial \hat{f}^0_1}{\partial \mathbf{v}_{1b}} \right) \right] \mathrm{d}\hat{\Omega}_{j\in N_\delta} \\ &+ \int \hat{f}^0_1 \frac{\partial \hat{f}^0_{j\in N_\delta}}{\partial \hat{t}_b} \mathrm{d}\hat{\Omega}_j + \alpha \int \frac{\partial}{\partial \mathbf{v}_{j\in N_\delta}} \cdot \left(\hat{\mathbf{F}}_{j\in N_\delta} \hat{f}^0_1 \hat{f}^0_{j\in N_\delta} \right) \mathrm{d}\hat{\Omega}_{j\in N_\delta}. \end{aligned} \tag{1.3.52}$$

In expression (1.3.52), the first integral on the right is zero, because of relation (1.3.43) and the third integral is zero for the same reasons as in Eq. (1.3.50). The second integral is equal to zero only if the influence of self-consistent field can be ignored in comparison with the external forces acting in the physical system. This problem is investigated in following sections devoted to applications of generalized Boltzmann physical kinetics in plasma physics.

Integral term in Eq. (1.3.50) is written as

$$\begin{aligned} &-\int \hat{\mathbf{F}}_{1j\in N_\delta} \cdot \frac{\partial}{\partial \hat{\mathbf{v}}_{1b}} \hat{f}^0_{2,j\in N_\delta} \mathrm{d}\hat{\Omega}_{j\in N_\delta} \\ &= \int \left(\hat{\mathbf{v}}_{1b} - \hat{\mathbf{v}}_{j\in N_\delta,b}\right) \cdot \frac{\partial \hat{f}^0_2}{\partial \hat{\mathbf{x}}_{1,j\in N_\delta}} \mathrm{d}\hat{\Omega}_{j\in N_\delta} \\ &= \int \left(\hat{\mathbf{v}}_{1b} - \hat{\mathbf{v}}_{j\in N_\delta,b}\right) \cdot \frac{\partial \hat{f}^0_2}{\partial \hat{\mathbf{x}}_{1,j\in N_\delta}} \mathrm{d}\hat{\mathbf{r}}_{j\in N_\delta,b} \mathrm{d}\hat{\mathbf{v}}_{j\in N_\delta,b}. \end{aligned} \tag{1.3.53}$$

Using $\hat{\mathbf{x}}_{1,j\in N_\delta}$ as an integrating variable we arrive at the expression for the collision term $\hat{J}^{st,0}$

$$\begin{aligned}\hat{J}^{\mathrm{st},0} &= -\sum_{\delta=1}^{\eta}\frac{N_\delta}{N}\int \hat{\mathbf{F}}_{1j\in N_\delta}\cdot\frac{\partial}{\partial\hat{\mathbf{v}}_{1b}}\hat{f}^0_{2,j\in N_\delta}\mathrm{d}\hat{\Omega}_{j\in N_\delta}\\ &=\sum_{\delta=1}^{\eta}\frac{N_\delta}{N}\int \hat{\mathbf{g}}_{1,j\in N_\delta}\cdot\frac{\partial \hat{f}^0_2}{\partial\hat{\mathbf{x}}_{1,j\in N_\delta}}\mathrm{d}\hat{\mathbf{x}}_{1,j\in N_\delta,b}\mathrm{d}\hat{\mathbf{v}}_{j\in N_\delta,b},\end{aligned} \tag{1.3.54}$$

where relative velocity of particles 1 and j is introduced:

$$\hat{\mathbf{g}}_{j\in N_\delta,1}=\hat{\mathbf{v}}_{j\in N_\delta,b}-\hat{\mathbf{v}}_{1b}. \tag{1.3.55}$$

Introduce the cylindrical coordinate system $\hat{l},\hat{b},\varphi$ with the origin of coordinates in point $\hat{\mathbf{r}}_{1b}$ and the axis $\hat{l}$, parallel to the relative velocity of encountered particles 1 and j; corresponding dimensionless impact parameter $\hat{b}$ and azimuth angle φ,

$$\begin{aligned}\hat{J}^{\mathrm{st},0} &=\sum_{\delta=1}^{\eta}\frac{N_\delta}{N}\int \hat{g}_{j\in N_\delta,1}\left[\int_{-\infty}^{+\infty}\frac{\partial \hat{f}^0_2}{\partial \hat{l}}\mathrm{d}\hat{l}\right]\hat{b}\mathrm{d}\hat{b}\mathrm{d}\varphi\ \mathrm{d}\hat{\mathbf{v}}_{j\in N_\delta,b}\\ &=\sum_{\delta=1}^{\eta}\frac{N_\delta}{N}\int\left[\hat{f}^0_2(+\infty)-\hat{f}^0_2(-\infty)\right]\hat{g}_{j\in N_\delta,1}\hat{b}\mathrm{d}\hat{b}\mathrm{d}\varphi\mathrm{d}\hat{\mathbf{v}}_{j\in N_\delta,b},\end{aligned} \tag{1.3.56}$$

where integrating is realized on the r_b-scale, i.e. the functions $\hat{f}^0_2(+\infty)$ and $\hat{f}^0_2(-\infty)$ are calculated for velocities $\hat{\mathbf{v}}'_{j\in N_\delta},\hat{\mathbf{v}}'_1$ and $\hat{\mathbf{v}}_{j\in N_\delta},\hat{\mathbf{v}}_1$, when particles are placed out of interaction zone. As usual the particle's velocities after collision are indicated by prime. Using assumption that condition of molecular chaos is valid in λ-scale for encountering particles—and, as consequence, the relation (1.3.51)—we obtain

$$\hat{J}^{\mathrm{st},0}=\sum_{\delta=1}^{\eta}\frac{N_\delta}{N}\int\left[\hat{f}^{0\prime}_1\hat{f}^{0\prime}_{j\in N_\delta}-\hat{f}^0_1\hat{f}^0_{j\in N_\delta}\right]\hat{g}_{j\in N_\delta,1}\hat{b}\mathrm{d}\hat{b}\mathrm{d}\varphi\mathrm{d}\hat{\mathbf{v}}_{j\in N_\delta} \tag{1.3.57}$$

the Boltzmann collision integral written for the multi-component gas.

By manipulating Eq. (1.3.44), we obtain

$$\frac{\mathrm{D}_1\hat{f}^1_1}{\mathrm{D}\hat{t}_b}+\frac{\mathrm{d}_1\hat{f}^0_1}{\mathrm{d}\hat{t}_{\lambda,L}}=\hat{J}^{\mathrm{st},0}, \tag{1.3.58}$$

where we have introduced the notation

$$\frac{\mathrm{D}_1\hat{f}^1_1}{\mathrm{D}\hat{t}_b}=\frac{\partial\hat{f}^1_1}{\partial\hat{t}_b}+\hat{\mathbf{v}}_{1b}\cdot\frac{\partial\hat{f}^1_1}{\partial\hat{\mathbf{r}}_{1b}}+\alpha\hat{\mathbf{F}}_1\cdot\frac{\partial\hat{f}^1_1}{\partial\hat{\mathbf{v}}_{1b}}, \tag{1.3.59}$$

$$\begin{aligned}\frac{\mathrm{d}_1\hat{f}^0_1}{\mathrm{d}\hat{t}_{\lambda,L}}&=\varepsilon_2\frac{\partial\hat{f}^0_1}{\partial\hat{t}_\lambda}+\hat{\mathbf{v}}_{1b}\cdot\frac{\partial\hat{f}^0_1}{\partial\hat{\mathbf{r}}_{1\lambda}}+\varepsilon_2\hat{\mathbf{F}}_1\cdot\frac{\partial\hat{f}^0_1}{\partial\hat{\mathbf{v}}_{1\lambda}}+\varepsilon_1\varepsilon_2\varepsilon_3\frac{\partial\hat{f}^0_1}{\partial\hat{t}_L}\\ &\quad+\varepsilon_1\hat{\mathbf{v}}_{1b}\cdot\frac{\partial\hat{f}^0_1}{\partial\hat{\mathbf{r}}_{1L}}+\frac{\varepsilon_2}{\varepsilon_3}\hat{\mathbf{F}}_1\cdot\frac{\partial\hat{f}^0_1}{\partial\hat{\mathbf{v}}_{1L}}.\end{aligned} \tag{1.3.60}$$

The following remarks are of fundamental importance in connection with the theory being developed.

(1) Until now no restrictions are imposed on the values of $\varepsilon_1,\varepsilon_2,\varepsilon_3$, including the Knudsen number ε_1.

(2) Equation (1.3.58) contains linking not only with respect to the lower but also to the upper index, implying that in order to employ the kinetic equation, additional assumptions should be made to reduce the equation to one dependent variable.

(3) The collision integral $\hat{J}^{\mathrm{st},0}$ transforms to the Boltzmann collision integral if the pair correlation functions in the zero-order ε-expansion vanish and if one can ignore, at the r_b scale, the explicit effect, on a given trial particle, of the self-consistent force of internal origin. We shall address this point in more detail below, when discussing the relationship between the generalized Boltzmann equations and alternative derivations of kinetic equations. The zero-order two-particle distribution function entering the Boltzmann collision integral is calculated at the λ scale and is presented, as usual, as a product of zero-order one-particle functions; this means that interacting particles are not correlated prior to a collision.

(4) The use of this representation makes it possible to express the collision integral $\hat{J}^{\mathrm{st},0}$ in the Boltzmannian form. The presence of superscript "0" in $\hat{J}^{\mathrm{st},0}$ is physically meant that even though the variation of the distribution function on the r_b scale is taken into account [the first term on the left-hand side of Eq. (1.3.58)], the form of the Boltzmann collision integral containing the function f_1^0 remains unchanged.

(5) It is crucial that the term $D_1\hat{f}_1^1/D\hat{t}_B$ in Eq. (1.3.58), accounting for the variation of the distribution function on the r_b scale, is of *the same order* of magnitude as the λ- and L-scale terms. This has nothing to do with whatever approximations for $D_1\hat{f}_1^1/D\hat{t}_B$ may later be made to break the Bogolyubov chain. The (unjustified) formal neglect of the term $D_1\hat{f}_1^1/D\hat{t}_B$ reduces Eq. (1.3.58) to the Boltzmann equation. This means, in turn, that the r_b-scale distribution function is left out of consideration in the Boltzmann kinetic theory; particles featuring in the Boltzmann kinetic theory are point-like and structureless. The system can be described in terms of the independent variables **r**, **p**, t, and the change in the distribution function due to collisions is instantaneous and is accounted for by the source term $\hat{J}^{\mathrm{st},0}$.

We intend to employ Eq. (1.3.58) for describing the evolution of the distribution function $\hat{f}_1^0$ on λ- and L-scales. But kinetic equation (1.3.58) contains the linking term $D_1\hat{f}_1^1/D\hat{t}_b$ with respect to the upper index, implying that in order to employ the kinetic equation. The problem arises concerning of approximation this term, in a definite sense analogous to the problem, which led us to approximations (1.3.9) or (1.3.51).

We now proceed to break the Bogolyubov chain at the r_b – scale with respect to the superscript in $D_1\hat{f}_1^1/D\hat{t}_b$. This term allows the exact representation using the series (1.3.26).

$$\frac{D_1\hat{f}_1^1}{D\hat{t}_b}=\frac{D_1}{D\hat{t}_b}\left[\frac{\partial\hat{f}_1}{\partial\varepsilon}\right]_{\varepsilon=0}. \tag{1.3.61}$$

Note, however, that in the "field" description the distribution function f_1 in the interaction on the r_b-scale depends on ε through the dynamical variables **r**, **v**, t, interrelated by the laws of classical mechanics. We can therefore use the approximation

$$\begin{aligned}\frac{D_1}{D\hat{t}_b}\left[\left(\frac{\partial\hat{f}_1}{\partial\varepsilon}\right)_{\varepsilon=0}\right]&\cong\frac{D_1}{D(-\hat{t}_b)}\left[\frac{\partial\hat{f}_1}{\partial(-\hat{t}_b)}\left(\frac{\partial(-\hat{t}_b)}{\partial\varepsilon}\right)_{\varepsilon=0}\right.\\&\left.+\frac{\partial\hat{f}_1}{\partial\hat{\mathbf{r}}_b}\cdot\frac{\partial\hat{\mathbf{r}}_b}{\partial(-\hat{t}_b)}\left(\frac{\partial(-\hat{t}_b)}{\partial\varepsilon}\right)_{\varepsilon=0}+\frac{\partial\hat{f}_1}{\partial\hat{\mathbf{v}}_b}\cdot\frac{\partial\hat{\mathbf{v}}_b}{\partial(-\hat{t}_b)}\left(\frac{\partial(-\hat{t}_b)}{\partial\varepsilon}\right)_{\varepsilon=0}\right]\\&=-\frac{D_1}{D\hat{t}_b}\left[\left(\frac{\partial\hat{t}_b}{\partial\varepsilon}\right)_{\varepsilon=0}\frac{D_1\hat{f}_1}{D\hat{t}_b}\right]\cong-\frac{D_1}{D\hat{t}_b}\left[\left(\frac{\partial\hat{t}_b}{\partial\varepsilon}\right)_{\varepsilon=0}\frac{D_1\hat{f}_1^0}{D\hat{t}_b}\right].\end{aligned} \tag{1.3.62}$$

The approximation introduced here proceeds *against* the course of time and corresponds to the condition that there be no correlations as to $t_0\rightarrow-\infty$, where is some instant of time t_0 on the r_b-scale at which the particles start to interact with each other. In the Boltzmann kinetic theory, the condition of correlation weakening has the form [16]

$$\lim_{t_0\rightarrow-\infty}W_2[\mathbf{r}_1-\mathbf{v}_1(t-t_0),\mathbf{v}_1;\mathbf{r}_2-\mathbf{v}_2(t-t_0),\mathbf{v}_2;t_0]=0, \tag{1.3.63}$$

where W_2 is the pair correlation function. For $t_0\rightarrow-\infty$ (but not for $t_0\rightarrow+\infty$!), the condition (1.3.63) of correlation weakening, together with the approximation (1.3.62), single out a time direction and lead to the time irreversibility in the real physical processes [31]. The next sections of this chapter discuss this point in detail in connection with the proof of the generalized H-theorem. In the following we will turn back to the chain of relations (1.3.62) discussing this approximation from disparate positions.

Return to the dimensional form of Eq. (1.3.44) taking into account new normalization condition for DF f_1, i.e.

$$\tilde{\tilde{f}}_\delta=\tilde{f}_1\frac{N_\delta}{N}. \tag{1.3.64}$$

Then

$$\int\tilde{\tilde{f}}_\delta d\mathbf{v}_\delta d\mathbf{r}=\int\tilde{f}_1 d\mathbf{v}\,d\mathbf{r}\,\frac{N_\delta}{N}=N_\delta, \tag{1.3.65}$$

or

$$\int\tilde{\tilde{f}}_\delta d\mathbf{v}_\delta=n_\delta, \tag{1.3.66}$$

and

$$\int n_\delta \mathrm{d}\mathbf{r} = N_\delta. \tag{1.3.67}$$

The DF normalization introduced for multi-component mixture allows us to treat this function $\overset{\approx}{f}_\delta$ as mathematical expectation of a number of particles (species δ) for which centers of mass at time moment t are placed the in the unit volume in the vicinity of $\mathbf{r}$, and velocities belong to the unit interval in the vicinity of $\mathbf{v}_\delta$. This condition will be used in what follows as a rule, and as a consequence of this fact we omit the upper sign $\approx$. But the lower index "1" in the DF f_1 corresponds not only to the notation of the one-particle DF but also the number of particle among all the numbered N-particles of the physical multi-component system considered. This lower index also will be omitted, as result, for the η-species mixture we have

$$\int f_\alpha \mathrm{d}\mathbf{v}_\alpha = n_\alpha, \quad \alpha = 1, \ldots, \eta. \tag{1.3.68}$$

Let the particle, indicated as number one in the system, belongs to species α in the mixture, it should be reflected by investigation of the physical sense of the parameter $(\partial \hat{t}_b / \partial \varepsilon)_{\varepsilon=0}$ in approximation (1.3.62).

The parameter

$$\hat{\tau} = \left[\frac{\partial \hat{t}}{\partial \varepsilon}\right]_{\varepsilon=0} \tag{1.3.69}$$

is the non-local parameter introduced in the *limit of the local* approximation (1.3.62) for *non-local* effects.

Let us return now to the dimension form of the kinetic equation (1.3.58). It is not the trivial procedure and we indicate the main transformations. In the corresponding equation

$$\begin{aligned} &\frac{\mathrm{D}_1 \hat{f}_1^1}{\mathrm{D}\hat{t}_b} + \frac{\mathrm{d}_1 \hat{f}_1^0}{\mathrm{d}\hat{t}_{\lambda,L}} \\ &= \sum_{\delta=1}^{\eta} \frac{N_\delta}{N} \int \left[\hat{f}_1^{0'} \hat{f}_{j\in N_\delta}^{0'} - \hat{f}_1^0 \hat{f}_{j\in N_\delta}^0\right] \hat{g}_{j\in N_\delta,1} \hat{b} \mathrm{d}\hat{b} \mathrm{d}\varphi \mathrm{d}\hat{\mathbf{v}}_{j\in N_\delta}, \end{aligned} \tag{1.3.70}$$

the scales $\hat{t}_b = \frac{t}{r_b v_{ob}^{-1}}, \hat{t}_\lambda = n_0 r_b^2 v_0 t, \hat{\mathbf{r}}_{ib} = \frac{\mathbf{r}_i}{r_b}$, $\hat{f}_1 = f_1 (v_0^3 / n_0)$, $\hat{F}_{ij} = F_{ij} / F_0$, and $\hat{\mathbf{F}}_i^b = \mathbf{F}_i^b / F_{0\lambda}$ are used for the dimensionless values in Eq. (1.3.70). We have the following chain of the transformations (1.3.71)–(1.3.75):

$$\begin{aligned} &\frac{\mathrm{D}_1 f_1^1}{\mathrm{D}\hat{t}_b} + \left(\frac{1}{v_0 n_0 r_b^2}\right) \frac{\mathrm{D}_1 f_1^0}{\mathrm{D}t} \\ &= \sum_{\delta=1}^{\eta} \frac{N_\delta}{N} \int \left[f_1^{0'} \hat{f}_{j\in N_\delta}^{0'} - f_1^0 \hat{f}_{j\in N_\delta}^0\right] \hat{g}_{j\in N_\delta,1} \hat{b} \mathrm{d}\hat{b} \mathrm{d}\varphi \mathrm{d}\hat{\mathbf{v}}_{j\in N_\delta}, \end{aligned} \tag{1.3.71}$$

$$\begin{aligned} &\frac{\mathrm{D}_1 f_1^1}{\mathrm{D}\hat{t}_b} + \left(\frac{1}{v_0 n_0 r_b^2}\right) \frac{\mathrm{D}_1 f_1^0}{\mathrm{D}t} \\ &= \frac{v_0^3}{n_0} \sum_{\delta=1}^{\eta} \frac{N_\delta}{N} \int \left[f_1^{0'} f_{j\in N_\delta}^{0'} - f_1^0 f_{j\in N_\delta}^0\right] \hat{g}_{j\in N_\delta,1} \hat{b} \mathrm{d}\hat{b} \mathrm{d}\varphi \mathrm{d}\hat{\mathbf{v}}_{j\in N_\delta}, \end{aligned} \tag{1.3.72}$$

$$\begin{aligned} &\left(\frac{r_b}{v_0}\right) \frac{\mathrm{D}_1 f_1^1}{\mathrm{D}t} + \left(\frac{1}{v_0 n_0 r_b^2}\right) \frac{\mathrm{D}_1 f_1^0}{\mathrm{D}t} \\ &= \frac{1}{n_0} \sum_{\delta=1}^{\eta} \frac{N_\delta}{N} \int \left[f_1^{0'} f_{j\in N_\delta}^{0'} - f_1^0 f_{j\in N_\delta}^0\right] \hat{g}_{j\in N_\delta,1} \hat{b} \mathrm{d}\hat{b} \mathrm{d}\varphi \mathrm{d}\mathbf{v}_{j\in N_\delta}, \end{aligned} \tag{1.3.73}$$

$$\begin{aligned} &r_b^3 \frac{\mathrm{D}_1 f_1^1}{\mathrm{D}t} + \left(\frac{1}{n_0}\right) \frac{\mathrm{D}_1 f_1^0}{\mathrm{D}t} \\ &= \frac{1}{n_0} \sum_{\delta=1}^{\eta} \frac{N_\delta}{N} \int \left[f_1^{0'} f_{j\in N_\delta}^{0'} - f_1^0 f_{j\in N_\delta}^0\right] g_{j\in N_\delta,1} b \mathrm{d}b \mathrm{d}\varphi \mathrm{d}\mathbf{v}_{j\in N_\delta}, \end{aligned} \tag{1.3.74}$$

$$\frac{\mathrm{D}_1}{\mathrm{D}t}\left(n_0 r_b^3 f_1^1\right) + \frac{\mathrm{D}_1 f_1^0}{\mathrm{D}t} = \sum_{\delta=1}^{\eta} \frac{N_\delta}{N} \int \left[f_1^{0'} f_{j\in N_\delta}^{0'} - f_1^0 f_{j\in N_\delta}^0 \right] g_{j\in N_\delta,1} b\mathrm{d}b\mathrm{d}\varphi \mathrm{d}\mathbf{v}_{j\in N_\delta}. \tag{1.3.75}$$

Now we can use the approximation (1.3.62) in the dimension variant for the function f_1^1, linked on the upper index. It should be taken into account that

$$\varepsilon_{\mathrm{eq}} = n_0 r_b^3 \tag{1.3.76}$$

is the equilibrium value (the "equilibrium" particle density in the interaction volume) for the ε parameter. It follows from Eqs. (1.3.75) and (1.3.76)

$$\frac{\mathrm{D}_1}{\mathrm{D}t}\left(\varepsilon_{\mathrm{eq}} \left(\frac{\partial t}{\partial \varepsilon}\right)_{\varepsilon=0} \frac{\mathrm{D}_1 f_1^0}{\mathrm{D}t}\right) + \frac{\mathrm{D}_1 f_1^0}{\mathrm{D}t} = \sum_{\delta=1}^{\eta} \frac{N_\delta}{N} \int \left[f_1^{0'} f_{j\in N_\delta}^{0'} - f_1^0 f_{j\in N_\delta}^0 \right] g_{j\in N_\delta,1} b\mathrm{d}b\mathrm{d}\varphi \mathrm{d}\mathbf{v}_{j\in N_\delta} \tag{1.3.77}$$

and Eq. (1.3.77) is written as follows

$$\frac{\mathrm{D}_1}{\mathrm{D}t}\left(\tau_{1\in N_\alpha} \frac{\mathrm{D}_1 f_1^0}{\mathrm{D}t}\right) + \frac{\mathrm{D}_1 f_1^0}{\mathrm{D}t} = \sum_{\delta=1}^{\eta} \frac{N_\delta}{N} \int \left[f_1^{0'} f_{j\in N_\delta}^{0'} - f_1^0 f_{j\in N_\delta}^0 \right] g_{j\in N_\delta,1} b\mathrm{d}b\mathrm{d}\varphi \mathrm{d}\mathbf{v}_{j\in N_\delta} \tag{1.3.78}$$

where

$$\tau_{1\in N_\alpha} = \frac{\varepsilon_{\mathrm{eq}}}{[\partial\varepsilon/\partial t]_{\varepsilon=0}}. \tag{1.3.79}$$

The relation (1.3.79) can be written in the typical relaxation form

$$\frac{\partial \varepsilon}{\partial t} = -\frac{\varepsilon(t) - \varepsilon^{\mathrm{eq}}}{\tau_\alpha}. \tag{1.3.80}$$

Consider in detail the relation (1.3.79) for the one-component gas of the hard spheres. The number N^{st} of collisions between particles per unit volume in a unit time is calculated using the DF f, for the Maxwellian function we have [35]:

$$N^{\mathrm{st}} = 2n^2\sigma^2 \left(\frac{2\pi k_{\mathrm{B}} T}{m}\right)^{1/2}, \tag{1.3.81}$$

where σ is the diameter of the particle and m is the particle mass. According to the physical meaning one obtains

$$\left[\frac{\partial \varepsilon}{\partial t}\right]_{\varepsilon=0} = \frac{N^{\mathrm{st}}}{n} n r_b^3, \tag{1.3.82}$$

or

$$\tau = \frac{\varepsilon_{eq}}{[\partial\varepsilon/\partial t]_{\varepsilon=0}} = \frac{n_0}{n} \frac{n}{N^{\mathrm{st}}}. \tag{1.3.83}$$

However, the ratio $n/N^{\mathrm{st}} = \tau_{\mathrm{mt}}$ is the mean time between collisions [35] and n_0/n is the scale factor which defines the relaxation time. It is sufficient in many cases to reason that τ is the mean free time between collisions and to use the model of the hard spheres for calculations for which in the Maxwellian approximation [35]

$$\tau_{\mathrm{mt}}^{(0)} p = \Pi \mu, \tag{1.3.84}$$

where p—static pressure, μ—dynamic viscosity. The Π parameter is defined not only by the model of the particle interaction. In the theory of the Botzmann equation in the mentioned approximation $\Pi \approx 0.8$. For example, in so-called "elementary kinetic theory" $\Pi = 1$. Successive approximations connected with Sonine's polynomials lead to the small correction for coefficient, $\Pi=0.786$. Other values of the Π parameter can be met in physics of the ionized gases. In the practical calculations the scale relaxation factor can be included in the Π parameter, if this factor admits the const numerical interpretation.

Therefore, the definition of the non-local parameters for the concrete physical systems is the separate problem. In the generalized hydrodynamics the non-local parameter τ plays, in the definite sense, the same role as the transport coefficients in the local hydrodynamics.

Reverting now to the dimensional form of the equation (1.3.44) which becomes for the multi-species mixture of gases

$$\frac{Df_\alpha}{Dt}-\frac{D}{Dt}\left(\tau_\alpha\frac{Df_\alpha}{Dt}\right)=\sum_{\beta=1}^{\eta}\int\left[f'_\alpha f'_\beta-f_\alpha f_\beta\right]g_{\beta\alpha}bdbd\varphi d\mathbf{v}_\beta,\ \alpha=,\ldots,\eta, \tag{1.3.85}$$

where

$$\frac{D}{Dt}=\frac{\partial}{\partial t}+\mathbf{v}_\alpha\cdot\frac{\partial}{\partial\mathbf{r}}+\mathbf{F}_\alpha\cdot\frac{\partial}{\partial\mathbf{v}_\alpha}. \tag{1.3.86}$$

Here $g_{\beta\alpha}$ is the relative velocity of the colliding particles (α and β), b is the impact parameter, and φ is the azimuth angle.

It is extremely important that the non-local parameter τ_α is proportional to the mean time τ_{mt} *between* the successive collisions of the α particle with particles of all kinds, defined by

$$\tau_{\alpha,\mathrm{mt}}=\frac{n_\alpha}{\sum_{\beta=1}^{\eta}N_{\alpha\beta}}. \tag{1.3.87}$$

The number $N_{\alpha\beta}$ of collisions between particles of α and β sorts per unit volume in a unit time is calculated using the functions f_α and f_β For the Maxwellian distribution functions [35, 36]

$$N_{\alpha\beta}=2n_\alpha n_\beta\sigma_{\alpha\beta}^2\left(\frac{2\pi k_BT}{m_{\alpha\beta}}\right)^{1/2}, \tag{1.3.88}$$

where σ_α is the diameter of the particle α and $m_{\alpha\beta}$ the reduced mass $m_{\alpha\beta}=m_\alpha m_\beta/(m_\alpha+m_\beta)$. For model of rigid spheres

$$\sigma_{\alpha\beta}=\frac{1}{2}\left(\sigma_\alpha+\sigma_\beta\right). \tag{1.3.89}$$

For an arbitrary DF, the total number of collisions per units of volume and time for molecules of α and β species can be find as

$$N_{\alpha\beta}=\int f_\alpha f_\beta g_{\alpha\beta}bdbd\varphi d\mathbf{v}_\alpha d\mathbf{v}_\beta. \tag{1.3.90}$$

Then the *generalized Boltzmann kinetic equation* (GBE) (1.3.85) involves the additional integral parameter $\tau\sim\tau_{\mathrm{mt}}$ defined by the same DF.

In the hydrodynamic limit when Knudsen numbers $Kn_\alpha=l_\alpha/L$ (l_α is the mean free path between collisions for αth particles) are small the mean time of free path $\tau_{\alpha,\mathrm{mt}}$ can be expressed as function of dynamical viscosity μ_α of species α.

In multi-species gas the approximate relations can be used

$$N_{\alpha\beta}=k_BT\frac{n_\alpha n_\beta}{\mu_{\alpha\beta}}, \tag{1.3.91}$$

$$\tau_{\alpha,\mathrm{mt}}^{-1}=k_BT\sum_\beta n_\beta\mu_{\alpha\beta}^{-1}, \tag{1.3.92}$$

$$\mu=\sum_\alpha\frac{n_\alpha}{n_1\mu_{\alpha1}^{-1}+n_2\mu_{\alpha2}^{-1}+\cdots}, \tag{1.3.93}$$

$$\mu=k_BT\sum_\alpha n_\alpha\tau_{\alpha,\mathrm{mt}}=\sum_\alpha\frac{n_\alpha}{n_1\mu_{\alpha1}^{-1}+n_2\mu_{\alpha2}^{-1}+\cdots}. \tag{1.3.94}$$

Considered theory related to multi-species non-reacting gases. The additional problem arises for adequate description of the inelastic particle's collisions. First of all, the conception of "species" should be scrutinized. In chemical reactions redistribution of atoms is realized. Then if, in the frame of the Liouville description, the motion of every such particle is traced, then total number of particles in ensemble of particles is not changing and Liouville equation can be written in standard form. This approach is wide applied in statistical physics. But in kinetic theory, this method leads to difficulties connected with expressions for collision integrals. As a result, another approach is used based on the classical concept of a chemical component. In this case, the inclusion of inelastic collision is realized with the aid of approximate collision integrals satisfying the law of mass conservation in non-relativistic chemical reactions. In fact we do not need more, because the exactness of those cross-sections of inelastic processes usually is not high.

Elastic collision integral $\hat{J}^{st,0}$ in GBE (1.3.85) contains only a DF of the zeroth order in series by the density parameter ε. It means that, in the generalized Boltzmann kinetic theory (GBKT), the elastic collision integral can be used in the same form as in the classical BKT. This is also true with regard to forms of inelastic collision integrals. It is significant to note that these affirmations are connected with local collision integrals. Below the effects of non-locality in space and time will be discussed separately.

Several important remarks, which also will be discussed in details in the following sections from different points of view:

(1) The generalized Boltzmann equation contains not only second derivatives with respect to time but also mixed (time-velocity and time-coordinate) partial derivatives. Introducing the "averaged" distribution function

$$f^{\alpha} = f - \tau \frac{Df}{Dt} \tag{1.3.95}$$

it assumes the form

$$\frac{Df^{\alpha}}{Dt} = J^{st}(f) \tag{1.3.96}$$

similar to the Boltzmann equation (I.1). Now it becomes clear that the Boltzmann equation, which does not contain fluctuation terms, is not a closed one, and there is no rigorous solution (to put it mildly) to the closure problem for the system of moment equations in the theory of turbulence, based on hydrodynamic equations derived from the Boltzmann equations.

(2) The parameter τ in the generalized Boltzmann equation can be assigned a clear physical meaning and, unlike the so-called kinetically consistent difference schemes [49, 70], to be discussed later, does not lead to secular terms.

(3) The GBE in the dimensionless form is written as

$$\frac{D\hat{f}_{\alpha}}{D\hat{t}} - \frac{D}{D\hat{t}}\left(Kn\hat{\tau}_{\alpha}\frac{D\hat{f}_{\alpha}}{D\hat{t}}\right) = \frac{1}{Kn}\hat{J}^{st}. \tag{1.3.97}$$

From this it follows that the second term is of the order of the Knudsen number (Kn) and turns out to dominate the left-hand side of this equation as the Knudsen number increases. Needless to say, this is not going beyond the free molecular limit of the equation because

$$\frac{D}{Dt}\frac{D\hat{f}_{\alpha}}{D\hat{t}} = 0, \quad \text{for } Kn \to \infty. \tag{1.3.98}$$

The solution of Eq. (1.3.98) is the equation of Knudsen flow

$$\frac{D\hat{f}_{\alpha}}{Dt} = 0 \tag{1.3.99}$$

i.e. the analog of the Liouville equation for a one-particle distribution function.

(4) Note, however, that the second term in Eq. (1.3.97) cannot be ignored even for small Knudsen numbers because, in that case, Kn acts as a small coefficient of higher derivatives, with an unavoidable consequence that the effect of this term will be strong in some regions. The neglect of formally small terms is equivalent, in particular, to dropping the (small-scale) Kolmogorov turbulence from consideration.

Consider now possibilities for simplification of GBE.

If cross-sections $\sigma_{\alpha\beta}$ and reduced masses $m_{\alpha\beta}$ are not too different, reasonable to suppose that the mean free times $\tau_{\alpha,mt}$ do not depend on number of species α and to use the mixture viscosity for calculation of τ.

In this case one obtains from equation

$$\frac{\mathrm{D}f_\alpha}{\mathrm{D}t}-\frac{\mathrm{D}}{\mathrm{D}t}\left(\tau_\alpha\frac{\mathrm{D}f_\alpha}{\mathrm{D}t}\right)=J_\alpha^{\mathrm{st,el}}+J_\alpha^{\mathrm{st,inel}} \tag{1.3.100}$$

the simplified form

$$\frac{\mathrm{D}f_\alpha}{\mathrm{D}t}-\frac{\mathrm{D}}{\mathrm{D}t}\left(\tau\frac{\mathrm{D}f_\alpha}{\mathrm{D}t}\right)=J_\alpha^{\mathrm{st,el}}+J_\alpha^{\mathrm{st,inel}} \tag{1.3.101}$$

The right-hand side of (1.3.100) contains integrals of elastic and inelastic collisions written in a symbolic form.

This equation can be simplified after omitting the terms, which are proportional to the logarithm of hydrodynamic quantities. Really

$$\begin{aligned}\frac{\mathrm{D}f_\alpha}{\mathrm{D}t}-\frac{\mathrm{D}}{\mathrm{D}t}\left(\tau\frac{\mathrm{D}f_\alpha}{\mathrm{D}t}\right)&=\frac{\mathrm{D}f_\alpha}{\mathrm{D}t}-\tau\frac{\mathrm{D}}{\mathrm{D}t}\frac{\mathrm{D}f_\alpha}{\mathrm{D}t}-\frac{\mathrm{D}\tau}{\mathrm{D}t}\frac{\mathrm{D}f_\alpha}{\mathrm{D}t}\\ &=\frac{\mathrm{D}f_\alpha}{\mathrm{D}t}-\tau\left(\frac{\mathrm{D}}{\mathrm{D}t}\frac{\mathrm{D}f_\alpha}{\mathrm{D}t}+\frac{\mathrm{D}\ln\tau}{\mathrm{D}t}\frac{\mathrm{D}f_\alpha}{\mathrm{D}t}\right)\\ &=\frac{\mathrm{D}f_\alpha}{\mathrm{D}t}-\tau\left(\frac{\mathrm{D}}{\mathrm{D}t}\frac{\mathrm{D}f_\alpha}{\mathrm{D}t}+\frac{\mathrm{D}\ln(\mu/p)}{\mathrm{D}t}\frac{\mathrm{D}f_\alpha}{\mathrm{D}t}\right).\end{aligned} \tag{1.3.102}$$

Usually, the derivative of the logarithm of hydrodynamic quantities is a close-to-zero quantity, and one can ignore the second term in brackets in (1.3.102) as compared to the first term and write the GBE in the form

$$\frac{\mathrm{D}f_\alpha}{\mathrm{D}t}-\tau\frac{\mathrm{D}}{\mathrm{D}t}\frac{\mathrm{D}f_\alpha}{\mathrm{D}t}=J_\alpha^{\mathrm{st,el}}+J_\alpha^{\mathrm{st,inel}}, \tag{1.3.103}$$

especially, if the mean collision times for different components do not differ too much from one another. These are the possibilities for simplifying GBE that stand out and may be used in applications.

There exists a possibility of analytical GBE integrating if the model collision integral can be used. For this aim we choose the BGK approximation for collision integral:

$$J^{\mathrm{st,el}}=\frac{f^{(0)}-f}{\tau}, \tag{1.3.104}$$

where $f^{(0)}$ is local equilibrium DF, and τ is a time of relaxation. Approach (1.3.89) is widely used in physics of ionized gases for a number of years through the work of Bhatnagar, Gross and Krook [71].

Let us consider a one-dimensional case in the absence of external forces, $\tau=\mathrm{const}$. GBE leads to the equation

$$\frac{\partial f}{\partial t}+v\frac{\partial f}{\partial x}-\tau\left(\frac{\partial^2 f}{\partial t^2}+2v\frac{\partial^2 f}{\partial x\partial t}+v^2\frac{\partial^2 f}{\partial x^2}\right)=\frac{f^{(0)}-f}{\tau}. \tag{1.3.105}$$

Write down (1.3.105) in the form

$$a_{11}\frac{\partial^2 f}{\partial x^2}+2a_{12}\frac{\partial^2 f}{\partial x\partial t}+a_{22}\frac{\partial^2 f}{\partial t^2}+b_1\frac{\partial f}{\partial x}+b_2\frac{\partial f}{\partial t}=\frac{f-f^{(0)}}{\tau}, \tag{1.3.106}$$

where

$$a_{11}=\tau v^2, a_{12}=\tau v, a_{22}=\tau, b_1=-v, b_2=-1,$$

Eq. (1.3.106) can be brought to the canonical form. Since $\Delta=a_{12}^2-a_{11}a_{22}=0$, we obtain equation of *parabolic* type and the following characteristic equation

$$a_{11}(\mathrm{d}t)^2-2a_{12}\mathrm{d}t\mathrm{d}x+a_{22}(\mathrm{d}x)^2=0. \tag{1.3.107}$$

This equation splits into two equations

$$\begin{aligned}\left(\frac{\mathrm{d}t}{\mathrm{d}x}\right)_1&=\frac{a_{12}+\sqrt{a_{12}^2-a_{11}a_{12}}}{a_{11}}\\ \left(\frac{\mathrm{d}t}{\mathrm{d}x}\right)_2&=\frac{a_{12}-\sqrt{a_{12}^2-a_{11}a_{12}}}{a_{11}}.\end{aligned} \tag{1.3.108}$$

Because of $\Delta=0$, one obtains only one characteristic equation for parabolic equation

$$\frac{dt}{dx}=\frac{a_{12}}{a_{11}}=\frac{1}{v}. \tag{1.3.109}$$

As result, the following independent variables can be taken for mentioned transformation:

$$\xi=x-vt, \zeta=t. \tag{1.3.110}$$

Using (1.3.110) we find:

$$\frac{\partial f}{\partial x}=\frac{\partial f}{\partial \xi}\frac{\partial \xi}{\partial x}+\frac{\partial f}{\partial \zeta}\frac{\partial \zeta}{\partial x}=\frac{\partial f}{\partial \xi} \tag{1.3.111}$$

and analogously the second derivatives

$$\begin{gathered}\frac{\partial^2 f}{\partial x^2}=\frac{\partial^2 f}{\partial \xi^2}, \frac{\partial f}{\partial t}=-v\frac{\partial f}{\partial \xi}+\frac{\partial f}{\partial \zeta}, \\ \frac{\partial^2 f}{\partial t^2}=v^2\frac{\partial^2 f}{\partial \xi^2}-2v\frac{\partial^2 f}{\partial \xi \partial \zeta}+\frac{\partial^2 f}{\partial \zeta^2}, \\ \frac{\partial^2 f}{\partial x \partial t}=-v\frac{\partial^2 f}{\partial \xi^2}+\frac{\partial^2 f}{\partial \xi \partial \zeta}.\end{gathered} \tag{1.3.112}$$

After substituting these derivatives in (1.3.106) one obtains

$$\tau\frac{\partial^2 f}{\partial \zeta^2}-\frac{\partial f}{\partial \zeta}=\frac{f-f^{(0)}}{\tau}. \tag{1.3.113}$$

Introducing a deviation $\hat{f}=f-f^{(0)}$, we find

$$\tau\hat{f''}-\hat{f'}-\frac{1}{\tau}\hat{f}=0. \tag{1.3.114}$$

Equation (1.3.114) has the characteristic equation

$$\tau k^2-k-\frac{1}{\tau}=0, \tag{1.3.115}$$

with the roots

$$k_{1;2}=\frac{1\pm\sqrt{5}}{2\tau}.$$

If, from the physical sense of solution, the root $k=\left(1-\sqrt{5}\right)/(2\tau)$ is chosen, we have

$$f-f_0=C(\xi)\exp\left(-\frac{\sqrt{5}-1}{2\tau}t\right). \tag{1.3.116}$$

Obviously value $C(\xi)$ is defined for the time moment $t=0$, and

$$f=f_0+C(\xi)\exp\left(-0.618\frac{t}{\tau}\right). \tag{1.3.117}$$

Analogically, for the Boltzmann equation we find the solution indicated by upper index BE

$$f^{BE}=f_0+C(\xi)\exp\left(-\frac{t}{\tau}\right). \tag{1.3.118}$$

The second solution of Eq. (1.3.113) is written as follows

$$f=f_0+C(\xi)\exp\left(1.618\frac{t}{\tau}\right). \tag{1.3.119}$$

Here the exponential growth of one of possible solution is eliminating by choosing an arbitrary factor—which is equal to zero—in corresponding solution. This situation will be discussed from position of the generalized H-theorem in the next section.

Relation (1.3.117) rules the process of gas relaxation in the form of a traveling decreasing wave to the local equilibrium, defined by the DF $f^{(0)}$.

For the space homogeneous relaxation corresponding to the equation

$$\frac{\partial f}{\partial t}-\tau\frac{\partial^2 f}{\partial t^2}=\frac{f^{(0)}-f}{\tau}, \tag{1.3.120}$$

one obtains

$$f=f_0+C(v)\exp\left(-0.618\frac{t}{\tau}\right). \tag{1.3.121}$$

An analogous equation in the Boltzmann physical kinetics is

$$\frac{\partial f}{\partial t}=\frac{f^{(0)}-f}{\tau}, \tag{1.3.122}$$

which has the solution

$$f^{\mathrm{BE}}=f_0+C(v)\exp\left(-\frac{t}{\tau}\right) \tag{1.3.123}$$

As we see, in the considered case GBE and BE solutions have the difference of the principal significance. Namely, the GBE contains the possibility of the exponential growth of one of the possible solutions.

1.4 GENERALIZED BOLTZMANN H-THEOREM AND THE PROBLEM OF IRREVERSIBILITY OF TIME

Boltzmann's H-theorem is of principal importance for kinetic theory and it provides, in fact, the kinetic substantiation of the theory. The generalized H-theorem was proven in 1992 (see, for example, [31]); we present below the main fragments of the derivation, discuss it from the standpoint of the problem of irreversibility of time, and generalize for the case of multi-component reacting gas.

Consider first a simple gas consisting of spherical molecules. The state of the gas is assumed to be constant, with external forces absent. Then, the GB-equation is reduced to the form

$$\frac{\partial f}{\partial t}-\frac{\partial}{\partial t}\left(\tau\frac{\partial f}{\partial t}\right)=J^{st}. \tag{1.4.1}$$

We will introduce Boltzmann's H-function,

$$H=\int f\,\ln f\,\mathrm{d}\mathbf{v}, \tag{1.4.2}$$

multiply both parts of equation (1.4.1) by ln f, and transform the equation to

$$\frac{\partial f}{\partial t}\ln f=\frac{\partial}{\partial t}(f\ln f)-\frac{\partial f}{\partial t}, \tag{1.4.3}$$

$$\frac{\partial^2 f}{\partial t^2}\ln f=\frac{\partial^2}{\partial t^2}(f\ln f)-\frac{1}{f}\left(\frac{\partial f}{\partial t}\right)^2-\frac{\partial^2 f}{\partial t^2}. \tag{1.4.4}$$

From GBE (1.4.1) and Eqs. (1.4.3) and (1.4.4) follow:

$$\frac{\partial}{\partial t}(f\ln f)-\frac{\partial f}{\partial t}-\tau\frac{\partial^2}{\partial t^2}(f\ln f)+\tau\frac{1}{f}\left(\frac{\partial f}{\partial t}\right)^2+\tau\frac{\partial^2 f}{\partial t^2}$$
$$-\ln f\frac{\partial\tau}{\partial t}\frac{\partial f}{\partial t}=J^{st}\ln f \tag{1.4.5}$$

or

$$\frac{\partial}{\partial t}(f\ \ln f)-\tau\frac{\partial^2}{\partial t^2}(f\ \ln f)+\tau\frac{1}{f}\left(\frac{\partial f}{\partial t}\right)^2-\frac{\partial\tau}{\partial t}\frac{\partial}{\partial t}(f\ \ln f) \\ =J^{st}\ \ln f+\frac{\partial f}{\partial t}-\tau\frac{\partial^2 f}{\partial t^2}-\frac{\partial\tau}{\partial t}\frac{\partial f}{\partial t}. \tag{1.4.6}$$

Using once more Eq. (1.4.1), we find

$$\frac{\partial}{\partial t}(f\ \ln f)-\tau\frac{\partial^2}{\partial t^2}(f\ \ln f)+\tau\frac{1}{f}\left(\frac{\partial f}{\partial t}\right)^2-\frac{\partial\tau}{\partial t}\frac{\partial}{\partial t}(f\ \ln f) \\ =(1+\ \ln f)J^{st} \tag{1.4.7}$$

We now integrate Eq. (1.4.7) term-by-term with respect to all the values of velocities and use the definition of the H-function,

$$\frac{\mathrm{d}H}{\mathrm{d}t}-\tau\frac{\mathrm{d}^2H}{\mathrm{d}t^2}-\frac{\mathrm{d}\tau}{\mathrm{d}t}\frac{\mathrm{d}H}{\mathrm{d}t}=-\tau\int\frac{1}{f}\left(\frac{\partial f}{\partial t}\right)^2\mathrm{d}\mathbf{v}+\int(1+\ \ln f)J^{st}\mathrm{d}\mathbf{v} \tag{1.4.8}$$

But the following inequalities is valid

$$-\tau\int\frac{1}{f}\left(\frac{\partial f}{\partial t}\right)^2\mathrm{d}\mathbf{v}+\int(1+\ \ln f)J^{st}\mathrm{d}\mathbf{v}\le 0. \tag{1.4.9}$$

Really, the first integral in Eq. (1.4.9) obviously is not positive, for the second integral Boltzmann's transformation can be applied

$$\int(1+\ \ln f)J^{st}\mathrm{d}\mathbf{v}=\int(1+\ \ln f)\left(f'f_1'-ff_1\right)gb\mathrm{d}b\mathrm{d}\varphi\mathrm{d}\mathbf{v}\mathrm{d}\mathbf{v}_1 \\ =\frac{1}{4}\int(1+\ \ln f+1+\ \ln f_1-1-\ \ln f'-1-\ \ln f)\left(f'f_1'\right. \\ \left.-ff_1\right)gb\mathrm{d}b\mathrm{d}\varphi\mathrm{d}\mathbf{v}\mathrm{d}\mathbf{v}_1=\frac{1}{4}\int\ \ln\frac{ff_1}{f'f_1'}\left(f'f_1'-ff_1\right)gb\mathrm{d}b\mathrm{d}\varphi\mathrm{d}\mathbf{v}\mathrm{d}\mathbf{v}_1. \tag{1.4.10}$$

The second relation in the chain of equalities (1.4.10) is obtained by formal re-notation of forward and backward collisions with using of the principle of microscopic reversibility which can be written, for this case, in the form

$$\mathrm{d}\mathbf{v}\mathrm{d}\mathbf{v}_1=\mathrm{d}\mathbf{v}'\mathrm{d}\mathbf{v}_1'. \tag{1.4.11}$$

Following Boltzmann we notice that the value $\ln(ff_1/f'f_1')$ is positive or negative depending on whether ff_1 are larger than $f'f_1'$ or smaller. In either case, the sign of $\ln(ff_1/f'f_1')$ is opposite to the sign of the difference $f'f_1'-ff_1$.

Then, we obtain

$$\frac{d}{\mathrm{d}t}\left(H-\tau\frac{\mathrm{d}H}{\mathrm{d}t}\right)\le 0. \tag{1.4.12}$$

We will introduce the H^a-function in accordance with the definition

$$H^a=H-\tau\frac{\mathrm{d}H}{\mathrm{d}t}. \tag{1.4.13}$$

Then the inequality is valid that yields the conclusion of the generalized H-theorem,

$$\frac{\mathrm{d}H^a}{\mathrm{d}t}\le 0. \tag{1.4.14}$$

If we suppose that τ is constant, not depending on time, then inequality (1.4.14) can be considered as combination of two principles—Boltzmann's principle

$$\frac{\mathrm{d}H}{\mathrm{d}t}\le 0 \tag{1.4.15}$$

and Prigogine's principle [72, 73]

$$\frac{d^2H}{dt^2} \geq 0. \tag{1.4.16}$$

For closed physical systems, H-function is a limited function, in particular, this function will be restricted from below. In other words, the integral $\int f \ln f d\mathbf{v}$ does not tends to $-\infty$ if $v \to \infty$, i.e. integral converges.

For this aim, consider the integral

$$\int f \frac{1}{2} mv^2 d\mathbf{v} = \frac{1}{2} \rho \overline{v^2}, \tag{1.4.17}$$

where ρ is the density and upper line is connected with averaging of the corresponding value.

Integral (1.4.17) is the value of kinetic energy for a unit volume and therefore it is finite. Suppose that function f decreases with $v \to \infty$ faster than $\exp(-(mv^2/2k_B T))$, i.e.,

$$f < \exp\left(-\frac{mv^2}{2k_B T}\right), \quad \ln f < -\frac{mv^2}{2k_B T}.$$

Then $\int f \ln f d\mathbf{v} > -\infty$.

If f decreases with $v \to \infty$ slower, than $e^{-(mv^2/2k_B T)}$, i.e.,

$$-\ln f < \frac{mv^2}{2k_B T},$$

the convergence of integral $\int f \ln f d\mathbf{v} > -\infty$ is defines by means of (1.4.17). Really, in this case $-f \ln f < \frac{1}{2k_B T} mv^2 f$ and integral of this value should be limited by virtue of (1.4.14), because the kinetic energy of physical system is limited. In all cases integral in (1.4.17)—connected with kinetic energy of closed system—is limited function.

During time evolution the decreasing H-function—and therefore H^a-function—is limited when $v \to \infty$.

At this stage of investigation we can state that generalized H-function (H^a) is not an increasing function for all hypothetical manners of the H-function behavior.

Let us consider from the possible evolution of H-function from this point of view.

If H-function is decreasing function, this function is restricted from below. Consider the time evolution of H- and H^a functions.

Be H-function is a monotonically decreasing function, as it is shown in Fig 1.7. Figure 1.7 also shows the dependence dH/dt on time—as result of graphical differentiating—and the time evolution of H^a as consequence of these two mentioned graphs and definition (1.4.13). As we see in this case, H^a is also monotonically decreasing function.

Maybe another situation shown in Fig. 1.8 could be realized? This dependence corresponds to the local growth of H-function. It leads, as result of graphical constructions, to the growth of H^a, but it is forbidden by inequality (1.4.14).

One proves [31] that, if

(I) $\frac{dH}{dt} \leq 0$, then $\frac{dH^a}{dt} \leq 0$ as well;

If in some portion of the evolution curve

(II) $\frac{dH}{dt} > 0$, then $\frac{dH^a}{dt} > 0$, as well, which is forbidden by inequality (1.4.14).

The system of inequalities (I), (II) does not forbid the evolution shown in Fig 1.9. In this case $d^2H/dt^2 \geq 0$ only in "linear" region and entropy production ($\sim dH/dt$) evolve by non-monotonic manner.

The possibility of appearance in equilibrium state (if $H^a = 0$) of the exponentially increasing in time H-function—and infinite growth of energy of closed system—should be exclude by turning into zero of the constant of integrating. In this case the model of the space homogeneous physical system is not correct.

Now what happens to the fluctuations that develop in the system? To see this, consider the generalized equation of continuity, which was "guessed" in Introduction and which—as will be shown in the next chapter—is a direct consequence of GBE in the hydrodynamic limit

$$\begin{aligned} &\frac{\partial}{\partial t}\left\{\rho - \tau\left[\frac{\partial \rho}{\partial t} + \frac{\partial}{\partial \mathbf{r}} \cdot (\rho \mathbf{v}_0)\right]\right\} \\ &+ \frac{\partial}{\partial \mathbf{r}} \cdot \left\{\rho \mathbf{v}_0 - \tau\left[\frac{\partial}{\partial t}(\rho \mathbf{v}_0) + \frac{\partial}{\partial \mathbf{r}} \cdot \rho \mathbf{v}_0 \mathbf{v}_0 + \overleftrightarrow{\mathbf{I}} \cdot \frac{\partial p}{\partial \mathbf{r}} - \rho \mathbf{a}\right]\right\} = 0. \end{aligned} \tag{1.4.18}$$

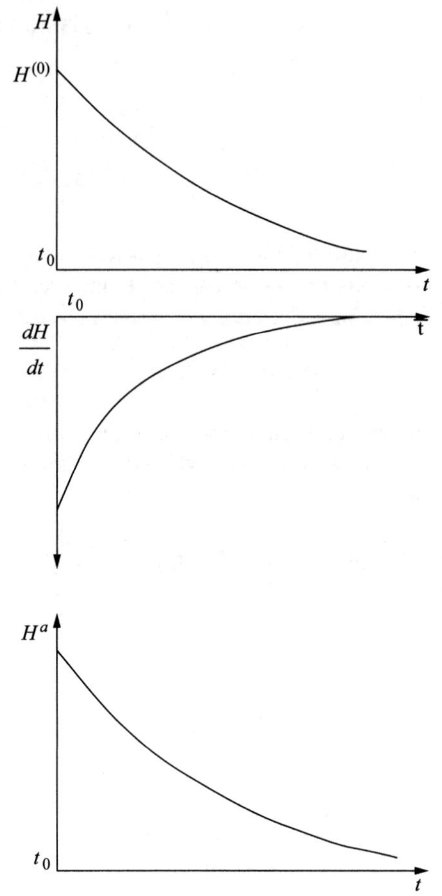

FIGURE 1.7 Hypothetical time evolution of the H- and H^a functions.

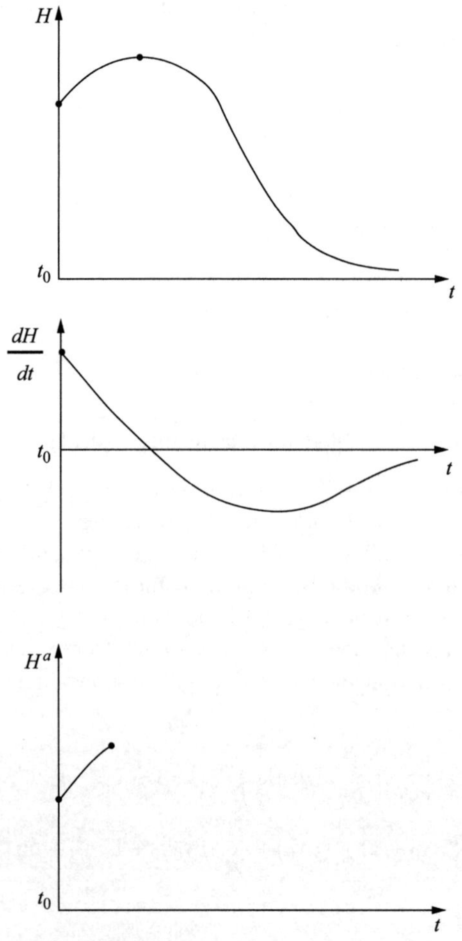

FIGURE 1.8 Hypothetical time evolution of the H- and H^a functions.

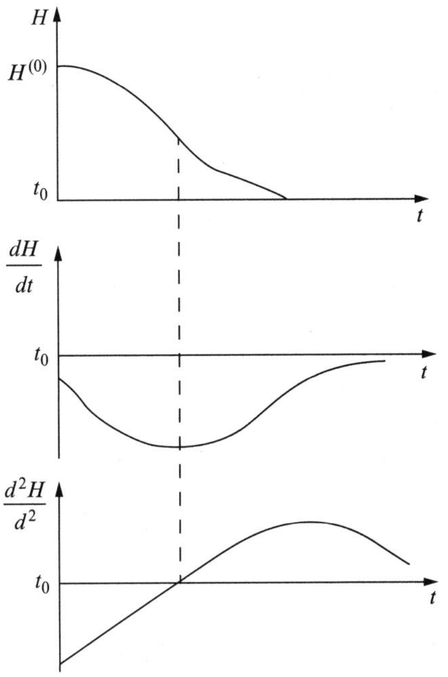

FIGURE 1.9 Hypothetical time evolution of the H- and H^a functions.

We write here this equation in the generalized Eulerian formulation under the assumption of no external forces and for the one-dimensional unsteady case:

$$\begin{aligned}&\frac{\partial}{\partial t}\left\{\rho-\tau^{(0)}\left[\frac{\partial \rho}{\partial t}+\frac{\partial}{\partial x}(\rho v_0)\right]\right\}+\frac{\partial}{\partial x}(\rho v_0)\\&=\frac{\partial}{\partial x}\left\{\tau^{(0)}\left[\frac{\partial}{\partial t}(\rho v_0)+\frac{\partial}{\partial x}\left(p+\rho v_0^2\right)\right]\right\},\end{aligned}\tag{1.4.19}$$

where $\tau^{(0)}$ is the mean time between collisions calculated in the locally Maxwellian approximation: $\tau^{(0)}p=\Pi\mu$ (as indicated, the factor Π being of order unity; for the hard-sphere model, $\Pi=0.8$ to first-order approximation in Sonine polynomials [35]).

We shall assume that except for shock-wave-type regions (to be discussed below within the framework of the generalized Boltzmann equation) hydrodynamical quantities vary not too rapidly on the scale of the order of the mean time between collisions:

$$\frac{\rho}{\tau^{(0)}}\gg\frac{\partial \rho}{\partial t},\quad \frac{\rho}{\tau^{(0)}}\gg\frac{\partial}{\partial x}(\rho v_0),$$

the temperature variations are small, the convective transfer is negligible, and the chaotic motion is highly energetic as compared to the kinetic energy of the flow, i.e. $\overline{V^2}/v_0^2\gg 1$(for example, for hydrogen at normal pressures and temperatures, we have $v_0=10\,\mathrm{cm\,s^{-1}}$, giving 3.4×10^8 for this ratio). Consequently, Eq. (1.4.19) becomes

$$\frac{\partial \rho}{\partial t}=\frac{\partial}{\partial x}\left(\frac{\tau^{(0)}p}{\rho}\frac{\partial \rho}{\partial x}\right)\tag{1.4.20}$$

or

$$\frac{\partial \rho}{\partial t}=\frac{\partial}{\partial x}\left(\mathrm{D}\frac{\partial \rho}{\partial x}\right),\tag{1.4.21}$$

where $\mathrm{D}=\Pi\mu/\rho$ is the self-diffusion coefficient [36]. Equation (1.4.21) is the diffusion equation, with the implication that (a) a locally increasing density fluctuation immediately activates the diffusion mechanism which smoothes it out, and (b) the generalized Boltzmann H-theorem proved above ensures that the smoothed fluctuations come to equilibrium.

Relation (1.4.8) can be rewritten as:

$$\frac{dH^a}{dt} = -\tau \int \frac{1}{f}\left(\frac{\partial f}{\partial t}\right)^2 d\mathbf{v} + \frac{1}{4}\int \ln\frac{ff_1}{f'f'_1}(f'f'_1 - ff_1)gbdbd\varphi d\mathbf{v}d\mathbf{v}_1, \tag{1.4.22}$$

The second integral in the right side of Eq. (1.4.15) is smaller or equal to zero because of the obvious inequality

$$(b-a)\ln\frac{a}{b} \le 0,$$

where $b = f'f'_1, a = ff_1$. Then, in the stationary state the equality is valid

$$f'_1 f' = ff_1, \tag{1.4.23}$$

which leads to the Maxwellian distribution function.

If the Boltzmann H-function defined by the distribution function f obtained from GBE, in hydrodynamic limit the value $\tau(dH/dt)$ should be considered as fluctuation of H-function on the Kolmogorov's level of the turbulence description; then H^a in the relation

$$H^a = H - \tau\frac{dH}{dt} \tag{1.4.24}$$

is an averaged value of H-function.

For multi-component gas, the analog of equation (1.4.1) has the form

$$\frac{\partial f_\alpha}{\partial t} - \frac{\partial}{\partial t}\left(\tau_\alpha \frac{\partial f_\alpha}{\partial t}\right) = J_\alpha^{\mathrm{st,el}}, \tag{1.4.25}$$

As a result, the H-function for the component α is written as

$$H_\alpha = \int f_\alpha \ln f_\alpha d\mathbf{v}_\alpha. \tag{1.4.26}$$

Subsequent mathematics is analogous, and therefore, the inequality assumes the following form:

$$\frac{d}{dt}\left(H_\alpha - \tau_\alpha \frac{dH_\alpha}{dt}\right) \le 0 \tag{1.4.27}$$

or

$$\frac{dH_\alpha^a}{dt} \le 0. \tag{1.4.14'}$$

The summation over all components leads to the H-function for the mixture,

$$\begin{gathered} H = \sum_\alpha \int f_\alpha \ln f_\alpha d\mathbf{v}_\alpha \\ H^a = \sum_\alpha \left(H_\alpha - \tau_\alpha \frac{dH_\alpha}{dt}\right) = H - \sum_{\alpha=1}^{\eta} \tau_\alpha \frac{dH_\alpha}{dt} \end{gathered} \tag{1.4.28}$$

and to the inequality

$$\frac{dH^a}{dt} \le 0,$$

which results from (1.4.14).

Let chemical reactions proceed in a multi-component gas mixture, for which the integral of bimolecular collisions has the form [21, 74] $(\beta, \gamma, \delta = 1, \ldots, \eta)$

$$J_\alpha^{\mathrm{st,nel}} = \frac{1}{2}\sum_r \sum_{\beta\gamma\delta} \int \left[\frac{s_\alpha s_\beta}{s_\gamma s_\delta}\left(\frac{m_\alpha m_\beta}{m_\gamma m_\delta}\right)^3 f'_\gamma f'_\delta - f_\alpha f_\beta\right] g_{\alpha\beta} d\sigma_{\alpha\chi}^{\gamma\delta\,r} d\mathbf{v}_\beta, \tag{1.4.29}$$

where $d\overset{r}{\sigma}_{\alpha\beta}^{\gamma\delta}$ is the differential cross section of inelastic collision in the r th reaction

$$A_\alpha + A_\beta \rightarrow A_\gamma + A_\delta,$$

and $s_\alpha, s_\beta, s_\gamma, s_\delta$ are statistical weights of the energy state of particles $A_\alpha, A_\beta, A_\gamma, A_\delta$, [44]. In this GBE has the form:

$$\frac{\partial f_\alpha}{\partial t} - \frac{\partial}{\partial t}\left(\tau_\alpha \frac{\partial f_\alpha}{\partial t}\right) = \sum_j \int \left[f'_\alpha f'_j - f_\alpha f_j\right] g_{\alpha j} d\sigma_{\alpha j}^{\alpha j} d\mathbf{v}_j$$
$$+\frac{1}{2}\sum_r \sum_{\beta\gamma\delta} \int \left[\xi_{\alpha\beta}^{\gamma\delta} f'_\gamma f'_\delta - f_\alpha f_\beta\right] g_{\alpha\beta} d\overset{r}{\sigma}_{\alpha\beta}^{\gamma\delta} d\mathbf{v}_\beta. \tag{1.4.30}$$

Now, the analog of Eq. (1.4.8) after summation over α is written as

$$\sum_\alpha \left[\frac{dH_\alpha}{dt} - \tau_\alpha \frac{d^2 H_\alpha}{dt^2} - \frac{d\tau_\alpha}{dt}\frac{dH_\alpha}{dt}\right]$$
$$= -\sum_\alpha \tau_\alpha \int \frac{1}{f_\alpha}\left(\frac{\partial f_\alpha}{\partial t}\right)^2 d\mathbf{v}_\alpha + \sum_\alpha \int (1 + \ln f_\alpha) J_\alpha^{st,el} d\mathbf{v}_\alpha$$
$$+\frac{1}{2}\sum_{r,\beta,\gamma,\delta,\alpha} \int (1 + \ln f_\alpha)\left(\xi_{\alpha\beta}^{\gamma\delta} f'_\gamma f'_\delta - f_\alpha f_\beta\right) g_{\alpha\beta} d\overset{r}{\sigma}_{\alpha\beta}^{\gamma\delta} d\mathbf{v}_\alpha d\mathbf{v}_\beta. \tag{1.4.31}$$

Obviously, in the integral sum in the right-hand side of Eq. (1.4.31) indexes α, β, γ, δ are dummy and the transformation is valid:

$$\sum_r \sum_{\alpha\beta\gamma\delta} \int (1 + \ln f_\alpha)\left(\xi_{\alpha\beta}^{\gamma\delta} f'_\gamma f'_\delta - f_\alpha f_\beta\right) g_{\alpha\beta} d\overset{r}{\sigma}_{\alpha\beta}^{\gamma\delta} d\mathbf{v}_\alpha d\mathbf{v}_\beta$$
$$= \frac{1}{4}\sum_r \sum_{\alpha\beta\gamma\delta} \int \ln\frac{f_\alpha f_\beta}{f_\gamma f_\delta}\left(\xi_{\alpha\beta}^{\gamma\delta} f'_\gamma f'_\delta - f_\alpha f_\beta\right) g_{\alpha\beta} d\overset{r}{\sigma}_{\alpha\beta}^{\gamma\delta} d\mathbf{v}_\alpha d\mathbf{v}_\beta, \tag{1.4.32}$$

We use the principle of microscopic reversibility

$$\xi_{\alpha\beta}^{\gamma\delta} g_{\alpha\beta} d\overset{r}{\sigma}_{\alpha\beta}^{\gamma\delta} d\mathbf{v}_\alpha d\mathbf{v}_\beta = g'_{\gamma\sigma} d\overset{r}{\sigma}_{\gamma\delta}^{\alpha\beta'} d\mathbf{v}'_\gamma d\mathbf{v}'_\delta. \tag{1.4.33}$$

and once again arrive at the formulation of the H-theorem

$$\frac{dH^a}{dt} \le 0, \quad H^a = H - \sum_{\alpha=1}^{\eta} \tau_\alpha \frac{dH_\alpha}{dt}, \tag{1.4.34}$$

although, as we can see, inequality (1.4.14) may prove invalid in the presence of chemical reactions.

In thermodynamics for equilibrium systems, entropy is introduced by relation

$$S = -k_B \int w(\Omega_1, \ldots, \Omega_N) \ln w(\Omega_1, \ldots, \Omega_N) d\Omega_1 \ldots d\Omega_N, \tag{1.4.35}$$

where $w(\Omega_1, \ldots, \Omega_N)$ is the probability density for canonical distribution. For ideal gas using the assumption about statistical independence of molecules, the simplification of Eq. (1.4.34) can be realized with the aid of probability density

$$W_s(\Omega_s) = \int w(\Omega_1, \ldots, \Omega_N) d\Omega_1 \ldots d\Omega_{s-1} d\Omega_{s+1} \ldots d\Omega_N, \tag{1.4.36}$$

defining the appearance of sth particle in the state Ω_s, when another particles occupy the arbitrary states admitted by physical system.

Then

$$w(\Omega_1, \ldots, \Omega_N) = \prod_{s=1}^{N} W_s(\Omega_s) \tag{1.4.37}$$

and after substitution in (1.4.35) we obtain

$$\begin{aligned} S &= -k_{\mathrm{B}}\sum_{s=1}^{N}\int\prod_{k=1}^{N}W_k(\Omega_k)\ln W_s(\Omega_s)\mathrm{d}\Omega_1\ldots\mathrm{d}\Omega_N \\ &= -k_{\mathrm{B}}\sum_{s=1}^{N}\int W_s(\Omega_s)\ln W_s(\Omega_s)\mathrm{d}\Omega_s. \end{aligned} \tag{1.4.38}$$

Relation (1.4.38) is written for unit normalization for W_k. If we pass over to the one-particle distribution functions $f(\mathbf{r},\mathbf{v},t)$ used by us, then, with an accuracy within the non-principal constant S_0 that is associated only with the level of entropy count, we have classical relation

$$S = -k_{\mathrm{B}}H + S_0, \tag{1.4.39}$$

which leads to the thermodynamic inequality

$$\frac{\mathrm{d}S}{\mathrm{d}t} \geq 0 \tag{1.4.40}$$

We will now investigate the thermodynamic inequality (1.4.40) from the standpoint of the existing causal relations and direction of time. To this end, one needs to answer the question of how it happened that inequality (1.4.34), which is the one leading to the increase in entropy (1.4.40) and to the existence of irreversible processes, appeared in our generalized Boltzmann physical kinetics?

This effect is a direct result of approximation (1.3.62), into which the motion in the direction opposite to the "time arrow" was introduced, so that the state of the system at the given moment of time is defined in a determinate manner by collisions that occurred in the past.

We will introduce the physical principle of causality as some operator which "cuts out," from all events possible at the present moment of time, only the certain event whose causes exist in the past, and which transfers the certain event under consideration in the present into the class of causal relations for some possible event in the future. Thereby, the irreversibility of time is introduced as well. In other words, one cannot speak of the principle of causality without using the concept of irreversibility of time.

What may be the result of formal rejection of the principle of causality in this particular case? If one abandons the additional statement that the cause precedes the effect, τ in relation (1.4.8) may be replaced by $(-\tau)$,

$$\frac{\mathrm{d}H^{a'}}{\mathrm{d}t} - \tau\int\frac{1}{f}\left(\frac{\partial f}{\partial t}\right)^2\mathrm{d}\mathbf{v} = \frac{1}{4}\int\ln\frac{ff_1}{f'f'_1}(f'f'_1 - ff_1)gb\mathrm{d}b\mathrm{d}\phi\mathrm{d}\mathbf{v}\mathrm{d}\mathbf{v}_1 \tag{1.4.41}$$

where

$$H^{a'} = H + \tau\frac{\mathrm{d}H}{\mathrm{d}t} \tag{1.4.42}$$

or, to put it differently,

$$\frac{\mathrm{d}H^{a'}}{\mathrm{d}t} - \tau\int\frac{1}{f}\left(\frac{\partial f}{\partial t}\right)^2\mathrm{d}\mathbf{v} \leq 0 \tag{1.4.43}$$

Nothing can now be said about the sign of derivative $\mathrm{d}H^{a'}/\mathrm{d}t$ in (1.4.43). Inequality (1.4.43) may also bold in case $\mathrm{d}H^{a'}/\mathrm{d}t > 0$, because from this value the non-negative integral $\tau\int 1/f(\partial f/\partial t)^2\mathrm{d}\mathbf{v}$ is subtracted, which does not vary when t is replaced by $(-t)$.

Therefore, the principle of entropy increase follows directly from the principle of irreversibility of time. After introducing the approximation of the two-particle distribution function via the product of one-particle functions and using the fact of correlation of the dynamic variables on the r_b-scale, the reversibility remains or is eliminated from treatment, depending on the approximation of the DF with respect to the hypothetical future or determinate past and, on the formal side, depending on the choice of sign before τ. The probability of reversible processes in closed dissipative physical systems is vanishingly small, and one must use GBE in the form of Eq. (1.3.71) for investigation of the transport processes in applied problems. However, as regards gigantic (such as human life) time intervals, it is of interest to investigate dissipative and, nevertheless, reversible systems.

In order to understand how the evolution of such a system may look like, consider the particular case of an alternative generalized Boltzmann equation,

$$\frac{Df}{Dt}+\frac{D}{Dt}\left(\tau\frac{Df}{Dt}\right)=J^{st}(f) \tag{1.4.44}$$

which corresponds to the evolution of a one-dimensional non-stationary system with the collision integral in the BGK form. The appropriate equation has the form

$$\frac{\partial f}{\partial t}+v\frac{\partial f}{\partial x}+\tau\left(\frac{\partial^2 f}{\partial t^2}+2v\frac{\partial^2 f}{\partial x\partial t}+v^2\frac{\partial^2 f}{\partial x^2}\right)=\frac{f^{(0)}-f}{\tau}. \tag{1.4.45}$$

Equation (1.4.45), like analogous Eq. (1.3.105), is parabolic and, with the aid of transformation $\xi=x-vt$, $\zeta=t$ reduces to the equation

$$\tau\frac{\partial^2\hat{f}}{\partial\zeta^2}+\frac{\partial\hat{f}}{\partial\zeta}+\frac{\hat{f}}{\tau}=0, \tag{1.4.46}$$

where $\hat{f}=f-f^{(0)}$.

Its characteristic equation

$$\tau k^2+k+\frac{1}{\tau}=0, \tag{1.4.47}$$

with roots

$$k_{1;2}=\frac{-1\pm\sqrt{-3}}{2\tau}. \tag{1.4.48}$$

It follows from (1.4.48) that the characteristic equation permits of non-monotonic solutions. We will study one of them,

$$f^{GBE}_{(+\tau)}=f_0+C(\xi)e^{-0.5t/\tau}\cos 0.866\frac{t}{\tau}. \tag{1.4.49}$$

Obviously value $C(\xi)$ is defined for time moment $t=0$, and the corresponding solutions of the BE and (1.3.114) have the form

$$f=f_0+C(\xi)e^{-\frac{t}{\tau}}, \tag{1.4.50}$$

$$f=f_0+C(\xi)e^{-0.618\frac{t}{\tau}}. \tag{1.4.51}$$

Figure 1.10 illustrates the evolution of the time parts of distribution functions (1.4.49)–(1.4.51), namely, $e^{-0.5t/\tau}\cos 0.866\frac{t}{\tau}$, $e^{-t/\tau}$, and $e^{-0.618t/\tau}$. It follows from Fig. 1.10 that, given a global approximation and the state of thermodynamic equilibrium, the DF may repeatedly assume the same values and, consequently, the system may repeatedly pass through the same states. Moreover, in the process of evolution the system may find itself in the state of thermodynamic equilibrium and, nevertheless, leave the latter state. The respective "apocalyptic" points $A_1, A_2\ldots$ are indicated

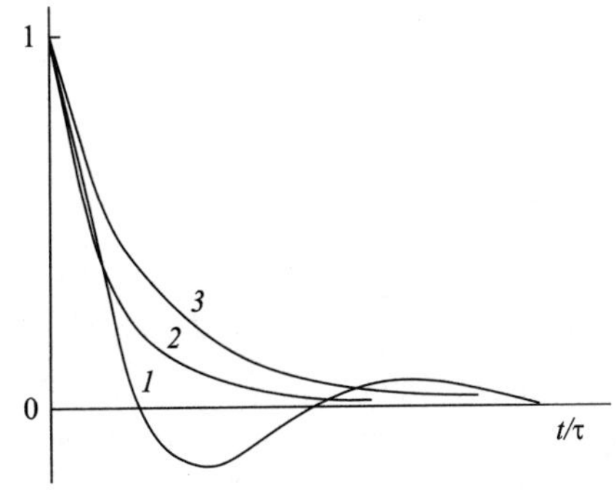

FIGURE 1.10 Evolution of the time parts of distribution function (1.4.49)–(1.4.51): (1) $e^{-0.5t/\tau}\cos 0.866\frac{t}{\tau}$, (2) $e^{-t/\tau}$, (3) $e^{-0.618t/\tau}$.

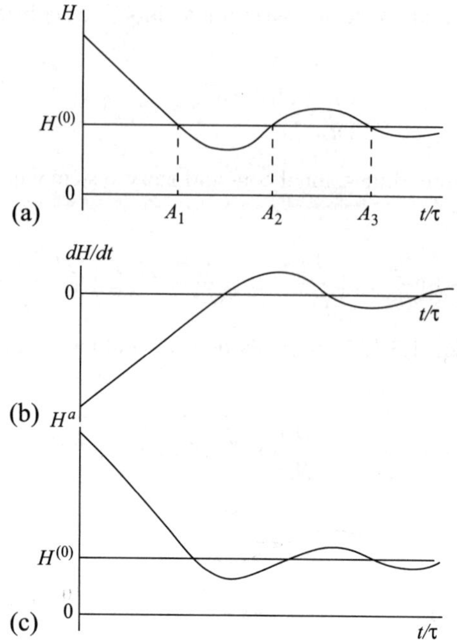

FIGURE 1.11 Vibrations of the (a) Boltzmann and (c) generalized Boltzmann H-functions when approaching the state of equilibrium, as well as (b) evolution of derivative Df/Dt for kinetic equation of the type of Eq. (1.4.44).

in Fig. 1.11. What happens in these points? The distribution function reaches its value corresponding to thermodynamic equilibrium,

$$H = H^{(0)} = \int f^{(0)} \ln f^{(0)} \mathrm{d}\mathbf{v}, \tag{1.4.52}$$

however, the derivative

$$\left(\frac{\partial H}{\partial t}\right)_{A_i} = \int \left(\frac{\partial f}{\partial t}\right)_{A_i} \left(\ln f^{(0)} + 1\right) \mathrm{d}\mathbf{v} \tag{1.4.53}$$

is other than zero, and the sign of the integral depends on the sign of derivative $(\partial f/\partial t)_{A_i}$ $(i=1, 2, \ldots)$.

Consequently, the Boltzmann entropy may experience vibrations, shown in Fig. 1.11, when approaching the state of equilibrium. Fig. 1.11 gives, as a result of qualitative graphic differentiation, the evolution of derivative $\partial H/\partial t$ and generalized entropy H^a, which also vibrationally approaches the global equilibrium. We use (1.4.49) to readily find the values of H^a in the apocalyptic points,

$$H^a_{A_i} = H^{(0)} - 0.866 \mathrm{e}^{-0.5\frac{t_i}{\tau}} (-1)^i \int C(v) \left(\ln f^{(0)} + 1\right) \mathrm{d}\mathbf{v}, \quad i = 1, 2, \ldots \tag{1.4.54}$$

because

$$\left(\frac{\partial f}{\partial t}\right)_{A_i} = C \frac{0.866}{\tau} \mathrm{e}^{-0.5\frac{t_i}{\tau}} (-1)^i, \tag{1.4.55}$$

$$t_i = \frac{\pi\tau}{0.866}\left(i - \frac{1}{2}\right). \tag{1.4.56}$$

Condition (1.4.56) follows from vanishing of $\cos(0.866t/\tau)$ in the apocalyptic points.

One can see now that the reversibility of processes shows up, on the level of one-particle description as well, as the reflection of reversible processes in a physical system, i.e. on the level of Liouville's equation written relative to the distribution function f_N.

By analogy with the notation of difference schemas with weight, one can replace Eqs. (1.3.71) and (1.4.45) by a generalized kinetic equation in the form

$$\frac{Df}{Dt}-\gamma_1\frac{D}{Dt}\left(\tau\frac{Df}{Dt}\right)+\gamma_2\frac{D}{Dt}\left(\tau\frac{Df}{Dt}\right)=J^{st}(f), \tag{1.4.57}$$

where γ_1, γ_2 are some functions proportional and correspondingly inversely proportional to T_{ret}, the time of return of the system to the initial state. For real dissipative processes, the time T_{ret} is so great, and the function γ_2 so small, as compared to unity that it is sufficient to use the GBE in the form of (1.3.71).

$$(1)\, e^{-0.5t/\tau}\cos 0.866\frac{t}{\tau},\quad (2) e^{-t/\tau},\quad (3) e^{-0.618t/\tau}.$$

approaching the state of equilibrium, as well as (b) evolution of derivative D*f*/D*t* for kinetic equation of the type of (1.4.44).

It is interesting to treat this problem also from the standpoint of the so-called "physical" derivation of the Boltzmann equation. Of course, subsequent reasoning will no longer be rigorous, but it will be nevertheless useful for understanding the situation. For this purpose, we will treat the variation of the number of particles of the species α, which were initially, at the moment of time t, found in the volume $d\mathbf{r}^t d\mathbf{v}_\alpha^t$ of the phase space. After the interval of time dt, the particles in the absence of collisions will be found in the volume $d\mathbf{r}^{t+dt} d\mathbf{v}_\alpha^{t+dt}$ and in so doing, the difference

$$f_\alpha[\mathbf{r}(t+dt), \mathbf{v}_\alpha(t+dt), t+dt]d\mathbf{r}^{t+dt}d\mathbf{v}_\alpha^{t+dt} - f_\alpha(\mathbf{r}, \mathbf{v}_\alpha, t)d\mathbf{r}^t d\mathbf{v}_\alpha^t$$

will go to zero. In the presence of external forces $\mathbf{F}_\alpha$ such as Lorentz forces, there is no reason, generally speaking, to assume that the elements of phase volume hold out in the course of time. It can be shown that within accuracy $O[(dt)^2]$ the elements of phase volumes can be transformed as

$$d\mathbf{r}^{t+dt} = \left[1+\frac{\partial}{\partial \mathbf{r}}\cdot \mathbf{F}_\alpha (dt)^2\right] d\mathbf{r}^t. \tag{1.4.58}$$

But, in the general case the six-dimensional Jacobian should be introduced $d[\mathbf{r}^{t+dt}, \mathbf{v}_\alpha^{t+dt}]/d[\mathbf{r}^t, \mathbf{v}_\alpha^t]$.

Write the balance equation:

$$\begin{aligned} &f_\alpha\left[\mathbf{r}+\mathbf{v}_\alpha dt+\frac{1}{2}\mathbf{F}_\alpha (dt)^2, \mathbf{v}_\alpha+\mathbf{F}_\alpha dt+\frac{1}{2}\frac{\partial \mathbf{F}_\alpha}{\partial t}(dt)^2, t+dt\right]\cdot\frac{d\left[\mathbf{r}^{t+dt}, \mathbf{v}_\alpha^{t+dt}\right]}{d\left[\mathbf{r}^t, \mathbf{v}_\alpha^t\right]} \\ &-f_\alpha(\mathbf{r}, \mathbf{v}_\alpha, t) = J_\alpha^{st} dt, \end{aligned} \tag{1.4.59}$$

containing the terms of $O[(dt)^2]$ in the left-hand side of Eq. (1.4.59).

Calculation of indicated Jacobian leads to the result

$$\frac{d\left[\mathbf{r}^{t+dt}, \mathbf{v}_\alpha^{t+dt}\right]}{d\left[\mathbf{r}^t, \mathbf{v}_\alpha^t\right]} = 1+\left[\left(\frac{q_\alpha}{m_\alpha}\right)^2 B^2 - \frac{1}{2}\frac{\partial}{\partial \mathbf{r}}\cdot \mathbf{F}_\alpha\right](dt)^2, \tag{1.4.60}$$

where q_α is charge of αth particle, **B** is magnetic induction, $\mathbf{F}_\alpha$ is the external force, acting on the particle of species α.

Expanding the distribution function as a power series and conserving terms of order $O[(dt)^2]$, the integro-differential difference equation is obtained [75, 76]

$$\begin{aligned} &\frac{\partial f_\alpha}{\partial t} + \mathbf{v}_\alpha\cdot\frac{\partial f_\alpha}{\partial \mathbf{r}} + \mathbf{F}_\alpha\cdot\frac{\partial f_\alpha}{\partial \mathbf{v}_\alpha} + \tau\left\{\mathbf{F}_\alpha\cdot\frac{\partial f_\alpha}{\partial \mathbf{r}} + \frac{\partial \mathbf{F}_\alpha}{\partial t}\cdot\frac{\partial f_\alpha}{\partial \mathbf{v}_\alpha}\right. \\ &+\frac{\partial f_\alpha}{\partial \mathbf{v}_\alpha}\mathbf{F}_\alpha : \frac{\partial}{\partial \mathbf{v}_\alpha}\mathbf{F}_\alpha + \frac{\partial f_\alpha}{\partial \mathbf{v}_\alpha}\mathbf{v}_\alpha : \frac{\partial}{\partial \mathbf{r}}\mathbf{F}_\alpha + \frac{\partial^2 f_\alpha}{\partial t^2} + 2\left(\frac{q_\alpha}{m_\alpha}\right)B^2 f_\alpha \\ &-f_\alpha\frac{\partial}{\partial \mathbf{r}}\cdot\mathbf{F}_\alpha + \frac{\partial^2 f_\alpha}{\partial \mathbf{r}\partial \mathbf{r}} : \mathbf{v}_\alpha\mathbf{v}_\alpha + \frac{\partial^2 f_\alpha}{\partial \mathbf{v}_\alpha \partial \mathbf{v}_\alpha} : \mathbf{F}_\alpha\mathbf{F}_\alpha \\ &\left. +2\frac{\partial^2 f_\alpha}{\partial \mathbf{v}_\alpha\partial \mathbf{r}} : \mathbf{v}_\alpha\mathbf{F}_\alpha + 2\frac{\partial^2 f_\alpha}{\partial \mathbf{v}_\alpha \partial t}\mathbf{F}_\alpha + 2\frac{\partial^2 f_\alpha}{\partial \mathbf{r}\partial t}\cdot\mathbf{v}_\alpha\right\} = J_\alpha^{st}, \end{aligned} \tag{1.4.61}$$

where $\tau = \Delta t/2$ is a half of time step in the finite-difference approximation.

Equation (1.4.61) must be treated as the source of difference-differential approximations for the left-hand side of the BE. Of course, the difference-differential operator in (1.4.61) does not coincide with the differential operator of GBE (1.3.85); τ is simply the time step of the difference schema, and with formal growth, secular terms appear in the left-hand side of (1.4.61).

Now, let $F_\alpha \equiv 0$, then Eq. (1.4.61) takes the form

$$\frac{Df_\alpha}{Dt} + \tau \frac{D}{Dt}\frac{Df_\alpha}{Dt} = J_\alpha^{st}, \tag{1.4.62}$$

where $\frac{D}{Dt} = \frac{\partial}{\partial t} + \mathbf{v}_\alpha \cdot \frac{\partial}{\partial \mathbf{r}}$ and the question of approximation of high accuracy for J_α^{st} formally remains open. In Eq. (1.4.62), τ is a constant of opposite sign as compared to the analogous term in (1.3.85). Is it possible to derive a difference-differential approximation with the negative sign at the second substantial derivative? No doubt, this is possible.

We will now treat the approximation "backward" in time in the form

$$\begin{aligned} &f_\alpha(t, \mathbf{r}, \mathbf{v}_\alpha) - f_\alpha(t - \Delta t, \mathbf{r} - \Delta \mathbf{r}, \mathbf{v}_\alpha - \Delta \mathbf{v}_\alpha) \\ &= \left[\frac{Df_\alpha}{Dt}\right]_t \Delta t - \frac{(\Delta t)^2}{2}\left[\frac{D}{Dt}\frac{Df_\alpha}{Dt}\right]_t + \cdots \end{aligned} \tag{1.4.63}$$

which leads to the balance equation:

$$\frac{Df_\alpha}{Dt} - \tau \frac{D}{Dt}\frac{Df_\alpha}{Dt} = J_\alpha^{st}, \tau = \frac{\Delta t}{2} \tag{1.4.64}$$

From the mathematical standpoint, both difference-differential approximations are completely equivalent, and one or the other approximation may be preferred only from the standpoint of stability of the difference schema employed. We will now use the assumptions, which have brought us to the approximate notation of the GBE in the form of Eq. (1.3.85). In this case, the difference-differential approximation is the same in these assumptions as in the GBE provided, naturally, that the time step $\Delta t = \text{const}$ of the difference schema from the GBKT is formally used as a doubled mean time between collisions.

Therefore, two mathematically equivalent difference approximations (1.4.59) and (1.4.63) have different physical meaning (see Eq. 1.3.62). One of them corresponds to the approximation to "predicted future," and the other one, to "determinate past" which is consistent with the principle of irreversibility of time.

It is significant that this fact is of no importance as regards the "physical derivation" of the Boltzmann equation, because no consideration at all is given to the DF evolution on the r_b-scale, and as a consequence, both approximations lead to one and the same result.

So, Boltzmann's result (treated as the effect of phenomenological derivation of BE) $(dH/dt) \leq 0$ may be derived without explicitly using the assumption on the irreversibility of time, while the analogous result in the theory of GBE calls for explicit use of the assumption on the irreversibility of time or, which is the same, of the principle of causality. Thereby, one of the main paradoxes, providing a subject for discussions in the Boltzmann physical kinetics, is resolved.

We will now treat the theory of kinetically consistent difference schemes (KCDS) (see, for example, [49, 70]). The ideas of this theory date back to the studies by Reitz (in particular, [77]), who used the splitting method to solve problems of the theory of transfer to the kinetic and hydrodynamic stages. As distinct from [77], the KCDS theory uses the DF expansion in Taylor series with respect to the parameter $v\tau$, where τ is some arbitrary parameter defined by the ratio h of the network step in space to the characteristic hydrodynamic velocity V_{hyd} with the accuracy within third-order terms

$$f^{j+1}(\mathbf{r}, \mathbf{v}, t^{j+1}) = f_0^j - \tau \sum_{\alpha=1}^{3} \frac{\partial f_0^j}{\partial r_\alpha} v_\alpha + \frac{\tau^2}{2} \sum_{\alpha,\beta=1}^{3} \frac{\partial^2 f_0^j}{\partial r_\alpha \partial r_\beta} v_\alpha v_\beta + \cdots \tag{1.4.65}$$

Here, f_0 is Maxwell's function, and j defines the number of step in time. In later studies dealing with the KCDS (see for example [78]), three parameters are introduced, $\tau^x = (h^x/2V_{\text{hyd}})$, $\tau^y = (h^y/2V_{\text{hyd}})$, $\tau^z = (h^z/2V_{\text{hyd}})$, which are defined by space steps on the coordinates x, y, and z.

In order to derive the values of gas-dynamic parameters on the new time layer $t = t^{j+1}$, the performed expansion is multiplied by the summator invariants and integrated with respect to the velocities of molecules of one-component gas. As a result, we derive a system of integro-differential equations with additional terms (on the right-hand sides, as distinct from the classical hydrodynamic equations), which present, by virtue of selected approximation, a combination of second space derivatives multiplied by the time step.

This approach does not lead to a new hydrodynamic description. Moreover, as follows from derivation (1.4.61) (see also [75, 76]), it does not even provide in the general case a second-order approximation for the Boltzmann equation or for generalized hydrodynamic equations, which in reality result directly from the GBE. Attempts at substantiating the KCDS on the basis of modified BE with an additional relaxation term are inadequate, because the BE "works" at times of the order of the relaxation time. In particular, with this approximation, as compared to the generalized hydrodynamic equations and GBE,

(a) all cross derivatives with respect to space and time are absent, as well as second time derivatives, and, as a result, the KCDS cannot be used correctly to simulate turbulent flows;
(b) the KCDS cannot be used to construct the generalized Navier-Stokes approximation;
(c) external forces cannot be introduced in the KCDS;
(d) it is impossible to clarify the physical meaning of the parameter τ, this leading to the appearance of secular terms in the equations;
(e) it is impossible to estimate the contribution to hydrodynamic equations due to inevitable modification of the collision term;
(f) as noted in [79], "the common drawback of [70, 78, 80] consists, in particular, in that the introduced additional terms disturb the invariance of the kinetic equation relative to Galilean transformations. In this case as well, the additional terms were introduced without adequate substantiation."

The following analogy may be drawn: let the finite-difference approximation of Newton's second law $\ddot{x} = F/m$ be written; depending on the accuracy of the scheme used, finite-difference increments of the third and higher orders may appear in the finite-difference approximation. This does not mean, naturally, that the return to differential notation gives a new law of nature of the type of $\ddot{x} + \tau \ldots x = F/m$. The reason for this situation is quite obvious: it is impossible to obtain a qualitatively new physical description by using a formally higher difference approximation for the classical equation.

In the approach developed by Klimontovich [41, 50], Liouville's equation is replaced by another kinetic equation with a source term (or "priming" term, using the terminology of [41, p. 319]), that differs from Liouville's equation by the source term [41] written in the τ form as

$$\frac{f_N(r, v, t) - \widetilde{f}_N(r', v', t)}{\tau_{\text{ph}}}$$

According to Klimontovich, this term describes the "adjustment" of microscopic distribution of particles to the appropriate smoothed distribution. Henceforward, the value of τ_{ph} is selected to be equal to that of τ during transition to the one-particle description. The resultant equation proves to be a combination of the Boltzmann and Fokker-Planck description (the differential part of the BE remains unchanged) with an additional "collision integral" [41, p. 251],

$$I_{(R)}(R, v, t) = \frac{\partial}{\partial R}\left[D\frac{\partial f}{\partial R} - bF(R)f\right] \tag{1.4.66}$$

with due regard for smoothing over the "point" dimensions, where D is one of three kinetic coefficients (kinematic viscosity υ_k, thermal diffusivity χ, and self-diffusion D), and b is the mobility.

It is assumed that all three coefficients are identical, and the difference between them may be taken into account by using some other, more complex, smoothing function. One can perceive an analogy between the continuity equations of Slezkin (I.20) and Klimontovich [41]. We see little point in discussing the remaining analogous HE.

In fact, the source term in Liouville's equation may only appear in the case of incomplete statistical description of the reacting system, or in the presence of special non-holonomic links and radiation, while the size of the "point" (using Klimontovich's terminology) is defined by the r_b-scale in the BE that was previously unaccounted for.

In closing this section we want to emphasize the fundamental point that the introduction of the third scale, which describes the distribution function variations on a time scale of the order of the collision time, leads to the single-order terms in the Boltzmann equation prior to Bogolyubov-chain-decoupling approximations, and to terms proportional to the mean time between collisions after these approximations. It follows that the Boltzmann equation requires a radical modification—which, in our opinion, is exactly what the generalized Boltzmann equation provides.

1.5 GENERALIZED BOLTZMANN EQUATION AND ITERATIVE CONSTRUCTION OF HIGHER-ORDER EQUATIONS IN THE BOLTZMANN KINETIC THEORY

Let us consider the relation between the generalized Boltzmann equation and the iterative construction of higher-order equations in the Boltzmann kinetic theory. Neglecting external forces, the Boltzmann equation for a spatially homogeneous case, with the right-hand side taken in the Bhatnagar-Gross-Krook (BGK) form, is given by

$$\frac{\partial f}{\partial t} = -\frac{f - f_0}{\tau_{\text{rel}}}, \tag{1.5.1}$$

where τ_{rel} is the relaxation time, and f_0 the equilibrium distribution function. From Eq. (1.5.1) it follows that

$$f = f_0 - \tau_{rel}\frac{\partial f}{\partial t} \cong f_0 - \tau_{rel}\frac{\partial f_0}{\partial t}. \tag{1.5.2}$$

The second iteration is constructed in a similar fashion giving

$$f = f_0 - \tau_{rel}\frac{\partial}{\partial t}\left(f_0 - \tau_{rel}\frac{\partial f}{\partial t}\right) \cong f_0 - \tau_{rel}\frac{\partial f_0}{\partial t} + \tau_{rel}^2\frac{\partial^2 f_0}{\partial t^2}. \tag{1.5.3}$$

Thus we obtain for the distribution function the series representation

$$f = \sum_{i=0}^{\infty}(-1)^i\frac{\partial^i f_0}{\partial t^i}\tau_{rel}^i \tag{1.5.4}$$

where the zero-order derivative operator corresponds to the distribution function f_0.

From Eq. (1.5.3) there follows an analog of the kinetic equation (1.5.1) for the second approximation:

$$-\tau_{rel}\frac{\partial^2 f_0}{\partial t^2} + \frac{\partial f_0}{\partial t} = -\frac{f - f_0}{\tau_{rel}} \tag{1.5.5}$$

It is important to note that the second time derivative of the distribution function in Eq. (1.5.5) occurs with a minus sign. In the general case we have the expansion

$$\sum_{i=1}^{\infty}\tau_{rel}^{i-1}(-1)^{i-1}\frac{\partial^i f_0}{\partial t^i} = -\frac{f - f_0}{\tau_{rel}}. \tag{1.5.6}$$

We now proceed to show that the generalized Boltzmann equation permits an iterative procedure similar to that just described. To this end we can write the second approximation in the form (using notations of approximation (1.3.62))

$$\begin{aligned}\frac{\mathrm{D}_1\hat{f}_1^1}{\mathrm{D}\hat{t}_b} &= -\frac{\mathrm{D}_1}{\mathrm{D}\hat{t}_b}\left(\hat{\tau}\frac{\mathrm{D}_1\hat{f}_1}{\mathrm{D}\hat{t}_b}\right) \cong -\frac{\mathrm{D}_1}{\mathrm{D}\hat{t}_b}\left[\hat{\tau}\frac{\mathrm{D}_1}{\mathrm{D}\hat{t}_b}(\hat{f}_1^0 + \hat{f}_1^1)\right] \\ &= -\frac{\mathrm{D}_1}{\mathrm{D}\hat{t}_b}\left(\hat{\tau}\frac{\mathrm{D}_1\hat{f}_1^0}{\mathrm{D}\hat{t}_b} + \hat{\tau}\frac{\mathrm{D}_1\hat{f}_1^1}{\mathrm{D}\hat{t}_b}\right) = -\frac{\mathrm{D}_1}{\mathrm{D}\hat{t}_b}\left(\hat{\tau}\frac{\mathrm{D}_1\hat{f}_1^0}{\mathrm{D}\hat{t}_b}\right) + \frac{\mathrm{D}_1}{\mathrm{D}\hat{t}_b}\left[\hat{\tau}\frac{\mathrm{D}_1}{\mathrm{D}\hat{t}_b}\left(\hat{\tau}\frac{\mathrm{D}_1\hat{f}_1^0}{\mathrm{D}\hat{t}_b}\right)\right].\end{aligned} \tag{1.5.7}$$

Higher approximations follow the same pattern. Thus, with the notation of relation (1.3.62) we obtain

$$\begin{aligned}\frac{\mathrm{D}_1\hat{f}_1^1}{\mathrm{D}\hat{t}_b} &= -\frac{\mathrm{D}_1}{\mathrm{D}\hat{t}_b}\left(\hat{\tau}\frac{\mathrm{D}_1\hat{f}_1^0}{\mathrm{D}\hat{t}_b}\right) + \frac{\mathrm{D}_1}{\mathrm{D}\hat{t}_b}\left[\hat{\tau}\frac{\mathrm{D}_1}{\mathrm{D}\hat{t}_b}\left(\hat{\tau}\frac{\mathrm{D}_1\hat{f}_1^0}{\mathrm{D}\hat{t}_b}\right)\right] \\ &\quad -\frac{\mathrm{D}_1}{\mathrm{D}\hat{t}_b}\left\{\hat{\tau}\frac{\mathrm{D}_1}{\mathrm{D}\hat{t}_b}\left[\hat{\tau}\frac{\mathrm{D}_1}{\mathrm{D}\hat{t}_b}\left(\hat{\tau}\frac{\mathrm{D}_1\hat{f}_1^0}{\mathrm{D}\hat{t}_b}\right)\right]\right\} + \cdots\end{aligned} \tag{1.5.8}$$

Interesting to estimate the accuracy of substitution of the zeroth-order term of the series f_1^0 instead of f_1 in (1.3.62)

$$\frac{\mathrm{D}_1}{\mathrm{D}\hat{t}_b}\left[\hat{\tau}\frac{\mathrm{D}_1\hat{f}_1}{\mathrm{D}\hat{t}_b}\right] \cong \frac{\mathrm{D}_1}{\mathrm{D}\hat{t}_b}\left[\hat{\tau}\frac{\mathrm{D}_1\hat{f}_1^0}{\mathrm{D}\hat{t}_b}\right]. \tag{1.5.9}$$

For this aim we use the model BGK collision integral:

$$\frac{\mathrm{D}f_1}{\mathrm{D}t} = -\frac{f_1 - f_1^0}{\tau_{rel}}. \tag{1.5.10}$$

The exact solution of Eq. (1.5.10) has the form of a non-local integral with the time delay:

$$f_1(\mathbf{v},\mathbf{r},t) = \int_0^{\infty} f_1^0(\mathbf{v} - \mathbf{F}\delta t, \mathbf{r} - \mathbf{v}\delta t, t - \delta t)\frac{1}{\tau_{rel}}\exp\left(-\frac{\delta t}{\tau_{rel}}\right)\mathrm{d}\delta t, \tag{1.5.11}$$

where δt is corresponding to the time delay.

This result can be easily proved by using the integration in (1.5.11) by parts. Relation (1.5.11) contains a convergent infinite integral as containing exponential function $\exp(-\delta t/\tau_{\text{rel}})$, which tends to zero as $\delta t \to \infty$.

In a linear approximation, expanding $f_1^0(\mathbf{v}-\mathbf{F}\delta\mathbf{t},\mathbf{r}-\mathbf{v}\delta t,t-\delta t)$ in powers of δt and neglecting $(\delta t)^2$ and higher powers, we obtain

$$f_1(\mathbf{v},\mathbf{r},t)=f_1^0(\mathbf{v},\mathbf{r},t)-(\delta t)\frac{\mathrm{D}f_1^0(\mathbf{v},\mathbf{r},t)}{\mathrm{D}t} \tag{1.5.12}$$

Comparing Eq. (1.5.2) with Eq. (1.5.12), we see that in linear approximation $\delta t \sim t_{\text{rel}}$. In the next Section these results will be discussed in the general theory of correlation functions.

Return now to a spatially homogeneous system free from forces. We obtain from Eq. (1.5.8) with $\tau=\text{const}$ that

$$\frac{\partial f_1^1}{\partial t}=\sum_{i=2}^{\infty}\tau^{i-1}(-1)^{i-1}\frac{\partial^i f_1^0}{\partial t^i}. \tag{1.5.13}$$

It follows that, in this particular case, the generalized Boltzmann equation takes the form

$$\sum_{i=2}^{\infty}\tau^{i-1}(-1)^{i-1}\frac{\partial^i f_1^0}{\partial t^i}+\frac{\partial f_1^0}{\partial t}=J^{\text{st},0} \tag{1.5.14}$$

or, collecting terms on the left, one finds

$$\sum_{i=1}^{\infty}\tau^{i-1}(-1)^{i-1}\frac{\partial^i f_1^0}{\partial t^i}=J^{\text{st},0}. \tag{1.5.15}$$

The analogy between Eqs. (1.5.6) and (1.5.15) is clearly seen.

In solid-state problems, concerning, for example, charge and energy transfer in non-degenerate semiconductors, one solves the Boltzmann equation by iterations for a spatially homogeneous system in the presence of an external electromagnetic field. For the BGK-approximated collision integral, the Boltzmann equation becomes

$$F\frac{\partial f}{\partial v_z}=-\frac{f-f_0}{\tau_{\text{rel}}} \tag{1.5.16}$$

(for a z-directed external force F), and the distribution function is written as

$$f=f_0-\tau_{\text{rel}}F\frac{\partial f}{\partial v_z}. \tag{1.5.17}$$

In the first approximation, we obtain

$$f=f_0-\tau_{\text{rel}}F\frac{\partial f_0}{\partial v_z}. \tag{1.5.18}$$

The substitution of Eq. (1.5.18) into the left-hand side of Eq. (1.5.17) yields the second-order approximation,

$$f=f_0-\tau_{\text{rel}}F\frac{\partial f_0}{\partial v_z}+\tau_{\text{rel}}^2F^2\frac{\partial^2 f_0}{\partial v_z^2}+\cdots, \tag{1.5.19}$$

provided the external force F acting on the particle is velocity-independent. The dots in this equation indicate that the procedure of constructing the series may be continued by this algorithm. From Eq. (1.5.19), the second-order accurate equation is

$$F\frac{\partial f_0}{\partial v_z}-\tau_{\text{rel}}F^2\frac{\partial^2 f_0}{\partial v_z^2}=-\frac{f-f_0}{\tau_{\text{rel}}}. \tag{1.5.20}$$

This equation turns out to be a particular case of the generalized Boltzmann equation if the system under study is stationary, spatially homogeneous, and if the applied field is sufficiently weak, giving hope for the fast convergence of the mentioned series, in which the corresponding derivatives are taken of the equilibrium distribution function. The representation of the distribution function in a series form, Eqs. (1.5.4) or (1.5.19), is only possible when one uses the BGK model for the Boltzmann collision integral.

Thus, the generalized Boltzmann equation automatically captures the second iteration in the Boltzmann theory for $\tau=\tau_{rel}$, but it does not, of course, presuppose the fulfillment of all the conditions listed. Note also that the appearance of the minus sign in the right-hand sides of Eqs. (1.5.1) and (1.5.16) in the BGK approximation has a deep physical meaning: this sign makes it possible to prove the H-theorem for the BGK-approximated Boltzmann equation and is related directly to the approximation proceeded against the course of time.

1.6 GENERALIZED BOLTZMANN EQUATION AND THE THEORY OF NON-LOCAL KINETIC EQUATIONS WITH TIME DELAY

It is of interest to examine the relation between the Boltzmann equation and the theory of kinetic equations accounting for time delay effects. We resort to the Bogolyubov Eq. (1.3.1) for determining the evolution of the s-particle distribution function in a one-component gas:

$$\frac{\partial f_s}{\partial t}+\sum_{i=1}^{s}\mathbf{v}_i\cdot\frac{\partial f_s}{\partial \mathbf{r}_i}+\sum_{i=1}^{s}\mathbf{F}_i\cdot\frac{\partial f_s}{\partial \mathbf{v}_i}+\sum_{ij=1}^{s}\mathbf{F}_{ij}\cdot\frac{\partial f_s}{\partial \mathbf{v}_i} = -\frac{1}{N}\sum_{i=1}^{s}\sum_{j=s+1}^{N}\int \mathbf{F}_{ij}\cdot\frac{\partial f_{s+1}}{\partial \mathbf{v}_i}d\Omega_{s+1}. \tag{1.6.1}$$

In writing Eq. (1.6.1) the normalization condition

$$\int f_s \mathrm{d}\Omega_1\ldots \mathrm{d}\Omega_s=N^s \tag{1.6.2}$$

is used and it also been assumed that the dynamic state of the system is fully described by the phase variables Ω_i.

Introducing the correlation functions W, the two-particle distribution function may be written as

$$f_2(\Omega_1,\Omega_2,t)=f_1(\Omega_1,t)f_1(\Omega_2,t)+W_2(\Omega_1,\Omega_2,t). \tag{1.6.3}$$

On the r_b scale, variables Ω_1 and Ω_2 turn out to be correlated, but because of definition (1.6.3) this effect is accounted for by the function W_2. Consequently, in this approach, it is the integral term containing W_2 which must lead to the Boltzmann (or a more general) collision integral. The BBGKY-I equation has the form

$$\frac{\partial f_1}{\partial t}+\mathbf{v}_1\cdot\frac{\partial f_1}{\partial \mathbf{r}_1}+\mathbf{F}_1\cdot\frac{\partial f_1}{\partial \mathbf{v}_1}+\frac{1}{N}\sum_{j=2}^{N}\frac{\partial f_1}{\partial \mathbf{v}_1}\cdot\int \mathbf{F}_{1j}f_1(2)\mathrm{d}\Omega_2 = -\frac{1}{N}\sum_{j=2}^{s}\int \mathbf{F}_{1j}\cdot\frac{\partial W_2}{\partial \mathbf{v}_1}\mathrm{d}\Omega_2 \tag{1.6.4}$$

The internal force $\mathbf{F}_1^{(\mathrm{in})}$ exerted on a given particle 1 from the side of particle 2 at its arbitrary location in phase space may be written as

$$\frac{1}{N}\sum_{j=2}^{N}\int \mathbf{F}_{1j}f_1(2)\mathrm{d}\Omega_2=\mathbf{F}_1^{(\mathrm{in})}. \tag{1.6.5}$$

Here, as usual, the symbol "2", the argument of the one-particle distribution function f_1 (2), denotes the phase variables of the particle 2. For identical particles, one finds

$$\mathbf{F}_1^{(in)}=\frac{N-1}{N}\int \mathbf{F}_{12}f_1(2)\mathrm{d}\Omega_2\cong\int \mathbf{F}_{12}f_1(2)\mathrm{d}\Omega_2 =\int \mathbf{F}_{1j}f_1(j)\mathrm{d}\Omega_j\,,\ j=2,3,\ldots. \tag{1.6.6}$$

If the self-consistent force $\mathbf{F}_1^{(\mathrm{sc},1)}$ acting on a probe particle in the one-particle picture, is introduced as the sum

$$\mathbf{F}_1^{(\mathrm{sc},\,1)}=\mathbf{F}_1+\mathbf{F}_1^{(\mathrm{in})}, \tag{1.6.7}$$

of the external force $\mathbf{F}_1$ and the internal force $\mathbf{F}_1^{(\mathrm{in})}$ defined by Eq. (1.6.5), then we arrive at the equation

$$\frac{\partial f_1}{\partial t}+\mathbf{v}_1\cdot\frac{\partial f_1}{\partial \mathbf{r}_1}+\mathbf{F}_1^{(\mathrm{sc},1)}\cdot\frac{\partial f_1}{\partial \mathbf{v}_1}=-\frac{1}{N}\sum_{j=2}^{N}\int \mathbf{F}_{1j}\cdot\frac{\partial W_2}{\partial \mathbf{v}_1}\mathrm{d}\Omega_2. \tag{1.6.8}$$

The BBGKY-2 equation has the form

$$\begin{aligned}&\frac{\partial f_2}{\partial t}+\mathbf{v}_1\cdot\frac{\partial f_2}{\partial \mathbf{r}_1}+\mathbf{v}_2\cdot\frac{\partial f_2}{\partial \mathbf{r}_2}+\mathbf{F}_1\cdot\frac{\partial f_2}{\partial \mathbf{v}_1}+\mathbf{F}_2\cdot\frac{\partial f_2}{\partial \mathbf{v}_2}+\mathbf{F}_{12}\cdot\frac{\partial f_2}{\partial \mathbf{v}_1}+\\&+\mathbf{F}_{21}\cdot\frac{\partial f_2}{\partial \mathbf{v}_2}=-\frac{1}{N}\sum_{j=3}^{N}\int\left[\mathbf{F}_{1j}\cdot\frac{\partial f_3}{\partial \mathbf{v}_1}+\mathbf{F}_{2j}\cdot\frac{\partial f_3}{\partial \mathbf{v}_2}\right]\mathrm{d}\Omega_j.\end{aligned} \tag{1.6.9}$$

We next express the distribution function f_3 in terms of the correlation functions as

$$\begin{aligned}f_3(\Omega_1,\Omega_2,\Omega_3,t)=&f_1(\Omega_1,t)f_1(\Omega_2,t)f_1(\Omega_3,t)+f_1(\Omega_1,t)W_2(\Omega_2,\Omega_3,t)\\&+f_1(\Omega_2,t)W_2(\Omega_1,\Omega_3,t)+f_1(\Omega_3,t)W_2(\Omega_1,\Omega_2,t)\\&+W_3(\Omega_1,\Omega_2,\Omega_3,t).\end{aligned} \tag{1.6.10}$$

and apply the theory of correlation functions to obtain an approximation for collision integrals.

Assumption 1
The correlation function W_3 may be neglected.
Using Eq. (1.6.2), Eq. (1.6.10) can be put into the form

$$f_3(1,2,3)=f_1(3)f_2(1,2)+f_1(2)W_2(1,3)+f_1(1)W_2(2,3). \tag{1.6.11}$$

From Eq. (1.6.9), Eq. (1.6.11) follow

$$\begin{aligned}&\frac{\partial f_2}{\partial t}+\mathbf{v}_1\cdot\frac{\partial f_2}{\partial \mathbf{r}_1}+\mathbf{v}_2\cdot\frac{\partial f_2}{\partial \mathbf{r}_2}+\mathbf{F}_1\cdot\frac{\partial f_2}{\partial \mathbf{v}_1}+\mathbf{F}_2\cdot\frac{\partial f_2}{\partial \mathbf{v}_2}+\mathbf{F}_{12}\cdot\frac{\partial f_2}{\partial \mathbf{v}_1}+\mathbf{F}_{21}\cdot\frac{\partial f_2}{\partial \mathbf{v}_2}\\&=-\frac{1}{N}\sum_{j=3}^{N}\int\left\{\begin{aligned}&\mathbf{F}_{1j}\cdot\frac{\partial}{\partial \mathbf{v}_1}[f_1(3)f_2(1,2)+f_1(2)W_2(1,3)+f_1(1)W_2(2,3)]\\&+\mathbf{F}_{2j}\cdot\frac{\partial}{\partial \mathbf{v}_2}[f_1(3)f_2(1,2)+f_1(2)W_2(1,3)+f_1(1)W_2(2,3)]\end{aligned}\right\}\mathrm{d}\Omega_j.\end{aligned} \tag{1.6.12}$$

Assumption 2
The polarization effects leading to the integrals

$$-\frac{1}{N}\sum_{j=3}^{N}\int\left[\mathbf{F}_{1j}\cdot\frac{\partial}{\partial \mathbf{v}_1}(f_1(1)W_2(2,3))\right]\mathrm{d}\Omega_j,\ -\frac{1}{N}\sum_{j=3}^{N}\int\left[\mathbf{F}_{2j}\cdot\frac{\partial}{\partial \mathbf{v}_2}(f_1(2)W_2(1,3))\right]\mathrm{d}\Omega_j$$

may be ignored.
Using Assumption 2 we find from Eq. (1.6.12)

$$\begin{aligned}&\frac{\partial f_2}{\partial t}+\mathbf{v}_1\cdot\frac{\partial f_2}{\partial \mathbf{r}_1}+\mathbf{v}_2\cdot\frac{\partial f_2}{\partial \mathbf{r}_2}+\mathbf{F}_1\cdot\frac{\partial f_2}{\partial \mathbf{v}_1}+\mathbf{F}_2\cdot\frac{\partial f_2}{\partial \mathbf{v}_2}+\mathbf{F}_{12}\cdot\frac{\partial f_2}{\partial \mathbf{v}_1}+\mathbf{F}_{21}\cdot\frac{\partial f_2}{\partial \mathbf{v}_2}\\&=-\frac{1}{N}\sum_{j=3}^{N}\int\left\{\begin{aligned}&\mathbf{F}_{1j}\cdot\frac{\partial}{\partial \mathbf{v}_1}[f_1(3)f_2(1,2)+f_1(2)W_2(1,3)]\\&+\mathbf{F}_{2j}\cdot\frac{\partial}{\partial \mathbf{v}_2}[f_1(3)f_2(1,2)+f_1(1)W_2(2,3)]\end{aligned}\right\}\mathrm{d}\Omega_j\end{aligned} \tag{1.6.13}$$

or

$$\begin{aligned}&\frac{\partial f_2}{\partial t}+\mathbf{v}_1\cdot\frac{\partial f_2}{\partial \mathbf{r}_1}+\mathbf{v}_2\cdot\frac{\partial f_2}{\partial \mathbf{r}_2}+\mathbf{F}_1\cdot\frac{\partial f_2}{\partial \mathbf{v}_1}+\mathbf{F}_2\cdot\frac{\partial f_2}{\partial \mathbf{v}_2}+\mathbf{F}_{12}\cdot\frac{\partial f_2}{\partial \mathbf{v}_1}+\mathbf{F}_{21}\cdot\frac{\partial f_2}{\partial \mathbf{v}_2}\\&+\frac{1}{N}\sum_{j=3}^{N}\int\left\{\begin{aligned}&\mathbf{F}_{1j}\cdot\frac{\partial}{\partial \mathbf{v}_1}[f_1(3)f_2(1,2)]\\&+\mathbf{F}_{2j}\cdot\frac{\partial}{\partial \mathbf{v}_2}[f_1(3)f_2(1,2)]\end{aligned}\right\}\mathrm{d}\Omega_j\\&=-\frac{1}{N}\sum_{j=3}^{N}\int\left\{\begin{aligned}&\mathbf{F}_{1j}\cdot\frac{\partial}{\partial \mathbf{v}_1}[f_1(2)W_2(1,3)]\\&+\mathbf{F}_{2j}\cdot\frac{\partial}{\partial \mathbf{v}_2}[f_1(1)W_2(2,3)]\end{aligned}\right\}\mathrm{d}\Omega_j.\end{aligned} \tag{1.6.14}$$

Introducing self-consistent forces in the framework of a two-particle description ($j=3, 4, 5\ldots$), viz.

$$\mathbf{F}_1^{(sc,2)}=\mathbf{F}_1+\mathbf{F}_{12}+\int \mathbf{F}_{1j}f_1(j)\mathrm{d}\Omega_j,\quad j=3,4,5\ldots, \tag{1.6.15}$$

analogously

$$\mathbf{F}_2^{(sc,2)}=\mathbf{F}_2+\mathbf{F}_{21}+\int \mathbf{F}_{2j}f_1(j)\mathrm{d}\Omega_j,\quad j=3,4,5\ldots, \tag{1.6.16}$$

we find from Eqs (1.6.14) to (1.6.16)

$$\begin{aligned}&\frac{\partial f_2}{\partial t}+\mathbf{v}_1\cdot\frac{\partial f_2}{\partial \mathbf{r}_1}+\mathbf{v}_2\cdot\frac{\partial f_2}{\partial \mathbf{r}_2}+\mathbf{F}_1^{(sc,2)}\cdot\frac{\partial f_2}{\partial \mathbf{v}_1}+\mathbf{F}_2^{(sc,2)}\cdot\frac{\partial f_2}{\partial \mathbf{v}_2}\\&=-\frac{1}{N}\sum_{j=3}^{N}\int\left\{\begin{aligned}&\mathbf{F}_{1j}\cdot\frac{\partial}{\partial \mathbf{v}_1}[f_1(2)W_2(1,3)]\\&+\mathbf{F}_{2j}\cdot\frac{\partial}{\partial \mathbf{v}_2}[f_1(1)W_2(2,3)]\end{aligned}\right\}\mathrm{d}\Omega_j.\end{aligned} \tag{1.6.17}$$

Making use of results

$$\frac{\partial f_1(1)}{\partial t}+\mathbf{v}_1\cdot\frac{\partial f_1(1)}{\partial \mathbf{r}_1}+\mathbf{F}_1^{(sc,1)}\cdot\frac{\partial f_1(1)}{\partial \mathbf{v}_1}=-\int \mathbf{F}_{13}\cdot\frac{\partial W_2(1,3)}{\partial \mathbf{v}_1}\mathrm{d}\Omega_3, \tag{1.6.18}$$

$$\frac{\partial f_1(2)}{\partial t}+\mathbf{v}_2\cdot\frac{\partial f_1(2)}{\partial \mathbf{r}_2}+\mathbf{F}_2^{(sc,1)}\cdot\frac{\partial f_1(2)}{\partial \mathbf{v}_2}=-\int \mathbf{F}_{23}\cdot\frac{\partial W_2(2,3)}{\partial \mathbf{v}_2}\mathrm{d}\Omega_3, \tag{1.6.19}$$

we arrive at the equation for $f_2(1, 2)$:

$$\begin{aligned}&\frac{\partial f_2}{\partial t}+\mathbf{v}_1\cdot\frac{\partial f_2}{\partial \mathbf{r}_1}+\mathbf{v}_2\cdot\frac{\partial f_2}{\partial \mathbf{r}_2}+\mathbf{F}_1^{(sc,2)}\cdot\frac{\partial f_2}{\partial \mathbf{v}_1}+\mathbf{F}_2^{(sc,2)}\cdot\frac{\partial f_2}{\partial \mathbf{v}_2}\\&=f_1(2)\left[\frac{\partial f_1(1)}{\partial t}+\mathbf{v}_1\cdot\frac{\partial f_1(1)}{\partial \mathbf{r}_1}+\mathbf{F}_1^{(sc,1)}\cdot\frac{\partial f_1(1)}{\partial \mathbf{v}_1}\right]\\&+f_1(1)\left[\frac{\partial f_1(2)}{\partial t}+\mathbf{v}_2\cdot\frac{\partial f_1(2)}{\partial \mathbf{r}_2}+\mathbf{F}_2^{(sc,1)}\cdot\frac{\partial f_1(2)}{\partial \mathbf{v}_2}\right].\end{aligned} \tag{1.6.20}$$

We next introduce the substantial derivatives

$$\frac{\mathrm{D}f_2(1,2)}{\mathrm{D}t}=\frac{\partial f_2}{\partial t}+\mathbf{v}_1\cdot\frac{\partial f_2}{\partial \mathbf{r}_1}+\mathbf{v}_2\cdot\frac{\partial f_2}{\partial \mathbf{r}_2}+\mathbf{F}_1^{(sc,2)}\cdot\frac{\partial f_2}{\partial \mathbf{v}_1}+\mathbf{F}_2^{(sc,2)}\cdot\frac{\partial f_2}{\partial \mathbf{v}_2} \tag{1.6.21}$$

$$\frac{\mathrm{D}_1 f_1(1)}{\mathrm{D}t}=\frac{\partial f_1(1)}{\partial t}+\mathbf{v}_1\cdot\frac{\partial f_1(1)}{\partial \mathbf{r}_1}+\mathbf{F}_1^{(sc,1)}\cdot\frac{\partial f_1(1)}{\partial \mathbf{v}_1}, \tag{1.6.22}$$

$$\frac{\mathrm{D}_2 f_1(2)}{\mathrm{D}t}=\frac{\partial f_1(2)}{\partial t}+\mathbf{v}_2\cdot\frac{\partial f_1(2)}{\partial \mathbf{r}_2}+\mathbf{F}_2^{(sc,1)}\cdot\frac{\partial f_1(2)}{\partial \mathbf{v}_2}, \tag{1.6.23}$$

which when substituted into Eq. (1.6.20) yield

$$\frac{\mathrm{D}f_2(1,2)}{\mathrm{D}t}=f_1(2)\frac{\mathrm{D}_1 f_1(1)}{\mathrm{D}t}+f_1(1)\frac{\mathrm{D}_2 f_1(2)}{\mathrm{D}t}. \tag{1.6.24}$$

Let us now integrate with respect to time along the phase trajectory in a six-dimensional space:

$$f_2(1,2)=f_{2,0}(1,2)+\int_{t_0}^{t_0+\tau} f_1(2)\frac{\mathrm{D}_1 f_1(1)}{\mathrm{D}t}\mathrm{d}\tau+\int_{t_0}^{t_0+\tau} f_1(1)\frac{\mathrm{D}_2 f_1(2)}{\mathrm{D}t}\mathrm{d}t, \tag{1.6.25}$$

where $f_{2,0}(1,2)$ denotes the initial value of the two-particle distribution function.

Assumption 3

We resort to the Bogolyubov condition of the weakening of initial correlations corresponding to a certain initial instant of time t_0 [see Eq. (1.3.62)]:

$$\lim_{t_0\to-\infty} W_2[\mathbf{r}_1(t_0-t),\mathbf{v}_1(t_0-t);\mathbf{r}_2(t_0-t),\mathbf{v}_2(t_0-t);t_0-t]=0. \tag{1.6.26}$$

This condition implies that

(a) *we are dealing with infinite motion in a two-body problem,*
(b) *we may speak of the condition of molecular chaos being fulfilled prior to the collision of the particles 1 and 2, which corresponds to the approximation in Eq. (1.3.62) proceeded against the course of time, and*
(c) *Eq. (1.6.22) is written at the r_b scale even though no scale is introduced explicit.*

Because of Assumption 3, Eq. (1.6.25) may be represented in the form

$$f_2(1,2) = f_1[\mathbf{r}_1(t_0),\mathbf{v}_1(t_0),t_0]f_1[\mathbf{r}_2(t_0),\mathbf{v}_2(t_0),t_0] + \int_{t_0}^{t_0+\tau} f_1(2)\frac{D_1 f_1(1)}{Dt}dt + \int_{t_0}^{t_0+\tau} f_1(1)\frac{D_2 f_1(2)}{Dt}dt. \tag{1.6.27}$$

Assumption 4
The collision of the probe particles, 1 and 2, is dominated by the forces of their internal interaction, so that [see Eqs. (1.6.7) and (1.6.15)] one obtains

$$\mathbf{F}_1^{(sc,2)} = \mathbf{F}_1^{(sc,1)}, \mathbf{F}_2^{(sc,2)} = \mathbf{F}_2^{(sc,1)}. \tag{1.6.28}$$

Equation (1.6.27) then becomes

$$f_2(1,2) = f_1[\mathbf{r}_1(t_0),\mathbf{v}_1(t_0),t_0]f_1[\mathbf{r}_2(t_0),\mathbf{v}_2(t_0),t_0] + \int_{t_0}^{t_0+\tau} \frac{D_{12}}{Dt}[f_1 f_2]dt. \tag{1.6.29}$$

Integrating by parts we find:

$$f_2(1,2) = f_1[\mathbf{r}_1(t_0),\mathbf{v}_1(t_0),t_0]f_1[\mathbf{r}_2(t_0),\mathbf{v}_2(t_0),t_0] + \tau\left[\frac{D_{12}}{Dt}[f_1(\mathbf{r}_1(t_0),\mathbf{v}_1(t_0),t_0)f_1(\mathbf{r}_2(t_0),\mathbf{v}_2(t_0),t_0)]\right]_{t=t_0} - \int_{t_0}^{t_0+\tau} t\frac{D_{12}}{Dt}\frac{D_{12}}{Dt}[f_1(1)f_1(2)]dt. \tag{1.6.30}$$

Assumption 5
Delay is sufficiently small that the linearization in the delay time can be used.

The sum of the first two terms in Eq. (1.6.30) determines the product $f_1(1)f_2(2)$ at the instant of time t in the linear approximation in τ, the velocities of particles 1 and 2 corresponding to their initial values at time t_0 (taken to be $t_0 = -\infty$ on the r_b scale).

If we now substitute $f_2(1,2)$ from Eq. (1.6.30) into the BBGKY-1 equation, we obtain:

$$\frac{\partial f_1(1)}{\partial t} + \mathbf{v}_1\cdot\frac{\partial f_1(1)}{\partial \mathbf{r}_1} + \mathbf{F}_1\cdot\frac{\partial f_1(1)}{\partial \mathbf{v}_1} = -\int \mathbf{F}_{12}\cdot\frac{\partial}{\partial \mathbf{v}_1}[f_1(\mathbf{r}_1,\mathbf{v}_1(-\infty),t)f_1(\mathbf{r}_2,\mathbf{v}_2(-\infty),t)]d\Omega_2 + \int \mathbf{F}_{12}\cdot\frac{\partial}{\partial \mathbf{v}_1}\left\{\int_{t_0}^{t_0+\tau} t\frac{D_{12}}{Dt}\frac{D_{12}}{Dt}[f_1(1)f_1(2)]dt\right\}d\Omega_2. \tag{1.6.31}$$

The first integral on the right corresponds to the classical form of the Bogolyubov collision integral and can be transformed in the usual manner to the Boltzmann collision integral [16]. The second collision integral accounts for the time delay effect and is amenable to a differential approximation analogous to Eq. (1.3.62). To obtain this approximation, the following assumption is made.

Assumption 6
For an arbitrary location of particle 2 in the phase space of interacting particles 1 and 2, the dependence on the integrand inside the braces in the time delay integral

$$J_2^{st} = \int \mathbf{F}_{12} \cdot \frac{\partial}{\partial \mathbf{v}_1} \left\{ \int_{t_0}^{t_0+\tau} t \frac{D_{12}}{Dt} \frac{D_{12}}{Dt} [f_1(1) f_1(2)] dt \right\} d\Omega_2 \tag{1.6.32}$$

is determined by the acting internal force $\mathbf{F}_{12}$ via *the change in the particle velocities. This assumption was used by Bogolyubov (see, for example, Ref. [81], p. 203).*

From Eq. (1.6.32) we have

$$\begin{aligned} J_2^{st} &= \int \mathbf{F}_{12} \cdot \frac{\partial}{\partial \mathbf{v}_1} \left\{ \int_{t_0}^{t} t' \frac{D_{12}}{Dt'} \frac{D_{12}}{Dt'} [f_1(1) f_1(2)] dt' \right\} d\Omega_2 \\ &= \int \left(\mathbf{F}_{12} \cdot \frac{\partial}{\partial \mathbf{v}_1} + \mathbf{F}_{21} \cdot \frac{\partial}{\partial \mathbf{v}_2} \right) \left\{ \int_{t_0}^{t} t' \frac{D_{12}}{Dt'} \frac{D_{12}}{Dt'} [f_1(1) f_1(2)] dt' \right\} d\Omega_2 \\ &\cong \tau_d \int \frac{D_{12}}{Dt} \frac{D_{12}}{Dt} [f_1(1) f_1(2)] d\Omega_2 = \tau_d \frac{D_1}{Dt} \frac{D_1 f_1(1)}{Dt}, \end{aligned} \tag{1.6.33}$$

where Assumption 5 has been used again and an effective delay time τ_d introduced.

Generally speaking, integration with respect to time in Eq. (1.6.33) is "eliminated" by the substantial derivative, which also contains spatial differentiation. However, to the linear approximation in the delay time this contribution is negligible. It can be seen from relations

$$\begin{aligned} J_2^{st} &= \int \left\{ \left[\mathbf{F}_{12} \cdot \frac{\partial}{\partial \mathbf{v}_1} + \mathbf{F}_{21} \cdot \frac{\partial}{\partial \mathbf{v}_2} + (\mathbf{v}_2 - \mathbf{v}_1) \cdot \frac{\partial}{\partial \mathbf{x}_{21}} \right] \int_{t_0}^{t} t' \frac{D_{12}}{Dt'} \frac{D_{12}}{Dt'} [f_1(1) f_1(2)] dt' \right\} d\Omega_2 \\ &\quad - \int \left\{ (\mathbf{v}_2 - \mathbf{v}_1) \cdot \frac{\partial}{\partial \mathbf{x}_{21}} \int_{t_0}^{t} t' \frac{D_{12}}{Dt'} \frac{D_{12}}{Dt'} [f_1(1) f_1(2)] dt' \right\} d\Omega_2 \\ &= \tau_d \frac{D_1}{Dt} \frac{D_1 f_1(1)}{Dt} - \int \left\{ (\mathbf{v}_2 - \mathbf{v}_1) \cdot \frac{\partial}{\partial \mathbf{x}_{21}} \int_{t_0}^{t} t' \frac{D_{12}}{Dt'} \frac{D_{12}}{Dt'} [f_1(1) f_1(2)] dt' \right\} d\Omega_2 \\ &\cong \tau_d \frac{D_1}{Dt} \frac{D_1 f_1(1)}{Dt} + \int (\mathbf{v}_2 - \mathbf{v}_1) \cdot \frac{\partial}{\partial \mathbf{x}_{21}} \left[[f_1(1) f_1(2)] - [f_1(1) f_1(2)]_{t_0} - \tau_d \frac{D_1}{Dt} [f_1(1) f_1(2)]_{t_0} \right] d\Omega_2 \\ &\cong \tau_d \frac{D_1}{Dt} \frac{D_1 f_1(1)}{Dt}, \end{aligned} \tag{1.6.34}$$

where $\mathbf{x}_{21} = \mathbf{r}_2 - \mathbf{r}_1$.

Thus, the appearance of the second substantial derivative with respect to time in the generalized Boltzmann equation may be considered as a differential approximation to the time delay integral that emerges in the theory of correlation functions for kinetic equations.

It would appear that the above theory does not require at all that we apply the method of many scales and expand the distribution function in a power series of a small parameter $\varepsilon = n r_b^3$. However, this is not the case. As we have seen above, the integration on the r_b-scale must be employed anyway, and giving up the ε-expansion of the distribution function, on the other hand, makes it impossible to estimate the value of τ_r. Each of the approaches outlined above actually complements one another and is interrelated with one another. The generalized Boltzmann equation can be treated both from the point of view of a higher-order Boltzmann theory and as a result of differential approximations to the collision integral accounting for time delay effects.

There is another point to be made. From Eq. (1.3.40), the equation for the distribution function f_2^0 accurate to the zeroth order in ε is

$$\begin{aligned} &\frac{\partial \hat{f}_2^0}{\partial \hat{t}_b} + \hat{\mathbf{v}}_{1b} \cdot \frac{\partial \hat{f}_2^0}{\partial \hat{\mathbf{r}}_{1b}} + \hat{\mathbf{v}}_{j \in N_\delta, b} \cdot \frac{\partial \hat{f}_2^0}{\partial \hat{\mathbf{r}}_{j \in N_\delta, b}} + \hat{\mathbf{F}}_{1, j \in N_\delta} \cdot \frac{\partial \hat{f}_2^0}{\partial \hat{\mathbf{v}}_{1b}} + \hat{\mathbf{F}}_{j \in N_\delta, 1} \cdot \frac{\partial \hat{f}_2^0}{\partial \hat{\mathbf{v}}_{j \in N_\delta, b}} \\ &+ \alpha \hat{\mathbf{F}}_1 \cdot \frac{\partial \hat{f}_2^0}{\partial \hat{\mathbf{v}}_{1b}} + \alpha \hat{\mathbf{F}}_{j \in N_\delta} \cdot \frac{\partial \hat{f}_2^0}{\partial \hat{\mathbf{v}}_{j \in N_\delta, b}} = 0 \end{aligned} \tag{1.6.35}$$

Comparing this with Eqs. (1.6.9) and (1.6.10) shows that the correlation functions accurate to zeroth order in ε are zero and that forces exerted on the colliding particles 1 and 2 from the side of other particles are not considered in the zero-order approximation at the r_b scale. This result is used for transforming the collision integral $\hat{J}^{st,0}$ to the Boltzmann form in the multi-scale method.

We may summarize then by saying that the derivation of the kinetic equation in the context of the theory of correlation functions for one-particle distribution functions leads to a kinetic equation of the form

$$\frac{\mathrm{D}f}{\mathrm{D}t} = J^{\mathrm{B}} + J^{\mathrm{td}}, \tag{1.6.36}$$

where J^{B} and J^{td} are the Boltzmann collision integral and the collision integral accounting for time delay effects, respectively.

The popularity of the BGK approximation to the Boltzmann collision integral:

$$J^{\mathrm{B}} = \frac{f^{(0)} - f}{\tau} \tag{1.6.37}$$

is due to the drastic simplifications it affords. Essentially, the generalized Boltzmann physical kinetics offer a local approximation for the second non-local collision integral

$$J^{\mathrm{nl}} = \frac{\mathrm{D}}{\mathrm{D}t}\left(\tau \frac{\mathrm{D}f}{\mathrm{D}t}\right). \tag{1.6.38}$$

Thus, Eq. (1.6.36) in its "simplest" version takes the form

$$\frac{\mathrm{D}f}{\mathrm{D}t} = \frac{f^{(0)} - f}{\tau} + \frac{\mathrm{D}}{\mathrm{D}t}\left(\tau \frac{\mathrm{D}f}{\mathrm{D}t}\right). \tag{1.6.39}$$

Since the ratio of the second to the first term on the right of this equation is $J^{\mathrm{nl}}/J^{\mathrm{B}} \approx \mathrm{O}(Kn^2)$, Kn being the Knudsen number, it would seem that the second term could be neglected for the small Knudsen numbers in hydrodynamic regime. However, in the transition to the hydrodynamic limit (after multiplying the kinetic equation by the collision invariants and subsequently integrating over velocities), the Boltzmann integral term vanishes, while the second term on the right-hand side of Eq. (1.6.36) gives a single-order contribution in the generalized Navier-Stokes description (let alone the effect of the small parameter of the higher derivative).

Finally we can state, that introduction of control volume by the reduced description for ensemble of particles of finite diameters leads to fluctuation of velocity moments in the volume. This fact can be considered in a definite sense as a classical analog of Heisenberg indeterminacy principle of quantum mechanics. Successive application of this consideration leads not only to non-local time-delay effects (connected particularly with molecules which centers of mass are inside of control volume) but also to "ghosts" particles, which are (at a time moment) partly inside in control volume without presence of their center of mass in this volume. Non-local effects lead to Johnson's flicker noise observed in 1925 for the first time by J.B. Johson by the measurement of current fluctuations of thermo-electron emission.

Chapter 2

Theory of Generalized Hydrodynamic Equations

ABSTRACT
The generalized Boltzmann equation (GBE) inevitably leads to formulation of new hydrodynamic equations, which are called generalized hydrodynamic equations (GHE). Classical hydrodynamic equations of Enskog, Euler, and Navier-Stokes are particular cases of these equations.
Keywords: Generalized hydrodynamic equations, Generalized Enskog, Euler and Navier, Stokes equations

2.1 TRANSPORT OF MOLECULAR CHARACTERISTICS

Consider a mixture of gases, which consists of η components. Note $\mathbf{v}_\alpha$ as molecule velocity in an immobile coordinate system. The mean velocity of molecules of α-species is defined by relation

$$\overline{\mathbf{v}}_\alpha = \frac{1}{n_\alpha}\int \mathbf{v}_\alpha f_\alpha(\mathbf{r}, \mathbf{v}_\alpha, t)\mathrm{d}\mathbf{v}_\alpha. \tag{2.1.1}$$

Mean mass-velocity $\mathbf{v}_0$ of the gas mixture is

$$\mathbf{v}_0 = \frac{1}{\rho}\sum_\alpha m_\alpha n_\alpha \overline{\mathbf{v}}_\alpha. \tag{2.1.2}$$

Thermal velocity $\mathbf{V}_\alpha$ of particle is the velocity of this particle in coordinate system moving with mean mass-velocity

$$\mathbf{V}_\alpha = \mathbf{v}_\alpha - \mathbf{v}_0. \tag{2.1.3}$$

Diffusive velocity $\overline{\mathbf{V}}_\alpha$ is mean molecule velocity of α-component in a coordinate system moving with mean mass-velocity

$$\overline{\mathbf{V}}_\alpha = \overline{\mathbf{v}}_\alpha - \mathbf{v}_0, \tag{2.1.4}$$

i.e. $\overline{\mathbf{V}}_\alpha$ is the mean thermal velocity of α-molecules.

On the whole, if $\psi_\alpha(\mathbf{r}, \mathbf{v}_\alpha, t)$ is arbitrary scalar, vector or tensor function then mean value of $\psi_\alpha(\mathbf{r}, \mathbf{v}_\alpha, t)$ is noted as $\overline{\psi}_\alpha$ and defines by relation

$$\overline{\psi}_\alpha(\mathbf{r}, t) = \frac{1}{n_\alpha}\int \psi_\alpha(\mathbf{r}, \mathbf{v}_\alpha, t) f_\alpha(\mathbf{r}, \mathbf{v}_\alpha, t)\mathrm{d}\mathbf{v}_\alpha. \tag{2.1.5}$$

Now we can introduce diffusive flux $\mathbf{J}_\alpha$ of α-species

$$\mathbf{J}_\alpha = m_\alpha n_\alpha \overline{\mathbf{V}}_\alpha, \alpha = 1, \ldots, \eta. \tag{2.1.6}$$

After summation of the left- and right-hand sides of Eq. (2.1.6) over all α ($\alpha = 1,.., \eta$) and using (2.1.2) and (2.1.3), one obtains

$$\sum_{\alpha=1}^{\eta} \mathbf{J}_\alpha = \sum_{\alpha=1}^{\eta} \rho_\alpha \overline{\mathbf{V}}_\alpha = \sum_{\alpha=1}^{\eta} \rho_\alpha (\overline{\mathbf{v}}_\alpha - \mathbf{v}_0) = 0. \tag{2.1.7}$$

Consider now the transport of molecular characteristics across an elementary surface $\mathrm{d}s$ moving in a gas with the mean mass-velocity $\mathbf{v}_0$ relatively of the chosen immovable coordinate system and introduce in $\mathrm{d}s$ a positive normal direction $\mathbf{n}$.

Unified non-local theory of transport processes. http://dx.doi.org/10.1016/B978-0-444-63478-8.00002-6

The transport of molecular characteristics across this surface can be found with the help of distribution function (DF) $f_\alpha(\mathbf{r}, \mathbf{v}_\alpha, t)$, which defines the probable position of mass center of molecules in phase space. Let us obtain the number of mass centers of α-molecules crossing this area $\mathrm{d}s$ in the positive direction $\mathbf{n}$ in time dt. Let Θ denote the angle between the positive direction of $\mathbf{n}$ and vector $\mathbf{V}_\alpha$. Because the own velocity of α-molecules relative to $\mathrm{d}s$ is $\mathbf{V}_\alpha$, in time $\mathrm{d}t$ the area $\mathrm{d}s$ will be crossed by all molecules belonging to the volume $\mathrm{d}\mathbf{r} = V_\alpha \cos \Theta \mathrm{d}s\mathrm{d}t$ ($\mathrm{d}\mathbf{r} \equiv \mathrm{d}x\mathrm{d}y\mathrm{d}z$). The number of those particles which velocities are belonging to the interval $\mathbf{V}_\alpha, \mathbf{V}_\alpha + \mathrm{d}\mathbf{v}_\alpha$ is equal to $f_\alpha(\mathbf{r}, \mathbf{v}_\alpha, t) V_\alpha \cos \Theta \mathrm{d}\mathbf{V}_\alpha \mathrm{d}t\mathrm{d}s$. Transportation of mass, momentum, and energy is realizing by flux of α-molecules across $\mathrm{d}s$. Arbitrary functions $\psi_\alpha(\mathbf{r}, \mathbf{v}_\alpha, t)$ of this kind can be titled as molecular markers. Flux $\mathrm{d}\Gamma_n^{(+)\psi_\alpha}$ of scalar molecular marker in $\mathbf{n}$-direction is given by

$$\mathrm{d}\Gamma_n^{(+)\psi_\alpha} = \psi_\alpha(\mathbf{r}, \mathbf{v}_\alpha, t) f_\alpha(\mathbf{r}, \mathbf{v}_\alpha, t) V_\alpha \cos \Theta \mathrm{d}\mathbf{V}_\alpha \mathrm{d}t\mathrm{d}s.$$

If $\psi_\alpha(\mathbf{r}, \mathbf{v}_\alpha, t)$ is a vector function (momentum, for example), it is convenient to consider the fluxes of scalar components of the mentioned marker ψ_α. Flux $\Gamma_n^{(+)\psi_\alpha}$ is given by integration over all velocities groups for which $V_{\alpha n} > 0$ ($V_{\alpha n}$ is projection of $\mathbf{V}_\alpha$ on normal direction $\mathbf{n}$):

$$\Gamma_n^{(+)\psi_\alpha} = \mathrm{d}s\mathrm{d}t \int\limits_{V_{\alpha n}>0} \psi_\alpha f_\alpha V_{\alpha n} \mathrm{d}\mathbf{V}_\alpha. \tag{2.1.8}$$

Similarly, the flux $\Gamma_n^{(-)\psi_\alpha}$ of markers from the positive side to the negative is

$$\Gamma_n^{(-)\psi_\alpha} = -\mathrm{d}s\mathrm{d}t \int\limits_{V_{\alpha n}<0} \psi_\alpha f_\alpha V_{\alpha n} \mathrm{d}\mathbf{V}_\alpha. \tag{2.1.9}$$

The full flux $\Gamma_n^{\psi_\alpha}$ of marker ψ_α in positive direction of $\mathbf{n}$ is

$$\Gamma_n^{\psi_\alpha} = \Gamma_n^{(+)\psi_\alpha} - \Gamma_n^{(-)\psi_\alpha} = \mathrm{d}s\mathrm{d}t \int \psi_\alpha f_\alpha V_{\alpha n} \mathrm{d}\mathbf{V}_\alpha, \tag{2.1.10}$$

where integrating is realizing over all $\mathbf{V}_\alpha$. The specific flux $\gamma_n^{\psi_\alpha}$ across the unit area in unit time is equal to

$$\gamma_n^{\psi_\alpha} = \int \psi_\alpha f_\alpha V_{\alpha n} \mathrm{d}\mathbf{V}_\alpha. \tag{2.1.11}$$

Consider transport of mass, momentum, and energy in a gas.

1. Let is $\psi_\alpha^{(1)} = m_\alpha$, then from (2.1.11) follows the relation for diffusive flux $\mathbf{J}_\alpha$ projection on the normal direction $\mathbf{n}$

$$J_{\alpha n} = m_\alpha \int f_\alpha V_{\alpha n} \mathrm{d}\mathbf{V}_\alpha = \rho_\alpha \overline{V}_{\alpha n}. \tag{2.1.12}$$

2. If $\psi_\alpha^{(2)} = mV_{\alpha i}$, $i = 1, 2, 3$, then relation (2.1.11) defines momentum transport in direction of $\mathbf{n}$. Let $\mathbf{n}$ be an alternate direction coinciding to positive directions of coordinate axes in physical space. Then obviously the momentum transport is defined by symmetric tensor of the second range

$$\overleftrightarrow{\mathrm{P}}_\alpha = \rho_\alpha \overline{\mathbf{V}_\alpha \mathbf{V}_\alpha}, \tag{2.1.13}$$

where $\overline{\mathbf{V}_\alpha \mathbf{V}_\alpha}$ is diada with nine components,

$$\overleftrightarrow{\mathrm{P}}_\alpha = \left\{ \begin{array}{c} \rho_\alpha \overline{V_{\alpha 1} V_{\alpha 1}}, \rho_\alpha \overline{V_{\alpha 1} V_{\alpha 2}}, \rho_\alpha \overline{V_{\alpha 1} V_{\alpha 3}} \\ \dots\dots\dots\dots\dots\dots\dots\dots \\ \rho_\alpha \overline{V_{\alpha 3} V_{\alpha 1}}, \rho_\alpha \overline{V_{\alpha 3} V_{\alpha 2}}, \rho_\alpha \overline{V_{\alpha 3} V_{\alpha 3}} \end{array} \right\}. \tag{2.1.14}$$

Sum of tensors of partial pressures gives pressure tensor for gas mixture

$$\overleftrightarrow{\mathrm{P}} = \sum_\alpha \rho_\alpha \overline{\mathbf{V}_\alpha \mathbf{V}_\alpha}. \tag{2.1.15}$$

Vector of pressure $\mathbf{p}$ for surface element cannot be coincides with normal direction $\mathbf{n}$ to this surface. Easy to check that normal component of pressure for an arbitrary oriented surface element in a gas, is essentially positive.

$$\mathbf{p} \cdot \mathbf{n} = \sum_\alpha \rho_\alpha \overline{\mathbf{V}_\alpha \mathbf{V}_{\alpha n}} \cdot \mathbf{n} = \sum_\alpha \rho_\alpha \overline{v_{\alpha n}^2}. \tag{2.1.16}$$

By definition, the mean value of normal pressure acting on three arbitrary but reciprocally orthogonal planes, gives the static pressure in gas

$$p = \frac{1}{3}\sum_{\alpha,n}\rho_\alpha \overline{V_{\alpha n}^2}. \tag{2.1.17}$$

3. Let $\psi_\alpha^{(3)} = m_\alpha V_\alpha^2/2 + \varepsilon_\alpha$, where ε_α is internal energy of particle of species α. Then relation (2.1.11) defines projection of heat flux $\mathbf{q}$ on the normal direction $\mathbf{n}$ to an element of surface, moving with the mean mass-velocity $\mathbf{v}_0$

$$q_n = \sum_\alpha q_{\alpha n} = \frac{1}{2}\sum_\alpha \rho_\alpha \overline{V_\alpha^2 V_{\alpha n}} + \sum_\alpha \varepsilon_\alpha. \tag{2.1.18}$$

Consider the quantity of kinetic energy E_k in unit volume for time moment t

$$E_k = \sum_\alpha \int \frac{1}{2} m_\alpha v_\alpha^2 f_\alpha \mathrm{d}\mathbf{v}_\alpha = \frac{1}{2}\sum_\alpha \rho_\alpha \overline{v_\alpha^2}. \tag{2.1.19}$$

Taking into account that

$$\mathbf{v}_\alpha = \mathbf{v}_0 + \mathbf{V}_\alpha \tag{2.1.20}$$

and using (2.1.7), one obtains

$$E_k = \sum_\alpha \frac{1}{2}\rho_\alpha \overline{V_\alpha^2} + \frac{1}{2}\rho v_0^2 = E_{\text{micro}} + E_{\text{macro}}. \tag{2.1.21}$$

Estimate the order of this value. Known from experiments that nitrogen density (temperature $T = 273.16$ K and mercury pressure $P = 760$ mm) $\rho = 1.25 \times 10^{-3}\,\text{g}\,\text{cm}^{-3}$. Then from (2.1.17) it follows, that for hydrodynamic velocity $v_0 \approx 1\,\text{m}\,\text{s}^{-1}$ the ratio $E_{\text{micro}}/E_{\text{macro}} \approx 2.4 \times 10^5$. It means that in this case practically all energy of molecules corresponds to energy of chaos movement, $\sqrt{\overline{V^2}} \approx 1.5 \times 10^2\,\text{m}\,\text{s}^{-1}$.

But a part of energy of translation movement, molecules have, generally speaking, vibration and rotation energy. Given chemical reactions, the transition of potential energy of molecules into kinetic energy of particles should be taken into account. All these kinds of energy including energy of chaotic movement identify as internal energy of gas. Temperature associated with translation movement can be defined as

$$\frac{3}{2}k_\mathrm{B} T n = \sum_\alpha \frac{\rho_\alpha \overline{V_\alpha^2}}{2}, \tag{2.1.22}$$

where k_B is Boltzmann constant, n—the number of particles per unit volume. It is natural to define temperature T_α of α-species as

$$m_\alpha \overline{V_\alpha^2} = 3k_\mathrm{B} T_\alpha. \tag{2.1.23}$$

In mixtures of chemical reacting gases, temperatures of species can differ significantly, [21].

2.2 HYDRODYNAMIC ENSKOG EQUATIONS

Recall derivation of the Enskog hydrodynamic equations. Multiplying Boltzmann equation by molecular marker ψ_α and integrating over all $\mathbf{v}_\alpha$, yields

$$\begin{aligned} &\int \psi_\alpha \left(\frac{\partial f_\alpha}{\partial t} + \mathbf{v}_\alpha \cdot \frac{\partial f_\alpha}{\partial \mathbf{r}} + \mathbf{F}_\alpha \cdot \frac{\partial f_\alpha}{\partial \mathbf{v}_\alpha} \right) \mathrm{d}\mathbf{v}_\alpha \\ &= \sum_{j=1}^{\eta} \int \psi_\alpha \left(f'_\alpha f'_j - f_\alpha f_j \right) P_{\alpha j}^{\alpha j} g_{\alpha j} b \mathrm{d}b \mathrm{d}\phi \mathrm{d}\mathbf{v}_\alpha \mathrm{d}\mathbf{v}_j + \int \psi_\alpha J_\alpha^{st,inel} \mathrm{d}\mathbf{v}_\alpha, \end{aligned} \tag{2.2.1}$$

where $P_{\alpha j}^{\alpha j}$ is probability density of elastic collisions of particles α and j, $J_\alpha^{st,inel}$ is integral of inelastic collisions. The use of notation for mean values leads to relations

$$\int \psi_\alpha \frac{\partial f_\alpha}{\partial t} d\mathbf{v}_\alpha = \frac{\partial}{\partial t}\int \psi_\alpha f_\alpha d\mathbf{v}_\alpha - \int f_\alpha \frac{\partial \psi_\alpha}{\partial t} d\mathbf{v}_\alpha = \frac{\partial n_\alpha \overline{\psi_\alpha}}{\partial t} - n_\alpha \overline{\frac{\partial \psi_\alpha}{\partial t}}, \tag{2.2.2}$$

$$\int \psi_\alpha \mathbf{v}_\alpha \cdot \frac{\partial f_\alpha}{\partial \mathbf{r}} d\mathbf{v}_\alpha = \frac{\partial}{\partial \mathbf{r}} \cdot \left(n_\alpha \overline{\psi_\alpha \mathbf{v}_\alpha}\right) - n_\alpha \overline{\mathbf{v}_\alpha \cdot \frac{\partial \psi_\alpha}{\partial \mathbf{r}}}. \tag{2.2.3}$$

If external forces do not depend on the particle's velocities or correspond to Lorentz forces, the following transformations are valid

$$\int \psi_\alpha F_{\alpha 1} \frac{\partial f_\alpha}{\partial v_{\alpha 1}} d\mathbf{v}_\alpha = \int \left[F_{\alpha 1}\psi_\alpha f_\alpha\right]_{v_{\alpha 1}=-\infty}^{v_{\alpha 1}=+\infty} dv_{\alpha 2} dv_{\alpha 3} - \int F_{\alpha 1} f_\alpha \frac{\partial \psi_\alpha}{\partial v_{\alpha 1}} d\mathbf{v}_\alpha = -n_\alpha \overline{F_{\alpha 1}\frac{\partial \psi_\alpha}{\partial v_{\alpha 1}}}. \tag{2.2.4}$$

Hence Enskog's equations are written as

$$\frac{\partial n_\alpha \overline{\psi_\alpha}}{\partial t} + \frac{\partial}{\partial \mathbf{r}} \cdot \left(n_\alpha \overline{\psi_\alpha \mathbf{v}_\alpha}\right) - n_\alpha \left[\overline{\frac{\partial \psi_\alpha}{\partial t}} + \overline{\mathbf{v}_\alpha \cdot \frac{\partial \psi_\alpha}{\partial \mathbf{r}}} + \overline{\mathbf{F}_\alpha \cdot \frac{\partial \psi_\alpha}{\partial \mathbf{v}_\alpha}}\right] = \sum_j \int \psi_\alpha \left(f'_\alpha f'_j - f_\alpha f_j\right) P^{\alpha j}_{\alpha j} g_{\alpha j} b db d\phi d\mathbf{v}_\alpha d\mathbf{v}_j + \int \psi_\alpha J^{st,inel}_\alpha d\mathbf{v}_\alpha. \tag{2.2.5}$$

Enskog's equations (2.2.5) are integro-differential equations and generally speaking, they are not more simple than initial Boltzmann equation. But for so-called summational invariants $(\psi^{(1)}_\alpha, \psi^{(2)}_\alpha, \psi^{(3)}_\alpha)$ integral terms in the right-hand-side of Eq. (2.2.5) could be significantly simplified. One proves [36], that for $\psi^{(1)}_\alpha = m_\alpha$, the first-mentioned term turns into zero and the second one leads to the mass rate of formation of α-species as a result of chemical reactions. For other invariants $\psi^{(2)}_\alpha, \psi^{(3)}_\alpha$, in view of conservation laws the integral terms are equal to zero after summation over all α $(\alpha = 1,\ldots,\eta)$.

In the generalized Boltzmann kinetic theory (GBKT), as it was shown, local collision integrals can be written in the same form like in classical Boltzmann theory. Therefore, GBKT does not create additional difficulties connected with collision integrals.

2.3 TRANSFORMATIONS OF THE GENERALIZED BOLTZMANN EQUATION

The generalized Boltzmann equation (GBE) inevitably leads to formulation of new hydrodynamic equations, which are called generalized hydrodynamic equations (GHE). Classical hydrodynamic equations of Enskog, Euler, and Navier-Stokes are particular cases of these equations. For the purpose of derivation of GHE let us transform GBE to the form convenient for further application. Write down the second term in the left of GBE

$$\frac{\mathrm{D}f_\alpha}{\mathrm{D}t} - \frac{\mathrm{D}}{\mathrm{D}t}\left(\tau_\alpha \frac{\mathrm{D}f_\alpha}{\mathrm{D}t}\right) = J^{st,el}_\alpha + J^{st,inel}_\alpha, \tag{2.3.1}$$

in explicit form

$$\frac{\mathrm{D}}{\mathrm{D}t}\left(\tau_\alpha \frac{\mathrm{D}f_\alpha}{\mathrm{D}t}\right) = \frac{\mathrm{D}\tau_\alpha}{\mathrm{D}t}\frac{\mathrm{D}f_\alpha}{\mathrm{D}t} + \tau_\alpha \frac{\mathrm{D}}{\mathrm{D}t}\frac{\mathrm{D}f_\alpha}{\mathrm{D}t},$$

where

$$\frac{\mathrm{D}}{\mathrm{D}t} = \frac{\partial}{\partial t} + \mathbf{v}_\alpha \cdot \frac{\partial}{\partial \mathbf{r}} + \mathbf{F}_\alpha \cdot \frac{\partial}{\partial \mathbf{v}_\alpha}, \tag{2.3.2}$$

$$
\begin{aligned}
\frac{\mathrm{D}}{\mathrm{D}t}\frac{\mathrm{D}f_\alpha}{\mathrm{D}t} = {} & \frac{\partial^2 f_\alpha}{\partial t^2} + 2\mathbf{v}_\alpha \cdot \frac{\partial^2 f_\alpha}{\partial \mathbf{r}\partial t} + 2\mathbf{F}_\alpha \cdot \frac{\partial^2 f_\alpha}{\partial \mathbf{v}_\alpha \partial t} \\
& + \mathbf{v}_\alpha \cdot \frac{\partial}{\partial \mathbf{r}}\left(\mathbf{v}_\alpha \cdot \frac{\partial f_\alpha}{\partial \mathbf{r}}\right) + \mathbf{F}_\alpha \cdot \frac{\partial}{\partial \mathbf{v}_\alpha}\left(\mathbf{v}_\alpha \cdot \frac{\partial f_\alpha}{\partial \mathbf{r}}\right) \\
& + \frac{\partial \mathbf{F}_\alpha}{\partial t} \cdot \frac{\partial f_\alpha}{\partial \mathbf{v}_\alpha} + \mathbf{v}_\alpha \cdot \frac{\partial}{\partial \mathbf{r}}\left(\mathbf{F}_\alpha \cdot \frac{\partial f_\alpha}{\partial \mathbf{v}_\alpha}\right) + \mathbf{F}_\alpha \cdot \frac{\partial}{\partial \mathbf{v}_\alpha}\left(\mathbf{F}_\alpha \cdot \frac{\partial f_\alpha}{\partial \mathbf{v}_\alpha}\right).
\end{aligned}
\tag{2.3.3}
$$

In doing so we should keep in mind that, generally speaking, non-local parameter τ_α depends on velocity, then

$$
\frac{\mathrm{D}\tau_\alpha}{\mathrm{D}t} = \frac{\partial \tau_\alpha}{\partial t} + \mathbf{v}_\alpha \cdot \frac{\partial \tau_\alpha}{\partial \mathbf{r}} + \mathbf{F}_\alpha \cdot \frac{\partial \tau_\alpha}{\partial \mathbf{v}_\alpha}.
\tag{2.3.4}
$$

Transform terms in (2.3.3):

$$
\mathbf{v}_\alpha \cdot \frac{\partial}{\partial \mathbf{r}}\left(\mathbf{v}_\alpha \cdot \frac{\partial f_\alpha}{\partial \mathbf{r}}\right) = \mathbf{v}_\alpha \mathbf{v}_\alpha : \frac{\partial^2 f_\alpha}{\partial \mathbf{r}\partial \mathbf{r}}.
\tag{2.3.5}
$$

$$
\mathbf{F}_\alpha \cdot \frac{\partial}{\partial \mathbf{v}_\alpha}\left(\mathbf{v}_\alpha \cdot \frac{\partial f_\alpha}{\partial \mathbf{r}}\right) = \mathbf{F}_\alpha \mathbf{v}_\alpha : \frac{\partial^2 f_\alpha}{\partial \mathbf{r}\partial \mathbf{v}_\alpha} + \mathbf{F}_\alpha \cdot \frac{\partial f_\alpha}{\partial \mathbf{r}},
\tag{2.3.6}
$$

$$
\mathbf{v}_\alpha \cdot \frac{\partial}{\partial \mathbf{r}}\left(\mathbf{F}_\alpha \cdot \frac{\partial f_\alpha}{\partial \mathbf{v}_\alpha}\right) = \mathbf{v}_\alpha \mathbf{F}_\alpha : \frac{\partial^2 f_\alpha}{\partial \mathbf{v}_\alpha \partial \mathbf{r}} + \frac{\partial f_\alpha}{\partial \mathbf{v}_\alpha}\mathbf{v}_\alpha : \frac{\partial}{\partial \mathbf{r}}\mathbf{F}_\alpha,
\tag{2.3.7}
$$

$$
\mathbf{F}_\alpha \cdot \frac{\partial}{\partial \mathbf{v}_\alpha}\left(\mathbf{F}_\alpha \cdot \frac{\partial f_\alpha}{\partial \mathbf{v}_\alpha}\right) = \mathbf{F}_\alpha \mathbf{F}_\alpha : \frac{\partial^2 f_\alpha}{\partial \mathbf{v}_\alpha \partial \mathbf{v}_\alpha} + \frac{\partial f_\alpha}{\partial \mathbf{v}_\alpha}\mathbf{F}_\alpha : \frac{\partial}{\partial \mathbf{v}_\alpha}\mathbf{F}_\alpha.
\tag{2.3.8}
$$

The sign ":" in (2.3.5)–(2.3.8) denotes as usually, the double tensor production. For example in (2.3.6)

$$
\mathbf{F}_\alpha \mathbf{v}_\alpha : \frac{\partial^2 f_\alpha}{\partial \mathbf{r}\partial \mathbf{v}_\alpha} = \sum_{ij=1}^{3} F_{\alpha i} v_{\alpha j} \frac{\partial^2 f_\alpha}{\partial r_j \partial v_{\alpha i}}.
\tag{2.3.9}
$$

The derivative of external forces $\mathbf{F}_\alpha$ $(\alpha = 1, \ldots, \eta)$ with respect to velocity appears on the right side of (2.3.8); force $\mathbf{F}_\alpha$—acting on the particle of species α—is related to the unit of mass of this particle. If $\mathbf{F}_\alpha$ does not depend on velocity, this derivative naturally turns into zero. In the following notation for the force independent of velocity, is $\mathbf{F}_\alpha^{(1)}$. If $\mathbf{F}_\alpha$ includes Lorentz force, noted as $\mathbf{F}_\alpha^B$ (**B**—magnetic induction),

$$
\mathbf{F}_\alpha = \mathbf{F}_\alpha^{(1)} + \mathbf{F}_\alpha^B
\tag{2.3.10}
$$

the subsequent transformations of (2.3.8) can be realized:

$$
\begin{aligned}
\frac{\partial f_\alpha}{\partial \mathbf{v}_\alpha}\mathbf{F}_\alpha : \frac{\partial}{\partial \mathbf{v}_\alpha}\mathbf{F}_\alpha &= \frac{\partial f_\alpha}{\partial \mathbf{v}_\alpha}\mathbf{F}_\alpha^{(1)} : \frac{\partial}{\partial \mathbf{v}_\alpha}\mathbf{F}_\alpha^B + \frac{\partial f_\alpha}{\partial \mathbf{v}_\alpha}\mathbf{F}_\alpha^B : \frac{\partial}{\partial \mathbf{v}_\alpha}\mathbf{F}_\alpha^B \\
&= \left(\frac{q_\alpha}{m_\alpha}\right)^2 \frac{\partial f_\alpha}{\partial \mathbf{v}_\alpha} \cdot \left\{\mathbf{B}(\mathbf{v}_\alpha \cdot \mathbf{B}) - B^2 \mathbf{v}_\alpha\right\} + \frac{\partial f_\alpha}{\partial \mathbf{v}_\alpha}\mathbf{F}_\alpha^{(1)} : \frac{\partial}{\partial \mathbf{v}_\alpha}\mathbf{F}_\alpha^B,
\end{aligned}
\tag{2.3.11}
$$

because

$$
\mathbf{F}_\alpha^B = \frac{q_\alpha}{m_\alpha}[\mathbf{v}_\alpha \times \mathbf{B}],
\tag{2.3.12}
$$

where q_α is the charge of the particle of species α.

The last term on the right side of relation (2.3.11) is written as:

$$
\frac{\partial f_\alpha}{\partial \mathbf{v}_\alpha}\mathbf{F}_\alpha^{(1)} : \frac{\partial}{\partial \mathbf{v}_\alpha}\mathbf{F}_\alpha^B = \frac{q_\alpha}{m_\alpha}\frac{\partial f_\alpha}{\partial \mathbf{v}_\alpha} \cdot \left(\mathbf{F}_\alpha^{(1)} \times \mathbf{B}\right).
\tag{2.3.13}
$$

Sign "×" corresponds to a vector product. GBE can contain Umov-Pointing vector $\mathbf{S} \sim [\mathbf{E}, \mathbf{H}]$ (**H** is magnetic intensity) in explicit form because the force of nonmagnetic origin $\mathbf{F}_\alpha^{(1)}$ can be connected with electric intensity **E**.

As a result

$$\frac{\partial f_\alpha}{\partial \mathbf{v}_\alpha}\mathbf{F}_\alpha : \frac{\partial}{\partial \mathbf{v}_\alpha}\mathbf{F}_\alpha = \left(\frac{q_\alpha}{m_\alpha}\right)^2 \frac{\partial f_\alpha}{\partial \mathbf{v}_\alpha}\cdot\left\{\mathbf{B}(\mathbf{v}_\alpha\cdot\mathbf{B}) - B^2\mathbf{v}_\alpha\right\} + \frac{q_\alpha}{m_\alpha}\frac{\partial f_\alpha}{\partial \mathbf{v}_\alpha}\cdot\left(\mathbf{F}_\alpha^{(1)}\times\mathbf{B}\right). \tag{2.3.14}$$

We reach the generalized Boltzmann equation in the form

$$\begin{aligned}
&\left(\frac{\partial f_\alpha}{\partial t} + \mathbf{v}_\alpha\cdot\frac{\partial f_\alpha}{\partial \mathbf{r}} + \mathbf{F}_\alpha\cdot\frac{\partial f_\alpha}{\partial \mathbf{v}_\alpha}\right)\left[1 - \left(\frac{\partial \tau_\alpha}{\partial t} + \mathbf{v}_\alpha\cdot\frac{\partial \tau_\alpha}{\partial \mathbf{r}} + \mathbf{F}_\alpha\cdot\frac{\partial \tau_\alpha}{\partial \mathbf{v}_\alpha}\right)\right]\\
&-\tau_\alpha\left[\frac{\partial^2 f_\alpha}{\partial t^2} + 2\frac{\partial^2 f_\alpha}{\partial \mathbf{r}\partial t}\cdot\mathbf{v}_\alpha + \frac{\partial^2 f_\alpha}{\partial \mathbf{r}\partial \mathbf{r}}:\mathbf{v}_\alpha\mathbf{v}_\alpha + 2\frac{\partial^2 f_\alpha}{\partial \mathbf{v}_\alpha\partial t}\cdot\mathbf{F}_\alpha\right.\\
&+\frac{\partial \mathbf{F}_\alpha}{\partial t}\cdot\frac{\partial f_\alpha}{\partial \mathbf{v}_\alpha} + \mathbf{F}_\alpha\cdot\frac{\partial f_\alpha}{\partial \mathbf{r}} + \frac{q_\alpha}{m_\alpha}\frac{\partial f_\alpha}{\partial \mathbf{v}_\alpha}\cdot\left[\mathbf{F}_\alpha^{(1)}\times\mathbf{B}\right]\\
&+\left(\frac{q_\alpha}{m_\alpha}\right)^2\frac{\partial f_\alpha}{\partial \mathbf{v}_\alpha}\cdot\left[\mathbf{B}(\mathbf{v}_\alpha\cdot\mathbf{B}) - B^2\mathbf{v}_\alpha\right] + \frac{\partial f_\alpha}{\partial \mathbf{v}_\alpha}\mathbf{v}_\alpha : \frac{\partial}{\partial \mathbf{r}}\mathbf{F}_\alpha\\
&\left.+\frac{\partial^2 f_\alpha}{\partial \mathbf{v}_\alpha\partial \mathbf{v}_\alpha}:\mathbf{F}_\alpha\mathbf{F}_\alpha + 2\frac{\partial^2 f_\alpha}{\partial \mathbf{v}_\alpha\partial \mathbf{r}}:\mathbf{v}_\alpha\mathbf{F}_\alpha\right] = J_\alpha^{st,el} + J_\alpha^{st,inel}
\end{aligned} \tag{2.3.15}$$

As is seen, the explicit form of differential part of GBE is much more complicated in comparison to Boltzmann equation. As a result the transition to generalized hydrodynamic equations (GHE) requires more effort.

Let us go to this work following the standard procedure of obtaining of the hydrodynamic description: multiply GBE by the particle collision invariants $\psi_\alpha^{(i)}$ $(i=1, 2, 3)$ and integrate over all $\mathbf{v}_\alpha$. Result of this procedure leads to the generalized Enskog equations of continuity, momentum, and energy.

Remark.

As it is indicated, in principle, non-local parameter τ_α is a function of time t, space variable $\mathbf{r}$, and velocity $\mathbf{v}_\alpha$, $\tau_\alpha=\tau_\alpha(t,\mathbf{r},\mathbf{v}_\alpha)$. The $\mathbf{v}_\alpha$-velocity dependence leads to the appearance in the GHE additional terms demanding the explicit expression for $\tau_\alpha(\mathbf{v}_\alpha)$. For example, in the generalized continuity equation we find the terms like

$$\begin{aligned}
&\int\frac{\partial f_\alpha}{\partial t}\mathbf{F}_\alpha\cdot\frac{\partial \tau_\alpha}{\partial \mathbf{v}_\alpha}d\mathbf{v}_\alpha = \frac{\partial}{\partial t}\int f_\alpha\mathbf{F}_\alpha\cdot\frac{\partial \tau_\alpha}{\partial \mathbf{v}_\alpha}d\mathbf{v}_\alpha - \int f_\alpha\frac{\partial}{\partial t}\left(\mathbf{F}_\alpha\cdot\frac{\partial \tau_\alpha}{\partial \mathbf{v}_\alpha}\right)d\mathbf{v}_\alpha\\
&=\frac{\partial}{\partial t}\left[n_\alpha\overline{\frac{\partial}{\partial \mathbf{v}_\alpha}\cdot(\mathbf{F}_\alpha\tau_\alpha)}\right] - n_\alpha\overline{\frac{\partial}{\partial t}\frac{\partial}{\partial \mathbf{v}_\alpha}\cdot(\mathbf{F}_\alpha\tau_\alpha)}.
\end{aligned} \tag{2.3.16}$$

It is sufficient, in all practical applications (including quantum hydrodynamics), to use the more restricted dependence $\tau_\alpha=\tau_\alpha(t,\mathbf{r},\overline{\mathbf{v}}_\alpha)$, where $\overline{\mathbf{v}}_\alpha$ is the mean hydrodynamic velocity of α-species.

2.4 GENERALIZED CONTINUITY EQUATION

Multiply GBE (2.3.15) by $\psi_\alpha^{(1)}=m_\alpha$, and realize term-by-term integrating of left and right sides of equation. Formally 22 terms should be transformed, here and in the following we of necessity demonstrate only character and complicated elements of these transformations and restrict our consideration to some comments in more simple cases.

Consider integrals of which transformation is realized by integration in parts, in doing so, the integrated part turns into zero because distribution function is equal to zero if $v_\alpha=\pm\infty$.

$$\begin{aligned}
&\int m_\alpha\frac{\partial f_\alpha}{\partial \mathbf{v}_\alpha}\mathbf{v}_\alpha : \frac{\partial}{\partial \mathbf{r}}\mathbf{F}_\alpha d\mathbf{v}_\alpha\\
&=-\rho_\alpha\frac{\partial}{\partial \mathbf{r}}\cdot\mathbf{F}_\alpha^{(1)} + \int m_\alpha\frac{\partial f_\alpha}{\partial \mathbf{v}_\alpha}\mathbf{v}_\alpha : \frac{\partial}{\partial \mathbf{r}}\mathbf{F}_\alpha^B d\mathbf{v}_\alpha\\
&=-\rho_\alpha\frac{\partial}{\partial \mathbf{r}}\cdot\mathbf{F}_\alpha^{(1)} + q_\alpha\int\left\{\frac{\partial f_\alpha}{\partial \mathbf{v}_\alpha}\cdot\left[\mathbf{v}_\alpha\times\left[\left(\mathbf{v}_\alpha\cdot\frac{\partial}{\partial \mathbf{r}}\right)\mathbf{B}\right]\right]\right\}d\mathbf{v}_\alpha\\
&=-\rho_\alpha\frac{\partial}{\partial \mathbf{r}}\cdot\mathbf{F}_\alpha^{(1)} + \frac{q_\alpha}{m_\alpha}\rho_\alpha\mathbf{v}_\alpha\cdot\mathrm{rot}\mathbf{B},
\end{aligned} \tag{2.4.1}$$

$$\int m_\alpha \frac{\partial f_\alpha}{\partial \mathbf{v}_\alpha} \cdot \{\mathbf{B}(\mathbf{v}_\alpha \cdot \mathbf{B}) - B^2 \mathbf{v}_\alpha\} d\mathbf{v}_\alpha = 2B^2 \rho_\alpha. \tag{2.4.2}$$

Consider also:

$$\begin{aligned} m_\alpha \int \left(\mathbf{F}_\alpha \cdot \frac{\partial f_\alpha}{\partial \mathbf{v}_\alpha} \right) \mathbf{v}_\alpha \cdot \frac{\partial \tau_\alpha}{\partial \mathbf{r}} d\mathbf{v}_\alpha &= -m_\alpha \frac{\partial \tau_\alpha}{\partial \mathbf{r}} \cdot \int f_\alpha \frac{\partial}{\partial \mathbf{v}_\alpha} \cdot (\mathbf{F}_\alpha \mathbf{v}_\alpha) d\mathbf{v}_\alpha \\ &= -m_\alpha \frac{\partial \tau_\alpha}{\partial \mathbf{r}} \cdot \int f_\alpha \left(\mathbf{F}_\alpha \cdot \frac{\partial}{\partial \mathbf{v}_\alpha} \right) \mathbf{v}_\alpha d\mathbf{v}_\alpha = -m_\alpha \frac{\partial \tau_\alpha}{\partial \mathbf{r}} \cdot \int \mathbf{F}_\alpha f_\alpha d\mathbf{v}_\alpha \\ &= -\rho_\alpha \frac{\partial \tau_\alpha}{\partial \mathbf{r}} \cdot \mathbf{F}_\alpha^{(1)} - \frac{q_\alpha}{m_\alpha} \rho_\alpha \frac{\partial \tau_\alpha}{\partial \mathbf{r}} \cdot [\overline{\mathbf{v}}_\alpha \times \mathbf{B}], \end{aligned} \tag{2.4.3}$$

Collecting all—transformed by this way—terms one obtains the generalized continuity equation:

$$\begin{aligned} &\frac{\partial}{\partial t} \left\{ \rho_\alpha - \tau_\alpha \left[\frac{\partial \rho_\alpha}{\partial t} + \frac{\partial}{\partial \mathbf{r}} \cdot (\rho_\alpha \overline{\mathbf{v}}_\alpha) \right] \right\} \\ &+ \frac{\partial}{\partial \mathbf{r}} \cdot \left\{ \rho_\alpha \overline{\mathbf{v}}_\alpha - \tau_\alpha \left[\frac{\partial}{\partial t} (\rho_\alpha \overline{\mathbf{v}}_\alpha) + \frac{\partial}{\partial \mathbf{r}} \cdot (\rho_\alpha \overline{\mathbf{v}_\alpha \mathbf{v}_\alpha}) \right. \right. \\ &\left. \left. - \rho_\alpha \mathbf{F}_\alpha^{(1)} - \frac{q_\alpha}{m_\alpha} \rho_\alpha \overline{\mathbf{v}}_\alpha \times \mathbf{B} \right] \right\} = R_\alpha, \end{aligned} \tag{2.4.4}$$

where R_α is mass rate α-particles formation in presence of inelastic (also chemical) processes

$$R_\alpha = m_\alpha \int J_\alpha^{st,inel} d\mathbf{v}_\alpha. \tag{2.4.5}$$

Generalized continuity equation differs in radical way from the classical continuity equation. The origin of this distinction—for more simple case of one-component nonreacting gas—was discussed in section "Historical Introduction" from qualitative positions. Nevertheless we repeat the main thoughts of this consideration because of their importance.

Recall the phenomenological derivation of continuity equation. Control volume is defined in an area occupied by gas. The character lengths of this control volume is much bigger than mean free path between collisions but much smaller than hydrodynamic scale. Write down the mass balance for this volume with transparent boundary surface. In another way, variation of mass in control volume could be connected only with fluxes of particles directed inside or outside of reference surface. This procedure leads to the well-known equation

$$\frac{\partial \rho}{\partial t} + \frac{\partial}{\partial \mathbf{r}} \cdot \rho \mathbf{v}_0 = 0 \tag{2.4.6}$$

Obviously, in the derivation of Eq. (2.4.6) implicit presumption was done that particles can be placed inside or outside of control volume. It means that only point-like particles are involved into consideration. This fact—as it was stated in Chapter 1 in the course of GBE derivation from the Bogolyubov chain—is the principal restriction of Boltzmann kinetic theory. In the generalized Boltzmann kinetic theory (GBKT) considering particles have finite sizes and then at some instant of time can be placed partly inside, partly outside of reference surface. This fact along with the use of the DF form oriented for describing of the point structureless particles leads to appearance of fluctuation terms, in particular in (2.4.4). For example the term $\tau_\alpha \left[\frac{\partial \rho_\alpha}{\partial t} + \frac{\partial}{\partial \mathbf{r}} \cdot \rho_\alpha \overline{\mathbf{v}}_\alpha \right]$ reflects the fluctuation of density ρ_α, and the term

$$\tau_\alpha \left[\frac{\partial}{\partial t} (\rho_\alpha \overline{\mathbf{v}}_\alpha) + \frac{\partial}{\partial \mathbf{r}} (\rho_\alpha \overline{\mathbf{v}_\alpha \mathbf{v}_\alpha}) - \rho_\alpha \mathbf{F}_\alpha^{(1)} - \frac{q_\alpha}{m_\alpha} \rho_\alpha \overline{\mathbf{v}}_\alpha \times \mathbf{B} \right].$$

corresponds to the fluctuation of momentum $\rho_\alpha \overline{\mathbf{v}}_\alpha$. For the clarity let us write down the generalized continuity equation in dimensionless form for the particular case of one-component gas:

$$\begin{aligned} &\frac{\partial}{\partial t} \left\{ \rho - \tau \left[\frac{\partial \rho}{\partial t} + \frac{\partial}{\partial \mathbf{r}} \cdot (\rho \mathbf{v}_0) \right] \right\} \\ &+ \frac{\partial}{\partial \mathbf{r}} \cdot \left\{ \rho \mathbf{v}_0 - \tau \left[\frac{\partial}{\partial t} (\rho \mathbf{v}_0) + \frac{\partial}{\partial \mathbf{r}} \cdot (\rho \overline{\mathbf{v}\mathbf{v}}) \right] \right\} = 0. \end{aligned} \tag{2.4.7}$$

Introduce the density scale ρ_{∞}, hydrodynamic lengths scale L, mean thermal velocity as the scale for particle velocity, and $v_{0\infty}$ as scale for hydrodynamic velocity $\mathbf{v}_0$. Hydrodynamic time scale is defined as $t_H = L v_{0\infty}^{-1}$, and scale for τ is l_λ / v_T. In dimensionless form Eq. (2.4.7) is written as

$$\begin{aligned}&\frac{\partial}{\partial \hat{t}}\left\{\hat{\rho} - A_v \hat{\tau}\frac{l}{L}\left[\frac{\partial \hat{\rho}}{\partial \hat{t}} + \frac{\partial}{\partial \hat{\mathbf{r}}}\cdot(\hat{\rho}\hat{\mathbf{v}}_0)\right]\right\}\\ &+\frac{\partial}{\partial \hat{\mathbf{r}}}\cdot\left\{\hat{\rho}\hat{\mathbf{v}}_0 - A_v\hat{\tau}\frac{l}{L}\left[\frac{\partial}{\partial \hat{t}}(\hat{\rho}\hat{\mathbf{v}}_0) + \frac{\partial}{\partial \hat{\mathbf{r}}}\cdot(\overline{\hat{\rho}\hat{\mathbf{v}}\hat{\mathbf{v}}})\right]\right\} = 0,\end{aligned} \tag{2.4.8}$$

where $A_v = v_{0\infty}/v_T$, $l/L = \mathrm{Kn}$ and l is the mean free path between collisions

Therefore, indicated fluctuation terms are proportional to Knudsen number and are really small in the case of small Knudsen numbers in the regime of continuum media. As is seen, for $\mathrm{Kn} \sim 1$ fluctuations of mass in control volume are of the same order as basic terms and introduction of additional terms becomes more and more significant. But no reason to think that for $\mathrm{Kn} \ll 1$ fluctuations terms could be omitted. As will be shown, the consideration of these terms is of principal significance for turbulence description on the Kolmogorov micro (or sub-grid) scale. Also should be noticed that for the case of $\mathrm{Kn} \ll 1$ hydrodynamic equations (including the generalized continuity equation) belong to the class of differential equations with small parameter in view of senior derivatives. This fact could lead to effects of "boundary layers" for solutions of these equations.

2.5 GENERALIZED MOMENTUM EQUATION FOR COMPONENT

Generalized momentum equation for α component is obtained by multiplying GBE (2.3.15) by collision invariant $\psi_\alpha^{(2)} = m_\alpha \mathbf{v}_\alpha$ and succeeding integration over all $\mathbf{v}_\alpha$.

$$\begin{aligned}&\int m_\alpha \mathbf{v}_\alpha\left(\frac{\partial f_\alpha}{\partial t} + \mathbf{v}_\alpha\cdot\frac{\partial f_\alpha}{\partial \mathbf{r}} + \mathbf{F}_\alpha\cdot\frac{\partial f_\alpha}{\partial \mathbf{v}_\alpha}\right)\left[1 - \left(\frac{\partial \tau_\alpha}{\partial t}\right.\right.\\ &\left.\left.+\mathbf{v}_\alpha\cdot\frac{\partial \tau_\alpha}{\partial \mathbf{r}}\right)\right]d\mathbf{v}_\alpha - \tau_\alpha\int m_\alpha\mathbf{v}_\alpha\left[\frac{\partial^2 f_\alpha}{\partial t^2} + 2\frac{\partial^2 f_\alpha}{\partial \mathbf{r}\partial t}\cdot\mathbf{v}_\alpha\right.\\ &+\frac{\partial^2 f_\alpha}{\partial \mathbf{r}\partial \mathbf{r}}:\mathbf{v}_\alpha\mathbf{v}_\alpha + 2\frac{\partial^2 f_\alpha}{\partial \mathbf{v}_\alpha\partial t}\mathbf{F}_\alpha + \frac{\partial \mathbf{F}_\alpha}{\partial t}\cdot\frac{\partial f_\alpha}{\partial \mathbf{v}_\alpha}\\ &+\mathbf{F}_\alpha\cdot\frac{\partial f_\alpha}{\partial \mathbf{r}} + \frac{q_\alpha}{m_\alpha}\frac{\partial f_\alpha}{\partial \mathbf{v}_\alpha}\cdot\left(\mathbf{F}_\alpha^{(1)}\times\mathbf{B}\right)\\ &+\left(\frac{q_\alpha}{m_\alpha}\right)^2\frac{\partial f_\alpha}{\partial \mathbf{v}_\alpha}\cdot\left[\mathbf{B}(\mathbf{v}_\alpha\cdot\mathbf{B}) - B^2\mathbf{v}_\alpha\right] + \frac{\partial f_\alpha}{\partial \mathbf{v}_\alpha}\mathbf{v}_\alpha : \frac{\partial}{\partial \mathbf{r}}\mathbf{F}_\alpha\\ &\left.+\frac{\partial^2 f_\alpha}{\partial \mathbf{v}_\alpha\partial \mathbf{v}_\alpha}:\mathbf{F}_\alpha\mathbf{F}_\alpha + 2\frac{\partial^2 f_\alpha}{\partial \mathbf{v}_\alpha\partial \mathbf{r}}:\mathbf{v}_\alpha\mathbf{F}_\alpha\right]d\mathbf{v}_\alpha =\\ &=\int m_\alpha\mathbf{v}_\alpha J_\alpha^{st,el}d\mathbf{v}_\alpha + \int m_\alpha\mathbf{v}_\alpha J_\alpha^{st,inel}d\mathbf{v}_\alpha.\end{aligned} \tag{2.5.1}$$

Transformation of the first integral on the left side of (2.5.1) follows the main features the standard transformations used by derivation of the Enskog momentum equation. For example ($i = 1,2,3$)

$$\begin{aligned}&\int m_\alpha v_{\alpha i}\left(\mathbf{F}_\alpha\cdot\frac{\partial f_\alpha}{\partial \mathbf{v}_\alpha}\right)\left(\mathbf{v}_\alpha\cdot\frac{\partial \tau_\alpha}{\partial \mathbf{r}}\right)d\mathbf{v}_\alpha\\ &= m_\alpha\sum_{k,l=1}^{3}\frac{\partial \tau_\alpha}{\partial r_k}\int v_{\alpha i}F_{\alpha l}v_{\alpha k}\frac{\partial f_\alpha}{\partial v_{\alpha l}}d\mathbf{v}_\alpha\\ &= -m_\alpha\sum_{kl}\frac{\partial \tau_\alpha}{\partial r_k}\int F_{\alpha l}f_\alpha\frac{\partial}{\partial v_{\alpha l}}(v_{\alpha i}v_{\alpha k})d\mathbf{v}_\alpha\\ &= -m_\alpha\sum_{k}\frac{\partial \tau_\alpha}{\partial r_k}\int F_{\alpha i}f_\alpha v_{\alpha k}d\mathbf{v}_\alpha\\ &\quad -m_\alpha\sum_{k}\frac{\partial \tau_\alpha}{\partial r_k}\int F_{\alpha k}f_\alpha v_{\alpha i}d\mathbf{v}_\alpha.\end{aligned} \tag{2.5.2}$$

By derivation of (2.5.2) is taken into account that possible dependence on velocity of the external force correspond to the Lorentz force; using also (2.3.10), one obtains

$$\begin{aligned}\int m_\alpha \mathbf{v}_\alpha \left(\mathbf{F}_\alpha \cdot \frac{\partial f_\alpha}{\partial \mathbf{v}_\alpha}\right)\left(\mathbf{v}_\alpha \cdot \frac{\partial \tau_\alpha}{\partial \mathbf{r}}\right) d\mathbf{v}_\alpha &= -\rho_\alpha \overline{\mathbf{F}_\alpha^{(1)} \frac{\partial \tau_\alpha}{\partial \mathbf{r}} \cdot \mathbf{v}_\alpha} \\ -\rho_\alpha \overline{\mathbf{v}}_\alpha \mathbf{F}_\alpha^{(1)} \cdot \frac{\partial \tau_\alpha}{\partial \mathbf{r}} - \frac{q_\alpha}{m_\alpha} &\overline{\left(\frac{\partial \tau_\alpha}{\partial \mathbf{r}} \cdot \rho_\alpha \mathbf{v}_\alpha\right)[\mathbf{v}_\alpha \times \mathbf{B}]} \\ -\frac{q_\alpha}{m_\alpha}\rho_\alpha &\overline{\left\{[\mathbf{v}_\alpha \times \mathbf{B}] \cdot \frac{\partial \tau_\alpha}{\partial \mathbf{r}}\right\}\mathbf{v}_\alpha}.\end{aligned} \tag{2.5.3}$$

As usual, top line in (2.5.3) denotes averaging over velocity with taking into account the rule (2.1.5). Consider now transformations of the second integral in (2.5.1). Below are given examples of mentioned transformations.

$$\begin{aligned}\int m_\alpha \mathbf{v}_\alpha \frac{\partial f_\alpha}{\partial \mathbf{r}} \cdot \mathbf{F}_\alpha d\mathbf{v}_\alpha &= \rho_\alpha \overline{\mathbf{v}}_\alpha \frac{\partial}{\partial \mathbf{r}} \cdot \mathbf{F}_\alpha^{(1)} + q_\alpha \int f_\alpha \mathbf{v}_\alpha \frac{\partial}{\partial \mathbf{r}} \cdot (\mathbf{v}_\alpha \times \mathbf{B}) d\mathbf{v}_\alpha \\ &= \rho_\alpha \overline{\mathbf{v}}_\alpha \frac{\partial}{\partial \mathbf{r}} \cdot \mathbf{F}_\alpha^{(1)} - \frac{q_\alpha}{m_\alpha} \rho_\alpha \overline{\mathbf{v}_\alpha \mathbf{v}_\alpha} \cdot \mathrm{rot}\mathbf{B}\end{aligned} \tag{2.5.4}$$

$$\begin{aligned}\int m_\alpha \mathbf{v}_\alpha \frac{\partial^2 f_\alpha}{\partial \mathbf{v}_\alpha \partial t} \cdot \mathbf{F}_\alpha d\mathbf{v}_\alpha &= -m_\alpha \mathbf{F}_\alpha^{(1)} \frac{\partial n_\alpha}{\partial t} \\ &+ q_\alpha \int \mathbf{v}_\alpha \frac{\partial}{\partial \mathbf{v}_\alpha} \cdot \left\{[\mathbf{v}_\alpha \times \mathbf{B}] \frac{\partial f_\alpha}{\partial t}\right\} d\mathbf{v}_\alpha \\ &= -\mathbf{F}_\alpha^{(1)} \frac{\partial \rho_\alpha}{\partial t} + \frac{q_\alpha}{m_\alpha} \mathbf{B} \times \frac{\partial}{\partial t} (\rho_\alpha \overline{\mathbf{v}}_\alpha).\end{aligned} \tag{2.5.5}$$

$$\begin{aligned}\int m_\alpha \mathbf{v}_\alpha \frac{\partial^2 f_\alpha}{\partial \mathbf{v}_\alpha \partial \mathbf{v}_\alpha} : \mathbf{F}_\alpha \mathbf{F}_\alpha d\mathbf{v}_\alpha &= 2m_\alpha \sum_{ij} F_{\alpha i}^{(1)} \int \mathbf{v}_\alpha \frac{\partial^2 f_\alpha}{\partial v_{\alpha i} \partial v_{\alpha j}} F_{\alpha j}^B d\mathbf{v}_\alpha \\ + \int m_\alpha \mathbf{v}_\alpha \frac{\partial^2 f_\alpha}{\partial \mathbf{v}_\alpha \partial \mathbf{v}_\alpha} : \mathbf{F}_\alpha^B \mathbf{F}_\alpha^B d\mathbf{v}_\alpha &= 2m_\alpha \sum_{ij} F_{\alpha i}^{(1)} \int \mathbf{v}_\alpha \frac{\partial^2 f_\alpha}{\partial v_{\alpha i} \partial v_{\alpha j}} F_{\alpha j}^B d\mathbf{v}_\alpha \\ + 2\left(\frac{q_\alpha}{m_\alpha}\right)^2 \rho_\alpha \left[\mathbf{B}(\mathbf{B} \cdot \overline{\mathbf{v}}_\alpha) - 2B^2 \overline{\mathbf{v}}_\alpha\right]\end{aligned} \tag{2.5.6}$$

Consider in details the evaluation of integral on the right-hand-side of (2.5.6) for velocity component $v_{\alpha 1}$; use the integration by parts

$$\begin{aligned}&\sum_{ij} m_\alpha F_{\alpha i}^{(1)} \int v_{\alpha 1} \frac{\partial^2 f_\alpha}{\partial v_{\alpha i} \partial v_{\alpha j}} F_{\alpha j}^B d\mathbf{v}_\alpha \\ &= \sum_{\substack{ij \\ i \neq j,\, i \neq 1}} m_\alpha F_{\alpha i}^{(1)} \int v_{\alpha 1} \frac{\partial^2 f_\alpha}{\partial v_{\alpha i} \partial v_{\alpha j}} F_{\alpha j}^B d\mathbf{v}_\alpha = q_\alpha m_\alpha \left(F_{\alpha 2}^{(1)} B_3 - F_{\alpha 3}^{(1)} B_2\right)\end{aligned} \tag{2.5.7}$$

Similarly transformed can be the terms related to other components of $\mathbf{v}_\alpha$, and (2.5.6) is written as:

$$\begin{aligned}\int m_\alpha \mathbf{v}_\alpha \frac{\partial^2 f_\alpha}{\partial \mathbf{v}_\alpha \partial \mathbf{v}_\alpha} : \mathbf{F}_\alpha \mathbf{F}_\alpha d\mathbf{v}_\alpha &= 2\rho_\alpha \left(\frac{q_\alpha}{m_\alpha}\right)^2 \left[\mathbf{B}(\mathbf{B} \cdot \overline{\mathbf{v}}_\alpha) - 2B^2 \overline{\mathbf{v}}_\alpha\right] \\ + 2\frac{q_\alpha}{m_\alpha} \rho_\alpha \mathbf{F}_\alpha^{(1)} \times \mathbf{B}.\end{aligned} \tag{2.5.8}$$

Consider now the integral

$$\begin{aligned}\int m_\alpha \mathbf{v}_\alpha \frac{\partial^2 f_\alpha}{\partial \mathbf{v}_\alpha \partial \mathbf{r}} : \mathbf{v}_\alpha \mathbf{F}_\alpha d\mathbf{v}_\alpha &= -m_\alpha \int \mathbf{F}_\alpha \left(\mathbf{v}_\alpha \cdot \frac{\partial f_\alpha}{\partial \mathbf{r}}\right) d\mathbf{v}_\alpha \\ -m_\alpha \int \mathbf{v}_\alpha \left(\mathbf{F}_\alpha \cdot \frac{\partial f_\alpha}{\partial \mathbf{r}}\right) d\mathbf{v}_\alpha &= -\mathbf{F}_\alpha^{(1)} \frac{\partial}{\partial \mathbf{r}} \cdot (\rho_\alpha \overline{\mathbf{v}}_\alpha) - \left(\mathbf{F}_\alpha^{(1)} \frac{\partial}{\partial \mathbf{r}}\right)(\rho_\alpha \overline{\mathbf{v}}_\alpha) \\ -m_\alpha \int \mathbf{F}_\alpha^B \left(\mathbf{v}_\alpha \cdot \frac{\partial f_\alpha}{\partial \mathbf{r}}\right) d\mathbf{v}_\alpha &- m_\alpha \int \mathbf{v}_\alpha \left(\mathbf{F}_\alpha^B \cdot \frac{\partial f_\alpha}{\partial \mathbf{r}}\right) d\mathbf{v}_\alpha.\end{aligned} \tag{2.5.9}$$

Integrals of (2.5.9) associated with magnetic field, can be transformed using formula for triple vector product

$$[\mathbf{v}_\alpha \times \mathbf{B}] \cdot \frac{\partial f_\alpha}{\partial \mathbf{r}} = \left[\mathbf{B} \times \frac{\partial}{\partial \mathbf{r}}\right] \cdot f_\alpha \mathbf{v}_\alpha. \tag{2.5.10}$$

We have:

$$\begin{aligned} m_\alpha \int \mathbf{F}_\alpha^B \left(\mathbf{v}_\alpha \cdot \frac{\partial f_\alpha}{\partial \mathbf{r}}\right) d\mathbf{v}_\alpha &= q_\alpha \int [\mathbf{v}_\alpha \times \mathbf{B}] \left(\mathbf{v}_\alpha \cdot \frac{\partial f_\alpha}{\partial \mathbf{r}}\right) d\mathbf{v}_\alpha \\ &= \frac{q_\alpha}{m_\alpha} \left[\frac{\partial}{\partial \mathbf{r}} \cdot \rho_\alpha \overline{\mathbf{v}_\alpha \mathbf{v}_\alpha}\right] \times \mathbf{B}, \end{aligned} \tag{2.5.11}$$

$$\begin{aligned} m_\alpha \int \mathbf{v}_\alpha \left(\mathbf{F}_\alpha^B \cdot \frac{\partial f_\alpha}{\partial \mathbf{r}}\right) d\mathbf{v}_\alpha &= q_\alpha \int \mathbf{v}_\alpha \left[(\mathbf{v}_\alpha \times \mathbf{B}) \cdot \frac{\partial f_\alpha}{\partial \mathbf{r}}\right] d\mathbf{v}_\alpha \\ &= \left[\mathbf{B} \times \frac{\partial}{\partial \mathbf{r}}\right] \cdot \frac{q_\alpha}{m_\alpha} \rho_\alpha \overline{\mathbf{v}_\alpha \mathbf{v}_\alpha}. \end{aligned} \tag{2.5.12}$$

As a result, (2.5.9) will be brought into form:

$$\begin{aligned} \int m_\alpha \mathbf{v}_\alpha \frac{\partial^2 f_\alpha}{\partial \mathbf{v}_\alpha \partial \mathbf{r}} : \mathbf{v}_\alpha \mathbf{F}_\alpha d\mathbf{v}_\alpha &= -\mathbf{F}_\alpha^{(1)} \frac{\partial}{\partial \mathbf{r}} \cdot (\rho_\alpha \overline{\mathbf{v}}_\alpha) - \left(\mathbf{F}_\alpha^{(1)} \cdot \frac{\partial}{\partial \mathbf{r}}\right)(\rho_\alpha \overline{\mathbf{v}}_\alpha) \\ &- \left[\frac{\partial}{\partial \mathbf{r}} \cdot \frac{q_\alpha}{m_\alpha} \rho_\alpha \overline{\mathbf{v}_\alpha \mathbf{v}_\alpha}\right] \times \mathbf{B} - \left(\mathbf{B} \times \frac{\partial}{\partial \mathbf{r}}\right) \cdot \frac{q_\alpha}{m_\alpha} \rho_\alpha \overline{\mathbf{v}_\alpha \mathbf{v}_\alpha}. \end{aligned} \tag{2.5.13}$$

Let us introduce the vector product of operator $\partial/\partial \mathbf{r}$ and diada $\overline{\mathbf{v}_\alpha \mathbf{v}_\alpha}$ as a tensor defined by vectors $\frac{\partial}{\partial \mathbf{r}} \times \overline{\mathbf{v}_\alpha v_{\alpha i}}$, in particular it means that

$$\left(\mathbf{B} \times \frac{\partial}{\partial \mathbf{r}}\right) \cdot \rho_\alpha \overline{\mathbf{v}_\alpha \mathbf{v}_\alpha} = \mathbf{B} \cdot \frac{\partial}{\partial \mathbf{r}} \times \rho_\alpha \overline{\mathbf{v}_\alpha \mathbf{v}_\alpha}. \tag{2.5.14}$$

We need also for the following transformation the relation

$$\mathbf{B}^2 \overline{\mathbf{v}_\alpha} - \mathbf{B}(\mathbf{B} \cdot \overline{\mathbf{v}_\alpha}) = -(\overline{\mathbf{v}_\alpha} \times \mathbf{B}) \times \mathbf{B}, \tag{2.5.15}$$

After substitution of all transformed integrals into (2.5.1) one obtains the form of the generalized momentum equation

$$\begin{aligned} &\frac{\partial}{\partial t}\left\{\rho_\alpha \overline{\mathbf{v}}_\alpha - \tau_\alpha \left[\frac{\partial}{\partial t}(\rho_\alpha \overline{\mathbf{v}}_\alpha) + \frac{\partial}{\partial \mathbf{r}} \cdot \rho_\alpha \overline{\mathbf{v}_\alpha \mathbf{v}_\alpha} - \rho_\alpha \mathbf{F}_\alpha^{(1)}\right.\right. \\ &\left.\left. - \frac{q_\alpha}{m_\alpha} \rho_\alpha \overline{\mathbf{v}}_\alpha \times \mathbf{B}\right]\right\} - \mathbf{F}_\alpha^{(1)} \left[\rho_\alpha - \tau_\alpha \left(\frac{\partial \rho_\alpha}{\partial t} + \frac{\partial}{\partial \mathbf{r}}(\rho_\alpha \overline{\mathbf{v}}_\alpha)\right)\right] \\ &- \frac{q_\alpha}{m_\alpha}\left\{\rho_\alpha \overline{\mathbf{v}}_\alpha - \tau_\alpha \left[\frac{\partial}{\partial t}(\rho_\alpha \overline{\mathbf{v}}_\alpha) + \frac{\partial}{\partial \mathbf{r}} \cdot \rho_\alpha \overline{\mathbf{v}_\alpha \mathbf{v}_\alpha} - \rho_\alpha \mathbf{F}_\alpha^{(1)}\right.\right. \\ &\left.\left. - \frac{q_\alpha}{m_\alpha} \rho_\alpha \overline{\mathbf{v}}_\alpha \times \mathbf{B}\right]\right\} \times \mathbf{B} + \frac{\partial}{\partial \mathbf{r}} \cdot \left\{\rho_\alpha \overline{\mathbf{v}_\alpha \mathbf{v}_\alpha} - \tau_\alpha \left[\frac{\partial}{\partial t} \rho_\alpha \overline{\mathbf{v}_\alpha \mathbf{v}_\alpha}\right.\right. \\ &+ \frac{\partial}{\partial \mathbf{r}} \cdot \rho_\alpha \overline{(\mathbf{v}_\alpha \mathbf{v}_\alpha) \mathbf{v}_\alpha} - \mathbf{F}_\alpha^{(1)} \rho_\alpha \overline{\mathbf{v}}_\alpha - \rho_\alpha \overline{\mathbf{v}}_\alpha \mathbf{F}_\alpha^{(1)} \\ &\left.\left. - \frac{q_\alpha}{m_\alpha} \rho_\alpha \overline{[\mathbf{v}_\alpha \times \mathbf{B}] \mathbf{v}_\alpha} - \frac{q_\alpha}{m_\alpha} \rho_\alpha \overline{\mathbf{v}_\alpha [\mathbf{v}_\alpha \times \mathbf{B}]}\right]\right\} \\ &= \int m_\alpha \mathbf{v}_\alpha J_\alpha^{st,el} d\mathbf{v}_\alpha + \int m_\alpha \mathbf{v}_\alpha J_\alpha^{st,inel} d\mathbf{v}_\alpha. \end{aligned} \tag{2.5.16}$$

By virtue of momentum conservation law, after summation the integral terms on the right side of (2.5.16) turn into zero

$$\sum_\alpha m_\alpha \int \mathbf{v}_\alpha J_\alpha^{st,el} d\mathbf{v}_\alpha = 0, \tag{2.5.17}$$

$$\sum_\alpha m_\alpha \int \mathbf{v}_\alpha J_\alpha^{st,inel} d\mathbf{v}_\alpha = 0. \tag{2.5.18}$$

Boltzmann collision integral satisfies these demands, moreover it can be proved [36, 54], that

$$\int \psi_\alpha^{(1)} J_\alpha^{st,el} d\mathbf{v}_\alpha = 0, \int J_\alpha^{st,inel} \psi_\alpha^{(2)} d\mathbf{v}_\alpha = 0, \tag{2.5.19}$$

$$\sum_\alpha \int \psi_\alpha^{(2)} J_\alpha^{st,el} d\mathbf{v}_\alpha = 0, \sum_\alpha \int \psi_\alpha^{(2)} J_\alpha^{st,inel} d\mathbf{v}_\alpha = 0 \tag{2.5.20}$$

$$\sum_\alpha \int \psi_\alpha^{(3)} J_\alpha^{st,el} d\mathbf{v}_\alpha = 0, \sum_\alpha \int \psi_\alpha^{(3)} J_\alpha^{st,inel} d\mathbf{v}_\alpha = 0 \tag{2.5.21}$$

where

$$\psi_\alpha^{(1)} = m_\alpha, \quad \psi_\alpha^{(2)} = m_\alpha \mathbf{v}_\alpha, \quad \psi_\alpha^{(3)} = \frac{m_\alpha v_\alpha^2}{2} + \varepsilon_\alpha.$$

The proof is based on inversion of forward and backward collisions and use of the mass, momentum, and energy conservation laws for nonrelativistic particle's collision. But in plasma physics—in particular in strong electric fields—the use of hydrodynamic equations for gas mixtures is not sufficient for adequate description of physical system. In this case one uses transport equations for components introducing approximations for integrals of types $\int \psi_\alpha^{(2,3)} J_\alpha^{st,el} d\mathbf{v}_\alpha$, $\int \psi_\alpha^{(2,3)} J_\alpha^{st,inel} d\mathbf{v}_\alpha$, like the BGK-approximation, [71].

2.6 GENERALIZED ENERGY EQUATION FOR COMPONENT

Generalized energy equation for α-component ($\alpha = 1,\ldots,\eta$) is result of a term-by-term multiplication of left and right sides of GBE (2.3.15) by collision invariant $\psi_\alpha^{(3)} = (m_\alpha v_\alpha^2)/2 + \varepsilon_\alpha$ and following integration over all velocities:

$$\begin{aligned}
&\int \left(\frac{m_\alpha {v_\alpha}^2}{2} + \varepsilon_\alpha\right)\left(\frac{\partial f_\alpha}{\partial t} + \mathbf{v}_\alpha \cdot \frac{\partial f_\alpha}{\partial \mathbf{r}} + \mathbf{F}_\alpha \cdot \frac{\partial f_\alpha}{\partial \mathbf{v}_\alpha}\right)\left[1 - \left(\frac{\partial \tau_\alpha}{\partial t}\right.\right.\\
&\left.\left. + \mathbf{v}_\alpha \cdot \frac{\partial \tau_\alpha}{\partial \mathbf{r}}\right)\right] d\mathbf{v}_\alpha - \tau_\alpha \int \left(\frac{m_\alpha {v_\alpha}^2}{2} + \varepsilon_\alpha\right)\left[\frac{\partial^2 f_\alpha}{\partial t^2}\right.\\
&+ 2\frac{\partial^2 f_\alpha}{\partial \mathbf{r} \partial t} \cdot \mathbf{v}_\alpha + \frac{\partial^2 f_\alpha}{\partial \mathbf{r} \partial \mathbf{r}} : \mathbf{v}_\alpha \mathbf{v}_\alpha + 2\frac{\partial^2 f_\alpha}{\partial \mathbf{v}_\alpha \partial t} \cdot \mathbf{F}_\alpha + \frac{\partial \mathbf{F}_\alpha}{\partial t} \cdot \frac{\partial f_\alpha}{\partial \mathbf{v}_\alpha}\\
&+ \mathbf{F}_\alpha \cdot \frac{\partial f_\alpha}{\partial \mathbf{r}} + \frac{q_\alpha}{m_\alpha} \frac{\partial f_\alpha}{\partial \mathbf{v}_\alpha} \cdot \left(\mathbf{F}_\alpha^{(1)} \times \mathbf{B}\right) + \left(\frac{q_\alpha}{m_\alpha}\right)^2 \frac{\partial f_\alpha}{\partial \mathbf{v}_\alpha} \cdot \left[\mathbf{B}(\mathbf{v}_\alpha \cdot \mathbf{B}) - B^2 \mathbf{v}_\alpha\right]\\
&+ \frac{\partial f_\alpha}{\partial \mathbf{v}_\alpha} \mathbf{v}_\alpha : \frac{\partial}{\partial \mathbf{r}} \mathbf{F}_\alpha + \frac{\partial^2 f_\alpha}{\partial \mathbf{v}_\alpha \partial \mathbf{v}_\alpha} : \mathbf{F}_\alpha \mathbf{F}_\alpha\\
&\left. + 2\frac{\partial^2 f_\alpha}{\partial \mathbf{v}_\alpha \partial \mathbf{r}} : \mathbf{v}_\alpha \mathbf{F}_\alpha\right] d\mathbf{v}_\alpha = \int \left(\frac{m_\alpha {v_\alpha}^2}{2} + \varepsilon_\alpha\right) J_\alpha^{st,el} d\mathbf{v}_\alpha\\
&+ \int \left(\frac{m_\alpha {v_\alpha}^2}{2} + \varepsilon_\alpha\right) J_\alpha^{st,inel} d\mathbf{v}_\alpha.
\end{aligned} \tag{2.6.1}$$

As reference material we present the result of transformation of all terms in the generalized energy equation (2.6.1):

$$\int \frac{\partial f_\alpha}{\partial t}\left(1 - \frac{\partial \tau_\alpha}{\partial t}\right)\left(\frac{m_\alpha v_\alpha^2}{2} + \varepsilon_\alpha\right) d\mathbf{v}_\alpha = \left(1 - \frac{\partial \tau_\alpha}{\partial t}\right)\frac{\partial}{\partial t}\left(\frac{\rho_\alpha \overline{v_\alpha^2}}{2} + n_\alpha \varepsilon_\alpha\right), \tag{2.6.2}$$

$$\int \frac{\partial f_\alpha}{\partial t}\left(\mathbf{v}_\alpha \cdot \frac{\partial \tau_\alpha}{\partial \mathbf{r}}\right)\left(\frac{m_\alpha v_\alpha^2}{2} + \varepsilon_\alpha\right) d\mathbf{v}_\alpha = \frac{\partial \tau_\alpha}{\partial \mathbf{r}} \cdot \left[\frac{1}{2}\rho_\alpha \overline{\mathbf{v}_\alpha v_\alpha^2} + \varepsilon_\alpha n_\alpha \overline{\mathbf{v}}_\alpha\right], \tag{2.6.3}$$

$$\int \left(\frac{m_\alpha v_\alpha^2}{2} + \varepsilon_\alpha\right)\left(\mathbf{v}_\alpha \cdot \frac{\partial f_\alpha}{\partial \mathbf{r}}\right) d\mathbf{v}_\alpha = \frac{\partial}{\partial \mathbf{r}} \cdot \left\{\frac{1}{2}\rho_\alpha \overline{\mathbf{v}_\alpha v_\alpha^2} + \varepsilon_\alpha n_\alpha \overline{\mathbf{v}}_\alpha\right\}, \tag{2.6.4}$$

$$\begin{aligned}
&\int \left(\frac{m_\alpha v_\alpha^2}{2} + \varepsilon_\alpha\right)\left(\mathbf{v}_\alpha \cdot \frac{\partial f_\alpha}{\partial \mathbf{r}}\right)\left(\mathbf{v}_\alpha \cdot \frac{\partial \tau_\alpha}{\partial \mathbf{r}}\right) d\mathbf{v}_\alpha\\
&= \frac{1}{2}\frac{\partial \tau_\alpha}{\partial \mathbf{r}} \frac{\partial}{\partial \mathbf{r}} : \rho_\alpha \overline{v_\alpha^2 \mathbf{v}_\alpha \mathbf{v}_\alpha} + \frac{\partial \tau_\alpha}{\partial \mathbf{r}} \frac{\partial}{\partial \mathbf{r}} : \varepsilon_\alpha n_\alpha \overline{\mathbf{v}_\alpha \mathbf{v}_\alpha},
\end{aligned} \tag{2.6.5}$$

$$\int\left(\frac{m_\alpha v_\alpha^2}{2}+\varepsilon_\alpha\right)\left(\mathbf{F}_\alpha\cdot\frac{\partial f_\alpha}{\partial\mathbf{v}_\alpha}\right)\mathrm{d}\mathbf{v}_\alpha=-\rho_\alpha\overline{\mathbf{F}_\alpha\mathbf{v}_\alpha},\tag{2.6.6}$$

$$\begin{aligned}&\int\left(\frac{m_\alpha v_\alpha^2}{2}+\varepsilon_\alpha\right)\left(\mathbf{F}_\alpha\cdot\frac{\partial f_\alpha}{\partial\mathbf{v}_\alpha}\right)\left(\mathbf{v}_\alpha\cdot\frac{\partial\tau_\alpha}{\partial\mathbf{r}}\right)\mathrm{d}\mathbf{v}_\alpha=-\rho_\alpha\frac{\partial\tau_\alpha}{\partial\mathbf{r}}\cdot\left(\overline{\mathbf{F}_\alpha\cdot\mathbf{v}_\alpha\mathbf{v}_\alpha}\right)\\&-\frac{1}{2}\frac{\partial\tau_\alpha}{\partial\mathbf{r}}\cdot\left(\rho_\alpha\overline{v_\alpha^2\mathbf{F}_\alpha}\right)-\frac{\partial\tau_\alpha}{\partial\mathbf{r}}\cdot\left(\varepsilon_\alpha n_\alpha\overline{\mathbf{F}}_\alpha\right),\end{aligned}\tag{2.6.7}$$

$$\int\left(\frac{m_\alpha v_\alpha^2}{2}+\varepsilon_\alpha\right)\frac{\partial^2 f_\alpha}{\partial t^2}\mathrm{d}\mathbf{v}_\alpha=\frac{\partial^2}{\partial t^2}\left(\frac{\rho_\alpha\overline{v_\alpha^2}}{2}+\varepsilon_\alpha n_\alpha\right),\tag{2.6.8}$$

$$\int\left(\frac{m_\alpha v_\alpha^2}{2}+\varepsilon_\alpha\right)\left(\mathbf{v}_\alpha\cdot\frac{\partial f_\alpha}{\partial\mathbf{r}}\right)\mathrm{d}\mathbf{v}_\alpha=\frac{\partial}{\partial\mathbf{r}}\cdot\left\{\frac{1}{2}\rho_\alpha\overline{\mathbf{v}_\alpha v_\alpha^2}+\varepsilon_\alpha n_\alpha\overline{\mathbf{v}}_\alpha\right\},\tag{2.6.9}$$

$$\begin{aligned}&\int\left(\frac{m_\alpha v_\alpha^2}{2}+\varepsilon_\alpha\right)\frac{\partial^2 f_\alpha}{\partial\mathbf{r}\partial\mathbf{r}}:\mathbf{v}_\alpha\mathbf{v}_\alpha\mathrm{d}\mathbf{v}_\alpha\\&=\frac{\partial^2}{\partial\mathbf{r}\partial\mathbf{r}}:\left[\frac{1}{2}\rho_\alpha\overline{\mathbf{v}_\alpha\mathbf{v}_\alpha v_\alpha^2}+\varepsilon_\alpha n_\alpha\overline{\mathbf{v}_\alpha\mathbf{v}_\alpha}\right],\end{aligned}\tag{2.6.10}$$

$$\int\left(\frac{m_\alpha v_\alpha^2}{2}+\varepsilon_\alpha\right)\frac{\partial^2 f_\alpha}{\partial\mathbf{v}_\alpha\partial t}\cdot\mathbf{F}_\alpha\mathrm{d}\mathbf{v}_\alpha=-\mathbf{F}_\alpha^{(1)}\cdot\frac{\partial}{\partial t}\left(\rho_\alpha\overline{\mathbf{v}_\alpha}\right),\tag{2.6.11}$$

$$\int\left(\frac{m_\alpha v_\alpha^2}{2}+\varepsilon_\alpha\right)\frac{\partial\mathbf{F}_\alpha}{\partial t}\cdot\frac{\partial f_\alpha}{\partial\mathbf{v}_\alpha}\mathrm{d}\mathbf{v}_\alpha=-\frac{\partial\mathbf{F}_\alpha^{(1)}}{\partial t}\cdot\rho_\alpha\overline{\mathbf{v}_\alpha},\tag{2.6.12}$$

$$\begin{aligned}&\int\left(\frac{m_\alpha v_\alpha^2}{2}+\varepsilon_\alpha\right)\mathbf{F}_\alpha\cdot\frac{\partial f_\alpha}{\partial\mathbf{r}}\mathrm{d}\mathbf{v}_\alpha\\&=\frac{1}{2}\frac{\partial}{\partial\mathbf{r}}\cdot\left(\rho_\alpha\overline{\mathbf{F}_\alpha v_\alpha^2}\right)-\frac{1}{2}\rho_\alpha\overline{v_\alpha^2\frac{\partial}{\partial\mathbf{r}}\cdot\mathbf{F}_\alpha}+\varepsilon_\alpha\left(\overline{\mathbf{F}_\alpha}\cdot\frac{\partial}{\partial\mathbf{r}}\right)n_\alpha\end{aligned}\tag{2.6.13}$$

$$\int\left(\frac{m_\alpha v_\alpha^2}{2}+\varepsilon_\alpha\right)\frac{\partial f_\alpha}{\partial\mathbf{v}_\alpha}\cdot\left[\mathbf{F}_\alpha^{(1)}\times\mathbf{B}\right]\mathrm{d}\mathbf{v}_\alpha=-\rho_\alpha\overline{\mathbf{v}}_\alpha\cdot\left[\mathbf{F}_\alpha^{(1)}\times\mathbf{B}\right],\tag{2.6.14}$$

$$\begin{aligned}&\int\left(\frac{m_\alpha v_\alpha^2}{2}+\varepsilon_\alpha\right)\frac{\partial f_\alpha}{\partial\mathbf{v}_\alpha}\cdot\left[\mathbf{B}(\mathbf{v}_\alpha\cdot\mathbf{B})-B^2\mathbf{v}_\alpha\right]d\mathbf{v}_\alpha=\\&=\frac{m_\alpha}{2}\int\frac{\partial f_\alpha}{\partial\mathbf{v}_\alpha}v_\alpha^2\cdot\left[\mathbf{B}(\mathbf{v}_\alpha\cdot\mathbf{B})-B^2\mathbf{v}_\alpha\right]d\mathbf{v}_\alpha+2\varepsilon_\alpha n_\alpha B^2\\&=-\rho_\alpha\overline{\mathbf{v}_\alpha\mathbf{v}_\alpha}:\mathbf{BB}-B^2\frac{\rho_\alpha\overline{v_\alpha^2}}{2}+2\varepsilon_\alpha n_\alpha B^2-\frac{m_\alpha}{2}B^2\int\frac{\partial f_\alpha}{\partial\mathbf{v}_\alpha}\cdot v_\alpha^2\mathbf{v}_\alpha d\mathbf{v}_\alpha\\&=-\rho_\alpha\overline{\mathbf{v}_\alpha\mathbf{v}_\alpha}:\mathbf{BB}+2\varepsilon_\alpha n_\alpha B^2+4B^2\frac{\rho_\alpha\overline{v_\alpha^2}}{2},\end{aligned}\tag{2.6.15}$$

$$\begin{aligned}&\int\left(\frac{m_\alpha v_\alpha^2}{2}+\varepsilon_\alpha\right)\frac{\partial f_\alpha}{\partial\mathbf{v}_\alpha}\mathbf{v}_\alpha:\frac{\partial}{\partial\mathbf{r}}\mathbf{F}_\alpha\mathrm{d}\mathbf{v}_\alpha\\&=\int\left(\frac{m_\alpha v_\alpha^2}{2}+\varepsilon_\alpha\right)\frac{\partial f_\alpha}{\partial\mathbf{v}_\alpha}\mathbf{v}_\alpha:\frac{\partial}{\partial\mathbf{r}}\mathbf{F}_\alpha^{(1)}\mathrm{d}\mathbf{v}_\alpha\\&+\int\left(\frac{m_\alpha v_\alpha^2}{2}+\varepsilon_\alpha\right)\frac{\partial f_\alpha}{\partial\mathbf{v}_\alpha}\mathbf{v}_\alpha:\frac{\partial}{\partial\mathbf{r}}\mathbf{F}_\alpha^{B}\mathrm{d}\mathbf{v}_\alpha.\end{aligned}\tag{2.6.16}$$

Transformation of integrals in the right side of (2.6.16):

$$\begin{aligned}&\int\left(\frac{m_\alpha v_\alpha^2}{2}+\varepsilon_\alpha\right)\frac{\partial f_\alpha}{\partial\mathbf{v}_\alpha}\mathbf{v}_\alpha:\frac{\partial}{\partial\mathbf{r}}\mathbf{F}_\alpha^{(1)}\mathrm{d}\mathbf{v}_\alpha\\&=-\left(\frac{\rho_\alpha\overline{v_\alpha^2}}{2}+n_\alpha\varepsilon_\alpha\right)\frac{\partial}{\partial\mathbf{r}}\cdot\mathbf{F}_\alpha^{(1)}-\rho_\alpha\overline{\mathbf{v}_\alpha\mathbf{v}_\alpha}:\frac{\partial}{\partial\mathbf{r}}\mathbf{F}_\alpha^{(1)},\end{aligned}\tag{2.6.17}$$

$$\begin{aligned}
&\int\left(\frac{m_\alpha v_\alpha^2}{2}+\varepsilon_\alpha\right)\frac{\partial f_\alpha}{\partial \mathbf{v}_\alpha}\mathbf{v}_\alpha : \frac{\partial}{\partial \mathbf{r}}\mathbf{F}_\alpha^B d\mathbf{v}_\alpha \\
&= \int \frac{m_\alpha v_\alpha^2}{2}\frac{\partial f_\alpha}{\partial \mathbf{v}_\alpha} : \frac{\partial}{\partial \mathbf{r}}\mathbf{F}_\alpha^B d\mathbf{v}_\alpha + \frac{q_\alpha}{m_\alpha} n_\alpha \varepsilon_\alpha \overline{\mathbf{v}}_\alpha \cdot \mathrm{rot}\mathbf{B} \\
&= -\sum_{i=1}^{3}\frac{m_\alpha}{2}\int v_\alpha^2 f_\alpha \frac{\partial}{\partial r_i}F_{\alpha i}^B d\mathbf{v}_\alpha - \sum_{ij} m_\alpha \int v_{\alpha i} v_{\alpha j} f_\alpha \frac{\partial}{\partial r_j} F_{\alpha i}^B d\mathbf{v}_\alpha \\
&+ \frac{q_\alpha}{m_\alpha} n_\alpha \varepsilon_\alpha \overline{\mathbf{v}}_\alpha \cdot \mathrm{rot}\mathbf{B} = \frac{1}{2} n_\alpha q_\alpha \overline{v_\alpha^2 \mathbf{v}_\alpha} \cdot \mathrm{rot}\mathbf{B} + \frac{q_\alpha}{m_\alpha} n_\alpha \varepsilon_\alpha \overline{\mathbf{v}}_\alpha \cdot \mathrm{rot}\mathbf{B}.
\end{aligned} \tag{2.6.18}$$

Using (2.6.17) and (2.6.18) we finish the transformation of (2.6.16):

$$\begin{aligned}
&\int\left(\frac{m_\alpha v_\alpha^2}{2}+\varepsilon_\alpha\right)\frac{\partial f_\alpha}{\partial \mathbf{v}_\alpha}\mathbf{v}_\alpha : \frac{\partial}{\partial \mathbf{r}}\mathbf{F}_\alpha d\mathbf{v}_\alpha = -\rho_\alpha \overline{\mathbf{v}_\alpha \mathbf{v}_\alpha} : \frac{\partial}{\partial \mathbf{r}}\mathbf{F}_\alpha^{(1)} \\
&- n_\alpha\left(\frac{m_\alpha \overline{v_\alpha^2}}{2}+\varepsilon_\alpha\right)\frac{\partial}{\partial \mathbf{r}}\cdot \mathbf{F}_\alpha^{(1)} + \frac{1}{2} n_\alpha q_\alpha \overline{v_\alpha^2 \mathbf{v}_\alpha} \cdot \mathrm{rot}\mathbf{B} \\
&+ n_\alpha \frac{q_\alpha}{m_\alpha}\varepsilon_\alpha \overline{\mathbf{v}}_\alpha \cdot \mathrm{rot}\mathbf{B},
\end{aligned} \tag{2.6.19}$$

$$\begin{aligned}
&\int\left(\frac{m_\alpha v_\alpha^2}{2}+\varepsilon_\alpha\right)\frac{\partial^2 f_\alpha}{\partial \mathbf{v}_\alpha \partial \mathbf{v}_\alpha} : \mathbf{F}_\alpha \mathbf{F}_\alpha d\mathbf{v}_\alpha \\
&= \int\left(\frac{m_\alpha v_\alpha^2}{2}+\varepsilon_\alpha\right)\frac{\partial^2 f_\alpha}{\partial \mathbf{v}_\alpha \partial \mathbf{v}_\alpha} : \mathbf{F}_\alpha^{(1)} \mathbf{F}_\alpha^{(1)} d\mathbf{v}_\alpha \\
&+ 2\int\left(\frac{m_\alpha v_\alpha^2}{2}+\varepsilon_\alpha\right)\frac{\partial^2 f_\alpha}{\partial \mathbf{v}_\alpha \partial \mathbf{v}_\alpha} : \mathbf{F}_\alpha^{(1)} \mathbf{F}_\alpha^{B} d\mathbf{v}_\alpha \\
&+ \int\left(\frac{m_\alpha v_\alpha^2}{2}+\varepsilon_\alpha\right)\frac{\partial^2 f_\alpha}{\partial \mathbf{v}_\alpha \partial \mathbf{v}_\alpha} : \mathbf{F}_\alpha^{B} \mathbf{F}_\alpha^{B} d\mathbf{v}_\alpha.
\end{aligned} \tag{2.6.20}$$

Consider transformation of every of three integrals on the right side of (2.6.20)

$$\int\left(\frac{m_\alpha v_\alpha^2}{2}+\varepsilon_\alpha\right)\frac{\partial^2 f_\alpha}{\partial \mathbf{v}_\alpha \partial \mathbf{v}_\alpha} : \mathbf{F}_\alpha^{(1)} \mathbf{F}_\alpha^{(1)} d\mathbf{v}_\alpha = \rho_\alpha F_\alpha^{(1)2}, \tag{2.6.21}$$

$$\begin{aligned}
&\int\left(\frac{m_\alpha v_\alpha^2}{2}+\varepsilon_\alpha\right)\frac{\partial^2 f_\alpha}{\partial \mathbf{v}_\alpha \partial \mathbf{v}_\alpha} : \mathbf{F}_\alpha^{(1)} \mathbf{F}_\alpha^{B} d\mathbf{v}_\alpha \\
&= m_\alpha \sum_{\substack{ij=1 \\ i\neq j}}^{3} \int f_\alpha F_{\alpha i}^{(1)} \frac{\partial F_{\alpha j}^B}{\partial v_{\alpha i}} v_{\alpha j} d\mathbf{v}_\alpha + m_\alpha \sum_{i=1}^{3} \int f_\alpha F_{\alpha i}^{(1)} F_{\alpha i}^B d\mathbf{v}_\alpha.
\end{aligned} \tag{2.6.22}$$

Using the explicit expression for the Lorentz force we find, that

$$\sum_{\substack{ij \\ i\neq j}} \int f_\alpha F_{\alpha i}^{(1)} \frac{\partial F_{\alpha j}^B}{\partial v_{\alpha i}} v_{\alpha j} d\mathbf{v}_\alpha = -\frac{q_\alpha}{m_\alpha} n_\alpha \mathbf{F}_\alpha^{(1)} \cdot [\overline{\mathbf{v}}_\alpha \times \mathbf{B}], \tag{2.6.23}$$

then the right-hand side of relation (2.6.22) is equal to zero. The last from three mentioned integrals can be transformed as follows

$$\begin{aligned}
&\int\left(\frac{m_\alpha v_\alpha^2}{2}+\varepsilon_\alpha\right)\frac{\partial^2 f_\alpha}{\partial \mathbf{v}_\alpha \partial \mathbf{v}_\alpha} : \mathbf{F}_\alpha^{B} \mathbf{F}_\alpha^{B} d\mathbf{v}_\alpha = \rho_\alpha \overline{\left(F_\alpha^B\right)^2} \\
&- 2\left(\frac{\rho_\alpha \overline{v_\alpha^2}}{2}+\varepsilon_\alpha n_\alpha\right)\left(\frac{q_\alpha}{m_\alpha}\right)^2 B^2 + 2\left(\frac{q_\alpha}{m_\alpha}\right)^2 \rho_\alpha \overline{\left[(\mathbf{B}\cdot\mathbf{v}_\alpha)^2 - v_\alpha^2 B^2\right]} \\
&= \rho_\alpha \left(\frac{q_\alpha}{m_\alpha}\right)^2 \overline{\left[(\mathbf{B}\cdot\mathbf{v}_\alpha)^2 - 2 v_\alpha^2 B^2\right]} - 2B^2 \left(\frac{q_\alpha}{m_\alpha}\right)^2 \varepsilon_\alpha n_\alpha.
\end{aligned} \tag{2.6.24}$$

Using (2.6.21)–(2.6.24), one obtains

$$\int\left(\frac{m_\alpha v_\alpha^2}{2}+\varepsilon_\alpha\right)\frac{\partial^2 f_\alpha}{\partial\mathbf{v}_\alpha\partial\mathbf{v}_\alpha}:\mathbf{F}_\alpha\mathbf{F}_\alpha d\mathbf{v}_\alpha=\rho_\alpha F_\alpha^{(1)2}$$
$$+\left(\frac{q_\alpha}{m_\alpha}\right)^2\rho_\alpha\left\{\overline{(\mathbf{B}\cdot\mathbf{v}_\alpha)^2-2v_\alpha^2B^2}\right\}-2B^2\left(\frac{q_\alpha}{m_\alpha}\right)^2\varepsilon_\alpha n_\alpha. \tag{2.6.25}$$

Calculate now the last integral of the left side of (2.6.1)

$$\int\left(\frac{m_\alpha v_\alpha^2}{2}+\varepsilon_\alpha\right)\frac{\partial^2 f_\alpha}{\partial\mathbf{v}_\alpha\partial\mathbf{r}}:\mathbf{v}_\alpha\mathbf{F}_\alpha d\mathbf{v}_\alpha$$
$$=\int\left(\frac{m_\alpha v_\alpha^2}{2}+\varepsilon_\alpha\right)\frac{\partial^2 f_\alpha}{\partial\mathbf{v}_\alpha\partial\mathbf{r}}:\mathbf{v}_\alpha\mathbf{F}_\alpha^{(1)}d\mathbf{v}_\alpha \tag{2.6.26}$$
$$+\int\left(\frac{m_\alpha v_\alpha^2}{2}+\varepsilon_\alpha\right)\frac{\partial^2 f_\alpha}{\partial\mathbf{v}_\alpha\partial\mathbf{r}}:\mathbf{v}_\alpha\mathbf{F}_\alpha^B d\mathbf{v}_\alpha,$$

The first integral can be written in the form

$$\int\left(\frac{m_\alpha v_\alpha^2}{2}+\varepsilon_\alpha\right)\frac{\partial^2 f_\alpha}{\partial\mathbf{v}_\alpha\partial\mathbf{r}}:\mathbf{v}_\alpha\mathbf{F}_\alpha^{(1)}d\mathbf{v}_\alpha$$
$$=-\mathbf{F}_\alpha^{(1)}\cdot\left\{\frac{\partial}{\partial\mathbf{r}}\cdot\rho_\alpha\overline{\mathbf{v}_\alpha\mathbf{v}_\alpha}+\frac{\partial}{\partial\mathbf{r}}\left(\frac{\rho_\alpha\overline{v_\alpha^2}}{2}+\varepsilon_\alpha n_\alpha\right)\right\}. \tag{2.6.27}$$

The second integral is transformed using explicit form of the Lorentz force

$$\int\left(\frac{m_\alpha v_\alpha^2}{2}+\varepsilon_\alpha\right)\frac{\partial^2 f_\alpha}{\partial\mathbf{v}_\alpha\partial\mathbf{r}}:\mathbf{v}_\alpha\mathbf{F}_\alpha^B d\mathbf{v}_\alpha$$
$$=\frac{q_\alpha}{m_\alpha}\left\{\int f_\alpha\left(\frac{m_\alpha v_\alpha^2}{2}+\varepsilon_\alpha\right)\frac{\partial}{\partial\mathbf{r}}\cdot[\mathbf{v}_\alpha\times\mathbf{B}]d\mathbf{v}_\alpha\right. \tag{2.6.28}$$
$$\left.-\frac{\partial}{\partial\mathbf{r}}\cdot\int f_\alpha[\mathbf{v}_\alpha\times\mathbf{B}]\left(\frac{m_\alpha v_\alpha^2}{2}+\varepsilon_\alpha\right)d\mathbf{v}_\alpha\right\}.$$

The identity holds

$$\frac{\partial}{\partial\mathbf{r}}\cdot[\mathbf{v}_\alpha\times\mathbf{B}]=-\mathbf{v}_\alpha\cdot\mathrm{rot}\mathbf{B}. \tag{2.6.29}$$

Then (2.6.28) is written as

$$\int\left(\frac{m_\alpha v_\alpha^2}{2}+\varepsilon_\alpha\right)\frac{\partial^2 f_\alpha}{\partial\mathbf{v}_\alpha\partial\mathbf{r}}:\mathbf{v}_\alpha\mathbf{F}_\alpha^B d\mathbf{v}_\alpha$$
$$=-\frac{1}{2}\mathrm{rot}\mathbf{B}\cdot\frac{q_\alpha}{m_\alpha}\rho_\alpha\overline{v_\alpha^2\mathbf{v}_\alpha}-\mathrm{rot}\mathbf{B}\cdot\frac{q_\alpha}{m_\alpha}\varepsilon_\alpha n_\alpha\overline{\mathbf{v}}_\alpha \tag{2.6.30}$$
$$-\frac{1}{2}\frac{\partial}{\partial\mathbf{r}}\cdot\rho_\alpha\frac{q_\alpha}{m_\alpha}\overline{[\mathbf{v}_\alpha\times\mathbf{B}]v_\alpha^2}-\frac{\partial}{\partial\mathbf{r}}\cdot\frac{q_\alpha}{m_\alpha}n_\alpha\varepsilon_\alpha[\overline{\mathbf{v}}_\alpha\times\mathbf{B}].$$

For integral (2.6.26) we have

$$\int\left(\frac{m_\alpha v_\alpha^2}{2}+\varepsilon_\alpha\right)\frac{\partial^2 f_\alpha}{\partial\mathbf{v}_\alpha\partial\mathbf{r}}:\mathbf{v}_\alpha\mathbf{F}_\alpha d\mathbf{v}_\alpha=-\mathbf{F}_\alpha^{(1)}\cdot\left\{\frac{\partial}{\partial\mathbf{r}}\cdot\rho_\alpha\overline{\mathbf{v}_\alpha\mathbf{v}_\alpha}\right.$$
$$\left.+\frac{\partial}{\partial\mathbf{r}}\left(\frac{\rho_\alpha\overline{v_\alpha^2}}{2}+\varepsilon_\alpha n_\alpha\right)\right\}-\frac{1}{2}\mathrm{rot}\mathbf{B}\cdot\frac{q_\alpha}{m_\alpha}\rho_\alpha\overline{v_\alpha^2\mathbf{v}_\alpha}$$
$$-\mathrm{rot}\mathbf{B}\cdot\frac{q_\alpha}{m_\alpha}\varepsilon_\alpha n_\alpha\overline{\mathbf{v}}_\alpha-\frac{1}{2}\frac{\partial}{\partial\mathbf{r}}\cdot\rho_\alpha\frac{q_\alpha}{m_\alpha}\overline{[\mathbf{v}_\alpha\times\mathbf{B}]v_\alpha^2} \tag{2.6.31}$$
$$-\frac{\partial}{\partial\mathbf{r}}\cdot\frac{q_\alpha}{m_\alpha}n_\alpha\varepsilon_\alpha[\overline{\mathbf{v}}_\alpha\times\mathbf{B}].$$

Combining all results of integral transformation in (2.6.1), we find the following form of the generalized energy equation:

$$\begin{aligned}
&\frac{\partial}{\partial t}\left\{\frac{\rho_\alpha \overline{v_\alpha^2}}{2}+\varepsilon_\alpha n_\alpha-\tau_\alpha\left[\frac{\partial}{\partial t}\left(\frac{\rho_\alpha \overline{v_\alpha^2}}{2}+\varepsilon_\alpha n_\alpha\right)\right.\right.\\
&\left.\left.+\frac{\partial}{\partial \mathbf{r}}\cdot\left(\frac{1}{2}\rho_\alpha\overline{v_\alpha^2\mathbf{v}_\alpha}+\varepsilon_\alpha n_\alpha\overline{\mathbf{v}_\alpha}\right)-\mathbf{F}_\alpha^{(1)}\cdot\rho_\alpha\overline{\mathbf{v}_\alpha}\right]\right\}\\
&+\frac{\partial}{\partial \mathbf{r}}\cdot\left\{\frac{1}{2}\rho_\alpha\overline{v_\alpha^2\mathbf{v}_\alpha}+\varepsilon_\alpha n_\alpha\overline{\mathbf{v}_\alpha}-\tau_\alpha\left[\frac{\partial}{\partial t}\left(\frac{1}{2}\rho_\alpha\overline{v_\alpha^2\mathbf{v}_\alpha}+\varepsilon_\alpha n_\alpha\overline{\mathbf{v}_\alpha}\right)\right.\right.\\
&+\frac{\partial}{\partial \mathbf{r}}\cdot\left(\frac{1}{2}\rho_\alpha\overline{v_\alpha^2\mathbf{v}_\alpha\mathbf{v}_\alpha}+\varepsilon_\alpha n_\alpha\overline{\mathbf{v}_\alpha\mathbf{v}_\alpha}\right)-\rho_\alpha\mathbf{F}_\alpha^{(1)}\cdot\overline{\mathbf{v}_\alpha\mathbf{v}_\alpha}\\
&\left.\left.-\frac{1}{2}\rho_\alpha\overline{v_\alpha^2\mathbf{F}_\alpha}-\varepsilon_\alpha n_\alpha\overline{\mathbf{F}_\alpha}\right]\right\}-\left\{\rho_\alpha\mathbf{F}_\alpha^{(1)}\cdot\overline{\mathbf{v}_\alpha}-\tau_\alpha\left[\mathbf{F}_\alpha^{(1)}\cdot\left(\frac{\partial}{\partial t}(\rho_\alpha\overline{\mathbf{v}_\alpha})\right.\right.\right.\\
&\left.\left.\left.+\frac{\partial}{\partial \mathbf{r}}\cdot\rho_\alpha\overline{\mathbf{v}_\alpha\mathbf{v}_\alpha}-\rho_\alpha\mathbf{F}_\alpha^{(1)}-q_\alpha n_\alpha\overline{\mathbf{v}_\alpha}\times\mathbf{B}\right)\right]\right\}\\
&=\int\left(\frac{m_\alpha v_\alpha^2}{2}+\varepsilon_\alpha\right)J_\alpha^{st,el}\mathbf{d}\mathbf{v}_\alpha+\int\left(\frac{m_\alpha v_\alpha^2}{2}+\varepsilon_\alpha\right)J_\alpha^{st,inel}\mathbf{d}\mathbf{v}_\alpha.
\end{aligned}\tag{2.6.32}$$

2.7 SUMMARY OF THE GENERALIZED ENSKOG EQUATIONS AND DERIVATION OF THE GENERALIZED HYDRODYNAMIC EULER EQUATIONS

Formulate now the summary of the generalized Enskog hydrodynamic equations for components.

Continuity equation

$$\begin{aligned}
&\frac{\partial}{\partial t}\left\{\rho_\alpha-\tau_\alpha\left[\frac{\partial\rho_\alpha}{\partial t}+\frac{\partial}{\partial \mathbf{r}}\cdot(\rho_\alpha\overline{\mathbf{v}_\alpha})\right]\right\}\\
&+\frac{\partial}{\partial \mathbf{r}}\cdot\left\{\rho_\alpha\overline{\mathbf{v}_\alpha}-\tau_\alpha\left[\frac{\partial}{\partial t}(\rho_\alpha\overline{\mathbf{v}_\alpha})+\frac{\partial}{\partial \mathbf{r}}\cdot(\rho_\alpha\overline{\mathbf{v}_\alpha\mathbf{v}_\alpha})-\rho_\alpha\mathbf{F}_\alpha^{(1)}\right.\right.\\
&\left.\left.-\frac{q_\alpha}{m_\alpha}\rho_\alpha\overline{\mathbf{v}_\alpha}\times\mathbf{B}\right]\right\}=R_\alpha.
\end{aligned}\tag{2.7.1}$$

Momentum equation

$$\begin{aligned}
&\frac{\partial}{\partial t}\left\{\rho_\alpha\overline{\mathbf{v}_\alpha}-\tau_\alpha\left[\frac{\partial}{\partial t}(\rho_\alpha\overline{\mathbf{v}_\alpha})+\frac{\partial}{\partial \mathbf{r}}\cdot\rho_\alpha\overline{\mathbf{v}_\alpha\mathbf{v}_\alpha}-\rho_\alpha\mathbf{F}_\alpha^{(1)}\right.\right.\\
&\left.\left.-\frac{q_\alpha}{m_\alpha}\rho_\alpha\overline{\mathbf{v}_\alpha}\times\mathbf{B}\right]\right\}-\mathbf{F}_\alpha^{(1)}\left[\rho_\alpha-\tau_\alpha\left(\frac{\partial\rho_\alpha}{\partial t}+\frac{\partial}{\partial \mathbf{r}}\cdot\rho_\alpha\overline{\mathbf{v}_\alpha}\right)\right]\\
&-\frac{q_\alpha}{m_\alpha}\left\{\rho_\alpha\overline{\mathbf{v}_\alpha}-\tau_\alpha\left[\frac{\partial}{\partial t}(\rho_\alpha\overline{\mathbf{v}_\alpha})+\frac{\partial}{\partial \mathbf{r}}\cdot\rho_\alpha\overline{\mathbf{v}_\alpha\mathbf{v}_\alpha}-\rho_\alpha\mathbf{F}_\alpha^{(1)}\right.\right.\\
&\left.\left.-\frac{q_\alpha}{m_\alpha}\rho_\alpha\overline{\mathbf{v}_\alpha}\times\mathbf{B}\right]\right\}\times\mathbf{B}+\frac{\partial}{\partial \mathbf{r}}\cdot\left\{\rho_\alpha\overline{\mathbf{v}_\alpha\mathbf{v}_\alpha}-\tau_\alpha\left[\frac{\partial}{\partial t}(\rho_\alpha\overline{\mathbf{v}_\alpha\mathbf{v}_\alpha})\right.\right.\\
&+\frac{\partial}{\partial \mathbf{r}}\cdot\rho_\alpha\overline{(\mathbf{v}_\alpha\mathbf{v}_\alpha)\mathbf{v}_\alpha}-\mathbf{F}_\alpha^{(1)}\rho_\alpha\overline{\mathbf{v}_\alpha}-\rho_\alpha\overline{\mathbf{v}_\alpha}\mathbf{F}_\alpha^{(1)}\\
&\left.\left.-\frac{q_\alpha}{m_\alpha}\rho_\alpha\overline{[\mathbf{v}_\alpha\times\mathbf{B}]\mathbf{v}_\alpha}-\frac{q_\alpha}{m_\alpha}\rho_\alpha\overline{\mathbf{v}_\alpha[\mathbf{v}_\alpha\times\mathbf{B}]}\right]\right\}\\
&=\int m_\alpha\mathbf{v}_\alpha J_\alpha^{st,el}\mathbf{d}\mathbf{v}_\alpha+\int m_\alpha\mathbf{v}_\alpha J_\alpha^{st,inel}\mathbf{d}\mathbf{v}_\alpha,
\end{aligned}\tag{2.7.2}$$

Energy equation

$$\frac{\partial}{\partial t}\left\{\frac{\rho_\alpha\overline{v_\alpha^2}}{2}+\varepsilon_\alpha n_\alpha-\tau_\alpha\left[\frac{\partial}{\partial t}\left(\frac{\rho_\alpha\overline{v_\alpha^2}}{2}+\varepsilon_\alpha n_\alpha\right)\right.\right.$$
$$\left.\left.+\frac{\partial}{\partial \mathbf{r}}\cdot\left(\frac{1}{2}\rho_\alpha\overline{v_\alpha^2\mathbf{v}_\alpha}+\varepsilon_\alpha n_\alpha\overline{\mathbf{v}_\alpha}\right)-\mathbf{F}_\alpha^{(1)}\cdot\rho_\alpha\overline{\mathbf{v}_\alpha}\right]\right\}$$
$$+\frac{\partial}{\partial \mathbf{r}}\cdot\left\{\frac{1}{2}\rho_\alpha\overline{\mathbf{v}_\alpha v_\alpha^2}+\varepsilon_\alpha n_\alpha\overline{\mathbf{v}_\alpha}-\tau_\alpha\left[\frac{\partial}{\partial t}\left(\frac{1}{2}\rho_\alpha\overline{v_\alpha^2\mathbf{v}_\alpha}\right.\right.\right.$$
$$\left.+\varepsilon_\alpha n_\alpha\overline{\mathbf{v}_\alpha}\right)+\frac{\partial}{\partial \mathbf{r}}\cdot\left(\frac{1}{2}\rho_\alpha\overline{v_\alpha^2\mathbf{v}_\alpha\mathbf{v}_\alpha}+\varepsilon_\alpha n_\alpha\overline{\mathbf{v}_\alpha\mathbf{v}_\alpha}\right)$$
$$\left.\left.-\rho_\alpha\mathbf{F}_\alpha^{(1)}\cdot\overline{\mathbf{v}_\alpha\mathbf{v}_\alpha}-\frac{1}{2}\rho_\alpha\overline{v_\alpha^2}\mathbf{F}_\alpha-\varepsilon_\alpha n_\alpha\mathbf{F}_\alpha\right]\right\} \tag{2.7.3}$$
$$-\left\{\rho_\alpha\mathbf{F}_\alpha^{(1)}\cdot\overline{\mathbf{v}_\alpha}-\tau_\alpha\left[\mathbf{F}_\alpha^{(1)}\cdot\left(\frac{\partial}{\partial t}(\rho_\alpha\overline{\mathbf{v}_\alpha})\right.\right.\right.$$
$$\left.\left.\left.+\frac{\partial}{\partial \mathbf{r}}\cdot\rho_\alpha\overline{\mathbf{v}_\alpha\mathbf{v}_\alpha}-\rho_\alpha\mathbf{F}_\alpha^{(1)}-q_\alpha n_\alpha\overline{\mathbf{v}_\alpha}\times\mathbf{B}\right)\right]\right\}$$
$$=\int\left(\frac{m_\alpha v_\alpha^2}{2}+\varepsilon_\alpha\right)J_\alpha^{st,el}d\mathbf{v}_\alpha+\int\left(\frac{m_\alpha v_\alpha^2}{2}+\varepsilon_\alpha\right)J_\alpha^{st,inel}d\mathbf{v}_\alpha.$$

GHE for mixture of gases are significantly simpler because after summation of the left and right sides of (2.7.1)–(2.7.3) over all species α ($\alpha=1,\ldots,\eta$), the right integral parts of these equations reduce to zero on the strength of conservation laws. Then we have

$$\frac{\partial}{\partial t}\left\{\rho-\sum_\alpha\tau_\alpha\left[\frac{\partial\rho_\alpha}{\partial t}+\frac{\partial}{\partial \mathbf{r}}\cdot(\rho_\alpha\overline{\mathbf{v}_\alpha})\right]\right\}+$$
$$+\frac{\partial}{\partial \mathbf{r}}\cdot\left\{\rho\mathbf{v}_0-\sum_\alpha\tau_\alpha\left[\frac{\partial}{\partial t}(\rho_\alpha\overline{\mathbf{v}_\alpha})+\frac{\partial}{\partial \mathbf{r}}\cdot(\rho_\alpha\overline{\mathbf{v}_\alpha\mathbf{v}_\alpha})-\rho_\alpha\mathbf{F}_\alpha^{(1)}\right.\right. \tag{2.7.4}$$
$$\left.\left.-\frac{q_\alpha}{m_\alpha}\rho_\alpha\overline{\mathbf{v}_\alpha}\times\mathbf{B}\right]\right\}=0,$$

$$\frac{\partial}{\partial t}\left\{\rho\mathbf{v}_0-\sum_\alpha\tau_\alpha\left[\frac{\partial}{\partial t}(\rho_\alpha\overline{\mathbf{v}_\alpha})+\frac{\partial}{\partial \mathbf{r}}\cdot\rho_\alpha\overline{\mathbf{v}_\alpha\mathbf{v}_\alpha}-\rho_\alpha\mathbf{F}_\alpha^{(1)}\right.\right.$$
$$\left.\left.-\frac{q_\alpha}{m_\alpha}\rho_\alpha\overline{\mathbf{v}_\alpha}\times\mathbf{B}\right]\right\}-\sum_\alpha\mathbf{F}_\alpha^{(1)}\left[\rho_\alpha-\tau_\alpha\left(\frac{\partial\rho_\alpha}{\partial t}+\frac{\partial}{\partial \mathbf{r}}\cdot\rho_\alpha\overline{\mathbf{v}_\alpha}\right)\right]$$
$$-\sum_\alpha\frac{q_\alpha}{m_\alpha}\left\{\rho_\alpha\overline{\mathbf{v}_\alpha}-\tau_\alpha\left[\frac{\partial}{\partial t}(\rho_\alpha\overline{\mathbf{v}_\alpha})+\frac{\partial}{\partial \mathbf{r}}\cdot\rho_\alpha\overline{\mathbf{v}_\alpha\mathbf{v}_\alpha}-\rho_\alpha\mathbf{F}_\alpha^{(1)}\right.\right.$$
$$\left.\left.-\frac{q_\alpha}{m_\alpha}\rho_\alpha\overline{\mathbf{v}_\alpha}\times\mathbf{B}\right]\right\}\times\mathbf{B}+\frac{\partial}{\partial \mathbf{r}}\cdot\left\{\sum_\alpha\rho_\alpha\overline{\mathbf{v}_\alpha\mathbf{v}_\alpha}-\sum_\alpha\tau_\alpha\left[\frac{\partial}{\partial t}(\rho_\alpha\overline{\mathbf{v}_\alpha\mathbf{v}_\alpha})\right.\right. \tag{2.7.5}$$
$$+\frac{\partial}{\partial \mathbf{r}}\cdot\rho_\alpha\overline{(\mathbf{v}_\alpha\mathbf{v}_\alpha)\mathbf{v}_\alpha}-\mathbf{F}_\alpha^{(1)}\rho_\alpha\overline{\mathbf{v}_\alpha}-\rho_\alpha\overline{\mathbf{v}_\alpha}\mathbf{F}_\alpha^{(1)}$$
$$\left.\left.-\frac{q_\alpha}{m_\alpha}\rho_\alpha\overline{[\mathbf{v}_\alpha\times\mathbf{B}]\mathbf{v}_\alpha}-\frac{q_\alpha}{m_\alpha}\rho_\alpha\overline{\mathbf{v}_\alpha[\mathbf{v}_\alpha\times\mathbf{B}]}\right]\right\}=0,$$

$$
\begin{aligned}
&\frac{\partial}{\partial t}\left\{\sum_{\alpha}\left(\frac{\rho_\alpha \overline{v_\alpha^2}}{2}+\varepsilon_\alpha n_\alpha\right)-\sum_{\alpha}\tau_\alpha\left[\frac{\partial}{\partial t}\left(\frac{\rho_\alpha \overline{v_\alpha^2}}{2}+\varepsilon_\alpha n_\alpha\right)\right.\right.\\
&\left.\left.+\frac{\partial}{\partial \mathbf{r}}\cdot\left(\frac{1}{2}\rho_\alpha \overline{v_\alpha^2\mathbf{v}_\alpha}+\varepsilon_\alpha n_\alpha \overline{\mathbf{v}_\alpha}\right)-\mathbf{F}_\alpha^{(1)}\cdot\rho_\alpha\overline{\mathbf{v}_\alpha}\right]\right\}\\
&+\frac{\partial}{\partial \mathbf{r}}\cdot\left\{\sum_{\alpha}\left(\frac{1}{2}\rho_\alpha \overline{\mathbf{v}_\alpha v_\alpha^2}+\varepsilon_\alpha n_\alpha\overline{\mathbf{v}_\alpha}\right)-\sum_{\alpha}\tau_\alpha\left[\frac{\partial}{\partial t}\left(\frac{1}{2}\rho_\alpha \overline{\mathbf{v}_\alpha v_\alpha^2}\right.\right.\right.\\
&\left.+\varepsilon_\alpha n_\alpha\overline{\mathbf{v}_\alpha}\right)+\frac{\partial}{\partial \mathbf{r}}\cdot\left(\frac{1}{2}\rho_\alpha \overline{v_\alpha^2\mathbf{v}_\alpha\mathbf{v}_\alpha}+\varepsilon_\alpha n_\alpha\overline{\mathbf{v}_\alpha\mathbf{v}_\alpha}\right)\\
&\left.\left.-\rho_\alpha\mathbf{F}_\alpha^{(1)}\cdot\overline{\mathbf{v}_\alpha\mathbf{v}_\alpha}-\frac{1}{2}\rho_\alpha\overline{v_\alpha^2}\mathbf{F}_\alpha-\varepsilon_\alpha n_\alpha\overline{\mathbf{F}_\alpha}\right]\right\}\\
&-\left\{\sum_{\alpha}\rho_\alpha\mathbf{F}_\alpha^{(1)}\cdot\overline{\mathbf{v}_\alpha}-\sum_{\alpha}\tau_\alpha\left[\mathbf{F}_\alpha^{(1)}\cdot\left(\frac{\partial}{\partial t}(\rho_\alpha\overline{\mathbf{v}_\alpha})+\frac{\partial}{\partial \mathbf{r}}\cdot\rho_\alpha\overline{\mathbf{v}_\alpha\mathbf{v}_\alpha}\right.\right.\right.\\
&\left.\left.\left.-\rho_\alpha\mathbf{F}_\alpha^{(1)}-q_\alpha n_\alpha\overline{\mathbf{v}_\alpha}\times\mathbf{B}\right)\right]\right\}=0.
\end{aligned}
\tag{2.7.6}
$$

Generalized hydrodynamic Enskog equations (GHEnE) are very complicated. In particular GEnE are of higher order than classical Enskog equations. This brings up immediately two problems: (1) is it possible to simplify these equations? (2) what about the additional boundary and initial conditions for these equations?

Let us begin by looking at the first question. Extraordinary features of GHEnE consist in appearance of terms proportional to non-local parameter τ_α in all equations (2.7.1)–(2.7.6). In due course for this case $\tau_\alpha \sim \tau_{\alpha,mt}$. Mean time between collisions $\tau_{\alpha,mt}$ of particles belonging to α-species can be calculated if DF f_α is known. For the model of hard spheres in one-component gas

$$\tau_{\alpha,mt}^{(0)}=\frac{1}{\pi\sqrt{2}\overline{v_\alpha}n_\alpha\sigma_\alpha^2},\tag{2.7.7}$$

where $\overline{v_\alpha}$ is mean velocity for α-molecules, σ_α is their diameter, upper index (0) underlines that calculation of $\tau_{\alpha,mt}^{(0)}$ is realized for local Maxwellian function,

$$\overline{v_\alpha}=\sqrt{\frac{8k_{\mathrm{B}}T}{\pi m_\alpha}}.\tag{2.7.8}$$

Then

$$\tau_{\alpha,mt}^{(0)}=\frac{\sqrt{m_\alpha}}{4\sqrt{\pi}n_\alpha\sigma_\alpha^2\sqrt{k_{\mathrm{B}}T}}.\tag{2.7.9}$$

Multiplying term-by-term (2.7.9) by $p_\alpha=n_\alpha k_{\mathrm{B}}T$, one obtains

$$\tau_{\alpha,mt}^{(0)}p_\alpha=\frac{\sqrt{m_\alpha k_{\mathrm{B}}T}}{4\sqrt{\pi}\sigma_\alpha^2}.\tag{2.7.10}$$

But for simple gas in the first ("Navier-Stokes") approximation dynamical viscosity for the hard spheres model is written as [36]

$$[\mu]_1=\frac{5}{16}\frac{\sqrt{m_\alpha k_{\mathrm{B}}T}}{\sqrt{\pi}\sigma_\alpha^2}.\tag{2.7.11}$$

Of course, the subject under discussion is changing of the transport coefficients calculation in GBKT but this question we leave to the next sections. Now we state that mean time between collisions in local Maxwellian approximation $\tau_{\alpha,mt}^{(0)}$ can be presented using viscosity $[\mu_\alpha]_1$, calculated in the next successive approximation. In other words,

$$\tau_\alpha^{(0)}p_\alpha=0.8[\mu_\alpha]_1.\tag{2.7.12}$$

This fact leads to very interesting conclusions in the theory of turbulence considered from positions of GBKT.

The use in BKT instead of the first approximation, the convergence series in Sonine polynomials for hard spheres leads to slightly changing of mentioned coefficient. One obtains 0.786 instead of 0.8. Consider a simple gas under no forces in BGK approximation for classical Boltzmann equation (BE)

$$\frac{\partial f}{\partial t}+\mathbf{v}\cdot\frac{\partial f}{\partial \mathbf{r}}=-\frac{f-f^{(0)}}{\tau}. \tag{2.7.13}$$

From Eq. (2.7.13) follows the expression for DF f

$$f=f^{(0)}-\tau\frac{\partial f}{\partial t}-\tau\mathbf{v}\cdot\frac{\partial f}{\partial \mathbf{r}}. \tag{2.7.14}$$

For a gas consisting of point-like particles in the case of slow changing of parameters of state, $f-f^{(0)}$ must be small and Eq. (2.7.14) becomes

$$f=f^{(0)}-\tau^{(0)}\frac{\partial f^{(0)}}{\partial t}-\tau^{(0)}\mathbf{v}\cdot\frac{\partial f^{(0)}}{\partial \mathbf{r}}. \tag{2.7.15}$$

In one-dimensional stationary approximation

$$f=f^{(0)}-\tau^{(0)}\mathbf{v}\cdot\frac{\partial f^{(0)}}{\partial \mathbf{r}}, \tag{2.7.16}$$

and for a simple gas of uniform density and temperature, streaming parallel to Ox, but with hydrodynamic velocity changing along the axis Oy it is reasonable to suppose that changing of local Maxwellian DF $f^{(0)}$ is governed by evolution of hydrodynamic velocity v_0. As result

$$f=f^{(0)}-\tau^{(0)}v_y\frac{\partial v_0}{\partial y}\frac{\partial f^{(0)}}{\partial v_0}. \tag{2.7.17}$$

The viscous stress in the y-direction across a plane $x=\text{const}$ is

$$p_{xy}=\int m(v_x-v_0)v_y f\,d\mathbf{v}. \tag{2.7.18}$$

Substitute f from (2.7.17) into Eq. (2.7.18)

$$p_{xy}=\int m(v_x-v_0)v_y f^{(0)}d\mathbf{v}-\tau^{(0)}\frac{\partial v_0}{\partial y}\int\frac{\partial f^{(0)}}{\partial v_0}m(v_x-v_0)v_y^2 d\mathbf{v}. \tag{2.7.19}$$

Because the first integrand is an odd function of v_x-v_0, the first integral vanishes and

$$p_{xy}=-\tau^{(0)}\int\frac{\partial f^{(0)}}{\partial v_0}m(v_x-v_0)v_y^2 d\mathbf{v}\frac{\partial v_0}{\partial y}, \tag{2.7.20}$$

or

$$p_{xy}=-\mu\frac{\partial v_0}{\partial y}, \tag{2.7.21}$$

where coefficient of viscosity μ is equal to

$$\mu=\tau^{(0)}\int\frac{\partial f^{(0)}}{\partial v_0}m(v_x-v_0)v_y^2 d\mathbf{v}, \tag{2.7.22}$$

or after integrating by parts

$$\mu=\tau^{(0)}\frac{\partial}{\partial v_0}\int mf^{(0)}(v_x-v_0)v_y^2 d\mathbf{v}+\tau^{(0)}\int mf^{(0)}v_y^2 d\mathbf{v}. \tag{2.7.23}$$

The first integral in (2.7.23) vanishes as containing an odd function in the integrand, but

$$p = \int m f^{(0)} v_y^2 \mathrm{d}\mathbf{v}. \tag{2.7.24}$$

It means that in considered stiff restrictions the formula is valid

$$\tau^{(0)} p = \mu. \tag{2.7.25}$$

Relation (2.7.25) is also consequence of so-called elementary theory of gases.

Finally we can state that

$$\tau^{(0)}_{\alpha,mt} p_\alpha = \Pi \mu_\alpha, \quad \alpha = 1, .., \eta. \tag{2.7.26}$$

where numerical coefficient Π reflects the character features of particles interactions.

In the theory of generalized hydrodynamic equations relations (2.7.26) allow to close GHE leaving all calculations—in the definite sense—on the macroscopic level of physical system's description.

If a mixture of gases contains particles which masses are not too different, it is possible to set in Eqs. (2.7.1)–(2.7.6) τ_α as independent of number of species. Therefore, in the simplest case, if $\tau_\alpha \cong \tau_{\alpha,mt}$ the problem of the τ_α calculation is not difficult or precisely the same as in classical BKT.

The difficulties of principal character ensue due the calculation of averaged values in (2.7.1)–(2.7.6) for which we need the explicit expression for DF. But for local Maxwellian DF all averaged values in (2.7.1)–(2.7.6) can be calculated in the finite form and lead to generalized Euler equations (GEuE).

Calculate averaged values for the velocities moments of DF $f_\alpha^{(0)}$. Summary of results follows.

For continuity equation:

$$\begin{aligned}\overline{\mathbf{v}_\alpha} &= \left(\frac{m_\alpha}{2\pi k_B T}\right)^{3/2} \int \mathbf{v}_\alpha \exp\left(-\frac{m_\alpha V_\alpha^2}{2k_B T}\right) \mathrm{d}\mathbf{v}_\alpha \\ &= \left(\frac{m_\alpha}{2\pi k_B T}\right)^{3/2} \int \mathbf{v}_0 \exp\left(-\frac{m_\alpha V_\alpha^2}{2k_B T}\right) \mathrm{d}\mathbf{v}_\alpha = \mathbf{v}_0\end{aligned} \tag{2.7.27}$$

$$\begin{aligned}\overline{\mathbf{v}_\alpha \mathbf{v}_\alpha} &= \left(\frac{m_\alpha}{2\pi k_B T}\right)^{3/2} \int \mathbf{v}_\alpha \mathbf{v}_\alpha \exp\left(-\frac{m_\alpha V_\alpha^2}{2k_B T}\right) \mathrm{d}\mathbf{v}_\alpha \\ &= \mathbf{v}_0 \mathbf{v}_0 + \frac{1}{n_\alpha} \int f_\alpha^{(0)} \mathbf{V}_\alpha \mathbf{V}_\alpha \mathrm{d}\mathbf{v}_\alpha = \mathbf{v}_0 \mathbf{v}_0 + \frac{p_\alpha}{\rho_\alpha} \overset{\leftrightarrow}{\mathbf{I}},\end{aligned} \tag{2.7.28}$$

where p_α is static pressure of α-species, $\overset{\leftrightarrow}{\mathbf{I}}$ is unit tensor. In calculations realized in (2.7.27), (2.7.28) take into account vanishing of integrals containing odd functions as integrand.

We have following generalized Euler continuity equation (compare with (2.7.1) and (2.7.4)):

$$\begin{aligned}&\frac{\partial}{\partial t}\left\{\rho_\alpha - \tau_\alpha^{(0)}\left[\frac{\partial \rho_\alpha}{\partial t} + \frac{\partial}{\partial \mathbf{r}} \cdot (\rho_\alpha \mathbf{v}_0)\right]\right\} \\ &+ \frac{\partial}{\partial \mathbf{r}} \cdot \left\{\rho_\alpha \mathbf{v}_0 - \tau_\alpha^{(0)}\left[\frac{\partial}{\partial t}(\rho_\alpha \mathbf{v}_0) + \right.\right. \\ &\left.\left. + \frac{\partial}{\partial \mathbf{r}} \cdot (\rho_\alpha \mathbf{v}_0 \mathbf{v}_0) + \overset{\leftrightarrow}{\mathbf{I}} \cdot \frac{\partial p_\alpha}{\partial \mathbf{r}} - \rho_\alpha \mathbf{F}_\alpha^{(1)} - \frac{q_\alpha}{m_\alpha} \rho_\alpha \mathbf{v}_0 \times \mathbf{B}\right]\right\} = R_\alpha,\end{aligned} \tag{2.7.29}$$

and for a mixture

$$\begin{aligned}&\frac{\partial}{\partial t}\left\{\rho - \sum_\alpha \tau_\alpha^{(0)}\left[\frac{\partial \rho_\alpha}{\partial t} + \frac{\partial}{\partial \mathbf{r}} \cdot (\rho_\alpha \mathbf{v}_0)\right]\right\} \\ &+ \frac{\partial}{\partial \mathbf{r}} \cdot \left\{\rho \mathbf{v}_0 - \sum_\alpha \tau_\alpha^{(0)}\left[\frac{\partial}{\partial t}(\rho_\alpha \mathbf{v}_0) + \frac{\partial}{\partial \mathbf{r}} \cdot (\rho_\alpha \mathbf{v}_0 \mathbf{v}_0) + \overset{\leftrightarrow}{\mathbf{I}} \cdot \frac{\partial p_\alpha}{\partial \mathbf{r}}\right.\right. \\ &\left.\left. - \rho_\alpha \mathbf{F}_\alpha^{(1)} - \frac{q_\alpha}{m_\alpha} \rho_\alpha \mathbf{v}_0 \times \mathbf{B}\right]\right\} = 0,\end{aligned} \tag{2.7.30}$$

Generalized Euler momentum equations for species can be obtain in a similar way:

$$\begin{aligned}
&\frac{\partial}{\partial t}\left\{\rho_\alpha \mathbf{v}_0-\tau_\alpha^{(0)}\left[\frac{\partial}{\partial t}(\rho_\alpha \mathbf{v}_0)+\frac{\partial}{\partial \mathbf{r}}\cdot\rho_\alpha \mathbf{v}_0\mathbf{v}_0+\frac{\partial p_\alpha}{\partial \mathbf{r}}-\rho_\alpha \mathbf{F}_\alpha^{(1)}\right.\right.\\
&\left.\left.-\left(\frac{q_\alpha}{m_\alpha}\right)\rho_\alpha \mathbf{v}_0\times\mathbf{B}\right]\right\}-\mathbf{F}_\alpha^{(1)}\left[\rho_\alpha-\tau_\alpha^{(0)}\left(\frac{\partial\rho_\alpha}{\partial t}+\frac{\partial}{\partial \mathbf{r}}\cdot(\rho_\alpha \mathbf{v}_0)\right)\right]\\
&-\frac{q_\alpha}{m_\alpha}\left\{\rho_\alpha \mathbf{v}_0-\tau_\alpha^{(0)}\left[\frac{\partial}{\partial t}(\rho_\alpha \mathbf{v}_0)+\frac{\partial}{\partial \mathbf{r}}\cdot\rho_\alpha \mathbf{v}_0\mathbf{v}_0+\frac{\partial p_\alpha}{\partial \mathbf{r}}-\rho_\alpha \mathbf{F}_\alpha^{(1)}\right.\right.\\
&\left.\left.-\frac{q_\alpha}{m_\alpha}\rho_\alpha \mathbf{v}_0\times\mathbf{B}\right]\right\}\times\mathbf{B}+\frac{\partial}{\partial \mathbf{r}}\cdot\left\{\rho_\alpha \mathbf{v}_0\mathbf{v}_0+p_\alpha\overleftrightarrow{\mathbf{I}}-\tau_\alpha^{(0)}\left[\frac{\partial}{\partial t}(\rho_\alpha \mathbf{v}_0\mathbf{v}_0+p_\alpha\overleftrightarrow{\mathbf{I}})\right.\right.\\
&+\frac{\partial}{\partial \mathbf{r}}\cdot\left(\rho_\alpha(\mathbf{v}_0\mathbf{v}_0)\mathbf{v}_0+\rho_\alpha\overline{(\mathbf{v}_0\mathbf{V}_\alpha)\mathbf{V}_\alpha}+\rho_\alpha\overline{(\mathbf{V}_\alpha\mathbf{v}_0)\mathbf{V}_\alpha}\right.\\
&\left.+\rho_\alpha\overline{(\mathbf{V}_\alpha\mathbf{V}_\alpha)}\mathbf{v}_0\right)-\mathbf{F}_\alpha^{(1)}\rho_\alpha\mathbf{v}_0-\rho_\alpha\mathbf{v}_0\mathbf{F}_\alpha^{(1)}\\
&-\frac{q_\alpha}{m_\alpha}\rho_\alpha[\mathbf{v}_0\times\mathbf{B}]\mathbf{v}_0-\frac{q_\alpha}{m_\alpha}\rho_\alpha\overline{[\mathbf{V}_\alpha\times\mathbf{B}]\mathbf{V}_\alpha}\\
&\left.\left.-\frac{q_\alpha}{m_\alpha}\rho_\alpha\mathbf{v}_0[\mathbf{v}_0\times\mathbf{B}]-\frac{q_\alpha}{m_\alpha}\rho_\alpha\overline{\mathbf{V}_\alpha[\mathbf{V}_\alpha\times\mathbf{B}]}\right]\right\}\\
&=\int m_\alpha\mathbf{v}_\alpha J_\alpha^{st,el}d\mathbf{v}_\alpha+\int m_\alpha\mathbf{v}_\alpha J_\alpha^{st,inel}d\mathbf{v}_\alpha.
\end{aligned} \tag{2.7.31}$$

The following averaged expressions should be calculated in local Maxwellian approximation:

$$\frac{\partial^2}{\partial\mathbf{r}\partial\mathbf{r}}:\rho_\alpha\overline{(\mathbf{V}_\alpha\mathbf{v}_0)\mathbf{V}_\alpha}=\frac{\partial}{\partial\mathbf{r}}\left[\frac{\partial}{\partial\mathbf{r}}\cdot(p_\alpha\mathbf{v}_0)\right], \tag{2.7.32}$$

$$\begin{aligned}
&\frac{\partial^2}{\partial\mathbf{r}\partial\mathbf{r}}:\rho_\alpha\overline{(\mathbf{V}_\alpha\mathbf{V}_\alpha)}\mathbf{v}_0=\frac{\partial^2}{\partial\mathbf{r}\partial\mathbf{r}}:\overleftrightarrow{\mathbf{P}}_\alpha^{(0)}\mathbf{v}_0\\
&=\frac{\partial^2}{\partial\mathbf{r}\partial\mathbf{r}}:p_\alpha\overleftrightarrow{\mathbf{I}}\mathbf{v}_0=\Delta(p_\alpha\mathbf{v}_0),
\end{aligned} \tag{2.7.33}$$

where Δ is Laplacian,

$$\frac{\partial}{\partial\mathbf{r}}\cdot\rho_\alpha\overline{[\mathbf{V}_\alpha\times\mathbf{B}]\mathbf{V}_\alpha}=-\frac{\partial}{\partial\mathbf{r}}\times(p_\alpha\mathbf{B}), \tag{2.7.34}$$

$$\frac{\partial}{\partial\mathbf{r}}\cdot\rho_\alpha\overline{\mathbf{V}_\alpha[\mathbf{V}_\alpha\times\mathbf{B}]}=\frac{\partial}{\partial\mathbf{r}}\times(p_\alpha\mathbf{B}). \tag{2.7.35}$$

Then summation of two last expressions leads to its canceling each other.

Write down the generalized Euler momentum equation for α-species:

$$\begin{aligned}
&\frac{\partial}{\partial t}\left\{\rho_\alpha \mathbf{v}_0-\tau_\alpha^{(0)}\left[\frac{\partial}{\partial t}(\rho_\alpha \mathbf{v}_0)+\frac{\partial}{\partial \mathbf{r}}\cdot\rho_\alpha \mathbf{v}_0\mathbf{v}_0+\frac{\partial p_\alpha}{\partial \mathbf{r}}-\rho_\alpha \mathbf{F}_\alpha^{(1)}\right.\right.\\
&\left.\left.-\frac{q_\alpha}{m_\alpha}\rho_\alpha \mathbf{v}_0\times\mathbf{B}\right]\right\}-\mathbf{F}_\alpha^{(1)}\left[\rho_\alpha-\tau_\alpha^{(0)}\left(\frac{\partial\rho_\alpha}{\partial t}+\frac{\partial}{\partial \mathbf{r}}(\rho_\alpha \mathbf{v}_0)\right)\right]\\
&-\frac{q_\alpha}{m_\alpha}\left\{\rho_\alpha \mathbf{v}_0-\tau_\alpha^{(0)}\left[\frac{\partial}{\partial t}(\rho_\alpha \mathbf{v}_0)+\frac{\partial}{\partial \mathbf{r}}\cdot\rho_\alpha \mathbf{v}_0\mathbf{v}_0+\frac{\partial p_\alpha}{\partial \mathbf{r}}-\rho_\alpha \mathbf{F}_\alpha^{(1)}\right.\right.\\
&\left.\left.-\frac{q_\alpha}{m_\alpha}\rho_\alpha \mathbf{v}_0\times\mathbf{B}\right]\right\}\times\mathbf{B}+\frac{\partial}{\partial \mathbf{r}}\cdot\left\{\rho_\alpha \mathbf{v}_0\mathbf{v}_0+p_\alpha\overleftrightarrow{\mathbf{I}}-\tau_\alpha^{(0)}\left[\frac{\partial}{\partial t}(\rho_\alpha \mathbf{v}_0\mathbf{v}_0\right.\right.\\
&+p_\alpha\overleftrightarrow{\mathbf{I}})+\frac{\partial}{\partial \mathbf{r}}\cdot\rho_\alpha(\mathbf{v}_0\mathbf{v}_0)\mathbf{v}_0+2\overleftrightarrow{\mathbf{I}}\left(\frac{\partial}{\partial \mathbf{r}}\cdot(p_\alpha\mathbf{v}_0)\right)+\frac{\partial}{\partial \mathbf{r}}\cdot(\overleftrightarrow{\mathbf{I}}p_\alpha\mathbf{v}_0)\\
&\left.\left.-\mathbf{F}_\alpha^{(1)}\rho_\alpha\mathbf{v}_0-\rho_\alpha\mathbf{v}_0\mathbf{F}_\alpha^{(1)}-\frac{q_\alpha}{m_\alpha}\rho_\alpha[\mathbf{v}_0\times\mathbf{B}]\mathbf{v}_0-\frac{q_\alpha}{m_\alpha}\rho_\alpha\mathbf{v}_0[\mathbf{v}_0\times\mathbf{B}]\right]\right\}\\
&=\int m_\alpha\mathbf{v}_\alpha J_\alpha^{st,el}d\mathbf{v}_\alpha+\int m_\alpha\mathbf{v}_\alpha J_\alpha^{st,inel}d\mathbf{v}_\alpha.
\end{aligned} \tag{2.7.36}$$

After summation (2.7.36) over all species, one obtains the generalized Euler momentum equation for mixture of gases

$$
\begin{aligned}
&\frac{\partial}{\partial t}\left\{\rho\mathbf{v}_0-\sum_\alpha \tau_\alpha^{(0)}\left[\frac{\partial}{\partial t}(\rho_\alpha\mathbf{v}_0)+\frac{\partial}{\partial \mathbf{r}}\cdot\rho_\alpha\mathbf{v}_0\mathbf{v}_0+\frac{\partial p_\alpha}{\partial \mathbf{r}}-\rho_\alpha\mathbf{F}_\alpha^{(1)}\right.\right.\\
&\left.\left.-\frac{q_\alpha}{m_\alpha}\rho_\alpha\mathbf{v}_0\times\mathbf{B}\right]\right\}-\sum_\alpha \mathbf{F}_\alpha^{(1)}\left[\rho_\alpha-\tau_\alpha^{(0)}\left(\frac{\partial\rho_\alpha}{\partial t}+\frac{\partial}{\partial \mathbf{r}}(\rho_\alpha\mathbf{v}_0)\right)\right]\\
&-\sum_\alpha\frac{q_\alpha}{m_\alpha}\left\{\rho_\alpha\mathbf{v}_0-\tau_\alpha^{(0)}\left[\frac{\partial}{\partial t}(\rho_\alpha\mathbf{v}_0)+\frac{\partial}{\partial \mathbf{r}}\cdot\rho_\alpha\mathbf{v}_0\mathbf{v}_0+\frac{\partial p_\alpha}{\partial \mathbf{r}}-\rho_\alpha\mathbf{F}_\alpha^{(1)}\right.\right.\\
&\left.\left.-\frac{q_\alpha}{m_\alpha}\rho_\alpha\mathbf{v}_0\times\mathbf{B}\right]\right\}\times\mathbf{B}+\frac{\partial}{\partial \mathbf{r}}\cdot\left\{\rho\mathbf{v}_0\mathbf{v}_0+p\overleftrightarrow{\mathbf{I}}-\sum_\alpha\tau_\alpha^{(0)}\left[\frac{\partial}{\partial t}\left(\rho_\alpha\mathbf{v}_0\mathbf{v}_0\right.\right.\right.\\
&\left.+p_\alpha\overleftrightarrow{\mathbf{I}}\right)+\frac{\partial}{\partial \mathbf{r}}\cdot\rho_\alpha(\mathbf{v}_0\mathbf{v}_0)\mathbf{v}_0+2\overleftrightarrow{\mathbf{I}}\left(\frac{\partial}{\partial \mathbf{r}}\cdot(p_\alpha\mathbf{v}_0)\right)+\frac{\partial}{\partial \mathbf{r}}\cdot\left(\overleftrightarrow{\mathbf{I}}p_\alpha\mathbf{v}_0\right)\\
&\left.\left.-\mathbf{F}_\alpha^{(1)}\rho_\alpha\mathbf{v}_0-\rho_\alpha\mathbf{v}_0\mathbf{F}_\alpha^{(1)}-\frac{q_\alpha}{m_\alpha}\rho_\alpha[\mathbf{v}_0\times\mathbf{B}]\mathbf{v}_0-\frac{q_\alpha}{m_\alpha}\rho_\alpha\mathbf{v}_0[\mathbf{v}_0\times\mathbf{B}]\right]\right\}=0
\end{aligned}
\tag{2.7.37}
$$

Let us go to the derivation of the generalized Euler energy equation. Write down this equation omitting all integral terms, which are obviously equal to zero as containing odd integrands of the thermal velocities V_α.

$$
\begin{aligned}
&\frac{\partial}{\partial t}\left\{\frac{\rho_\alpha v_0^2}{2}+\frac{\rho_\alpha\overline{V_\alpha^2}}{2}+\varepsilon_\alpha n_\alpha-\tau_\alpha^{(0)}\left[\frac{\partial}{\partial t}\left(\frac{\rho_\alpha v_0^2}{2}\right.\right.\right.\\
&\left.+\frac{\rho_\alpha\overline{V_\alpha^2}}{2}+\varepsilon_\alpha n_\alpha\right)+\frac{\partial}{\partial \mathbf{r}}\cdot\left(\frac{1}{2}\rho_\alpha v_0^2\mathbf{v}_0\right.\\
&\left.\left.\left.+\frac{1}{2}\rho_\alpha\overline{V_\alpha^2}\mathbf{v}_0+\rho_\alpha\overline{(\mathbf{v}_0\cdot\mathbf{V}_\alpha)\mathbf{V}_\alpha}+\varepsilon_\alpha n_\alpha\mathbf{v}_0\right)-\mathbf{F}_\alpha^{(1)}\cdot\rho_\alpha\mathbf{v}_0\right]\right\}\\
&+\frac{\partial}{\partial \mathbf{r}}\cdot\left\{\frac{1}{2}\rho_\alpha v_0^2\mathbf{v}_0+\rho_\alpha\overline{(\mathbf{v}_0\cdot\mathbf{V}_\alpha)\mathbf{V}_\alpha}+\frac{1}{2}\rho_\alpha\overline{V_\alpha^2}\mathbf{v}_0\right.\\
&+\varepsilon_\alpha n_\alpha\mathbf{v}_0-\tau_\alpha^{(0)}\left[\frac{\partial}{\partial t}\left(\frac{1}{2}\rho_\alpha v_0^2\mathbf{v}_0+\rho_\alpha\overline{(\mathbf{v}_0\cdot\mathbf{V}_\alpha)\mathbf{V}_\alpha}\right.\right.\\
&\left.+\frac{1}{2}\rho_\alpha\overline{V_\alpha^2}\mathbf{v}_0+\varepsilon_\alpha n_\alpha\mathbf{v}_0\right)+\frac{\partial}{\partial \mathbf{r}}\cdot\left[\frac{1}{2}\rho_\alpha v_0^2\mathbf{v}_0\mathbf{v}_0+\frac{1}{2}\rho_\alpha\overline{V_\alpha^2}\mathbf{v}_0\mathbf{v}_0\right.\\
&+\frac{1}{2}\rho_\alpha v_0^2\overline{\mathbf{V}_\alpha\mathbf{V}_\alpha}+\rho_\alpha\overline{(\mathbf{v}_0\cdot\mathbf{V}_\alpha)\mathbf{V}_\alpha}\mathbf{v}_0+\rho_\alpha\overline{(\mathbf{v}_0\cdot\mathbf{V}_\alpha)\mathbf{v}_0\mathbf{V}_\alpha}\\
&\left.+\frac{1}{2}\rho_\alpha\overline{V_\alpha^2\mathbf{V}_\alpha\mathbf{V}_\alpha}+\varepsilon_\alpha n_\alpha\mathbf{v}_0\mathbf{v}_0+\varepsilon_\alpha n_\alpha\overline{\mathbf{V}_\alpha\mathbf{V}_\alpha}\right]-\rho_\alpha\mathbf{F}_\alpha^{(1)}\cdot\mathbf{v}_0\mathbf{v}_0\\
&\left.\left.-\rho_\alpha\mathbf{F}_\alpha^{(1)}\cdot\overline{\mathbf{V}_\alpha\mathbf{V}_\alpha}-\frac{1}{2}\rho_\alpha\overline{v_\alpha^2}\mathbf{F}_\alpha-\varepsilon_\alpha n_\alpha\overline{\mathbf{F}}_\alpha\right]\right\}\\
&-\left\{\rho_\alpha\mathbf{F}_\alpha^{(1)}\cdot\mathbf{v}_0-\tau_\alpha^{(0)}\left[\mathbf{F}_\alpha^{(1)}\cdot\left(\frac{\partial}{\partial t}(\rho_\alpha\mathbf{v}_0)+\frac{\partial}{\partial \mathbf{r}}\cdot\rho_\alpha\mathbf{v}_0\mathbf{v}_0\right.\right.\right.\\
&\left.\left.\left.+\frac{\partial}{\partial \mathbf{r}}\cdot\rho_\alpha\overline{\mathbf{V}_\alpha\mathbf{V}_\alpha}-\rho_\alpha\mathbf{F}_\alpha^{(1)}-q_\alpha n_\alpha\mathbf{v}_0\times\mathbf{B}\right)\right]\right\}\\
&=\int\left(\frac{m_\alpha v_\alpha^2}{2}+\varepsilon_\alpha\right)J_\alpha^{st,el}d\mathbf{v}_\alpha+\int\left(\frac{m_\alpha v_\alpha^2}{2}+\varepsilon_\alpha\right)J_\alpha^{st,inel}d\mathbf{v}_\alpha.
\end{aligned}
\tag{2.7.38}
$$

Let us calculate all average values connected with local Maxwellian DF

$$
f_\alpha^{(0)}=n_\alpha\left(\frac{m_\alpha}{2\pi k_B T}\right)^{3/2}\exp\left(-\frac{m_\alpha V_\alpha^2}{2k_B T}\right).
\tag{2.7.39}
$$

We have

$$\frac{\rho_\alpha \overline{V_\alpha^2}}{2} = \frac{3}{2} n_\alpha k_\mathrm{B} T, \tag{2.7.40}$$

$$\begin{aligned}
&\frac{\partial}{\partial \mathbf{r}} \cdot \rho_\alpha \overline{(\mathbf{v}_0 \cdot \mathbf{V}_\alpha)\mathbf{V}_\alpha} \\
&= m_\alpha \frac{\partial}{\partial \mathbf{r}} \cdot \left\{ \mathbf{v}_0 \cdot \int \mathbf{V}_\alpha \mathbf{V}_\alpha e^{-W_\alpha^2} \mathrm{d}\mathbf{V}_\alpha n_\alpha \left(\frac{m_\alpha}{2\pi k_\mathrm{B} T} \right)^{3/2} \right\} \\
&= \frac{4\pi}{\pi^{3/2}} \frac{2k_B}{m_\alpha} m_\alpha \sum_{ij=1}^{3} \frac{\partial}{\partial r_i} \left\{ n_\alpha v_{0j} T \int W_{\alpha j} W_{\alpha i} W_\alpha^2 e^{-W_\alpha^2} \mathrm{d}W_\alpha \right\} \\
&= \frac{8k_B}{\sqrt{\pi}} \sum_i \frac{\partial}{\partial r_i} \left\{ n_\alpha v_{0i} T \int W_{\alpha i}^2 W_\alpha^2 e^{-W_\alpha^2} \mathrm{d}W_\alpha \right\} \\
&= \frac{8k_B}{3\sqrt{\pi}} \sum_i \frac{\partial}{\partial r_i} \left\{ n_\alpha v_{0i} T \int_0^\infty W_\alpha^4 e^{-W_\alpha^2} \mathrm{d}W_\alpha \right\} \\
&= \frac{8}{3\sqrt{\pi}} \frac{\sqrt{\pi}}{2} \frac{1}{2} \frac{3}{2} \frac{\partial}{\partial \mathbf{r}} \cdot (\mathbf{v}_0 p_\alpha) = \frac{\partial}{\partial \mathbf{r}} \cdot (p_\alpha \mathbf{v}_0),
\end{aligned} \tag{2.7.41}$$

where $W_\alpha = \sqrt{\frac{m_\alpha V_\alpha^2}{2k_\mathrm{B} T}}$.

Other integrals in (2.7.38) can be evaluated in a similar way:

$$\frac{\partial}{\partial \mathbf{r}} \cdot \{\overline{\mathbf{V}_\alpha \mathbf{V}_\alpha} \rho_\alpha v_0^2\} = \frac{\partial}{\partial \mathbf{r}} \cdot \left(p_\alpha v_0^2 \overset{\leftrightarrow}{\mathrm{I}} \right), \tag{2.7.42}$$

$$\begin{aligned}
&\frac{\partial}{\partial \mathbf{r}} \cdot \left\{ \rho_\alpha \overline{(\mathbf{v}_0 \cdot \mathbf{V}_\alpha)\mathbf{V}_\alpha} \mathbf{v}_0 \right\} \\
&= \frac{\partial}{\partial \mathbf{r}} \cdot \left\{ \rho_\alpha \overline{(\mathbf{v}_0 \cdot \mathbf{V}_\alpha) \mathbf{v}_0 \mathbf{V}_\alpha} \right\} = \frac{\partial}{\partial \mathbf{r}} \cdot p_\alpha \mathbf{v}_0 \mathbf{v}_0,
\end{aligned} \tag{2.7.43}$$

$$\frac{\partial}{\partial \mathbf{r}} \cdot \rho_\alpha \overline{V_\alpha^2 \mathbf{V}_\alpha \mathbf{V}_\alpha} = 5 \frac{\partial}{\partial \mathbf{r}} \cdot \left(\frac{p_\alpha^2}{\rho_\alpha} \overset{\leftrightarrow}{\mathrm{I}} \right), \tag{2.7.44}$$

$$\varepsilon_\alpha n_\alpha \overline{\mathbf{F}}_\alpha = \varepsilon_\alpha \frac{q_\alpha}{m_\alpha} n_\alpha [\mathbf{v}_0 \times \mathbf{B}] + \varepsilon_\alpha n_\alpha \mathbf{F}_\alpha^{(1)}. \tag{2.7.45}$$

Calculate also in Euler approximation

$$\begin{aligned}
&\frac{\partial}{\partial \mathbf{r}} \cdot \rho_\alpha \overline{v_\alpha^2 \mathbf{F}_\alpha} \\
&= \frac{\partial}{\partial \mathbf{r}} \cdot \left\{ m_\alpha \int v_\alpha^2 \mathbf{F}_\alpha^{(1)} f_\alpha^{(0)} \mathrm{d}\mathbf{v}_\alpha + q_\alpha \int v_\alpha^2 [\mathbf{v}_\alpha \times \mathbf{B}] f_\alpha^{(0)} \mathrm{d}\mathbf{v}_\alpha \right\} \\
&= \frac{\partial}{\partial \mathbf{r}} \cdot \left\{ \rho_\alpha \mathbf{F}_\alpha^{(1)} v_0^2 + \mathbf{F}_\alpha^{(1)} m_\alpha \int V_\alpha^2 f_\alpha^{(0)} \mathrm{d}\mathbf{V}_\alpha + \frac{q_\alpha}{m_\alpha} \rho_\alpha [\mathbf{v}_0 \times \mathbf{B}] v_0^2 \right. \\
&\left. + 2q_\alpha \mathbf{v}_0 \cdot \int \mathbf{V}_\alpha [\mathbf{V}_\alpha \times \mathbf{B}] f_\alpha^{(0)} \mathrm{d}\mathbf{V}_\alpha + q_\alpha [\mathbf{v}_0 \times \mathbf{B}] \int V_\alpha^2 f_\alpha^{(0)} \mathrm{d}\mathbf{V}_\alpha \right\} \\
&= \frac{\partial}{\partial \mathbf{r}} \cdot \left\{ \rho_\alpha \mathbf{F}_\alpha^{(1)} v_0^2 + 3\mathbf{F}_\alpha^{(1)} p_\alpha + \rho_\alpha v_0^2 \frac{q_\alpha}{m_\alpha} [\mathbf{v}_0 \times \mathbf{B}] \right. \\
&\left. + 5p_\alpha \frac{q_\alpha}{m_\alpha} [\mathbf{v}_0 \times \mathbf{B}] \right\}.
\end{aligned} \tag{2.7.46}$$

As a result, the following form of the generalized Euler energy equation for species can be stated:

$$\begin{aligned}
&\frac{\partial}{\partial t}\left\{\frac{\rho_\alpha v_0^2}{2}+\frac{3}{2}p_\alpha+\varepsilon_\alpha n_\alpha-\tau_\alpha^{(0)}\left[\frac{\partial}{\partial t}\left(\frac{\rho_\alpha v_0^2}{2}+\frac{3}{2}p_\alpha+\varepsilon_\alpha n_\alpha\right)\right.\right.\\
&\left.\left.+\frac{\partial}{\partial \mathbf{r}}\cdot\left(\frac{1}{2}\rho_\alpha v_0^2\mathbf{v}_0+\frac{5}{2}p_\alpha\mathbf{v}_0+\varepsilon_\alpha n_\alpha\mathbf{v}_0\right)-\mathbf{F}_\alpha^{(1)}\cdot\rho_\alpha\mathbf{v}_0\right]\right\}\\
&+\frac{\partial}{\partial \mathbf{r}}\cdot\left\{\frac{1}{2}\rho_\alpha v_0^2\mathbf{v}_0+\frac{5}{2}p_\alpha\mathbf{v}_0+\varepsilon_\alpha n_\alpha\mathbf{v}_0-\tau_\alpha^{(0)}\left[\frac{\partial}{\partial t}\left(\frac{1}{2}\rho_\alpha v_0^2\mathbf{v}_0\right.\right.\right.\\
&\left.+\frac{5}{2}p_\alpha\mathbf{v}_0+\varepsilon_\alpha n_\alpha\mathbf{v}_0\right)+\frac{\partial}{\partial \mathbf{r}}\cdot\left(\frac{1}{2}\rho_\alpha v_0^2\mathbf{v}_0\mathbf{v}_0+\frac{7}{2}p_\alpha\mathbf{v}_0\mathbf{v}_0+\frac{1}{2}p_\alpha v_0^2\overleftrightarrow{\mathrm{I}}\right.\\
&\left.+\frac{5p_\alpha^2}{2\rho_\alpha}\overleftrightarrow{\mathrm{I}}+\varepsilon_\alpha n_\alpha\mathbf{v}_0\mathbf{v}_0+\varepsilon_\alpha\frac{p_\alpha}{m_\alpha}\overleftrightarrow{\mathrm{I}}\right)-\rho_\alpha\mathbf{F}_\alpha^{(1)}\cdot\mathbf{v}_0\mathbf{v}_0-p_\alpha\mathbf{F}_\alpha^{(1)}\cdot\overleftrightarrow{\mathrm{I}}\\
&-\frac{1}{2}\rho_\alpha v_0^2\mathbf{F}_\alpha^{(1)}-\frac{3}{2}\mathbf{F}_\alpha^{(1)}p_\alpha-\frac{\rho_\alpha v_0^2}{2}\frac{q_\alpha}{m_\alpha}[\mathbf{v}_0\times\mathbf{B}]-\frac{5}{2}p_\alpha\frac{q_\alpha}{m_\alpha}[\mathbf{v}_0\times\mathbf{B}]\\
&\left.\left.-\varepsilon_\alpha n_\alpha\frac{q_\alpha}{m_\alpha}[\mathbf{v}_0\times\mathbf{B}]-\varepsilon_\alpha n_\alpha\mathbf{F}_\alpha^{(1)}\right]\right\}-\left\{\rho_\alpha\mathbf{F}_\alpha^{(1)}\cdot\mathbf{v}_0\right.\\
&\left.-\tau_\alpha^{(0)}\left[\mathbf{F}_\alpha^{(1)}\cdot\left(\frac{\partial}{\partial t}(\rho_\alpha\mathbf{v}_0)+\frac{\partial}{\partial \mathbf{r}}\cdot\rho_\alpha\mathbf{v}_0\mathbf{v}_0+\frac{\partial}{\partial \mathbf{r}}\cdot p_\alpha\overleftrightarrow{\mathrm{I}}-\rho_\alpha\mathbf{F}_\alpha^{(1)}-q_\alpha n_\alpha[\mathbf{v}_0\times\mathbf{B}]\right)\right]\right\}\\
&=\int\left(\frac{m_\alpha v_\alpha^2}{2}+\varepsilon_\alpha\right)J_\alpha^{st,el}d\mathbf{v}_\alpha+\int\left(\frac{m_\alpha v_\alpha^2}{2}+\varepsilon_\alpha\right)J_\alpha^{st,inel}d\mathbf{v}_\alpha.
\end{aligned} \tag{2.7.47}$$

Finally, after summation of Eq. (2.7.47) over all species we reach the generalized Euler energy equation for mixture of gases:

$$\begin{aligned}
&\frac{\partial}{\partial t}\left\{\frac{\rho v_0^2}{2}+\frac{3}{2}p+\sum_\alpha\varepsilon_\alpha n_\alpha-\sum_\alpha\tau_\alpha^{(0)}\left[\frac{\partial}{\partial t}\left(\frac{\rho_\alpha v_0^2}{2}+\frac{3}{2}p_\alpha+\varepsilon_\alpha n_\alpha\right)\right.\right.\\
&\left.\left.+\frac{\partial}{\partial \mathbf{r}}\cdot\left(\frac{1}{2}\rho_\alpha v_0^2\mathbf{v}_0+\frac{5}{2}p_\alpha\mathbf{v}_0+\varepsilon_\alpha n_\alpha\mathbf{v}_0\right)-\mathbf{F}_\alpha^{(1)}\cdot\rho_\alpha\mathbf{v}_0\right]\right\}\\
&+\frac{\partial}{\partial \mathbf{r}}\cdot\left\{\frac{1}{2}\rho v_0^2\mathbf{v}_0+\frac{5}{2}p\mathbf{v}_0+\mathbf{v}_0\sum_\alpha\varepsilon_\alpha n_\alpha-\sum_\alpha\tau_\alpha^{(0)}\left[\frac{\partial}{\partial t}\left(\frac{1}{2}\rho_\alpha v_0^2\mathbf{v}_0\right.\right.\right.\\
&\left.+\frac{5}{2}p_\alpha\mathbf{v}_0+\varepsilon_\alpha n_\alpha\mathbf{v}_0\right)+\frac{\partial}{\partial \mathbf{r}}\cdot\left(\frac{1}{2}\rho_\alpha v_0^2\mathbf{v}_0\mathbf{v}_0+\frac{7}{2}p_\alpha\mathbf{v}_0\mathbf{v}_0+\frac{1}{2}p_\alpha v_0^2\overleftrightarrow{\mathrm{I}}\right.\\
&\left.+\frac{5p_\alpha^2}{2\rho_\alpha}\overleftrightarrow{\mathrm{I}}+\varepsilon_\alpha n_\alpha\mathbf{v}_0\mathbf{v}_0+\varepsilon_\alpha\frac{p_\alpha}{m_\alpha}\overleftrightarrow{\mathrm{I}}\right)-\rho_\alpha\mathbf{F}_\alpha^{(1)}\cdot\mathbf{v}_0\mathbf{v}_0-p_\alpha\mathbf{F}_\alpha^{(1)}\cdot\overleftrightarrow{\mathrm{I}}\\
&-\frac{1}{2}\rho_\alpha v_0^2\mathbf{F}_\alpha^{(1)}-\frac{3}{2}\mathbf{F}_\alpha^{(1)}p_\alpha-\frac{\rho_\alpha v_0^2}{2}\frac{q_\alpha}{m_\alpha}[\mathbf{v}_0\times\mathbf{B}]-\frac{5}{2}p_\alpha\frac{q_\alpha}{m_\alpha}[\mathbf{v}_0\times\mathbf{B}]\\
&\left.\left.-\varepsilon_\alpha n_\alpha\frac{q_\alpha}{m_\alpha}[\mathbf{v}_0\times\mathbf{B}]-\varepsilon_\alpha n_\alpha\mathbf{F}_\alpha^{(1)}\right]\right\}-\mathbf{v}_0\cdot\sum_\alpha\rho_\alpha\mathbf{F}_\alpha^{(1)}\\
&+\sum_\alpha\tau_\alpha^{(0)}\mathbf{F}_\alpha^{(1)}\cdot\left[\frac{\partial}{\partial t}(\rho_\alpha\mathbf{v}_0)+\frac{\partial}{\partial \mathbf{r}}\cdot\rho_\alpha\mathbf{v}_0\mathbf{v}_0+\frac{\partial}{\partial \mathbf{r}}\cdot p_\alpha\overleftrightarrow{\mathrm{I}}-\rho_\alpha\mathbf{F}_\alpha^{(1)}-q_\alpha n_\alpha[\mathbf{v}_0\times\mathbf{B}]\right]=0.
\end{aligned} \tag{2.7.48}$$

Write down the system of GHE in the Euler approach taking into account the force of gravitation:

continuity equation

$$\begin{aligned}
&\frac{\partial}{\partial t}\left\{\rho-\Pi\frac{\mu}{p}\left[\frac{\partial\rho}{\partial t}+\frac{\partial}{\partial \mathbf{r}}\cdot(\rho\mathbf{v}_0)\right]\right\}+\frac{\partial}{\partial \mathbf{r}}\cdot\left\{\rho\mathbf{v}_0-\Pi\frac{\mu}{p}\left[\frac{\partial}{\partial t}(\rho\mathbf{v}_0)\right.\right.\\
&\left.\left.+\frac{\partial}{\partial \mathbf{r}}\cdot\rho\mathbf{v}_0\mathbf{v}_0+\overleftrightarrow{\mathrm{I}}\cdot\frac{\partial p}{\partial \mathbf{r}}-\rho\mathbf{g}\right]\right\}=0,
\end{aligned} \tag{2.7.49}$$

momentum equation

$$\begin{aligned}&\frac{\partial}{\partial t}\left\{\rho\mathbf{v}_0-\Pi\frac{\mu}{p}\left[\frac{\partial}{\partial t}(\rho\mathbf{v}_0)+\frac{\partial}{\partial\mathbf{r}}\cdot\rho\mathbf{v}_0\mathbf{v}_0+\frac{\partial p}{\partial\mathbf{r}}-\rho\mathbf{g}\right]\right\}\\&-\mathbf{g}\left[\rho-\Pi\frac{\mu}{p}\left(\frac{\partial\rho}{\partial t}+\frac{\partial}{\partial\mathbf{r}}\cdot(\rho\mathbf{v}_0)\right)\right]+\frac{\partial}{\partial\mathbf{r}}\cdot\left\{\rho\mathbf{v}_0\mathbf{v}_0\right.\\&+p\overset{\leftrightarrow}{I}-\Pi\frac{\mu}{p}\left[\frac{\partial}{\partial t}(\rho\mathbf{v}_0\mathbf{v}_0+p\overset{\leftrightarrow}{I})+\frac{\partial}{\partial\mathbf{r}}\cdot\rho(\mathbf{v}_0\mathbf{v}_0)\mathbf{v}_0\right.\\&\left.\left.+2\overset{\leftrightarrow}{I}\left[\frac{\partial}{\partial\mathbf{r}}\cdot(p\mathbf{v}_0)\right]+\frac{\partial}{\partial\mathbf{r}}\cdot\left(\overset{\leftrightarrow}{I}p\mathbf{v}_0\right)-\mathbf{g}\rho\mathbf{v}_0-\mathbf{v}_0\mathbf{g}\rho\right]\right\}=0,\end{aligned}\tag{2.7.50}$$

energy equation

$$\begin{aligned}&\frac{\partial}{\partial t}\left\{\frac{\rho v_0^2}{2}+\frac{3}{2}p-\Pi\frac{\mu}{p}\left[\frac{\partial}{\partial t}\left(\frac{\rho v_0^2}{2}+\frac{3}{2}p\right)\right.\right.\\&\left.\left.+\frac{\partial}{\partial\mathbf{r}}\cdot\left(\frac{1}{2}\rho v_0^2\mathbf{v}_0+\frac{5}{2}p\mathbf{v}_0\right)-\mathbf{g}\cdot\rho\mathbf{v}_0\right]\right\}\\&+\frac{\partial}{\partial\mathbf{r}}\cdot\left\{\frac{1}{2}\rho v_0^2\mathbf{v}_0+\frac{5}{2}p\mathbf{v}_0-\Pi\frac{\mu}{p}\left[\frac{\partial}{\partial t}\left(\frac{1}{2}\rho v_0^2\mathbf{v}_0\right.\right.\right.\\&\left.+\frac{5}{2}p\mathbf{v}_0\right)+\frac{\partial}{\partial\mathbf{r}}\cdot\left(\frac{1}{2}\rho v_0^2\mathbf{v}_0\mathbf{v}_0+\frac{7}{2}p\mathbf{v}_0\mathbf{v}_0+\frac{1}{2}pv_0^2\overset{\leftrightarrow}{I}\right.\\&\left.\left.\left.+\frac{5p^2}{2\rho}\overset{\leftrightarrow}{I}\right)-\rho\mathbf{g}\cdot\mathbf{v}_0\mathbf{v}_0-p\mathbf{g}\cdot\overset{\leftrightarrow}{I}-\frac{1}{2}\rho v_0^2\mathbf{g}-\frac{3}{2}\mathbf{g}p\right]\right\}\\&-\left\{\rho\mathbf{g}\cdot\mathbf{v}_0-\Pi\frac{\mu}{p}\left[\mathbf{g}\cdot\left(\frac{\partial}{\partial t}(\rho\mathbf{v}_0)+\frac{\partial}{\partial\mathbf{r}}\cdot\rho\mathbf{v}_0\mathbf{v}_0\right.\right.\right.\\&\left.\left.\left.+\frac{\partial}{\partial\mathbf{r}}\cdot p\overset{\leftrightarrow}{I}-\rho\mathbf{g}\right)\right]\right\}=0,\end{aligned}\tag{2.7.51}$$

where $\mathbf{g}$ is acceleration in gravitational field.

Let us summarize generalized Euler hydrodynamic equations for the simplest case of one-dimensional motion in absence of external forces.

continuity equation:

$$\begin{aligned}&\frac{\partial}{\partial t}\left\{\rho-\tau^{(0)}\left[\frac{\partial\rho}{\partial t}+\frac{\partial}{\partial x}(\rho v_0)\right]\right\}+\frac{\partial}{\partial x}\left\{\rho v_0-\tau^{(0)}\left[\frac{\partial}{\partial t}(\rho v_0)\right.\right.\\&\left.\left.+\frac{\partial}{\partial x}(\rho v_0^2)+\frac{\partial p}{\partial x}\right]\right\}=0,\end{aligned}\tag{2.7.52}$$

momentum equation

$$\begin{aligned}&\frac{\partial}{\partial t}\left\{\rho v_0-\tau^{(0)}\left[\frac{\partial}{\partial t}(\rho v_0)+\frac{\partial}{\partial x}(\rho v_0^2)+\frac{\partial p}{\partial x}\right]\right\}+\\&+\frac{\partial}{\partial x}\left\{\rho v_0^2+p-\tau^{(0)}\left[\frac{\partial}{\partial t}(\rho v_0^2+p)+\frac{\partial}{\partial x}(\rho v_0^3+3pv_0)\right]\right\}=0,\end{aligned}\tag{2.7.53}$$

energy equation:

$$\begin{aligned}&\frac{\partial}{\partial t}\left\{\rho v_0^2+3p-\tau^{(0)}\left[\frac{\partial}{\partial t}(\rho v_0^2+3p)+\frac{\partial}{\partial x}(\rho v_0^3+5pv_0)\right]\right\}+\\&+\frac{\partial}{\partial x}\left\{\rho v_0^3+5pv_0-\tau^{(0)}\left[\frac{\partial}{\partial t}(\rho v_0^3+5pv_0)+\frac{\partial}{\partial x}\left(\rho v_0^4+8pv_0^2+5\frac{p^2}{\rho}\right)\right]\right\}=0.\end{aligned}\tag{2.7.54}$$

Chapter 3

Quantum Non-Local Hydrodynamics

ABSTRACT

The goal of the following consideration in Chapter 3 consists in the inclusion of quantum mechanics (in the form of quantum non-local hydrodynamics) in the general construction of the unified theory of transport processes.

Main ideas which were realized:

1. Madelung wrote [67]: "It is shown that the Schrödinger equation for one-electron problems can be transformed into the form of hydrodynamic equations."
2. It means that even the *single*-particle evolution can be presented in the hydrodynamic form.
3. Madelung's quantum hydrodynamics is equivalent to the Schrödinger equation (SE) and leads to the description of the quantum particle evolution in the form of Euler equation and continuity equation. Quantum Euler equation contains additional potential of non-local origin which can be written, for example, in the Bohm form.
4. SE is consequence of the Liouville equation as a result of the *local* approximation of non-local equations.
5. Generalized Boltzmann physical kinetics leads to the strict approximation of non-local effects in space and time and *in the local limit* leads to parameter τ, which on the quantum level corresponds to the uncertainty principle "time-energy."
6. Generalized hydrodynamic equations (GHEs) lead to SE as a deep particular case of the generalized Boltzmann physical kinetics and therefore of non-local hydrodynamics. The result—quantum mechanics is the particular case of the unified theory of transport processes.

Keywords: Generalized Boltzmann physical kinetics, Quantum non-local hydrodynamics, Shnoll's effect

3.1 GENERALIZED HYDRODYNAMIC EQUATIONS AND QUANTUM MECHANICS

It is well known that basic equation of quantum mechanics—SE—cannot be strictly derived. SE could be "guessed" from reasonable physical considerations and (after comparison the SE solutions with some experimental data) declared as one of postulates of quantum mechanics. The main steps of this guess can be characterized as follows:

(a) The complex function $\psi(x,t)$ is introduced as characteristics of physical objects with corpuscular and wave features. The simplest wave form which could be imagined is:

$$\psi = \mathrm{e}^{-\mathrm{i}(\omega t - kx)}, \tag{3.1.1}$$

with additional conditions:

$$\omega = E_{\mathrm{K}}/\hbar, \;\; k = 2\pi/\lambda = p/\hbar, \tag{3.1.2}$$

where traditional nomenclatures are used for frequency ω, kinetic energy E_{K}, wave number k, and impulse p. Substitution of Eq. (3.1.2) in Eq. (3.1.1) and following differentiation (once in time and twice in space) leads to the relations

$$\mathrm{i}\hbar\frac{\partial\psi}{\partial t} = E_{\mathrm{K}}\psi, \tag{3.1.3}$$

$$-\frac{\hbar^2}{2m}\frac{\partial^2\psi}{\partial x^2} = E_{\mathrm{K}}\psi, \tag{3.1.4}$$

because for individual free particle of mass m

$$E_{\mathrm{K}} = \frac{p^2}{2m}. \tag{3.1.5}$$

Unified non-local theory of transport processes. http://dx.doi.org/10.1016/B978-0-444-63478-8.00003-8

As a result, the one-dimensional (1D) quantum equation (E. Schrödinger, 1926) takes the form

$$\mathrm{i}\hbar\frac{\partial\psi}{\partial t}=-\frac{\hbar^2}{2m}\frac{\partial^2\psi}{\partial x^2}. \tag{3.1.6}$$

(b) After obvious generalizations for the 3D situation with potential external energy $U(x,y,z,t)$, Eq. (3.1.6) takes the form

$$\mathrm{i}\hbar\frac{\partial\psi}{\partial t}=-\frac{\hbar^2}{2m}\left(\frac{\partial^2\psi}{\partial x^2}+\frac{\partial^2\psi}{\partial y^2}+\frac{\partial^2\psi}{\partial z^2}\right)+U\psi, \tag{3.1.7}$$

as a quantum mechanical postulate. Another manner of differentiating leads to another basic quantum equations (see, for example, [82]).

The obvious next step should be done and was realized by E. Madelung in 1927—the derivation of the special form of SE after the introduction of the wave function ψ as

$$\psi(x,y,z,t)=\alpha(x,y,z,t)\mathrm{e}^{\mathrm{i}\beta(x,y,z,t)}. \tag{3.1.8}$$

Write down the SE in the form

$$\mathrm{i}\hbar\frac{\partial\psi}{\partial t}=\widehat{H}\psi, \tag{3.1.9}$$

where nonrelativistic Hamiltonian $\widehat{H}$ for a quantum mechanical system placed in electromagnetic field, is written as

$$\widehat{H}=\left(\widehat{\vec{p}}-e\widehat{\vec{A}}\right)^2/(2m)+\widehat{U}, \tag{3.1.10}$$

where **A** is vector potential, $\widehat{\vec{A}}$ is corresponding operator, which reflects the magnet field influence on the quantum system. We use the standard notations: ψ—wave function, $\widehat{U}$—operator of potential energy, $\widehat{\vec{p}}=\frac{\hbar}{\mathrm{i}}\nabla\equiv\frac{\hbar}{\mathrm{i}}\frac{\partial}{\partial\mathbf{r}}\equiv\frac{\hbar}{\mathrm{i}}\mathbf{grad}$—momentum vector operator; m, e—mass and charge of particle, and $h=2\pi\hbar$—Planck constant.

In these notations,

$$\widehat{H}=\frac{\widehat{\vec{p}}^{\,2}}{2m}+\widehat{U}-\frac{e}{m}\left(\widehat{\vec{A}},\widehat{\vec{p}}\right)+\frac{e^2}{2m}\widehat{\vec{A}}^{\,2}. \tag{3.1.11}$$

Use the complex form (3.1.8) of the wave function ψ and introduce Laplacian Δ defined as the usual

$$\Delta=\frac{\partial^2}{\partial x^2}+\frac{\partial^2}{\partial y^2}+\frac{\partial^2}{\partial z^2}=\frac{\partial}{\partial\mathbf{r}}\cdot\frac{\partial}{\partial\mathbf{r}}. \tag{3.1.12}$$

After differentiating Eq. (3.1.8) one obtains

$$\frac{1}{\psi}\frac{\partial\psi}{\partial t}=\frac{1}{\alpha}\frac{\partial\alpha}{\partial t}+\mathrm{i}\frac{\partial\beta}{\partial t}, \tag{3.1.13}$$

$$\frac{1}{\psi}\frac{\partial\psi}{\partial\mathbf{r}}=\frac{1}{\alpha}\frac{\partial\alpha}{\partial\mathbf{r}}+\mathrm{i}\frac{\partial\beta}{\partial\mathbf{r}}, \tag{3.1.14}$$

$$\frac{\Delta\psi}{\psi}=\frac{\Delta\alpha}{\alpha}-\left(\frac{\partial\beta}{\partial\mathbf{r}}\right)^2+\mathrm{i}\left(\Delta\beta+\frac{2}{\alpha}\frac{\partial\alpha}{\partial\mathbf{r}}\cdot\frac{\partial\beta}{\partial\mathbf{r}}\right). \tag{3.1.15}$$

Using relations (3.1.8) and (3.1.13)–(3.1.15) and separate real and imaginary parts of complex numbers in Eq. (3.1.9), we find

$$\Delta\alpha-\alpha\left(\frac{\partial\beta}{\partial\mathbf{r}}\right)^2-\frac{2m}{\hbar^2}\alpha\widetilde{U}-\frac{2m}{\hbar}\frac{\partial\beta}{\partial t}\alpha+\frac{2e}{\hbar}\alpha\mathbf{A}\cdot\frac{\partial\beta}{\partial\mathbf{r}}=0, \tag{3.1.16}$$

$$\alpha\Delta\beta+2\frac{\partial\alpha}{\partial\mathbf{r}}\cdot\frac{\partial\beta}{\partial\mathbf{r}}+\frac{2m}{\hbar}\frac{\partial\alpha}{\partial t}-\frac{2e}{\hbar}\mathbf{A}\cdot\frac{\partial\alpha}{\partial\mathbf{r}}=0, \tag{3.1.17}$$

where

$$\widetilde{U} = U + \frac{e^2 A^2}{2m}. \tag{3.1.18}$$

By direct differentiating, the following identity can be proved

$$\frac{\partial}{\partial \mathbf{r}} \cdot \left(\alpha^2 \frac{\partial \beta}{\partial \mathbf{r}}\right) \equiv \operatorname{div}\left(\alpha^2 \frac{\partial \beta}{\partial \mathbf{r}}\right) = \alpha^2 \Delta \beta + 2\alpha \frac{\partial \alpha}{\partial \mathbf{r}} \cdot \frac{\partial \beta}{\partial \mathbf{r}}, \tag{3.1.19}$$

Then Eq. (3.1.19) is written as

$$\frac{\partial \alpha^2}{\partial t} + \operatorname{div}\left(\frac{\alpha^2 \hbar}{m} \frac{\partial \beta}{\partial \mathbf{r}}\right) = \frac{e}{m} \mathbf{A} \cdot \frac{\partial \alpha^2}{\partial \mathbf{r}}. \tag{3.1.20}$$

Introduce notations:

$$\rho = \alpha^2, \tag{3.1.21}$$

$$\mathbf{v} = \frac{\partial}{\partial \mathbf{r}} (\beta \hbar / m). \tag{3.1.22}$$

Following Madelung's idea, we identify ρ, having the sense of probability density, with density of a hydrodynamic flow, and $\mathbf{v}$—with velocity of this flow.

Notice immediately, that existence of relation (3.1.22) means that the considered flow belongs to the class of potential flows with potential

$$\phi = \beta \hbar / m. \tag{3.1.23}$$

Equation (3.1.23) can be treated as hydrodynamic continuity equation with source term

$$R = \frac{e}{m} \mathbf{A} \cdot \frac{\partial \rho}{\partial \mathbf{r}}, \tag{3.1.24}$$

connected with fictitious rising and vanishing of hypothetical particles:

$$\frac{\partial \rho}{\partial t} + \frac{\partial}{\partial \mathbf{r}} \cdot (\rho \mathbf{v}) = R. \tag{3.1.25}$$

Continuity Eq. (3.1.25) has a typical form for hydrodynamics of reacting gases, (see, for example, [21]).

Transform now Eq. (3.1.19) to the hydrodynamic form. With this aim divide left and right sides of (3.1.16) by $2\alpha m^2/\hbar^2$ and apply gradient operator to the terms of obtaining equation:

$$\frac{\hbar^2}{2m^2} \frac{\partial}{\partial \mathbf{r}} \frac{\Delta \alpha}{\alpha} - \frac{\hbar^2}{2m^2} \frac{\partial}{\partial \mathbf{r}} \left(\frac{\partial \beta}{\partial \mathbf{r}}\right)^2 - \frac{1}{m} \frac{\partial}{\partial \mathbf{r}} \widetilde{U} - \frac{\hbar}{m} \frac{\partial}{\partial t} \frac{\partial \beta}{\partial \mathbf{r}} + \frac{e\hbar}{m^2} \frac{\partial}{\partial \mathbf{r}} \left(\mathbf{A} \cdot \frac{\partial \beta}{\partial \mathbf{r}}\right) = 0. \tag{3.1.26}$$

Using definition of velocity Eq. (3.1.22), it is found

$$\frac{\partial \mathbf{v}}{\partial t} + \frac{1}{2} \frac{\partial}{\partial \mathbf{r}} v^2 = -\frac{1}{m} \frac{\partial}{\partial \mathbf{r}} \left(\widetilde{U} - e\mathbf{A} \cdot \mathbf{v} - \frac{\hbar^2}{2m} \frac{\Delta \alpha}{\alpha}\right). \tag{3.1.27}$$

Notice that

$$\frac{\Delta \alpha}{\alpha} = \frac{\Delta \alpha^2}{2\alpha^2} - \frac{1}{\alpha^2} \left(\frac{\partial \alpha}{\partial \mathbf{r}}\right)^2, \tag{3.1.28}$$

and—for the potential flow:

$$\frac{1}{2} \frac{\partial}{\partial \mathbf{r}} v^2 = \left(\mathbf{v} \cdot \frac{\partial}{\partial \mathbf{r}}\right) \mathbf{v}, \tag{3.1.29}$$

then Eq. (3.1.27) becomes

$$\frac{\partial \mathbf{v}}{\partial t} + \left(\mathbf{v} \cdot \frac{\partial}{\partial \mathbf{r}}\right) \mathbf{v} = -\frac{1}{m} \frac{\partial}{\partial \mathbf{r}} U^*, \tag{3.1.30}$$

where

$$U^* = \widetilde{U} - e\mathbf{A}\cdot\mathbf{v} - \frac{\hbar^2}{4m\rho}\left[\Delta\rho - \frac{1}{2\rho}\left(\frac{\partial\rho}{\partial\mathbf{r}}\right)^2\right]. \tag{3.1.31}$$

The additive part of potential in the right side of Eq. (3.1.31) can be written in more compact form taking into account, that

$$\frac{\hbar^2}{2m\sqrt{\rho}}\Delta\sqrt{\rho} = \frac{\hbar^2}{4m\rho}\left[\Delta\rho - \frac{1}{2\rho}\left(\frac{\partial\rho}{\partial\mathbf{r}}\right)^2\right]. \tag{3.1.32}$$

Then effective potential energy U^* is sum of potential energy $\widetilde{\widetilde{U}}$

$$\widetilde{\widetilde{U}} = \widetilde{U} - e\mathbf{A}\cdot\mathbf{v}, \tag{3.1.33}$$

which is not equal to zero in classical limit, and quantum part U_{qu},

$$U_{\mathrm{qu}} = -\frac{\hbar^2}{2m\sqrt{\rho}}\Delta\sqrt{\rho} = -\frac{\hbar^2}{4m\rho}\left[\Delta\rho - \frac{1}{2\rho}\left(\frac{\partial\rho}{\partial\mathbf{r}}\right)^2\right], \tag{3.1.34}$$

which contains squared Planck constant as coefficient. Then

$$U^* = \widetilde{\widetilde{U}} + U_{\mathrm{qu}} = \widetilde{\widetilde{U}} - \frac{\hbar^2}{2m\sqrt{\rho}}\Delta\sqrt{\rho} = \widetilde{\widetilde{U}} - \frac{\hbar^2}{4m\rho}\left[\Delta\rho - \frac{1}{2\rho}\left(\frac{\partial\rho}{\partial\mathbf{r}}\right)^2\right]. \tag{3.1.35}$$

Usually, transition to the classical description can be realized using $\hbar\to 0$. Obviously, in classical limit U_{qu} turns to zero.

For this potential flow, hydrodynamic Cauchy integral exists

$$\frac{\partial\phi}{\partial t} + \frac{v^2}{2} + \frac{U^*}{m} = \mathrm{const.} \tag{3.1.36}$$

Boundary and initial conditions for the hydrodynamic Eqs. (3.1.25) and (3.1.30) should be formulated separately taking into account the specific features of concrete quantum mechanical problem. One of these problems is considered as an example below.

Consider quantum mechanical analogue of stabilized flow. If potential U does not depend on time, wave function has the form

$$\psi = \alpha(x, y, z)\exp\left\{\mathrm{i}\left[f(x, y, z) - \frac{Et}{\hbar}\right]\right\}, \tag{3.1.37}$$

where E is total energy. Then from Eqs. (3.1.23) and (3.1.36) follows Bernoulli equation

$$\frac{mv^2}{2} + U^* = E. \tag{3.1.38}$$

Incompressible flow corresponds—in quantum mechanical interpretation—to the motion of free particle.

Therefore, the SE can be treated in terms of the potential flow of a compressible ideal liquid with rising and vanishing of fictitious particles, therewith the rate of "particles" formation is proportional to scalar product of vector potential by gradient of density. However, vector potential $\mathbf{A}$ is proportional to electrokinetic momentum $\mathbf{p}_{\mathrm{ek}}$ which is defined as

$$\mathbf{p}_{\mathrm{ek}} = \frac{1}{c}q\mathbf{A}, \tag{3.1.39}$$

where q is particle charge. Appearance of the additional electrokinetic momentum leads to the change of probability density; quantum hydrodynamics reflects this fact by the rate of fictitious particles formation.

If magnetic field is absent, the usual continuity equation can be written as

$$\frac{\partial\rho}{\partial t} + \frac{\partial}{\partial\mathbf{r}}\cdot\rho\mathbf{v} = 0. \tag{3.1.40}$$

And potential energy of the flow has reduced form

$$U^* = U - \frac{\hbar^2}{4m\rho}\left[\Delta\rho - \frac{1}{2\rho}\left(\frac{\partial\rho}{\partial\mathbf{r}}\right)^2\right]. \tag{3.1.41}$$

Cauchy and Bernoulli integrals should be used for a choice of energy levels.

The outlined theory can be applied for investigation of many quantum mechanical problems from position of hydrodynamics, and first of all in the context of numerical hydrodynamics. Difficulties, arising by numerical investigations of quantum mechanics problems, are well known. From this point of view, it is interesting to apply good developed methods of numerical hydrodynamics (see, for example, [83–85]) for solution of quantum mechanical tasks.

Let us consider, for example, [21, 86] the calculation of differential cross-sections by elastic scattering of electron bunch in spherically symmetric potential field of an atom ($\hbar = m = c = 1$):

$$U(r) = -\frac{2Z}{r}\sum_{i=1}^{n}\gamma_i e^{-\lambda_i r}, \tag{3.1.42}$$

where n is number of terms in potential function, γ_i, λ_i are constants calculated in [86].

Numerical calculation of system of hydrodynamic Eqs. (3.1.30), (3.1.31), and (3.1.40) was realized in a spherical coordinate system taking into account the space symmetry of differential cross-section and initial and boundary conditions

$$\rho(r, \vartheta, 0) = 1, \tag{3.1.43}$$

$$v_r(r, \vartheta, 0) = \sqrt{2E_{\mathrm{k}}}\sin\vartheta, \tag{3.1.44}$$

$$v_\vartheta(r, \vartheta, 0) = \sqrt{2E_{\mathrm{k}}}\cos\vartheta, \tag{3.1.45}$$

$$U^*(r, \vartheta, 0) = 0, \tag{3.1.46}$$

$$\rho(0, \vartheta, t) = 0, \tag{3.1.47}$$

$$v_r(r_0, \vartheta, t) = \sqrt{2E_{\mathrm{k}}}, \tag{3.1.48}$$

$$v_\vartheta(r_0, \vartheta, t) = 0, \tag{3.1.49}$$

$$U^*(r_0, \vartheta, t) = \frac{1}{4\rho r_0^2}\left[\frac{1}{2\rho}\left(\frac{\partial\rho}{\partial\vartheta}\right)^2 - \frac{1}{\sin\vartheta}\frac{\partial}{\partial\vartheta}\left(\sin\vartheta\frac{\partial\rho}{\partial\vartheta}\right)\right], \tag{3.1.50}$$

where E_{k} is kinetic energy of scattered electron bunch, r_0 is parameter of cutting of sphere of the atom action. Relation (3.1.50) is obtained with the help of asymptotic of scattered spherical wave.

Figure 3.1 presents results of calculations of differential cross-section $\sigma(\vartheta)$ of electrons with energy 10 keV scattering by krypton atom (solid curve). The results are in good coincidence with analogous results obtained by the method of partial waves [87] (a dashed curve).

For comparison, the results the Spencer's theory with Moliere's screening are presented (dot-dashed curve in Fig. 3.1) [88, 89]. This method of quantum hydrodynamics was applied later to calculations with taking into account the external force fields and nonspherical atom potentials.

Flux of probability density,

$$\mathbf{j} = \frac{i\hbar}{2m}\left(\psi\frac{\partial\psi^\bullet}{\partial\mathbf{r}} - \psi^\bullet\frac{\partial\psi}{\partial\mathbf{r}}\right), \tag{3.1.51}$$

also can be calculated in terms of Madelung hydrodynamics using (3.1.8), (3.1.14), (3.1.21), and (3.1.22):

$$\mathbf{j} = \rho\mathbf{v}. \tag{3.1.52}$$

If the quantum mechanics can be treated in terms of hydrodynamics, then the backward affirmation is also true. Hydrodynamics equations can be reduced (as minimum for ideal liquids) to SE, equations of quantum mechanics.

From this point of view, no surprise comes in appearance of discrete structures in flow investigations like strange attractors [90]. Quantum mechanical technique of quantization can be propagated in physics of continuum media on the whole and in particular in hydrodynamics. It is important to note that the SE is treated as Euler equation for ideal gas more exact as Euler equation for probabilistic liquid which is considered from position of mechanics of continuum media. It means that outgoing SE is a non-dissipative equation reflecting reversible processes in closed system.

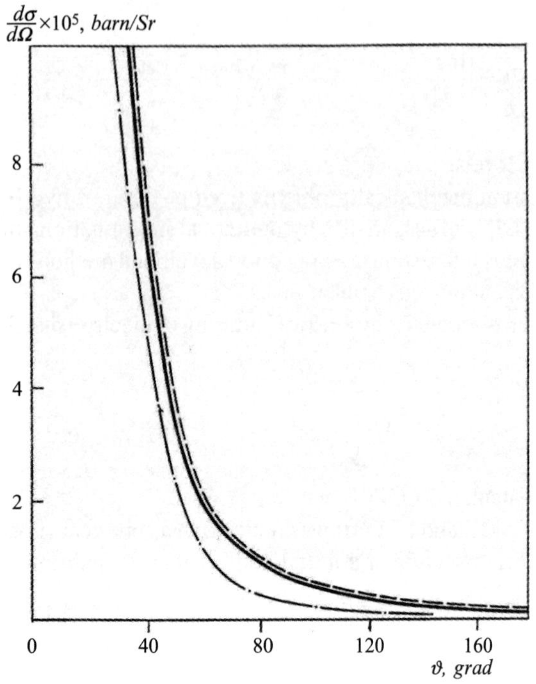

FIGURE 3.1 Results of calculations of differential cross-section $\sigma(\vartheta)$ of electrons with energy 10 keV scattering by a krypton atom (a solid curve); the results obtained by the method of partial waves shown by dashed curve, Molier's screening—dot-dashed curve.

In spite of Euler form, SE leads to reversibility by the change $t \to -t$ only by passing in SE to the complex conjugate values. The SE conserves its form by substitutions $t \to -t, \psi \to \psi^*$. In other words, "the derivation" of SE from Eq. (3.1.1) written as $\psi = e^{i(\omega t - kx)}$ leads to other hydrodynamic equations:

$$-\frac{\partial \rho}{\partial t} + \frac{\partial}{\partial \mathbf{r}} \cdot \rho \mathbf{v} = 0, \tag{3.1.53}$$

$$-\frac{\partial \mathbf{v}}{\partial t} + (\mathbf{v}\nabla)\mathbf{v} = -\frac{1}{m}\nabla U^*. \tag{3.1.54}$$

It means that SE contains in implicit form approximation against of the time arrow. In other words, the theory of irreversible processes denies the existence of such processes but the Poincare-Zermelo theorem admits in principal the return in the previous state of physical systems obeyed to Newton's dynamics.

In the general case of open quantum system, the energy equation should be used in consideration. From the first glance, it forced us to introduce hypercomplex numbers description for wave function instead of complex ψ-function, real and imaginary parts of which lead only to two hydrodynamic equations: continuity and Euler momentum equations. However, a much more effective way exists as it is shown in the next section. GHEs lead to the SE as a deep particular case of the generalized Boltzmann physical kinetics and therefore of non-local hydrodynamics. The result—quantum hydrodynamics can be integrated into the unified theory of transport processes

3.2 GHEs, QUANTUM HYDRODYNAMICS. SE AS THE CONSEQUENCE OF GHE

In the following, we intend to formulate in explicit form all assumptions which should be done for obtaining of SE from GHE. On the finalized step for simplicity, we shall use the nonstationary 1D model without external forces.

Let us write down equations of the Madelung's hydrodynamics in the form

$$\frac{\partial \rho}{\partial t} + \frac{\partial}{\partial \mathbf{r}} \cdot \rho \mathbf{v}_0 = 0 \tag{3.2.1}$$

and

$$\frac{\partial}{\partial t}(\rho \mathbf{v}_0) + \frac{\partial}{\partial \mathbf{r}} \cdot \rho \mathbf{v}_0 \mathbf{v}_0 = -\frac{\rho}{m}\nabla U^*, \tag{3.2.2}$$

where $\mathbf{v}_0$ is hydrodynamic velocity and potential

$$U^* = U - \frac{\hbar^2}{2m\sqrt{\rho}}\Delta\sqrt{\rho} = U - \frac{\hbar^2}{4m\rho}\left[\Delta\rho - \frac{1}{2\rho}\left(\frac{\partial \rho}{\partial \mathbf{r}}\right)^2\right]. \tag{3.2.3}$$

This form is more convenient for following transformations of GHE.

Step 1. Generalized Euler equations can be written in the form (2.7.29), (2.7.30), (2.7.36), (2.7.37), and (2.7.48);

- Continuity equation for species α:

$$\begin{aligned}&\frac{\partial}{\partial t}\left\{\rho_\alpha-\tau_\alpha\left[\frac{\partial \rho_\alpha}{\partial t}+\frac{\partial}{\partial \mathbf{r}}\cdot(\rho_\alpha \mathbf{v}_0)\right]\right\}\\&+\frac{\partial}{\partial \mathbf{r}}\cdot\left\{\rho_\alpha \mathbf{v}_0-\tau_\alpha\left[\frac{\partial}{\partial t}(\rho_\alpha \mathbf{v}_0)\right.\right.\\&\left.\left.+\frac{\partial}{\partial \mathbf{r}}\cdot(\rho_\alpha \mathbf{v}_0\mathbf{v}_0)+\overleftrightarrow{\mathbf{I}}\cdot\frac{\partial p_\alpha}{\partial \mathbf{r}}-\rho_\alpha \mathbf{F}_\alpha^{(1)}-\frac{q_\alpha}{m_\alpha}\rho_\alpha \mathbf{v}_0\times\mathbf{B}\right]\right\}=R_\alpha.\end{aligned} \tag{3.2.4}$$

- Continuity equation for a mixture:

$$\begin{aligned}&\frac{\partial}{\partial t}\left\{\rho-\sum_\alpha\tau_\alpha\left[\frac{\partial \rho_\alpha}{\partial t}+\frac{\partial}{\partial \mathbf{r}}\cdot(\rho_\alpha \mathbf{v}_0)\right]\right\}\\&+\frac{\partial}{\partial \mathbf{r}}\cdot\left\{\rho \mathbf{v}_0-\sum_\alpha\tau_\alpha\left[\frac{\partial}{\partial t}(\rho_\alpha \mathbf{v}_0)+\frac{\partial}{\partial \mathbf{r}}\cdot(\rho_\alpha \mathbf{v}_0\mathbf{v}_0)+\overleftrightarrow{\mathbf{I}}\cdot\frac{\partial p_\alpha}{\partial \mathbf{r}}\right.\right.\\&\left.\left.-\rho_\alpha \mathbf{F}_\alpha^{(1)}-\frac{q_\alpha}{m_\alpha}\rho_\alpha \mathbf{v}_0\times\mathbf{B}\right]\right\}=0.\end{aligned} \tag{3.2.5}$$

- Momentum equation for species α:

$$\begin{aligned}&\frac{\partial}{\partial t}\left\{\rho_\alpha \mathbf{v}_0-\tau_\alpha\left[\frac{\partial}{\partial t}(\rho_\alpha \mathbf{v}_0)+\frac{\partial}{\partial \mathbf{r}}\cdot\rho_\alpha \mathbf{v}_0\mathbf{v}_0+\frac{\partial p_\alpha}{\partial \mathbf{r}}-\rho_\alpha \mathbf{F}_\alpha^{(1)}\right.\right.\\&\left.\left.-\frac{q_\alpha}{m_\alpha}\rho_\alpha \mathbf{v}_0\times\mathbf{B}\right]\right\}-\mathbf{F}_\alpha^{(1)}\left[\rho_\alpha-\tau_\alpha\left(\frac{\partial \rho_\alpha}{\partial t}+\frac{\partial}{\partial \mathbf{r}}(\rho_\alpha \mathbf{v}_0)\right)\right]\\&-\frac{q_\alpha}{m_\alpha}\left\{\rho_\alpha \mathbf{v}_0-\tau_\alpha\left[\frac{\partial}{\partial t}(\rho_\alpha \mathbf{v}_0)+\frac{\partial}{\partial \mathbf{r}}\cdot\rho_\alpha \mathbf{v}_0\mathbf{v}_0+\frac{\partial p_\alpha}{\partial \mathbf{r}}-\rho_\alpha \mathbf{F}_\alpha^{(1)}\right.\right.\\&\left.\left.-\frac{q_\alpha}{m_\alpha}\rho_\alpha \mathbf{v}_0\times\mathbf{B}\right]\right\}\times\mathbf{B}+\frac{\partial}{\partial \mathbf{r}}\cdot\left\{\rho_\alpha \mathbf{v}_0\mathbf{v}_0+p_\alpha\overleftrightarrow{\mathbf{I}}-\tau_\alpha\left[\frac{\partial}{\partial t}(\rho_\alpha \mathbf{v}_0\mathbf{v}_0\right.\right.\\&+p_\alpha\overleftrightarrow{\mathbf{I}})+\frac{\partial}{\partial \mathbf{r}}\cdot\rho_\alpha(\mathbf{v}_0\mathbf{v}_0)\mathbf{v}_0+2\overleftrightarrow{\mathbf{I}}\left(\frac{\partial}{\partial \mathbf{r}}\cdot(\rho_\alpha \mathbf{v}_0)\right)+\frac{\partial}{\partial \mathbf{r}}\cdot\left(\overleftrightarrow{\mathbf{I}}\rho_\alpha \mathbf{v}_0\right)\\&\left.\left.-\mathbf{F}_\alpha^{(1)}\rho_\alpha \mathbf{v}_0-\rho_\alpha \mathbf{v}_0\mathbf{F}_\alpha^{(1)}-\frac{q_\alpha}{m_\alpha}\rho_\alpha[\mathbf{v}_0\times\mathbf{B}]\mathbf{v}_0-\frac{q_\alpha}{m_\alpha}\rho_\alpha \mathbf{v}_0[\mathbf{v}_0\times\mathbf{B}]\right]\right\}\\&=\int m_\alpha \mathbf{v}_\alpha J_\alpha^{\mathrm{st,el}}d\mathbf{v}_\alpha+\int m_\alpha \mathbf{v}_\alpha J_\alpha^{\mathrm{st,inel}}d\mathbf{v}_\alpha.\end{aligned} \tag{3.2.6}$$

- Momentum equation for mixture:

$$\begin{aligned}&\frac{\partial}{\partial t}\left\{\rho \mathbf{v}_0-\sum_\alpha\tau_\alpha^{(0)}\left[\frac{\partial}{\partial t}(\rho_\alpha \mathbf{v}_0)+\frac{\partial}{\partial \mathbf{r}}\cdot\rho_\alpha \mathbf{v}_0\mathbf{v}_0+\frac{\partial p_\alpha}{\partial \mathbf{r}}-\rho_\alpha \mathbf{F}_\alpha^{(1)}\right.\right.\\&\left.\left.-\frac{q_\alpha}{m_\alpha}\rho_\alpha \mathbf{v}_0\times\mathbf{B}\right]\right\}-\sum_\alpha\mathbf{F}_\alpha^{(1)}\left[\rho_\alpha-\tau_\alpha^{(0)}\left(\frac{\partial \rho_\alpha}{\partial t}+\frac{\partial}{\partial \mathbf{r}}(\rho_\alpha \mathbf{v}_0)\right)\right]\\&-\sum_\alpha\frac{q_\alpha}{m_\alpha}\left\{\rho_\alpha \mathbf{v}_0-\tau_\alpha\left[\frac{\partial}{\partial t}(\rho_\alpha \mathbf{v}_0)+\frac{\partial}{\partial \mathbf{r}}\cdot\rho_\alpha \mathbf{v}_0\mathbf{v}_0+\frac{\partial p_\alpha}{\partial \mathbf{r}}-\rho_\alpha \mathbf{F}_\alpha^{(1)}\right.\right.\\&\left.\left.-\frac{q_\alpha}{m_\alpha}\rho_\alpha \mathbf{v}_0\times\mathbf{B}\right]\right\}\times\mathbf{B}+\frac{\partial}{\partial \mathbf{r}}\cdot\left\{\rho \mathbf{v}_0\mathbf{v}_0+p\overleftrightarrow{\mathbf{I}}-\sum_\alpha\tau_\alpha\left[\frac{\partial}{\partial t}(\rho_\alpha \mathbf{v}_0\mathbf{v}_0\right.\right.\\&+p_\alpha\overleftrightarrow{\mathbf{I}})+\frac{\partial}{\partial \mathbf{r}}\cdot\rho_\alpha(\mathbf{v}_0\mathbf{v}_0)\mathbf{v}_0+2\overleftrightarrow{\mathbf{I}}\left(\frac{\partial}{\partial \mathbf{r}}\cdot(p_\alpha \mathbf{v}_0)\right)+\frac{\partial}{\partial \mathbf{r}}\cdot\left(\overleftrightarrow{\mathbf{I}}p_\alpha \mathbf{v}_0\right)\\&\left.\left.-\mathbf{F}_\alpha^{(1)}\rho_\alpha \mathbf{v}_0-\rho_\alpha \mathbf{v}_0\mathbf{F}_\alpha^{(1)}-\frac{q_\alpha}{m_\alpha}\rho_\alpha[\mathbf{v}_0\times\mathbf{B}]\mathbf{v}_0-\frac{q_\alpha}{m_\alpha}\rho_\alpha \mathbf{v}_0[\mathbf{v}_0\times\mathbf{B}]\right]\right\}=0.\end{aligned} \tag{3.2.7}$$

• Energy equation:

$$\begin{aligned}
&\frac{\partial}{\partial t}\left\{\frac{\rho v_0^2}{2}+\frac{3}{2}p+\sum_\alpha \varepsilon_\alpha n_\alpha-\sum_\alpha \tau_\alpha\left[\frac{\partial}{\partial t}\left(\frac{\rho_\alpha v_0^2}{2}+\frac{3}{2}p_\alpha+\varepsilon_\alpha n_\alpha\right)\right.\right.\\
&\left.\left.+\frac{\partial}{\partial \mathbf{r}}\cdot\left(\frac{1}{2}\rho_\alpha v_0^2\mathbf{v}_0+\frac{5}{2}p_\alpha\mathbf{v}_0+\varepsilon_\alpha n_\alpha\mathbf{v}_0\right)-\mathbf{F}_\alpha^{(1)}\cdot\rho_\alpha\mathbf{v}_0\right]\right\}\\
&+\frac{\partial}{\partial \mathbf{r}}\cdot\left\{\frac{1}{2}\rho v_0^2\mathbf{v}_0+\frac{5}{2}p\mathbf{v}_0+\mathbf{v}_0\sum_\alpha\varepsilon_\alpha n_\alpha-\sum_\alpha\tau_\alpha\left[\frac{\partial}{\partial t}\left(\frac{1}{2}\rho_\alpha v_0^2\mathbf{v}_0\right.\right.\right.\\
&\left.+\frac{5}{2}p_\alpha\mathbf{v}_0+\varepsilon_\alpha n_\alpha\mathbf{v}_0\right)+\frac{\partial}{\partial \mathbf{r}}\cdot\left(\frac{1}{2}\rho_\alpha v_0^2\mathbf{v}_0\mathbf{v}_0+\frac{7}{2}p_\alpha\mathbf{v}_0\mathbf{v}_0+\frac{1}{2}p_\alpha v_0^2\overleftrightarrow{\mathbf{I}}\right.\\
&\left.+\frac{5}{2}\frac{p_\alpha^2}{\rho_\alpha}\overleftrightarrow{\mathbf{I}}+\varepsilon_\alpha n_\alpha\mathbf{v}_0\mathbf{v}_0+\varepsilon_\alpha\frac{p_\alpha}{m_\alpha}\overleftrightarrow{\mathbf{I}}\right)-\rho_\alpha\mathbf{F}_\alpha^{(1)}\cdot\mathbf{v}_0\mathbf{v}_0-p_\alpha\mathbf{F}_\alpha^{(1)}\cdot\overleftrightarrow{\mathbf{I}}\\
&-\frac{1}{2}\rho_\alpha v_0^2\mathbf{F}_\alpha^{(1)}-\frac{3}{2}\mathbf{F}_\alpha^{(1)}p_\alpha-\frac{\rho_\alpha v_0^2}{2}\frac{q_\alpha}{m_\alpha}[\mathbf{v}_0\times\mathbf{B}]-\frac{5}{2}p_\alpha\frac{q_\alpha}{m_\alpha}[\mathbf{v}_0\times\mathbf{B}]\\
&\left.\left.-\varepsilon_\alpha n_\alpha\frac{q_\alpha}{m_\alpha}[\mathbf{v}_0\times\mathbf{B}]-\varepsilon_\alpha n_\alpha\mathbf{F}_\alpha^{(1)}\right]\right\}-\mathbf{v}_0\cdot\sum_\alpha\rho_\alpha\mathbf{F}_\alpha^{(1)}\\
&+\sum_\alpha\tau_\alpha\mathbf{F}_\alpha^{(1)}\cdot\left[\frac{\partial}{\partial t}(\rho_\alpha\mathbf{v}_0)+\frac{\partial}{\partial \mathbf{r}}\cdot\rho_\alpha\mathbf{v}_0\mathbf{v}_0+\frac{\partial}{\partial \mathbf{r}}\cdot p_\alpha\overleftrightarrow{\mathbf{I}}-\rho_\alpha\mathbf{F}_\alpha^{(1)}-q_\alpha n_\alpha[\mathbf{v}_0\times\mathbf{B}]\right]=0.
\end{aligned}\tag{3.2.8}$$

Here, $\mathbf{F}_\alpha^{(1)}$ are the forces of the nonmagnetic origin, $\mathbf{B}$—magnetic induction, $\overleftrightarrow{\mathbf{I}}$—unit tensor, q_α—charge of the α-component particle, p_α—static pressure for α-component, $\mathbf{V}_\alpha$—thermal velocity, ε_α—internal energy for the particles of α-component.

Step 2. Pass on to one component physical system without external forces.

Continuity equation

$$\begin{aligned}
&\frac{\partial}{\partial t}\left\{\rho-\tau\left(\frac{\partial \rho}{\partial t}+\frac{\partial}{\partial \mathbf{r}}\cdot(\rho\mathbf{v}_0)\right)\right\}+\frac{\partial}{\partial \mathbf{r}}\cdot\left\{\rho\mathbf{v}_0-\tau\left(\frac{\partial}{\partial t}(\rho\mathbf{v}_0)\right.\right.\\
&\left.\left.+\frac{\partial}{\partial \mathbf{r}}\cdot(\rho\mathbf{v}_0\mathbf{v}_0)+\frac{\partial p}{\partial \mathbf{r}}\right)\right\}=0.
\end{aligned}\tag{3.2.9}$$

Motion equation

$$\begin{aligned}
&\frac{\partial}{\partial t}\left\{\rho v_{0\beta}-\tau\left[\frac{\partial}{\partial t}(\rho v_{0\beta})+\frac{\partial}{\partial r_\alpha}(p\delta_{\alpha\beta}+\rho v^2{}_{0\alpha})\right]\right\}+\frac{\partial}{\partial r_\alpha}\left\{p\delta_{\alpha\beta}+\rho v_{0\alpha}v_{0\beta}\right.\\
&\left.-\tau\left[\frac{\partial}{\partial t}(p\delta_{\alpha\beta}+\rho v_{0\alpha}v_{0\beta})+\frac{\partial}{\partial r_\gamma}(p\delta_{\alpha\gamma}v_{0\beta}+pv_{0\alpha}\delta_{\beta\gamma}+pv_{0\gamma}\delta_{\alpha\beta}+\rho v_{0\alpha}v_{0\beta}v_{0\gamma})\right]\right\}=0.
\end{aligned}\tag{3.2.10}$$

In Eq. (3.2.10), Einstein's rule is used for summation with index $\alpha,\beta,\gamma=1,2,3$.

Energy equation

$$\begin{aligned}
&\frac{\partial}{\partial t}\left\{3p+\rho v_0^2-\tau\left[\frac{\partial}{\partial t}(3p+\rho v_0^2)+\frac{\partial}{\partial \mathbf{r}}\cdot\left(\mathbf{v}_0(\rho v_0^2+5p)\right)\right]\right\}\\
&+\frac{\partial}{\partial \mathbf{r}}\cdot\left\{\begin{array}{l}\mathbf{v}_0(\rho v_0^2+5p)\\ -\tau\left[\frac{\partial}{\partial t}\left(\mathbf{v}_0(\rho v_0^2+5p)\right)+\frac{\partial}{\partial \mathbf{r}}\cdot\left(\overleftrightarrow{\mathbf{I}}pv_0^2+\rho v_0^2\mathbf{v}_0\mathbf{v}_0+7p\mathbf{v}_0\mathbf{v}_0+5\overleftrightarrow{\mathbf{I}}\frac{p^2}{\rho}\right)\right]\end{array}\right\}=0.
\end{aligned}\tag{3.2.11}$$

Step 3. Pass on to the nonstationary 1D model for the generalized Euler equations.

The following inscription $\tau=\tau^{(qu)}$ corresponds to choice of the scale on the quantum level.

Continuity equation:

$$\frac{\partial}{\partial t}\left\{\rho-\tau^{(\mathrm{qu})}\left[\frac{\partial \rho}{\partial t}+\frac{\partial}{\partial x}(\rho v_0)\right]\right\}+\frac{\partial}{\partial x}\left\{\rho v_0-\tau^{(\mathrm{qu})}\left[\frac{\partial}{\partial t}(\rho v_0)+\frac{\partial}{\partial x}(\rho v_0^2)+\frac{\partial p}{\partial x}\right]\right\}=0. \tag{3.2.12}$$

Motion equation:

$$\frac{\partial}{\partial t}\left\{\rho v_0-\tau^{(\mathrm{qu})}\left[\frac{\partial}{\partial t}(\rho v_0)+\frac{\partial}{\partial x}(\rho v_0^2)+\frac{\partial p}{\partial x}\right]\right\}+\frac{\partial}{\partial x}\left\{\rho v_0^2+p-\tau^{(\mathrm{qu})}\left[\frac{\partial}{\partial t}(\rho v_0^2+p)+\frac{\partial}{\partial x}(\rho v_0^3+3pv_0)\right]\right\}=0. \tag{3.2.13}$$

Energy equation:

$$\frac{\partial}{\partial t}\left\{\rho v_0^2+3p-\tau^{(\mathrm{qu})}\left[\frac{\partial}{\partial t}(\rho v_0^2+3p)+\frac{\partial}{\partial x}(\rho v_0^3+5pv_0)\right]\right\}+\frac{\partial}{\partial x}\left\{\rho v_0^3+5pv_0-\tau^{(\mathrm{qu})}\left[\frac{\partial}{\partial t}(\rho v_0^3+5pv_0)+\frac{\partial}{\partial x}\left(\rho v_0^4+8pv_0^2+5\frac{p^2}{\rho}\right)\right]\right\}=0. \tag{3.2.14}$$

Step 4. All the time non-local terms are omitted following the Schrödinger-Madelung model.
Continuity equation:

$$\frac{\partial \rho}{\partial t}+\frac{\partial}{\partial x}\left\{\rho v_0-\tau^{(\mathrm{qu})}\left[\frac{\partial}{\partial t}(\rho v_0)+\frac{\partial}{\partial x}(\rho v_0^2)+\frac{\partial p}{\partial x}\right]\right\}=0. \tag{3.2.15}$$

Momentum equation:

$$\frac{\partial}{\partial t}(\rho v_0)+\frac{\partial}{\partial x}\left\{\rho v_0^2+p-\tau^{(\mathrm{qu})}\left[\frac{\partial}{\partial t}(\rho v_0^2+p)+\frac{\partial}{\partial x}(\rho v_0^3+3pv_0)\right]\right\}=0. \tag{3.2.16}$$

Energy equation:

$$\frac{\partial}{\partial t}(\rho v_0^2+3p)+\frac{\partial}{\partial x}\left\{\rho v_0^3+5pv_0-\tau^{(\mathrm{qu})}\left[\frac{\partial}{\partial t}(\rho v_0^3+5pv_0)+\frac{\partial}{\partial x}\left(\rho v_0^4+8pv_0^2+5\frac{p^2}{\rho}\right)\right]\right\}=0. \tag{3.2.17}$$

Step 5. All terms containing the static pressure are omitted following the Schrödinger-Madelung model.
Continuity equation:

$$\frac{\partial \rho}{\partial t}+\frac{\partial}{\partial x}\left\{\rho v_0-\tau^{(\mathrm{qu})}\left[\frac{\partial}{\partial t}(\rho v_0)+\frac{\partial}{\partial x}(\rho v_0^2)\right]\right\}=0. \tag{3.2.18}$$

Momentum equation:

$$\frac{\partial}{\partial t}(\rho v_0)+\frac{\partial}{\partial x}\left\{\rho v_0^2-\tau^{(\mathrm{qu})}\left[\frac{\partial}{\partial t}(\rho v_0^2)+\frac{\partial}{\partial x}(\rho v_0^3)\right]\right\}=0. \tag{3.2.19}$$

Energy equation:

$$\frac{\partial}{\partial t}(\rho v_0^2)+\frac{\partial}{\partial x}\left\{\rho v_0^3-\tau^{(\mathrm{qu})}\left[\frac{\partial}{\partial t}(\rho v_0^3)+\frac{\partial}{\partial x}(\rho v_0^4)\right]\right\}=0. \tag{3.2.20}$$

Step 6. Following the Schrödinger-Madelung model, we omit all explicit time dependence of non-local terms.
Continuity equation:

$$\frac{\partial \rho}{\partial t}+\frac{\partial}{\partial x}\left[\rho u-\tau^{(\mathrm{qu})}\frac{\partial}{\partial x}(\rho u^2)\right]=0. \tag{3.2.21}$$

Motion equation:

$$\frac{\partial}{\partial t}(\rho u)+\frac{\partial}{\partial x}\left[\rho u^2-\tau^{(\mathrm{qu})}\frac{\partial}{\partial x}(\rho u^3)\right]=0. \tag{3.2.22}$$

Energy equation:

$$\frac{\partial}{\partial t}(\rho u^2)+\frac{\partial}{\partial x}\left[\rho u^3-\tau^{(\mathrm{qu})}\frac{\partial}{\partial x}(\rho u^4)\right]=0, \tag{3.2.23}$$

where u is velocity along the x axes. Now we formulate the next step.

Step 7. Following the Schrödinger-Madelung model, we reduce the system of quantum hydrodynamic equations to the continuity and motion equations.

This assumption leads to the condition following from the energy equation (3.2.23):

$$\rho u^3=\tau^{(\mathrm{qu})}\frac{\partial}{\partial x}(\rho u^4). \tag{3.2.24}$$

In this case, the energy equation (3.2.23) can be transformed into the relation

$$\frac{\partial}{\partial t}(\rho u^2)=0, \tag{3.2.25}$$

or

$$\rho u^2=C(x). \tag{3.2.26}$$

This energy conservation law $\rho u^2=C(x)$ does not lead out of the limits of approximations for the formulated assumptions. Moreover, the energy space dependence remains in Eqs. (3.2.21) and (3.2.22).

It is shown below that step 7 and condition (3.2.24) *lead to the Madelung's type potential.*

After substitution of the condition (3.2.24) in Eq. (3.2.22), we reach the system of two hydrodynamic equations

$$\frac{\partial \rho}{\partial t}+\frac{\partial}{\partial x}\left[\rho u-\tau^{(\mathrm{qu})}\frac{\partial}{\partial x}(\rho u^2)\right]=0 \tag{3.2.27}$$

and

$$\frac{\partial}{\partial t}(\rho u)+\frac{\partial}{\partial x}(\rho u^2)=\frac{\partial}{\partial x}\left\{\tau^{(\mathrm{qu})}\frac{\partial}{\partial x}\left[\tau^{(\mathrm{qu})}\frac{\partial}{\partial x}(\rho u^4)\right]\right\}. \tag{3.2.28}$$

Step 8. Following the Schrödinger-Madelung model, we should exclude non-local terms in the continuity equation.

$$\frac{\partial \rho}{\partial t}+\frac{\partial}{\partial x}(\rho u)=0. \tag{3.2.29}$$

It means that estimation can be used in the following transformations of momentum equation (3.2.28):

$$\frac{1}{u}\frac{\partial u}{\partial x}\cong-\frac{1}{2\rho}\frac{\partial \rho}{\partial x}. \tag{3.2.30}$$

From (I.53) follows approximation for non-local parameter $\tau^{(\mathrm{qu})}$

$$\tau^{(\mathrm{qu})}=H/(mu^2), \tag{3.2.31}$$

where H is a coefficient of proportionality, which reflects the state of physical system. In the simplest case, H is equal to Plank constant $\hbar/2$ and relation (3.2.31) becomes compatible with the Heisenberg relation.

Generally speaking, the expression

$$U^{(\mathrm{qu})}=\tau^{(\mathrm{qu})}\frac{\partial}{\partial x}\left[\tau^{(\mathrm{qu})}\frac{\partial}{\partial x}(\rho u^4)\right] \tag{3.2.32}$$

can be considered as the quantum potential, but it contains a dependence on velocity u. This dependence can be excluded by different ways. Let us estimate potential (3.2.32) conserving only the senior powers in velocity

$$U^{(\mathrm{qu})} = \tau^{(\mathrm{qu})}\frac{\partial}{\partial x}\left[\tau^{(\mathrm{qu})}\frac{\partial}{\partial x}(\rho u^4)\right] \approx \tau^{(\mathrm{qu})^2}\left[u^4\frac{\partial^2\rho}{\partial x^2} + \gamma u^3\frac{\partial u}{\partial x}\frac{\partial\rho}{\partial x}\right] \tag{3.2.33}$$

and introducing γ as a numerical parameter for possible corrections.

In this case (see also Eqs. (3.2.30) and (3.2.31)),

$$U^{(\mathrm{qu})} \approx \frac{H^2}{m^2}\left[\frac{\partial^2\rho}{\partial x^2} + \gamma\frac{1}{u}\frac{\partial u}{\partial x}\frac{\partial\rho}{\partial x}\right] = \frac{H^2}{m^2}\left[\frac{\partial^2\rho}{\partial x^2} - \gamma\frac{1}{2\rho}\left(\frac{\partial\rho}{\partial x}\right)^2\right]. \tag{3.2.34}$$

Obviously, potential (3.2.34) coincides with the Madelung potential, if $\gamma = 1$. If the case requires, the Bohm transformation can be involved in consideration

$$\frac{1}{2\rho}\left(\frac{\partial^2\rho}{\partial x^2} - \frac{1}{2\rho}\left(\frac{\partial\rho}{\partial x}\right)^2\right) = \frac{1}{\sqrt{\rho}}\frac{\partial^2}{\partial x^2}\sqrt{\rho}. \tag{3.2.35}$$

As a result, we have the quantum hydrodynamic equation

$$\frac{\partial}{\partial t}(\rho u) + \frac{\partial}{\partial x}(\rho u^2) = \frac{H^2}{m^2}\frac{\partial}{\partial x}\left\{\frac{\partial^2\rho}{\partial x^2} - \frac{1}{2\rho}\left(\frac{\partial\rho}{\partial x}\right)^2\right\}. \tag{3.2.36}$$

Madelung equation has the form

$$\frac{\partial}{\partial t}(\rho u) + \frac{\partial}{\partial x}(\rho u^2) = \frac{\hbar^2}{4m^2}\rho\frac{\partial}{\partial x}\left\{\frac{1}{\rho}\left[\frac{\partial^2\rho}{\partial x^2} - \frac{1}{2\rho}\left(\frac{\partial\rho}{\partial x}\right)^2\right]\right\}. \tag{3.2.37}$$

Transform Eq. (3.2.36) to the Madelung form:

$$\begin{aligned}
\frac{\partial}{\partial t}(\rho u) + \frac{\partial}{\partial x}(\rho u^2) &= \frac{H^2}{m^2}\frac{\partial}{\partial x}\left\{\rho\frac{1}{\rho}\left[\frac{\partial^2\rho}{\partial x^2} - \frac{1}{2\rho}\left(\frac{\partial\rho}{\partial x}\right)^2\right]\right\} \\
&= \frac{H^2}{m^2}\rho\frac{\partial}{\partial x}\left\{\frac{1}{\rho}\left[\frac{\partial^2\rho}{\partial x^2} - \frac{1}{2\rho}\left(\frac{\partial\rho}{\partial x}\right)^2\right]\right\} \\
&+ \frac{H^2}{m^2}\left[\frac{\partial^2\rho}{\partial x^2} - \frac{\rho}{2}\left(\frac{\partial\ln\rho}{\partial x}\right)^2\right]\frac{\partial\ln\rho}{\partial x}.
\end{aligned} \tag{3.2.38}$$

Using the relation

$$\frac{\partial^2\rho}{\partial x^2} = \rho\frac{\partial^2\ln\rho}{\partial x^2} + \rho\left(\frac{\partial\ln\rho}{\partial x}\right)^2, \tag{3.2.39}$$

one obtains

$$\begin{aligned}
\frac{\partial}{\partial t}(\rho u) + \frac{\partial}{\partial x}(\rho u^2) &= \frac{H^2}{m^2}\rho\frac{\partial}{\partial x}\left\{\frac{1}{\rho}\left[\frac{\partial^2\rho}{\partial x^2} - \frac{1}{2\rho}\left(\frac{\partial\rho}{\partial x}\right)^2\right]\right\} \\
&+ \frac{H^2}{m^2}\rho\left[\frac{\partial^2\ln\rho}{\partial x^2} + \frac{1}{2}\left(\frac{\partial\ln\rho}{\partial x}\right)^2\right]\frac{\partial\ln\rho}{\partial x}.
\end{aligned} \tag{3.2.40}$$

The last term on the left-hand side of Eq. (3.2.40) contains the derivatives on the logarithmic functions and can be omitted. After introducing the definition for H

$$H = \frac{\hbar}{2}, \tag{3.2.41}$$

we reach the Madelung hydrodynamic Eq. (3.2.37).

Quantum hydrodynamic equation can be written in the unified form

$$\frac{\partial}{\partial t}(\rho u)+\frac{\partial}{\partial x}(\rho u^2)=\frac{\hbar^2}{4m^2}\rho\frac{\partial}{\partial x}\left\{\frac{1}{\rho}\left[\gamma_1\frac{\partial^2\rho}{\partial x^2}-\gamma_2\frac{1}{\rho}\left(\frac{\partial\rho}{\partial x}\right)^2\right]\right\}, \tag{3.2.42}$$

the numerical coefficients $\gamma_1=1$, $\gamma_2=0.5$ correspond to SE.

Therefore, the condition (3.2.24) $\rho u^3=\tau^{(qu)}\frac{\partial}{\partial x}(\rho u^4)$ leads by $\gamma_1=1, \gamma_2=0.5$ to the Bohm potential, reflecting "the last traces" of the omitted energy equation. In the other words, Bohm potential reflects the condition of the dissipation absence in the particular case, when $\gamma_1=1$, $\gamma_2=0.5$.

Rejection of some assumptions formulated above could change the coefficients γ_1, γ_2 in Eq. (3.2.42).

Conclusion: SE is a deep particular case of the GHEs.

However, some remarks of the principal significance should be done concerning the quantization methods in quantum hydrodynamics. These problems are considered in Section 5.3. It is shown that the adiabatic theorem and consequences of this theory deliver the general quantization conditions for non-local quantum hydrodynamics.

3.3 SE AND ITS DERIVATION FROM LIOUVILLE EQUATION

Let us consider now from positions of non-local physics the main steps of the SE derivation directly from Liouville equation. This derivation of classical SE can be found in [63, 64, 91]. Starting point is the Liouville equation written for one-particle distribution function $f(x,p,t)$

$$\frac{\partial f}{\partial t}+\frac{p}{m}\frac{\partial f}{\partial x}+F(x)\frac{\partial f}{\partial p}=0. \tag{3.3.1}$$

Equation (3.3.1) is a collisionless Boltzmann equation, which cannot describe dissipation. The external force $F(x)=-\partial U(x)/\partial x$ is acting on the particle with mass m having impulse p. Equation (3.3.1) is the *local* equation, it means that *no chance* to obtain the SE (and therefore Madelung equation) from Eq. (3.3.1) without additional assumptions. As it is shown in Section 3.3, it is sufficient to introduce (artificial) space nonlocality into consideration. One introduces:

(a) The classical probability amplitude $\Psi(x,t)$ for which

$$|\Psi(x,t)|^2=\int\limits_{-\infty}^{+\infty}f(x,p,t)\mathrm{d}p, \tag{3.3.2}$$

generally speaking $\Psi(x,t)$ is a complex function;

(b) The Wigner type of Fourier transform defined by

$$T[f](x,y,t)=\int\limits_{-\infty}^{+\infty}f(x,p,t)\mathrm{e}^{\frac{2ipy}{\alpha}}\mathrm{d}p \tag{3.3.3}$$

with the a real parameter α.

Transform $T[f](x,y,t)$ introduces the artificial space nonlocality in physical system leaving without account the nonlocality in time. Transform $T[f](x,y,t)$ has physical meaning if $y\approx 0$.

Transform $T[f](x,y,t)$ by $y\neq 0$ can be written as

$$T[f](x,y,t)=\Psi^*(t,x-y)\Psi(t,x+y). \tag{3.3.4}$$

The next step is to consider the derivative

$$\frac{\partial}{\partial t}T[f](x,y,t)=\int\limits_{-\infty}^{+\infty}\frac{\partial f(x,p,t)}{\partial t}\mathrm{e}^{\frac{2ipy}{\alpha}}\mathrm{d}p \tag{3.3.5}$$

using the Liouville equation:

$$\frac{\partial}{\partial t}T[f](x,y,t)=-\int\limits_{-\infty}^{+\infty}\left[\frac{p}{m}\frac{\partial f}{\partial x}+F(x)\frac{\partial f}{\partial p}\right]\mathrm{e}^{\frac{2ipy}{\alpha}}\mathrm{d}p \tag{3.3.6}$$

or

$$\frac{\partial}{\partial t}T[f](x,y,t)=\frac{\mathrm{i}\alpha}{2m}\int\limits_{-\infty}^{+\infty}\frac{\partial^2}{\partial x\partial y}\left[f\mathrm{e}^{\frac{2ipy}{\alpha}}\right]\mathrm{d}p+F(x)\frac{2\mathrm{i}y}{\alpha}T[f](x,y,t). \tag{3.3.7}$$

Equation (3.3.7) transforms into

$$\mathrm{i}\alpha\frac{\partial}{\partial t}T[f](x,y,t)=-\frac{\alpha^2}{2m}\frac{\partial^2}{\partial x\partial y}T[f](x,y,t)-2yF(x)T[f](x,y,t) \tag{3.3.8}$$

and its solution can be obtained by the perturbation method (see Appendix 1)

$$T[f](t,x,y)=\sum_{n=0}^{\infty}T_n[f](t,x)y^n. \tag{3.3.9}$$

Authors [91] transform Eq. (3.3.8) using the substitution:

$$s=x-y,\quad r=x+y, \tag{3.3.10}$$

$$\mathrm{i}\alpha\frac{\partial}{\partial t}T[f](r,s,t)=\left[-\frac{\alpha^2}{2m}\left(\frac{\partial^2}{\partial r^2}-\frac{\partial^2}{\partial s^2}\right)-(r-s)F\left(\frac{r+s}{2}\right)\right]T[f](r,s,t). \tag{3.3.11}$$

Using Eq. (3.3.4) in the form

$$T[f](s,r,t)=\Psi^*(t,s)\Psi(t,r), \tag{3.3.12}$$

one obtains

$$\begin{aligned}&\Psi^*(t,s)\left[\mathrm{i}\alpha\frac{\partial\Psi(t,r)}{\partial t}+\frac{\alpha^2}{2m}\frac{\partial^2\Psi(t,r)}{\partial r^2}-V(r)\Psi(t,r)\right]\\&=\Psi(t,r)\left[-\mathrm{i}\alpha\frac{\partial\Psi^*(t,s)}{\partial t}+\frac{\alpha^2}{2m}\frac{\partial^2\Psi^*(t,s)}{\partial s^2}-V(s)\Psi^*(t,s)\right].\end{aligned} \tag{3.3.13}$$

After introducing notation

$$K(t,r)=\mathrm{i}\alpha\frac{\partial\Psi(t,r)}{\partial t}+\frac{\alpha^2}{2m}\frac{\partial^2\Psi(t,r)}{\partial r^2}-V(r)\Psi(t,r). \tag{3.3.14}$$

Equation (3.3.13) is rewritten as

$$\Psi^*(t,s)K(t,r)=\Psi(t,r)K^*(t,s) \tag{3.3.15}$$

and can be satisfied identically if

$$K(t,r)=\mathrm{i}\alpha\frac{\partial\Psi(t,r)}{\partial t}+\frac{\alpha^2}{2m}\frac{\partial^2\Psi(t,r)}{\partial r^2}-V(r)\Psi(t,r)=0. \tag{3.3.16}$$

The assumption $K(t,r)=i\alpha\frac{\partial\Psi(t,r)}{\partial t}+\frac{\alpha^2}{2m}\frac{\partial^2\Psi(t,r)}{\partial r^2}-V(r)\Psi(t,r)=0$ *means from physical point of view the transition in the finalized results to the local space approximation of non-local equations.*

Obviously Eq. (3.3.16) is the SE if arbitrary parameter α coincides with Plank constant $\hbar$ and amplitude $\Psi(x,t)$ transforms into the wave function $\psi(x,t)$.

3.4 DIRECT EXPERIMENTAL CONFIRMATIONS OF THE NON-LOCAL EFFECTS

It is well known that by the end of the nineteenth century, the physical picture of the world comprised of two basic entities, particles and fields. The charged particles (like electrons, which are considered as the structureless point-like particles) were the source of fields and were also moved by fields. The quantum revolution after the year 1900, contains the idea of the principle significance—waves acquired particle properties and particles behaved like waves.

Erwin Schrödinger noted a peculiar property of this new theory which he called "entanglement." As it was shown by Madelung, even a single electron can be "smeared" in hydrodynamics. This fact leads in the theory to the curious feature of this new quantum connection—if two quantum systems are brought together and then separated, they remain still connected.

From the entanglement idea follows:

1. The notion of "locality" should be revised. Local description holds that one particle influences another only by direct contact or via some intermediary field; this influence can travel no faster than light.
2. "Nonlocality," on the other hand, would mean that one particle could influence another distant particle without anything passing between them, in an instantaneous manner, faster than light. One particle influences another, not via a conventional force field, but simply because they touched one another sometime in the distant past.
3. Such understanding of the nonlocality leads to "spooky actions at a distance" (A. Einstein). Quantum theory is nonlocal; quantum facts are always local.

It would seem that if the facts are always local, then the statistical description of the physical reality must be local as well.

As it was shown in the previous consideration—it is not the case. In other words, even though all the quantum facts are local, these facts cannot be simulated by an underlying local statistical theory of nonequilibrium processes.

It is the main sense of the Bell's theorem. Bell's proof does not depend on the details of quantum theory but only on simple logic applied to a few experimental facts. Furthermore, Bell's results remain true no matter what kind of theory explained these facts. Bell's theorem shows that local statistical theories place strong restrictions.

One way of describing a Bell test is as follows. Two observers, Alice and Bob, receive (entangled) particles emitted by some source; they each choose a measurement setting, a and b, respectively, and then record their measurement outcome values, A and B. Many Bell tests have been performed to date; only two possibilities were removed separately:

1. The locality "loophole" (the possibility that the outcome on one side is causally influenced by the setting choice or outcome on the other side)
2. The fair-sampling or detection loophole (the possibility that only a nonrepresentative sub-ensemble of particles is measured).

A loophole-free test must close all loopholes; otherwise, the measured data can still be explained in terms of local realism. The work [92] tries to organize an experiment which simultaneously closes the locality and the freedom-of-choice loophole. For the Bell test, they used the Clauser-Horne-Shimony-Holt (CHSH) form of Bell's inequality:

$$\mathrm{S}(a_1, a_2, b_1, b_2) = |E(a_1, b_1) + E(a_1, b_2) + E(a_2, b_1) - E(a_2, b_2)| \leq 2, \tag{3.4.1}$$

a_1, a_2, b_1, b_2 are Alice's (Bob's) possible polarizer settings and $E(a_i, b_j)$ $i,j=1,2$ is the expectation value of the correlation between Alice's and Bob's local (dichotomic) polarization measurement outcomes. For the used singlet state, quantum mechanics predicts a violation of this inequality with a maximum value of $\mathrm{S}^{\mathrm{qm}}_{\max} = 2\sqrt{2}$ when Alice and Bob make their measurement choices between appropriate mutually unbiased bases, e.g., with polarization analyzer $a_1 = 0°$, $b_1 = 22.5°$; $a_1 = 0°$, $b_2 = 67.5°$; $a_2 = 45°$, $b_1 = 22.5°$; and $a_2 = 45°$, $b_2 = 67.5°$.

During four 600-s-long measurement runs, the authors of [92] detected 19,917 photon-pair coincidences and violated the CHSH inequality, with $\mathrm{S}^{\exp} = 2.37 \pm 0.02$ (no background subtraction), by 16 standard deviations above the local realistic bound of 2. This result represents a clear violation of local realism in an experimental arrangement which explicitly closes both the locality and the freedom-of-choice loopholes, while only relying on the fair-sampling assumption. Assuming fair sampling, these results significantly reduce the set of possible local hidden variable theories.

Is it possible to present the experimental evidence of the nonlocality effect existence on the level of distribution functions? The experimental results (obtained under Shnoll's leadership) of the mentioned kind really exist (see, for example, [93–96]. Until nowadays, these results were not discussed from the positions of non-local physics and the theoretical conclusions of [93–96] should be used with caution.

The following results were obtained (omitting, in my mind, the wrong theoretical interpretations) in experiments starting with 1958:

1. Fine structure of the spectrum of amplitude variations in the results of measurements of the processes of different nature (in other words, the fine structure of the dispersion of results or the pattern of the corresponding histograms) is subject to "macroscopic fluctuations," changing regularly with time.
2. These changes indicate that the "dispersion of results" (that remains after all artifacts are excluded) inevitably accompanies any measurements (from the biochemical reactions and noise in the gravitational antenna to the α-decay) and reflects very basic features of our world.
3. The most general conclusion is the evidence that the fine structure of stochastic distributions is not accidental.
4. The corresponding histograms have much the same shape at any given time and for processes of a different nature and are very likely to change shape simultaneously for various processes and in widely distant laboratories. For a series of

successive histograms, any given one is highly probably similar to its nearest neighbors and occurs repeatedly with a period of 24 hours, 27 days, and about 365 days, thus implying that the phenomenon has a very profound cosmophysical origin.

S.E. Shnoll points out in detail the following effects:

1. **The "effect of near zone."**
The first evidence of the histogram pattern changing regularly in time is the "effect of near zone." This effect means that *similar histograms are significantly more probable* to appear in the nearby (neighboring) intervals of the time series of the results of measurements.
2. **Measurements of processes of different nature.**
The second evidence comes from the *similarity of the pattern of histograms plotted from the results of simultaneous independent measurements of processes of different nature at the same geographical point.*
3. **Regular changes in the histogram patterns.**
The third evidence of noncasuality of the histogram patterns is *their regular changing with time*. The regularities are revealed in the existence of the following periods in the change of the probability of similar histograms to appear.
   3.1. **Near-daily periods**: these are well-resolvable "sidereal" (1436 min) and "solar" (1440 min) daily periods. These periods mean the dependence of the histogram pattern on the rotation of the Earth around its axis.
   3.2. **Approximately 27-day periods**: these period scan be considered as an indication of the dependence of the histogram pattern on the disposition relative to the nearby celestial bodies: the Sun, the Moon and, probably, the planets.
   3.3. **Yearly periods**: these are well-resolvable "calendar" (365 solar days) and "sidereal" (365 solar days plus 6 h and 9 min) yearly periods. All these periods imply the dependence of the histogram pattern on the (1) rotation of the Earth around its axis and (2) movement of the Earth along its circumsolar orbit.
4. **The observed local-time synchronism.**
The dependence of the histogram pattern *on the Earth rotation* around its axis is clearly revealed in a phenomenon of "synchronization at the local time," when similar histograms are highly probable to appear at different geographical points (from Arctic to Antarctic, in the Western and Eastern hemispheres) at the same local time. It is astonishing that the local-time synchronism with the precision of 1 min is observed independently of the *regional latitude* at the most extreme distances—as extreme as possible on the Earth (about 15,000 km).
5. **The synchronism observed at different latitudes.**
The dependence of the histogram pattern on the Earth rotation around its axis is also revealed in the *disappearance of the near-daily periods close to the North Pole.*
6. **The collimator directed at the polestar.**
Measurements were taken *with the collimator directed at the polestar*. In the analysis of histograms plotted from the results of counting α-particles that were traveling North (in the direction of the polestar), *the near-daily periods were not observed* nor was the near-zone effect. The measurements were made in Pushchino (54. latitude North), but the effect is as would be expected at 90. North, i. e. at the North Pole. This means that the histogram pattern depends on the spatial direction of the process measured.
7. **The east and west-directed collimators.**
This effect was confirmed in the experiments with two collimators, directed east and west correspondingly. In those experiments, two important effects were discovered.
   7.1. The histograms registered in the experiments with the east-directed collimator ("east histograms") are similar to those "west histograms" that are delayed by 718 min, i.e., *by a half of the sidereal day.*
   7.2. **No similar histograms were observed at the simultaneous measurements with the "east" and "west" collimators.** Without collimators, similar histograms are highly probable to appear at the same place and time.

Now, we are ready to discuss these observations from positions of non-local physics. Let us write down the GBE once more

$$\frac{\mathrm{D}f_\alpha}{\mathrm{D}t}-\frac{\mathrm{D}}{\mathrm{D}t}\left[\tau_\alpha\frac{\mathrm{D}f_\alpha}{\mathrm{D}t}\right]=J^B, \tag{3.4.2}$$

where

$$\frac{\mathrm{D}}{\mathrm{D}t}=\frac{\partial}{\partial t}+\mathbf{v}_\alpha\cdot\frac{\partial}{\partial\mathbf{r}}+\mathbf{F}_\alpha^{\mathrm{sc}}\cdot\frac{\partial}{\partial\mathbf{v}_\alpha}. \tag{3.4.3}$$

The relation

$$\frac{\mathrm{D}f_\alpha^a}{\mathrm{D}t} = J^B, \tag{3.4.4}$$

where

$$f_\alpha^a = f_\alpha - \tau_\alpha \frac{\mathrm{D}f_\alpha}{\mathrm{D}t}, \tag{3.4.5}$$

means that the inclusion in consideration the non-local effects leads to inevitable (in the definite sense—noncasual) "macroscopic fluctuations,"

$$f_\alpha^{fl} = \tau_\alpha \frac{\mathrm{D}f_\alpha}{\mathrm{D}t}. \tag{3.4.6}$$

The normal distribution is

$$f(x) = \frac{1}{\sigma\sqrt{2\pi}} \mathrm{e}^{-\frac{(x-\mu)^2}{2\sigma^2}}. \tag{3.4.7}$$

The parameter μ in this formula is the mean or expectation of the distribution. The parameter σ is its standard deviation; its variance is therefore σ^2. A random variable with a Gaussian distribution is said to be normally distributed and is called a normal deviate. If $\mu = 0$ and $\sigma = 1$, the distribution is called the standard normal distribution or the unit normal distribution, and a random variable with that distribution is a standard normal deviate. For the normal (Maxwellian) distribution function (DF) $f_\alpha^a = f_\alpha^{(0)}$:

$$f_\alpha^{(0)} = n_\alpha \left(\frac{m_\alpha}{2\pi k_\mathrm{B} T}\right)^{3/2} \exp\left(\frac{-m_\alpha(\mathbf{v}_\alpha - \mathbf{v_0})^2}{2k_\mathrm{B}T}\right), \tag{3.4.8}$$

variance σ^2 is written as

$$\sigma^2 = k_\mathrm{B} T / m_\alpha, \tag{3.4.9}$$

and the variation coefficient is $V_\alpha^{\mathrm{var}} = \dfrac{\sigma}{v_0} = \dfrac{1}{v_0}\sqrt{k_\mathrm{B}T/m_\alpha}$.

Write down the non-local DF

$$f_\alpha = f_\alpha^{(0)} + \tau_\alpha \frac{\mathrm{D}f_\alpha^{(0)}}{\mathrm{D}t}, \tag{3.4.10}$$

where

$$\frac{\mathrm{D}f_\alpha^{(0)}}{\mathrm{D}t} = \frac{\partial f_\alpha^{(0)}}{\partial t} + \mathbf{v}_\alpha \cdot \frac{\partial f_\alpha^{(0)}}{\partial \mathbf{r}} + \mathbf{F}_\alpha \cdot \frac{\partial f_\alpha^{(0)}}{\partial \mathbf{v}_\alpha}. \tag{3.4.11}$$

Let us calculate f_α,

$$f_\alpha = f_\alpha^{(0)} + \tau_\alpha \left[\frac{\partial f_\alpha^{(0)}}{\partial t} + \mathbf{v}_\alpha \cdot \frac{\partial f_\alpha^{(0)}}{\partial \mathbf{r}} + \mathbf{F}_\alpha \cdot \frac{\partial f_\alpha^{(0)}}{\partial \mathbf{v}_\alpha}\right], \tag{3.4.12}$$

using (3.4.8). We thence derive, $\dfrac{\partial f_\alpha^{(0)}}{\partial t}$:

$$\frac{\partial f_\alpha^{(0)}}{\partial t} = f_\alpha^{(0)} \left[\frac{\partial \ln n_\alpha}{\partial t} - \frac{3}{2}\frac{\partial \ln T}{\partial t} + \frac{m_\alpha \mathbf{V}_\alpha^2}{2k_\mathrm{B}T}\frac{\partial \ln T}{\partial t} + \frac{m_\alpha}{k_\mathrm{B}T}\frac{\partial \mathbf{v_0}}{\partial t} \cdot (\mathbf{v}_\alpha - \mathbf{v}_0)\right]. \tag{3.4.13}$$

After introducing the kinetic temperature for species $\alpha \cdot \frac{3}{2} k_\mathrm{B} T_\alpha = \frac{1}{2} m_\alpha \overline{V_\alpha^2}$ and the heat velocity

$$\mathbf{V}_\alpha = \mathbf{v}_\alpha - \mathbf{v}_0, \tag{3.4.14}$$

we find

$$\frac{\partial f_\alpha^{(0)}}{\partial t} = f_\alpha^{(0)} \left[\frac{\partial \ln n_\alpha}{\partial t} + \left(\frac{m_\alpha \mathbf{V}_\alpha^2}{2k_\mathrm{B}T} - \frac{3}{2}\right)\frac{\partial \ln T}{\partial t} + \frac{m_\alpha}{k_\mathrm{B}T}\frac{\partial \mathbf{v_0}}{\partial t} \cdot \mathbf{V}_\alpha\right]. \tag{3.4.15}$$

We calculate $\mathbf{v}_\alpha \cdot \dfrac{\partial f_\alpha^{(0)}}{\partial \mathbf{r}}$ and $\mathbf{F}_\alpha \cdot \dfrac{\partial f_\alpha^{(0)}}{\partial \mathbf{v}_\alpha}$:

$$\mathbf{v}_\alpha \cdot \frac{\partial f_\alpha^{(0)}}{\partial \mathbf{r}} = f_\alpha^{(0)} \left[\mathbf{v}_\alpha \cdot \frac{\partial \ln n_\alpha}{\partial \mathbf{r}} + \left(\frac{m_\alpha \mathbf{V}_\alpha^2}{2k_\mathrm{B}T} - \frac{3}{2} \right) \mathbf{v}_\alpha \cdot \frac{\partial \ln T}{\partial \mathbf{r}} + \frac{m_\alpha}{k_\mathrm{B}T} \mathbf{v}_\alpha \cdot \sum_i V_{\alpha i} \frac{\partial v_{0i}}{\partial \mathbf{r}} \right]. \tag{3.4.16}$$

$$\mathbf{F}_\alpha \cdot \frac{\partial f_\alpha^{(0)}}{\partial \mathbf{v}_\alpha} = f_\alpha^{(0)} \frac{m_\alpha}{k_\mathrm{B}T} \mathbf{F}_\alpha \cdot (\mathbf{v}_\alpha - \mathbf{v}_0). \tag{3.4.17}$$

The force $\mathbf{F}_\alpha$ acting on the mass unit of α-particle is

$$\mathbf{F}_\alpha = \mathbf{g} + \frac{q_\alpha}{m_\alpha} \mathbf{E} + \frac{q_\alpha}{m_\alpha} \mathbf{v}_\alpha \times \mathbf{B}, \tag{3.4.18}$$

where $\mathbf{g}$ is the gravitational acceleration, q_α is the charge of the particle of the α species, $\mathbf{E}$ is the intensity of the electric field, and $\mathbf{B}$ is magnetic induction. In this case, $\mathbf{g}$, $\mathbf{E}$, and $\mathbf{B}$ can be considered as known functions. Then,

$$\mathbf{F}_\alpha \cdot \frac{\partial f_\alpha^{(0)}}{\partial \mathbf{v}_\alpha} = f_\alpha^{(0)} \frac{m_\alpha}{k_\mathrm{B}T} \left\{ \mathbf{g} + \frac{q_\alpha}{m_\alpha} \mathbf{E} + \frac{q_\alpha}{m_\alpha} \mathbf{v}_\alpha \times \mathbf{B} \right\} \cdot (\mathbf{v}_\alpha - \mathbf{v}_0), \tag{3.4.19}$$

$$\mathbf{F}_\alpha \cdot \frac{\partial f_\alpha^{(0)}}{\partial \mathbf{v}_\alpha} = f_\alpha^{(0)} \frac{m_\alpha}{k_\mathrm{B}T} \left[\mathbf{g} + \frac{q_\alpha}{m_\alpha} \mathbf{E} \right] \cdot \mathbf{V}_\alpha + f_\alpha^{(0)} \frac{m_\alpha}{k_\mathrm{B}T} \frac{q_\alpha}{m_\alpha} [\mathbf{v}_\alpha \times \mathbf{B}] \cdot \mathbf{V}_\alpha. \tag{3.4.20}$$

The scalar triple product can be written as follows:

$$(\mathbf{a} \times \mathbf{b}) \cdot \mathbf{c} = (\mathbf{b} \times \mathbf{c}) \cdot \mathbf{a} = (\mathbf{c} \times \mathbf{a}) \cdot \mathbf{b} = -(\mathbf{a} \times \mathbf{c}) \cdot \mathbf{b} = -(\mathbf{b} \times \mathbf{a}) \cdot \mathbf{c} = -(\mathbf{c} \times \mathbf{b}) \cdot \mathbf{a}, \tag{3.4.21}$$

then

$$\mathbf{F}_\alpha \cdot \frac{\partial f_\alpha^{(0)}}{\partial \mathbf{v}_\alpha} = f_\alpha^{(0)} \frac{m_\alpha}{k_\mathrm{B}T} \left[\mathbf{g} + \frac{q_\alpha}{m_\alpha} \mathbf{E} \right] \cdot \mathbf{V}_\alpha + f_\alpha^{(0)} \frac{m_\alpha}{k_\mathrm{B}T} \frac{q_\alpha}{m_\alpha} [\mathbf{v}_0 \times \mathbf{B}] \cdot \mathbf{V}_\alpha, \tag{3.4.22}$$

or

$$\mathbf{F}_\alpha \cdot \frac{\partial f_\alpha^{(0)}}{\partial \mathbf{v}_\alpha} = f_\alpha^{(0)} \frac{m_\alpha}{k_\mathrm{B}T} \left\{ \mathbf{g} + \frac{q_\alpha}{m_\alpha} \mathbf{E} \right\} \cdot \mathbf{V}_\alpha - f_\alpha^{(0)} \frac{m_\alpha}{k_\mathrm{B}T} \frac{q_\alpha}{m_\alpha} \{ \mathbf{V}_\alpha \times \mathbf{B} \} \cdot \mathbf{v}_0, \tag{3.4.23}$$

or

$$\mathbf{F}_\alpha \cdot \frac{\partial f_\alpha^{(0)}}{\partial \mathbf{v}_\alpha} = f_\alpha^{(0)} \frac{m_\alpha}{k_\mathrm{B}T} \left\{ \mathbf{g} + \frac{q_\alpha}{m_\alpha} \mathbf{E} \right\} \cdot \mathbf{V}_\alpha + f_\alpha^{(0)} \frac{m_\alpha}{k_\mathrm{B}T} \frac{q_\alpha}{m_\alpha} \{ \mathbf{V}_\alpha \times \mathbf{v}_0 \} \cdot \mathbf{B}. \tag{3.4.24}$$

From relations (3.4.15)–(3.4.17) follow

$$\begin{aligned} f_\alpha &= f_\alpha^{(0)} + \tau_\alpha \frac{\mathrm{D} f_\alpha^{(0)}}{\mathrm{D}t} \\ &= f_\alpha^{(0)} \left\{ 1 + \tau_\alpha \left[\begin{aligned} & \frac{\partial \ln n_\alpha}{\partial t} + \left(\frac{m_\alpha \mathbf{V}_\alpha^2}{2k_\mathrm{B}T} - \frac{3}{2} \right) \frac{\partial \ln T}{\partial t} + \frac{m_\alpha}{k_\mathrm{B}T} \frac{\partial \mathbf{v}_0}{\partial t} \cdot \mathbf{V}_\alpha \\ & + \mathbf{v}_\alpha \cdot \frac{\partial \ln n_\alpha}{\partial \mathbf{r}} + \left(\frac{m_\alpha \mathbf{V}_\alpha^2}{2k_\mathrm{B}T} - \frac{3}{2} \right) \mathbf{v}_\alpha \cdot \frac{\partial \ln T}{\partial \mathbf{r}} + \frac{m_\alpha}{k_\mathrm{B}T} \mathbf{v}_\alpha \cdot \sum_i V_{\alpha i} \frac{\partial v_{0i}}{\partial \mathbf{r}} \\ & + \frac{m_\alpha}{k_\mathrm{B}T} \left\{ \mathbf{g} + \frac{q_\alpha}{m_\alpha} \mathbf{E} \right\} \cdot \mathbf{V}_\alpha + \frac{m_\alpha}{k_\mathrm{B}T} \frac{q_\alpha}{m_\alpha} \{ \mathbf{V}_\alpha \times \mathbf{v}_0 \} \cdot \mathbf{B} \end{aligned} \right] \right\} \end{aligned} \tag{3.4.25}$$

It is reasonable to suppose that the derivatives of the logarithmic terms in Eq. (3.4.25) can be omitted as the small values. In this case,

$$f_\alpha = f_\alpha^{(0)} + \tau_\alpha \frac{\mathrm{D} f_\alpha^{(0)}}{\mathrm{D}t} = f_\alpha^{(0)} \left\{ 1 + \tau_\alpha \frac{m_\alpha}{k_\mathrm{B}T} \left[\begin{array}{l} \dfrac{\partial \mathbf{v_0}}{\partial t} \cdot \mathbf{V}_\alpha + (\mathbf{V}_\alpha + \mathbf{v_0}) \cdot \sum_i V_{\alpha i} \dfrac{\partial v_{0i}}{\partial \mathbf{r}} \\ + \left\{ \mathbf{g} + \dfrac{q_\alpha}{m_\alpha} \mathbf{E} \right\} \cdot \mathbf{V}_\alpha + \dfrac{q_\alpha}{m_\alpha} \{ \mathbf{V}_\alpha \times \mathbf{v_0} \} \cdot \mathbf{B} \end{array} \right] \right\}, \tag{3.4.26}$$

or

$$f_\alpha = f_\alpha^{(0)} \left\{ 1 + \tau_\alpha \frac{m_\alpha}{k_\mathrm{B}T} \left[\begin{array}{l} \mathbf{V}_\alpha \cdot \left(\dfrac{\partial \mathbf{v_0}}{\partial t} + \sum_{i=1}^{3} V_{\alpha i} \dfrac{\partial v_{0i}}{\partial \mathbf{r}} \right) + \mathbf{v_0} \cdot \sum_{i=1}^{3} V_{\alpha i} \dfrac{\partial v_{0i}}{\partial \mathbf{r}} \\ + \left\{ \mathbf{g} + \dfrac{q_\alpha}{m_\alpha} \mathbf{E} \right\} \cdot \mathbf{V}_\alpha + \dfrac{q_\alpha}{m_\alpha} \{ \mathbf{V}_\alpha \times \mathbf{v_0} \} \cdot \mathbf{B} \end{array} \right] \right\} \tag{3.4.27}$$

$$f_\alpha = f_\alpha^{(0)} \left\{ 1 + \tau_\alpha \frac{m_\alpha}{k_\mathrm{B}T} \left[\begin{array}{l} \mathbf{V}_\alpha \cdot \left(\dfrac{\partial \mathbf{v_0}}{\partial t} + \sum_{i=1}^{3} V_{\alpha i} \dfrac{\partial v_{0i}}{\partial \mathbf{r}} \right) + \mathbf{v_0} \cdot \sum_{i=1}^{3} V_{\alpha i} \dfrac{\partial v_{0i}}{\partial \mathbf{r}} \\ + \left(\mathbf{g} + \dfrac{q_\alpha}{m_\alpha} \mathbf{E} \right) \cdot \mathbf{V}_\alpha + \dfrac{q_\alpha}{m_\alpha} (\mathbf{B} \times \mathbf{V}_\alpha) \cdot \mathbf{v_0} \end{array} \right] \right\}. \tag{3.4.28}$$

Introducing the variance $\sigma_\alpha^2 = k_\mathrm{B}T/m_\alpha$ for species α, we obtain from Eq. (3.4.28)

$$f_\alpha = f_\alpha^{(0)} \left\{ 1 + \frac{\tau_\alpha}{\sigma_\alpha^2} \left[\begin{array}{l} \mathbf{V}_\alpha \cdot \left(\dfrac{\partial \mathbf{v_0}}{\partial t} + \sum_{i=1}^{3} V_{\alpha i} \dfrac{\partial v_{0i}}{\partial \mathbf{r}} \right) + \mathbf{v_0} \cdot \sum_{i=1}^{3} V_{\alpha i} \dfrac{\partial v_{0i}}{\partial \mathbf{r}} \\ + \left(\mathbf{g} + \dfrac{q_\alpha}{m_\alpha} \mathbf{E} \right) \cdot \mathbf{V}_\alpha + \dfrac{q_\alpha}{m_\alpha} (\mathbf{B} \times \mathbf{V}_\alpha) \cdot \mathbf{v_0} \end{array} \right] \right\}. \tag{3.4.29}$$

The DF (3.4.29) can be written in the form

$$f_\alpha = n_\alpha \left(\frac{m_\alpha}{2\pi k_\mathrm{B} T} \right)^{3/2} \mathrm{e}^{-\left(\frac{m_\alpha V_\alpha^2}{2 k_\mathrm{B} T} \right)} \left\{ 1 + \tau_\alpha \frac{m_\alpha}{k_\mathrm{B}T} \left[\begin{array}{l} \left(\dfrac{\partial}{\partial t} + \mathbf{V}_\alpha \cdot \dfrac{\partial}{\partial \mathbf{r}} + \mathbf{v_0} \cdot \dfrac{\partial}{\partial \mathbf{r}} \right) (\mathbf{V}_\alpha \cdot \mathbf{v_0}) \\ + \mathbf{g} \cdot \mathbf{V}_\alpha + \dfrac{q_\alpha}{m_\alpha} \mathbf{E} \cdot \mathbf{V}_\alpha + \dfrac{q_\alpha}{m_\alpha} (\mathbf{B} \times \mathbf{V}_\alpha) \cdot \mathbf{v_0} \end{array} \right] \right\}. \tag{3.4.30}$$

Then non-local physics forecasts the effect of the fluctuation appearance (non-local noncasual macroscopic fluctuations) which should be taking into account by the data processing. In relations like (3.4.30), $\mathbf{V}_\alpha$ is the velocity relative $\mathbf{v}_0$, and $\mathbf{v}_0$ is not the set velocity relative the observer—it is the effective velocity of the physical system movement in an inertial coordinate system. *Obviously, the coordinate system associated with the Earth surface does not belong to the inertial systems.*

Let us give some explanations for the mentioned Shnoll's data using the developed theory.

1. The "effect of near zone" means that *similar histograms are significantly more probable* to appear in the nearby (neighboring) intervals of the time series of the results of measurements. This effect is connected with the continuous dependence of the external parameters (like $\mathbf{v}_0$) on independent variables. In particular, this effect manifests itself via the term $\mathbf{V}_\alpha \cdot \left(\frac{\partial \mathbf{v_0}}{\partial t} + \sum_{i=1}^{3} V_{\alpha i} \frac{\partial v_{0i}}{\partial \mathbf{r}} \right)$ in Eq. (3.4.29).
2. *Similarity* (of the pattern of histograms plotted from the results of simultaneous independent measurements of processes of different nature at the same geographical point) follows from the structure of the relation (3.4.29).
3. *Regular changes in the histogram patterns* including near-daily periods; approximately 27-day periods and early periods, follows from the structure (3.4.29) and Fig. 3.2 (taken from [97]) which explains the periodical velocity dependence $\mathbf{v}_0$ on the Earth rotation and revolution around a moving Sun. Skewed lines on Fig. 3.2 corresponds to the movement direction of Sun in Milky Way.
4. The effects of *synchronism* observed at different latitudes and effects of the mirror similarity can be explained by the follow way.

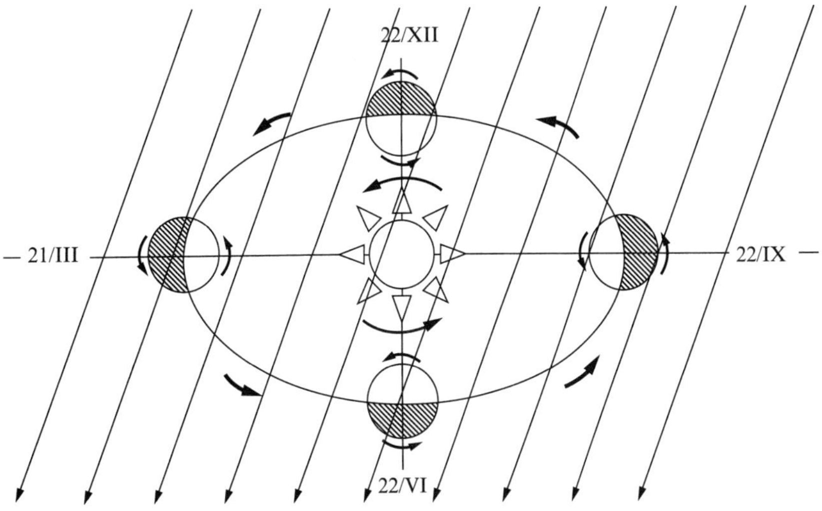

FIGURE 3.2 Earth rotation and revolution around a moving Sun. (See monograph [97].)

Let us introduce the character linear velocity of the Earth rotation $\mathbf{v}_0^{c}$ and the velocity $\mathbf{v}_0^{or}$ of the revolution around a Sun, $\mathbf{v}_0^{or} \gg \mathbf{v}_0^{c}$. As we see from Fig. 3.2, two cases can be observed related to $\mathbf{v}_0 = \mathbf{v}_0^{or} + \mathbf{v}_0^{c}$ and $\mathbf{v}_0 = \mathbf{v}_0^{or} - \mathbf{v}_0^{c}$. For the first case, we find from

$$f_\alpha = f_\alpha^{(0)}\left\{1 + \frac{\tau_\alpha}{\sigma_\alpha^2}\left[\begin{array}{l} \mathbf{V}_\alpha\cdot\left(\dfrac{\partial \mathbf{v}_0^{or}}{\partial t} + \displaystyle\sum_{i=1}^{3} V_{\alpha i}\frac{\partial v_{oi}^{or}}{\partial \mathbf{r}}\right) + \mathbf{V}_\alpha\cdot\left(\dfrac{\partial \mathbf{v}_0^{c}}{\partial t} + \displaystyle\sum_{i=1}^{3} V_{\alpha i}\frac{\partial v_{0i}^{c}}{\partial \mathbf{r}}\right) \\ + \mathbf{v}_0^{or}\cdot\displaystyle\sum_{i=1}^{3} V_{\alpha i}\frac{\partial v_{oi}^{or}}{\partial \mathbf{r}} + \mathbf{v}_0^{or}\cdot\displaystyle\sum_{i=1}^{3} V_{\alpha i}\frac{\partial v_{0i}^{c}}{\partial \mathbf{r}} \\ + \mathbf{v}_0^{c}\cdot\displaystyle\sum_{i=1}^{3} V_{\alpha i}\frac{\partial v_{0i}^{or}}{\partial \mathbf{r}} + \mathbf{v}_0^{c}\cdot\displaystyle\sum_{i=1}^{3} V_{\alpha i}\frac{\partial v_{0i}^{c}}{\partial \mathbf{r}} + \left(\mathbf{g} + \dfrac{q_\alpha}{m_\alpha}\mathbf{E}\right)\cdot\mathbf{V}_\alpha \\ + \dfrac{q_\alpha}{m_\alpha}(\mathbf{B}\times\mathbf{V}_\alpha)\cdot\left(\mathbf{v}_0^{or} + \mathbf{v}_0^{c}\right) \end{array}\right]\right\} \tag{3.4.31}$$

or

$$f_\alpha = f_\alpha^{(0)}\left\{1 + \frac{\tau_\alpha}{\sigma_\alpha^2}\left[\begin{array}{l} \mathbf{V}_\alpha\cdot\left(\dfrac{\partial \mathbf{v}_0^{or}}{\partial t} + \displaystyle\sum_{i=1}^{3} V_{\alpha i}\frac{\partial v_{oi}^{or}}{\partial \mathbf{r}}\right) - \mathbf{V}_\alpha\cdot\left(\dfrac{\partial \mathbf{v}_0^{c}}{\partial t} + \displaystyle\sum_{i=1}^{3} V_{\alpha i}\frac{\partial v_{0i}^{c}}{\partial \mathbf{r}}\right) \\ + \mathbf{v}_0^{or}\cdot\displaystyle\sum_{i=1}^{3} V_{\alpha i}\frac{\partial v_{oi}^{or}}{\partial \mathbf{r}} - \mathbf{v}_0^{or}\cdot\displaystyle\sum_{i=1}^{3} V_{\alpha i}\frac{\partial v_{0i}^{c}}{\partial \mathbf{r}} \\ - \mathbf{v}_0^{c}\cdot\displaystyle\sum_{i=1}^{3} V_{\alpha i}\frac{\partial v_{0i}^{or}}{\partial \mathbf{r}} + \mathbf{v}_0^{c}\cdot\displaystyle\sum_{i=1}^{3} V_{\alpha i}\frac{\partial v_{0i}^{c}}{\partial \mathbf{r}} + \left(\mathbf{g} + \dfrac{q_\alpha}{m_\alpha}\mathbf{E}\right)\cdot\mathbf{V}_\alpha \\ + \dfrac{q_\alpha}{m_\alpha}(\mathbf{B}\times\mathbf{V}_\alpha)\cdot\left(\mathbf{v}_0^{or} - \mathbf{v}_0^{c}\right) \end{array}\right]\right\} \tag{3.4.32}$$

if the velocities $\mathbf{v}_0^{or}$ and $\mathbf{v}_0^{c}$ are directed in the opposite directions. As a result, we have the effects of the mirror similarity. Disappearance of the near-daily periods close to the North Pole follows from Eqs. (3.4.31) and (3.4.32) because on the North Pole $\mathbf{v}_0^{c} = 0$.

5. In the analysis of histograms plotted from the results of counting α-particles that were traveling North (in the direction of the polestar), the near-daily periods were not observed, nor was the near-zone effect.

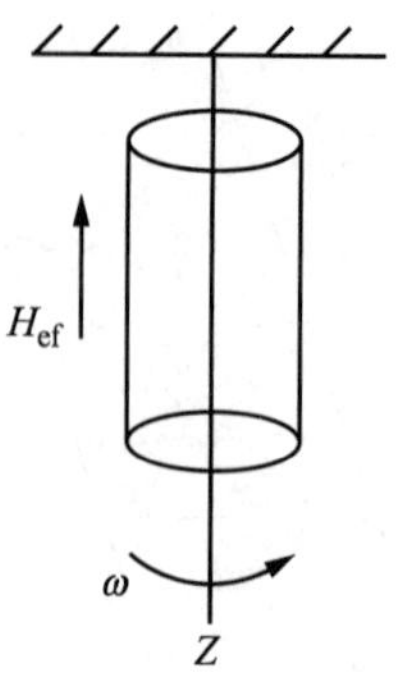

FIGURE 3.3 Scheme of the Barnett effect.

The polestar direction coincides practically with the north direction. It means that in the relation

$$f_\alpha = f_\alpha^{(0)}\left\{1 + \frac{\tau_\alpha}{\sigma_\alpha^2}\left[\mathbf{V}_\alpha \cdot \left(\frac{\partial \mathbf{v}_0}{\partial t} + \sum_{i=1}^{3} V_{\alpha i}\frac{\partial \mathbf{v}_{0i}}{\partial \mathbf{r}}\right) + \mathbf{v}_0 \cdot \sum_{i=1}^{3} V_{\alpha i}\frac{\partial \mathbf{v}_{0i}}{\partial \mathbf{r}} + \left(\mathbf{g} + \frac{q_\alpha}{m_\alpha}\mathbf{E}\right)\cdot \mathbf{V}_\alpha + \frac{q_\alpha}{m_\alpha}(\mathbf{B}\times\mathbf{V}_\alpha)\cdot\mathbf{v}_0\right]\right\}$$

the last term should be omitted

$$\frac{q_\alpha}{m_\alpha}\{\mathbf{B}\times\mathbf{V}_\alpha\}\cdot \mathbf{v}_0 = 0, \tag{3.4.33}$$

because $\mathbf{V}_\alpha \parallel \mathbf{B}$. As a result, the Earth rotation practically does not influence on the measurements.

6. An interesting effect is observed during measurements of the alpha particles activity of ^{239}Pu near the rotating centrifuge, (see [97], p. 352). The rotating centrifuge influences on the histogram form. The explanation consists in the influence of the term $(q_\alpha/m_\alpha)\{\mathbf{B}\times\mathbf{V}_\alpha\}\cdot\mathbf{v}_0$ as a result of variation of the magnetic induction because of the Barnett effect in ferromagnetic (see Fig. 3.3).

The difference of the relative amplitude of fluctuations in the processes of the different physical nature defines of the τ_α change (see Eq. 3.4.29). In the local theory, the non-local parameter $\tau_\alpha = 0$, then conclusion can be formulated in the form of the proved theorem:

"Shnoll's effect cannot be explained in the frame of local physics. Shnoll's data are the direct experimental evidence of the non-local effect existence on the level of the DF observations."

Chapter 4

Application of Unified Non-Local Theory to the Calculation of the Electron and Proton Inner Structures

ABSTRACT

The proton and electron inner structures are considered in the frame of unified non-local theory. From calculations follow that proton and electron can be considered like charged balls (shortly CB-model) in which charges are concentrated mainly in the shell of these balls. The proton-electron collision in the frame of the CB-model should be considered as *collision of two resonators*. Application of the CB – model allows to explain a number of character effects observed by inelastic electron – proton collisions (including the initial and final electron energies and the scattering angles) without application of the quark theory.

Keywords: Electron and proton inner structures, Quantum hydrodynamics

4.1 GENERALIZED QUANTUM HYDRODYNAMIC EQUATIONS

Strict consideration leads to the following system of generalized hydrodynamic equations (GHE) written in the generalized Euler form:

continuity equation for species α

$$\frac{\partial}{\partial t}\left\{\rho_\alpha - \tau_\alpha\left[\frac{\partial \rho_\alpha}{\partial t} + \frac{\partial}{\partial \mathbf{r}}\cdot(\rho_\alpha \mathbf{v}_0)\right]\right\} + \frac{\partial}{\partial \mathbf{r}}\cdot\left\{\rho_\alpha \mathbf{v}_0 - \tau_\alpha\left[\frac{\partial}{\partial t}(\rho_\alpha \mathbf{v}_0) + \frac{\partial}{\partial \mathbf{r}}\cdot(\rho_\alpha \mathbf{v}_0\mathbf{v}_0) + \overleftrightarrow{I}\cdot\frac{\partial p_\alpha}{\partial \mathbf{r}} - \rho_\alpha \mathbf{F}_\alpha^{(1)} - \frac{q_\alpha}{m_\alpha}\rho_\alpha \mathbf{v}_0 \times \mathbf{B}\right]\right\} = R_\alpha, \tag{4.1.1}$$

and continuity equation for mixture

$$\frac{\partial}{\partial t}\left\{\rho - \sum_\alpha \tau_\alpha\left[\frac{\partial \rho_\alpha}{\partial t} + \frac{\partial}{\partial \mathbf{r}}\cdot(\rho_\alpha \mathbf{v}_0)\right]\right\} + \frac{\partial}{\partial \mathbf{r}}\cdot\left\{\rho \mathbf{v}_0 - \sum_\alpha \tau_\alpha\left[\frac{\partial}{\partial t}(\rho_\alpha \mathbf{v}_0) + \frac{\partial}{\partial \mathbf{r}}\cdot(\rho_\alpha \mathbf{v}_0\mathbf{v}_0) + \overleftrightarrow{I}\cdot\frac{\partial p_\alpha}{\partial \mathbf{r}} - \rho_\alpha \mathbf{F}_\alpha^{(1)} - \frac{q_\alpha}{m_\alpha}\rho_\alpha \mathbf{v}_0 \times \mathbf{B}\right]\right\} = 0. \tag{4.1.2}$$

Momentum equation for species

$$\begin{aligned}
&\frac{\partial}{\partial t}\left\{\rho_\alpha \mathbf{v}_0 - \tau_\alpha\left[\frac{\partial}{\partial t}(\rho_\alpha \mathbf{v}_0) + \frac{\partial}{\partial \mathbf{r}}\cdot\rho_\alpha \mathbf{v}_0\mathbf{v}_0 + \frac{\partial p_\alpha}{\partial \mathbf{r}} - \rho_\alpha \mathbf{F}_\alpha^{(1)}\right.\right.\\
&\left.\left. - \frac{q_\alpha}{m_\alpha}\rho_\alpha \mathbf{v}_0 \times \mathbf{B}\right]\right\} - \mathbf{F}_\alpha^{(1)}\left[\rho_\alpha - \tau_\alpha\left(\frac{\partial \rho_\alpha}{\partial t} + \frac{\partial}{\partial \mathbf{r}}(\rho_\alpha \mathbf{v}_0)\right)\right]\\
&- \frac{q_\alpha}{m_\alpha}\left\{\rho_\alpha \mathbf{v}_0 - \tau_\alpha\left[\frac{\partial}{\partial t}(\rho_\alpha \mathbf{v}_0) + \frac{\partial}{\partial \mathbf{r}}\cdot\rho_\alpha \mathbf{v}_0\mathbf{v}_0 + \frac{\partial p_\alpha}{\partial \mathbf{r}} - \rho_\alpha \mathbf{F}_\alpha^{(1)}\right.\right.\\
&\left.\left. - \frac{q_\alpha}{m_\alpha}\rho_\alpha \mathbf{v}_0 \times \mathbf{B}\right]\right\} \times \mathbf{B} + \frac{\partial}{\partial \mathbf{r}}\cdot\left\{\rho_\alpha \mathbf{v}_0\mathbf{v}_0 + p_\alpha \overleftrightarrow{I} - \tau_\alpha\left[\frac{\partial}{\partial t}\left(\rho_\alpha \mathbf{v}_0\mathbf{v}_0 + p_\alpha \overleftrightarrow{I}\right)\right.\right.\\
&+ \frac{\partial}{\partial \mathbf{r}}\cdot\rho_\alpha(\mathbf{v}_0\mathbf{v}_0)\mathbf{v}_0 + 2\overleftrightarrow{I}\left(\frac{\partial}{\partial \mathbf{r}}\cdot(p_\alpha \mathbf{v}_0)\right) + \frac{\partial}{\partial \mathbf{r}}\cdot\left(\overleftrightarrow{I}p_\alpha \mathbf{v}_0\right)
\end{aligned}$$

Unified non-local theory of transport processes. http://dx.doi.org/10.1016/B978-0-444-63478-8.00004-X

$$
-\mathbf{F}_\alpha^{(1)}\rho_\alpha\mathbf{v}_0-\rho_\alpha\mathbf{v}_0\mathbf{F}_\alpha^{(1)}-\frac{q_\alpha}{m_\alpha}\rho_\alpha[\mathbf{v}_0\times\mathbf{B}]\mathbf{v}_0-\frac{q_\alpha}{m_\alpha}\rho_\alpha\mathbf{v}_0[\mathbf{v}_0\times\mathbf{B}]\Big]\Big\}
$$

$$
=\int m_\alpha\mathbf{v}_\alpha J_\alpha^{\mathrm{st,el}}d\mathbf{v}_\alpha+\int m_\alpha\mathbf{v}_\alpha J_\alpha^{\mathrm{st,inel}}d\mathbf{v}_\alpha. \tag{4.1.3}
$$

Generalized moment equation for mixture

$$
\begin{aligned}
&\frac{\partial}{\partial t}\left\{\rho\mathbf{v}_0-\sum_\alpha\tau_\alpha\left[\frac{\partial}{\partial t}(\rho_\alpha\mathbf{v}_0)+\frac{\partial}{\partial\mathbf{r}}\cdot\rho_\alpha\mathbf{v}_0\mathbf{v}_0+\frac{\partial p_\alpha}{\partial\mathbf{r}}-\rho_\alpha\mathbf{F}_\alpha^{(1)}\right.\right.\\
&\left.\left.-\frac{q_\alpha}{m_\alpha}\rho_\alpha\mathbf{v}_0\times\mathbf{B}\right]\right\}-\sum_\alpha\mathbf{F}_\alpha^{(1)}\left[\rho_\alpha-\tau_\alpha\left(\frac{\partial\rho_\alpha}{\partial t}+\frac{\partial}{\partial\mathbf{r}}(\rho_\alpha\mathbf{v}_0)\right)\right]\\
&-\sum_\alpha\frac{q_\alpha}{m_\alpha}\left\{\rho_\alpha\mathbf{v}_0-\tau_\alpha^{(0)}\left[\frac{\partial}{\partial t}(\rho_\alpha\mathbf{v}_0)+\frac{\partial}{\partial\mathbf{r}}\cdot\rho_\alpha\mathbf{v}_0\mathbf{v}_0+\frac{\partial p_\alpha}{\partial\mathbf{r}}-\rho_\alpha\mathbf{F}_\alpha^{(1)}\right.\right.\\
&\left.\left.-\frac{q_\alpha}{m_\alpha}\rho_\alpha\mathbf{v}_0\times\mathbf{B}\right]\right\}\times\mathbf{B}+\frac{\partial}{\partial\mathbf{r}}\cdot\left\{\rho\mathbf{v}_0\mathbf{v}_0+p\overleftrightarrow{\mathrm{I}}-\sum_\alpha\tau_\alpha\left[\frac{\partial}{\partial t}\left(\rho_\alpha\mathbf{v}_0\mathbf{v}_0+p_\alpha\overleftrightarrow{\mathrm{I}}\right)\right.\right.\\
&+\frac{\partial}{\partial\mathbf{r}}\cdot\rho_\alpha(\mathbf{v}_0\mathbf{v}_0)\mathbf{v}_0+2\overleftrightarrow{\mathrm{I}}\left(\frac{\partial}{\partial\mathbf{r}}\cdot(p_\alpha\mathbf{v}_0)\right)+\frac{\partial}{\partial\mathbf{r}}\cdot\left(\overleftrightarrow{\mathrm{I}}p_\alpha\mathbf{v}_0\right)\\
&\left.\left.-\mathbf{F}_\alpha^{(1)}\rho_\alpha\mathbf{v}_0-\rho_\alpha\mathbf{v}_0\mathbf{F}_\alpha^{(1)}-\frac{q_\alpha}{m_\alpha}\rho_\alpha[\mathbf{v}_0\times\mathbf{B}]\mathbf{v}_0-\frac{q_\alpha}{m_\alpha}\rho_\alpha\mathbf{v}_0[\mathbf{v}_0\times\mathbf{B}]\right]\right\}=0.
\end{aligned}\tag{4.1.4}
$$

Energy equation for component

$$
\begin{aligned}
&\frac{\partial}{\partial t}\left\{\frac{\rho_\alpha v_0^2}{2}+\frac{3}{2}p_\alpha+\varepsilon_\alpha n_\alpha-\tau_\alpha\left[\frac{\partial}{\partial t}\left(\frac{\rho_\alpha v_0^2}{2}+\frac{3}{2}p_\alpha+\varepsilon_\alpha n_\alpha\right)\right.\right.\\
&\left.\left.+\frac{\partial}{\partial\mathbf{r}}\cdot\left(\frac{1}{2}\rho_\alpha v_0^2\mathbf{v}_0+\frac{5}{2}p_\alpha\mathbf{v}_0+\varepsilon_\alpha n_\alpha\mathbf{v}_0\right)-\mathbf{F}_\alpha^{(1)}\cdot\rho_\alpha\mathbf{v}_0\right]\right\}\\
&+\frac{\partial}{\partial\mathbf{r}}\cdot\left\{\frac{1}{2}\rho_\alpha v_0^2\mathbf{v}_0+\frac{5}{2}p_\alpha\mathbf{v}_0+\varepsilon_\alpha n_\alpha\mathbf{v}_0-\tau_\alpha\left[\frac{\partial}{\partial t}\left(\frac{1}{2}\rho_\alpha v_0^2\mathbf{v}_0\right.\right.\right.\\
&\left.+\frac{5}{2}p_\alpha\mathbf{v}_0+\varepsilon_\alpha n_\alpha\mathbf{v}_0\right)+\frac{\partial}{\partial\mathbf{r}}\cdot\left(\frac{1}{2}\rho_\alpha v_0^2\mathbf{v}_0\mathbf{v}_0+\frac{7}{2}p_\alpha\mathbf{v}_0\mathbf{v}_0+\frac{1}{2}p_\alpha v_0^2\overleftrightarrow{\mathrm{I}}\right.\\
&\left.+\frac{5}{2}\frac{p_\alpha^2}{\rho_\alpha}\overleftrightarrow{\mathrm{I}}+\varepsilon_\alpha n_\alpha\mathbf{v}_0\mathbf{v}_0+\varepsilon_\alpha\frac{p_\alpha}{m_\alpha}\overleftrightarrow{\mathrm{I}}\right)-\rho_\alpha\mathbf{F}_\alpha^{(1)}\cdot\mathbf{v}_0\mathbf{v}_0-p_\alpha\mathbf{F}_\alpha^{(1)}\cdot\overleftrightarrow{\mathrm{I}}\\
&-\frac{1}{2}\rho_\alpha v_0^2\mathbf{F}_\alpha^{(1)}-\frac{3}{2}\mathbf{F}_\alpha^{(1)}p_\alpha-\frac{\rho_\alpha v_0^2}{2}\frac{q_\alpha}{m_\alpha}[\mathbf{v}_0\times\mathbf{B}]-\frac{5}{2}p_\alpha\frac{q_\alpha}{m_\alpha}[\mathbf{v}_0\times\mathbf{B}]\\
&\left.\left.-\varepsilon_\alpha n_\alpha\frac{q_\alpha}{m_\alpha}[\mathbf{v}_0\times\mathbf{B}]-\varepsilon_\alpha n_\alpha\mathbf{F}_\alpha^{(1)}\right]\right\}-\rho_\alpha\mathbf{F}_\alpha^{(1)}\cdot\mathbf{v}_0\\
&+\tau_\alpha\mathbf{F}_\alpha^{(1)}\cdot\left(\frac{\partial}{\partial t}(\rho_\alpha\mathbf{v}_0)+\frac{\partial}{\partial\mathbf{r}}\cdot\rho_\alpha\mathbf{v}_0\mathbf{v}_0+\frac{\partial}{\partial\mathbf{r}}\cdot p_\alpha\overleftrightarrow{\mathrm{I}}-\rho_\alpha\mathbf{F}_\alpha^{(1)}-q_\alpha n_\alpha[\mathbf{v}_0\times\mathbf{B}]\right)\\
&=\int\left(\frac{m_\alpha v_\alpha^2}{2}+\varepsilon_\alpha\right)J_\alpha^{\mathrm{st,el}}d\mathbf{v}_\alpha+\int\left(\frac{m_\alpha v_\alpha^2}{2}+\varepsilon_\alpha\right)J_\alpha^{\mathrm{st,inel}}d\mathbf{v}_\alpha.
\end{aligned}\tag{4.1.5}
$$

and after summation the generalized energy equation for mixture

$$
\begin{aligned}
&\frac{\partial}{\partial t}\left\{\frac{\rho v_0^2}{2}+\frac{3}{2}p+\sum_\alpha \varepsilon_\alpha n_\alpha-\sum_\alpha \tau_\alpha\left[\frac{\partial}{\partial t}\left(\frac{\rho_\alpha v_0^2}{2}+\frac{3}{2}p_\alpha+\varepsilon_\alpha n_\alpha\right)\right.\right.\\
&\left.\left.+\frac{\partial}{\partial \mathbf{r}}\cdot\left(\frac{1}{2}\rho_\alpha v_0^2\mathbf{v}_0+\frac{5}{2}p_\alpha\mathbf{v}_0+\varepsilon_\alpha n_\alpha\mathbf{v}_0\right)-\mathbf{F}_\alpha^{(1)}\cdot\rho_\alpha\mathbf{v}_0\right]\right\}\\
&+\frac{\partial}{\partial \mathbf{r}}\cdot\left\{\frac{1}{2}\rho v_0^2\mathbf{v}_0+\frac{5}{2}p\mathbf{v}_0+\mathbf{v}_0\sum_\alpha\varepsilon_\alpha n_\alpha-\sum_\alpha\tau_\alpha\left[\frac{\partial}{\partial t}\left(\frac{1}{2}\rho_\alpha v_0^2\mathbf{v}_0\right.\right.\right.\\
&\left.+\frac{5}{2}p_\alpha\mathbf{v}_0+\varepsilon_\alpha n_\alpha\mathbf{v}_0\right)+\frac{\partial}{\partial \mathbf{r}}\cdot\left(\frac{1}{2}\rho_\alpha v_0^2\mathbf{v}_0\mathbf{v}_0+\frac{7}{2}p_\alpha\mathbf{v}_0\mathbf{v}_0+\frac{1}{2}p_\alpha v_0^2\overleftrightarrow{\mathrm{I}}\right.\\
&\left.+\frac{5p_\alpha^2}{2\rho_\alpha}\overleftrightarrow{\mathrm{I}}+\varepsilon_\alpha n_\alpha\mathbf{v}_0\mathbf{v}_0+\varepsilon_\alpha\frac{p_\alpha}{m_\alpha}\overleftrightarrow{\mathrm{I}}\right)-\rho_\alpha\mathbf{F}_\alpha^{(1)}\cdot\mathbf{v}_0\mathbf{v}_0-p_\alpha\mathbf{F}_\alpha^{(1)}\cdot\overleftrightarrow{\mathrm{I}}\\
&-\frac{1}{2}\rho_\alpha v_0^2\mathbf{F}_\alpha^{(1)}-\frac{3}{2}\mathbf{F}_\alpha^{(1)}p_\alpha-\frac{\rho_\alpha v_0^2}{2}\frac{q_\alpha}{m_\alpha}[\mathbf{v}_0\times\mathbf{B}]-\frac{5}{2}p_\alpha\frac{q_\alpha}{m_\alpha}[\mathbf{v}_0\times\mathbf{B}]\\
&\left.\left.-\varepsilon_\alpha n_\alpha\frac{q_\alpha}{m_\alpha}[\mathbf{v}_0\times\mathbf{B}]-\varepsilon_\alpha n_\alpha\mathbf{F}_\alpha^{(1)}\right]\right\}-\mathbf{v}_0\cdot\sum_\alpha\rho_\alpha\mathbf{F}_\alpha^{(1)}\\
&+\sum_\alpha\tau_\alpha\mathbf{F}_\alpha^{(1)}\cdot\left[\frac{\partial}{\partial t}(\rho_\alpha\mathbf{v}_0)+\frac{\partial}{\partial \mathbf{r}}\cdot\rho_\alpha\mathbf{v}_0\mathbf{v}_0+\frac{\partial}{\partial \mathbf{r}}\cdot p_\alpha\overleftrightarrow{\mathrm{I}}-\rho_\alpha\mathbf{F}_\alpha^{(1)}-q_\alpha n_\alpha[\mathbf{v}_0\times\mathbf{B}]\right]=0.
\end{aligned}
\tag{4.1.6}
$$

Here, $\mathbf{F}_\alpha^{(1)}$ are the forces of the non-magnetic origin, $\mathbf{B}$—magnetic induction, $\overleftrightarrow{\mathrm{I}}$—unit tensor, q_α—charge of the α component particle, p_α—static pressure for α species, ε_α—internal energy for the particles of α component, $\mathbf{v}_0$—hydrodynamic velocity for mixture.

Some remarks to the system of Eqs. (4.1.1)–(4.1.6):

(1) For calculations in the self-consistent electromagnetic field, the system of non-local Maxwell equations should be added (see (I.50), (I.51)).

(2) The system of the generalized quantum hydrodynamic equations contains energy equation written for an unknown dependent value, which can be specified as quantum pressure p_α of non-local origin.

(3) In chemical reaction systems, the internal energies ε_α define the reaction's heat Q. For example, for bimolecular reaction $A_a+A_b\rightarrow A_c+A_d$, the reaction heat $Q=\varepsilon_c+\varepsilon_d-\varepsilon_a-\varepsilon_b$.

(4) For so-called "elementary particles," the internal energy can contain the spin and magnetic parts. For example, electron has the internal energy ε

$$\varepsilon_e=\varepsilon_{el,sp}+\varepsilon_{el,m}, \tag{4.1.7}$$

with the spin and magnetic parts, namely

$$\varepsilon_{el,sp}=\hbar\omega/2,\quad \varepsilon_{el,m}=-\mathbf{p_m}\cdot\mathbf{B}; \tag{4.1.8}$$

$\mathbf{p_m}$—electron magnetic moment, $\mathbf{B}$—magnetic induction. But, $\mathbf{p_m}=-\frac{e}{m_e}\frac{\hbar}{2c}$, then $\varepsilon_e=\frac{\hbar}{2}\omega_{eff}$.

The effective frequencies ω_{eff} can be altered in the process of the interaction with the surrounding environment. In this case, additional equations defining the change of internal energies should be added to Eqs. (4.1.1)–(4.1.6).

Let us consider this situation in detail beginning with the case when the particle internal energy is constant.

After dividing both the sides of the continuity Eq. (4.1.1) by m_α and multiplying by ε_α, this equation takes the form

$$\frac{\partial}{\partial t}\left\{\varepsilon_\alpha n_\alpha - \tau_\alpha\left[\frac{\partial \varepsilon_\alpha n_\alpha}{\partial t} + \frac{\partial}{\partial \mathbf{r}}\cdot(\varepsilon_\alpha n_\alpha \mathbf{v}_0)\right]\right\}$$
$$+\frac{\partial}{\partial \mathbf{r}}\cdot\left\{\varepsilon_\alpha n_\alpha \mathbf{v}_0 - \tau_\alpha\left[\frac{\partial}{\partial t}(\varepsilon_\alpha n_\alpha \mathbf{v}_0) + \frac{\partial}{\partial \mathbf{r}}\cdot(\varepsilon_\alpha n_\alpha \mathbf{v}_0\mathbf{v}_0)\right.\right. \tag{4.1.9}$$
$$\left.\left.+\frac{1}{m_\alpha}\varepsilon_\alpha \overleftrightarrow{I}\cdot\frac{\partial p_\alpha}{\partial \mathbf{r}} - \varepsilon_\alpha n_\alpha \mathbf{F}_\alpha^{(1)} - \frac{q_\alpha}{m_\alpha}\varepsilon_\alpha n_\alpha \mathbf{v}_0\times\mathbf{B}\right]\right\} = \frac{1}{m_\alpha}\varepsilon_\alpha R_\alpha.$$

In the general case, if $\varepsilon_\alpha \neq \text{const}$, Eq. (4.1.9) is the internal energy equation in which the right hand side of equation $\frac{1}{m_\alpha}\varepsilon_\alpha R_\alpha$ transforms into function $E_\alpha(\varepsilon_\alpha)$. After subtraction of the both sides of Eq. (4.1.9) from the corresponding parts of Eq. (4.1.5), one obtains

$$\frac{\partial}{\partial t}\left\{\frac{\rho_\alpha v_0^2}{2} + \frac{3}{2}p_\alpha - \tau_\alpha\left[\frac{\partial}{\partial t}\left(\frac{\rho_\alpha v_0^2}{2} + \frac{3}{2}p_\alpha\right) + \frac{\partial}{\partial \mathbf{r}}\cdot\left(\frac{1}{2}\rho_\alpha v_0^2\mathbf{v}_0 + \frac{5}{2}p_\alpha \mathbf{v}_0\right) - \mathbf{F}_\alpha^{(1)}\cdot\rho_\alpha \mathbf{v}_0\right]\right\}$$
$$+\frac{\partial}{\partial \mathbf{r}}\cdot\left\{\frac{1}{2}\rho_\alpha v_0^2\mathbf{v}_0 + \frac{5}{2}p_\alpha \mathbf{v}_0 - \tau_\alpha\left[\frac{\partial}{\partial t}\left(\frac{1}{2}\rho_\alpha v_0^2\mathbf{v}_0 + \frac{5}{2}p_\alpha \mathbf{v}_0\right)\right.\right.$$
$$+\frac{\partial}{\partial \mathbf{r}}\cdot\left(\frac{1}{2}\rho_\alpha v_0^2\mathbf{v}_0\mathbf{v}_0 + \frac{7}{2}p_\alpha \mathbf{v}_0\mathbf{v}_0 + \frac{1}{2}p_\alpha v_0^2\overleftrightarrow{I} + \frac{5}{2}p_\alpha \mathbf{v}_0\right)$$
$$+\frac{\partial}{\partial \mathbf{r}}\cdot\left(\frac{1}{2}\rho_\alpha v_0^2\mathbf{v}_0\mathbf{v}_0 + \frac{7}{2}p_\alpha \mathbf{v}_0\mathbf{v}_0 + \frac{1}{2}p_\alpha v_0^2\overleftrightarrow{I} + \frac{5p_\alpha^2}{2\rho_\alpha}\overleftrightarrow{I}\right) - \rho_\alpha \mathbf{F}_\alpha^{(1)}\cdot\mathbf{v}_0\mathbf{v}_0 - p_\alpha \mathbf{F}_\alpha^{(1)}\cdot\overleftrightarrow{I} \tag{4.1.10}$$
$$\left.\left.-\frac{1}{2}\rho_\alpha v_0^2\mathbf{F}_\alpha^{(1)} - \frac{3}{2}\mathbf{F}_\alpha^{(1)}p_\alpha - \frac{\rho_\alpha v_0^2}{2}\frac{q_\alpha}{m_\alpha}[\mathbf{v}_0\times\mathbf{B}] - \frac{5}{2}p_\alpha\frac{q_\alpha}{m_\alpha}[\mathbf{v}_0\times\mathbf{B}]\right]\right\} - \rho_\alpha \mathbf{F}_\alpha^{(1)}\cdot\mathbf{v}_0$$
$$+\tau_\alpha \mathbf{F}_\alpha^{(1)}\cdot\left(\frac{\partial}{\partial t}(\rho_\alpha \mathbf{v}_0) + \frac{\partial}{\partial \mathbf{r}}\cdot\rho_\alpha \mathbf{v}_0\mathbf{v}_0 + \frac{\partial}{\partial \mathbf{r}}\cdot p_\alpha\overleftrightarrow{I} - \rho_\alpha \mathbf{F}_\alpha^{(1)} - q_\alpha n_\alpha[\mathbf{v}_0\times\mathbf{B}]\right)$$
$$= \int\frac{m_\alpha v_\alpha^2}{2}J_\alpha^{\mathrm{st,el}}d\mathbf{v}_\alpha + \int\frac{m_\alpha v_\alpha^2}{2}J_\alpha^{\mathrm{st,inel}}d\mathbf{v}_\alpha.$$

taking into account that

$$\int\varepsilon_\alpha J_\alpha^{\mathrm{st,el}}d\mathbf{v}_\alpha + \int\varepsilon_\alpha J_\alpha^{\mathrm{st,inel}}d\mathbf{v}_\alpha = \varepsilon_\alpha\frac{R_\alpha}{m_\alpha}. \tag{4.1.11}$$

Conclusion: In general, the change of the species internal energies takes place as result of the interaction of the physical system with environment. If the change of the species internal energies is absent, the solution of the full system of equations (4.1.1)–(4.1.6) can be reduced to the system (4.1.1)–(4.1.5), (4.1.10).

4.2 THE CHARGE INTERNAL STRUCTURE OF ELECTRON

Let us consider a negative charged physical system placed in a bounded region of a space. Internal energy ε_α of this one species object and a possible influence of the magnetic field are taken into account. The character linear scale of this region will be defined as the result of the self-consistent solution of the generalized non-local quantum hydrodynamic Eqs. (4.1.1)–(4.1.6). Let us suppose also that the mentioned physical object for simplicity has the spherical form and therefore the system (4.1.1)–(4.1.6) is reasonable to write in the spherical coordinate system (see Appendix 2). Remark also that the terms ρg_r, ρg_θ, and ρg_φ correspond to the components of the mass forces acting on the unit of volume. For example, for the potential forces of the electrical origin, $\rho g_r = m_e n g_r \to -m_e n\frac{eE}{m_e} = -neE = q\frac{\partial\psi}{\partial r}$. It also means that in the following, q is the absolute value of the negative charge per unit of volume.

We have:
Non-local continuity equation

$$
\begin{aligned}
&\frac{\partial}{\partial t}\left\{\rho-\tau\left[\frac{\partial \rho}{\partial t}+\frac{1}{r^2}\frac{\partial(r^2\rho v_{0r})}{\partial r}+\frac{1}{r\sin\theta}\frac{\partial(\rho v_{0\varphi})}{\partial\varphi}+\frac{1}{r\sin\theta}\frac{\partial(\rho v_{0\theta}\sin\theta)}{\partial\theta}\right]\right\}\\
&+\frac{1}{r^2}\frac{\partial}{\partial r}\left\{r^2\left\{\rho v_{0r}-\tau\left[\frac{\partial}{\partial t}(\rho v_{0r})+\frac{1}{r^2}\frac{\partial(r^2\rho v_{0r}^2)}{\partial r}+\frac{1}{r\sin\theta}\frac{\partial(\rho v_{0\varphi}v_{0r})}{\partial\varphi}\right.\right.\right.\\
&\left.\left.\left.+\frac{1}{r\sin\theta}\frac{\partial(\rho v_{0\theta}v_{0r}\sin\theta)}{\partial\theta}-\rho g_r-\frac{q}{m}\rho\left(v_{0\varphi}B_\theta-v_{0\theta}B_\varphi\right)\right]\right\}\right\}+\frac{1}{r\sin\theta}\frac{\partial}{\partial\varphi}\left\{\rho v_{0\varphi}-\tau\left[\frac{\partial}{\partial t}(\rho v_{0\varphi})\right.\right.\\
&\left.\left.+\frac{1}{r^2}\frac{\partial(r^2\rho v_{0r}v_{0\varphi})}{\partial r}+\frac{1}{r\sin\theta}\frac{\partial\left(\rho v_{0\varphi}^2\right)}{\partial\varphi}+\frac{1}{r\sin\theta}\frac{\partial(\rho v_{0\theta}v_{0\varphi}\sin\theta)}{\partial\theta}-\rho g_\varphi-\frac{q}{m}\rho\left(v_{0\theta}B_r-v_{0r}B_\theta\right)\right]\right\}\\
&+\frac{1}{r\sin\theta}\frac{\partial}{\partial\theta}\left\{\sin\theta\left\{\rho v_{0\theta}-\tau\left[\frac{\partial}{\partial t}(\rho v_{0\theta})+\frac{1}{r^2}\frac{\partial(r^2\rho v_{0r}v_{0\theta})}{\partial r}+\frac{1}{r\sin\theta}\frac{\partial(\rho v_{0\varphi}v_{0\theta})}{\partial\varphi}\right.\right.\right.\\
&\left.\left.\left.+\frac{1}{r\sin\theta}\frac{\partial(\rho v_{0\theta}^2\sin\theta)}{\partial\theta}-\rho g_\theta-\frac{q}{m}\rho\left(v_{0r}B_\varphi-v_{0\varphi}B_r\right)\right]\right\}\right\}\\
&-\frac{1}{r^2}\frac{\partial}{\partial r}\left(\tau r^2\frac{\partial p}{\partial r}\right)-\frac{1}{r^2\sin\theta}\frac{\partial}{\partial\theta}\left(\tau\sin\theta\frac{\partial p}{\partial\theta}\right)-\frac{1}{r^2\sin^2\theta}\frac{\partial}{\partial\varphi}\left(\tau\frac{\partial p}{\partial\varphi}\right)=0.
\end{aligned}
\tag{4.2.1}
$$

Non-local momentum equation ($\mathbf{e}_r$ projection)

$$
\begin{aligned}
&\frac{\partial}{\partial t}\left\{\rho v_{0r}-\tau\left[\frac{\partial}{\partial t}(\rho v_{0r})+\frac{1}{r^2}\frac{\partial(r^2\rho v_{0r}^2)}{\partial r}+\frac{1}{r\sin\theta}\frac{\partial(\rho v_{0\varphi}v_{0r})}{\partial\varphi}+\frac{1}{r\sin\theta}\frac{\partial(\rho v_{0\theta}v_{0r}\sin\theta)}{\partial\theta}\right.\right.\\
&\left.\left.+\frac{\partial p}{\partial r}-\rho g_r-\frac{q}{m}\rho\left(v_{0\varphi}B_\theta-v_{0\theta}B_\varphi\right)\right]\right\}\\
&-g_r\left[\rho-\tau\left(\frac{\partial\rho}{\partial t}+\frac{1}{r^2}\frac{\partial(r^2\rho v_{0r})}{\partial r}+\frac{1}{r\sin\theta}\frac{\partial(\rho v_{0\phi})}{\partial\varphi}+\frac{1}{r\sin\theta}\frac{\partial(\rho v_{0\theta}\sin\theta)}{\partial\theta}\right)\right]\\
&-\frac{q}{m}\left(\rho v_{0\varphi}-\tau\left[\frac{\partial}{\partial t}(\rho v_{0\varphi})+\frac{1}{r^2}\frac{\partial(r^2\rho v_{0r}v_{0\varphi})}{\partial r}+\frac{1}{r\sin\theta}\frac{\partial\left(\rho v_{0\varphi}^2\right)}{\partial\varphi}+\frac{1}{r\sin\theta}\frac{\partial(\rho v_{0\theta}v_{0\varphi}\sin\theta)}{\partial\theta}\right.\right.\\
&\left.\left.+\frac{1}{r\sin\theta}\frac{\partial p}{\partial\varphi}-\rho g_\varphi-\frac{q}{m}\rho(v_{0\theta}B_r-v_{0r}B_\theta)\right]\right)B_\theta\\
&+\frac{q}{m}\left(\rho v_{0\theta}-\tau\left[\frac{\partial}{\partial t}(\rho v_{0\theta})+\frac{1}{r^2}\frac{\partial(r^2\rho v_{0r}v_{0\theta})}{\partial r}+\frac{1}{r\sin\theta}\frac{\partial(\rho v_{0\varphi}v_{0\theta})}{\partial\varphi}+\frac{1}{r\sin\theta}\frac{\partial(\rho v_{0\theta}^2\sin\theta)}{\partial\theta}\right.\right.\\
&\left.\left.+\frac{1}{r}\frac{\partial p}{\partial\theta}-\rho g_\theta-\frac{q}{m}\rho\left(v_{0r}B_\varphi-v_{0\varphi}B_r\right)\right]\right)B_\varphi\\
&+\frac{1}{r^2}\frac{\partial}{\partial r}\left\{r^2\left\{\rho v_{0r}^2-\tau\left[\frac{\partial}{\partial t}(\rho v_{0r}^2)+\frac{1}{r^2}\frac{\partial(r^2\rho v_{0r}^3)}{\partial r}+\frac{1}{r\sin\theta}\frac{\partial(\rho v_{0\varphi}v_{0r}^2)}{\partial\varphi}\right.\right.\right.\\
&\left.\left.\left.+\frac{1}{r\sin\theta}\frac{\partial(\rho v_{0\theta}v_{0r}^2\sin\theta)}{\partial\theta}-2g_r\rho v_{0r}-2\frac{q}{m}\rho\left(v_{0\varphi}B_\theta-v_{0\theta}B_\varphi\right)\right]\right\}\right\}\\
&+\frac{1}{r\sin\theta}\frac{\partial}{\partial\varphi}\left\{\rho v_{0\varphi}v_{0r}-\tau\left[\frac{\partial}{\partial t}(\rho v_{0\varphi}v_{0r})+\frac{1}{r^2}\frac{\partial(r^2\rho v_{0\varphi}v_{0r}^2)}{\partial r}+\frac{1}{r\sin\theta}\frac{\partial\left(\rho v_{0\varphi}^2v_{0r}\right)}{\partial\varphi}\right.\right.\\
&+\frac{1}{r\sin\theta}\frac{\partial(\rho v_{0\theta}v_{0\varphi}v_{0r}\sin\theta)}{\partial\theta}-g_\varphi\rho v_{0r}-\frac{q}{m}\rho(v_{0\theta}B_r-v_{0r}B_\theta)v_{0r}\\
&\left.\left.-\frac{q}{m}\rho v_{0\varphi}\left(v_{0\varphi}B_\theta-v_{0\theta}B_\varphi\right)-v_{0\varphi}\rho g_r\right]\right\}\\
&+\frac{1}{r\sin\theta}\frac{\partial}{\partial\theta}\left\{\sin\theta\left\{\rho v_{0\theta}v_{0r}-\tau\left[\frac{\partial}{\partial t}(\rho v_{0\theta}v_{0r})+\frac{1}{r^2}\frac{\partial(r^2\rho v_{0\theta}v_{0r}^2)}{\partial r}+\frac{1}{r\sin\theta}\frac{\partial(\rho v_{0\varphi}v_{0\theta}v_{0r})}{\partial\varphi}\right.\right.\right.\\
&+\frac{1}{r\sin\theta}\frac{\partial(\rho v_{0\theta}^2v_{0r}\sin\theta)}{\partial\theta}-g_\theta\rho v_{0r}-\frac{q}{m}\rho\left(v_{0r}B_\varphi-v_{0\varphi}B_r\right)v_{0r}\\
&\left.\left.\left.-v_{0\theta}\rho g_r-\frac{q}{m}\rho\left(v_{0\varphi}B_\theta-v_{0\theta}B_\varphi\right)v_{0\theta}\right]\right\}\right\}\\
&+\frac{\partial p}{\partial r}-\frac{\partial}{\partial r}\left(\tau\frac{\partial p}{\partial t}\right)-2\frac{\partial}{\partial r}\left(\tau\left(\frac{1}{r^2}\frac{\partial(r^2pv_{0r})}{\partial r}+\frac{1}{r\sin\theta}\frac{\partial(pv_{0\phi})}{\partial\varphi}+\frac{1}{r\sin\theta}\frac{\partial(pv_{0\theta}\sin\theta)}{\partial\theta}\right)\right)\\
&-\frac{1}{r^2}\frac{\partial}{\partial r}\left(\tau r^2\frac{\partial(pv_{0r})}{\partial r}\right)-\frac{1}{r^2\sin\theta}\frac{\partial}{\partial\theta}\left(\tau\sin\theta\frac{\partial(pv_{0r})}{\partial\theta}\right)-\frac{1}{r^2\sin^2\theta}\frac{\partial}{\partial\varphi}\left(\tau\frac{\partial(pv_{0r})}{\partial\varphi}\right)=0.
\end{aligned}
\tag{4.2.2}
$$

Non-local momentum equation ($\mathbf{e}_\varphi$ projection)

$$
\begin{aligned}
&\frac{\partial}{\partial t}\left\{\rho v_{0\varphi}-\tau\left[\frac{\partial}{\partial t}(\rho v_{0\varphi})+\frac{1}{r^2}\frac{\partial(r^2\rho v_{0r}v_{0\varphi})}{\partial r}+\frac{1}{r\sin\theta}\frac{\partial\left(\rho v_{0\phi}^2\right)}{\partial\varphi}+\frac{1}{r\sin\theta}\frac{\partial(\rho v_{0\theta}v_{0\varphi}\sin\theta)}{\partial\theta}\right.\right.\\
&\left.\left.+\frac{1}{r\sin\theta}\frac{\partial p}{\partial\varphi}-\rho g_\varphi-\frac{q}{m}\rho(v_{0\theta}B_r-v_{0r}B_\theta)\right]\right\}\\
&-g_\varphi\left[\rho-\tau\left(\frac{\partial\rho}{\partial t}+\frac{1}{r^2}\frac{\partial(r^2\rho v_{0r})}{\partial r}+\frac{1}{r\sin\theta}\frac{\partial(\rho v_{0\varphi})}{\partial\varphi}+\frac{1}{r\sin\theta}\frac{\partial(\rho v_{0\theta}\sin\theta)}{\partial\theta}\right)\right]\\
&-\frac{q}{m}\left(\rho v_{0\theta}-\tau\left[\frac{\partial}{\partial t}(\rho v_{0\theta})+\frac{1}{r^2}\frac{\partial(r^2\rho v_{0r}v_{0\theta})}{\partial r}+\frac{1}{r\sin\theta}\frac{\partial(\rho v_{0\varphi}v_{0\theta})}{\partial\varphi}+\frac{1}{r\sin\theta}\frac{\partial(\rho v_{0\theta}^2\sin\theta)}{\partial\theta}\right.\right.\\
&\left.\left.+\frac{1}{r}\frac{\partial p}{\partial\theta}-\rho g_\theta-\frac{q}{m}\rho\left(v_{0r}B_\varphi-v_{0\varphi}B_r\right)\right]\right)B_r\\
&+\frac{q}{m}\left(\rho v_{0r}-\tau\left[\frac{\partial}{\partial t}(\rho v_{0r})+\frac{1}{r^2}\frac{\partial(r^2\rho v_{0r}^2)}{\partial r}+\frac{1}{r\sin\theta}\frac{\partial(\rho v_{0\varphi}v_{0r})}{\partial\varphi}+\frac{1}{r\sin\theta}\frac{\partial(\rho v_{0\theta}v_{0r}\sin\theta)}{\partial\theta}\right.\right.\\
&\left.\left.+\frac{\partial p}{\partial r}-\rho g_r-\frac{q}{m}\rho\left(v_{0\varphi}B_\theta-v_{0\theta}B_\varphi\right)\right]\right)B_\theta+\frac{1}{r^2}\frac{\partial}{\partial r}\left\{r^2\left\{\rho v_{0r}v_{0\varphi}-\tau\left[\frac{\partial}{\partial t}(\rho v_{0r}v_{0\varphi})\right.\right.\right.\\
&+\frac{1}{r^2}\frac{\partial(r^2\rho v_{0r}^2v_{0\varphi})}{\partial r}+\frac{1}{r\sin\theta}\frac{\partial\left(\rho v_{0\varphi}^2v_{0r}\right)}{\partial\varphi}+\frac{1}{r\sin\theta}\frac{\partial(\rho v_{0\theta}v_{0r}v_{0\varphi}\sin\theta)}{\partial\theta}\\
&\left.\left.\left.-g_r\rho v_{0\varphi}-v_{0r}\rho g_\varphi-\frac{q}{m}\rho(v_{0\varphi}B_\theta-v_{0\theta}B_\varphi)v_{0\varphi}-v_{0r}\frac{q}{m}\rho(v_{0\theta}B_r-v_{0r}B_\theta)\right]\right\}\right\}\\
&+\frac{1}{r\sin\theta}\frac{\partial}{\partial\varphi}\left\{\rho v_{0\varphi}^2-\tau\left[\frac{\partial}{\partial t}\left(\rho v_{0\varphi}^2\right)+\frac{1}{r^2}\frac{\partial\left(r^2\rho v_{0r}v_{0\varphi}^2\right)}{\partial r}+\frac{1}{r\sin\theta}\frac{\partial\left(\rho v_{0\varphi}^3\right)}{\partial\varphi}+\frac{1}{r\sin\theta}\frac{\partial\left(\rho v_{0\theta}v_{0\varphi}^2\sin\theta\right)}{\partial\theta}\right.\right.\\
&\left.\left.-2g_\varphi\rho v_{0\varphi}-2\frac{q}{m}\rho(v_{0\theta}B_r-v_{0r}B_\theta)v_{0\varphi}\right]\right\}\\
&+\frac{1}{r\sin\theta}\frac{\partial}{\partial\theta}\left\{\sin\theta\left\{\rho v_{0\theta}v_{0\varphi}-\tau\left[\frac{\partial}{\partial t}(\rho v_{0\theta}v_{0\varphi})+\frac{1}{r^2}\frac{\partial(r^2\rho v_{0r}v_{0\theta}v_{0\varphi})}{\partial r}+\frac{1}{r\sin\theta}\frac{\partial\left(\rho v_{0\varphi}^2v_{0\theta}\right)}{\partial\varphi}\right.\right.\right.\\
&+\frac{1}{r\sin\theta}\frac{\partial(\rho v_{0\theta}^2v_{0\varphi}\sin\theta)}{\partial\theta}-g_\theta\rho v_{0\varphi}-\frac{q}{m}\rho\left(v_{0r}B_\varphi-v_{0\varphi}B_r\right)v_{0\varphi}\\
&\left.\left.\left.-v_{0\theta}\rho g_\varphi-\frac{q}{m}\rho(v_{0\theta}B_r-v_{0r}B_\theta)v_{0\theta}\right]\right\}\right\}+\frac{1}{r\sin\theta}\frac{\partial p}{\partial\varphi}-\frac{1}{r\sin\theta}\frac{\partial}{\partial\varphi}\left(\tau\frac{\partial p}{\partial t}\right)\\
&-\frac{2}{r\sin\theta}\frac{\partial}{\partial\varphi}\left(\tau\left(\frac{1}{r^2}\frac{\partial(r^2pv_{0r})}{\partial r}+\frac{1}{r\sin\theta}\frac{\partial(pv_{0\varphi})}{\partial\varphi}+\frac{1}{r\sin\theta}\frac{\partial(pv_{0\theta}\sin\theta)}{\partial\theta}\right)\right)\\
&-\frac{1}{r^2}\frac{\partial}{\partial r}\left(\tau r^2\frac{\partial(pv_{0\varphi})}{\partial r}\right)-\frac{1}{r^2\sin\theta}\frac{\partial}{\partial\theta}\left(\tau\sin\theta\frac{\partial(pv_{0\varphi})}{\partial\theta}\right)-\frac{1}{r^2\sin^2\theta}\frac{\partial}{\partial\varphi}\left(\tau\frac{\partial(pv_{0\varphi})}{\partial\varphi}\right)=0.
\end{aligned}
\tag{4.2.3}
$$

Non-local momentum equation ($\mathbf{e}_\theta$ projection)

$$
\begin{aligned}
&\frac{\partial}{\partial t}\Big\{\rho v_{0\theta}-\tau\Big[\frac{\partial}{\partial t}(\rho v_{0\theta})+\frac{1}{r^2}\frac{\partial(r^2\rho v_{0r}v_{0\theta})}{\partial r}+\frac{1}{r\sin\theta}\frac{\partial(\rho v_{0\varphi}v_{0\theta})}{\partial\varphi}+\frac{1}{r\sin\theta}\frac{\partial(\rho v_{0\theta}^2\sin\theta)}{\partial\theta}\\
&+\frac{1}{r}\frac{\partial p}{\partial\theta}-\rho g_\theta-\frac{q}{m}\rho\left(v_{0r}B_\varphi-v_{0\varphi}B_r\right)\Big]\Big\}\\
&-g_\theta\Big[\rho-\tau\Big(\frac{\partial\rho}{\partial t}+\frac{1}{r^2}\frac{\partial(r^2\rho v_{0r})}{\partial r}+\frac{1}{r\sin\theta}\frac{\partial(\rho v_{0\varphi})}{\partial\varphi}+\frac{1}{r\sin\theta}\frac{\partial(\rho v_{0\theta}\sin\theta)}{\partial\theta}\Big)\Big]\\
&-\frac{q}{m}\Big(\rho v_{0r}-\tau\Big[\frac{\partial}{\partial t}(\rho v_{0r})+\frac{1}{r^2}\frac{\partial(r^2\rho v_{0r}^2)}{\partial r}+\frac{1}{r\sin\theta}\frac{\partial(\rho v_{0\varphi}v_{0r})}{\partial\varphi}+\frac{1}{r\sin\theta}\frac{\partial(\rho v_{0\theta}v_{0r}\sin\theta)}{\partial\theta}\\
&+\frac{\partial p}{\partial r}-\rho g_r-\frac{q}{m}\rho\left(v_{0\varphi}B_\theta-v_{0\theta}B_\varphi\right)\Big]\Big)B_\varphi\\
&+\frac{q}{m}\Big(\rho v_{0\varphi}-\tau\Big[\frac{\partial}{\partial t}(\rho v_{0\varphi})+\frac{1}{r^2}\frac{\partial(r^2\rho v_{0r}v_{0\varphi})}{\partial r}+\frac{1}{r\sin\theta}\frac{\partial\left(\rho v_{0\varphi}^2\right)}{\partial\varphi}+\frac{1}{r\sin\theta}\frac{\partial(\rho v_{0\theta}v_{0\varphi}\sin\theta)}{\partial\theta}\\
&+\frac{1}{r\sin\theta}\frac{\partial p}{\partial\varphi}-\rho g_\varphi-\frac{q}{m}\rho(v_{0\theta}B_r-v_{0r}B_\theta)\Big]\Big)B_r+\frac{1}{r^2}\frac{\partial}{\partial r}\Big\{r^2\Big\{\rho v_{0r}v_{0\theta}-\tau\Big[\frac{\partial}{\partial t}(\rho v_{0r}v_{0\theta})\\
&+\frac{1}{r^2}\frac{\partial(r^2\rho v_{0r}^2v_{0\theta})}{\partial r}+\frac{1}{r\sin\theta}\frac{\partial(\rho v_{0\varphi}v_{0r}v_{0\theta})}{\partial\varphi}+\frac{1}{r\sin\theta}\frac{\partial(\rho v_{0\theta}^2v_{0r}\sin\theta)}{\partial\theta}\\
&-g_r\rho v_{0\theta}-v_{0r}\rho g_\theta-\frac{q}{m}\rho\left(v_{0\varphi}B_\theta-v_{0\theta}B_\varphi\right)v_{0\theta}-v_{0r}\frac{q}{m}\rho\left(v_{0r}B_\varphi-v_{0\varphi}B_r\right)\Big]\Big\}\Big\}\\
&+\frac{1}{r\sin\theta}\frac{\partial}{\partial\varphi}\Big\{\rho v_{0\varphi}v_{0\theta}-\tau\Big[\frac{\partial}{\partial t}(\rho v_{0\varphi}v_{0\theta})+\frac{1}{r^2}\frac{\partial(r^2\rho v_{0r}v_{0\varphi}v_{0\theta})}{\partial r}+\frac{1}{r\sin\theta}\frac{\partial\left(\rho v_{0\varphi}^2v_{0\theta}\right)}{\partial\varphi}\\
&+\frac{1}{r\sin\theta}\frac{\partial(\rho v_{0\theta}^2v_{0\varphi}\sin\theta)}{\partial\theta}-g_\varphi\rho v_{0\theta}-\frac{q}{m}\rho(v_{0\theta}B_r-v_{0r}B_\theta)v_{0\theta}-v_{0\varphi}\frac{q}{m}\rho\left(v_{0r}B_\varphi-v_{0\varphi}B_r\right)-v_{0\varphi}\rho g_\theta\Big]\Big\}\\
&+\frac{1}{r\sin\theta}\frac{\partial}{\partial\theta}\Big\{\sin\theta\Big\{\rho v_{0\theta}^2-\tau\Big[\frac{\partial}{\partial t}(\rho v_{0\theta}^2)+\frac{1}{r^2}\frac{\partial(r^2\rho v_{0r}v_{0\theta}^2)}{\partial r}+\frac{1}{r\sin\theta}\frac{\partial(\rho v_{0\varphi}v_{0\theta}^2)}{\partial\varphi}\\
&+\frac{1}{r\sin\theta}\frac{\partial(\rho v_{0\theta}^3\sin\theta)}{\partial\theta}-2g_\theta\rho v_{0\theta}-\frac{q}{m}\rho\left(v_{0r}B_\varphi-v_{0\varphi}B_r\right)v_{0\theta}\Big]\Big\}\Big\}+\frac{1}{r}\frac{\partial p}{\partial\theta}-\frac{1}{r}\frac{\partial}{\partial\theta}\left(\tau\frac{\partial p}{\partial t}\right)\\
&-\frac{2}{r}\frac{\partial}{\partial\theta}\left(\tau\left(\frac{1}{r^2}\frac{\partial(r^2pv_{0r})}{\partial r}+\frac{1}{r\sin\theta}\frac{\partial(pv_{0\varphi})}{\partial\varphi}+\frac{1}{r\sin\theta}\frac{\partial(pv_{0\theta}\sin\theta)}{\partial\theta}\right)\right)\\
&-\frac{1}{r^2}\frac{\partial}{\partial r}\left(\tau r^2\frac{\partial(pv_{0\theta})}{\partial r}\right)-\frac{1}{r^2\sin\theta}\frac{\partial}{\partial\theta}\left(\tau\sin\theta\frac{\partial(pv_{0\theta})}{\partial\theta}\right)-\frac{1}{r^2\sin^2\theta}\frac{\partial}{\partial\varphi}\left(\tau\frac{\partial(pv_{0\theta})}{\partial\varphi}\right)=0.
\end{aligned}
\tag{4.2.4}
$$

Energy equation

$$
\begin{aligned}
&\frac{\partial}{\partial t}\bigg\{\frac{1}{2}\rho v_0^2+\varepsilon n+\frac{3}{2}p-\tau\bigg[\frac{\partial}{\partial t}\left(\frac{1}{2}\rho v_0^2+\varepsilon n+\frac{3}{2}p\right)+\frac{1}{r^2}\frac{\partial}{\partial r}\bigg(r^2 v_{0r}\left(\frac{1}{2}\rho v_0^2+\varepsilon n+\frac{5}{2}p\right)\bigg)\\
&+\frac{1}{r\sin\theta}\frac{\partial}{\partial\varphi}\bigg(v_{0\varphi}\left(\frac{1}{2}\rho v_0^2+\varepsilon n+\frac{5}{2}p\right)\bigg)+\frac{1}{r\sin\theta}\frac{\partial}{\partial\theta}\bigg(\sin\theta\, v_{0\theta}\left(\frac{1}{2}\rho v_0^2+\varepsilon n+\frac{5}{2}p\right)\bigg)\\
&-\rho\left(g_r v_{0r}+g_\varphi v_{0\varphi}+g_\theta v_{0\theta}\right)\bigg]\bigg\}\\
&+\frac{1}{r^2}\frac{\partial}{\partial r}\bigg\{r^2\bigg\{\left(\frac{1}{2}\rho v_0^2+\varepsilon n+\frac{5}{2}p\right)v_{0r}-\tau\bigg[\frac{\partial}{\partial t}\bigg(\left(\frac{1}{2}\rho v_0^2+\varepsilon n+\frac{5}{2}p\right)v_{0r}\bigg)\\
&+\frac{1}{r^2}\frac{\partial}{\partial r}\bigg(r^2\left(\frac{1}{2}\rho v_0^2+\varepsilon n+\frac{7}{2}p\right)v_{0r}^2\bigg)+\frac{1}{r\sin\theta}\frac{\partial}{\partial\varphi}\bigg(\left(\frac{1}{2}\rho v_0^2+\varepsilon n+\frac{7}{2}p\right)v_{0\varphi}v_{0r}\bigg)\\
&+\frac{1}{r\sin\theta}\frac{\partial}{\partial\theta}\bigg(\sin\theta\left(\frac{1}{2}\rho v_0^2+\varepsilon n+\frac{7}{2}p\right)v_{0\theta}v_{0r}\bigg)\\
&-\rho\left(g_r v_{0r}+g_\varphi v_{0\varphi}+g_\theta v_{0\theta}\right)v_{0r}-\left(\frac{1}{2}\rho v_0^2+\varepsilon n+\frac{3}{2}p\right)g_r\\
&-\left(\frac{1}{2}\rho v_0^2+\varepsilon n+\frac{5}{2}p\right)\frac{q}{m}\left(v_{0\varphi}B_\theta-v_{0\theta}B_\varphi\right)\bigg]\bigg\}\bigg\}+\frac{1}{r\sin\theta}\frac{\partial}{\partial\varphi}\bigg\{\left(\frac{1}{2}\rho v_0^2+\varepsilon n+\frac{5}{2}p\right)v_{0\varphi}\\
&-\tau\bigg[\frac{\partial}{\partial t}\bigg(\left(\frac{1}{2}\rho v_0^2+\varepsilon n+\frac{5}{2}p\right)v_{0\varphi}+\frac{1}{r^2}\frac{\partial}{\partial r}\bigg(r^2\left(\frac{1}{2}\rho v_0^2+\varepsilon n+\frac{7}{2}p\right)v_{0r}v_{0\varphi}\bigg)\\
&+\frac{1}{r\sin\theta}\frac{\partial}{\partial\varphi}\bigg(\left(\frac{1}{2}\rho v_0^2+\varepsilon n+\frac{7}{2}p\right)v_{0\varphi}^2\bigg)+\frac{1}{r\sin\theta}\frac{\partial}{\partial\theta}\bigg(\sin\theta\left(\frac{1}{2}\rho v_0^2+\varepsilon n+\frac{7}{2}p\right)v_{0\theta}v_{0\varphi}\bigg)\\
&-\rho\left(g_r v_{0r}+g_\varphi v_{0\varphi}+g_\theta v_{0\theta}\right)v_{0\varphi}-\left(\frac{1}{2}\rho v_0^2+\varepsilon n+\frac{3}{2}p\right)g_\varphi-\left(\frac{1}{2}\rho v_0^2+\varepsilon n+\frac{5}{2}p\right)\frac{q}{m}\left(v_{0\theta}B_r-v_{0r}B_\theta\right)\bigg]\bigg\}\\
&+\frac{1}{r\sin\theta}\frac{\partial}{\partial\theta}\bigg\{\sin\theta\bigg\{\left(\frac{1}{2}\rho v_0^2+\varepsilon n+\frac{5}{2}p\right)v_{0\theta}-\tau\bigg[\frac{\partial}{\partial t}\bigg(\left(\frac{1}{2}\rho v_0^2+\varepsilon n+\frac{5}{2}p\right)v_{0\theta}\\
&+\frac{1}{r^2}\frac{\partial}{\partial r}\bigg(r^2\left(\frac{1}{2}\rho v_0^2+\varepsilon n+\frac{7}{2}p\right)v_{0r}v_{0\theta}\bigg)+\frac{1}{r\sin\theta}\frac{\partial}{\partial\varphi}\bigg(\left(\frac{1}{2}\rho v_0^2+\varepsilon n+\frac{7}{2}p\right)v_{0\varphi}v_{0\theta}\bigg)\\
&+\frac{1}{r\sin\theta}\frac{\partial}{\partial\theta}\bigg(\sin\theta\left(\frac{1}{2}\rho v_0^2+\varepsilon n+\frac{7}{2}p\right)v_{0\theta}^2\bigg)-\rho\left(g_r v_{0r}+g_\varphi v_{0\varphi}+g_\theta v_{0\theta}\right)v_{0\theta}-\left(\frac{1}{2}\rho v_0^2+\varepsilon n+\frac{3}{2}p\right)g_\theta\\
&-\left(\frac{1}{2}\rho v_0^2+\varepsilon n+\frac{5}{2}p\right)\frac{q}{m}\left(v_{0r}B_\varphi-v_{0\varphi}B_r\right)\bigg]\bigg\}\bigg\}-\bigg\{\rho\left(g_r v_{0r}+g_\phi v_{0\phi}+g_\theta v_{0\theta}\right)\\
&-\tau\bigg[g_r\bigg(\frac{\partial}{\partial t}(\rho v_{0r})+\frac{1}{r^2}\frac{\partial}{\partial r}(r^2\rho v_{0r}^2)+\frac{1}{r\sin\theta}\frac{\partial}{\partial\varphi}(\rho v_{0\varphi}v_{0r})\\
&+\frac{1}{r\sin\theta}\frac{\partial}{\partial\theta}(\rho v_{0\theta}v_{0r}\sin\theta)+\frac{\partial p}{\partial r}-\rho g_r-qn\left(v_{0\varphi}B_\theta-v_{0\theta}B_\varphi\right)\bigg)\\
&+g_\varphi\bigg(\frac{\partial}{\partial t}(\rho v_{0\varphi})+\frac{1}{r^2}\frac{\partial}{\partial r}(r^2\rho v_{0r}v_{0\varphi})+\frac{1}{r\sin\theta}\frac{\partial}{\partial\varphi}\left(\rho v_{0\varphi}^2\right)+\frac{1}{r\sin\theta}\frac{\partial}{\partial\theta}(\rho v_{0\theta}v_{0\varphi}\sin\theta)\\
&+\frac{1}{r\sin\theta}\frac{\partial p}{\partial\varphi}-\rho g_\varphi-qn(v_{0\theta}B_r-v_{0r}B_\theta)\bigg)+g_\theta\bigg(\frac{\partial}{\partial t}(\rho v_{0\theta})+\frac{1}{r^2}\frac{\partial}{\partial r}(r^2\rho v_{0r}v_{0\theta})+\frac{1}{r\sin\theta}\frac{\partial}{\partial\varphi}(\rho v_{0\varphi}v_{0\theta})\\
&+\frac{1}{r\sin\theta}\frac{\partial}{\partial\theta}(\rho v_{0\theta}^2\sin\theta)+\frac{1}{r}\frac{\partial p}{\partial\theta}-\rho g_\theta-qn\left(v_{0r}B_\varphi-v_{0\varphi}B_r\right)\bigg)\bigg]\bigg\}\\
&-\frac{1}{r^2}\frac{\partial}{\partial r}\bigg(\tau r^2\frac{\partial}{\partial r}\left(\frac{1}{2}pv_0^2+\varepsilon\frac{p}{m}+\frac{5}{2}\frac{p^2}{\rho}\right)\bigg)\\
&-\frac{1}{r^2\sin\theta}\frac{\partial}{\partial\theta}\bigg(\tau\sin\theta\frac{\partial}{\partial\theta}\left(\frac{1}{2}pv_0^2+\varepsilon\frac{p}{m}+\frac{5}{2}\frac{p^2}{\rho}\right)\bigg)-\frac{1}{r^2\sin^2\theta}\frac{\partial}{\partial\varphi}\bigg(\tau\frac{\partial}{\partial\varphi}\left(\frac{1}{2}pv_0^2+\varepsilon\frac{p}{m}+\frac{5}{2}\frac{p^2}{\rho}\right)\bigg)\\
&+\frac{1}{r^2}\frac{\partial}{\partial r}(r^2\tau p g_r)+\frac{1}{r\sin\theta}\frac{\partial}{\partial\varphi}\left(\tau p g_\varphi\right)+\frac{1}{r\sin\theta}\frac{\partial}{\partial\theta}(\tau p g_\theta\sin\theta)=0
\end{aligned}
\tag{4.2.5}
$$

Let us point out the particular non-stationary one-dimensional case corresponding to the negative charged system evolution in the potential electric field:

– continuity equation

$$\frac{\partial}{\partial t}\left\{\rho-\tau\left[\frac{\partial \rho}{\partial t}+\frac{1}{r^2}\frac{\partial(r^2\rho v_{0r})}{\partial r}\right]\right\}+\frac{1}{r^2}\frac{\partial}{\partial r}\left\{r^2\left\{\rho v_{0r}-\tau\left[\frac{\partial}{\partial t}(\rho v_{0r})\right.\right.\right.$$
$$\left.\left.\left.+\frac{1}{r^2}\frac{\partial(r^2\rho v_{0r}^2)}{\partial r}-q\frac{\partial \psi}{\partial r}\right]\right\}\right\}-\frac{1}{r^2}\frac{\partial}{\partial r}\left(\tau r^2\frac{\partial p}{\partial r}\right)=0, \tag{4.2.6}$$

– momentum equation

$$\frac{\partial}{\partial t}\left\{\rho v_{0r}-\tau\left[\frac{\partial}{\partial t}(\rho v_{0r})+\frac{1}{r^2}\frac{\partial(r^2\rho v_{0r}^2)}{\partial r}+\frac{\partial p}{\partial r}-q\frac{\partial \psi}{\partial r}\right]\right\}-\frac{q}{\rho}\frac{\partial \psi}{\partial r}\left[\rho-\tau\left(\frac{\partial \rho}{\partial t}+\frac{1}{r^2}\frac{\partial(r^2\rho v_{0r})}{\partial r}\right)\right]$$
$$+\frac{1}{r^2}\frac{\partial}{\partial r}\left\{r^2\left\{\rho v_{0r}^2-\tau\left[\frac{\partial}{\partial t}(\rho v_{0r}^2)+\frac{1}{r^2}\frac{\partial(r^2\rho v_{0r}^3)}{\partial r}-2q\frac{\partial \psi}{\partial r}v_{0r}\right]\right\}\right\}$$
$$+\frac{\partial p}{\partial r}-\frac{\partial}{\partial r}\left(\tau\frac{\partial p}{\partial t}\right)-2\frac{\partial}{\partial r}\left(\frac{\tau}{r^2}\frac{\partial(r^2 p v_{0r})}{\partial r}\right)-\frac{1}{r^2}\frac{\partial}{\partial r}\left(\tau r^2\frac{\partial(p v_{0r})}{\partial r}\right)=0, \tag{4.2.7}$$

– energy equation

$$\frac{\partial}{\partial t}\left\{\frac{1}{2}\rho v_{0r}^2+\frac{3}{2}p-\tau\left[\frac{\partial}{\partial t}\left(\frac{1}{2}\rho v_{0r}^2+\frac{3}{2}p\right)+\frac{1}{r^2}\frac{\partial}{\partial r}\left(r^2 v_{0r}\left(\frac{1}{2}\rho v_{0r}^2+\frac{5}{2}p\right)\right)-q\frac{\partial \psi}{\partial r}v_{0r}\right]\right\}$$
$$+\frac{1}{r^2}\frac{\partial}{\partial r}\left\{r^2\left\{\left(\frac{1}{2}\rho v_{0r}^2+\frac{5}{2}p\right)v_{0r}-\tau\left[\frac{\partial}{\partial t}\left(\left(\frac{1}{2}\rho v_{0r}^2+\frac{5}{2}p\right)v_{0r}+\frac{1}{r^2}\frac{\partial}{\partial r}\left(r^2\left(\frac{1}{2}\rho v_{0r}^2+\frac{7}{2}p\right)v_{0r}^2\right)\right.\right.\right.\right.$$
$$\left.\left.\left.\left.-q\frac{\partial \psi}{\partial r}v_{0r}^2-\frac{q}{\rho}\frac{\partial \psi}{\partial r}\left(\frac{1}{2}\rho v_{0r}^2+\frac{3}{2}p\right)\right)\right]\right\}\right\}-\left\{q\frac{\partial \psi}{\partial r}v_{0r}\right.$$
$$\left.-\tau\left[\frac{q}{\rho}\frac{\partial \psi}{\partial r}\left(\frac{\partial}{\partial t}(\rho v_{0r})+\frac{1}{r^2}\frac{\partial}{\partial r}(r^2\rho v_{0r}^2)+\frac{\partial p}{\partial r}-q\frac{\partial \psi}{\partial r}\right)\right]\right\}-\frac{1}{r^2}\frac{\partial}{\partial r}\left(\tau r^2\frac{\partial}{\partial r}\left(\frac{1}{2}p v_{0r}^2+\frac{5p^2}{2\rho}\right)\right)$$
$$+\frac{1}{r^2}\frac{\partial}{\partial r}\left(r^2\tau p\frac{q}{\rho}\frac{\partial \psi}{\partial r}\right)=0. \tag{4.2.8}$$

Assuming that non-stationary physical system is at rest, namely $v_{0r}=0$. Taking into account also the forces of the magnetic origin, one obtains from the system of Eqs. (4.2.1)–(4.2.5) for the non-stationary one-dimensional (1D; along r) case:

– continuity equation

$$\frac{\partial}{\partial t}\left\{\rho-\tau\frac{\partial \rho}{\partial t}\right\}+\frac{1}{r^2}\frac{\partial}{\partial r}\left[r^2\tau\left(q\frac{\partial \psi}{\partial r}-\frac{\partial p}{\partial r}\right)\right]=0, \tag{4.2.9}$$

– momentum equation, $\mathbf{e}_r$ projection

$$\frac{\partial}{\partial t}\left[\tau q\frac{\partial \psi}{\partial r}\right]-\frac{q}{\rho}\frac{\partial \psi}{\partial r}\left[\rho-\tau\frac{\partial \rho}{\partial t}\right]+\frac{\partial}{\partial r}\left[p-\tau\frac{\partial p}{\partial t}\right]=0, \tag{4.2.10}$$

– momentum equation, $\mathbf{e}_\varphi$ projection

$$\frac{q}{m}\tau\left[\frac{\partial p}{\partial r}-q\frac{\partial \psi}{\partial r}\right]B_\theta=0, \tag{4.2.11}$$

– momentum equation, $\mathbf{e}_\theta$ projection

$$\tau\frac{q}{m}\left[\frac{\partial p}{\partial r}-q\frac{\partial\psi}{\partial r}\right]B_\varphi=0. \tag{4.2.12}$$

– energy equation

$$\begin{aligned}&\frac{\partial}{\partial t}\left\{\varepsilon n+\frac{3}{2}p-\tau\frac{\partial}{\partial t}\left(\varepsilon n+\frac{3}{2}p\right)\right\}\\&=\frac{1}{r^2}\frac{\partial}{\partial r}\left(\tau r^2\frac{\partial}{\partial r}\left[\frac{p}{\rho}\left(\varepsilon n+\frac{5}{2}p\right)\right]\right)-\frac{1}{r^2}\frac{\partial}{\partial r}\left\{\tau r^2\left(\varepsilon n+\frac{5}{2}p\right)\frac{q}{\rho}\frac{\partial\psi}{\partial r}\right\}\\&-\tau\frac{q}{\rho}\frac{\partial\psi}{\partial r}\left(\frac{\partial p}{\partial r}-q\frac{\partial\psi}{\partial r}\right),\end{aligned} \tag{4.2.13}$$

where ε is the internal particle energy. To the system of Eqs. (4.2.9), (4.2.10), and (4.2.13) the Poisson equation should be added:

$$\frac{1}{r^2}\frac{\partial}{\partial r}\left(r^2\frac{\partial\psi}{\partial r}\right)=4\pi q, \tag{4.2.14}$$

where ψ is the scalar electric potential and q is the absolute value of the negative charge (per unit of volume) of the one species quantum object.

4.3 THE DERIVATION OF THE ANGLE RELAXATION EQUATION

Let us consider an electron which is at rest at the time moment $t=0$. This electron has the internal energy ε (see also (4.1.7), (4.1.8))

$$\varepsilon=\varepsilon_{\mathrm{el,sp}}+\varepsilon_{\mathrm{el,m}}, \tag{4.3.1}$$

containing the spin and magnetic parts, namely

$$\varepsilon_{\mathrm{el,sp}}=\hbar\omega/2,\quad \varepsilon_{\mathrm{el,m}}=-\mathbf{p_m}\cdot\mathbf{B}; \tag{4.3.2}$$

$\mathbf{p_m}$—electron magnetic moment, $\mathbf{B}$—magnetic induction. But, $p_{\mathrm{m}}=-\frac{e}{m_e}\frac{\hbar}{2c}$ and relation (4.3.1) can be written as

$$\varepsilon=\frac{\hbar}{2}\left[\omega+\frac{e}{m_e c}B\cos\vartheta\right], \tag{4.3.3}$$

where the angle ϑ reflects the possible deviation between a separated direction of the spin at the initial time moment and the direction of magnetic momentum after an external perturbation. For example, this perturbation can be considered as the result of the approach of the second electron to the previous one at the distance r_{in} with appearance of the virtual photon with the wavelength

$$\lambda_{\mathrm{ph}}=2\pi r_{\mathrm{in}}. \tag{4.3.4}$$

The fine-structure constant α has physical interpretations as the ratio of two energies:

(1) the energy E_c needed to overcome the electrostatic repulsion between two electrons with a distance of r_{in} apart, and
(2) the energy of a single photon of wavelength $\lambda_{\mathrm{ph}}=2\pi r_{\mathrm{in}}$.

Taking into account the previous remarks, let us consider the charge time evolution inside of the first electron. In principle, we need to solve the general complicated system (4.2.1)–(4.2.5). It is reasonable to a obtain much more simple solution using the perturbation method. Namely, all unknown functions can be expanded in a Taylor series like

$$\rho = \rho_0 + \left[\frac{\partial \rho}{\partial t}\right]_{t=t_0} \delta t + \cdots \tag{4.3.5}$$

In particular, we need to find the time derivation of the value $\varepsilon_{\mathrm{el,m}} = -\mathbf{p_m} \cdot \mathbf{B}$ and therefore the derivative with signs reflecting two possible projection orientations $\pm \frac{\partial}{\partial t} \cos \vartheta = \mp \sin \vartheta \frac{\partial \vartheta}{\partial t}$. The derivative $\frac{\partial \vartheta}{\partial t}$ is written in the relaxation form

$$\frac{\partial \vartheta}{\partial t} = \frac{\pi}{\tau}. \tag{4.3.6}$$

As it was supposed, the deviation of the magnetic moment from the spin orientation is the result of the approach of the second electron with impulse p to the first electron at the distance r_{in}. In this case

$$\frac{1}{\tau} = \frac{p}{m_{\mathrm{e}} \lambda}, \tag{4.3.7}$$

where $\lambda \sim r_{\mathrm{in}}$. After introduction of the coefficient s, we have $r_{\mathrm{in}} = \lambda s$ and

$$\frac{1}{\tau} = s \frac{p}{m_{\mathrm{e}} r_{\mathrm{in}}}. \tag{4.3.8}$$

It means

$$\frac{\partial \vartheta}{\partial t} = s \frac{\pi p}{m_{\mathrm{e}} r_{\mathrm{in}}} \tag{4.3.9}$$

or

$$\frac{\partial \vartheta}{\partial t} = s \frac{2\pi}{p r_{\mathrm{in}}} \frac{p^2}{2 m_{\mathrm{e}}} = s \frac{2\pi}{p r_{\mathrm{in}}} E_{\mathrm{c}}. \tag{4.3.10}$$

Let us introduce now the fine-structure constant α, where

$$\alpha = \frac{E_{\mathrm{c}}}{E_{\mathrm{ph}}} \tag{4.3.11}$$

and transform (4.3.10)

$$\frac{\partial \vartheta}{\partial t} = \frac{2\pi}{p r_{\mathrm{in}}} s \alpha E_{\mathrm{ph}}, \tag{4.3.12}$$

$$\frac{\partial \vartheta}{\partial t} = \frac{2\pi}{h} s \alpha E_{\mathrm{ph}} \frac{\lambda}{r_{\mathrm{in}}}; \tag{4.3.13}$$

using $r_{\mathrm{in}} = \lambda s$ one obtains

$$\pm \sin \vartheta \frac{\partial \vartheta}{\partial t} = \pm \frac{2\pi}{h} \alpha E_{\mathrm{ph}} \sin \vartheta = \pm \alpha \omega_{\mathrm{in}} \sin \vartheta \tag{4.3.14}$$

or

$$\frac{\partial \vartheta}{\partial t} = \alpha \omega_{\mathrm{in}}, \tag{4.3.15}$$

where ω_{in} is the photon frequency which the wave length is $2\pi r_{\mathrm{in}}$. But

$$\alpha = \frac{e^2}{\hbar c}. \tag{4.3.16}$$

It means that Eq. (4.3.15) takes the transparent physical form

$$\frac{\partial \vartheta}{\partial t} = \frac{e^2}{\hbar r_{\mathrm{in}}}. \tag{4.3.17}$$

4.4 THE MATHEMATICAL MODELING OF THE CHARGE DISTRIBUTION IN ELECTRON AND PROTON

Let us deliver the derivation of the non-local equations in the first approximation. From (4.2.9) to (4.2.12)

$$\frac{\partial p}{\partial r}-q\frac{\partial \psi}{\partial r}=0. \tag{4.4.1}$$

Transform the energy Eq. (4.2.13) using (4.2.14) and (4.4.1), (see also (4.3.2), (4.3.15))

$$\pm\frac{\hbar}{2}n\frac{e}{m_e c}B\alpha\omega_{\text{in}}\sin\vartheta=\frac{1}{r^2}\frac{\partial}{\partial r}\left(\tau r^2\frac{\partial}{\partial r}\left[\frac{p}{\rho}\left(\varepsilon n+\frac{5}{2}p\right)\right]\right)-\frac{1}{r^2}\frac{\partial}{\partial r}\left\{\tau r^2\left(\varepsilon n+\frac{5}{2}p\right)\frac{q}{\rho}\frac{\partial \psi}{\partial r}\right\}. \tag{4.4.2}$$

or

$$\pm\frac{\hbar}{2}n\frac{e}{m_e c}B\alpha\omega_{\text{in}}r^2\sin\vartheta=\frac{\partial}{\partial r}\left(\tau r^2\frac{p}{m}\frac{\partial \varepsilon}{\partial r}\right)+\frac{5}{2}\frac{\partial}{\partial r}\left(\tau r^2 p\frac{\partial}{\partial r}\left[\frac{p}{\rho}\right]\right). \tag{4.4.3}$$

Naturally, to suppose that $\frac{\partial \varepsilon}{\partial r}=0$ and non-local parameter τ does not depend on r

$$\pm\frac{\hbar}{2}nB\alpha\omega_{\text{in}}r^2\sin\vartheta=\frac{5}{2}\tau\frac{\partial}{\partial r}\left(r^2 p\frac{\partial}{\partial r}\left[\frac{p}{q}\right]\right). \tag{4.4.4}$$

Using the relation $\left(\rho=m\frac{q}{e}\right)$, the scales r_0, p_0, q_0 for the values r, p, q and denoting by tilde the dimensionless values, one obtains

$$\pm\frac{\hbar}{5}\frac{1}{e}B\frac{\omega_{\text{in}}}{\tau}\frac{r_0^2 q_0^2}{p_0^2}\alpha\sin\vartheta\,\tilde{r}^2\tilde{q}=\frac{\partial}{\partial\tilde{r}}\left(\tilde{r}^2\tilde{p}\frac{\partial}{\partial\tilde{r}}\left[\frac{\tilde{p}}{\tilde{q}}\right]\right). \tag{4.4.5}$$

Introduce the notation $\widetilde{B}$ for the dimensionless coefficient

$$\widetilde{B}=\frac{\hbar}{5}\frac{1}{e}B\frac{\omega_{\text{in}}}{\tau}\frac{r_0^2 q_0^2}{p_0^2}\alpha\sin\vartheta, \tag{4.4.6}$$

we have

$$\frac{\partial}{\partial\tilde{r}}\left(\tilde{r}^2\tilde{p}\frac{\partial}{\partial\tilde{r}}\left[\frac{\tilde{p}}{\tilde{q}}\right]\right)=\pm\widetilde{B}\tilde{r}^2\tilde{q}. \tag{4.4.7}$$

The Poisson Eq. (4.2.14) takes the dimensionless form

$$A\frac{\partial}{\partial\tilde{r}}\left(\tilde{r}^2\frac{\partial\tilde{\psi}}{\partial\tilde{r}}\right)=\tilde{r}^2\tilde{q}, \tag{4.4.8}$$

where the dimensionless coefficient A is introduced

$$A=\frac{\psi_0}{4\pi r_0^2 q_0}, \tag{4.4.9}$$

ψ_0 is the scale for the potential ψ.

In the absence of perturbations, $\widetilde{B}=0$ and from (4.4.7) one obtains

$$p=Cq. \tag{4.4.10}$$

From (4.4.7), (4.4.8) follow also

$$\frac{\partial}{\partial\tilde{r}}\left(\tilde{r}^2\tilde{p}\frac{\partial}{\partial\tilde{r}}\left[\frac{\tilde{p}}{\tilde{q}}\right]\right)=\pm\widetilde{B}A\frac{\partial}{\partial\tilde{r}}\left(\tilde{r}^2\frac{\partial\tilde{\psi}}{\partial\tilde{r}}\right). \tag{4.4.11}$$

Writing the Eq. (4.4.1) in the dimensionless form

$$\frac{p_0}{q_0\psi_0}\frac{\partial\tilde{p}}{\partial\tilde{r}}-\tilde{q}\frac{\partial\tilde{\psi}}{\partial\tilde{r}}=0. \tag{4.4.12}$$

and introducing the obvious relation between scales for the simplification

$$\frac{p_0}{q_0\psi_0}=1, \tag{4.4.13}$$

we have

$$\frac{\partial \widetilde{p}}{\partial \widetilde{r}} - \widetilde{q}\frac{\partial \widetilde{\psi}}{\partial \widetilde{r}} = 0 \tag{4.4.14}$$

and

$$\widetilde{B} = B\frac{\hbar\omega_{\text{in}}}{\tau}\frac{r_0^2}{5e\psi_0^2}\alpha\sin\vartheta. \tag{4.4.15}$$

Using Eqs. (4.4.11) and (4.4.14), it turns out that

$$\frac{\partial \widetilde{p}}{\partial \widetilde{r}} \pm \widetilde{B}A\frac{\partial \ln \widetilde{p}}{\partial \widetilde{r}} = \frac{\widetilde{p}}{\widetilde{q}}\frac{\partial \widetilde{q}}{\partial \widetilde{r}}, \tag{4.4.16}$$

$$\frac{\partial}{\partial \widetilde{r}}\left(\ln\frac{\widetilde{p}}{\widetilde{q}}\right) \pm \widetilde{B}A\frac{1}{p}\frac{\partial \ln \widetilde{p}}{\partial \widetilde{r}} = 0, \tag{4.4.17}$$

then the second term on the left hand side of Eq. (4.4.16) reflects the influence of perturbation. Omitting this term, we return to the relation (4.4.10).

Write down once more the system of equation which was used in the mathematical modeling (SYSTEM I)

$$\frac{\partial}{\partial \widetilde{r}}\left(\widetilde{r}^{\,2}\widetilde{p}\frac{\partial}{\partial \widetilde{r}}\left[\frac{\widetilde{p}}{\widetilde{q}}\right]\right) = \widetilde{B}\widetilde{r}^{\,2}\widetilde{q}\,,$$

$$A\frac{\partial}{\partial \widetilde{r}}\left(\widetilde{r}^{\,2}\frac{\partial \widetilde{\psi}}{\partial \widetilde{r}}\right) = \widetilde{r}^{\,2}\widetilde{q}\,,$$

$$\widetilde{q}\frac{\partial \widetilde{\psi}}{\partial \widetilde{r}} - \frac{\partial \widetilde{p}}{\partial \widetilde{r}} = 0\,,$$

where

$$A = \frac{\psi_0}{4\pi r_0^2 q_0}, \quad \widetilde{B} = \pm B\frac{\hbar\omega_{\text{in}}}{\tau c}\frac{r_0^2}{5e\psi_0^2}\alpha\sin\vartheta.$$

Some significant remarks:

1. Solving of the set of equations SYSTEM I belongs to the class of Cauchy problems and need not in introduction the strictly defined electron radius beforehand.
2. From here on, for convenience, the different signs were included in $\widetilde{B}$.
3. The classical electron radius is only one from the possible scales r_0.

Really, from (4.4.8) follows that the absolute electron charge q_{el} is equal to

$$q_{\text{el}} = |e| = \int_0^{r_{\text{el}}} 4\pi r^2 q(r)\text{d}r = 4\pi r_0^3 q_0 \int_0^{\widetilde{r}_{el}} \widetilde{r}^2\widetilde{q}\,\text{d}\widetilde{r} = 4\pi r_0^3 q_0 A \int_0^{\widetilde{r}_{el}} \frac{\partial}{\partial \widetilde{r}}\left(\widetilde{r}^2\frac{\partial \widetilde{\psi}}{\partial \widetilde{r}}\right)\text{d}\widetilde{r}$$

$$= r_0\psi_0 \int_0^{\widetilde{r}_{\text{el}}} \frac{\partial}{\partial \widetilde{r}}\left(\widetilde{r}^2\frac{\partial \widetilde{\psi}}{\partial \widetilde{r}}\right)\text{d}\widetilde{r} = r_0\psi_0\widetilde{r}_{\text{el}}^2\left[\frac{\partial \widetilde{\psi}}{\partial \widetilde{r}}\right]_{r=r_{\text{el}}} \tag{4.4.18}$$

or

$$\left[\frac{\partial \widetilde{\psi}}{\partial \widetilde{r}}\right]_{r=r_{\text{el}}} = 1 \tag{4.4.19}$$

for the scales chosen as

$$\psi_0 = |e|/r_0, \quad r_0 = r_{\text{el}}, \quad q_0 = |e|/r_{el}^3, \quad p_0 = q_0\psi_0 = \frac{e^2}{r_{el}^4}. \tag{4.4.20}$$

In this case

$$\left[\frac{\partial\psi}{\partial r}\right]_{r=r_{el}}=\frac{|e|}{r_0^2},\tag{4.4.21}$$

or

$$[F]_{r=r_{el}}=\frac{e^2}{r_{el}^2}.\tag{4.4.22}$$

But, in the definition of the fine-structure constant α, the energy E_c was introduced as the energy needed to overcome the electrostatic repulsion between two electrons with a distance of r_{in} apart (see also (4.3.4)). It means that for this problem, naturally to put the scale $r_0=r_{in}$. In this case (system of conditions SYSTEM II):

$$\psi_0=|e|/r_0,\quad r_0=r_{in},\quad q_0=|e|/r_{in}^3,\quad p_0=q_0\psi_0=e^2/r_{in}^4.$$

$$A=\frac{\psi_0}{4\pi r_0^2 q_0}=\frac{1}{4\pi},\quad \widetilde{B}=\pm B\frac{\hbar\omega_{in}}{\tau c}\frac{r_{in}^4}{5|e|^3}\alpha\sin\vartheta.$$

Parameter (4.4.15) can be rewritten as

$$\widetilde{B}=\pm B\frac{\hbar\omega_{in}}{\tau c}\frac{r_{in}^4}{5|e|^3}\alpha\sin\vartheta=\pm B\frac{r_{in}^3}{5|e|\tau c}\sin\vartheta.\tag{4.4.23}$$

Let us introduce the character magnetic force

$$F_{mag}=\frac{|e|}{c}\frac{r_{in}}{5\tau}B\tag{4.4.24}$$

and the character electrostatic force

$$F_{elect}=\frac{e^2}{r_{in}^2}.\tag{4.4.25}$$

It means that parameter $\widetilde{B}$ can be written in the transparent physical form

$$\widetilde{B}=\frac{F_{mag}}{F_{elect}}\sin\vartheta.\tag{4.4.26}$$

Is it possible to obtain the soliton type solution for this object under these conditions? Let us show that the SYSTEM I admits such kind of solutions.

Calculations reflected on Figs. 4.1–4.18 are realized under conditions SYSTEM II (in particular by $A=\frac{1}{4\pi}$ and different $\widetilde{B}$). The influence $\widetilde{B}$ is investigated from zero up to value $|\widetilde{B}|=10$. Maple notations are used ($v=\widetilde{\psi}$, $D(v)(t)=\frac{\partial\widetilde{\psi}}{\partial\widetilde{r}}$, $q=\widetilde{q}$, $t=\widetilde{r}$, $B=\widetilde{B}$).

Cauchy conditions for SYSTEM II: $v(0)=\widetilde{\psi}(0)=1$, $D(v)(0)=\frac{\partial\widetilde{\psi}}{\partial\widetilde{r}}(0)=0$; $q(0)=\widetilde{q}(0)=1$, $D(q)(0)=\frac{\partial\widetilde{q}}{\partial\widetilde{r}}(0)=0$; $p(0)=\widetilde{p}(0)=1$, $D(p)(0)=\frac{\partial\widetilde{p}}{\partial\widetilde{r}}(0)=0$.

Figs. 4.1–4.3 correspond to the case when the angle ϑ is nil and then $\widetilde{B}=0$. Solutions in all calculations exist only in a bounded region of the 1D space. The size of this region r_{lim} defines the electron radius. For the case $\widetilde{B}=0$, one obtains $\widetilde{r}_{lim}=0.9235$.

For the case $B=\widetilde{B}=0.001$, one obtains also $\widetilde{r}_{lim}=0.9235$.

For the case $B=\widetilde{B}=0.01$, one obtains $\widetilde{r}_{lim}=0.9239$.

For the case $B=\widetilde{B}=0.1$, one obtains $\widetilde{r}_{lim}=0.9272$.

For the case $B=\widetilde{B}=1$, one obtains $\widetilde{r}_{lim}=0.9614$.

For the case $B=\widetilde{B}=10$, one obtains $\widetilde{r}_{lim}=1.4397$. Calculations reflected on Figs. 4.19–4.21 are realized by conditions SYSTEM III: $B=\widetilde{B}=0.1$, $v(0)=\widetilde{\psi}(0)=1$, $D(v)(0)=\frac{\partial\widetilde{\psi}}{\partial\widetilde{r}}(0)=0$; $q(0)=\widetilde{q}(0)=0.1$, $D(q)(0)=\frac{\partial\widetilde{q}}{\partial\widetilde{r}}(0)=0$; $p(0)=\widetilde{p}(0)=0.01$, $D(p)(0)=\frac{\partial\widetilde{p}}{\partial\widetilde{r}}(0)=0$.

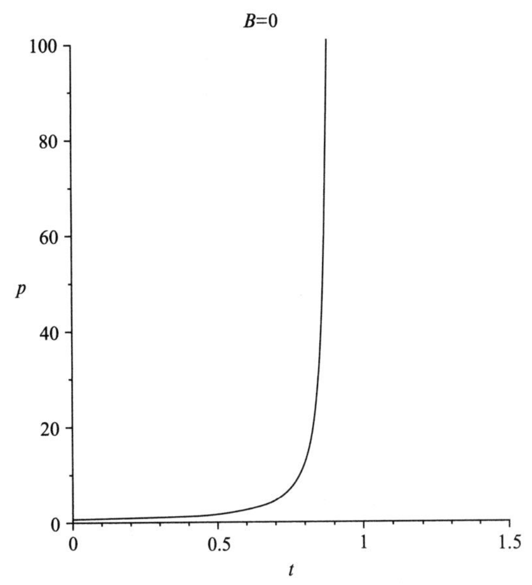

FIGURE 4.1 $p = \widetilde{p}(\widetilde{r}), \widetilde{B} = 0.$

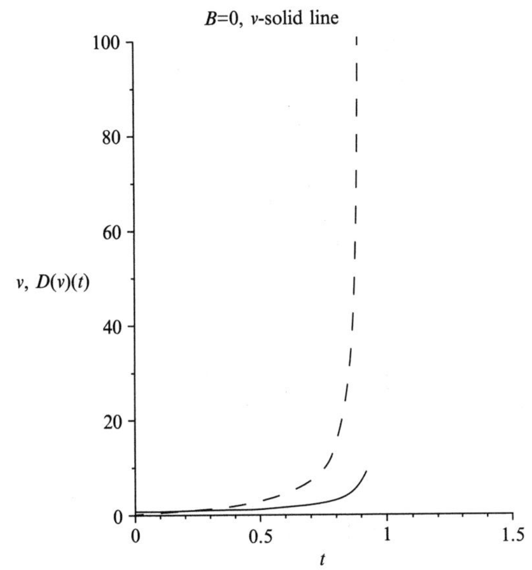

FIGURE 4.2 $v = \widetilde{\psi}(\widetilde{r}), D(v)(t) = \frac{\partial \widetilde{\psi}}{\partial \widetilde{r}}(\widetilde{r})$, solid line $v = \widetilde{\psi}(\widetilde{r}), \widetilde{B} = 0.$

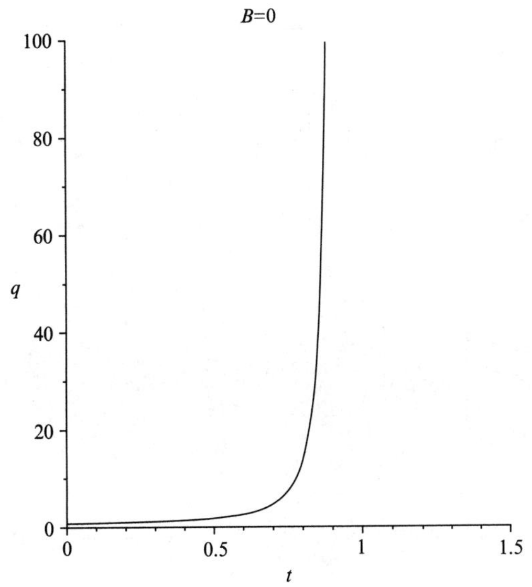

FIGURE 4.3 $q = \widetilde{q}(\widetilde{r}), \widetilde{B} = 0.$

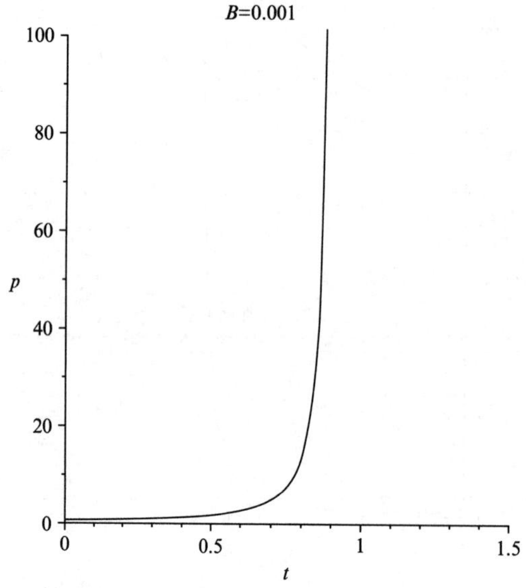

FIGURE 4.4 $p=\widetilde{p}(\widetilde{r}), \widetilde{B}=0.001.$

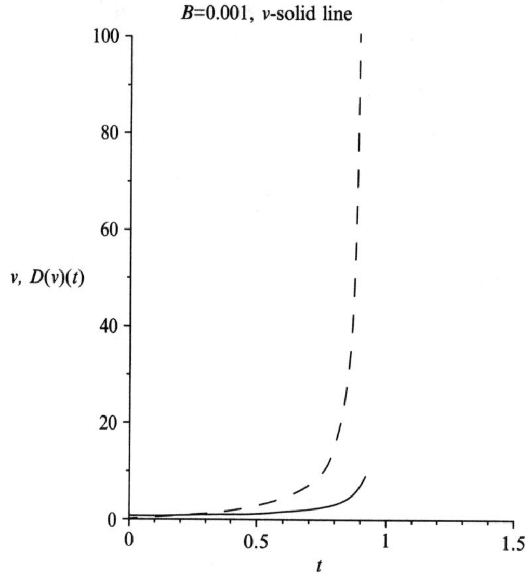

FIGURE 4.5 $v=\widetilde{\psi}(\widetilde{r}), D(v)(t)=\frac{\partial\widetilde{\psi}}{\partial\widetilde{r}}(\widetilde{r})$, solid line $v=\widetilde{\psi}(\widetilde{r}), \widetilde{B}=0.001.$

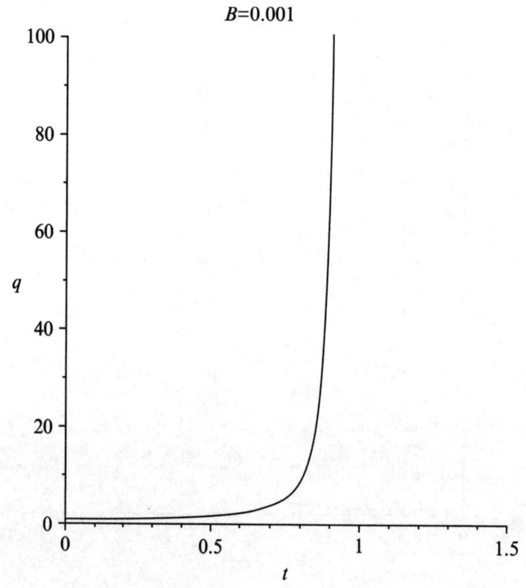

FIGURE 4.6 $q=\widetilde{q}(\widetilde{r}), \widetilde{B}=0.001.$

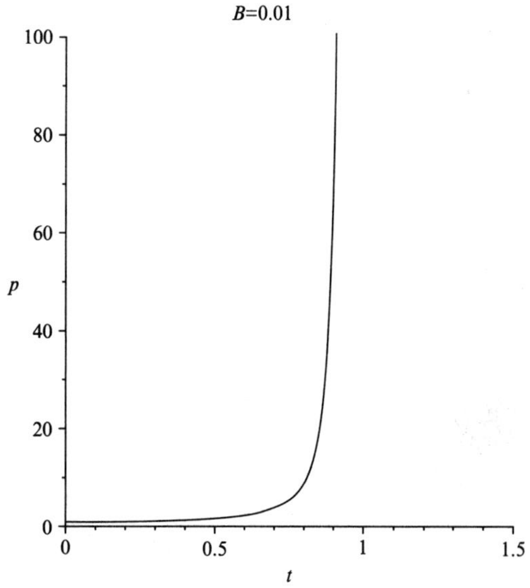

FIGURE 4.7 $p=\widetilde{p}(\widetilde{r}),\ \widetilde{B}=0.01.$

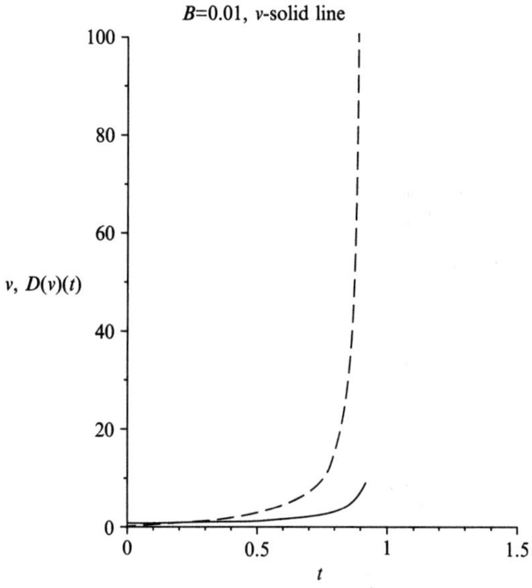

FIGURE 4.8 $v=\widetilde{\psi}(\widetilde{r}),\ D(v)(t)=\frac{\partial\widetilde{\psi}}{\partial\widetilde{r}}(\widetilde{r})$, solid line $v=\widetilde{\psi}(\widetilde{r}),\ \widetilde{B}=0.01.$

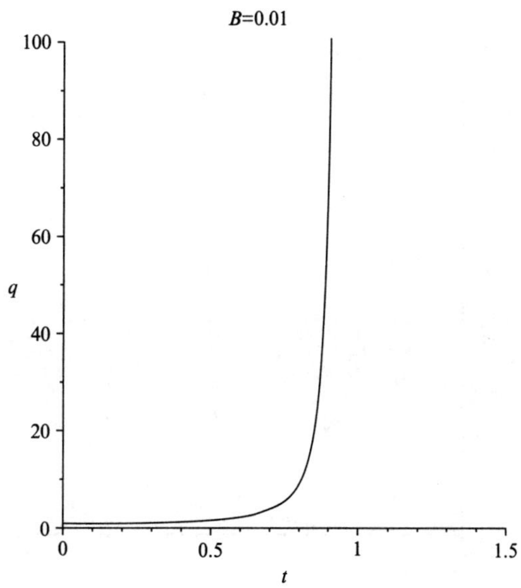

FIGURE 4.9 $q=\widetilde{q}(\widetilde{r}),\ \widetilde{B}=0.01.$

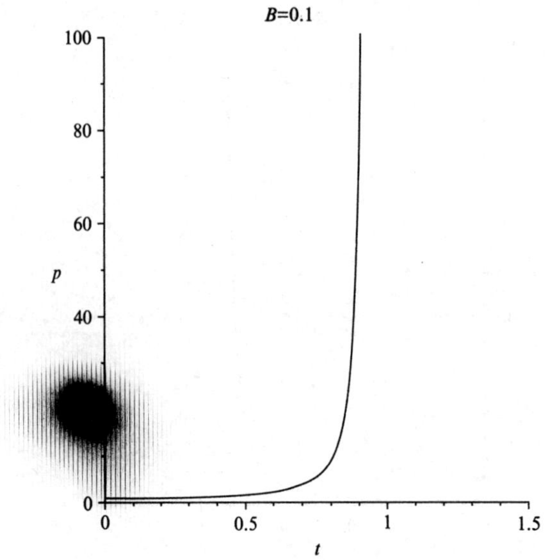

FIGURE 4.10 $p=\widetilde{p}(\widetilde{r}), \widetilde{B}=0.1.$

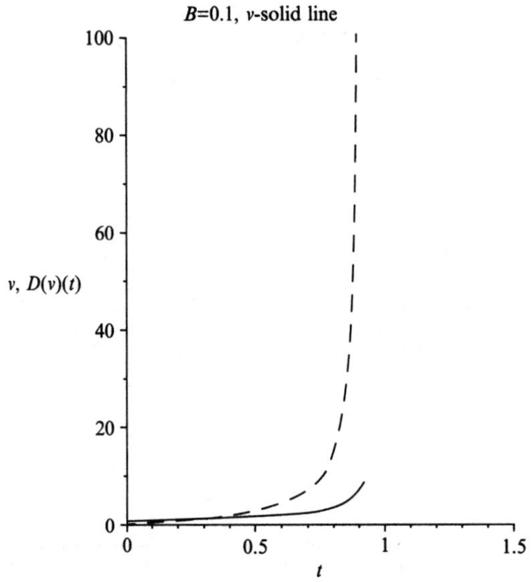

FIGURE 4.11 $v=\widetilde{\psi}(\widetilde{r}), D(v)(t)=\frac{\partial\widetilde{\psi}}{\partial\widetilde{r}}(\widetilde{r})$, solid line $v=\widetilde{\psi}(\widetilde{r}), \widetilde{B}=0.1.$

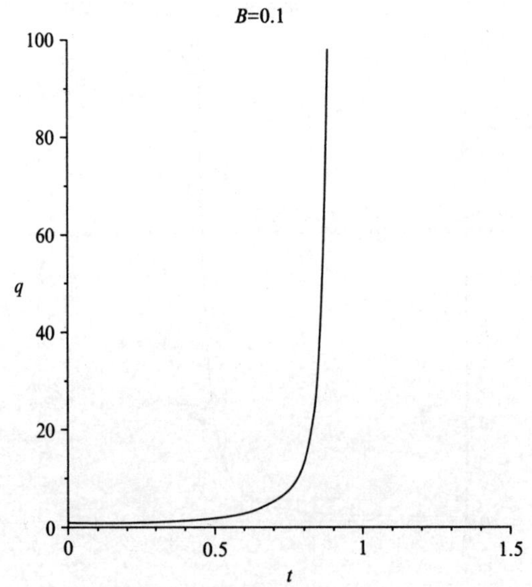

FIGURE 4.12 $q=\widetilde{q}(\widetilde{r}), \widetilde{B}=0.1.$

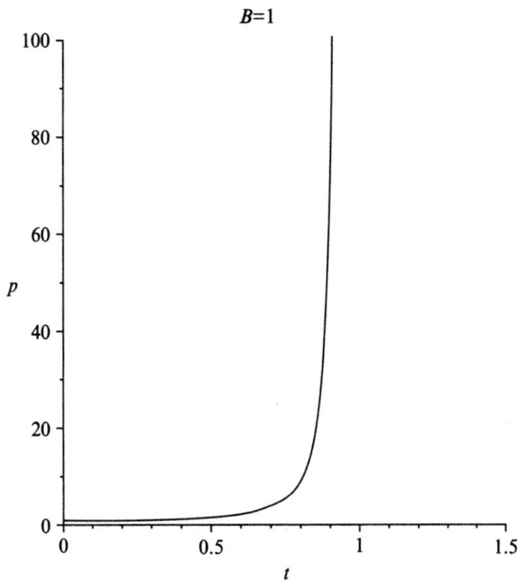

FIGURE 4.13 $p=\widetilde{p}(\widetilde{r}),\ \widetilde{B}=1.$

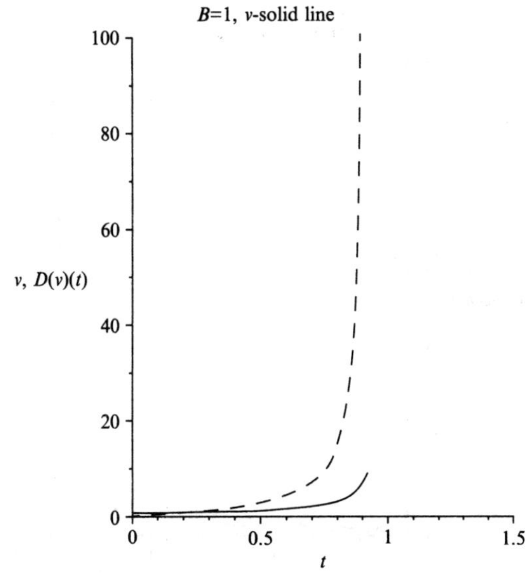

FIGURE 4.14 $v=\widetilde{\psi}(\widetilde{r}),\ D(v)(t)=\frac{\partial\widetilde{\psi}}{\partial\widetilde{r}}(\widetilde{r})$, solid line $v=\widetilde{\psi}(\widetilde{r}),\ \widetilde{B}=1.$

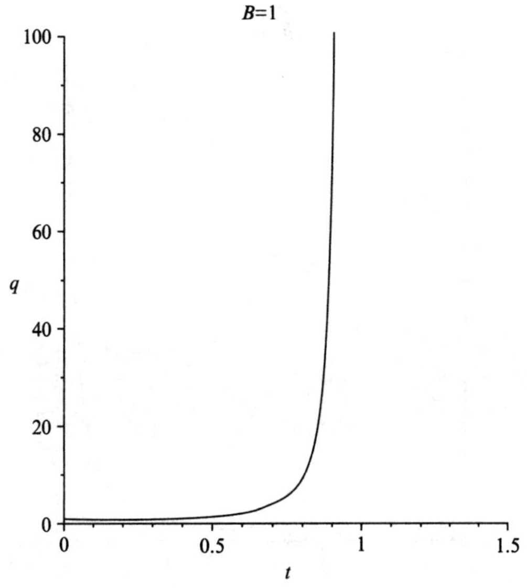

FIGURE 4.15 $q=\widetilde{q}(\widetilde{r}),\ \widetilde{B}=1.$

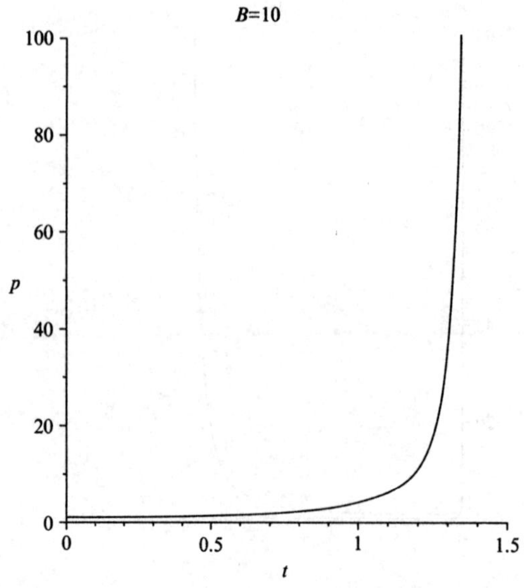

FIGURE 4.16 $p = \widetilde{p}(\widetilde{r}), \widetilde{B} = 10.$

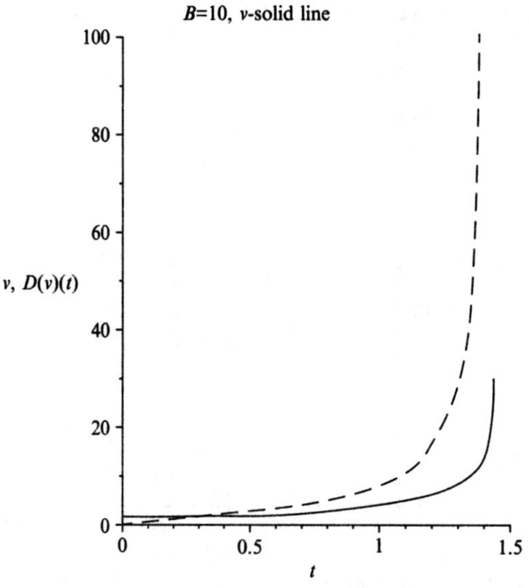

FIGURE 4.17 $v = \widetilde{\psi}(\widetilde{r}), D(v)(t) = \frac{\partial \widetilde{\psi}}{\partial \widetilde{r}}(\widetilde{r})$, solid line $v = \widetilde{\psi}(\widetilde{r}), \widetilde{B} = 10.$

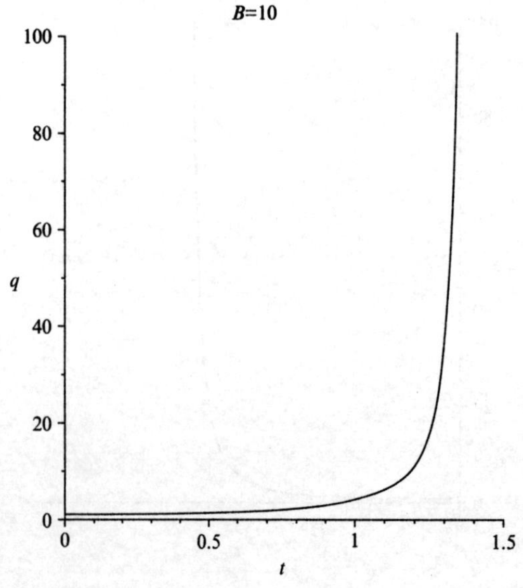

FIGURE 4.18 $q = \widetilde{q}(\widetilde{r}), \widetilde{B} = 10.$

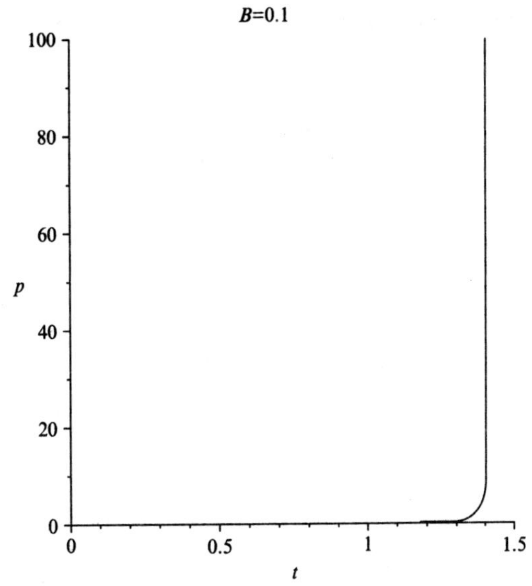

FIGURE 4.19 $p=\widetilde{p}(\widetilde{r}),\ \widetilde{B}=0.1.$

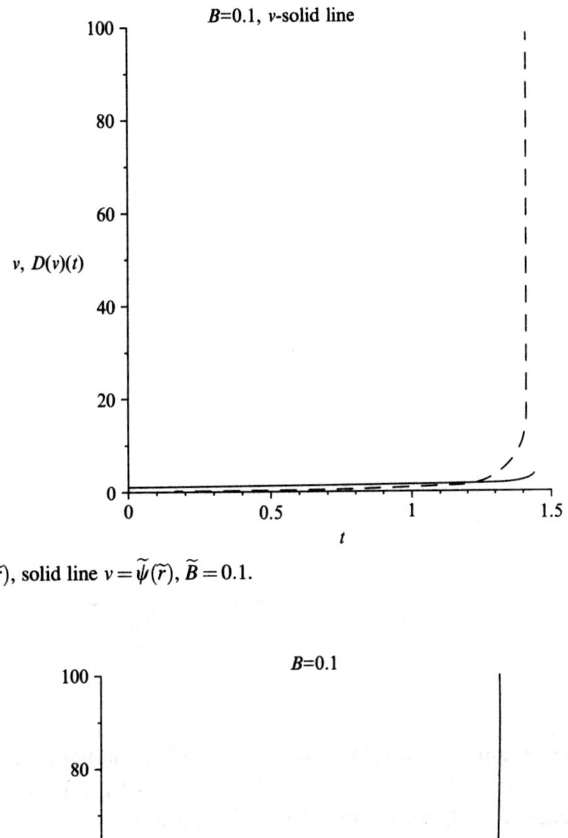

FIGURE 4.20 $v=\widetilde{\psi}(\widetilde{r}),\ D(v)(t)=\frac{\partial\widetilde{\psi}}{\partial\widetilde{r}}(\widetilde{r})$, solid line $v=\widetilde{\psi}(\widetilde{r}),\ \widetilde{B}=0.1.$

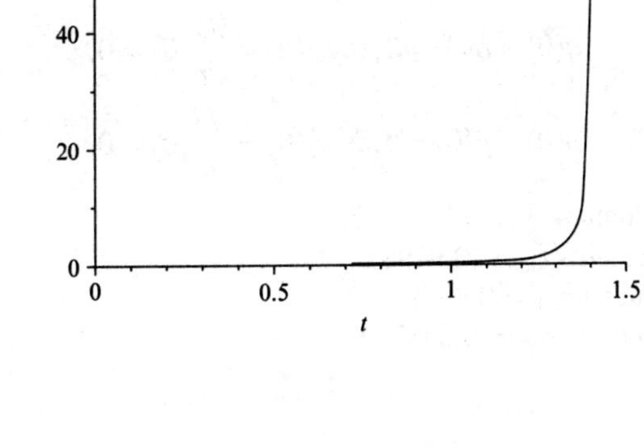

FIGURE 4.21 $q=\widetilde{q}(\widetilde{r}),\ \widetilde{B}=0.1.$

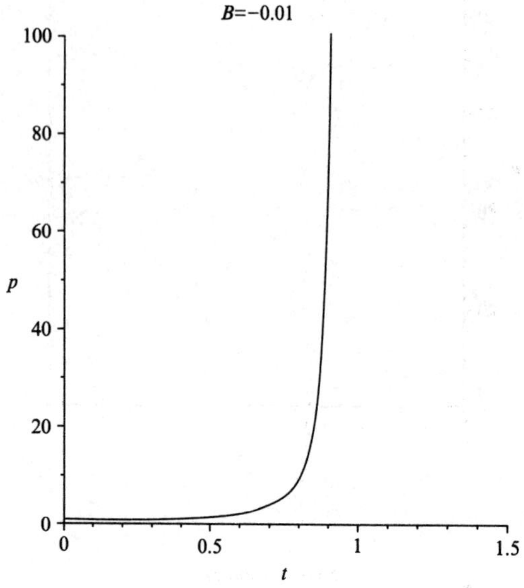

FIGURE 4.22 $p=\widetilde{p}(\widetilde{r}), \widetilde{B}=-0.01$.

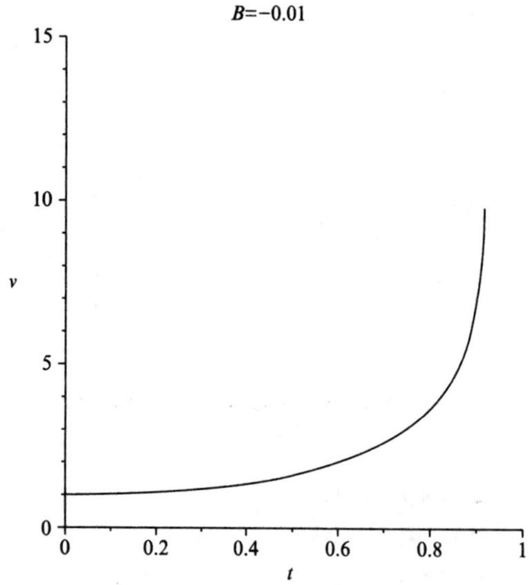

FIGURE 4.23 $v=\widetilde{\psi}(\widetilde{r}), B=\widetilde{B}=-0.01$.

For the case SYSTEM III one obtains $\widetilde{r}_{\rm lim}=1.44$. Figs. 4.21–4.22 demonstrate the results of calculations for the negative values $B=\widetilde{B}$ and the Cauchy condition chosen in the form SYSTEM 2. Figs. 4.23–4.36 demonstrate the results of calculations for the negative values $B=\widetilde{B}$, but for the Cauchy conditions:

$$v(0)=\widetilde{\psi}(0)=1, D(v)(0)=\frac{\partial\widetilde{\psi}}{\partial\widetilde{r}}(0)=0;$$

$$q(0)=\widetilde{q}(0)=1, D(q)(0)=\frac{\partial\widetilde{q}}{\partial\widetilde{r}}(0)=0;$$

$$p(0)=\widetilde{p}(0)=1, D(p)(0)=\frac{\partial\widetilde{p}}{\partial\widetilde{r}}(0)=0.$$

For the case $B=\widetilde{B}=-0.01$, one obtains $\widetilde{r}_{\rm lim}=0.92312$.
For the case $B=\widetilde{B}=-0.1$, one obtains $\widetilde{r}_{\rm lim}=0.9198$.
For the case $B=\widetilde{B}=-1$, one obtains $\widetilde{r}_{\rm lim}=0.8979$.
For the case $B=\widetilde{B}=-10$, one obtains $\widetilde{r}_{\rm lim}=0.6487$.

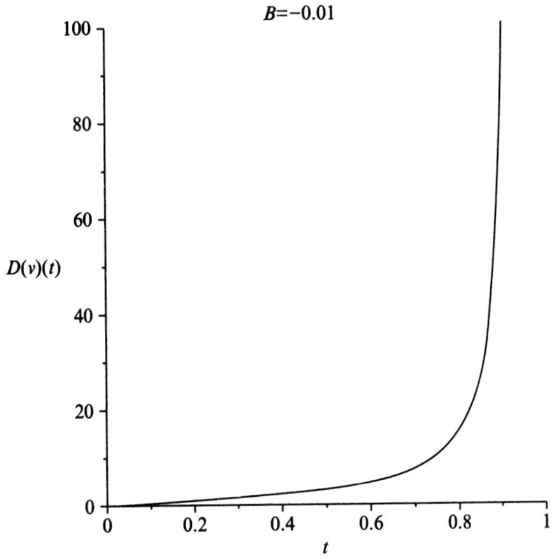

FIGURE 4.24 $D(v)(t) = \frac{\partial\widetilde{\psi}}{\partial\widetilde{r}}(\widetilde{r}),\ B = \widetilde{B} = -0.01.$

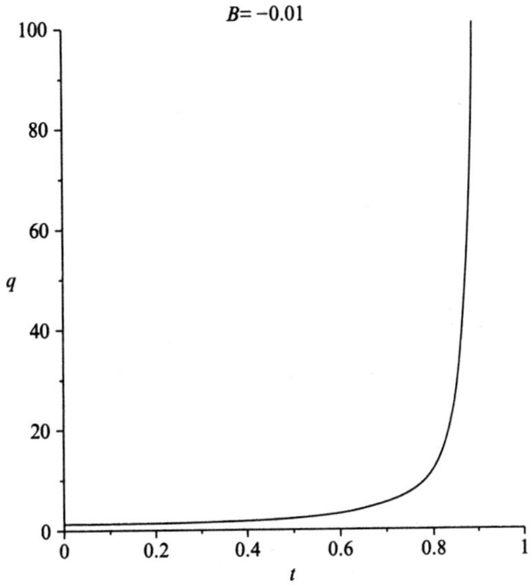

FIGURE 4.25 $q = \widetilde{q}(\widetilde{r}),\ B = \widetilde{B} = -0.01.$

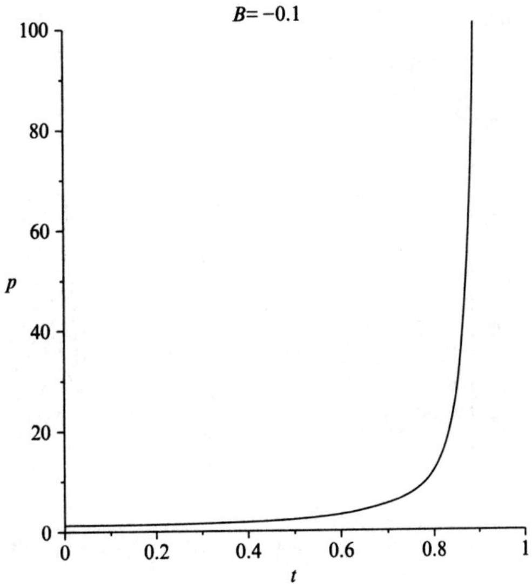

FIGURE 4.26 $p = \widetilde{p}(\widetilde{r}),\ \widetilde{B} = -0.1.$

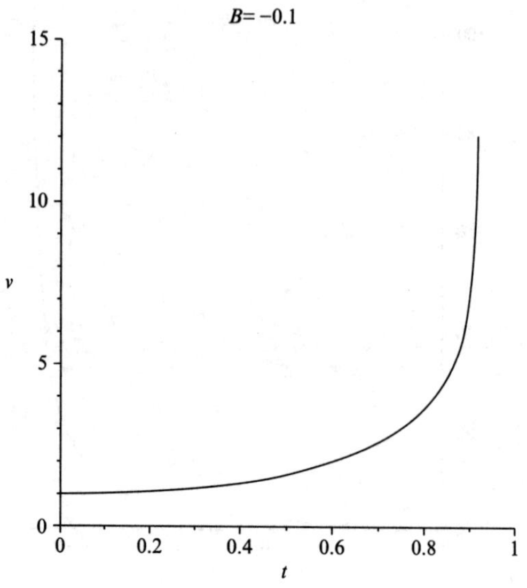

FIGURE 4.27 $v=\widetilde{\psi}(\widetilde{r}),\ B=\widetilde{B}=-0.1.$

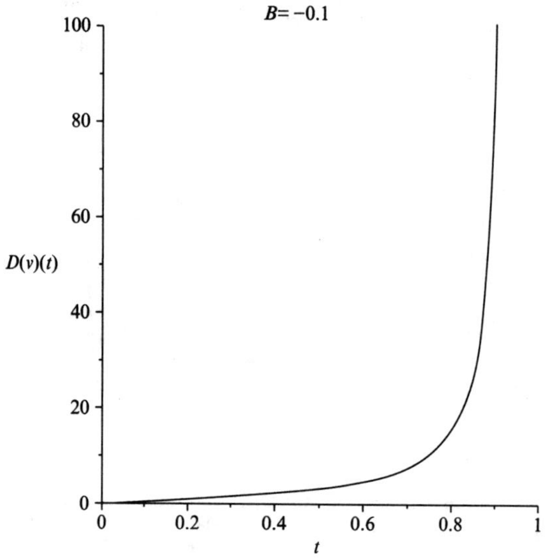

FIGURE 4.28 $D(v)(t)=\frac{\partial\widetilde{\psi}}{\partial\widetilde{r}}(\widetilde{r}),\ B=\widetilde{B}=-0.1.$

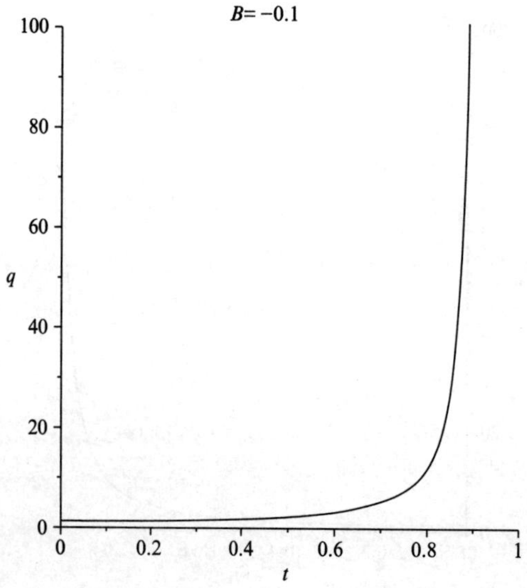

FIGURE 4.29 $q=\widetilde{q}(\widetilde{r}),\ B=\widetilde{B}=-0.1.$

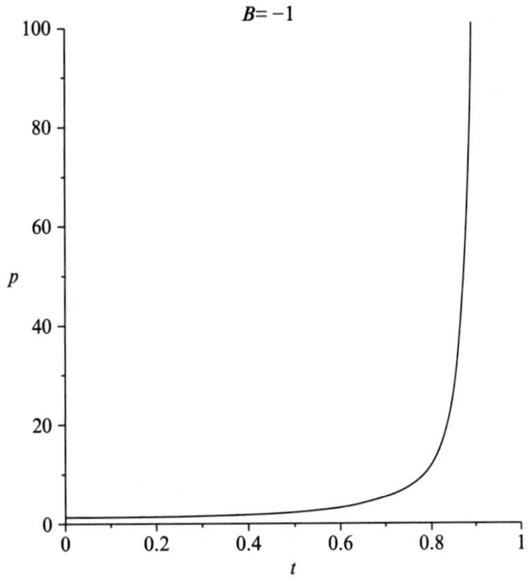

FIGURE 4.30 $p=\widetilde{p}(\widetilde{r}),\ \widetilde{B}=-1.$

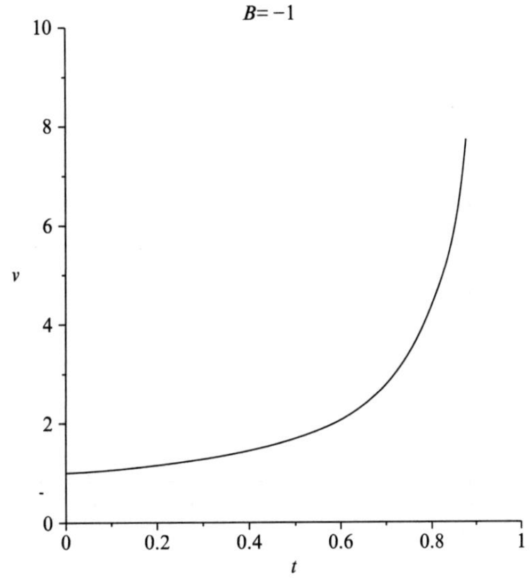

FIGURE 4.31 $v=\widetilde{\psi}(\widetilde{r}),\ B=\widetilde{B}=-1.$

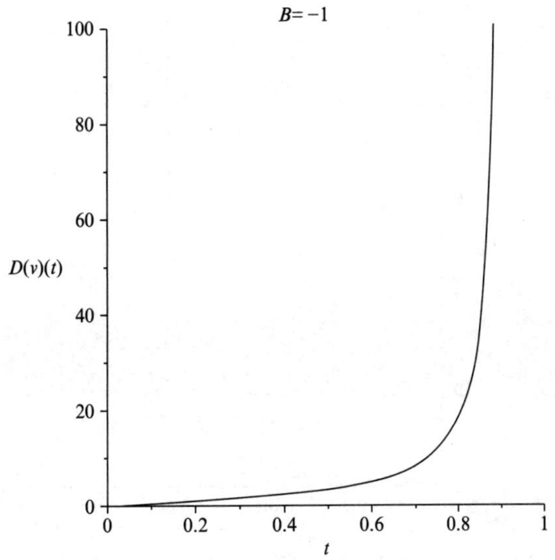

FIGURE 4.32 $D(v)(t)=\frac{\partial\widetilde{\psi}}{\partial\widetilde{r}}(\widetilde{r}),\ B=\widetilde{B}=-1.$

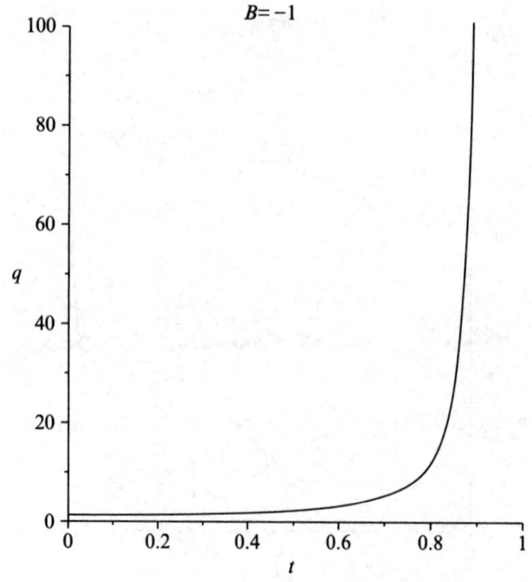

FIGURE 4.33 $q = \widetilde{q}(\widetilde{r}),\ B = \widetilde{B} = -1.$

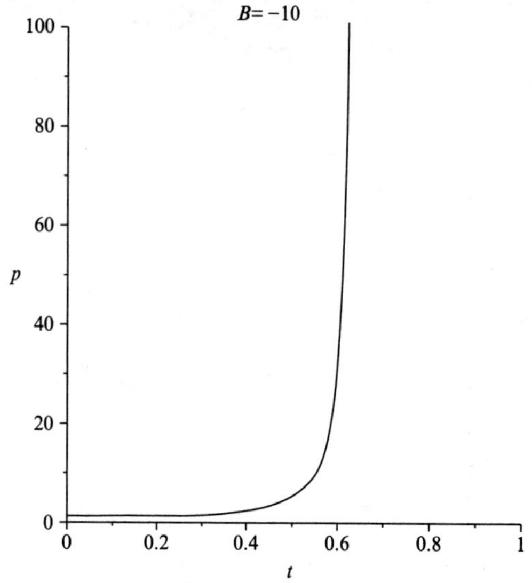

FIGURE 4.34 $p = \widetilde{p}(\widetilde{r}),\ \widetilde{B} = -10.$

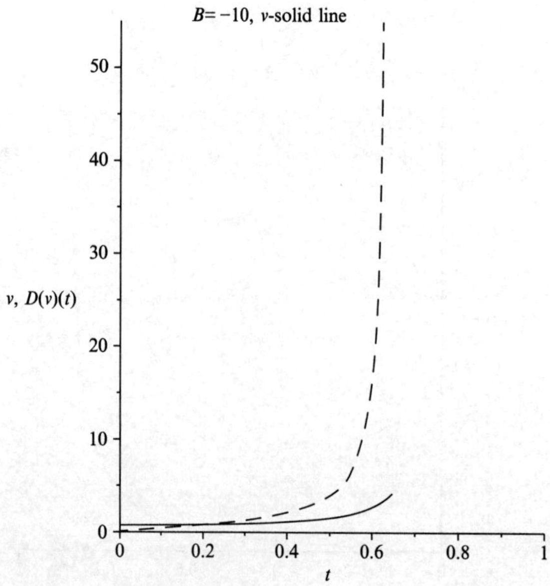

FIGURE 4.35 $v = \widetilde{\psi}(\widetilde{r}),\ D(v)(t) = \frac{\partial \widetilde{\psi}}{\partial \widetilde{r}}(\widetilde{r})$, solid line $v = \widetilde{\psi}(\widetilde{r}),\ B = \widetilde{B} = -10.$

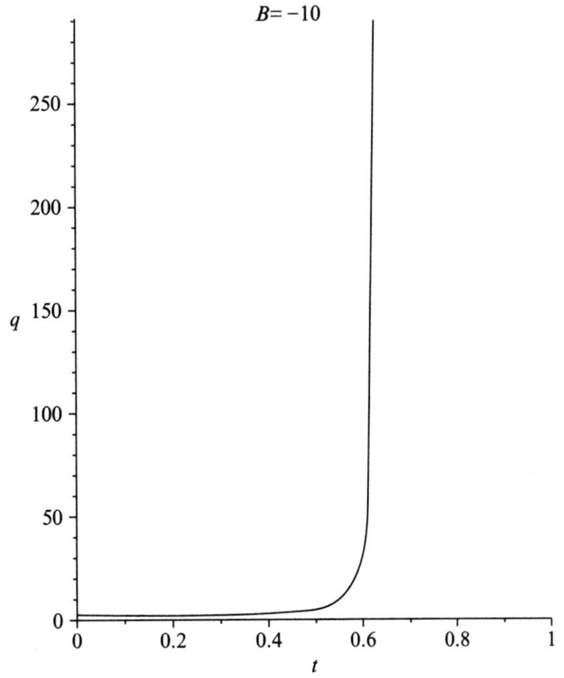

FIGURE 4.36 $q=\widetilde{q}(\widetilde{r}),\ B=\widetilde{B}=-10.$

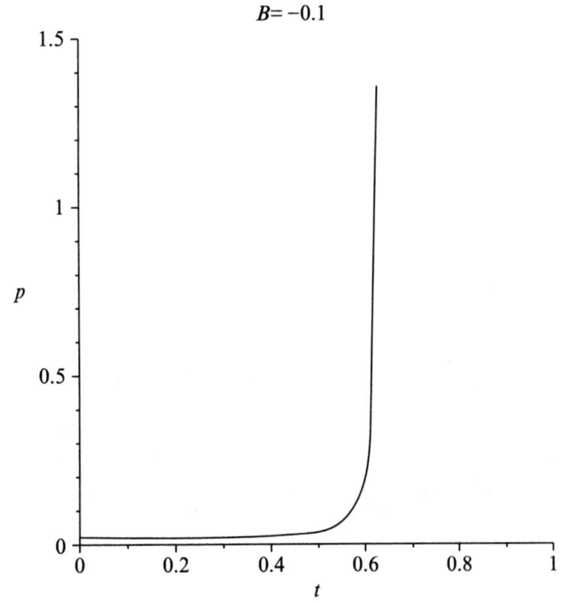

FIGURE 4.37 $p=\widetilde{p}(\widetilde{r}),\ B=\widetilde{B}=-0.1.$

Finally, I show some results obtained for the case $v(0)=\widetilde{\psi}(0)=1,\ D(v)(0)=\frac{\partial\widetilde{\psi}}{\partial\widetilde{r}}(0)=0;\ q(0)=\widetilde{q}(0)=0.1,$ $D(q)(0)=\frac{\partial\widetilde{q}}{\partial\widetilde{r}}(0)=0;\ p(0)=\widetilde{p}(0)=0.01,\ D(p)(0)=\frac{\partial\widetilde{p}}{\partial\widetilde{r}}(0)=0$, but for the negative value $B=\widetilde{B}=-0.1$; compare the corresponding Figs. 4.37–4.40 with Figs. 4.19–4.21.

For the last case $B=\widetilde{B}=-0.1$, one obtains $\widetilde{r}_{\lim}=0.6487$.

Obviously, the theory for the positive charged objects (like proton) can be constructed by the analogous way. In the absence of the magnetic field, we should take into account changing of the sign in the right side of Eq. (4.2.14). After denoting the dimensionless values by wave $\widetilde{q}=q/q_0,\ \widetilde{\psi}=\psi/\psi_0$, Eq. (4.2.14) can be transformed into the dimensionless equation

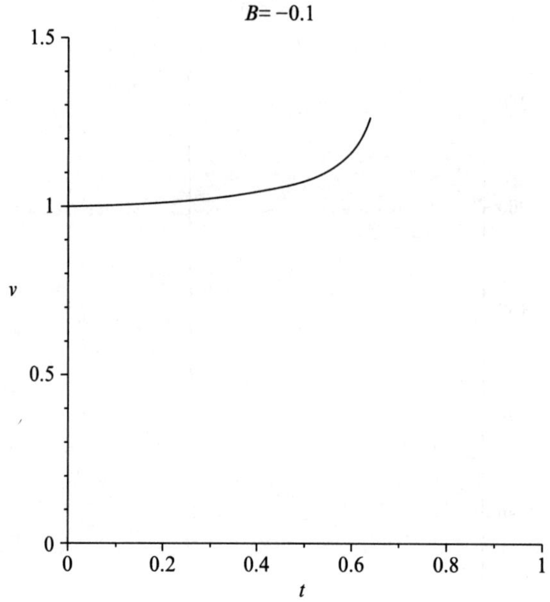

FIGURE 4.38 $v = \widetilde{\psi}(\widetilde{r}),\ B = \widetilde{B} = -0.1.$

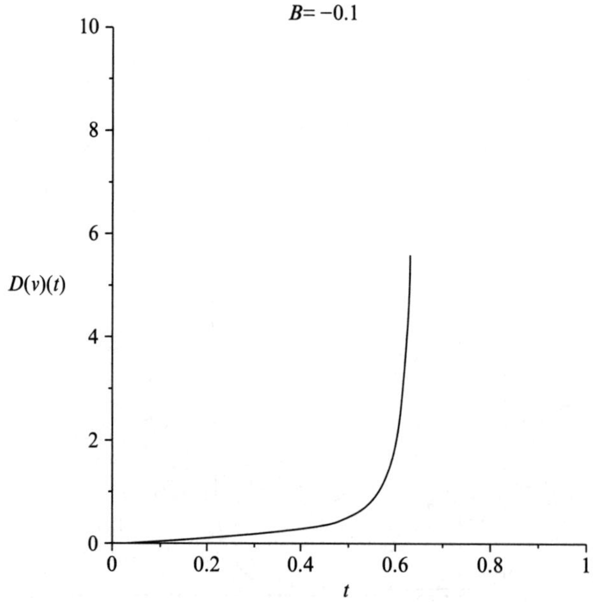

FIGURE 4.39 $D(v)(t) = \frac{\partial \widetilde{\psi}}{\partial \widetilde{r}}(\widetilde{r}),\ B = \widetilde{B} = -0.1.$

$$A\frac{\partial}{\partial \widetilde{r}}\left(\widetilde{r}^2\frac{\partial \widetilde{\psi}}{\partial \widetilde{r}}\right) = -\widetilde{r}^2\exp\left(-\widetilde{\psi}\right), \tag{4.4.27}$$

where the form-factor $A = \frac{\psi_0}{4\pi r_0^2 q_0}$ (see (4.4.9)) is introduced. For other equations, one obtains

$$A\frac{\partial}{\partial \widetilde{r}}\left(\widetilde{r}^2\frac{1}{\widetilde{q}}\frac{\partial \widetilde{q}}{\partial \widetilde{r}}\right) = \widetilde{r}^2\widetilde{q} \tag{4.4.28}$$

$$A\frac{\partial}{\partial \widetilde{r}}\left(\widetilde{r}^2\frac{1}{\widetilde{p}}\frac{\partial \widetilde{p}}{\partial \widetilde{r}}\right) = \widetilde{r}^2\widetilde{p}, \tag{4.4.29}$$

where the scale for pressure is $p_0 = \psi_0 q_0$. Equations (4.4.28) and (4.4.29) have the same dimensionless solutions. Definition (4.4.9) for dimensionless factor A can be written as

$$q_0 = \frac{1}{4\pi r_0^2 A}\psi_0. \tag{4.4.30}$$

It means that $C_{0,\mathrm{cap}} = \frac{1}{4\pi r_0^2 A}$ can be considered as the scale of proton capacity per unit of volume and for scale of volume $V_0 = \frac{4}{3}\pi r_0^3$, the scale of proton capacity is equal r_0 if $A = 1/3$.

Figs. 4.41 and 4.42 reflect the solutions of (4.4.27) and (4.4.29) correspondingly for $A = 1/3$; Maple notations are used ($v = \widetilde{\psi}$, $D(v)(t) = \frac{\partial \widetilde{\psi}}{\partial \widetilde{r}}$, $q = \widetilde{q}$). Cauchy conditions for these calculations: $v(0) = \widetilde{\psi}(0) = 1$, $D(v)(0) = \frac{\partial \widetilde{\psi}}{\partial \widetilde{r}}(0) = 0$; $\widetilde{q}(0) = e^{-1}$, $D(q)(0) = \frac{\partial \widetilde{q}}{\partial \widetilde{r}}(0) = 0$. From Figs. 4.41 and 4.42 follow that solutions exist for this case in the domain less than $r/r_0 \cong 3$.

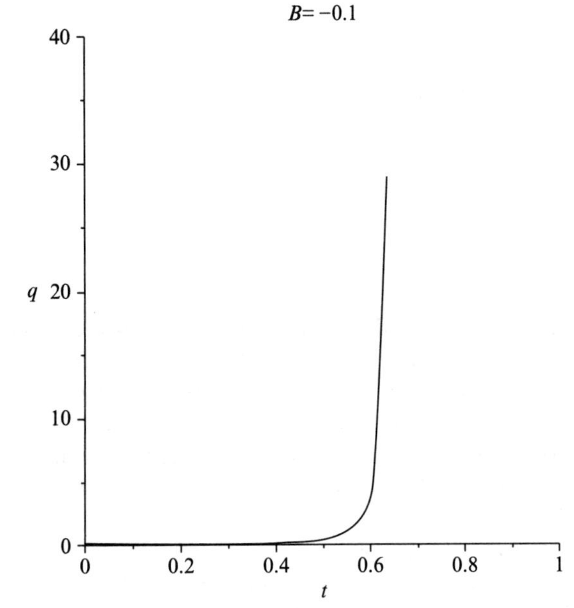

FIGURE 4.40 $q = \widetilde{q}(\widetilde{r})$, $B = \widetilde{B} = -0.1$.

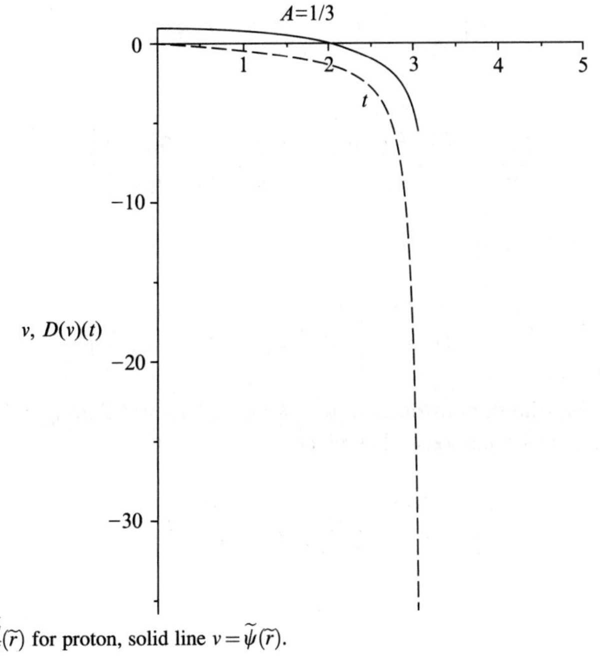

FIGURE 4.41 $v = \widetilde{\psi}(\widetilde{r})$, $D(v)(t) = \frac{\partial \widetilde{\psi}}{\partial \widetilde{r}}(\widetilde{r})$ for proton, solid line $v = \widetilde{\psi}(\widetilde{r})$.

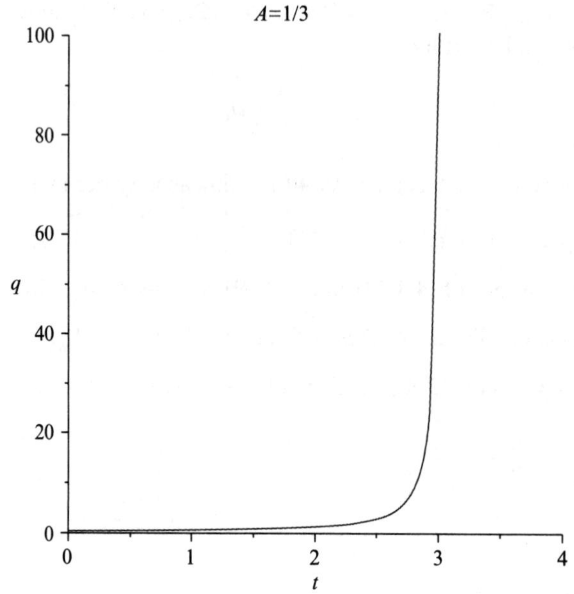

FIGURE 4.42 $q=\tilde{q}(\tilde{r})$ for proton.

Before going further, some points need to be made about the so-called "classical electron radius." This is a calculated radius based on an assumption that the electron is the empty charged sphere with a certain radius. It has a value of $r_0=2.82\times10^{-15}$ m obtained as result of calculation by equating the potential electrostatic energy e^2/r_0 to the energy of rest $m_e c^2$. Now, compare this radius with the measured radius of a proton, which is 1.11×10^{-15} m. There are several sources with different values, but they appear to be around 10^{-15} m. According to this, an electron has a radius 2.5 times larger than a proton. Given that a proton is 1836 heavier, it is difficult to know if we should take this "classical radius" seriously.

As it was shown, the Schrödinger equation (SE) is a deep particular case of the developed theory. But, SE is also a non-local equation. Can SE be used for the calculations of the electron inner structure? Let us consider this question in detail.

Writing down the stationary Madelung equation in the absence of the magnetic field for the case $\mathbf{v}_0=0$

$$\frac{\partial}{\partial \mathbf{r}}\left[U-\frac{\hbar^2}{4m\rho}\left[\Delta\rho-\frac{1}{2\rho}\left(\frac{\partial\rho}{\partial\mathbf{r}}\right)^2\right]\right]=0, \tag{4.4.31}$$

or in the spherical coordinate system

$$\frac{\partial U}{\partial r}=\frac{\partial}{\partial r}\left[\frac{\hbar^2}{4m}\frac{1}{\rho}\left[\frac{\partial^2\rho}{\partial r^2}+\frac{2}{r}\frac{\partial\rho}{\partial r}-\frac{1}{2\rho}\left(\frac{\partial\rho}{\partial r}\right)^2\right]\right]. \tag{4.4.32}$$

For the potential forces of the electrical origin

$$e\frac{\partial\psi}{\partial r}=\frac{\partial}{\partial r}\left[\frac{\hbar^2}{4m}\frac{1}{q}\left[\frac{\partial^2 q}{\partial r^2}+\frac{2}{r}\frac{\partial q}{\partial r}-\frac{1}{2q}\left(\frac{\partial q}{\partial r}\right)^2\right]\right], \tag{4.4.33}$$

where scalar electric potential ψ has the dimension $\mathrm{Dim}[\psi]=\mathrm{CGSE_e}/cm$ and $\mathrm{Dim}[q]=\mathrm{CGSE_e}/cm^3$.

As a result, the following system of equations takes place

$$\frac{\partial^2 q}{\partial r^2}+\frac{2}{r}\frac{\partial q}{\partial r}-\frac{1}{2q}\left(\frac{\partial q}{\partial r}\right)^2-\frac{4me}{\hbar^2}\psi q=0, \tag{4.4.34}$$

$$\frac{\partial^2\psi}{\partial r^2}+\frac{2}{r}\frac{\partial\psi}{\partial r}=4\pi q. \tag{4.4.35}$$

Introducing the scales q_0, r_0, ψ_0 in the dimensionless form we have

$$\frac{\partial^2\tilde{q}}{\partial\tilde{r}^2}+\frac{2}{\tilde{r}}\frac{\partial\tilde{q}}{\partial\tilde{r}}-\frac{1}{2\tilde{q}}\left(\frac{\partial\tilde{q}}{\partial\tilde{r}}\right)^2-\frac{4me\psi_0 r_0^2}{\hbar^2}\tilde{\psi}\tilde{q}=0, \tag{4.4.36}$$

$$\frac{\partial^2\tilde{\psi}}{\partial\tilde{r}^2}+\frac{2}{\tilde{r}}\frac{\partial\tilde{\psi}}{\partial\tilde{r}}=\frac{4\pi q_0 r_0^2}{\psi_0}\tilde{q}. \tag{4.4.37}$$

If the scale system $q_0=\frac{e}{r_0^3}, r_0, \psi_0=\frac{e}{r_0}$ is used, then

$$\begin{aligned}&\frac{\partial^2\tilde{q}}{\partial\tilde{r}^2}+\frac{2}{\tilde{r}}\frac{\partial\tilde{q}}{\partial\tilde{r}}-\frac{1}{2\tilde{q}}\left(\frac{\partial\tilde{q}}{\partial\tilde{r}}\right)^2-\frac{4me^2 r_0}{\hbar^2}\tilde{\psi}\tilde{q}=0,\\&\frac{\partial^2\tilde{\psi}}{\partial\tilde{r}^2}+\frac{2}{\tilde{r}}\frac{\partial\tilde{\psi}}{\partial\tilde{r}}=4\pi\tilde{q}.\end{aligned} \tag{4.4.38}$$

From Eq. (4.4.37), follows that $r_0=\frac{\hbar^2}{4me^2}=r_\tau$ is the character non-local length in the SE. For electron, $r_0=\frac{\hbar^2}{4me^2}=1.32294\times10^{-9}$ cm. We reach the following system of equations

$$\frac{\partial^2\tilde{q}}{\partial\tilde{r}^2}+\frac{2}{\tilde{r}}\frac{\partial\tilde{q}}{\partial\tilde{r}}-\frac{1}{2\tilde{q}}\left(\frac{\partial\tilde{q}}{\partial\tilde{r}}\right)^2-\tilde{\psi}\tilde{q}=0, \tag{4.4.39}$$

$$\frac{\partial^2\tilde{\psi}}{\partial\tilde{r}^2}+\frac{2}{\tilde{r}}\frac{\partial\tilde{\psi}}{\partial\tilde{r}}=4\pi\tilde{q}. \tag{4.4.40}$$

Figs. 4.43 and 4.44 show the dependences $\tilde{\psi}(\tilde{r})$, $\tilde{q}(\tilde{r})$; in the Maple notations, it corresponds to $v(t)$, $q(t)$. Cauchy conditions for the case Fig. 4.43: $v(0)=1, q(0)=1, D(v)(0)=0, D(q)(0)=0$.

Cauchy conditions for the case Fig. 4.44: $v(0)=0.001, q(0)=0.001, D(v)(0)=0, D(q)(0)=0$.

Interesting to notice is that all calculations of the domain existing of the SE solution in the considered cases demonstrate weakly dependence on the initial conditions. Really,

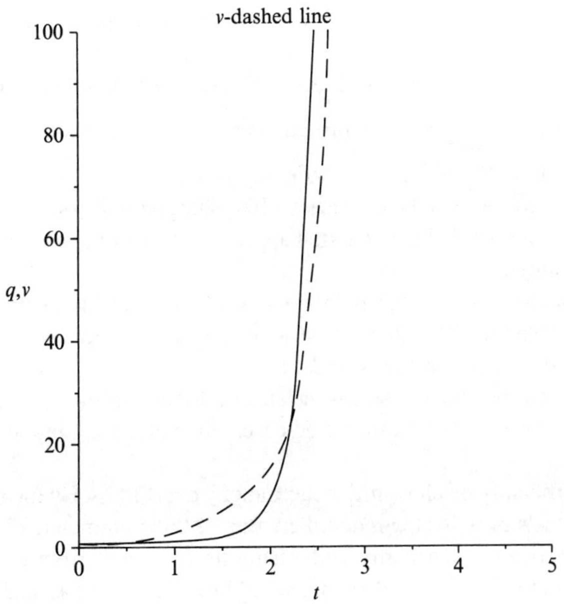

FIGURE 4.43 $v(0)=1, q(0)=1, D(v)(0)=0, D(q)(0)=0$. $\tilde{r}_{\lim}=2.9679941$, $v(t)$—dashed line.

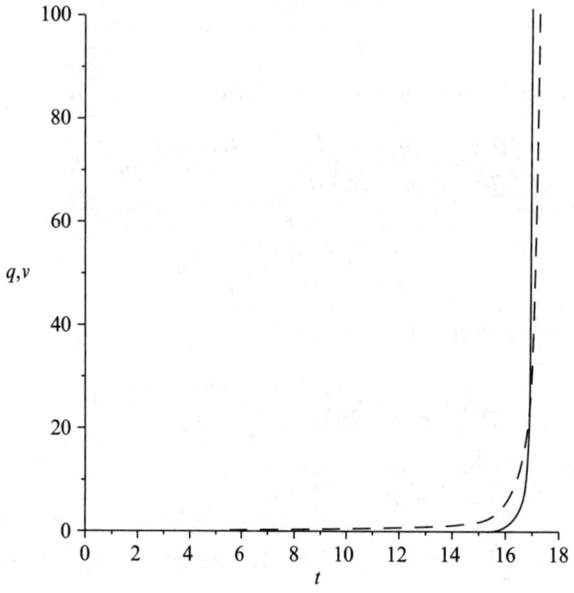

FIGURE. 4.44 Cauchy conditions: $v(0)=0.001, q(0)=0.001, D(v)(0)=0$, $D(q)(0)=0$. $\tilde{r}_{\lim}=17.699269$, $v(t)$—dashed line.

$$v(0)=1, q(0)=1, D(v)(0)=0, D(q)(0)=0, \tilde{r}_{\lim}=2.9679941$$

$$v(0)=10^{-1}, q(0)=10^{-1}, D(v)(0)=0, D(q)(0)=0, \tilde{r}_{\lim}=5.4939122$$

$$v(0)=10^{-3}, q(0)=10^{-3}, D(v)(0)=0, D(q)(0)=0, \tilde{r}_{\lim}=17.699269$$

$$v(0)=10^{-4}, q(0)=10^{-4}, D(v)(0)=0, D(q)(0)=0, \tilde{r}_{\lim}=31.508655$$

$$v(0)=10^{-5}, q(0)=10^{-5}, D(v)(0)=0, D(q)(0)=0, \tilde{r}_{\lim}=56.056471$$

$$v(0)=10^{-7}, q(0)=10^{-7}, D(v)(0)=0, D(q)(0)=0, \tilde{r}_{\lim}=177.30168.$$

Some conclusions from the delivered calculations:

1. The Schrödinger—Madelung model gives the qualitative results akin to the results obtained in the frame of the non-local unified transport theory. But, the quantitative results are inadequate—the linear electron size (estimated via Eq. (4.4.19) or the domain of the solution existing) leads to value more than 10^{-15}cm. Moreover, the electron linear size becomes much more than $r_0 = \frac{\hbar^2}{4m_p e^2} = r_\tau$ for proton. The origin of this situation comes from the absence of the independent energy equation in the Schrödinger—Madelung model.
2. From calculations follow that electrons can be considered like charged balls (shortly CB-model) in which charges are concentrated mainly in the shell of these balls. In the first approximation (when $\vartheta = 0$), this result does not depend on the choice of the non-locality parameter.
3. Electron radius cannot be indicated exactly in principle; its radius depends on the physical system where an electron is placed. It is possible to speak about the different electron shells connected with evolution of the charge density, quantum pressure, electric potential, and forces near the boundary.
4. From the theoretical point of view the electron size is the size of domain of the existence of the corresponding solution. The mentioned sizes $\tilde{r}_{\lim}$ are indicated for all considered cases; the values $\tilde{r}_{\lim}$ practically do not depend on the chosen numerical method.
5. The value of $\tilde{r}_{\lim}$ depends significantly on choosing of the Cauchy conditions. By the same Cauchy conditions, the weak dependence on parameter $\tilde{B}$ exists only for the moderate value of this parameter. If $|\tilde{B}|$ is of the unit order or of more value, $\tilde{r}_{\lim}$ may vary very significantly especially with changing of sign in front of $\tilde{B}$.
6. The proton-electron collision in the frame of CB-model should be considered as *collision of two resonators*. Curves of the equal amplitudes of the intensity of electric field create domains in proton in the form of many "islands"—*caustic surfaces of electromagnetic field*, which can serve as additional scattering centers. It can open a new way for the

explanation of a number of character collisional features depending on the initial and final electron energies without considering partons or quarks as scattering centers, [98].

7. Items 1–6 of conclusion should be taken into account in the theory of the single floating electron been isolated in a Penning trap (see for example [99, 100]).

4.5 TO THE THEORY OF PROTON AND ELECTRON AS BALL-LIKE CHARGED OBJECTS

The affirmation that protons and electrons can be considered as ball-like charged objects radically changes the theoretical results of e^-p scattering. The wave length λ is correlated with the particle impulse p_p as $\lambda = \frac{h}{p_p}$; this relation leads to condition of the particle localization. At very low electron energies when the wavelength is much more than the proton radius r_p, the scattering is equivalent to that from "point-like" spin-less object. The localization begins when $\lambda \sim r_p$ and impulse is about 1 GeV/c. At high electron energies $\lambda < r_p$, the wavelengths are sufficiently short to resolve substructure.

It is reasonable to write down the Rosenbluth formula for elastic e^-p scattering expressed as

$$\frac{d\sigma}{d\Omega} = \left(\frac{d\sigma}{d\Omega}\right)_{\text{Mott}} \left(\frac{G_E^2 + \tau_p G_M^2}{1+\tau_p} + 2\tau_p G_M^2 \tan^2\frac{\theta}{2}\right) \tag{4.5.1}$$

where $\left(\frac{d\sigma}{d\Omega}\right)_{\text{Mott}}$ is the Mott differential cross-section, which includes the proton recoil. It corresponds to scattering from a spin-0 proton. Formula (4.5.1) formally is valid for extended charged object. With this aim, Rosenbluth introduced the dimensionless electric G_E and magnetic G_M form factors. Formula (4.5.1) contains also the Lorentz invariant quantity $\tau_p = -\frac{q_p^2}{4M^2c^2} > 0$. The form factors are a function of q_p^2 rather than scalar production $(\mathbf{p_p} - \mathbf{p_p}')^2$ in the three-dimensional space and, generally speaking, cannot be considered in terms of the Fourier transformation of the charge and magnetic moment distribution. If E and E' are initial and final electron energies then

$$q_p^2 = \frac{1}{c^2}(E-E')^2 - \left(\mathbf{p_p} - \mathbf{p_p'}\right)^2, \tag{4.5.2}$$

or

$$-\left(\mathbf{p_p} - \mathbf{p_p'}\right)^2 = q_p^2\left[1 - \left(\frac{q_p}{2Mc}\right)^2\right]; \tag{4.5.3}$$

only if

$$\frac{\left|q_p^2\right|}{4M^2c^2} \ll 1, \tag{4.5.4}$$

one obtains $q_p^2 \approx -(\mathbf{p_p} - \mathbf{p_p}')^2$ with typical approximation $G_E = G_E(\mathbf{p_p} - \mathbf{p_p}')^2$, $G_M = G_M(\mathbf{p_p} - \mathbf{p_p}')^2$. In the limit (4.5.4), the Fourier transforms are used like

$$G_E\left(q_p^2\right) = \int d^3\mathbf{r}\{q(\mathbf{r})\exp\left[i(\tilde{\mathbf{p}} - \tilde{\mathbf{p}}')\cdot\tilde{\mathbf{r}}\right]\}, \tag{4.5.5}$$

$$G_M\left(q_p^2\right) = \int d^3\mathbf{r}\{\mu(\mathbf{r})\exp\left[i(\tilde{\mathbf{p}} - \tilde{\mathbf{p}}')\cdot\tilde{\mathbf{r}}\right]\}. \tag{4.5.6}$$

Rosenbluth formula is derived for a spin-half Dirac particle, then for magnetic moment $\vec{\mu} = \frac{e}{M}\vec{S}$. The typical experimental correction leads to additional coefficient.

$$\vec{\mu} = 2.79\frac{e}{M}\vec{S}. \tag{4.5.7}$$

All calculations depend significantly on the choice of the approximation for the charge density $q(\mathbf{r})$. Until now, the following $q(\mathbf{r})$ approximations are considered: point-like, exponential dependence which grows smaller with the distance from the proton center, Gaussian, uniform sphere, and Fermi function. In general, the calculations are sensitive to the choice of $\tau_p = -\frac{q_p^2}{4M^2c^2} > 0$. For low $|q_p^2|(\tau_p \approx 0)$, we have

$$\frac{d\sigma}{d\Omega}=\left(\frac{d\sigma}{d\Omega}\right)_{\text{Mott}}G_{\text{E}}^{2}. \tag{4.5.8}$$

For high $|q_{\text{p}}^{2}|(\tau_{p}\gg 1)$,

$$\frac{d\sigma}{d\Omega}=\left(\frac{d\sigma}{d\Omega}\right)_{\text{Mott}}G_{\text{M}}^{2}\left(1+2\tau_{\text{p}}\tan^{2}\frac{\theta}{2}\right). \tag{4.5.9}$$

From the first glance it seems that the theory of elastic scattering needs only in the recalculation (4.5.5), (4.5.6) with the use of new models for the internal proton and electron structures. But situation is more complicated.

The proton-electron collision in the frame of CB-model should be considered as *collision of two resonators.*

Application of the CB – model allows to explain a number of character effects observed by inelastic electron – proton collisions (including the initial and final electron energies and the scattering angles) without application of the quark theory. At low $-q_{\text{p}}^{2}\approx(\mathbf{p_p}-\mathbf{p_p}')^{2}$, one observes not only the elastic peak but also the resonance curves typical for the excitations in resonators including discrete and continuum spectra. Resonance curves disappear when the cavity cannot serve as a resonator. It is the situation which is well known from radio physics. For example, the usual resonance band corresponds to approximately 8 GeV and the wavelength to $\sim 1.5\times 10^{-14}$ cm, [101]. It leads to the relation $\lambda/r_{\text{pr}}\sim 0.1$. This relation is typical for axially symmetric resonators (see for example [102]). This paper contains calculations for the complex shaped cavity with axial symmetry; the resonance frequency is about 70 GHz with the character length ~35 mm. In the definite sense, it is the similar situation; the cavity contains very complicated topology of the electromagnetic field. Curves of equal amplitudes of the intensity of electric field create domains in the form of many "islands"—*caustic surfaces of electromagnetic field.* These "islands" could be the origin of specific features of the electron scattering usually related with partons or quarks. It can be indicated the results [103] in support of this conjecture. In [103], M. Popescu realizes the analyses of the cross-sections by collisions π^{+}p and π^{-}p. It was noticed that the experimental data converge to show that the curves giving the total cross-section versus energy are smoothed curves, with large peaks (resonance) for medium energies and a slow variation with some waving on the high-energy side. Popescu points out that the curves exhibit striking resemblance to scattering curves obtained by electron, neutron, or X-ray scattering in liquid and amorphous materials. Starting from this observation, Popescu developed a method to extract structural information on the proton internal structure in the hypothesis that ions are scattered by unknown internal centers when they knock the proton. The result of this analysis was very interesting because the total number of scattering centers was estimated as 20-30 with the severe variance with the number of quarks now supposed to be integrated in the proton. It should be added that mentioned number of scattering centers is the typical quantity of "islands" in resonators for the mentioned conditions.

In this connection, another interesting problem arises. Is it possible to obtain the experimental confirmation of the resonator model for the electron? In this case, it is reasonable to remind one old Blokhintsev paper published in Physics-Uspekhi as the letter to Editor [104]. He considered the process of the interaction neutrino υ and electron e with transformation of electron in μ—meson: $\upsilon+\text{e}\rightarrow\mu+\upsilon'$. In this case, the energy density W can be estimated as

$$W=g^{*}\overline{\psi}_{\text{e}}\psi_{\mu}\overline{\psi}_{\upsilon}\psi_{\upsilon'}, \tag{4.5.10}$$

where g^{*} is the Fermi constant, $\psi_{\text{e}},\psi_{\mu},\psi_{\upsilon}$ are wave functions for electron, μ—meson and neutrino correspondingly. Following I.S. Shapiro, Blokhintsev estimated g^{*} as

$$\frac{g^{*}}{\hbar c}=\Lambda_{0}^{2}, \tag{4.5.11}$$

with $\Lambda_{0}\sim 10^{-16}$ cm. His conclusion consists in affirmation that the strong interaction of electron and neutrino takes place when the wave length $\lambda\!\!\!^{-}$ of the neutrino wave packet is less than Λ_{0}.

$$\lambda\!\!\!^{-}<\Lambda_{0}. \tag{4.5.12}$$

The inequality (4.5.12) can be considered as estimation for revealing of the resonance electron properties. Blohintzev supposes that fulfilling of (4.5.12) leads to the significant changes in the Compton effect and other changes in electromagnetic interaction of electrons. It is also possible to wait for the influence of the resonance electron effects on investigation of hypothetical neutrino oscillations.

Chapter 5

Non-Local Quantum Hydrodynamics in the Theory of Plasmoids and the Atom Structure

ABSTRACT

Quantum solitons are discovered with the help of generalized quantum hydrodynamics. The solitons have the character of the stable quantum objects in the self-consistent electric field. Particularly these effects can be considered as explanation of the existence of the stable plasmoids, lightning balls, and atoms with the separated electronic shell and the positive kernel. The delivered theory demonstrates the great possibilities of the generalized quantum hydrodynamics in investigation of the quantum solitons. The theory leads to solitons as typical formations in the generalized quantum hydrodynamics. The self-consistent theory of plasmoids cannot be constructed in the frame of local physics.

Keywords: Quantum hydrodynamics, Theory of plasmoids, Atom structure

5.1 THE STATIONARY SINGLE SPHERICAL PLASMOID

Ball lightning is an atmospheric electrical phenomenon. The properties of a "typical" ball lightning are associated with:

1. Thunderstorms, but lasts considerably longer than the split-second flash of a lightning bolt.
2. Shapes that vary between spheres, ovals, tear-drops, rods, or disks.
3. Its capability to change form, split into fragments, and penetrate through chinks.
4. Peculiar character of its movement (absence of convection, movement against the wind, floating along conductors).
5. The lifetime of each event is from 1 s to over a minute with the brightness remaining fairly constant during that time.
6. Quiet dying or destruction with explosion.
7. Absence of heat emission, and burns at close contact.
8. Its ability to penetrate through obstacles (glasses, nets) with or without damaging them.
9. The presence or absence of noise and odor, accompanying its appearance.

Ball lightning is often erroneously identified as St. Elmo's fire. St. Elmo's fire is named after St. Erasmus, the patron saint of sailors. The phenomenon sometimes appeared on ships at sea during thunderstorms. St. Elmo's light is a weather phenomenon in which luminous plasma is created by a coronal discharge from a sharp or pointed object in a strong electric field in the atmosphere.

Given the wide range of physical conditions under which events have been reported in nature (including UFO like foo-fighters), it is quite likely that ball lightning is not a single phenomenon but a collection of physical phenomena that gives rise to similar observables [105–109]. Over the last century, there have been numerous attempts to produce an atmospheric ball lightning. The first reproducible experimental production of ball lightning-like phenomena is attributed to Nicola Tesla during his infamous year at the Colorado Springs laboratory in 1899-1900 [110]. In January 1900, Tesla noted that "the phenomenon of the 'fireball' is produced by the sudden heating, to high incandescence, of a mass of air or other gas as the case may be, by the passage of a powerful discharge."

As you see "the ball lightning" is not aptly called. More preferable name is plasmoid. The word *plasmoid* was coined in 1956 by Winston H. Bostick to mean a "plasma-magnetic entity." Hereafter we intend to use "plasmoid" in the extended sense for an object with the separated positive and negative charges—it does not matter whether the magnetic field is existing or not.

Unified non-local theory of transport processes. http://dx.doi.org/10.1016/B978-0-444-63478-8.00005-1

Plasmoid phenomenon has attracted the attention of researchers for more than 200 years. Considerable number of papers is published in this area including monographs and review articles. Many efforts have been made for theoretical explanation generation, structure, and long lifetime of ball lightning. A number of models for the ball lightning have been developed. But all theoretical models have the same character features—they are developed in the frame of the local physics. Further is shown that local models have no chance for success.

Moreover, the creation of the plasmoid theory means also the creation of theory of the atom structure with the simultaneous description of the electronic shell and the positive nucleus. Let us consider in the frame of the non-local hydrodynamic description the charge system placed in a space domain. The character linear size of the object will be defined as a result of the self-consistent solution of the generalized hydrodynamic equations (GHE). The internal species energy and the possible influence magnetic field here are not interesting for us. Moreover we suppose that this single charged object has the spherical symmetry. For the case under consideration we have (see also (4.1.1)–(4.1.6), (4.2.1)–(4.2.5)):

- continuity equation for species α:

$$\begin{aligned}&\frac{\partial}{\partial t}\left\{\rho_\alpha-\tau_\alpha\left[\frac{\partial\rho_\alpha}{\partial t}+\frac{1}{r^2}\frac{\partial(r^2\rho_\alpha v_{0r})}{\partial r}\right]\right\}+\frac{1}{r^2}\frac{\partial}{\partial r}\left\{r^2\left\{\rho_\alpha v_{0r}-\tau_\alpha\left[\frac{\partial}{\partial t}(\rho_\alpha v_{0r})\right.\right.\right.\\&\left.\left.\left.+\frac{1}{r^2}\frac{\partial(r^2\rho_\alpha v_{0r}^2)}{\partial r}+n_\alpha q_\alpha\frac{\partial\psi}{\partial r}\right]\right\}\right\}-\frac{1}{r^2}\frac{\partial}{\partial r}\left(\tau_\alpha r^2\frac{\partial p_\alpha}{\partial r}\right)=R_\alpha,\end{aligned}\tag{5.1.1}$$

- momentum equation for mixture

$$\begin{aligned}&\frac{\partial}{\partial t}\left\{\rho v_{0r}-\sum_\alpha\tau_\alpha\left[\frac{\partial}{\partial t}(\rho_\alpha v_{0r})+\frac{1}{r^2}\frac{\partial(r^2\rho_\alpha v_{0r}^2)}{\partial r}+\frac{\partial p_\alpha}{\partial r}+n_\alpha q_\alpha\frac{\partial\psi}{\partial r}\right]\right\}\\&+\frac{\partial\psi}{\partial r}\sum_\alpha\left[n_\alpha q_\alpha-\tau_\alpha\left(q_\alpha\frac{\partial n_\alpha}{\partial t}+\frac{1}{r^2}q_\alpha\frac{\partial(r^2n_\alpha v_{0r})}{\partial r}\right)\right]\\&+\frac{1}{r^2}\frac{\partial}{\partial r}\left\{r^2\left\{\rho v_{0r}^2-\sum_\alpha\tau_\alpha\left[\frac{\partial}{\partial t}(\rho_\alpha v_{0r}^2)+\frac{1}{r^2}\frac{\partial(r^2\rho_\alpha v_{0r}^3)}{\partial r}+2q_\alpha n_\alpha\frac{\partial\psi}{\partial r}v_{0r}\right]\right\}\right\}\\&+\frac{\partial p}{\partial r}-\frac{\partial}{\partial t}\sum_\alpha\left(\tau_\alpha\frac{\partial p_\alpha}{\partial r}\right)-2\frac{\partial}{\partial r}\sum_\alpha\left(\frac{\tau_\alpha}{r^2}\frac{\partial(r^2p_\alpha v_{0r})}{\partial r}\right)-\frac{1}{r^2}\sum_\alpha\frac{\partial}{\partial r}\left(\tau_\alpha r^2\frac{\partial(p_\alpha v_{0r})}{\partial r}\right)=0.\end{aligned}\tag{5.1.2}$$

- energy equation for species α

$$\begin{aligned}&\frac{\partial}{\partial t}\left\{\frac{1}{2}\rho_\alpha v_{0r}^2+\frac{3}{2}p_\alpha-\tau_\alpha\left[\frac{\partial}{\partial t}\left(\frac{1}{2}\rho_\alpha v_{0r}^2+\frac{3}{2}p_\alpha\right)+\frac{1}{r^2}\frac{\partial}{\partial r}\left(r^2v_{0r}\left(\frac{1}{2}\rho_\alpha v_{0r}^2+\frac{5}{2}p_\alpha\right)\right)\right.\right.\\&\left.\left.+q_\alpha n_\alpha\frac{\partial\psi}{\partial r}v_{0r}\right]\right\}+\frac{1}{r^2}\frac{\partial}{\partial r}\left\{r^2\left\{\left(\frac{1}{2}\rho_\alpha v_{0r}^2+\frac{5}{2}p_\alpha\right)v_{0r}-\tau_\alpha\left[\frac{\partial}{\partial t}\left(\left(\frac{1}{2}\rho_\alpha v_{0r}^2+\frac{5}{2}p_\alpha\right)v_{0r}\right)\right.\right.\right.\\&\left.\left.\left.+\frac{1}{r^2}\frac{\partial}{\partial r}\left(r^2\left(\frac{1}{2}\rho_\alpha v_{0r}^2+\frac{7}{2}p_\alpha\right)v_{0r}^2\right)+q_\alpha n_\alpha\frac{\partial\psi}{\partial r}v_{0r}^2+\frac{\partial\psi}{\partial r}\left(\frac{1}{2}n_\alpha q_\alpha v_{0r}^2+\frac{n_\alpha q_\alpha}{\rho_\alpha}\frac{3}{2}p_\alpha\right)\right]\right\}\right\}\\&+\left\{q_\alpha n_\alpha\frac{\partial\psi}{\partial r}v_{0r}-\tau_\alpha\left[\frac{q_\alpha n_\alpha}{\rho_\alpha}\frac{\partial\psi}{\partial r}\left(\frac{\partial}{\partial t}(\rho_\alpha v_{0r})+\frac{1}{r^2}\frac{\partial}{\partial r}(r^2\rho_\alpha v_{0r}^2)+\frac{\partial p_\alpha}{\partial r}+q_\alpha n_\alpha\frac{\partial\psi}{\partial r}\right)\right]\right\}\\&-\frac{1}{r^2}\frac{\partial}{\partial r}\left(\tau_\alpha r^2\frac{\partial}{\partial r}\left(\frac{1}{2}p_\alpha v_{0r}^2+\frac{5}{2}\frac{p_\alpha^2}{\rho_\alpha}\right)\right)-\frac{1}{r^2}\frac{\partial}{\partial r}\left(r^2\tau_\alpha p_\alpha\frac{q_\alpha n_\alpha}{\rho_\alpha}\frac{\partial\psi}{\partial r}\right)\\&=\int\left(\frac{m_\alpha v_\alpha^2}{2}+\varepsilon_\alpha\right)J_\alpha^{st,el}\mathbf{d}\mathbf{v}_\alpha+\int\left(\frac{m_\alpha v_\alpha^2}{2}+\varepsilon_\alpha\right)J_\alpha^{st,inel}\mathbf{d}\mathbf{v}_\alpha.\end{aligned}\tag{5.1.3}$$

Here q_α—charge of the α-component particle, p_α—static pressure for α-component, $\mathbf{v}_0$—hydrodynamic velocity for mixture, τ_α—non-local parameter. Suppose also that the stationary physical system is in rest; it means

$$v_{0r}=0,\partial/\partial t\equiv 0.\tag{5.1.4}$$

The main problem from the mathematical point of view—is it possible to find the soliton kind solution from Eqs. (5.1.1)–(5.1.3) as result of the Cauchy problem consideration? This question has the positive answer.

5.2 RESULTS OF THE MATHEMATICAL MODELING OF THE REST SOLITONS

Let us write down Eqs. (5.1.1)–(5.1.3) using the conditions (5.1.4). By the conditions (5.1.4) Poisson equation can be written in the usual form

$$\frac{1}{r^2}\frac{\partial}{\partial r}\left(r^2\frac{\partial\psi}{\partial r}\right)=-4\pi e[n_i-n_e]. \tag{5.2.1}$$

Continuity equation for species α:

$$\frac{\partial p_\alpha}{\partial r}+n_\alpha q_\alpha\frac{\partial\psi}{\partial r}=0. \tag{5.2.2}$$

Momentum equation for mixture:

$$\sum_\alpha\left\{\frac{\partial\psi}{\partial r}n_\alpha q_\alpha+\frac{\partial p_\alpha}{\partial r}\right\}=0. \tag{5.2.3}$$

This equation can be satisfied identically by the condition (5.2.2). Using (5.2.4) we find from the energy equation

$$\begin{aligned}&-\frac{5}{2}\frac{1}{r^2}\frac{\partial}{\partial r}\left[\tau_\alpha r^2\frac{n_\alpha q_\alpha}{\rho_\alpha}p_\alpha\frac{\partial\psi}{\partial r}\right]-\tau_\alpha\frac{q_\alpha n_\alpha}{\rho_\alpha}\frac{\partial\psi}{\partial r}\left(\frac{\partial p_\alpha}{\partial r}+q_\alpha n_\alpha\frac{\partial\psi}{\partial r}\right)\\&-\frac{5}{2}\frac{1}{r^2}\frac{\partial}{\partial r}\left(\tau_\alpha r^2\frac{\partial}{\partial r}\left(\frac{p_\alpha^2}{\rho_\alpha}\right)\right)=\int\left(\frac{m_\alpha v_\alpha^2}{2}+\varepsilon_\alpha\right)J_\alpha^{st,el}d\mathbf{v}_\alpha+\int\left(\frac{m_\alpha v_\alpha^2}{2}+\varepsilon_\alpha\right)J_\alpha^{st,inel}d\mathbf{v}_\alpha;\end{aligned} \tag{5.2.4}$$

taking into account (5.2.2):

$$-\frac{5}{2}\frac{1}{r^2}\frac{\partial}{\partial r}\left[\tau_\alpha r^2\left(\frac{n_\alpha q_\alpha}{\rho_\alpha}p_\alpha\frac{\partial\psi}{\partial r}+\frac{\partial}{\partial r}\left(\frac{p_\alpha^2}{\rho_\alpha}\right)\right)\right]=\int\left(\frac{m_\alpha v_\alpha^2}{2}+\varepsilon_\alpha\right)J_\alpha^{st,el}d\mathbf{v}_\alpha+\int\left(\frac{m_\alpha v_\alpha^2}{2}+\varepsilon_\alpha\right)J_\alpha^{st,inel}d\mathbf{v}_\alpha$$

and after the repeated (5.2.2) application we reach the equation

$$-\frac{1}{r^2}\frac{5}{2}\frac{\partial}{\partial r}\left[r^2\tau_\alpha p_\alpha\frac{\partial}{\partial r}\left(\frac{p_\alpha}{\rho_\alpha}\right)\right]=\int\left(\frac{m_\alpha v_\alpha^2}{2}+\varepsilon_\alpha\right)J_\alpha^{st,el}d\mathbf{v}_\alpha+\int\left(\frac{m_\alpha v_\alpha^2}{2}+\varepsilon_\alpha\right)J_\alpha^{st,inel}d\mathbf{v}_\alpha, \tag{5.2.5}$$

where index $\alpha=e^-,i^+$ corresponds to the negative and positive species. Local collision integrals of the right hand side of equation (5.2.5) can be written in the relaxation form.

$$\frac{1}{r^2}\frac{5}{2}\frac{\partial}{\partial r}\left[r^2\tau_i p_i\frac{\partial}{\partial r}\left(\frac{p_i}{\rho_i}\right)\right]=\frac{p_i-p_e}{\tau_{ei}}, \tag{5.2.6}$$

$$\frac{1}{r^2}\frac{5}{2}\frac{\partial}{\partial r}\left[r^2\tau_e p_e\frac{\partial}{\partial r}\left(\frac{p_e}{\rho_e}\right)\right]=\frac{p_e-p_i}{\tau_{ei}}. \tag{5.2.7}$$

Non-local parameter τ_{ei} of the interaction of the positive and negative particles is written as

$$\tau_{ei}^{-1}=\tau_e^{-1}+\tau_i^{-1}. \tag{5.2.8}$$

In this case τ_{ei} is the relaxation time in the process of the particle interaction of different species. The relation (5.2.8) is in keeping with the Heisenberg principle of uncertainty. No reason to discuss τ_e, τ_i, τ_{ei} separately because in the following all these parameters will be introduced in equations as a combination. Further the dependent variables $q_i=en_i$, $q_e=en_e$ are used (instead of being introduced before q_α as the particle charges), where e is the absolute value of the electron charge, n_i, n_e are the numerical densities of the positive and negative species. Transform Eqs. (5.2.1), (5.2.2), (5.2.6), and (5.2.7) to the dimensionless forms using the sign tilde for the dimensionless values, the scales $\rho_0, \psi_0, r_0, p_0=q_0\psi_0$ are used.

$$A\frac{\partial}{\partial\tilde r}\left(\tilde r^2\frac{\partial\tilde\psi}{\partial\tilde r}\right)=\tilde r^2(\tilde q_e-\tilde q_i), \tag{5.2.9}$$

$$\frac{5}{2}\frac{e}{m_i}\tau_i\tau_{ei}\frac{\psi_0}{r_0^2}\frac{\partial}{\partial\tilde r}\left[\tilde r^2\tilde p_i\frac{\partial}{\partial\tilde r}\left(\frac{\tilde p_i}{\tilde q_i}\right)\right]=\tilde r^2(\tilde p_i-\tilde p_e), \tag{5.2.10}$$

$$\frac{5}{2}\frac{e}{m_e}\tau_e\tau_{ei}\frac{\psi_0}{r_0^2}\frac{\partial}{\partial\tilde r}\left[\tilde r^2\tilde p_e\frac{\partial}{\partial\tilde r}\left(\frac{\tilde p_e}{\tilde q_e}\right)\right]=\tilde r^2(\tilde p_e-\tilde p_i), \tag{5.2.11}$$

$$\frac{\partial\tilde p_i}{\partial\tilde r}+\tilde q_i\frac{\partial\tilde\psi}{\partial\tilde r}=0, \tag{5.2.12}$$

$$\frac{\partial \tilde{p}_e}{\partial \tilde{r}} - \tilde{q}_e \frac{\partial \tilde{\psi}}{\partial \tilde{r}} = 0, \tag{5.2.13}$$

where the factor

$$A = \psi_0 / \left(4\pi r_0^2 q_0\right). \tag{5.2.14}$$

is used. Introduce the definition of non-local Reynolds number

$$\mathrm{Re}_i = LV_i / \nu_i, \tag{5.2.15}$$

where the linear size $L = r_0$, the character velocity $V_i = r_0/\tau_i$, and the kinematic viscosity has the following definition

$$\nu_i = \frac{5}{2}\frac{e}{m_i}\psi_0 \tau_{ei} = \frac{5}{2}\frac{e}{m_i}\tau_{ei}\frac{p_0}{q_0} = \frac{5}{2}\frac{e}{m_i}\tau_{ei}\frac{p_0}{en_0} = \frac{5}{2}\tau_{ei}\frac{p_0}{m_i n_0}. \tag{5.2.16}$$

In this case

$$\frac{1}{\mathrm{Re}_i} = \frac{5}{2}\frac{e}{m_i}\tau_i \tau_{ei}\frac{\psi_0}{r_0^2}. \tag{5.2.17}$$

Analogically

$$\nu_e = \frac{5}{2}\tau_{ei}\frac{p_0}{m_e n_0}, \tag{5.2.18}$$

$$\frac{1}{\mathrm{Re}_e} = \frac{5}{2}\frac{e}{m_e}\tau_e \tau_{ei}\frac{\psi_0}{r_0^2}. \tag{5.2.19}$$

Equations (5.2.10) and (5.2.11) takes the form

$$\frac{1}{\mathrm{Re}_i}\frac{\partial}{\partial \tilde{r}}\left[\tilde{r}^2 \tilde{p}_i \frac{\partial}{\partial \tilde{r}}\left(\frac{\tilde{p}_i}{\tilde{q}_i}\right)\right] = \tilde{r}^2(\tilde{p}_i - \tilde{p}_e), \tag{5.2.20}$$

$$\frac{1}{\mathrm{Re}_e}\frac{\partial}{\partial \tilde{r}}\left[\tilde{r}^2 \tilde{p}_e \frac{\partial}{\partial \tilde{r}}\left(\frac{\tilde{p}_e}{\tilde{q}_e}\right)\right] = \tilde{r}^2(\tilde{p}_e - \tilde{p}_i). \tag{5.2.21}$$

From (5.2.20) and (5.2.21) follow

$$\frac{1}{\mathrm{Re}_i}\frac{\partial}{\partial \tilde{r}}\left[\tilde{r}^2 \tilde{p}_i \frac{\partial}{\partial \tilde{r}}\left(\frac{\tilde{p}_i}{\tilde{q}_i}\right)\right] + \frac{1}{\mathrm{Re}_e}\frac{\partial}{\partial \tilde{r}}\left[\tilde{r}^2 \tilde{p}_e \frac{\partial}{\partial \tilde{r}}\left(\frac{\tilde{p}_e}{\tilde{q}_e}\right)\right] = 0. \tag{5.2.22}$$

One integration can be realized with taking into account that constant of integration is equal to zero.

$$\frac{\mathrm{Re}_e}{\mathrm{Re}_i}\tilde{p}_i \frac{\partial}{\partial \tilde{r}}\left(\frac{\tilde{p}_i}{\tilde{q}_i}\right) + \tilde{p}_e \frac{\partial}{\partial \tilde{r}}\left(\frac{\tilde{p}_e}{\tilde{q}_e}\right) = 0. \tag{5.2.23}$$

Using (5.2.12) and (5.2.13), transform (5.2.23)

$$\frac{\partial \tilde{\psi}}{\partial \tilde{r}}\left(\tilde{p}_e - \frac{\mathrm{Re}_e}{\mathrm{Re}_i}\tilde{p}_i\right) = \left(\frac{\tilde{p}_i}{\tilde{q}_i}\right)^2 \frac{\partial \tilde{q}_i}{\partial \tilde{r}} + \left(\frac{\tilde{p}_e}{\tilde{q}_e}\right)^2 \frac{\partial \tilde{q}_e}{\partial \tilde{r}}, \tag{5.2.24}$$

and

$$\frac{\mathrm{Re}_e}{\mathrm{Re}_i} = \frac{\tau_i/m_i}{\tau_e/m_e}. \tag{5.2.25}$$

Relations (5.2.24) and (5.2.25) are useful in particular for the qualitative analysis of the calculation results.

Some conclusions from the developed theory:

1. The basic system of Eqs. (5.2.9)–(5.2.13) contain the similarity criteria Re_i, Re_e, and A, defining the interrelation of the scales of the electrostatic and hydrostatic origin.
2. Non-local description is of the principal significance. Really, from Eqs. (5.2.9)–(5.2.13) follow only trivial solutions of the equilibrium plasma if the non-locality parameters τ_i, τ_e turn into zero, namely $\tilde{p}_i = \tilde{p}_e$, $\tilde{q}_i = \tilde{q}_e$, and $\tilde{\psi} = const.$
3. Therefore the plasmoid theory (Ball lighting theory) can be constructed *only* in the frame of non-local physics.
4. The criteria variations Re_i, Re_e, A, Cauchy conditions and the possible chemical composition lead to the tremendous class of the possible solutions.

Take into account the formulated remarks and demonstrate the character features of the numerical solutions. The following figures reflect the results of calculations realized with the help of the Maple program. The notations on figures are used: r—$\widetilde{q}_i$ (dimensionless charge for the positive particles); s—$\widetilde{q}_e$ (absolute dimensionless charge for the negative particles); p—pressure $\widetilde{p}_i$; q—pressure $\widetilde{p}_e$ and v—self-consistent potential $\widetilde{\psi}$. Explanations placed under all following figures, Maple program contains Maple's notations—for example the expression $D(v)(0)=0$ means in usual notations $\left(\partial\widetilde{\psi}/\partial\widetilde{r}\right)(0)=0$, independent variable t responds to $\widetilde{r}$.

We investigate the problem of the principal significance—is it possible to obtain the soliton solution after perturbations defined the Cauchy conditions? With this aim introduce the Cauchy perturbations (VARIANT 1):

p(0) = 0.9, q(0) = 1, v(0) = 1, r(0) = 1, s(0) = 1, D(p)(0) = 0, D(q)(0) = 0, D(v)(0) = 0, D(r)(0) = 0, D(s)(0) = 0; $A=\mathrm{Re}_e=\mathrm{Re}_i=1$. Figures 5.1–5.3 reflect the calculation results for VARIANT 1.

Figures 5.4 and 5.5 correspond to the Cauchy perturbations (VARIANT 2):

p(0) = 1, q(0) = 0.9, v(0) = 1, r(0) = 1, s(0) = 1, D(p)(0) = 0, D(q)(0) = 0, D(v)(0) = 0, D(r)(0) = 0, D(s)(0) = 0; $A=\mathrm{Re}_e=\mathrm{Re}_i=1$.

The soliton configurations can remind the atomic structures with the separated positive and negative shells (Fig. 5.6).

Figure 5.6 reflects the results of calculations for VARIANT 3: **p(0) = 1, q(0) = 0.9, v(0) = 1, r(0) = 1, s(0) = 0.1, D(p)(0) = 0, D(q)(0) = 0, D(v)(0) = 0, D(r)(0) = 0, D(s)(0) = 0;** $A=1, B^{-1}=\mathrm{Re}_i=10^6, C^{-1}=\mathrm{Re}_e=10^3$; $r_{\mathrm{lim}}=0.699$.

Figure 5.7 reflects the results of calculations for VARIANT 4: **p(0) = 0.9, q(0) = 1, v(0) = 1, r(0) = 0.1, s(0) = 1, D(p)(0) = 0, D(q)(0) = 0, D(v)(0) = 0, D(r)(0) = 0, D(s)(0) = 0;** $A=1, B^{-1}=\mathrm{Re}_i=10^3, C^{-1}=\mathrm{Re}_e=10^6$, $r_{\mathrm{lim}}=0.699$.

Figures 5.8 and 5.9 reflect the results of calculations for VARIANT 5: **p(0) = 1, q(0) = 0.9, v(0) = 1, r(0) = 1, s(0) = 0.1, D(p)(0) = 0, D(q)(0) = 0, D(v)(0) = 0, D(r)(0) = 0, D(s)(0) = 0;** $A=1, B^{-1}=\mathrm{Re}_i=10^3, C^{-1}=\mathrm{Re}_e=10^6$, $r_{\mathrm{lim}}=0.220$.

Figures 5.10, 5.11 reflects the results of calculations for VARIANT 6: **p(0) = 0.9, q(0) = 1, v(0) = 1, r(0) = 1, s(0) = 0.1, D(p)(0) = 0, D(q)(0) = 0, D(v)(0) = 0, D(r)(0) = 0, D(s)(0) = 0;** $A=1, B^{-1}=\mathrm{Re}_i=10^6, C^{-1}=\mathrm{Re}_e=10^3$, $r_{\mathrm{lim}}=0.00697$.

Important conclusions follow from mathematical modeling:

1. Plasmoid (Ball lightning) is the non-equilibrium product of the matter self-organization, placed in the finite domain of space. This non-equilibrium object has the excess charge (in comparison with the equilibrium state) of one sign along the radial direction and a deficit of the charge of another sign. Figures 5.1 and 5.4 demonstrate the existence of two spherical layers of the significantly different linear sizes. For example one of these domains has the excess positive charge near the kernel (VARIANT 1), but the excess negative charge moves at the periphery. The stability of the plasma object has been reached as the result of the equilibrium of forces of the electrostatic origin and kinetic pressure of the non-local origin. In this case the external spherical layer can have the negative charge if $\widetilde{p}_i(0)<\widetilde{p}_e(0)$, or the positive charge if $\widetilde{p}_i(0)>\widetilde{p}_e(0)$. From the experimental data follow that both these cases can be realized in actual practice [111, 112].

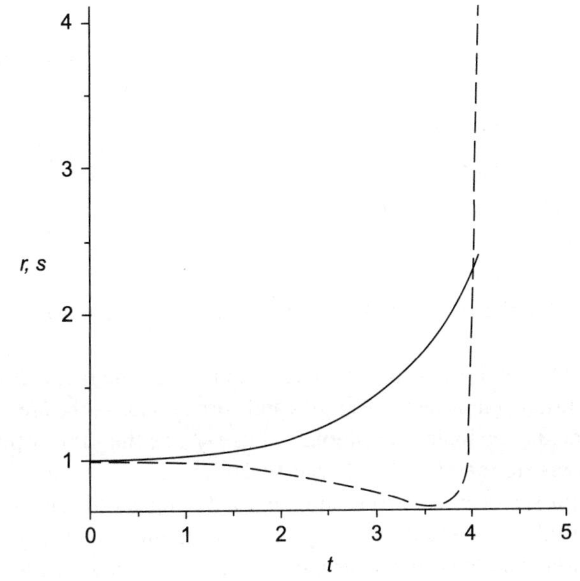

FIGURE 5.1 r—$\widetilde{q}_i(\widetilde{r})$ and s—$\widetilde{q}_e(\widetilde{r})$, VARIANT 1. ($s$—dashed line.)

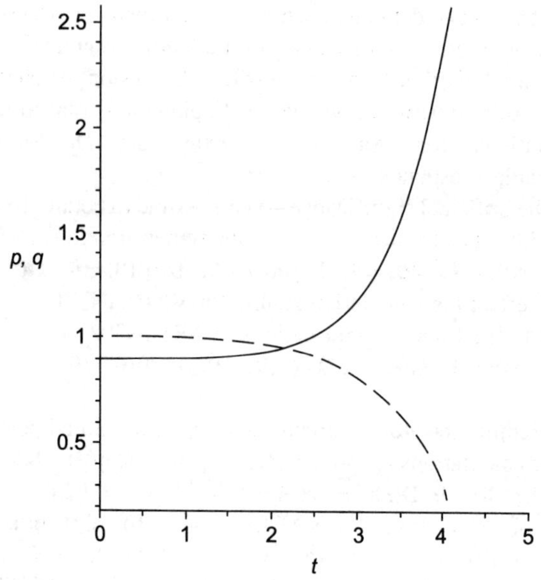

FIGURE 5.2 p—$\widetilde{p}_i(\widetilde{r})$ and q—$\widetilde{p}_e(\widetilde{r})$, VARIANT 1. ($q$—dashed line.)

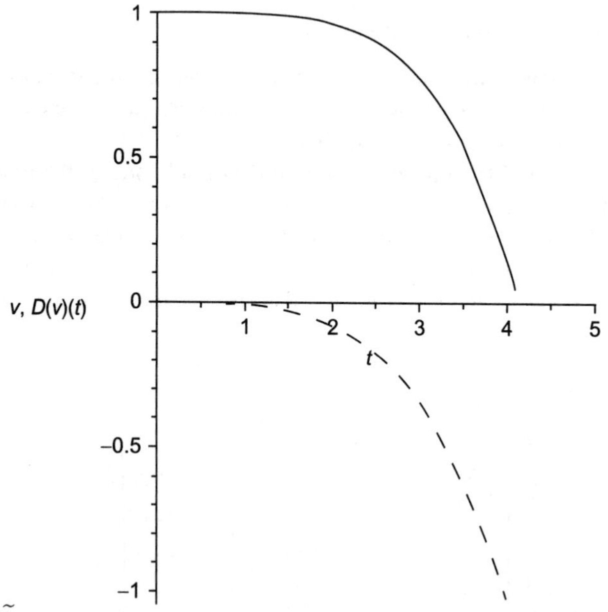

FIGURE 5.3 v—$\widetilde{\psi}(\widetilde{r})$ and $D(v)(t)$—$\frac{\partial\widetilde{\psi}}{\partial r}(\widetilde{r})$, VARIANT 1. ($D(v)(t)$—dashed line.)

2. Mathematical modeling realized in the frame of non-local physics, leads to existence of the stable objects even in the absence of magnetic fields.
3. In the delivered theory no need to use the external boundary conditions. The radial dimensionless size of plasmoid is a result of the self-consistent solution of non-local equations and corresponds to the area of the solution existence. For the considered variants 1 and 2 the dimensionless plasmoid radius $\widetilde{r}_{BL}$ has the same value, $\widetilde{r}_{BL} \sim 4.09$.
4. The theory does not contain restrictions for the charge scales or the object sizes. No need to introduce the convoying magnetic field. It is no surprise—the Schrödinger-Madelung atom theory is the theory of plasmoid with the separated charges (as postulate) without the magnetic confinement of the physical system. From this point of view the Tunguska explosion and Gagarin catastrophe can have the same physical origin—plasmoid appearance in the Earth atmosphere [113].

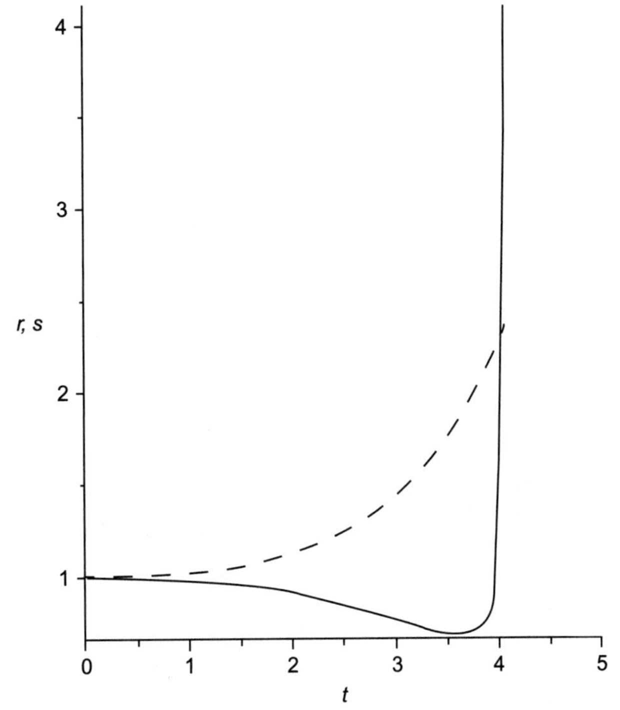

FIGURE 5.4 r—$\tilde{q}_i(\tilde{r})$ and s—$\tilde{q}_e(\tilde{r})$, VARIANT 2. (s—dashed line.)

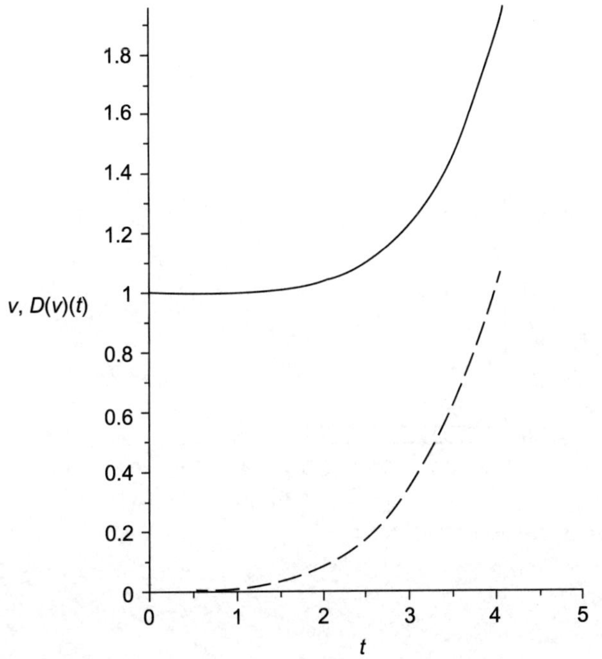

FIGURE 5.5 v—$\tilde{\psi}(\tilde{r})$ and $D(v)(t)$—$\frac{\partial\tilde{\psi}}{\partial\tilde{r}}(\tilde{r})$, ($D(v)(t)$—dashed line). VARIANT 2.

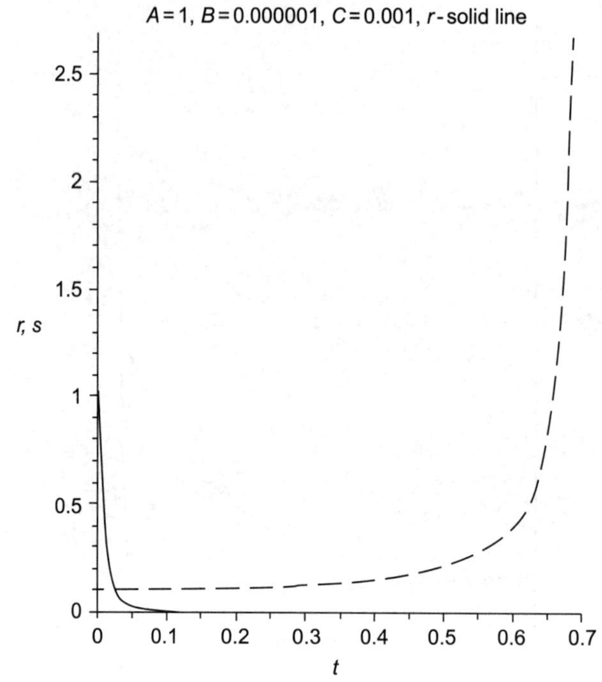

FIGURE 5.6 r—$\widetilde{q}_i(\widetilde{r})$ and s—$\widetilde{q}_e(\widetilde{r})$, ($s$—dashed line). VARIANT 3.

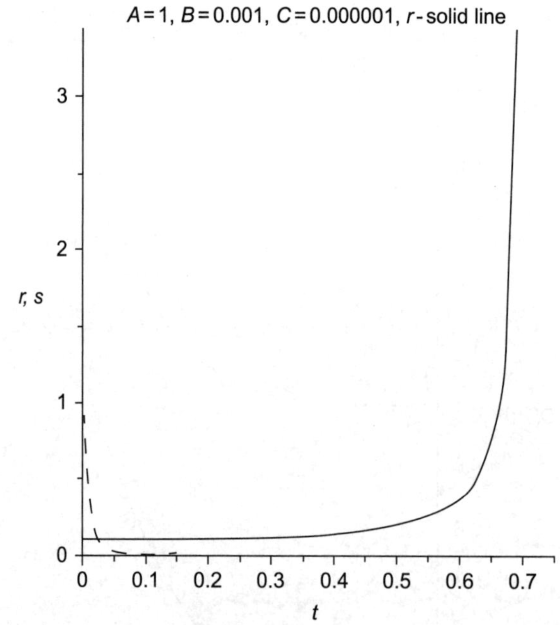

FIGURE 5.7 r—$\widetilde{q}_i(\widetilde{r})$ and s—$\widetilde{q}_e(\widetilde{r})$, ($s$—dashed line). VARIANT 4.

5. As follows from calculations (as minimum) two kinds of plasmoids can exist. Plasmoids as product of plasma polarization (see Figs. 5.1–5.5) and plasmoids with atomic structures (like Fig. 5.6). Obviously the theory of the second type plasmoids describes (in the frame generalized quantum hydrodynamics) the atom structure with the coincident description of nucleus and the electron shell.

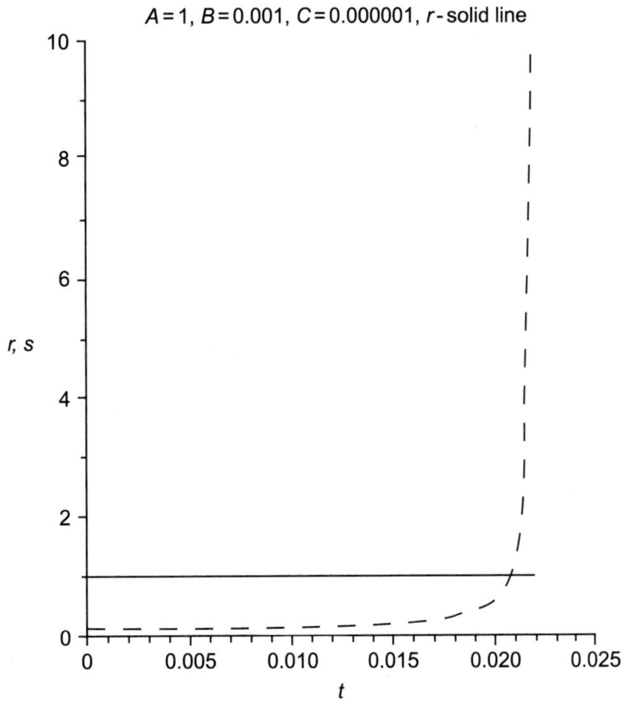

FIGURE 5.8 $r—\widetilde{q}_i(\widetilde{r})$ and $s—\widetilde{q}_e(\widetilde{r})$, ($s$—dashed line). VARIANT 5.

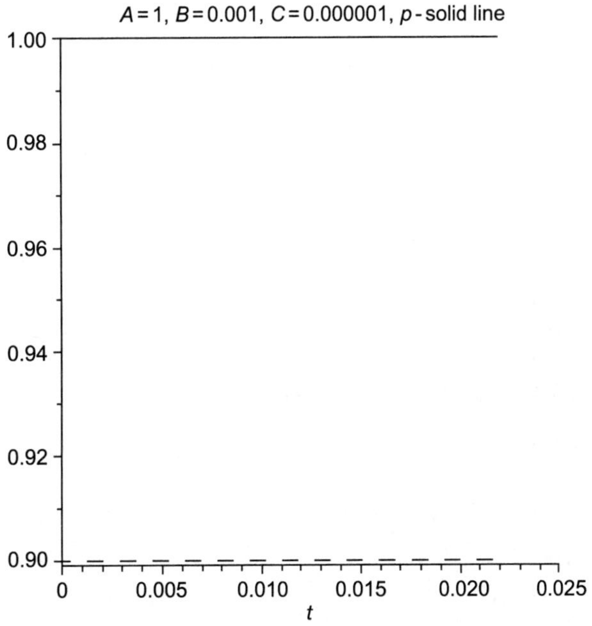

FIGURE 5.9 $p—\widetilde{p}_i(\widetilde{r})$ and $q—\widetilde{p}_e(\widetilde{r})$, (dashed line). VARIANT 5.

6. The controlled discharge should serve for the plasmoid production; this charge should follow the solution of the *non-stationary* non-local equations which leads to the stationary (considered here) charge separation.
7. The main result can be formulated as the proved theorem—the plasmoid theory cannot be constructed in the frame of local physics.

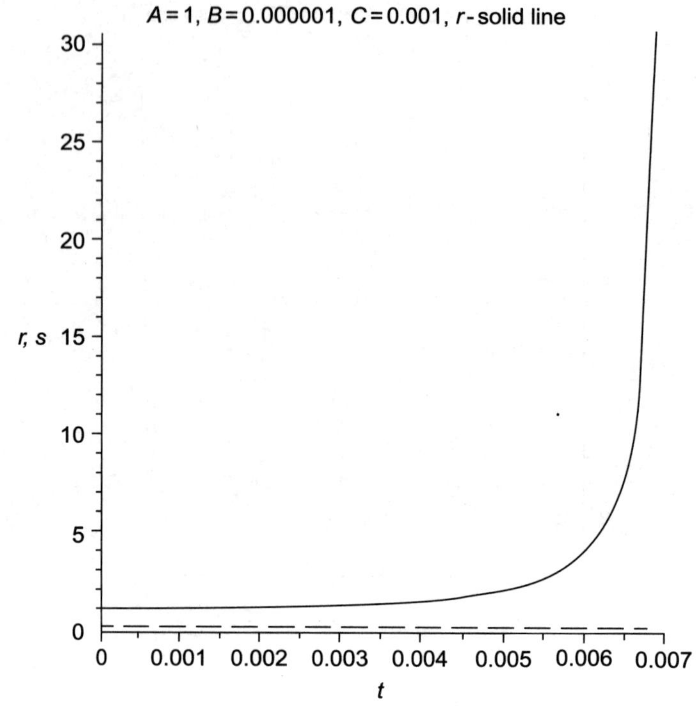

FIGURE 5.10 r—$\widetilde{q}_i(\widetilde{r})$ and s—$\widetilde{q}_e(\widetilde{r})$, ($s$—dashed line). VARIANT 6.

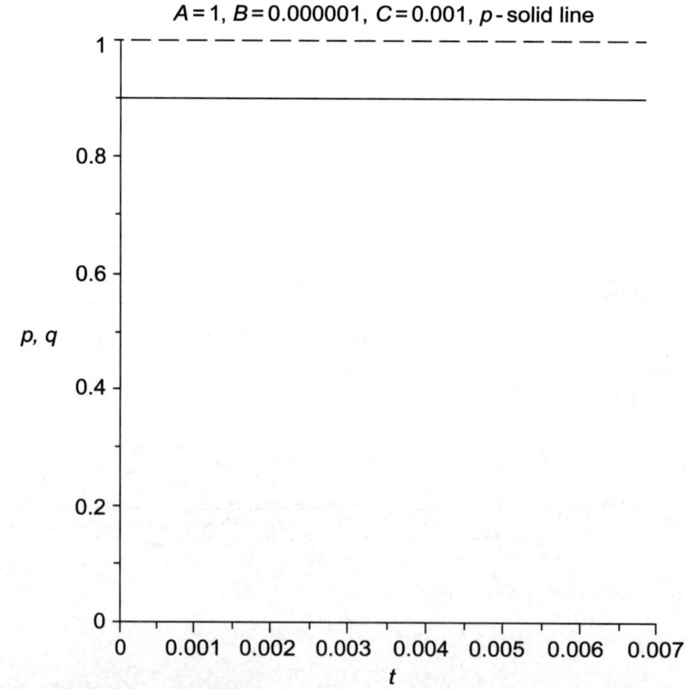

FIGURE 5.11 p—$\widetilde{p}_i(\widetilde{r})$ and q—$\widetilde{p}_e(\widetilde{r})$, (dashed line). VARIANT 6.

5.3 NONSTATIONARY 1D GENERALIZED HYDRODYNAMIC EQUATIONS IN THE SELF-CONSISTENT ELECTRICAL FIELD. QUANTIZATION IN THE GENERALIZED QUANTUM HYDRODYNAMICS

In the following we intend to obtain the soliton's type of solution of the generalized hydrodynamic equations for plasma in the self-consistent electrical field. All elements of possible formation like quantum soliton should move with the same translational velocity. Then the system of GHE consist from the generalized Poisson equation reflecting the effects of the charge and the charge flux perturbations, two continuity equations for positive and negative species (in particular, for ion and electron components), one motion equation and two energy equations for ion and electron components. This system of six equations for nonstationary 1D case can be written in the form (see (4.1.1)–(4.1.6)):

– Poisson equation

$$\frac{\partial^2\psi}{\partial x^2}=-4\pi e\left\{\left[n_i-\tau_i\left(\frac{\partial n_i}{\partial t}+\frac{\partial}{\partial x}(n_i u)\right)\right]-\left[n_e-\tau_e\left(\frac{\partial n_e}{\partial t}+\frac{\partial}{\partial x}(n_e u)\right)\right]\right\}, \tag{5.3.1}$$

– continuity equation for positive ion component

$$\frac{\partial}{\partial t}\left\{\rho_i-\tau_i\left[\frac{\partial\rho_i}{\partial t}+\frac{\partial}{\partial x}(\rho_i u)\right]\right\}+\frac{\partial}{\partial x}\left\{\rho_i u-\tau_i\left[\frac{\partial}{\partial t}(\rho_i u)+\frac{\partial}{\partial x}(\rho_i u^2)+\frac{\partial p_i}{\partial x}-\rho_i F_i\right]\right\}=0, \tag{5.3.2}$$

– continuity equation for electron component

$$\frac{\partial}{\partial t}\left\{\rho_e-\tau_e\left[\frac{\partial\rho_e}{\partial t}+\frac{\partial}{\partial x}(\rho_e u)\right]\right\}+\frac{\partial}{\partial x}\left\{\rho_e u-\tau_e\left[\frac{\partial}{\partial t}(\rho_e u)+\frac{\partial}{\partial x}(\rho_e u^2)+\frac{\partial p_e}{\partial x}-\rho_e F_e\right]\right\}=0, \tag{5.3.3}$$

– momentum equation

$$\begin{aligned}&\frac{\partial}{\partial t}\left\{\rho u-\tau_i\left[\frac{\partial}{\partial t}(\rho_i u)+\frac{\partial}{\partial x}(p_i+\rho_i u^2)-\rho_i F_i\right]-\tau_e\left[\frac{\partial}{\partial t}(\rho_e u)+\frac{\partial}{\partial x}(p_e+\rho_e u^2)-\rho_e F_e\right]\right\}\\&-\rho_i F_i-\rho_e F_e+F_i\tau_i\left(\frac{\partial\rho_i}{\partial t}+\frac{\partial}{\partial x}(\rho_i u)\right)+F_e\tau_e\left(\frac{\partial\rho_e}{\partial t}+\frac{\partial}{\partial x}(\rho_e u)\right)\\&+\frac{\partial}{\partial x}\left\{\begin{array}{l}\rho u^2+p-\tau_i\left[\frac{\partial}{\partial t}(\rho_i u^2+p_i)+\frac{\partial}{\partial x}(\rho_i u^3+3p_i u)-2\rho_i u F_i\right]\\-\tau_e\left[\frac{\partial}{\partial t}(\rho_e u^2+p_e)+\frac{\partial}{\partial x}(\rho_e u^3+3p_e u)\right]-2\rho_e u F_e\end{array}\right\}=0;\end{aligned} \tag{5.3.4}$$

– energy equation for positive ion component

$$\begin{aligned}&\frac{\partial}{\partial t}\left\{\rho_i u^2+3p_i-\tau_i\left[\frac{\partial}{\partial t}(\rho_i u^2+3p_i)+\frac{\partial}{\partial x}(\rho_i u^3+5p_i u)-2\rho_i F_i u\right]\right\}\\&+\frac{\partial}{\partial x}\left\{\rho_i u^3+5p_i u-\tau_i\left[\frac{\partial}{\partial t}(\rho_i u^3+5p_i u)+\frac{\partial}{\partial x}\left(\rho_i u^4+8p_i u^2+5\frac{p_i^2}{\rho_i}\right)-F_i(3\rho_i u^2+5p_i)\right]\right\}\\&-2u\rho_i F_i+2\tau_i F_i\left[\frac{\partial}{\partial t}(\rho_i u)+\frac{\partial}{\partial x}(\rho_i u^2+p_i)-\rho_i F_i\right]=-\frac{p_i-p_e}{\tau_{ei}},\end{aligned} \tag{5.3.5}$$

– energy equation for electron component

$$\begin{aligned}&\frac{\partial}{\partial t}\left\{\rho_e u^2+3p_e-\tau_e\left[\frac{\partial}{\partial t}(\rho_e u^2+3p_e)+\frac{\partial}{\partial x}(\rho_e u^3+5p_e u)-2\rho_e F_e u\right]\right\}\\&+\frac{\partial}{\partial x}\left\{\rho_e u^3+5p_e u-\tau_e\left[\frac{\partial}{\partial t}(\rho_e u^3+5p_e u)+\frac{\partial}{\partial x}\left(\rho_e u^4+8p_e u^2+5\frac{p_e^2}{\rho_e}\right)-F_e(3\rho_e u^2+5p_e)\right]\right\}\\&-2u\rho_e F_e+2\tau_e F_e\left[\frac{\partial}{\partial t}(\rho_e u)+\frac{\partial}{\partial x}(\rho_e u^2+p_e)-\rho_e F_e\right]=-\frac{p_e-p_i}{\tau_{ei}},\end{aligned} \tag{5.3.6}$$

where u is the translational velocity of the quantum object, ψ—scalar potential, n_i and n_e are the number density of the charged species, F_i and F_e are the forces acting on the unit mass of ion and electron.

Approximations for non-local parameters τ_i, τ_e, and τ_{ei} need special consideration. In principal GHE need not in using of the "time-energy" uncertainty relation for estimation of the value of the non-locality parameter τ. Moreover the "time-energy" uncertainty relation does not produce the exact relations and from position of non-local physics is only the simplest estimation of the non-local effects. Really, let us consider two neighboring physically infinitely small volumes $PhSV_1$ and $PhSV_2$ in a non-equilibrium system. Obviously the time τ should tend to diminish with increasing of the velocities u of particles invading the nearest neighboring physically infinitely small volume ($PhSV_1$ or $PhSV_2$):

$$\tau = H/u^n. \tag{5.3.7}$$

But the value τ cannot depend on the velocity direction and naturally to tie τ with the particle kinetic energy, then

$$\tau = \frac{H}{mu^2}, \tag{5.3.8}$$

where H is a coefficient of proportionality, which reflects the state of physical system. In the simplest case H is equal to Plank constant $\hbar$ and relation (5.3.8) became compatible with the Heisenberg relation. For the τ_i and τ_i approximation, the relation (5.3.8) is used in the forms

$$\tau_i = \frac{H}{m_i u^2}, \quad \tau_e = \frac{H}{m_e u^2}. \tag{5.3.9}$$

For non-local parameter of electron-ion interaction τ_{ei} is applicable the relation

$$\frac{1}{\tau_{ei}} = \frac{1}{\tau_e} + \frac{1}{\tau_i}. \tag{5.3.10}$$

In this case parameter τ_{ei} serves as relaxation time in the process of the particle interaction of different kinds. Transformation (5.3.10) for the case $H = \hbar$ leads to the obvious compatibility with the Heisenberg principle

$$\frac{1}{\tau_{ei}} = \frac{\tau_e + \tau_i}{\tau_e \tau_i} = \frac{\hbar/m_e u^2 + \hbar/m_i u^2}{\dfrac{\hbar^2}{u^4}\dfrac{1}{m_e m_i}} = \frac{u^2}{\hbar}(m_e + m_i). \tag{5.3.11}$$

Then

$$u^2(m_e + m_i)\tau_{ei} = \hbar. \tag{5.3.12}$$

Equality (5.3.12) is consequence of "time-energy" uncertainty relation for combined particle with mass $m_i + m_e$.

In principal the time values τ_i and τ_e should be considered as sums of mean times between collisions (τ_i^{tr}, τ_e^{tr}) and discussed above non-local quantum values (τ_i^{qu}, τ_e^{qu}), namely for example

$$\tau_i = \tau_i^{tr} + \tau_i^{qu}. \tag{5.3.13}$$

For molecular hydrogen in standard conditions mean time between collisions is equal to 6.6×10^{-11} s. For quantum objects moving with velocities typical for plasmoids or lightning balls τ^{qu} is much more than τ^{tr} and the usual static pressure p transforms in the pressure which can be named as the rest non-local pressure. In the definite sense this kind of pressure can be considered as analog of the Bose condensate pressure.

The following formulae are valid for acting forces

$$F_i = -\frac{e}{m_i}\frac{\partial \psi}{\partial x}, \quad F_e = \frac{e}{m_e}\frac{\partial \psi}{\partial x}. \tag{5.3.14}$$

Let us consider now the introduction of quantization in quantum hydrodynamics. With this aim write down the expression for the total energy E of a particle moving along the positive direction of x -axis with velocity u in the attractive field of Coulomb forces

$$E = \frac{mu^2}{2} - \frac{Ze^2}{x}, \tag{5.3.15}$$

where Z is the charge number and x is the distance from the center of forces. If this movement obeys to the condition of non-locality $mu^2 = H/\tau$ and $x = u\tau$, then

$$E = \frac{H^2}{2mx^2} - \frac{Ze^2}{x}. \tag{5.3.16}$$

Minimal total energy corresponds to the condition $(\partial E/\partial x)_{x=x_B}=0$ and

$$\frac{H^2}{mx_B^3}=\frac{Ze^2}{x_B^3}. \tag{5.3.17}$$

From (5.3.17) follows

$$x_B=\frac{H^2}{Zme^2}. \tag{5.3.18}$$

and from (5.3.16) and (5.3.18)

$$E=-\frac{Z^2me^4}{2H^2}. \tag{5.3.19}$$

The comparison of Eq. (5.3.19) with the Balmer's relation leads to condition

$$H=n\hbar \tag{5.3.20}$$

with integer $n=1,2,\ldots$ known as principal quantum number and to well known relation

$$E=-\frac{Z^2m_e e^4}{2\hbar^2}\frac{1}{n^2}. \tag{5.3.21}$$

For atom with sole electron in the shell ($Z=1$) moving on the Bohr's orbit, the length x_B (from the relation (5.3.18)) is equal to the Bohr radius $r_B=\frac{\hbar^2}{m_e e^2}=5.2917720859(36)\times 10^{-11}$ m.

Equation (5.3.17) with taking into account the relation

$$\frac{\hbar^2}{m_e r_B{}^3}=\frac{m_e u^2}{r_B} \tag{5.3.22}$$

leads to equality of Coulomb and inertial forces for the orbit electron

$$\frac{m_e u^2}{r_B}=\frac{e^2}{r_B{}^2} \tag{5.3.23}$$

and to the character velocity for this obviously model problem

$$u=\frac{e^2}{\hbar} \tag{5.3.24}$$

with the velocity which is equal to 2.187×10^8 cm/s.

As we see the mentioned simple considerations allow in principal to introduce quantization in the quantum hydrodynamics without direct application of Schrödinger equation. Important to notice that conditions of quantization are not the intrinsic feature of Schrödinger equation, for example the appearance of quantization in Schrödinger's theory is connected with the truncation of infinite series and transformation in polynomials with the finite quantity of terms.

It is known that Ehrenfest adiabatic theorem is one of the most important and widely studied theorems in Schrödinger quantum mechanics. It states that if we have a slowly changing Hamiltonian that depends on time, and the system is prepared in one of the instantaneous eigenstates of the Hamiltonian then the state of the system at any time is given by an the instantaneous eigenfunction of the Hamiltonian up to multiplicative phase factors.

The adiabatic theory can be naturally incorporated in generalized quantum hydrodynamics based on local approximations of non-local terms. In the simplest case if ΔQ is the elementary heat quantity delivered for a system executing the transfer from one state (the corresponding time moment is t_{in}) to the next one (the time moment t_e) then

$$\Delta Q=\frac{1}{\tau}2\delta\left(\overline{T}\tau\right), \tag{5.3.25}$$

where $\tau=t_e-t_{in}$ and $\overline{T}$ is the average kinetic energy. For adiabatic case Ehrenfest supposes that

$$2\overline{T}\tau=\Omega_1,\Omega_2,\ldots \tag{5.3.26}$$

where $\Omega_1,\Omega_2,\ldots$ are adiabatic invariants. Obviously for Plank's oscillator (compare with (5.3.8))

$$2\overline{T}\tau=nh. \tag{5.3.27}$$

Then the adiabatic theorem and consequences of this theory deliver the general quantization conditions for non-local quantum hydrodynamics.

5.4 MOVING QUANTUM SOLITONS IN SELF-CONSISTENT ELECTRIC FIELD

Let us introduce the coordinate system moving along the positive direction of x -axis in ID space with velocity $C=u_0$ equal to phase velocity of considering quantum object

$$\xi = x - Ct. \tag{5.4.1}$$

Taking into account the De Broglie relation we should wait that the group velocity u_g is equal $2u_0$. Really, the energy of the relativistic particle

$$E = mc^2, \tag{5.4.2}$$

where

$$m = \frac{m_0}{\sqrt{1-\left(v_g^2/c^2\right)}}, \tag{5.4.3}$$

c is the light velocity, m_0 is the particle rest mass, v_g is the group velocity, can be written as

$$E = p\frac{c^2}{v_g}. \tag{5.4.4}$$

In (5.4.4) $p=mv_g$ is the particle impulse. In the non-relativistic case (5.4.4) takes the form

$$E = \frac{1}{2}m_0 v_g^2. \tag{5.4.5}$$

Using the De Broglie principle of the wave-particle duality, writing down the article energy

$$E = \hbar\omega = \hbar k v_{ph}, \tag{5.4.6}$$

whereas usual ω is the frequency, $v_{ph}=\omega/\kappa$ is the phase velocity and $k=2\pi/\lambda$ is the wave number. The particle impulse is $p=\hbar k$ and

$$E = p v_{ph}. \tag{5.4.7}$$

From the relation (5.4.5) follows

$$E = \frac{1}{2}m_0 v_g^2 = \frac{1}{2}p v_g. \tag{5.4.8}$$

Comparing (5.4.7) and (5.4.8) we find in the non-relativistic approximation

$$v_g = 2v_{ph}. \tag{5.4.9}$$

In the moving coordinate system all dependent hydrodynamic values are functions of (ξ, t). We investigate the possibility of the quantum object formation of the soliton type. For this solution there is no explicit dependence on time for coordinate system moving with the phase velocity u_0. Write down the system of equations (5.3.1)–(5.3.6) for the two component mixture of charged particles without taking into account the component's internal energy in the dimensionless form, where dimensionless symbols are marked by tildes. We begin with introduction of the scales for velocity

$$[u] = u_0 \tag{5.4.10}$$

and for coordinate x

$$\frac{\hbar}{m_e u_0} = x_0. \tag{5.4.11}$$

Generalized Poisson equation (5.3.1) now is written as

$$\frac{\partial^2 \tilde{\psi}}{\partial \tilde{\xi}^2} = -\left\{\frac{m_e}{m_i}\left[\tilde{\rho}_i - \frac{1}{\tilde{u}^2}\frac{m_e}{m_i}\left(-\frac{\partial \tilde{p}_i}{\partial \tilde{\xi}} + \frac{\partial}{\partial \tilde{\xi}}(\tilde{\rho}_i \tilde{u})\right)\right] - \left[\tilde{\rho}_e - \frac{1}{\tilde{u}^2}\left(-\frac{\partial \tilde{p}_e}{\partial \tilde{\xi}} + \frac{\partial}{\partial \tilde{\xi}}(\tilde{\rho}_e \tilde{u})\right)\right]\right\}. \tag{5.4.12}$$

if the potential scale ψ_0 and the density scale ρ_0 are chosen as

$$\psi_0 = \frac{m_e}{e} u_0^2, \tag{5.4.13}$$

$$\rho_0 = \frac{m_e^4}{4\pi\hbar^2 e^2} u_0^4. \tag{5.4.14}$$

Scaled forces will be described by (e—absolute electron charge) relations

$$\rho_i F_i = -\frac{u_0^2}{x_0} \rho_0 \frac{m_e}{m_i} \frac{\partial \tilde{\psi}}{\partial \tilde{\xi}} \tilde{\rho}_i, \tag{5.4.15}$$

$$\rho_e F_e = \frac{u_0^2}{x_0} \rho_0 \frac{\partial \tilde{\psi}}{\partial \tilde{\xi}} \tilde{\rho}_e. \tag{5.4.16}$$

Analogical transformations should be applied to the other equations of the system (5.3.1)–(5.3.6). We have the following system of six ordinary non-linear equations:

$$\frac{\partial^2 \tilde{\psi}}{\partial \tilde{\xi}^2} = -\left\{ \frac{m_e}{m_i} \left[\tilde{\rho}_i - \frac{1}{\tilde{u}^2} \frac{m_e}{m_i} \left(-\frac{\partial \tilde{\rho}_i}{\partial \tilde{\xi}} + \frac{\partial}{\partial \tilde{\xi}} (\tilde{\rho}_i \tilde{u}) \right) \right] - \left[\tilde{\rho}_e - \frac{1}{\tilde{u}^2} \left(-\frac{\partial \tilde{\rho}_e}{\partial \tilde{\xi}} + \frac{\partial}{\partial \tilde{\xi}} (\tilde{\rho}_e \tilde{u}) \right) \right] \right\}, \tag{5.4.17}$$

$$\frac{\partial \tilde{\rho}_i}{\partial \tilde{\xi}} - \frac{\partial \tilde{\rho}_i \tilde{u}}{\partial \tilde{\xi}} + \frac{m_e}{m_i} \frac{\partial}{\partial \tilde{\xi}} \left\{ \frac{1}{\tilde{u}^2} \left[\frac{\partial}{\partial \tilde{\xi}} (\tilde{p}_i + \tilde{\rho}_i + \tilde{\rho}_i \tilde{u}^2 - 2\tilde{\rho}_i \tilde{u}) + \frac{m_e}{m_i} \tilde{\rho}_i \frac{\partial \tilde{\psi}}{\partial \tilde{\xi}} \right] \right\} = 0, \tag{5.4.18}$$

$$\frac{\partial \tilde{\rho}_e}{\partial \tilde{\xi}} - \frac{\partial \tilde{\rho}_e \tilde{u}}{\partial \tilde{\xi}} + \frac{\partial}{\partial \tilde{\xi}} \left\{ \frac{1}{\tilde{u}^2} \left[\frac{\partial}{\partial \tilde{\xi}} (\tilde{p}_e + \tilde{\rho}_e + \tilde{\rho}_e \tilde{u}^2 - 2\tilde{\rho}_e \tilde{u}) - \tilde{\rho}_e \frac{\partial \tilde{\psi}}{\partial \tilde{\xi}} \right] \right\} = 0, \tag{5.4.19}$$

$$\begin{aligned}
&\frac{\partial}{\partial \tilde{\xi}} \{ (\tilde{\rho}_i + \tilde{\rho}_e) \tilde{u}^2 + (\tilde{p}_i + \tilde{p}_e) - (\tilde{\rho}_i + \tilde{\rho}_e) \tilde{u} \} \\
&+ \frac{\partial}{\partial \tilde{\xi}} \left\{ \begin{array}{l} \dfrac{1}{\tilde{u}^2} \dfrac{m_e}{m_i} \left[\dfrac{\partial}{\partial \tilde{\xi}} (2\tilde{p}_i + 2\tilde{\rho}_i \tilde{u} - \tilde{\rho}_i \tilde{u} - \tilde{\rho}_i \tilde{u} - 3\tilde{p}_i \tilde{u}) + \tilde{\rho}_i \dfrac{m_e}{m_i} \dfrac{\partial \tilde{\psi}}{\partial \tilde{\xi}} \right] \\ + \dfrac{1}{\tilde{u}^2} \left[\dfrac{\partial}{\partial \tilde{\xi}} (2\tilde{p}_e + 2\tilde{\rho}_e \tilde{u} - \tilde{\rho}_e \tilde{u} - \tilde{\rho}_e \tilde{u} - 3\tilde{p}_e \tilde{u}) - \tilde{\rho}_e \dfrac{\partial \tilde{\psi}}{\partial \tilde{\xi}} \right] \end{array} \right\} \\
&+ \tilde{\rho}_i \frac{m_e}{m_i} \frac{\partial \tilde{\psi}}{\partial \tilde{\xi}} - \tilde{\rho}_e \frac{\partial \tilde{\psi}}{\partial \tilde{\xi}} - \frac{\partial \tilde{\psi}}{\partial \tilde{\xi}} \frac{1}{\tilde{u}^2} \left(\frac{m_e}{m_i} \right)^2 \left(-\frac{\partial \tilde{\rho}_i}{\partial \tilde{\xi}} + \frac{\partial \tilde{\rho}_i \tilde{u}}{\partial \tilde{\xi}} \right) \\
&+ \frac{\partial \tilde{\psi}}{\partial \tilde{\xi}} \frac{1}{\tilde{u}^2} \left(-\frac{\partial \tilde{\rho}_e}{\partial \tilde{\xi}} + \frac{\partial \tilde{\rho}_e \tilde{u}}{\partial \tilde{\xi}} \right) - 2 \frac{\partial}{\partial \tilde{\xi}} \left\{ \frac{1}{\tilde{u}} \frac{\partial \tilde{\psi}}{\partial \tilde{\xi}} \left[\left(\frac{m_e}{m_i} \right)^2 \tilde{\rho}_i - \tilde{\rho}_e \right] \right\},
\end{aligned} \tag{5.4.20}$$

$$\begin{aligned}
&\frac{\partial}{\partial \tilde{\xi}} \{ \tilde{\rho}_i \tilde{u} + 5\tilde{p}_i \tilde{u} - \tilde{\rho}_i \tilde{u} - 3\tilde{p}_i \} \\
&+ \frac{\partial}{\partial \tilde{\xi}} \left\{ \frac{1}{\tilde{u}^2} \frac{m_e}{m_i} \left[\begin{array}{l} \dfrac{\partial}{\partial \tilde{\xi}} \left(2\tilde{\rho}_i \tilde{u} + 10\tilde{p}_i \tilde{u} - \tilde{\rho}_i \tilde{u} - 8\tilde{p}_i \tilde{u}^2 - 5\dfrac{\tilde{p}_i^2}{\tilde{\rho}_i} - \tilde{\rho}_i \tilde{u} - 3\tilde{p}_i \right) \\ + \dfrac{m_e}{m_i} \dfrac{\partial \tilde{\psi}}{\partial \tilde{\xi}} (2\tilde{\rho}_i \tilde{u} - 3\tilde{\rho}_i \tilde{u} - 5\tilde{p}_i) \end{array} \right] \right\} \\
&+ 2\tilde{\rho}_i \tilde{u} \frac{m_e}{m_i} \frac{\partial \tilde{\psi}}{\partial \tilde{\xi}} - 2 \frac{\partial \tilde{\psi}}{\partial \tilde{\xi}} \frac{1}{\tilde{u}^2} \left(\frac{m_e}{m_i} \right)^2 \left[\frac{\partial}{\partial \tilde{\xi}} (\tilde{\rho}_i \tilde{u} + \tilde{p}_i - \tilde{\rho}_i \tilde{u}) + \tilde{\rho}_i \frac{m_e}{m_i} \frac{\partial \tilde{\psi}}{\partial \tilde{\xi}} \right] \\
&= -(\tilde{p}_i - \tilde{p}_e) \tilde{u}^2 \left(1 + \frac{m_i}{m_e} \right),
\end{aligned} \tag{5.4.21}$$

$$
\begin{aligned}
&\frac{\partial}{\partial\tilde{\xi}}\{\tilde{\rho}_e\tilde{u}+5\tilde{p}_e\tilde{u}-\tilde{\rho}_e\tilde{u}-3\tilde{p}_e\}\\
&+\frac{\partial}{\partial\tilde{\xi}}\left\{\frac{1}{\tilde{u}^2}\left[\begin{array}{l}\frac{\partial}{\partial\tilde{\xi}}\left(2\tilde{\rho}_e\tilde{u}+10\tilde{p}_e\tilde{u}-\tilde{\rho}_e\tilde{u}-8\tilde{p}_e\tilde{u}^2-5\frac{\tilde{p}_e^2}{\tilde{\rho}_e}-\tilde{\rho}_e\tilde{u}-3\tilde{p}_e\right)\\ -\frac{\partial\tilde{\psi}}{\partial\tilde{\xi}}(2\tilde{\rho}_e\tilde{u}-3\tilde{\rho}_e\tilde{u}-5\tilde{p}_e)\end{array}\right]\right\}\\
&-2\tilde{\rho}_e\tilde{u}\frac{\partial\tilde{\psi}}{\partial\tilde{\xi}}+2\frac{\partial\tilde{\psi}}{\partial\tilde{\xi}}\frac{1}{\tilde{u}^2}\left[\frac{\partial}{\partial\tilde{\xi}}(\tilde{\rho}_e\tilde{u}+\tilde{p}_e-\tilde{\rho}_e\tilde{u})-\tilde{p}_e\frac{\partial\tilde{\psi}}{\partial\tilde{\xi}}\right]\\
&=-(\tilde{p}_e-\tilde{p}_i)\tilde{u}^2\left(1+\frac{m_i}{m_e}\right),
\end{aligned}
\tag{5.4.22}
$$

Some comments to Eqs. (5.4.17)–(5.4.22):

1. Every equation from the system is of the second order and needs two conditions. The problem belongs to the class of Cauchy problems.
2. In comparison with the Schrödinger theory connected with behavior of the wave function, no special conditions are applied for dependent variables including the domain of the solution existing. This domain is defined automatically in the process of the numerical solution of the concrete variant of calculations.
3. From the introduced scales

$$u_0,\ x_0=\frac{\hbar}{m_e}\frac{1}{u_0},\ \psi_0=\frac{m_e}{e}u_0^2,\ \rho_0=\frac{m_e^4}{4\pi\hbar^2e^2}u_0^4,\ p_0=\rho_0u_0^2=\frac{m_e^4}{4\pi\hbar^2e^2}u_0^6$$

only two parameters are independent—the phase velocity u_0 of the quantum object, and external parameter H, which is proportional to Plank constant $\hbar$ and in general case should be inserted in the scale relation as $x_0=(H/m_eu_0)=(n\hbar/m_eu_0)$. It leads to exchange in all scales $\hbar\leftrightarrow H$. But the value $\nu^{qu}=\hbar/m_e$ has the dimension [cm^2/s] and can be titled as quantum viscosity, $\nu^{qu}=1.1577$ cm^2/s. Of course in principal the electron mass can be replaced in scales by mass of other particles with the negative charge. From this point of view the obtained solutions which will be discussed below have the universal character defined only by the Cauchy conditions.

5.5 MATHEMATICAL MODELING OF MOVING SOLITONS

The system of generalized quantum hydrodynamic equations (5.4.17)–(5.4.22) has the great possibilities of mathematical modeling as result of changing of twelve Cauchy conditions describing the character features of initial perturbations which lead to the soliton formation.

On this step of investigation I intend to demonstrate the influence of different conditions on the soliton formation. The following figures reflect some results of calculations realized according to the system of equations (5.4.17)–(5.4.22) with the help of Maple 9.

The following notations on figures are used: r—density $\tilde{\rho}_i$, s—density $\tilde{\rho}_e$, u—velocity $\tilde{u}$, p—pressure $\tilde{p}_i$, q—pressure $\tilde{p}_e$, and v—self-consistent potential $\tilde{\psi}$. Explanations placed under all following figures, Maple program contains Maple's notations—for example the expression $D(u)(0)=0$ means in usual notations $\left(\partial\tilde{u}/\partial\tilde{\xi}\right)(0)=0$, independent variable t responds to $\tilde{\xi}$.

We begin with investigation of the problem of principle significance—is it possible after a perturbation (defined by Cauchy conditions) to obtain the quantum object of the soliton's kind as result of the self-organization of ionized matter? In the case of the positive answer, what is the origin of the existence of this stable object? By the way the mentioned questions belong to the typical problem in the theory of the ball lightning. With this aim let us consider the initial perturbations

v(0) = 1,r(0) = 1,s(0) = 1,u(0) = 1,p(0) = 1,q(0) = 0.95,
D(v)(0) = 0,D(r)(0) = 0,D(s)(0) = 0,D(u)(0) = 0,D(p)(0) = 0,D(q)(0) = 0

in the mixture of positive and negative ions of equal masses if the pressure $\tilde{p}_i(0)$ of positive particles is larger than $\tilde{p}_e(0)$ of the negative ones (for the variant under consideration **p(0) = 1,q(0) = 0.95**). The following Figs. 5.12–5.15 reflect the result of solution of Eqs. (5.4.17)–(5.4.22).

Figure 5.12 displays the quantum object placed in bounded region of 1D space, all parts of this object are moving with the same velocity. Important to underline that no special boundary conditions were used for this and all following cases. Then this soliton is product of the self-organization of ionized matter. Figures 5.13 and 5.14 contain the answer for formulated above questions about stability of the object. Really the object is restricted by negative shell. The derivative $\partial\tilde{\psi}/\partial\tilde{\xi}$ is proportional to the self-consistent forces acting on the positive and negative parts of the soliton. Consider for example the right side of soliton. The self-consistent force of the electrical origin compresses the positive part of this soliton and

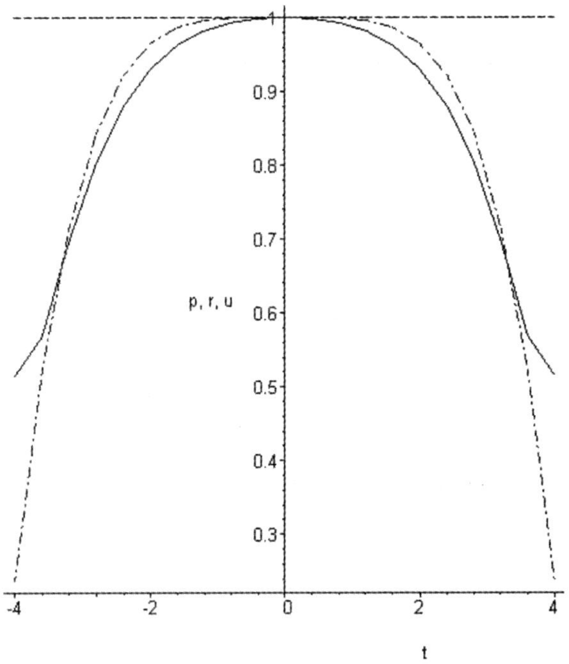

FIGURE 5.12 *r*—density $\tilde{\rho}_i$ (solid line), *u*—velocity $\tilde{u}$ (dashed line), *p*—pressure $\tilde{p}_i$ (dash dotted line).

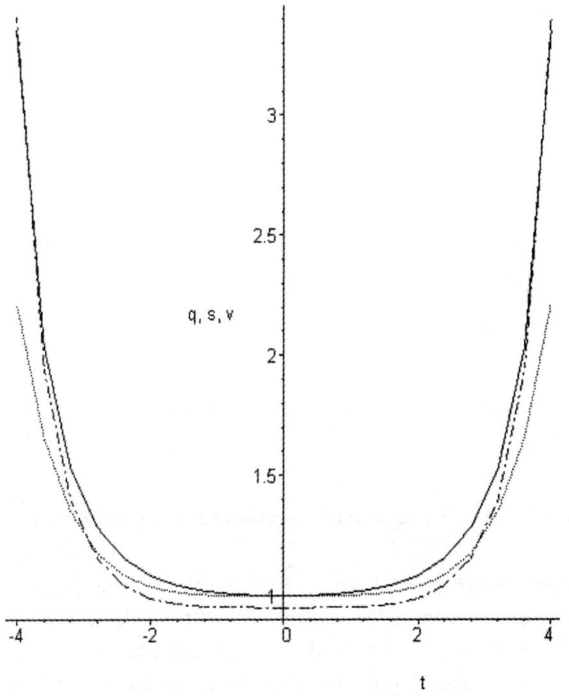

FIGURE 5.13 *s*—density $\tilde{\rho}_e$ (solid line), *q*—pressure $\tilde{p}_e$ (dash dotted line), *v*—self-consistent potential $\tilde{\psi}$ (dotted line) in quantum soliton.

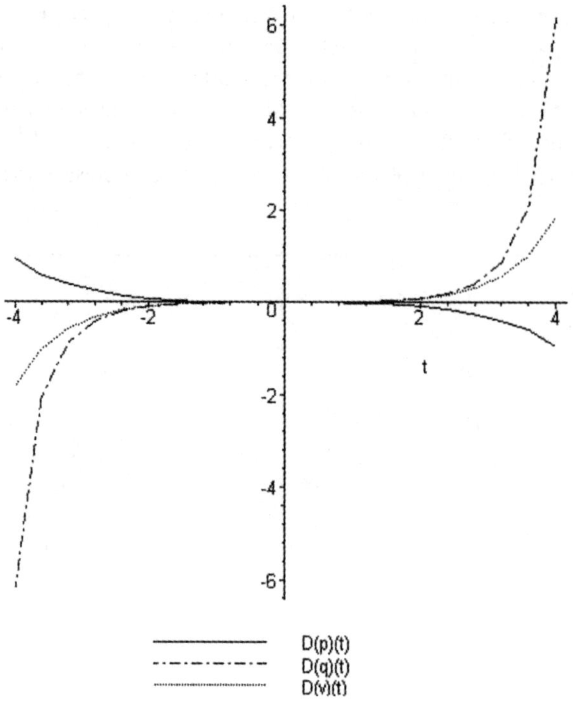

FIGURE 5.14 The derivative of pressure of the positive component $\partial\widetilde{p}_i/\partial\widetilde{\xi}$, the derivative of pressure of negative component $\partial\widetilde{p}_e/\partial\widetilde{\xi}$, the derivative of the self-consistent potential $\partial\widetilde{\psi}/\partial\widetilde{\xi}$ in quantum soliton.

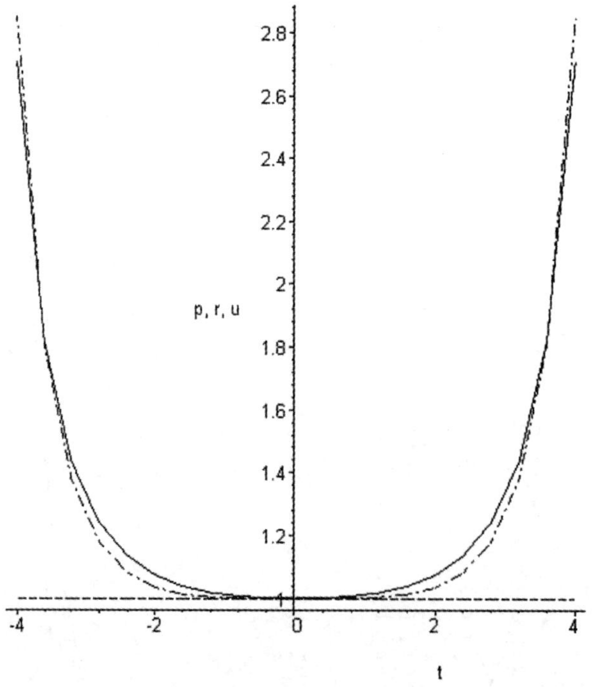

FIGURE 5.15 r—density $\widetilde{\rho}_i$ (solid line), u—velocity $\widetilde{u}$ (dashed line), p—pressure $\widetilde{p}_i$ (dash dotted line).

provokes the movement of the negative part along the positive direction of the $\widetilde{\xi}$-axis (t-axis in the notation of Fig. 5.14). But the increasing of quantum pressure prevent to destruction of soliton. Therefore the stability of the quantum object is result of the self-consistent influence of electric potential and quantum pressures.

Interesting to notice that stability can be also achieved if soliton has *positive* shell and *negative* kernel but $\widetilde{p}_i(0) < \widetilde{p}_e(0)$, see Figs. 5.15–5.17 obtained as result of mathematical modeling for the case

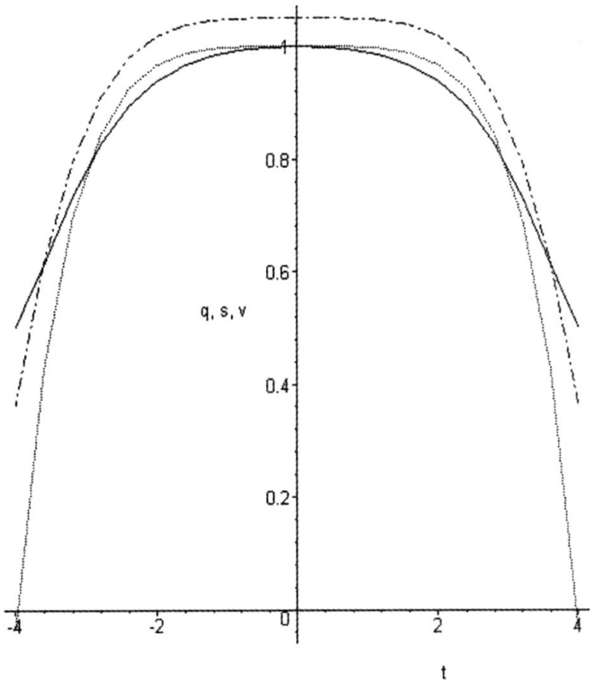

FIGURE 5.16 *s*—density $\widetilde{\rho}_e$ (solid line), *q*—pressure $\widetilde{p}_e$ (dash dotted line), *v*—self-consistent potential $\widetilde{\psi}$ (dotted line) in quantum soliton.

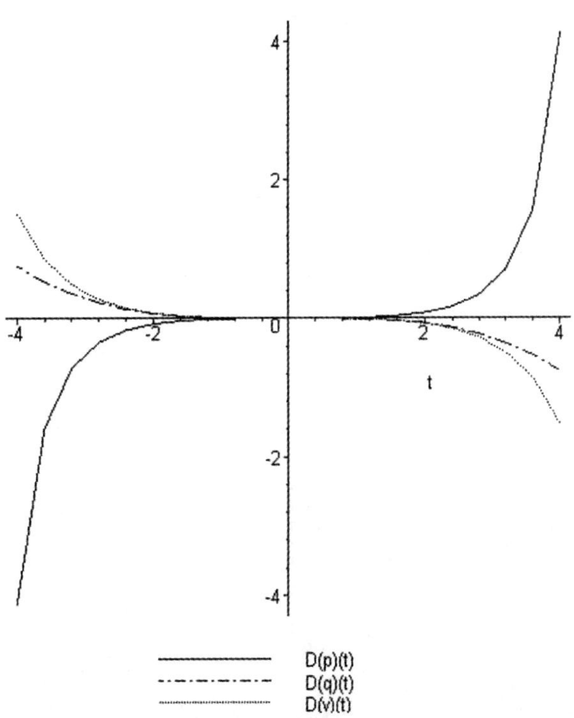

FIGURE 5.17 The derivative of pressure of the positive component $\partial\widetilde{p}_i/\partial\widetilde{\xi}$, the derivative of pressure of negative component $\partial\widetilde{p}_e/\partial\widetilde{\xi}$, the derivative of the self-consistent potential $\partial\widetilde{\psi}/\partial\widetilde{\xi}$ in quantum soliton.

v(0) = 1,r(0) = 1,s(0) = 1,u(0) = 1,p(0) = 1,q(0) = 1.05,
D(v)(0) = 0,D(r)(0) = 0,D(s)(0) = 0,D(u)(0) = 0,D(p)(0) = 0,D(q)(0) = 0.

The explanation for this case has practically the same character as in the previous case but positive and negative species change their roles.

Everywhere in following calculations we use the typical ratio of masses $m_i/m_e = 1838$. The initial perturbations in the mixture of heavy positive particles and electrons produce the soliton formation if the pressure $\widetilde{p}_i(0)$ of the positive particles is larger than $\widetilde{p}_e(0)$ of the negative ones (in Figs. 5.18 and 5.19 for the following variant under consideration **p(0) = 1,q(0) = 0.95**):

v(0) = 1,r(0) = 1,s(0) = 1/1838,u(0) = 1,p(0) = 1,q(0) = 0.95,
D(v)(0) = 0,D(r)(0) = 0,D(s)(0) = 0,D(u)(0) = 0,D(p)(0) = 0,D(q)(0) = 0.

In comparison with Figs. 5.12 and 5.13 we observe the explicit positive kernel which is typical for the atom structures.

Now can be demonstrated the influence of the significant difference in mass of particles $m_i/m_e = 1838$ for the case $\widetilde{p}_i(0) < \widetilde{p}_e(0)$ in Figs. 5.20–5.22. We use the Cauchy conditions

v(0) = 1,r(0) = 1,s(0) = 1/1838,u(0) = 1,p(0) = 1,q(0) = 1.05,
D(v)(0) = 0,D(r)(0) = 0,D(s)(0) = 0,D(u)(0) = 0,D(p)(0) = 0,D(q)(0) = 0

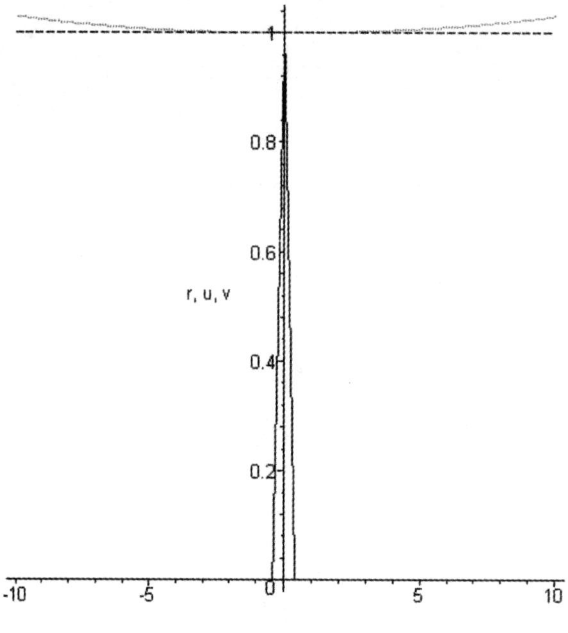

FIGURE 5.18 r—density $\widetilde{\rho}_i$ (solid bold black line), u—velocity $\widetilde{u}$ (dashed line), v—self-consistent potential $\widetilde{\psi}$ in quantum soliton (solid line).

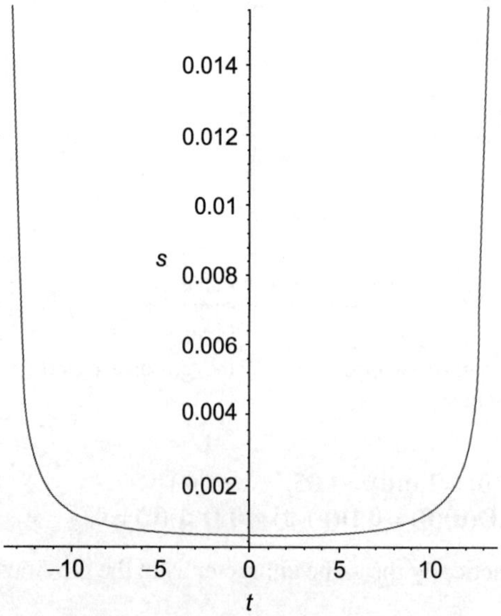

FIGURE 5.19 s—density $\widetilde{\rho}_e$ in quantum soliton.

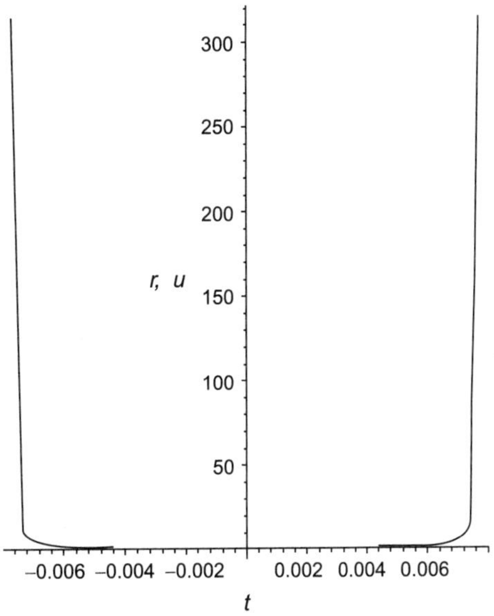

FIGURE 5.20 r—density $\widetilde{\rho}_i$, u—velocity $\widetilde{u}$.

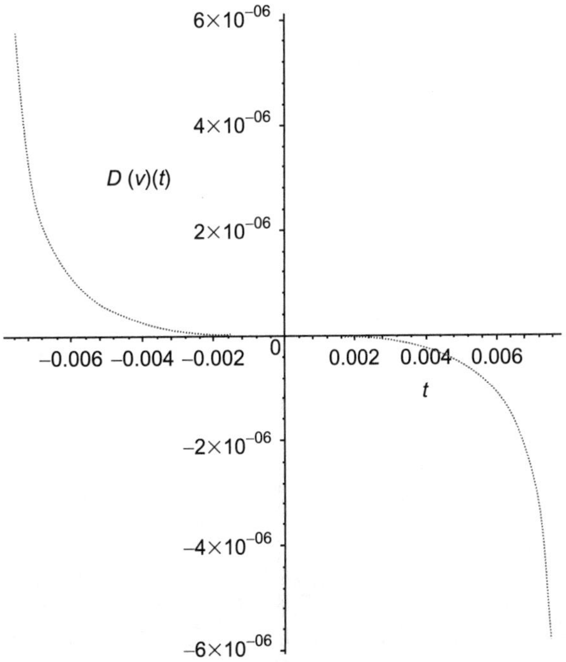

FIGURE 5.21 The derivative of the self-consistent potential $\partial \widetilde{\psi}/\partial \widetilde{\xi}$ in quantum soliton.

Consider the influence of changing of the rest non-local pressures $\widetilde{p}_i(0), \widetilde{p}_e(0)$. Figures 5.23–5.25 reflect the following Cauchy conditions:

$$v(0)=1, r(0)=1, s(0)=1/1838, u(0)=1, p(0)=1, q(0)=0.999,$$
$$D(v)(0)=0, D(r)(0)=0, D(s)(0)=0, D(u)(0)=0, D(p)(0)=0, D(q)(0)=0$$

Figures 5.26 and 5.27 reflect calculations realized for the same Cauchy conditions but in spherical coordinate system, 1D case.

Figure 5.28 contains the detailed image of a carbon atom as it was obtained in Kharkov Institute for Physics and Technology (open access Internet resources). The intensity of the spherical slice color corresponds to the density $\widetilde{p}_e$ in the cloud of electrons. The most interesting experimental result reflects the relative ratio of the thickness of the electron shell and the

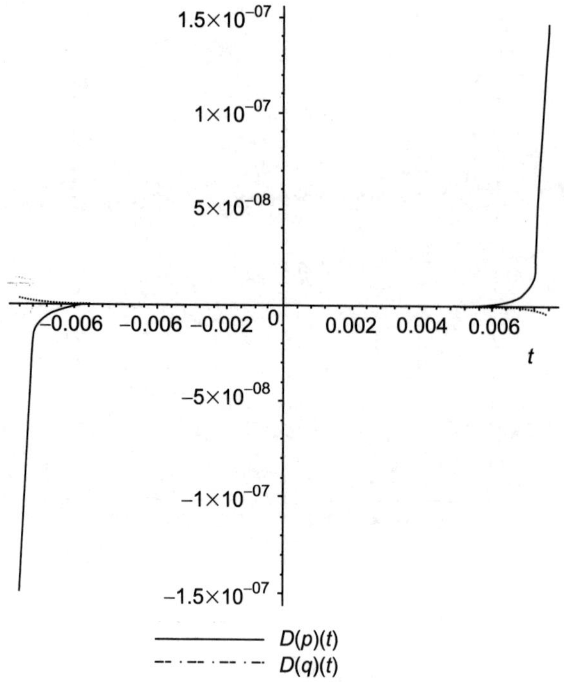

FIGURE 5.22 The derivative of pressures $\partial\widetilde{p}_i/\partial\widetilde{\xi}$ and $\partial\widetilde{p}_e/\partial\widetilde{\xi}$ in quantum soliton.

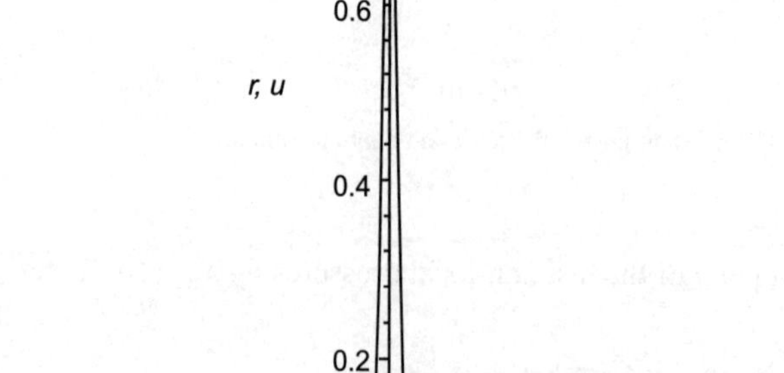

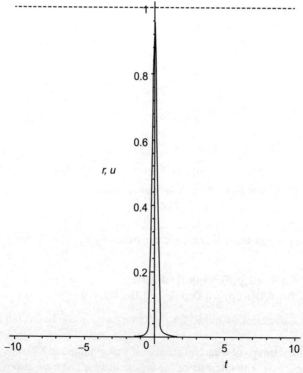

FIGURE 5.23 r—density $\widetilde{\rho}_i$, u—velocity $\widetilde{u}$ in quantum soliton (dashed line).

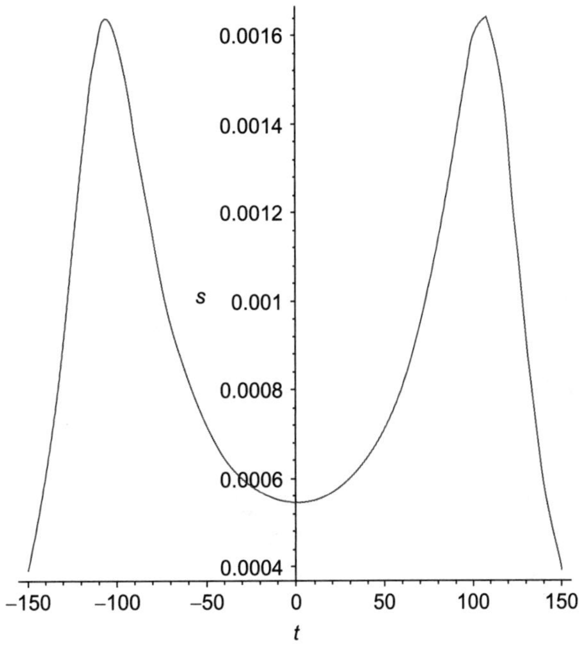

FIGURE 5.24 *s*—density $\widetilde{\rho}_e$ in quantum soliton.

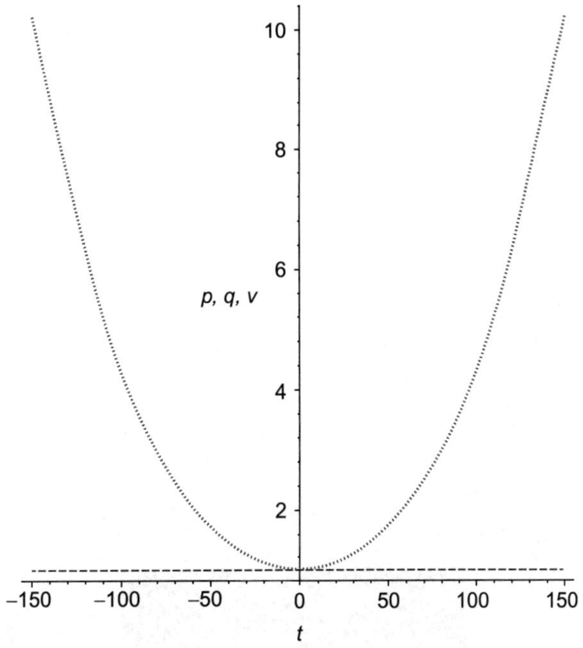

FIGURE 5.25 *v*—self-consistent potential $\widetilde{\psi}$ (dotted line) in quantum soliton and pressures $\widetilde{p}_i$, $\widetilde{p}_e$.

atom radius—it is not the small value. As we see the $\widetilde{\rho}_e$ distribution in Fig. 5.27 (obtained before the experimental results) resembles the Fig. 5.28.

From Figs. 5.18, 5.19, and 5.23–5.25 follow that increasing the difference $p_i(0) - p_e(0)$ brings to diminishing of the character domain occupied by soliton. The classical construction with the positive kernel and negative shell is existing if $\widetilde{p}_i(0) > \widetilde{p}_e(0)$. But in the opposite case $\widetilde{p}_i(0) < \widetilde{p}_e(0)$ mathematical modeling leads to construction with negative kernel and positive shell for soliton. Let us demonstrate the possibility to calculate the soliton formations with very significant difference from the used scales. In following Figs. 5.29 and 5.30 this difference is in 10^5 times more than the corresponding scales.

v(0) = 10^5,r(0) = 10^5,s(0) = (10^5)/1838,u(0) = 1,p(0) = 10^5,q(0) = 0.95 × 10^5,D(v)(0) = 0,D(r)(0) = 0,D(s)(0) = 0,D(u)(0) = 0,D(p)(0) = 0,D(q)(0) = 0

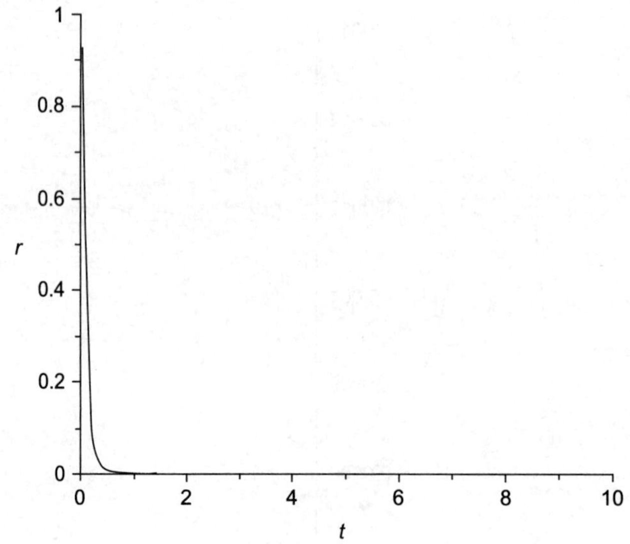

FIGURE 5.26 r—density $\widetilde{\rho}_i$ in quantum soliton.

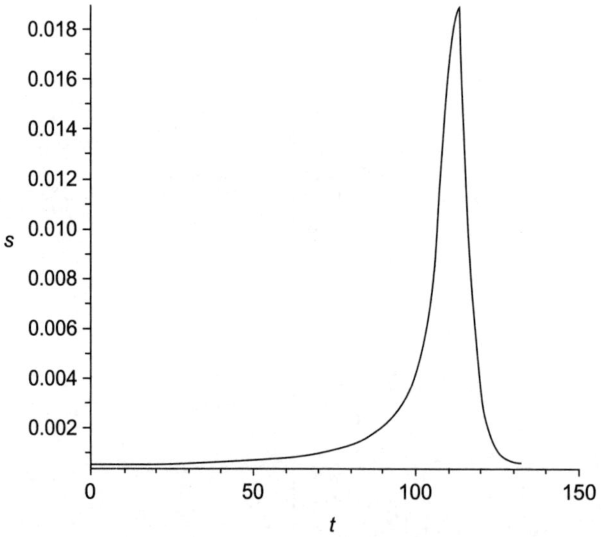

FIGURE 5.27 s—density $\widetilde{\rho}_e$ in quantum soliton.

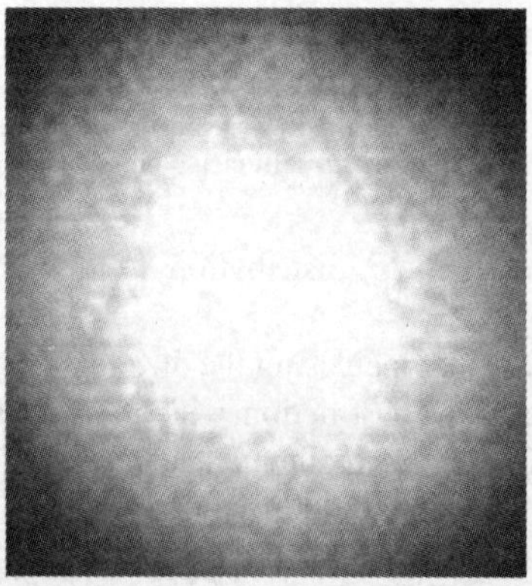

FIGURE 5.28 Detailed image of a carbon atom.

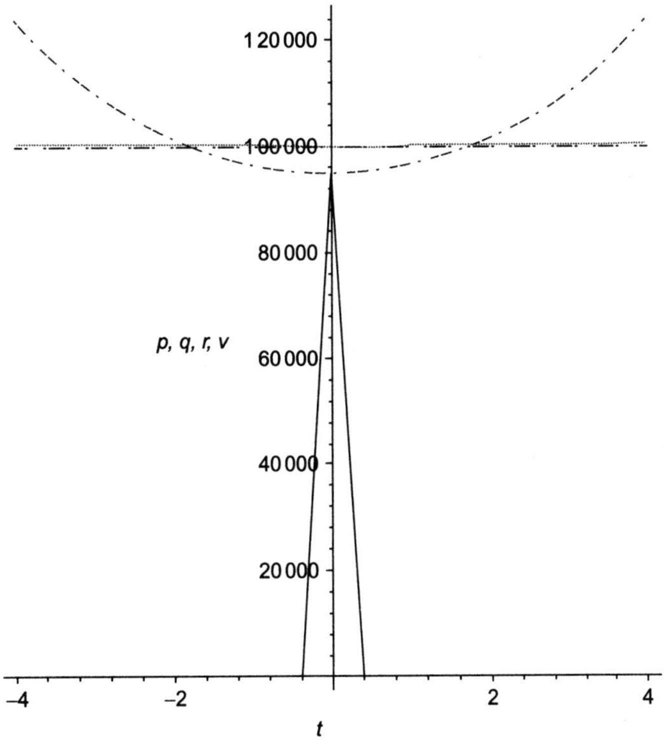

FIGURE 5.29 *r*—density $\widetilde{\rho}_i$ (solid line), *p*—pressure $\widetilde{p}_i$, *q*—pressure $\widetilde{p}_e$ (dash dotted line), *v*—self-consistent potential $\widetilde{\psi}$ (dotted line) in quantum soliton.

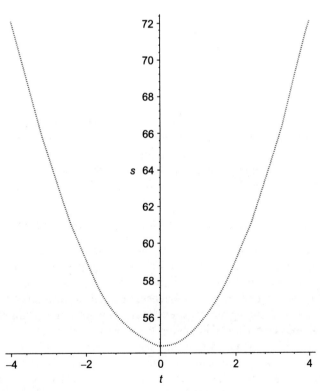

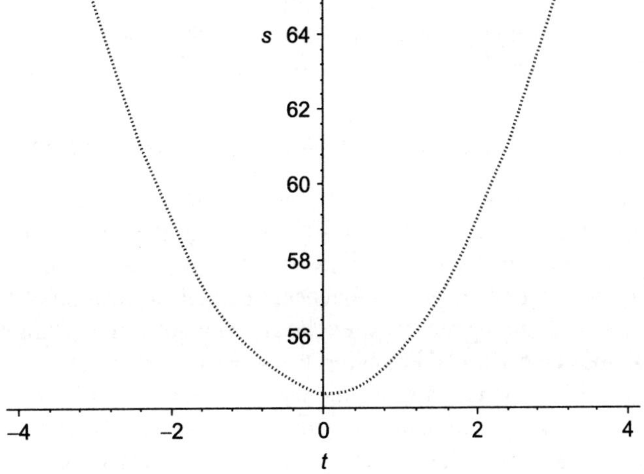

FIGURE 5.30 *s*—density $\widetilde{\rho}_e$ in quantum soliton.

Let us demonstrate the influence of initial perturbation connected with gradient of potential and compare the case $\partial\widetilde{\psi}/\partial\widetilde{\xi}(0)=0$ (Figs. 5.18 and 5.19) after increasing this gradient in the following calculations to the value $\partial\widetilde{\psi}/\partial\widetilde{\xi}(0)=1$:

v(0) = 1,r(0) = 1,s(0) = 1/1838,u(0) = 1,p(0) = 1,q(0) = 0.95,
D(v)(0) = 1,D(r)(0) = 0,D(s)(0) = 0,D(u)(0) = 0,D(p)(0) = 0,D(q)(0) = 0

As we see from Figs. 5.18, 5.19, and 5.31–5.33 this very significant changing of the initial gradient of potential does not provoke the soliton destruction.

Important to notice that all elements of soliton are moving with the same self-consistent constant velocity if initial perturbation $\widetilde{u}(0)=1$ corresponds to phase velocity. The next calculation (Fig. 5.34) demonstrates the soliton destruction if the initial perturbation brought another velocity of the object ($\widetilde{u}(0)=0.5$) into being.

v(0) = 1,r(0) = 1,s(0) = 1/1838,u(0) = 0.5,p(0) = 1,q(0) = 0.95,
D(v)(0) = 0,D(r)(0) = 0,D(s)(0) = 0,D(u)(0) = 0,D(p)(0) = 0,D(q)(0) = 0

Let us consider now the situation when the soliton is catched by the external periodical longitudinal electric field $F_i^{(npot)}=eE/m_i\cos(kx-\omega t)$ for which phase velocity is equal $\omega/k=u_0$. In this case

$$\rho_i F_i^{(npot)}=\frac{eE}{m_i}\rho_0\widetilde{\rho}_i\cos\left[2\pi\frac{\hbar}{\lambda m_e u_0}\widetilde{\xi}\right] \tag{5.5.1}$$

and effective forces acting on positive and negative charges can be written as

$$\rho_i F_i^{(pot)}+\rho_i F_i^{(npot)}=-\frac{u_0^2 m_e}{x_0\, m_i}\rho_0\widetilde{\rho}_i\left[\frac{\partial\widetilde{\psi}}{\partial\widetilde{\xi}}-\widetilde{E}\cos\left(2\pi\frac{\hbar}{\lambda m_e u_0}\widetilde{\xi}\right)\right], \tag{5.5.2}$$

$$\rho_e F_e^{(pot)}+\rho_e F_e^{(npot)}=\frac{u_0^2}{x_0}\rho_0\widetilde{\rho}_e\left[\frac{\partial\widetilde{\psi}}{\partial\widetilde{\xi}}-\widetilde{E}\cos\left(2\pi\frac{\hbar}{\lambda m_e u_0}\widetilde{\xi}\right)\right]. \tag{5.5.3}$$

The expressions (5.5.2) and (5.5.3) should be introduced in general system of quantum hydrodynamic equations (5.4.17)–(5.4.22). The amplitude $\widetilde{E}$ and coefficient $\widetilde{\Lambda}=2\pi(\hbar/\lambda m_e u_0)$ are the parameters of calculations. Let us show the typical result of calculations in the external resonance electric field using the following Cauchy conditions with $\widetilde{E}=1$, $\widetilde{\Lambda}=1$, and $\widetilde{\rho}_e(0)=\widetilde{\rho}_i(0)$:

v(0) = 1,r(0) = 1,s(0) = 1/1838,u(0) = 1,p(0) = 1,q(0) = 1,
D(v)(0) = 0,D(r)(0) = 0,D(s)(0) = 0,D(u)(0) = 0,D(p)(0) = 0,D(q)(0) = 0

Figures 5.35–5.38 reflect the results of calculations

Quantum solitons are discovered with the help of generalized quantum hydrodynamics. The solitons have the character of quantum objects (with positive or negative shells) which reach stability as result of equalizing of corresponding pressure of the non-local origin and the self-consistent electric forces. These effects can be considered also as explanation of the existence of lightning balls. If the initial rest pressures of non-local origin for the positive and negative components are equal each other but the quantum object is moving in the periodic resonance electric field, the mentioned quantum object is the stable soliton. This effect can be significant for nano-electronics. In this case disappearing of the mentioned field leads to the blow up destruction of soliton (Figs. 5.35–5.38).

5.6 SOME REMARKS CONCERNING CPT (CHARGE-PARITY-TIME) PRINCIPLE

One of the fundamental symmetries of today's theoretical physics consists in the CPT-principle. From position of a quantum field theory describing a universe, it means that after reversing the parity, charges, and the arrow of time in the theory, the resulting theory still describes the same universe. The Big Bang should have produced equal amounts of ordinary matter and antimatter with the obvious consequence, the total cancelation of both. This would have resulted in a sea of photons in the universe with no matter. Most explanations involve modifying the standard model of particle physics, to allow for some reactions (specifically involving the weak nuclear force) to proceed more easily than their opposite. This is called "violating CP symmetry" in weak interactions. Such a violation could allow matter to be produced more commonly than antimatter in conditions immediately after the Big Bang. However, as yet, no theoretical consensus has been reached regarding this, and there is no experimental evidence of an imbalance in the creation rates of matter and antimatter. Since this is evidently not the case, after the Big Bang, some physical explanations should be done in the frame of the unified non-local theory of transport processes.

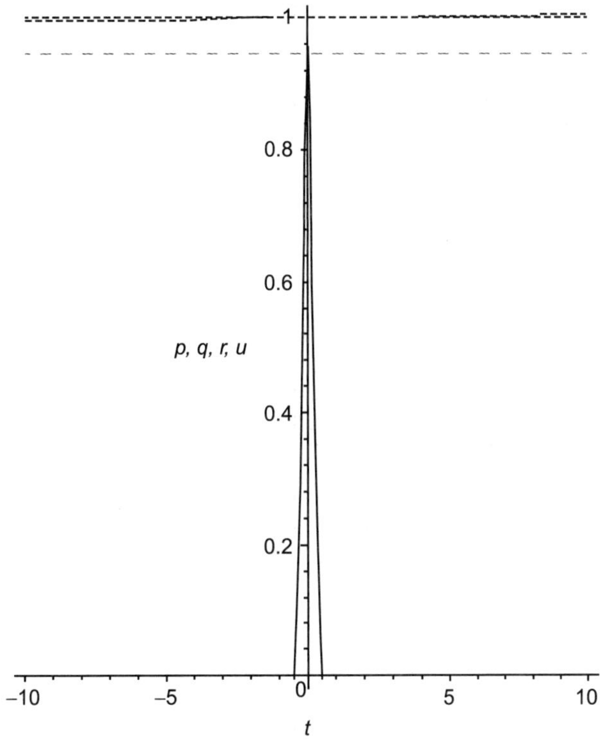

FIGURE 5.31 *r*—density $\widetilde{\rho}_i$ (solid line), *u*—velocity $\widetilde{u}$ (dashed line), *p*—pressure $\widetilde{p}_i$ (dashed inclined line), *q*—pressure $\widetilde{p}_e$ (low dashed line) in quantum soliton.

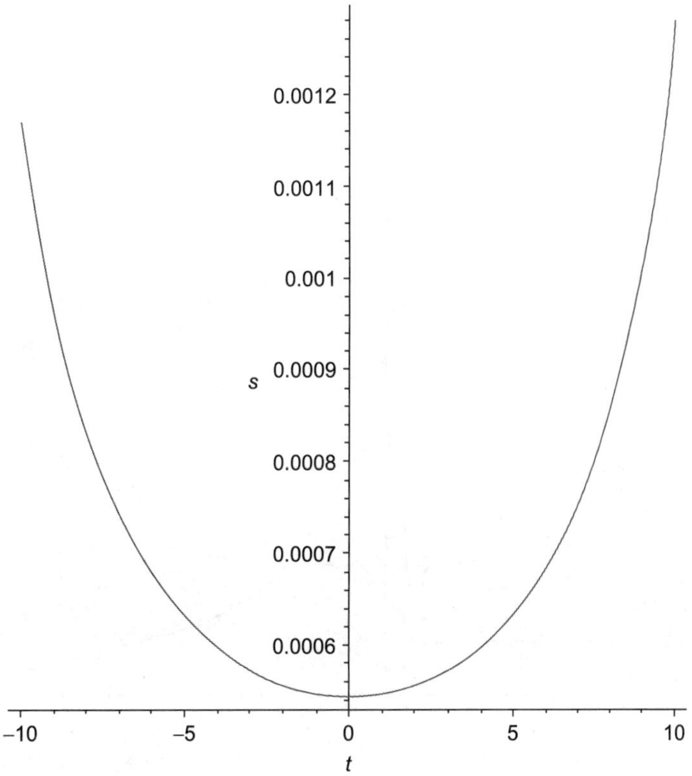

FIGURE 5.32 *s*—density $\widetilde{\rho}_e$ in quantum soliton.

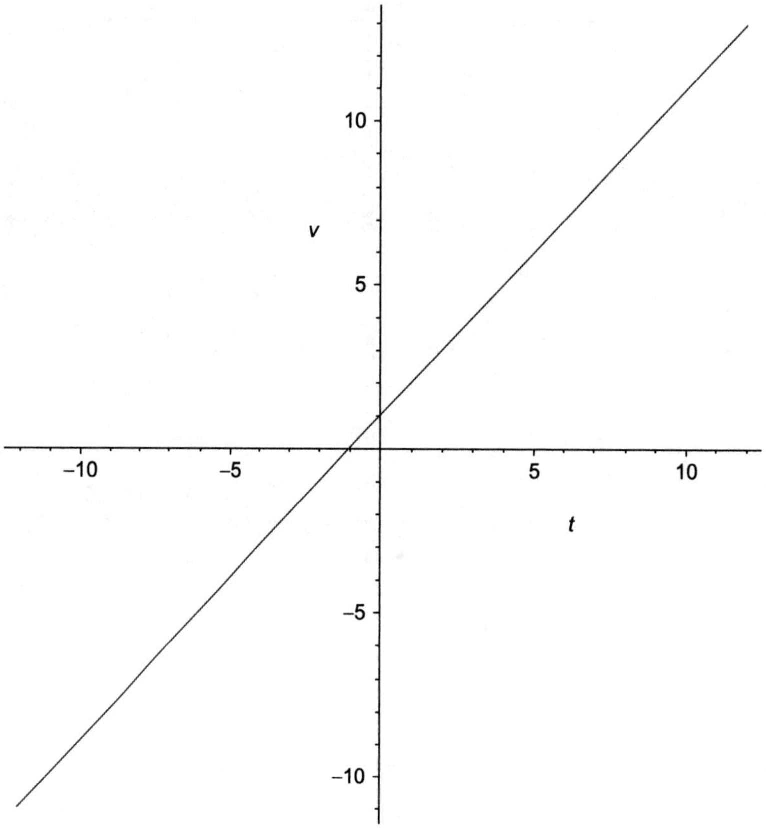

FIGURE 5.33 v—self-consistent potential $\widetilde{\psi}$ in quantum soliton.

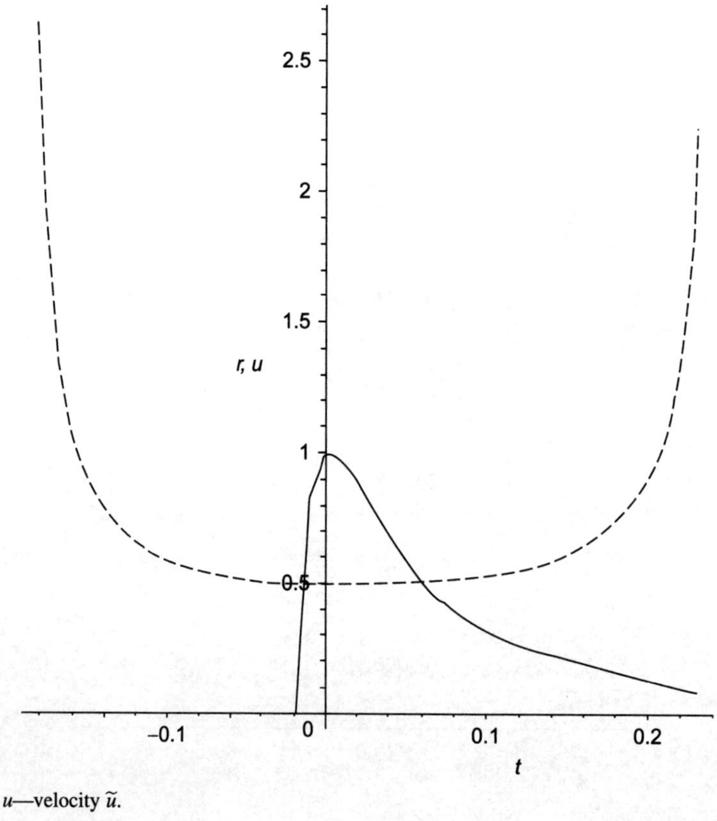

FIGURE 5.34 r—density $\widetilde{\rho}_i$, u—velocity $\widetilde{u}$.

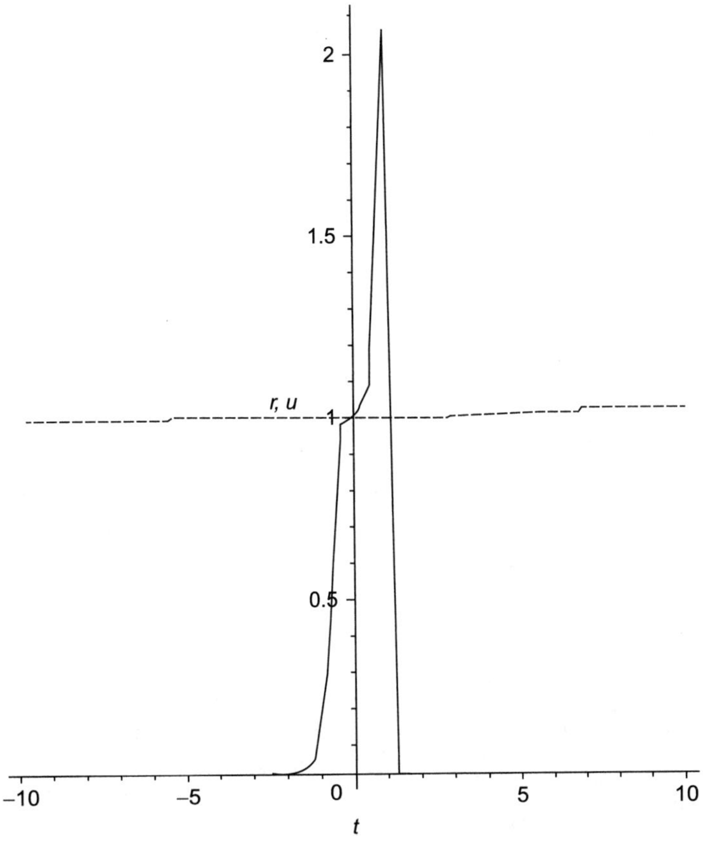

FIGURE 5.35 r—density $\widetilde{\rho}_i$, u—velocity $\widetilde{u}$ (dashed line).

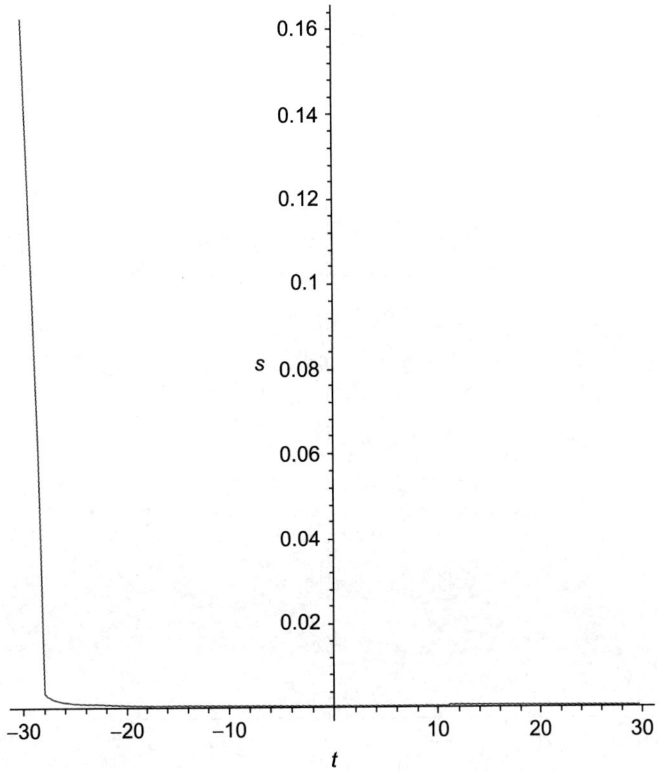

FIGURE 5.36 s—density $\widetilde{\rho}_s$.

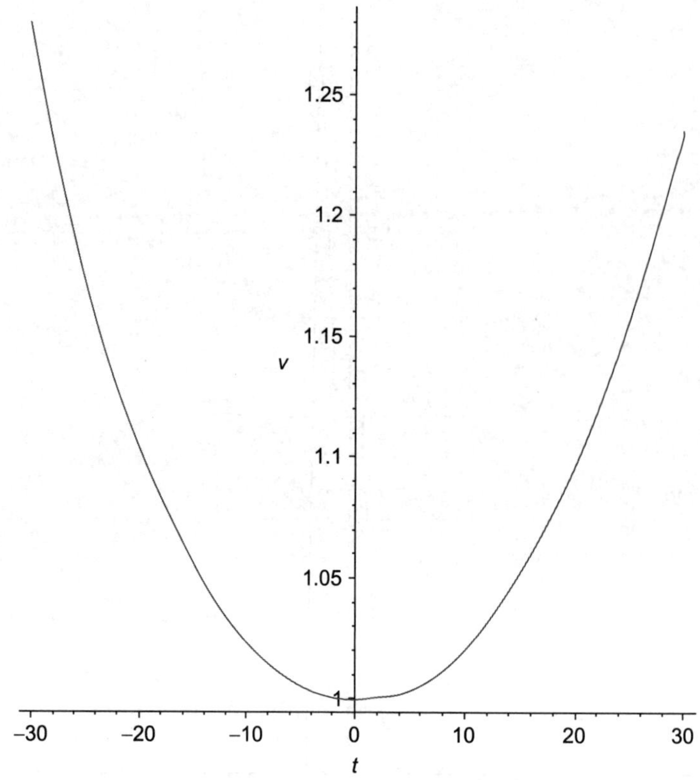

FIGURE 5.37 v—self-consistent potential $\widetilde{\psi}$.

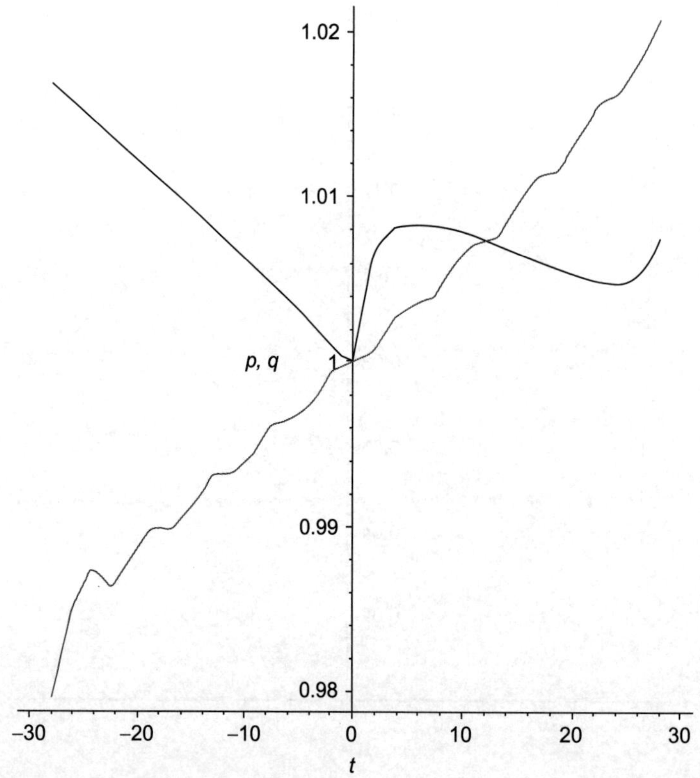

FIGURE 5.38 p—pressure $\widetilde{p}_i$ (black line), q—pressure $\widetilde{p}_e$.

We will now return to the calculations realized in Section 5.3 from this standpoint. Figure 5.6 reflects the results of calculations for VARIANT 3 written in the usual notations:

$$\tilde{p}_i(0)=1,\ \tilde{p}_e(0)=0.9,\ \tilde{\psi}(0)=1,\ \tilde{q}_i(0)=1,\ \tilde{q}_e(0)=0.1,\ \frac{\partial \tilde{p}_i}{\partial \tilde{r}}(0)=0,\ \frac{\partial \tilde{p}_e}{\partial \tilde{r}}(0)=0,\ \frac{\partial \tilde{\psi}}{\partial \tilde{r}}(0)=0,\ \frac{\partial \tilde{q}_i}{\partial \tilde{r}}(0)=0,\ \frac{\partial \tilde{q}_e}{\partial \tilde{r}}(0)=0;$$
$$A=1, B^{-1}=\mathrm{Re}_i=10^6, C^{-1}=\mathrm{Re}_e=10^3;\ r_{\lim}=0.699.$$

Figure 5.7 reflects the results of calculations for VARIANT 4:

$$\tilde{p}_i(0)=0.9,\ \tilde{p}_e(0)=1,\ \tilde{\psi}(0)=1,\ \tilde{q}_i(0)=0.1,\ \tilde{q}_e(0)=1,\ \frac{\partial \tilde{p}_i}{\partial \tilde{r}}(0)=0,\ \frac{\partial \tilde{p}_e}{\partial \tilde{r}}(0)=0,\ \frac{\partial \tilde{\psi}}{\partial \tilde{r}}(0)=0,\ \frac{\partial \tilde{q}_i}{\partial \tilde{r}}(0)=0,\ \frac{\partial \tilde{q}_e}{\partial \tilde{r}}(0)=0;$$
$$A=1, B^{-1}=\mathrm{Re}_i=10^3, C^{-1}=\mathrm{Re}_e=10^6;\ r_{\lim}=0.699.$$

As we see from Figs. 5.6 and 5.7 the result of calculations are absolutely identical, but the dashed and solid lines changed their places. This motivation is twofold:

1. It is the typical appearance of the CP symmetry in the concrete calculations.
2. It does not matter whether we specify the positive or negative species by index "*e*" or "*i*"; but important to introduce changing in all parameters and Cauchy conditions.

What was happened with the antimatter after the Big Bang? The role of the principal significance plays *the difference of the quantum pressures* in this case. Returning to Figs. 5.8 and 5.9, we see strong diminishing of the quantum object.

Figures 5.10 and 5.11 reflect the results of calculations for VARIANT 6:

$$\tilde{p}_i(0)=0.9,\ \tilde{p}_e(0)=1,\ \tilde{\psi}(0)=1,\ \tilde{q}_i(0)=1,\ \tilde{q}_e(0)=0.1,\ \frac{\partial \tilde{p}_i}{\partial \tilde{r}}(0)=0,\ \frac{\partial \tilde{p}_e}{\partial \tilde{r}}(0)=0,\ \frac{\partial \tilde{\psi}}{\partial \tilde{r}}(0)=0,\ \frac{\partial \tilde{q}_i}{\partial \tilde{r}}(0)=0,\ \frac{\partial \tilde{q}_e}{\partial \tilde{r}}(0)=0;$$
$$A=1, B^{-1}=\mathrm{Re}_i=10^6, C^{-1}=\mathrm{Re}_e=10^3;\ r_{\lim}=0.00697.$$

Calculations for VARIANT 3 and VARIANT 6 correspond to atoms of the ordinary matter and antimatter. As we see from Figs. 5.6 and 5.10, 5.11 the character volume of the antimatter atom for these calculations is less than the ordinary atom volume by a factor of $\sim 10^6$. The corresponding results for moving solitons are discussed in [58] and can lead to the neutron charge structure with the negative central charge density and the positive shell. It means that antimatter atoms (because of their small cross-sections) are placed now on the periphery of Universe after the Big Bang. Then we should observe now the effects of the matter-antimatter annihilations on the periphery of our Universe.

Therefore the explanation of the apparent baryon asymmetry is that there are regions of the universe in which ordinary matter is dominant, and other regions of the universe in which antimatter is dominant. These regions are separated and the process of the matter-antimatter annihilations in this region should appear (roughly speaking) as the discharge in the spherical capacitor by the breakdown voltage.

From this model immediately follows:

1. The antimatter's presence should be detectable in the form of the extremely energetic explosions that have been observed in distant galaxies.
2. Matter and antimatter annihilation (particles and anti-particles, stars and "anti-stars", galaxies and "anti-galaxies)" leads to the subsequent production of the gamma radiation flashes. No two gamma-ray burst light curves are identical, with the large variation observed in almost every property.
3. These brightest electromagnetic events followed by a longer-lived "afterglow" emitted at longer wavelengths including X-ray, ultraviolet, optical, infrared, microwave, and radio emission.
4. The events distribution has the *isotropic* character with no concentration toward the plane of the Milky Way.
5. Gamma-ray bursts are thought to be highly focused explosions.

All these consequences of fundamental importance are observed really in the form of the so-called Gamma-ray bursts (GRBs) which were first detected in 1967 by the Vela satellites, a series of satellites designed to detect covert nuclear weapons tests. Results of observations of the GRBs of cosmic origin are reflected on Fig. 5.39; the plane of the Milky Way runs horizontally through the center of the image.

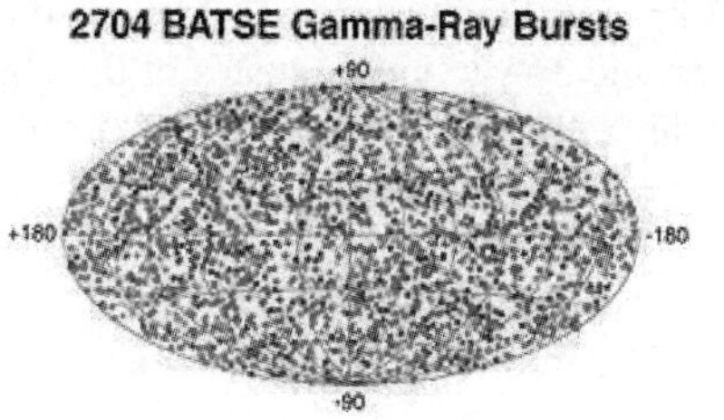

FIGURE 5.39 Positions on the sky of all gamma-ray bursts detected during the BATSE mission.

Gamma-ray bursts are very bright as observed from Earth despite their typically immense distances. An average long GRB has a bolometric flux comparable to a bright star of our galaxy despite a distance of billions of light years. Most of this energy is released in gamma-rays, although some GRBs have extremely luminous optical counterparts as well. GRB 080319B, for example, was accompanied by an optical counterpart that peaked at a visible magnitude of 5.8, comparable to that of the dimmest naked-eye stars despite the burst's distance of 7.5 billion light years. No known process (apart the matter-antimatter annihilation) in the Universe can produce this much energy in such a short time.

Gamma-ray bursts are highly focused explosions, with most of the explosion energy collimated into a narrow jet traveling at speeds exceeding 99.995% of the speed of light. Observations suggest a variation in the jet angle from between 2° and 20°. Because their energy is strongly beamed, the gamma-rays emitted by most bursts are expected to miss the Earth and never be detected.

The most distant known GRB, GRB 090429B, is now the most distant known object in the universe. On November 21, 2013, NASA released detailed data about a gamma-ray burst, designated GRB 130427A, that was observed on April 27, 2013. Originating billions of light years away, this explosion of light has now been confirmed as the brightest gamma-ray burst ever observed; it was also one of the most energetic. Its light traveled 3.6 billion years before arriving at Earth, about one-third the travel time for light from typical bursts. The record-setting 20 h was longer than any other observed GRB.

The delivered theory demonstrates the great possibilities of the generalized quantum hydrodynamics in the investigation of the quantum solitons, which creation and following evolution demands the non-steady 3D consideration in the frame of the unified non-local theory of transport processes. The usual Schrödinger' quantum mechanics cannot be useful in this situation—at least but not at last—because Schrödinger-Madelung quantum theory does not contain the energy equation in principal.

5.7 ABOUT SOME MYSTERIOUS EVENTS OF THE LAST HUNDRED YEARS

5.7.1 Tunguska Event (TE)

The Tunguska event was an enormously powerful explosion that occurred near the Podkamennaya Tunguska River in what is now Krasnoyarsk Krai, Russia, at about 07:14 (Krasnoyarsk local time, 00:14 Universal time) on June 30, 1908. The explosion had the epicenter 60°55′N101°57′E. Tunguska explosion caused the felling of 80 million trees over area of over 2000 km^2.

The explosion registered at seismic stations across Eurasia. The resulting shock wave was equivalent to an earthquake measuring 5.0 on the Richter scale. It also produced fluctuations in atmospheric pressure strong enough to be detected in Great Britain. Over the next few days, night skies in Asia and Europe were aglow. An explosion of this magnitude is capable of destroying a large metropolitan area; a tremendous sound wave traveled twice around the globe. Since then, dozens of research expeditions have visited the area, hundreds of scientific papers (mainly in Russian) have been written and several hundred hypotheses put forward about the causes of the event. Not one of them, however, has been able to explain fully the complex phenomena that preceded and accompanied the Tunguska explosion. Many scientists have participated in Tunguska studies; the best known are Leonid Kulik, Yevgeny Krinov, Kirill Florensky, and N.V. Vasiliev. The results of their investigations are well known and have the free access in Internet. From the first glance the simple explanation can be used for the TE—impact of the celestial bodies. The chief difficulty in the celestial impact hypothesis is that a stony object should have produced a large crater where it struck the ground, but no such crater has been found. Many people believe that the crater lies under the water of Lake Checko in Western Siberia. Yet, the scientists have found no object or material from this cosmic body itself.

Maybe we have with the probable airburst of small asteroid or comet? But a body composed of cometary material, traveling through the atmosphere along such a shallow trajectory, ought to have disintegrated, whereas the Tunguska object (TO) apparently remained intact into the lower atmosphere. The leading scientific explanation for the explosion is the air burst of an asteroid 6-10 km above Earth's surface.

Practically all energy estimations are based on the asteroid version. Meteoroids enter Earth's atmosphere from outer space every day, traveling at a speed of at least 11 km/s. In literature we have tremendous differences in estimations of size and mass of the object. Different studies have yielded widely varying estimates of the object's size, on the order of 60–190 m.

If we have the stone spherical object which diameter are roughly 30 m and the mass of about 10^8 kg moving with the velocity of 15 km/s, the kinetic energy of the object as large as $\sim 10^{16}$ J. Obviously it is only the rough estimation biased to the concrete (maybe wrong model). The "megaton of TNT" is a unit of energy equal to 4.184 PJ. The Hiroshima bomb represented only 8×10^{13} J of energy. Thus, our estimate is that the Tunguska had an explosive energy on order of 2 MT of TNT. It was closer in effect to a very large H-bomb. Most likely estimates are between 10 and 15 Mton of TNT (42-63 PJ).

By the way the Tsar Bomba (the nickname for the AN602 hydrogen bomb) developed by the Soviet Union, the bomb had the yield of 50-58 Mton of TNT (210-240 PJ). Only one bomb of this type was ever officially built and it was tested on October 30, 1961, in the Novaya Zemlya archipelago, at Sukhoy Nos. Many unusual effects convoyed this event; these effects cannot be explained from positions of the celestial impact. Really,

1. Many meteorological factors point toward the possibility of a meteorological event occurring. There was evidence of strong cyclones near Siberia that summer. Significant increases in air pressure were associated with the area at that time. Increased thunderstorm activity and intensity inundated Siberia. Witness accounts detail hearing thunder and seeing lightning as the event occurred. Perhaps there is some credit to the theory that the Tunguska event transpired as a meteorological occurrence.
2. It was established that the zone of leveled forest occupied an area of some 2150 km^2 with the shape resembling a gigantic spread-eagled butterfly with a "wingspan" of 70 km and a "body length" of 55 km. Upon closer examination it was found that several explosions took place. Siberian Life newspaper (July 27, 1908) reported about some kind of artillery barrage, that repeated in intervals of 15 min at least 10 times.
3. In the 10 days before the explosion, in many countries of Europe as well as western Siberia, the darkness of night was replaced by an unusual illumination as if those areas were experiencing the "white nights" phenomenon of high-latitude summers. Everywhere there appeared, shining brightly in the twilight of dawn and dusk, silvery clouds stretching east to west that formed like along "the lines of force". Professor Weber about a powerful geo-magnetic disturbance observed in a laboratory at Kiel University in Germany for 3 days before the intrusion of the Tunguska object, and which ended at the very hour after the explosion in the Central Siberian Plateau. There was a sense of the approach of some unusual natural phenomenon.
4. Some climatologists and scientists concur that the Tunguska event caused major damage to the air layer of the mesosphere. These atmospheric changes resulted in an ozone depletion lasting up to four years after the event. A cooling trend in the years following the 1908 event was recorded in weather records around the Earth.
5. The TO followed a trajectory from southeast to northwest. It was the discrepancies in the accounts of eyewitnesses—who at one and the same time observed objects above areas of Siberia far remote from one another, moving on different courses but toward a single point—that confused researchers, prompting the hypothesis that it was probably a spaceship that had been maneuvering above the Siberian taiga. Meteorites and comets do not fly like that!
6. The reports contain information about objects moving slowly, parallel to the Earth's surface, sometimes stopping, changing course, and speed. Thousands of observers could not have mistaken what they saw, as the sky was cloudless that morning. People living within a radius of over 800 km from the place where the cosmic intruder fell observed the unusual flight of enormous fiery bodies giving off sparks and leaving rainbow trails behind them. As a result, one of hypothesis sounds that they did not all see one and the same object, but several different bodies.

One other possible cause of the Tunguska event which can explain all main character features of the TE, is plasmoid (ball lightning). It can move horizontally, hover or in a zigzag motion. It is not a new idea, but until now, ball lightning was a phenomenon not consensually understood in the scientific world. The non-local theory of plasmoids gives grounds to solve the TO problem. Energy content of plasmoids have no restrictions in comparison with the chemical models. The energy density is defined by the initial conditions of the plasmoid creation and calls for the application of the nonstationary models. It has been known about the very large plasmoids with diameter up to 260 m.

As it follows from the calculations, the separated charges in plasmoid can correspond to the model of the spherical capacitor. The maximum energy that can be stored in a capacitor is limited by the breakdown voltage. But the breakdown process can have rather lengthy character realized in the several stages. This fact can explain the anomalies in the forest felling.

The spherical capacitor energy W is written as

$$W = 2\pi\varepsilon_0\varepsilon \frac{R_1 R_2}{R_2 - R_1}(\Delta\psi)^2, \tag{5.7.1}$$

where $\Delta\psi = \psi_1 - \psi_2$ is the potential difference between the conductors for a given charge q on each. The voltage between the spheres can be found by integrating the electric field along a radial line:

$$\Delta\psi = \psi_1 - \psi_2 = \frac{q}{4\pi\varepsilon_0\varepsilon}\int_{R_1}^{R_2}\frac{\mathrm{d}r}{r^2} = \frac{q}{4\pi\varepsilon_0\varepsilon}\left(\frac{1}{R_1} - \frac{1}{R_2}\right), \tag{5.7.2}$$

If $R_2 \gg R_1$, then

$$W = 2\pi\varepsilon_0\varepsilon R_1(\Delta\psi)^2. \tag{5.7.3}$$

The force $F = (-dW_p/dr)(\mathbf{r}/r)$, acting on the internal conductor

$$F_{in} = -\frac{\partial W_p}{\partial R_1} \cong 2\pi\varepsilon_0\varepsilon(\Delta\psi)^2, \tag{5.7.4}$$

does not depend in the first approximation on the radius of the internal sphere. For the external sphere, the force acts in the opposite direction.

$$F_{ex} = -\frac{\partial W_p}{\partial R_2} = -2\pi\varepsilon_0\varepsilon_1(\Delta\psi)^2\frac{R_1^2}{(R_2 - R_1)^2}. \tag{5.7.5}$$

If $R_2 \gg R_1$, then

$$F_{ex} = -\frac{\partial W_p}{\partial R_2} = -2\pi\varepsilon_0\varepsilon_1(\Delta\psi)^2\left(\frac{R_1}{R_2}\right)^2, \tag{5.7.6}$$

If for the TO $W = 10^{16}$ J, radius of the internal sphere is 100 m, $\varepsilon = 1$, then $\Delta\psi = 1.34 \times 10^{12}$ V.

Electrostatic generator (which uses a moving belt to accumulate very high amounts of electrical potential on a hollow metal globe on the top of the stand) was invented by American physicist Robert J. Van de Graaff in 1929. The potential difference achieved in Van de Graaff generators reaches 7×10^6 V in the 30th of the last century.

A Marx generator (Arkadyev-Marks generator in the Russian scientific literature) generates a high-voltage pulse. The circuit generates a high-voltage pulse by charging a number of capacitors in parallel, then suddenly connecting them in series. Marx generators are used in high energy physics experiments, as well as to simulate the effects of lightning on power line gear and aviation equipment. The high-voltage pulse can reach up to 10^7 V. The mega-joule estimates are known for the ball lightings.

It is stated that the ball lighting explosion damages the plane navigation equipment, but it is the theme of the next section.

5.7.2 Gagarin and Seryogin Air Crash

In 1960, after much searching and a selection process, Yuri Gagarin was chosen with many other pilots for the Soviet space program. A Soviet Air Force doctor evaluated his personality as follows:

"Modest; embarrasses when his humor gets a little too racy; high degree of intellectual development evident in Yuri; fantastic memory; distinguishes himself from his colleagues by his sharp and far-ranging sense of attention to his surroundings; a well-developed imagination; quick reactions; persevering, prepares himself painstakingly for his activities and training exercises, handles celestial mechanics and mathematical formulae with ease as well as excels in higher mathematics; does not feel constrained when he has to defend his point of view if he considers himself right; appears that he understands life better than a lot of his friends."

Gagarin was also a favored candidate by his peers. When the 20 candidates were asked to anonymously vote for which other candidate they would like to see as the first to fly, all but three chose Gagarin. On 12 April 1961, aboard the Vostok 1, Gagarin became both the first human to travel into space, and the first to orbit the earth.

On 27 March 1968, while on a routine training flight from Chkalovsky Air Base, he and flight instructor Vladimir Seryogin died in a MiG-15UTI crash near the town of Kirzhach. The bodies of Gagarin and Seryogin were cremated and the ashes were buried in the walls of the Kremlin on Red Square. It was the tragedy of the national scale. (Read more for example: [113–121].) The cause of the crash that killed Gagarin is not entirely certain, and has been subject to speculation about conspiracy theories over the ensuing decades.

In April 2011, documents from a 1968 commission set up by the Central Committee of the Communist Party to investigate the accident were declassified. Those documents revealed that the commission's original conclusion was that Gagarin or Seryogin had maneuvered sharply either to avoid a weather balloon, leading the jet into a "super-critical flight regime and to its stalling in complex meteorological conditions," or to avoid "entry into the upper limit of the first layer of cloud cover."

Soviet documents declassified in March 2003 showed that the KGB had conducted their own investigation of the accident, in addition to one government and two military investigations. The KGB's report dismissed various conspiracy theories.

In the years and decades that followed, rumors swirled about Gagarin's death. No reason to discuss fantastic hypotheses on the level of the provocation like "Had Gagarin been drinking?" or "Was he distracted, taking pictures of birds from the air when he should have been paying attention to his aircraft?"

About the aim of the Gagarin airplane flight, pilot-cosmonaut Vladimir Aksenov wrote in his book "The Roads of Tests":

"Gagarin and Yevgeny Khrunov were supposed to be the first to go through check flights. According to flight rules, check flights, prior to independent flights, could be conducted by the heads of flight departments, rather than instructor pilots. They could be squadron commanders, deputy commanders, and commanders of regiments. So it was Vladimir Seryogin, the regiment commander, who joined Yuri Gagarin in the check flight. Another important peculiarity of that check flight was as follows: it was a flight in the area for the execution of complex aerobatics stunts. In classical training programs, the check flight, and the first solo flight are performed on the so-called "box" that is, takeoff, height gain, flying around the airfield, landing approach, and landing. Prior to solo flights in the area to perform aerobatic maneuvers, another check flight should be made." It should be added that V. Seryogin was the leading test pilot for the plane MiG-15UTI.

The KGB report states that an air traffic controller provided Gagarin with outdated weather information, and that by the time of his flight, conditions had deteriorated significantly. Vladimir Aksenov writes: "On that day clouds were unusual. The lower edge of almost continuous clouds was about 600 m above the ground. Then, 4000 m above, there were only dense clouds. The upper edge was flat, and there were no clouds above that—there was a clear sky and very good visibility." The last message from the MiG-15UTI contains information (without unusual emotions in the voice), that the check flight is finished and they return to landing. Further on the height less than 4000 m the plane entered in the clouds.

Here is an extract from the book of distinguished test pilot of the USSR Stepan Mikoyan "We Are Children of War. Memoirs of a Military Test Pilot.":

"The time determined by imprints of the hands of the remains of aircraft clock and Gagarin's watch differed by about 15 s. That moment occurred only in *45-60 s after the last broadcast* from Gagarin that was recorded on the magnetic tape." The investigation concluded that Gagarin's aircraft executes the maneuver trying to avoid the collision with unknown object. The hypotheses about possible objects like balloons or flocks of birds should be ruled out—too high for birds and no traces of the balloon on the place of the crash.

The investigation concluded that

1. The maneuver led to the aircraft going into a tailspin and crashing, killing both men.
2. Gagarin and Seryogin have the control until the end.
3. The crew believed their altitude to be higher than it actually was, and could not react properly to bring the MiG-15 out of its spin. It was discovered that altitude sensor was out of order but the crew believed their altitude to be higher than it actually was.
4. The plane was not destroyed in the air. It means that the plane with outboard tanks had the overloads less than eight which were not unusual for the crew.
5. The reading of the pressure sensor scale displayed that the glass cockpit was destroyed. *Only* 2/3 of the glass splits were discovered on the crash place, *for other parts* ~96%. It means that cockpit was destroyed in air.

Hypotheses that a cabin air vent was accidentally left open by the crew or the previous pilot, leading to oxygen deprivation, and leaving the crew incapable of controlling the aircraft, cannot be true. The height of about or even less than 4000 m is usual for alpinists. For example, the "Shelter of 11" (4130 m) was a hotel near Elbrus. Large groups of climbers would usually leave this base camp at 2-3 a.m. to challenge the summit.

In his 2004 book Two Sides of the Moon, Alexey Leonov, who was part of a State Commission established to investigate the death in 1968, recounts that he was flying a helicopter in the same area that day when he heard "two loud booms in the distance." Corroborating other theories, his conclusion is that a Sukhoi jet (which he identifies as a Su-15) was flying below its minimum allowed altitude, and "without realizing it because of the terrible weather conditions, he passed within 10 or 20 m of Yuri and Seregin's plane while breaking the sound barrier." The resulting turbulence would have sent the MiG into an uncontrolled spin. Leonov believes the first boom he heard was that of the jet breaking the sound barrier, and the second was Gagarin's plane crashing. In a June 2013 interview with Russian television network RT, Leonov said that a declassified report on the incident revealed the presence of a second, "unauthorized" Su-15 flying in the area. Leonov states that "the aircraft reduced its echelon at a distance of 10–15 m in the clouds, passing close to Gagarin, turning his plane and thus sending it into a tailspin—a deep spiral, to be precise—at a speed of 750 km/h."

It is the very significant evidence which was checked by cosmonaut Tolboyev. He said (for example during the television interview on January, 7 (2013)) that the special experiments were organized; during the Su-15 flight two MiG-15 UTI entered in the turbulent wake of Su-15. In all cases both MiG-15 UTI were pushed out from the stream without going into a tailspin.

Interesting information from cosmonaut Tolboyev during this interview—he retailed about the aviation accident in the Russian Ahtuba aviation division. The pilot broadcasted about the UFO (unknown flying object). He was commanded to return immediately for landing, but the pilot tried to close to this object. The result—he was landing with the tremendous difficulties without cabin electronics.

I believe that the cause of the Gagarin accident consists in the impact of the MiG 15 UTI with plasmoid.

5.7.3 Accident with Malaysia Airlines Flight MH370

Let us consider other mystery accidents from this point of view. For example, the Malaysia Airlines flight MH370 with 239 people onboard. It "lost all contact" with Subang Air Traffic Control at 2:40 a.m., 2 h into the flight. The plane was expected to land in Beijing at 6:30 a.m. Saturday (on March 8, 2014). Known facts:

1. Around the time the plane vanished, the weather was fine and the plane was already at cruising altitude, making its disappearance all the more mysterious. Just 9% of fatal accidents happen when a plane is at cruising altitude, according to a statistical summary of commercial jet accidents done by Boeing.
2. Military radar indicated that the plane may have turned from its flight route before losing contact. Aviation sources in China report that radar data suggest a steep and sudden descent of the aircraft, during which the track of the aircraft changed from 024° to 333°.
3. A Malaysia Airlines plane sent signals to a satellite for 4 h after the aircraft went missing, an indication that it was still flying for hundreds of miles or more. Boeing offers a satellite service that can receive a stream of data during flight on how the aircraft is functioning and relay the information to the plane's home base. Malaysia Airlines not a subscriber to Boeing service but still automatically sent pings to satellite. If the plane had disintegrated during flight or had suffered some other catastrophic failure, all signals—the pings to the satellite, the data messages, and the transponder—would be expected to stop at the same time.
4. There was no distress signal. The lack of a radio call suggests something *very sudden* and *very violent happened.*
5. The plane had enough fuel for *four* more hours of flight. The plane lost all contact and radar signal 1 m before it entered Vietnam's air traffic control.
6. Officials said two men, later identified as Iranians, boarded the plane with stolen passports. It was later reported that they were unlikely to be linked to terrorist groups.
7. The plane was last inspected 10 days before the accident and found to be in proper condition.

Investigators have not ruled out any possible cause for the plane's disappearance. As a result, experts say one possibility that could explain why the transponders were not working is that the pilot, or a passenger, likely one with some technical knowledge, switched off the transponders in the hope of flying undetected.

It is known that the appearance of the ball lightning in the airplane is dangerous, because it can cause a short circuit and hence lead to crash of the airplane. Plasmoids were really observed on board the airplane [122]. Taking into account the plasmoid theory created by me, we can make the preliminary conclusions:

(a) The accident has very sudden and very violent character. The aircraft was partly disintegrated, as result—the loss of pressure and practically of all electronic equipment.
(b) The loss of pressure was so severe that it knocked passengers and crew out.
(c) In this case, the pilots should have been able to react quickly and connect to oxygen masks, but did not. The plane transformed into, so to speak, "flying Dutchman." The aircraft flew for the rest hours until it ran out of fuel and crashed. Really, the aircraft has fuel for approximately 4 h for flight, and plane sent signals to a satellite for 4 h after the aircraft went missing. This fact indicates the possible area of the crash. But this area has no site for landing. In its turn it excludes the version of hijacking.
(d) It should be added that the area of the plane crash contains the boundary between two tectonic plates, the Burma plate and the Sunda Plate. The boundary between two major tectonic plates results in high seismic activity, *anomalous atmospheric and ocean events* in the region. Numerous earthquakes have been recorded, and at least six, in 1797, 1833, 1861, 2004, 2005, and 2007, had the magnitude of 8.4 or higher. On December 26, 2004, a large portion of the boundary between the Burma Plate and the Indo-Australian Plate slipped, causing the 2004 Indian Ocean earthquake.

This earthquake had a magnitude of 9.3. Between 1300 and 1600 km of the boundary underwent thrust faulting and shifted by about 20 m, with the sea floor being uplifted several meters. This rise in the sea floor generated a massive tsunami with an estimated height of 28 m that killed approximately 280,000 people along the coast of the Indian Ocean.

Now the final reasonable conclusions could be done from the position of the previous theory:

1. *Malaysia Airlines flight MH370 met the atmospheric plasmoid.*
2. This is not the only time a plane has disappeared without a trace or sparks an investigation surrounded by confusion. It is reasonable to look back at other baffling aviation disasters (this quantity may as much as 14%) from the formulated point of view.
3. The special program should be developed for avoiding this class of accidents.
4. From the point of view of non-local physics the Tunguska explosion, Gagarin catastrophe and accident with Malaysia Airlines flight MH370 can have the same physical origin—plasmoid appearance in the Earth atmosphere [113, 123].

Chapter 6

Quantum Solitons in Solid Matter

ABSTRACT
The quantum oscillator (QO) is the quantum-mechanical analog of the classical harmonic oscillator. An arbitrary potential can usually be approximated as a harmonic potential at the vicinity of a stable equilibrium point. Then the QO can be used as the important model systems in quantum mechanics. It is one of the few quantum-mechanical systems for which an exact, analytical solution is known from the Schrödinger equation (SE). Here, we consider the QO behavior on the basement the unified non-local theory (UNLT), compare results with Madelung's hydrodynamics and apply the UNLT to transport processes in the much more complicated physical systems like graphene. The origin of the charge density waves (CDWs) is a long-standing problem relevant to a number of important issues in condensed matter physics. Mathematical modeling of the CDW expansion as well as the problem of the high-temperature superconducting can be solved only on the basement of the non-local quantum hydrodynamics.
Keywords: Quantum solitons, Quantum oscillators, Charged density waves in the graphene crystal lattice

6.1 QUANTUM OSCILLATORS IN THE UNIFIED NON-LOCAL THEORY

Let us consider the physical object moving along x-axis in one-dimensional (1D) space and introduce the coordinate system moving along the positive direction of x-axis with velocity $C = u_0$ equal to phase velocity of considering quantum object

$$\xi = x - Ct. \tag{6.1.1}$$

Taking into account the De Broglie relation, we expect that the group velocity u_g is equal 2 u_0 (see Section 5.4).

In the moving coordinate system, all dependent hydrodynamic values are functions of (ξ, t). We investigate the possibility of the quantum object formation of the soliton type. For this solution, there is no explicit dependence on time for coordinate system moving with the phase velocity u_0. Write down the system of Eqs. (5.3.1)–(5.3.6) for the one-species system. As before, the dimensionless symbols are marked by tildes. The next scales are used

$$\rho_0,\ u_0,\ t_0 = \tau^{(qu)},\ x_0 = u_0 t_0,\ U_0 = u_0^2,\ p_0 = \rho_0 u_0^2,\ x_0 = u_0 \tau^{(qu)} = u_0 \frac{\hbar}{m u_0^2},$$

The subsequent system of equation is the particular case of Eqs. (5.4.17)–(5.4.22).

$$\frac{\partial \tilde{\rho}}{\partial \tilde{\xi}} - \frac{\partial \tilde{\rho}\tilde{u}}{\partial \tilde{\xi}} + \frac{\partial}{\partial \tilde{\xi}} \left\{ \frac{1}{\tilde{u}^2} \left[\frac{\partial}{\partial \tilde{\xi}} \left(\tilde{p} + \tilde{\rho} + \tilde{\rho}\tilde{u}^2 - 2\tilde{\rho}\tilde{u} \right) + \tilde{\rho} \frac{\partial \tilde{U}}{\partial \tilde{\xi}} \right] \right\} = 0, \tag{6.1.2}$$

$$\begin{aligned} &\frac{\partial}{\partial \tilde{\xi}} \left(\tilde{\rho}\tilde{u}^2 + \tilde{p} - \tilde{\rho}\tilde{u} \right) \\ &+ \frac{\partial}{\partial \tilde{\xi}} \left\{ \frac{1}{\tilde{u}^2} \left[\frac{\partial}{\partial \tilde{\xi}} \left(2\tilde{\rho}\tilde{u}^2 - \tilde{\rho}\tilde{u} + 2\tilde{p} - \tilde{\rho}\tilde{u}^3 - 3\tilde{p}\tilde{u} \right) + \tilde{\rho} \frac{\partial \tilde{U}}{\partial \tilde{\xi}} \right] \right\} \\ &+ \frac{\partial \tilde{U}}{\partial \tilde{\xi}} \left\{ \tilde{\rho} - \frac{1}{\tilde{u}^2} \left[-\frac{\partial \tilde{\rho}}{\partial \tilde{\xi}} + \frac{\partial}{\partial \tilde{\xi}} \left(\tilde{\rho}\tilde{u} \right) \right] \right\} - 2 \frac{\partial}{\partial \tilde{\xi}} \left\{ \frac{\tilde{\rho}}{\tilde{u}} \frac{\partial \tilde{U}}{\partial \tilde{\xi}} \right\} = 0, \end{aligned} \tag{6.1.3}$$

Unified non-local theory of transport processes. http://dx.doi.org/10.1016/B978-0-444-63478-8.00006-3

$$\begin{aligned}&\frac{\partial}{\partial\tilde{\xi}}\left(\widetilde{\rho}\widetilde{u}^2+3\widetilde{p}-\widetilde{\rho}\widetilde{u}^3-5\widetilde{p}\widetilde{u}\right)\\&-\frac{\partial}{\partial\tilde{\xi}}\left\{\frac{1}{\tilde{u}^2}\frac{\partial}{\partial\tilde{\xi}}\left(2\widetilde{\rho}\widetilde{u}^3+10\widetilde{p}\widetilde{u}-\widetilde{\rho}\widetilde{u}^2-3\widetilde{p}-\widetilde{\rho}\widetilde{u}^4-8\widetilde{p}\widetilde{u}^2-5\frac{\widetilde{p}^2}{\widetilde{\rho}}\right)\right\}\\&+\frac{\partial}{\partial\tilde{\xi}}\left\{\frac{1}{\tilde{u}^2}\left(3\widetilde{\rho}\widetilde{u}^2+5\widetilde{p}\right)\frac{\partial\widetilde{U}}{\partial\tilde{\xi}}\right\}-2\widetilde{\rho}\widetilde{u}\frac{\partial\widetilde{U}}{\partial\tilde{\xi}}-2\frac{\partial}{\partial\tilde{\xi}}\left\{\frac{\widetilde{p}}{\widetilde{u}}\frac{\partial\widetilde{U}}{\partial\tilde{\xi}}\right\}\\&+\frac{2}{\tilde{u}^2}\frac{\partial\widetilde{U}}{\partial\tilde{\xi}}\left[-\frac{\partial}{\partial\tilde{\xi}}\left(\widetilde{\rho}\widetilde{u}\right)+\frac{\partial}{\partial\tilde{\xi}}\left(\widetilde{\rho}\widetilde{u}^2+\widetilde{p}\right)+\widetilde{\rho}\frac{\partial\widetilde{U}}{\partial\tilde{\xi}}\right]=0.\end{aligned}\tag{6.1.4}$$

Some comments to Eqs. (6.1.2)–(6.1.4):

1. Every equation from the system is of the second order and needs two conditions. The problem belongs to the class of Cauchy problems.
2. In comparison with the Schrödinger theory connected with behavior of the wave function, no special conditions are applied for dependent variables including the domain of the solution existing. This domain is defined automatically in the process of the numerical solution of the concrete variant of calculations.
3. From the introduced scales

$$\rho_0,\ u_0,\ p_0,\ t_0=\frac{\hbar}{mu_0^2},\ x_0=u_0t_0,\ U_0=u_0^2,$$

only two parameters are independent—the phase velocity u_0 of the quantum object and the density ρ_0. It is important to underline that the value $\nu^{qu}=\hbar/m_e$ has the dimension [cm^2/s] and can be denoted as quantum viscosity, $\nu^{qu}=1.1577\text{cm}^2/\text{s}$. Of course, in principal the electron mass can be replaced in scales by mass of other particles. From this point of view, the obtained solutions that will be discussed below have the universal character defined only by the Cauchy conditions.

The system of generalized quantum hydrodynamic equations (6.1.2)–(6.1.4) have the great possibilities of mathematical modeling as result of changing of twelve Cauchy conditions describing the character features of initial perturbations which lead to the soliton formation.

On this step of investigation, I intend to demonstrate the influence of difference conditions on the soliton formation. The following figures reflect some results of calculations realized according to the system of Eqs. (6.1.2)–(6.1.4) with the help of Maple. We use the typical dimensionless potential in the form $\widetilde{U}=\widetilde{\xi}^2$ (Figs. 6.1–6.6).

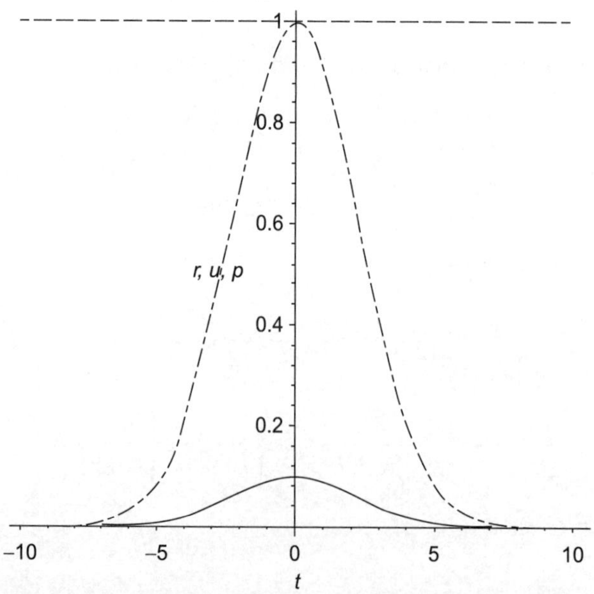

FIGURE 6.1 Solution of Eqs. (6.1.2)–(6.1.4). $u(0)=1$, $p(0)=1$, $r(0)=.1$, D(u)(0)=0, D(p)(0)=0, D(r)(0)=0, r—quantum density (solid line), u—velocity (dashed line), p—quantum pressure (dash-dotted line).

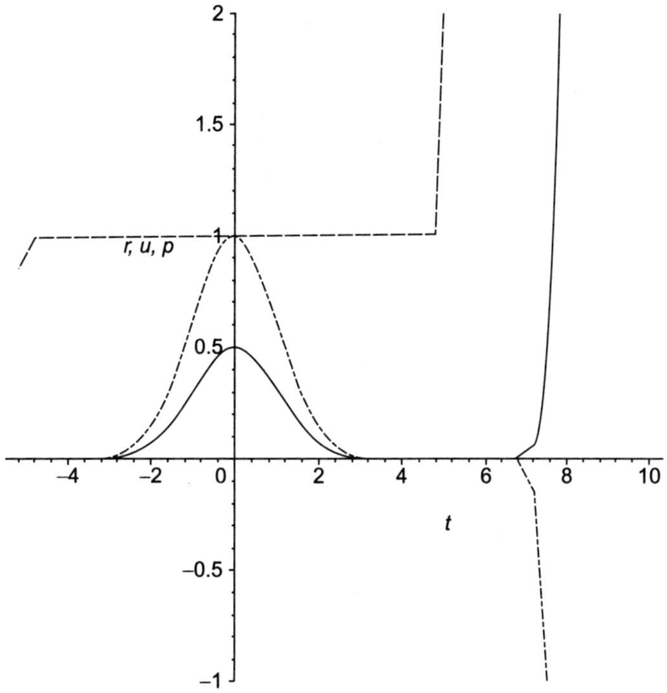

FIGURE 6.2 Solution of Eqs. (6.1.2)–(6.1.4). $u(0)=1$, $p(0)=1$, $r(0)=0.5$, $D(u)(0)=0$, $D(p)(0)=0$, $D(r)(0)=0$, r—quantum density (solid line), u—velocity (dashed line), p—quantum pressure (dash-dotted line).

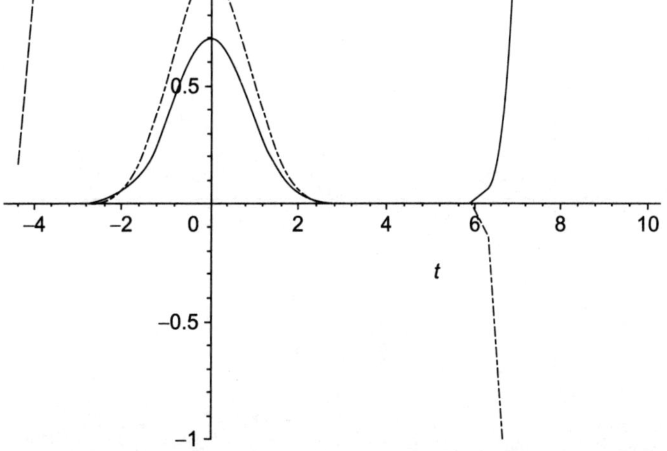

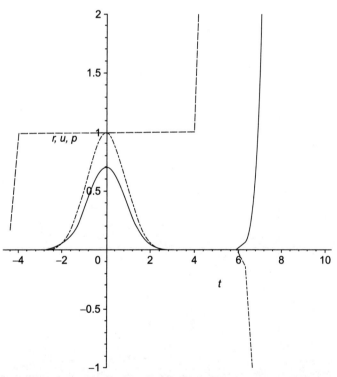

FIGURE 6.3 Solution of Eqs. (6.1.2)–(6.1.4). $u(0)=1$, $p(0)=1$, $r(0)=0.7$, $D(u)(0)=0$, $D(p)(0)=0$, $D(r)(0)=0$, r—quantum density (solid line), u—velocity (dashed line), p—quantum pressure (dash-dotted line).

The following notations on figures are used: r—density $\tilde{\rho}$, u—velocity $\tilde{u}$, and p—pressure $\tilde{p}$. Explanations placed under all following figures, Maple program contains Maple's notations—for example, the expression $D(u)(0)=0$ means in usual notations $\frac{\partial \tilde{u}}{\partial \tilde{\xi}}(0)=0$, independent variable t responds to $\tilde{\xi}$. We begin with investigation of the problem of principle significance—is it possible after a perturbation (defined by Cauchy conditions) to obtain the quantum object of the soliton's

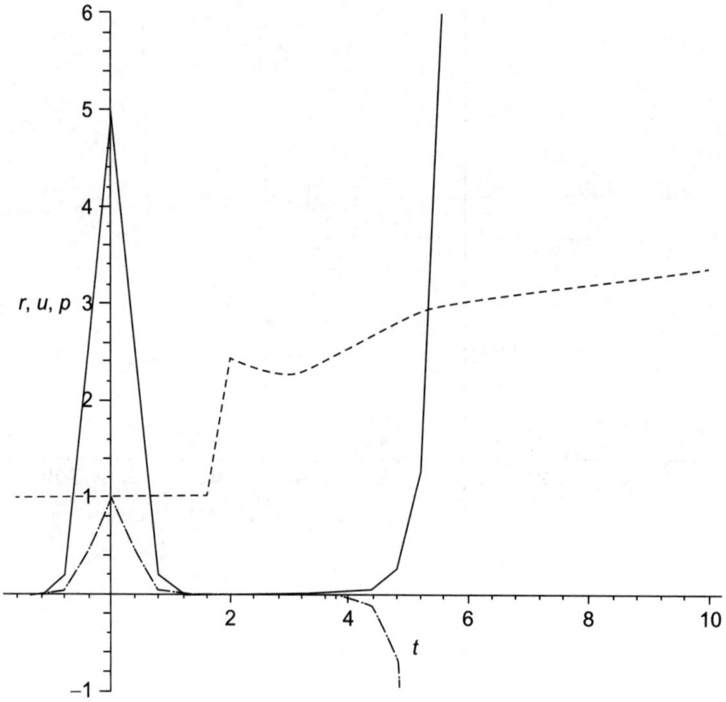

FIGURE 6.4 Solution of Eqs. (6.1.2)–(6.1.4). $u(0)=1$, $p(0)=1$, $r(0)=5$, $D(u)(0)=0$, $D(p)(0)=0$, $D(r)(0)=0$, r—quantum density (solid line), u—velocity (dashed line), p—quantum pressure (dash-dotted line).

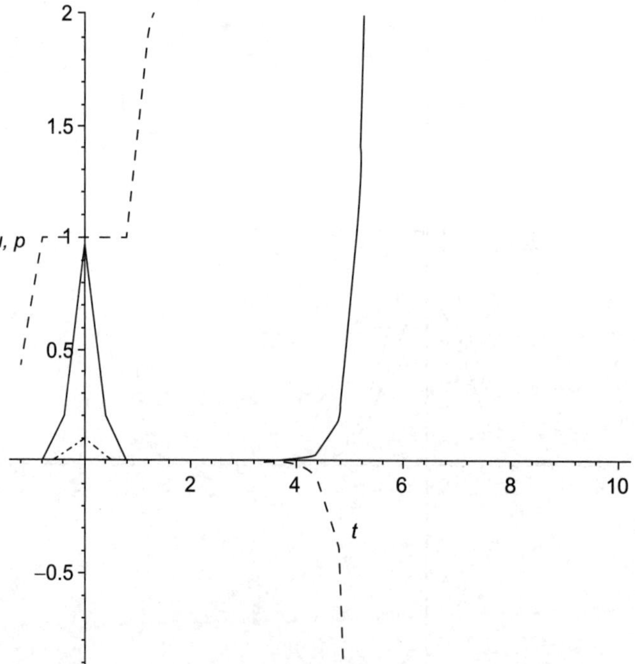

FIGURE 6.5 Solution of Eqs. (6.1.2)–(6.1.4), $u(0)=1$, $p(0)=0.1$, $r(0)=1$, $D(u)(0)=0$, $D(p)(0)=0$, $D(r)(0)=0$, r—quantum density (solid line), u—velocity (dashed line), p—quantum pressure (dash-dotted line).

kind as result of the self-organization of matter without boundary conditions? What is the influence of the quantum density and pressure on the soliton construction?

Figures 6.1–6.4 demonstrate the influence of the quantum density change.

The following increase of the initial quantum density leads to appearance of quantization of the first derivatives of density and pressure (see Figs. 6.3 and 6.4).

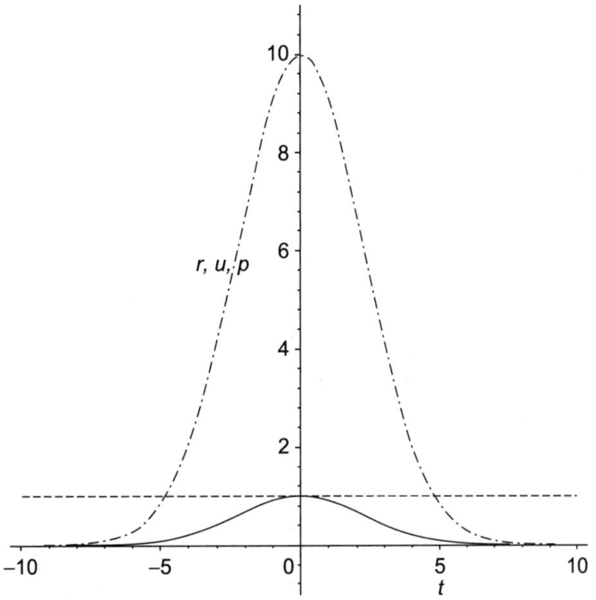

FIGURE 6.6 Solution of Eqs. (6.1.2)–(6.1.4). $u(0)=1$, $p(0)=10$, $r(0)=1$, $D(u)(0)=0$, $D(p)(0)=0$, $D(r)(0)=0$, r—quantum density (solid line), u—velocity (dashed line), p—quantum pressure (dash-dotted line).

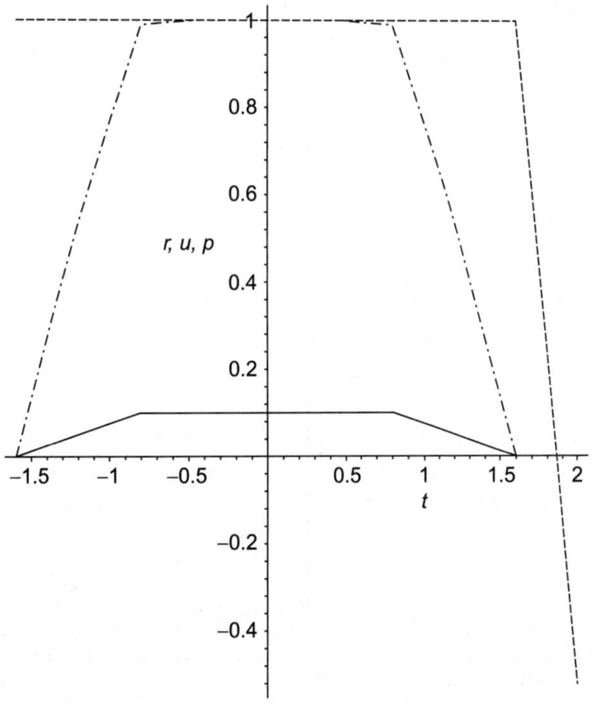

FIGURE 6.7 Solution of Eqs. (6.1.2)–(6.1.4). $u(0)=1$, $p(0)=1$, $r(0)=0.1$, $D(u)(0)=0$, $D(p)(0)=0$, $D(r)(0)=0$, $\widetilde{U}=\widetilde{\xi}^{10}$.

As we see from Figs. 6.1–6.4, the $\widetilde{\rho}(0)$ increasing leads to narrowing of the domain of the soliton existence.

Let us demonstrate now the influence of the quantum pressure change (from $\widetilde{p}(0)=0.1$ to $\widetilde{p}(0)=10$, Figs. 6.5 and 6.6).

The ensuing figures demonstrate results of calculations for nonharmonic oscillators with $\widetilde{U}=\widetilde{\xi}^{10}$ (Fig. 6.7) and $\widetilde{U}=\widetilde{\xi}^{2}+\widetilde{\xi}^{10}$ (Fig. 6.8).

The curves of Figs. 6.7 and 6.8 have come to light the typical effects of quantization—the curves transform into the short straight lines.

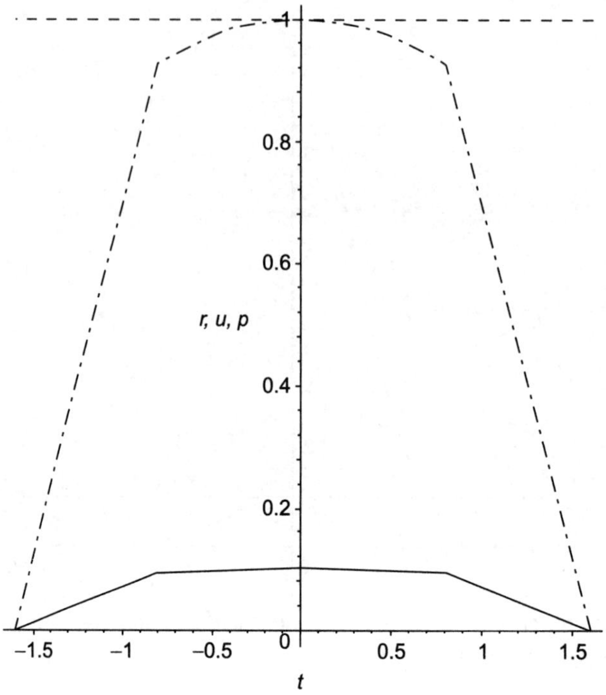

FIGURE 6.8 Solution of Eqs. (6.1.2)–(6.1.4) for the same Cauchy conditions but $\widetilde{U} = \widetilde{\xi}^2 + \widetilde{\xi}^{10}$, r—quantum density (solid line), u—velocity (dashed line), p—quantum pressure (dash-dotted line.

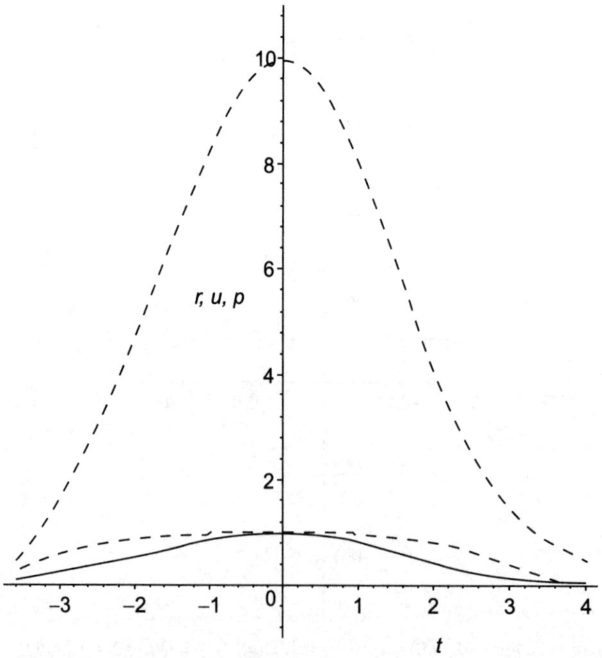

FIGURE 6.9 Solution of Eqs. (6.1.2)–(6.1.4). $u(0) = 1$, $p(0) = 10$, $r(0) = 1$, $D(u)(0) = 0$, $D(p)(0) = 0$, $D(r)(0) = 0$, $\widetilde{U} = \widetilde{\xi}^2 + \left(\widetilde{\xi} - 1\right)^2$.

It is interesting to investigate the behavior of two displaced oscillators with $\widetilde{U} = \widetilde{\xi}^2 + \left(\widetilde{\xi} - 1\right)^2$. It means, that the expressions $\widetilde{U} = \widetilde{\xi}^2 + \left(\widetilde{\xi} - 1\right)^2$ and $\left(\frac{\partial \widetilde{U}}{\partial \widetilde{\xi}}\right) = 2\left(2\widetilde{\xi} - 1\right)$ in the system (6.1.2)–(6.1.4) should be introduced. We have results presented on Figs. 6.9 and 6.10.

Figure 6.10 reflects the same calculation as before, but the Cauchy conditions displaced in the point $\widetilde{\xi} = 0.5$. In all cases, numerical solution cannot be obtained in all points and exists only in the soliton's area.

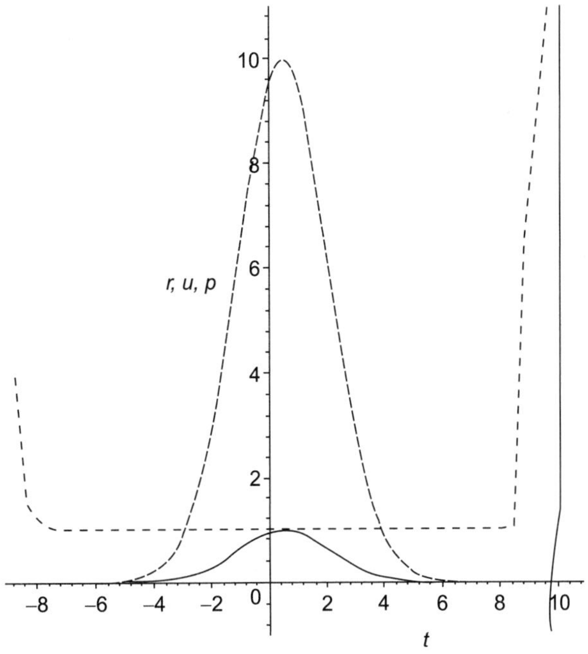

FIGURE 6.10 Solution of Eqs. (6.1.2)–(6.1.4). $u(0.5)=1$, $p(0.5)=10$, $r(0.5)=1$, $D(u)(0.5)=0$, $D(p)(0.5)=0$, $D(r)(0.5)=0$, $\tilde{U}=\tilde{\xi}^2+\left(\tilde{\xi}-1\right)^2$. r—quantum density (solid line), u—velocity (dashed line), p—quantum pressure (dash-dotted line).

The harmonic oscillator is a good example of how different quantums and classical results can be. The usual comparison consists in the calculation of the probability of finding the object which is oscillating at a given distance x from the equilibrium position. For the classical case, the probability is greatest at the ends of the motion since it is moving more slowly and comes to the rest instantaneously at the extremes of the motion. The relative probability of finding it in any interval Δx is just the inverse of its average velocity over that interval. For the Schrödinger quantum-mechanical case, the probability of finding the oscillator in an interval Δx is the square of the wave function, and that is very different for the lower energy states. The maximum probability for the ground state is at the equilibrium point $x=0$. As we see at this section, the UNLT for oscillators leads analogically to the soliton creation without additional boundary conditions. In the frame of the classical harmonic oscillator, the particle spends most of its time (and is therefore most likely to be found) at the turning points, where it is the slowest.

Let us look more thoroughly on theory of the Schrödinger quantum harmonic oscillator.

To find the ground state solution of the Schrödinger equation (SE) for the quantum 1D harmonic oscillator, we try to use the following form for the wave function ψ

$$\psi(x)=C\exp\left(-\alpha x^2/2\right) \tag{6.1.5}$$

for the SE equation

$$-\frac{\hbar^2}{2m}\frac{d^2\psi}{dx^2}+\frac{1}{2}m\omega^2x^2=E\psi, \tag{6.1.6}$$

where ω is the angular frequency and C is constant. The following implicit assumptions are introduced:

1. The form (6.1.5) means the introduction of the boundary conditions on the infinity $x=\pm\infty$.
2. The solution does not contain the explicit dependence on time.
3. After the substitution, the obtained equation should be satisfied for all independent x.

Realization of these principles leads to the equation

$$-\frac{\hbar^2}{2m}\left(-\alpha+\alpha^2x^2\right)\psi+\frac{1}{2}m\omega^2x^2=E\psi \tag{6.1.7}$$

and relations

$$-\frac{\hbar^2}{m}\alpha^2=m\omega^2,\quad \frac{\hbar^2}{2m}\alpha=E. \tag{6.1.8}$$

It means that

$$\alpha = \frac{m\omega}{\hbar}, \; E_0 = \frac{\hbar\omega}{2}, \tag{6.1.9}$$

where $E_0 = \hbar\omega/2$ is the oscillator energy of the ground state. The same method is applied in the theory of the Schrödinger quantum harmonic oscillator for more general form of the wave function ψ and the item 3.

$$\psi(x) = f(x)\exp(-\alpha x^2/2). \tag{6.1.10}$$

Realization of items 1–3 permits to avoid solving of the explicit time dependent problem; of course, the obtained solution cannot be satisfied for all x in the class of the continuous functions.

As we know, the Madelung equations are the hydrodynamic equivalent of the SE. Let us solve the Madelung equations using only the mentioned assumption 3. Then the Madelung equations (see (3.1.30) and (3.1.40))

$$\frac{\partial \rho}{\partial t} + \frac{\partial}{\partial \mathbf{r}} \cdot \rho \mathbf{v} = 0, \tag{6.1.11}$$

$$\frac{\partial \mathbf{v}}{\partial t} + (\mathbf{v}\nabla)\mathbf{v} = -\frac{1}{m}\nabla U^*, \tag{6.1.12}$$

where

$$U^* = \widetilde{U} - \frac{\hbar^2}{4m\rho}\left[\Delta\rho - \frac{1}{2\rho}\left(\frac{\partial \rho}{\partial \mathbf{r}}\right)^2\right], \tag{6.1.13}$$

transform in the following system of 1D equations (6.1.14) and (6.1.15).

$$\frac{\partial}{\partial x}(\rho u) = 0, \tag{6.1.14}$$

$$u\frac{\partial u}{\partial x} = -\frac{1}{m}\frac{\partial}{\partial x}\left[U - \frac{\hbar^2}{4m\rho}\left[\frac{\partial^2 \rho}{\partial x^2} - \frac{1}{2\rho}\left(\frac{\partial \rho}{\partial x}\right)^2\right]\right]. \tag{6.1.15}$$

Equation (6.1.14) allows to exclude velocity u from the consideration because $\rho u = C$, where C does not depend on x.

After the obvious transformation, we have

$$\rho^2\frac{\partial^3 \rho}{\partial x^3} - 2\rho\frac{\partial^2 \rho}{\partial x^2}\left(\frac{\partial \rho}{\partial x}\right) + \left(\frac{\partial \rho}{\partial x}\right)^3 + \frac{4C^2m^2}{\hbar^2}\frac{\partial \rho}{\partial x} - \frac{8km}{\hbar^2}\rho^3 x = 0. \tag{6.1.16}$$

For the derivation of Eq. (6.1.16), the expression for the potential energy is used

$$U = \frac{1}{2}m\omega^2 x^2 = kx^2. \tag{6.1.17}$$

As usual, Eq. (6.1.16) may be written in the nondimensional form:

$$\widetilde{\rho}^2\frac{\partial^3 \widetilde{\rho}}{\partial \widetilde{x}^3} - 2\widetilde{\rho}\frac{\partial^2 \widetilde{\rho}}{\partial \widetilde{x}^2}\left(\frac{\partial \widetilde{\rho}}{\partial \widetilde{x}}\right) + \left(\frac{\partial \widetilde{\rho}}{\partial \widetilde{x}}\right)^3 + \widetilde{A}\frac{\partial \hat{\rho}}{\partial \hat{x}} - \widetilde{B}\widetilde{\rho}^3\widetilde{x} = 0, \tag{6.1.18}$$

which contains the dimensionless parameters $\widetilde{A} = \frac{4C^2m^2}{\hbar^2}\frac{x_0^2}{\rho_0^2}$ and $\widetilde{B} = \frac{8km}{\hbar^2}x_0^4$.

Figures 6.11 and 6.12 reflect the results of solution of Eq. (6.1.18) for the Cauchy conditions $\widetilde{\rho}(0) = 1, \; \frac{d\widetilde{\rho}}{d\widetilde{x}}(0) = 0, \; \frac{d^2\widetilde{\rho}}{d\widetilde{x}^2}(0) = 0$ and different parameters $\widetilde{A}$ and $\widetilde{B}$.

As we see from Figs. 6.11 and 6.12, the quantum density distribution corresponds to the "classical" dependence in the extremely large variations of parameters $\widetilde{A}$ and $\widetilde{B}$. Important to notice that we need not to use the boundary conditions. The solution exists only in the restricted region of the 1D space.Important conclusions should be done:

1. The discovery of solutions in mathematics is considered as a rare and outstanding event. As you see in the UNLT, it is the typical often recurring fact. In many cases, the theory leads to the discrete levels for the derivatives of the dependent values.
2. It is the commonplace to meet the questions like "Why do wave functions spread out over time? Where in the math does quantum mechanics state this?" The answer consists in shortcomings of the local theory of transport processes on the whole and in particular, in the absence of the independent energy equation in the Schrödinger theory.

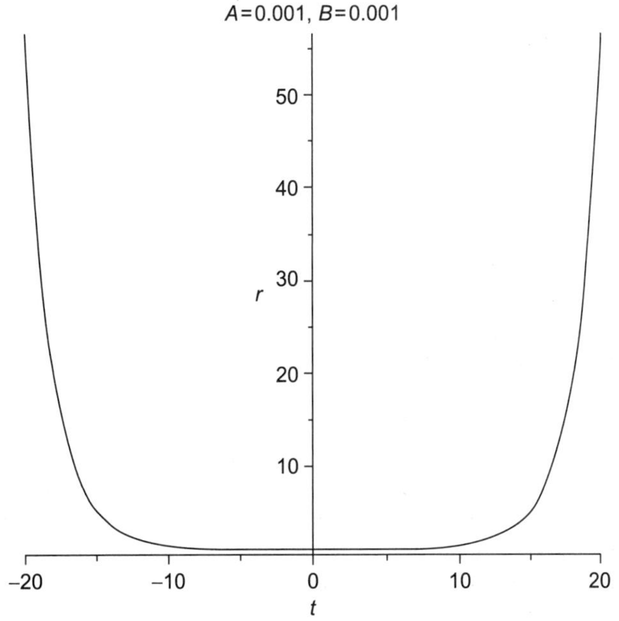

FIGURE 6.11 Solution of Eq. (6.1.18) for the case $\widetilde{A}=0.001$, $\widetilde{B}=0.001$.

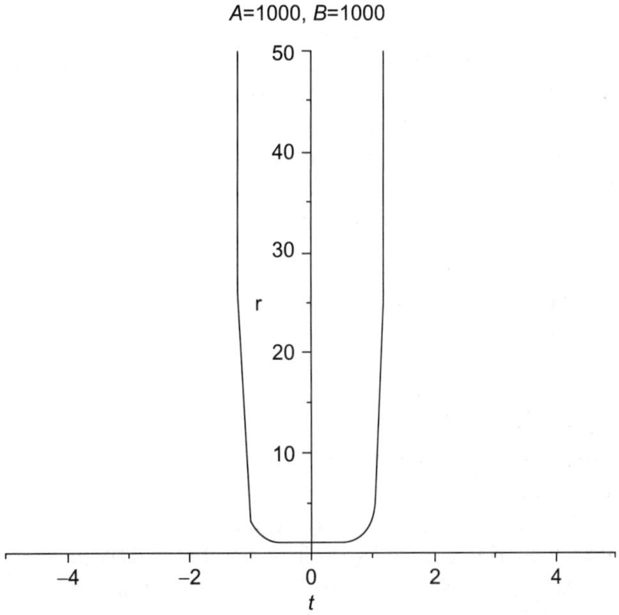

FIGURE 6.12 Solution of Eq. (6.1.18) for the case $\widetilde{A}=1000$, $\widetilde{B}=1000$.

6.2 APPLICATION OF NON-LOCAL QUANTUM HYDRODYNAMICS TO THE DESCRIPTION OF THE CHARGED DENSITY WAVES IN THE GRAPHENE CRYSTAL LATTICE

The motion of the charged particles in graphene in the frame of the quantum non-local hydrodynamic description is considered. It is shown as results of the mathematical modeling that the mentioned motion is realized in the soliton forms. The dependence of the size and structure of solitons on the different physical parameters is investigated.

Here, we investigate applications of the non-local quantum hydrodynamics in the theory of transport processes in graphene including the effects of the charge density waves (CDW). Is known that graphene, a single-atom-thick sheet of graphite, is a new material which combines aspects of semiconductors and metals. For example, the mobility, a measure of how well a material conducts electricity, is higher than for other known material at room temperature. In graphene, a resistivity is of about 1.0 μΩ-cm (resistivity defined as a specific measure of resistance; the resistance of a piece material is its resistivity times its length and divided by its cross-sectional area). This is about 35% less than the resistivity of copper, the lowest resistivity material known at room temperature.

Measurements lead to conclusion that the influence of thermal vibrations on the conduction of electrons in graphene is extraordinarily small. From the other side the typical reasoning exists:

"In any material, the energy associated with the temperature of the material causes the atoms of the material to vibrate in place. As electrons travel through the material, they can bounce off these vibrating atoms, giving rise to electrical resistance. This electrical resistance is 'intrinsic' to the material: it cannot be eliminated unless the material is cooled to absolute zero temperature, and hence sets the upper limit to how well a material can conduct electricity."

Obviously, this point of view leads to the principal elimination of effects of the high-temperature SC. From the mentioned point of view, the restrictions in mobility of known semiconductors can be explained as the influence of the thermal vibration of the atoms. The limit to mobility of electrons in graphene is about 200,000 $cm^2/(Vs)$ at room temperature, compared to about 1400 $cm^2/(Vs)$ in silicon, and 77,000 $cm^2/(Vs)$ in indium antimonide, the highest mobility conventional semiconductor known. The opinion of a part of investigators can be formulated as follows: "Other extrinsic sources in today's fairly dirty graphene samples add some extra resistivity to graphene," (see for example [124]) "so the overall resistivity isn't quite as low as copper's at room temperature yet. However, graphene has far fewer electrons than copper, so in graphene the electrical current is carried by only a few electrons moving much faster than the electrons in copper." Mobility determines the speed at which an electronic device (for instance, a field-effect transistor, which forms the basis of modern computer chips) can turn on and off. The very high mobility makes graphene promising for applications in which transistors much switch extremely fast, such as in processing extremely high-frequency signals. The low resistivity and extremely thin nature of graphene also promises applications in thin, mechanically tough, electrically conducting, transparent films. Such films are sorely needed in a variety of electronics applications from touch screens to photovoltaic cells.

In the last years, the direct observation of the atomic structures of superconducting materials (as usual superconducting materials in the cuprate family like $YBa_2Cu_3O_{6.67}$ ($T_c = 67$ K)) was realized with the scanning tunneling microscope (STM) and other instruments, STMs scan a surface in steps smaller than an atom.

Superconductivity (SC), in which an electric current flows with zero resistance, was first discovered in metals cooled very close to absolute zero. New materials called cuprates—copper oxides "doped" with other atoms—superconduct as "high" as minus 123 C. Some conclusions from direct observations [125, 126]:

1. Observations of high-temperature superconductors show an "energy gap" where electronic states are missing. Sometimes this energy gap appears but the material still does not superconduct—a so-called "pseudogap" phase. The pseudogap appears at higher temperatures than any SC, offering the promise of someday developing materials that would superconduct at or near room temperature.
2. STM image of a partially doped cuprate superconductor shows regions with an electronic "pseudogap." As doping increases, pseudogap regions spread and connect, making the whole sample a superconductor.
3. High-temperature SC in layered cuprates can develop from an electronically ordered state called a CDW. The results of observation can be interpreted as the creation of the "checkerboard pattern" due to the modulation of the atomic positions in the CuO_2 layers of $YBa_2Cu_3O_{6+x}$ caused by the CDW.
4. Application of the method of high-energy X-ray diffraction shows that a CDW develops at zero field in the normal state of superconducting $YBa_2Cu_3O_{6.67}$ ($T_c = 67$ K). Below T_c, the application of a magnetic field suppresses SC and enhances the CDW. It means that the high-T_c SC forms from a preexisting CDW environment.

Important conclusion: high-temperature superconductors demonstrate new type of electronic order and modulation of atomic positions. The above mentioned graphene properties can be explained only in the frame of the self-consistent non-local quantum theory which leads to appearance of the soliton waves moving in graphene.

The transport properties in graphene can be described at low energies approximately by a massless Dirac-fermion model with chiral quasiparticles [127, 128]. The Boltzmann and Schrödinger approaches are used also [129], [130].

Some remarks should be remained:

(a) SE transforms in hydrodynamic form without additional assumptions. But numerical methods of hydrodynamics are very good developed. SE reduces to the system of continuity equation and particular case of the Euler equation with the additional potential proportional to $\hbar^2$. Hydrodynamic form of the Dirac equation (DE) is considered in Chapter 12.

(b) SE (obtained in the frame of the theory of classical complex variables) and DE cannot contain the independent energy equation in principle.

(c) The system of the generalized quantum hydrodynamic equations contains energy equation written for unknown dependent value which can be specified as quantum pressure p_α of non-local origin.

Applications of these SE sand DE approaches are directed on the calculation of kinetic coefficients. The non-local kinetic equations also are used for calculation of graphene electrical conductivity. Here we intend to investigate the possibilities of non-local quantum hydrodynamics for modeling of the CDWs in graphene. In non-local quantum hydrodynamics, the many particles correlations manifest themselves in equations in the terms proportional to non locality parameter τ.

The influence of spin and magnetic moment of particles can be taken into account by the natural elegant way via the internal energy of particles. On this stage of investigations, we omit the influence of the internal energy of particles, therefore spin waves will be investigated separately (see Section 6.6).

6.3 GENERALIZED QUANTUM HYDRODYNAMIC EQUATIONS DESCRIBING THE SOLITON MOVEMENT IN THE CRYSTAL LATTICE

Let us consider the CDWs which are periodic modulation of conduction electron density. From the direct observations of CDWs, follow that CDW develop at zero external fields. In the following we suppose that the effective charge movement was created in graphene lattice as result of an initial fluctuation.

The movement of the soliton waves at the presence of the external electrical potential difference will be considered also.

This effective charge is created due to interference of the induced electron waves and correlating potentials as result of the polarized modulation of atomic positions. Therefore, in this approach the conduction in graphene convoys the transfer of the positive ($+e$, m_p) and negative ($-e$, m_e) charges. Let us formulate the problem in detail. The nonstationary 1D motion of the combined soliton is considered under influence of the self-consistent electric forces of the potential and non-potential origin. The mentioned soliton can exists without a chemical bond formation. For better understanding, let us investigate the situation for the case when the external forces are absent. Introduce the coordinate system ($\xi = x - Ct$) moving along the positive direction of the x axis with the velocity $C = u_0$, which is equal to the phase velocity of this quantum object.

Let us find the soliton type solutions for the system of the generalized quantum equations for two-species mixture. The graphene crystal lattice is 2D flat structure which is considered in the moving coordinate system with the independent variables $\xi = x - u_0 t$, y.

In the following, we intend (without taking into account the component's internal energy) to apply generalized non-local quantum hydrodynamic Eqs. (4.1.1)–(4.1.6) to the investigation of the CDWs in the frame of two-species model which led to the following dimensional equations:

(Poisson equation for the self-consistent electric field)

$$\frac{\partial^2 \psi}{\partial \xi^2} + \frac{\partial^2 \psi}{\partial y^2} = -4\pi e \left\{ \left[n_p - \tau_p \frac{\partial}{\partial \xi}(n_p(u - u_0)) \right] - \left[n_e - \tau_e \frac{\partial}{\partial \xi}(n_e(u - u_0)) \right] \right\}. \tag{6.3.1}$$

(Continuity equation for the positive particles)

$$\begin{aligned} &\frac{\partial}{\partial \xi}\left[\rho_p(u_0 - u)\right] + \frac{\partial}{\partial \xi}\left\{ \tau_p \frac{\partial}{\partial \xi}\left[\rho_p (u - u_0)^2\right] \right\} \\ &+ \frac{\partial}{\partial \xi}\left\{ \tau_p \left[\frac{\partial}{\partial \xi} p_p - \rho_p F_{p\xi} \right] \right\} + \frac{\partial}{\partial y}\left\{ \tau_p \left[\frac{\partial}{\partial y} p_p - \rho_p F_{py} \right] \right\} = 0. \end{aligned} \tag{6.3.2}$$

(Continuity equation for electrons)

$$\begin{aligned} &\frac{\partial}{\partial \xi}\left[\rho_e(u_0 - u)\right] + \frac{\partial}{\partial \xi}\left\{ \tau_e \frac{\partial}{\partial \xi}\left[\rho_e (u - u_0)^2\right] \right\} \\ &+ \frac{\partial}{\partial \xi}\left\{ \tau_e \left[\frac{\partial}{\partial \xi} p_e - \rho_e F_{e\xi} \right] \right\} + \frac{\partial}{\partial y}\left\{ \tau_e \left[\frac{\partial}{\partial y} p_e - \rho_e F_{ey} \right] \right\} = 0. \end{aligned} \tag{6.3.3}$$

(Momentum equation for the x direction)

$$\begin{aligned}
&\frac{\partial}{\partial \xi}\{\rho u(u-u_0)+p\}-\rho_p F_{p\xi}-\rho_e F_{e\xi} \\
&+\frac{\partial}{\partial \xi}\left\{\tau_p\left[\frac{\partial}{\partial \xi}\left(2p_p(u_0-u)-\rho_p u(u_0-u)^2\right)-\rho_p F_{p\xi}(u_0-u)\right]\right\} \\
&+\frac{\partial}{\partial \xi}\left\{\tau_e\left[\frac{\partial}{\partial \xi}\left(2p_e(u_0-u)-\rho_e u(u_0-u)^2\right)-\rho_e F_{e\xi}(u_0-u)\right]\right\} \\
&+\tau_p F_{p\xi}\left(\frac{\partial}{\partial \xi}(\rho_p(u-u_0))\right)+\tau_e F_{e\xi}\left(\frac{\partial}{\partial \xi}(\rho_e(u-u_0))\right) \\
&-\frac{\partial}{\partial \xi}\left\{\tau_p\frac{\partial}{\partial \xi}(p_p u)\right\}-\frac{\partial}{\partial \xi}\left\{\tau_e\frac{\partial}{\partial \xi}(p_e u)\right\}-\frac{\partial}{\partial y}\left\{\tau_p\frac{\partial}{\partial y}(p_p u)\right\}-\frac{\partial}{\partial y}\left\{\tau_e\frac{\partial}{\partial y}(p_e u)\right\} \\
&+\frac{\partial}{\partial \xi}\{\tau_p[F_{p\xi}\rho_p u]\}+\frac{\partial}{\partial \xi}\{\tau_e[F_{e\xi}\rho_e u]\}+\frac{\partial}{\partial y}\{\tau_p[F_{py}\rho_p u]\}+\frac{\partial}{\partial y}\{\tau_e[F_{ey}\rho_e u]\}=0
\end{aligned} \tag{6.3.4}$$

(Energy equation for the positive particles)

$$\begin{aligned}
&\frac{\partial}{\partial \xi}\left[\rho_p u^2(u-u_0)+5p_p u-3p_p u_0\right]-2\rho_p F_{p\xi}u \\
&+\frac{\partial}{\partial \xi}\left\{\tau_p\left[\begin{array}{l}\frac{\partial}{\partial \xi}\left(-\rho_p u^2(u_0-u)^2+7p_p u(u_0-u)+3p_p u_0(u-u_0)-p_p u^2-5\frac{p_p^2}{\rho_p}\right) \\ -2F_{p\xi}\rho_p u(u_0-u)+\rho_p u^2 F_{p\xi}+5p_p F_{p\xi}\end{array}\right]\right\} \\
&-\frac{\partial}{\partial y}\left\{\tau_p\left[\frac{\partial}{\partial y}\left(p_p u^2+5\frac{p_p^2}{\rho_p}\right)-\rho_p F_{py}u^2-5p_p F_{py}\right]\right\} \\
&-2\tau_p F_{p\xi}\left[\frac{\partial}{\partial \xi}(\rho_p u(u_0-u))\right]-2\tau_p\rho_p\left[(F_{p\xi})^2+(F_{py})^2\right] \\
&+2\tau_p F_{p\xi}\left[\frac{\partial}{\partial \xi}p_p\right]+2\tau_p F_{py}\left[\frac{\partial}{\partial y}p_p\right]=-\frac{p_p-p_e}{\tau_{ep}}
\end{aligned} \tag{6.3.5}$$

(Energy equation for electrons)

$$\begin{aligned}
&\frac{\partial}{\partial \xi}\left[\rho_e u^2(u-u_0)+5p_e u-3p_e u_0\right]-2\rho_e F_{e\xi}u \\
&+\frac{\partial}{\partial \xi}\left\{\tau_e\left[\begin{array}{l}\frac{\partial}{\partial \xi}\left(-\rho_e u^2(u_0-u)^2+7p_e u(u_0-u)+3p_e u_0(u-u_0)-p_e u^2-5\frac{p_e^2}{\rho_e}\right) \\ -2F_{e\xi}\rho_e u(u_0-u)+\rho_e u^2 F_{e\xi}+5p_e F_{e\xi}\end{array}\right]\right\} \\
&-\frac{\partial}{\partial y}\left\{\tau_e\left[\frac{\partial}{\partial y}\left(p_e u^2+5\frac{p_e^2}{\rho_e}\right)-\rho_e F_{ey}u^2-5p_e F_{ey}\right]\right\} \\
&-2\tau_e F_{e\xi}\left[\frac{\partial}{\partial \xi}(\rho_e u(u_0-u))\right]-2\tau_e\rho_e\left[(F_{e\xi})^2+(F_{ey})^2\right] \\
&+2\tau_e F_{e\xi}\left[\frac{\partial}{\partial \xi}p_e\right]+2\tau_e F_{ey}\left[\frac{\partial}{\partial y}p_e\right]=-\frac{p_e-p_p}{\tau_{ep}}
\end{aligned} \tag{6.3.6}$$

Here, u—hydrodynamic velocity; ψ—self-consistent electric potential; ρ_e, ρ_p—densities for the electron and positive species; p_e, p_p—quantum electron pressure and the pressure of positive species; F_e, F_p—the forces acting on the mass unit of electrons and the positive particles. The right-hand sides of the energy equations are written in the relaxation forms following from BGK (Bhatnagar – Gross – Krook operator) kinetic approximation

Non-local parameters can be written in the form (see Eqs. (5.3.9) and (5.3.10))

$$\tau_p=\frac{N_R\hbar}{m_p u^2},\ \tau_e=\frac{N_R\hbar}{m_e u^2},\ \frac{1}{\tau_{ep}}=\frac{1}{\tau_e}+\frac{1}{\tau_p}, \tag{6.3.7}$$

where N_R—integer.

Acting forces are the sum of three terms: the self-consistent potential force (scalar potential ψ), connected with the displacement of positive and negative charges, potential forces originated by the graphene crystal lattice (potential U), and the external electrical field creating the intensity E. As result, the following relations are valid

$$F_{p\xi}=\frac{e}{m_p}\left(-\frac{\partial\psi}{\partial\xi}-\frac{\partial U}{\partial\xi}+E_{0\xi}\right),\ F_{e\xi}=\frac{e}{m_e}\left(\frac{\partial\psi}{\partial\xi}+\frac{\partial U}{\partial\xi}-E_{0\xi}\right),\ F_{py}=\frac{e}{m_p}\left(-\frac{\partial\psi}{\partial y}-\frac{\partial U}{\partial y}+E_{0y}\right),\ F_{ey}=\frac{e}{m_e}\left(\frac{\partial\psi}{\partial y}+\frac{\partial U}{\partial y}-E_{0y}\right). \tag{6.3.8}$$

Let us write down these equations in the dimensionless form, where dimensionless symbols are marked by tildes; introduce the scales:

$$u=u_0\tilde{u},\ \xi=x_0\tilde{\xi},\ y=x_0\tilde{y},\ \psi=\psi_0\tilde{\psi},\ \rho_e=\rho_0\tilde{\rho}_e,\ \rho_p=\rho_0\tilde{\rho}_p,$$

where u_0, x_0, ψ_0, ρ_0—scales for velocity, distance, potential, and density. Let there be also $p_p=\rho_0V_{0p}^2\tilde{p}_p$, $p_e=\rho_0V_{0e}^2\tilde{p}_e$, where V_{0p} and V_{0e}—the scales for thermal velocities for the electron and positive species; $F_p=\tilde{\mathrm{F}}_{\mathrm{p}}\frac{e\psi_0}{m_px_0}$, $F_e=\tilde{\mathrm{F}}_{\mathrm{e}}\frac{e\psi_0}{m_ex_0}$; $\tau_p=\frac{m_ex_0H}{m_pu_0\tilde{u}^2}$, $\tau_e=\frac{x_0H}{u_0\tilde{u}^2}$, where dimensionless parameter $H=\frac{N_R\hbar}{m_ex_0u_0}$ is introduced. Then, $\frac{1}{\tau_{ep}}=\frac{u_0\tilde{u}^2}{x_0H}\left(1+\frac{m_p}{m_e}\right)$.

Introduce also the following dimensionless parameters

$$R=\frac{e\rho_0x_0^2}{m_e\psi_0},\ E=\frac{e\psi_0}{m_eu_0^2}. \tag{6.3.9}$$

Taking into account the introduced values, the following system of dimensionless non-local hydrodynamic equations for the 2D soliton description can be written:

(Poisson equation for the self-consistent electric field)

$$\frac{\partial^2\tilde{\psi}}{\partial\tilde{\xi}^2}+\frac{\partial^2\tilde{\psi}}{\partial\tilde{y}^2}=-4\pi R\left\{\frac{m_e}{m_p}\left[\tilde{\rho}_p-\frac{m_eH}{m_p\tilde{u}^2}\frac{\partial}{\partial\tilde{\xi}}(\tilde{\rho}_p(\tilde{u}-1))\right]-\left[\tilde{\rho}_e-\frac{H}{\tilde{u}^2}\frac{\partial}{\partial\tilde{\xi}}(\tilde{\rho}_e(\tilde{u}-1))\right]\right\}. \tag{6.3.10}$$

(Continuity equation for the positive particles)

$$\begin{aligned}&\frac{\partial}{\partial\tilde{\xi}}\left[\tilde{\rho}_p(1-\tilde{u})\right]+\frac{m_e}{m_p}\frac{\partial}{\partial\tilde{\xi}}\left\{\frac{H}{\tilde{u}^2}\frac{\partial}{\partial\tilde{\xi}}\left[\tilde{\rho}_p(\tilde{u}-1)^2\right]\right\}\\&+\frac{m_e}{m_p}\frac{\partial}{\partial\tilde{\xi}}\left\{\frac{H}{\tilde{u}^2}\left[\frac{V_{0p}^2}{u_0^2}\frac{\partial}{\partial\tilde{\xi}}\tilde{p}_p-\frac{m_e}{m_p}E\tilde{\rho}_p\tilde{\mathrm{F}}_{\mathrm{p}\xi}\right]\right\}+\frac{m_e}{m_p}\frac{\partial}{\partial\tilde{y}}\left\{\frac{H}{\tilde{u}^2}\left[\frac{V_{0p}^2}{u_0^2}\frac{\partial}{\partial\tilde{y}}\tilde{p}_p-\frac{m_e}{m_p}E\tilde{\rho}_p\tilde{\mathrm{F}}_{\mathrm{py}}\right]\right\}=0.\end{aligned} \tag{6.3.11}$$

Continuity equation for electrons:

$$\begin{aligned}&\frac{\partial}{\partial\tilde{\xi}}\left[\tilde{\rho}_e(1-\tilde{u})\right]+\frac{\partial}{\partial\tilde{\xi}}\left\{\frac{H}{\tilde{u}^2}\frac{\partial}{\partial\tilde{\xi}}\left[\tilde{\rho}_e(\tilde{u}-1)^2\right]\right\}\\&+\frac{\partial}{\partial\tilde{\xi}}\left\{\frac{H}{\tilde{u}^2}\left[\frac{V_{0e}^2}{u_0^2}\frac{\partial}{\partial\tilde{\xi}}\tilde{p}_e-\tilde{\rho}_eE\tilde{\mathrm{F}}_{\mathrm{e}\xi}\right]\right\}+\frac{\partial}{\partial\tilde{y}}\left\{\frac{H}{\tilde{u}^2}\left[\frac{V_{0e}^2}{u_0^2}\frac{\partial}{\partial\tilde{y}}\tilde{p}_e-\tilde{\rho}_eE\tilde{\mathrm{F}}_{\mathrm{ey}}\right]\right\}=0.\end{aligned} \tag{6.3.12}$$

Momentum equation for the x direction:

$$\begin{aligned}&\frac{\partial}{\partial\tilde{\xi}}\left\{(\tilde{\rho}_p+\tilde{\rho}_e)\tilde{u}(\tilde{u}-1)+\frac{V_{0p}^2}{u_0^2}\tilde{p}_p+\frac{V_{0e}^2}{u_0^2}\tilde{p}_e\right\}-\frac{m_e}{m_p}\tilde{\rho}_pE\tilde{\mathrm{F}}_{p\xi}-\tilde{\rho}_eE\tilde{\mathrm{F}}_{e\xi}\\&+\frac{m_e}{m_p}\frac{\partial}{\partial\tilde{\xi}}\left\{\frac{H}{\tilde{u}^2}\left[\frac{\partial}{\partial\tilde{\xi}}\left(2\frac{V_{0p}^2}{u_0^2}\tilde{p}_p(1-\tilde{u})-\tilde{\rho}_p\tilde{u}(1-\tilde{u})^2\right)-\frac{m_e}{m_p}\tilde{\rho}_pE\tilde{\mathrm{F}}_{p\xi}(1-\tilde{u})\right]\right\}\\&+\frac{\partial}{\partial\tilde{\xi}}\left\{\frac{H}{\tilde{u}^2}\left[\frac{\partial}{\partial\tilde{\xi}}\left(2\frac{V_{0e}^2}{u_0^2}\tilde{p}_e(1-\tilde{u})-\tilde{\rho}_e\tilde{u}(1-\tilde{u})^2\right)-\tilde{\rho}_eE\tilde{\mathrm{F}}_{e\xi}(1-\tilde{u})\right]\right\}\\&+\frac{H}{\tilde{u}^2}E\left(\frac{m_e}{m_p}\right)^2\tilde{\mathrm{F}}_{p\xi}\left(\frac{\partial}{\partial\tilde{\xi}}(\tilde{\rho}_p(\tilde{u}-1))\right)+\frac{H}{\tilde{u}^2}E\tilde{\mathrm{F}}_{e\xi}\left(\frac{\partial}{\partial\tilde{\xi}}(\tilde{\rho}_e(\tilde{u}-1))\right)\\&-\frac{m_e}{m_p}\frac{\partial}{\partial\tilde{\xi}}\left\{\frac{HV_{0p}^2}{\tilde{u}^2u_0^2}\frac{\partial}{\partial\tilde{\xi}}(\tilde{p}_p\tilde{u})\right\}-\frac{\partial}{\partial\tilde{\xi}}\left\{\frac{HV_{0e}^2}{\tilde{u}^2u_0^2}\frac{\partial}{\partial\tilde{\xi}}(\tilde{p}_e\tilde{u})\right\}-\frac{m_e}{m_p}\frac{\partial}{\partial\tilde{y}}\left\{\frac{HV_{0p}^2}{\tilde{u}^2u_0^2}\frac{\partial}{\partial\tilde{y}}(\tilde{p}_p\tilde{u})\right\}-\frac{\partial}{\partial\tilde{y}}\left\{\frac{HV_{0e}^2}{\tilde{u}^2u_0^2}\frac{\partial}{\partial\tilde{y}}(\tilde{p}_e\tilde{u})\right\}\\&+\left(\frac{m_e}{m_p}\right)^2\frac{\partial}{\partial\tilde{\xi}}\left\{\frac{H}{\tilde{u}^2}E\left[\tilde{\mathrm{F}}_{p\xi}\tilde{\rho}_p\tilde{u}\right]\right\}+\frac{\partial}{\partial\tilde{\xi}}\left\{\frac{H}{\tilde{u}^2}E\left[\tilde{\mathrm{F}}_{e\xi}\tilde{\rho}_e\tilde{u}\right]\right\}+\left(\frac{m_e}{m_p}\right)^2\frac{\partial}{\partial\tilde{y}}\left\{\frac{H}{\tilde{u}^2}E\left[\tilde{\mathrm{F}}_{py}\tilde{\rho}_p\tilde{u}\right]\right\}+\frac{\partial}{\partial\tilde{y}}\left\{\frac{H}{\tilde{u}^2}E\left[\tilde{\mathrm{F}}_{ey}\tilde{\rho}_e\tilde{u}\right]\right\}=0.\end{aligned} \tag{6.3.13}$$

(Energy equation for the positive particles)

$$
\begin{aligned}
&\frac{\partial}{\partial\tilde{\xi}}\left[\tilde{\rho}_p\tilde{u}^2(\tilde{u}-1)+5\frac{V_{0p}^2}{u_0^2}\tilde{p}_p\tilde{u}-3\frac{V_{0p}^2}{u_0^2}\tilde{p}_p\right]-2\frac{m_e}{m_p}\tilde{\rho}_pE\tilde{\mathrm{F}}_{p\xi}\tilde{u}\\
&+\frac{\partial}{\partial\tilde{\xi}}\left\{\frac{H\,m_e}{\tilde{u}^2m_p}\left[\begin{array}{l}\dfrac{\partial}{\partial\tilde{\xi}}\left(-\tilde{\rho}_p\tilde{u}^2(1-\tilde{u})^2+7\dfrac{V_{0p}^2}{u_0^2}\tilde{p}_p\tilde{u}(1-\tilde{u})+3\dfrac{V_{0p}^2}{u_0^2}\tilde{p}_p(\tilde{u}-1)-\dfrac{V_{0p}^2}{u_0^2}\tilde{p}_p\tilde{u}^2-5\dfrac{V_{0p}^4}{u_0^4}\dfrac{\tilde{p}_p^2}{\tilde{\rho}_p}\right)\\[2ex] -2\dfrac{m_e}{m_p}E\tilde{\mathrm{F}}_{p\xi}\tilde{\rho}_p\tilde{u}(1-\tilde{u})+\dfrac{m_e}{m_p}\tilde{\rho}_p\tilde{u}^2E\tilde{\mathrm{F}}_{p\xi}+5\dfrac{m_e}{m_p}\dfrac{V_{0p}^2}{u_0^2}\tilde{p}_pE\tilde{\mathrm{F}}_{p\xi}\end{array}\right]\right\}\\
&-\frac{\partial}{\partial\tilde{y}}\left\{\frac{H\,m_e}{\tilde{u}^2m_p}\left[\frac{\partial}{\partial\tilde{y}}\left(\frac{V_{0p}^2}{u_0^2}\tilde{p}_p\tilde{u}^2+5\frac{V_{0p}^4}{u_0^4}\frac{\tilde{p}_p^2}{\tilde{\rho}_p}\right)-\frac{m_e}{m_p}\tilde{\rho}_pE\tilde{\mathrm{F}}_{py}\tilde{u}^2-5\frac{m_e}{m_p}\frac{V_{0p}^2}{u_0^2}\tilde{p}_pE\tilde{\mathrm{F}}_{py}\right]\right\}\\
&-2\frac{H}{\tilde{u}^2}\left(\frac{m_e}{m_p}\right)^2E\tilde{\mathrm{F}}_{p\xi}\left[\frac{\partial}{\partial\tilde{\xi}}\left(\tilde{\rho}_p\tilde{u}(1-\tilde{u})\right)\right]-2\frac{H}{\tilde{u}^2}\left(\frac{m_e}{m_p}\right)^3\tilde{\rho}_pE^2\left[\left(\tilde{\mathrm{F}}_{p\xi}\right)^2+\left(\tilde{\mathrm{F}}_{py}\right)^2\right]\\
&+2\frac{H}{\tilde{u}^2}\left(\frac{m_e}{m_p}\right)^2E\tilde{\mathrm{F}}_{p\xi}\left[\frac{V_{0p}^2}{u_0^2}\frac{\partial}{\partial\tilde{\xi}}\tilde{p}_p\right]+2\frac{H}{\tilde{u}^2}\left(\frac{m_e}{m_p}\right)^2E\tilde{\mathrm{F}}_{py}\left[\frac{V_{0p}^2}{u_0^2}\frac{\partial}{\partial\tilde{y}}\tilde{p}_p\right]=-\frac{\tilde{u}^2}{Hu_0^2}\left(V_{0p}^2\tilde{p}_p-\tilde{p}_eV_{0e}^2\right)\left(1+\frac{m_p}{m_e}\right)
\end{aligned}
\tag{6.3.14}
$$

(Energy equation for electrons)

$$
\begin{aligned}
&\frac{\partial}{\partial\tilde{\xi}}\left[\tilde{\rho}_e\tilde{u}^2(\tilde{u}-1)+5\frac{V_{0e}^2}{u_0^2}\tilde{p}_e\tilde{u}-3\frac{V_{0e}^2}{u_0^2}\tilde{p}_e\right]-2\tilde{\rho}_eE\tilde{\mathrm{F}}_{e\xi}\tilde{u}\\
&+\frac{\partial}{\partial\tilde{\xi}}\left\{\frac{H}{\tilde{u}^2}\left[\begin{array}{l}\dfrac{\partial}{\partial\tilde{\xi}}\left(-\tilde{\rho}_e\tilde{u}^2(1-\tilde{u})^2+7\dfrac{V_{0e}^2}{u_0^2}\tilde{p}_e\tilde{u}(1-\tilde{u})+3\dfrac{V_{0e}^2}{u_0^2}\tilde{p}_e(\tilde{u}-1)-\dfrac{V_{0e}^2}{u_0^2}\tilde{p}_e\tilde{u}^2-5\dfrac{V_{0e}^4}{u_0^4}\dfrac{\tilde{p}_e^2}{\tilde{\rho}_e}\right)\\[2ex] -2E\tilde{\mathrm{F}}_{e\xi}\tilde{\rho}_e\tilde{u}(1-\tilde{u})+\tilde{\rho}_e\tilde{u}^2E\tilde{\mathrm{F}}_{e\xi}+5\dfrac{V_{0e}^2}{u_0^2}\tilde{p}_eE\tilde{\mathrm{F}}_{e\xi}\end{array}\right]\right\}\\
&-\frac{\partial}{\partial\tilde{y}}\left\{\frac{H}{\tilde{u}^2}\left[\frac{\partial}{\partial\tilde{y}}\left(\frac{V_{0e}^2}{u_0^2}\tilde{p}_e\tilde{u}^2+5\frac{V_{0e}^4}{u_0^4}\frac{\tilde{p}_e^2}{\tilde{\rho}_e}\right)-\tilde{\rho}_eE\tilde{\mathrm{F}}_{ey}\tilde{u}^2-5\frac{V_{0e}^2}{u_0^2}\tilde{p}_eE\tilde{\mathrm{F}}_{ey}\right]\right\}\\
&-2\frac{H}{\tilde{u}^2}E\tilde{\mathrm{F}}_{e\xi}\left[\frac{\partial}{\partial\tilde{\xi}}\left(\tilde{\rho}_e\tilde{u}(1-\tilde{u})\right)\right]-2\frac{H}{\tilde{u}^2}\tilde{\rho}_eE^2\left[\left(\tilde{\mathrm{F}}_{e\xi}\right)^2+\left(\tilde{\mathrm{F}}_{ey}\right)^2\right]\\
&+2\frac{H}{\tilde{u}^2}E\tilde{\mathrm{F}}_{e\xi}\left[\frac{V_{0e}^2}{u_0^2}\frac{\partial}{\partial\tilde{\xi}}\tilde{p}_e\right]+2\frac{H}{\tilde{u}^2}E\tilde{\mathrm{F}}_{ey}\left[\frac{V_{0e}^2}{u_0^2}\frac{\partial}{\partial\tilde{y}}\tilde{p}_e\right]=-\frac{\tilde{u}^2}{Hu_0^2}\left(V_{0e}^2\tilde{p}_e-V_{0p}^2\tilde{p}_p\right)\left(1+\frac{m_p}{m_e}\right).
\end{aligned}
\tag{6.3.15}
$$

We have the following dimensionless relations for forces:

$$
\tilde{\mathrm{F}}_{p\xi}=-\frac{\partial\tilde{\psi}}{\partial\tilde{\xi}}-\frac{\partial\tilde{U}}{\partial\tilde{\xi}}+\tilde{E}_\xi,\ \tilde{\mathrm{F}}_{e\xi}=\frac{\partial\tilde{\psi}}{\partial\tilde{\xi}}+\frac{\partial\tilde{U}}{\partial\tilde{\xi}}-\tilde{E}_\xi,\ \tilde{\mathrm{F}}_{py}=-\frac{\partial\tilde{\psi}}{\partial\tilde{y}}-\frac{\partial\tilde{U}}{\partial\tilde{y}}+\tilde{E}_y,\ \tilde{\mathrm{F}}_{ey}=\frac{\partial\tilde{\psi}}{\partial\tilde{y}}+\frac{\partial\tilde{U}}{\partial\tilde{y}}-\tilde{E}_y. \tag{6.3.16}
$$

Graphene is a single layer of carbon atoms densely packed in a honeycomb lattice. Figure 6.13 reflects the structure of graphene as the 2D hexagonal carbon crystal, the distance a between the nearest atoms is equal to $a=0.142\,\mathrm{nm}$.

Elementary cell contains two atoms (for example A and B, Fig. 6.13) and the primitive lattice vectors are given by

$$
\mathbf{a}_1=\frac{a}{2}\left(3;\sqrt{3}\right),\ \mathbf{a}_2=\frac{a}{2}\left(3;-\sqrt{3}\right).
$$

Coordinates of the nearest atoms to the given atom define by vectors

$$
\boldsymbol{\delta}_1=\frac{a}{2}\left(1;\sqrt{3}\right),\ \boldsymbol{\delta}_2=\frac{a}{2}\left(1;-\sqrt{3}\right),\ \boldsymbol{\delta}_3=-a(1;0).
$$

Six neighboring atoms of the second order are placed in knots defined by vectors

$$
\boldsymbol{\delta}_1'=\pm\mathbf{a}_1,\ \boldsymbol{\delta}_2'=\pm\mathbf{a}_2,\ \boldsymbol{\delta}_3'=\pm(\mathbf{a}_2-\mathbf{a}_1).
$$

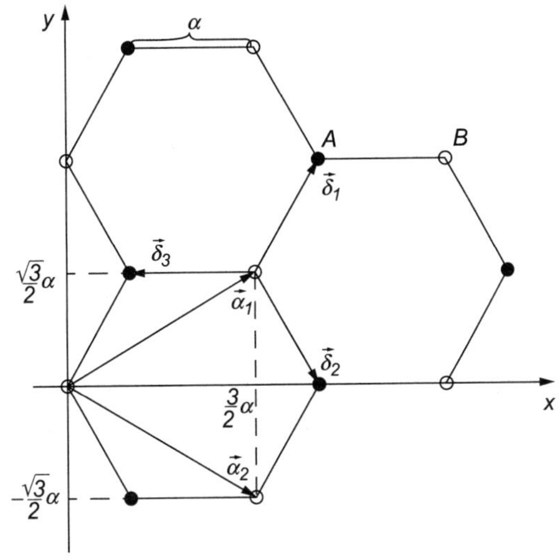

FIGURE 6.13 Crystal graphene lattice.

Let us take the first atom of the elementary cell in the origin of the coordinate system (Fig. 6.13) and compose the radii-vector of the second atom with respect to the basis $\mathbf{a}_1$ and $\mathbf{a}_2$:

$$\mathbf{r}_1 = u\mathbf{a}_1 + v\mathbf{a}_2 = u\left(3\frac{a}{2}\mathbf{e}_x + \sqrt{3}\frac{a}{2}\mathbf{e}_y\right) + v\left(3\frac{a}{2}\mathbf{e}_x - \sqrt{3}\frac{a}{2}\mathbf{e}_y\right). \tag{6.3.17}$$

Let us find u and v, taking into account that

$$\mathbf{r}_1 = \boldsymbol{\delta}_1 = \frac{a}{2}\left(1;\sqrt{3}\right) = \frac{a}{2}\mathbf{e}_x + \frac{a}{2}\sqrt{3}\mathbf{e}_y. \tag{6.3.18}$$

Equalizing (6.3.17) and (6.3.18), we have $u = \frac{2}{3}$, $v = -\frac{1}{3}$, then

$$\mathbf{r}_1 = \frac{2}{3}\mathbf{a}_1 - \frac{1}{3}\mathbf{a}_2. \tag{6.3.19}$$

Assume that $V_1(\mathbf{r})$ is the periodical potential created by one sublattice. Then potential of crystal is

$$V(\mathbf{r}) = V_1(\mathbf{r}) + V_1(\mathbf{r} - \mathbf{r}_1) = \sum_{n=0}^{1} V_1(\mathbf{r} - \mathbf{r}_n). \tag{6.3.20}$$

Atoms in crystal form the periodic structure and as the consequence, the corresponding potential is periodic function

$$V_1(\mathbf{r}) = V_1(\mathbf{r} + \mathbf{a}_m),$$

where for 2D structure

$$\mathbf{a}_m = m_1\mathbf{a}_1 + m_2\mathbf{a}_2,$$

and m_1 and m_2 are arbitrary entire numbers. Expanding $V_1(\mathbf{r})$ in the Fourier series one obtains

$$V_1(\mathbf{r} - \mathbf{r}_n) = \sum_{\mathbf{b}} V_{\mathbf{b}} e^{i\mathbf{b}\cdot(\mathbf{r} - \mathbf{r}_n)}. \tag{6.3.21}$$

In our case, both the basis atoms ($n = 0,1$) are the same. Here,

$$\mathbf{b} = g_1\mathbf{b}_1 + g_2\mathbf{b}_2,$$

$\mathbf{b}_1$ and $\mathbf{b}_2$ are the translational vectors of the reciprocal lattice. For graphene,

$$\mathbf{b}_1 = \frac{2\pi}{3a}\left(1;\sqrt{3}\right),\ \mathbf{b}_2 = \frac{2\pi}{3a}\left(1;-\sqrt{3}\right). \tag{6.3.22}$$

Then,

$$V(\mathbf{r})=\sum_{\mathbf{b}}\sum_{n=0}^{1}V_{1\mathbf{b}}e^{i\mathbf{b}\cdot(\mathbf{r}-\mathbf{r}_n)}=\sum_{\mathbf{b}}V_{\mathbf{b}}e^{i\mathbf{b}\cdot\mathbf{r}}, \tag{6.3.23}$$

where $V_{\mathbf{b}}=V_{1\mathbf{b}}\cdot\sum_{n}e^{-i\mathbf{b}\cdot\mathbf{r}_n}=V_{1\mathbf{b}}\cdot S_{\mathbf{b}}$. The structure factor $S_{\mathbf{b}}$ for graphene:

$$S_{\mathbf{b}}=e^{-i\mathbf{b}\cdot 0}+\exp\left(-i\mathbf{b}\cdot\left(\frac{2}{3}\mathbf{a}_1-\frac{1}{3}\mathbf{a}_2\right)\right)=1+\exp\left(i\frac{2\pi}{3}(g_2-2g_1)\right). \tag{6.3.24}$$

$$V(\mathbf{r})=\sum_{g_1,g_2}V_{1g_1,g_2}e^{i(g_1\mathbf{b}_1+g_2\mathbf{b}_2)\cdot\mathbf{r}}\left(1+\exp\left(i\frac{2\pi}{3}(g_2-2g_1)\right)\right). \tag{6.3.25}$$

For the approximate calculation, we use the terms of the series with $|g_1|\le 2,\ |g_2|\le 2$. Therefore

$$\begin{aligned}
V(\mathbf{r})&=2V_{1,(00)}\\
&+4V_{1,(10)}\left(\cos\left(\frac{1}{2}(\mathbf{b}_1+\mathbf{b}_2)\cdot\mathbf{r}\right)\cos\left(\frac{1}{2}(\mathbf{b}_1-\mathbf{b}_2)\cdot\mathbf{r}\right)+\cos\left(\frac{1}{2}(\mathbf{b}_1+\mathbf{b}_2)\cdot\mathbf{r}+\frac{2\pi}{3}\right)\cos\left(\frac{1}{2}(\mathbf{b}_1-\mathbf{b}_2)\cdot\mathbf{r}\right)\right)\\
&+2V_{1,(11)}\left(\cos((\mathbf{b}_1+\mathbf{b}_2)\cdot\mathbf{r})+\cos\left((\mathbf{b}_1+\mathbf{b}_2)\cdot\mathbf{r}-\frac{2\pi}{3}\right)+2\cos((\mathbf{b}_1-\mathbf{b}_2)\cdot\mathbf{r})\right)\\
&-4V_{1,(20)}\cos((\mathbf{b}_2-\mathbf{b}_1)\cdot\mathbf{r})\cos\left((\mathbf{b}_2+\mathbf{b}_1)\cdot\mathbf{r}+\frac{2\pi}{3}\right)+2V_{1,(12)}(2\cos((\mathbf{b}_1+2\mathbf{b}_2)\cdot\mathbf{r})+2\cos((2\mathbf{b}_1+\mathbf{b}_2)\cdot\mathbf{r})\\
&+\cos\left((\mathbf{b}_1-2\mathbf{b}_2)\cdot\mathbf{r}-\frac{\pi}{3}\right)-\cos\left((2\mathbf{b}_1-\mathbf{b}_2)\cdot\mathbf{r}-\frac{2\pi}{3}\right)\Big)\\
&+2V_{1,(22)}\left(2\cos(2(\mathbf{b}_1-\mathbf{b}_2)\cdot\mathbf{r})-\cos\left(2(\mathbf{b}_1+\mathbf{b}_2)\cdot\mathbf{r}-\frac{2\pi}{3}\right)\right).
\end{aligned} \tag{6.3.26}$$

Using the vectors $\mathbf{b}_1$ and $\mathbf{b}_2$ of the reciprocal lattice from Eq. (6.3.22) and coordinates x and y one obtains from Eq. (6.3.26):

$$\begin{aligned}
V(x,y)&=2V_{1,(00)}+4V_{1,(10)}\cos\left(\frac{2\pi}{3a}x+\frac{\pi}{3}\right)\cos\left(\frac{2\pi}{3a}\sqrt{3}y\right)\\
&+2V_{1,(11)}\left(\cos\left(\frac{4\pi}{3a}x-\frac{\pi}{3}\right)+2\cos\left(\frac{4\pi}{3a}\sqrt{3}y\right)\right)-4V_{1,(20)}\cos\left(\frac{4\pi}{3a}\sqrt{3}y\right)\cos\left(\frac{4\pi}{3a}x+\frac{2\pi}{3}\right)\\
&+4V_{1,(12)}\left(2\cos\left(\frac{2\pi}{a}x\right)\cos\left(\frac{2\pi}{3a}\sqrt{3}y\right)-\sin\left(\frac{2\pi}{3a}x-\frac{\pi}{6}\right)\cos\left(\frac{2\pi}{a}\sqrt{3}y\right)\right)\\
&+2V_{1,(22)}\left(2\cos\left(\frac{8\pi}{3a}\sqrt{3}y\right)-\cos\left(\frac{8\pi}{3a}x-\frac{2\pi}{3}\right)\right).
\end{aligned} \tag{6.3.27}$$

We need the derivatives for the forces components in dimensionless form

$$\begin{aligned}
-\frac{\partial\tilde{U}}{\partial\tilde{\xi}}&=\tilde{U}'_{10}\sin\left(\frac{2\pi}{3\tilde{a}}\tilde{x}+\frac{\pi}{3}\right)\cos\left(\frac{2\pi}{3\tilde{a}}\sqrt{3}\tilde{y}\right)+\tilde{U}'_{11}\sin\left(\frac{4\pi}{3\tilde{a}}\tilde{x}-\frac{\pi}{3}\right)\\
&-\tilde{U}'_{20}\cos\left(\frac{4\pi}{3\tilde{a}}\sqrt{3}\tilde{y}\right)\sin\left(\frac{4\pi}{3\tilde{a}}\tilde{x}+\frac{2\pi}{3}\right)+\tilde{U}'_{12}\left(6\sin\left(\frac{2\pi}{\tilde{a}}\tilde{x}\right)\cos\left(\frac{2\pi}{3\tilde{a}}\sqrt{3}\tilde{y}\right)\right.\\
&\left.+\cos\left(\frac{2\pi}{3\tilde{a}}\tilde{x}-\frac{\pi}{6}\right)\cos\left(\frac{2\pi}{\tilde{a}}\sqrt{3}\tilde{y}\right)\right)-\tilde{U}'_{22}\sin\left(\frac{8\pi}{3\tilde{a}}\tilde{x}-\frac{2\pi}{3}\right),
\end{aligned} \tag{6.3.28}$$

$$\begin{aligned}
-\frac{\partial\tilde{U}}{\partial\tilde{y}}&=\tilde{U}'_{10}\sqrt{3}\cos\left(\frac{2\pi}{3\tilde{a}}\tilde{x}+\frac{\pi}{3}\right)\sin\left(\frac{2\pi}{3\tilde{a}}\sqrt{3}\tilde{y}\right)+\tilde{U}'_{11}2\sqrt{3}\sin\left(\frac{4\pi}{3\tilde{a}}\sqrt{3}\tilde{y}\right)\\
&-\sqrt{3}\tilde{U}'_{20}\sin\left(\frac{4\pi}{3\tilde{a}}\sqrt{3}\tilde{y}\right)\cos\left(\frac{4\pi}{3\tilde{a}}\tilde{x}+\frac{2\pi}{3}\right)+\tilde{U}'_{12}\left(2\sqrt{3}\cos\left(\frac{2\pi}{\tilde{a}}\tilde{x}\right)\sin\left(\frac{2\pi}{3\tilde{a}}\sqrt{3}\tilde{y}\right)\right.\\
&\left.-3\sqrt{3}\sin\left(\frac{2\pi}{3\tilde{a}}\tilde{x}-\frac{\pi}{6}\right)\sin\left(\frac{2\pi}{\tilde{a}}\sqrt{3}\tilde{y}\right)\right)+2\sqrt{3}\tilde{U}'_{22}\sin\left(\frac{8\pi}{3\tilde{a}}\sqrt{3}\tilde{y}\right),
\end{aligned} \tag{6.3.29}$$

where the notations are introduced:

$$\widetilde{U}'_{10}=\frac{8\pi}{3\widetilde{a}}\widetilde{V}_{1,(10)},\ \widetilde{U}'_{11}=\frac{8\pi}{3\widetilde{a}}\widetilde{V}_{1,(11)},\ \widetilde{U}'_{20}=\frac{16\pi}{3\widetilde{a}}\widetilde{V}_{1,(20)},\ \widetilde{U}'_{12}=\frac{8\pi}{3\widetilde{a}}\widetilde{V}_{1,(12)},\ \widetilde{U}'_{22}=\frac{16\pi}{3\widetilde{a}}\widetilde{V}_{1,(22)}. \tag{6.3.30}$$

Consider as the approximation the acting forces by $\widetilde{t}=0$, when $\widetilde{\xi}=\widetilde{x}$. After substitution of Eqs. (6.3.28) and (6.3.29) in Eq. (6.3.16), one obtains the expressions for the dimensionless forces acting on the unit of mass of particles:

$$\begin{aligned}\widetilde{F}_{p\xi}=&-\frac{\partial\widetilde{\psi}}{\partial\widetilde{\xi}}+\widetilde{U}'_{10}\sin\left(\frac{2\pi}{3\widetilde{a}}\widetilde{\xi}+\frac{\pi}{3}\right)\cos\left(\frac{2\pi}{3\widetilde{a}}\sqrt{3}\widetilde{y}\right)+\widetilde{U}'_{11}\sin\left(\frac{4\pi}{3\widetilde{a}}\widetilde{\xi}-\frac{\pi}{3}\right)\\&-\widetilde{U}'_{20}\cos\left(\frac{4\pi}{3\widetilde{a}}\sqrt{3}\widetilde{y}\right)\sin\left(\frac{4\pi}{3\widetilde{a}}\widetilde{\xi}+\frac{2\pi}{3}\right)+\widetilde{U}'_{12}\left(6\sin\left(\frac{2\pi}{\widetilde{a}}\widetilde{\xi}\right)\cos\left(\frac{2\pi}{3\widetilde{a}}\sqrt{3}\widetilde{y}\right)\right.\\&\left.+\cos\left(\frac{2\pi}{3\widetilde{a}}\widetilde{\xi}-\frac{\pi}{6}\right)\cos\left(\frac{2\pi}{\widetilde{a}}\sqrt{3}\widetilde{y}\right)\right)-\widetilde{U}'_{22}\sin\left(\frac{8\pi}{3\widetilde{a}}\widetilde{\xi}-\frac{2\pi}{3}\right)+\widetilde{E}_{\xi},\end{aligned} \tag{6.3.31}$$

$$\begin{aligned}\widetilde{F}_{py}=&-\frac{\partial\widetilde{\psi}}{\partial\widetilde{y}}+\widetilde{U}'_{10}\sqrt{3}\cos\left(\frac{2\pi}{3\widetilde{a}}\widetilde{\xi}+\frac{\pi}{3}\right)\sin\left(\frac{2\pi}{3\widetilde{a}}\sqrt{3}\widetilde{y}\right)+\widetilde{U}'_{11}2\sqrt{3}\sin\left(\frac{4\pi}{3\widetilde{a}}\sqrt{3}\widetilde{y}\right)\\&-\sqrt{3}\widetilde{U}'_{20}\sin\left(\frac{4\pi}{3\widetilde{a}}\sqrt{3}\widetilde{y}\right)\cos\left(\frac{4\pi}{3\widetilde{a}}\widetilde{\xi}+\frac{2\pi}{3}\right)+\widetilde{U}'_{12}\left(2\sqrt{3}\cos\left(\frac{2\pi}{\widetilde{a}}\widetilde{\xi}\right)\sin\left(\frac{2\pi}{3\widetilde{a}}\sqrt{3}\widetilde{y}\right)\right.\\&\left.-3\sqrt{3}\sin\left(\frac{2\pi}{3\widetilde{a}}\widetilde{\xi}-\frac{\pi}{6}\right)\sin\left(\frac{2\pi}{\widetilde{a}}\sqrt{3}\widetilde{y}\right)\right)+2\sqrt{3}\widetilde{U}'_{22}\sin\left(\frac{8\pi}{3\widetilde{a}}\sqrt{3}\widetilde{y}\right)+\widetilde{E}_{y}.\end{aligned} \tag{6.3.32}$$

Analogically,

$$\widetilde{F}_{e\xi}=-\widetilde{F}_{p\xi},\ \widetilde{F}_{ey}=-\widetilde{F}_{py}. \tag{6.3.33}$$

The forces (6.3.31)–(6.3.33) should be introduced in the system of the hydrodynamic equations (6.3.10)–(6.3.15).

Suppose that the external field intensity E is equal to zero. Average on $\widetilde{y}$ the obtained system of quantum hydrodynamic equations taking into account that effective hydrodynamic velocity is directed along x axis. The averaging will be realized in the limit of one hexagonal crystal cell. Let us carry out the integration of the left and right hand sides of the hydrodynamic equations calculating the integral of the type $\frac{1}{\sqrt{3}\widetilde{a}}\int_{-\frac{\sqrt{3}}{2}\widetilde{a}}^{\frac{\sqrt{3}}{2}\widetilde{a}}d\widetilde{y}$ (see Fig. 6.13). We should take into account that $\frac{1}{\sqrt{3}\widetilde{a}}\int_{-\frac{\sqrt{3}}{2}\widetilde{a}}^{\frac{\sqrt{3}}{2}\widetilde{a}}\frac{\partial\Psi}{\partial\widetilde{y}}d\widetilde{y}=0$ because of system symmetry for an arbitrary function ψ, characterizing the state of the physical system. We suppose therefore that by averaging all physical values, characterizing the state of the physical system do not depend on $\widetilde{y}$.

As a result, we have the following system of equations:

Dimensionless Poisson equation for the self-consistent potential $\widetilde{\psi}$ of the electric field:

$$\frac{\partial^2\widetilde{\psi}}{\partial\widetilde{\xi}^2}=-4\pi R\left\{\frac{m_e}{m_p}\left[\widetilde{\rho}_p-\frac{m_eH}{m_p\widetilde{u}^2}\frac{\partial}{\partial\widetilde{\xi}}(\widetilde{\rho}_p(\widetilde{u}-1))\right]-\left[\widetilde{\rho}_e-\frac{H}{\widetilde{u}^2}\frac{\partial}{\partial\widetilde{\xi}}(\widetilde{\rho}_e(\widetilde{u}-1))\right]\right\}. \tag{6.3.34}$$

Continuity equation for the positive particles:

$$\begin{aligned}&\frac{\partial}{\partial\widetilde{\xi}}\left[\widetilde{\rho}_p(1-\widetilde{u})\right]+\frac{m_e}{m_p}\frac{\partial}{\partial\widetilde{\xi}}\left\{\frac{H}{\widetilde{u}^2}\frac{\partial}{\partial\widetilde{\xi}}\left[\widetilde{\rho}_p(\widetilde{u}-1)^2\right]\right\}+\frac{m_e}{m_p}\frac{\partial}{\partial\widetilde{\xi}}\left\{\frac{H}{\widetilde{u}^2}\left[\frac{V_{0p}^2}{u_0^2}\frac{\partial}{\partial\widetilde{\xi}}\widetilde{p}_p\right.\right.\\&\left.\left.-\frac{m_e}{m_p}\widetilde{\rho}_pE\left(-\frac{\partial\widetilde{\psi}}{\partial\widetilde{\xi}}+\widetilde{U}'_{11}\sin\left(\frac{4\pi}{3\widetilde{a}}\widetilde{\xi}-\frac{\pi}{3}\right)-\widetilde{U}'_{22}\sin\left(\frac{8\pi}{3\widetilde{a}}\widetilde{\xi}-\frac{2\pi}{3}\right)\right)\right]\right\}=0\end{aligned} \tag{6.3.35}$$

Continuity equation for electrons:

$$\begin{aligned}&\frac{\partial}{\partial\widetilde{\xi}}\left[\widetilde{\rho}_e(1-\widetilde{u})\right]+\frac{\partial}{\partial\widetilde{\xi}}\left\{\frac{H}{\widetilde{u}^2}\frac{\partial}{\partial\widetilde{\xi}}\left[\widetilde{\rho}_e(\widetilde{u}-1)^2\right]\right\}+\frac{\partial}{\partial\widetilde{\xi}}\left\{\frac{H}{\widetilde{u}^2}\left[\frac{V_{0e}^2}{u_0^2}\frac{\partial}{\partial\widetilde{\xi}}\widetilde{p}_e\right.\right.\\&\left.\left.-\widetilde{\rho}_eE\left(\frac{\partial\widetilde{\psi}}{\partial\widetilde{\xi}}-\widetilde{U}'_{11}\sin\left(\frac{4\pi}{3\widetilde{a}}\widetilde{\xi}-\frac{\pi}{3}\right)+\widetilde{U}'_{22}\sin\left(\frac{8\pi}{3\widetilde{a}}\widetilde{\xi}-\frac{2\pi}{3}\right)\right)\right]\right\}=0\end{aligned} \tag{6.3.36}$$

Momentum equation for the movement along the x direction:

$$\begin{aligned}
&\frac{\partial}{\partial\widetilde{\xi}}\left\{(\widetilde{\rho}_p+\widetilde{\rho}_e)\widetilde{u}(\widetilde{u}-1)+\frac{V_{0p}^2}{u_0^2}\widetilde{p}_p+\frac{V_{0e}^2}{u_0^2}\widetilde{p}_e\right\}\\
&-\frac{m_e}{m_p}\widetilde{\rho}_pE\left(-\frac{\partial\widetilde{\psi}}{\partial\widetilde{\xi}}+\widetilde{U}'_{11}\sin\left(\frac{4\pi}{3\widetilde{a}}\widetilde{\xi}-\frac{\pi}{3}\right)-\widetilde{U}'_{22}\sin\left(\frac{8\pi}{3\widetilde{a}}\widetilde{\xi}-\frac{2\pi}{3}\right)\right)\\
&-\widetilde{\rho}_eE\left(\frac{\partial\widetilde{\psi}}{\partial\widetilde{\xi}}-\widetilde{U}'_{11}\sin\left(\frac{4\pi}{3\widetilde{a}}\widetilde{\xi}-\frac{\pi}{3}\right)+\widetilde{U}'_{22}\sin\left(\frac{8\pi}{3\widetilde{a}}\widetilde{\xi}-\frac{2\pi}{3}\right)\right)\\
&+\frac{m_e}{m_p}\frac{\partial}{\partial\widetilde{\xi}}\left\{\frac{H}{\widetilde{u}^2}\left[\frac{\partial}{\partial\widetilde{\xi}}\left(2\frac{V_{0p}^2}{u_0^2}\widetilde{p}_p(1-\widetilde{u})-\widetilde{\rho}_p\widetilde{u}(1-\widetilde{u})^2\right)\right.\right.\\
&\left.\left.-\frac{m_e}{m_p}\widetilde{\rho}_p(1-\widetilde{u})E\left(-\frac{\partial\widetilde{\psi}}{\partial\widetilde{\xi}}+\widetilde{U}'_{11}\sin\left(\frac{4\pi}{3\widetilde{a}}\widetilde{\xi}-\frac{\pi}{3}\right)-\widetilde{U}'_{22}\sin\left(\frac{8\pi}{3\widetilde{a}}\widetilde{\xi}-\frac{2\pi}{3}\right)\right)\right]\right\}\\
&+\frac{\partial}{\partial\widetilde{\xi}}\left\{\frac{H}{\widetilde{u}^2}\left[\frac{\partial}{\partial\widetilde{\xi}}\left(2\frac{V_{0e}^2}{u_0^2}\widetilde{p}_e(1-\widetilde{u})-\widetilde{\rho}_e\widetilde{u}(1-\widetilde{u})^2\right)\right.\right.\\
&\left.\left.-\widetilde{\rho}_e(1-\widetilde{u})E\left(\frac{\partial\widetilde{\psi}}{\partial\widetilde{\xi}}-\widetilde{U}'_{11}\sin\left(\frac{4\pi}{3\widetilde{a}}\widetilde{\xi}-\frac{\pi}{3}\right)+\widetilde{U}'_{22}\sin\left(\frac{8\pi}{3\widetilde{a}}\widetilde{\xi}-\frac{2\pi}{3}\right)\right)\right]\right\}\\
&+\frac{H}{\widetilde{u}^2}E\left(\frac{m_e}{m_p}\right)^2\left(-\frac{\partial\widetilde{\psi}}{\partial\widetilde{\xi}}+\widetilde{U}'_{11}\sin\left(\frac{4\pi}{3\widetilde{a}}\widetilde{\xi}-\frac{\pi}{3}\right)-\widetilde{U}'_{22}\sin\left(\frac{8\pi}{3\widetilde{a}}\widetilde{\xi}-\frac{2\pi}{3}\right)\right)\left(\frac{\partial}{\partial\widetilde{\xi}}(\widetilde{\rho}_p(\widetilde{u}-1))\right)\\
&+\frac{H}{\widetilde{u}^2}E\left(\frac{\partial\widetilde{\psi}}{\partial\widetilde{\xi}}-\widetilde{U}'_{11}\sin\left(\frac{4\pi}{3\widetilde{a}}\widetilde{\xi}-\frac{\pi}{3}\right)+\widetilde{U}'_{22}\sin\left(\frac{8\pi}{3\widetilde{a}}\widetilde{\xi}-\frac{2\pi}{3}\right)\right)\left(\frac{\partial}{\partial\widetilde{\xi}}(\widetilde{\rho}_e(\widetilde{u}-1))\right)\\
&-\frac{m_e}{m_p}\frac{\partial}{\partial\widetilde{\xi}}\left\{\frac{H}{\widetilde{u}^2}\frac{V_{0p}^2}{u_0^2}\frac{\partial}{\partial\widetilde{\xi}}(\widetilde{p}_p\widetilde{u})\right\}-\frac{\partial}{\partial\widetilde{\xi}}\left\{\frac{H}{\widetilde{u}^2}\frac{V_{0e}^2}{u_0^2}\frac{\partial}{\partial\widetilde{\xi}}(\widetilde{p}_e\widetilde{u})\right\}\\
&+\left(\frac{m_e}{m_p}\right)^2E\frac{\partial}{\partial\widetilde{\xi}}\left\{\frac{H}{\widetilde{u}^2}\left[\left(-\frac{\partial\widetilde{\psi}}{\partial\widetilde{\xi}}+\widetilde{U}'_{11}\sin\left(\frac{4\pi}{3\widetilde{a}}\widetilde{\xi}-\frac{\pi}{3}\right)-\widetilde{U}'_{22}\sin\left(\frac{8\pi}{3\widetilde{a}}\widetilde{\xi}-\frac{2\pi}{3}\right)\right)\widetilde{\rho}_p\widetilde{u}\right]\right\}\\
&+E\frac{\partial}{\partial\widetilde{\xi}}\left\{\frac{H}{\widetilde{u}^2}\left[\left(\frac{\partial\widetilde{\psi}}{\partial\widetilde{\xi}}-\widetilde{U}'_{11}\sin\left(\frac{4\pi}{3\widetilde{a}}\widetilde{\xi}-\frac{\pi}{3}\right)+\widetilde{U}'_{22}\sin\left(\frac{8\pi}{3\widetilde{a}}\widetilde{\xi}-\frac{2\pi}{3}\right)\right)\widetilde{\rho}_e\widetilde{u}\right]\right\}=0
\end{aligned}\tag{6.3.37}$$

Energy equation for the positive particles:

$$\begin{aligned}
&\frac{\partial}{\partial\widetilde{\xi}}\left[\widetilde{\rho}_p\widetilde{u}^2(\widetilde{u}-1)+5\frac{V_{0p}^2}{u_0^2}\widetilde{p}_p\widetilde{u}-3\frac{V_{0p}^2}{u_0^2}\widetilde{p}_p\right]\\
&-2\frac{m_e}{m_p}\widetilde{\rho}_pE\left(-\frac{\partial\widetilde{\psi}}{\partial\widetilde{\xi}}+\widetilde{U}'_{11}\sin\left(\frac{4\pi}{3\widetilde{a}}\widetilde{\xi}-\frac{\pi}{3}\right)-\widetilde{U}'_{22}\sin\left(\frac{8\pi}{3\widetilde{a}}\widetilde{\xi}-\frac{2\pi}{3}\right)\right)\widetilde{u}\\
&+\frac{\partial}{\partial\widetilde{\xi}}\left\{\frac{H}{\widetilde{u}^2}\frac{m_e}{m_p}\left[\frac{\partial}{\partial\widetilde{\xi}}\left(-\widetilde{\rho}_p\widetilde{u}^2(1-\widetilde{u})^2+7\frac{V_{0p}^2}{u_0^2}\widetilde{p}_p\widetilde{u}(1-\widetilde{u})+3\frac{V_{0p}^2}{u_0^2}\widetilde{p}_p(\widetilde{u}-1)-\frac{V_{0p}^2}{u_0^2}\widetilde{p}_p\widetilde{u}^2-5\frac{V_{0p}^4}{u_0^4}\frac{\widetilde{p}_p^2}{\widetilde{\rho}_p}\right)\right.\right.\\
&+E\left(-2\frac{m_e}{m_p}\widetilde{\rho}_p\widetilde{u}(1-\widetilde{u})+\frac{m_e}{m_p}\widetilde{\rho}_p\widetilde{u}^2+5\frac{m_e}{m_p}\frac{V_{0p}^2}{u_0^2}\widetilde{p}_p)\right)\left(-\frac{\partial\widetilde{\psi}}{\partial\widetilde{\xi}}\right.\\
&\left.\left.\left.+\widetilde{U}'_{11}\sin\left(\frac{4\pi}{3\widetilde{a}}\widetilde{\xi}-\frac{\pi}{3}\right)-\widetilde{U}'_{22}\sin\left(\frac{8\pi}{3\widetilde{a}}\widetilde{\xi}-\frac{2\pi}{3}\right)\right)\right]\right\}+2\frac{H}{\widetilde{u}^2}E\left(\frac{m_e}{m_p}\right)^2\left[-\frac{\partial}{\partial\widetilde{\xi}}(\widetilde{\rho}_p\widetilde{u}(1-\widetilde{u}))\right.\\
&\left.+\frac{V_{0p}^2}{u_0^2}\frac{\partial}{\partial\widetilde{\xi}}\widetilde{p}_p\right]\left(-\frac{\partial\widetilde{\psi}}{\partial\widetilde{\xi}}+\widetilde{U}'_{11}\sin\left(\frac{4\pi}{3\widetilde{a}}\widetilde{\xi}-\frac{\pi}{3}\right)-\widetilde{U}'_{22}\sin\left(\frac{8\pi}{3\widetilde{a}}\widetilde{\xi}-\frac{2\pi}{3}\right)\right)
\end{aligned}$$

$$
\begin{aligned}
&-2\frac{H}{\widetilde{u}^2}E^2\left(\frac{m_e}{m_p}\right)^3\widetilde{\rho}_p\left[\left(-\frac{\partial\widetilde{\psi}}{\partial\widetilde{\xi}}+\widetilde{U}'_{11}\sin\left(\frac{4\pi}{3\widetilde{a}}\widetilde{\xi}-\frac{\pi}{3}\right)-\widetilde{U}'_{22}\sin\left(\frac{8\pi}{3\widetilde{a}}\widetilde{\xi}-\frac{2\pi}{3}\right)\right)^2\right.\\
&+\frac{1}{2}\left(\widetilde{U}'_{10}\sin\left(\frac{2\pi}{3\widetilde{a}}\widetilde{\xi}+\frac{\pi}{3}\right)+6\widetilde{U}'_{12}\sin\left(\frac{2\pi}{\widetilde{a}}\widetilde{\xi}\right)\right)^2+\frac{1}{2}\left(\widetilde{U}'_{12}\right)^2\cos^2\left(\frac{2\pi}{3\widetilde{a}}\widetilde{\xi}-\frac{\pi}{6}\right)\\
&+\frac{1}{2}\left(\widetilde{U}'_{02}\right)^2\sin^2\left(\frac{4\pi}{3\widetilde{a}}\widetilde{\xi}+\frac{2\pi}{3}\right)-\frac{4}{3\pi}\widetilde{U}'_{02}\sin\left(\frac{4\pi}{3\widetilde{a}}\widetilde{\xi}+\frac{2\pi}{3}\right)\left(\widetilde{U}'_{10}\sin\left(\frac{2\pi}{3\widetilde{a}}\widetilde{\xi}+\frac{\pi}{3}\right)+6\widetilde{U}'_{12}\sin\left(\frac{2\pi}{\widetilde{a}}\widetilde{\xi}\right)\right)\\
&-\frac{12}{5\pi}\widetilde{U}'_{02}\widetilde{U}'_{12}\sin\left(\frac{4\pi}{3\widetilde{a}}\widetilde{\xi}+\frac{2\pi}{3}\right)\cos\left(\frac{2\pi}{3\widetilde{a}}\widetilde{\xi}-\frac{\pi}{6}\right)+\frac{3}{2}\left(\widetilde{U}'_{10}\cos\left(\frac{2\pi}{3\widetilde{a}}\widetilde{\xi}+\frac{\pi}{3}\right)+2\widetilde{U}'_{12}\cos\left(\frac{2\pi}{\widetilde{a}}\widetilde{\xi}\right)\right)^2\\
&+\frac{3}{2}\left(2\widetilde{U}'_{11}-\widetilde{U}'_{02}\cos\left(\frac{4\pi}{3\widetilde{a}}\widetilde{\xi}+\frac{2\pi}{3}\right)\right)^2+\frac{27}{2}\left(\widetilde{U}'_{12}\right)^2\sin^2\left(\frac{2\pi}{3\widetilde{a}}\widetilde{\xi}-\frac{\pi}{6}\right)+6\left(\widetilde{U}'_{22}\right)^2\\
&+\frac{8}{\pi}\left(\widetilde{U}'_{10}\cos\left(\frac{2\pi}{3\widetilde{a}}\widetilde{\xi}+\frac{\pi}{3}\right)+2\widetilde{U}'_{12}\cos\left(\frac{2\pi}{\widetilde{a}}\widetilde{\xi}\right)\right)\left(2\widetilde{U}'_{11}-\widetilde{U}'_{02}\cos\left(\frac{4\pi}{3\widetilde{a}}\widetilde{\xi}+\frac{2\pi}{3}\right)\right)\\
&-\frac{96}{15\pi}\left(\widetilde{U}'_{10}\cos\left(\frac{2\pi}{3\widetilde{a}}\widetilde{\xi}+\frac{\pi}{3}\right)+2\widetilde{U}'_{12}\cos\left(\frac{2\pi}{\widetilde{a}}\widetilde{\xi}\right)\right)\widetilde{U}'_{22}\\
&\left.-\frac{72}{5\pi}\widetilde{U}'_{12}\left(2\widetilde{U}'_{11}-\widetilde{U}'_{02}\cos\left(\frac{4\pi}{3\widetilde{a}}\widetilde{\xi}+\frac{2\pi}{3}\right)\right)\sin\left(\frac{2\pi}{3\widetilde{a}}\widetilde{\xi}-\frac{\pi}{6}\right)-\frac{288}{7\pi}\widetilde{U}'_{12}\widetilde{U}'_{22}\sin\left(\frac{2\pi}{3\widetilde{a}}\widetilde{\xi}-\frac{\pi}{6}\right)\right]\\
&=-\frac{\widetilde{u}^2}{Hu_0^2}\left(V_{0p}^2\widetilde{p}_p-\widetilde{p}_eV_{0e}^2\right)\left(1+\frac{m_p}{m_e}\right)
\end{aligned}
\tag{6.3.38}
$$

Energy equation for electrons:

$$
\begin{aligned}
&\frac{\partial}{\partial\widetilde{\xi}}\left[\widetilde{\rho}_e\widetilde{u}^2(\widetilde{u}-1)+5\frac{V_{0e}^2}{u_0^2}\widetilde{p}_e\widetilde{u}-3\frac{V_{0e}^2}{u_0^2}\widetilde{p}_e\right]\\
&-2\widetilde{\rho}_e\widetilde{u}E\left(\frac{\partial\widetilde{\phi}}{\partial\widetilde{\xi}}-\widetilde{U}'_{11}\sin\left(\frac{4\pi}{3\widetilde{a}}\widetilde{\xi}-\frac{\pi}{3}\right)+\widetilde{U}'_{22}\sin\left(\frac{8\pi}{3\widetilde{a}}\widetilde{\xi}-\frac{2\pi}{3}\right)\right)\\
&+\frac{\partial}{\partial\widetilde{\xi}}\left\{\frac{H}{\widetilde{u}^2}\left[\frac{\partial}{\partial\widetilde{\xi}}\left(-\widetilde{\rho}_e\widetilde{u}^2(1-\widetilde{u})^2+7\frac{V_{0e}^2}{u_0^2}\widetilde{p}_e\widetilde{u}(1-\widetilde{u})+3\frac{V_{0e}^2}{u_0^2}\widetilde{p}_e(\widetilde{u}-1)-\frac{V_{0e}^2}{u_0^2}\widetilde{p}_e\widetilde{u}^2-5\frac{V_{0e}^4}{u_0^4}\frac{\widetilde{p}_e^2}{\widetilde{\rho}_e}\right)\right.\right.\\
&\left.\left.+E\left(-2\widetilde{\rho}_e\widetilde{u}(1-\widetilde{u})+\widetilde{\rho}_e\widetilde{u}^2+5\frac{V_{0e}^2}{u_0^2}\widetilde{p}_e\right)\left(\frac{\partial\widetilde{\psi}}{\partial\widetilde{\xi}}-\widetilde{U}'_{11}\sin\left(\frac{4\pi}{3\widetilde{a}}\widetilde{\xi}-\frac{\pi}{3}\right)+\widetilde{U}'_{22}\sin\left(\frac{8\pi}{3\widetilde{a}}\widetilde{\xi}-\frac{2\pi}{3}\right)\right)\right]\right\}\\
&+E\left(-2\frac{H}{\widetilde{u}^2}\frac{\partial}{\partial\widetilde{\xi}}\left(\widetilde{\rho}_e\widetilde{u}(1-\widetilde{u})+2\frac{H}{\widetilde{u}^2}\frac{V_{0e}^2}{u_0^2}\frac{\partial}{\partial\widetilde{\xi}}\widetilde{p}_e\right)\right)\left(\frac{\partial\widetilde{\psi}}{\partial\widetilde{\xi}}-\widetilde{U}'_{11}\sin\left(\frac{4\pi}{3\widetilde{a}}\widetilde{\xi}-\frac{\pi}{3}\right)+\widetilde{U}'_{22}\sin\left(\frac{8\pi}{3\widetilde{a}}\widetilde{\xi}-\frac{2\pi}{3}\right)\right)-\\
&-2E^2\frac{H}{\widetilde{u}^2}\widetilde{\rho}_e\left[\left(-\frac{\partial\widetilde{\psi}}{\partial\widetilde{\xi}}+\widetilde{U}'_{11}\sin\left(\frac{4\pi}{3\widetilde{a}}\widetilde{\xi}-\frac{\pi}{3}\right)-\widetilde{U}'_{22}\sin\left(\frac{8\pi}{3\widetilde{a}}\widetilde{\xi}-\frac{2\pi}{3}\right)\right)^2\right.\\
&+\frac{1}{2}\left(\widetilde{U}'_{10}\sin\left(\frac{2\pi}{3\widetilde{a}}\widetilde{\xi}+\frac{\pi}{3}\right)+6\widetilde{U}'_{12}\sin\left(\frac{2\pi}{\widetilde{a}}\widetilde{\xi}\right)\right)^2+\frac{1}{2}\left(\widetilde{U}'_{12}\right)^2\cos^2\left(\frac{2\pi}{3\widetilde{a}}\widetilde{\xi}-\frac{\pi}{6}\right)\\
&+\frac{1}{2}\left(\widetilde{U}'_{02}\right)^2\sin^2\left(\frac{4\pi}{3\widetilde{a}}\widetilde{\xi}+\frac{2\pi}{3}\right)-\frac{4}{3\pi}\widetilde{U}'_{02}\sin\left(\frac{4\pi}{3\widetilde{a}}\widetilde{\xi}+\frac{2\pi}{3}\right)\left(\widetilde{U}'_{10}\sin\left(\frac{2\pi}{3\widetilde{a}}\widetilde{\xi}+\frac{\pi}{3}\right)+6\widetilde{U}'_{12}\sin\left(\frac{2\pi}{\widetilde{a}}\widetilde{\xi}\right)\right)\\
&-\frac{12}{5\pi}\widetilde{U}'_{02}\widetilde{U}'_{12}\sin\left(\frac{4\pi}{3\widetilde{a}}\widetilde{\xi}+\frac{2\pi}{3}\right)\cos\left(\frac{2\pi}{3\widetilde{a}}\widetilde{\xi}-\frac{\pi}{6}\right)+\frac{3}{2}\left(\widetilde{U}'_{10}\cos\left(\frac{2\pi}{3\widetilde{a}}\widetilde{\xi}+\frac{\pi}{3}\right)+2\widetilde{U}'_{12}\cos\left(\frac{2\pi}{\widetilde{a}}\widetilde{\xi}\right)\right)^2\\
&+\frac{3}{2}\left(2\widetilde{U}'_{11}-\widetilde{U}'_{02}\cos\left(\frac{4\pi}{3\widetilde{a}}\widetilde{\xi}+\frac{2\pi}{3}\right)\right)^2+\frac{27}{2}\left(\widetilde{U}'_{12}\right)^2\sin^2\left(\frac{2\pi}{3\widetilde{a}}\widetilde{\xi}-\frac{\pi}{6}\right)+6\left(\widetilde{U}'_{22}\right)^2\\
&+\frac{8}{\pi}\left(\widetilde{U}'_{10}\cos\left(\frac{2\pi}{3\widetilde{a}}\widetilde{\xi}+\frac{\pi}{3}\right)+2\widetilde{U}'_{12}\cos\left(\frac{2\pi}{\widetilde{a}}\widetilde{\xi}\right)\right)\left(2\widetilde{U}'_{11}-\widetilde{U}'_{02}\cos\left(\frac{4\pi}{3\widetilde{a}}\widetilde{\xi}+\frac{2\pi}{3}\right)\right)
\end{aligned}
$$

$$-\frac{96}{15\pi}\left(\widetilde{U}'_{10}\cos\left(\frac{2\pi}{3\widetilde{a}}\widetilde{\xi}+\frac{\pi}{3}\right)+2\widetilde{U}'_{12}\cos\left(\frac{2\pi}{\widetilde{a}}\widetilde{\xi}\right)\right)\widetilde{U}'_{22}$$

$$\left.-\frac{72}{5\pi}\widetilde{U}'_{12}\left(2\widetilde{U}'_{11}-\widetilde{U}'_{02}\cos\left(\frac{4\pi}{3\widetilde{a}}\widetilde{\xi}+\frac{2\pi}{3}\right)\right)\sin\left(\frac{2\pi}{3\widetilde{a}}\widetilde{\xi}-\frac{\pi}{6}\right)-\frac{288}{7\pi}\widetilde{U}'_{12}\widetilde{U}'_{22}\sin\left(\frac{2\pi}{3\widetilde{a}}\widetilde{\xi}-\frac{\pi}{6}\right)\right]$$

$$=-\frac{\widetilde{u}^2}{Hu_0^2}\left(V_{0e}^2\widetilde{\rho}_e-V_{0p}^2\widetilde{\rho}_p\right)\left(1+\frac{m_p}{m_e}\right) \tag{6.3.39}$$

We need estimations for the numerical values of dimensionless parameters for solutions of the hydrodynamic Eqs. (6.3.34)–(6.3.39). In turn, these parameters depend on choosing of the independent scales of physical values. Analyze the independent scales for the physical problem under consideration. It should be underlined that we choose just scales but not real physical values which may differ significantly from scale values. Real physical values will be obtained as a result of numerical self-consistent calculations.

Assume that the surface electron density in graphene is about $\breve{n}_e \approx 10^{10}\,\mathrm{cm}^{-2}$ (such value is typical for many experiments (see [131–133]), the thickness of the graphene layer is equal to ~ 1 *nm*. Then the electron concentration consists $n_e \approx 10^{17}\,\mathrm{cm}^{-3}$, and the density for the electron species $\rho_e = m_e n_e \approx 10^{-10}\,\mathrm{g/cm^3}$ which leads to the scale $\rho_0 = 10^{-10}\,\mathrm{g/cm^3}$. For numerical solutions of the hydrodynamic Eqs. (6.3.27)–(6.3.32), we need Cauchy conditions, obviously in the typical for grapheme conditions the estimation $\widetilde{\rho}_e \sim 1$ is valid which can be used as the condition by $\widetilde{\xi} = 0$.

The process of the carbon atoms polarization leads to displacement of the atoms from the regular chain and to the creation of the "effective" positive particles which concentration $n_p \approx n_e$. Masses of these particles are about the mass of the carbon atom $m_p \approx 2\times 10^{-23}\,г$. Then, $\frac{L}{T} = \frac{m_e}{m_p} \approx 5\times 10^{-5}$; $\rho_p = m_p n_p \approx 2\times 10^{-6}\,\mathrm{g/cm^3}$ and by the chosen scale for the density ρ_0 we have $\widetilde{\rho}_p \sim 2\times 10^4$.

Going to the scales for thermal velocities for electrons and the positive particles we have by $T = 300°\mathrm{K}$:
$V_{0e} \sim \sqrt{\frac{k_B T}{m_e}} \approx 6.4 \mp 10^6\,\mathrm{см/с}$, take the scale $V_{0e} = 5\times 10^6\,\mathrm{cm/s}$;
$V_{0p} \sim \sqrt{\frac{k_B T}{m_p}} \approx 4.5 \mp 10^4\,\mathrm{см/с}$, take the scale $V_{0p} = 5\times 10^4\,\mathrm{cm/s}$.

The theoretical mobility in graphene reaches up to $10^6\,\mathrm{cm^2/Vs}$ [134]. Let us use the scale $u_0 = 5\times 10^6\,\mathrm{cm/s}$. Then, $N = \frac{V_{0e}^2}{u_0^2} = 1$, $P = \frac{V_{0p}^2}{u_0^2} = 10^{-4}$.

Let us estimate the parameters E and R. For this estimation, we need the scale ψ_0. Admit $\psi_0 \approx \delta\frac{e}{a}$, where δ is a "shielding coefficient." Naturally, to take $x_0 = a = 0.142\,\mathrm{nm}$ (see Fig. 6.13) as the length scale, then $\widetilde{a} = 1$. In the situation of an uncertainty in ψ_0 choosing let us consider two limit cases:

1. $\delta \sim 1$.
Then, $E = \frac{e\psi_0}{m_e u_0^2} \sim 1000$, $R = \frac{e\rho_0 x_0^2}{m_e \psi_0} \sim 3\times 10^{-7}$.

2. $\delta = 0.0001$.
Then, $E = \frac{e\psi_0}{m_e u_0^2} \sim 0.1$, $R = \frac{e\rho_0 x_0^2}{m_e \psi_0} \sim 3\times 10^{-3}$.

Consider the terms describing the lattice influence. We should estimate the coefficients (6.3.23) using ψ_0 as the scale for the potential V, $V = \psi\psi_0\widetilde{V}$. Three possible cases under consideration:

(1) $V \sim \psi_0$
We choose $U = \widetilde{U}'_{10} \sim 10$, $F = \widetilde{U}'_{11} \sim 10$, $J = \widetilde{U}'_{20} \sim \pm 5$, $B = \widetilde{U}'_{12} \sim \pm 2.5$, $G = \widetilde{U}'_{22} \sim \pm 5$.
In this case, the coefficients of "the second order" are less than the coefficients of "the first order."

(2) $V \ll \psi_0$ (the small influence of the lattice),
we choose $U = \widetilde{U}'_{10} \sim 0.1$, $F = \widetilde{U}'_{11} \sim 0.1$, $J = \widetilde{U}'_{20} \sim 0.05$, $B = \widetilde{U}'_{12} \sim 0.025$, $G = \widetilde{U}'_{22} \sim 0.05$.

(3) $V \gg \psi_0$ (the great influence of the lattice),
we choose $U = \widetilde{U}'_{10} \sim 1000$, $F = \widetilde{U}'_{11} \sim 1000$, $J = \widetilde{U}'_{20} \sim 500$, $B = \widetilde{U}'_{12} \sim 250$, $G = \widetilde{U}'_{22} \sim 500$.

Estimate parameter $H = \frac{N_R \hbar}{m_e x_0 u_0}$ for two limit cases:

(1) $N_R = 1$, then $H \sim 15$.

(2) $N_R = 100$, then $H \sim 1500$.

TABLE 6.1 Initial conditions

$\widetilde{\rho}_e(0)$	$\widetilde{\rho}_p(0)$	$\widetilde{\psi}(0)$	$\widetilde{p}_e(0)$	$\widetilde{p}_p(0)$	$\frac{\partial \widetilde{\rho}_e}{\partial \widetilde{\xi}}(0)$	$\frac{\partial \widetilde{\rho}_p}{\partial \widetilde{\xi}}(0)$	$\frac{\partial \widetilde{\psi}}{\partial \widetilde{\xi}}(0)$	$\frac{\partial \widetilde{p}_e}{\partial \widetilde{\xi}}(0)$	$\frac{\partial \widetilde{p}_p}{\partial \widetilde{\xi}}(0)$
1	2×10^4	1	1	2×10^4	0	0	0	0	0

TABLE 6.2 Constant parameters

$\widetilde{a}$	L	T	N	P
1	1	20,000	1	10^{-4}

Initial conditions demand also the estimations for the quantum electron pressure and the pressure for the positive species. For the electron pressure, we have $p_e=\rho_0 V_{0e}^2 \widetilde{p}_e$ and using for the scale estimation $p_e=n_e k_B T \sim n_e m_e V_{oe}^2=\rho_e V_{oe}^2 \sim \rho_0 V_{oe}^2$, one obtains $\widetilde{p}_e \sim 1$. Analogically for the positive particles $p_p=\rho_0 V_{0p}^2 \widetilde{p}_p$, and using $p_p=n_p k_B T \sim n_p m_p V_{op}^2=\rho_p V_{0p}^2$, we have $p_p \sim 2\times 10^4 \rho_0 V_{0p}^2$, $\widetilde{p}_p \sim 2\times 10^4$.

Tables 6.1 and 6.2 contain the initial conditions and parameters which were not varied by the numerical modeling.

In the present time there are no the foolproof methods of the calculations of the potential lattice forces in graphene. In the following mathematical modeling, the strategy is taken consisting in the vast variation of the parameters defining the evolution of the physical system.

6.4 RESULTS OF THE MATHEMATICAL MODELING WITHOUT THE EXTERNAL ELECTRIC FIELD

The calculations are realized on the basement of equations (6.3.34)–(6.3.39) by the initial conditions and parameters containing in the Tables 6.1–6.3. Now we are ready to display the results of the mathematical modeling realized with the help of Maple (the versions Maple 9 or more can be used). The system of generalized hydrodynamic Eqs. (6.3.34)–(6.3.39) have the great possibilities of mathematical modeling as result of changing of Cauchy conditions and parameters describing the character features of initial perturbations which lead to the soliton formation.

The following Maple notations on figures are used: r—density $\widetilde{\rho}_p$, s—density $\widetilde{\rho}_e$, u—velocity $\widetilde{u}$ (solid black line), p—pressure $\widetilde{p}_p$ (black dashed line), q—pressure $\widetilde{p}_e$, and v—self-consistent potential $\widetilde{\psi}$. Explanations placed under all following figures, Maple program contains Maple's notations—for example, the expression $D(u)(0)=0$ means in the usual notations $\left(\partial \widetilde{u}/\partial \widetilde{\xi}\right)(0)=0$, independent variable t responds to $\widetilde{\xi}$.

Important to underline that no special boundary conditions were used for all following cases. The aim of the numerical investigation consists in the discovery of the soliton waves as a product of the self-organization of matter in graphene. It means that the solution should exist only in the restricted domain of the 1D space and the obtained object in the moving coordinate system ($\widetilde{\xi}=\widetilde{x}-\widetilde{t}$) has the constant velocity $\widetilde{u}=1$ for all parts of the object. In this case, the domain of the solution existence defines the character soliton size. The following numerical results demonstrate the realization of mentioned principles.

Figures 6.14–6.21 reflect the result of calculations for Variant 1 (Table 6.3) in the first and the second approximations. In the first approximation the terms of series (6.3.25) with $|g_1|\leq 1$, $|g_2|\leq 1$ (then coefficients U and F) were taken into account. The second approximation contains all terms of the series (6.3.25) with $|g_1|\leq 2$, $|g_2|\leq 2$ (then coefficients U, F, J, B, and G).

From Figs. 6.14–6.21, follow that the size of the created soliton is about $0.5a$, where $a=0.142$ nm. The domain size occupied by the polarized positive charge is about $0.025a$ (see Figs. 6.15 and 6.19). However, the negative charge distributes over the entire soliton domain (Figs. 6.14–6.18), but the negative charge density increases to the edges of the soliton. Therefore, the soliton structure reminds the 1D atom with the positive nuclei and the negative shell.

The self-consistent potential $\widetilde{\psi}$ is practically constant in the soliton boundaries, (Figs. 6.16–6.20). The small grows of the positive particles pressure exists in the x direction. This effect can be connected with the hydrodynamic movement along x and "the reconstruction" of the polarized particles in the soliton front.

Comparing the Figs. 6.14–6.17 and 6.18–6.21, we conclude that the calculation results in the first and the second approximation do not vary significantly. Seemingly significant difference of Figs. 6.14 and 6.18 on the edges of the domain

TABLE 6.3 Varied parameters

Variant No.	E	R	H	U	F	J	B	G
1	0.1	0.003	15	10	10	5	2.5	5
2	0.1	0.003	15	0.1	0.1	0.05	0.025	0.05
3	0.1	0.003	15	10	10	−5	−2.5	−5
4	1000	3×10^{-7}	15	10	10	5	2.5	5
5	0.1	0.003	1500	10	10	5	2.5	5
6	0.1	0.003	15	1000	1000	500	250	500

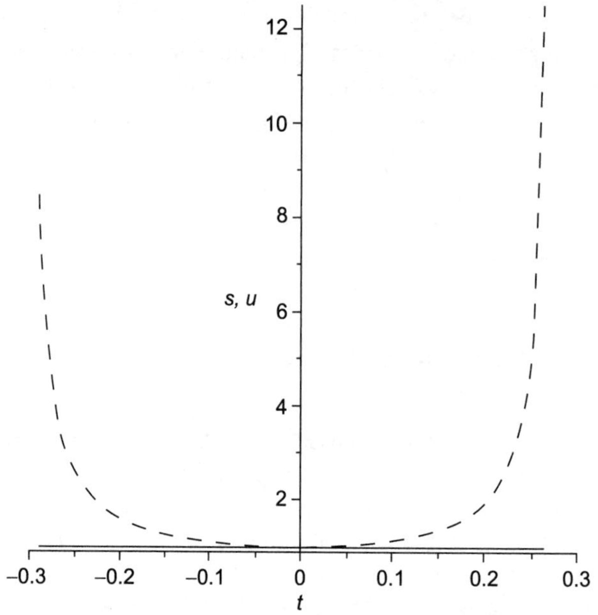

FIGURE 6.14 s—the electron density $\widetilde{\rho}_e$, u—velocity $\widetilde{u}$ (solid line) (first approximation, Variant 1).

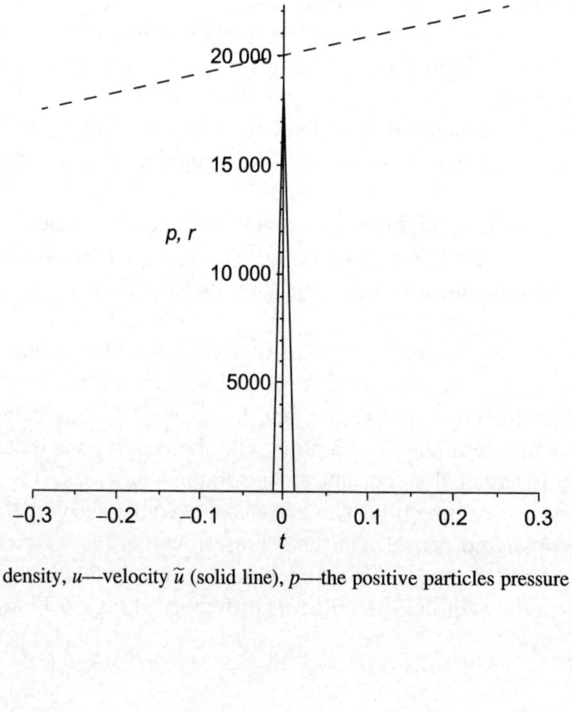

FIGURE 6.15 r—the positive particles density, u—velocity $\widetilde{u}$ (solid line), p—the positive particles pressure (first approximation, Variant 1).

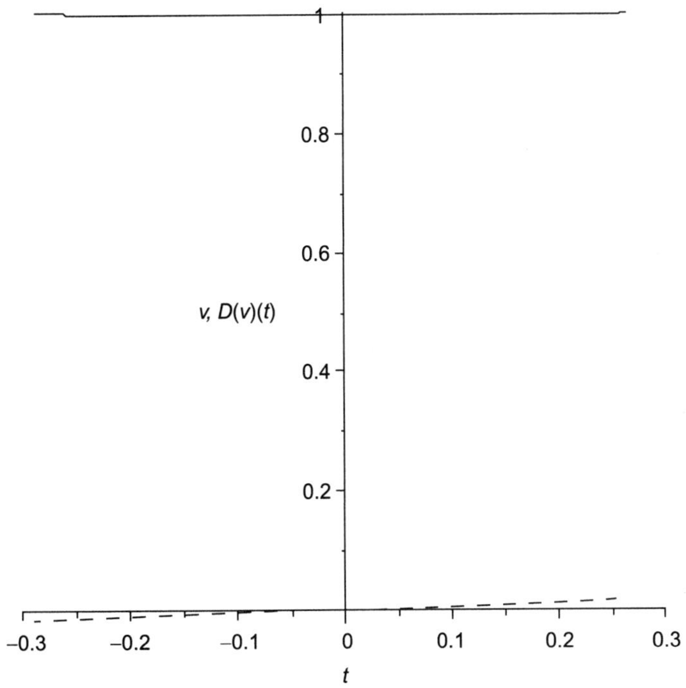

FIGURE 6.16 *v*—potential $\tilde{\psi}$ (solid line) and derivative *D(v)(t)* (first approximation, Variant 1).

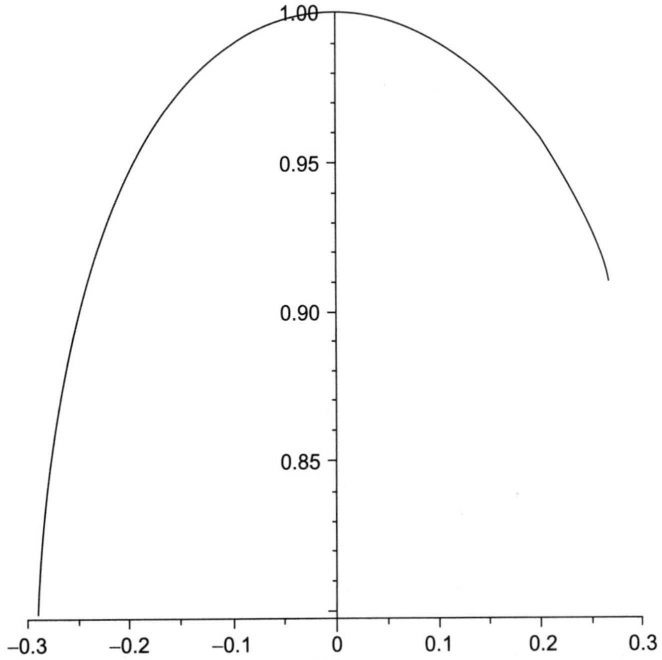

FIGURE 6.17 *q*—pressure of the negative particles (first approximation, Variant 1).

has not the physical sense because corresponds to the regions where $u \neq \text{const}$. Then the restriction of two successive approximations is justified. The question about the convergence of the series (like (6.3.25)) lives open because the first and the second approximations include only the restricted quantity of terms of the infinite series with the coefficients known with the small accuracy.

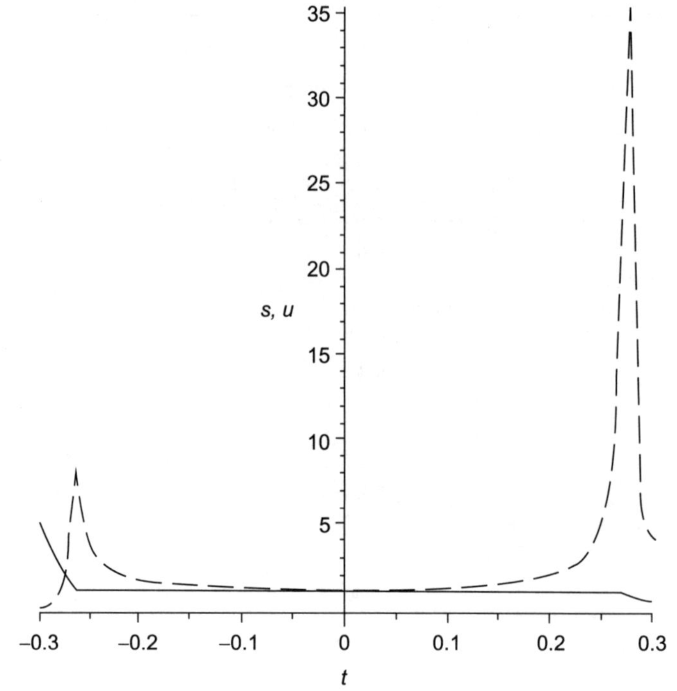

FIGURE 6.18 s—electron density $\tilde{\rho}_e$, u—velocity $\tilde{u}$ (solid line), (the second approximation, Variant 1).

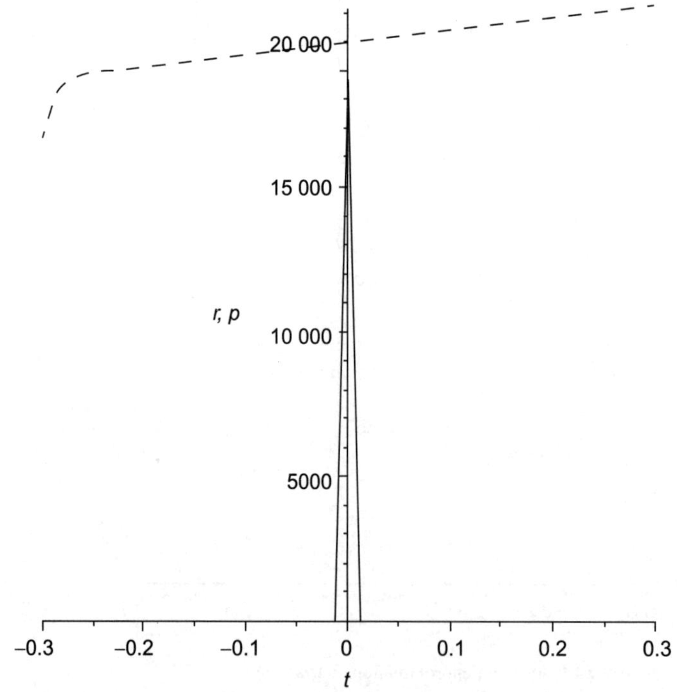

FIGURE 6.19 r—the positive particles density (solid line) p—the positive particles pressure (the second approximation, Variant 1).

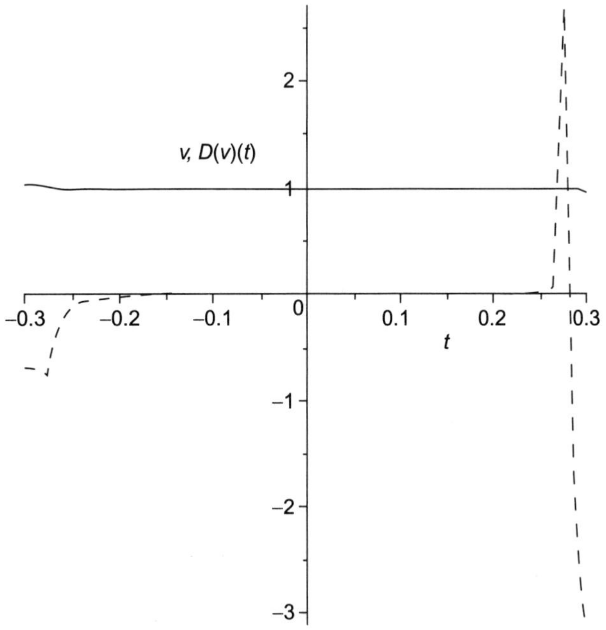

FIGURE 6.20 v—potential $\tilde{\psi}$ (solid line), and derivative $D(v)(t)$ (first approximation, Variant 1).

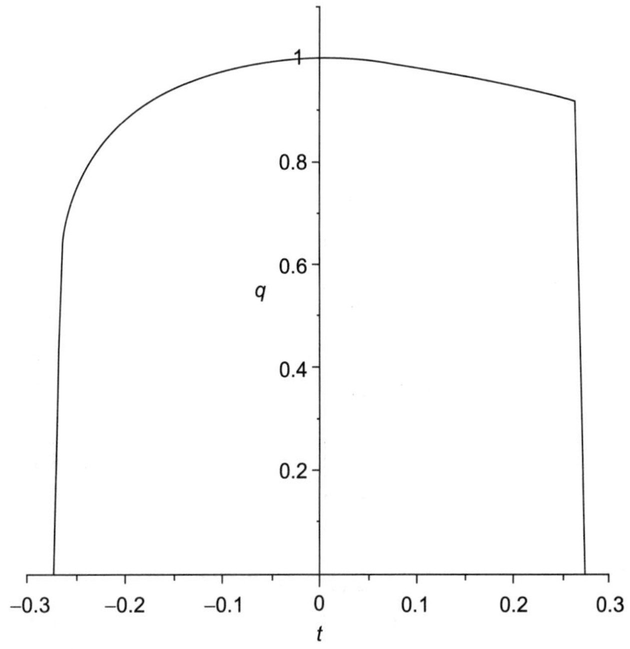

FIGURE 6.21 q—the negative particles pressure (the second approximation, Variant 1).

Figures 6.22–6.27 show the results of calculations responding to Variant 3 (Table 6.3). In the first approximation, Variant 3 is identical to Variant 1 (coefficients $J=B=G=0$) and only the results of the second approximation are delivered. These calculations are more complicated in the numerical realization and all curves are imaged separately, (Figs. 6.22–6.27).

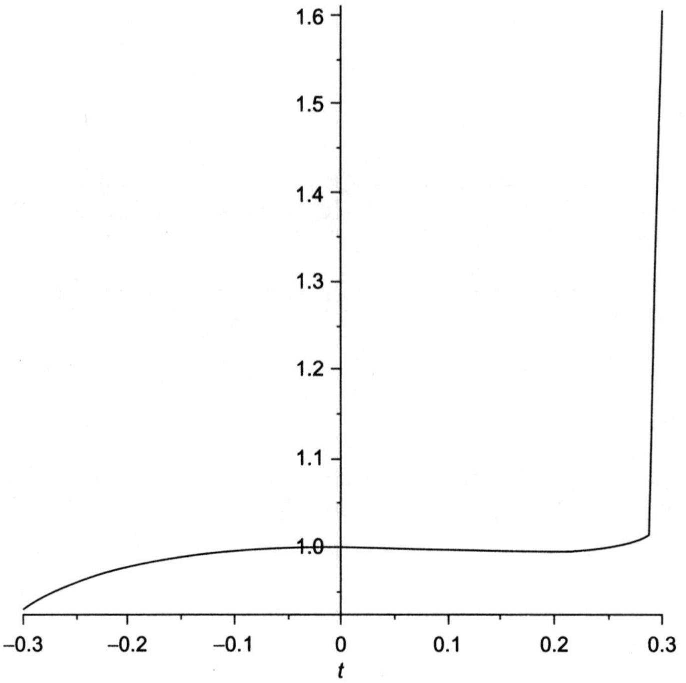

FIGURE 6.22 u—velocity $\widetilde{u}$ (the second approximation, Variant 3).

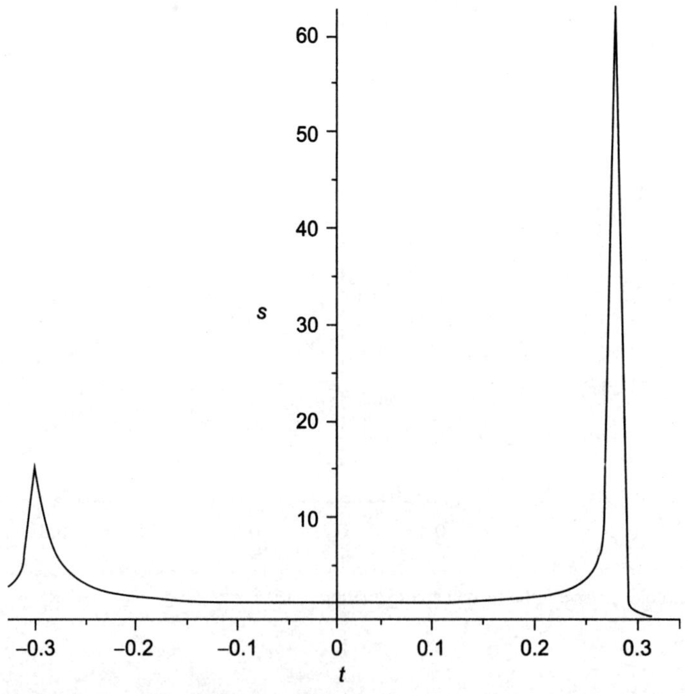

FIGURE 6.23 s—electron density $\widetilde{\rho}_e$, (the second approximation, Variant 3).

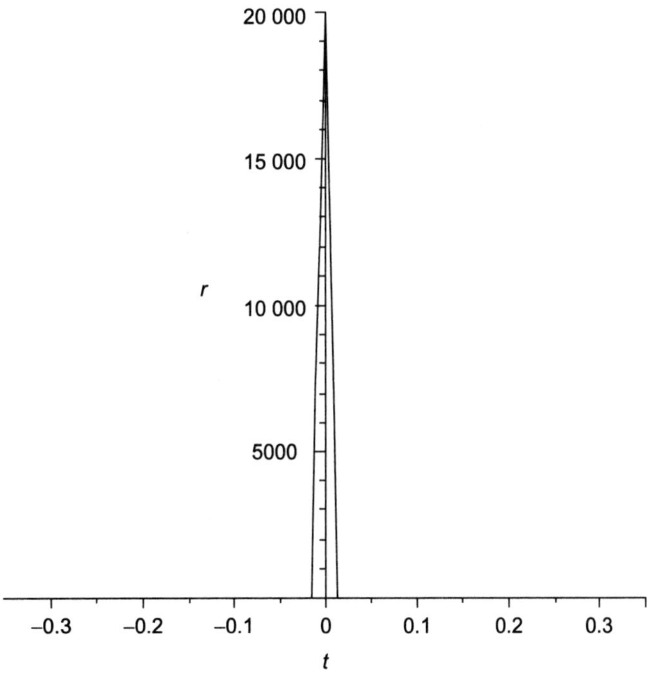

FIGURE 6.24 r—the positive particles density (the second approximation, Variant 3).

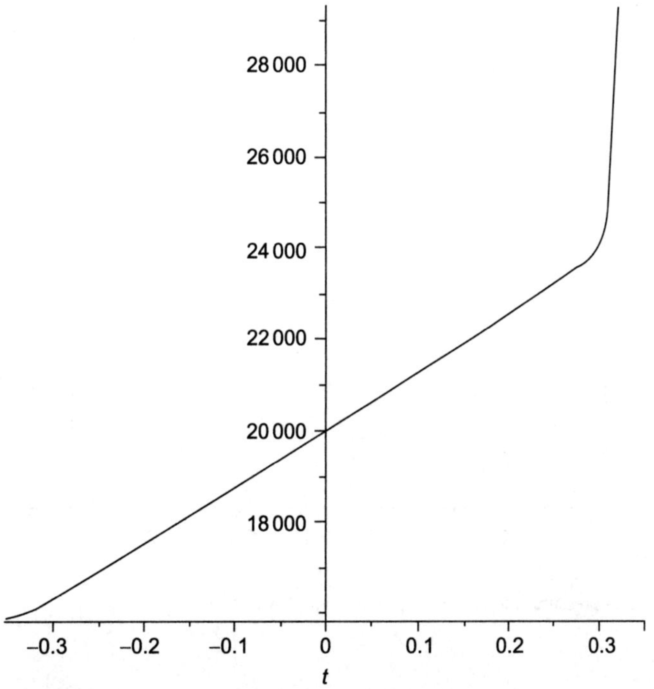

FIGURE 6.25 p—the positive particles pressure, (the second approximation, Variant 3).

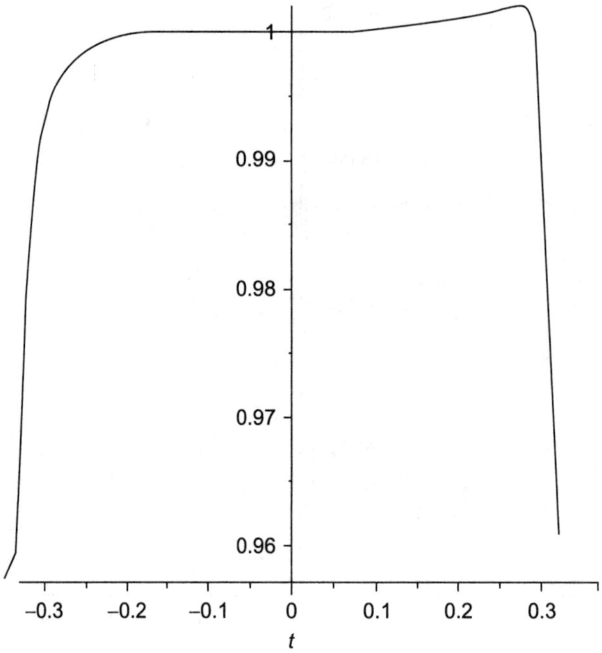

FIGURE 6.26 v—potential $\widetilde{\psi}$ (the second approximation, Variant 3).

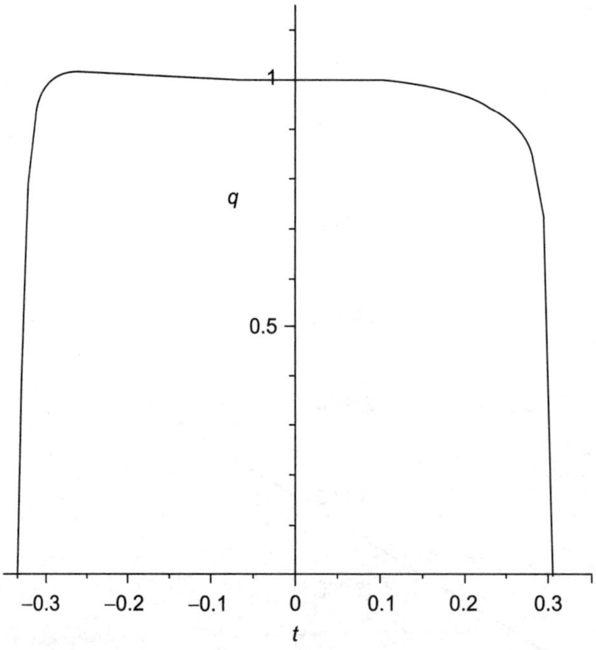

FIGURE 6.27 q—the negative particles pressure (the second approximation, Variant 3).

In the comparison with Variant 1, the calculations in Variant 3 are realized for the case with opposite signs in front of the coefficients of second order. In this case, the distortion of the left side of soliton is observed because by $\widetilde{\xi} < 0$ the velocity $\widetilde{u}$ is not constant. Then this kind of potential for lattice is not favorable for creation of the superconducting structures.

Variant 2 (Table 6.3) corresponds to diminishing of the lattice potential in 100 times by the same practically self-consistent potential, (see Figs. 6.28–6.35).

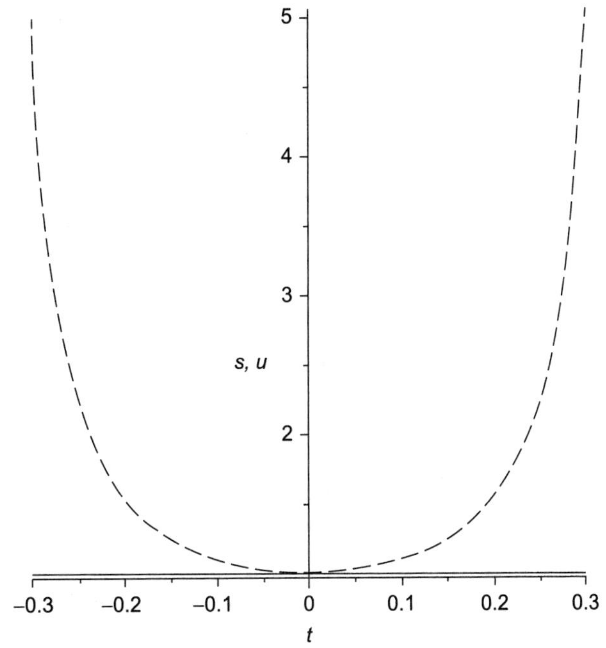

FIGURE 6.28 *s*—electron density $\widetilde{\rho}_e$, *u*—velocity $\widetilde{u}$ (solid line) (the first approximation, Variant 2).

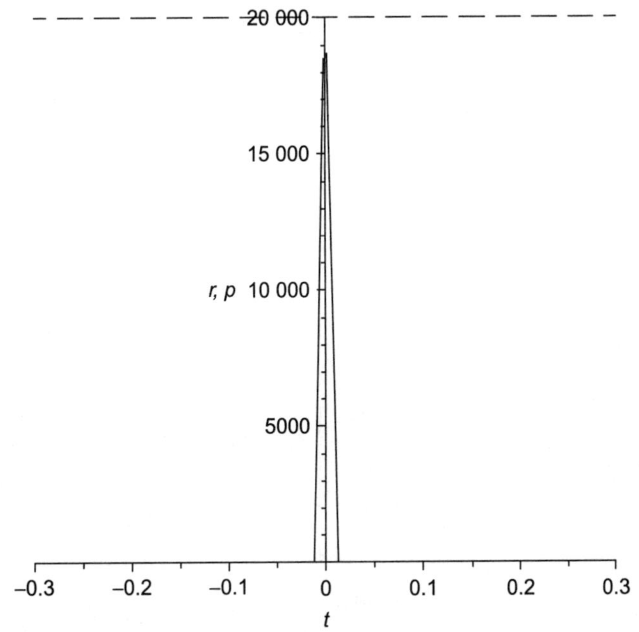

FIGURE 6.29 *r*—the positive particles density, (solid line); *p*—the positive particles pressure (the first approximation, Variant 2).

From comparison of Figs. 6.14–6.21 and 6.28–6.35, follow that numerical diminishing of the lattice potential (by the practically the same value of the self-consistent potential) does not influence on soliton size. However, at the same time the solitons gain the more symmetrical forms. Therefore, namely the self-consistent potential plays the basic role in the soliton formation.

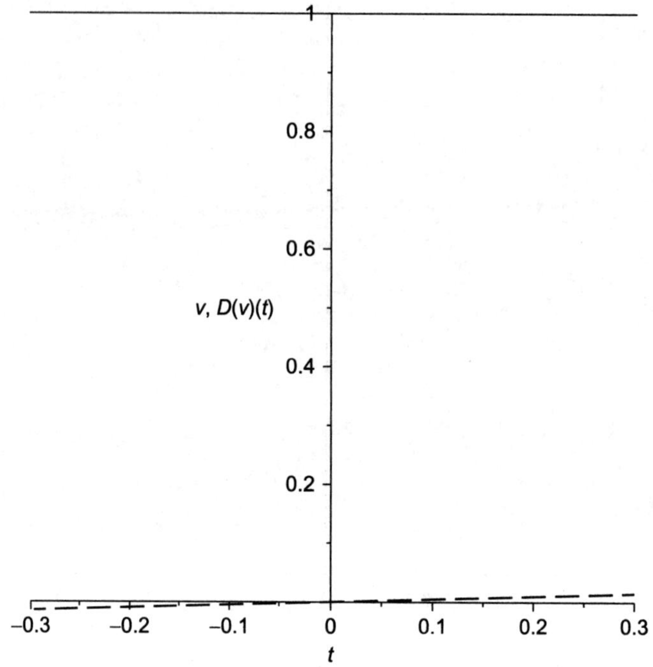

FIGURE 6.30 v—potential $\widetilde{\psi}$ (solid line), $D(v)(t)$, (the first approximation, Variant 2).

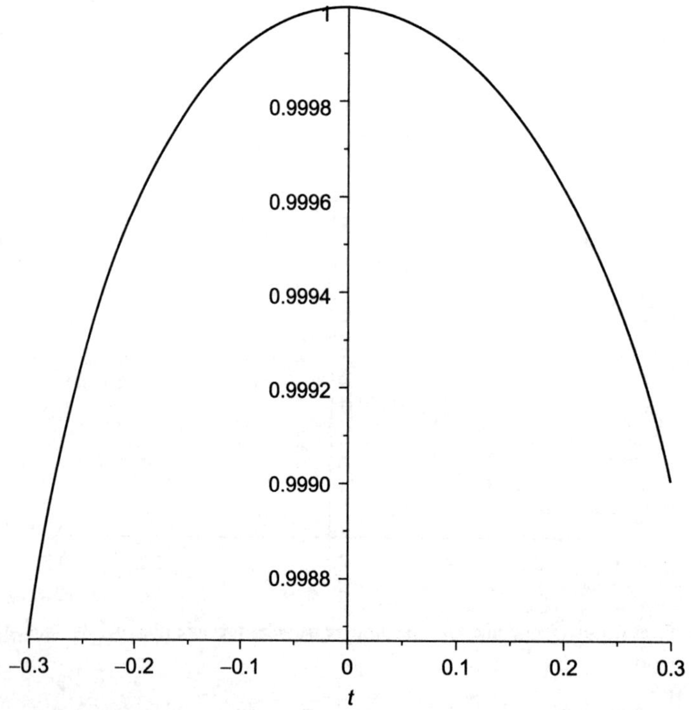

FIGURE 6.31 q—the negative particles pressure (the first approximation, Variant 2).

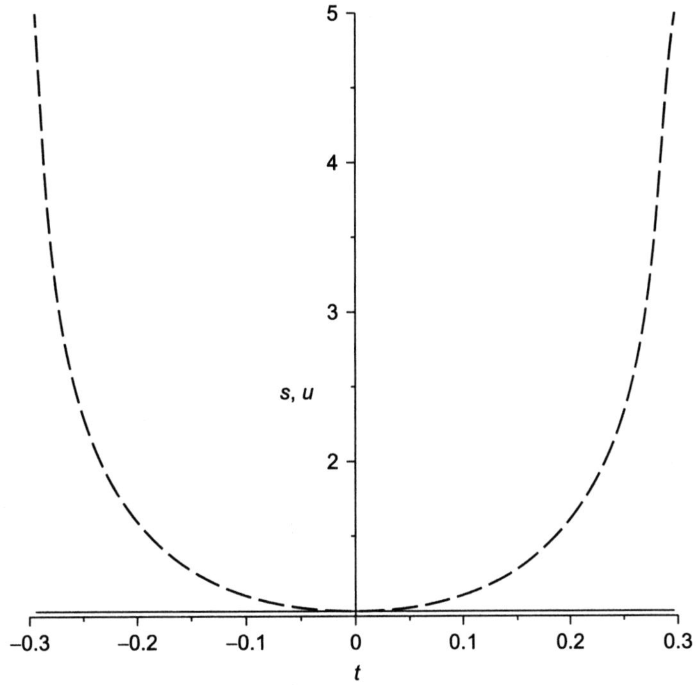

FIGURE 6.32 *s*—electron density $\widetilde{\rho}_e$, *u*—velocity $\widetilde{u}$ (solid line) (the second approximation, Variant 2).

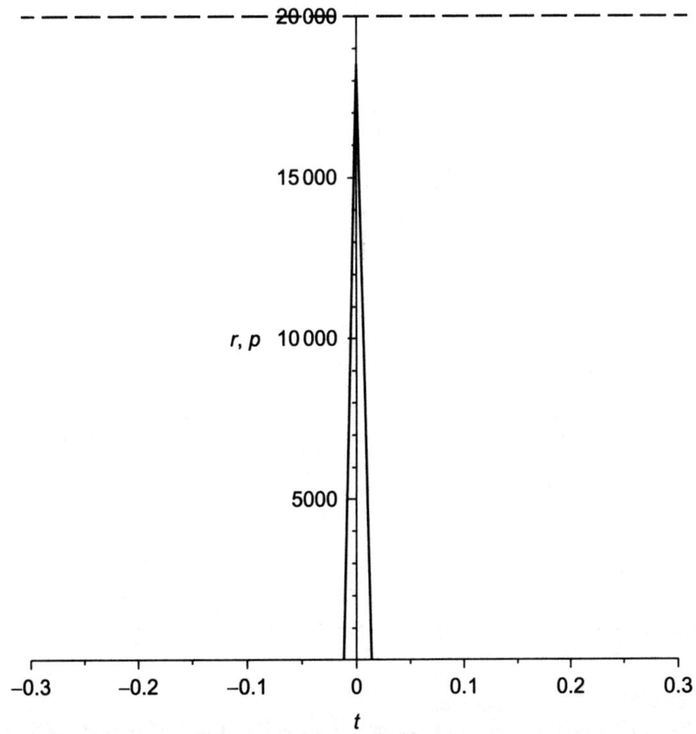

FIGURE 6.33 *r*—the positive particles density, (solid line); *p*—the positive particles pressure (the second approximation, Variant 2).

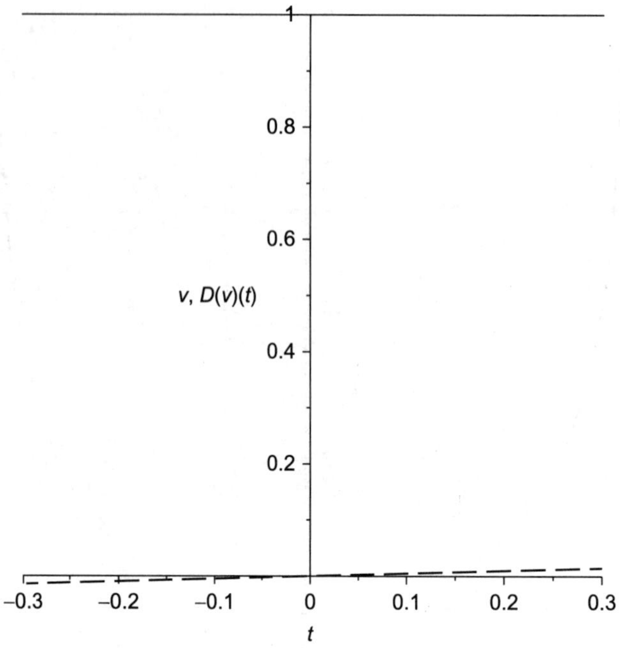

FIGURE 6.34 v—potential $\widetilde{\psi}$ (solid line), $D(v)(t)$ (the second approximation, Variant 2).

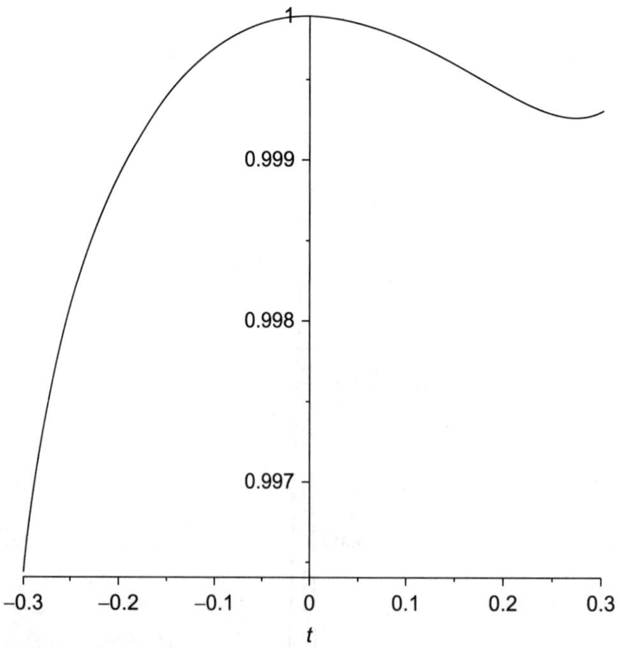

FIGURE 6.35 q—the negative particles pressure (the second approximation, Variant 2).

Let us analyze now the influence of H-parameter, practically the influence of the non locality parameter. Figures 6.36–6.43 (Variant 5) correspond to increasing of the parameter H in 100 times in comparison with Variant 1.

The comparison of Figs. 6.14–6.17 and 6.36–6.39 indicates that in the first approximation the very significant increasing in of the H value in 100 times leads to increasing of the soliton size only in two times without significant changing of the soliton structure. The comparison of calculations (see Figs. 6.20 and 6.42) in the second approximation leads to conclusion that the region (where the velocity $\widetilde{u}$ is constant) has practically the same size.

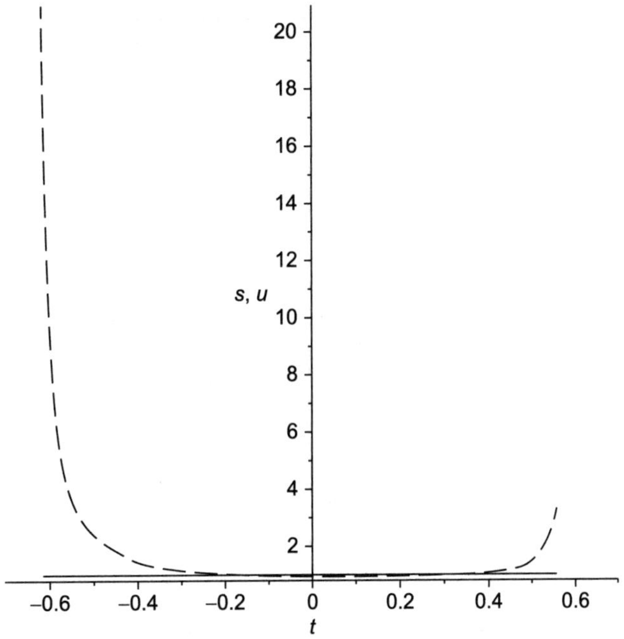

FIGURE 6.36 *s*—electron density $\widetilde{\rho}_e$, *u*—velocity $\widetilde{u}$ (solid line) (the first approximation, Variant 5).

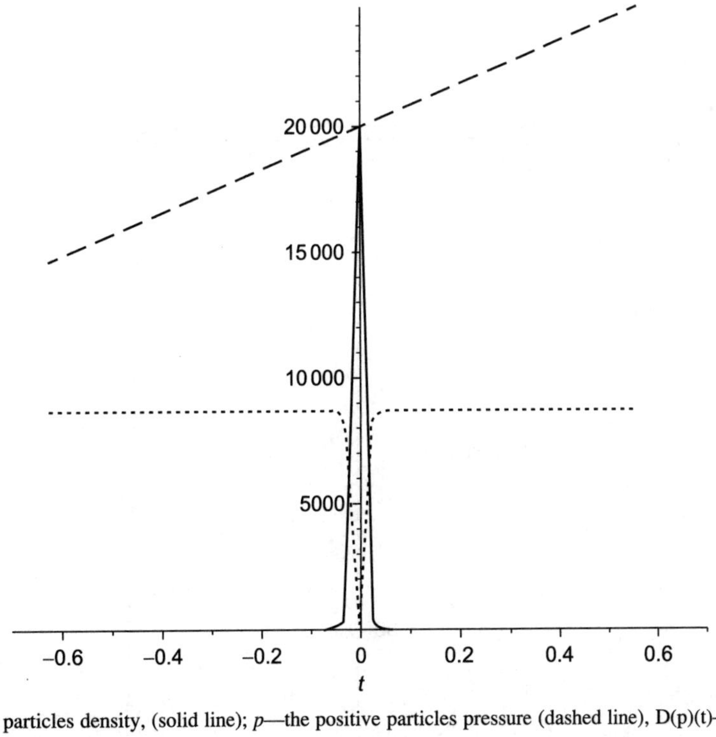

FIGURE 6.37 *r*—the positive particles density, (solid line); *p*—the positive particles pressure (dashed line), D(p)(t)—dotted line. (The first approximation, Variant 5).

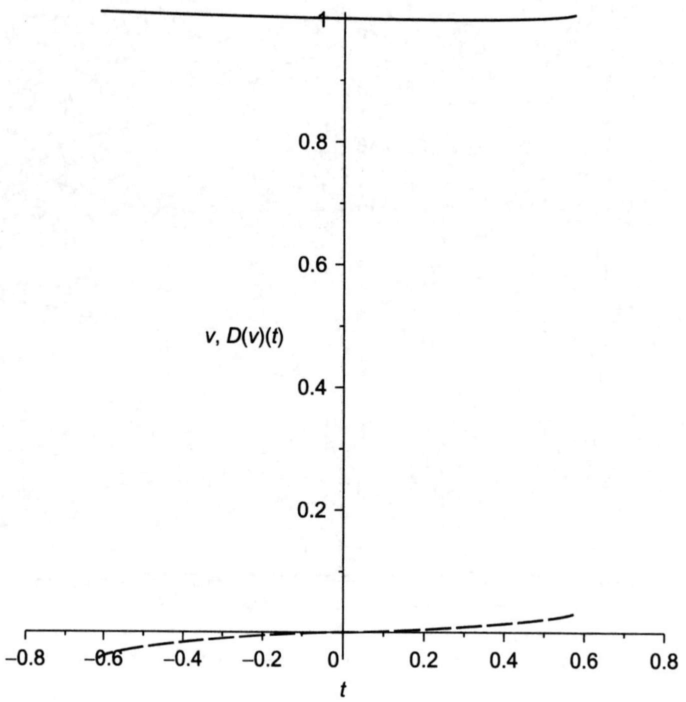

FIGURE 6.38 *v*—potential $\widetilde{\psi}$ (solid line); *D(v)(t)* (the first approximation, Variant 5).

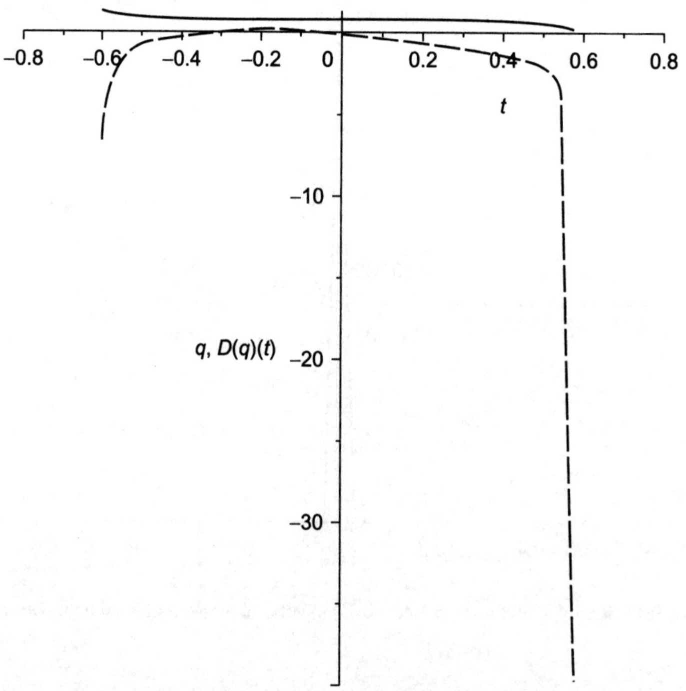

FIGURE 6.39 *q*—the negative particles pressure (solid line), *D(q)(t)*, (the first approximation, Variant 5).

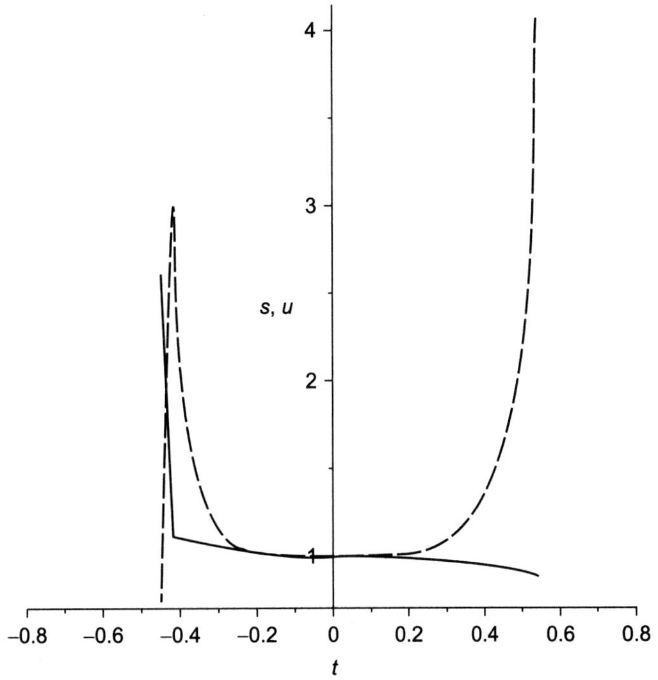

FIGURE 6.40 *s*—electron density $\widetilde{\rho}_e$, *u*—velocity $\widetilde{u}$ (solid line) (the second approximation, Variant 5).

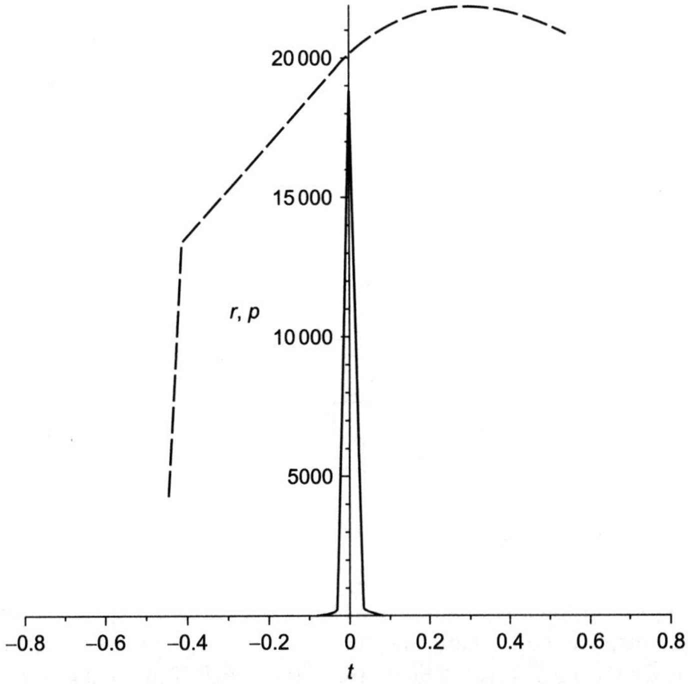

FIGURE 6.41 *r*—the positive particles density, (solid line); *p*—the positive particles pressure (the second approximation, Variant 5).

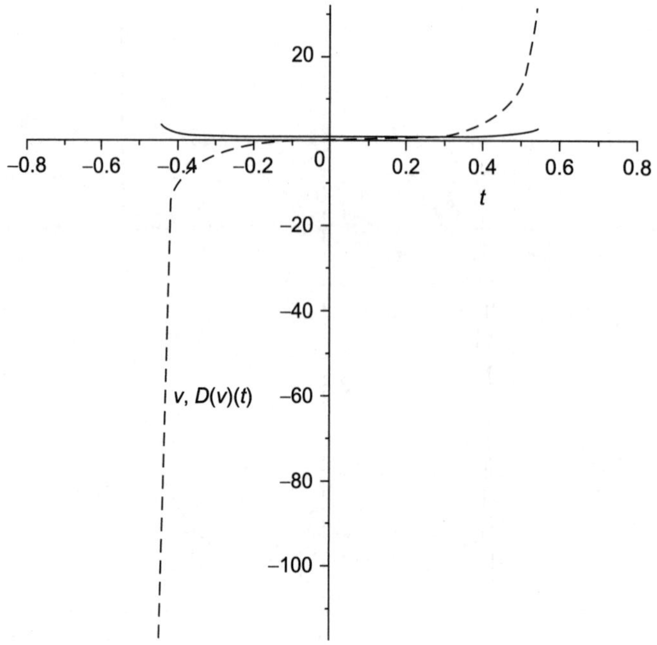

FIGURE 6.42 *v*—potential $\tilde{\psi}$ (solid line); *D(v)(t)*, (the second approximation, Variant 5).

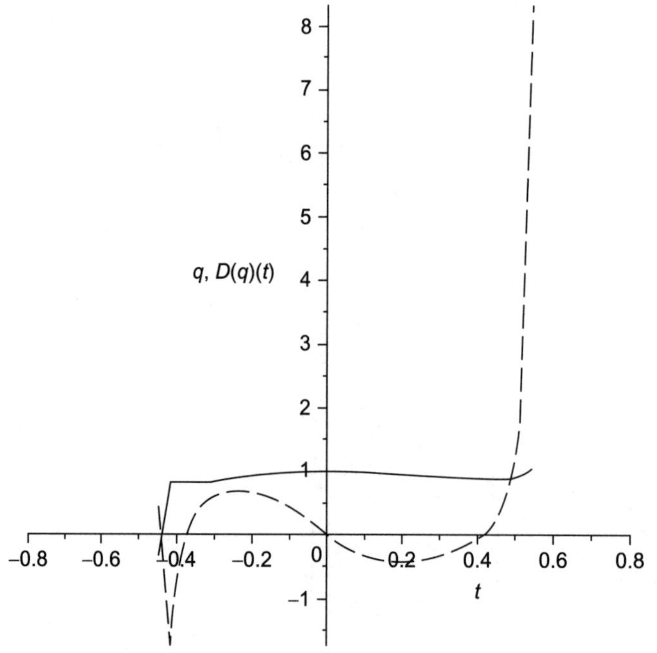

FIGURE 6.43 *q*—the negative particles pressure (solid line), *D(q)(t)*, (the second approximation, Variant 5).

Consider now the calculations responding to Variant 4 (Table 6.3). Increasing in 10^4 times of the scale ψ_0 denotes increasing the self-consistent potential and the lattice potential introduced in the process of the mathematical modeling. This case leads to the drastic diminishing of the soliton size. Figures 6.44–6.47 demonstrate that in the calculations of the first approximation the soliton size is $\sim 10^{-4}a = 1.42 \times 10^{-12}$ cm and exceeds the nuclei size only in several times. The positive kernel of the soliton decreasing in the less degree and occupies now the half of the soliton size. It is no surprise because the low boundary of this kernel size is the character size of the nuclei. Application of the second approximation for the lattice potential function in the mathematical modeling leads to the significant soliton deformation but the same soliton size (see Figs. 6.48–6.51).

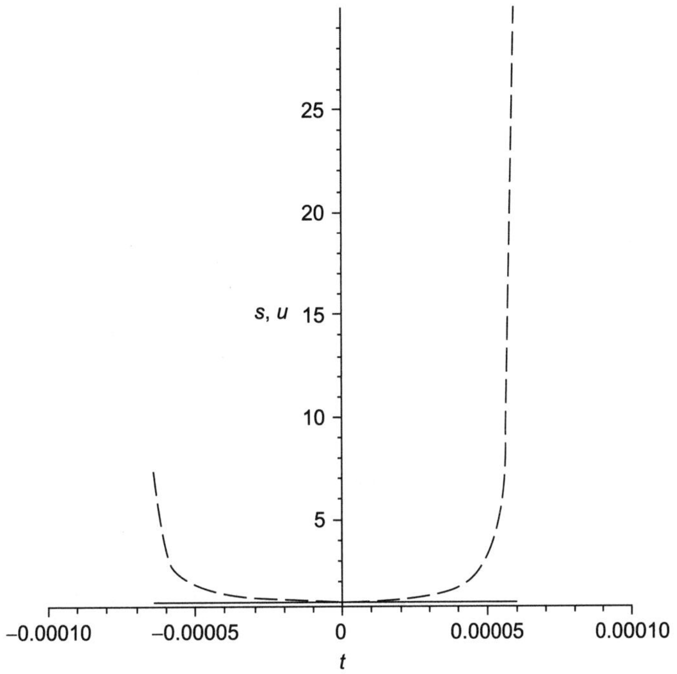

FIGURE 6.44 s—electron density $\widetilde{\rho}_e$, u—velocity $\widetilde{u}$ (solid line) (the first approximation, Variant 4).

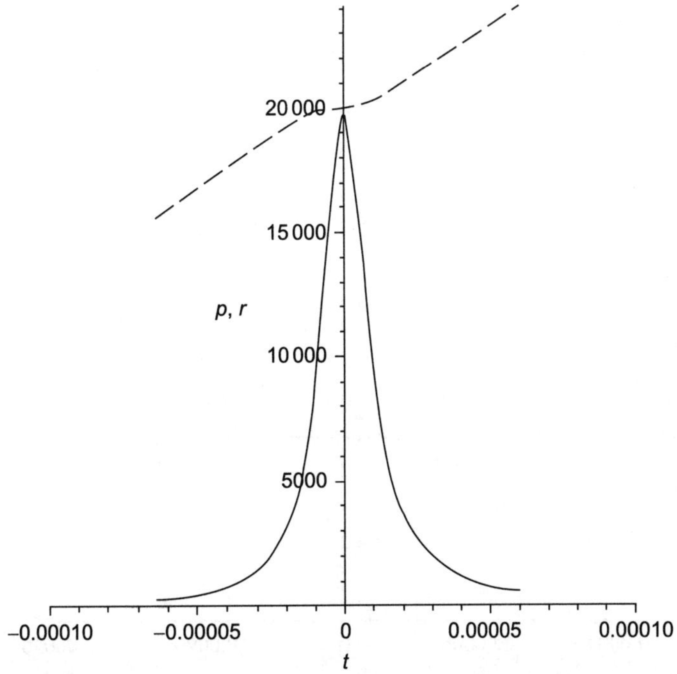

FIGURE 6.45 r—the positive particles density, (solid line); p—the positive particles pressure (the first approximation, Variant 4).

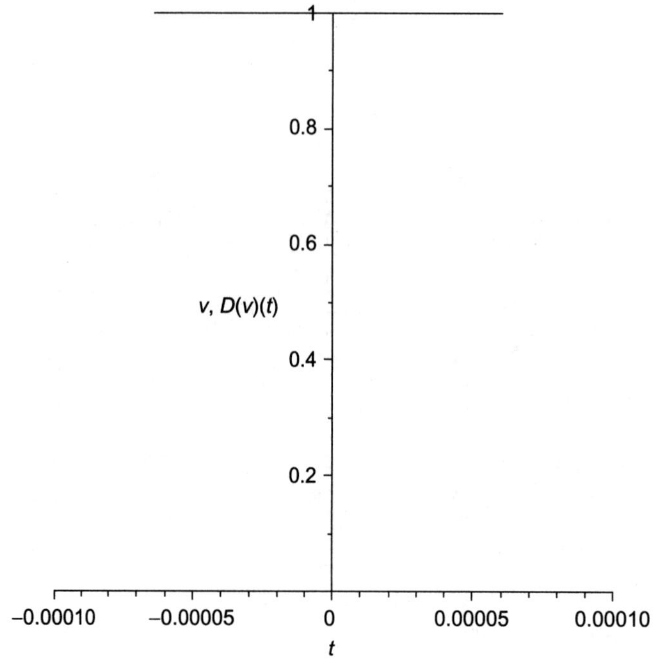

FIGURE 6.46 v—potential $\widetilde{\psi}$ (solid line). (The first approximation, Variant 4).

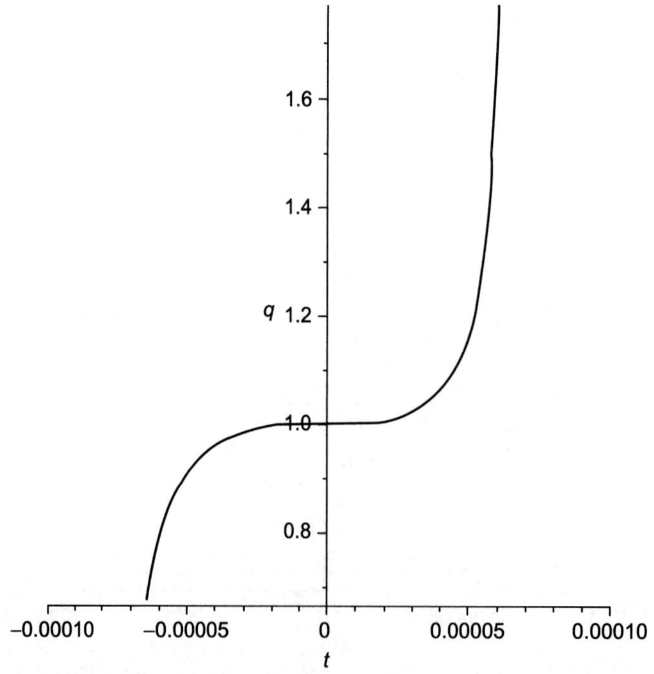

FIGURE 6.47 q—the negative particles pressure. (The first approximation, Variant 4).

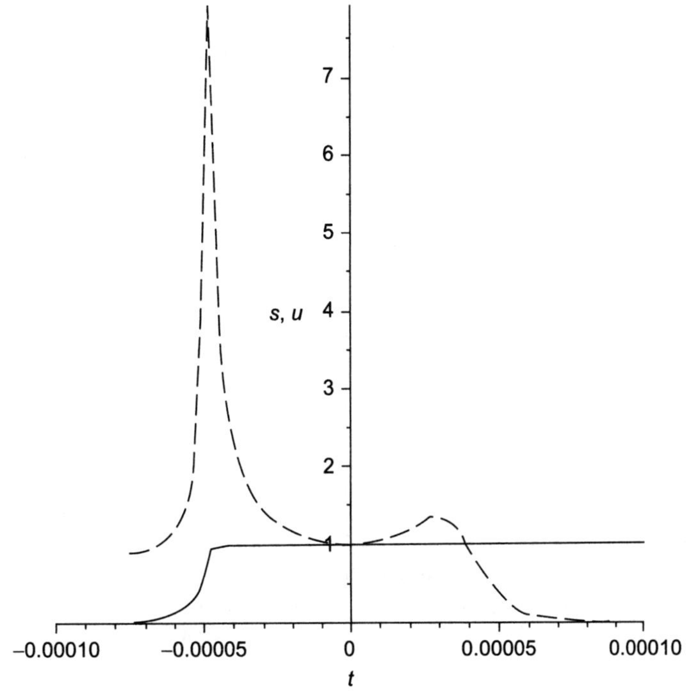

FIGURE 6.48 *s*—electron density $\widetilde{\rho}_e$, *u*—velocity $\widetilde{u}$ (solid line). (The second approximation, Variant 4).

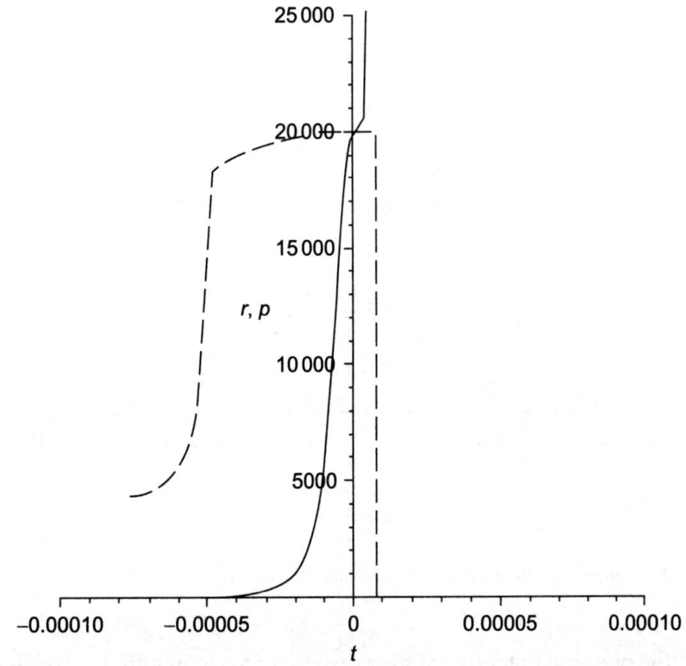

FIGURE 6.49 *r*—the positive particles density, (solid line); *p*—the positive particles pressure, (the second approximation, Variant 4).

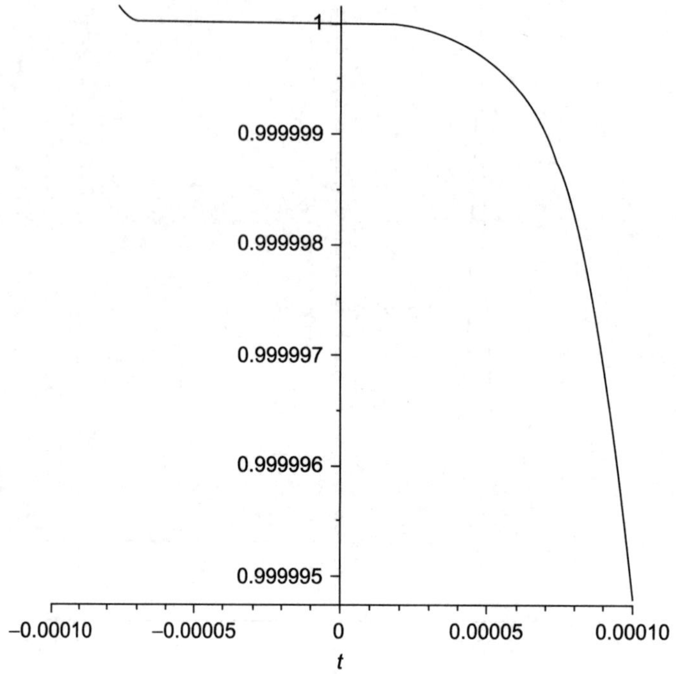

FIGURE 6.50 v—potential $\tilde{\psi}$ (solid line). (The second approximation, Variant 4).

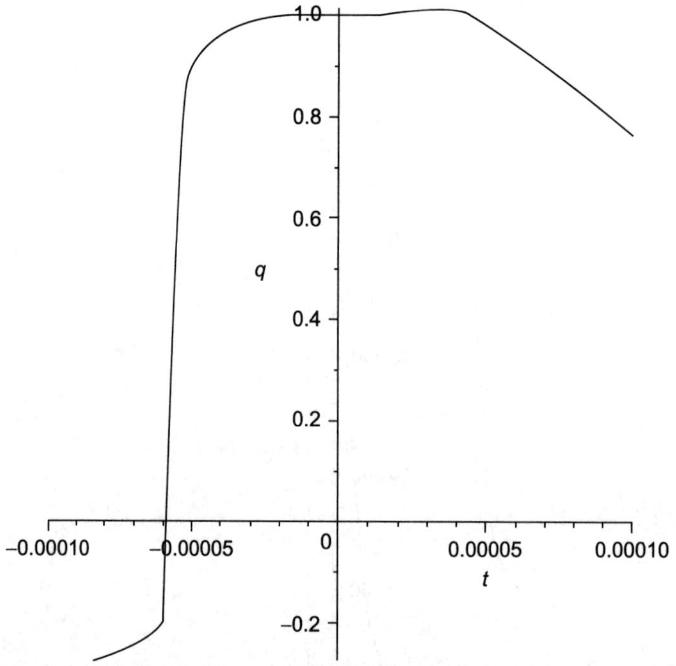

FIGURE 6.51 q—the negative particles pressure. (The second approximation, Variant 4).

The drastic increasing of the periodic potential of the crystal lattice (in hundred times, see Figs. 6.52–6.60) in comparison with the self-consistent potential also leads to diminishing of the soliton size. For the case Variant 6, Table 6.3 this size consists only $\sim 10^{-2}a$. However, this increasing does not lead to the relative increasing of the soliton kernel and to the mentioned above the soliton deformation in the second approximation (see Figs. 6.45–6.48). Figure 6.41 demonstrate the extremely high accuracy of the soliton stability, the velocity fluctuation inside the soliton is only $\sim 10^{-16}\tilde{u}$.

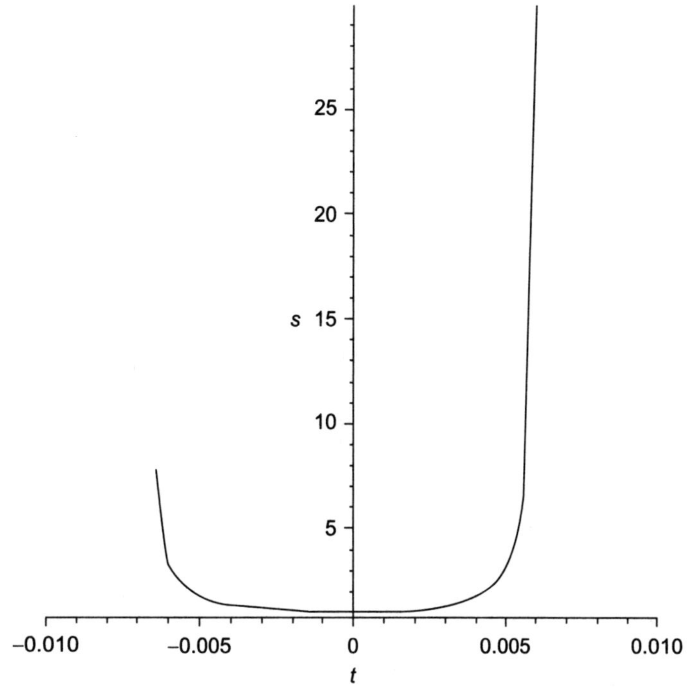

FIGURE 6.52 *s*—electron density $\widetilde{\rho}_e$, (the first approximation, Variant 6).

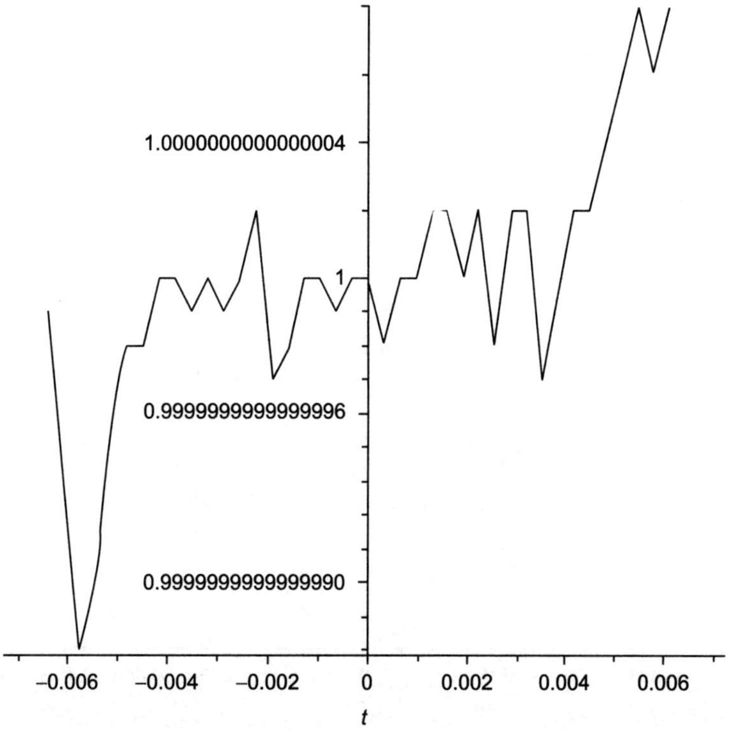

FIGURE 6.53 *u*—velocity $\widetilde{u}$. (The first approximation, Variant 6).

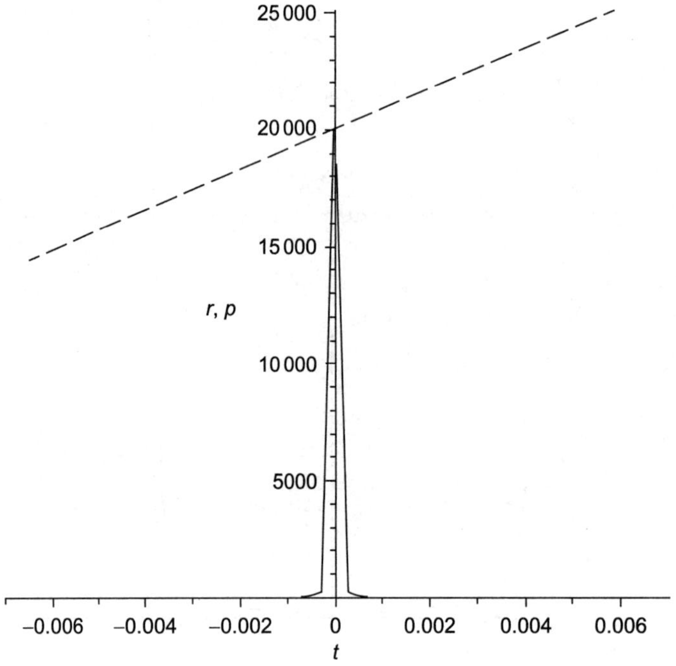

FIGURE 6.54 *r*—the positive particles density, *p*—the positive particles pressure (the first approximation, Variant 6).

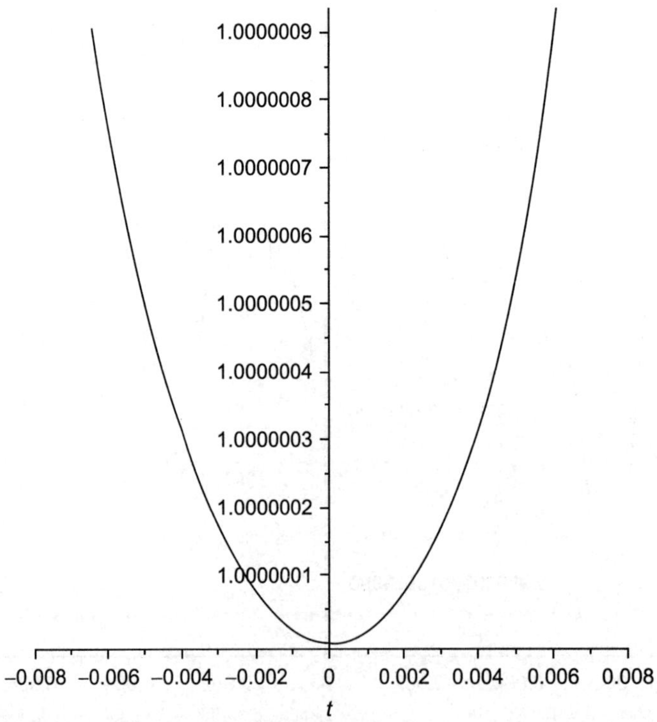

FIGURE 6.55 *v*—potential $\tilde{\psi}$ (solid line); (the first approximation, Variant 6).

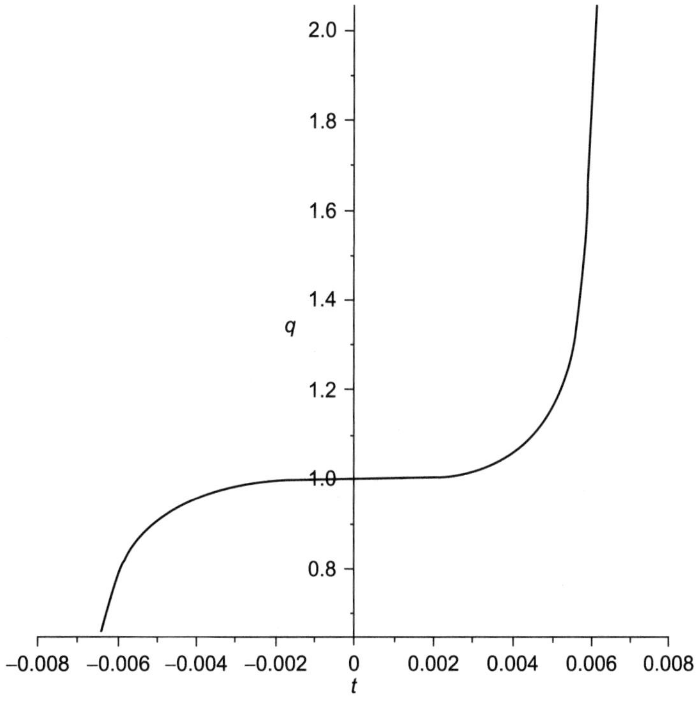

FIGURE 6.56 q—the negative particles pressure (the first approximation, Variant 6).

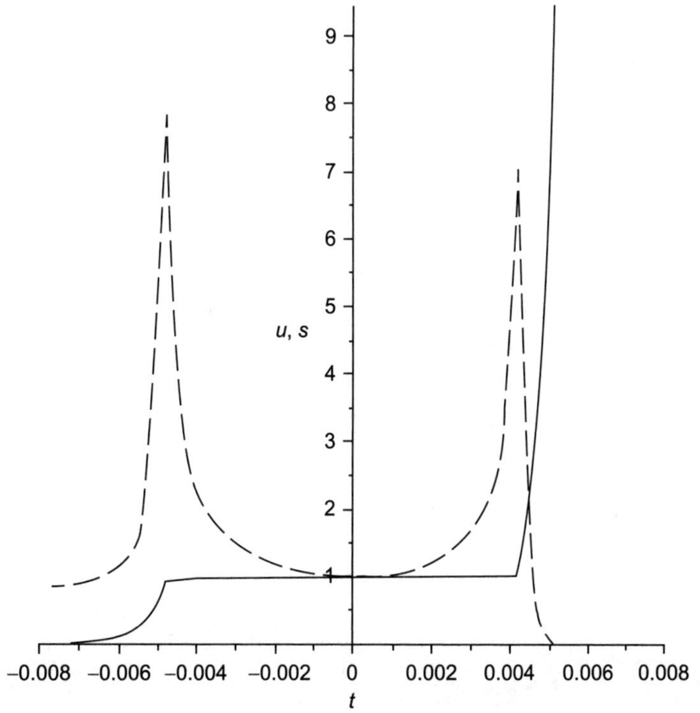

FIGURE 6.57 s—electron density $\widetilde{\rho}_e$, u—velocity $\widetilde{u}$ (solid line) (the second approximation, Variant 6).

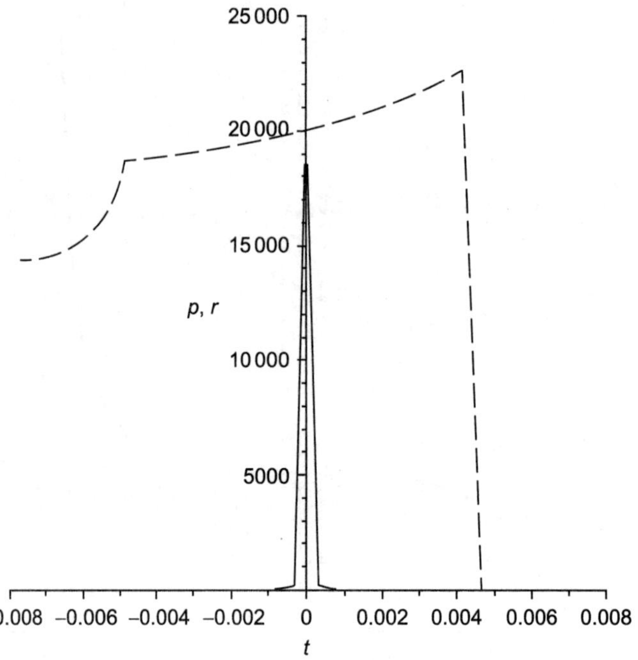

FIGURE 6.58 *r*—the positive particles density (solid line); *p*—the positive particles, (the second approximation, Variant 6).

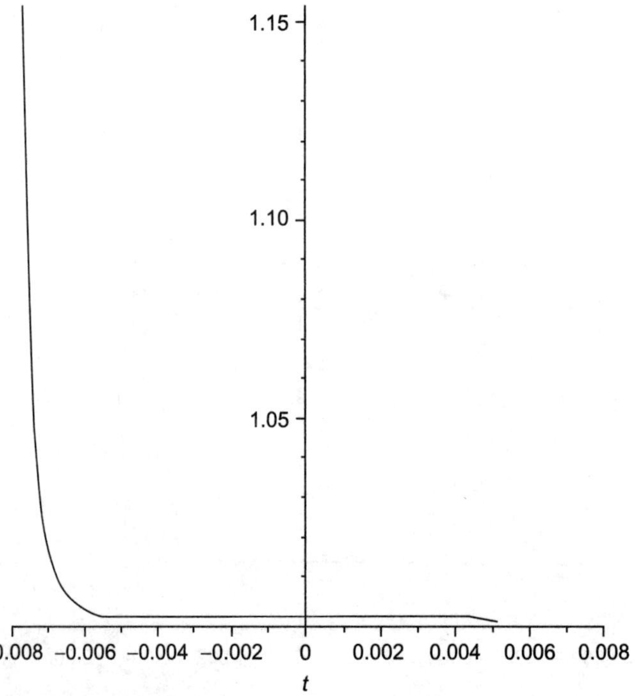

FIGURE 6.59 *v*—potential $\tilde{\psi}$ pressure (the second approximation, Variant 6).

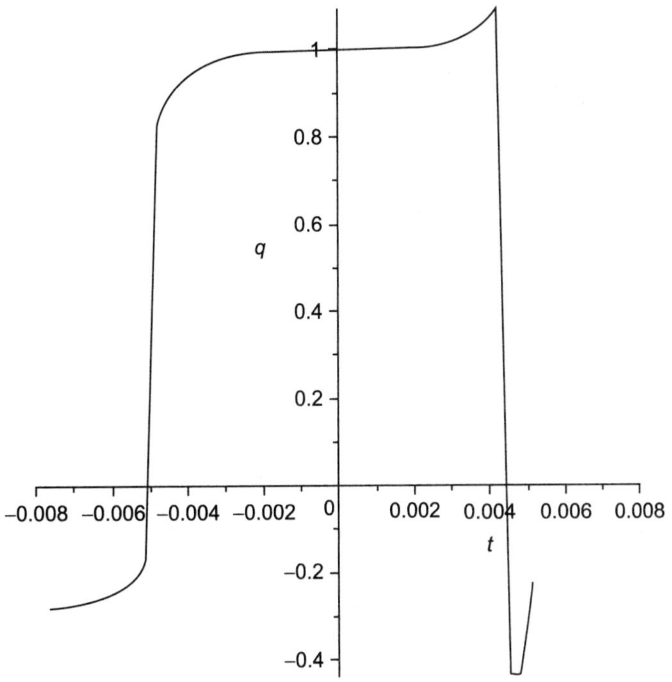

FIGURE 6.60 q—the negative particles pressure (the second approximation, Variant 6).

6.5 RESULTS OF THE MATHEMATICAL MODELING WITH THE EXTERNAL ELECTRIC FIELD

Let us consider now the results of the mathematical modeling with taking into account the intensity of the external electric field which does not depend on y. In this case, the solution of the hydrodynamic system (6.3.10)–(6.3.15) should be found. After averaging and in the moving coordinate system it leads to the following equations written in the first approximation (compare with the system (6.3.34)–(6.3.39)):

Dimensionless Poisson equation for the self-consistent electric field:

$$\frac{\partial^2 \tilde{\psi}}{\partial \tilde{\xi}^2} = -4\pi R \left\{ \frac{m_e}{m_p} \left[\tilde{\rho}_p - \frac{m_e H}{m_p \tilde{u}^2} \frac{\partial}{\partial \tilde{\xi}} (\tilde{\rho}_p (\tilde{u} - 1)) \right] - \left[\tilde{\rho}_e - \frac{H}{\tilde{u}^2} \frac{\partial}{\partial \tilde{\xi}} (\tilde{\rho}_e (\tilde{u} - 1)) \right] \right\}. \tag{6.5.1}$$

Continuity equation for the positive particles:

$$\begin{aligned} &\frac{\partial}{\partial \tilde{\xi}} [\tilde{\rho}_p (1 - \tilde{u})] + \frac{m_e}{m_p} \frac{\partial}{\partial \tilde{\xi}} \left\{ \frac{H}{\tilde{u}^2} \frac{\partial}{\partial \tilde{\xi}} \left[\tilde{\rho}_p (\tilde{u} - 1)^2 \right] \right\} + \frac{m_e}{m_p} \frac{\partial}{\partial \tilde{\xi}} \left\{ \frac{H}{\tilde{u}^2} \left[\frac{V_{0p}^2}{u_0^2} \frac{\partial}{\partial \tilde{\xi}} \tilde{p}_p \right. \right. \\ &\left. \left. - \frac{m_e}{m_p} \tilde{\rho}_p E \left(- \frac{\partial \tilde{\psi}}{\partial \tilde{\xi}} + \tilde{U}'_{11} \sin \left(\frac{4\pi}{3\tilde{a}} \tilde{\xi} - \frac{\pi}{3} \right) + \tilde{E}_0 \right) \right] \right\} = 0 \end{aligned} \tag{6.5.2}$$

Continuity equation for electrons:

$$\begin{aligned} &\frac{\partial}{\partial \tilde{\xi}} [\tilde{\rho}_e (1 - \tilde{u})] + \frac{\partial}{\partial \tilde{\xi}} \left\{ \frac{H}{\tilde{u}^2} \frac{\partial}{\partial \tilde{\xi}} \left[\tilde{\rho}_e (\tilde{u} - 1)^2 \right] \right\} + \frac{\partial}{\partial \tilde{\xi}} \left\{ \frac{H}{\tilde{u}^2} \left[\frac{V_{0e}^2}{u_0^2} \frac{\partial}{\partial \tilde{\xi}} \tilde{p}_e \right. \right. \\ &\left. \left. - \tilde{\rho}_e E \left(\frac{\partial \tilde{\psi}}{\partial \tilde{\xi}} - \tilde{U}'_{11} \sin \left(\frac{4\pi}{3\tilde{a}} \tilde{\xi} - \frac{\pi}{3} \right) - \tilde{E}_0 \right) \right] \right\} = 0 \end{aligned} \tag{6.5.3}$$

Momentum equation for the x direction:

$$
\begin{aligned}
&\frac{\partial}{\partial\tilde{\xi}}\left\{(\tilde{\rho}_p+\tilde{\rho}_e)\tilde{u}(\tilde{u}-1)+\frac{V_{0p}^2}{u_0^2}\tilde{p}_p+\frac{V_{0e}^2}{u_0^2}\tilde{p}_e\right\}\\
&-\frac{m_e}{m_p}\tilde{\rho}_pE\left(-\frac{\partial\tilde{\phi}}{\partial\tilde{\xi}}+\tilde{U}_{11}'\sin\left(\frac{4\pi}{3\tilde{a}}\tilde{\xi}-\frac{\pi}{3}\right)+\tilde{E}_0\right)\\
&-\tilde{\rho}_eE\left(\frac{\partial\tilde{\psi}}{\partial\tilde{\xi}}-\tilde{U}_{11}'\sin\left(\frac{4\pi}{3\tilde{a}}\tilde{\xi}-\frac{\pi}{3}\right)-\tilde{E}_0\right)\\
&+\frac{m_e}{m_p}\frac{\partial}{\partial\tilde{\xi}}\left\{\frac{H}{\tilde{u}^2}\left[\frac{\partial}{\partial\tilde{\xi}}\left(2\frac{V_{0p}^2}{u_0^2}\tilde{p}_p(1-\tilde{u})-\tilde{\rho}_p\tilde{u}(1-\tilde{u})^2\right)\right.\right.\\
&\left.\left.-\frac{m_e}{m_p}\tilde{\rho}_p(1-\tilde{u})E\left(-\frac{\partial\tilde{\psi}}{\partial\tilde{\xi}}+\tilde{U}_{11}'\sin\left(\frac{4\pi}{3\tilde{a}}\tilde{\xi}-\frac{\pi}{3}\right)+\tilde{E}_0\right)\right]\right\}\\
&+\frac{\partial}{\partial\tilde{\xi}}\left\{\frac{H}{\tilde{u}^2}\left[\frac{\partial}{\partial\tilde{\xi}}\left(2\frac{V_{0e}^2}{u_0^2}\tilde{p}_e(1-\tilde{u})-\tilde{\rho}_e\tilde{u}(1-\tilde{u})^2\right)-\tilde{\rho}_e(1-\tilde{u})E\left(\frac{\partial\tilde{\psi}}{\partial\tilde{\xi}}-\tilde{U}_{11}'\sin\left(\frac{4\pi}{3\tilde{a}}\tilde{\xi}-\frac{\pi}{3}\right)-\tilde{E}_0\right)\right]\right\}\\
&+\frac{H}{\tilde{u}^2}E\left(\frac{m_e}{m_p}\right)^2\left(-\frac{\partial\tilde{\psi}}{\partial\tilde{\xi}}+\tilde{U}_{11}'\sin\left(\frac{4\pi}{3\tilde{a}}\tilde{\xi}-\frac{\pi}{3}\right)+\tilde{E}_0\right)\left(\frac{\partial}{\partial\tilde{\xi}}(\tilde{\rho}_p(\tilde{u}-1))\right)\\
&+\frac{H}{\tilde{u}^2}E\left(\frac{\partial\tilde{\psi}}{\partial\tilde{\xi}}-\tilde{U}_{11}'\sin\left(\frac{4\pi}{3\tilde{a}}\tilde{\xi}-\frac{\pi}{3}\right)-\tilde{E}_0\right)\left(\frac{\partial}{\partial\tilde{\xi}}(\tilde{\rho}_e(\tilde{u}-1))\right)\\
&-\frac{m_e}{m_p}\frac{\partial}{\partial\tilde{\xi}}\left\{\frac{H}{\tilde{u}^2}\frac{V_{0p}^2}{u_0^2}\frac{\partial}{\partial\tilde{\xi}}(\tilde{p}_p\tilde{u})\right\}-\frac{\partial}{\partial\tilde{\xi}}\left\{\frac{H}{\tilde{u}^2}\frac{V_{0e}^2}{u_0^2}\frac{\partial}{\partial\tilde{\xi}}(\tilde{p}_e\tilde{u})\right\}\\
&+\left(\frac{m_e}{m_p}\right)^2E\frac{\partial}{\partial\tilde{\xi}}\left\{\frac{H}{\tilde{u}^2}\left[\left(-\frac{\partial\tilde{\psi}}{\partial\tilde{\xi}}+\tilde{U}_{11}'\sin\left(\frac{4\pi}{3\tilde{a}}\tilde{\xi}-\frac{\pi}{3}\right)+\tilde{E}_0\right)\tilde{\rho}_p\tilde{u}\right]\right\}\\
&+E\frac{\partial}{\partial\tilde{\xi}}\left\{\frac{H}{\tilde{u}^2}\left[\left(\frac{\partial\tilde{\psi}}{\partial\tilde{\xi}}-\tilde{U}_{11}'\sin\left(\frac{4\pi}{3\tilde{a}}\tilde{\xi}-\frac{\pi}{3}\right)-\tilde{E}_0\right)\tilde{\rho}_e\tilde{u}\right]\right\}=0
\end{aligned}
\tag{6.5.4}
$$

Energy equation for the positive particles:

$$
\begin{aligned}
&\frac{\partial}{\partial\tilde{\xi}}\left[\tilde{\rho}_p\tilde{u}^2(\tilde{u}-1)+5\frac{V_{0p}^2}{u_0^2}\tilde{p}_p\tilde{u}-3\frac{V_{0p}^2}{u_0^2}\tilde{p}_p\right]-2\frac{m_e}{m_p}\tilde{\rho}_pE\left(-\frac{\partial\tilde{\psi}}{\partial\tilde{\xi}}+\tilde{U}_{11}'\sin\left(\frac{4\pi}{3\tilde{a}}\tilde{\xi}-\frac{\pi}{3}\right)+\tilde{E}_0\right)\tilde{u}\\
&+\frac{\partial}{\partial\tilde{\xi}}\left\{\frac{H}{\tilde{u}^2}\frac{m_e}{m_p}\left[\frac{\partial}{\partial\tilde{\xi}}\left(-\tilde{\rho}_p\tilde{u}^2(1-\tilde{u})^2+7\frac{V_{0p}^2}{u_0^2}\tilde{p}_p\tilde{u}(1-\tilde{u})+3\frac{V_{0p}^2}{u_0^2}\tilde{p}_p(\tilde{u}-1)-\frac{V_{0p}^2}{u_0^2}\tilde{p}_p\tilde{u}^2-5\frac{V_{0p}^4}{u_0^4}\frac{\tilde{p}_p^2}{\tilde{\rho}_p}\right)\right.\right.\\
&+E\left(-2\frac{m_e}{m_p}\tilde{\rho}_p\tilde{u}(1-\tilde{u})+\frac{m_e}{m_p}\tilde{\rho}_p\tilde{u}^2+5\frac{m_e}{m_p}\frac{V_{0p}^2}{u_0^2}\tilde{p}_p\right)\left(-\frac{\partial\tilde{\psi}}{\partial\tilde{\xi}}\right.\\
&\left.\left.\left.+\tilde{U}_{11}'\sin\left(\frac{4\pi}{3\tilde{a}}\tilde{\xi}-\frac{\pi}{3}\right)+\tilde{E}_0\right)\right]\right\}+2\frac{H}{\tilde{u}^2}E\left(\frac{m_e}{m_p}\right)^2\left[-\frac{\partial}{\partial\tilde{\xi}}(\tilde{\rho}_p\tilde{u}(1-\tilde{u}))\right.\\
&\left.+\frac{V_{0p}^2}{u_0^2}\frac{\partial}{\partial\tilde{\xi}}\tilde{p}_p\right]\left(-\frac{\partial\tilde{\psi}}{\partial\tilde{\xi}}+\tilde{U}_{11}'\sin\left(\frac{4\pi}{3\tilde{a}}\tilde{\xi}-\frac{\pi}{3}\right)+\tilde{E}_0\right)\\
&-2\frac{H}{\tilde{u}^2}E^2\left(\frac{m_e}{m_p}\right)^3\tilde{\rho}_p\left[\left(-\frac{\partial\tilde{\psi}}{\partial\tilde{\xi}}+\tilde{U}_{11}'\sin\left(\frac{4\pi}{3\tilde{a}}\tilde{\xi}-\frac{\pi}{3}\right)+\tilde{E}_0\right)^2+\frac{1}{2}\left(\tilde{U}_{10}'\sin\left(\frac{2\pi}{3\tilde{a}}\tilde{\xi}+\frac{\pi}{3}\right)\right)^2\right.\\
&\left.+\frac{3}{2}\left(\tilde{U}_{10}'\cos\left(\frac{2\pi}{3\tilde{a}}\tilde{\xi}+\frac{\pi}{3}\right)\right)^2+6\left(\tilde{U}_{11}'\right)^2+\frac{16}{\pi}\left(\tilde{U}_{10}'\tilde{U}_{11}'\right)\cos\left(\frac{2\pi}{3\tilde{a}}\tilde{\xi}+\frac{\pi}{3}\right)\right]\\
&=-\frac{\tilde{u}^2}{Hu_0^2}\left(V_{0p}^2\tilde{p}_p-\tilde{p}_eV_{0e}^2\right)\left(1+\frac{m_p}{m_e}\right)
\end{aligned}
\tag{6.5.5}
$$

Energy equation for electrons:

$$
\begin{aligned}
&\frac{\partial}{\partial\tilde{\xi}}\left[\tilde{\rho}_e\tilde{u}^2(\tilde{u}-1)+5\frac{V_{0e}^2}{u_0^2}\tilde{p}_e\tilde{u}-3\frac{V_{0e}^2}{u_0^2}\tilde{p}_e\right]-2\tilde{\rho}_e\tilde{u}E\left(\frac{\partial\tilde{\psi}}{\partial\tilde{\xi}}-\tilde{U}'_{11}\sin\left(\frac{4\pi}{3\tilde{a}}\tilde{\xi}-\frac{\pi}{3}\right)-\tilde{E}_0\right)\\
&+\frac{\partial}{\partial\tilde{\xi}}\left\{\frac{H}{\tilde{u}^2}\left[\frac{\partial}{\partial\tilde{\xi}}\left(-\tilde{\rho}_e\tilde{u}^2(1-\tilde{u})^2+7\frac{V_{0e}^2}{u_0^2}\tilde{p}_e\tilde{u}(1-\tilde{u})+3\frac{V_{0e}^2}{u_0^2}\tilde{p}_e(\tilde{u}-1)-\frac{V_{0e}^2}{u_0^2}\tilde{p}_e\tilde{u}^2-5\frac{V_{0e}^4}{u_0^4}\frac{\tilde{p}_e^2}{\tilde{\rho}_e}\right)\right.\right.\\
&\left.\left.+E\left(-2\tilde{\rho}_e\tilde{u}(1-\tilde{u})+\tilde{\rho}_e\tilde{u}^2+5\frac{V_{0e}^2}{u_0^2}\tilde{p}_e\right)\left(\frac{\partial\tilde{\psi}}{\partial\tilde{\xi}}-\tilde{U}'_{11}\sin\left(\frac{4\pi}{3\tilde{a}}\tilde{\xi}-\frac{\pi}{3}\right)-\tilde{E}_0\right)\right]\right\}\\
&+E\left(-2\frac{H}{\tilde{u}^2}\frac{\partial}{\partial\tilde{\xi}}(\tilde{\rho}_e\tilde{u}(1-\tilde{u}))+2\frac{H}{\tilde{u}^2}\frac{V_{0e}^2}{u_0^2}\frac{\partial}{\partial\tilde{\xi}}\tilde{p}_e\right)\left(\frac{\partial\tilde{\psi}}{\partial\tilde{\xi}}-\tilde{U}'_{11}\sin\left(\frac{4\pi}{3\tilde{a}}\tilde{\xi}-\frac{\pi}{3}\right)-\tilde{E}_0\right)\\
&-2E^2\frac{H}{\tilde{u}^2}\tilde{\rho}_e\left[\left(-\frac{\partial\tilde{\psi}}{\partial\tilde{\xi}}+\tilde{U}'_{11}\sin\left(\frac{4\pi}{3\tilde{a}}\tilde{\xi}-\frac{\pi}{3}\right)+\tilde{E}_0\right)^2\right.\\
&\left.+\frac{1}{2}\left(\tilde{U}'_{10}\sin\left(\frac{2\pi}{3\tilde{a}}\tilde{\xi}+\frac{\pi}{3}\right)\right)^2+\frac{3}{2}\left(\tilde{U}'_{10}\cos\left(\frac{2\pi}{3\tilde{a}}\tilde{\xi}+\frac{\pi}{3}\right)\right)^2+6\left(\tilde{U}'_{11}\right)^2+\frac{16}{\pi}\left(\tilde{U}'_{10}\tilde{U}'_{11}\right)\cos\left(\frac{2\pi}{3\tilde{a}}\tilde{\xi}+\frac{\pi}{3}\right)\right]\\
&=-\frac{\tilde{u}^2}{Hu_0^2}\left(V_{0e}^2\tilde{p}_e-V_{0p}^2\tilde{p}_p\right)\left(1+\frac{m_p}{m_e}\right)
\end{aligned}
\tag{6.5.6}
$$

Two classes of parameters were used by the mathematical modeling—parameters and scales which were not changed during calculations and varied parameters indicated in Table 6.4.

Parameters, scales, and Cauchy conditions which are common for modeling with the external field: $\frac{m_e}{m_p}=5\times10^{-5}$, the scales $\rho_0=10^{-10}\,\mathrm{g/cm^3}$, $u_0=5\times10^6\,\mathrm{cm/s}$, $V_{0e}=5\times10^6\,\mathrm{cm/s}$, $V_{0p}=5\times10^4\,\mathrm{cm/s}$, $x_0=a=0.142\,\mathrm{nm}$, $\psi_0=10^{-4}\frac{e}{a}=3.4\times10^{-6}\,CGSE_\psi$.

Dimensionless parameters $R=3\times10^{-3}$, $E=0.1$, $H=15$ (by $N_R=1$). Admit that for the lattice $U\sim V_{1,(10)}\sim V_{1,(11)}\sim\psi_0$ and choose $\tilde{U}'_{10}=10$, $\tilde{U}'_{11}=10$.

Cauchy conditions $\tilde{\rho}_e(0)=1$, $\tilde{\rho}_p(0)=2\times10^4$, $\tilde{p}_e(0)=1$, $\tilde{p}_p(0)=2\times10^4$, $\tilde{\psi}(0)=1$, $\frac{\partial\tilde{\rho}_e}{\partial\tilde{\xi}}(0)=0$, $\frac{\partial\tilde{\rho}_p}{\partial\tilde{\xi}}(0)=0$.

The external intensity of the electric field is written as $E_0=\frac{\psi_0}{x_0}\tilde{E}_0=10^{-4}\frac{e}{a^2}\tilde{E}_0=238CGSE_E\tilde{E}_0=7.14\times10^6\frac{V}{m}\tilde{E}_0$. It means that even by $\tilde{E}_0=1$ we are dealing with the rather strong fields. But namely, strong external fields can exert the influence on the soliton structures compared with the Coulomb forces in the lattice. For example, in [135] the influence of the external electric field in graphene up to 10^7-10^8 V/m. The values $\tilde{E}_0$ are indicated in Table 6.4, variants 9.0 and 9.1 respond to the extremely strong external field.

Table 6.4 contains in the first line the reminder about the first variant of calculations reflected on Figs. 6.14–6.17. These data (in the absence of the external field, $\tilde{E}_0=0$) are convenient for the following result comparison. The variants of calculations in Table 6.4 are grouped on principle of the $\tilde{E}_0$ increasing. In more details: Figs. 6.61–6.70 correspond to $\tilde{E}_0=10$, Figs. 6.71–6.80 correspond to $\tilde{E}_0=100$, Figs. 6.81–6.92 correspond to $\tilde{E}_0=10{,}000$.

TABLE 6.4 Varied parameters in calculations with the external electric field

Variant No	$\tilde{E}_0$	$\frac{\partial\tilde{\psi}}{\partial\tilde{\xi}}(0)$	$\frac{\partial\tilde{p}_p}{\partial\tilde{\xi}}(0)$	$\frac{\partial\tilde{p}_e}{\partial\tilde{\xi}}(0)$
1	0	0	0	0
7.0	10	10	0	0
7.1	10	10	10	−1
8.0	100	100	0	0
8.1	100	100	10	0
9.0	10,000	10,000	0	0
9.1	10,000	10,000	10	−1

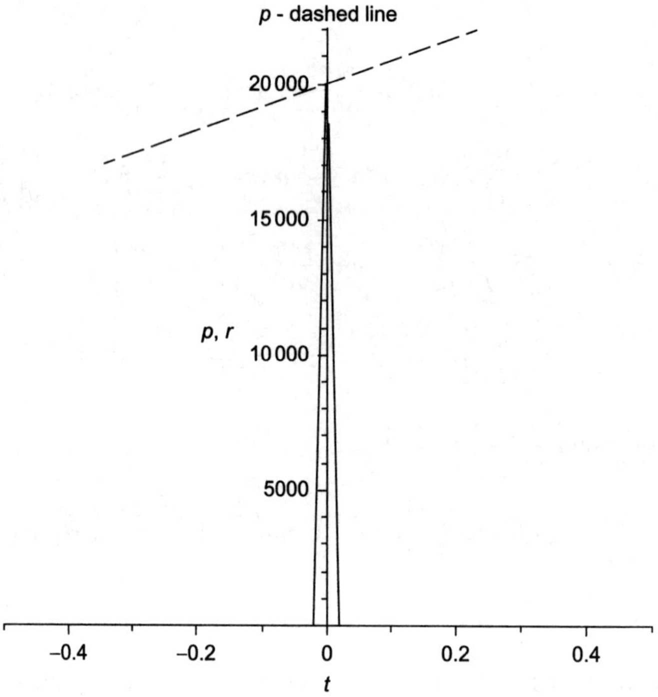

FIGURE 6.61 *r*—the positive particles density, (solid line); *p*—the positive particles pressure. (Variant 7.0).

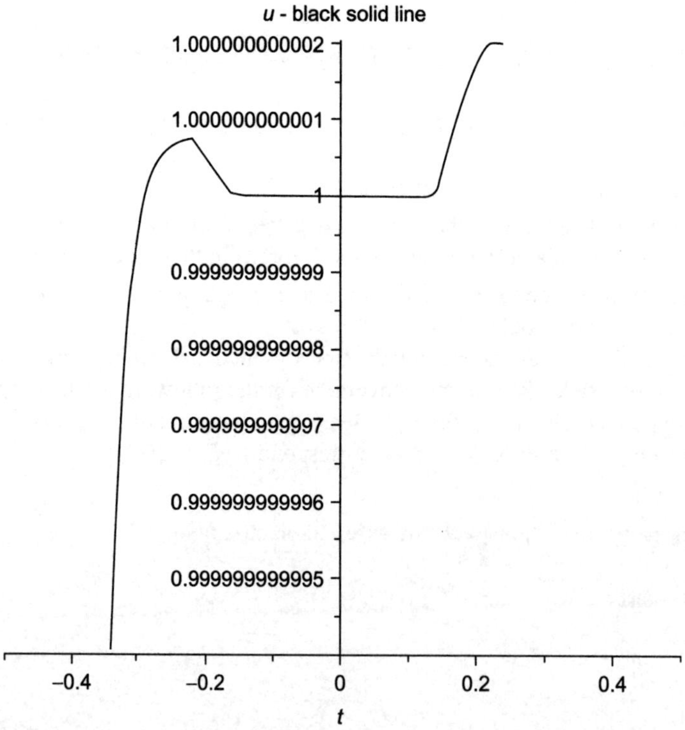

FIGURE 6.62 *u*—velocity $\tilde{u}$. (Variant 7.0).

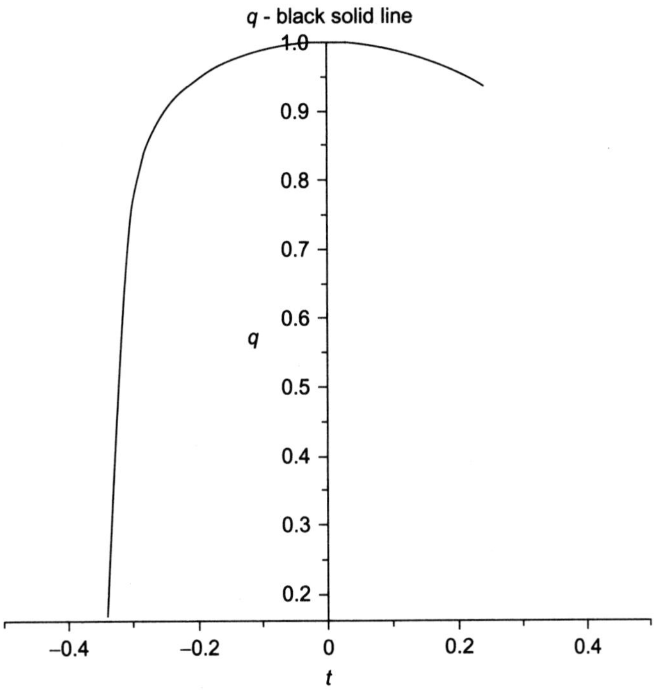

FIGURE 6.63 q—the negative particles pressure. (Variant 7.0).

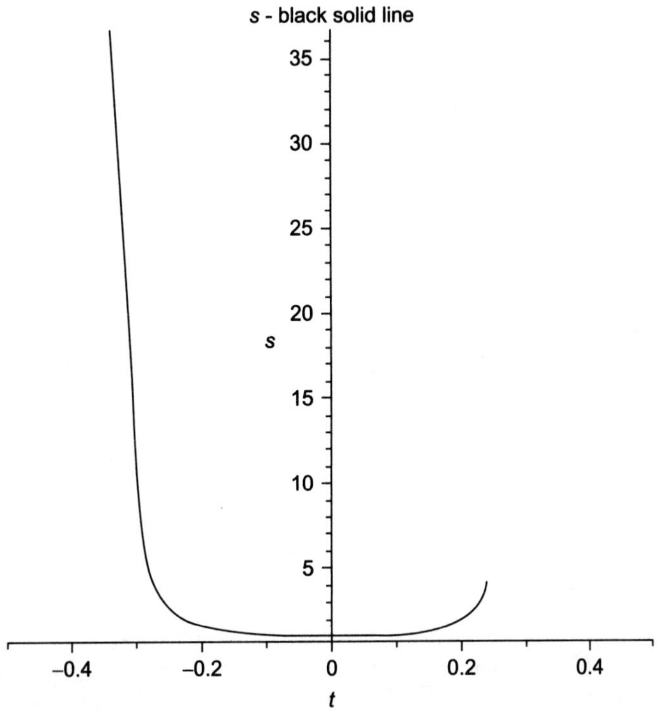

FIGURE 6.64 s—electron density $\widetilde{\rho}_e$, (Variant 7.0).

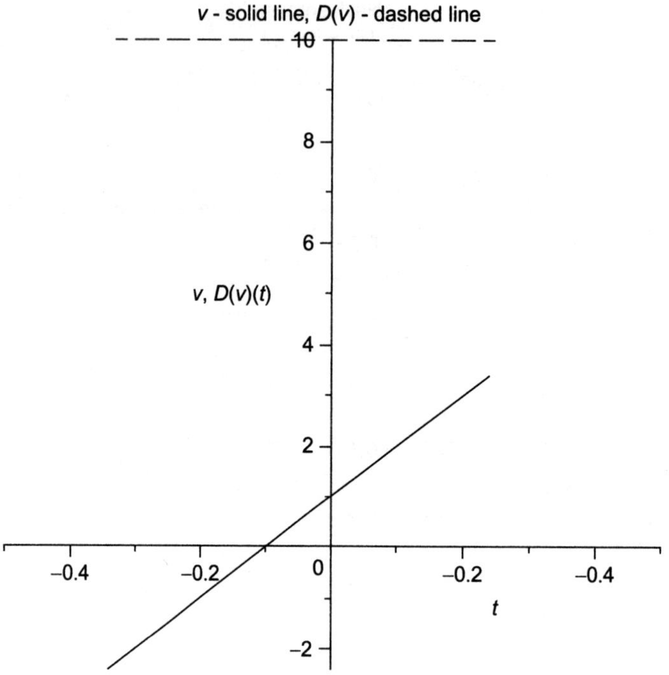

FIGURE 6.65 v—potential $\widetilde{\psi}$ (solid line); $D(v)(t)$ (Variant 7.0).

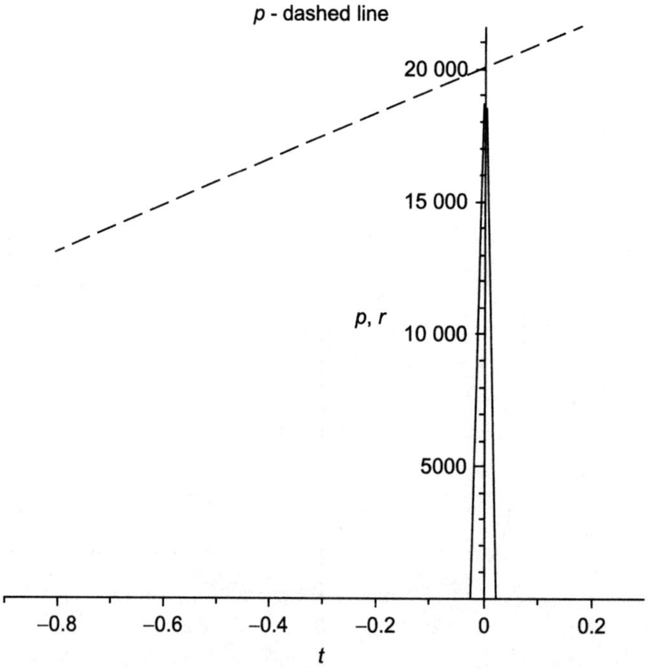

FIGURE 6.66 r—the positive particles density, (solid line); p—the positive particles pressure. (Variant 7.1).

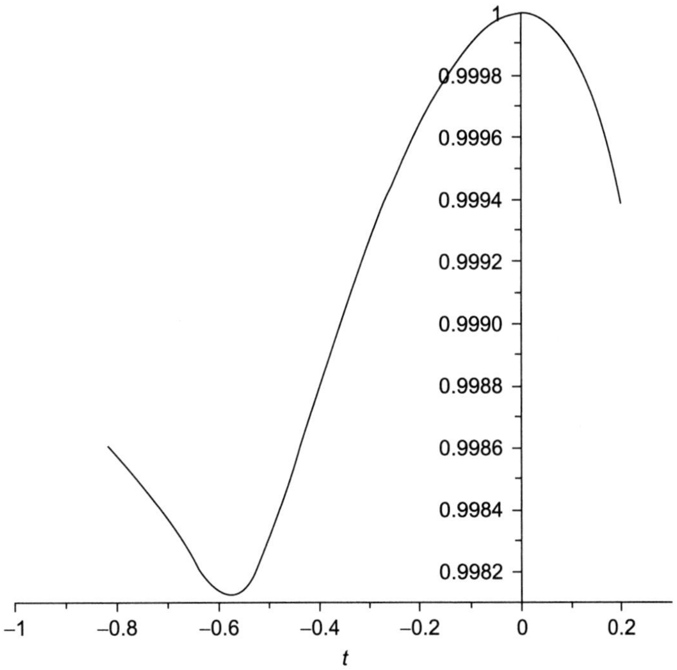

FIGURE 6.67 *u*—velocity $\widetilde{u}$. (Variant 7.1).

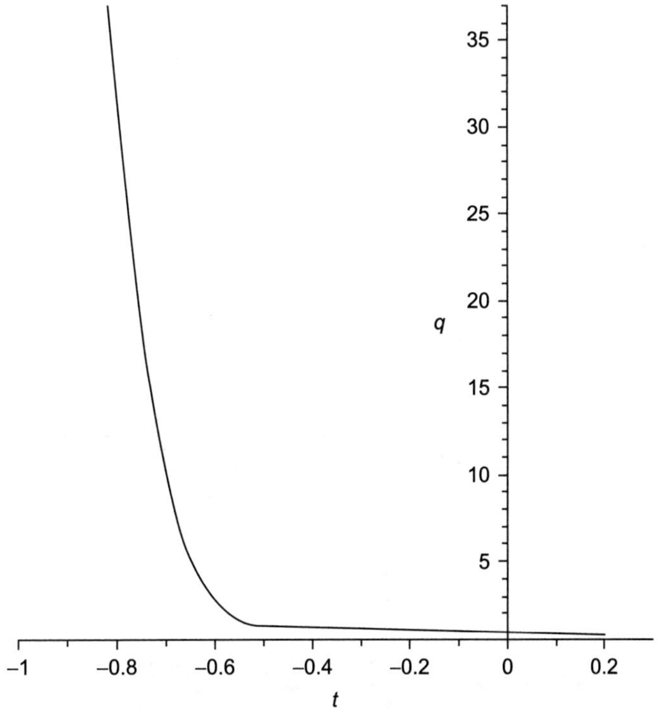

FIGURE 6.68 *q*—the negative particles pressure. (Variant 7.1).

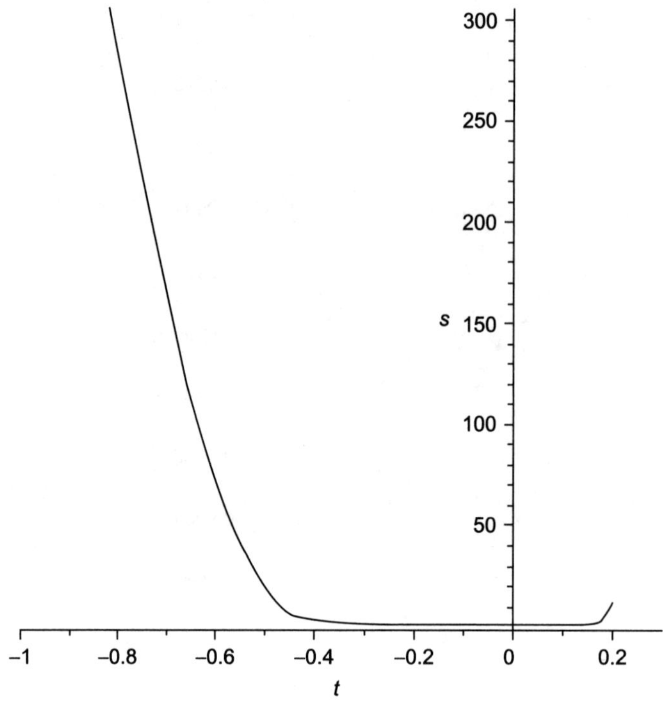

FIGURE 6.69 s—electron density $\widetilde{\rho}_e$, (Variant 7.1).

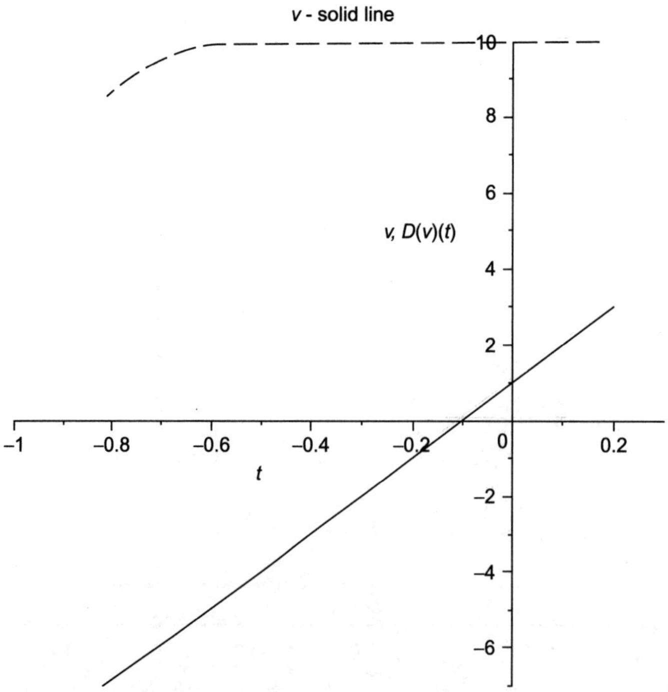

FIGURE 6.70 v—potential $\widetilde{\psi}$ (solid line); D(v)(t), (Variant 7.1).

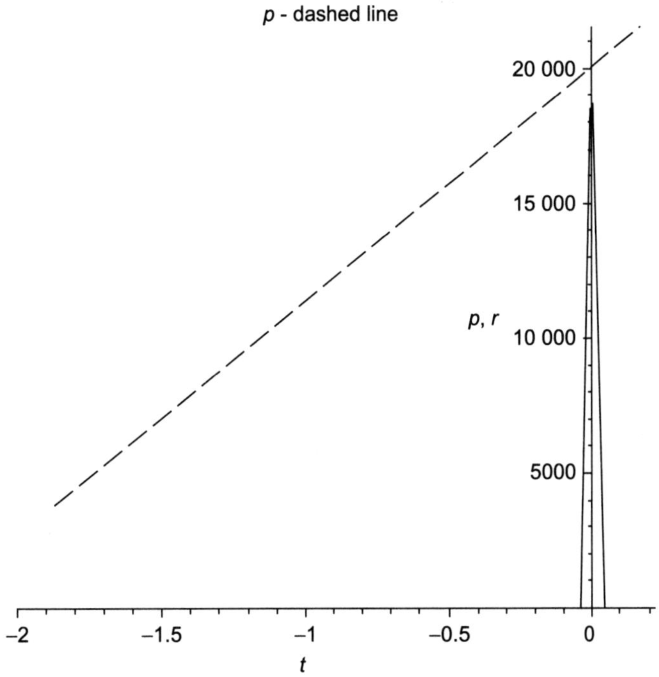

FIGURE 6.71 *r*—the positive particles density, (solid line); *p*—the positive particles pressure. (Variant 8.0).

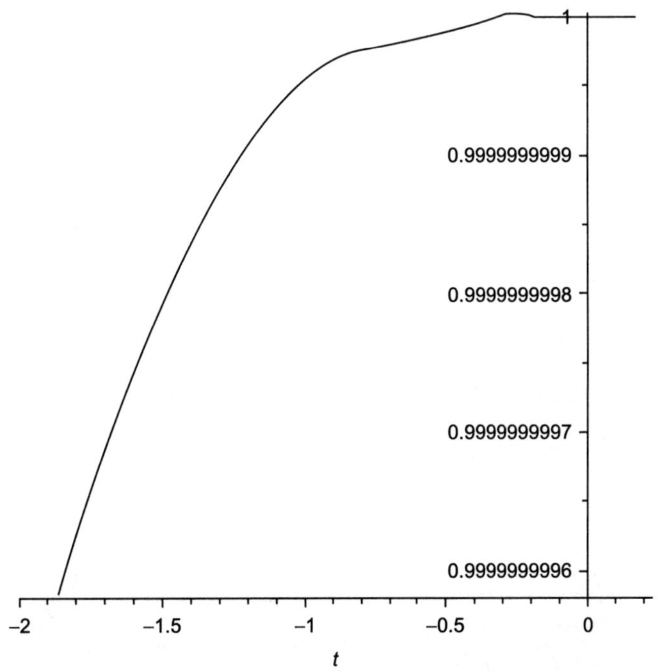

FIGURE 6.72 *u*—velocity $\widetilde{u}$. (Variant 8.0).

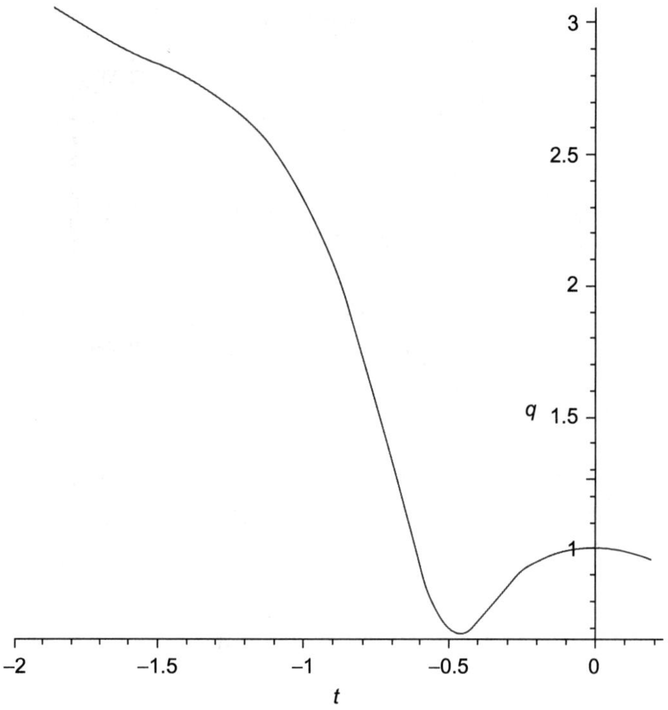

FIGURE 6.73 *q*—the negative particles pressure. (Variant 8.0).

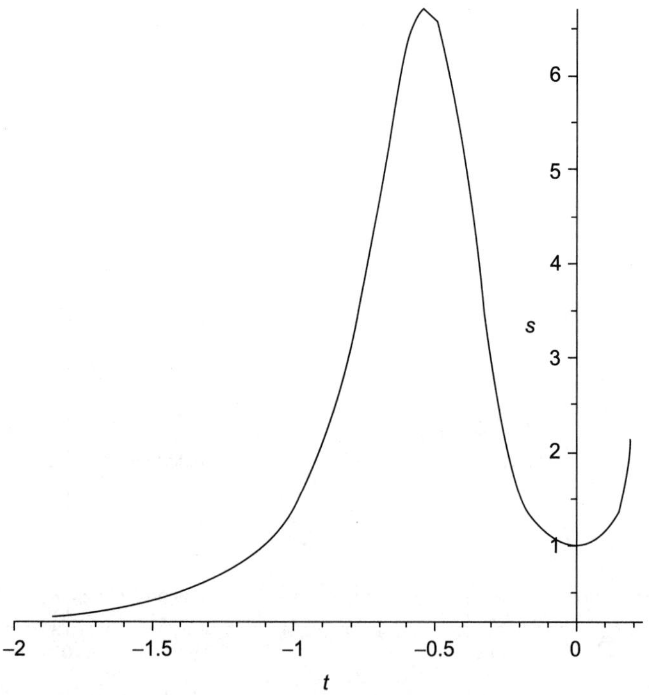

FIGURE 6.74 *s*—electron density $\widetilde{p}_e$, (Variant 8.0).

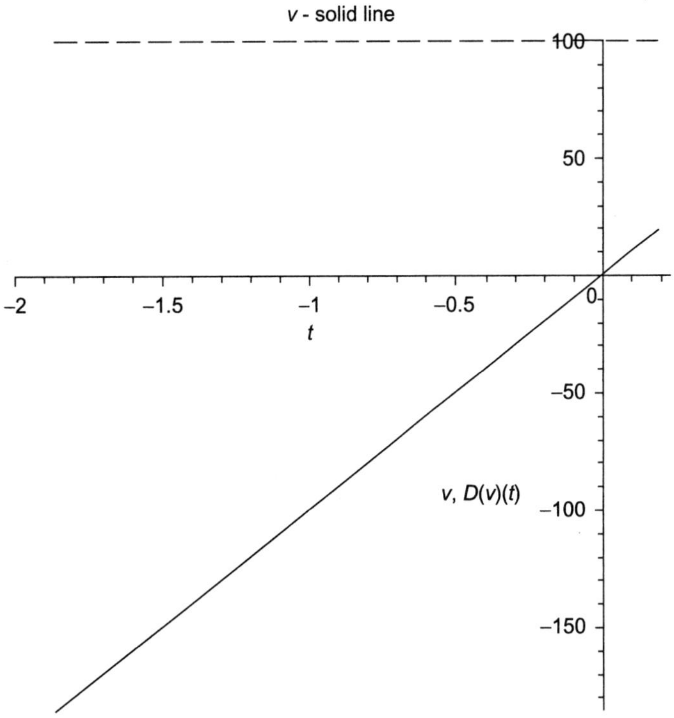

FIGURE 6.75 v—potential $\widetilde{\psi}$ (solid line); $D(v)(t)$, (Variant 8.0).

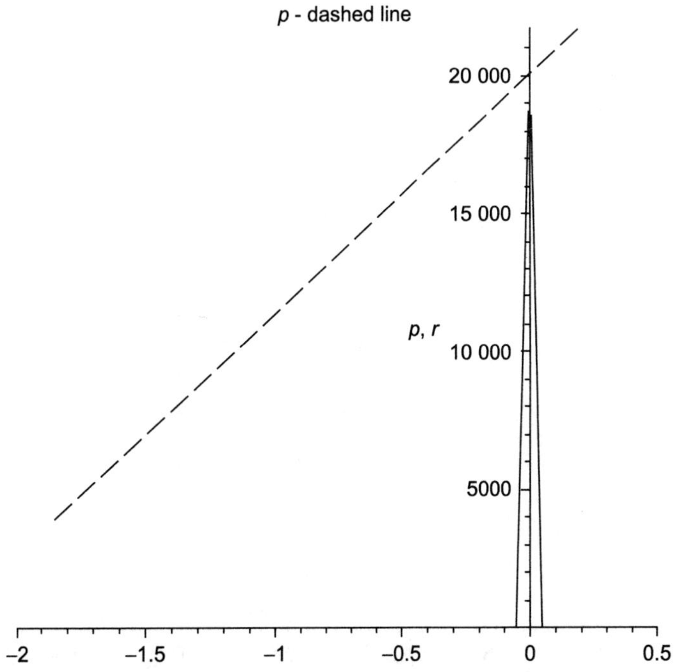

FIGURE 6.76 r—the positive particles density, (solid line); p—the positive particles pressure. (Variant 8.1).

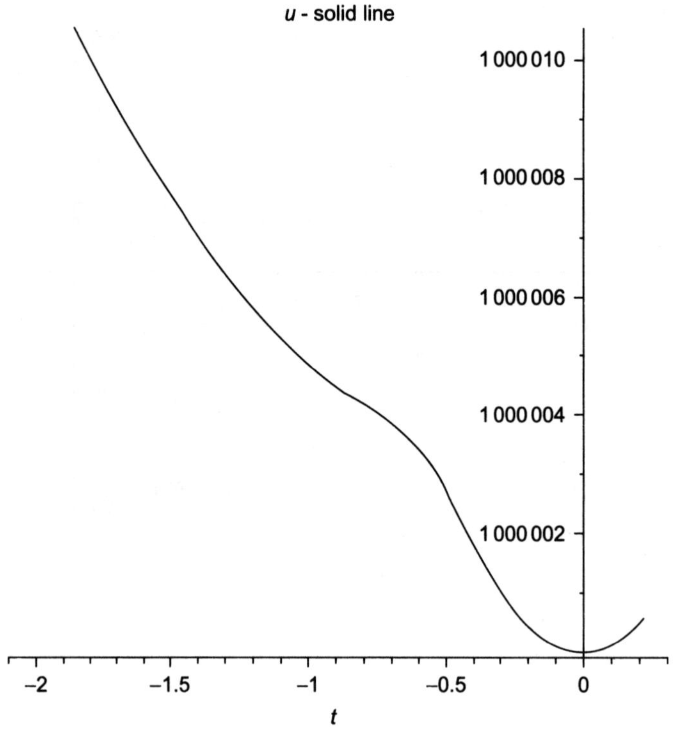

FIGURE 6.77 *u*—velocity $\widetilde{u}$. (Variant 8.1).

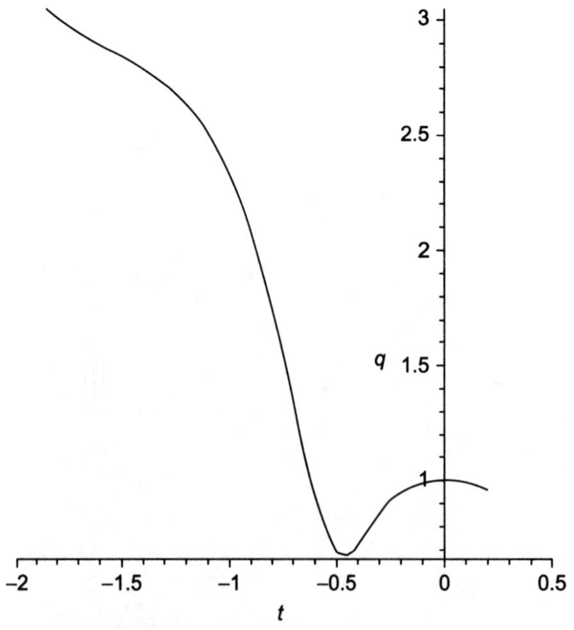

FIGURE 6.78 *q*—the negative particles pressure. (Variant 8.1).

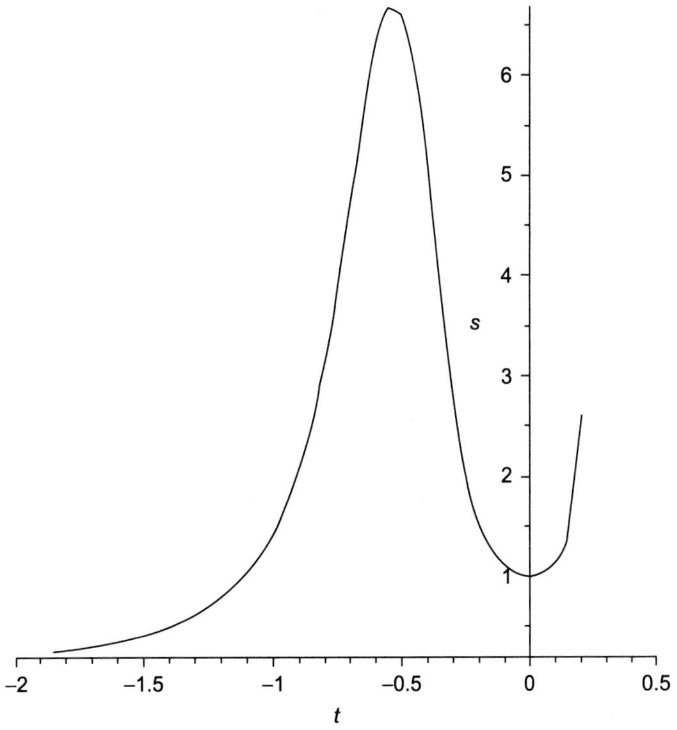

FIGURE 6.79 *s*—electron density $\widetilde{\rho}_e$, (Variant 8.1).

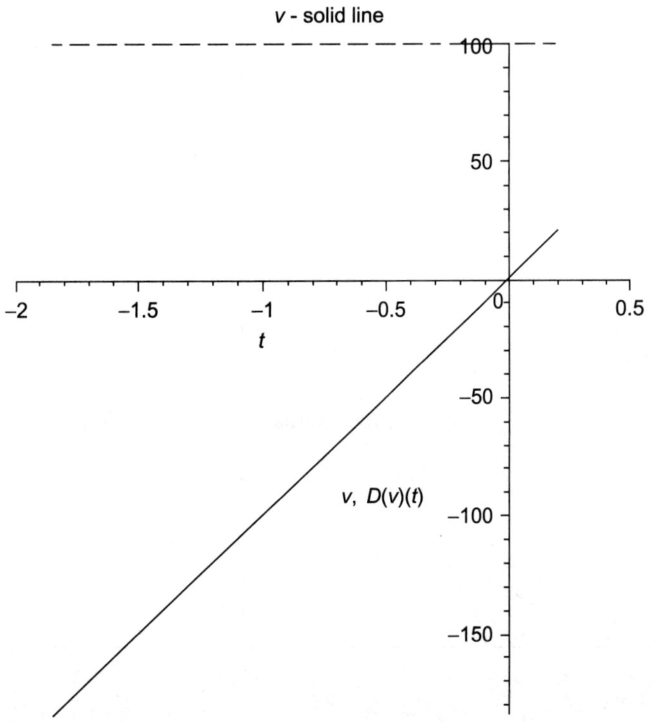

FIGURE 6.80 *v*—potential $\widetilde{\psi}$ (solid line); *D(v)(t)*, (Variant 8.1).

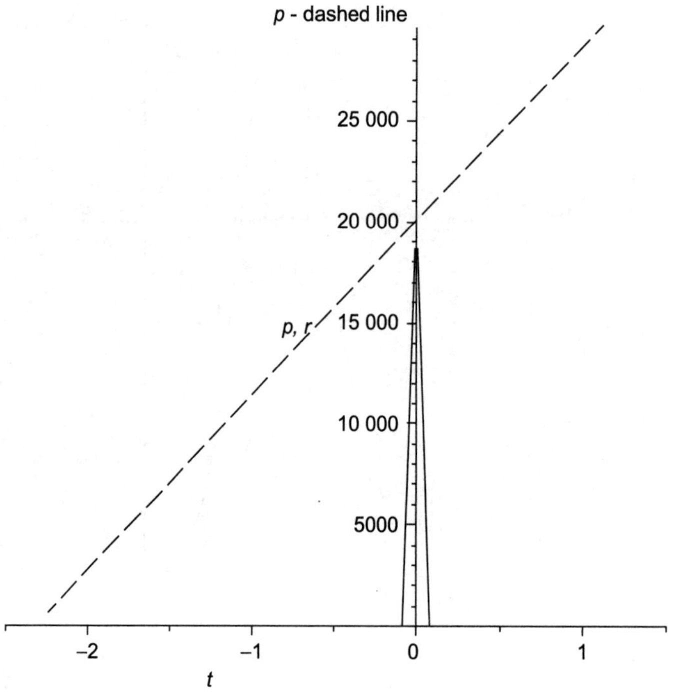

FIGURE 6.81 *r*—the positive particles density, (solid line); *p*—the positive particles pressure. (Variant 9.0).

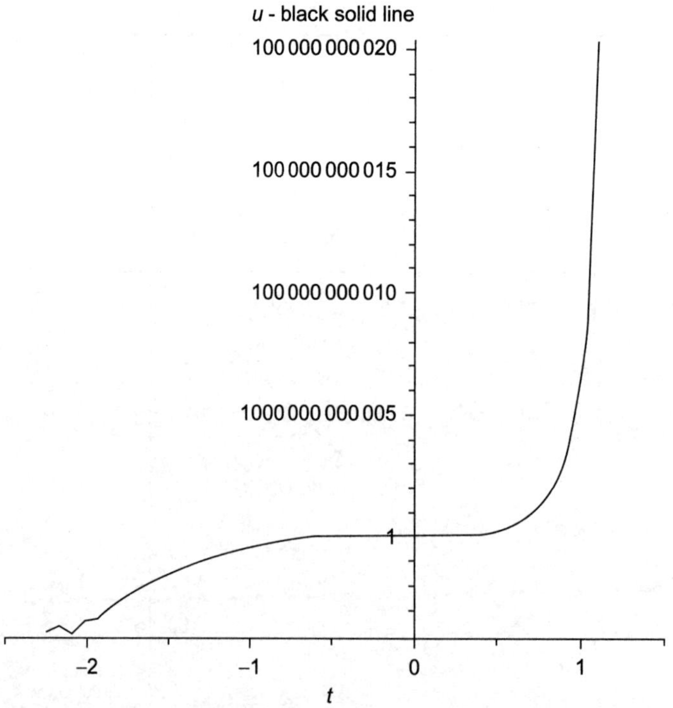

FIGURE 6.82 *u*—velocity $\tilde{u}$. (Variant 9.0).

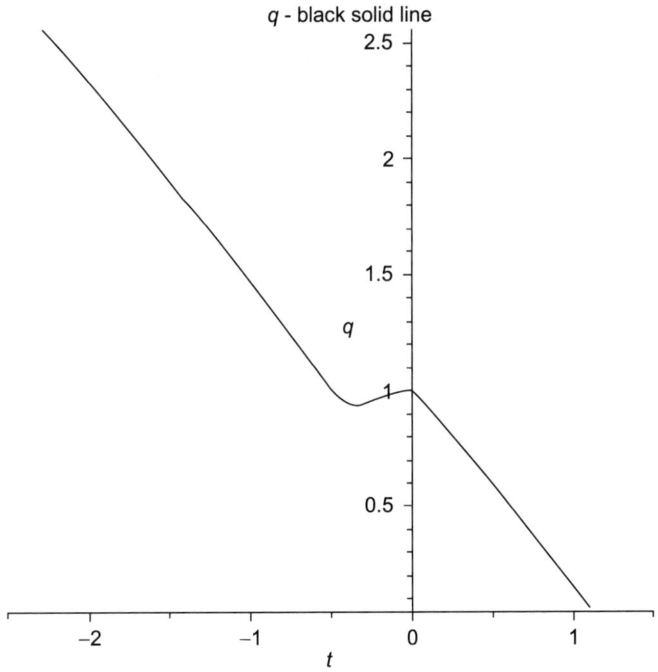

FIGURE 6.83 q—the negative particles pressure. (Variant 9.0).

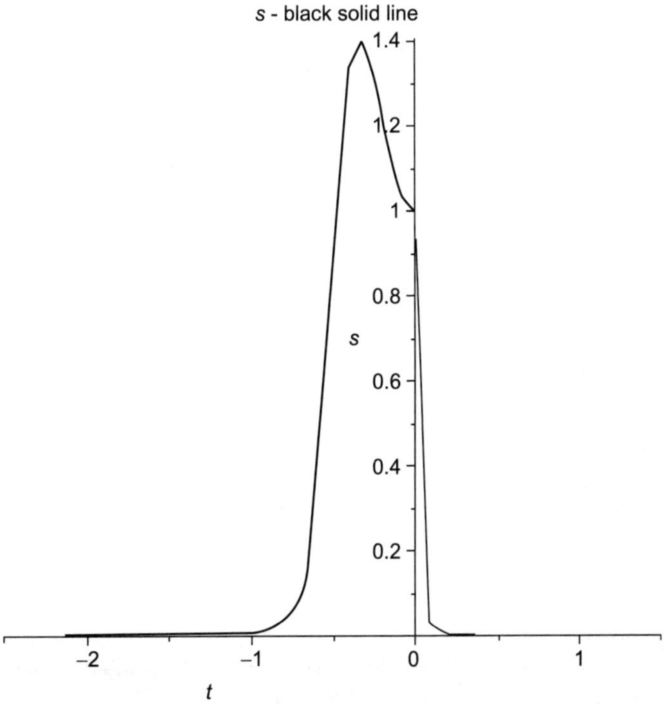

FIGURE 6.84 s—electron density $\widetilde{\rho}_e$, (Variant 9.0).

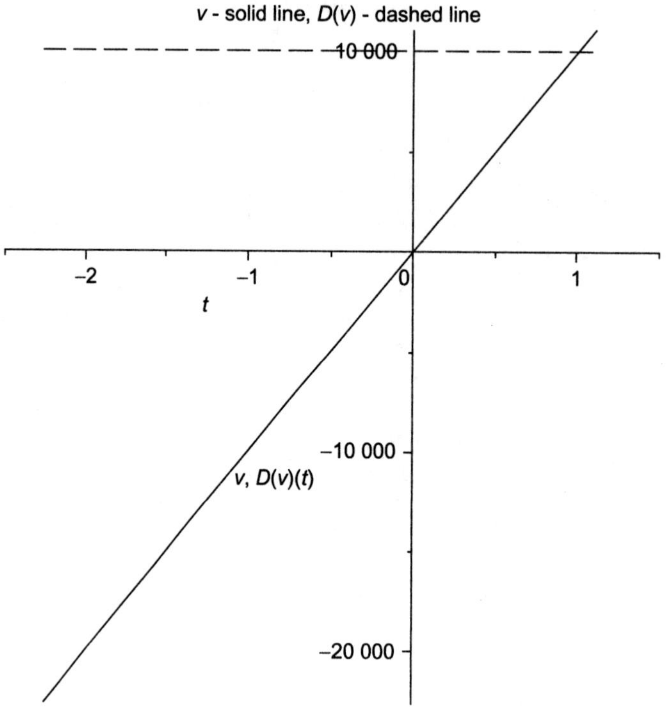

FIGURE 6.85 v—potential $\widetilde{\psi}$ (solid line); $D(v)(t)$, (Variant 9.0).

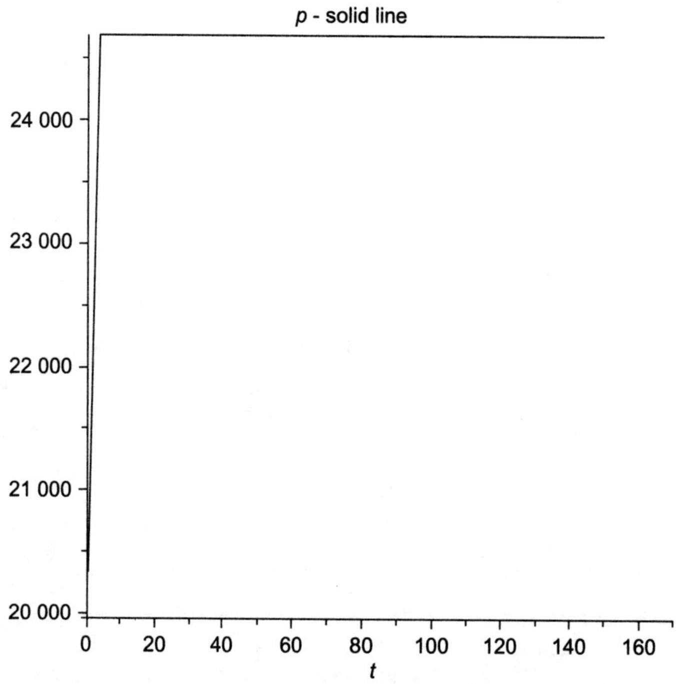

FIGURE 6.86 p—the positive particles pressure. (Variant 9.1).

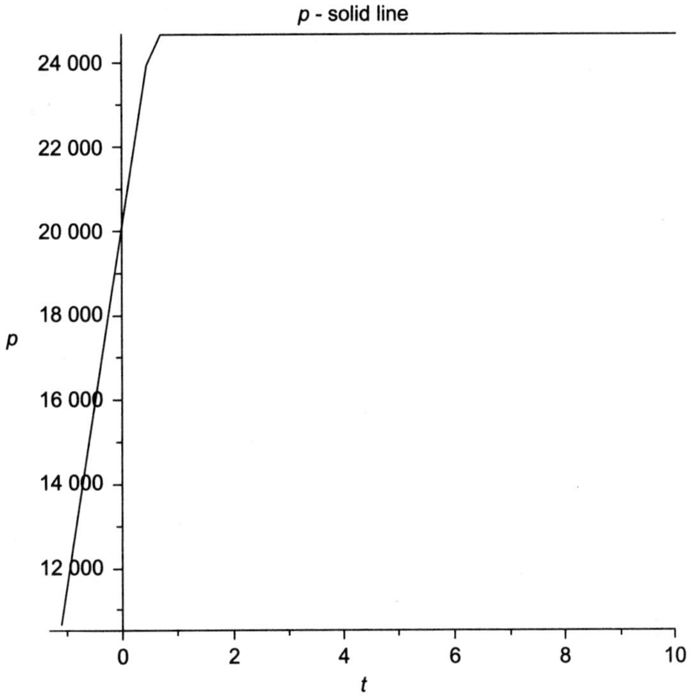

FIGURE 6.87 p—the positive particles pressure. (Variant 9.1).

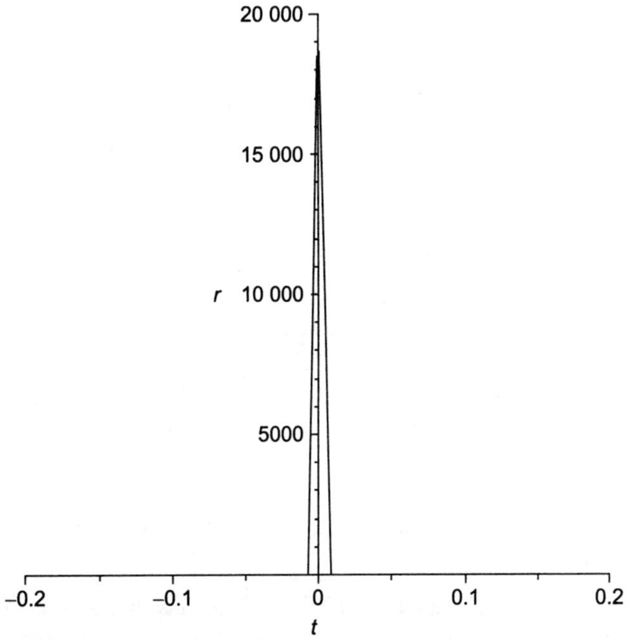

FIGURE 6.88 r—the positive particles density, (Variant 9.1).

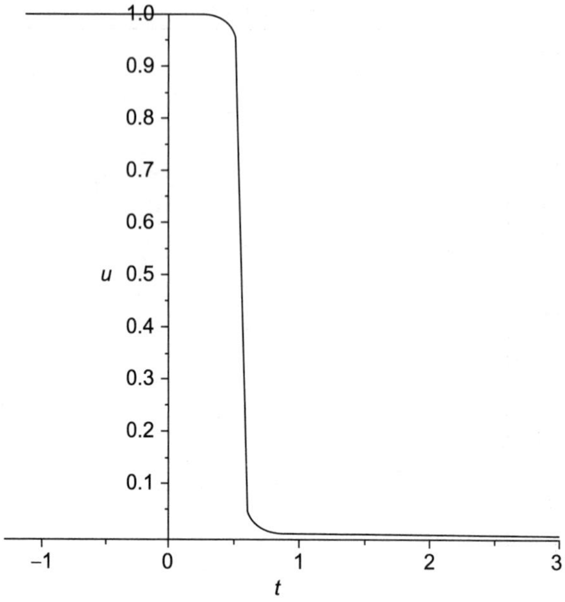

FIGURE 6.89 *u*—velocity $\tilde{u}$. (Variant 9.1).

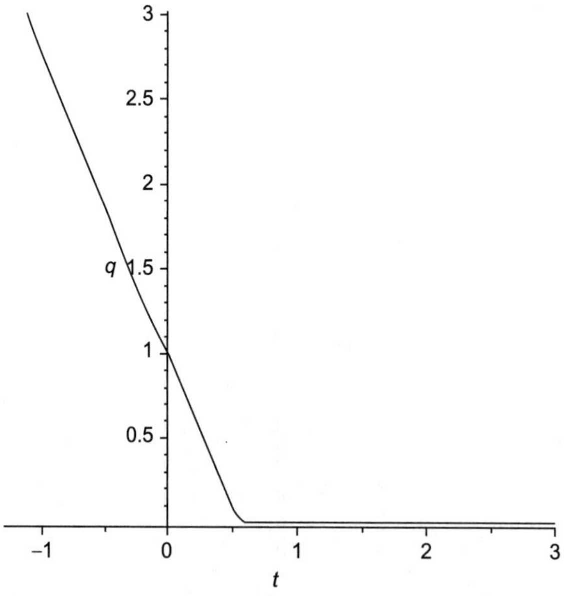

FIGURE 6.90 *q*—the negative particles pressure. (Variant 9.1).

Consider now the character features of the soliton evolution and the change of the charge distribution in solitons with growing of the external field intensity:

1. The character soliton size is defined by the area where $\tilde{u}=1$. It means that all part of the soliton wave are moving without destruction. The size of this area is practically independent on choosing of the numerical method of calculations.
2. Figures 6.87–6.89 demonstrate the typical situation when the area of possible numerical calculations for a physical variable does not coincide with area $\tilde{u}=1$ where the soliton regime exists.
3. In the area of the soliton existence, the condition $\tilde{u}=1$ is fulfilled with the high accuracy defined practically by accuracy of the chosen numerical method (see Figs. 6.62, 6.67, 6.72, 6.77, 6.82, and 6.89).

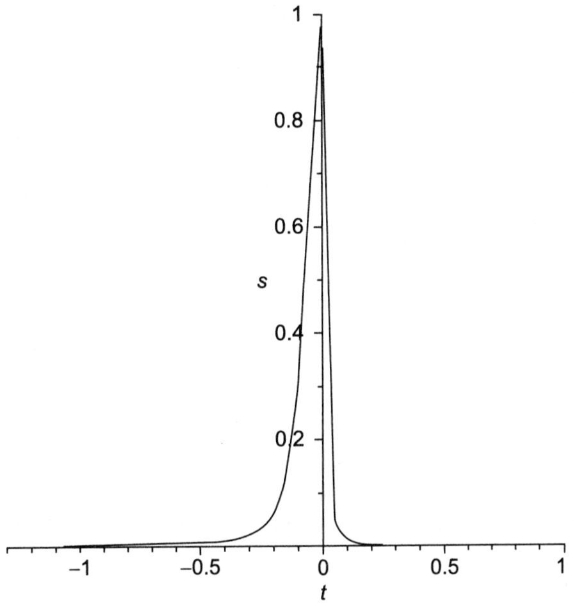

FIGURE 6.91 *s*—electron density $\widetilde{\rho}_e$ (Variant 9.1).

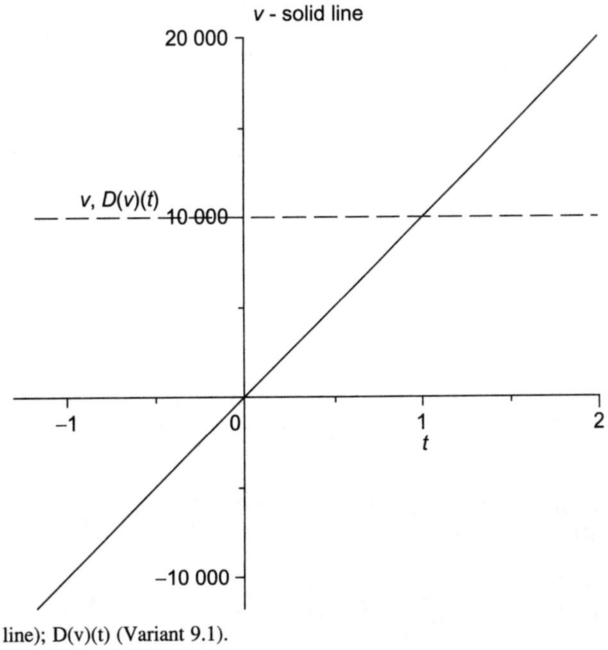

FIGURE 6.92 *v*—potential $\widetilde{\psi}$ (solid line); D(v)(t) (Variant 9.1).

4. As a rule for the chosen topology of the electric field the size of the soliton existence is growing with increasing of the electric field intensity.
5. Under the influence of the external electric field the captured electron cloud is displacing in the opposite direction (of the negative variable $\widetilde{\xi}$). The soliton kernel is loosing its symmetry.
6. The redistribution of the self-consistent effective charge creates the self-consistence field with the opposite (to the external field) direction, (see Figs. 6.65, 6.70, 6.75, 6.80, 6.85, and 6.92).
7. The quantum pressure of the positive particle is growing with the $\widetilde{\xi}$ increase. On the whole, the specific features of the $\widetilde{p},\widetilde{q}$ pressures are defined by the process of the soliton formation.

6.6 SPIN EFFECTS IN THE GENERALIZED QUANTUM HYDRODYNAMIC EQUATIONS

The influence of spin and magnetic moment of particles can be taken into account by the natural elegant way via the internal energy of particles. For the so-called "elementary particles" the internal energy can contain the spin and magnetic parts. For example electron has the internal energy ε (see also (4.3.1)–(4.3.3))

$$\varepsilon = \varepsilon_{el,sp} + \varepsilon_{el,m}, \tag{6.6.1}$$

which contains the spin and magnetic parts, namely

$$\varepsilon_{el,sp} = \hbar\omega/2, \quad \varepsilon_{el,m} = -\mathbf{p_m}\cdot\mathbf{B}, \tag{6.6.2}$$

where $\mathbf{p}_{\mathrm{m}}$ – electron magnetic moment, $\mathbf{B}$ – magnetic induction. But $p_m = -\frac{e}{m_e}\frac{\hbar}{2c}$, then $\varepsilon_{\mathrm{el}} = \frac{\hbar}{2}\omega_{\mathrm{eff}}$. Relation (6.6.1) can be written as

$$\varepsilon = \frac{\hbar}{2}\left[\omega \pm \frac{e}{m_e c}B\right], \tag{6.6.3}$$

if $\mathbf{B}$ is directed along the spin direction. On this stage of investigations, we omit the influence of the internal energy of particles, therefore spin waves will be investigated separately.

Let us rewrite the system of the generalized hydrodynamic equations (GHEs) taking into account the spin effects: (continuity equation for species α)

$$\begin{aligned}&\frac{\partial}{\partial t}\left\{\rho_\alpha - \tau_\alpha\left[\frac{\partial\rho_\alpha}{\partial t} + \frac{\partial}{\partial\mathbf{r}}\cdot(\rho_\alpha\mathbf{v}_0)\right]\right\} + \frac{\partial}{\partial\mathbf{r}}\cdot\left\{\rho_\alpha\mathbf{v}_0 - \tau_\alpha\left[\frac{\partial}{\partial t}(\rho_\alpha\mathbf{v}_0) + \frac{\partial}{\partial\mathbf{r}}\cdot(\rho_\alpha\mathbf{v}_0\mathbf{v}_0) + \overleftrightarrow{\mathbf{I}}\cdot\frac{\partial p_\alpha}{\partial\mathbf{r}}\right.\right.\\&\left.\left. - \rho_\alpha\mathbf{F}_\alpha^{(1)} - \frac{q_\alpha}{m_\alpha}\rho_\alpha\mathbf{v}_0\times\mathbf{B}\right]\right\} = R_\alpha,\end{aligned} \tag{6.6.4}$$

(continuity equation for mixture)

$$\begin{aligned}&\frac{\partial}{\partial t}\left\{\rho - \sum_\alpha\tau_\alpha\left[\frac{\partial\rho_\alpha}{\partial t} + \frac{\partial}{\partial\mathbf{r}}\cdot(\rho_\alpha\mathbf{v}_0)\right]\right\} + \frac{\partial}{\partial\mathbf{r}}\cdot\left\{\rho\mathbf{v}_0 - \sum_\alpha\tau_\alpha\left[\frac{\partial}{\partial t}(\rho_\alpha\mathbf{v}_0) + \frac{\partial}{\partial\mathbf{r}}\cdot(\rho_\alpha\mathbf{v}_0\mathbf{v}_0)\right.\right.\\&\left.\left. + \overleftrightarrow{\mathbf{I}}\cdot\frac{\partial p_\alpha}{\partial\mathbf{r}} - \rho_\alpha\mathbf{F}_\alpha^{(1)} - \frac{q_\alpha}{m_\alpha}\rho_\alpha\mathbf{v}_0\times\mathbf{B}\right]\right\} = 0,\end{aligned} \tag{6.6.5}$$

(momentum equation for species)

$$\begin{aligned}&\frac{\partial}{\partial t}\left\{\rho_\alpha\mathbf{v}_0 - \tau_\alpha\left[\frac{\partial}{\partial t}(\rho_\alpha\mathbf{v}_0) + \frac{\partial}{\partial\mathbf{r}}\cdot\rho_\alpha\mathbf{v}_0\mathbf{v}_0 + \frac{\partial p_\alpha}{\partial\mathbf{r}} - \rho_\alpha\mathbf{F}_\alpha^{(1)}\right.\right.\\&\left.\left. - \frac{q_\alpha}{m_\alpha}\rho_\alpha\mathbf{v}_0\times\mathbf{B}\right]\right\} - \mathbf{F}_\alpha^{(1)}\left[\rho_\alpha - \tau_\alpha\left(\frac{\partial\rho_\alpha}{\partial t} + \frac{\partial}{\partial\mathbf{r}}\cdot(\rho_\alpha\mathbf{v}_0)\right)\right]\\&- \frac{q_\alpha}{m_\alpha}\left\{\rho_\alpha\mathbf{v}_0 - \tau_\alpha\left[\frac{\partial}{\partial t}(\rho_\alpha\mathbf{v}_0) + \frac{\partial}{\partial\mathbf{r}}\cdot\rho_\alpha\mathbf{v}_0\mathbf{v}_0 + \frac{\partial p_\alpha}{\partial\mathbf{r}} - \rho_\alpha\mathbf{F}_\alpha^{(1)}\right.\right.\\&\left.\left. - \frac{q_\alpha}{m_\alpha}\rho_\alpha\mathbf{v}_0\times\mathbf{B}\right]\right\}\times\mathbf{B} + \frac{\partial}{\partial\mathbf{r}}\cdot\left\{\rho_\alpha\mathbf{v}_0\mathbf{v}_0 + p_\alpha\overleftrightarrow{\mathbf{I}} - \tau_\alpha\left[\frac{\partial}{\partial t}\left(\rho_\alpha\mathbf{v}_0\mathbf{v}_0\right.\right.\right.\\&\left. + p_\alpha\overleftrightarrow{\mathbf{I}}\right) + \frac{\partial}{\partial\mathbf{r}}\cdot\rho_\alpha(\mathbf{v}_0\mathbf{v}_0)\mathbf{v}_0 + 2\overleftrightarrow{\mathbf{I}}\left(\frac{\partial}{\partial\mathbf{r}}\cdot(p_\alpha\mathbf{v}_0)\right) + \frac{\partial}{\partial\mathbf{r}}\cdot\left(\overleftrightarrow{\mathbf{I}}p_\alpha\mathbf{v}_0\right)\\&\left.\left. - \mathbf{F}_\alpha^{(1)}\rho_\alpha\mathbf{v}_0 - \rho_\alpha\mathbf{v}_0\mathbf{F}_\alpha^{(1)} - \frac{q_\alpha}{m_\alpha}\rho_\alpha[\mathbf{v}_0\times\mathbf{B}]\mathbf{v}_0 - \frac{q_\alpha}{m_\alpha}\rho_\alpha\mathbf{v}_0[\mathbf{v}_0\times\mathbf{B}]\right]\right\}\\&= \int m_\alpha\mathbf{v}_\alpha J_\alpha^{st,el}d\mathbf{v}_\alpha + \int m_\alpha\mathbf{v}_\alpha J_\alpha^{st,inel}d\mathbf{v}_\alpha,\end{aligned} \tag{6.6.6}$$

(generalized moment equation for mixture)

$$\begin{aligned}
&\frac{\partial}{\partial t}\left\{\rho\mathbf{v}_0-\sum_\alpha\tau_\alpha\left[\frac{\partial}{\partial t}(\rho_\alpha\mathbf{v}_0)+\frac{\partial}{\partial\mathbf{r}}\cdot\rho_\alpha\mathbf{v}_0\mathbf{v}_0+\frac{\partial p_\alpha}{\partial\mathbf{r}}-\rho_\alpha\mathbf{F}_\alpha^{(1)}\right.\right.\\
&\left.\left.-\frac{q_\alpha}{m_\alpha}\rho_\alpha\mathbf{v}_0\times\mathbf{B}\right]\right\}-\sum_\alpha\mathbf{F}_\alpha^{(1)}\left[\rho_\alpha-\tau_\alpha\left(\frac{\partial\rho_\alpha}{\partial t}+\frac{\partial}{\partial\mathbf{r}}(\rho_\alpha\mathbf{v}_0)\right)\right]\\
&-\sum_\alpha\frac{q_\alpha}{m_\alpha}\left\{\rho_\alpha\mathbf{v}_0-\tau_\alpha^{(0)}\left[\frac{\partial}{\partial t}(\rho_\alpha\mathbf{v}_0)+\frac{\partial}{\partial\mathbf{r}}\cdot\rho_\alpha\mathbf{v}_0\mathbf{v}_0+\frac{\partial p_\alpha}{\partial\mathbf{r}}-\rho_\alpha\mathbf{F}_\alpha^{(1)}\right.\right.\\
&\left.\left.-\frac{q_\alpha}{m_\alpha}\rho_\alpha\mathbf{v}_0\times\mathbf{B}\right]\right\}\times\mathbf{B}+\frac{\partial}{\partial\mathbf{r}}\cdot\left\{\rho\mathbf{v}_0\mathbf{v}_0+p\overleftrightarrow{\mathrm{I}}-\sum_\alpha\tau_\alpha\left[\frac{\partial}{\partial t}\left(\rho_\alpha\mathbf{v}_0\mathbf{v}_0\right.\right.\right.\\
&\left.+p_\alpha\overleftrightarrow{\mathrm{I}}\right)+\frac{\partial}{\partial\mathbf{r}}\cdot\rho_\alpha(\mathbf{v}_0\mathbf{v}_0)\mathbf{v}_0+2\overleftrightarrow{\mathrm{I}}\left(\frac{\partial}{\partial\mathbf{r}}\cdot(p_\alpha\mathbf{v}_0)\right)+\frac{\partial}{\partial\mathbf{r}}\cdot\left(\overleftrightarrow{\mathrm{I}}p_\alpha\mathbf{v}_0\right)\\
&\left.\left.-\mathbf{F}_\alpha^{(1)}\rho_\alpha\mathbf{v}_0-\rho_\alpha\mathbf{v}_0\mathbf{F}_\alpha^{(1)}-\frac{q_\alpha}{m_\alpha}\rho_\alpha[\mathbf{v}_0\times\mathbf{B}]\mathbf{v}_0-\frac{q_\alpha}{m_\alpha}\rho_\alpha\mathbf{v}_0[\mathbf{v}_0\times\mathbf{B}]\right]\right\}=0,
\end{aligned}\tag{6.6.7}$$

(energy equation for component)

$$\begin{aligned}
&\frac{\partial}{\partial t}\left\{\frac{\rho_\alpha v_0^2}{2}+\frac{3}{2}p_\alpha+\varepsilon_\alpha n_\alpha-\tau_\alpha\left[\frac{\partial}{\partial t}\left(\frac{\rho_\alpha v_0^2}{2}+\frac{3}{2}p_\alpha+\varepsilon_\alpha n_\alpha\right)\right.\right.\\
&\left.\left.+\frac{\partial}{\partial\mathbf{r}}\cdot\left(\frac{1}{2}\rho_\alpha v_0^2\mathbf{v}_0+\frac{5}{2}p_\alpha\mathbf{v}_0+\varepsilon_\alpha n_\alpha\mathbf{v}_0\right)-\mathbf{F}_\alpha^{(1)}\cdot\rho_\alpha\mathbf{v}_0\right]\right\}\\
&+\frac{\partial}{\partial\mathbf{r}}\cdot\left\{\frac{1}{2}\rho_\alpha v_0^2\mathbf{v}_0+\frac{5}{2}p_\alpha\mathbf{v}_0+\varepsilon_\alpha n_\alpha\mathbf{v}_0-\tau_\alpha\left[\frac{\partial}{\partial t}\left(\frac{1}{2}\rho_\alpha v_0^2\mathbf{v}_0\right.\right.\right.\\
&\left.+\frac{5}{2}p_\alpha\mathbf{v}_0+\varepsilon_\alpha n_\alpha\mathbf{v}_0\right)+\frac{\partial}{\partial\mathbf{r}}\cdot\left(\frac{1}{2}\rho_\alpha v_0^2\mathbf{v}_0\mathbf{v}_0+\frac{7}{2}p_\alpha\mathbf{v}_0\mathbf{v}_0+\frac{1}{2}p_\alpha v_0^2\overleftrightarrow{\mathrm{I}}\right.\\
&\left.+\frac{5p_\alpha^2}{2\rho_\alpha}\overleftrightarrow{\mathrm{I}}+\varepsilon_\alpha n_\alpha\mathbf{v}_0\mathbf{v}_0+\varepsilon_\alpha\frac{p_\alpha}{m_\alpha}\overleftrightarrow{\mathrm{I}}\right)-\rho_\alpha\mathbf{F}_\alpha^{(1)}\cdot\mathbf{v}_0\mathbf{v}_0-p_\alpha\mathbf{F}_\alpha^{(1)}\cdot\overleftrightarrow{\mathrm{I}}\\
&-\frac{1}{2}\rho_\alpha v_0^2\mathbf{F}_\alpha^{(1)}-\frac{3}{2}\mathbf{F}_\alpha^{(1)}p_\alpha-\frac{\rho_\alpha v_0^2}{2}\frac{q_\alpha}{m_\alpha}[\mathbf{v}_0\times\mathbf{B}]-\frac{5}{2}p_\alpha\frac{q_\alpha}{m_\alpha}[\mathbf{v}_0\times\mathbf{B}]\\
&\left.\left.-\varepsilon_\alpha n_\alpha\frac{q_\alpha}{m_\alpha}[\mathbf{v}_0\times\mathbf{B}]-\varepsilon_\alpha n_\alpha\mathbf{F}_\alpha^{(1)}\right]\right\}-\left\{\rho_\alpha\mathbf{F}_\alpha^{(1)}\cdot\mathbf{v}_0\right.\\
&\left.-\tau_\alpha\left[\mathbf{F}_\alpha^{(1)}\cdot\left(\frac{\partial}{\partial t}(\rho_\alpha\mathbf{v}_0)+\frac{\partial}{\partial\mathbf{r}}\cdot\rho_\alpha\mathbf{v}_0\mathbf{v}_0+\frac{\partial}{\partial\mathbf{r}}\cdot p_\alpha\overleftrightarrow{\mathrm{I}}-\rho_\alpha\mathbf{F}_\alpha^{(1)}-q_\alpha n_\alpha[\mathbf{v}_0\times\mathbf{B}]\right)\right]\right\}\\
&=\int\left(\frac{m_\alpha v_\alpha^2}{2}+\varepsilon_\alpha\right)J_\alpha^{\mathrm{st,el}}\mathrm{d}\mathbf{v}_\alpha+\int\left(\frac{m_\alpha v_\alpha^2}{2}+\varepsilon_\alpha\right)J_\alpha^{\mathrm{st,inel}}\mathrm{d}\mathbf{v}_\alpha,
\end{aligned}\tag{6.6.8}$$

and after summation the generalized energy equation for mixture

$$\begin{aligned}
&\frac{\partial}{\partial t}\left\{\frac{\rho v_0^2}{2}+\frac{3}{2}p+\sum_\alpha\varepsilon_\alpha n_\alpha-\sum_\alpha\tau_\alpha\left[\frac{\partial}{\partial t}\left(\frac{\rho_\alpha v_0^2}{2}+\frac{3}{2}p_\alpha+\varepsilon_\alpha n_\alpha\right)\right.\right.\\
&\left.\left.+\frac{\partial}{\partial\mathbf{r}}\cdot\left(\frac{1}{2}\rho_\alpha v_0^2\mathbf{v}_0+\frac{5}{2}p_\alpha\mathbf{v}_0+\varepsilon_\alpha n_\alpha\mathbf{v}_0\right)-\mathbf{F}_\alpha^{(1)}\cdot\rho_\alpha\mathbf{v}_0\right]\right\}\\
&+\frac{\partial}{\partial\mathbf{r}}\cdot\left\{\frac{1}{2}\rho v_0^2\mathbf{v}_0+\frac{5}{2}p\mathbf{v}_0+\mathbf{v}_0\sum_\alpha\varepsilon_\alpha n_\alpha-\sum_\alpha\tau_\alpha\left[\frac{\partial}{\partial t}\left(\frac{1}{2}\rho_\alpha v_0^2\mathbf{v}_0\right.\right.\right.\\
&\left.+\frac{5}{2}p_\alpha\mathbf{v}_0+\varepsilon_\alpha n_\alpha\mathbf{v}_0\right)+\frac{\partial}{\partial\mathbf{r}}\cdot\left(\frac{1}{2}\rho_\alpha v_0^2\mathbf{v}_0\mathbf{v}_0+\frac{7}{2}p_\alpha\mathbf{v}_0\mathbf{v}_0+\frac{1}{2}p_\alpha v_0^2\overleftrightarrow{\mathrm{I}}\right.
\end{aligned}$$

$$+\frac{5p_\alpha^2}{2\rho_\alpha}\overleftrightarrow{I}+\varepsilon_\alpha n_\alpha \mathbf{v}_0\mathbf{v}_0+\varepsilon_\alpha\frac{p_\alpha}{m_\alpha}\overleftrightarrow{I}\Bigg)-\rho_\alpha\mathbf{F}_\alpha^{(1)}\cdot\mathbf{v}_0\mathbf{v}_0-p_\alpha\mathbf{F}_\alpha^{(1)}\cdot\overleftrightarrow{I}$$

$$-\frac{1}{2}\rho_\alpha v_0^2\mathbf{F}_\alpha^{(1)}-\frac{3}{2}\mathbf{F}_\alpha^{(1)}p_\alpha-\frac{\rho_\alpha v_0^2}{2}\frac{q_\alpha}{m_\alpha}[\mathbf{v}_0\times\mathbf{B}]-\frac{5}{2}p_\alpha\frac{q_\alpha}{m_\alpha}[\mathbf{v}_0\times\mathbf{B}]$$

$$-\varepsilon_\alpha n_\alpha\frac{q_\alpha}{m_\alpha}[\mathbf{v}_0\times\mathbf{B}]-\varepsilon_\alpha n_\alpha\mathbf{F}_\alpha^{(1)}\Bigg]\Bigg\}-\mathbf{v}_0\cdot\sum_\alpha\rho_\alpha\mathbf{F}_\alpha^{(1)}$$

$$+\sum_\alpha\tau_\alpha\mathbf{F}_\alpha^{(1)}\cdot\left[\frac{\partial}{\partial t}(\rho_\alpha\mathbf{v}_0)+\frac{\partial}{\partial\mathbf{r}}\cdot\rho_\alpha\mathbf{v}_0\mathbf{v}_0+\frac{\partial}{\partial\mathbf{r}}\cdot p_\alpha\overleftrightarrow{I}-\rho_\alpha\mathbf{F}_\alpha^{(1)}-q_\alpha n_\alpha[\mathbf{v}_0\times\mathbf{B}]\right]=0. \quad (6.6.9)$$

Here, $\mathbf{F}_\alpha^{(1)}$ are the forces of the nonmagnetic origin, $\mathbf{B}$—magnetic induction, $\overleftrightarrow{I}$—unit tensor, q_α—charge of the α-component particle, p_α—static pressure for α-component, ε_α— internal energy for the particles of α-component, $\mathbf{v}_0$—hydrodynamic velocity for mixture. For calculations in the self-consistent electromagnetic field, the system of non-local Maxwell equations should be added.

In chemical reaction systems, the internal energies ε_α define the reactions heat Q. For example for bimolecular reaction $A_a+A_b\rightarrow A_c+A_d$ the reaction heat $Q=\varepsilon_c+\varepsilon_d-\varepsilon_a-\varepsilon_b$.

The effective frequencies ω_{eff} (see Eqs. (6.6.1)–(6.6.3)) can be altered in the process of the interaction with the surrounding environment. In this case, the additional equations defining the change of the internal energies should be added to Eqs. (6.6.4)–(6.6.9). Let us consider this situation in detail. I begin with case when the particle internal energy is constant.

After dividing the both sides of the continuity Eq. (6.6.4) by m_α and multiplying by ε_α this equation takes the form

$$\frac{\partial}{\partial t}\left\{\varepsilon_\alpha n_\alpha-\tau_\alpha\left[\frac{\partial\varepsilon_\alpha n_\alpha}{\partial t}+\frac{\partial}{\partial\mathbf{r}}\cdot(\varepsilon_\alpha n_\alpha\mathbf{v}_0)\right]\right\}$$
$$+\frac{\partial}{\partial\mathbf{r}}\cdot\left\{\varepsilon_\alpha n_\alpha\mathbf{v}_0-\tau_\alpha\left[\frac{\partial}{\partial t}(\varepsilon_\alpha n_\alpha\mathbf{v}_0)+\frac{\partial}{\partial\mathbf{r}}\cdot(\varepsilon_\alpha n_\alpha\mathbf{v}_0\mathbf{v}_0)\right.\right.$$
$$\left.\left.+\frac{1}{m_\alpha}\varepsilon_\alpha\overleftrightarrow{I}\cdot\frac{\partial p_\alpha}{\partial\mathbf{r}}-\varepsilon_\alpha n_\alpha\mathbf{F}_\alpha^{(1)}-\frac{q_\alpha}{m_\alpha}\varepsilon_\alpha n_\alpha\mathbf{v}_0\times\mathbf{B}\right]\right\} \quad (6.6.10)$$
$$=\frac{1}{m_\alpha}\varepsilon_\alpha R_\alpha,$$

In general case if $\varepsilon_\alpha\neq$const Eq. (6.6.10) is the internal energy equation in which the right-hand side of equation $\frac{1}{m_\alpha}\varepsilon_\alpha R_\alpha$ transforms into function $E_\alpha(\varepsilon_\alpha)$. After subtraction of the both sides of Eq. (6.6.10) from the corresponding parts of Eq. (6.6.8) one obtains

$$\frac{\partial}{\partial t}\left\{\frac{\rho_\alpha v_0^2}{2}+\frac{3}{2}p_\alpha-\tau_\alpha\left[\frac{\partial}{\partial t}\left(\frac{\rho_\alpha v_0^2}{2}+\frac{3}{2}p_\alpha\right)+\frac{\partial}{\partial\mathbf{r}}\cdot\left(\frac{1}{2}\rho_\alpha v_0^2\mathbf{v}_0+\frac{5}{2}p_\alpha\mathbf{v}_0\right)-\mathbf{F}_\alpha^{(1)}\cdot\rho_\alpha\mathbf{v}_0\right]\right\}$$
$$+\frac{\partial}{\partial\mathbf{r}}\cdot\left\{\frac{1}{2}\rho_\alpha v_0^2\mathbf{v}_0+\frac{5}{2}p_\alpha\mathbf{v}_0-\tau_\alpha\left[\frac{\partial}{\partial t}\left(\frac{1}{2}\rho_\alpha v_0^2\mathbf{v}_0+\frac{5}{2}p_\alpha\mathbf{v}_0\right)\right.\right.$$
$$+\frac{\partial}{\partial\mathbf{r}}\cdot\left(\frac{1}{2}\rho_\alpha v_0^2\mathbf{v}_0\mathbf{v}_0+\frac{7}{2}p_\alpha\mathbf{v}_0\mathbf{v}_0+\frac{1}{2}p_\alpha v_0^2\overleftrightarrow{I}+\frac{5}{2}p_\alpha\mathbf{v}_0\right)$$
$$+\frac{\partial}{\partial\mathbf{r}}\cdot\left(\frac{1}{2}\rho_\alpha v_0^2\mathbf{v}_0\mathbf{v}_0+\frac{7}{2}p_\alpha\mathbf{v}_0\mathbf{v}_0+\frac{1}{2}p_\alpha v_0^2\overleftrightarrow{I}+\frac{5p_\alpha^2}{2\rho_\alpha}\overleftrightarrow{I}\right)-\rho_\alpha\mathbf{F}_\alpha^{(1)}\cdot\mathbf{v}_0\mathbf{v}_0-p_\alpha\mathbf{F}_\alpha^{(1)}\cdot\overleftrightarrow{I} \quad (6.6.11)$$
$$\left.\left.-\frac{1}{2}\rho_\alpha v_0^2\mathbf{F}_\alpha^{(1)}-\frac{3}{2}\mathbf{F}_\alpha^{(1)}p_\alpha-\frac{\rho_\alpha v_0^2}{2}\frac{q_\alpha}{m_\alpha}[\mathbf{v}_0\times\mathbf{B}]-\frac{5}{2}p_\alpha\frac{q_\alpha}{m_\alpha}[\mathbf{v}_0\times\mathbf{B}]\right]\right\}$$
$$-\left\{\rho_\alpha\mathbf{F}_\alpha^{(1)}\cdot\mathbf{v}_0-\tau_\alpha\left[\mathbf{F}_\alpha^{(1)}\cdot\left(\frac{\partial}{\partial t}(\rho_\alpha\mathbf{v}_0)+\frac{\partial}{\partial\mathbf{r}}\cdot\rho_\alpha\mathbf{v}_0\mathbf{v}_0+\frac{\partial}{\partial\mathbf{r}}\cdot p_\alpha\overleftrightarrow{I}-\rho_\alpha\mathbf{F}_\alpha^{(1)}-q_\alpha n_\alpha[\mathbf{v}_0\times\mathbf{B}]\right)\right]\right\}$$
$$=\int\frac{m_\alpha v_\alpha^2}{2}J_\alpha^{\text{st,el}}d\mathbf{v}_\alpha+\int\frac{m_\alpha v_\alpha^2}{2}J_\alpha^{\text{st,inel}}d\mathbf{v}_\alpha,$$

taking into account that

$$\int \varepsilon_\alpha J_\alpha^{\text{st,el}} d\mathbf{v}_\alpha + \int \varepsilon_\alpha J_\alpha^{\text{st,inel}} d\mathbf{v}_\alpha = \varepsilon_\alpha \frac{R_\alpha}{m_\alpha}. \tag{6.6.12}$$

Conclusion:

In the case when the change of the species internal energies is absent as result of interaction of the physical system with environment, the solution of the full system of equations (6.6.4)–(6.6.9) can be reduced to the system (6.6.4)–(6.6.7), (6.6.11).

It is interesting to confirm this conclusion by the direct numerical calculation. With this aim, let us consider the CDWs which are periodic modulation of the conduction electron density. The movement of the soliton waves in graphene was considered in the previous section. It was shown that the effective charge is created due to interference of the induced electron waves and correlating potentials as result of the polarized modulation of atomic positions.

The effective charge is created due to interference of the induced electron waves and correlating potentials as result of the polarized modulation of atomic positions. Therefore in this approach the conduction in graphene convoys the transfer of the positive ($+e$, m_p) and negative ($-e$, m_e) charges. The nonstationary 1D motion of the combined soliton is considered under influence of the self-consistent electric forces of the potential and non-potential origin. The mentioned soliton can exists without a chemical bond formation.

Let us find the soliton type solutions for the system of the generalized quantum equations for two-species mixture. The graphene crystal lattice is 2D flat structure which is considered in the moving coordinate system ($\xi = x - u_0 t$, y). In the following, we intend to apply generalized non-local quantum hydrodynamic Eqs. (6.6.4)–(6.6.9) to the investigation of the CDWs in the frame of two-species model which lead to the following set of the dimensional equations: (Poisson equation for the self-consistent electric field)

$$\frac{\partial^2 \psi}{\partial \xi^2} + \frac{\partial^2 \psi}{\partial y^2} = -4\pi e\left\{\left[n_p - \tau_p \frac{\partial}{\partial \xi}\left(n_p(u-u_0)\right)\right] - \left[n_e - \tau_e \frac{\partial}{\partial \xi}\left(n_e(u-u_0)\right)\right]\right\}. \tag{6.6.13}$$

(Continuity equation for the positive particles)

$$\begin{gathered}\frac{\partial}{\partial \xi}\left[\rho_p(u_0-u)\right] + \frac{\partial}{\partial \xi}\left\{\tau_p \frac{\partial}{\partial \xi}\left[\rho_p(u-u_0)^2\right]\right\} \\ + \frac{\partial}{\partial \xi}\left\{\tau_p\left[\frac{\partial}{\partial \xi}p_p - \rho_p F_{p\xi}\right]\right\} + \frac{\partial}{\partial y}\left\{\tau_p\left[\frac{\partial}{\partial y}p_p - \rho_p F_{py}\right]\right\} = 0.\end{gathered} \tag{6.6.14}$$

(Continuity equation for electrons)

$$\begin{gathered}\frac{\partial}{\partial \xi}\left[\rho_e(u_0-u)\right] + \frac{\partial}{\partial \xi}\left\{\tau_e \frac{\partial}{\partial \xi}\left[\rho_e(u-u_0)^2\right]\right\} \\ \frac{\partial}{\partial \xi}\left\{\tau_e\left[\frac{\partial}{\partial \xi}p_e - \rho_e F_{e\xi}\right]\right\} + \frac{\partial}{\partial y}\left\{\tau_e\left[\frac{\partial}{\partial y}p_e - \rho_e F_{ey}\right]\right\} = 0.\end{gathered} \tag{6.6.15}$$

(Momentum equation for the x direction)

$$\begin{aligned}&\frac{\partial}{\partial \xi}\left\{\rho u(u-u_0) + p\right\} - \rho_p F_{p\xi} - \rho_e F_{e\xi} \\ &+ \frac{\partial}{\partial \xi}\left\{\tau_p\left[\frac{\partial}{\partial \xi}\left(2p_p(u_0-u) - \rho_p u(u_0-u)^2\right) - \rho_p F_{p\xi}(u_0-u)\right]\right\} \\ &+ \frac{\partial}{\partial \xi}\left\{\tau_e\left[\frac{\partial}{\partial \xi}\left(2p_e(u_0-u) - \rho_e u(u_0-u)^2\right) - \rho_e F_{e\xi}(u_0-u)\right]\right\} \\ &+ \tau_p F_{p\xi}\left(\frac{\partial}{\partial \xi}\left(\rho_p(u-u_0)\right)\right) + \tau_e F_{e\xi}\left(\frac{\partial}{\partial \xi}\left(\rho_e(u-u_0)\right)\right) \\ &- \frac{\partial}{\partial \xi}\left\{\tau_p \frac{\partial}{\partial \xi}(p_p u)\right\} - \frac{\partial}{\partial \xi}\left\{\tau_e \frac{\partial}{\partial \xi}(p_e u)\right\} - \frac{\partial}{\partial y}\left\{\tau_p \frac{\partial}{\partial y}(p_p u)\right\} - \frac{\partial}{\partial y}\left\{\tau_e \frac{\partial}{\partial y}(p_e u)\right\} \\ &+ \frac{\partial}{\partial \xi}\left\{\tau_p\left[F_{p\xi}\rho_p u\right]\right\} + \frac{\partial}{\partial \xi}\left\{\tau_e\left[F_{e\xi}\rho_e u\right]\right\} + \frac{\partial}{\partial y}\left\{\tau_p\left[F_{py}\rho_p u\right]\right\} + \frac{\partial}{\partial y}\left\{\tau_e\left[F_{ey}\rho_e u\right]\right\} = 0.\end{aligned} \tag{6.6.16}$$

(Energy equation for the positive particles)

$$
\begin{aligned}
&\frac{\partial}{\partial \xi}\left[\rho_p u^2(u-u_0)+2\varepsilon_p n_p(u-u_0)+5p_p u-3p_p u_0\right]-2\rho_p F_{p\xi} u \\
&+\frac{\partial}{\partial \xi}\left\{\tau_p\left[\begin{array}{l}\frac{\partial}{\partial \xi}\left(-\rho_p u^2(u_0-u)^2-2\varepsilon_p n_p(u_0-u)^2+7p_p u(u_0-u)+3p_p u_0(u-u_0)\right. \\ \left.-p_p u^2-2\varepsilon_p \frac{p_p}{m_p}-5\frac{p_p^2}{\rho_p}\right)-2F_{p\xi}\rho_p u(u_0-u)+\rho_p u^2 F_{p\xi}+2\varepsilon_p n_p F_{p\xi}+5p_p F_{p\xi}\end{array}\right]\right\} \\
&-\frac{\partial}{\partial y}\left\{\tau_p\left[\frac{\partial}{\partial y}\left(p_p u^2+2\varepsilon_p\frac{p_p}{m_p}+5\frac{p_p^2}{\rho_p}\right)-\rho_p F_{py}u^2-2\varepsilon_p n_p F_{py}-5p_p F_{py}\right]\right\} \\
&-2\tau_p F_{p\xi}\left[\frac{\partial}{\partial \xi}(\rho_p u(u_0-u))\right]-2\tau_p\rho_p\left[(F_{p\xi})^2+(F_{py})^2\right] \\
&+2\tau_p F_{p\xi}\left[\frac{\partial}{\partial \xi}p_p\right]+2\tau_p F_{py}\left[\frac{\partial}{\partial y}p_p\right]=-\frac{p_p-p_e}{\tau_{ep}}.
\end{aligned}
\tag{6.6.17}
$$

(Energy equation for electrons)

$$
\begin{aligned}
&\frac{\partial}{\partial \xi}\left[\rho_e u^2(u-u_0)+2\varepsilon_e n_e(u-u_0)+5p_e u-3p_e u_0\right]-2\rho_e F_{e\xi} u \\
&+\frac{\partial}{\partial \xi}\left\{\tau_e\left[\begin{array}{l}\frac{\partial}{\partial \xi}\left(-\rho_e u^2(u_0-u)^2-2\varepsilon_e n_e(u_0-u)^2+7p_e u(u_0-u)+3p_e u_0(u-u_0)\right. \\ \left.-p_e u^2-2\varepsilon_e \frac{p_e}{m_e}-5\frac{p_e^2}{\rho_e}\right)-2F_{e\xi}\rho_e u(u_0-u)+\rho_e u^2 F_{e\xi}+2\varepsilon_e n_e F_{e\xi}+5p_e F_{e\xi}\end{array}\right]\right\} \\
&-\frac{\partial}{\partial y}\left\{\tau_e\left[\frac{\partial}{\partial y}\left(p_e u^2+2\varepsilon_e\frac{p_e}{m_e}+5\frac{p_e^2}{\rho_e}\right)-\rho_e F_{ey}u^2-2\varepsilon_e n_e F_{ey}-5p_e F_{ey}\right]\right\} \\
&-2\tau_e F_{e\xi}\left[\frac{\partial}{\partial \xi}(\rho_e u(u_0-u))\right]-2\tau_e\rho_e\left[(F_{e\xi})^2+(F_{ey})^2\right] \\
&+2\tau_e F_{e\xi}\left[\frac{\partial}{\partial \xi}p_e\right]+2\tau_e F_{ey}\left[\frac{\partial}{\partial y}p_e\right]=-\frac{p_e-p_p}{\tau_{ep}}
\end{aligned}
\tag{6.6.18}
$$

As before, write down these equations in the dimensionless form, where dimensionless symbols are marked by tildes; introduce the scales:

$$u=u_0\tilde{u},\ \xi=x_0\tilde{\xi},\ y=x_0\tilde{y},\ \psi=\psi_0\tilde{\psi},\ \rho_e=\rho_0\tilde{\rho}_e,\ \rho_p=\rho_0\tilde{\rho}_p$$

where u_0, x_0, ψ_0, ρ_0—scales for velocity, distance, potential, and density. Let us introduce also $p_p=\rho_0 V_{0p}^2\tilde{p}_p, p_e=\rho_0 V_{0e}^2\tilde{p}_e$, where V_{0p} and V_{0e}—the scales for thermal velocities for the electron and positive species;

$$F_p=\tilde{F}_p\frac{e\psi_0}{m_p x_0},\ F_e=\tilde{F}_e\frac{e\psi_0}{m_e x_0};\ \tau_p=\frac{m_e x_0 H}{m_p u_0\tilde{u}^2},\ \tau_e=\frac{x_0 H}{u_0\tilde{u}^2}$$

where $H=\dfrac{N_R\hbar}{m_e x_0 u_0}$ is dimensionless parameter. Then, $\dfrac{1}{\tau_{ep}}=\dfrac{u_0\tilde{u}^2}{x_0 H}\left(1+\dfrac{m_p}{m_e}\right)$.

Let us introduce also the following dimensionless parameters

$$R=\frac{e\rho_0 x_0^2}{m_e\phi_0},\ E=\frac{e\psi_0}{m_e u_0^2}$$

and dimensionless spin parameters characterizing the internal particles energy:

$$S_e=\frac{2\varepsilon_e}{m_e u_0^2},\ S_p=\frac{2\varepsilon_p}{m_p u_0^2} \tag{6.6.19}$$

Acting forces are the sum of three terms: the self-consistent potential force (scalar potential ψ), connected with the displacement of positive and negative charges, potential forces originated by the graphene crystal lattice (potential U), and the external electrical field creating the intensity **E**. As a result, the following relations are valid

$$F_{p\xi}=\frac{e}{m_p}\left(-\frac{\partial\psi}{\partial\xi}-\frac{\partial U}{\partial\xi}+E_{0\xi}\right),\ F_{e\xi}=\frac{e}{m_e}\left(\frac{\partial\psi}{\partial\xi}+\frac{\partial U}{\partial\xi}-E_{0\xi}\right),$$

$$F_{py}=\frac{e}{m_p}\left(-\frac{\partial\psi}{\partial y}-\frac{\partial U}{\partial y}+E_{0y}\right),\ F_{ey}=\frac{e}{m_e}\left(\frac{\partial\psi}{\partial y}+\frac{\partial U}{\partial y}-E_{0y}\right).$$

Graphene is a single layer of carbon atoms densely packed in a honeycomb lattice. Taking into account the introduced values and approximations acting forces along y-direction for graphene (all details of the corresponding approximations are delivered in Section 6.5) the following set of the dimensionless non-local hydrodynamic equations for the 2D soliton description can be written in the first approximation:

(Poisson equation for the self-consistent electric field)

$$\frac{\partial^2\widetilde{\psi}}{\partial\widetilde{\xi}^2}=-4\pi R\left\{\frac{m_e}{m_p}\left[\widetilde{\rho}_p-\frac{m_eH}{m_p\widetilde{u}^2}\frac{\partial}{\partial\widetilde{\xi}}(\widetilde{\rho}_p(\widetilde{u}-1))\right]-\left[\widetilde{\rho}_e-\frac{H}{\widetilde{u}^2}\frac{\partial}{\partial\widetilde{\xi}}(\widetilde{\rho}_e(\widetilde{u}-1))\right]\right\}. \tag{6.6.20}$$

(Continuity equation for the positive particles)

$$\begin{aligned}&\frac{\partial}{\partial\widetilde{\xi}}[\widetilde{\rho}_p(1-\widetilde{u})]+\frac{m_e}{m_p}\frac{\partial}{\partial\widetilde{\xi}}\left\{\frac{H}{\widetilde{u}^2}\frac{\partial}{\partial\widetilde{\xi}}\left[\widetilde{\rho}_p(\widetilde{u}-1)^2\right]\right\}+\frac{m_e}{m_p}\frac{\partial}{\partial\widetilde{\xi}}\left\{\frac{H}{\widetilde{u}^2}\left[\frac{V_{0p}^2}{u_0^2}\frac{\partial}{\partial\widetilde{\xi}}\widetilde{p}_p\right.\right.\\&\left.\left.-\frac{m_e}{m_p}\widetilde{\rho}_pE\left(-\frac{\partial\widetilde{\phi}}{\partial\widetilde{\xi}}+\widetilde{U}_{11}'\sin\left(\frac{4\pi}{3\widetilde{a}}\widetilde{\xi}-\frac{\pi}{3}\right)\right)\right]\right\}=0\end{aligned} \tag{6.6.21}$$

Continuity equation for electrons:

$$\begin{aligned}&\frac{\partial}{\partial\widetilde{\xi}}[\widetilde{\rho}_e(1-\widetilde{u})]+\frac{\partial}{\partial\widetilde{\xi}}\left\{\frac{H}{\widetilde{u}^2}\frac{\partial}{\partial\widetilde{\xi}}\left[\widetilde{\rho}_e(\widetilde{u}-1)^2\right]\right\}+\frac{\partial}{\partial\widetilde{\xi}}\left\{\frac{H}{\widetilde{u}^2}\left[\frac{V_{0e}^2}{u_0^2}\frac{\partial}{\partial\widetilde{\xi}}\widetilde{p}_e\right.\right.\\&\left.\left.-\widetilde{\rho}_eE\left(\frac{\partial\widetilde{\psi}}{\partial\widetilde{\xi}}-\widetilde{U}_{11}'\sin\left(\frac{4\pi}{3\widetilde{a}}\widetilde{\xi}-\frac{\pi}{3}\right)\right)\right]\right\}=0\end{aligned} \tag{6.6.22}$$

Momentum equation for the x direction:

$$\begin{aligned}&\frac{\partial}{\partial\widetilde{\xi}}\left\{(\widetilde{\rho}_p+\widetilde{\rho}_e)\widetilde{u}(\widetilde{u}-1)+\frac{V_{0p}^2}{u_0^2}\widetilde{p}_p+\frac{V_{0e}^2}{u_0^2}\widetilde{p}_e\right\}-\frac{m_e}{m_p}\widetilde{\rho}_pE\left(-\frac{\partial\widetilde{\psi}}{\partial\widetilde{\xi}}+\widetilde{U}_{11}'\sin\left(\frac{4\pi}{3\widetilde{a}}\widetilde{\xi}-\frac{\pi}{3}\right)\right)\\&-\widetilde{\rho}_eE\left(\frac{\partial\widetilde{\psi}}{\partial\widetilde{\xi}}-\widetilde{U}_{11}'\sin\left(\frac{4\pi}{3\widetilde{a}}\widetilde{\xi}-\frac{\pi}{3}\right)\right)+\frac{m_e}{m_p}\frac{\partial}{\partial\widetilde{\xi}}\left\{\frac{H}{\widetilde{u}^2}\left[\frac{\partial}{\partial\widetilde{\xi}}\left(2\frac{V_{0p}^2}{u_0^2}\widetilde{p}_p(1-\widetilde{u})-\widetilde{\rho}_p\widetilde{u}(1-\widetilde{u})^2\right)-\right.\right.\\&\left.\left.-\frac{m_e}{m_p}\widetilde{\rho}_p(1-\widetilde{u})E\left(-\frac{\partial\widetilde{\psi}}{\partial\widetilde{\xi}}+\widetilde{U}_{11}'\sin\left(\frac{4\pi}{3\widetilde{a}}\widetilde{\xi}-\frac{\pi}{3}\right)\right)\right]\right\}\\&+\frac{\partial}{\partial\widetilde{\xi}}\left\{\frac{H}{\widetilde{u}^2}\left[\frac{\partial}{\partial\widetilde{\xi}}\left(2\frac{V_{0e}^2}{u_0^2}\widetilde{p}_e(1-\widetilde{u})-\widetilde{\rho}_e\widetilde{u}(1-\widetilde{u})^2\right)-\widetilde{\rho}_e(1-\widetilde{u})E\left(\frac{\partial\widetilde{\psi}}{\partial\widetilde{\xi}}-\widetilde{U}_{11}'\sin\left(\frac{4\pi}{3\widetilde{a}}\widetilde{\xi}-\frac{\pi}{3}\right)\right)\right]\right\}\\&+\frac{H}{\widetilde{u}^2}E\left(\frac{m_e}{m_p}\right)^2\left(-\frac{\partial\widetilde{\psi}}{\partial\widetilde{\xi}}+\widetilde{U}_{11}'\sin\left(\frac{4\pi}{3\widetilde{a}}\widetilde{\xi}-\frac{\pi}{3}\right)\right)\left(\frac{\partial}{\partial\widetilde{\xi}}(\widetilde{\rho}_p(\widetilde{u}-1))\right)\\&+\frac{H}{\widetilde{u}^2}E\left(\frac{\partial\widetilde{\psi}}{\partial\widetilde{\xi}}-\widetilde{U}_{11}'\sin\left(\frac{4\pi}{3\widetilde{a}}\widetilde{\xi}-\frac{\pi}{3}\right)\right)\left(\frac{\partial}{\partial\widetilde{\xi}}(\widetilde{\rho}_e(\widetilde{u}-1))\right)\\&-\frac{m_e}{m_p}\frac{\partial}{\partial\widetilde{\xi}}\left\{\frac{H}{\widetilde{u}^2}\frac{V_{0p}^2}{u_0^2}\frac{\partial}{\partial\widetilde{\xi}}(\widetilde{p}_p\widetilde{u})\right\}-\frac{\partial}{\partial\widetilde{\xi}}\left\{\frac{H}{\widetilde{u}^2}\frac{V_{0e}^2}{u_0^2}\frac{\partial}{\partial\widetilde{\xi}}(\widetilde{p}_e\widetilde{u})\right\}\\&+\left(\frac{m_e}{m_p}\right)^2E\frac{\partial}{\partial\widetilde{\xi}}\left\{\frac{H}{\widetilde{u}^2}\left[\left(-\frac{\partial\widetilde{\psi}}{\partial\widetilde{\xi}}+\widetilde{U}_{11}'\sin\left(\frac{4\pi}{3\widetilde{a}}\widetilde{\xi}-\frac{\pi}{3}\right)\right)\widetilde{\rho}_p\widetilde{u}\right]\right\}\\&+E\frac{\partial}{\partial\widetilde{\xi}}\left\{\frac{H}{\widetilde{u}^2}\left[\left(\frac{\partial\widetilde{\psi}}{\partial\widetilde{\xi}}-\widetilde{U}_{11}'\sin\left(\frac{4\pi}{3\widetilde{a}}\widetilde{\xi}-\frac{\pi}{3}\right)\right)\widetilde{\rho}_e\widetilde{u}\right]\right\}=0\end{aligned} \tag{6.6.23}$$

Energy equation for the positive particles:

$$\begin{aligned}
&\frac{\partial}{\partial\tilde{\xi}}\left[\tilde{\rho}_p\tilde{u}^2(\tilde{u}-1)+S_p\tilde{\rho}_p(\tilde{u}-1)+5\frac{V_{0p}^2}{u_0^2}\tilde{p}_p\tilde{u}-3\frac{V_{0p}^2}{u_0^2}\tilde{p}_p\right]\\
&-2\frac{m_e}{m_p}\tilde{\rho}_pE\left(-\frac{\partial\tilde{\psi}}{\partial\tilde{\xi}}+\tilde{U}_{11}'\sin\left(\frac{4\pi}{3\tilde{a}}\tilde{\xi}-\frac{\pi}{3}\right)\right)\tilde{u}\\
&+\frac{\partial}{\partial\tilde{\xi}}\left\{\frac{H}{\tilde{u}^2}\frac{m_e}{m_p}\left[\frac{\partial}{\partial\tilde{\xi}}\left(-\tilde{\rho}_p\tilde{u}^2(1-\tilde{u})^2-S_p\tilde{\rho}_p(\tilde{u}-1)^2+7\frac{V_{0p}^2}{u_0^2}\tilde{p}_p\tilde{u}(1-\tilde{u})\right.\right.\right.\\
&\left.+3\frac{V_{0p}^2}{u_0^2}\tilde{p}_p(\tilde{u}-1)-\frac{V_{0p}^2}{u_0^2}\tilde{p}_p\tilde{u}^2-\frac{V_{0p}^2}{u_0^2}S_p\tilde{p}_p-5\frac{V_{0p}^4}{u_0^4}\frac{\tilde{p}_p^2}{\tilde{\rho}_p}\right)\\
&+E\left(-2\frac{m_e}{m_p}\tilde{\rho}_p\tilde{u}(1-\tilde{u})+\frac{m_e}{m_p}\tilde{\rho}_p\tilde{u}^2+\frac{m_e}{m_p}S_p\tilde{\rho}_{pp}+5\frac{m_e}{m_p}\frac{V_{0p}^2}{u_0^2}\tilde{p}_p\right)\left(-\frac{\partial\tilde{\psi}}{\partial\tilde{\xi}}\right.\\
&\left.\left.\left.+\tilde{U}_{11}'\sin\left(\frac{4\pi}{3\tilde{a}}\tilde{\xi}-\frac{\pi}{3}\right)\right)\right]\right\}+2\frac{H}{\tilde{u}^2}E\left(\frac{m_e}{m_p}\right)^2\left[-\frac{\partial}{\partial\tilde{\xi}}(\tilde{\rho}_p\tilde{u}(1-\tilde{u}))\right.\\
&\left.+\frac{V_{0p}^2}{u_0^2}\frac{\partial}{\partial\tilde{\xi}}\tilde{p}_p\right]\left(-\frac{\partial\tilde{\psi}}{\partial\tilde{\xi}}+\tilde{U}_{11}'\sin\left(\frac{4\pi}{3\tilde{a}}\tilde{\xi}-\frac{\pi}{3}\right)\right)\\
&-2\frac{H}{\tilde{u}^2}E^2\left(\frac{m_e}{m_p}\right)^3\tilde{\rho}_p\left[\left(-\frac{\partial\tilde{\psi}}{\partial\tilde{\xi}}+\tilde{U}_{11}'\sin\left(\frac{4\pi}{3\tilde{a}}\tilde{\xi}-\frac{\pi}{3}\right)\right)^2+\frac{1}{2}\left(\tilde{U}_{10}'\sin\left(\frac{2\pi}{3\tilde{a}}\tilde{\xi}+\frac{\pi}{3}\right)\right)^2\right.\\
&\left.+\frac{3}{2}\left(\tilde{U}_{10}'\cos\left(\frac{2\pi}{3\tilde{a}}\tilde{\xi}+\frac{\pi}{3}\right)\right)^2+6\left(\tilde{U}_{11}'\right)^2+\frac{16}{\pi}\left(\tilde{U}_{10}'\tilde{U}_{11}'\right)\cos\left(\frac{2\pi}{3\tilde{a}}\tilde{\xi}+\frac{\pi}{3}\right)\right]\\
&=-\frac{\tilde{u}^2}{Hu_0^2}\left(V_{0p}^2\tilde{p}_p-\tilde{p}_eV_{0e}^2\right)\left(1+\frac{m_p}{m_e}\right).
\end{aligned}\tag{6.6.24}$$

(Energy equation for electrons)

$$\begin{aligned}
&\frac{\partial}{\partial\tilde{\xi}}\left[\tilde{\rho}_e\tilde{u}^2(\tilde{u}-1)+S_e\tilde{\rho}_e(\tilde{u}-1)+5\frac{V_{0e}^2}{u_0^2}\tilde{p}_e\tilde{u}-3\frac{V_{0e}^2}{u_0^2}\tilde{p}_e\right]-2\tilde{\rho}_e\tilde{u}E\left(\frac{\partial\tilde{\psi}}{\partial\tilde{\xi}}-\tilde{U}_{11}'\sin\left(\frac{4\pi}{3\tilde{a}}\tilde{\xi}-\frac{\pi}{3}\right)\right)\\
&+\frac{\partial}{\partial\tilde{\xi}}\left\{\frac{H}{\tilde{u}^2}\left[\frac{\partial}{\partial\tilde{\xi}}\left(-\tilde{\rho}_e\tilde{u}^2(1-\tilde{u})^2-S_e\tilde{\rho}_e(\tilde{u}-1)^2+7\frac{V_{0e}^2}{u_0^2}\tilde{p}_e\tilde{u}(1-\tilde{u})+3\frac{V_{0e}^2}{u_0^2}\tilde{p}_e(\tilde{u}-1)-\frac{V_{0e}^2}{u_0^2}\tilde{p}_e\tilde{u}^2\right.\right.\right.\\
&\left.\left.\left.-\frac{V_{0e}^2}{u_0^2}S_e\tilde{p}_e-5\frac{V_{0e}^4}{u_0^4}\frac{\tilde{p}_e^2}{\tilde{\rho}_e}\right)+E\left(-2\tilde{\rho}_e\tilde{u}(1-\tilde{u})+\tilde{\rho}_e\tilde{u}^2+S_e\tilde{\rho}_e+5\frac{V_{0e}^2}{u_0^2}\tilde{p}_e\right)\left(\frac{\partial\tilde{\psi}}{\partial\tilde{\xi}}-\tilde{U}_{11}'\sin\left(\frac{4\pi}{3\tilde{a}}\tilde{\xi}-\frac{\pi}{3}\right)\right)\right]\right\}\\
&+E\left(-2\frac{H}{\tilde{u}^2}\frac{\partial}{\partial\tilde{\xi}}(\tilde{\rho}_e\tilde{u}(1-\tilde{u}))+2\frac{H}{\tilde{u}^2}\frac{V_{0e}^2}{u_0^2}\frac{\partial}{\partial\tilde{\xi}}\tilde{p}_e\right)\left(\frac{\partial\tilde{\psi}}{\partial\tilde{\xi}}-\tilde{U}_{11}'\sin\left(\frac{4\pi}{3\tilde{a}}\tilde{\xi}-\frac{\pi}{3}\right)\right)\\
&-2E^2\frac{H}{\tilde{u}^2}\tilde{\rho}_e\left[\left(-\frac{\partial\tilde{\psi}}{\partial\tilde{\xi}}+\tilde{U}_{11}'\sin\left(\frac{4\pi}{3\tilde{a}}\tilde{\xi}-\frac{\pi}{3}\right)\right)^2+\frac{1}{2}\left(\tilde{U}_{10}'\sin\left(\frac{2\pi}{3\tilde{a}}\tilde{\xi}+\frac{\pi}{3}\right)\right)^2+6\left(\tilde{U}_{11}'\right)^2\right.\\
&\left.+\frac{3}{2}\left(\tilde{U}_{10}'\cos\left(\frac{2\pi}{3\tilde{a}}\tilde{\xi}+\frac{\pi}{3}\right)\right)^2+\frac{16}{\pi}\left(\tilde{U}_{10}'\tilde{U}_{11}'\right)\cos\left(\frac{2\pi}{3\tilde{a}}\tilde{\xi}+\frac{\pi}{3}\right)\right]=-\frac{\tilde{u}^2}{Hu_0^2}\left(V_{0e}^2\tilde{p}_e-V_{0p}^2\tilde{p}_p\right)\left(1+\frac{m_p}{m_e}\right).
\end{aligned}\tag{6.6.25}$$

The calculations are realized on the basement of Eqs. (6.6.20)–(6.6.25) by the initial conditions and parameters containing in the Table 6.5 and corresponding to Variant 1 (see Section 6.4). The following numerical data (Table 6.6) demonstrate some results of calculations.Now I can formulate some principal conclusions:

1. All calculations realized as Variant 1 and containing in Table 6.6 correspond to spin variables $S_e = S_p = 0$. The domain of the soliton existence is equal to $\tilde{\xi}$ varying in interval (−0.305, 0.274).

TABLE 6.5 Initial conditions and parameters of calculations for variant 1

$\tilde{a}$	1
L	1
T	2×10^4
$\tilde{\rho}_e(0)$	1
$\tilde{\rho}_p(0)$	2×10^4
$N=\dfrac{V_{0e}^2}{u_0^2}$	1
$P=\dfrac{V_{0p}^2}{u_0^2}$	10^{-4}
$\tilde{p}_e(0)$	1
$\tilde{p}_p(0)$	2×10^4
$\tilde{\varphi}(0)$	1
$E=\dfrac{e\varphi_0}{m_e u_0^2}$	0.1
$R=\dfrac{e\rho_0 x_0^2}{m_e \varphi_0}$	0.003
H	15
$\dfrac{\partial\tilde{\rho}_e}{\partial\tilde{\xi}}(0)$	0
$\dfrac{\partial\tilde{\rho}_p}{\partial\tilde{\xi}}(0)$	0
$\dfrac{\partial\tilde{p}_e}{\partial\tilde{\xi}}(0)$	0
$\dfrac{\partial\tilde{p}_p}{\partial\tilde{\xi}}(0)$	0
$K=\dfrac{\partial\tilde{\varphi}}{\partial\tilde{\xi}}(0)$	0
$\tilde{U}'_{10}$	10
$\tilde{U}'_{11}$	10

2. All calculations realized as Variant 1 corresponding to the constant spin variables S_e, S_p (varying from $S_e = S_p = 0$ to $S_e = S_p = 10^9$) lead to the absolutely the same results shown in Table 6.6. The domain of the soliton existence is also equal to $(-0.305, 0.274)$.
3. This fact confirms the previous theoretical result—in the case when the change of the species internal energies is absent as result of interaction with external media, the solution of the full system of Eqs. (6.6.4)–(6.6.9) can be reduced to the system (6.6.4)–(6.6.7) and (6.6.11).
4. These calculations realized by several numerical methods are the direct evidence in favor of the high accuracy of numerical methods in the interactive Maple system for solution of the ordinary differential equations.

6.7 TO THE THEORY OF THE SC

Real following progress in the SC can be reached only on the way of the non-local theory development. Then no reason here go into details of existing theories based on the Schrödinger (SE) apart of the facts of the principal significance leading to the non-local model of SC.

TABLE 6.6 Numerical results of calculations for variant 1

	$t=\tilde{\xi}=0.2$	$t=\tilde{\xi}=0.25$
$p=\tilde{p}_p$	21,731.595	22,164.607
$p'=\partial\tilde{p}_p/\partial\tilde{\xi}$	8660.254	8660.254
$q=\tilde{p}_e$	0.956424	0.925592
$q'=\partial\tilde{p}_e/\partial\tilde{\xi}$	−0.476250	−0.859321
$r=\tilde{\rho}_p$	0.622976×10^{-3}	0.407662×10^{-3}
$r'=\partial\tilde{\rho}_p/\partial\tilde{\xi}$	−0.593701	-0.308647×10^{-2}
$s=\tilde{\rho}_e$	1.866551	4.681384
$s'=\partial\tilde{\rho}_e/\partial\tilde{\xi}$	18.453042	170.620851
$u=\tilde{u}$	1.000000	1.000000
$v=\tilde{\psi}$	1.000819	1.0013853
$v'=\partial\tilde{\psi}/\partial\tilde{\xi}$	0.909275×10^{-2}	0.143210×10^{-1}
$S_e=S_p=0$		

It is well known that SC was discovered more than 100 years ago in the laboratory of Heike Kamerlingh Onnes, at Leiden University in the Netherlands. It was done on 8 April 1911, after testing for electrical resistance in a sample of mercury at 3 K.

Quantum mechanics created in the 1920s provided an underlying model for the structure of ordinary metals. Metal atoms form a regular crystalline lattice with tightly bound inner core of electrons. However, their loosely attached outer electrons become unbound, collecting into a mobile "electron cloud." Under the influence of an electric field, these free electrons will drift throughout the lattice, forming the basis of conductivity. However, random thermal fluctuations scatter the electrons, interrupting their forward motion and dissipating energy—thereby producing electrical resistance. If some metals are cooled to temperatures close to absolute zero, the electrons suddenly shift into a highly ordered state and travel collectively without deviating from their path. Below a critical temperature T_c (or better to say a narrow temperature diapason ΔT) the electrical resistance falls to zero, and they become superconductors. In the case of low temperature SC ΔT is about $0.001 \div 0.1$ K, for high temperature SC ΔT can be more than 1 K. From position of non-local physics it means that the solution of the generalized non-local quantum equations transforms into soliton's type without destruction in space and time.

In the May issue of Physical Review, two experimental papers [136,137] (both received by the journal on March 24th, 1950) were describing measurements of the critical temperature of mercury for different isotopes, reporting that "there is a systematic decrease of transition temperature with increasing mass" [136], and that [137] "From these results one may infer that the transition temperature of a superconductor is a function of the nuclear mass, the lighter the mass the higher the transition temperature." From experiments follow that for crystal lattices of the different isotopes the relation takes place

$$T_c\sqrt{M}=\text{const}, \tag{6.7.1}$$

where M is isotopic mass and const is the same for all isotopes. The frequency of the lattice vibration ω is connected with the square root of M

$$\omega \sim 1/\sqrt{M}. \tag{6.7.2}$$

It means that the interaction between electrons and lattice vibrations (phonons) was responsible for SC.

It should be noticed that on May 16th, 1950, Herbert Fröhlich's paper entitled "Theory of the Superconducting State. I. The Ground State at the Absolute Zero of Temperature" was received by the Physical Review [138]. This theoretical paper also proposed that the interaction between electrons and lattice vibrations (phonons) was responsible for SC. The paper made no mention of the experimental papers [136, 137], moreover later Herbert Fröhlich stated that the isotope effect experiments "have just come to my notice" and pointed out that the formalism in his May 16th paper [138] in fact predicted the effect. The question of priority is discussed until now [139].

Significant step in explanation of the behavior of superconducting materials consists in creation of BCS theory (John Bardeen, Leon N. Cooper, and John R. Schrieffer, 1957, [140]). From the qualitative point of view the physical model looks as follows. Two forces of the electric origin influence on the electrons behavior in a metal—the repulsion between electrons and attraction between electrons and positive ions that make up the rigid lattice of the metal.

Let us estimate the character scale Δr of the phonon interaction. Phonon energy is $\hbar\omega_D \sim \hbar v_s/a$, where ω_D is Debye frequency, v_s—sound speed, a—the lattice constant. For example, the lattice constant for a common carbon diamond is $a = 3.57$ Å at 300 K. The character impulse is $\Delta p \sim \hbar\omega_D/v_F$, where v_F ($\sim 10^6$ m/s)—electron velocity near Fermi surface. Then the scale Δr can be found using the Heisenberg uncertainty relation

$$\Delta r \sim \hbar/\Delta p \sim v_F/\omega_D \sim \frac{v_F}{v_s} a \sim \sqrt{\frac{M}{m}} a, \tag{6.7.3}$$

where M, m—ion and electron masses, respectively. Usually, $\Delta r \sim 10^{-5} \div 10^{-6}$ cm.

The mentioned attraction distorts the ion lattice, increasing the local positive charge density of the lattice. This perturbation of the positive charge can attract other electrons. At long distances, this attraction between electrons due to the displaced ions can overcome the electrons' repulsion due to their negative charge, and cause them to pair up. A Cooper pair (described in 1956 by Leon Cooper) has special construction:

1. Electrons are fermions, but a Cooper pair is a composite boson as its total spin is integer. As result the wave functions are symmetric under particle interchange, and they are allowed to be in the same state. The Cooper pairs "condense" in a body into the same ground quantum state.
2. If the electric current is absent, the combined impulse of a Cooper pair is equal to zero. After application of the external electric field a Cooper pair receive the additional impulse

$$(\mathbf{p} + \mathbf{p}') + (-\mathbf{p} + \mathbf{p}') = 2\mathbf{p}', \tag{6.7.4}$$

if the initial impulse of the first electron was $\mathbf{p}$ and the second one $(-\mathbf{p})$. The impulse of the Cooper pair is

$$p_C = 2mv \tag{6.7.5}$$

The electrons in a pair are not necessarily close together; because the interaction is long range. The typical estimation is $\sim 10^{-4}$ cm. This distance is usually greater than the average inter-electron distance. As a result, a Cooper pair begins to move as a single object under influence of the self-consistent electric field. As a superconductor is warmed, its Cooper pairs separate into individual electrons, and the material becomes no superconducting. In other words, thermal energy can break the pairs. On the Internet, the animations can be found for illustration of the mentioned motion. The following remarks should be taken into account by using such kind of animations: (a) the distance between electrons in a Cooper pair practically much more than the distance between neighboring ions in the crystal lattice and (b) by the concurrent motion of many Cooper pairs the lattice ions perform the oscillations along the directed motion of electrons.

From the previous considerations follows that we have quantum hydrodynamic non-local effects. Later in June 1986, physicists Georg Bednorz and Alex Müller at the IBM Laboratory in Zurich, Switzerland, reported that they had created a material that became superconducting at 35 K. Extremely important that they were looking not at metals, but at insulating materials called copper oxides. Ceramics $YBa_2Cu_3O_7$ and TlBaCaCuO became superconducting at 93 K. For high temperature SC the classic BCS theory can't be applied. Finishing the introduction is reasonable to cite the beginning of the BCS-paper [140]: "Since the discovery of the isotope effect, it has been known that SC arises from the interaction between electrons and lattice vibrations, but it has proved difficult to construct an adequate theory based on this concept."

Research of superconductors is curried out very actively. But in spite of obvious success the following conclusion could be established:

1. Contemporary theories of SC based on the SE, practically exhaust their arguments and have no possibility to explain effects of the high temperature SC.
2. Contemporary theories of the high temperature SC (including BCS) based on the SE, cannot propose the principles of search and creation of superconducting materials.
3. The necessity of creation of principal new non-local quantum theories of SC exists.

The main aim consists in construction "an adequate theory based on this concept," the theory distinguished from BCS. Practically the main features of this new theory of SC (including the high temperature SC) were formulated in the previous sections of Chapter 6.

It is known that the Schrödinger-Madelung quantum physics leads to the destruction of the wave packets and cannot be used for the solution of this kind of problems. The appearance in mathematics the soliton solutions is the rare and remarkable effect. As we see the soliton's appearance in the generalized hydrodynamics created by Alexeev is an "ordinary" oft-recurring fact. The realized here mathematical modeling CDW expansion support established mechanism of the relay ("estafette") motion of the soliton' system ("lattice ion—electron") which is realizing by the absence of chemical bonds.

Important to underline that the soliton mechanism of CDW expansion in graphene (and other substances such as $NbSe_2$) takes place in the extremely large diapason of physical parameters. However, CDW existence belongs to effects convoying the high temperature SC. It means that the high temperature SC can be explained in the frame of the non-local soliton quantum hydrodynamics.

From position of the quantum hydrodynamics the problem of search and creation of superconductive materials come to the search of materials which lattices ensure the soliton movement in the self-consistent fields without destruction. In my opinion, the mentioned materials can be created artificially using the technology of the special introduction of quantum dots in matrices on the basement of proposed quantum hydrodynamics. It is known that technology of material creation with special quantum dots exists now in other applications.

The problem of existing and propagation of solitons in graphene and in the perspective high superconducting materials belong to the class of significantly non-local nonlinear problems which can be sold only in the frame of vast numerical modeling.

Chapter 7

Generalized Boltzmann Physical Kinetics in Physics of Plasma

ABSTRACT

We now proceed to apply the generalized Boltzmann kinetic theory to plasmas, where self-consistent forces with appropriately cut-off radius of their action are introduced to expand the capabilities of the generalized Boltzmann equation (GBE).

Keywords: Generalized Boltzmann physical kinetics, Dispersion equations of plasma, Damping of plasma waves, Linear theory of waves propagation, Landau damping

7.1 EXTENSION OF GENERALIZED BOLTZMANN PHYSICAL KINETICS FOR THE TRANSPORT PROCESSES DESCRIPTION IN PLASMA

Let us remind, that the dimensionless equation of the Bogolyubov-Born-Green-Kirkwood-Yvon (BBGKY) hierarchy for the s-particle distribution function (DF) f_s ($s=1,\ldots, N$, N—number of particles in the system) has the form

$$\begin{aligned}&\frac{\partial \hat{f}_s}{\partial \hat{t}_b}+\sum_{i=1}^{s}\hat{\mathbf{v}}_{ib}\cdot\frac{\partial \hat{f}_s}{\partial \hat{\mathbf{r}}_b}+\sum_{ij=1}^{s}\hat{\mathbf{F}}_{ij}\cdot\frac{\partial \hat{f}_s}{\partial \hat{\mathbf{v}}_{ib}}+\alpha\sum_{i=1}^{s}\hat{\mathbf{F}}_i\cdot\frac{\partial \hat{f}_s}{\partial \hat{\mathbf{v}}_{ib}}\\&=-\varepsilon\frac{1}{N}\sum_{i=1}^{s}\sum_{j=s+1}^{N}\int\hat{\mathbf{F}}_{ij}\cdot\frac{\partial}{\partial \hat{\mathbf{v}}_{ib}}\hat{f}_{s+1}\left(\hat{t},\hat{\Omega}_1,\ldots,\hat{\Omega}_s,\hat{\Omega}_j\right)\mathrm{d}\hat{\Omega}_j,\end{aligned}\tag{7.1.1}$$

where $\hat{f}_s=f_s v_{0b}^{3s}n^{-s}$; v_{0b} is the characteristic collision velocity; n is the number density of particles; $\alpha=F_{0\lambda}/F_0$ is the ratio of the scales of the internal and external forces; $\mathrm{d}\Omega_j=\mathrm{d}\mathbf{r}_j\mathrm{d}\mathbf{v}_j$ is an elementary phase volume of the particle j, whose position is determined by the radius-vector r_j and whose velocity v_j. We employ dot notation for a scalar product. A spatial variable is non-dimensionalized by introducing the interaction length r_b, and characteristic timescale is set by $r_b v_{0b}^{-1}$; ε corresponds to the number of particles which is contained in the interaction volume υ_{int} and serves as a small parameter in the kinetic theory of rarefied neutral gases.

There are actually at least three groups of scales to consider in a rarefied gas. Apart from $r_b, v_{0b}, t_{0b}=r_b/v_{0b}$, there exist 'mean free path' λ-scales (the mean free path λ, the mean free-flight velocity $v_{0\lambda}$, and the characteristic time scale $\lambda/v_{0\lambda}$) and L-parameters corresponding to hydrodynamic flow parameters (the characteristic hydrodynamic dimension L, the hydrodynamic velocity v_{0L}, and the hydrodynamic time L/v_{0L}).

The fundamental aspect of plasma physics is the presence of multi-particle interaction. The choice of the characteristic scales which determine the evolution of a plasma volume and are used in the method of many scales below is discussed in Appendix 3. Let us introduce a small parameter $\varepsilon=nr_b^3=\upsilon_{\mathrm{int}}$ assuming that the interaction energy per particle is much less than the particle's kinetic energy. We also assume that the plasma is non-degenerate and employ the multi-scale approach. In the discussion to follow we shall concern ourselves with describing a physical system at the level of one-particle DF f_1 on the scales $r_b\equiv l$, λ, L (l is the Landau length, λ is the mean free path of a probe particle between two "close" collisions, and L is the hydrodynamic scale). Note that the mean free path of a plasma particle is introduced as

$$\lambda_n=\Lambda^{-1}\lambda,\tag{7.1.2}$$

with Λ being the Coulomb logarithm. The mean free path λ_n or the corresponding mean time between the collisions are involved in the definition of kinetic coefficients. In the multi-scale method, $\hat{f}_s$ is expressed in the form of an asymptotic series

Unified non-local theory of transport processes. http://dx.doi.org/10.1016/B978-0-444-63478-8.00007-5

$$\hat{f}_s=\sum_{\upsilon=0}^{\infty}\hat{f}_s^{\upsilon}(\hat{t}_b,\hat{\mathbf{r}}_{ib},\hat{\mathbf{v}}_{ib};\hat{t}_\lambda,\hat{\mathbf{r}}_{i\lambda},\hat{\mathbf{v}}_{i\lambda};\hat{t}_L,\hat{\mathbf{r}}_{iL},\hat{\mathbf{v}}_{iL})\varepsilon^{\upsilon}, \tag{7.1.3}$$

in which the functions $\hat{f}_s^{\upsilon}$ depend on all the three types of variables.

From the above-written BBGKY equation, taking the derivatives of the composite functions on the left-hand side of this equation and then equating the coefficients of ε^0 and ε^1, we find that

$$\frac{\partial \hat{f}_1^0}{\partial \hat{t}_b}+\hat{\mathbf{v}}_{ib}\cdot\frac{\partial \hat{f}_1^0}{\partial \hat{\mathbf{r}}_{1b}}+\alpha\hat{\mathbf{F}}_1\cdot\frac{\partial \hat{f}_1^0}{\partial \hat{\mathbf{v}}_{1b}}=0, \tag{7.1.4}$$

$$\begin{aligned}&\frac{\partial \hat{f}_1^1}{\partial \hat{t}_b}+\hat{\mathbf{v}}_{ib}\cdot\frac{\partial \hat{f}_1^1}{\partial \hat{\mathbf{r}}_{1b}}+\alpha\hat{\mathbf{F}}_1\cdot\frac{\partial \hat{f}_1^1}{\partial \hat{\mathbf{v}}_{1b}}+\varepsilon_2\frac{\partial \hat{f}_1^0}{\partial \hat{t}_\lambda}+\hat{\mathbf{v}}_{ib}\cdot\frac{\partial \hat{f}_1^0}{\partial \hat{\mathbf{r}}_{1\lambda}}+\varepsilon_2\hat{\mathbf{F}}_1\cdot\frac{\partial \hat{f}_1^0}{\partial \hat{\mathbf{v}}_{1\lambda}}\\&+\varepsilon_1\varepsilon_2\varepsilon_3\frac{\partial \hat{f}_1^0}{\partial \hat{t}_L}+\varepsilon_1\hat{\mathbf{v}}_{ib}\cdot\frac{\partial \hat{f}_1^0}{\partial \hat{\mathbf{r}}_{1L}}+\frac{\varepsilon_2}{\varepsilon_3}\hat{\mathbf{F}}_1\cdot\frac{\partial \hat{f}_1^0}{\partial \hat{\mathbf{v}}_{1L}}\\&=-\sum_{\delta=1}^{\eta}\frac{N_\delta}{N}\int\hat{\mathbf{F}}_{1,j\in N_\delta}\cdot\frac{\partial}{\partial \hat{\mathbf{v}}_{1b}}\hat{f}^0_{2,j\in N_\delta}\mathrm{d}\hat{\Omega}_{j\in N_\delta},\end{aligned} \tag{7.1.5}$$

where η is the number of components in the mixture, N_δ is the number of particles of the δth kind, $\varepsilon_1=\lambda/L$ (the Knudsen number), $\varepsilon_2=v_{0\lambda}/v_{0b}$, $\varepsilon_3=v_{0L}/v_{0\lambda}$. The integration in Eq. (7.1.5) is performed on the r_b scale. Importantly, no restriction is placed on the value of the Knudsen number. Equation (7.1.4) shows that the function $\hat{f}_1^0$ does not change along the phase trajectory on the r_b-scale—in other words, following the integration on the r_b-scale we have

$$\hat{f}_1^0=\hat{f}_1^0(\hat{t}_\lambda,\hat{\mathbf{v}}_{1\lambda},\hat{\mathbf{r}}_{1\lambda};\ \hat{t}_L,\hat{\mathbf{v}}_{1L},\hat{\mathbf{r}}_{1L}). \tag{7.1.6}$$

If the last function is known, $\hat{f}_1^1$ needs to be found from Eq. (7.1.5). This is possible if certain additional assumptions are posed on the function $\hat{f}_2^0$ entering the right-hand side, integral part of the expression (7.1.5). Thus we see that the system of equations contains linked terms. In real life, the dependence (Eq. 7.1.6) is unknown beforehand. Then Eq. (7.1.5) can serve to determine $\hat{f}_1^0$ on the λ- and L-scales, but in this case it becomes doubly linked, with respect to both the lower index '2' and the upper index '1'. As a result, the problem of breaking the equations arises.

Let us now write the analog of Eq. (7.1.4) for the two-particle function $\hat{f}_2^0$ dependent on time and on the dynamic variables for the particles 1 and j:

$$\begin{aligned}&\frac{\partial \hat{f}_2^0}{\partial \hat{t}_b}+\hat{\mathbf{v}}_{1b}\cdot\frac{\partial \hat{f}_2^0}{\partial \hat{\mathbf{r}}_{1b}}+\hat{\mathbf{v}}_{j\in N_\delta,b}\cdot\frac{\partial \hat{f}_2^0}{\partial \hat{\mathbf{r}}_{j\in N_\delta,b}}+\hat{\mathbf{F}}_{1,j\in N_\delta}\cdot\frac{\partial \hat{f}_2^0}{\partial \hat{\mathbf{v}}_{1b}}\\&+\hat{\mathbf{F}}_{j\in N_\delta,1}\cdot\frac{\partial \hat{f}_2^0}{\partial \hat{\mathbf{v}}_{j\in N_\delta}}+\alpha\hat{\mathbf{F}}_1\cdot\frac{\partial \hat{f}_2^0}{\partial \hat{\mathbf{v}}_{1b}}+\alpha\hat{\mathbf{F}}_{j\in N_\delta}\cdot\frac{\partial \hat{f}_2^0}{\partial \hat{\mathbf{v}}_{j\in N_\delta,1}}=0.\end{aligned} \tag{7.1.7}$$

Introducing the new variable $\hat{\mathbf{x}}_{1,j\in N_\delta}=\hat{\mathbf{r}}_{1,b}-\hat{\mathbf{r}}_{j\in N_\delta,b}$, we find from Eq. (7.1.7) that:

$$\begin{aligned}-\hat{\mathbf{F}}_{1,j\in N_\delta}\cdot\frac{\partial \hat{f}_2^0}{\partial \hat{\mathbf{v}}_{1b}}&=\frac{\partial \hat{f}_2^0}{\partial \hat{t}_b}+\hat{\mathbf{v}}_{1b}\cdot\frac{\partial \hat{f}_2^0}{\partial \hat{\mathbf{r}}_{1b}}+\left(\hat{\mathbf{v}}_{1b}-\hat{\mathbf{v}}_{j\in N_\delta,b}\right)\cdot\frac{\partial \hat{f}_2^0}{\partial \hat{\mathbf{x}}_{1,j\in N_\delta}}\\&+\hat{\mathbf{F}}_{j\in N_\delta,1}\cdot\frac{\partial \hat{f}_2^0}{\partial \hat{\mathbf{v}}_{j\in N_\delta,b}}+\alpha\hat{\mathbf{F}}_1\cdot\frac{\partial \hat{f}_2^0}{\partial \hat{\mathbf{v}}_{1b}}+\alpha\hat{\mathbf{F}}_{j\in N_\delta}\cdot\frac{\partial \hat{f}_2^0}{\partial \hat{\mathbf{v}}_{j\in N_\delta,b}}.\end{aligned} \tag{7.1.8}$$

Using the last equation, we obtain the following representation for the integral in Eq. (7.1.5):

$$\begin{aligned}&-\int\hat{\mathbf{F}}_{1,j\in N_\delta}\cdot\frac{\partial}{\partial \hat{\mathbf{v}}_{1b}}\hat{f}_2^0\left(\hat{t},\hat{\Omega}_1,\hat{\Omega}_{j\in N_\delta}\right)\mathrm{d}\hat{\Omega}_{j\in N_\delta}\\&=\int\left(\hat{\mathbf{v}}_{1b}-\hat{\mathbf{v}}_{j\in N_\delta,b}\right)\cdot\frac{\partial \hat{f}_2^0}{\partial \hat{\mathbf{x}}_{1,j\in N_\delta}}\mathrm{d}\hat{\Omega}_{j\in N_\delta}\\&+\int\left(\frac{\partial \hat{f}_2^0}{\partial \hat{t}_b}+\hat{\mathbf{v}}_{1b}\cdot\frac{\partial \hat{f}_2^0}{\partial \hat{\mathbf{r}}_{1b}}+\alpha\hat{\mathbf{F}}_1\cdot\frac{\partial \hat{f}_2^0}{\partial \hat{\mathbf{v}}_{1b}}+\alpha\hat{\mathbf{F}}_{j\in N_\delta}\cdot\frac{\partial \hat{f}_2^0}{\partial \hat{\mathbf{v}}_{j\in N_\delta,b}}\right)\mathrm{d}\hat{\Omega}_{j\in N_\delta}\\&+\int\hat{\mathbf{F}}_{j\in N_\delta,1}\cdot\frac{\partial \hat{f}_2^0}{\partial \hat{\mathbf{v}}_{j\in N_\delta,b}}\mathrm{d}\hat{\Omega}_{j\in N_\delta}.\end{aligned} \tag{7.1.9}$$

The last integral on the right-hand side of Eq. (7.1.9) can be written in the form

$$\int \hat{\mathbf{F}}_{j\in N_\delta,1} \cdot \frac{\partial \hat{f}_2^0}{\partial \hat{\mathbf{v}}_{j\in N_\delta,b}} \mathrm{d}\hat{\Omega}_{j\in N_\delta} = \int \left[\int \frac{\partial}{\partial \hat{\mathbf{v}}_{j\in N_\delta,b}} \cdot \left(\hat{\mathbf{F}}_{j\in N_\delta,1} \hat{f}_2^0 \right) \mathrm{d}\hat{\mathbf{v}}_{j\in N_\delta} \right] \mathrm{d}\hat{\mathbf{r}}_{j\in N_\delta}. \tag{7.1.10}$$

But the inner integral can be transformed by the Gauss theorem into an integral taken over an infinitely distant surface in the velocity space, which vanishes because $\hat{f}_2^0 \to 0$ for $\hat{v}_j \to \infty$. Let us now introduce two-particle correlation functions $\hat{W}_2\left(\hat{t}, \hat{\Omega}_1, \hat{\Omega}_{j\in N_\delta}\right)$ (hereinafter f_j is the one-particle function corresponding to the particles N_j):

$$\hat{f}_2^0\left(\hat{t}, \hat{\Omega}_1, \hat{\Omega}_j\right) = \hat{f}_1^0\left(\hat{t}, \hat{\Omega}_1\right) \hat{f}_{j\in N_\delta}\left(\hat{t}, \hat{\Omega}_{j\in N_\delta}\right) + \hat{W}_2^0\left(\hat{t}, \hat{\Omega}_1, \hat{\Omega}_{j\in N_\delta}\right). \tag{7.1.11}$$

The next to last integral in Eq. (7.1.9) then becomes

$$\begin{aligned}
&\int \left(\frac{\partial \hat{f}_2^0}{\partial \hat{t}_b} + \hat{\mathbf{v}}_{1b} \cdot \frac{\partial \hat{f}_2^0}{\partial \hat{\mathbf{r}}_{1b}} + \alpha \hat{\mathbf{F}}_1 \cdot \frac{\partial \hat{f}_2^0}{\partial \hat{\mathbf{v}}_{1b}} + \alpha \hat{\mathbf{F}}_{j\in N_\delta} \cdot \frac{\partial \hat{f}_2^0}{\partial \hat{\mathbf{v}}_{j\in N_\delta}} \right) \mathrm{d}\hat{\Omega}_{j\in N_\delta} \\
&= \int \left[\hat{f}_{j\in N_\delta}^0 \left(\frac{\partial \hat{f}_1^0}{\partial \hat{t}_b} + \hat{\mathbf{v}}_{1b} \cdot \frac{\partial \hat{f}_1^0}{\partial \hat{\mathbf{r}}_{1b}} + \alpha \hat{\mathbf{F}}_1 \cdot \frac{\partial \hat{f}_1^0}{\partial \hat{\mathbf{v}}_{1b}} \right) \right] \mathrm{d}\hat{\Omega}_{j\in N_\delta} \\
&+ \int \hat{f}_1^0 \frac{\partial \hat{f}_{j\in N_\delta}^0}{\partial \hat{t}_b} \mathrm{d}\hat{\Omega}_{j\in N_\delta} + \alpha \int \frac{\partial}{\partial \hat{\mathbf{v}}_{j\in N_\delta,b}} \cdot \left(\hat{\mathbf{F}}_{j\in N_\delta} \hat{f}_1^0 \hat{f}_{j\in N_\delta}^0 \right) \mathrm{d}\hat{\Omega}_{j\in N_\delta} \\
&+ \int \left(\frac{\partial \hat{W}_2^0}{\partial \hat{t}_b} + \hat{\mathbf{v}}_{1b} \cdot \frac{\partial \hat{W}_2^0}{\partial \hat{\mathbf{r}}_{1b}} + \alpha \hat{\mathbf{F}}_1 \cdot \frac{\partial \hat{W}_2^0}{\partial \hat{\mathbf{v}}_{1b}} + \alpha \hat{\mathbf{F}}_{j\in N_\delta} \cdot \frac{\partial \hat{W}_2^0}{\partial \hat{\mathbf{v}}_{j\in N_\delta}} \right) \mathrm{d}\hat{\Omega}_{j\in N_\delta}.
\end{aligned} \tag{7.1.12}$$

In expression (7.1.12), the first integral on the right is zero because of the relation (7.1.4), and the third integral is zero for the same reasons as in Eq. (7.1.10). The situation with the second and fourth integrals, however, requires a more detailed treatment. Consider first the integral

$$A = \int \hat{f}_1^0 \frac{\partial \hat{f}_{j\in N_\delta}^0}{\partial \hat{t}_b} \mathrm{d}\hat{\Omega}_{j\in N_\delta}. \tag{7.1.13}$$

The dynamic variables determining the motion of the given trial particles 1 and j are correlated with one another in the collision of the particles, i.e., on the r_b-scale. In the center-of-mass system, the equations of motion for these particles are written as

$$\dot{\mathbf{v}}_{1b} = \mathbf{F}_{1j}; \quad \dot{\mathbf{v}}_{jb} = \mathbf{F}_{j1}; \quad \mathbf{p}_1 = -\mathbf{p}_j, \tag{7.1.14}$$

where a dot over denotes differentiation with respect to time, and $\mathbf{p}$ is the particle momentum.

Using Eq. (7.1.14) and integrating by parts, we arrive at the relation

$$A \cong -\hat{\mathbf{F}}_{1\delta}^{\mathrm{a}} \cdot \frac{\partial \hat{f}_1}{\partial \hat{\mathbf{v}}_{1b}}, \tag{7.1.15}$$

where $\hat{\mathbf{F}}_{1\delta}^{\mathrm{a}}$ is the average force acting on particle 1 during its collision with particle j which has an arbitrary velocity and an arbitrary position on the r_b-scale (particles j belong to the species δ):

$$\hat{\mathbf{F}}_{1\delta}^{\mathrm{a}} = \int \hat{f}_j \hat{\mathbf{F}}_{1j} \mathrm{d}\hat{\mathbf{v}}_{j\in N_\delta} \mathrm{d}\hat{\mathbf{r}}_{j\in N_\delta}. \tag{7.1.16}$$

Thus, the integral A vanishes provided that the self-consistent force of internal nature can be neglected, in particular, in comparison with the external force acting on particle 1. We next transform the integral A further by using the series (7.1.2), to obtain

$$A \cong -\hat{\mathbf{F}}_{1\delta}^{\mathrm{a}} \cdot \frac{\partial \hat{f}_1^0}{\partial \hat{\mathbf{v}}_{1b}} - \varepsilon \hat{\mathbf{F}}_{1\delta}^{\mathrm{a}} \cdot \frac{\partial \hat{f}_1^1}{\partial \hat{\mathbf{v}}_{1b}}. \tag{7.1.17}$$

The last term in Eq. (7.1.17) ensures, as we shall see below, that the generalized kinetic equation is written in a symmetrical form. Now consider the fourth—the last—integral on the right-hand side of Eq. (7.1.12). To do this, we write down an equation for the two-particle function f_2 of the Bogolyubov chain, in which, in this case, we do not separate the groups of particles belonging to a certain chemical component. The two-particle function f_2 corresponds to the dynamical variables of particles N_1, N_2 and is written in the form $f_2 = f_2(1,2)$.

Thus, one finds

$$\frac{\partial f_2}{\partial t} + \mathbf{v}_1 \cdot \frac{\partial f_2}{\partial \mathbf{r}_1} + \mathbf{v}_2 \cdot \frac{\partial f_2}{\partial \mathbf{r}_2} + \mathbf{F}_{12} \cdot \frac{\partial f_2}{\partial \mathbf{v}_1} + \mathbf{F}_{21} \cdot \frac{\partial f_2}{\partial \mathbf{v}_2} + \mathbf{F}_1 \cdot \frac{\partial f_2}{\partial \mathbf{v}_1} + \mathbf{F}_2 \cdot \frac{\partial f_2}{\partial \mathbf{v}_2}$$
$$= \int \left\{ \begin{array}{l} \mathbf{F}_{13} \cdot \dfrac{\partial}{\partial \mathbf{v}_1} [f_1(1) f_1(2) f_1(3) + f_1(1) W_2(2,3) + f_1(2) W_2(1,3) \\ f_1(3) W_2(1,2)] + \mathbf{F}_{23} \cdot \dfrac{\partial}{\partial \mathbf{v}_2} [f_1(1) f_1(2) f_1(3) + f_1(1) W_2(2,3) \\ + f_1(2) W_2(1,3) + f_1(3) W_2(1,2)] \end{array} \right\} \tag{7.1.18}$$

where the three-particle function f_3 has been approximated by using correlation functions as follows:

$$\begin{aligned} f_3(\Omega_1,\Omega_2,\Omega_3,t) &= f_1(\Omega_1,t) f_1(\Omega_2,t) f_1(\Omega_3,t) + f_1(\Omega_1,t) W_2(\Omega_2,\Omega_3,t) \\ &+ f_1(\Omega_2,t) W_2(\Omega_1,\Omega_3,t) + f_1(\Omega_3,t) W_2(\Omega_1,\Omega_2,t) + W_3(\Omega_1,\Omega_2,\Omega_3,t) \end{aligned} \tag{7.1.19}$$

The effect of the correlation function $W_3(\Omega_1,\Omega_2,\Omega_3,t)$ is here neglected.

Equation (7.1.18) written in the zeroth approximation in ε as an equation for finding f_2^0 is identical with Eq. (7.1.7) only when the zero-order correlation functions are zero, viz.

$$\begin{aligned} &W_2^0(\Omega_2,\Omega_3,t) = 0, \quad W_2^0(\Omega_1,\Omega_3,t) = 0, \\ &W_2^0(\Omega_1,\Omega_2,t) = 0, \quad W_3^0(\Omega_1,\Omega_2,\Omega_3,t) = 0, \end{aligned} \tag{7.1.20}$$

and when the interaction forces determining the effect of the third particle on the first and the second ones during their "close" collision are small, i.e. $F_{13} \approx 0$, $F_{23} \approx 0$. Hence, in the multi-scale approach, polarization terms on the right-hand side of Eq. (7.1.18) appear in the next, of order small $O(\varepsilon^2)$ approximation. Thus, in the multi-scale approach, the last integral on the right-hand side of Eq. (7.1.12) vanishes because of the condition (7.1.20). The integral relation (7.1.9) can be written in the following form:

$$\begin{aligned} &\int \hat{\mathbf{F}}_{1,j\in N_\delta} \cdot \frac{\partial}{\partial \hat{\mathbf{v}}_{1b}} \hat{f}_2^0 \left(\hat{t}, \hat{\Omega}_1, \hat{\Omega}_{j\in N_\delta}\right) \mathrm{d}\hat{\Omega}_{j\in N_\delta} \\ &= \int \left(\hat{\mathbf{v}}_{1b} - \hat{\mathbf{v}}_{j\in N_\delta}\right) \cdot \frac{\partial \hat{f}_2^0}{\partial \hat{\mathbf{x}}_{1,j\in N_\delta}} \mathrm{d}\hat{\Omega}_{j\in N_\delta} - \hat{\mathbf{F}}_{1\delta}^a \cdot \frac{\partial \hat{f}_1^0}{\partial \hat{\mathbf{v}}_{1b}} - \varepsilon \hat{\mathbf{F}}_{1\delta}^a \cdot \frac{\partial \hat{f}_1^1}{\partial \hat{\mathbf{v}}_{1b}}. \end{aligned} \tag{7.1.21}$$

We now introduce the cylindrical coordinate system $\hat{l}, \hat{b}, \varphi$ with the origin at point $\mathbf{r}_i$ and $\hat{l}$-axis parallel to the vector of the relative velocity of the colliding particles 1 and j. Then, in terms of $\hat{b}$ (dimensionless impact parameter) and φ (azimuth angle), the first term on the right-hand side of Eq. (7.1.21) is written as

$$\begin{aligned} \hat{J}^{st,0} &= \sum_{\delta=1}^{\eta} \frac{N_\delta}{N} \int \hat{g}_{j\in N_\delta,1} \left[\int_{-\infty}^{+\infty} \frac{\partial \hat{f}_2^0}{\partial \hat{l}} \mathrm{d}\hat{l} \right] \hat{b}\,\mathrm{d}\hat{b}\,\mathrm{d}\varphi\,\mathrm{d}\hat{\mathbf{v}}_{j\in N_\delta,b} \\ &= \sum_{\delta=1}^{\eta} \frac{N_\delta}{N} \int \left[\hat{f}_2^0(+\infty) - \hat{f}_2^0(-\infty) \right] \hat{g}_{j\in N_\delta,1} \hat{b}\,\mathrm{d}\hat{b}\,\mathrm{d}\varphi\,\mathrm{d}\hat{\mathbf{v}}_{j\in N_\delta,b}. \end{aligned} \tag{7.1.22}$$

The integration in Eq. (7.1.22) was performed on the r_b-scale, i.e., the DFs $\hat{f}_2^0(+\infty), \hat{f}_2^0(-\infty)$ are calculated for the velocities $\hat{\mathbf{v}}_1', \hat{\mathbf{v}}_{j\in N_\delta}'$ and $\hat{\mathbf{v}}_1, \hat{\mathbf{v}}_{j\in N_\delta}$ in the situation where the particles are found outside of their region of interaction—in other words, before or after the collision (with primed velocities in the latter case). If before the collision the conditions of molecular chaos are fulfilled on the λ-scale, then the two-particle DFs can be expressed as a product of one-particle DFs. In this case $\hat{J}^{st,0}$ is the Boltzmann collision integral:

$$\hat{J}^{st,0} = \sum_{\delta=1}^{\eta} \frac{N_\delta}{N} \int \left[\hat{f}_1'^0 \hat{f}'^0_{j\in N_\delta} - \hat{f}_1^0 \hat{f}^0_{j\in N_\delta} \right] \hat{g}_{j\in N_\delta,1} \hat{b}\,\mathrm{d}\hat{b}\,\mathrm{d}\varphi\,d\hat{\mathbf{v}}_{j\in N_\delta}. \tag{7.1.23}$$

Lenard [141] and Balescu [142] solved the equation for the correlation function W_2 under the assumptions of a weakened initial correlation, no time delay, and spatially uniform DF f_1. The corresponding collision integral (the Balescu-Lenard collision integral) incorporates the polarization of the plasma and allows elimination of the logarithmic divergence of the Boltzmann collision integral for a Coulomb plasma [142–144]. If, however, the Boltzmann collision integral is still used in plasma description, a cut-off procedure involving Debye screening must be employed.

Using expressions (7.1.20), the kinetic equation (7.1.5) is written down in the form

$$\begin{aligned}&\frac{\partial\hat{f}_1^1}{\partial\hat{t}_b}+\hat{\mathbf{v}}_{1b}\cdot\frac{\partial\hat{f}_1^1}{\partial\hat{\mathbf{r}}_{1b}}+\left(\alpha\hat{\mathbf{F}}_1+\varepsilon\hat{\mathbf{F}}_1^{\mathrm{a}}\right)\cdot\frac{\partial\hat{f}_1^1}{\partial\hat{\mathbf{v}}_{1b}}+\varepsilon_2\frac{\partial\hat{f}_1^0}{\partial\hat{t}_\lambda}+\hat{\mathbf{v}}_{1b}\cdot\frac{\partial\hat{f}_1^0}{\partial\hat{\mathbf{r}}_{1\lambda}}\\&+\varepsilon_1\varepsilon_2\varepsilon_3\frac{\partial\hat{f}_1^0}{\partial\hat{t}_L}+\varepsilon_1\hat{\mathbf{v}}_{1b}\cdot\frac{\partial\hat{f}_1^0}{\partial\hat{\mathbf{r}}_{1L}}+\hat{\mathbf{F}}_1^{\mathrm{a}}\cdot\frac{\partial\hat{f}_1^0}{\partial\hat{\mathbf{v}}_{1b}}+\varepsilon_2\hat{\mathbf{F}}_1\cdot\frac{\partial\hat{f}_1^0}{\partial\hat{\mathbf{v}}_{1\lambda}}+\frac{\varepsilon_2}{\varepsilon_3}\hat{\mathbf{F}}_1\cdot\frac{\partial\hat{f}_1^0}{\partial\hat{\mathbf{v}}_{1L}}=\hat{J}^{\mathrm{st},0},\end{aligned}\tag{7.1.24}$$

where

$$\hat{\mathbf{F}}_1^{\mathrm{a}}=\sum_{\delta=1}^{\eta}\frac{N_\delta}{N}\hat{\mathbf{F}}_{1\delta}^{\mathrm{a}}.\tag{7.1.25}$$

It should be emphasized that in its dimensional form the factor $\varepsilon\hat{\mathbf{F}}_{1\delta}^{\mathrm{a}}$ can be written in the form

$$\varepsilon\mathbf{F}_{1\delta}^{\mathrm{a}}=\frac{1}{\left(v_{0b}^2/r_b\right)}\int f_{j\in N_\delta}\mathbf{F}_{1,j\in N_\delta}\mathrm{d}\mathbf{v}_{j\in N_\delta}\mathrm{d}\mathbf{r}_{j\in N_\delta},\tag{7.1.26}$$

if it is remembered that

$$\varepsilon=nr_b^3,\ \hat{f}=fv_{0b}^3n^{-1},\ \hat{v}=v/v_{0b},\ \hat{r}=r/r_b,\ \hat{F}_{1j}=F_{1j}/\left(v_{0b}^2/r_b\right).$$

The scale of the internal force F_{1j} corresponds to choosing the Landau length l as r_b. Let us now write Eq. (7.1.24) in the form (cf. Eq. 1.3.58)

$$\frac{\mathrm{D}_1\hat{f}_1^1}{\mathrm{D}\hat{t}_b}+\frac{\mathrm{d}_1\hat{f}_1^0}{\mathrm{d}\hat{t}_{b,\lambda,L}}=\hat{J}^{\mathrm{st},0},\tag{7.1.27}$$

where we have introduced the notation

$$\frac{\mathrm{D}_1\hat{f}_1^1}{\mathrm{D}\hat{t}_b}=\frac{\partial\hat{f}_1^1}{\partial\hat{t}_b}+\hat{\mathbf{v}}_{1b}\cdot\frac{\partial\hat{f}_1^1}{\partial\hat{\mathbf{r}}_{1b}}+\left(\alpha\hat{\mathbf{F}}_1+\varepsilon\hat{\mathbf{F}}_1^{\mathrm{a}}\right)\cdot\frac{\partial\hat{f}_1^1}{\partial\hat{\mathbf{v}}_{1b}}\tag{7.1.28}$$

$$\begin{aligned}\frac{\mathrm{d}_1\hat{f}_1^0}{\mathrm{d}\hat{t}_{b,\lambda,L}}&=\varepsilon_2\frac{\partial\hat{f}_1^0}{\partial\hat{t}_\lambda}+\hat{\mathbf{v}}_{1b}\cdot\frac{\partial\hat{f}_1^0}{\partial\hat{\mathbf{r}}_{1\lambda}}+\varepsilon_1\varepsilon_2\varepsilon_3\frac{\partial\hat{f}_1^0}{\partial\hat{t}_L}+\varepsilon_1\hat{\mathbf{v}}_{1b}\cdot\frac{\partial\hat{f}_1^0}{\partial\hat{\mathbf{r}}_{1L}}\\&+\hat{\mathbf{F}}_1^{\mathrm{a}}\cdot\frac{\partial\hat{f}_1^0}{\partial\hat{\mathbf{v}}_{1b}}+\varepsilon_2\hat{\mathbf{F}}_1\cdot\frac{\partial\hat{f}_1^0}{\partial\hat{\mathbf{v}}_{1\lambda}}+\frac{\varepsilon_2}{\varepsilon_3}\hat{\mathbf{F}}_1\cdot\frac{\partial\hat{f}_1^0}{\partial\hat{\mathbf{v}}_{1L}}\end{aligned}\tag{7.1.29}$$

We now wish to use Eq. (7.1.27) for describing the evolution of the DF $\hat{f}_1^0$—but the trouble is, this equation involves a single-order term $\mathrm{D}_1\hat{f}_1^1/\mathrm{D}_1\hat{t}_b$ linked with respect to upper index. So we are faced with the problem of how to approximate this term—a problem which is similar in a sense to that of approximating the two-particle DF in the collision integral. Using the series (7.1.3) allows an exact representation for the term of interest:

$$\frac{\mathrm{D}_1\hat{f}_1^1}{\mathrm{D}\hat{t}_b}=\frac{\mathrm{D}}{\mathrm{D}\hat{t}_b}\left[\frac{\partial\hat{f}_1}{\partial\varepsilon}\right]_{\varepsilon=0}.\tag{7.1.30}$$

The term $\dfrac{\mathrm{D}_1\hat{f}_1^1}{\mathrm{D}_1\hat{t}_b}$ describes how the DF varies over times of the order of the collision time, or equivalently on the r_b-scale. If this term is left out of account then, from the viewpoint of the derivation of the hierarchy of kinetic equations, this means that

(1) the DF does not vary on the r_b-scale (provided we also neglect the average internal force that gives rise to the second and third terms on the right in Eq. (7.1.21);

(2) the particles are point-like and structureless;
(3) changes in DF due to collisions take place instantaneously and are described by the source term $\hat{J}^{st,0}$.

In the field description, however, the DF f_1 on the interaction scale (r_b-scale) depends on ε through the dynamic variables $\mathbf{r},\mathbf{v},t$ related by the laws of classical mechanics, thus allowing the approximation

$$\begin{aligned}
&\frac{\mathrm{D}_1\hat{f}_1^0}{\mathrm{D}\hat{t}_b}\left[\left(\frac{\partial \hat{f}_1}{\partial \varepsilon}\right)_{\varepsilon=0}\right] \\
&\cong \frac{\mathrm{D}_1}{\mathrm{D}(-\hat{t}_b)}\left[\frac{\partial \hat{f}_1^0}{\partial(-\hat{t}_b)}\left(\frac{\partial(-\hat{t}_b)}{\partial \varepsilon}\right)_{\varepsilon=0}+\frac{\partial \hat{f}_1}{\partial \hat{\mathbf{r}}_{1b}}\cdot\frac{\partial \hat{\mathbf{r}}_b}{\partial(-\hat{t}_b)}\left(\frac{\partial(-\hat{t}_b)}{\partial \varepsilon}\right)_{\varepsilon=0}+\frac{\partial \hat{f}_1}{\partial \hat{\mathbf{v}}_{1b}}\cdot\frac{\partial \hat{\mathbf{v}}_b}{\partial(-\hat{t}_b)}\left(\frac{\partial(-\hat{t}_b)}{\partial \varepsilon}\right)_{\varepsilon=0}\right] \\
&= -\frac{\mathrm{D}_1}{\mathrm{D}\hat{t}_b}\left[\left(\frac{\partial \hat{t}_b}{\partial \varepsilon}\right)_{\varepsilon=0}\frac{D_1\hat{f}_1}{D\hat{t}_b}\right] \\
&\cong -\frac{\mathrm{D}_1}{\mathrm{D}\hat{t}_b}\left[\left(\frac{\partial \hat{t}_b}{\partial \varepsilon}\right)_{\varepsilon=0}\frac{\mathrm{D}_1\hat{f}_1^0}{\mathrm{D}\hat{t}_b}\right].
\end{aligned} \tag{7.1.31}$$

In expression (7.1.31), we have introduced an approximation proceeded against the flying direction of an arrow of time, which corresponds to the condition of there being no correlations for $t_0 \to -\infty$, with t_0 being a certain instant of time on the r_b-scale at which the particles start to interact.

In this way, Markov processes are separated out from all stochastic processes possible in the system.

For the particles of the chemical sort α ($\alpha=1,\ldots,\eta$) we employ the following normalized DF:

$$f_\alpha = f_1 N_\alpha/N, \quad \int f_\alpha \mathrm{d}\mathbf{v}_\alpha = n_\alpha, \quad \int n_\alpha \mathrm{d}\mathbf{r} = N_\alpha. \tag{7.1.32}$$

In Eq. (7.1.32), f_1 is a one-particle DF. Returning to the expression (7.1.27) written in the dimensional form, we convolute the multi-scale substantial derivatives to obtain

$$\frac{\mathrm{D}f_\alpha}{\mathrm{D}t}-\frac{\mathrm{D}}{\mathrm{D}t}\left[\tau_\alpha\frac{\mathrm{D}f_\alpha}{\mathrm{D}t}\right]=\sum_{\beta=1}^{\eta}\int\left[f'_\alpha f'_\beta - f_\alpha f_\beta\right]g_{\beta\alpha}b\mathrm{d}b\mathrm{d}\varphi\mathrm{d}\mathbf{v}_\beta, \tag{7.1.33}$$

where

$$\frac{\mathrm{D}}{\mathrm{D}t}=\frac{\partial}{\partial t}+\mathbf{v}_\alpha\cdot\frac{\partial}{\partial \mathbf{r}}+\mathbf{F}_\alpha^{\mathrm{sc}}\cdot\frac{\partial}{\partial \mathbf{v}_\alpha}, \quad \mathbf{F}_\alpha^{\mathrm{sc}}=\mathbf{F}_\alpha+\mathbf{F}_\alpha^{\mathrm{a}}. \tag{7.1.34}$$

Let us comment on Eq. (7.1.33).

(1) We consider that the particle numbered 1 in the multi-component mixture belongs to a component α, which is exactly what the subscript α on the symbol of the DF indicates. Note also that we dropped the superscript 0 from this symbol: carrying it no longer makes sense because all the equations hereinafter already contain only functions of zero-order (in the sense of the series expansion in terms of the density parameter ε).
(2) The non-local parameter τ_α is written in the form (see also Eq. 1.3.69)

$$\tau_{1\in N_\alpha}=\frac{\varepsilon_{\mathrm{eq}}}{\left[\frac{\partial \varepsilon}{\partial t}\right]_{\varepsilon=0}}, \tag{7.1.35}$$

where ε is the number of particles of all kinds that find themselves in the interaction volume of an a particle by the instant of time t; introducing $\varepsilon^{\mathrm{eq}}$ (the "equilibrium" particle density in the close interaction volume), Eq. (7.1.35) is written in a typical relaxation form (compare with (1.3.80))

$$\frac{\partial \varepsilon}{\partial t}=-\frac{\varepsilon(t)-\varepsilon^{\mathrm{eq}}}{\tau_\alpha}. \tag{7.1.36}$$

The denominator in Eq. (7.1.35) is interpreted as the number of particles that find themselves within the interaction volume of a certain particle belonging to the αth component per unit time; the derivative is calculated under the additional condition $\varepsilon = 0$.

Clearly, this number is equal to the number of collisions occurring in the interaction volume per unit time. Hence, τ_α is the relaxation time proportional to the mean time between close collisions of a particle of the αth sort $\tau_{\alpha,\mathrm{mt}}$ with particles of all other sorts. The procedure includes the action, during the collision of the particles, of the self-consistent force $\mathbf{F}^{\mathrm{sc}}$ being the sum of the external force and the force $\mathbf{F}^{\mathrm{a}}$ of internal origin.

As the derivation of formula (7.1.35) suggests, τ_α is determined by close collisions occurring in the plasma.

By analogy with expression (7.1.2) we have

$$\tau^n_{\alpha,\mathrm{mt}} = \Lambda^{-1}\tau_\alpha, \tag{7.1.37}$$

where $\tau^n_{\alpha,\mathrm{mt}}$ is the mean time between collisions.

In the hydrodynamic approximation, the time τ_α can be expressed in terms of the viscosity μ_α of the component α [145]; for example, for ions one has

$$\tau_\alpha = \Lambda \Pi \mu_\alpha p_\alpha^{-1}. \tag{7.1.38}$$

Equation (7.1.38) involves the coefficient Π, which is determined by the interaction model (for ions $\Pi = 1.04$ [144,145]), and the static pressure

$$p_\alpha = n_\alpha k_{\mathrm{B}} T_\alpha. \tag{7.1.39}$$

The relaxation correction can be included in the parameter Π.

The generalized Boltzmann equation (GBE) is invariant under the Galileo transformation and has a correct free-molecular and Maxwellian asymptotic behavior.

We shall now write down the system of generalized hydrodynamic equations. These equations have been obtained previously for gaseous systems in an external field of forces. The distinguishing feature of the generalized Enskog equations we display below is the inclusion of the self-consistent forces $\mathbf{F}^{\mathrm{sc}}$ [see formulas 7.1.34].

The continuity equation for the component α is given by

$$\begin{aligned} &\frac{\partial}{\partial t}\left\{\rho_\alpha - \tau_\alpha\left[\frac{\partial \rho_\alpha}{\partial t} + \frac{\partial}{\partial \mathbf{r}}\cdot(\rho_\alpha \overline{\mathbf{v}}_\alpha)\right]\right\} \\ &+ \frac{\partial}{\partial \mathbf{r}}\cdot\left\{\rho_\alpha \overline{\mathbf{v}}_\alpha - \tau_\alpha\left[\frac{\partial}{\partial t}(\rho_\alpha \overline{\mathbf{v}}_\alpha) + \frac{\partial}{\partial \mathbf{r}}\cdot \rho_\alpha \overline{\mathbf{v}_\alpha \mathbf{v}_\alpha} - \rho_\alpha \mathbf{F}_\alpha^{(1)\mathrm{sc}} - \frac{q_\alpha}{m_\alpha}\rho_\alpha \overline{\mathbf{v}}_\alpha \mathbf{B}^{\mathrm{sc}}\right]\right\} = R_\alpha, \end{aligned} \tag{7.1.40}$$

the equation of motion is written as

$$\begin{aligned} &\frac{\partial}{\partial t}\left\{\rho_\alpha \overline{\mathbf{v}}_\alpha - \tau_\alpha\left[\frac{\partial}{\partial t}(\rho_\alpha \overline{\mathbf{v}}_\alpha) + \frac{\partial}{\partial \mathbf{r}}\cdot \rho_\alpha \overline{\mathbf{v}_\alpha \mathbf{v}_\alpha} - \rho_\alpha \mathbf{F}_\alpha^{(1)\mathrm{sc}} - \frac{q_\alpha}{m_\alpha}\rho_\alpha \overline{\mathbf{v}}_\alpha \times \mathbf{B}^{\mathrm{sc}}\right]\right\} \\ &- \mathbf{F}_\alpha^{(1)\mathrm{sc}}\left[\rho_\alpha - \tau_\alpha\left(\frac{\partial \rho_\alpha}{\partial t} + \frac{\partial}{\partial \mathbf{r}}\cdot(\rho_\alpha \overline{\mathbf{v}}_\alpha)\right)\right] \\ &- \frac{q_\alpha}{m_\alpha}\left\{\rho_\alpha \overline{\mathbf{v}}_\alpha - \tau_\alpha\left[\frac{\partial}{\partial t}(\rho_\alpha \overline{\mathbf{v}}_\alpha) + \frac{\partial}{\partial \mathbf{r}}\cdot \rho_\alpha \overline{\mathbf{v}_\alpha \mathbf{v}_\alpha} - \rho_\alpha \mathbf{F}_\alpha^{(1)\mathrm{sc}} - \frac{q_\alpha}{m_\alpha}\rho_\alpha \overline{\mathbf{v}}_\alpha \times \mathbf{B}^{\mathrm{sc}}\right]\right\} \times \mathbf{B}^{\mathrm{sc}} \\ &+ \frac{\partial}{\partial \mathbf{r}}\cdot\left\{\rho_\alpha \overline{\mathbf{v}_\alpha \mathbf{v}_\alpha} - \tau_\alpha\left[\frac{\partial}{\partial t}(\rho_\alpha \overline{\mathbf{v}_\alpha \mathbf{v}_\alpha}) + \frac{\partial}{\partial \mathbf{r}}\cdot \rho_\alpha \overline{(\mathbf{v}_\alpha \mathbf{v}_\alpha)\mathbf{v}_\alpha} - \rho_\alpha \mathbf{F}_\alpha^{(1)\mathrm{sc}} \overline{\mathbf{v}}_\alpha - \rho_\alpha \overline{\mathbf{v}}_\alpha \mathbf{F}_\alpha^{(1)\mathrm{sc}}\right.\right. \\ &\left.\left. - \frac{q_\alpha}{m_\alpha}\rho_\alpha \overline{(\mathbf{v}_\alpha \times \mathbf{B}^{\mathrm{sc}})\mathbf{v}_\alpha} - \frac{q_\alpha}{m_\alpha}\rho_\alpha \overline{\mathbf{v}_\alpha(\mathbf{v}_\alpha \times \mathbf{B}^{\mathrm{sc}})}\right]\right\} = \mathbf{I}_{\alpha,\mathrm{mot}} \end{aligned} \tag{7.1.41}$$

and the equation of energy has the form

$$\begin{aligned}
&\frac{\partial}{\partial t}\left\{\frac{\rho_\alpha\overline{v_\alpha^2}}{2}+\varepsilon_\alpha n_\alpha-\tau_\alpha\left[\frac{\partial}{\partial t}\left(\frac{\rho_\alpha\overline{v_\alpha^2}}{2}+\varepsilon_\alpha n_\alpha\right)+\frac{\partial}{\partial\mathbf{r}}\cdot\left(\frac{1}{2}\rho_\alpha\overline{v_\alpha^2\mathbf{v}_\alpha}+\varepsilon_\alpha n_\alpha\overline{\mathbf{v}}_\alpha\right)-\rho_\alpha\mathbf{F}_\alpha^{(1)\mathrm{sc}}\cdot\overline{\mathbf{v}}_\alpha\right]\right\}\\
&+\frac{\partial}{\partial\mathbf{r}}\cdot\left\{\frac{1}{2}\rho_\alpha\overline{v_\alpha^2\mathbf{v}_\alpha}+\varepsilon_\alpha n_\alpha\overline{\mathbf{v}}_\alpha-\tau_\alpha\left[\frac{\partial}{\partial t}\left(\frac{1}{2}\rho_\alpha\overline{v_\alpha^2\mathbf{v}_\alpha}+\varepsilon_\alpha n_\alpha\overline{\mathbf{v}}_\alpha\right)\right.\right.\\
&\left.\left.+\frac{\partial}{\partial\mathbf{r}}\cdot\left(\frac{1}{2}\rho_\alpha\overline{v_\alpha^2\mathbf{v}_\alpha\mathbf{v}_\alpha}+\varepsilon_\alpha n_\alpha\overline{\mathbf{v}_\alpha\mathbf{v}_\alpha}\right)-\rho_\alpha\mathbf{F}_\alpha^{(1)\mathrm{sc}}\cdot\overline{\mathbf{v}_\alpha\mathbf{v}_\alpha}-\frac{1}{2}\rho_\alpha\overline{v_\alpha^2\mathbf{F}_\alpha^{\mathrm{sc}}}-\varepsilon_\alpha n_\alpha\overline{\mathbf{F}_\alpha^{\mathrm{sc}}}\right]\right\}\\
&-\left\{\rho_\alpha\mathbf{F}_\alpha^{(1)\mathrm{sc}}\cdot\overline{\mathbf{v}}_\alpha-\tau_\alpha\left[\mathbf{F}_\alpha^{(1)\mathrm{sc}}\cdot\left(\frac{\partial}{\partial t}(\rho_\alpha\overline{\mathbf{v}}_\alpha)+\frac{\partial}{\partial\mathbf{r}}\cdot\rho_\alpha\overline{\mathbf{v}_\alpha\mathbf{v}_\alpha}-\rho_\alpha\mathbf{F}_\alpha^{(1)\mathrm{sc}}-q_\alpha n_\alpha\overline{\mathbf{v}}_\alpha\times\mathbf{B}^{\mathrm{sc}}\right)\right]\right\}\\
&=\overline{I}_{\alpha,\mathrm{en}},\quad(\alpha=1,\ldots,\eta).
\end{aligned}\tag{7.1.42}$$

where $\mathbf{F}_\alpha^{\mathrm{sc}}$ is the total self-consistent force acting on the unit mass of species of the αth kind, $\mathbf{F}_\alpha^{(1)\mathrm{sc}}$ is the component of the self-consistent force independent of the velocity of the charged particle, $\mathbf{B}^{\mathrm{sc}}$ is the magnetic induction, q_α is the charge of the particle α, ε_α its internal energy, and ρ_α is the density of component α; the bar indicates an average over the velocities.

Thus, the generalized Enskog hydrodynamic equations involve self-consistent forces due to the collective nature of plasma particle interactions. In the following sections, we discuss the applicability of the above theory to plasma physics problems.

7.2 DISPERSION EQUATIONS OF PLASMA IN GENERALIZED BOLTZMANN THEORY

The GBE describes how the one-particle DF $f_\alpha(\alpha=1,\ldots,\eta)$ in a η-component gas mixture changes over times of the order of the time between collisions, of the order of the hydrodynamic flow time, and, unlike the conventional Boltzmann equation (BE), over a time of the order of the collision time. The GBE for a plasma medium has the form

$$\frac{\mathrm{D}f_\alpha}{\mathrm{D}t}-\frac{\mathrm{D}}{\mathrm{D}t}\left(\tau_\alpha\frac{\mathrm{D}f_\alpha}{\mathrm{D}t}\right)=J_\alpha,\tag{7.2.1}$$

where

$$\frac{\mathrm{D}}{\mathrm{D}t}=\frac{\partial}{\partial t}+\mathbf{v}_\alpha\cdot\frac{\partial}{\partial\mathbf{r}}+\mathbf{F}_\alpha\cdot\frac{\partial}{\partial\mathbf{v}_\alpha}\tag{7.2.2}$$

is the substantial (particle) derivative containing the self-consistent force $\mathbf{F}_\alpha$, J_α is the classical (Boltzmann) collision integral, and τ_α is non-local parameter proportional to the mean time between the close particle collisions. In the hydrodynamic regime τ_α can be expressed in terms of the Coulomb logarithm Λ, viscosity μ_α, static pressure p_α, and the coefficient Π dependent on the particle interaction model and the relaxation corrections [see Eq. 7.1.38].

The GBE in general and that for plasma in particular have a fundamentally important feature that the additional GBE terms prove to be of the order of the Knudsen number. This does not mean that in the hydrodynamic (small Kn) limit these terms may be neglected: the Knudsen number in this case appears as a small parameter of the higher derivative in the GBE. Consequently, the additional GBE terms (as compared to the BE) are significant for any Kn, and the order of magnitude of the difference between the BE and GBE solutions is impossible to tell beforehand.

In this connection, it is of interest to apply the GBE model to obtain the dispersion relation for a plasma in the absence of a magnetic field. In doing so, we will make the same assumptions that were used in the BE-based derivation, namely:

(a) the integral collision term is neglected;
(b) the evolution of electrons and ions in a self-consistent electric field corresponds to a nonstationary one-dimensional model;
(c) the DFs for ions, f_i, and for electrons, f_e, deviate little from their equilibrium counterparts $f_{0\mathrm{i}}$ and $f_{0\mathrm{e}}$;
(d) a wave number k and a complex frequency ω are appropriate to the wave mode considered;
(e) the quadratic GBE terms determining the deviation from the equilibrium DF are neglected, and
(f) the self-consistent forces F_i and F_e are small.

Results of the calculations done under these assumptions are given in Appendix 4. Proceeding now to the dispersion relation, we lift one of these assumptions, the first, by introducing the Bhatnagar-Gross-Krook (BGK) collision term

$$J_{\alpha} = -\frac{f_{\alpha} - f_{0\alpha}}{\nu_{\alpha}^{-1}}. \tag{7.2.3}$$

into the right-hand side of the GBE. Here, $f_{0\alpha}$ and $\nu_{\alpha}^{-1} = \tau_{p\alpha}$ are respectively a certain equilibrium DF and the relaxation time for species of the αth kind. Using Eqs. (A.4.9) and (7.2.1), we arrive at the dispersion relation

$$\begin{aligned} 1 = -\frac{e^2}{\varepsilon_0 k}\Bigg\{ & \frac{1}{m_{\mathrm{e}}}\int_{-\infty}^{+\infty} \frac{(\partial f_{0\mathrm{e}}/\partial u)[i - \tau_{\mathrm{e}}(\omega - ku)]}{i(\omega - ku) - \tau_{\mathrm{e}}(\omega - ku)^2 - \nu_{\mathrm{e}}} du \\ & + \frac{1}{m_{\mathrm{i}}}\int_{-\infty}^{+\infty} \frac{(\partial f_{0\mathrm{i}}/\partial u)[i - \tau_{\mathrm{i}}(\omega - ku)]}{i(\omega - ku) - \tau_{\mathrm{i}}(\omega - ku)^2 - \nu_{\mathrm{i}}} du \Bigg\}. \end{aligned} \tag{7.2.4}$$

In the Boltzmann kinetic theory, the analog of Eq. (7.2.4) is the equation [146]

$$1 = -\frac{e^2}{\varepsilon_0 k}\left\{ \frac{1}{m_{\mathrm{e}}}\int_{-\infty}^{+\infty} \frac{(\partial f_{0\mathrm{e}}/\partial u)}{\omega - ku + i\nu_{\mathrm{e}}} du + \frac{1}{m_{\mathrm{i}}}\int_{-\infty}^{+\infty} \frac{(\partial f_{0\mathrm{i}}/\partial u)}{\omega - ku + i\nu_{\mathrm{i}}} du \right\}. \tag{7.2.5}$$

To solve Eq. (7.2.4), we take advantage of the additional conjectures. Let us assume that the ions are at rest and that both the temperature and average velocity of the electrons are zero. Then the electron DF can be expressed in terms of the delta function:

$$f_{0\mathrm{e}}(u) = n_{\mathrm{e}}\delta(u). \tag{7.2.6}$$

Upon integration by parts in Eq. (7.2.4), we arrive at the equation (the subscript "e" on ν_{e}, τ_{e} is dropped for brevity)

$$1 + \frac{e^2 n_{\mathrm{e}}}{\varepsilon_0 m_{\mathrm{e}}}\int_{-\infty}^{+\infty} \frac{\delta(u)\left\{[1 + \mathrm{i}\tau(\omega - ku)]^2 + \nu\tau\right\}}{\left[\mathrm{i}(\omega - ku) - \tau(\omega - ku)^2 - \nu\right]^2} \mathrm{d}u = 0. \tag{7.2.7}$$

In the special case of Boltzmann collisionless plasma, Eq. (7.2.7) leads to the classical formula

$$1 - \frac{e^2 n_{\mathrm{e}}}{\varepsilon_0 m_{\mathrm{e}}}\int_{-\infty}^{+\infty} \frac{\delta(u)}{(\omega - ku)^2} \mathrm{d}u = 0. \tag{7.2.8}$$

Using the properties of the delta function and performing the integration in Eq. (7.2.7), it is found that

$$\omega_{\mathrm{e}}^2 = -\frac{(\nu\tau + \omega^2\tau^2 - \mathrm{i}\omega\tau)^2}{\tau^2(1 + \nu\tau - \omega^2\tau^2 + 2\mathrm{i}\omega\tau)}, \tag{7.2.9}$$

with $\omega_{\mathrm{e}} = \sqrt{e^2 n_{\mathrm{e}}/\varepsilon_0 m_{\mathrm{e}}}$ —being the plasma frequency.

Let us consider the limiting cases inherent in Eq. (7.2.9), namely:

(a) In collisionless limit when $\nu = 0$, $\tau = 0$ we have

$$\omega^2 = \omega_{\mathrm{e}}^2, \tag{7.2.10}$$

or

$$\omega' = \omega_{\mathrm{e}}, \quad \omega'' = 0. \tag{7.2.11}$$

(b) If $\nu \sim \omega' \sim |\omega''|$, $\omega'\tau \ll 1$, then

$$\omega_{\mathrm{e}}^2 = \omega'^2 + 2\mathrm{i}\omega'(\omega'' + \nu) - (\omega'' + \nu)^2 \tag{7.2.12}$$

and separating the real and imaginary parts of relation (7.2.12) leads to the result

$$\omega' = \omega_{\mathrm{e}}, \quad \omega'' = -\nu. \tag{7.2.13}$$

Let us introduce now in consideration the generalized Maxwell equations (GME) (see, for example (I.50) and (I.51))

$$\frac{\partial}{\partial \mathbf{r}}\cdot\mathbf{B}=0,\ \frac{\partial}{\partial \mathbf{r}}\cdot\mathbf{D}=\rho^{a},\ \frac{\partial}{\partial \mathbf{r}}\times\mathbf{E}=-\frac{\partial \mathbf{B}}{\partial t},\ \frac{\partial}{\partial \mathbf{r}}\times\mathbf{H}=\mathbf{j}^{a}+\frac{\partial \mathbf{B}}{\partial t}, \tag{7.2.14}$$

where

$$\rho^{a}=\rho-\rho^{fl},\mathbf{j}^{a}=\mathbf{j}-\mathbf{j}^{fl}, \tag{7.2.15}$$

Fluctuations ρ^{fl}, $\mathbf{j}^{fl}$—which were calculated in the frame of GBE—can be taken from Eq. (2.4.4) and Table 10.1. For example, for one-dimensional (1D) case, we have

$$\delta n_{e}^{fl}=\tau_{e}\left(\frac{\partial \delta n_{e}}{\partial t}+\overline{u}_{e}\frac{\partial \delta n_{e}}{\partial x}\right), \tag{7.2.16}$$

$$\delta n_{i}^{fl}=\tau_{e}\left(\frac{\partial \delta n_{i}}{\partial t}+\overline{u}_{i}\frac{\partial \delta n_{i}}{\partial x}\right), \tag{7.2.17}$$

where $\overline{u}_{e}$ and $\overline{u}_{i}$ are the hydrodynamic velocities of electrons and ions. Equations (7.2.14)–(7.2.17) lead to modification of Eq. (7.2.4):

$$\begin{aligned}1=-\frac{e^{2}}{\varepsilon_{0}k}\Bigg\{&\frac{1}{m_{e}}\int_{-\infty}^{+\infty}\frac{(\partial f_{0e}/\partial u)[i-\tau_{e}(\omega-ku)]}{i(\omega-ku)-\tau_{e}(\omega-ku)^{2}-\nu_{e}}du[1+i\tau_{e}(\omega-k\overline{u}_{e})]\\&+\frac{1}{m_{i}}\int_{-\infty}^{+\infty}\frac{(\partial f_{0i}/\partial u)[i-\tau_{i}(\omega-ku)]}{i(\omega-ku)-\tau_{i}(\omega-ku)^{2}-\nu_{i}}du[1+i\tau_{i}(\omega-k\overline{u}_{i})]\Bigg\}\end{aligned} \tag{7.2.18}$$

Assuming again that the ions are at rest and that both the temperature and average velocity of the electrons are zero, we obtain

$$\omega_{e}^{2}=-\frac{(\nu+\omega^{2}\tau-i\omega)^{2}}{(1+\nu\tau-\omega^{2}\tau^{2}+2i\omega\tau)(1+i\omega\tau)}. \tag{7.2.19}$$

For considered above limit case $\nu\sim\omega'\sim|\omega''|$, $\omega'\tau\ll 1$ the dispersion equation (7.2.19) have the same solution as Eqs. (7.2.9) (see 7.2.13).

Let us show now that GME can lead to reasonable solutions of dispersion equations when in the classical case the corresponding dispersion equation has no physical sense.

One obtains for the limit case

$$\omega'\tau\ll 1,\ |\omega''|\tau\gg 1,\ \nu\sim\omega', \tag{7.2.20}$$

from Eq. (7.2.19):

$$\omega_{e}^{2}=-\frac{\left(\tau\omega''^{2}-2i\tau\omega'\omega''-\nu\right)^{2}}{\tau^{2}\left(\tau\omega''^{2}-2i\tau\omega'\omega''+\nu\right)(i\omega'-\omega'')} \tag{7.2.21}$$

and therefore

$$\omega'=0,\ \omega''=\tau\omega_{e}^{2}. \tag{7.2.22}$$

Then conditions (7.2.20) and by the way conditions

$$\omega'\tau\ll 1,\ |\omega''|\tau\gg 1,\ \nu\sim\omega'' \tag{7.2.23}$$

lead to arising of instability. Interesting to notice that for the mentioned conditions (7.2.20) and (7.2.23) dispersion equation (7.2.9) has no solutions.

Now let the ions be at rest, and the electron component have the following velocity distribution:

$$f_{0e}(u)=n_{e}\delta(u-u_{1}). \tag{7.2.24}$$

The generalized dispersion equation found with taking into account GME has the form

$$1+\frac{e^2 n_e}{\varepsilon_0 m_e}[1+i\tau(\omega-\bar{u}k)]\int_{-\infty}^{+\infty}\frac{\delta(u-u_1)\left\{[1+i\tau(\omega-ku)]^2+\nu\tau\right\}}{\left[i(\omega-ku)-\tau(\omega-ku)^2-\nu\right]^2}du=0. \tag{7.2.25}$$

and after integration in Eq. (7.2.25) one obtains

$$1+\frac{e^2 n_e}{\varepsilon_0 m_e}A\frac{[1+i\tau(\omega-ku_1)]^2+\nu\tau}{\left[i(\omega-ku_1)-\tau(\omega-ku_1)^2-\nu\right]^2}=0, \tag{7.2.26}$$

where

$$A=1-\tau\omega''+i\tau(\omega'-\bar{u}k). \tag{7.2.27}$$

Dispersion equation written with the help of classical Maxwell equations has the form

$$1+\frac{e^2 n_e}{\varepsilon_0 m_e}\frac{[1+i\tau(\omega-ku_1)]^2+\nu\tau}{\left[i(\omega-ku_1)-\tau(\omega-ku_1)^2-\nu\right]^2}=0. \tag{7.2.28}$$

It can be shown that (7.2.26) have physical solutions when dispersion equation (7.2.28) has no solutions at all. With this aim we consider the case when the velocity u_1 of electron beam is equal to hydrodynamic velocity $\bar{u}$ and electrons are trapped by the wave of electrical field:

$$u_1=\bar{u}, \quad \omega'=k\bar{u}. \tag{7.2.29}$$

In this case it follows correspondingly from Eqs. (7.2.26) and (7.2.28):

$$1+\omega_e^2\tau^2\frac{x(x^2+\nu\tau)}{[x(1-x)+\nu\tau]^2}=0, \tag{7.2.30}$$

$$1+\omega_e^2\tau^2\frac{x^2+\nu\tau}{[x(1-x)+\nu\tau]^2}=0, \tag{7.2.31}$$

where

$$\omega_e^2=\frac{e^2 n_e}{\varepsilon_0 m_e}, \quad x=1-\omega''\tau. \tag{7.2.32}$$

Because x is a real number then (7.2.31) has no solutions having physical sense, but Eq. (7.2.30)—by the negative values of x—have solutions which correspond of developing of instability in the system. In particular, if $|x|\ll 1$, $|x|\ll \nu\tau$, then it follows from (7.2.30)

$$x=-(\nu/\omega_e)^2 \tag{7.2.33}$$

or

$$\omega''\tau=1+(\nu/\omega_e)^2. \tag{7.2.34}$$

This solution has a transparent physical meaning.

Now let the ions be at rest, and the electron component has a Maxwellian velocity distribution:

$$f_{0e}=n_e\left(\frac{m_e}{2\pi k_B T}\right)^{3/2}e^{-m_e V^2/2k_B T}, \tag{7.2.35}$$

where k_B is the Boltzmann constant. Then Eq. (7.2.4) becomes

$$1+\frac{e^2 n_e}{\varepsilon_0 k m_e}\left(\frac{m_e}{2\pi k_B T}\right)^{1/2}\int_{-\infty}^{+\infty}\frac{[i-\tau(\omega-ku)]\frac{\partial}{\partial u}\mathrm{e}^{-m_e u^2/2k_B T}}{i(\omega-ku)-\tau(\omega-ku)^2-\nu}\mathrm{d}u=0, \tag{7.2.36}$$

where we have reintroduced the notation ($u\equiv V_x$) for the velocity of the 1D, unsteady wave motion.

From the above equation one derives the expression

$$1+\frac{1}{r_D^2 k^2}\left[1-\sqrt{\frac{m_e}{2\pi k_B T}}\int_{-\infty}^{+\infty}\frac{\{[i-\tau(\omega-ku)]\omega-\nu\}\mathrm{e}^{-m_e u^2/2k_B T}}{i(\omega-ku)-\tau(\omega-ku)^2-\nu}\mathrm{d}u\right]=0, \tag{7.2.37}$$

where $r_D=\sqrt{\varepsilon_0 k_B T/n_e e^2}$ is the Debye-Hückel radius.

Introducing now the dimensionless variables

$$\widehat{u}=u\sqrt{\frac{m_e}{2k_B T}},\quad \widehat{\omega}=\omega\frac{1}{k}\sqrt{\frac{m_e}{2k_B T}}, \tag{7.2.38}$$

$$\widehat{\nu}=\nu\frac{1}{k}\sqrt{\frac{m_e}{2k_B T}},\quad \widehat{\tau}=\tau k\sqrt{\frac{2k_B T}{m_e}}, \tag{7.2.39}$$

we can rewrite Eq. (7.2.37) in the form

$$1+\frac{1}{r_D^2 k^2}\left[1-\sqrt{\frac{1}{\pi}}\int_{-\infty}^{+\infty}\frac{\{[i-\widehat{\tau}(\widehat{\omega}-\widehat{u})]\widehat{\omega}-\widehat{\nu}\}\mathrm{e}^{-\widehat{u}^2}}{i(\widehat{\omega}-\widehat{u})-\widehat{\tau}(\widehat{\omega}-\widehat{u})^2-\widehat{\nu}}\mathrm{d}\widehat{u}\right]=0. \tag{7.2.40}$$

Equation (7.2.40) has the fundamental significance in the following considerations.

7.3 THE GENERALIZED THEORY OF LANDAU DAMPING

The collisionless damping of electron plasma waves was predicted by Landau in 1946 [147–149] and later was confirmed experimentally. Landau damping plays a significant role in many electronics experiments and belongs to the most well known phenomenon in statistical physics of ionized gases.

The physical origin of the collisionless Landau wave damping is simple. Really, if individual electron of mass m_e moves in the periodic electric field, this electron can diminish its energy (electron velocity larger than phase velocity of wave) or receive additional energy from the wave (electron velocity less than phase velocity of wave). Then the total energy balance for a swarm of electrons depends on quantity of "cold" and "hot" electrons. For the Maxwellian DF, the quantity of "cold" electrons is more than quantity of "hot" electrons. This fact leads to, so-called, the collisionless Landau damping of the electric field perturbation.

In spite of transparent physical sense, the effect of Landau damping has continued to be of great interest to theorist as well. Much of this interest is connected with counterintuitive nature of result itself coupled with the rather abstruse mathematical nature of Landau's original derivation (including so-called Landau's rule of complex integral calculation). Moreover, for these reason there were publications containing some controversy over the reality of the phenomenon (see, for example, [150–152]).

In this chapter, I hope to clarify difficulties originated by Landau's derivative. The following consideration leads to another solution of Vlasov-Landau equation, these ones in agreement with data of experiments. The problem Landau damping can be considered from the positions of Generalized Boltzmann Physical Kinetics [153, 154] as the corresponding asymptotic solution of GBE. But here on purpose all consideration will be based on the classical BE written for collisionless case.

Let us remind the classical formulation of the problem. The usual derivation of Landau's damping begins by linearizing the Vlasov equation for the infinite homogeneous collisionless plasma. In doing so, we will make the same assumptions that were used in the Landau derivation namely:

(a) The integral collision term in BE is neglected.
(b) The evolution of electrons in a self-consistent electric field corresponds to a nonstationary 1D model.

(c) Ions are in rest, the DFs for electrons f_e deviates small from the Maxwellian function f_{0e}.

$$f_e = f_{0e}(u) + \delta f_e(x, u, t), \tag{7.3.1}$$

(d) A wave number k and complex frequency ω are appropriate to the wave mode considered.

$$\delta f_e = \langle \delta f_e \rangle e^{i(kx-\omega t)}, \tag{7.3.2}$$

(e) The quadratic terms determining the deviation from the equilibrium DF in kinetic equation

$$\frac{\partial f_e}{\partial t} + u\frac{\partial f_e}{\partial x} + F_e\frac{\partial f_e}{\partial u} = 0, \tag{7.3.3}$$

are neglected.

(f) The change of the electrical potential corresponds to the same spatial-time dependence as the perturbation of DF

$$\psi = \langle \psi \rangle e^{i(kx-\omega t)}, \tag{7.3.4}$$

(g) The self-consistent force F_e is not too large

$$F_e = \frac{e}{m_e}\frac{\partial \psi}{\partial x}, \tag{7.3.5}$$

where e—absolute electron charge. It means that for calculation of function $F_e(\partial f_e/\partial u)$ the equilibrium DF is sufficient:

$$F_e\frac{\partial f_e}{\partial u} = \frac{e}{m_e}\frac{\partial \psi}{\partial x}\frac{\partial f_{0e}}{\partial u} \tag{7.3.6}$$

Let us write down the complex frequency ω in the form

$$\omega = \omega' + i\omega'' \tag{7.3.7}$$

It means that

$$\delta f_e = \langle \delta f_e \rangle e^{i(kx-\omega' t)} e^{\omega'' t}. \tag{7.3.8}$$

As result the Landau's question can be formulated as follows—is it possible to find the solution of Eq. (7.3.3) by the all formulated restrictions, if $\omega'' < 0$? It is well known that the answer of this question leads to the necessity of solution of the dispersion equation (7.2.37) by the conditions $\tau = 0, v = 0$:

$$1 + \frac{1}{r_D^2 k^2}\left[1 - \sqrt{\frac{m_e}{2\pi k_B T}}\int_{-\infty}^{+\infty}\frac{e^{-m_e u^2/2k_B T}}{\omega - ku}du\right] = 0. \tag{7.3.9}$$

After introducing the dimensionless values

$$t = \frac{u\sqrt{m_e}}{\sqrt{2k_B T}}, \quad z_0 = \frac{\omega\sqrt{m_e}}{k\sqrt{2k_B T}}, \tag{7.3.10}$$

the dispersion equation (7.3.9) takes the standard form

$$\int_{-\infty}^{+\infty}\frac{e^{-t^2}}{z_0 - t}dt - \frac{\sqrt{\pi}}{z_0} = r_D^2 k^2 \frac{\sqrt{\pi}}{z_0}, \tag{7.3.11}$$

where

$$r_D = \sqrt{\frac{k_B T}{4\pi n_e e^2}} \tag{7.3.12}$$

k is the wave number, r_D is Debye radius. Let us introduce also the useful notations

$$z_0 \equiv x + iy \equiv \widehat{\omega}' + \widehat{\omega}'', \tag{7.3.13}$$

where

$$\widehat{\omega}' = \frac{\omega'\sqrt{m_e}}{k\sqrt{2k_B T}}, \quad \widehat{\omega}'' = \frac{\omega''\sqrt{m_e}}{k\sqrt{2k_B T}}. \tag{7.3.14}$$

As we see the solution of the dispersion equation (7.3.11) depends significantly on parameter $r_D^2 k^2$ or $r_D^2(2\pi)^2/\lambda^2$. In the so-called long wave approximation when $r_D^2 k^2 \ll 1$, Eq. (7.3.11) can be additionally simplified:

$$\int_{-\infty}^{+\infty} \frac{e^{-t^2}}{z_0 - t} dt = \frac{\sqrt{\pi}}{z_0}. \tag{7.3.15}$$

It seems that after all these restrictions and simplifications the quest of solution could not lead to troubles but it is very far from reality. The main problem consists in evaluation of the Landau integral

$$L(z_0) = \int_{-\infty}^{+\infty} \frac{e^{-t^2}}{z_0 - t} dt. \tag{7.3.16}$$

In the worst case this complex singular integral can be evaluated numerically, but numerical estimation of (7.3.16) leads to numerical solution of (7.3.11). It is reasonable, but un-convenient way because in many cases the problem of Landau damping is only a part of more complicated analytical problem. In the following, I intend to estimate the accuracy of the Landau's original proposition for the $L(z_0)$ evaluation and therefore the accuracy of so-called Landau's rule, which can be found practically in all plasma physics textbooks. It will be shown that Landau rule contains the additional implicit condition implied for physical system.

7.4 EVALUATION OF LANDAU INTEGRAL

The separation of the real and imaginary parts of the integral $L(z_0)$ leads to the following formulas:

$$L(z_0) = U + iV = \int_{-\infty}^{+\infty} \frac{e^{-t^2}}{z_0 - t} dt, \tag{7.4.1}$$

$$U = \operatorname{Re} L(z_0) = \int_{-\infty}^{+\infty} \frac{(x-t)e^{-t^2}}{(x-t)^2 + y^2} dt, \quad V = \operatorname{Im} L(z_0) = -y \int_{-\infty}^{+\infty} \frac{(x-t)e^{-t^2}}{(x-t)^2 + y^2} dt. \tag{7.4.2}$$

Integral $L(z_0)$ can be evaluated by numerical way, the results of corresponding calculations for real part $\operatorname{Re} L(z_0)$ and $\operatorname{Im} L(z_0)$ are shown in Figs. 7.1 and 7.2. As we see the integral surfaces have very complicated character. But the problem is not only in finding of numerical solution of (7.3.11). As it was mentioned above, the ideology

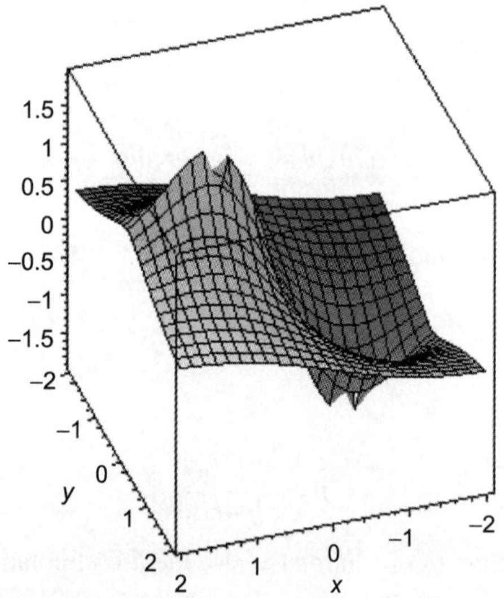

FIGURE 7.1 $\operatorname{Re} L(z_0) = \int_{-\infty}^{+\infty} \frac{(x-t)e^{-t^2}}{(x-t)^2 + y^2} dt$ in domain ($x = -2, \ldots, 2$; $y = -2, \ldots, 2$).

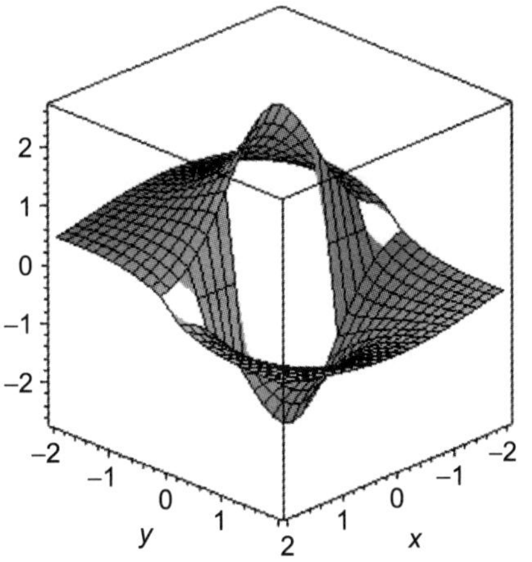

FIGURE 7.2 $\operatorname{Im} L(z_0) = -y\int_{-\infty}^{+\infty} \frac{(x-t)e^{-t^2}}{(x-t)^2+y^2}dt$ in domain ($x=-2,\ldots,2$; $y=-2,\ldots,2$).

of Landau damping penetrates in all physics (not only in electronics) and hundreds thousands of references connected with the consideration of this problem. This fact defines also the importance of analytical solution of the Landau problem.

For the $L(z_0)$ evaluation the well known function (see, for example, the corresponding tables [155–157])

$$w(z_0) = e^{-z_0^2}\left(1 + \frac{2i}{\sqrt{\pi}}\int_0^{z_0} e^{t^2}dt\right) \tag{7.4.3}$$

of the complex argument z_0 could be useful. Let us consider in details the connection of these functions $L(z_0)$ and $w(z_0)$. Introduce the substitution $t=z_0+u$ in (7.4.1). The imaginary parts of z and u cancel each other because t—real variable. But z_0 is fixed complex number ($z_0=x+iy$), then for complex variable u the imaginary part will be also constant and integration should be realized along a line N, which is parallel to the real axis:

$$L(z_0) = -e^{-z_0^2}\int_N \frac{e^{-2z_0u-u^2}}{u}du. \tag{7.4.4}$$

After introduction the function

$$f(z_0) = \int_N \frac{e^{-2z_0u-u^2}}{u}du \tag{7.4.5}$$

we have

$$\int_{-\infty}^{+\infty} \frac{e^{-t^2}}{z_0 - t}dt = -e^{-z_0^2}f(z_0). \tag{7.4.6}$$

Differentiation (7.4.5) with respect to z_0 leads to result

$$f'(z_0) = -2\int_N e^{-2z_0u-u^2}du \tag{7.4.7}$$

or, returning to the variable t:

$$f'(z_0) = -2\int_N e^{z_0^2-t^2}dt. \tag{7.4.8}$$

But integral in the right-hand side of Eq. (7.4.8) contains Poisson integral, then

$$f'(z_0) = -2\sqrt{\pi}e^{z_0^2}. \tag{7.4.9}$$

Upon integrating Eq. (7.4.9) with respect to complex z_0:

$$f(z_0) = -2\sqrt{\pi}\int_0^{z_0} e^{s^2}\,ds + C, \tag{7.4.10}$$

where C is constant, which should be defined. After comparison of (7.4.6) and (7.4.10), one obtains

$$\int_{-\infty}^{+\infty} \frac{e^{-t^2}}{z_0 - t}\,dt = e^{-z_0^2}\left(2\sqrt{\pi}\int_0^{z_0} e^{s^2}\,ds - C\right). \tag{7.4.11}$$

Let us find C. With this aim Eq. (7.4.11) is written by the conditions $x=0, y\to+\infty$. In this case, left-hand side tends to zero and

$$C = 2\sqrt{\pi}\int_0^{iy} e^{s^2}\,ds, \quad y\to+\infty. \tag{7.4.12}$$

Introduce the variable $v=is$. Because the integration is realized along imaginary axis, then v is real value and

$$C = -2i\sqrt{\pi}\int_0^{-\infty} e^{-v^2}\,dv = \pi i, \tag{7.4.13}$$

After substitution of Eq. (7.4.13) in Eq. (7.4.11), one obtains for upper half plane ($y>0$)

$$\int_{-\infty}^{+\infty} \frac{e^{-t^2}}{z_0 - t}\,dt = e^{-z_0^2}\left(2\sqrt{\pi}\int_0^{z_0} e^{s^2}\,ds - \pi i\right) \tag{7.4.14}$$

or (for $y>0$)

$$\int_{-\infty}^{+\infty} \frac{e^{-t^2}}{z_0 - t}\,dt = -\pi i w(z_0). \tag{7.4.15}$$

For lower half plane the formula (7.4.11) should transformed by another way. Let us return to the definition of constant C and calculate this constant by the condition $x=0, y\to-\infty$. In this case

$$C = 2\sqrt{\pi}\int_0^{iy} e^{s^2}\,ds, \text{ by } y\to-\infty; \tag{7.4.16}$$

$$C = -2i\sqrt{\pi}\int_0^{+\infty} e^{-v^2}\,dv = -\pi i. \tag{7.4.17}$$

As result for lower half plane ($y<0$)

$$\int_{-\infty}^{+\infty} \frac{e^{-t^2}}{z_0 - t}\,dt = e^{-z_0^2}\left(2\sqrt{\pi}\int_0^{z_0} e^{s^2}\,ds + \pi i\right). \tag{7.4.18}$$

It means that for the case of Landau damping when we try to find solution of Eq. (7.3.11) in lower half plane $\left(y<0 \text{ or } \widehat{\omega}''<0\right)$, Landau integral $L(z_0)$ cannot be written via function $w(z_0)$.The comparison of Eqs. (7.4.14) and (7.4.18) leads to conclusion that Landau integral function $L(z_0)$ has jump discontinuity in the vicinity of real axis. Let us consider this problem in details using "the approximation of small y." Let us return to evaluation of the function $L(z_0)$ near the real axis with the help of (7.4.14). In the first approximation of the small $y\,(z_0=x+iy,\ 0<y\ll 1)$ we have for $L(z_0)=U(z_0)+iV(z_0)$

$$U = -2\pi xye^{-x^2} + 2\sqrt{\pi}e^{-x^2}\int_0^x e^{s^2}\,ds, \tag{7.4.19}$$

$$V = -\pi e^{-x^2} + 2\sqrt{\pi}y - 4\sqrt{\pi}xye^{-x^2}\int_0^x e^{s^2}\,ds. \tag{7.4.20}$$

For the case $y<0,\ |y|\ll 1$, from Eq. (7.4.18) follows

$$U = 2\pi xye^{-x^2} + 2\sqrt{\pi}e^{-x^2}\int_0^x e^{s^2}\,ds, \tag{7.4.21}$$

$$V = \pi e^{-x^2} + 2\sqrt{\pi}y - 4\sqrt{\pi}xye^{-x^2}\int_0^x e^{s^2}\,ds. \tag{7.4.22}$$

From relations (7.4.19)–(7.4.22) follow:

(A) Function U by crossing real axis from the domain $y<0$ in the domain where $y>0$ has no jump discontinuity in the vicinity of real axis.

(B) Function V by crossing real axis from the domain $y<0$ in the domain where $y>0$ has jump discontinuity in the vicinity of real axis, namely $V(x,y=-0)=\pi e^{-x^2}$, $V(x,y=+0)=-\pi e^{-x^2}$. In particular, $V(x=0,y=-0)=\pi$, $V(x=0,y=+0)=-\pi$.

Let us consider now another possible approximation of the Landau integral $L(z_0)=\int_{-\infty}^{+\infty}\frac{e^{-t^2}}{z_0-t}dt$, namely "the approximation of large z_0." The formal expansion of function $\left(1-\frac{t}{z_0}\right)^{-1}$ in complex series can be written as

$$\left(1-\frac{t}{z_0}\right)^{-1} = 1-\frac{t}{z_0}+\left(\frac{t}{z_0}\right)^2-\left(\frac{t}{z_0}\right)^3+\left(\frac{t}{z_0}\right)^4-\cdots. \tag{7.4.23}$$

After substitution of Eq. (7.4.23) in Landau integral one obtains

$$L(z_0)=\int_{-\infty}^{+\infty}\frac{e^{-t^2}}{z_0-t}dt=\frac{1}{z_0}\sum_{n=0}^{\infty}\int_{-\infty}^{+\infty}e^{-t^2}(-1)^n\left(\frac{t}{z_0}\right)^n dt. \tag{7.4.24}$$

But integrals in the right-hand side of Eq. (7.4.24) containing odd powers of $\left(\frac{t}{z_0}\right)^{2n+1}$ turning into zero then

$$L(z_0)=\int_{-\infty}^{+\infty}\frac{e^{-t^2}}{z_0-t}dt=\frac{1}{z_0}\int_{-\infty}^{+\infty}e^{-t^2}\left(1+\left(\frac{t}{z_0}\right)^2+\left(\frac{t}{z_0}\right)^4+\cdots\right)dt. \tag{7.4.25}$$

For convenience of the series convergence, write down the integral series (7.4.24) as

$$\int_{-\infty}^{+\infty}\frac{e^{-t^2}}{z_0-t}dt=\sum_{k=0}^{\infty}S_{2k+1}. \tag{7.4.26}$$

In this case the subscript in the sum corresponds to maximum reversed power of z_0 in partial sum of infinite series. The first terms of the series have the form

$$S_1=\frac{\sqrt{\pi}}{z_0},\ S_3=\frac{\sqrt{\pi}}{z_0}\left(1+\frac{1}{2}\frac{1}{z_0^2}\right),\ S_5=\frac{\sqrt{\pi}}{z_0}\left(1+\frac{1}{2}\frac{1}{z_0^2}+\frac{3}{4}\frac{1}{z_0^4}\right),\ S_7=\frac{\sqrt{\pi}}{z_0}\left(1+\frac{1}{2}\frac{1}{z_0^2}+\frac{3}{4}\frac{1}{z_0^4}+\frac{15}{8}\frac{1}{z_0^6}\right),$$

$$S_9=\frac{\sqrt{\pi}}{z_0}\left(1+\frac{1}{2}\frac{1}{z_0^2}+\frac{3}{4}\frac{1}{z_0^4}+\frac{15}{8}\frac{1}{z_0^6}+\frac{105}{16}\frac{1}{z_0^8}\right),\ S_{11}=\frac{\sqrt{\pi}}{z_0}\left(1+\frac{1}{2}\frac{1}{z_0^2}+\frac{3}{4}\frac{1}{z_0^4}+\frac{15}{8}\frac{1}{z_0^6}+\frac{105}{16}\frac{1}{z_0^8}+\frac{945}{32}\frac{1}{z_0^{10}}\right),$$

$$S_{13}=\frac{\sqrt{\pi}}{z_0}\left(1+\frac{1}{2}\frac{1}{z_0^2}+\frac{3}{4}\frac{1}{z_0^4}+\frac{15}{8}\frac{1}{z_0^6}+\frac{105}{16}\frac{1}{z_0^8}+\frac{945}{32}\frac{1}{z_0^{10}}+\frac{10395}{64}\frac{1}{z_0^{12}}\right),$$

$$S_{15}=\frac{\sqrt{\pi}}{z_0}\left(1+\frac{1}{2}\frac{1}{z_0^2}+\frac{3}{4}\frac{1}{z_0^4}+\frac{15}{8}\frac{1}{z_0^6}+\frac{105}{16}\frac{1}{z_0^8}+\frac{945}{32}\frac{1}{z_0^{10}}+\frac{10395}{64}\frac{1}{z_0^{12}}+\frac{135135}{128}\frac{1}{z_0^{14}}\right),$$

$$S_{17}=\frac{\sqrt{\pi}}{z_0}\left(1+\frac{1}{2}\frac{1}{z_0^2}+\frac{3}{4}\frac{1}{z_0^4}+\frac{15}{8}\frac{1}{z_0^6}+\frac{105}{16}\frac{1}{z_0^8}+\frac{945}{32}\frac{1}{z_0^{10}}+\frac{10395}{64}\frac{1}{z_0^{12}}+\frac{135135}{128}\frac{1}{z_0^{14}}+\frac{135135\cdot 15}{256}\frac{1}{z_0^{16}}\right),$$

$$S_{19}=\frac{\sqrt{\pi}}{z_0}\left(\begin{array}{l}1+\frac{1}{2}\frac{1}{z_0^2}+\frac{3}{4}\frac{1}{z_0^4}+\frac{15}{8}\frac{1}{z_0^6}+\frac{105}{16}\frac{1}{z_0^8}+\frac{945}{32}\frac{1}{z_0^{10}}+\frac{10395}{64}\frac{1}{z_0^{12}}+\frac{135135}{128}\frac{1}{z_0^{14}}+\frac{135135\cdot 15}{256}\frac{1}{z_0^{16}}\\ +\frac{135135\cdot 15\cdot 17}{512}\frac{1}{z_0^{18}}\end{array}\right)$$

Consider the peculiar features of convergence of the series (7.4.26). Figures 7.3–7.12 contain the results of the $L(z_0)$ calculation for three successive approximations in the method of "large z_0" for real and imagine parts of the corresponding complex functions and sectional views for line $x=y$. As we see the convergence of successive approximations exist in domain $x,y\geq 3$.

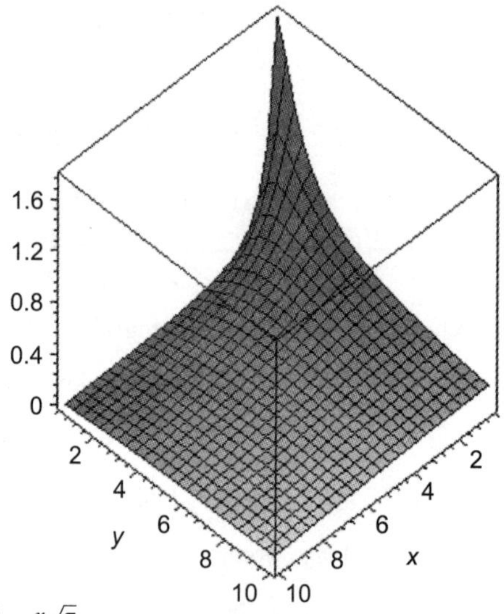

FIGURE 7.3 The 3D image of function $\mathrm{Re}\dfrac{\sqrt{\pi}}{z_0}=\dfrac{x\sqrt{\pi}}{x^2+y^2}$, $(x=0.5,\ldots,10;\, y=0.5,\ldots,10)$.

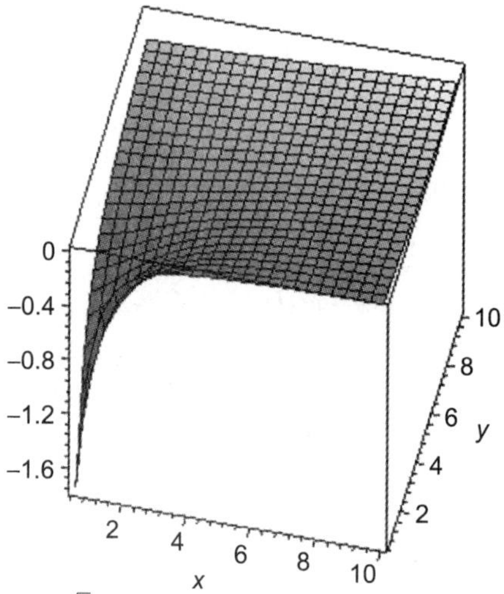

FIGURE 7.4 The 3D image of function $\mathrm{Im}\dfrac{\sqrt{\pi}}{z_0}=-\dfrac{y\sqrt{\pi}}{x^2+y^2}$ $(x=0.5,\ldots,10;\, y=0.5,\ldots,10)$.

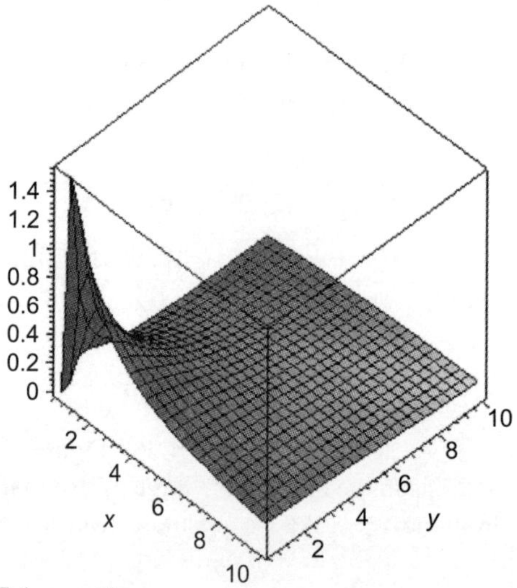

FIGURE 7.5 The 3D image of function $\mathrm{Re}\dfrac{\sqrt{\pi}}{z_0}\left(1+0.5\dfrac{1}{z_0^2}\right)$, $(x=0.5,\ldots,10;\, y=0.5,\ldots,10)$.

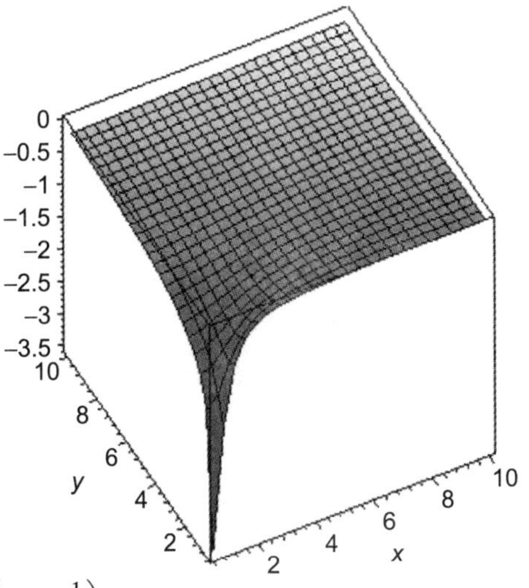

FIGURE 7.6 The 3D image function Im $\frac{\sqrt{\pi}}{z_0}\left(1+0.5\frac{1}{z_0^2}\right)$, $(x=0.5,\ldots,10;\ y=0.5,\ldots,10)$.

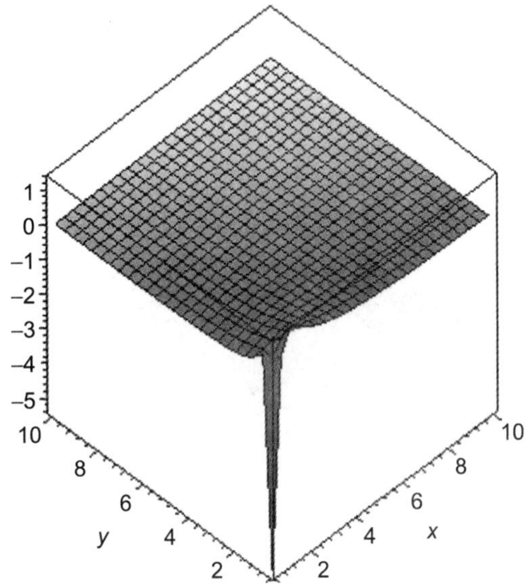

FIGURE 7.7 The 3D image of function Re $\frac{\sqrt{\pi}}{z_0}\left(1+0.5\frac{1}{z_0^2}+0.75\frac{1}{z_0^4}\right)$, $(x=0.5,\ldots,10;\ y=0.5,\ldots,10)$.

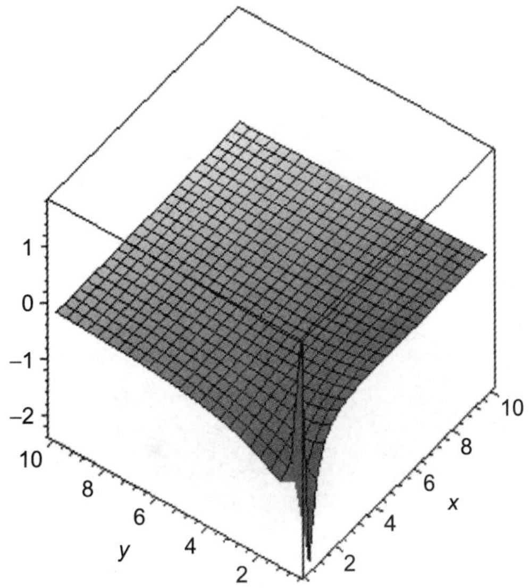

FIGURE 7.8 The 3D image of function Im $\frac{\sqrt{\pi}}{z_0}\left(1+0.5\frac{1}{z_0^2}+0.75\frac{1}{z_0^4}\right)$, $(x=0.5,\ldots,10;\ y=0.5,\ldots,10)$.

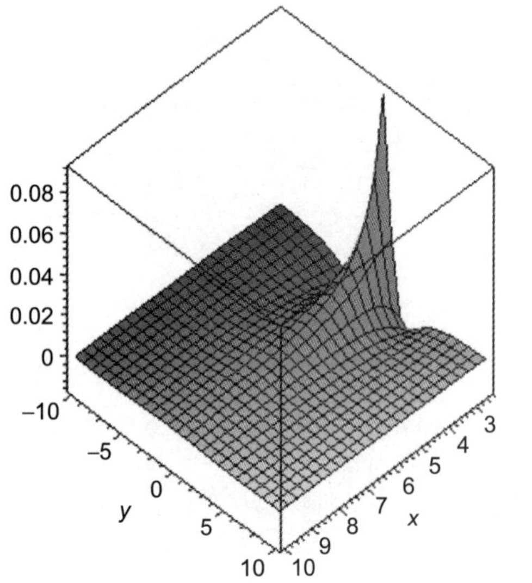

FIGURE 7.9 The 3D image for the difference Re $H_7 = \text{Re}(S_7 - S_1)$.

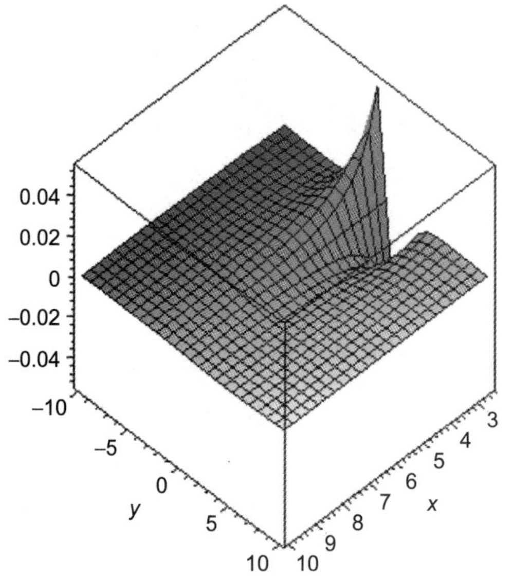

FIGURE 7.10 The 3D image for the difference Im $H_7 = \text{Im}(S_7 - S_1)$.

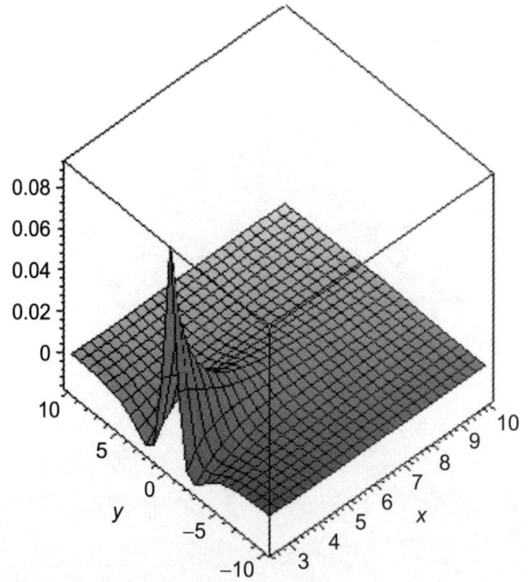

FIGURE 7.11 The 3D image for the difference Re $H_{19} = \text{Re}(S_{19} - S_1)$.

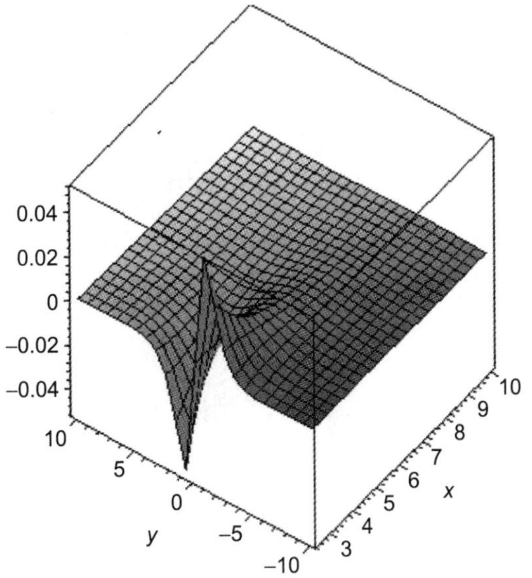

FIGURE 7.12 The 3D image for the difference Im $H_{19}=\mathrm{Im}(S_{19}-S_1)$; $x=2.5,\ldots,10$; $y=-10,\ldots,+10$.

What can be said about the accuracy of the first approximation $S_1=\sqrt{\pi}/z_0$ in comparison with the others S_m? The answer for this question can be received with the help of difference $H_m=S_m-S_1$. The following Figs. 7.9–7.12 correspond to difference $H_7=S_7-S_1$ and $H_{19}=S_{19}-S_1$.

For the better observation of the differences Re $H_{19}=\mathrm{Re}(S_{19}-S_1)$ and Im $H_{19}=\mathrm{Im}(S_{19}-S_1)$ the region of heavy change of the functions are turned around the vertical axis.

The mathematical modeling attests about good convergence of successive approximations of the method of "large z_0" apart of the region near the image axes. From this point of view for accuracy investigation is sufficient to investigate the difference between the numerical evaluation of $L(z_0)$ and its presentation in the form of partial sum S_1. The corresponding difference is written as

$$H_{\mathrm{ex},1}=L(z_0)-S_1, \tag{7.4.27}$$

or

$$H_{\mathrm{ex},1}=\int_{-\infty}^{+\infty}\frac{\mathrm{e}^{-t^2}}{z_0-t}\mathrm{d}t-\frac{\sqrt{\pi}}{z_0}. \tag{7.4.28}$$

(see Figs. 7.13 and 7.14)

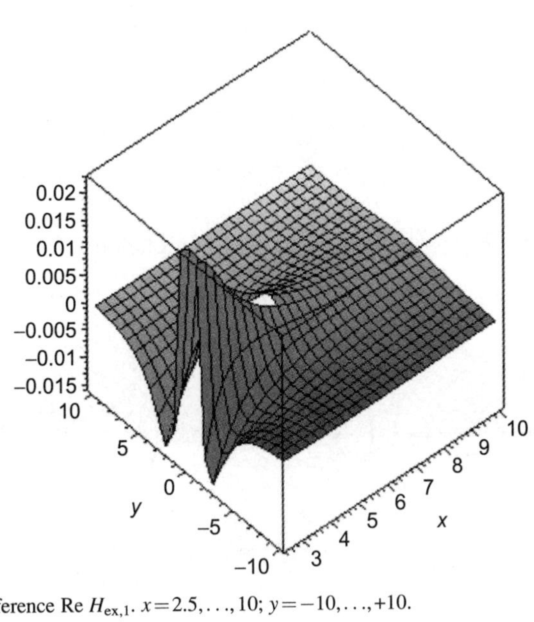

FIGURE 7.13 The 3D image for the difference Re $H_{\mathrm{ex},1}$. $x=2.5,\ldots,10$; $y=-10,\ldots,+10$.

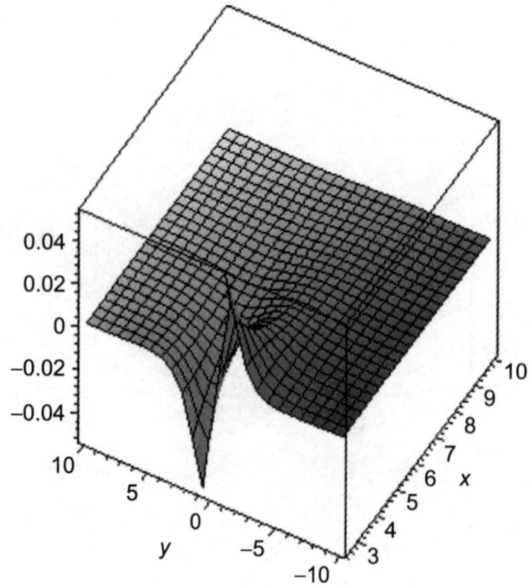

FIGURE 7.14 The 3D image for the difference Im $H_{ex,1}$; $x=2.5,\ldots,10$; $y=-10,\ldots,+10$.

7.5 ESTIMATION OF THE ACCURACY OF LANDAU APPROXIMATION

Let us remind the main steps of derivation of the Landau damping solution for dispersion equation (7.3.11) for the long wave limit case when $r_D^2k^2 \ll 1$. Landau proposal for $L(z_0)$ approximation consists in combination of the sum of semi-residual (written for the upper plane) and three terms from the $L(z_0)$ series of the method of "large z_0":

$$L(z_0)=\int_{-\infty}^{+\infty}\frac{e^{-t^2}}{z_0-t}dt=-i\pi e^{-z_0^2}+\frac{\sqrt{\pi}}{z_0}\left(1+0.5\frac{1}{z_0^2}+0.75\frac{1}{z_0^4}+\cdots\right). \tag{7.5.1}$$

Immediately should be noted that the first term in the right-hand side of (7.5.1) coincides with

$$V=-\pi e^{-x^2}+2\sqrt{\pi}y-4\sqrt{\pi}xye^{-x^2}\int_0^x e^{s^2}ds \tag{7.5.2}$$

from (7.4.20) for small y, but for upper side in spite of the solution is searching in the lower half plane.

After substitution of Eq. (7.5.1) in Eq. (7.3.11) and transformation of Eq. (7.3.11) with the help of the nomenclature (7.3.10), (7.3.12)–(7.3.14) used above one obtains

$$i\omega\sqrt{\pi}\left(\frac{m_e}{2k_BT}\right)^{1.2}\frac{1}{r_D^2k^3}e^{-\widehat{\omega}^2}=-1+\left(0.5\frac{1}{\widehat{\omega}^2}+0.75\frac{1}{\widehat{\omega}^4}+\cdots\right)\frac{1}{r_D^2k^2}. \tag{7.5.3}$$

or

$$i\omega\sqrt{\frac{\pi}{2}}\frac{\omega_e^2}{k^3\left(\frac{k_BT}{m_e}\right)^{3/2}}e^{-\widehat{\omega}^2}=-1+\frac{\omega_e^2}{\omega^2}+\frac{3}{4\widehat{\omega}^4}\frac{1}{r_D^2k^2}, \tag{7.5.4}$$

where plasma frequency is introduced $\omega_e=\sqrt{k_BT/m_e}/r_D$. Using the relation

$$r_D^2k^2\widehat{\omega}^4=\frac{1}{k^2}\omega^4\omega_e^{-2}\frac{m_e}{4k_BT} \tag{7.5.5}$$

transform (7.5.4)

$$1-\frac{\omega_e^2}{(\omega'+i\omega'')^2}-\frac{3\omega_e^2}{(\omega'+i\omega'')^4}k^2\left(\frac{k_BT}{m_e}\right)+i(\omega'+i\omega'')\sqrt{\frac{\pi}{2}}\frac{\omega_e^2}{k^3\left(\frac{k_BT}{m_e}\right)^{3/2}}e^{-(\widehat{\omega}'+i\widehat{\omega}'')^2}=0. \tag{7.5.6}$$

If

$$|\omega''|\ll\omega', \tag{7.5.7}$$

(it is valid for Landau solution), then separation of real and imaginary parts leads to relations

$$1-\frac{\omega_e^2}{\omega'^2}=0, \tag{7.5.8}$$

(real part of (7.5.6) in the first approximation) or

$$\omega'=\omega_e. \tag{7.5.9}$$

Formally from Landau solution follows also for real part

$$1-\frac{\omega_e^2}{\omega'^2}-\frac{3\omega_e^2}{\omega'^4}k^2\left(\frac{k_BT}{m_e}\right)=0 \tag{7.5.10}$$

or

$$\omega'^2=\omega_e^2+3k^2\left(\frac{k_BT}{m_e}\right). \tag{7.5.11}$$

Imaginary part of Eq. (7.5.6) has the form

$$\sqrt{\frac{\pi}{2}}\frac{\omega'}{k^3\left(\frac{k_BT}{m_e}\right)^{3/2}}e^{-\widehat{\omega}'^2}=-\frac{2\omega''}{\omega'^3}. \tag{7.5.12}$$

For damping oscillations the decrement γ can be introduced as

$$\gamma=-\omega'' \tag{7.5.13}$$

and

$$\gamma=\sqrt{\frac{\pi}{8}\frac{\omega_e^4}{k^3}}\left(\frac{k_BT}{m_e}\right)^{-3/2}e^{-\widehat{\omega}_e^{\,2}}, \tag{7.5.14}$$

where

$$\widehat{\omega}_e^{\,2}=\frac{1}{2k^2r_D^2} \tag{7.5.15}$$

It leads to the standard form for the Landau formula

$$\gamma=\sqrt{\frac{\pi}{8}\frac{\omega_e^4}{k^3}}\left(\frac{k_BT}{m_e}\right)^{-3/2}\exp\left[-\frac{1}{2k^2r_D^2}\right]. \tag{7.5.16}$$

In the long wave case Landau formula leads to very small γ, this fact (see Eq. 7.5.16) was used beforehand by the transformations in Landau approximation.

Realize now the direct numerical integration for $L(z_0)$ near the real axis for analysis of the Landau solution. Figure 7.15 contains the integral surface of Re $L(z_0)$ near real axis ($x=-10,\ldots,+10$; $y=-0.01,\ldots,-0.001$), and Fig. 7.16—the integral surface Im $L(z_0)$ also near real axis ($x=-10,\ldots,+10$; $y=-0.01,\ldots,-0.001$). What is the difference between these results and the Landau approximation for $L(z_0)$ (see also 7.5.1)?

The answers for these questions can be found from the Figs. 7.17 and 7.18, which reproduce the difference $\mathrm{Im}\left[L(z_0)+\pi i e^{-z_0^2}-\frac{\sqrt{\pi}}{z_0}\right]$ near the real axis for the domains ($x=0,\ldots,+5$; $y=-0.01,\ldots,-0.001$) and ($x=2,\ldots,+5$; $y=-0.01,\ldots,-0.001$) correspondingly:

1. The results for small x are absolutely inadequate.
2. For $x=2$ the imaginary part of difference between the exact solution and the Landau approximation is about of 0.12. It seems not too bad—the approximation of "large z_0" begins to work. But strict application of combination of semi-residual and the approximation of "large z_0" for lower half plane leads to changing sign in front of the exponential term in the right-hand-side of Eq. (7.5.1) (see strict results (7.4.21) and (7.4.22)).
3. The correction of sign corresponds to Fig. 7.19. The correction of sign (which leads to the integral surface $\mathrm{Im}\left[L(z_0)-\pi i e^{-z_0^2}-\frac{\sqrt{\pi}}{z_0}\right]$ gives the much better approximation near the real axis ($x=2,\ldots,+5$; $y=-0.01,\ldots,-0.001$). The corresponding difference for $x=2$ does not exceed 0.0005.

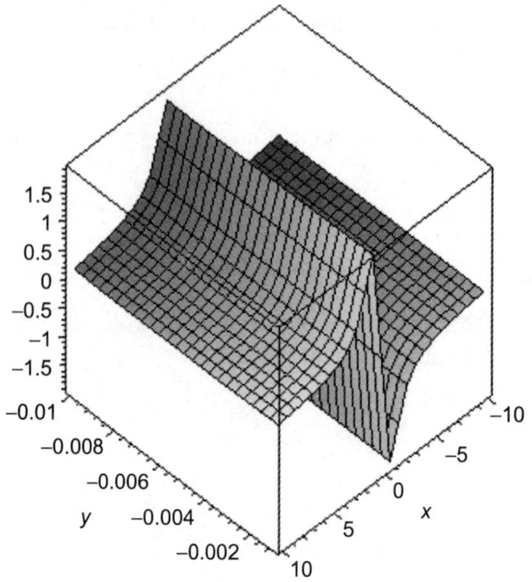

FIGURE 7.15 The integral surface of Re $L(z_0)$ near real axis ($x=-10,\ldots,+10$; $y=-0.01,\ldots,-0.001$).

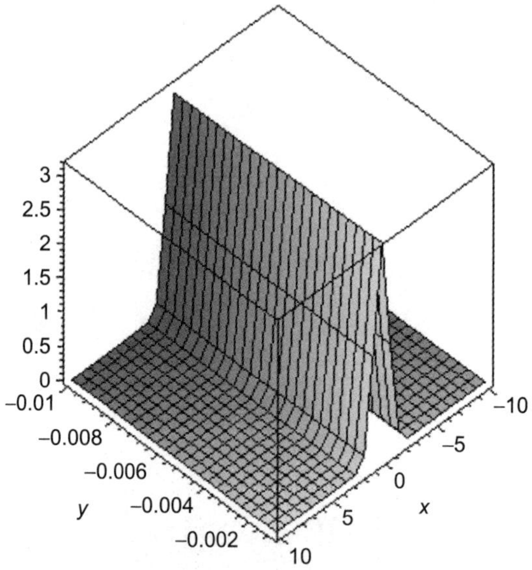

FIGURE 7.16 The integral surface of Im $L(z_0)$ near real axis ($x=-10,\ldots,+10$; $y=-0.01,\ldots,-0.001$).

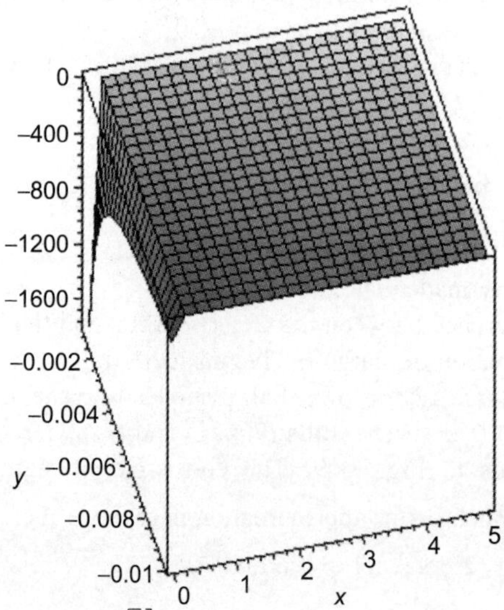

FIGURE 7.17 The integral surface of $\mathrm{Im}\left[L(z_0)+\pi i e^{-z_0^2}-\dfrac{\sqrt{\pi}}{z_0}\right]$ near real axis ($x=0,\ldots,+5$; $y=-0.01,\ldots,-0.001$).

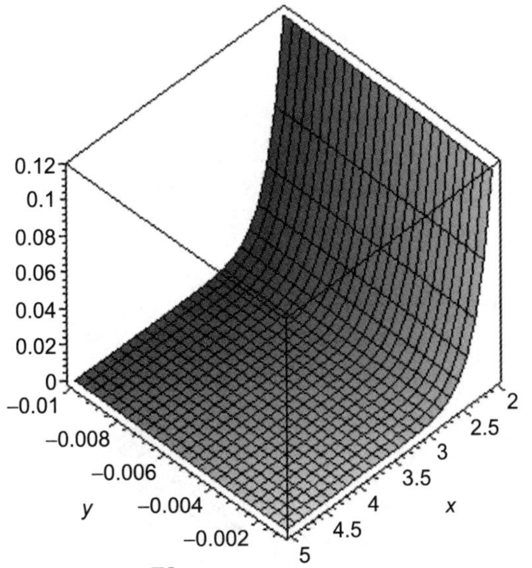

FIGURE 7.18 The integral surface of $\mathrm{Im}\left[L(z_0)+\pi i e^{-z_0^2}-\frac{\sqrt{\pi}}{z_0}\right]$ near real axis ($x=2,\ldots,+5; y=-0.01,\ldots,-0.001$).

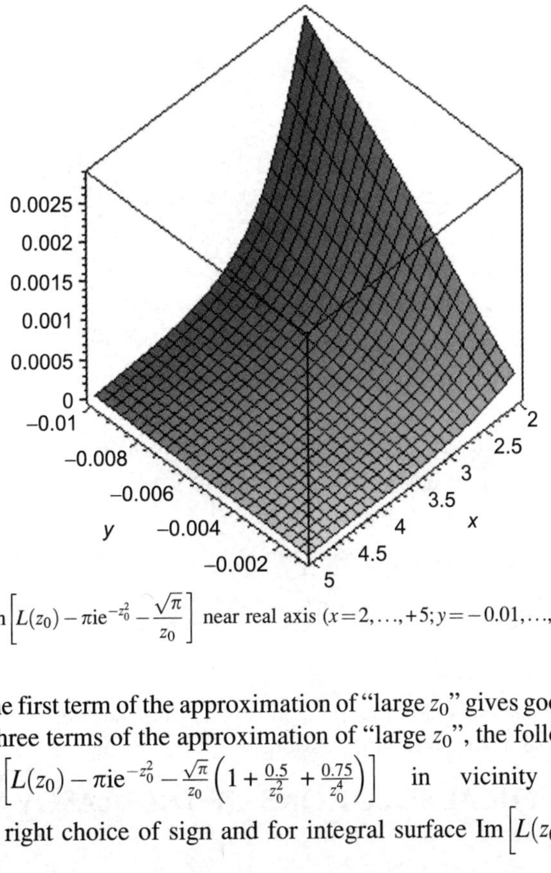

FIGURE 7.19 The integral surface of $\mathrm{Im}\left[L(z_0)-\pi i e^{-z_0^2}-\frac{\sqrt{\pi}}{z_0}\right]$ near real axis ($x=2,\ldots,+5; y=-0.01,\ldots,-0.001$).

4. As it was demonstrated even the first term of the approximation of "large z_0" gives good approximation of $L(z_0)$ for large x. But, because Landau used three terms of the approximation of "large z_0", the following Figs. 7.20 and 7.21 demonstrate the difference $\mathrm{Im}\left[L(z_0)-\pi i e^{-z_0^2}-\frac{\sqrt{\pi}}{z_0}\left(1+\frac{0.5}{z_0^2}+\frac{0.75}{z_0^4}\right)\right]$ in vicinity of real axis ($x=2,\ldots,+5$; $y=-0.01,\ldots,-0.001$) by the right choice of sign and for integral surface $\mathrm{Im}\left[L(z_0)+-\pi i e^{-z_0^2}-\frac{\sqrt{\pi}}{z_0}\left(1+\frac{0.5}{z_0^2}+\frac{0.75}{z_0^4}\right)\right]$ for the Landau approximation.
5. The right correction of the sign leads to liquidation the Landau damping effect, moreover to $\widehat{\omega}''>0$ and to reinforcement of oscillations.

As we see the Landau approximation for the complex integral function $L(z)$ leads to the mathematical contradictions in spite of clear physical sense of effect and the experimental confirmations.

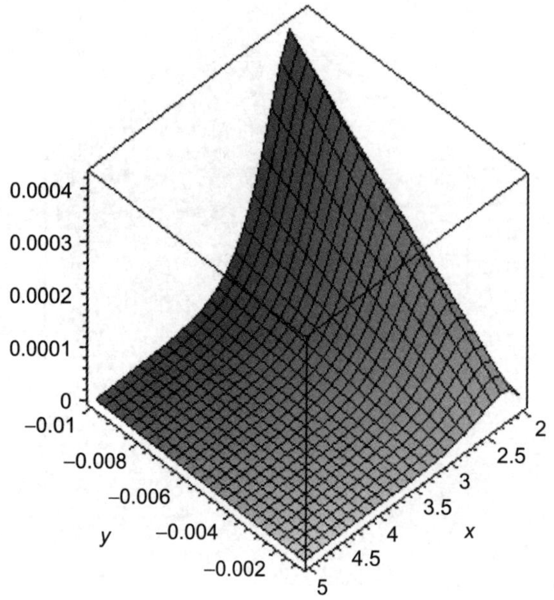

FIGURE 7.20 The integral surface of $\mathrm{Im}\left[L(z_0)-\pi i e^{-z_0^2}-\frac{\sqrt{\pi}}{z_0}\left(1+\frac{0.5}{z_0^2}+\frac{0.75}{z_0^4}\right)\right]$ near real axis ($x=2,\ldots,+5$; $y=-0.01,\ldots,-0.001$).

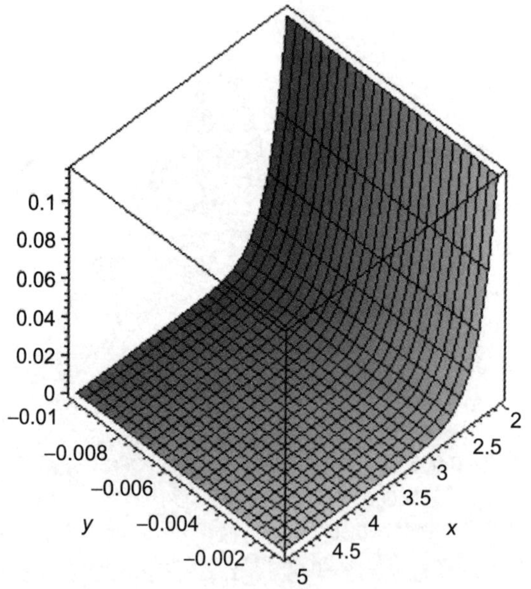

FIGURE 7.21 The integral surface of $\mathrm{Im}\left[L(z_0)+\pi i e^{-z_0^2}-\frac{\sqrt{\pi}}{z_0}\left(1+\frac{0.5}{z_0^2}+\frac{0.75}{z_0^4}\right)\right]$ near real axis ($x=2,\ldots,+5$; $y=-0.01,\ldots,-0.001$). The difference for the Landau approximation.

So called "Landau rule" should be considered as the additional condition implied for physical system. The physical sense of this condition is considered in the following section.

7.6 ALTERNATIVE ANALYTICAL SOLUTIONS OF THE VLASOV-LANDAU DISPERSION EQUATION

From the theory of complex variables is known Cauchy's integral formula: if the function $f(z)$ is analytic inside and on a simple closed curve C, and z_0 is any point inside C, then

$$f(z_0) = -\frac{1}{2\pi i}\oint_C \frac{f(z)}{z_0 - z}\mathrm{d}z \tag{7.6.1}$$

where C is traversed in the positive (counterclockwise) sense.

Let C be the boundary of a simple closed curve placed in lower half plane (for example a semicircle of radius R) with the corresponding element of real axis, z_0 is an interior point. As usual after adding to this semicircle a cross-cut connecting semicircle C with the interior circle (surrounding z_0) of the infinite small radius for analytic $f(z)$ the formula takes place

$$\oint_C \frac{f(z)}{z_0 - z}\mathrm{d}z = -\int_{-R}^{R} \frac{f(\widetilde{x})}{z_0 - \widetilde{x}}\mathrm{d}\widetilde{x} + \int_{C_R} \frac{f(z)}{z_0 - z}\mathrm{d}z + 2\pi i f(z_0), \tag{7.6.2}$$

because the integrals along cross-cut cancel each other $(z = \widetilde{x} + \mathrm{i}\widetilde{y})$.

Analogically for upper half plane

$$\oint_C \frac{f(z)}{z_0 - z}\mathrm{d}z = \int_{-R}^{R} \frac{f(\widetilde{x})}{z_0 - \widetilde{x}}\mathrm{d}\widetilde{x} + \int_{C_R} \frac{f(z)}{z_0 - z}\mathrm{d}z + 2\pi i f(z_0). \tag{7.6.3}$$

The formulas (7.6.2) and (7.6.3) could be used for calculation (including the case $R \to \infty$) of the integrals along the real axis with the help of the residual theory for arbitrary z_0 if analytical function $f(z)$ satisfies the special conditions of decreasing by $R \to \infty$.

Let us consider now integral $\int_{C_R} \frac{\mathrm{e}^{-z^2}}{z_0 - z}\mathrm{d}z$. Generally speaking for function $f(z) = \mathrm{e}^{-z^2}$ Cauchy's conditions are not satisfied. Really for a point $z = \widetilde{x} + \mathrm{i}\widetilde{y}$ this function $f(z) = \mathrm{e}^{\widetilde{y}^2} - \widetilde{x}^2[\cos(2\widetilde{x}\widetilde{y}) - \mathrm{i}\sin(2\widetilde{x}\widetilde{y})]$ and by $|\widetilde{y}| > |\widetilde{x}|$ the function is growing for this part of C_R.

But from physical point of view in the linear problem of interaction of individual electrons only with waves of potential electric field the natural assumption can be introduced that solution depends only of concrete $z_0 = \widehat{\omega}' + \mathrm{i}\widehat{\omega}''$, but does not depend of another possible modes of oscillations in physical system.

It can be realized only if the calculations do not depend of choosing of contour C_R. This fact leads to the additional conditions, for lower half plane

$$\int_{-\infty}^{+\infty} \frac{f(\widetilde{x})}{z_0 - \widetilde{x}}\mathrm{d}\widetilde{x} = 2\pi i f(z_0), \tag{7.6.4}$$

and for upper half plane

$$\int_{-\infty}^{+\infty} \frac{f(\widetilde{x})}{z_0 - \widetilde{x}}\mathrm{d}\widetilde{x} = -2\pi i f(z_0). \tag{7.6.5}$$

From Eqs. (7.6.2) and (7.5.1) follow that Landau approximation contains in implicit form restrictions (valid only for close vicinity of $\widetilde{x}$-axis) for the contour C choosing which leads to the continuous spectrum.

The question arises, is it possible to find solutions of Eq. (7.3.11) by the restriction (7.6.4)? In the following will be shown that the conditions (7.6.4) and (7.6.5) together with (7.3.11) lead to the discrete spectrum of $z_0 = \widehat{\omega}' + \mathrm{i}\widehat{\omega}''$ and from physical point of view condition (7.6.4) can be considered as condition of quantization. Substitution, for example, Eq. (7.6.4) into Eq. (7.3.11) gives the result

$$2\sqrt{\pi}\mathrm{e}^{-z_0^2} = -\mathrm{i}\frac{\beta}{z_0}, \tag{7.6.6}$$

where $\beta = 1 + k^2 r_{\mathrm{D}}^2$. The condition (7.6.5) leads to the formal changing of the sign in front of β. The following construction of solution will depend on β^2, has the universal character but needs to corresponding choosing of solutions for the low and upper planes. Write down Eq. (7.6.6) via complex frequencies

$$2\sqrt{\pi}\,\exp\left(-\widehat{\omega}'^2 + \widehat{\omega}''^2 - 2\widehat{\omega}'\widehat{\omega}''\mathrm{i}\right) = -\mathrm{i}\frac{\beta}{\widehat{\omega}' + \mathrm{i}\widehat{\omega}''} \tag{7.6.7}$$

and separate the real and imaginary parts of equation. Real part

$$-\frac{\beta}{2\sqrt{\pi}}\mathrm{e}^{\widehat{\omega}'^2 - \widehat{\omega}''^2}\cos\left(2\widehat{\omega}'\widehat{\omega}''\right) = \widehat{\omega}'', \tag{7.6.8}$$

imaginary part

$$-\frac{\beta}{2\sqrt{\pi}}\mathrm{e}^{\widehat{\omega}'^2 - \widehat{\omega}''^2}\sin\left(2\widehat{\omega}'\widehat{\omega}''\right) = -\widehat{\omega}'. \tag{7.6.9}$$

After dividing of Eq. (7.6.8) on Eq. (7.6.9), one obtains

$$\widehat{\omega}'\cos\left(2\widehat{\omega}'\widehat{\omega}''\right)+\widehat{\omega}''\sin\left(2\widehat{\omega}'\widehat{\omega}''\right)=0. \tag{7.6.10}$$

After introducing notation

$$\alpha=2\widehat{\omega}'\widehat{\omega}'', \tag{7.6.11}$$

the following system of transcendent equations takes place

$$-\frac{\beta}{4\sqrt{\pi}}\mathrm{e}^{\widehat{\omega}'^2-\widehat{\omega}''^2}\sin 2\alpha=\widehat{\omega}''\sin\alpha, \tag{7.6.12}$$

$$\widehat{\omega}'\cos\alpha+\widehat{\omega}''\sin\alpha=0. \tag{7.6.13}$$

Using the relations

$$\widehat{\omega}'^2=-\frac{1}{2}\alpha\tan\alpha, \widehat{\omega}''^2=-\frac{1}{2}\alpha\cot\alpha, \widehat{\omega}'^2-\widehat{\omega}''^2=\alpha\cot 2\alpha. \tag{7.6.14}$$

we find

$$\mathrm{e}^{\alpha\cot 2\alpha}\sin 2\alpha=-\frac{4\sqrt{\pi}}{\beta}\widehat{\omega}''\sin\alpha, \tag{7.6.15}$$

and after squaring of both parts of (7.6.15)

$$\mathrm{e}^{2\alpha\cot 2\alpha}\sin^2 2\alpha=\frac{16\pi}{\beta^2}\widehat{\omega}''^2\sin^2\alpha \tag{7.6.16}$$

As it was mentioned, this equation does not depend on the sign in front of parameter β. Using once more (7.6.14), we find

$$\mathrm{e}^{2\alpha\cot 2\alpha}\sin 2\alpha=-\frac{4\pi}{\beta^2}\alpha. \tag{7.6.17}$$

After introducing notation

$$\sigma=-2\alpha \tag{7.6.18}$$

the dispersion equation takes the finalized form

$$\mathrm{e}^{\sigma\cot\sigma}\sin\sigma=-\frac{2\pi}{\beta^2}\sigma. \tag{7.6.19}$$

The exact solution of equation (7.6.19) can be found with the help of the W-function of Lambert

$$\sigma_n=\mathrm{Im}\left[W_n\left(\frac{\beta^2}{2\pi}\right)\right], \tag{7.6.20}$$

frequencies $\widehat{\omega}'_n, \widehat{\omega}''_n$ are (see Eqs. 7.3.14, 7.6.14, and 7.6.18)

$$\omega'_n=k\sqrt{-\frac{k_\mathrm{B}T}{2m_\mathrm{e}}\sigma_n\tan\frac{\sigma_n}{2}}, \omega''_n=-k\sqrt{-\frac{k_\mathrm{B}T}{2m_\mathrm{e}}\sigma_n\cot\frac{\sigma_n}{2}}. \tag{7.6.21}$$

In asymptotic for large entire positive n

$$\sigma_n=\left(n+\frac{1}{2}\right)\pi, \widehat{\omega}'_n=\frac{\sqrt{\sigma_n}}{2}=\frac{1}{2}\sqrt{\pi\left(n+\frac{1}{2}\right)}, \widehat{\omega}''_n=-\frac{\sqrt{\sigma_n}}{2}=-\frac{1}{2}\sqrt{\pi\left(n+\frac{1}{2}\right)}. \tag{7.6.22}$$

The exact solution for the nth discrete solution from the spectrum of oscillations:

$$\begin{aligned}\widehat{\omega}_n=&\frac{1}{2}\sqrt{-\mathrm{Im}\left[W_n\left(\frac{\left(1+r_\mathrm{D}^2k^2\right)^2}{2\pi}\right)\right]\tan\left[\frac{1}{2}\mathrm{Im}\left[W_n\left(\frac{\left(1+r_\mathrm{D}^2k^2\right)^2}{2\pi}\right)\right]\right]}\\&-\frac{\mathrm{i}}{2}\sqrt{-\mathrm{Im}\left[W_n\left(\frac{\left(1+r_\mathrm{D}^2k^2\right)^2}{2\pi}\right)\right]\cot\left[\frac{1}{2}\mathrm{Im}\left[W_n\left(\frac{\left(1+r_\mathrm{D}^2k^2\right)^2}{2\pi}\right)\right]\right]}.\end{aligned} \tag{7.6.23}$$

The square of the oscillation frequency of plasma waves ω'^2_n is proportional to the wave energy. Hence, the energy of plasma waves is quantized, and as n grows we have the asymptotic expression

$$\widehat{\omega}'^2_n = \frac{\pi}{4}\left(n + \frac{1}{2}\right) \tag{7.6.24}$$

and the squares of possible dimensionless frequencies become equally spaced:

$$\widehat{\omega}'^2_{n+1} - \widehat{\omega}'^2_n = \frac{\pi}{4}. \tag{7.6.25}$$

Figures 7.22 and 7.23 reflect the result of calculations for 200 discrete levels for low complex half plane. For high levels this spectrum contains many very close practically straight lines, which human eyes can perceive as background. Moreover plotter from the technical point of view has no possibility to reflect the small curvature of lines approximating this curvature as a long step. My suggestion is to turn this shortcoming into merit in explication of topology of high quantum levels in quantum systems.

Really, extremely interesting that this (from the first glance) grave shortcoming of plotters lead to the automatic construction of approximation for derivatives $d(r_D k)/d\widehat{\omega}'$ and $d(r_D k)/d\widehat{\omega}''$. This effect has no attitude to the mathematical programming. You can see this very complicated topology of curves including the spectrum of the bell-like dispersion curves in Figs. 7.22 and 7.23, which also form the discrete spectrum.

But mentioned derivatives can be written in the form

$$\frac{d(r_D k)}{d\widehat{\omega}'} = \frac{dk}{d(\omega'/k)}\frac{k_B T\sqrt{2}}{m_e \omega_e} = \frac{dk}{dv_\phi}\frac{k_B T\sqrt{2}}{m_e \omega_e}, \tag{7.6.26}$$

where v_ϕ is the wave phase velocity. Then

$$\frac{dv_\phi}{d\lambda} = -\frac{2\pi k_B T\sqrt{2}}{\lambda^2 m_e \omega_e}\left[\frac{d(r_D k)}{d\widehat{\omega}'}\right]^{-1} \tag{7.6.27}$$

or

$$\frac{d(r_D k)}{d\widehat{\omega}'} = \frac{dk}{dv_\phi}\frac{k_B T}{e\sqrt{2\pi \rho_e}}, \tag{7.6.28}$$

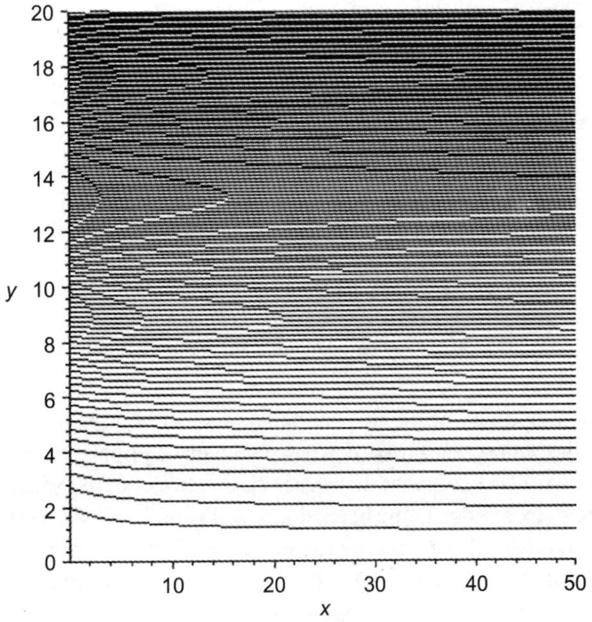

FIGURE 7.22 The dimensionless frequency $\widehat{\omega}'$ (y axes) versus parameter $r_D k$ (x axes).

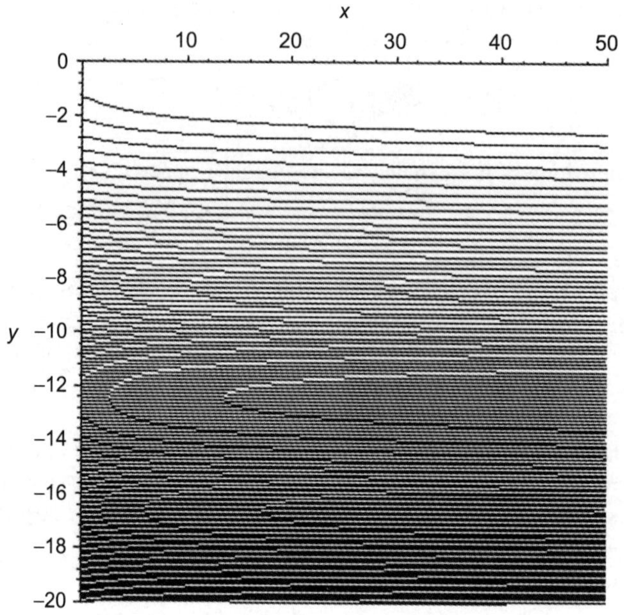

FIGURE 7.23 The dimensionless frequency $\widehat{\omega}''$ (y axes) versus parameter $r_{\mathrm{D}}k$ (x axes).

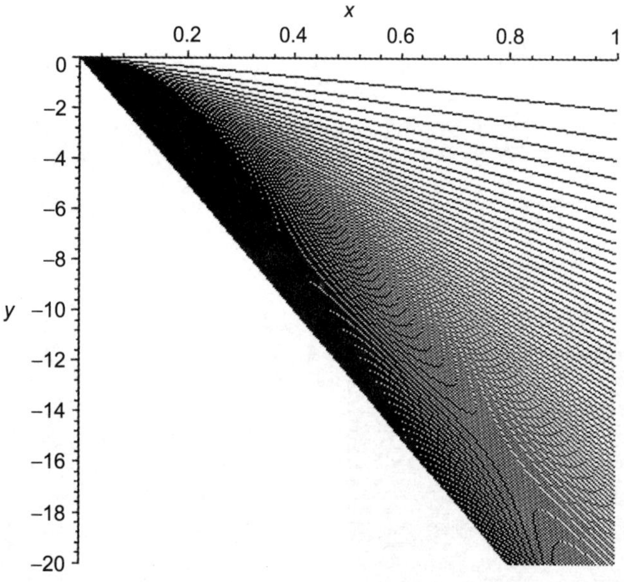

FIGURE 7.24 Dependence $\widehat{\omega}''/\widehat{\omega}_{\mathrm{e}}$ (y axes) on $r_{\mathrm{D}}k$ (x axes).

$$\frac{\mathrm{d}v_{\phi}}{\mathrm{d}\lambda}=-\frac{1}{\lambda^2}\frac{k_{\mathrm{B}}T}{e}\sqrt{\frac{2\pi}{\rho_{\mathrm{e}}}}\left[\frac{\mathrm{d}(r_{\mathrm{D}}k)}{\mathrm{d}\widehat{\omega}'}\right]^{-1}, \tag{7.6.29}$$

where $\rho_{\mathrm{e}}=m_{\mathrm{e}}n_{\mathrm{e}}$. Very complicated topology of the dispersion curves can be revealed by the construction of dependence $\widehat{\omega}''/\widehat{\omega}_{\mathrm{e}}$ versus $r_{\mathrm{D}}k$ (Fig. 7.24). By the curves compression we can see very complicated topology until the intermediate area passes over the black domain where (in chosen scale) the curves cannot be observed separately. Enlarging of scaling shows that the complicated curves topology exists also in the black domain. Then Figs. 7.22–7.24 can be used for understanding of the future development of events in physical system after the initial linear stage. For example, Fig. 7.22 shows the discrete set of frequencies which vicinity corresponds to passing over from abnormal to normal dispersion (for example, by $\widehat{\omega}'\sim 9$) for discrete systems of $r_{\mathrm{D}}k$. Of course the non-linear stage needs the special investigation with using of another methods including the method of direct mathematical modeling [158–161].

We now proceed to compare the theoretical results with those of the computer experiment. Extensive simulations of the attenuation of Langmuir waves in plasma have been performed at the SB RAS Institute of Nuclear Physics (Novosibirsk) in the 1970s and 1980s (see, for example, Refs. [158–161]). Of interest to us here is the formulation of the problem close to the classical Landau formulation. The problem involves a 1D plasma system subject to periodic boundary conditions. The velocity distribution of plasma electrons is taken to be Maxwellian, and ions are assumed to be at rest $(m_i/m_e)=10^4$ and distributed uniformly over the length of the system. It is also assumed that at some initial point in time the system is subjected to small electron velocity and electron density perturbations of the form

$$\frac{\delta n}{n_0}=\frac{k_0E_0}{4\pi en_0}\sin(\omega_0 t-k_0x),\ \delta v=\frac{\omega_0E_0}{4\pi en_0}\sin(\omega_0 t-k_0x), \tag{7.6.30}$$

corresponding to the linear wave $E(x,t)=E_0\sin(\omega_0 t-k_0x)$, where $\omega_0^2=\omega_e^2+\frac{3}{2}k_0^2v_T^2$, $k_0=2\pi/\lambda_0$, v_T is the thermal electron velocity. The quantities E_0, ψ_0, λ_0, ω_0, and k_0 are the initial values of the field amplitude, potential, wavelength, frequency, and wave number, respectively. The numerical integration is performed using the 'particles-in-cells' method. The number of particles is not large (in Refs. [158–161], the authors usually put $N=10^4$, with about 10^2 particles per cell). To reduce the initial noise level, the 'easy start' method is used, in which neither the coordinate nor velocity DFs of the electrons change from one cell to another. In this case, it was noted in Ref. [158] that the noise level is determined by computation errors but increases for the computation scheme chosen; the noise level increases with increasing E_0 and with decreasing λ_0. The computation only makes sense until the noise level remains small compared to the harmonics of the effect under study that arise in the calculation.

The calculations in Refs. [158–161] were performed over a wide range of initial wave parameters. The time dependence of the field strength is quite complicated, but the initial stage always corresponds to the wave damping regime in which an increase in the amplitude and the phase velocity v_ϕ in the range $e\psi_0/(k_BT)>1$ (and the corresponding decrease of the parameter k_0r_D) dramatically increases the damping decrement compared to the Landau value (7.5.16).

In theory and mathematical experiments the question can be put about the number of trapped electrons by traveling wave electric field. But the mathematical expectation of the number of electrons moving with the phase velocity is zero. Then it is possible to speak only about a definite velocity interval, which contains phase velocity of wave, and about electrons (or probe particles in mathematical experiments) belonging to this interval.

In mentioned mathematical experiments they use also another possibility of calculation of the trapped particles. The mathematical experiments were realized with the initial number of particles in cell, which was not more than 100 and maximum initial velocity, which was less than $2.15v_T$, where $v_T=\sqrt{k_BT/m_e}$ is the thermal electron velocity. In experiments, the phase velocity was changed in diapason from $2.46v_T$ to $22.4v_T$. Then it is possible to calculate the number of trapped probe particles which velocity—in the process of evolution of the physical system—become more than phase velocity v_ϕ. At the beginning time moment all high energetic electrons in the tail of the Maxwell DF are cutting off. But the number of these "lost" electrons (estimation of authors of experiments) is not more than $\sim$1.6%.

From discussed numerical experiments follow that the number of trapped electrons is not more than 20%. On the macroscopic level of the system description we can speak only about average (hydrodynamic) velocity $\bar{u}$ of electrons and about difference between $\bar{u}$ and v_ϕ.

Figure 7.25 summarizes the results of the numerical simulation in series experiments (3-1,. . .,3-6), [158]. Table 7.1 explains the used nomenclature.

Black squares in Fig. 7.25 correspond to the data of the proposed analytical theory. From formulated numerical results of the mathematical modeling follow:

1. Reasonable coincidence of the damping decrement γ with the Landau decrement γ_L in series 3-1,. . .,3-6 is observed when potential energy of the electric field is of order of the heat energy and by rather large (but not small) values r_Dk.
2. The results of the γ/γ_L calculations depends only slightly on the ratio $\sqrt{e\langle\psi\rangle/m_e}/v_T$ (in [161] $v_T=\sqrt{k_BT/m_e}$), but the difference between γ and γ_L catastrophically increases in spite of decreasing of the parameter r_Dk. It is of important that in series of the mathematical experiments (1-1,. . .,1-8), authors [161] come to a halt by the ratio $\gamma/\gamma_L\sim10^{104}$ by $r_Dk=0.045$.
3. The results of the proposed analytical theory are in good coincidence with the data of the direct mathematical modeling. By the way the linear theory does not contain the explicit dependence on the electrical field energy.

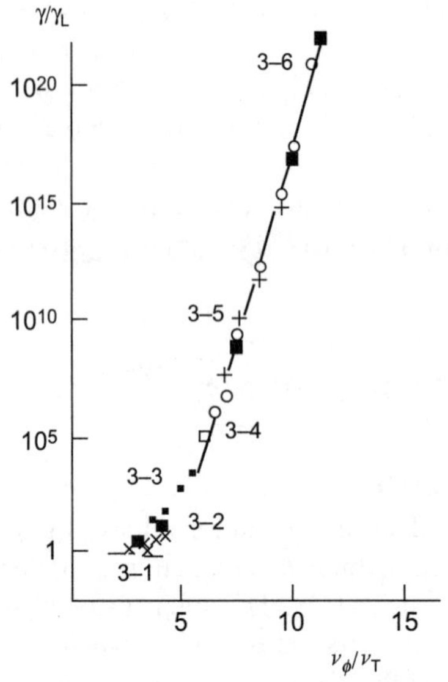

FIGURE 7.25 Dependence γ/γ_L (y axes) on v_ϕ/v_T, (x axes).

TABLE 7.1 Nomenclature and Data of the Mathematical Experiments by $\sqrt{e\langle\Psi\rangle/m_e} = \text{const}$.

Nomenclature Fig. 7.25	Oblique Cross	Black Points	Light Points	Straight Cross
Series	3-2	3-3	3-4	3-5
$\sqrt{e\langle\psi\rangle/m_e}/v_T$	1.6	2.6	4.2	5.4
$r_D k$	0.42-0.26	0.3-0.19	0.17-0.14	0.15-0.11

This disagreement is easy to explain from the computational point of view. Let us consider the Landau formula (7.5.16). For $k_0 r_D \ll 1$, the damping decrement calculated by Eq. (7.5.16) becomes very small, whereas its simulation counterpart does not differ much from the plasma frequency. Application of developed theory and relations (7.6.23) to the solution of classical problem of Landau damping makes it possible, even in Landau's linear formulation, to obtain a quite satisfactory agreement with mathematical experiments.

As it was shown the Landau rule for analytical estimation of the corresponding singular integral should be considered as additional condition imposed on the physical system. Another condition with the transparent physical sense (connected with calculation of independent oscillations in physical system) leads to appearance of spectrum of oscillations.

The results of calculations are in good agreement with results of direct mathematical experiment even in Landau's linear formulation, when the Landau formulae for decrement calculation leads to strong disagreement with results of direct mathematical experiments.

In spite of linear formulation of the problem the discussed complicated topology of curves (including the spectrum of the bell-like dispersion curves in Figs. 7.22 and 7.23) permits to give some conclusions about possible scenario of non-linear stage of process. The discrete spectrum of $\mathrm{d}v_\phi/\mathrm{d}\lambda$ and the discrete spectrum of $\mathrm{d}(r_D k)/\mathrm{d}\widehat{\omega}''$ can lead to series of many successive components, which have some features of solitary waves. Moreover, the Landau formulation contains two restrictions of principal significance—potential of the force field and Maxwellian function as approximation for DF. Then the modified Landau formulation can be imposed in consideration of another physical system. From this point of view no surprise that appearance of successive solitary waves with velocities decreasing by about a factor of 2 was discovered in experiments in the principal different physical systems (see, for example, [162, 163]). But of course these effects need special thorough theoretical investigations.

7.7 THE GENERALIZED THEORY OF LANDAU DAMPING IN COLLISIONAL MEDIA

Let us remind the derivation the dispersion equations of plasma in the generalized Boltzmann theory (see also Section 7.2) using the assumptions that were used by Landau in the BE-based derivation and taking into account the close collisions, namely:

(a) the evolution of electrons and ions in a self-consistent electric field corresponds to a nonstationary 1D model;
(b) the DFs for ions, f_i, and for electrons, f_e, deviate little from their equilibrium counterparts f_{0i} and f_{0e};

$$f_e = f_{0e}(u) + \delta f_e(x, u, t), \tag{7.7.1}$$

$$f_i = f_{0i}(u) + \delta f_i(x, u, t); \tag{7.7.2}$$

(c) a wave number k and a complex frequency ω are appropriate to the wave mode considered; for example

$$\delta f_e = \langle \delta f_e \rangle e^{i(kx - \omega t)}, \tag{7.7.3}$$

(d) the quadratic GBE terms determining the deviation from the equilibrium DF are neglected;
(e) the self-consistent forces F_i and F_e are small. It particular for electron species the self-consistent force F_e is not too large

$$F_e = \frac{e}{m_e}\frac{\partial \psi}{\partial x}, \tag{7.7.4}$$

where e—absolute electron charge. It means that the equilibrium DF is sufficient for calculation of function $F_e\left(\frac{\partial f_e}{\partial u}\right)$:

$$F_e \frac{\partial f_e}{\partial u} = \frac{e}{m_e}\frac{\partial \psi}{\partial x}\frac{\partial f_{0e}}{\partial u} \tag{7.7.5}$$

(f) the change of the electrical potential corresponds to the same spatial-time dependence as the perturbation of DF

$$\psi = \langle \psi \rangle e^{i(kx - \omega t)}. \tag{7.7.6}$$

Considering the more general case we take into account the collisional term written in the BGK form

$$J_\alpha = -\frac{f_\alpha - f_{0\alpha}}{v_{r\alpha}^{-1}} \tag{7.7.7}$$

in the right-hand side of the GBE. Here, $f_{0\alpha}$ and $v_\alpha^{-1} = \tau_{r\alpha}$ are respectively a certain equilibrium DF and the relaxation time for species of the αth kind ($\alpha = e, i$).

Strengths and weaknesses of the BGK method are known. BGK—approximation for the local collision integral conserves (for point-like particles) the number of particles in physical system, leads to H-theorem, but violates the laws of impulse and energy conservation. The way for overcoming of these shortages is well known and connected with application of S-method [164]. But accuracy of BGK approximation is sufficient for following investigation of asymptotic solutions. The quadratic terms determining the deviation from the equilibrium DF in kinetic equations

$$\begin{aligned}\frac{\partial f_\alpha}{\partial t} + u\frac{\partial f_\alpha}{\partial x} + F_\alpha \frac{\partial f_\alpha}{\partial u} - \tau_\alpha \Bigg\{ &\frac{\partial^2 f_\alpha}{\partial t^2} + 2u\frac{\partial^2 f_\alpha}{\partial t \partial x} + u^2 \frac{\partial^2 f_\alpha}{\partial x^2} + 2F_\alpha \frac{\partial^2 f_\alpha}{\partial t \partial u} \\ + \frac{\partial F_\alpha}{\partial t}\frac{\partial f_\alpha}{\partial u} + F_\alpha \frac{\partial f_\alpha}{\partial x} + u\frac{\partial F_\alpha}{\partial x}\frac{\partial f_\alpha}{\partial u} + F_\alpha^2 \frac{\partial^2 f_\alpha}{\partial u^2} &+ 2uF_\alpha \frac{\partial^2 f_\alpha}{\partial u \partial x}\Bigg\} = v_{r\alpha}\delta f_\alpha.\end{aligned} \tag{7.7.8}$$

are neglected. Let us write down the complex frequency ω in the form $\omega = \omega' + i\omega''$. It means that

$$\delta f_e = \langle \delta f_e \rangle e^{i(kx - \omega' t)} e^{\omega'' t}. \tag{7.7.9}$$

Using assumptions formulated above, one obtains from Eq. (7.7.8) for electron species

$$i(ku - \omega)\langle \delta f_e \rangle + i\frac{e}{m_e}k\langle \psi \rangle \frac{\partial f_{e0}}{\partial u} - (ku - \omega)\tau_e \left\{ -(ku - \omega)\langle \delta f_e \rangle - \langle \psi \rangle \frac{ek}{m_e}\frac{\partial f_{e0}}{\partial u} \right\} = v_{re}\delta f_e. \tag{7.7.10}$$

or

$$\langle \delta f_{\mathrm{e}} \rangle = -\frac{ek\langle \psi \rangle}{m_{\mathrm{e}}} \frac{\partial f_{\mathrm{e0}}}{\partial u} \frac{i + (ku - \omega)\tau_{\mathrm{e}}}{i(ku - \omega) + (ku - \omega)^2 \tau_{\mathrm{e}} + \nu_{\mathrm{re}}}. \tag{7.7.11}$$

The next step consists in application of Poisson equation for exclusion of $\langle \psi \rangle$ from Eq. (7.7.11) and analogical equation for i-species. Poisson equation will be used in the form

$$\Delta \psi = -\frac{1}{\varepsilon_0} \rho, \tag{7.7.12}$$

then

$$\varepsilon_0 k^2 \psi = e(\delta n_{\mathrm{i}} - \delta n_{\mathrm{e}}), \tag{7.7.13}$$

After integration over velocity u we find the dispersion equation

$$1 = -\frac{e^2}{\varepsilon_0 k} \left\{ \frac{1}{m_{\mathrm{e}}} \int_{-\infty}^{+\infty} \frac{(\partial f_{\mathrm{e0}}/\partial u)[i - \tau_{\mathrm{e}}(\omega - ku)]}{i(\omega - ku) - \tau_{\mathrm{e}}(\omega - ku)^2 - \nu_{\mathrm{re}}} du + \frac{1}{m_i} \int_{-\infty}^{+\infty} \frac{(\partial f_{\mathrm{i0}}/\partial u)[i - \tau_{\mathrm{i}}(\omega - ku)]}{i(\omega - ku) - \tau_{\mathrm{i}}(\omega - ku)^2 - \nu_{\mathrm{ri}}} du \right\}, \tag{7.7.14}$$

originated by e, i—concentrations

$$\langle \delta n_{\mathrm{e}} \rangle = -\frac{ek\langle \psi \rangle}{m_{\mathrm{e}}} \int_{-\infty}^{\infty} \frac{\partial f_{\mathrm{e0}}}{\partial u} \frac{i + (ku - \omega)\tau_{\mathrm{e}}}{i(ku - \omega) + (ku - \omega)^2 \tau_{\mathrm{e}} + \nu_{\mathrm{re}}} du, \tag{7.7.15}$$

$$\langle \delta n_{\mathrm{i}} \rangle = \frac{ek\langle \psi \rangle}{m_{\mathrm{i}}} \int_{-\infty}^{\infty} \frac{\partial f_{\mathrm{i0}}}{\partial u} \frac{i + (ku - \omega)\tau_{\mathrm{i}}}{i(ku - \omega) + (ku - \omega)^2 \tau_{\mathrm{i}} + \nu_{\mathrm{ri}}} du. \tag{7.7.16}$$

Obviously analog of Eq. (7.7.14) in the Boltzmann kinetic theory looks like

$$1 = -\frac{e^2}{\varepsilon_0 k} \left\{ \frac{1}{m_{\mathrm{e}}} \int_{-\infty}^{+\infty} \frac{(\partial f_{0\mathrm{e}}/\partial u)}{\omega - ku + i\nu_{\mathrm{re}}} du + \frac{1}{m_{\mathrm{i}}} \int_{-\infty}^{+\infty} \frac{(\partial f_{0\mathrm{i}}/\partial u)}{\omega - ku + i\nu_{\mathrm{ri}}} du \right\}. \tag{7.7.17}$$

Now we introduce the next Landau assumption: ions are in rest, the DFs for electrons f_{e} corresponds to the Maxwellian function $f_{0\mathrm{e}}$ written for 1D case. As result from Eq. (7.7.14)

$$1 - \frac{1}{r_{\mathrm{D}}^2 k^2} \left(\frac{m_{\mathrm{e}}}{2\pi k_{\mathrm{B}} T} \right)^{1/2} \int_{-\infty}^{+\infty} \frac{[i - \tau(\omega - ku)] k u e^{-m_{\mathrm{e}} u^2/2k_{\mathrm{B}} T}}{i(\omega - ku) - \tau(\omega - ku)^2 - \nu_{\mathrm{re}}} du = 0, \tag{7.7.18}$$

or after transformations

$$1 + \frac{1}{r_{\mathrm{D}}^2 k^2} \left[1 - \sqrt{\frac{m_{\mathrm{e}}}{2\pi k_{\mathrm{B}} T}} \int_{-\infty}^{+\infty} \frac{\{[i - \tau(\omega - ku)]\omega - \nu_{\mathrm{re}}\} e^{-m_{\mathrm{e}} u^2/2k_{\mathrm{B}} T}}{i(\omega - ku) - \tau(\omega - ku)^2 - \nu_{\mathrm{re}}} du \right] = 0, \tag{7.7.19}$$

where $r_{\mathrm{D}} = \sqrt{\varepsilon_0 k_{\mathrm{B}} T / n_{\mathrm{e}} e^2}$ is Debye-Hückel radius, k_{B}—Boltzmann constant. Eq. (7.7.19) is convenient to write in the dimensionless form

$$1 + \frac{1}{r_{\mathrm{D}}^2 k^2} \left\{ \begin{array}{l} 1 + \dfrac{1}{\sqrt{\pi}} \left[\dfrac{i\widehat{\nu}_{\mathrm{re}} + 0.5\widehat{\omega}}{\sqrt{1 + 4\widehat{\tau}\widehat{\nu}_{\mathrm{re}}}} - 0.5\widehat{\omega} \right] \displaystyle\int_{-\infty}^{+\infty} \frac{e^{-\widehat{u}^2}}{\widehat{u}_1 - \widehat{u}} d\widehat{u} \\ -\dfrac{1}{\sqrt{\pi}} \left[\dfrac{i\widehat{\nu}_{\mathrm{re}} + 0.5\widehat{\omega}}{\sqrt{1 + 4\widehat{\tau}\widehat{\nu}_{\mathrm{re}}}} + 0.5\widehat{\omega} \right] \displaystyle\int_{-\infty}^{+\infty} \frac{e^{-\widehat{u}^2}}{\widehat{u}_2 - \widehat{u}} d\widehat{u} \end{array} \right\} = 0, \tag{7.7.20}$$

where

$$\widehat{u} = u\sqrt{\frac{m_{\mathrm{e}}}{2k_{\mathrm{B}} T}}, \quad \widehat{\omega} = \omega \frac{1}{k} \sqrt{\frac{m_{\mathrm{e}}}{2k_{\mathrm{B}} T}}, \quad \widehat{\nu}_{\mathrm{re}} = \nu_{\mathrm{re}} \frac{1}{k} \sqrt{\frac{m_{\mathrm{e}}}{2k_{\mathrm{B}} T}}, \quad \widehat{\tau} = \tau k \sqrt{\frac{2k_{\mathrm{B}} T}{m_{\mathrm{e}}}}, \tag{7.7.21}$$

$$\widehat{u}_1 = \widehat{\omega} - \frac{i}{2\widehat{\tau}} \left(1 + \sqrt{1 + 4\widehat{\tau}\widehat{\nu}_{\mathrm{re}}} \right), \tag{7.7.22}$$

$$\widehat{u}_2=\widehat{\omega}-\frac{i}{2\widehat{\tau}}\left(1-\sqrt{1+4\widehat{\tau}\widehat{\nu}_{\mathrm{re}}}\right). \tag{7.7.23}$$

Equation (7.7.20) can be transformed for the collisionless case $\left(\widehat{\nu}_{\mathrm{re}}=0\right)$ in the Landau dispersion equation

$$\widehat{\omega}\int_{-\infty}^{+\infty}\frac{\mathrm{e}^{-\widehat{u}^2}}{\widehat{u}_2-\widehat{u}}\mathrm{d}\widehat{u}=\sqrt{\pi}\left(1+r_{\mathrm{D}}^2k^2\right). \tag{7.7.24}$$

Equation (7.2.20) contains improper Cauchy type integrals. From the theory of complex variables is known Cauchy's integral formula: if the function $f(z)$ is analytic inside and on a simple closed curve C, and z_0 is any point inside C, then

$$f(z_0)=-\frac{1}{2\pi\mathrm{i}}\oint_C\frac{f(z)}{z_0-z}\mathrm{d}z, \tag{7.7.25}$$

where C is traversed in the positive (counterclockwise) sense.

Let C be the boundary of a simple closed curve placed in lower half plane (for example, a semicircle of radius R) with the corresponding element of real axis, z_0 is an interior point. As usual after adding to this semicircle a cross-cut connecting semicircle C with the interior circle (surrounding z_0) of the infinite small radius for analytic $f(z)$ the formula takes place

$$\oint_C\frac{f(z)}{z_0-z}\mathrm{d}z=-\int_{-R}^{R}\frac{f(\widetilde{x})}{z_0-\widetilde{x}}\mathrm{d}\widetilde{x}+\int_{C_R}\frac{f(z)}{z_0-z}\mathrm{d}z+2\pi\mathrm{i}f(z_0), \tag{7.7.26}$$

because the integrals along cross-cut cancel each other, $(z=\widetilde{x}+\mathrm{i}\widetilde{y})$.

Analogically for upper half plane

$$\oint_C\frac{f(z)}{z_0-z}\mathrm{d}z=\int_{-R}^{R}\frac{f(\widetilde{x})}{z_0-\widetilde{x}}\mathrm{d}\widetilde{x}+\int_{C_R}\frac{f(z)}{z_0-z}\mathrm{d}z+2\pi\mathrm{i}f(z_0), \tag{7.7.27}$$

The formulas (7.2.26) and (7.2.27) could be used for calculation (including the case $R\to\infty$) of the integrals along the real axis with the help of the residual theory for arbitrary z_0 if analytical function $f(z)$ satisfies the special conditions of decreasing by $R\to\infty$.

Let us consider now integral $\int_{C_R}\frac{\mathrm{e}^{-z^2}}{z_0-z}\mathrm{d}z$. Generally speaking for function $f(z)=\mathrm{e}^{-z^2}$ Cauchy's conditions are not satisfied. Really for a point $z=\widetilde{x}+\mathrm{i}\widetilde{y}$ this function is $f(z)=\mathrm{e}^{\widetilde{y}^2-\widetilde{x}^2}[\cos(2\widetilde{x}\widetilde{y})-\mathrm{i}\sin(2\widetilde{x}\widetilde{y})]$ and by $|\widetilde{y}|>|\widetilde{x}|$ the function is growing for this part of C_R.

*But from physical point of view in **the linear problem** of interaction of individual electrons **only** with waves of potential electric field the natural assumption can be introduced that solution depends **only** of concrete $z_0=\widehat{\omega}'+\mathrm{i}\widehat{\omega}''$, but does not depend of another possible modes of oscillations in physical system.*

It can be realized only if the calculations do not depend of choosing of contour C_R. This fact leads to the additional conditions, for lower half plane

$$\int_{-\infty}^{\infty}\frac{f(\widetilde{x})}{z_0-\widetilde{x}}\mathrm{d}\widetilde{x}=2\pi\mathrm{i}f(z_0), \tag{7.7.28}$$

and for upper half plane

$$\int_{-\infty}^{\infty}\frac{f(\widetilde{x})}{z_0-\widetilde{x}}\mathrm{d}\widetilde{x}=-2\pi\mathrm{i}f(z_0). \tag{7.7.29}$$

As it was shown, Landau approximation for improper integral in Eq. (7.7.29) contains in implicit form restrictions for the contour C choosing which leads to the continuous spectrum.

The question arises, is it possible to find solutions of Eq. (7.7.20) by the restrictions (7.7.28) and (7.7.29)? In the following will be shown that the conditions (7.7.28) and (7.7.29) together with (7.7.20) lead to the discrete spectrum of $z_0=\widehat{\omega}'+\mathrm{i}\widehat{\omega}''$ and from physical point of view conditions (7.7.28) and (7.7.29) can be considered as condition of quantization.

Let us first analyze the conditions under which plasma waves can be damped. This requires, that (see Eq. 7.7.9) the imaginary part of the complex frequency fulfil the condition

$$\omega''<0. \tag{7.7.30}$$

If the collisional relaxation is not significant and $\widehat{\nu}_{\mathrm{re}}\to 0$, we have two equal integrals in relation (7.7.20), $\widehat{u}_1=\widehat{u}_2$; coefficient in front of the integral containing $\widehat{u}_1$ turns into zero by $\widehat{\nu}_{\mathrm{re}}=0$.

Let us investigate the positions of the poles involved in the calculation of the integrals in Eq. (7.7.20) taking into account the condition (7.7.28):

(A) The pole $\widehat{u}_2$ lies in the lower half plane

$$\widehat{\omega}''<\frac{1-\sqrt{1+4\widehat{\tau}\,\widehat{v}_{\mathrm{re}}}}{2\widehat{\tau}}. \tag{7.7.31}$$

Then it can be realized if only

$$\left|\widehat{\omega}''\right|>\left|\frac{1-\sqrt{1+4\widehat{\tau}\,\widehat{v}_{\mathrm{re}}}}{2\widehat{\tau}}\right|, \tag{7.7.32}$$

or for small $\widehat{v}_{\mathrm{re}}$, if only

$$\left|\widehat{\omega}''\right|>\widehat{v}_{\mathrm{re}}. \tag{7.7.33}$$

(B) The pole $\widehat{u}_2$ lies in the upper half plane. Analogically for this case we have

$$\left|\widehat{\omega}''\right|<\left|\frac{1-\sqrt{1+4\widehat{\tau}\,\widehat{v}_{\mathrm{re}}}}{2\widehat{\tau}}\right| \tag{7.7.34}$$

or for small $\widehat{v}_{\mathrm{re}}$

$$\left|\widehat{\omega}''\right|<\widehat{v}_{\mathrm{re}}. \tag{7.7.35}$$

Conditions (7.7.32)–(7.7.35) have physical sense, but lead to the different expressions for conditions of quantization (7.7.28) and (7.7.29), because for lower half plane we have

$$I_2=\int_{-\infty}^{+\infty}\frac{\mathrm{e}^{-\widehat{u}^2}}{\widehat{u}_2-\widehat{u}}\,\mathrm{d}\widehat{u}=2\pi\mathrm{i}\mathrm{e}^{-\widehat{u}_2^2} \tag{7.7.36}$$

and for the upper half plane

$$I_2=\int_{-\infty}^{+\infty}\frac{\mathrm{e}^{-\widehat{u}^2}}{\widehat{u}_2-\widehat{u}}\,\mathrm{d}\widehat{u}=-2\pi\mathrm{i}\mathrm{e}^{-\widehat{u}_2^2} \tag{7.7.37}$$

If $\widehat{\omega}''<0$ the pole $\widehat{u}_1=\widehat{\omega}'+\mathrm{i}\left(\widehat{\omega}''-\frac{1+\sqrt{1+4\widehat{\tau}\,\widehat{v}_{\mathrm{re}}}}{2\widehat{\tau}}\right)$ is placed in the lower half plane and

$$I_1=\int_{-\infty}^{+\infty}\frac{\mathrm{e}^{-\widehat{u}^2}}{\widehat{u}_1-\widehat{u}}\,\mathrm{d}\widehat{u}=2\pi\mathrm{i}\mathrm{e}^{-\widehat{u}_1^2}. \tag{7.7.38}$$

In many cases of the practical significance this integral can be excluded from consideration. Really, let us write down the dispersion equation where the both poles are in the lower half-plane

$$\frac{-\widehat{v}_{\mathrm{re}}+0.5\mathrm{i}\widehat{\omega}}{\sqrt{1+4\widehat{\tau}\,\widehat{v}_{\mathrm{re}}}}+0.5\mathrm{i}\widehat{\omega}-\left[\frac{-\widehat{v}_{\mathrm{re}}+0.5\mathrm{i}\widehat{\omega}}{\sqrt{1+4\widehat{\tau}\,\widehat{v}_{\mathrm{re}}}}-0.5\mathrm{i}\widehat{\omega}\right]\mathrm{e}^{\widehat{u}_2^2-\widehat{u}_1^2}=\frac{1+r_{\mathrm{D}}^2k^2}{2\sqrt{\pi}}\mathrm{e}^{\widehat{u}_2^2}, \tag{7.7.39}$$

where

$$\widehat{u}_2^2 - \widehat{u}_1^2 = \frac{\sqrt{1+4\widehat{\tau}\,\widehat{\nu}_{\mathrm{re}}}}{\widehat{\tau}}\left[\frac{1}{\widehat{\tau}} + 2\mathrm{i}\widehat{\omega}' - 2\widehat{\omega}''\right]. \tag{7.7.40}$$

If $\widehat{\nu}_{\mathrm{re}} \ll 1$ or $\nu_{\mathrm{re}} \ll \frac{2\pi}{\lambda}\sqrt{\frac{2k_{\mathrm{B}}T}{m_{\mathrm{e}}}}$ the term in square brackets close to zero and exponential term $\exp\left(\widehat{u}_2^2 - \widehat{u}_1^2\right) \sim 1$ for large mean time between close collisions.

Equation (7.7.39) is written for poles in the lower half plane. If the poles are placed in the upper half plane (in spite of $\widehat{\omega}'' < 0$) the sign before the exponential term should be changed and we have unified formula in the considering case of small $\widehat{\nu}_{\mathrm{re}}$.

$$\mp \mathrm{e}^{\widehat{u}_2^2}\frac{1+r_{\mathrm{D}}^2k^2}{2\sqrt{\pi}} = \frac{\widehat{\nu}_{\mathrm{re}}}{\sqrt{1+4\widehat{\tau}\,\widehat{\nu}_{\mathrm{re}}}} - \frac{\mathrm{i}\widehat{\omega}}{2}\left(1+\frac{1}{\sqrt{1+4\widehat{\tau}\,\widehat{\nu}_{\mathrm{re}}}}\right). \tag{7.7.41}$$

where

$$\widehat{u}_2^2 = \widehat{\omega}'^2 - \widehat{\omega}''^2 - \widehat{\omega}''\frac{\sqrt{1+4\widehat{\tau}\,\widehat{\nu}_{\mathrm{re}}}-1}{\widehat{\tau}} - \frac{1+2\widehat{\tau}\,\widehat{\nu}_{\mathrm{re}} - \sqrt{1+4\widehat{\tau}\,\widehat{\nu}_{\mathrm{re}}}}{2\widehat{\tau}^2} + \mathrm{i}\left(\frac{2\widehat{\omega}'' + \sqrt{1+4\widehat{\tau}\,\widehat{\nu}_{\mathrm{re}}}-1}{\widehat{\tau}}\right)\widehat{\omega}'. \tag{7.7.42}$$

As we will see lower, Eq. (7.7.41) admits the exact solution.

The relaxation time τ_{re} can be estimated in terms of the mean time τ between close collisions and the Coulomb logarithm [145]:

$$\tau_{\mathrm{re}} = \tau\Lambda^{-1} \tag{7.7.43}$$

We can then write

$$\tau\nu_{\mathrm{re}} = \Lambda, \quad \widehat{\tau}\,\widehat{\nu}_{\mathrm{re}} = \Lambda, \tag{7.7.44}$$

Now, in Eq. (7.7.41) we separate the real and imaginary parts. One obtains for the real part using Eq. (7.7.44)

$$\begin{aligned}
&\mp\frac{1+r_{\mathrm{D}}^2k^2}{2\sqrt{\pi}}\exp\left\{\widehat{\omega}'^2 - \widehat{\omega}''^2 - \widehat{\omega}''\widehat{\nu}_{\mathrm{re}}\frac{\sqrt{1+4\Lambda}-1}{\Lambda} - \widehat{\nu}_{\mathrm{re}}^2\frac{1+2\Lambda-\sqrt{1+4\Lambda}}{2\Lambda^2}\right\} \\
&= \left[\frac{\widehat{\nu}_{\mathrm{re}}}{\sqrt{1+4\Lambda}} + 0.5\widehat{\omega}'' + \frac{0.5\widehat{\omega}''}{\sqrt{1+4\Lambda}}\right]\cos\left[\widehat{\omega}'\left(2\widehat{\omega}'' + \widehat{\nu}_{\mathrm{re}}\frac{\sqrt{1+4\Lambda}-1}{\Lambda}\right)\right] \\
&-0.5\widehat{\omega}'\left[1+\frac{1}{\sqrt{1+4\Lambda}}\right]\sin\left[\widehat{\omega}'\left(2\widehat{\omega}'' + \widehat{\nu}_{\mathrm{re}}\frac{\sqrt{1+4\Lambda}-1}{\Lambda}\right)\right].
\end{aligned} \tag{7.7.45}$$

Similarly, for the imaginary part we find

$$\begin{aligned}
&0.5\widehat{\omega}'\left[1+\frac{1}{\sqrt{1+4\Lambda}}\right]\cos\left[\widehat{\omega}'\left(2\widehat{\omega}'' + \widehat{\nu}_{\mathrm{re}}\frac{\sqrt{1+4\Lambda}-1}{\Lambda}\right)\right] \\
&+\left[\frac{\widehat{\nu}_{\mathrm{re}}}{\sqrt{1+4\Lambda}} + 0.5\widehat{\omega}'' + \frac{0.5\widehat{\omega}''}{\sqrt{1+4\Lambda}}\right]\sin\left[\widehat{\omega}'\left(2\widehat{\omega}'' + \widehat{\nu}_{\mathrm{re}}\frac{\sqrt{1+4\Lambda}-1}{\Lambda}\right)\right] = 0.
\end{aligned} \tag{7.7.46}$$

The system of complicated transcendent Eqs. (7.7.45) and (7.7.46) can generally be solved only on a computer. If, however, the Coulomb logarithm Λ is large enough for terms of order $\Lambda^{-1/2}$ to be negligible or for small Coulomb logarithm Λ, the exact solution can be obtained. The system of Eqs. (7.7.45) and (7.7.46) for the large Coulomb logarithm Λ simplifies to

$$\mp\frac{1+r_{\mathrm{D}}^2k^2}{\sqrt{\pi}}\mathrm{e}^{\widehat{\omega}'^2 - \widehat{\omega}''^2} = \widehat{\omega}''\cos\left(2\widehat{\omega}'\widehat{\omega}''\right) - \widehat{\omega}'\sin\left(2\widehat{\omega}'\widehat{\omega}''\right), \tag{7.7.47}$$

$$\widehat{\omega}'\cos\left(2\widehat{\omega}'\widehat{\omega}''\right) + \widehat{\omega}''\sin\left(2\widehat{\omega}'\widehat{\omega}''\right) = 0. \tag{7.7.48}$$

Let us introduce the notation

$$\alpha = 2\widehat{\omega}'\widehat{\omega}'', \quad \beta = 1 + r_{\mathrm{D}}^2 k^2, \tag{7.7.49}$$

and note that

$$\widehat{\omega}'^2 = -\frac{1}{2}\alpha \tan\alpha, \quad \widehat{\omega}''^2 = -\frac{1}{2}\alpha \cot\alpha, \quad \widehat{\omega}'^2 - \widehat{\omega}''^2 = \alpha \cot 2\alpha. \tag{7.7.50}$$

From relations (7.7.47)–(7.7.50) follow

$$\mp \frac{\beta}{\sqrt{\pi}} \mathrm{e}^{\alpha \cot 2\alpha} = \widehat{\omega}'' \cos\alpha - \widehat{\omega}' \sin\alpha. \tag{7.7.51}$$

Taking the square of both sides of Eq. (7.7.52) we obtain the universal equation

$$-\mathrm{e}^{\sigma \cot \sigma} \sin\sigma = \frac{\pi}{2\beta^2}\sigma, \tag{7.7.52}$$

where the notation is introduced $\sigma = -2\alpha = -4\widehat{\omega}'\widehat{\omega}''$. As it was mentioned, this equation does not depend on the sign in front of parameter β.

The exact solution of Eq. (7.7.52) can be found with the help of the W-function of Lambert

$$\sigma_n = \mathrm{Im}\left[W_n\left(\frac{2\beta^2}{\pi}\right)\right], \tag{7.7.53}$$

frequencies $\widehat{\omega}'_n, \widehat{\omega}''_n$ are (see Eq. 7.7.50)

$$\omega'_n = k\sqrt{-\frac{k_{\mathrm{B}}T}{2m_{\mathrm{e}}}\sigma_n \tan\frac{\sigma_n}{2}}, \quad \omega''_n = -k\sqrt{-\frac{k_{\mathrm{B}}T}{2m_{\mathrm{e}}}\sigma_n \cot\frac{\sigma_n}{2}}. \tag{7.7.54}$$

In asymptotic for large entire positive n

$$\sigma_n = \left(n + \frac{1}{2}\right)\pi, \quad \widehat{\omega}'_n = \frac{\sqrt{\sigma_n}}{2} = \frac{1}{2}\sqrt{\pi\left(n + \frac{1}{2}\right)}, \quad \widehat{\omega}''_n = -\frac{\sqrt{\sigma_n}}{2} = -\frac{1}{2}\sqrt{\pi\left(n + \frac{1}{2}\right)}. \tag{7.7.55}$$

The exact solution for the nth discrete solution from the spectrum of oscillations follows from Eqs. (7.7.53) and (7.7.54):

$$\begin{aligned}\widehat{\omega}_n = {} & \frac{1}{2}\sqrt{-\mathrm{Im}\left[W_n\left(\frac{2(1 + r_{\mathrm{D}}^2 k^2)^2}{\pi}\right)\right] \tan\left[\frac{1}{2}\mathrm{Im}\left[W_n\left(\frac{2(1 + r_{\mathrm{D}}^2 k^2)^2}{\pi}\right)\right]\right]} \\ & - \frac{\mathrm{i}}{2}\sqrt{-\mathrm{Im}\left[W_n\left(\frac{2(1 + r_{\mathrm{D}}^2 k^2)^2}{\pi}\right)\right] \cot\left[\frac{1}{2}\mathrm{Im}\left[W_n\left(\frac{2(1 + r_{\mathrm{D}}^2 k^2)^2}{\pi}\right)\right]\right]}.\end{aligned} \tag{7.7.56}$$

The square of the oscillation frequency of plasma waves ω'^2_n is proportional to the wave energy. Hence, the energy of plasma waves is quantized, and as n grows we have the asymptotic expression

$$\widehat{\omega}'^2_n = \frac{\pi}{4}\left(n + \frac{1}{2}\right), \tag{7.7.57}$$

the squares of possible dimensionless frequencies become equally spaced:

$$\widehat{\omega}'^2_{n+1} - \widehat{\omega}'^2_n = \frac{\pi}{4}. \tag{7.7.58}$$

Figures 7.26 and 7.27 reflect the result of calculations for 200 discrete levels for the case of the large Coulomb logarithm Λ.

Plotter from the technical point of view has no possibility to reflect the small curvature of lines and approximates this curvature as a long step with shifting to the next pick cell. For high levels this spectrum contains many very close practically straight lines with small but long steps, which human eyes can perceive as background. As result plotters are realizing the discrete constriction of derivatives $\mathrm{d}(r_{\mathrm{D}}k)/\mathrm{d}\widehat{\omega}'$ and $\mathrm{d}(r_{\mathrm{D}}k)/\mathrm{d}\widehat{\omega}''$ for discontinuous functions. This effect has no attitude to the mathematical programming. You can see this very complicated topology of curves including the spectrum of the bell-like dispersion curves in Figs. 7.26 and 7.27, which also form the discrete spectrum.

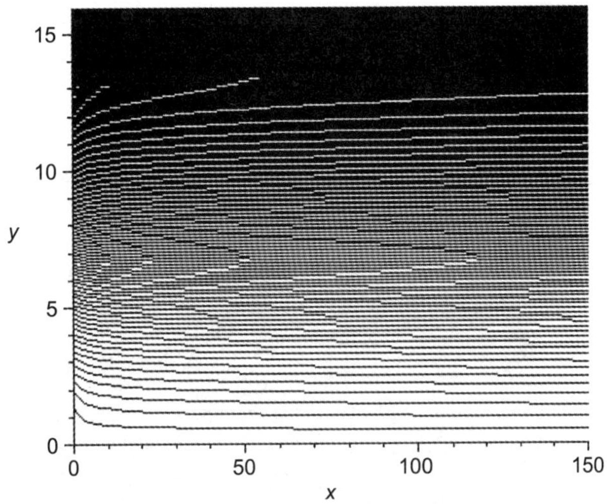

FIGURE 7.26 The dimensionless frequency $\widehat{\omega}'$ (y axes) versus parameter $r_{\mathrm{D}}k$ (x axes).

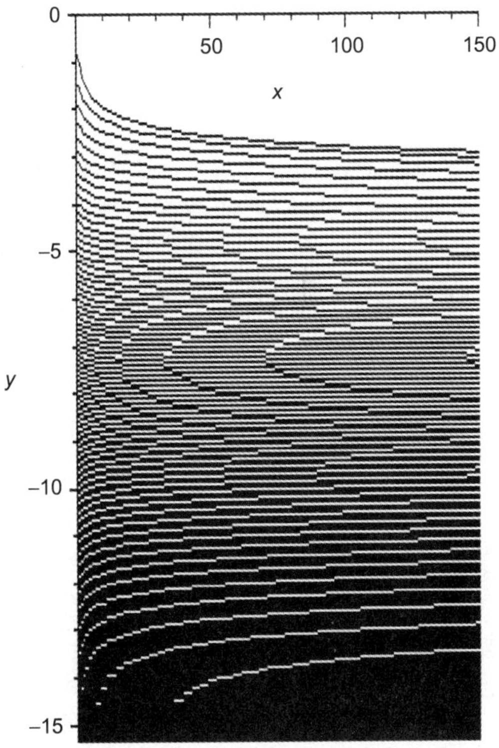

FIGURE 7.27 The dimensionless frequency $\widehat{\omega}''$ (y axes) versus parameter $r_{\mathrm{D}}k$ (x axes).

But mentioned derivatives can be written as before (see Section 7.3) in the form

$$\frac{\mathrm{d}(r_{\mathrm{D}}k)}{\mathrm{d}\widehat{\omega}'} = \frac{\mathrm{d}k}{\mathrm{d}(\omega'/k)} \frac{k_{\mathrm{B}}T\sqrt{2}}{m_{\mathrm{e}}\omega_{\mathrm{e}}} = \frac{\mathrm{d}k}{\mathrm{d}v_{\phi}} \frac{k_{\mathrm{B}}T\sqrt{2}}{m_{\mathrm{e}}\omega_{\mathrm{e}}}, \tag{7.7.59}$$

where v_{ϕ} is the wave phase velocity. Then

$$\frac{\mathrm{d}v_{\phi}}{\mathrm{d}\lambda} = -\frac{1}{\lambda^2}\frac{k_{\mathrm{B}}T}{e}\sqrt{\frac{2\pi}{\rho_{\mathrm{e}}}}\left[\frac{\mathrm{d}(r_{\mathrm{D}}k)}{\mathrm{d}\widehat{\omega}'}\right]^{-1}, \tag{7.7.60}$$

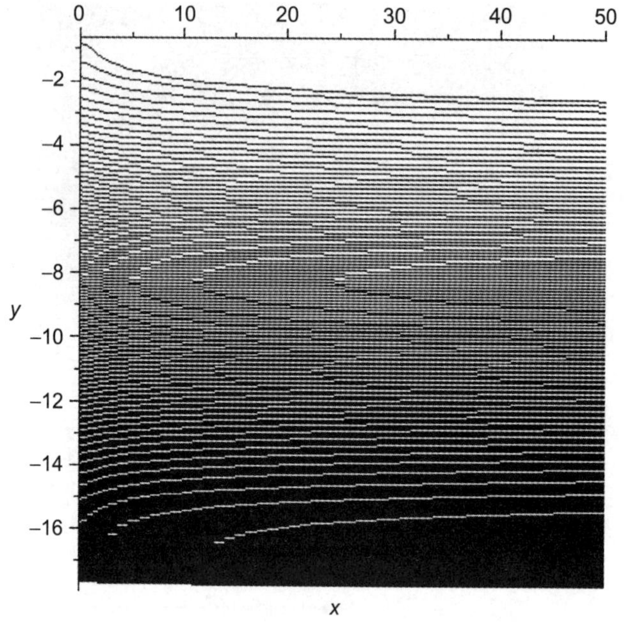

FIGURE 7.28 The dimensionless frequency $\widehat{\omega}''$ (y axes) versus parameter $r_{\mathrm{D}}k$ (x axes).

where $\rho_{\mathrm{e}}=m_{\mathrm{e}}n_{\mathrm{e}}$. Very complicated topology of the dispersion curves exist also in the black domains of Figs. 7.26 and 7.27 where (in chosen scale) the curves cannot be observed separately. Then Figs. 7.26 and 7.27 can be used for understanding of the future development of events in physical system after the initial linear stage. For example, Fig. 7.26 shows the discrete set of frequencies which vicinity corresponds to passing over from abnormal to normal dispersion (for example, by $\widehat{\omega}'\sim 7$) for discrete systems of $r_{\mathrm{D}}k$. This discrete systems of $r_{\mathrm{D}}k$ has regular characters with factor of about two. Figure 7.28 demonstrates the close up view of the mentioned discrete system of passing over from abnormal to normal dispersion in the range between 0 and 50 $r_{\mathrm{D}}k$. Of course the non-linear stage needs the special investigation with using of other methods including the method of direct mathematical modeling.

Let us investigate now the case of small Coulomb logarithm Λ using Eqs. (7.7.45) and (7.7.46). One obtains for real and imaginary parts correspondingly

$$\begin{aligned}&\mp\frac{1+r_{\mathrm{D}}^2k^2}{2\sqrt{\pi}}\exp\left\{\widehat{\omega}'^2-\left[\widehat{\omega}''+\widehat{v}_{\mathrm{re}}\right]^2\right\}\\&=\left[\widehat{\omega}''+\widehat{v}_{\mathrm{re}}\right]\cos\left[2\widehat{\omega}'\left(\widehat{\omega}''+\widehat{v}_{\mathrm{re}}\right)\right]-\widehat{\omega}'\sin\left[2\widehat{\omega}'\left(\widehat{\omega}''+\widehat{v}_{\mathrm{re}}\right)\right].\end{aligned}\tag{7.7.61}$$

$$\widehat{\omega}'\cos\left[2\widehat{\omega}'\left(\widehat{\omega}''+\widehat{v}_{\mathrm{re}}\right)\right]+\left(\widehat{\omega}''+\widehat{v}_{\mathrm{re}}\right)\sin\left[2\widehat{\omega}'\left(\widehat{\omega}''+\widehat{v}_{\mathrm{re}}\right)\right]=0.\tag{7.7.62}$$

Equations (7.7.61) and (7.7.62) have practically the same form as the system of Eqs. (7.7.47) and (7.7.48) after replacing $\widehat{\omega}''$ with the variable $\widehat{\omega}''_1=\widehat{\omega}''+\widehat{v}_{\mathrm{re}}$. It should also be noted that, in the asymptotics we are considering, an additional factor 0.5 appears on the left-hand side of Eq. (7.7.61). Equations (7.7.61) and (7.7.62) are then solved in exactly the same manner to give (for large n):

$$\widehat{\omega}'_n=\frac{1}{2}\sqrt{\pi\left(n+\frac{1}{2}\right)},\quad \widehat{\omega}''_n=-\frac{1}{2}\sqrt{\pi\left(n+\frac{1}{2}\right)}-\widehat{v}_{\mathrm{re}}.\tag{7.7.63}$$

Then the problem reduces to that of solving the transcendent equation

$$-\mathrm{e}^{\sigma_1\cot\sigma_1}\sin\sigma_1=\frac{2\pi}{\beta^2}\sigma_1,\tag{7.7.64}$$

where

$$\sigma_1 = -4\widehat{\omega}'\widehat{\omega}_1'', \quad \widehat{\omega}_1'' = \widehat{\omega}'' + \widehat{\nu}_{\mathrm{re}}. \tag{7.7.65}$$

After introducing $\widehat{\nu}_{\mathrm{re}} \to 0$ in Eqs. (7.7.64) and (7.7.65) we realize the passage to the generalized dispersion equation considered before for the non-collisional case.

The exact solution of Eq. (7.7.64) can be found with the help of the W-function of Lambert

$$\sigma_{1n} = \mathrm{Im}\left[W_n \left|\frac{\beta^2}{2\pi}\right|\right], \tag{7.7.66}$$

frequencies $\widehat{\omega}_n', \widehat{\omega}_n''$ are

$$\omega_n' = k\sqrt{-\frac{k_{\mathrm{B}}T}{2m_{\mathrm{e}}}\sigma_{1n}\tan\frac{\sigma_{1n}}{2}}, \tag{7.7.67}$$

$$\omega_n'' = k\sqrt{-\frac{k_{\mathrm{B}}T}{2m_{\mathrm{e}}}\sigma_{1n}\cot\frac{\sigma_{1n}}{2}}. \tag{7.7.68}$$

Figures 7.29 and 7.30 reflect the result of calculations for 200 discrete levels for the case of the small Coulomb logarithm Λ. We see the same character features of the topology as in the case of large Coulomb logarithm Λ reflected on Figs. 7.26–7.28, but of course with another quantitative characteristics.

Dispersion equations (7.7.52) and (7.7.64) can be written in the unified form:

$$-e^{\sigma\cot\gamma}\sin\sigma = \xi\sigma, \tag{7.7.69}$$

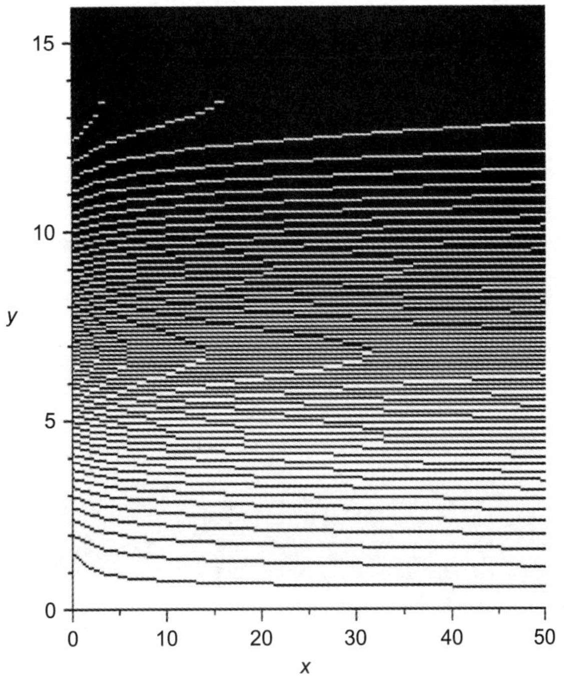

FIGURE 7.29 The dimensionless frequency $\widehat{\omega}'$ (y axes) versus parameter $r_{\mathrm{D}}k$ (x axes).

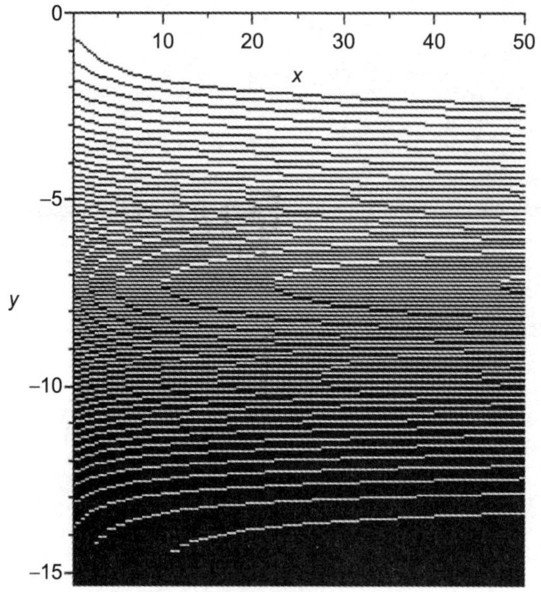

FIGURE 7.30 The dimensionless frequency $\widehat{\omega}_1''$ (y axes) versus parameter $r_D k$ (x axes).

which solution is

$$\sigma_n = \mathrm{Im}\left[W_n\left(\frac{1}{\xi}\right)\right], \tag{7.7.70}$$

where $\xi = \pi/2\beta^2$ for large Coulomb logarithm Λ and $\xi = 2\pi/\beta^2$ for small Coulomb logarithm Λ, $\beta = 1 + r_D^2 k^2$. Both considered cases correspond to assumption of small values ν_{re}, in other words—to small influence of local collision integral on solution of kinetic equation. If this influence is not small, dispersion equation should contains omitted integral I_2 (see relations (7.7.36) and (7.7.37)) and should be written with taking into account the fluctuation effects in Poisson equation. This consideration cannot be realized in the analytical form.

Chapter 8

Physics of a Weakly Ionized Gas

ABSTRACT

Physics of weakly ionized gas has very vast area of technical applications. Investigations of transport processes in gas discharge, quantum generators, magneto-hydrodynamic generators, and so on are based on physics of weakly ionized gases and, in particular on the Boltzmann kinetic theory (BKT). Methods of physics of weakly ionized gas are significant of course not only for technical applications but also for fundamental physics for example for physics of ionosphere and astrophysics.

The traditional area of application of the BKT is the physics of a weakly ionized gas. It is interesting to see what the generalized Boltzmann equation (GBE) yields in this case and how its results differ from those of the classical theory. To answer this fundamental question, let us consider the classical Lorentz formulation of the problem. We consider a spatially homogeneous, weakly ionized gas, for which it is assumed that collisions between charged particles may be ignored

$$\upsilon_{\rm e} \ll \delta \upsilon_{\rm ea},$$

where $\upsilon_{\rm e}$ is the collision rate between charged particles; $\upsilon_{\rm ea}$, is the collision rate between charged and neutral particles, and δ is the relative amount of energy which a charged particle loses in one collision with a neutral particle. We assume that the magnetic field is either absent or has a static component B_z, and that the electric field is along the x-axis; all inelastic interactions are neglected. Topology of the external electromagnetic field and models of particle interactions will be discussed in the following for every considered case separately.

Keywords: Weakly ionized gas, Charged particles relaxation, Conductivity of a weakly ionized gas

8.1 CHARGED PARTICLES RELAXATION IN "MAXWELLIAN" GAS AND THE HYDRODYNAMIC ASPECTS OF THE THEORY

The classical Boltzmann equation (BE) in this case takes the form

$$\frac{{\rm D}f_{\rm e}}{{\rm D}t} \equiv \frac{\partial f_{\rm e}}{\partial t} + \mathbf{F}_{\rm e}\cdot\frac{\partial f_{\rm e}}{\partial \mathbf{v}_{\rm e}} = J_{\rm ea}, \tag{8.1.1}$$

where $\mathbf{F}_{\rm e} = q_{\rm e}\mathbf{E}/m_{\rm e}$ is a force acting on a unit mass of the charged particle, and $q_{\rm e}$ is the particle charge. The generalized Boltzmann equation (GBE) is written in the following way:

$$\frac{{\rm D}f_{\rm e}}{{\rm D}t} - \frac{\rm D}{{\rm D}t}\left(\tau_{\rm ea}\frac{{\rm D}f_{\rm e}}{{\rm D}t}\right) = J_{\rm ea}. \tag{8.1.2}$$

In Eq. (8.1.2), $\tau_{\rm ea}$ is a non-local parameter proportional to the mean time between the collisions of neutral and charged particles. As the GBE theory suggests, the collision integral can be taken in the Boltzmann form. Let us compare the results obtained in frames of Boltzmann kinetic theory (BKT) and generalized Boltzmann kinetic theory (GBKT). Multiplying Eqs. (8.1.1) and (8.1.2) by the collision invariants $m_{\rm e}, m_{\rm e}\mathbf{v}_{\rm e}, m_{\rm e}v_{\rm e}^2/2$ and then integrating them over the velocities $\mathbf{v}_{\rm e}$, we arrive at the classical hydrodynamic equations (HEs) and the generalized hydrodynamic equations (GHEs), which assume a closed form provided we know how to evaluate the moments of the collision integrals involved. Note that in this case the following relation holds:

$$\int J_{\rm ea} m_{\rm e} {\rm d}\mathbf{v}_{\rm e} = 0, \tag{8.1.3}$$

owing to the law of conservation of mass in the elastic non-relativistic collisions. But the integrals $\int J_{\rm ea} m_{\rm e}\mathbf{v}_{\rm e}{\rm d}\mathbf{v}_{\rm e}$, and $\int J_{\rm ea}\frac{m_{\rm e}v_{\rm e}^2}{2}{\rm d}\mathbf{v}_{\rm e}$ can be taken explicitly only for special models of particle interaction. Let us adopt the Maxwell model, in which the force $F_{\rm ea}$ of the intermolecular interaction depends on the inverse fifth power of the inter-particle spacing

Unified non-local theory of transport processes. http://dx.doi.org/10.1016/B978-0-444-63478-8.00008-7

$$F_{\mathrm{ea}}=\frac{\chi_{\mathrm{ea}}}{r^5}. \tag{8.1.4}$$

For this model, the integrals mentioned above are well known and the quantity τ_{ea} (hereinafter the subscript ea is dropped) was calculated to be

$$\tau=\left[2\pi 0.422\left(\frac{\chi(m_{\mathrm{e}}+m_{\mathrm{a}})}{m_{\mathrm{e}}m_{\mathrm{a}}}\right)^{1/2}n_{\mathrm{a}}\right]^{-1}. \tag{8.1.5}$$

Introducing the quantities

$$A=\frac{8\sqrt{\pi}}{3}\Gamma\left(\frac{5}{2}\right)0.422(m_{\mathrm{e}}+m_{\mathrm{a}})^{-1}\left(\frac{\chi}{m_{\mathrm{a}}M_{\mathrm{e}}}\right)^{1/2} \tag{8.1.6}$$

and

$$M_{\mathrm{a}}=m_{\mathrm{a}}/(m_{\mathrm{a}}+m_{\mathrm{e}}),\quad M_{\mathrm{e}}=m_{\mathrm{e}}/(m_{\mathrm{a}}+m_{\mathrm{e}}), \tag{8.1.7}$$

the rate υ of collisions between charged and neutral particles can be written in the form

$$\upsilon=n_{\mathrm{a}}(m_{\mathrm{a}}+m_{\mathrm{e}})A, \tag{8.1.8}$$

where $\tau=\upsilon^{-1}$, and n_{a} is the number density of neutral particles.

The continuity equations obtained from Eq. (8.1.2) yield the condition $n_{\mathrm{e}}=\mathrm{const}$, and the GBE in this case becomes

$$\frac{\partial f_{\mathrm{e}}}{\partial t}+\mathbf{F}_{\mathrm{e}}\cdot\frac{\partial f_{\mathrm{e}}}{\partial \mathbf{v}_{\mathrm{e}}}-\tau\left\{\frac{\partial^2 f_{\mathrm{e}}}{\partial t^2}+2\mathbf{F}_{\mathrm{e}}\cdot\frac{\partial^2 f_{\mathrm{e}}}{\partial \mathbf{v}_{\mathrm{e}}\partial t}+\frac{\partial^2 f_{\mathrm{e}}}{\partial \mathbf{v}_{\mathrm{e}}\partial \mathbf{v}_{\mathrm{e}}}:\mathbf{F}_{\mathrm{e}}\mathbf{F}_{\mathrm{e}}\right\}=J_{\mathrm{ea}}. \tag{8.1.9}$$

As before, a colon denotes the double scalar product of tensors.

We now introduce the drift velocity $\overline{v}_{\mathrm{ex}}$ defined by the expression

$$\overline{v}_{\mathrm{ex}}=\frac{1}{n_{\mathrm{e}}}\int f_{\mathrm{e}}v_{\mathrm{ex}}\mathrm{d}\mathbf{v}_{\mathrm{e}}. \tag{8.1.10}$$

Then the equation of motion entering the system of GHEs takes the form

$$\tau\frac{\mathrm{d}^2\overline{v}_{\mathrm{ex}}}{\mathrm{d}t^2}-\frac{\mathrm{d}\overline{v}_{\mathrm{ex}}}{\mathrm{d}t}-Am_{\mathrm{a}}n_{\mathrm{a}}\overline{v}_{\mathrm{ex}}+q_{\mathrm{e}}Em_{\mathrm{e}}^{-1}=0. \tag{8.1.11}$$

In writing Eq. (8.1.11) we have used the result

$$\int m_{\mathrm{e}}v_{\mathrm{ex}}J_{\mathrm{ea}}\mathrm{d}\mathbf{v}_{\mathrm{e}}=-Am_{\mathrm{a}}n_{\mathrm{a}}\overline{v}_{\mathrm{ea}}n_{\mathrm{a}}n_{\mathrm{e}}.$$

The solution of Eq. (8.1.11) takes the form

$$\overline{v}_{\mathrm{ex}}^{\mathrm{BAE}}=\left(\overline{v}_{\mathrm{ex}}^{0}-\frac{q_{\mathrm{e}}E}{m_{\mathrm{e}}m_{\mathrm{a}}n_{\mathrm{a}}A}\right)\mathrm{e}^{-(t/2\tau)\left[\sqrt{4M_{\mathrm{a}}+1}-1\right]}+\frac{q_{\mathrm{e}}E}{m_{\mathrm{e}}m_{\mathrm{a}}n_{\mathrm{a}}A}. \tag{8.1.12}$$

The superscript 0 here refers to the initial instant of time.

The problem of the time relaxation of Maxwell particles in an electric field is known to be amenable to a BE solution [21] giving for the drift velocity the result

$$\overline{v}_{\mathrm{ex}}^{\mathrm{BE}}=\left(\overline{v}_{\mathrm{ea}}^{0}-\frac{q_{\mathrm{e}}E}{m_{\mathrm{e}}m_{\mathrm{a}}n_{\mathrm{a}}A}\right)\mathrm{e}^{-tAn_{\mathrm{a}}m_{\mathrm{a}}}+\frac{q_{\mathrm{e}}E}{m_{\mathrm{e}}m_{\mathrm{a}}n_{\mathrm{a}}A}. \tag{8.1.13}$$

For example, let us assume that $m_{\mathrm{e}}\ll m_{\mathrm{a}}$ Then from Eqs. (8.1.12) and (8.1.13), it follows that

$$\overline{v}_{\mathrm{ex}}^{\mathrm{GBE}}=(\overline{v}_{\mathrm{ex}}^{0}-F_{\mathrm{ex}}\tau)\mathrm{e}^{-0.618(t/\tau)}+F_{\mathrm{ex}}\tau, \tag{8.1.14}$$

$$\overline{v}_{\mathrm{ex}}^{\mathrm{BE}}=(\overline{v}_{\mathrm{ex}}^{0}-F_{\mathrm{ex}}\tau)\mathrm{e}^{-(t/\tau)}+F_{\mathrm{ex}}\tau. \tag{8.1.15}$$

Thus, all other things being equal, the relaxation of the drift velocity $\overline{v}_{\mathrm{ex}}$ in the framework of BKT proceeds faster than in the GBKT, whereas the steady-state drift velocities are found to be the same. We now turn our attention to the equation of energy and introduce the energy temperatures $\hat{T}_{\mathrm{e}}$ and $\tilde{T}_{\mathrm{e}}$ in accord with the definitions:

$$\hat{T}_{\mathrm{e}}=\frac{m_{\mathrm{e}}}{3n_{\mathrm{e}}}\int f_{\mathrm{e}}v_{\mathrm{e}}^2\mathrm{d}\mathbf{v}_{\mathrm{e}}, \tag{8.1.16}$$

$$\widetilde{T}_{\mathrm{e}}=\frac{m_{\mathrm{e}}}{3n_{\mathrm{e}}}\int f_{\mathrm{e}}(\mathbf{v}_{\mathrm{e}}-\overline{\mathbf{v}}_{\mathrm{e}})^2 d\mathbf{v}_{\mathrm{e}}. \tag{8.1.17}$$

Clearly, which of these temperatures is used is a matter of convenience, and in our case one obtains

$$\hat{T}_{\mathrm{e}}=\widetilde{T}_{\mathrm{e}}+\frac{1}{3}m_{\mathrm{e}}\overline{v}_{\mathrm{ex}}^2. \tag{8.1.18}$$

We next evaluate the moments on the left-hand side of the kinetic equations. For example, the following relations hold true:

$$\frac{\partial}{\partial t}\int\frac{m_{\mathrm{e}}v_{\mathrm{e}}^2}{2}f_{\mathrm{e}}d\mathbf{v}_{\mathrm{e}}=\frac{3}{2}n_{\mathrm{e}}\frac{\partial\hat{T}_{\mathrm{e}}}{\partial t}, \tag{8.1.19}$$

$$\int F_{\mathrm{e}}\cdot\frac{\partial^2 f_{\mathrm{e}}}{\partial\mathbf{v}_{\mathrm{e}}\partial t}\frac{m_{\mathrm{e}}v_{\mathrm{e}}^2}{2}d\mathbf{v}_{\mathrm{e}}=-F_{\mathrm{ex}}m_{\mathrm{e}}n_{\mathrm{e}}\frac{\partial\overline{v}_{\mathrm{ex}}}{\partial t}, \tag{8.1.20}$$

$$\int\frac{m_{\mathrm{e}}v_{\mathrm{e}}^2}{2}\frac{\partial^2 f_{\mathrm{e}}}{\partial\mathbf{v}_{\mathrm{e}}\partial\mathbf{v}_{\mathrm{e}}}:\mathbf{F}_{\mathrm{e}}\mathbf{F}_{\mathrm{e}}d\mathbf{v}_{\mathrm{e}}=F_{\mathrm{ex}}^2m_{\mathrm{e}}n_{\mathrm{e}}. \tag{8.1.21}$$

The corresponding integral on the right-hand side was calculated in Ref. [165] and is found to be

$$\int J_{\mathrm{ea}}\frac{m_{\mathrm{e}}v_{\mathrm{e}}^2}{2}d\mathbf{v}_{\mathrm{e}}=-\frac{3(\hat{T}_{\mathrm{e}}-\hat{T}_{\mathrm{a}})}{m_{\mathrm{e}}+m_{\mathrm{a}}}Am_{\mathrm{e}}m_{\mathrm{a}}n_{\mathrm{e}}n_{\mathrm{a}}. \tag{8.1.22}$$

We have then the following inhomogeneous linear second-order differential equation:

$$\begin{aligned}&\frac{d^2\widetilde{T}_{\mathrm{e}}}{dt^2}-\frac{1}{\tau}\frac{d\widetilde{T}_{\mathrm{e}}}{dt}-2\frac{\hat{T}_{\mathrm{e}}-\hat{T}_{\mathrm{a}}}{m_{\mathrm{e}}+m_{\mathrm{a}}}\frac{A}{\tau}m_{\mathrm{a}}m_{\mathrm{e}}n_{\mathrm{a}}\\&=\frac{1}{3}\frac{m_{\mathrm{e}}}{\tau}\frac{d}{dt}\overline{v}_{\mathrm{ex}}^2-\frac{2}{3}\frac{m_{\mathrm{e}}}{\tau}F_{\mathrm{ex}}\overline{v}_{\mathrm{ex}}-\frac{1}{3}m_{\mathrm{e}}\frac{d^2}{dt^2}\overline{v}_{\mathrm{ex}}^2+\frac{4}{3}F_{\mathrm{ex}}m_{\mathrm{e}}\frac{d\overline{v}_{\mathrm{ex}}}{dt}-\frac{2}{3}m_{\mathrm{e}}F_{\mathrm{ex}}^2.\end{aligned} \tag{8.1.23}$$

Omitting the straightforward but tedious algebra we arrive at the following results. For example, by setting $m_{\mathrm{e}}\ll m_{\mathrm{a}}$, the GBE yields

$$\begin{aligned}\widetilde{T}_{\mathrm{e}}^{\mathrm{GBE}}&=\widetilde{T}_{\mathrm{a}}+C_2e^{-2(m_{\mathrm{e}}/m_{\mathrm{a}})(t/\tau)}-2.157m_{\mathrm{e}}\tau F_{\mathrm{ex}}\left(\overline{v}_{\mathrm{ex}}^0-F_{\mathrm{ex}}\tau\right)e^{-0.618(t/\tau)}\\&-\frac{1}{3}m_{\mathrm{e}}\left(\overline{v}_{\mathrm{ex}}^0-F_{\mathrm{ex}}\tau\right)^2e^{-1.236(t/\tau)}+\frac{2}{3}F_{\mathrm{ex}}^2\tau^2m_{\mathrm{a}},\end{aligned} \tag{8.1.24}$$

where the following notation was used:

$$\begin{aligned}C_2=&\widetilde{T}_{\mathrm{e}}^0-\widetilde{T}_{\mathrm{a}}+2.157m_{\mathrm{e}}\tau F_{\mathrm{ex}}\left(\overline{v}_{\mathrm{ex}}^0-F_{\mathrm{ex}}\tau\right)\\&+\frac{1}{3}m_{\mathrm{e}}\left(\overline{v}_{\mathrm{ex}}^0-F_{\mathrm{ex}}\tau\right)^2-\frac{2}{3}F_{\mathrm{ex}}^2\tau^2m_{\mathrm{a}}.\end{aligned}$$

Similar BE results are as follows:

$$\begin{aligned}\widetilde{T}_{\mathrm{e}}^{\mathrm{BE}}&=\widetilde{T}_{\mathrm{a}}+C_2e^{-2(m_{\mathrm{e}}/m_{\mathrm{a}})(t/\tau)}-\frac{4}{3}m_{\mathrm{e}}\tau F_{\mathrm{ex}}\left(\overline{v}_{\mathrm{ex}}^0-F_{\mathrm{ex}}\tau\right)e^{-(t/\tau)}\\&-\frac{1}{3}m_{\mathrm{e}}\left(\overline{v}_{\mathrm{ex}}^0-F_{\mathrm{ex}}\tau\right)^2e^{-2(t/\tau)}+\frac{1}{3}F_{\mathrm{ex}}^2\tau^2m_{\mathrm{a}}.\end{aligned} \tag{8.1.25}$$

Here, the notation was used:

$$\begin{aligned}C_2=&\widetilde{T}_{\mathrm{e}}^0-\widetilde{T}_{\mathrm{a}}+\frac{4}{3}m_{\mathrm{e}}\tau F_{\mathrm{ex}}\left(\overline{v}_{\mathrm{ex}}^0-F_{\mathrm{ex}}\tau\right)\\&+\frac{1}{3}m_{\mathrm{e}}\left(\overline{v}_{\mathrm{ex}}^0-F_{\mathrm{ex}}\tau\right)^2-\frac{1}{3}F_{\mathrm{ex}}^2\tau^2m_{\mathrm{a}}.\end{aligned}$$

In the steady-state regime, the above solutions are related by the expression

$$\left(\widetilde{T}_{\mathrm{e}}-\widetilde{T}_{\mathrm{a}}\right)_{\mathrm{st}}^{\mathrm{GBE}}=2\left(\widetilde{T}_{\mathrm{e}}-\widetilde{T}_{\mathrm{a}}\right)_{\mathrm{st}}^{\mathrm{BE}}. \tag{8.1.26}$$

As before, the superscripts on the energy temperature differences in Eq. (8.1.26) refer to the type of the solution. We are now in a position to write down the solutions for the energy temperatures $\hat{T}_{\mathrm{e}}$; in the GBE scheme we have

$$\hat{T}_{\mathrm{e}} = \hat{T}_{\mathrm{a}} + \left[\hat{T}^{0}_{\mathrm{ea}} - \frac{2}{3}F_{\mathrm{ex}}\tau\frac{m_{\mathrm{e}}}{M_{\mathrm{a}}(2M_{\mathrm{e}}-1)}\sqrt{4M_{\mathrm{a}}+1}\left(\bar{v}^{0}_{\mathrm{ex}} - \frac{F_{\mathrm{ex}}\tau}{M_{\mathrm{a}}}\right)\right.$$
$$\left. - \frac{\tau^2 F^2_{\mathrm{ex}}}{3\,M^2_{\mathrm{a}}}(1+M_{\mathrm{a}})(m_{\mathrm{e}}+m_{\mathrm{a}})\right]\mathrm{e}^{-(t/2\tau)\left(\sqrt{8M_{\mathrm{a}}M_{\mathrm{e}}+1}-1\right)}$$
$$+\frac{2}{3}F_{\mathrm{ex}}\tau\frac{m_{\mathrm{e}}}{M_{\mathrm{a}}(2M_{\mathrm{e}}-1)}\sqrt{4M_{\mathrm{a}}+1}\left(\bar{v}^{0}_{\mathrm{ex}} - \frac{F_{\mathrm{ex}}\tau}{M_{\mathrm{a}}}\right)\mathrm{e}^{-(t/2\tau)\left(\sqrt{4M_{\mathrm{a}}+1}-1\right)} \tag{8.1.27}$$
$$+\frac{\tau^2 F^2_{\mathrm{ex}}}{3\,M^2_{\mathrm{a}}}(m_{\mathrm{e}}+m_{\mathrm{a}})(1+M_{\mathrm{a}}),$$

with $\hat{T}^{0}_{\mathrm{ea}} = \hat{T}^{0}_{\mathrm{e}} - \hat{T}_{\mathrm{a}}$.

In the framework of the classical BE we find [21]

$$\hat{T}_{\mathrm{e}} = \hat{T}_{\mathrm{a}} + \left[\hat{T}^{0}_{\mathrm{ea}} - \frac{2}{3}F_{\mathrm{ex}}\tau\frac{m_{\mathrm{e}}}{M_{\mathrm{a}}(2M_{\mathrm{e}}-1)}\left(\bar{v}^{0}_{\mathrm{ex}} - \frac{F_{\mathrm{ex}}\tau}{M_{\mathrm{a}}}\right)\right.$$
$$\left. - \frac{\tau^2 F^2_{\mathrm{ex}}}{3\,M^2_{\mathrm{a}}}(m_{\mathrm{e}}+m_{\mathrm{a}})\right]\mathrm{e}^{-2M_{\mathrm{a}}M_{\mathrm{e}}(t/\tau)}$$
$$+\frac{2}{3}F_{\mathrm{ex}}\tau\frac{m_{\mathrm{e}}}{M_{\mathrm{a}}(2M_{e}-1)}\left(\bar{v}^{0}_{\mathrm{ex}} - \frac{F_{\mathrm{ex}}\tau}{M_{\mathrm{a}}}\right)\mathrm{e}^{-M_{\mathrm{a}}(t/\tau)} \tag{8.1.28}$$
$$+\frac{\tau^2 F^2_{\mathrm{ex}}}{3\,M^2_{\mathrm{a}}}(m_{\mathrm{e}}+m_{\mathrm{a}}).$$

Notice that the vanishing of the term $2M_{\mathrm{e}}-1$ in the denominators in Eqs. (8.1.27) and (8.1.28) does not actually lead to singularities at $M_{\mathrm{e}}=0.5$, because the corresponding terms cancel due to the exponential factors being equal. From Eqs. (8.1.27) and (8.1.28) it follows that:

$$\hat{T}^{\mathrm{GBE}}_{\mathrm{ea,st}} = (1+M_{\mathrm{a}})\hat{T}^{\mathrm{BE}}_{\mathrm{ea,st}}. \tag{8.1.29}$$

Thus, unlike drift velocity calculations, not only the GBE changes the trend of the relaxation curves but it also leads to different steady-state values of the energy temperatures. For a weakly ionized Lorentz gas, the effect of the self-consistent forces of electromagnetic origin can be neglected.

8.2 DISTRIBUTION FUNCTION (DF) OF THE CHARGED PARTICLES IN THE "LORENTZ" GAS

Calculating the DF for charged particles added as an impurity to a neutral gas in an external electric field is a classical problem in gas discharge physics, whose long history dates back to Pidduck's 1913 attempt to calculate the ion drift velocity in gases [166]. Mention should also be made of Compton's work concerned with computing the charge particle DF and its moments [167, 168]. Later on, Druyvesteyn [169, 170], and Davydov [171] obtained analytical expressions for the DF and transport coefficients for the special case of elastic collisions. More recent work (note, in particular, the monograph [172]) has been aimed principally at investigating the effect of inelastic collisions on the DF and transport processes within the BKT framework. It is important to note that the calculation of the DF depends heavily on what model of particle interaction is adopted - and hence ultimately on the collision cross sections involved. For example, the Davydov-Druyvesteyn distribution obtained under the assumption of a constant mean free path l for elastically colliding, charged gas particles significantly under predicts the number of "hot" particles on the tail of the DF and leads ultimately to unacceptable results when the theory is extended to calculating the kinetics of inelastic processes [172].

We apply the GBE

$$\mathbf{F}_{\mathrm{e}}\cdot\frac{\partial f_{\mathrm{e}}}{\partial \mathbf{v}_{\mathrm{e}}} - \tau\frac{\partial^2 f_{\mathrm{e}}}{\partial \mathbf{v}_{\mathrm{e}}\partial \mathbf{v}_{\mathrm{e}}}:\mathbf{F}_{\mathrm{e}}\mathbf{F}_{\mathrm{e}} = J_{\mathrm{ea}}, \tag{8.2.1}$$

to consider charged particles in a steady-state in a Lorentz gas subject to a stationary external electric field, where

$$\mathbf{F}_{\mathrm{e}} = eE/m_{\mathrm{e}}$$

The Boltzmann kinetic equation is usually solved by expanding the DF in a power series of zero-order solid spherical harmonics, i.e., of Legendre polynomials. The corresponding system of the linked equations was obtained elsewhere [173, 174]. The solution to the GBE (8.2.1) is conveniently sought as an expansion in terms of solid spherical harmonics:

$$f(v_e) = f_0(v_e) + \mathbf{F}_e\cdot\mathbf{v}_e f_1(v_e) + \mathbf{F}_e\mathbf{F}_e : \mathbf{v}_e^0\mathbf{v}_e f_2(v_e) + \cdots \tag{8.2.2}$$

Here $\mathbf{v}_e^0\mathbf{v}_e$ is the zero-trace tensor. For our further calculations in this section we assume that the force $\mathbf{F}_e$ is along the positive direction of a certain chosen coordinate axis. We now substitute expansion (8.2.2) into Eq. (8.2.1) and transform the corresponding terms; we have, for example, the following relations:

$$\begin{aligned}\mathbf{F}_e\cdot\frac{\partial}{\partial \mathbf{v}_e}\left(f_2\mathbf{F}_e\mathbf{F}_e : \mathbf{v}_e^0\mathbf{v}_e\right) &= \frac{(\mathbf{F}_e\cdot\mathbf{v}_e)^3}{v_e}\frac{\partial f_2}{\partial v_e} - \frac{1}{3}F_e^2 v_e(\mathbf{F}_e\cdot\mathbf{v}_e)\frac{\partial f_2}{\partial v_e}\\ &+\frac{4}{3}F_e^2 f_2(\mathbf{v}_e\cdot\mathbf{F}_e),\end{aligned} \tag{8.2.3}$$

$$\begin{aligned}\frac{\partial^2}{\partial \mathbf{v}_e\partial \mathbf{v}_e}:\left\{f_2\left(\mathbf{F}_e\mathbf{F}_e : \mathbf{v}_e^0\mathbf{v}_e\right)\mathbf{F}_e\mathbf{F}_e\right\} &= 5F_e^2\frac{(\mathbf{F}_e\cdot\mathbf{v}_e)^2}{v_e}\frac{\partial f_2}{\partial v_e} + 2f_2F_e^4\\ &+\frac{(\mathbf{F}_e\cdot\mathbf{v}_e)^4}{v_e^2}\frac{\partial^2 f_2}{\partial v_e^2} - \frac{(\mathbf{F}_e\cdot\mathbf{v}_e)^4}{v_e^3}\frac{\partial f_2}{\partial v_e}\\ &-\frac{1}{3}\sum_{i=1}^{3}F_{ei}^2F_e^2\frac{\partial^2}{\partial v_{ei}^2}\left(f_2v_e^2\right)\end{aligned} \tag{8.2.4}$$

The left-hand side Y of the GBE then takes the form

$$\begin{aligned}Y &= (\mathbf{F}_e\cdot\mathbf{v}_e)\frac{1}{v_e}\frac{\partial f_0}{\partial v_e} + F_e^2 f_1 + \frac{(\mathbf{F}_e\cdot\mathbf{v}_e)^2}{v_e}\frac{\partial f_1}{\partial v_e} + \frac{(\mathbf{F}_e\cdot\mathbf{v}_e)^3}{v_e}\frac{\partial f_2}{\partial v_e}\\ &-\frac{1}{3}F_e^2v_e(\mathbf{F}_e\cdot\mathbf{v}_e)\frac{\partial f_2}{\partial v_e} + \frac{4}{3}F_e^2 f_2(\mathbf{F}_e\cdot\mathbf{v}_e)\\ &-\tau\left\{\frac{(\mathbf{F}_e\cdot\mathbf{v}_e)^2}{v_e}\frac{\partial}{\partial v_e}\left(\frac{1}{v_e}\frac{\partial f_0}{\partial v_e}\right) + \frac{F_e^2}{v_e}\frac{\partial f_0}{\partial v_e} + 3F_e^2\frac{(\mathbf{F}_e\cdot\mathbf{v}_e)}{v_e}\frac{\partial f_1}{\partial v_e}\right.\\ &-\frac{(\mathbf{F}_e\cdot\mathbf{v}_e)^3}{v_e^3}\frac{\partial f_1}{\partial v_e} + \frac{(\mathbf{F}_e\cdot\mathbf{v}_e)^3}{v_e^2}\frac{\partial^2 f_1}{\partial v_e^2} + 4F_e^2\frac{(\mathbf{F}_e\cdot\mathbf{v}_e)^2}{v_e}\frac{\partial f_2}{\partial v_e}\\ &+\frac{4}{3}f_2F_e^2 + \frac{(\mathbf{F}_e\cdot\mathbf{v}_e)^4}{v_e^2}\frac{\partial^2 f_2}{\partial v_e^2} - \frac{1}{3}F_e^2(\mathbf{F}_e\cdot\mathbf{v}_e)^2\frac{\partial^2 f_2}{\partial v_e^2}\\ &\left.-\frac{(\mathbf{F}_e\cdot\mathbf{v}_e)^4}{v_e^3}\frac{\partial f_2}{\partial v_e} - \frac{1}{3}F_e^4 v_e\frac{\partial f_2}{\partial v_e}\right\}.\end{aligned} \tag{8.2.5}$$

Denoting the angle between the vectors $\mathbf{F}_e$ and $\mathbf{v}_e$ by ϑ ($0 \le \vartheta \le \pi$), multiplying the GBE by $d\cos\vartheta$, and integrating over the entire range of angles, we arrive at

$$\begin{aligned}f_1 &+ \frac{1}{3}v_e\frac{\partial f_1}{\partial v_e} - \frac{\tau}{3}\left\{\frac{2}{v_e}\frac{\partial f_0}{\partial v_e} + \frac{\partial^2 f_0}{\partial v_e^2} + \frac{12}{5}v_e\frac{\partial f_2}{\partial v_e}F_e^2\right.\\ &\left.+4f_2F_e^2 + \frac{4}{15}F_e^2v_e^2\frac{\partial^2 f_2}{\partial v_e^2}\right\} = \frac{1}{F_e^2}J_{ea}.\end{aligned} \tag{8.2.6}$$

Multiplying the GBE by $\cos\vartheta\, d\cos\vartheta$, a similar procedure leads to the result

$$\frac{\partial f_0}{\partial v_e} + \frac{4}{15}F_e^2v_e^2\frac{\partial f_2}{\partial v_e} + \frac{4}{3}F_e^2v_e f_2 - \frac{3}{5}\tau F_e^2\left\{4\frac{\partial f_1}{\partial v_e} + v_e\frac{\partial^2 f_1}{\partial v_e^2}\right\} = -\frac{3}{2}\frac{1}{F_e}J_1. \tag{8.2.7}$$

As has been indicated, the collision terms J_{ea}, and J_1 in the GBKT can be taken in the form in which they are usually written in the BE. In the case we consider below, assuming that the change in the electron energy due to an elastic collision [approximately equal to $\sim(m_e/m_a)^{1/2}\varepsilon$] is much less that the electron energy prior to the collision, in the Fokker-Planck approximation we have

$$J_{ea} = \frac{m_e \hat{T}_a}{m_a v_e^2}\frac{\partial}{\partial v_e}\left[v_e^3 \upsilon \left(\frac{f_0}{\hat{T}} + \frac{1}{m_e v_e}\frac{\partial f_0}{\partial v_e}\right)\right], \tag{8.2.8}$$

$$J_1 = \frac{2}{3} F_e \frac{v_e^2}{\ell} f_1, \tag{8.2.9}$$

where $\hat{T}_a$ is the energy temperature of the neutral gas $\hat{T}_a = k_B T_a$, υ is the collision rate which generally depends on the velocity, and ℓ is the mean free path for collisions of neutral and charged particles. It is relations (8.2.6)–(8.2.9) which provide the required basis for determining the DF and its moments. Traditionally, two limiting situations are considered in detail:

(1) a constant frequency rate, $\upsilon = \text{const}$, $\upsilon = \tau^{-1} = v_e \ell^{-1}$ and
(2) a constant mean free path between the collisions of charged and neutral particles, $\ell = \text{const}$.

We take up the former case first. Multiplying Eq. (8.2.6) through by $3v_e^2$ and using Eq. (4.2.8), we find, after some algebra, that

$$F_e^2 \frac{d}{dv_e}\left\{v_e^3 f_1 - \tau v_e^2 \frac{df_0}{dv_e}\right\} = 3\frac{\hat{T}_a m_e}{\tau m_a}\frac{d}{dv_e}\left\{v_e^3\left(\frac{f_0}{\hat{T}_a} + \frac{1}{m_e v_e}\frac{df_0}{dv_e}\right)\right\} \tag{8.2.10}$$

or upon integration over v_e

$$f_1 v_e = \left[\tau + \frac{3\hat{T}_a}{F_e^2 m_a \tau}\right]\frac{df_0}{dv_e} + \frac{3}{F_e^2}\frac{m_e}{m_a}\frac{v_e}{\tau} f_0, \tag{8.2.11}$$

because the constant of integration is zero due to the fact that both the left-hand and right-hand sides of Eq. (8.2.11) vanish for $v_e = 0$. Equation (8.2.11) was obtained under the condition (which will also be used in the following analysis) that small terms proportional to f_2 may be dropped. Substituting Eq. (8.2.11) into Eq. (8.2.9) and making use of the result produced to eliminate f_1 from Eq. (8.2.7), we arrive at the following equation in f_0:

$$\begin{aligned} &v_e^2\left(\tau + \frac{3\hat{T}_a}{F_e^2 m_a \tau}\right)\frac{d^3 f_0}{dv_e^3} + v_e\left(2\tau + \frac{3m_e}{F_e^2 m_a \tau}v_e^2 + \frac{6\hat{T}_a}{F_e^2 m_a \tau}\right)\frac{d^2 f_0}{dv_e^2} \\ &+\left(-2\tau - \frac{10}{3\tau F_e^2}v_e^2 - \frac{5\hat{T}_a}{\tau^3 F_e^4 m_a}v_e^2 + \frac{12 m_e}{F_e^2 m_a \tau}v_e^2 - \frac{6\hat{T}_a}{F_e^2 m_a \tau}\right)\frac{df_0}{dv_e} \\ &-\frac{5m_e}{F_e^4 \tau^3 m_a}v_e^3 f_0 = 0. \end{aligned} \tag{8.2.12}$$

To solve Eq. (8.2.12), three boundary conditions are needed. These are in fact quite obvious. Indeed, for $v_e = 0$, we can specify a certain value of f_0, determined only by the normalization of the function. From Eq. (8.2.12) it is also seen that $f_0' = 0$ for $v_e = 0$.

Finally, dividing the above equation by v_e^3 we find that $f_0 \to 0$ for $v_e \to \infty$.

Thus, Eq. (8.2.12) is easily solved by, for example, the sweep method. To do this, it is convenient to first bring the equation to the dimensionless form by introducing the following dimensionless quantities labeled with arcs over the symbols:

$$\check{v}_e = \frac{v_e}{F_e \tau},\ \check{\varepsilon} = \frac{m_e F_e^2 \tau^2}{\hat{T}_a},\ \check{f}_0 = \frac{f_0}{f_0(v_e = 0)}. \tag{8.2.13}$$

The procedure is realized to yield

$$\begin{aligned} &\check{v}_e^2\left[1 + \frac{3m_e}{m_a \check{\varepsilon}}\right]\frac{d^3\check{f}_0}{d\check{v}_e^3} + \check{v}_e\left[2 + \frac{6m_e}{m_a\check{\varepsilon}} + \frac{3m_e}{m_a}\check{v}_e^2\right]\frac{d^2\check{f}_0}{d\check{v}_e^2} \\ &-\left[2 + \frac{6m_e}{m_a\check{\varepsilon}} + \check{v}_e^2\left(\frac{10}{3} + \frac{5m_e}{m_a\check{\varepsilon}} - 12\frac{m_e}{m_a}\right)\right]\frac{d\check{f}_0}{d\check{v}_e} - 5\frac{m_e}{m_a}\check{v}_e^3\check{f}_0 = 0. \end{aligned} \tag{8.2.14}$$

Let us define the energy temperature of charged particles as follows:

$$\hat{T}_e = \frac{1}{3n_e}\int f_e m_e v_e^2 dv_e \cong \frac{1}{3n_e}\int f_0 m_e v_e^2 dv_e. \tag{8.2.15}$$

This means, for example, that, in terms of definitions (4.2.13), the Maxwellian function $\check{f}_{\mathrm{M}}$ has the form

$$\check{f}_{\mathrm{M}}=\exp\left[-\frac{\hat{T}_{\mathrm{a}}\check{\varepsilon}}{2\hat{T}_{\mathrm{e}}}\check{v}_{\mathrm{e}}^{2}\right]. \tag{8.2.16}$$

Let us examine the asymptotic of the function f_0 at the large velocities v_{e}. From Eq. (8.2.14) it follows that for $v_{\mathrm{e}}\to\infty$ the equation:

$$\frac{\mathrm{d}^2 f_0}{\mathrm{d}v_{\mathrm{e}}^2}-\frac{5}{3F_{\mathrm{e}}^2\tau^2}f_0=0, \tag{8.2.17}$$

holds, which has the solution

$$f_0\sim\exp\left[-\sqrt{\frac{5}{3}}\frac{v_{\mathrm{e}}}{F_{\mathrm{e}}\tau}\right]. \tag{8.2.18}$$

Note that, in the limiting case we are considering, the classical solution of the BE [175] leads to a Maxwellian distribution function (MDF) with a temperature $\hat{T}_{\mathrm{e}}$ different from the neutral gas temperature $\hat{T}_{\mathrm{a}}$. Thus, the solution of the GBE results in a large number of "hot" charged particles on the tail of the DF. Of course, the moments of the DF—the temperature $\hat{T}_{\mathrm{e}}$ and the drift velocity $\overline{v}_{\mathrm{ex}}$—can be found by properly integrating the DF after the solution of Eq. (8.2.14) has been found. There is no need to do this, however. Indeed, multiplying Eq. (8.2.14) by v_{e} and integrating term-by term we obtain

$$\left(\frac{3\hat{T}_{\mathrm{a}}}{m_{\mathrm{a}}\tau^2F_{\mathrm{e}}^2}+2\right)\int_0^{\infty}f_0v_{\mathrm{e}}^2\mathrm{d}v_{\mathrm{e}}=\frac{m_{\mathrm{e}}}{m_{\mathrm{a}}\tau^2F_{\mathrm{e}}^2}\int_0^{\infty}f_0v_{\mathrm{e}}^4\mathrm{d}v_{\mathrm{e}}. \tag{8.2.19}$$

Assuming that

$$\int f\mathrm{d}\mathbf{v}_{\mathrm{e}}\cong\int f_0\mathrm{d}\mathbf{v}_{\mathrm{e}}=4\pi\int_0^{\infty}f_0v_{\mathrm{e}}^2\mathrm{d}v_{\mathrm{e}}=n_{\mathrm{e}}, \tag{8.2.20}$$

as it was done in Eq. (8.2.15), it is found that

$$\hat{T}_{\mathrm{e}}=\hat{T}_{\mathrm{a}}+\frac{2}{3}m_{\mathrm{a}}\tau^2F_{\mathrm{e}}^2. \tag{8.2.21}$$

In a similar way, without explicitly solving Eqs. (8.2.11) and (8.2.12), we can determine the drift velocity. To accomplish this, we multiply Eq. (8.2.11) term-wise by v_{e}^3 and integrate the resulting expression to yield

$$\int_0^{\infty}f_1v_{\mathrm{e}}^4\mathrm{d}v_{\mathrm{e}}=\left[\tau+\frac{3\hat{T}_{\mathrm{a}}}{F_{\mathrm{e}}^2m_{\mathrm{a}}\tau}\right]\int_0^{\infty}v_{\mathrm{e}}^3\frac{\partial f_0}{\partial v_{\mathrm{e}}}\mathrm{d}v_{\mathrm{e}}+\frac{3}{F_{\mathrm{e}}^2}\frac{m_{\mathrm{e}}}{m_{\mathrm{a}}\tau}\int_0^{\infty}f_0v_{\mathrm{e}}^4\mathrm{d}v_{\mathrm{e}}, \tag{8.2.22}$$

leading to

$$\overline{v}_{\mathrm{ex}}=\frac{3\left(\hat{T}_{\mathrm{e}}-\hat{T}_{\mathrm{a}}\right)}{m_{\mathrm{a}}\tau F_{\mathrm{e}}}-\tau F_{\mathrm{e}}, \tag{8.2.23}$$

because, by definition, the following relations hold true:

$$\overline{v}_{\mathrm{ex}}=\frac{1}{n_{\mathrm{e}}}\int fv_{\mathrm{ex}}\mathrm{d}\mathbf{v}_{\mathrm{e}}=\frac{4\pi F_{\mathrm{e}}}{3n_{\mathrm{e}}}\int_0^{\infty}f_1v_{\mathrm{e}}^4\mathrm{d}v_{\mathrm{e}}. \tag{8.2.24}$$

Using expressions (8.2.21) and (8.2.23), we achieve the result sought:

$$\overline{v}_{\mathrm{ex}}=\tau F_{\mathrm{e}}. \tag{8.2.25}$$

Comparing relations (8.2.21) and (8.2.25) with known classical results (Ref. [175], p. 108) suggests that in the limiting case $\upsilon=\mathrm{const}$ the drift velocity remains unchanged and that $\hat{T}_{\mathrm{e}}$ increases [the classical analogue of Eq. (8.2.21) contains the numerical coefficient 1/2 instead of 2/3]. In concluding the discussion of this limiting case, we present the corresponding form of Eq. (8.2.14) $m_{\mathrm{e}}\ll m_{\mathrm{a}}$ for $\check{\varepsilon}\gtrsim 1$

$$\check{v}_{\mathrm{e}}^2\frac{\mathrm{d}^3\check{f}_0}{\mathrm{d}\check{v}_{\mathrm{e}}^3}+\left(2+3\frac{m_{\mathrm{e}}}{m_{\mathrm{a}}}\check{v}_{\mathrm{e}}^2\right)\check{v}_{\mathrm{e}}\frac{\mathrm{d}^2\check{f}_0}{\mathrm{d}\check{v}_{\mathrm{e}}^2}-\left(2+\frac{10}{3}\check{v}_{\mathrm{e}}^2\right)\frac{\mathrm{d}\check{f}_0}{\mathrm{d}\check{v}_{\mathrm{e}}}-5\frac{m_{\mathrm{e}}}{m_{\mathrm{a}}}\check{v}_{\mathrm{e}}^3\check{f}_0=0. \tag{8.2.26}$$

As a check on the correctness of the above results, note that if $F_e \equiv 0$ then Eq. (8.2.11) leads, as it should, to the MDF f_{0M}

$$\frac{df_0}{dv_e} = -\frac{m_e v_e}{\hat{T}} f_0, \tag{8.2.27}$$

$$f_0 = C\exp\left[-\frac{m_e v_e^2}{2\hat{T}}\right]. \tag{8.2.28}$$

We proceed now to the second limiting case, ℓ=const. In this case, the analog of Eq. (8.2.11) is as follows:

$$f_1 v_e = \left[\tau + \frac{3\hat{T} v_e}{F_e^2 m_a \ell}\right]\frac{df_0}{dv_e} + \frac{3}{F_e^2 m_a}\frac{m_e v_e^2}{\ell} f_0. \tag{8.2.29}$$

By the same procedure used in the limiting case υ=const, we arrive at the following equation in f_0:

$$\begin{aligned} & v_e^2\left[\tau + \frac{3\hat{T}_a v_e}{F_e^2 m_a \ell}\right]\frac{d^3 f_0}{dv_e^3} + \left[2\tau + \frac{12\hat{T}_a}{m_a \ell F_e^2} v_e + \frac{3 m_e v_e^3}{m_a \ell F_e^2}\right] v_e \frac{d^2 f_0}{dv_e^2} \\ & + \left[18\frac{m_e}{m_a \ell F_e^2} v_e^3 - 2\tau - \frac{5 v_e^2}{3\tau F_e^2} - \frac{5}{3F_e^2 \ell} v_e^3 - \frac{5\hat{T}_a}{\tau F_e^4 m_a \ell^2} v_e^4\right]\frac{df_0}{dv_e} \\ & + v_e^2\left[12\frac{m_e}{F_e^2 m_a \ell} - 5\frac{m_e}{\tau F_e^4 m_a \ell^2} v_e^3\right] f_0 = 0. \end{aligned} \tag{8.2.30}$$

Again, it is easily seen by multiplying Eq. (8.2.30) term-wise by F_e^4 that the vanishing of the external force F_e leads to Eq. (8.2.27) and then, upon integration, to the MDF (8.2.28). The boundary conditions for Eq. (8.2.30) are as follows: f_0 is specified for $v_e = 0$ in accord with the chosen normalization; for $v_e = 0$ as Eq. (8.2.30) suggests $f_0' = 0$, and, finally, $f_0 \to 0$, when $v_e \to \infty$.

The last result becomes evident if one first divides Eq. (8.2.30) through by v_e^5. In order to numerically integrate Eq. (8.2.30), it is convenient to bring this equation to the dimensionless form by using the dimensionless quantities

$$\check{v}_e = \frac{v_e}{(\ell/\tau)}, \quad \check{\varepsilon} = \frac{m_e F_e^2 \tau^2}{\hat{T}_a}, \quad \check{A} = \frac{F_e^2 \tau^4}{\ell^2}, \tag{8.2.31}$$

to give the ordinary differential equation

$$\begin{aligned} & \check{v}_e^2 \check{A}\left[1 + 3\frac{m_e}{m_a}\frac{\check{v}_e}{\check{\varepsilon}}\right]\frac{d^3 \check{f}_0}{d\check{v}_e^3} + \left[\check{A}\left(2 + 12\frac{m_e}{m_a}\frac{\check{v}_e}{\check{\varepsilon}}\right)\right. \\ & \left. + 3\frac{m_e}{m_a}\check{v}_e^3\right]\check{v}_e \frac{d^2 \check{f}_0}{d\check{v}_e^2} + \left[-2\check{A} - \frac{5}{3}\check{v}_e^2 - \left(\frac{5}{3} - 18\frac{m_e}{m_a}\right)\check{v}_e^3\right. \\ & \left. - 5\frac{m_e}{m_a}\frac{\check{v}_e^4}{\check{\varepsilon}}\right]\frac{d\check{f}_0}{d\check{v}_e} + \check{v}_e^2 \frac{m_e}{m_a}\left[12 - 5\frac{\check{v}_e^3}{\check{A}}\right]\check{f}_0 = 0, \end{aligned} \tag{8.2.32}$$

with the boundary conditions

$$\check{f}_0(0) = 1, \quad \check{f}_0'(0) = 0, \quad \check{f}_0(\infty) = 0. \tag{8.2.33}$$

Here the term-by-term integration no longer leads to elegant results like Eqs. (8.2.21) and (8.2.25). We can, however, give a useful formula for computing the drift velocity $\overline{v}_{ex}$, which is obtained from Eq. (8.2.32) by multiplying it by v_e^3 and then integrating, to yield

$$\overline{v}_{ex} = \frac{4\pi F_e}{3n_e}\left\{\frac{18\hat{T}_a \tau}{5 m_a}\int_0^\infty f_0 dv_e + C\frac{6}{5}\tau^2 \ell F_e^2 + 2\ell\int_0^\infty f_0 v_e dv_e - \frac{9}{10\pi}\frac{m_e}{m_a}\tau n_e\right\}, \tag{8.2.34}$$

with $C = f_0(v_e = 0)$.

Although Eq. (8.2.34) can of course be used only after numerically integrating Eq. (8.2.33), it is of interest to note that, unlike Eq. (8.2.25), the drift velocity is a non-linear function of F_e in this limiting case. Let us consider here some numerical results for the DF of charged particles in an external electric field, produced when employing the GBE. Numerical integration of corresponding differential equations was realized by the three-diagonal method of Gauss elimination techniques for the ordinary differential equations (see Appendix 5).

In Fig. 8.1, the dimensionless DF $\check{f}_0$, is plotted versus the dimensionless velocity $\check{v}_e$ for $\check{\varepsilon}= 10^{-3}$ and $\tau=\text{const}$. Line 1 corresponds to the MDF, and curve 2, to the DF obtained using the GBE. As $\check{\varepsilon}$ is decreased, the two distributions approach each other. Note that the function $\check{f}_0^{\,GBE}$ lies above the Maxwellian function.

Figures 8.2 and 8.3 present $\check{f}_0$, calculated in the case of $\ell=\text{const}$ under the conditions $\check{\varepsilon}= 10^{-2}, \check{A}=1$ and $\check{\varepsilon}= 10^{-2}, \check{A}= 10^{-1}$, respectively. Curves 1, 2, and 3 in Figs. 8.2 and 8.3 correspond to the MDF, the GBE, and the Druyvesteyn DF, respectively. It is interesting to note that the DF $\check{f}_0^{\,GBE}$ may lie both between the Maxwell and Druyvesteyn functions and above the two. In practical computations, to reemphasize, the DFs can be normalized to the number density of the charged particles involved.

The previous considerations can be applied in the theory of transport processes in the non-degenerative semiconductors. Let us consider the stationary 1D electron drift in an isotropic semiconductor. In this case for the elected x-direction we have from Eq. (8.2.1)

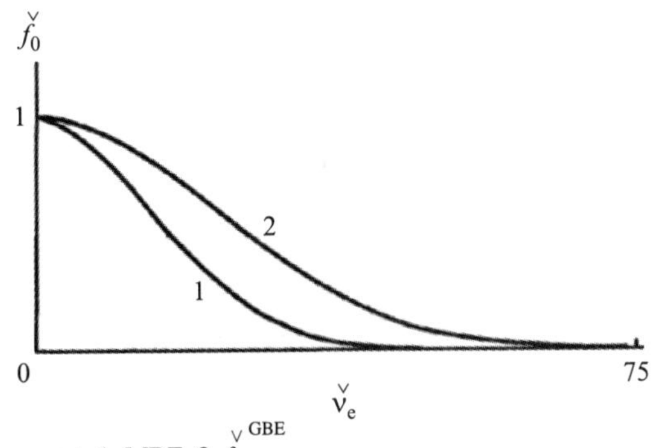

FIGURE 8.1 Dependence of $\check{f}_0$ on $\check{v}_e$ for $\tau=\text{const}$. 1, MDF; 2, $\check{f}_0^{\,GBE}$.

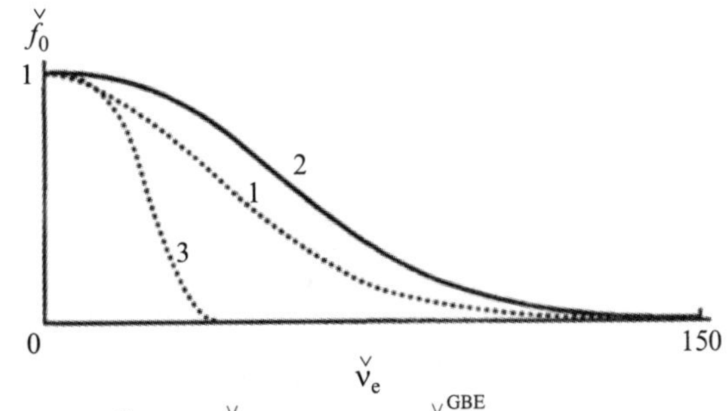

FIGURE 8.2 Dependence of $\check{f}_0$ on $\check{v}_e$ for $\ell=\text{const}$, $\check{\varepsilon}= 10^{-2}$, $\check{A}= 1$. 1, MDF; 2, $\check{f}_0^{\,GBE}$; 3, Druyvesteyn DF.

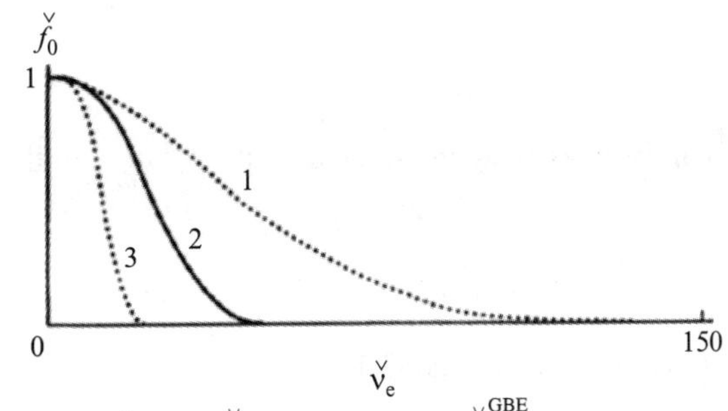

FIGURE 8.3 Dependence of $\check{f}_0$ on $\check{v}_e$ for $\ell=\text{const}$, $\check{\varepsilon}= 10^{-2}$, $\check{A}= 10^{-1}$. 1, MDF; 2, $\check{f}_0^{\,GBE}$; 3, Druyvesteyn DF.

$$-\frac{eE}{m_e}\frac{\partial f_e}{\partial v_{ex}} - \tau\left(\frac{eE}{m_e}\right)^2\frac{\partial^2 f_e}{\partial v_{ex}^2} = J_e^B, \tag{8.2.35}$$

e is the absolute electron charge. Obviously the corresponding BE has the form

$$-\frac{eE}{m_e}\frac{\partial f_e}{\partial v_{ex}} = J_e^B, \tag{8.2.36}$$

We introduce the momentum collision integrals

$$\mathbf{I}_{e,mot} = -\int m_e \mathbf{v}_e \frac{f_e}{\tau_p(\varepsilon)} d\mathbf{v}_e, \tag{8.2.37}$$

$$I_{e,en} = -\int \frac{m_e v_e^2}{2}\frac{f_e}{\tau_\varepsilon(\varepsilon)} d\mathbf{v}_e, \tag{8.2.38}$$

where $\varepsilon = m_e v_e^2/2$, τ_p, and τ_ε are the impulse relaxation time and the energy relaxation time correspondingly. The electron scattering by acoustic phonons for the non-piezoelectric crystals can be taken from [176]

$$\frac{1}{\tau_p(\varepsilon)} = \frac{1}{\tau_a}\sqrt{\frac{\varepsilon}{k_B T}}, \tag{8.2.39}$$

$$\frac{1}{\tau_\varepsilon(\varepsilon)} = \frac{1}{\tau_a}\frac{2m_e s^2}{k_B T}\left[\sqrt{\frac{\varepsilon}{k_B T}} - \sqrt{\frac{k_B T}{\varepsilon}}\right], \tag{8.2.40}$$

where

$$\frac{1}{\tau_a} = \frac{\sqrt{2}G^2(m_e k_B T)^{3/2}}{\pi \rho s^2 \hbar^4}, \tag{8.2.41}$$

T and ρ are the temperature and density of the lattice, s–the longitudinal sound velocity, G is the constant of the interaction, which has the energy dimension. Application of relations (8.2.37)–(8.2.41) needs in the explicit expression for the DF. Introduce the typical approximate expression the DF (compared with Eq. (8.2.2))

$$f_e = f_0(\varepsilon) + Ef_1(\varepsilon)v_{ex} + E^2 f_2(\varepsilon)v_e^2\left(\cos^2\theta - \frac{1}{3}\right) + \cdots \tag{8.2.42}$$

where θ is the angle between the velocity $\mathbf{v}_e$ direction and the x-axis. Using the Bhatnagar-Gross-Krook (BGK) model and relations (8.2.35), Eq. (8.2.42) leads in the first approximation

$$f_1(\varepsilon) = \tau_p(\varepsilon)\frac{e}{m_e v_{ex}}\frac{\partial f_0}{\partial v_{ex}}, \tag{8.2.43}$$

Then for the drift velocity

$$\bar{v}_{ex} = \frac{1}{n_e}\int f_e(\mathbf{v}_e)v_{ex}d\mathbf{v}_e \tag{8.2.44}$$

as for BE so GBE in the first approximation

$$\bar{v}_{ex} = \frac{eE}{m_e n_e}\int \tau_p(\varepsilon)\frac{\partial f_0}{\partial v_{ex}}v_{ex}d\mathbf{v}_e \tag{8.2.45}$$

The energy DF for the hot electrons can be usually approximated by the MDF with the effective electron temperature T_e

$$f_0 = n_e\left(\frac{m_e}{2\pi k_B T_e}\right)^{3/2}\exp\left(-\frac{\varepsilon}{k_B T_e}\right) \tag{8.2.46}$$

and as a result to the same drift velocities for the BE and GBE

$$\bar{v}_{ex} = v_{ex}^0 \left(\frac{T}{T_e}\right)^{1/2} \tag{8.2.47}$$

where

$$v_{ex}^0 = -\frac{4eE\tau_a}{3\sqrt{\pi}m_e}. \tag{8.2.48}$$

Estimate now the electron temperature as the result of the electron evolution in the electric field with the E intensity. From the hydrodynamic energy equation (the BE-model)

$$\bar{v}_{ex}en_eE = I_{e,en} \tag{8.2.49}$$

and for GBE

$$\bar{v}_{ex}en_eE - \frac{\tau e^2E^2}{m_e}n_e = I_{e,en}. \tag{8.2.50}$$

For the relaxation integral we find

$$I_{e,en} = -\int \frac{m_e v_e^2}{2}\frac{f_e}{\tau_\varepsilon(\varepsilon)}d\mathbf{v}_e = -\frac{8}{\sqrt{\pi}}\frac{m_e s^2}{\tau_a}\left[\left(\frac{T_e}{T}\right)^{3/2} - \left(\frac{T_e}{T}\right)^{1/2}\right]. \tag{8.2.51}$$

Rewrite the energy equations (8.2.49) and (8.2.50)
(for the Boltzmann model)

$$\frac{(eE)^2}{3m_e}\tau_a\left(\frac{T}{T_e}\right)^{1/2} = \frac{2m_e s^2}{\tau_a}\left[\left(\frac{T_e}{T}\right)^{3/2} - \left(\frac{T_e}{T}\right)^{1/2}\right], \tag{8.2.52}$$

(for the non-local model)

$$\frac{(eE)^2}{m_e}\tau + \frac{4(eE)^2}{3\sqrt{\pi}m_e}\tau_a\left(\frac{T}{T_e}\right)^{1/2} = \frac{8m_e s^2}{\tau_a\sqrt{\pi}}\left[\left(\frac{T_e}{T}\right)^{3/2} - \left(\frac{T_e}{T}\right)^{1/2}\right]. \tag{8.2.53}$$

The T_e—dependence T_e of T for the Boltzmann model follows from (8.2.51)

$$T_e = \frac{T}{2}\left[1 + \sqrt{1 + \frac{2\tau_a^2}{3s^2}\left(\frac{eE}{m_e}\right)^2}\right]. \tag{8.2.54}$$

Transform Eq. (8.2.53) using the substitution $y = (T_e/T)^{1/2}$.

$$y^4 - y^2 - \frac{\sqrt{\pi}}{8}\left(\frac{eE}{m_e}\right)^2\frac{\tau\tau_a}{s^2}y - \frac{1}{6}\left(\frac{eE}{m_e}\right)^2\frac{\tau_a^2}{s^2} = 0. \tag{8.2.55}$$

We find the "local" and "non-local" combinations in Eq. (8.2.55)

$$\frac{\tau_a}{s^2} = \frac{\pi\rho\hbar^4}{\sqrt{2}G^2(m_e k_B T)^{3/2}}, \tag{8.2.56}$$

$$\frac{\tau\tau_a}{s^2} = \frac{\pi\rho\hbar^4}{\sqrt{2}G(m_e k_B T)^{3/2}}, \tag{8.2.57}$$

where

$$G_1 = G/\sqrt{\tau}. \tag{8.2.58}$$

The solutions of the fourth degree equation (8.2.55) can be obtained in the analytical (but rather complicated) form. Much more simple the asymptotic solutions of Eq. (8.2.55). For the hot electrons ($T_e \gg T$) we find

(for the Boltzmann model)

$$T_e = T\frac{1}{\sqrt{6}m_e}\frac{eE\tau_a}{s}, \tag{8.2.59}$$

(for the non-local theory)

$$T_e = T\frac{\pi^{1/3}}{4}\left(\frac{eE}{m_e}\right)^{4/3}\left(\frac{\tau\tau_a}{s^2}\right)^{2/3} \tag{8.2.60}$$

or

$$T_e = T\frac{\pi^{1/3}}{4}\left(\frac{eE}{m_e}\right)^{4/3}\left(\frac{\pi\rho\hbar^4}{\sqrt{2}G(m_e k_B T)^{2/3}}\right)^{2/3}. \tag{8.2.61}$$

In contrast to the Boltzmann kinetics, non-local kinetic theory leads to the non-linear dependence of the electron effective temperature on the intensity of the electric field, $T_e \sim E^{4/3}$.

For so-called warm charge carriers ($T_e - T \ll T$) we have from relations (8.2.52) and (8.2.53) (BE-model)

$$T_e = T\left[1 + \frac{\tau_a^2}{6s^2}\left(\frac{eE}{m_e}\right)^2\right], \tag{8.2.62}$$

(non-local theory)

$$T_e = T\left[1 + \frac{\tau_a^2}{6s^2}\left(\frac{eE}{m_e}\right)^2\left(1 + \frac{3\sqrt{\pi}}{4}\frac{\tau}{\tau_a}\right)\right]. \tag{8.2.63}$$

As we see the temperature of the warm carriers T_e^{GBE} is higher than T_e^{BE}. Compare now the theoretical and experimental results. For this aim let us obtain the relation between the intensity E and the charge carrier mobility μ. One obtains from Eq. (8.2.47)

$$\mu = \mu^0\left(\frac{T}{T_e}\right)^{1/2}, \tag{8.2.64}$$

where

$$\mu^0 = \bar{v}_{ex}^0/E, \tag{8.2.65}$$

and from relations (8.2.52) and (8.2.53) correspondingly (BE)

$$E^2 = \frac{6s^2}{\tau_a^2}\left(\frac{m_e}{e}\right)^2\left(\frac{\mu_0}{\mu}\right)^2\left[\left(\frac{\mu_0}{\mu}\right)^2 - 1\right], \tag{8.2.66}$$

(GBE)

$$E^2 = \frac{6s^2}{\tau_a^2}\left(\frac{m_e}{e}\right)^2\left(\frac{\mu_0}{\mu}\right)^2\frac{(\mu_0/\mu)^2 - 1}{1 + (3\sqrt{\pi}/4)(\tau/\tau_a)(\mu_0/\mu)}. \tag{8.2.67}$$

We can introduce the non-local correction defined by the coefficient

$$\alpha = \frac{3\sqrt{\pi}}{4}\frac{\tau}{\tau_a}, \tag{8.2.68}$$

namely

$$E^2 = \frac{6s^2}{\tau_a^2}\left(\frac{m_e}{e}\right)^2\left(\frac{\mu_0}{\mu}\right)^2\frac{(\mu_0/\mu)^2 - 1}{1 + \alpha(\mu_0/\mu)}. \tag{8.2.69}$$

The electron scattering by the longitudinal acoustic phonons is the basic mechanism of scattering for the sufficiently high temperatures when we need not to take into account the scattering by the ionized dopant. At the same time the temperature should sufficiently low for the exclusion of scattering by the optical phonons [177]. This situation corresponds for example for the case of the electron scattering in the covalent crystals Ge and Si for the temperatures $20 < T < 100$ K [177–179]. The μ/μ_0—dependence of E in the logarithmic coordinates is shown in Fig. 8.4 for n-Ge ($T = 77$ K) [177, 179, 180]. Figures 8.5 and 8.6 contain the analogical dependences for n-Si obtained from the measurement of the drift velocity for 20 and 77 K [181]. All mentioned figures list the data obtained for $\alpha = 0$, $\alpha = 1$, $\alpha = 2$. The case $\alpha = 0$ responds to the local Boltzmann model. The energy constant G is practically unknown value for the local models and the comparison with the experimental data achieves after the parallel shift of the theoretical curve along E-axis.

From the Figs. 8.4–8.6 stem that the discrepancy between local models ($\alpha = 0$) is very significant, but "non-local" curves (calculated for $\alpha = 1$, and $\alpha = 2$) much better respond to the experiment.

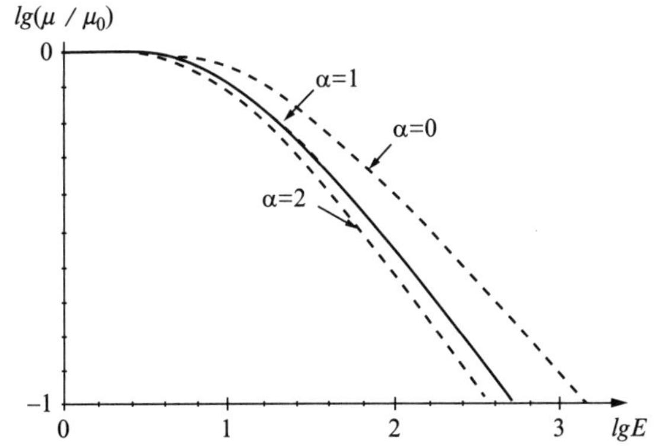

FIGURE 8.4 The lg(μ/μ_0)—dependence on lg E for n-Ge ($T = 77$ K, dim E=V/cm)

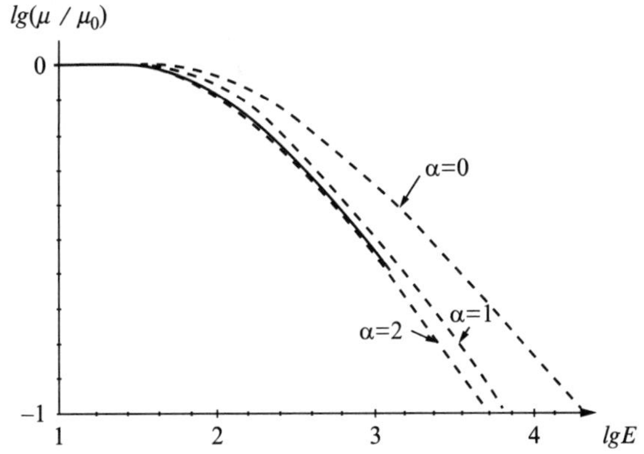

FIGURE 8.5 The μ/μ_0—dependence of E in the logarithmic coordinates for n-Si ($T = 20$ K), (left).

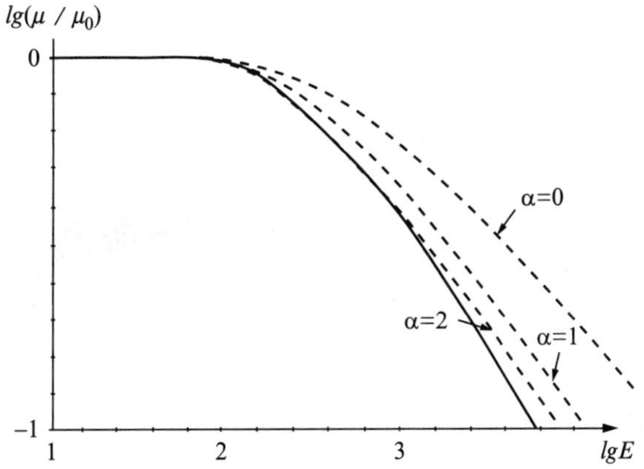

FIGURE 8.6 The μ/μ_0—dependence of E in the logarithmic coordinates for n-Si ($T = 77$ K), (right), (dim E = V/cm).

Figures 8.7 and 8.8 demonstrate the convergence of the μ/μ_0–dependence of E. Let us investigate this effect in detail from the mathematical point of view using the following form of Eq. (8.2.69) (see also Eqs. 8.2.41 and 8.2.68)

$$\widetilde{E}=\sqrt{\frac{1}{y^4}-\frac{1}{y^2}\frac{1}{\sqrt{1+(\alpha/y)}}}, \tag{8.2.70}$$

where $y=\mu/\mu_0, \widetilde{E}=E\left[\frac{\sqrt{6}sm_e}{\tau_a\ e}\right]^{-1}$, $0<y\leq 1$. Fig. 8.7 is plotted using the Maple notations, namely $y=A[i]=\ln\left(\mathrm{sqrt}\left(\frac{1}{x^4}-\frac{1}{x^2}\right)\cdot\frac{1}{\mathrm{sqrt}(1+((i-1)x))}\right), 1\leq i\leq 140, 0.001\leq x\leq 1$. It means that the α-parameter varies in relation (8.2.70) in vary vast limits $0\leq\alpha\leq 139$. Obviously the upper curve on Fig. 8.7 ($\alpha=0$) corresponds to the local theory.

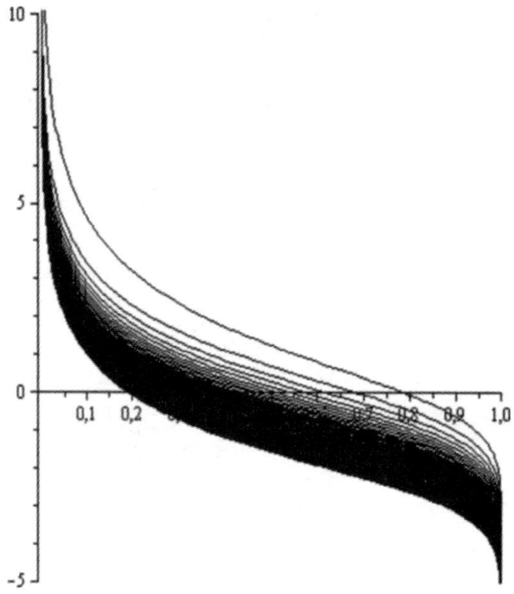

FIGURE 8.7 The dependence $\ln\widetilde{E}$ on μ/μ_0.

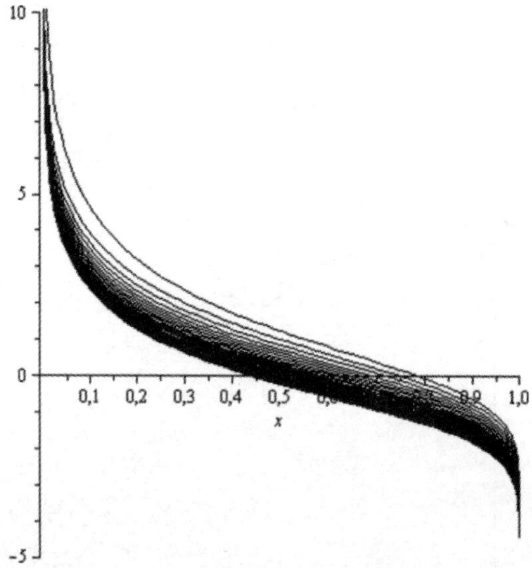

FIGURE 8.8 The dependence $\ln\widetilde{E}$ on μ/μ_0.

Figure 8.8 is plotted also using the Maple notations for the case

$$y = A[i] = \ln\left(\mathrm{sqrt}\left(\frac{1}{x^4} - \frac{1}{x^2}\right) \cdot \frac{1}{\mathrm{sqrt}(1 + (i-1)x0.3)}\right), 1 \le i \le 31, 0.001 \le x \le 1.$$

In this case the α-parameter varies in relation (8.2.70) in limits $0 \le \alpha \le 9$. As before the upper curve on Fig. 8.8 ($\alpha = 0$) corresponds to the local theory

The black areas on Figs. 8.7 and 8.8 reflect the convergence of "the family" of curves by enlarging of α-parameter.

8.3 CHARGED PARTICLES IN ALTERNATING ELECTRIC FIELD

As another example of the application of the GBE, let us consider the time evolution of the DF of charged particles moving in an alternating electric field. In this problem, only elastic collisions will be considered; the GBE takes the form

$$\begin{aligned} &\left(\frac{\partial f_e}{\partial t} + \mathbf{F}_e \cdot \frac{\partial f_e}{\partial \mathbf{v}_e}\right)\left(1 - \frac{\partial \tau}{\partial t}\right) - \tau\left\{\frac{\partial^2 f_e}{\partial t^2} + 2\mathbf{F}_e \cdot \frac{\partial^2 f_e}{\partial \mathbf{v}_e \partial t}\right. \\ &\left. + \frac{\partial \mathbf{F}_e}{\partial t} \cdot \frac{\partial f_e}{\partial \mathbf{v}_e} + \frac{\partial^2 f_e}{\partial \mathbf{v}_e \partial \mathbf{v}_e} : \mathbf{F}_e \mathbf{F}_e\right\} = J_{ea}. \end{aligned} \tag{8.3.1}$$

If we make use of the expansion

$$f_e(\mathbf{v}_e, t) = f_0(v_e, t) + \mathbf{F}_e \cdot \mathbf{v}_e f_1(v_e, t) + \mathbf{F}_e \mathbf{F}_e : \mathbf{v}_e^0 \mathbf{v}_e f_2(v_e, t), \tag{8.3.2}$$

then utilizing a procedure analogous to that described above we arrive at the following equations in the functions f_0 and f_1:

$$\begin{aligned} &\left(1 - \frac{\partial \tau}{\partial t}\right)\left[\frac{\partial f_0}{\partial t} + F_e^2 f_1 + \frac{1}{3} F_e^2 v_e \frac{\partial f_1}{\partial v_e}\right] \\ &-\tau\left[\frac{\partial^2 f_0}{\partial t^2} + \frac{1}{3} F_e^2 \frac{\partial^2 f_0}{\partial v_e^2} + \frac{2}{3} F_e^2 \frac{1}{v_e} \frac{\partial f_0}{\partial v_e}\right. \\ &+ \frac{1}{2} f_1 \frac{\partial F_e^2}{\partial t} + 2F_e^2 \frac{\partial f_1}{\partial t} + \frac{1}{2} \frac{\partial F_e^2}{\partial t} v_e \frac{\partial f_1}{\partial v_e} \\ &\left. + \frac{2}{3} F_e^2 v_e \frac{\partial^2 f_1}{\partial v_e \partial t}\right] = J_{ea}, \end{aligned} \tag{8.3.3}$$

$$\begin{aligned} &\left(1 - \frac{\partial \tau}{\partial t}\right)\left[F_e \frac{\partial f_0}{\partial v_e} + v_e \frac{\partial F_e}{\partial t} f_1 + F_e v_e \frac{\partial f_1}{\partial t}\right] - \tau\left[2F_e \frac{\partial^2 f_0}{\partial v_e \partial t}\right. \\ &+ \frac{\partial F_e}{\partial t} \frac{\partial f_0}{\partial v_e} + 2v_e \frac{\partial F_e}{\partial t} \frac{\partial f_1}{\partial t} + v_e \frac{\partial^2 F_e}{\partial t^2} f_1 + F_e v_e \frac{\partial^2 f_1}{\partial t^2} \\ &\left. + \frac{12}{5} F_e^3 \frac{\partial f_1}{\partial v_e} + \frac{3}{5} F_e^3 v_e \frac{\partial^2 f_1}{\partial v_e^2}\right] = -\frac{3}{2} J_1, \end{aligned} \tag{8.3.4}$$

where, for example,

$$J_1 = \frac{2}{3} F_e f_1 \frac{v_e^2}{\ell}$$

Now consider a case in which the mean time between collisions $\upsilon = v_e/\ell$, (which is used for J_{ea} and J_1 approximations), is independent of the velocity. It proves possible to determine the DF moments $\overline{v}_{ex}$ and $\hat{T}_e$ without directly solving Eqs. (8.3.3) and (8.3.4). Multiply Eq. (8.3.4) term-wise by v_e^3 and integrate the resulting expression over all absolute velocities. Using the additional conditions

$$\int_0^\infty f_1 v_e^4 \mathrm{d}v_e = \frac{3n_e}{4\pi F_e} \overline{v}_{ex}, \quad \int_0^\infty f_0 v_e^2 \mathrm{d}v_e = \frac{n_e}{4\pi}, \quad \int_0^\infty f_0 v_e^4 \mathrm{d}v_e = \frac{3}{2} n_e \frac{\hat{T}_e}{m_e},$$

we derive the following equation:

$$\tau\frac{\mathrm{d}^2\overline{v}_{\mathrm{ex}}}{\mathrm{d}t^2}-\frac{\mathrm{d}\overline{v}_{\mathrm{ex}}}{\mathrm{d}t}-\frac{1}{\tau}\overline{v}_{\mathrm{ex}}+F_{\mathrm{e}}-\tau\frac{\mathrm{d}F_{\mathrm{e}}}{\mathrm{d}t}=0. \tag{8.3.5}$$

Suppose that the time dependence of the external force can be represented as $F_{\mathrm{e}}=\frac{eE_0}{m_{\mathrm{e}}}\cos\omega t$, where the frequency ω is related to the external electric field strength. The solution of the inhomogeneous differential Eq. (8.3.5) is written as

$$\overline{v}_{ex}^{\mathrm{GBE}}=C_1e^{-a(t/\tau)}+\frac{b\tau}{(\omega\tau)^4+3(\omega\tau)^2+1}\left[\cos\omega t+\omega\tau\left(2+\omega^2\tau^2\right)\sin\omega t\right], \tag{8.3.6}$$

where $b=eE_0/m_{\mathrm{e}}$, and C_1 is the constant of integration, which is determined by the initial conditions of the problem. The classical result which can be obtained from the BE for the quasi-stationary case is given by

$$\overline{v}_{\mathrm{ex}}^{\mathrm{BE}}=\frac{b\tau}{(\omega\tau)^2+1}[\cos\omega t+\omega t\sin\omega t], \tag{8.3.7}$$

The introduction of the mobility K usually defined through the expression

$$\overline{v}_{\mathrm{ex}}=K\frac{m_{\mathrm{e}}}{e}F_{\mathrm{e}}$$

would serve no purpose due to singularities that can appear for $F_{\mathrm{e}}=0$.

We now turn our attention to Eq. (8.3.3). We multiply this equation term-wise by v_{e}^4 and integrate the resulting expression over all v_{e}.

$$\begin{aligned}\frac{\mathrm{d}^2\hat{T}_{\mathrm{e}}}{\mathrm{d}t^2}-\frac{1}{\tau}\frac{\mathrm{d}\hat{T}_{\mathrm{e}}}{\mathrm{d}t}-\frac{2m_{\mathrm{e}}}{m_{\mathrm{a}}\tau^2}\left(\hat{T}_{\mathrm{e}}-\hat{T}_{\mathrm{a}}\right)&=-\frac{2}{3}m_{\mathrm{e}}\frac{F_{\mathrm{e}}}{\tau}\overline{v}_{\mathrm{ex}}\\ -\frac{2}{3}m_{\mathrm{e}}F_{\mathrm{e}}^2+\frac{2}{3}m_{\mathrm{e}}\overline{v}_{\mathrm{ex}}\frac{\mathrm{d}F_{\mathrm{e}}}{\mathrm{d}t}&+\frac{4}{3}m_{\mathrm{e}}F_{\mathrm{e}}\frac{\mathrm{d}\overline{v}_{\mathrm{ex}}}{\mathrm{d}t},\end{aligned} \tag{8.3.8}$$

where $F_{\mathrm{e}}=b\sin\omega t$ and consequently,

$$\overline{v}_{\mathrm{ex}}=C_1\mathrm{e}^{-a(t/\tau)}+\frac{b\tau}{(\omega\tau)^4+3(\omega\tau)^2+1}\left\{\sin\omega t-\left(2+(\omega\tau)^2\right)\omega\tau\cos\omega t\right\}. \tag{8.3.9}$$

The differential equation (8.3.8) integrates in the finite form to the following expression:

$$\begin{aligned}\hat{T}_{\mathrm{ea}}&=C_1\mathrm{e}^{-d(t/\tau)}\\ &+\mathrm{e}^{-a(t/\tau)}\frac{\tau^2Z}{\left[(d+a+1)^2+\omega^2\tau^2\right]\left[(d-a)^2+\omega^2\tau^2\right]}\left\{\sin\omega t\left[\omega^2\tau^2\left(2\sqrt{5}\right.\right.\right.\\ &-2(d+a+1)+\omega^2\tau^2(\sqrt{5}-1-d-a))+(d-a)(2\omega^2\tau^2+\omega^4\tau^4\\ &+\omega^2\tau^2\sqrt{5}(d+a+1)+2\sqrt{5}(d+a+1))\big]+\cos\omega t\left[(d-a)\omega\tau(2\sqrt{5}\right.\\ &+\omega^2\tau^2(\sqrt{5}-1-d-a)-2(d+a+1))-\omega\tau(2\omega^2\tau^2+\omega^4\tau^4\\ &\left.\left.+\omega^2\tau^2\sqrt{5}(d+a+1)+2\sqrt{5}(d+a+1))\right]\right\}\\ &+Z\tau^2\left\{\frac{\sin^2\omega t}{2(d^2+4\omega^2\tau^2)\left[(d+1)^2+4\omega^2\tau^2\right]}[4d(d+1)\right.\\ &-2d(d+1)\omega^2\tau^2-4\omega^2\tau^2-2d(d+1)\omega^4\tau^4-28\omega^4\tau^4-16\omega^6\tau^6]\\ &+\cos^2\omega t\frac{\omega^2\tau^2(2+\omega^2\tau^2)d(d+1)}{\left[(d+1)^2+4\omega^2\tau^2\right]\left[d^2+4\omega^2\tau^2\right]}\\ &-\omega\tau\cos\omega t\sin\omega t\frac{\omega^2\tau^2(d^2+d+14)+4d+4}{(d^2+4\omega^2\tau^2)((d+1)^2+4\omega^2\tau^2)}\\ &\left.+\omega^2\tau^2\frac{9d^2+9d+4+\omega^2\tau^2(12d^2+10d+18)+4\omega^4\tau^4(d^2+d+2)}{d(d+1)\left[(d+1)+4\tau^2\omega^2\right]\left[d^2+4\tau^2\omega^2\right]}\right\},\end{aligned} \tag{8.3.10}$$

where the notation was used

$$a=\frac{\sqrt{5}-1}{2},\ d=\frac{1}{2}\left[\sqrt{1+8\frac{m_e}{m_a}}-1\right]\cong 2\frac{m_e}{m_a}\ll 1,\ Z=\frac{2}{3}\frac{m_e b^2}{(\omega^4\tau^4+3\omega^2\tau^2-1)}. \tag{8.3.5}$$

In the quasi-stationary limiting case, under the condition $\omega\tau\gg 1$, one finds

$$\hat{T}_{ea}=-\tau^2 Z\frac{\omega^2\tau^2}{2}\sin^2\omega t+\tau^2 Z\frac{\omega^2\tau^2}{2d}, \tag{8.3.11}$$

or, taking into account that the time average

$$\overline{\sin^2\omega t}=\frac{1}{2}\ll d^{-1},$$

we arrive at

$$\hat{T}_e=\frac{m_a}{6}\left(\frac{eE_0}{m_e\omega}\right)^2. \tag{8.3.12}$$

Thus, in the limiting case $l=\text{const}$, the following equality holds true:

$$\hat{T}_e^{BE}=\hat{T}_e^{GBE}. \tag{8.3.13}$$

In the opposite limit of $\omega\tau\ll d\ll 1$, one has

$$\hat{T}_{ea}=\tau^2 Z\frac{2}{d}\sin^2\omega t$$

or, computing the average over the time, we obtain

$$\hat{T}_{ea}=\frac{m_a\tau^2}{3}\left(\frac{eE_0}{m_e}\right)^2, \tag{8.3.14}$$

i.e., a result which corresponds to the solution of the classical BE [182].

8.4 CONDUCTIVITY OF A WEAKLY IONIZED GAS IN THE CROSSED ELECTRIC AND MAGNETIC FIELDS

In this section, the conductivity of a weakly ionized gas subject to crossed magnetic and (alternating) electric field will be examined using the GBE with the BGK approximation for the elastic collision integral. The BGK-approximated kinetic equation takes the form

$$\begin{aligned}&\frac{\partial f_e}{\partial t}+\mathbf{F}_e\cdot\frac{\partial f_e}{\partial \mathbf{v}_e}-\tau\left[\frac{\partial^2 f_e}{\partial t^2}+2\mathbf{F}_e\cdot\frac{\partial^2 f_e}{\partial \mathbf{v}_e\partial t}+\frac{\partial \mathbf{F}_e}{\partial t}\cdot\frac{\partial f_e}{\partial \mathbf{v}_e}\right.\\&\left.+\frac{\partial^2 f_e}{\partial \mathbf{v}_e\partial \mathbf{v}_e}:\mathbf{F}_e\mathbf{F}_e+\frac{\partial f_e}{\partial \mathbf{v}_e}\mathbf{F}_e:\frac{\partial}{\partial \mathbf{v}_e}\mathbf{F}_e\right]=-\frac{f_e-f_e^{(0)}}{\tau},\end{aligned} \tag{8.4.1}$$

where $\mathbf{F}_e=\mathbf{F}_e^{(1)}+\mathbf{F}_B$ is the Lorentz force which, in our case, includes the effect of the alternating electric field

$$\mathbf{F}_e^{(1)}=\frac{e\mathbf{E}^0}{m_e}e^{i\omega t},$$

directed along the x-axis, and of a static magnetic field, whose induction is along the z-axis. The equation of motion (4.1.32) reduces to the form

$$\begin{aligned}&\frac{\partial}{\partial t}\left\{\overline{\mathbf{v}}_e-\tau\left[\frac{\partial \overline{\mathbf{v}}_e}{\partial t}-\frac{e\mathbf{E}^0}{m_e}e^{i\omega t}-\frac{e}{m_e}\overline{\mathbf{v}}_e\times\mathbf{B}\right]\right\}-\frac{e\mathbf{E}^0}{m_e}e^{i\omega t}\\&-\frac{e}{m_e}\left\{\overline{\mathbf{v}}_e-\tau\left[\frac{\partial \overline{\mathbf{v}}_e}{\partial t}-\frac{e\mathbf{E}^0}{m_e}e^{i\omega t}-\frac{e}{m_e}\overline{\mathbf{v}}_e\times\mathbf{B}\right]\right\}\times\mathbf{B}=-\frac{\overline{\mathbf{v}}_e}{\tau}.\end{aligned} \tag{8.4.2}$$

The components of the drift velocity $\overline{\mathbf{v}}_e$ along the axes x and y are determined by the following set of equations ($\overline{v}_{ez}=0$):

$$\frac{\partial}{\partial t}\left\{\overline{v}_{ex}-\tau\left[\frac{\partial \overline{v}_{ex}}{\partial t}-\frac{eE^0}{m_e}e^{i\omega t}-\frac{e}{m_e}\overline{v}_{ey}B\right]\right\}-\frac{eE^0}{m_e}e^{i\omega t} \\ -\frac{e}{m_e}\left\{B\overline{v}_{ey}-\tau B\frac{\partial \overline{v}_{ey}}{\partial t}\right\}=-\frac{\overline{v}_{ex}}{\tau}-\frac{e^2\tau}{m_e^2}B^2\overline{v}_{ex}, \tag{8.4.3}$$

$$\frac{\partial}{\partial t}\left\{\overline{v}_{ey}-\tau\left[\frac{\partial \overline{v}_{ey}}{\partial t}+\frac{e}{m_e}\overline{v}_{ex}B\right]\right\}-\frac{e}{m_e}\left\{-\overline{v}_{ex}B+\tau B\frac{\partial \overline{v}_{ex}}{\partial t} \\ -\frac{eE^0}{m_e}e^{i\omega t}\tau B\right\}=-\frac{\overline{v}_{ey}}{\tau}-\frac{e^2\tau}{m_e^2}B^2\overline{v}_{ey}, \tag{8.4.4}$$

The solution to Eq. (8.4.2) is naturally sought in the form $\overline{v}=\overline{v}^0 e^{i\omega t}$, thus leading to the following system of algebraic equations:

$$\overline{v}^0_{ex}\left[i\omega+\tau\left(\omega^2+\omega_B^2\right)+\frac{1}{\tau}\right]=\frac{eE^0}{m_e}+\overline{v}^0_{ey}\omega_B[1-2i\omega\tau]-i\omega\tau\frac{eE^0}{m_e}, \\ \omega_B=eB/m_e, \tag{8.4.5}$$

$$\overline{v}^0_{ey}\left[i\omega+\tau\left(\omega^2+\omega_B^2\right)+\frac{1}{\tau}\right]=\overline{v}^0_{ex}\omega_B[2i\omega\tau-1]-\omega_B\tau\frac{eE^0}{m_e}, \tag{8.4.6}$$

where $\omega_B=eB/m_e$.

From these equations it is not difficult to find the components $\overline{v}^0_{ex}$ and $\overline{v}^0_{ey}$ of the drift velocity, and hence to determine the components of the electrical conductivity tensor. In our case, the complex conductivity σ_x assumes the form

$$\sigma_x=\sigma_0\frac{1+2\omega^2\tau^2+i\left(\omega_B^2\tau^2-\omega^2\tau^2\right)\omega\tau}{1+\omega^2\tau^2+\omega^4\tau^4+3\omega_B^2\tau^2+\omega_B^4\tau^4-2\omega^2\tau^2\omega_B^2\tau^2+2i\omega\tau(1+\omega^2\tau^2-\omega_B^2\tau^2)}, \tag{8.4.7}$$

where we have used the notation:

$$\sigma_x=\frac{en_e\overline{v}^0_{ex}}{E^0},\sigma_0=\frac{n_e e^2\tau}{m_e}.$$

Separating the real part of σ_x now yields

$$\mathrm{Re}\,\sigma_x=\sigma_0\frac{1+3\omega^2\tau^2+\omega^4\tau^4+\omega_B^2\tau^2\left[3+6\omega^2\tau^2+\omega_B^2\tau^2\right]}{\left[1+\omega^2\tau^2+\omega^4\tau^4+\omega_B^2\tau^2(3+\omega_B^2\tau^2-2\omega^2\tau^2)\right]^2+4\omega^2\tau^2[1+\omega^2\tau^2-\omega_B^2\tau^2]^2}. \tag{8.4.8}$$

The Boltzmann theory, as is known, leads to the following results

$$\mathrm{Re}\,\sigma_x=\sigma_0\frac{1+\omega^2\tau^2+\omega_B^2\tau^2}{1+2\omega^2\tau^2+\omega^4\tau^4+\omega_B^2\tau^2[2-2\omega^2\tau^2+\omega_B^2\tau^2]}, \tag{8.4.9}$$

$$\sigma_x=\sigma_0\frac{1+i\omega\tau}{1+(\omega_B^2-\omega^2)\tau^2+2i\omega\tau}. \tag{8.4.10}$$

For the traditionally considered limiting cases, Eqs. (8.4.7)–(8.4.10) give the following results, [182]:

(a) $\omega=0$, a constant electric field:

$$\mathrm{Re}\,\sigma_x^{GBE}=\sigma_0\frac{1}{1+3\omega_B^2\tau^2+\omega_B^4\tau^4}, \tag{8.4.11}$$

$$\mathrm{Re}\,\sigma_x^{BE}=\sigma_0\frac{1}{1+\omega_B^2\tau^2}, \tag{8.4.12}$$

(b) $\omega_B = 0$, no magnetic field:

$$\mathrm{Re}\,\sigma_x^{\mathrm{GBE}} = \sigma_0 \frac{1+3\omega^2\tau^2+\omega^4\tau^4}{[1+\omega^2\tau^2+\omega^4\tau^4]^2+4\omega^2\tau^2[1+\omega^2\tau^2]^2} = \frac{1}{1+3\omega^2\tau^2+\omega^4\tau^4}, \tag{8.4.13}$$

$$\mathrm{Re}\,\sigma_x^{\mathrm{BE}} = \sigma_0 \frac{1}{1+\omega^2\tau^2}, \tag{8.4.14}$$

(c) $\omega = \omega_B$,

$$\mathrm{Re}\,\sigma_x^{\mathrm{GBE}} = \sigma_0 \frac{1+6\omega^2\tau^2+8\omega^4\tau^4}{1+12\omega^2\tau^2+16\omega^4\tau^4}, \tag{8.4.15}$$

$$\mathrm{Re}\,\sigma_x^{\mathrm{BE}} = \sigma_0 \frac{1+2\omega^2\tau^2}{1+4\omega^2\tau^2}; \tag{8.4.16}$$

(d) $\omega = \omega_B$, $\omega\tau \gg 1$, the cyclotron resonance condition:

$$\mathrm{Re}\,\sigma_x^{\mathrm{GBE}} = \mathrm{Re}\,\sigma_x^{\mathrm{BE}} = \frac{1}{2}\sigma_0. \tag{8.4.17}$$

Finally, from the system of Eqs. (8.4.5) and (8.4.6) the drift velocity along the y-axis is found as

$$\bar{v}_{ey}^0 = \frac{eE^0}{m_e}\omega_B\tau^2\left[\left(\omega^4-\omega_B^4\right)\tau^4 - 3\left(\omega^2+\omega_B^2\right)\tau^2 - 2 + \mathrm{i}\omega\tau\left(3\omega^2\tau^2+\omega_B^2\tau^2\right)\right]D^{-1}, \tag{8.4.18}$$

where

$$D = 1+\left(\omega^6+\omega_B^6\right)\tau^6 + 4\omega_B^2\tau^2 + 4\omega^2\tau^2\omega_B^2\tau^2 - \omega^4\tau^4\omega_B^2\tau^2 + 4\omega_B^4\tau^4 - \omega^2\tau^2\omega_B^4\tau^4$$
$$+\mathrm{i}\omega\tau\left[3\omega^4\tau^4 - 5\omega^2\tau^2 + 3 + 3\omega_B^2\tau^2 - 2\omega^2\tau^2\omega_B^2\tau^2 - \omega_B^4\tau^4\right].$$

Notice that the BE implies that

$$\bar{v}_{ey}^0 = -\frac{eE^0}{m_e}\omega_B\tau^2 \frac{1}{1+(\omega_B^2-\omega^2)\tau^2+2\mathrm{i}\omega\tau}. \tag{8.4.19}$$

In particular, it follows that

(a) $\omega = 0$,

$$\bar{v}_{ey}^{0\mathrm{GBE}} = -\frac{eE^0}{m_e}\omega_B\tau^2 \frac{\omega_B^4\tau^4+3\omega_B^2\tau^2+2}{\omega_B^6\tau^6+4\omega_B^4\tau^4+4\omega_B^2\tau^2+1}, \tag{8.4.20}$$

$$\bar{v}_{ey}^{0\mathrm{BE}} = -\frac{eE^0}{m_e}\omega_B\tau^2 \frac{1}{\omega_B^2\tau^2+1}, \tag{8.4.21}$$

(b) $\omega = \omega_B$,

$$\mathrm{Re}\,\bar{v}_{ey}^{0\mathrm{GBE}} = -\frac{eE^0}{m_e}\omega\tau^2 \frac{2+14\omega^2\tau^2+28\omega^4\tau^4+56\omega^6\tau^6}{[1+4\omega^2\tau^2+8\omega^4\tau^4]^2+\omega^2\tau^2[3-2\omega^2\tau^2]^2}, \tag{8.4.22}$$

$$\mathrm{Re}\,\bar{v}_{ey}^{0\mathrm{BE}} = -\frac{eE^0}{m_e}\omega\tau^2 \frac{1}{1+4\omega^2\tau^2}. \tag{8.4.23}$$

The calculations show that while the BE and GBE results may happen to be identical, they may also be significantly different, both qualitatively and quantitatively. The question of exactly how significantly can only be answered through the solution of concrete problems. In particular, the GBE has been applied successfully to transport processes in a partially ionized gas of inelastic colliding particles [183].

Table 4.1 contains the calculation of ratio of conductivities Re σ_x^{GBE}/Re σ_x^{BE} for the case "a", ($\omega=0$, a constant electric field).

$$\frac{\mathrm{Re}\,\sigma_x^{\mathrm{GBE}}}{\mathrm{Re}\,\sigma_x^{\mathrm{BE}}}=\frac{1+\omega_{\mathrm{B}}^2\tau^2}{1+3\omega_{\mathrm{B}}^2\tau^2+\omega_{\mathrm{B}}^4\tau^4}. \tag{8.4.24}$$

From Eqs. (8.4.13) and (8.4.14) follow that for case "b" (no magnetic field), the ratio of conductivities can be written as

$$\frac{\mathrm{Re}\,\sigma_x^{\mathrm{GBE}}}{\mathrm{Re}\,\sigma_x^{\mathrm{BE}}}=\frac{1+\omega^2\tau^2}{1+3\omega^2\tau^2+\omega^4\tau^4}, \tag{8.4.25}$$

and dependence $\frac{\mathrm{Re}\,\sigma_x^{\mathrm{GBE}}}{\mathrm{Re}\,\sigma_x^{\mathrm{BE}}}$ as function of $\omega\tau$ has the same character reflected in Table 8.1.

Table 8.2 corresponds the case "b", when exists the coincidence of frequencies $\omega=\omega_{\mathrm{B}}$.

In Tables 8.3 and 8.4, the calculation of ratio of drift velocities Re $\bar{v}_{ey}^{0\mathrm{GBE}}$/Re $\bar{v}_{ey}^{0\mathrm{BE}}$ across magnetic field is presented Re $\bar{v}_{ey}^{0\mathrm{GBE}}$/Re $\bar{v}_{ey}^{0\mathrm{BE}}$ in constant electric field ($\omega=0$) and in electromagnetic field by the condition $\omega=\omega_{\mathrm{B}}$. In these cases correspondingly

$$\frac{\bar{v}_{ey}^{0\mathrm{GBE}}}{\bar{v}_{ey}^{0\mathrm{BE}}}=\frac{\omega_{\mathrm{B}}^6\tau^6+4\omega_{\mathrm{B}}^4\tau^4+5\omega_{\mathrm{B}}^2\tau^2+2}{\omega_{\mathrm{B}}^6\tau^6+4\omega_{\mathrm{B}}^4\tau^4+4\omega_{\mathrm{B}}^2\tau^2+1}, \tag{8.4.26}$$

TABLE 8.1 Calculation of Re σ_x^{GBE}/Re σ_x^{BE} for $\omega=0$

$\omega_{\mathrm{B}}\tau$	0	1	2	3	4	5	6	7	8	9	10
$\frac{\mathrm{Re}\,\sigma_x^{\mathrm{GBE}}}{\mathrm{Re}\,\sigma_x^{\mathrm{BE}}}$	1	0.4	0.172	0.0917	0.0557	0.0371	0.0263	0.0196	0.0152	0.0120	0.0098

TABLE 8.2 Calculation of Re σ_x^{GBE}/Re σ_x^{BE} for the case $\omega=\omega_{\mathrm{B}}$

$\omega\tau$	0	0.1	0.2	0.3	0.4	0.5	0.6	0.7	0.8	0.9
$\frac{\mathrm{Re}\,\sigma_x^{\mathrm{GBE}}}{\mathrm{Re}\,\sigma_x^{\mathrm{BE}}}$	1	0.964	0.937	0.837	0.807	0.8	0.805	0.817	0.832	0.847
$\omega\tau$	1	2	3	4	5	6	7	8	9	10
$\frac{\mathrm{Re}\,\sigma_x^{\mathrm{GBE}}}{\mathrm{Re}\,\sigma_x^{\mathrm{BE}}}$	0.862	0.948	0.974	0.985	0.990	0.993	0.995	0.996	0.997	0.9975
$\omega\tau$	11	12	13	14	15	16	17	18	19	20
$\frac{\mathrm{Re}\,\sigma_x^{\mathrm{GBE}}}{\mathrm{Re}\,\sigma_x^{\mathrm{BE}}}$	0.998	0.9983	0.9985	0.9987	0.9989	0.999	0.9991	0.9992	0.9993	0.9994
$\omega\tau$	30	40	50							
$\frac{\mathrm{Re}\,\sigma_x^{\mathrm{GBE}}}{\mathrm{Re}\,\sigma_x^{\mathrm{BE}}}$	0.9997	0.9998	0.9999							

TABLE 8.3 Calculation of $\bar{v}_{ey}^{0\mathrm{GBE}}/\bar{v}_{ey}^{0\mathrm{BE}}$ in constant electric field ($\omega=0$)

$\omega_{\mathrm{B}}\tau$	0.1	0.2	0.3	0.4	0.5	0.6	0.7	0.8	0.9	
$\frac{\bar{v}_{ey}^{0\mathrm{GBE}}}{\bar{v}_{ey}^{0\mathrm{BE}}}$	1.971	1.891	1.782	1.664	1.552	1.453	1.369	1.300	1.245	
$\omega_{\mathrm{B}}\tau$	1	2	3	4	5	6	7	8	9	10
$\frac{\bar{v}_{ey}^{0\mathrm{GBE}}}{\bar{v}_{ey}^{0\mathrm{BE}}}$	1.2	1.034	1.009	1.003	1.001	1.001	1.0004	1.0002	1.0001	1.0001

TABLE 8.4 Calculation of $\text{Re}\,\bar{v}_{ey}^{0GBE}/\text{Re}\,\bar{v}_{ey}^{0BE}$ for the case $\omega=\omega_B$

$\omega\tau$	0.1	0.2	0.3	0.4	0.5	0.6	0.7	0.8	0.9
$\frac{\text{Re}\,\bar{v}_{ey}^{0GBE}}{\text{Re}\,\bar{v}_{ey}^{0BE}}$	1.901	1.763	1.747	1.868	2	2.321	2.542	2.716	2.846
$\omega\tau$	1	2	3	4	5	6	7	8	9
$\frac{\text{Re}\,\bar{v}_{ey}^{0GBE}}{\text{Re}\,\bar{v}_{ey}^{0BE}}$	2.941	3.291	3.393	3.437	3.458	3.471	3.478	3.483	3.487
$\omega\tau$	10		20		30		40		50
$\frac{\text{Re}\,\bar{v}_{ey}^{0GBE}}{\text{Re}\,\bar{v}_{ey}^{0BE}}$	3.489		3.497		3.499		3.4993		3.4996

$$\frac{\text{Re}\,\bar{v}_{\text{ey}}^{0\text{GBE}}}{\text{Re}\,\bar{v}_{\text{ey}}^{0\text{BE}}}=\frac{(2+14\omega^2\tau^2+28\omega^4\tau^4+56\omega^6\tau^6)(1+4\omega^2\tau^2)}{(1+4\omega^2\tau^2+8\omega^4\tau^4)^2+\omega^2\tau^2(3-2\omega^2\tau^2)^2}. \tag{8.4.27}$$

The drift velocity $\bar{v}_{\text{ey}}^0$ defines diffusion of charged particles in the perpendicular direction to vectors of intensity of electric field and magnetic induction and then the Hall effect. Usually, experimental data in the theory of the Hall effect are presented with the help of Hall constant. Let us introduce Hall constant using Eq. (8.4.21). After multiplying both parts of Eq. (8.4.21) on $n_e e$ and using current density $J_{\text{ey}}=n_e e v_{\text{ey}}^0$, (where e is particle charge) we have

$$J_{\text{ey}}\frac{m_e}{e^2 n}\omega_B\left(1+\frac{1}{\omega_B^2\tau^2}\right)=E^0. \tag{8.4.28}$$

Taking into account that $\omega_B=eB/m_e$, from Eq. (8.4.28) can be found

$$E^0=-\frac{1}{ne}\left(1+\frac{1}{\omega_B^2\tau^2}\right)J_{\text{ey}}B. \tag{8.4.29}$$

But Hall constant is the coefficient of proportionality between potential difference and production of current density and magnetic induction, then

$$R^{\text{BE}}=\frac{1}{ne}\left(1+\frac{1}{\omega_B^2\tau^2}\right). \tag{8.4.30}$$

Sign of R defines by the sign of the charge e. Obviously in this case

$$R^{\text{GBE}}=\frac{1}{ne}\frac{\omega_B^6\tau^6+4\omega_B^6\tau^6+4\omega_B^2\tau^2+1}{\omega_B^2\tau^2(\omega_B^4\tau^4+3\omega_B^2\tau^2+2)}. \tag{8.4.31}$$

From relations (8.4.20), (8.4.21), (8.4.26), (8.4.30), and (8.4.31) follow:

$$\frac{\bar{v}_{\text{ey}}^{0\text{GBE}}}{\bar{v}_{\text{ey}}^{0\text{BE}}}=\frac{R^{\text{BE}}}{R^{\text{GBE}}}. \tag{8.4.32}$$

Therefore for the case "a" Table 8.3 defines also the ratio of the Hall coefficients R^{BE} and R^{GBE}, obtained in frames of BKT and GBKT.

By the formulated assumptions the following conclusions can be drawn from Tables 8.1–8.4:

1. In constant electric field the diminution of conductivity with increase of $\omega_B\tau$ is realizing in the frame of GBKT much faster than it follows from BKT (more significant effect of confining of ionized gas in magnetic field).
2. In absence of magnetic field the increase of $\omega\tau$ in alternating electric field in the frame of GBKT leads to more significant decrease of conductivity than it follows from BKT.
3. In the case of equality of frequencies $\omega=\omega_B$ the ratio of conductivities is varying by non-monotonic way with grows of $\omega\tau$.
4. Transverse drift velocities in electromagnetic fields are larger than it follows from BKT.

The calculations show that while the BE and GBE results may happen to be identical, they may also be significantly different, both qualitatively and quantitatively. The question of exactly how significantly can only be answered through the solution of concrete problems.

8.5 INVESTIGATION OF THE GBE FOR ELECTRON ENERGY DISTRIBUTION IN A CONSTANT ELECTRIC FIELD WITH DUE REGARD FOR INELASTIC COLLISIONS

We begin with numerical solutions of the GBE, which are carried out for electrons of a weakly ionized plasma in a constant electric field with due regard for inelastic collisions for the case of constant collision frequency between electrons and heavy particles ($\upsilon_{ea}=1/\tau=\text{const}$). For this purpose, a set of cross sections is used of collisions between electrons and heavy particles, applied in simulation of processes in hydrogen plasma. The results are compared with those obtained by means of numerical solution of the "traditional" BE.

The aim of this study [182, 183] (important in solving numerous problems of physics and plasma chemistry) is to construct an algorithm for numerical solution of GBE for plasma electrons in a constant electric field with due regard for inelastic collisions between electrons and heavy particles.

In view of this, a nontrivial problem is that on the set of cross sections used in solution of the equation. Strictly speaking, this should be a set corresponding to a certain atom (molecule) obtained by calculation or experimentally. In practice, such a rigorous approach cannot be realized because of the fact that far from all the cross sections are known even for well-studied particles such as inert gases and hydrogen. Therefore, sets of cross sections are used which were constructed from the known ones in such a way that the calculated drift velocity and the first Townsend coefficient should coincide with experimental quantities in a wide range of values of reduced electric field E/N. In so doing, the cross sections entering the set may be changed in magnitude and shifted along the axis of electron energy (as compared to initial values). Thus, it is only a full set of cross sections that has the physical meaning, because it yields the electron DF in energies, corresponding to the gas being described. No cross-section may be added to or excluded from this set. The rate coefficients of the processes involving electrons are calculated with the help of electron energy distribution and cross sections of these processes.

All known solutions of the BE are obtained using such sets of cross sections. In this study, we restricted ourselves to the construction of the algorithm of numerical solution of the GBE for the case of constant collision frequency between electrons and heavy particles ($\upsilon_{ea}=\text{const}$). As is known, this condition is valid for helium and hydrogen [184, 185]. In solving the GBE, we used the set of cross sections for molecular hydrogen [186, 187].

The second objective of this study was to compare the results obtained by numerical solution of the GBE and BE for one and the same set of cross sections. The results of such comparison are of extraordinary importance in defining the trend of further treatment of the GBE. Indeed, if both equations yield coinciding results, this may be indicative of the fact that, for the problems being treated, additional terms appearing in the GBE make no contribution to the results. This conclusion must simply be tested for various sets of cross sections.

If the results are different, the role of additional terms is considerable. However, the physical interpretation of the results is hardly possible at this stage, because, as follows from the foregoing, the used set of cross sections corresponds to the BE rather than to GBE. Therefore, the further stage should be construction of a set of cross sections for the GBE (in the way described above) and physical interpretation of the obtained solutions. This is an independent problem falling beyond the scope of this study.

Consider the case of spatially homogeneous weakly ionized gas, assuming that collisions between charged particles may be ignored. This means that

$$\upsilon_{ee} \ll \delta\upsilon_{ea}, \tag{8.5.1}$$

here, υ_{ee} is the frequency of collisions between electrons; $\upsilon_{ea}=\overline{N\sigma_{ea}(v)v}$ is the frequency of collisions between electrons and neutral particles; N is the concentration of heavy particles; σ_{ea} is the cross-section of electron collisions with atoms (molecules); and δ is the fraction of energy lost by an electron in a single collision against a neutral particle. It is also assumed that the external electric field is stationary, and there is no magnetic field. In this case, the GBE is written in the form (see also Eqs. 8.2.1 and 8.2.2)

$$\mathbf{F}_e\cdot\frac{\partial f_e}{\partial \mathbf{v}_e}-\tau\frac{\partial^2 f_e}{\partial \mathbf{v}_e\partial \mathbf{v}_e}:\mathbf{F}_e\mathbf{F}_e=J_{ea}, \tag{8.5.2}$$

where $\mathbf{F}_e=e\mathbf{E}/m_e$. Solution (8.5.2) for the energy distribution of electrons is found using the traditional perturbation method with the help of expansion,

$$f_e(v_e)=f_0(v_e)+\mathbf{F}_e\cdot\mathbf{v}_e f_1(v_e)+\cdots \tag{8.5.3}$$

In the case of weak anisotropy, we may restrict our consideration to the first two terms of energy expansion, and, with the elastic electron collisions alone taken into account, the set of equation with respect to $f_0(v)$ and $f_1(v)$ has the form (see also Eq. 8.2.6)

$$f_1+\frac{1}{3}v_e\frac{\partial f_1}{\partial v_e}-\frac{\tau}{3}\left\{\frac{2}{v_e}\frac{\partial f_0}{\partial v_e}+\frac{\partial^2 f_0}{\partial v_e^2}\right\}=\frac{1}{F_e^2}J_0, \tag{8.5.4}$$

$$\frac{\partial f_0}{\partial v_e}-\frac{3}{5}\tau F_e^2\left\{4\frac{\partial f_1}{\partial v_e}+v_e\frac{\partial^2 f_1}{\partial v_e^2}\right\}=-\frac{3}{2}\frac{1}{F_e}J_1, \tag{8.5.5}$$

In Eqs. (8.5.4) and (8.5.5), J_0, J_1 are the collision integrals, which should be presented in the form conventional for the BE

$$J_0=\frac{m_e\hat{T}_a}{m_a v_e^2}\frac{\partial}{\partial v_e}\left[v_e^3\upsilon_{ea}^{(el)}\left(\frac{f_0}{\hat{T}_a}+\frac{1}{m_e v_e}\frac{\partial f_0}{\partial v_e}\right)\right], \tag{8.5.6}$$

$$J_1=\frac{2}{3}F_e\frac{v_e^2}{\ell}f_1, \tag{8.5.7}$$

where $\upsilon_{ea}^{(el)}$ is the frequency of elastic collisions; ℓ is the mean free path of electrons; and $\hat{T}_a=k_B T$ is the gas temperature in energy units. Further transformation of Eqs. (8.5.4) and (8.5.5) makes it possible to derive the equation for the case of $\upsilon_{ea}=\text{const}$

$$\begin{gathered}\check{v}_e^2\left[1+\frac{3m_e}{\check{\varepsilon}\, m_a}\right]\check{f}_0'''+\check{v}_e\left[2+\frac{6m_e}{\check{\varepsilon}\, m_a}+\frac{3m_e\check{v}_e^2}{m_a}\right]\check{f}_0''\\-\left[2+\frac{6m_e}{\check{\varepsilon}\, m_a}+\check{v}_e^2\left(\frac{10}{3}+\frac{5m_e}{m_a}-12\frac{m_e}{m_a}\right)\right]\check{f}_0'-5\frac{m_e}{m_a}\check{v}_e^3\check{f}_0=0,\end{gathered} \tag{8.5.8}$$

where

$$\check{v}_e=\frac{v_e}{\tau F_e},\ \check{\varepsilon}=\frac{m_e\tau^2F_e^2}{\hat{T}_a},\ \check{f}_0=\frac{f_0}{f_0(v_e=0)},$$

The next stage of transformation of the GBE is the inclusion of real cross sections of elastic and inelastic collisions. The BE with due regard for inelastic collisions is written in the form

$$\mathbf{F}_e\cdot\frac{\partial f_e}{\partial \mathbf{v}_e}=J_{ea}^{(el)}+\sum_m J_{ea}^{(m)}+\sum_n J_i^{(n)}+J_{ea}^{(r)}. \tag{8.5.9}$$

Here, $J_{ea}^{(el)}, J_{ea}^{(m)}, J_i^{(n)}, J_{ea}^{(r)}$ are the terms of the collision integral, taking into account elastic collisions between electrons and neutral particles, excitation of the mth level, n-fold ionization, and recombination, respectively. We introduce the following notation for the inelastic part of the collision integral:

$$J_{ea}^{(inel)}=\sum_m J_{ea}^{(m)}+\sum_n J_i^{(n)}+J_{ea}^{(r)}. \tag{8.5.10}$$

Now, write the GBE with due regard for inelastic collisions

$$\mathbf{F}_e\cdot\frac{\partial f_e}{\partial \mathbf{v}_e}-\tau\frac{\partial^2 f_e}{\partial \mathbf{v}_e\partial \mathbf{v}_e}:\mathbf{F}_e\mathbf{F}_e=J_{ea}^{(el)}+J_{ea}^{(inel)}, \tag{8.5.11}$$

which is reduced, with the help of Eq. (8.5.3), to the following set of equations:

$$f_1+\frac{1}{3}v_e\frac{\partial f_1}{\partial v_e}-\frac{\tau}{3}\left\{\frac{2}{v_e}\frac{\partial f_0}{\partial v_e}+\frac{\partial^2 f_0}{\partial v_e^2}\right\}=\frac{1}{F_e^2}J_0^{(el)}+\frac{1}{F_e^2}J_{ea}^{(inel)}(f_0), \tag{8.5.12}$$

$$\frac{\partial f_0}{\partial v_e}-\frac{3}{5}\tau F_e^2\left\{4\frac{\partial f_1}{\partial v_e}+v_e\frac{\partial^2 f_1}{\partial v_e^2}\right\}=-\frac{3}{2}\frac{1}{F_e}\left[J_1^{(el)}+J_{ea}^{(inel)}(f_1)\right]. \tag{8.5.13}$$

where J_0 and J_1, are defined by Eqs. (8.5.6) and (8.5.7), respectively.

We dwell in more detail on the choice of the inelastic part of the collision integral. We will use the following approximation for the component parts of $J_{\mathrm{ea}}^{(\mathrm{inel})}$ (the first term on the right-hand side allows for electron energy losses in collisions against heavy particles in direct processes, and the second one for the energy return to electrons upon collisions of the second kind) [183, 184]:

$$
\begin{aligned}
J_{\mathrm{ea}}^{(m)} &= -\upsilon_{\mathrm{e}}^{(m)}(v_{\mathrm{e}})f(v_{\mathrm{e}}) + \upsilon_{\mathrm{e}}^{(m)}\left(v_{\mathrm{e}}^{(m)}\right)\frac{v_{\mathrm{e}}^{(m)}}{v_{\mathrm{e}}}f\left(v_{\mathrm{e}}^{(m)}\right) \\
J_{\mathrm{i}}^{(n)}(v_{\mathrm{e}}) &= -\upsilon_{\mathrm{i}}^{(n)}(v_{\mathrm{e}})f(v_{\mathrm{e}}) + \eta v_{\mathrm{i}}^{(n)}\left(v_{\mathrm{e}}^{(n)}\right)\frac{v_{\mathrm{e}}^{(n)}}{v_{\mathrm{e}}}f\left(v_{\mathrm{e}}^{(n)}\right), \\
J_{\mathrm{ea}}^{(r)} &= -\upsilon_{r}(v_{\mathrm{e}})f(v_{\mathrm{e}})
\end{aligned}
\tag{8.5.14}
$$

where

$$
\upsilon_{\mathrm{e}}^{(m)} = Nv_{\mathrm{e}}\sigma_{\mathrm{e}}^{(m)}(v_{\mathrm{e}}), \upsilon_{\mathrm{i}}^{(n)} = Nv_{\mathrm{e}}\sigma_{\mathrm{i}}^{(n)}(v_{\mathrm{e}});
$$

$v_{\mathrm{e}}^{(m)}, v_{\mathrm{i}}^{(n)}$ are the velocities of electrons defined by the equations

$$
\begin{aligned}
J_{\mathrm{ea}}^{(m)} &= -\upsilon_{\mathrm{e}}^{(m)}(v_{\mathrm{e}})f(v_{\mathrm{e}}) + \upsilon_{\mathrm{e}}^{(m)}\left(v_{\mathrm{e}}^{(m)}\right)\frac{v_{\mathrm{e}}^{(m)}}{v_{\mathrm{e}}}f\left(v_{\mathrm{e}}^{(m)}\right) \\
J_{\mathrm{i}}^{(n)}(v_{\mathrm{e}}) &= -\upsilon_{\mathrm{i}}^{(n)}(v_{\mathrm{e}})f(v_{\mathrm{e}}) + \eta v_{\mathrm{i}}^{(n)}\left(v_{\mathrm{e}}^{(n)}\right)\frac{v_{\mathrm{e}}^{(n)}}{v_{\mathrm{e}}}f\left(v_{\mathrm{e}}^{(n)}\right), \\
J_{\mathrm{ea}}^{(r)} &= -\upsilon_{r}(v_{\mathrm{e}})f(v_{\mathrm{e}})
\end{aligned}
\tag{8.5.15}
$$

Here, $\varepsilon_{\mathrm{e}}^{(m)}, \varepsilon_{\mathrm{i}}^{(n)}, \sigma_{\mathrm{e}}^{(m)}, \sigma_{\mathrm{i}}^{(n)}$ are the threshold energies and cross sections of the mth and nth processes, respectively, which are defined by the equilibrium between the primary and secondary electrons $\eta \approx 1$. For a weakly ionized plasma, the terms $J_{\mathrm{ea}}^{(n)}, J_{\mathrm{ea}}^{(r)}$ in the collision integral may be ignored [188–190].

We will now turn to the transformation of Eqs. (8.5.12) and (8.5.13) for the case of $\tau = \mathrm{const}$. It follows from Eqs. (8.5.12) and (8.5.13) that

$$
\begin{aligned}
F_{\mathrm{e}}^{2}\left\{v_{\mathrm{e}}^{3}f_{1} - \tau v_{\mathrm{e}}^{2}\frac{\partial f_{0}}{\partial v_{\mathrm{e}}}\right\}' &= 3\hat{T}_{\mathrm{a}}\left\{v_{e}^{3}\upsilon_{\mathrm{ea}}^{(\mathrm{el})}\left(\frac{f_{0}}{\hat{T}_{\mathrm{a}}} + \frac{1}{m_{\mathrm{e}}v_{\mathrm{e}}}\frac{\partial f_{0}}{\partial v_{\mathrm{e}}}\right)\right\}' \\
&+ 3v_{\mathrm{e}}^{3}J_{\mathrm{ea}}^{(\mathrm{inel})}(f_{0}),
\end{aligned}
\tag{8.5.16}
$$

$$
\frac{\partial f_{0}}{\partial v_{\mathrm{e}}} - \frac{3}{5}\tau F_{\mathrm{e}}^{2}\left\{4\frac{\partial f_{1}}{\partial v_{\mathrm{e}}} + v_{\mathrm{e}}\frac{\partial^{2} f_{1}}{\partial v_{\mathrm{e}}^{2}}\right\} = -\upsilon_{\mathrm{ea}}v_{\mathrm{e}}f_{1},
\tag{8.5.17}
$$

where

$$
\upsilon_{\mathrm{ea}} = \upsilon_{\mathrm{ea}}^{(\mathrm{el})} + \upsilon_{\mathrm{ea}}^{(\mathrm{inel})}.
$$

Relations (8.5.16) and (8.5.17) make it possible to exclude f_1, from the set of equations, because an explicit representation for f_1, may be obtained from Eq. (8.5.17).

It follows directly from Eq. (8.5.16) that

$$
\begin{aligned}
\{v_{\mathrm{e}}^{3}f_{1}\}' &= 3\hat{T}_{\mathrm{a}}\frac{m_{\mathrm{e}}}{F_{\mathrm{e}}^{2}m_{\mathrm{a}}}\left\{v_{\mathrm{e}}^{3}\upsilon_{\mathrm{ea}}^{(\mathrm{el})}\left(\frac{f_{0}}{\hat{T}_{\mathrm{a}}} + \frac{1}{m_{\mathrm{e}}v_{\mathrm{e}}}\frac{\partial f_{0}}{\partial v_{\mathrm{e}}}\right)\right\}' + \left(\tau v_{\mathrm{e}}^{2}\frac{\partial f_{0}}{\partial v_{\mathrm{e}}}\right)' \\
&+ 3\frac{v_{\mathrm{e}}^{2}}{F_{\mathrm{e}}^{2}}J_{\mathrm{ea}}^{(\mathrm{inel})}(f_{0}).
\end{aligned}
\tag{8.5.18}
$$

On performing the integration of Eq. (8.5.18) in a way similar to that used for the expression for the case of elastic collisions in [182], we derive

$$
\begin{aligned}
v_{\mathrm{e}}^{3}f_{1} &= 3\hat{T}_{\mathrm{a}}\frac{m_{\mathrm{e}}}{F_{\mathrm{e}}^{2}m_{\mathrm{a}}}v_{\mathrm{e}}^{3}\upsilon_{\mathrm{ea}}^{(\mathrm{el})}\left(\frac{f_{0}}{\hat{T}_{\mathrm{a}}} + \frac{1}{m_{\mathrm{e}}v_{\mathrm{e}}}\frac{\partial f_{0}}{\partial v_{\mathrm{e}}}\right) + \tau v_{\mathrm{e}}^{2}\frac{\partial f_{0}}{\partial v_{\mathrm{e}}} \\
&+ 3\frac{1}{F_{\mathrm{e}}^{2}}\int_{0}^{v_{\mathrm{e}}} v_{\mathrm{e}}^{2}J_{\mathrm{ea}}^{(\mathrm{inel})}(f_{0})\mathrm{d}v_{\mathrm{e}}.
\end{aligned}
\tag{8.5.19}
$$

It follows directly from Eq. (8.5.19) that

$$f_1 = 3\hat{T}_a \frac{m_e}{F_e^2 m_a} \upsilon_{ea}^{(el)} \left(\frac{f_0}{\hat{T}_a} + \frac{1}{m_e v_e} \frac{\partial f_0}{\partial v_e} \right) + \frac{\tau}{v_e} \frac{\partial f_0}{\partial v_e} + \frac{3}{F_e^2 v_e^3} \int_0^{v_e} v_e^2 J_{ea}^{(inel)}(f_0) dv_e. \tag{8.5.20}$$

We perform double differentiation of Eq. (8.5.20) to derive

$$\begin{aligned} \frac{\partial f_1}{\partial v_e} &= 3\hat{T}_a \frac{m_e}{F_e^2 m_a} \frac{\partial \upsilon_{ea}^{(el)}}{\partial v_e} \left(\frac{f_0}{\hat{T}_a} + \frac{1}{m_e v_e} \frac{\partial f_0}{\partial v_e} \right) - \frac{\tau}{v_e^2} \frac{\partial f_0}{\partial v_e} \\ &+ 3\hat{T}_a \frac{m_e}{F_e^2 m_a} \upsilon_{ea}^{(el)} \left(\frac{1}{\hat{T}_a} \frac{\partial f_0}{\partial v_e} + \frac{1}{m_e v_e} \frac{\partial^2 f_0}{\partial v_e^2} - \frac{1}{m_e v_e^2} \frac{\partial f_0}{\partial v_e} \right) \\ &+ \frac{\tau}{v_e} \frac{\partial^2 f_0}{\partial v_e^2} - \frac{9}{F_e^2 v_e^4} \int_0^{v_e} v_e^2 J_{ea}^{(inel)}(f_0) dv_e + \frac{3}{F_e^2 v_e} J_{ea}^{(inel)}(f_0), \end{aligned} \tag{8.5.21}$$

$$\begin{aligned} \frac{\partial^2 f_1}{\partial v_e^2} &= \frac{3 m_e}{F_e^2 m_a} \frac{\partial^2 \upsilon_{ea}^{(el)}}{\partial v_e^2} f_0 + \left(\frac{6 m_e}{F_e^2 m_a} \frac{\partial \upsilon_{ea}^{(el)}}{\partial v_e} - \frac{6\hat{T}_a}{F_e^2 m_a v_e^2} \frac{\partial \upsilon_{ea}^{(el)}}{\partial v_e} \right. \\ &\left. + \frac{2\tau}{v_e^3} + \frac{6\hat{T}_a}{F_e^2 m_a v_e^3} \upsilon_{ea}^{(el)} + \frac{3\hat{T}_a}{F_e^2 m_a v_e} \frac{\partial^2 \upsilon_{ea}^{(el)}}{\partial v_e^2} \right) \frac{\partial f_0}{\partial v_e} + \left(\frac{6\hat{T}_a}{F_e^2 m_a v_e} \frac{\partial \upsilon_{ea}^{(el)}}{\partial v_e} \right. \\ &\left. + \frac{3 m_e}{F_e^2 m_a} \upsilon_{ea}^{(el)} - \frac{6\hat{T}_a}{F_e^2 m_a v_e^2} \upsilon_{ea}^{(el)} - \frac{2\tau}{v_e^2} \right) \frac{\partial^2 f_0}{\partial v_e^2} \\ &+ \left(\frac{\tau}{v_e} + \frac{3\hat{T}_a}{F_e^2 m_a v_e} \upsilon_{ea}^{(el)} \right) \frac{\partial^3 f_0}{\partial v_e^3} + \frac{36}{F_e^2 v_e^5} \int_0^{v_e} v_e^2 J_{ea}^{(inel)}(f_0) dv_e \\ &- \frac{12}{F_e^2 v_e^2} J_{ea}^{(inel)}(f_0) + \frac{3}{F_e^2 v_e} \frac{\partial J_{ea}^{(inel)}(f_0)}{\partial v_e}, \end{aligned} \tag{8.5.22}$$

accordingly,

$$J_{ea}^{(inel)}(f_0) = \sum_\alpha \left[\upsilon_{ea}^{(inel)}(v_e + v_e^\alpha) \frac{v_e^{(m)}}{v_e} f_0(v_e + v_e^\alpha) - \upsilon_{ea}^{(inel)}(v_e) f_0(v_e) \right],$$

where $\upsilon_{ea}^{(el)}$ is the frequency of elastic collisions; $\upsilon_{ea}^{(inel)}$ is the frequency of inelastic collisions; v_e^α is the threshold of the process a; and $v_e^{(m)}$ is calculated by Eq. (8.5.15).

The resultant equation is considerably simplified if collisions of the second kind are ignored, and the integral of inelastic collisions is written in the form

$$J_{ea}^{(inel)}(f_e) = -\sum_\alpha \upsilon_{ea_\alpha}^{(inel)} f_e(v_e) \equiv -\upsilon_{ea}^{(inel)} f_e(v_e).$$

On substituting Eqs. (8.5.20)–(8.5.22) into Eq. (8.5.17), we derive

$$
\begin{aligned}
& f_0'''\left(\tau+\frac{3\hat{T}_{\mathrm{a}}\upsilon_{\mathrm{a}}^{(\mathrm{el})}}{F_{\mathrm{e}}^2 m_{\mathrm{a}}}\right)+f_0''\left(2\frac{\tau}{v_{\mathrm{e}}}+\frac{6\hat{T}_{\mathrm{a}}}{F_{\mathrm{e}}^2 m_{\mathrm{a}}}\frac{\partial \upsilon_{\mathrm{ea}}^{(\mathrm{el})}}{\partial v_{\mathrm{e}}}+\frac{3m_{\mathrm{e}}v_{\mathrm{e}}}{F_{\mathrm{e}}^2 m_{\mathrm{a}}}\upsilon_{\mathrm{ea}}^{(\mathrm{el})}\right.\\
& \left.+\frac{6\hat{T}_{\mathrm{a}}}{F_{\mathrm{e}}^2 m_{\mathrm{a}} v_{\mathrm{e}}}\upsilon_{\mathrm{ea}}^{(\mathrm{el})}\right)+f_0'\left(-\frac{2\tau}{v_{\mathrm{e}}^2}-\frac{6\hat{T}_{\mathrm{a}}}{F_{\mathrm{e}}^2 m_{\mathrm{a}} v_{\mathrm{e}}^2}\upsilon_{\mathrm{ea}}^{(\mathrm{el})}\right.\\
& +\frac{6m_{\mathrm{e}}}{F_{\mathrm{e}}^2 m_{\mathrm{a}}}\frac{\partial \upsilon_{\mathrm{ea}}^{(\mathrm{el})}}{\partial v_{\mathrm{e}}}\left(\frac{\hat{T}_{\mathrm{a}}}{m_{\mathrm{a}} v_{\mathrm{e}}}+v_{\mathrm{e}}\right)\\
& +12\frac{m_{\mathrm{e}}}{F_{\mathrm{e}}^2 m_{\mathrm{a}}}\upsilon_{\mathrm{ea}}^{(\mathrm{el})}-\frac{5}{3F_{\mathrm{e}}^2}\left(\upsilon_{\mathrm{ea}}^{(\mathrm{el})}+\upsilon_{\mathrm{ea}}^{(\mathrm{inel})}\right)+\frac{3\hat{T}_{\mathrm{a}}}{F_{\mathrm{e}}^2 m_{\mathrm{a}}}\frac{\partial^2 \upsilon_{\mathrm{ea}}^{(\mathrm{el})}}{\partial v_{\mathrm{e}}^2}\\
& \left.-\frac{3}{F_{\mathrm{e}}^2}\upsilon_{\mathrm{ea}}^{(\mathrm{inel})}-\frac{5}{3}\frac{1}{\tau F_{\mathrm{e}}^2}-5\left(\upsilon_{\mathrm{ea}}^{(\mathrm{el})}+\upsilon_{\mathrm{ea}}^{(\mathrm{inel})}\right)\frac{\hat{T}_{\mathrm{a}}\upsilon_{\mathrm{ea}}^{(\mathrm{el})}}{F_{\mathrm{e}}^4 \tau m_{\mathrm{a}}}\right)\\
& +f_0\left(12\frac{m_{\mathrm{e}}}{F_{\mathrm{e}}^2 m_{\mathrm{a}}}\frac{\partial \upsilon_{\mathrm{ea}}^{(\mathrm{el})}}{\partial v_{\mathrm{e}}}+3\frac{m_{\mathrm{e}}}{F_{\mathrm{e}}^2 m_{\mathrm{a}}}\frac{\partial^2 \upsilon_{\mathrm{ea}}^{(\mathrm{el})}}{\partial v_{\mathrm{e}}^2}v_{\mathrm{e}}-\frac{3}{F_{\mathrm{e}}^2}\frac{\partial \upsilon_{\mathrm{ea}}^{(\mathrm{inel})}}{\partial v_{\mathrm{e}}}\right.\\
& \left.-5\left(\upsilon_{\mathrm{ea}}^{(\mathrm{el})}+\upsilon_{\mathrm{ea}}^{(\mathrm{inel})}\right)v_{\mathrm{e}}\frac{m_{\mathrm{e}}\upsilon_{\mathrm{ea}}^{(\mathrm{el})}}{F_{\mathrm{e}}^2 \tau m_{\mathrm{a}}}\right)\\
& +5\frac{\upsilon_{\mathrm{ea}}^{(\mathrm{el})}+\upsilon_{\mathrm{ea}}^{(\mathrm{inel})}}{F_{\mathrm{e}}^4 \tau v_{\mathrm{e}}^2}\int_0^{v_{\mathrm{e}}} v_{\mathrm{e}}^2 \upsilon_{\mathrm{ea}}^{(\mathrm{inel})} f_0 \mathrm{d}v_{\mathrm{e}}=0.
\end{aligned}
\tag{8.5.23}
$$

In dimensionless form, the equation for the isotropic part of energy distribution of electrons may be presented in the form

$$
\begin{aligned}
& \breve{f}_0'''\left(1+3\frac{\check{m}}{\breve{\varepsilon}}\breve{\upsilon}_{\mathrm{ea}}^{(\mathrm{el})}\right)\breve{v}_{\mathrm{e}}^2+\breve{f}_0''\left(2+6\frac{\check{m}}{\breve{\varepsilon}}\frac{\partial \breve{\upsilon}_{\mathrm{ea}}^{(\mathrm{el})}}{\partial \breve{v}_{\mathrm{e}}}\breve{v}_{\mathrm{e}}+3\check{m}\,\breve{v}_{\mathrm{e}}^2\breve{\upsilon}_{\mathrm{ea}}^{(\mathrm{el})}\right.\\
& \left.+6\frac{\check{m}}{\breve{\varepsilon}}\breve{\upsilon}_{\mathrm{ea}}^{(\mathrm{el})}\right)\breve{v}_{\mathrm{e}}+\breve{f}_0'\left(-2-6\frac{\check{m}}{\breve{\varepsilon}}\breve{\upsilon}_{\mathrm{ea}}^{(\mathrm{el})}+6\check{m}\left(\frac{\breve{v}_{\mathrm{e}}}{\breve{\varepsilon}}+\breve{v}_{\mathrm{e}}^3\right)\frac{\partial \breve{\upsilon}_{\mathrm{ea}}^{(\mathrm{el})}}{\partial \breve{v}_{\mathrm{e}}}\right.\\
& -\breve{v}_{\mathrm{e}}^2\left(\frac{5}{3}+\frac{5}{3}\breve{\upsilon}_{\mathrm{ea}}^{(\mathrm{el})}+\frac{14}{3}\breve{\upsilon}_{\mathrm{ea}}^{(\mathrm{inel})}-12\check{m}\,\breve{\upsilon}_{\mathrm{ea}}^{(\mathrm{el})}-3\frac{\check{m}}{\breve{\varepsilon}}\frac{\partial^2 \breve{\upsilon}_{\mathrm{ea}}^{(\mathrm{el})}}{\partial \breve{v}_{\mathrm{e}}^2}\right.\\
& \left.\left.+5\left(\breve{\upsilon}_{\mathrm{ea}}^{(\mathrm{el})}+\breve{\upsilon}_{\mathrm{ea}}^{(\mathrm{inel})}\right)\frac{\check{m}}{\breve{\varepsilon}}\breve{\upsilon}_{\mathrm{ea}}^{(\mathrm{el})}\right)\right)+\breve{f}_0\left(12\check{m}\,\breve{v}_{\mathrm{e}}^2\frac{\partial \breve{\upsilon}_{\mathrm{ea}}^{(\mathrm{el})}}{\partial \breve{v}_{\mathrm{e}}}\right.\\
& \left.+3\check{m}\,\breve{v}_{\mathrm{e}}^3\frac{\partial^2 \breve{\upsilon}_{\mathrm{ea}}^{(\mathrm{el})}}{\partial \breve{v}_{\mathrm{e}}^2}-3\breve{v}_{\mathrm{e}}^2\frac{\partial \breve{\upsilon}_{\mathrm{ea}}^{(\mathrm{inel})}}{\partial \breve{v}_{\mathrm{e}}}-5\left(\breve{\upsilon}_{\mathrm{ea}}^{(\mathrm{el})}+\breve{\upsilon}_{\mathrm{ea}}^{(\mathrm{inel})}\right)\breve{\upsilon}_{\mathrm{ea}}^{(\mathrm{inel})}\breve{v}_{\mathrm{ea}}^3\check{m}\right)\\
& +5\left(\breve{\upsilon}_{\mathrm{ea}}^{(\mathrm{el})}+\breve{\upsilon}_{\mathrm{ea}}^{(\mathrm{inel})}\right)\int_0^{\breve{v}_{\mathrm{e}}}\breve{v}_{\mathrm{e}}^2\breve{\upsilon}_{\mathrm{ea}}^{(\mathrm{inel})} f_0 \mathrm{d}\breve{v}_{\mathrm{e}}=0,
\end{aligned}
\tag{8.5.24}
$$

where

$$
\begin{aligned}
& \check{m}=\frac{m_{\mathrm{e}}}{m_{\mathrm{a}}},\quad \breve{v}_{\mathrm{e}}=\frac{v_{\mathrm{e}}}{\tau F_{\mathrm{e}}},\quad \breve{\varepsilon}=\frac{m_{\mathrm{e}}F_{\mathrm{e}}^2\tau}{\hat{T}_{\mathrm{a}}},\\
& \breve{\upsilon}_{\mathrm{ea}}^{(\mathrm{el})}=\frac{\upsilon_{\mathrm{ea}}^{(\mathrm{el})}}{1/\tau},\quad \breve{\upsilon}_{\mathrm{ea}}^{(\mathrm{inel})}=\frac{\upsilon_{\mathrm{ea}}^{(\mathrm{inel})}}{1/\tau},\quad \breve{f}_0=\frac{f_0}{f_0(v_{\mathrm{e}}=0)}.
\end{aligned}
\tag{8.5.25}
$$

Note that Eq. (8.5.24) transforms to Eq. (8.5.8) if elastic collisions are ignored ($\breve{\nu}_{ea}^{(el)} \to 1$ and $\breve{\nu}_{ea}^{(inel)} \to 0$, respectively). The boundary conditions for the solution of Eq. (8.5.24) have the form

$$\breve{f}_0(\breve{v}_e = 0) = 1, \quad \left(\frac{\partial \breve{f}_0}{\partial \breve{v}_e}\right)_{\breve{v}_e=0} = 0, \quad \breve{f}_0(\breve{v}_e = \infty) = 0. \tag{8.5.26}$$

Equation (8.5.24) was solved by the iterative three-diagonal method of Gauss elimination technique for the differential second-order equation (see Appendix 5). Note once again that the distribution functions of electrons (DFE) were found for conditions under which the electron-electron collisions and collisions of the second kind may be ignored. Besides the DF, some moments of the energy distribution are considered below, such as the average electron energy ($\bar{\varepsilon}$).

$$\bar{\varepsilon} = \int_0^\infty \varepsilon^{3/2} f_0(\varepsilon) d\varepsilon, \tag{8.5.27}$$

electron diffusion coefficient (D)

$$D = \frac{4\pi}{3} \int_0^\infty \frac{v_e^4}{\sum_\alpha \upsilon_{ea}^\alpha} f_0(v) dv, \tag{8.5.28}$$

and the ratio of mobility of electrons (drift velocity) calculated by the GBE to the mobility calculated by the BE (μ_{GBE}/μ_{BE}). The electron mobility was defined with due regard for normalization of the DF (8.5.27) and expansion in spherical harmonics by the formula

$$\mu_{GBE} = \frac{4\pi}{3} \frac{e}{m_e} \int_0^\infty v_e^4 f_1(v_e) dv_e. \tag{8.5.29}$$

Results are given that were derived within the GBKT (GBE) and BKT (BE), respectively.

Figure 8.9 shows the characteristic form of the electron DFs in hydrogen plasma obtained by the BE (curve 1) and GBE (curve 2). The energy distributions differ considerably, this is indicative of the fact that the role of additional terms in the GBE is great. It is of interest to compare the shapes of energy distributions. Because the average energies of electrons calculated by the DFE's obtained using two different methods are different, it is reasonable to perform comparison with the help of the functions f_0/f_m (f_m is the MDF at the same average electron energy as that defined by f_0). The abscissa is the quantity $\varepsilon/\bar{\varepsilon}$ rather than the electron energy. Thus, upon comparison, it is possible to exclude from consideration the differences caused by different average electron energies. Such curves are shown in Fig. 8.10. Table 8.5, and Figs. 8.11 and 8.12 give the dependencies of $\bar{\varepsilon}$, D and (μ_{GBE}/μ_{BE}) on the parameter E/N.

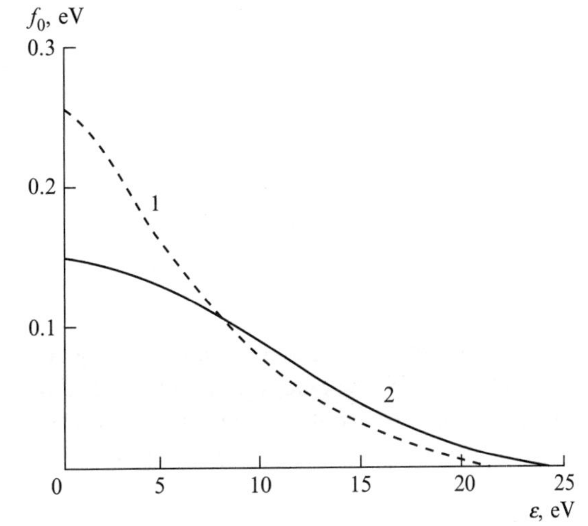

FIGURE 8.9 Electron DF in energies for $T = 1000°C$, $p = 1$ torr, $E = 500$ V/m. 1, The solution of the BE; 2, the solution of the GBE.

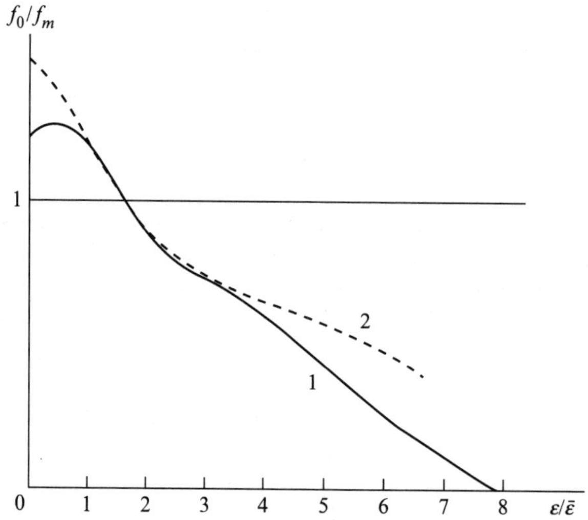

FIGURE 8.10 The ratio of the DFs to MDF. 1, The ratio of solution of the BE to MDF; 2, the ratio of solution of the GBE to MDF.

TABLE 8.5 The dependence of average energy, diffusion coefficient and mobility on the parameter E/N

$(E/N) \times 10^{16}$, (V cm^2)	Average Energy (eV)		Diffusion Coefficient $D \times 10^{-6}$ (cm^2/s)		Mobility $\mu \times 10^{-6}$ (cm^2/V s)	
	GBE	BE	GBE	BE	GBE	BE
2.07	1.85	1.17	1.67	1.24	1.36	1.46
3.1 1	2.88	1.80	2.29	1.59	1.23	1.25
4.14	3.91	2.59	2.97	2.07	1.14	1.15
5.18	4.81	3.27	3.59	2.50	1.13	1.12
6.21	5.56	3.78	4.11	2.84	1.12	1.10
7.25	6.19	4.21	4.56	3.11	1.11	1.09
8.02	6.59	4.48	4.84	3.35	1.10	1.08
9.06	7.05	4.78	5.18	3.55	1.10	1.07
10.10	7.45	5.06	5.46	3.73	1.09	1.07
11.10	7.79	5.31	5.69	3.90	1.09	1.06
12.20	8.09	5.54	5.93	4.06	1.09	1.05
13.20	8.35	5.75	6.12	4.20	1.08	1.05
14.20	8.58	5.94	6.29	4.33	1.09	1.04

The problems related to the used set of cross sections apart, and assuming this set to be model, we can draw the following conclusions. First, for one and the same value of E/N, the quantities $\bar{\varepsilon}$ and D calculated by the GBE are greater than those calculated using the BE. Second, although the ratio of f_0 to the Maxwellian DFE is similar (Fig. 8.10), the difference from the Maxwellian DFE is greater in the region of low energies and smaller in the region of high energies for the DFE calculated using the GBE rather than the BE. This means that the rate coefficients of high-threshold processes are higher in the first case than in the second one, and additional terms in the GBE result in an increase of the energy stored in electron gas per electron.

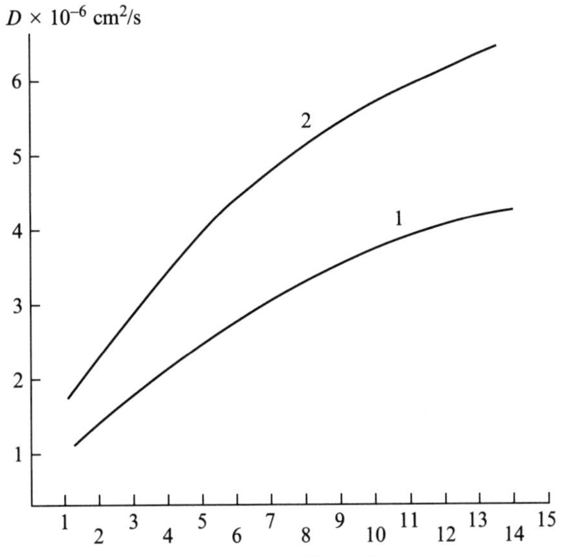

FIGURE 8.11 Dependence of the diffusion coefficient on the parameter E/N. 1, The BE; 2, The GBE.

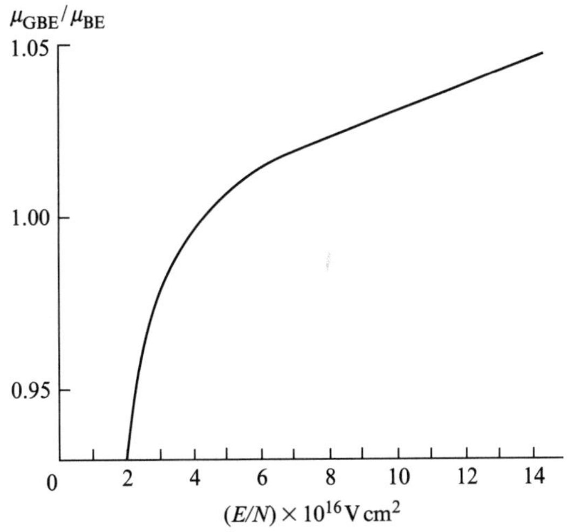

FIGURE 8.12 Dependence of the ratio of the electron mobilities on the parameter E/N.

These results are also indicative of the fact that the calculated values of the mobility and ionization coefficient are different from the measured ones, and the used set of cross sections should be changed as applied to the GBE. This should be the main trend of further studies into numerical solution of the GBE. Other trends include the construction of algorithms of solution of the GBE for the case of arbitrary dependence of the collision frequency ν_{ea} on the energy (velocity) of electrons, as well as treatment of the behavior of electron gas in variable electric fields. An analytical approach to solution of the latter problem is developed in [182].

Chapter 9

Generalized Boltzmann Equation in the Theory of the Rarefied Gases and Liquids

ABSTRACT

Discuss now the logic pattern of following investigations. The generalized Boltzmann equation (GBE) can be used for the investigation of transport processes in the hydrodynamic limit, when Knudsen numbers are small. The perturbation method of the GBE and BE (Boltzmann equation) solution is developed using expansion of distribution function f_α in a power series in Knudsen numbers. The important character features of theory should be underlined:

1. The usual perturbation theory applied to BE, leads in the minor approximations to Euler and Navier-Stokes hydrodynamic equations (HE). The qualitative and quantitative difference between these HE solutions so significant, then, so to speak, impossible to believe that these solutions are constructed in the frame of the same perturbation method.
2. What is the origin of this difference from the position of the non-local theory? Navier-Stokes equations contain the terms proportional to viscosity (to the mean time between collisions) and therefore to the non-local parameter τ. But these terms reflect only a part of all changes which should be introduced in the theory. In other words these "non-local" terms were extracted from the "local" theory and have no chance to improve radically the physical description.
3. Generalized non-local hydrodynamic equations written for the Maxwellian distribution function (Maxwellian DF), contain non-local parameter τ, which can be considered as a unified kinetic coefficient. Obviously the usual perturbation routine applied to GBE should lead to many additional complicated calculations. The question arises, is it reasonable to take into account as a rule the higher approximations? In other words, is the difference between generalized Euler and Navier-Stokes so significant, as between the classic Euler and Navier-Stokes solutions? Let us consider these problems.

Keywords: Kinetic theory of rarefied gas and liquids, Non-local hydrodynamics, Sound propagation

9.1 KINETIC COEFFICIENTS IN THE THEORY OF THE GENERALIZED KINETIC EQUATIONS. LINEARIZATION OF THE GENERALIZED BOLTZMANN EQUATION

Let us use the dimensionless form of the generalized Boltzmann equation (GBE) convenient for the investigation of transport processes in the hydrodynamic limit. We take into account only the pair collisions. Define for GBE

$$\frac{\mathrm{D}f_\alpha}{\mathrm{D}t}-\frac{\mathrm{D}}{\mathrm{D}t}\left(\tau_\alpha\frac{\mathrm{D}f_\alpha}{\mathrm{D}t}\right)=\sum_{j=1}^{\eta}\int\left[f'_\alpha f'_j-f_\alpha f_j\right]g_{\alpha j}b\mathrm{d}b\mathrm{d}\varphi\mathrm{d}\mathbf{v}_j, \tag{9.1.1}$$

where

$$\frac{\mathrm{D}}{\mathrm{D}t}=\frac{\partial}{\partial t}+\mathbf{v}_\alpha\cdot\frac{\partial}{\partial\mathbf{r}}+\mathbf{F}_\alpha\cdot\frac{\partial}{\partial\mathbf{v}_\alpha}, \tag{9.1.2}$$

are the corresponding scales. Namely, for radius-vector $\mathbf{r}$ the character hydrodynamic lengths L is used as scale, $(M_r=L)$; for velocity $\mathbf{v}_\alpha$ the scale is $M_{v_\alpha}=v_{0\lambda,\alpha}$, $\left(M_{v_\alpha}=\sqrt{\overline{v_\alpha^2}}\right)$ which is consistent with the root-mean-square velocity for particles belonging to species α ("α-particles"). In accordance with the definition

Unified non-local theory of transport processes. http://dx.doi.org/10.1016/B978-0-444-63478-8.00009-9

$$\overline{v_\alpha^2} = \frac{1}{n_\alpha}\int f_\alpha v_\alpha^2 \mathrm{d}\mathbf{v}_\alpha. \tag{9.1.3}$$

For distribution function f_α the scale is introduced compatible with its normalization

$$M_{f_\alpha} = n_{\alpha 0}/v_{0\lambda,\alpha}^3, \tag{9.1.4}$$

where $n_{\alpha 0}$ is character value of numerical concentration for α-particles. The scale $G_{\alpha j}$ for module of relative velocity $g_{\alpha j}$ defines as:

$$M_{g_{\alpha j}} = \frac{1}{n_\alpha n_j}\int |\mathbf{v}_\alpha - \mathbf{v}_j| f_\alpha f_j \mathrm{d}\mathbf{v}_\alpha \mathrm{d}\mathbf{v}_j = G_{\alpha j}, \tag{9.1.5}$$

and the scale M_b for impact parameter b of encountering particles α and j

$$M_b = \sigma_{\alpha j}\sqrt[4]{1+\frac{m_\alpha}{m_j}}, \quad \sigma_{\alpha j} = (\sigma_\alpha + \sigma_j)/2. \tag{9.1.6}$$

The rest scales: $M_{F_\alpha} = F_{0\alpha}$, $M_{\tau_\alpha} = l_\alpha / v_{0\lambda,\alpha}$, where l_α—mean free path for α-particles, $M_\varphi = \pi$. Rewrite now Eq. (9.1.1) in the dimensionless form, marking off the dimensionless values by sign^.

$$\begin{aligned}
&\frac{v_{0\lambda,\alpha}}{L}\frac{\partial \hat{f}_\alpha}{\partial \hat{t}} + \frac{v_{0\lambda,\alpha}}{L}\hat{\mathbf{v}}_\alpha\cdot\frac{\partial \hat{f}_\alpha}{\partial \hat{\mathbf{r}}} + \frac{F_{0\alpha}}{v_{0\lambda,\alpha}}\hat{\mathbf{F}}_\alpha\cdot\frac{\partial \hat{f}_\alpha}{\partial \hat{\mathbf{v}}_\alpha} \\
&-\left\{\left(\frac{v_{0\lambda,\alpha}}{L}\frac{\partial}{\partial \hat{t}} + \frac{v_{0\lambda,\alpha}}{L}\hat{\mathbf{v}}_\alpha\cdot\frac{\partial}{\partial \hat{\mathbf{r}}} + \frac{F_{0\alpha}}{v_{0\lambda,\alpha}}\hat{\mathbf{F}}_\alpha\cdot\frac{\partial}{\partial \hat{\mathbf{v}}_\alpha}\right)\left[\frac{\hat{\tau}_\alpha\lambda_\alpha}{v_{0\lambda,\alpha}}\left(\frac{v_{0\lambda,\alpha}}{L}\frac{\partial \hat{f}_\alpha}{\partial \hat{t}}\right.\right.\right. \\
&\left.\left.\left.+\frac{v_{0\lambda,\alpha}}{L}\hat{\mathbf{v}}_\alpha\cdot\frac{\partial \hat{f}_\alpha}{\partial \hat{\mathbf{r}}} + \frac{F_{0\alpha}}{v_{0\lambda,\alpha}}\hat{\mathbf{F}}_\alpha\cdot\frac{\partial \hat{f}_\alpha}{\partial \hat{\mathbf{v}}_\alpha}\right)\right]\right\} = \pi\sum_{j=1}^{\eta} n_{j0}\sigma_{\alpha j}^2 G_{\alpha j}\int\left(\hat{f}'_\alpha \hat{f}'_j\right. \\
&\left.-\hat{f}_\alpha \hat{f}_j\right)\hat{g}_{\alpha j}\hat{b}\,\mathrm{d}\hat{b}\,\mathrm{d}\hat{\varphi}\,\mathrm{d}\hat{\mathbf{v}}_j\sqrt{1+\frac{m_\alpha}{m_j}}.
\end{aligned} \tag{9.1.7}$$

After dividing Eq. (9.1.7) term by term into $v_{0\lambda,\alpha}/L$, we find

$$\begin{aligned}
&\frac{\partial \hat{f}_\alpha}{\partial \hat{t}} + \hat{\mathbf{v}}_\alpha\cdot\frac{\partial \hat{f}_\alpha}{\partial \hat{\mathbf{r}}} + \frac{F_{0\alpha}L}{v_{0\lambda,\alpha}^2}\hat{\mathbf{F}}_\alpha\cdot\frac{\partial \hat{f}_\alpha}{\partial \hat{\mathbf{v}}_\alpha} - \left\{\left(\frac{\partial}{\partial \hat{t}} + \mathbf{v}_\alpha\cdot\frac{\partial}{\partial \hat{\mathbf{r}}}\right.\right. \\
&\left.\left.+\frac{F_{0\alpha}L}{v_{0\lambda,\alpha}^2}\hat{\mathbf{F}}_\alpha\cdot\frac{\partial}{\partial \hat{\mathbf{v}}_\alpha}\right)\left[\hat{\tau}_\alpha\frac{l_\alpha}{L}\left(\frac{\partial \hat{f}_\alpha}{\partial \hat{t}} + \hat{\mathbf{v}}_\alpha\cdot\frac{\partial \hat{f}_\alpha}{\partial \hat{\mathbf{r}}} + \frac{F_{0\alpha}L}{v_{0\lambda,\alpha}^2}\hat{\mathbf{F}}_\alpha\cdot\frac{\partial \hat{f}_\alpha}{\partial \hat{\mathbf{v}}_\alpha}\right)\right]\right\} \\
&= \frac{\pi L}{v_{0\lambda,\alpha}}\sum_{j=1}^{\eta} n_{j0}\sqrt{1+\frac{m_\alpha}{m_j}}\sigma_{\alpha j}^2 G_{\alpha j}\int\left(\hat{f}'_\alpha \hat{f}'_j -\right. \\
&\left.-\hat{f}_\alpha \hat{f}_j\right)\hat{g}_{\alpha j}\hat{b}\,\mathrm{d}\hat{b}\,\mathrm{d}\hat{\varphi}\,\mathrm{d}\hat{\mathbf{v}}_j.
\end{aligned} \tag{9.1.8}$$

For the hard spheres model in multi-component mixture of gases the relation is valid:

$$l_\alpha^{-1} = \pi\sum_{j=1}^{\eta} n_j\sigma_{\alpha j}^2\sqrt{1+\frac{m_\alpha}{m_j}}. \tag{9.1.9}$$

Let us introduce Knudsen numbers

$$\mathrm{Kn}_{\alpha j}^{-1} = \frac{L}{l_{\alpha j}} = \pi L n_j \sigma_{\alpha j}^2\sqrt{1+\frac{m_\alpha}{m_j}}, \tag{9.1.10}$$

$$\mathrm{Kn}_\alpha^{-1} = \frac{L}{l_\alpha} = \pi L\sum_{j=1}^{\eta} n_j\sigma_{\alpha j}^2\sqrt{1+\frac{m_\alpha}{m_j}} = \sum_{j=1}^{\eta}\mathrm{Kn}_{\alpha j}^{-1}, \tag{9.1.11}$$

and rewrite Eq. (9.1.8) in the form

$$\frac{\mathrm{D}\hat{f}_\alpha}{\mathrm{D}\hat{t}} - \frac{\mathrm{D}}{\mathrm{D}\hat{t}}\left[\hat{\tau}_\alpha \mathrm{Kn}_\alpha\frac{\mathrm{D}\hat{f}_\alpha}{\mathrm{D}\hat{t}}\right] = \sum_j \mathrm{Kn}_{\alpha j}^{-1}\frac{G_{\alpha j}}{v_{0\lambda,\alpha}}\int\left[\hat{f}'_\alpha \hat{f}'_j - \hat{f}_\alpha \hat{f}_j\right]\hat{g}_{\alpha j}\hat{b}\,\mathrm{d}\hat{b}\,\mathrm{d}\hat{\varphi}\,\mathrm{d}\hat{\mathbf{v}}_j, \tag{9.1.12}$$

where

$$\frac{\mathrm{D}}{\mathrm{D}\hat{t}}=\frac{\partial}{\partial \hat{t}}+\hat{\mathbf{v}}_\alpha\cdot\frac{\partial}{\partial \hat{\mathbf{r}}}+\gamma\hat{\mathbf{F}}_\alpha\cdot\frac{\partial}{\partial \hat{\mathbf{v}}_\alpha},\quad \gamma=\frac{F_{0\alpha}}{\left(v_{0\lambda,\alpha}^2/L\right)}. \tag{9.1.13}$$

For mixture of neutral gases, which molecules are not strongly different in masses, the scale of relative velocity $G_{\alpha j}$ can be adopted as $v_{0\lambda,\alpha}$. In the hydrodynamic limit, when Knudsen numbers $\mathrm{Kn}_{\alpha j}$ $(\alpha,j=1,\ldots,\eta)$ are small (or, that is the same, the mean free path between collisions much less than character hydrodynamic length L), the perturbation method of the GBE solution can be developed using expansion of distribution function f_α in a power series in Knudsen numbers. Namely, introduce now the small parameter

$$\varepsilon=\max\left\{\mathrm{Kn}_{\alpha j}\right\}\ll 1,\ (\alpha,j=1,\ldots,\eta), \tag{9.1.14}$$

in this case $\mathrm{Kn}_\alpha^{-1}\geq\mu\varepsilon^{-1}$, then

$$\mathrm{Kn}_\alpha\leq\frac{\varepsilon}{\eta}, \tag{9.1.15}$$

where η is number of species in the gas mixture. The mentioned expansion for f_α has the form

$$f_\alpha=\sum_k f_\alpha^{(k)}\varepsilon^k. \tag{9.1.16}$$

Now is obvious that for obtaining of the successive approach, the model equation with the large parameter ε^{-1} in the right integral part of equation can be used

$$\frac{\mathrm{D}f_\alpha}{\mathrm{D}t}-\frac{\mathrm{D}}{\mathrm{D}t}\left(\tau_\alpha\frac{\mathrm{D}f_\alpha}{\mathrm{D}t}\right)=\frac{1}{\varepsilon}\sum_{j=1}^{\eta}\int\left(f'_\alpha f'_j-f_\alpha f_j\right)g_{\alpha j}b\,\mathrm{d}b\,\mathrm{d}\varphi\,\mathrm{d}\mathbf{v}_j. \tag{9.1.17}$$

If the inequality Eq. (9.1.14) is not valid for all components of mixture further modification of the Chapman-Enskog method for solution of Eq. (9.1.17) needs a correction which can lead to multi-velocities hydrodynamics.

Equating the coefficients of ε^{-1} on the both sides of Eq. (9.1.17) leads to the integral relation:

$$\sum_{j=1}^{\eta}\int\left(f'^{(0)}_\alpha f'^{(0)}_j-f^{(0)}_\alpha f^{(0)}_j\right)g_{\alpha j}b\mathrm{d}b\mathrm{d}\varphi\mathrm{d}\mathbf{v}_j=0, \tag{9.1.18}$$

where $f_\alpha^{(0)}$ is distribution function (DF) corresponding to the state of the local equilibrium, then to the Maxwellian DF

$$f_\alpha^{(0)}=n_\alpha\left(\frac{m_\alpha}{2\pi k_\mathrm{B}T}\right)^{3/2}\mathrm{e}^{-\left(m_\alpha V_\alpha^2/2k_\mathrm{B}T\right)}, \tag{9.1.19}$$

in Eq. (9.1.19) $\mathbf{V}_\alpha$ is thermal velocity of particles of α-component, T—temperature of gas mixture. The following approach corresponds to equating of coefficients in ε^0 of Eq. (9.1.17). In this case two alternative methods exist:

1. The second term in the left-hand-side of Eq. (9.1.12), containing the second substantive derivation, is proportional to Knudsen number (then to ε) and formally speaking this term is out of the Navier-Stokes approximation. In this case the linear equation which forms the basis for obtaining of the kinetic coefficients becomes the standard form

$$\frac{\mathrm{D}f_\alpha^{(0)}}{\mathrm{D}t}=\sum_j\int\left[f_\alpha^{(0)\prime}f_j^{(1)\prime}+f_j^{(0)\prime}f_\alpha^{(1)\prime}-f_\alpha^{(0)}f_j^{(1)}-f_\alpha^{(1)}f_j^{(0)}\right]\mathrm{d}\omega, \tag{9.1.20}$$

where

$$\mathrm{d}\omega\equiv g_{\alpha j}b\,\mathrm{d}b\mathrm{d}\varphi\mathrm{d}\mathbf{v}_j.$$

Application to Eq. (9.1.20) of the Chapman-Enskog procedure allows obtaining the classical expressions for kinetic coefficients [35, 36]. Differences in results from known results of the Chapman-Enskog method make their appearance on the level of $f_\alpha^{(2)}$ obtaining, i.e., in the generalized Barnett approach.

2. In the formulated first method in essence the turbulent fluctuations of kinetic coefficients on the Kolmogorov micro-scale (or statistical fluctuations of transport coefficients) are ignored. These remarks lead to equation:

$$\begin{aligned}&-\frac{\mathrm{D}}{\mathrm{D}t}\left(\tau_\alpha^{(0)}\left(\frac{\mathrm{D}f_\alpha^{(0)}}{\mathrm{D}t}\right)_{\tau=0}\right)+\left(\frac{\mathrm{D}f_\alpha^{(0)}}{\mathrm{D}t}\right)\\&=\sum_j\int\left[f_\alpha^{(0)\prime}f_j^{(1)\prime}+f_j^{(0)\prime}f_\alpha^{(1)\prime}-f_\alpha^{(0)}f_j^{(1)}-f_j^{(0)}f_\alpha^{(1)}\right]\mathrm{d}\omega.\end{aligned} \tag{9.1.21}$$

Important to notice, that additional term in the left-hand-side of the linearized equation does not go out from the limit of approximation and can be calculated using only Maxwellian DF. In particular, $\tau_\alpha^{(0)}$ is non-local parameter proportional to the mean time between collisions of α-particles, calculated for the state of local thermodynamic equilibrium. Moreover, introduction of this term leaves the description on the level of the linear Fredholm equations.

But for proof of this affirmation we need to show that the arising additional integral equation satisfies the corresponding solubility conditions. With this aim we consider for simplicity one-component gas and obtain the explicit form of $\mathrm{D}f_\alpha^{(0)}/\mathrm{D}t$ with taking into account τ-terms.

The next relation is valid in independent variables $\mathbf{r}, \mathbf{V}, t$ [35]:

$$\frac{\mathrm{D}f^{(0)}}{\mathrm{D}t} \equiv \frac{\breve{\mathrm{D}}f^{(0)}}{\mathrm{D}t} + \mathbf{V}\cdot\frac{\partial f^{(0)}}{\partial \mathbf{r}} + \left(\mathbf{F} - \frac{\mathrm{D}\mathbf{v}_0}{\mathrm{D}t}\right)\cdot\frac{\partial f^{(0)}}{\partial \mathbf{V}} - \frac{\partial f^{(0)}}{\partial \mathbf{V}}\mathbf{V} : \frac{\partial}{\partial \mathbf{r}}\mathbf{v}_0, \tag{9.1.22}$$

where operator is introduced

$$\frac{\breve{\mathrm{D}}}{\mathrm{D}t} \equiv \frac{\partial}{\partial t} + \mathbf{v}_0 \cdot \frac{\partial}{\partial \mathbf{r}}. \tag{9.1.23}$$

Generalized hydrodynamic Euler equations are written as:

$$\frac{\partial n}{\partial t} = -\frac{\partial}{\partial \mathbf{r}}\cdot(n\mathbf{v}_0) + \int \frac{\mathrm{D}}{\mathrm{D}t}\left(\tau^{(0)}\frac{\mathrm{D}f^{(0)}}{\mathrm{D}t}\right)\mathrm{d}\mathbf{v}, \tag{9.1.24}$$

$$\frac{\partial \mathbf{v}_0}{\partial t} = -\left(\mathbf{v}_0\cdot\frac{\partial}{\partial \mathbf{r}}\right)\mathbf{v}_0 + \mathbf{F} - \frac{1}{\rho}\frac{\partial p}{\partial \mathbf{r}} + \frac{1}{n}\int \mathbf{V}\frac{\mathrm{D}}{\mathrm{D}t}\left(\tau^{(0)}\frac{\mathrm{D}f^{(0)}}{\mathrm{D}t}\right)\mathrm{d}\mathbf{v}, \tag{9.1.25}$$

$$\begin{aligned}\frac{\partial}{\partial t}\left(\frac{3}{2}nk_\mathrm{B}T\right) &= -\mathbf{v}_0\cdot\frac{\partial}{\partial \mathbf{r}}\left(\frac{3}{2}nk_\mathrm{B}T\right) - \frac{3}{2}nk_\mathrm{B}T\frac{\partial}{\partial \mathbf{r}}\cdot\mathbf{v}_0 - p\frac{\partial}{\partial \mathbf{r}}\cdot\mathbf{v}_0 \\ &+ \int \frac{mV^2}{2}\frac{\mathrm{D}}{\mathrm{D}t}\left(\tau^{(0)}\frac{\mathrm{D}f^{(0)}}{\mathrm{D}t}\right)\mathrm{d}\mathbf{v}.\end{aligned} \tag{9.1.26}$$

Substantial derivative $\mathrm{D}f^{(0)}/\mathrm{D}t$ in integral terms (9.1.24)–(9.1.26) should be calculated in the Euler approach, because terms proportional to $\tau^{(0)2}$ are not taking into account. In Eq. (9.1.21) this fact is reflected by introduction of the symbolic condition $\tau=0$ for $[\mathrm{D}f^{(0)}/\mathrm{D}t]_{\tau=0}$. Rewrite hydrodynamic equations using operator $\breve{\mathrm{D}}/\mathrm{D}t$ from Eq. (9.1.23):

$$\frac{\breve{\mathrm{D}}n}{\mathrm{D}t} = -n\frac{\partial}{\partial \mathbf{r}}\cdot\mathbf{v}_0 + \int \frac{\mathrm{D}}{\mathrm{D}t}\left(\tau^{(0)}\frac{\mathrm{D}f^{(0)}}{\mathrm{D}t}\right)\mathrm{d}\mathbf{v}, \tag{9.1.27}$$

$$\frac{\breve{D}\mathbf{v}_0}{Dt} = \mathbf{F} - \frac{1}{\rho}\frac{\partial p}{\partial \mathbf{r}} + \frac{1}{n}\int \mathbf{V}\frac{D}{Dt}\left(\tau^{(0)}\frac{Df^{(0)}}{Dt}\right)\mathrm{d}\mathbf{v}, \tag{9.1.28}$$

$$\begin{aligned}\frac{\breve{\mathrm{D}}T}{\mathrm{D}t} &= -\frac{2}{3}T\frac{\partial}{\partial \mathbf{r}}\cdot\mathbf{v}_0 - \frac{T}{n}\int \frac{\mathrm{D}}{\mathrm{D}t}\left(\tau^{(0)}\frac{\mathrm{D}f^{(0)}}{\mathrm{D}t}\right)\mathrm{d}\mathbf{v} \\ &+ \frac{2}{3k_\mathrm{B}n}\int \frac{mV^2}{2}\frac{\mathrm{D}}{\mathrm{D}t}\left(\tau^{(0)}\frac{\mathrm{D}f^{(0)}}{\mathrm{D}t}\right)\mathrm{d}\mathbf{v}.\end{aligned} \tag{9.1.29}$$

Transform relation (9.1.22)

$$\begin{aligned}\frac{\mathrm{D}f^{(0)}}{\mathrm{D}t} = f^{(0)}\Bigg\{&\frac{\breve{\mathrm{D}}\ln f^{(0)}}{\mathrm{D}t} + \mathbf{V}\cdot\frac{\partial \ln f^{(0)}}{\partial \mathbf{r}} + \left(\mathbf{F} - \frac{\breve{\mathrm{D}}\mathbf{v}_0}{\mathrm{D}t}\right)\cdot\frac{\partial \ln f^{(0)}}{\partial \mathbf{V}} \\ &- \frac{\partial \ln f^{(0)}}{\partial \mathbf{V}}\mathbf{V} : \frac{\partial}{\partial \mathbf{r}}\mathbf{v}_0\Bigg\},\end{aligned} \tag{9.1.30}$$

and use set of Eqs. (9.1.27)–(9.1.29), the result is the relation

$$\frac{\breve{\mathrm{D}}\ln f^{(0)}}{\mathrm{D}t} = -\frac{mV^2}{3k_{\mathrm{B}}T}\frac{\partial}{\partial \mathbf{r}}\cdot\mathbf{v}_0 + \left(\frac{5}{2n}-\frac{mV^2}{2p}\right)\int\frac{\mathrm{D}}{\mathrm{D}t}\left(\tau^{(0)}\frac{\mathrm{D}f^{(0)}}{\mathrm{D}t}\right)\mathrm{d}\mathbf{v}$$
$$+\left(\frac{mV^2}{3nk_{\mathrm{B}}{}^2T^2}-\frac{1}{p}\right)\int\frac{mV^2}{2}\frac{\mathrm{D}}{\mathrm{D}t}\left(\tau^{(0)}\frac{\mathrm{D}f^{(0)}}{\mathrm{D}t}\right)\mathrm{d}\mathbf{v}. \tag{9.1.31}$$

We use Eq. (9.1.31) for further transformation of Eq. (9.1.30):

$$\frac{\mathrm{D}f^{(0)}}{\mathrm{D}t} = f^{(0)}\left\{\left(\frac{mV^2}{2k_{\mathrm{B}}T}-\frac{5}{2}\right)\mathbf{V}\cdot\frac{\partial\ln T}{\partial\mathbf{r}} + \frac{m}{k_{\mathrm{B}}T}\mathbf{V}^0\mathbf{V}:\frac{\partial}{\partial\mathbf{r}}\mathbf{v}_0\right.$$
$$+\left(\frac{5}{2n}-\frac{mV^2}{2p}\right)\int\frac{\mathrm{D}}{\mathrm{D}t}\left(\tau^{(0)}\frac{\mathrm{D}f^{(0)}}{\mathrm{D}t}\right)\mathrm{d}\mathbf{v} + \frac{m\mathbf{V}}{p}\cdot\int\mathbf{V}\frac{\mathrm{D}}{\mathrm{D}t}\left(\tau^{(0)}\frac{\mathrm{D}f^{(0)}}{\mathrm{D}t}\right)\mathrm{d}\mathbf{v} \tag{9.1.32}$$
$$\left.+\frac{1}{p}\left(\frac{mV^2}{3k_{\mathrm{B}}T}-1\right)\int\frac{mV^2}{2}\frac{\mathrm{D}}{\mathrm{D}t}\left(\tau^{(0)}\frac{\mathrm{D}f^{(0)}}{\mathrm{D}t}\right)\mathrm{d}\mathbf{v}\right\}.$$

Thence, in view of Eq. (9.1.32) the integral Eq. (9.1.21) is written as:

$$f^{(0)}\left\{\left(\frac{mV^2}{2k_{\mathrm{B}}T}-\frac{5}{2}\right)\mathbf{V}\cdot\frac{\partial\ln T}{\partial\mathbf{r}} + \frac{m}{k_{\mathrm{B}}T}\mathbf{V}^0\mathbf{V}:\frac{\partial}{\partial\mathbf{r}}\mathbf{v}_0\right.$$
$$+\left(\frac{5}{2n}-\frac{mV^2}{2p}\right)\int\frac{\mathrm{D}}{\mathrm{D}t}\left(\tau^{(0)}\left[\frac{\mathrm{D}f^{(0)}}{\mathrm{D}t}\right]_{\tau=0}\right)\mathrm{d}\mathbf{v}$$
$$+\frac{m\mathbf{V}}{p}\cdot\int\mathbf{V}\frac{\mathrm{D}}{\mathrm{D}t}\left(\tau^{(0)}\left[\frac{\mathrm{D}f^{(0)}}{\mathrm{D}t}\right]_{\tau=0}\right)\mathrm{d}\mathbf{v}$$
$$\left.+\frac{1}{p}\left(\frac{mV^2}{3k_{\mathrm{B}}T}-1\right)\int\frac{mV^2}{2}\frac{\mathrm{D}}{\mathrm{D}t}\left(\tau^{(0)}\left[\frac{\mathrm{D}f^{(0)}}{\mathrm{D}t}\right]_{\tau=0}\right)\mathrm{d}\mathbf{v}\right\} \tag{9.1.33}$$
$$-\frac{\mathrm{D}}{\mathrm{D}t}\left(\tau^{(0)}\left[\frac{\mathrm{D}f^{(0)}}{\mathrm{D}t}\right]_{\tau=0}\right)$$
$$=\int f^{(0)}f_1^{(0)}\left(\psi_1'+\psi'-\psi-\psi_1\right)gb\mathrm{d}b\mathrm{d}\varphi\mathrm{d}\mathbf{v}_1,$$

where

$$f=f^{(0)}(1+\psi),\ f^{(1)}=\psi f^{(0)}. \tag{9.1.34}$$

The solution ψ of Eq. (9.1.33) is the sum of two functions

$$\psi=\psi^E+\psi^\tau, \tag{9.1.35}$$

corresponding to equations

$$f^{(0)}\left\{\left(\frac{mV^2}{2k_{\mathrm{B}}T}-\frac{5}{2}\right)\mathbf{V}\cdot\frac{\partial\ln T}{\partial\mathbf{r}} + \frac{m}{k_{\mathrm{B}}T}\mathbf{V}^0\mathbf{V}:\frac{\partial}{\partial\mathbf{r}}\mathbf{v}_0\right\}$$
$$=\int f^{(0)}f_1^{(0)}\left(\psi_1'^E+\psi'^E-\psi_1^E-\psi^E\right)\mathrm{d}\omega, \tag{9.1.36}$$

$$f^{(0)}\left\{\left(\frac{5}{2n}-\frac{mV^2}{2p}\right)\int\frac{\mathrm{D}}{\mathrm{D}t}\left(\tau^{(0)}\left[\frac{\mathrm{D}f^{(0)}}{\mathrm{D}t}\right]_{\tau=0}\right)\mathrm{d}\mathbf{v}\right.$$
$$+\frac{m\mathbf{V}}{p}\cdot\int\mathbf{V}\frac{\mathrm{D}}{\mathrm{D}t}\left(\tau^{(0)}\left[\frac{\mathrm{D}f^{(0)}}{\mathrm{D}t}\right]_{\tau=0}\right)\mathrm{d}\mathbf{v}$$
$$\left.+\frac{1}{p}\left(\frac{mV^2}{3k_{\mathrm{B}}T}-1\right)\int\frac{mV^2}{2}\frac{\mathrm{D}}{\mathrm{D}t}\left(\tau^{(0)}\left[\frac{\mathrm{D}f^{(0)}}{\mathrm{D}t}\right]_{\tau=0}\right)\mathrm{d}\mathbf{v}\right\} \tag{9.1.37}$$
$$-\frac{\mathrm{D}}{\mathrm{D}t}\left(\tau^{(0)}\left[\frac{\mathrm{D}f^{(0)}}{\mathrm{D}t}\right]_{\tau=0}\right)$$
$$=\int f^{(0)}f_1^{(0)}\left(\psi_1'^\tau+\psi'^\tau-\psi^\tau-\psi_1^\tau\right)\mathrm{d}\omega.$$

Equation (9.1.36) is the classical Enskog integral equation, the solubility conditions for this equation are satisfied, and its solution is known, [35].

Let us investigate the solubility of integral Eq. (9.1.37). With this aim we multiply both (integral and differential) parts of Eq. (9.1.37) by additive invariants $\psi^i(i=1,\mathbf{V},(V^2/2))$ and integer with respect to **v**. Right-hand-sides of these relations are equal to zero because of conservation laws of mass, momentum, and energy for encountering particles. Now we show that integrated left-hand-sides of these relations also are equal to zero, therefore the solubility conditions for this Fredholm equation are fulfilled.

1. $\psi^{(1)}=1$.
For this invariant we have

$$\begin{aligned}&\int f^{(0)}\left(\frac{5}{2n}-\frac{mV^2}{2p}\right)\mathrm{d}\mathbf{v}\int\frac{\mathrm{D}}{\mathrm{D}t}\left(\tau^{(0)}\left[\frac{\mathrm{D}f^{(0)}}{\mathrm{D}t}\right]_{\tau=0}\right)\mathrm{d}\mathbf{v}\\&=\int\frac{\mathrm{D}}{\mathrm{D}t}\left(\tau^{(0)}\left[\frac{\mathrm{D}f^{(0)}}{\mathrm{D}t}\right]_{\tau=0}\right)\mathrm{d}\mathbf{v}.\end{aligned}\tag{9.1.38}$$

The second integrated term in Eq. (9.1.37) is equal to zero because the integrand is odd function in the thermal velocity, the third term is also equal to zero because of equality

$$\int\left(\frac{mV^2}{3k_BT}-1\right)f^{(0)}\mathrm{d}\mathbf{v}=0.\tag{9.1.39}$$

Notice that expression (9.1.38) and the rest term of Eq. (9.1.37) (integrated in **v**) of the left-hand-side cancel each other.

2. $\boldsymbol{\psi}^{(2)}=\mathbf{V}$.
Define R_i as

$$R_i=\int V_i\frac{\mathrm{D}}{\mathrm{D}t}\left(\tau^{(0)}\left[\frac{\mathrm{D}f^{(0)}}{\mathrm{D}t}\right]_{\tau=0}\right)\mathrm{d}\mathbf{v},\quad(i=1,2,3)$$

then

$$\frac{m}{p}\int f^{(0)}V_i(\mathbf{V}\cdot\mathbf{R})\mathrm{d}\mathbf{v}-R_i=0.\tag{9.1.40}$$

Relation Eq. (9.1.40) leads to accomplishing of the solubility condition for invariant $\boldsymbol{\psi}^{(2)}$.

3. $\psi^{(3)}=mV^2/2$.

$$\begin{aligned}&\int\frac{mV^2}{2}\left(\frac{5}{2n}-\frac{mV^2}{2p}\right)f^{(0)}\mathrm{d}\mathbf{v}\int\frac{\mathrm{D}}{\mathrm{D}t}\left(\tau^{(0)}\left[\frac{\mathrm{D}f^{(0)}}{\mathrm{D}t}\right]_{\tau=0}\right)\mathrm{d}\mathbf{v}\\&+\frac{1}{p}\int\left(\frac{mV^2}{3k_\mathrm{B}T}-1\right)\frac{mV^2}{2}f^{(0)}\mathrm{d}\mathbf{v}\int\frac{mV^2}{2}\frac{\mathrm{D}}{\mathrm{D}t}\left(\tau^{(0)}\left[\frac{\mathrm{D}f^{(0)}}{\mathrm{D}t}\right]_{\tau=0}\right)\mathrm{d}\mathbf{v}\\&-\int\frac{mV^2}{2}\frac{\mathrm{D}}{\mathrm{D}t}\left(\tau^{(0)}\left[\frac{\mathrm{D}f^{(0)}}{\mathrm{D}t}\right]_{\tau=0}\right)\mathrm{d}\mathbf{v}=0,\end{aligned}\tag{9.1.41}$$

because

$$\begin{aligned}&\int\frac{mV^2}{2}\left(\frac{5}{2n}-\frac{mV^2}{2p}\right)f^{(0)}\mathrm{d}\mathbf{v}=0\\&\frac{1}{p}\int\left(\frac{mV^2}{3k_\mathrm{B}T}-1\right)\frac{mV^2}{2}f^{(0)}\mathrm{d}\mathbf{v}=1.\end{aligned}\tag{9.1.42}$$

Then solubility conditions are fulfilled for Eq. (9.1.37). For inscription of the part ψ^τ of solution we need the explicit form of operator

$$\begin{aligned}\frac{\mathrm{D}}{\mathrm{D}t}\left(\tau^{(0)}\left[\frac{\mathrm{D}f^{(0)}}{\mathrm{D}t}\right]_{\tau=0}\right)&=\left\{\frac{\breve{\mathrm{D}}}{\mathrm{D}t}+\mathbf{V}\cdot\frac{\partial}{\partial\mathbf{r}}+\left(\mathbf{F}-\frac{\breve{\mathrm{D}}\mathbf{v}_0}{\mathrm{D}t}\right)\cdot\frac{\partial}{\partial\mathbf{V}}\right.\\&\left.-\frac{\partial}{\partial\mathbf{V}}\mathbf{V}:\frac{\partial}{\partial\mathbf{r}}\mathbf{v}_0\right\}\left\{\tau^{(0)}\left[\left(W^2-\frac{5}{2}\right)\mathbf{V}\cdot\frac{\partial\ln T}{\partial\mathbf{r}}+2\mathbf{W}^0\mathbf{W}:\frac{\partial}{\partial\mathbf{r}}\mathbf{v}_0\right]\right\}.\end{aligned}\tag{9.1.43}$$

Transformation of Eq. (9.1.43) leads to the result

$$\begin{aligned}
\frac{\mathrm{D}}{\mathrm{D}t}\left(\tau\left[\frac{\mathrm{D}f^{(0)}}{\mathrm{D}t}\right]_{\tau=0}\right) &= \left(W^2-\frac{5}{2}\right)\frac{\breve{\mathrm{D}}}{\mathrm{D}t}\left[\tau^{(0)}\frac{\partial \ln T}{\partial \mathbf{r}}\right] \\
&+2\mathbf{W}^0\mathbf{W}:\frac{\breve{\mathrm{D}}}{\mathrm{D}t}\left[\tau^{(0)}\frac{\partial}{\partial \mathbf{r}}\mathbf{v}_0\right]-\tau^{(0)}W^2\left(\frac{\partial \ln T}{\partial \mathbf{r}}\cdot\mathbf{V}\right)\frac{\partial \ln T}{\partial t} \\
&-\tau^{(0)}W^2\left[\mathbf{v}_0\cdot\frac{\partial \ln T}{\partial \mathbf{r}}\right]\left[\mathbf{V}\cdot\frac{\partial \ln T}{\partial \mathbf{r}}\right]-2\frac{\breve{\mathrm{D}}\ln T}{\mathrm{D}t}\tau^{(0)}\mathbf{W}^0\mathbf{W}:\frac{\partial}{\partial \mathbf{r}}\mathbf{v}_0 \\
&+\tau^{(0)}\left(\mathbf{F}-\frac{\breve{\mathrm{D}}\mathbf{v}_0}{\mathrm{D}t}\right)\cdot\frac{\partial \ln T}{\partial \mathbf{r}}\left(W^2-\frac{5}{2}\right)+2\tau^{(0)}\left(\mathbf{F}-\frac{\breve{\mathrm{D}}\mathbf{v}_0}{\mathrm{D}t}\right)\frac{\partial \ln T}{\partial \mathbf{r}}:\mathbf{W}\mathbf{W} \\
&+2\sqrt{\frac{m}{2k_\mathrm{B}T}}\left(\mathbf{F}-\frac{\breve{\mathrm{D}}\mathbf{v}_0}{\mathrm{D}t}\right)\mathbf{W}:\frac{\partial}{\partial \mathbf{r}}\mathbf{v}_0+2\sqrt{\frac{m}{2kT}}\mathbf{W}\left(\mathbf{F}-\frac{\breve{\mathrm{D}}\mathbf{v}_0}{\mathrm{D}t}\right):\frac{\partial}{\partial \mathbf{r}}\mathbf{v}_0 \\
&-\tau^{(0)}\left(W^2-\frac{5}{2}\right)\frac{\partial \ln T}{\partial \mathbf{r}}\mathbf{V}:\frac{\partial}{\partial \mathbf{r}}\mathbf{v}_0-2\tau^{(0)}\left(\frac{\partial \ln T}{\partial \mathbf{r}}\cdot\mathbf{W}\right)\mathbf{W}\mathbf{V}:\frac{\partial}{\partial \mathbf{r}}\mathbf{v}_0 \\
&-2\tau^{(0)}\left(\mathbf{W}\cdot\frac{\partial}{\partial \mathbf{r}}\right)\mathbf{v}_0\cdot\left(\mathbf{W}\cdot\frac{\partial}{\partial \mathbf{r}}\right)\mathbf{v}_0-2\tau^{(0)}\left[\left(\mathbf{W}\cdot\frac{\partial}{\partial \mathbf{r}}\right)\mathbf{v}_0\cdot\frac{\partial}{\partial \mathbf{r}}\right](\mathbf{v}_0\cdot\mathbf{W}).
\end{aligned} \tag{9.1.44}$$

Taking into account the explicit form of $\mathrm{D}f^{(0)}/\mathrm{D}t$ with the terms proportional to $\tau^{(0)}$ (see also Appendices 6 and 7, and corresponding generalized hydrodynamic equations), we state, that the structure of the left-hand-side of Eq. (9.1.33) is so complicated that in many cases more convenient to use self-consistent numerical schemes of solution of GHE and Eq. (9.1.33) (defining kinetic coefficients) or to use variants of the generalized Grad's method. Nevertheless we formulate approximate method of the Eq. (9.1.33) solution making possible to find explicit forms of the fluctuation terms for kinetic coefficients.

9.2 APPROXIMATE MODIFIED CHAPMAN-ENSKOG METHOD

We intend to develop in the theory of GBE the approximate modified Chapman-Enskog method. With this aim let us rewrite Eq. (9.1.33) in the form

$$\begin{aligned}
& f^{(0)}\left\{\left(\frac{mV^2}{2k_\mathrm{B}T}-\frac{5}{2}\right)\left[\mathbf{V}\cdot\frac{\partial \ln T}{\partial \mathbf{r}}-\frac{1}{n}\int\frac{\mathrm{D}}{\mathrm{D}t}\left(\tau^{(0)}\left(\frac{\mathrm{D}f^{(0)}}{\mathrm{D}t}\right)_{\tau=0}\right)\mathrm{d}\mathbf{v}\right]\right. \\
& +\frac{m}{k_\mathrm{B}T}\mathbf{V}\mathbf{V}:\frac{\partial}{\partial \mathbf{r}}\mathbf{v}_0-\frac{mV^2}{3k_\mathrm{B}T}\left[\frac{\partial}{\partial \mathbf{r}}\cdot\mathbf{v}_0-\frac{1}{p}\int\frac{mV^2}{2}\frac{\mathrm{D}}{\mathrm{D}t}\left(\tau^{(0)}\left(\frac{\mathrm{D}f^{(0)}}{\mathrm{D}t}\right)_{\tau=0}\right)\mathrm{d}\mathbf{v}\right] \\
& +\frac{m}{p}\left[\mathbf{V}\cdot\int\mathbf{V}\frac{\mathrm{D}}{\mathrm{D}t}\left(\tau^{(0)}\left(\frac{\mathrm{D}f^{(0)}}{\mathrm{D}t}\right)_{\tau=0}\right)\mathrm{d}\mathbf{v}-\int\frac{V^2}{2}\frac{\mathrm{D}}{\mathrm{D}t}\left(\tau^{(0)}\left(\frac{\mathrm{D}f^{(0)}}{\mathrm{D}t}\right)_{\tau=0}\right)\mathrm{d}\mathbf{v}\right] \\
& \left.-\frac{\mathrm{D}}{\mathrm{D}t}\left(\tau^{(0)}\left(\frac{\mathrm{D}f^{(0)}}{\mathrm{D}t}\right)_{\tau=0}\right)\right\} \\
& =\int f^{(0)}f_1^{(0)}(\psi_1'+\psi'-\psi-\psi_1)\mathrm{d}\omega.
\end{aligned} \tag{9.2.1}$$

In the following calculations of the kinetic coefficient's perturbations we neglect the small influence of τ-terms in square brackets of Eq. (9.2.1). It means that we intend to obtain solution of the shortened approximate integral equation

$$\left(\frac{\mathrm{D}f^{(0)}}{\mathrm{D}t}\right)_{\tau=0}-\frac{\mathrm{D}}{\mathrm{D}t}\left(\tau^{(0)}\left(\frac{\mathrm{D}f^{(0)}}{\mathrm{D}t}\right)_{\tau=0}\right)=\int f^{(0)}f_1^{(0)}(\psi_1'+\psi'-\psi-\psi_1)\mathrm{d}\omega, \tag{9.2.2}$$

where the substantial derivative $(\mathrm{D}f^{(0)}/\mathrm{D}t)_{\tau=0}$ exists. Let us consider more general case of the multi-component mixture of gases but neglecting the influence of the external forces $\mathbf{F}_\alpha$. We use also approximation $\tau_\alpha=\tau$. It leads to some simplification [30, 69, 191] the following calculations, which nevertheless remain complicated. Then,

$$\left(\frac{\mathrm{D}f_\alpha^{(0)}}{\mathrm{D}t}\right)_{\tau=0}-\frac{\mathrm{D}}{\mathrm{D}t}\left(\tau^{(0)}\left(\frac{\mathrm{D}f_\alpha^{(0)}}{\mathrm{D}t}\right)_{\tau=0}\right)$$
$$=\sum_{j=1}^{\eta}\int\left[f_\alpha^{(0)\prime}f_j^{(1)\prime}+f_\alpha^{(1)\prime}f_j^{(0)\prime}-f_\alpha^{(0)}f_j^{(1)}-f_\alpha^{(1)}f_j^{(0)}\right]\mathrm{d}\omega,\quad \mathrm{d}\omega=g_{\alpha j}b\mathrm{d}b\mathrm{d}\varphi\mathrm{d}\mathbf{v}_j. \tag{9.2.3}$$

Suppose ψ_α be the solution of equation

$$\left[\frac{\mathrm{D}f_\alpha^{(0)}}{\mathrm{D}t}\right]_{\tau=0}=\sum_j\int f_\alpha^{(0)}f_j^{(0)}\left(\psi_\alpha'+\psi_j'-\psi_\alpha-\psi_j\right)\mathrm{d}\omega, \tag{9.2.4}$$

$$\frac{\mathrm{D}}{\mathrm{D}t}=\frac{\partial}{\partial t}+\mathbf{v}_\alpha\cdot\frac{\partial}{\partial \mathbf{r}}. \tag{9.2.5}$$

We intend to find the solution of the linearized GBE Eq. (9.2.3) in the form

$$f_\alpha^{(1)}=f_\alpha^{(0)}\psi_\alpha-\frac{\mathrm{D}}{\mathrm{D}t}\left[\tau^{(0)}f_\alpha^{(0)}\psi_\alpha\right]. \tag{9.2.6}$$

After substituting of Eq. (9.2.6) in Eq. (9.2.3):

$$\left[\frac{\mathrm{D}f_\alpha^{(0)}}{\mathrm{D}t}\right]_{\tau=0}-\frac{\mathrm{D}}{\mathrm{D}t}\left[\tau^{(0)}\left(\frac{\mathrm{D}f_\alpha^{(0)}}{\mathrm{D}t}\right)_{\tau=0}\right]$$
$$=\sum_{j=1}^{\eta}\int f_\alpha^{(0)}f_j^{(0)}\left(\psi_\alpha'+\psi_j'-\psi_\alpha-\psi_j\right)\mathrm{d}\omega$$
$$-\sum_{j=1}^{\eta}\int\left\{f_\alpha^{(0)\prime}\frac{\mathrm{D}'}{\mathrm{D}t}\left(\tau^{(0)}f_j^{(0)\prime}\psi_j'\right)+f_j^{(0)\prime}\frac{\mathrm{D}'}{\mathrm{D}t}\left(\tau^{(0)}f_\alpha^{(0)\prime}\psi_\alpha'\right)\right.$$
$$\left.-f_\alpha^{(0)}\frac{\mathrm{D}}{\mathrm{D}t}\left(\tau^{(0)}f_j^{(0)}\psi_j\right)-f_j^{(0)}\frac{\mathrm{D}}{\mathrm{D}t}\left(\tau^{(0)}f_\alpha^{(0)}\psi_\alpha\right)\right\}\mathrm{d}\omega, \tag{9.2.7}$$

where

$$\frac{\mathrm{D}'}{\mathrm{D}t}=\frac{\partial}{\partial t}+\mathbf{V}_\alpha'\cdot\frac{\partial}{\partial \mathbf{r}}. \tag{9.2.8}$$

Equation (9.2.7) is simplified with the help of (9.2.4):

$$\frac{\mathrm{D}}{\mathrm{D}t}\left[\tau^{(0)}\left(\frac{\mathrm{D}f_\alpha^{(0)}}{\mathrm{D}t}\right)_{\tau=0}\right]=\sum_j\int\left\{f_\alpha^{(0)\prime}\frac{\mathrm{D}'}{\mathrm{D}t}\left(\tau^{(0)}f_j^{(0)\prime}\psi_j'\right)\right.$$
$$+f_j^{(0)\prime}\frac{\mathrm{D}'}{\mathrm{D}t}\left(\tau^{(0)}f_\alpha^{(0)\prime}\psi_\alpha'\right)-f_\alpha^{(0)}\frac{\mathrm{D}}{\mathrm{D}t}\left(\tau^{(0)}f_j^{(0)}\psi_j\right)$$
$$\left.-f_j^{(0)}\frac{\mathrm{D}}{\mathrm{D}t}\left(\tau^{(0)}f_\alpha^{(0)}\psi_\alpha\right)\right\}\mathrm{d}\omega. \tag{9.2.9}$$

Maxwellian functions in front of substantial derivatives in right-hand-sides of Eq. (9.2.9) can be introduced in the round squares after sign of derivatives with the accuracy $O\left[\psi^2\right]$. Let us consider for example the term

$$f_\alpha^{(0)}\frac{\mathrm{D}}{\mathrm{D}t}\left(\tau^{(0)}f_j^{(0)}\psi_j\right)=\frac{\mathrm{D}}{\mathrm{D}t}\left(\tau^{(0)}f_\alpha^{(0)}f_j^{(0)}\psi_j\right)-\tau^{(0)}f_j^{(0)}\psi_j\frac{\mathrm{D}f_\alpha^{(0)}}{\mathrm{D}t}. \tag{9.2.10}$$

The form (9.2.10) is obtained with the help of relation

$$\frac{\mathrm{D}}{\mathrm{D}t}\equiv\frac{\partial}{\partial t}+\mathbf{v}_\alpha\cdot\frac{\partial}{\partial \mathbf{r}}=\frac{\partial}{\partial t}+\frac{\partial}{\partial \mathbf{r}}\cdot\mathbf{v}_\alpha, \tag{9.2.11}$$

which is valid, because $\mathbf{r}, \mathbf{v}_\alpha, t$ are independent variables. But from Eq. (9.2.4) follows that $\mathrm{D}f_\alpha^{(0)}/\mathrm{D}t \sim \psi$, and therefore the formulated affirmation. Then

$$\frac{\mathrm{D}}{\mathrm{D}t}\left(\tau^{(0)}\left[\frac{\mathrm{D}f_\alpha^{(0)}}{\mathrm{D}t}\right]_{\tau=0}\right) = \sum_{j=1}^{\eta}\int\left\{\frac{\mathrm{D}'}{\mathrm{D}t}\left(f_\alpha^{(0)\prime}f_j^{(0)\prime}\tau^{(0)}\psi_j'\right)\right.$$
$$+\frac{\mathrm{D}'}{\mathrm{D}t}\left(f_\alpha^{(0)\prime}f_j^{(0)\prime}\tau^{(0)}\psi_\alpha'\right) - \frac{\mathrm{D}}{\mathrm{D}t}\left(f_\alpha^{(0)}f_j^{(0)}\tau^{(0)}\psi_j\right) \tag{9.2.12}$$
$$\left.-\frac{\mathrm{D}}{\mathrm{D}t}\left(f_\alpha^{(0)}f_j^{(0)}\tau^{(0)}\psi_\alpha\right)\right\}\mathrm{d}\omega.$$

This linear equation can be split in the following integral equations.

$$\frac{\partial}{\partial t}\left(\tau^{(0)}\left[\frac{\mathrm{D}f_\alpha^{(0)}}{\mathrm{D}t}\right]_{\tau=0}\right) = \frac{\partial}{\partial t}\left[\tau^{(0)}\sum_{j=1}^{\eta}\int f_\alpha^{(0)}f_j^{(0)}\left(\psi_\alpha'+\psi_j'-\psi_\alpha-\psi_j\right)\mathrm{d}\omega\right], \tag{9.2.13}$$

$$\frac{\partial}{\partial \mathbf{r}}\cdot\left(\mathbf{v}_0\tau^{(0)}\left[\frac{\mathrm{D}f_\alpha^{(0)}}{\mathrm{D}t}\right]_{\tau=0}\right) = \frac{\partial}{\partial \mathbf{r}}\cdot\left[\mathbf{v}_0\tau^{(0)}\sum_{j=1}^{\eta}\int f_\alpha^{(0)}f_j^{(0)}\left(\psi_\alpha'+\psi_j'-\psi_\alpha-\psi_j\right)\mathrm{d}\omega\right], \tag{9.2.14}$$

$$\frac{\partial}{\partial \mathbf{r}}\cdot\left(\mathbf{V}_\alpha\tau^{(0)}\left[\frac{\mathrm{D}f_\alpha^{(0)}}{\mathrm{D}t}\right]_{\tau=0}\right) = \frac{\partial}{\partial \mathbf{r}}\cdot\left[\tau^{(0)}\sum_{j=1}^{\eta}\int f_\alpha^{(0)}f_j^{(0)}\left(\psi_\alpha'\mathbf{V}_\alpha'\right.\right.$$
$$\left.\left.-\psi_j'\mathbf{V}_j'-\psi_\alpha\mathbf{V}_\alpha-\psi_j\mathbf{V}_j\right)\mathrm{d}\omega\right]. \tag{9.2.15}$$

Obviously, Eqs. (9.2.4), (9.2.13), and (9.2.14) are consistent. Let us show that Eq. (9.2.15) can be satisfied identically by solution of Eq. (9.2.16) in the frame of the moment Enskog method; in other words can be state that by the use of the moment method for the solution of equation

$$\mathbf{V}_\alpha\left[\frac{\mathrm{D}f_\alpha^{(0)}}{\mathrm{D}t}\right]_{\tau=0} = \sum_j\int f_\alpha^{(0)}f_j^{(0)}\left(\psi_\alpha'\mathbf{V}_\alpha'-\psi_j'\mathbf{V}_j'\right.$$
$$\left.-\psi_\alpha'\mathbf{V}_\alpha'-\psi_j'\mathbf{V}_j'\right)\mathrm{d}\omega, \tag{9.2.16}$$

where ψ_α—Enskog solution, the moment equations corresponding first approximation, are satisfying identically. In variables $\mathbf{r}$, $\mathbf{V}_\alpha$, t after exclusion of time derivatives using hydrodynamic equations [35, 36], the left-hand-side of Eq. (9.2.16) takes the form

$$\mathbf{V}_\alpha\left[\frac{\mathrm{D}f_\alpha^{(0)}}{\mathrm{D}t}\right]_{\tau=0} = f_\alpha^{(0)}\mathbf{V}_\alpha\left[\frac{n}{n_\alpha}(\mathbf{V}_\alpha\cdot\mathbf{d}_\alpha)+2\mathbf{W}_\alpha^0\mathbf{W}_\alpha:\frac{\partial}{\partial\mathbf{r}}\mathbf{v}_0\right.$$
$$\left.-\left(\frac{5}{2}-W_\alpha^2\right)\left(\mathbf{V}_\alpha\cdot\frac{\partial \ln T}{\partial \mathbf{r}}\right)\right], \tag{9.2.17}$$

where

$$\mathbf{d}_\alpha = \frac{\partial}{\partial \mathbf{r}}\left(\frac{n_\alpha}{n}\right)+\left(\frac{n_\alpha}{n}-\frac{\rho_\alpha}{\rho}\right)\frac{\partial \ln p}{\partial \mathbf{r}}, \tag{9.2.18}$$

$$\mathbf{W}_\alpha = \sqrt{\frac{m_\alpha}{2k_\mathrm{B}T}}\mathbf{V}_\alpha. \tag{9.2.19}$$

The function ψ_α is the linear function of derivatives and can be written as:

$$\psi_\alpha = -\mathbf{A}_\alpha\cdot\frac{\partial \ln T}{\partial \mathbf{r}} - \overleftrightarrow{B}_\alpha:\frac{\partial}{\partial \mathbf{r}}\mathbf{v}_0 + n\sum_j\mathbf{C}_\alpha^{(j)}\cdot\mathbf{d}_j, \tag{9.2.20}$$

$$\mathbf{C}_\alpha^{(j)} = C_\alpha^{(j)}(W_\alpha)\mathbf{W}_\alpha,\quad \mathbf{A}_\alpha = A_\alpha(W_\alpha)\mathbf{W}_\alpha,\quad \overleftrightarrow{B}_\alpha = \mathbf{W}_\alpha^0\mathbf{W}_\alpha B_\alpha(W_\alpha). \tag{9.2.21}$$

Because of Eq. (9.2.20), Eq. (9.2.16) splits into three equations

$$\begin{aligned}\frac{1}{n_\alpha}V_{\alpha i}\mathbf{V}_\alpha f_\alpha^{(0)}(\delta_{\alpha n}-\delta_{\alpha k})\mathbf{V}_\alpha = \sum_j\int\Big[&V'_{\alpha i}\mathbf{C}_\alpha^{(n)\prime}+V'_{ji}\mathbf{C}_j^{(n)\prime}-V'_{\alpha i}\mathbf{C}_\alpha^{(k)\prime}\\ &-V'_{ji}\mathbf{C}_j^{(k)\prime}-V_{\alpha i}\mathbf{C}_\alpha^{(n)}-V_{ji}\mathbf{C}_j^{(n)}+V_{\alpha i}\mathbf{C}_\alpha^{(k)}+V_{ji}\mathbf{C}_j^{(k)}\Big]\\ &\times f_\alpha^{(0)}f_j^{(0)}\mathrm{d}\omega,\ (i=1,2,3),\end{aligned} \tag{9.2.22}$$

$$\begin{aligned}2V_{\alpha i}f_\alpha^{(0)}\mathbf{W}_\alpha^0\mathbf{W}_\alpha = -\sum_j\int\Big[&V'_{\alpha i}\overleftrightarrow{B}'_\alpha+V_{ji}\overleftrightarrow{B}'_j-V_{\alpha i}\overleftrightarrow{B}_\alpha\\ &-V_{ji}\overleftrightarrow{B}_j\Big]f_\alpha^{(0)}f_j^{(0)}\mathrm{d}\omega,\end{aligned} \tag{9.2.23}$$

$$\begin{aligned}V_{\alpha i}f_\alpha^{(0)}\left(\frac{5}{2}-W_\alpha^2\right)\mathbf{V}_\alpha = \sum_j\int\big[&V'_{\alpha i}\mathbf{A}'_\alpha+V'_{ji}\mathbf{A}'_j-V_{\alpha i}\mathbf{A}_\alpha\\ &-V_{ji}\mathbf{A}_j\big]f_\alpha^{(0)}f_j^{(0)}\mathrm{d}\omega.\end{aligned} \tag{9.2.24}$$

By solution of Eq. (9.2.4), as probe functions $\overleftrightarrow{t}_\alpha^{(n,k)}$ (which correspond to functions $\mathbf{A}_\alpha$, $\overleftrightarrow{B}_\alpha$, $\mathbf{C}_\alpha^{(n)}-\mathbf{C}_\alpha^{(k)}$) can be used the finite linear combinations of Sonine polynomials ($\overleftrightarrow{W}_\alpha=\mathbf{W}_\alpha$, $\mathbf{W}_\alpha^0\mathbf{W}_\alpha$, $\mathbf{W}_\alpha$ relevant to Eqs. (9.2.22)–(9.2.24)).

$$\overleftrightarrow{t}_\alpha^{(n,k)}=\overleftrightarrow{W}_\alpha\sum_{m=0}^{\xi}t_{\alpha m}^{(n,k)}S_n^{(m)}\left(W_\alpha^2\right),\quad m=0,1,2\ldots \tag{9.2.25}$$

After multiplication of the left-hand-sides and right-hand-sides of Eqs. (9.2.22)–(9.2.24) into Sonine polynomials—correspondingly $\mathbf{W}_\alpha S_{3/2}^{(m)}(W_\alpha^2)$, $\mathbf{W}_\alpha^0\mathbf{W}_\alpha S_{5/2}^{(m)}(W_\alpha^2)$, $\mathbf{W}_\alpha S_{3/2}^{(m)}(W_\alpha^2)$—the following integrals appear in the left-hand sides of equations

$$I_1=\frac{1}{n_\alpha}\int f_\alpha^{(0)}V_{\alpha i}(\delta_{\alpha n}-\delta_{\alpha k})\mathbf{V}_\alpha\cdot\mathbf{W}_\alpha S_{3/2}^{(m)}\left(W_\alpha^2\right)\mathrm{d}\mathbf{V}_\alpha, \tag{9.2.26}$$

$$I_2=2\int f_\alpha^{(0)}V_{\alpha i}f_\alpha^{(0)}\mathbf{W}_\alpha^0\mathbf{W}_\alpha:\mathbf{W}_\alpha^0\mathbf{W}_\alpha S_{5/2}^{(m)}\left(W_\alpha^2\right)\mathrm{d}\mathbf{V}_\alpha, \tag{9.2.27}$$

$$I_3=\int f_\alpha^{(0)}V_{\alpha i}\left(\frac{5}{2}-W_\alpha^2\right)\mathbf{V}_\alpha\cdot\mathbf{W}_\alpha S_{3/2}^{(m)}\left(W_\alpha^2\right)\mathrm{d}\mathbf{V}_\alpha,\quad (i=1,\ldots,3). \tag{9.2.28}$$

Integrands in Eqs. (9.2.26)–(9.2.28) are odd functions of velocities components and then integrals I_1, I_2, I_3 overall $\mathbf{V}$-space are equal to zero. This affirmation is obvious for integrals I_1 and I_3. For proof that $I_2=0$ is sufficient to notice that

$$\mathbf{W}_\alpha^0\mathbf{W}_\alpha:\mathbf{W}_\alpha^0\mathbf{W}_\alpha=\frac{2}{3}W_\alpha^4. \tag{9.2.29}$$

Let us consider now the right-hand-sides of moment equations. By scalar multiplication of right-hand-sides of Eqs. (9.2.22) and (9.2.24) by $\mathbf{W}_\alpha S_{3/2}^{(m)}(W_\alpha^2)$ and following integration over all $\mathbf{V}_\alpha$, the bracket expressions appear of the type

$$\begin{aligned}&\left[V_{\alpha i}\mathbf{W}_\alpha S_{3/2}^{(p)}\left(W_\alpha^2\right),\mathbf{W}_j S_{3/2}^{(q)}\left(W_j^2\right)\right]_{\alpha j}\\ &=-\frac{1}{n_\alpha n_j}\int f_\alpha^{(0)}f_j^{(0)}S_{3/2}^{(q)}\left(W_j^2\right)\mathbf{W}_j\cdot\left[V'_{\alpha i}\mathbf{W}'_\alpha S_{3/2}^{(p)}\left(W_\alpha'^2\right)\right.\\ &\left.-V_{\alpha i}\mathbf{W}_\alpha S_{3/2}^{(p)}\left(W_\alpha^2\right)\right]g_{\alpha j}b\,\mathrm{d}b\,\mathrm{d}\varphi\,\mathrm{d}\mathbf{v}_\alpha\mathrm{d}\mathbf{v}_j,\end{aligned} \tag{9.2.30}$$

$$\begin{aligned}&\left[V_{\alpha i}\mathbf{W}_\alpha S_{3/2}^{(p)}(W_\alpha^2),\mathbf{W}_\alpha S_{3/2}^{(q)}(W_\alpha^2)\right]_{\alpha j}\\ &=-\frac{1}{n_\alpha n_j}\int f_\alpha^{(0)}f_j^{(0)}S_{3/2}^{(q)}(W_\alpha^2)\mathbf{W}_\alpha\cdot\left[V'_{\alpha i}\mathbf{W}'_\alpha S_{3/2}^{(p)}\left(W_\alpha'^2\right)\right.\\ &\left.-V_{\alpha i}\mathbf{W}_\alpha S_{3/2}^{(p)}(W_\alpha^2)\right]g_{\alpha j}b\,\mathrm{d}b\,\mathrm{d}\varphi\,\mathrm{d}\mathbf{v}_\alpha\mathrm{d}\mathbf{v}_j.\end{aligned} \tag{9.2.31}$$

Expression (9.2.30) is coefficient by $s^p t^q$ in the expansion of function

$$J = \frac{1}{n_\alpha n_j}\int f_\alpha^{(0)} f_j^{(0)} (1-s)^{-5/2}(1-t)^{-5/2}\left\{\mathbf{W}_\alpha \cdot \mathbf{W}_j V_{\alpha i} \exp\left[-\frac{W_\alpha^2 s}{1-s} - \frac{W_j^2 t}{1-t}\right]\right.$$
$$\left. -\mathbf{W}'_\alpha \cdot \mathbf{W}'_j V'_{\alpha i} \exp\left[-\frac{W'^2_\alpha s}{1-s} - \frac{W_j^2 t}{1-t}\right]\right\} g_{\alpha j} b\, \mathrm{d}b\, \mathrm{d}\varphi\, \mathrm{d}\mathbf{v}_\alpha \mathrm{d}\mathbf{v}_j. \tag{9.2.32}$$

Let us use now in relation (9.2.32) another variables $\mathbf{G}_0, \mathbf{g}_{\alpha j}$ using formulae

$$\mathbf{V}_\alpha = \mathbf{G}_0 - M_j \mathbf{g}_{j\alpha}. \tag{9.2.33}$$

$$\mathbf{V}_j = \mathbf{G}_0 + M_\alpha \mathbf{g}_{j\alpha}, \tag{9.2.34}$$

where $\mathbf{G}_0$ is velocity of center of mass of particles α, j relative coordinate axes moving with the mean mass velocity $\mathbf{v}_0$ of gas, ($M_j = m_j/m_0$, $m_0 = m_\alpha + m_j$). In this coordinate system

$$\frac{1}{2} m_\alpha V_\alpha^2 + \frac{1}{2} m_j V_j^2 = \frac{1}{2} m_0 \left(G_0^2 + M_\alpha M_j g_{\alpha j}^2\right). \tag{9.2.35}$$

Introduce notations

$$\widetilde{G}_0 = G_0 \sqrt{m_0/(2k_\mathrm{B}T)}, \quad \widetilde{g}_{\alpha j} = g_{\alpha j}\sqrt{m_0 M_\alpha M_j/(2k_\mathrm{B}T)}. \tag{9.2.36}$$

We use now in formula (9.2.32) the integration over variables $\mathbf{G}_0, \mathbf{g}_{\alpha j}$. Jacobian of transformation can be written as

$$\frac{\partial\left(\widetilde{\mathbf{G}}_0, \widetilde{\mathbf{g}}_{j\alpha}\right)}{\partial\left(\mathbf{v}_\alpha, \mathbf{v}_j\right)} = \frac{(m_\alpha m_j)^{3/2}}{(2k_\mathrm{B}T)^3}. \tag{9.2.37}$$

Formula (9.2.32) in variables $\mathbf{G}_0, \mathbf{g}_{\alpha j}$ has the form

$$J = (1-s)^{-5/2}(1-t)^{-5/2}\pi^{-3}\int e^{-\widetilde{G}_0^2 - \widetilde{g}_{\alpha j}^2}\left(e^{-\widetilde{S}W_\alpha^2}\mathbf{W}_\alpha V_{\alpha i}\right.$$
$$\left. - e^{-\widetilde{S}W'^2_\alpha}\mathbf{W}'_\alpha V'_{\alpha i}\right) \cdot \mathbf{W}_j e^{-\widetilde{T}W_j^2} g_{\alpha j} b\, \mathrm{d}b\, \mathrm{d}\varphi\, \mathrm{d}\widetilde{\mathbf{G}}_0 \mathrm{d}\widetilde{\mathbf{g}}_{j\alpha}, \tag{9.2.38}$$

where

$$\widetilde{S} = s/(1-s), \quad \widetilde{T} = t/(1-t). \tag{9.2.39}$$

Calculate

$${}^1H_{\alpha j, i} = \int e^{-\widetilde{G}_0^2 - \widetilde{g}_{\alpha j}^2 - \widetilde{S}W'^2_\alpha - \widetilde{T}W_j^2}\mathbf{W}'_\alpha \cdot \mathbf{W}_j V'_{\alpha i} \mathrm{d}\widetilde{\mathbf{G}}_0. \tag{9.2.40}$$

With this aim introduce the new variable

$$\mathbf{c} = \widetilde{\mathbf{G}}_0 + \frac{1}{i_{\alpha j}}\sqrt{M_\alpha M_j}\left(\widetilde{T}\widetilde{\mathbf{g}}_{j\alpha} - \widetilde{S}\widetilde{\mathbf{g}}'_{j\alpha}\right), \tag{9.2.41}$$

where

$$i_{\alpha j} = \frac{1 - sM_j - tM_\alpha}{(1-s)(1-t)}. \tag{9.2.42}$$

Then exponential index in Eq. (9.2.38) has the form

$$\widetilde{G}_0^2 + \widetilde{g}_{\alpha j}^2 + \widetilde{S}W'^2_\alpha + \widetilde{T}W_j^2 = i_{\alpha j}c^2 + j_{\alpha j}\widetilde{g}_{\alpha j}^2, \tag{9.2.43}$$

where

$$j_{\alpha j} = \frac{1 - 2M_\alpha M_j st(1-\cos\chi)}{1 - sM_j - tM_\alpha}, \quad \chi = \widehat{\mathbf{g}_{j\alpha}, \mathbf{g}'_{j\alpha}}. \tag{9.2.44}$$

Notice, that

$$\mathbf{W}'_{\alpha}\cdot\mathbf{W}_j=\left(\sqrt{M_\alpha}\widetilde{\mathbf{G}}_0-\sqrt{M_j}\,\widetilde{\mathbf{g}}_{j\alpha}\right)\cdot\left(\sqrt{M_j}\widetilde{\mathbf{G}}_0+\sqrt{M_\alpha}\,\widetilde{\mathbf{g}}'_{j\alpha}\right)$$
$$=\left(\sqrt{M_\alpha}\mathbf{c}-\sqrt{M_j}\mathbf{c}_\alpha\right)\cdot\left(\sqrt{M_j}\mathbf{c}-\sqrt{M_\alpha}\mathbf{c}_j\right)$$
$$=\sqrt{M_\alpha M_j}c^2-\mathbf{c}\cdot\left(M_j\mathbf{c}_\alpha+M_\alpha\mathbf{c}_j\right)+\sqrt{M_\alpha M_j}\mathbf{c}_\alpha\cdot\mathbf{c}_j, \tag{9.2.45}$$

$$W'_{\alpha i}=\sqrt{M_\alpha}c_i-\sqrt{M_j}c_{\alpha i}\quad (i=1,2,3). \tag{9.2.46}$$

Therefore

$$\mathbf{W}'_{\alpha}\cdot\mathbf{W}_jW'_{\alpha i}=M_\alpha\sqrt{M_j}c^2c_i-\sqrt{M_\alpha}c_i\mathbf{c}\cdot\left(M_j\mathbf{c}_\alpha+M_\alpha\mathbf{c}_j\right)$$
$$+M_\alpha\sqrt{M_j}c_i\mathbf{c}_\alpha\cdot\mathbf{c}_j-M_j\sqrt{M_\alpha}c_{\alpha i}c^2$$
$$+\sqrt{M_j}c_{\alpha i}\mathbf{c}\cdot\left(M_j\mathbf{c}_\alpha+M_\alpha\mathbf{c}_j\right)-M_j\sqrt{M_\alpha}c_{\alpha i}\mathbf{c}_\alpha\cdot\mathbf{c}_j, \tag{9.2.47}$$

where

$$\mathbf{c}_\alpha=\frac{M_\alpha}{i_{\alpha j}}\left(\widetilde{T\mathbf{g}}_{j\alpha}-\widetilde{S\mathbf{g}}'_{j\alpha}\right)+\widetilde{\mathbf{g}}'_{j\alpha}, \tag{9.2.48}$$

$$\mathbf{c}_j=\frac{M_j}{i_{\alpha j}}\left(\widetilde{T\mathbf{g}}'_{j\alpha}-\widetilde{S\mathbf{g}}_{j\alpha}\right)-\widetilde{\mathbf{g}}_{j\alpha}. \tag{9.2.49}$$

By integration overall **c** in **c**—space all integrals containing the odd in "c" integrand will be equal to zero. It means that after substitution of Eq. (9.2.47) in integrand of Eq. (9.2.40) first, second, and fifth terms in right-hand-side of (9.2.47) should be omitted. As result we have

$$^1H_{\alpha j,i}=4\pi\sqrt{\frac{2k_BT}{m_\alpha}}\int_0^\infty e^{-i_{\alpha j}c^2}c^2\left[-\frac{1}{3}c^2\sqrt{M_\alpha}\left(M_jc_{\alpha i}+M_\alpha c_{ji}\right)\right.$$
$$\left.-M_j\sqrt{M_\alpha}c_{\alpha i}c^2-M_j\sqrt{M_\alpha}c_{\alpha i}\mathbf{c}_\alpha\cdot\mathbf{c}_j\right]dc\,e^{-j_{\alpha j}\tilde{g}^2_{\alpha j}}. \tag{9.2.50}$$

Fulfill the integration over "c" in Eq. (9.2.50):

$$^1H_{\alpha j,i}=\frac{\pi^{3/2}}{2}\sqrt{\frac{2k_BT}{m_0}}\left\{\left(M_jc_{\alpha i}+M_\alpha c_{ji}\right)i_{\alpha j}^{-5/2}-3M_jc_{\alpha i}i_{\alpha j}^{-5/2}\right.$$
$$\left.-2M_j\mathbf{c}_\alpha\cdot\mathbf{c}_jc_{\alpha i}i_{\alpha j}^{-3/2}\right\}e^{-j_{\alpha j}\tilde{g}^2_{\alpha j}}, \tag{9.2.51}$$

or

$$^1H_{\alpha j,i}=-\frac{\pi^{3/2}}{2}\sqrt{\frac{2k_BT}{m_0}}\left\{2M_jc_{\alpha i}+2M_j\mathbf{c}_\alpha\cdot\mathbf{c}_jc_{\alpha i}i_{\alpha j}\right.$$
$$\left.-M_\alpha c_{ji}\right\}e^{-j_{\alpha j}\tilde{g}^2_{\alpha j}}i_{\alpha j}^{-5/2},\quad i=1,2,3. \tag{9.2.52}$$

Using also

$$i_{\alpha j}\mathbf{c}_\alpha\cdot\mathbf{c}_j=\tilde{g}^2_{\alpha j}\left(1-j_{\alpha j}-\cos\chi\right), \tag{9.2.53}$$

we find

$$^1H_{\alpha j,i}=-\frac{\pi^{3/2}}{2}\sqrt{\frac{2k_BT}{m_0}}\left\{2M_jc_{\alpha i}+2M_jc_{\alpha i}\tilde{g}^2_{\alpha j}\left(1-j_{\alpha j}-\cos\chi\right)\right.$$
$$\left.-M_\alpha c_{ji}\right\}e^{-j_{\alpha j}\tilde{g}^2_{\alpha j}}i_{\alpha j}^{-5/2}. \tag{9.2.54}$$

After substitutions $c_{\alpha i}, c_{ji}$ from relations (9.2.48) and (9.2.49) into integral $\int {}^1H_{\alpha j,i}d\widetilde{\mathbf{g}}_{j\alpha}$ the following integrals appear in calculations:

$$J_1=\int e^{-j_{\alpha j}\tilde{g}^2_{\alpha j}}Z_1\left(\tilde{g}^2_{\alpha j}\right)\tilde{g}_{\alpha j,i}d\widetilde{\mathbf{g}}_{\alpha j}, \tag{9.2.55}$$

$$J_2 = \int e^{-j_{\alpha j}\tilde{g}_{\alpha j}^2} Z_2\left(\tilde{g}_{\alpha j}^2\right)\tilde{g}'_{\alpha j,i}\,d\tilde{\mathbf{g}}_{\alpha j},\quad (i=1,2,3). \tag{9.2.56}$$

The first from mentioned above integrals is known to be equal to zero as having the odd $\tilde{g}_{\alpha j,i}$ in integrand. Consider now integral J_2. Because

$$d\tilde{\mathbf{G}}_0\, d\tilde{\mathbf{g}}_{\alpha j} = d\mathbf{v}_\alpha\, d\mathbf{v}_j \left[\frac{(m_\alpha m_j)^{1/2}}{2k_B T}\right]^3, \tag{9.2.57}$$

and

$$d\mathbf{v}_\alpha\, d\mathbf{v}_j = d\mathbf{v}'_\alpha\, d\mathbf{v}'_j, \tag{9.2.58}$$

then

$$d\tilde{\mathbf{G}}'_0\, d\tilde{\mathbf{g}}'_{\alpha j} = d\mathbf{v}_\alpha\, d\mathbf{v}_j \left[\frac{(m_\alpha m_j)^{1/2}}{2k_B T}\right]^3. \tag{9.2.59}$$

The mass center velocity is not changing during collision of particles also as module of the relative velocity by the elastic collisions, $\tilde{\mathbf{G}}_0 = \tilde{\mathbf{G}}'_0$, $\tilde{\mathbf{g}}_{\alpha j} = \tilde{\mathbf{g}}'_{\alpha j}$. Then the use of variables connected with the backward collisions leads to relation

$$d\tilde{\mathbf{G}}_0\, d\tilde{\mathbf{g}}'_{\alpha j} = d\mathbf{v}_\alpha\, d\mathbf{v}_j \left[\frac{(m_\alpha m_j)^{1/2}}{2k_B T}\right]^3. \tag{9.2.60}$$

and to possible transformation of integral J_2:

$$\int e^{-j_{\alpha j}\tilde{g}'^2_{\alpha j}} Z_2\left(\tilde{g}'^2_{\alpha j}\right)\tilde{g}'_{\alpha j,i}\,d\tilde{\mathbf{g}}_{\alpha j},$$

which also turns into zero because integrand contains the odd function.

Calculate now ${}^2H_{\alpha j,i}$

$${}^2H_{\alpha j,i} = \int e^{-\tilde{G}_0^2 - \tilde{g}_{\alpha j}^2 - \tilde{S}W_\alpha^2 - \tilde{T}W_j^2}\,\mathbf{W}_\alpha \cdot \mathbf{W}_j V_{\alpha i}\, d\tilde{\mathbf{G}}_0. \tag{9.2.61}$$

Notice, that

$$\mathbf{W}_\alpha = \sqrt{M_\alpha}\mathbf{c} - \sqrt{M_j}\mathbf{c}_\alpha + \sqrt{M_j}\left(\tilde{\mathbf{g}}'_{j\alpha} - \tilde{\mathbf{g}}_{j\alpha}\right). \tag{9.2.62}$$

Using also

$$\mathbf{W}_j = \sqrt{M_j}\mathbf{c} - \sqrt{M_\alpha}\mathbf{c}_j, \tag{9.2.63}$$

all velocities combinations we need for calculation of ${}^2H_{\alpha j,i}$ in Eq. (9.2.61) can be found. But we need not to do it because [35] the arbitrary function of $\mathbf{W}'_\alpha$ and $\mathbf{W}_j$ can be transformed into corresponding function of the dimensionless velocities $\mathbf{W}_\alpha$, $\mathbf{W}_j$, if suppose $\chi = 0$, (see Eq. (9.2.54)). Then all considerations, which were led to the conditions $J_1 = 0$, $J_2 = 0$, are valid and the bracket expression (9.2.30) turns into zero.

By calculation of bracket expression (9.2.31) the following function appear

$$H_{\alpha j,i}(\chi) = \int e^{-\tilde{G}_0^2 - \tilde{g}_{\alpha j}^2 - \tilde{S}W'^2_\alpha - \tilde{T}W_\alpha^2}\,\mathbf{W}_\alpha \cdot \mathbf{W}'_\alpha V'_{\alpha i}\, d\tilde{\mathbf{G}}_0, \tag{9.2.64}$$

because the bracket expression (9.2.31) is coefficient by $s^p t^q$ in expansion of function

$$\begin{aligned} N = \frac{1}{n_\alpha n_j}\int & f_\alpha^{(0)} f_j^{(0)} (1-s)^{-5/2}(1-t)^{-5/2}\left\{e^{-W_\alpha^2\tilde{S}} W_\alpha^2 V_{\alpha i}\right.\\ & \left. - e^{-W'^2_\alpha \tilde{S}}\mathbf{W}'_\alpha \cdot \mathbf{W}_\alpha V_{\alpha i}\right\} e^{-W_\alpha^2 \tilde{T}} g_{\alpha j}\, b\, db\, d\varphi\, d\mathbf{v}_\alpha\, d\mathbf{v}_j. \end{aligned} \tag{9.2.65}$$

Really in variables $\widetilde{\mathrm{G}}_0$, $\widetilde{\mathrm{g}}_{\alpha j}$ function N takes the form

$$N=(1-s)^{-5/2}(1-t)^{-5/2}\pi^{-3}\int\left[H_{\alpha j}(0)-H_{\alpha}(\chi)\right]g_{\alpha j}b\mathrm{d}b\mathrm{d}\varphi\mathrm{d}\widetilde{\mathbf{g}}_{j\alpha}, \tag{9.2.66}$$

where

$$H_{\alpha j,i}(0)=\int \mathrm{e}^{-\widetilde{G}_0^2-\widetilde{g}_{\alpha j}^2-SW_{\alpha}^2-TW_{\alpha}^2}W_{\alpha}^2V_{\alpha i}\mathrm{d}\widetilde{G}_0.$$

Let us go from variables $\widetilde{\mathbf{G}}_0, \widetilde{\mathbf{g}}_{j\alpha}$ to variables $\mathbf{c}, \widetilde{\mathbf{g}}_{j\alpha}$ (see for example [21], p. 167), supposing that

$$\mathbf{c}=\mathbf{G}_0+d_1\widetilde{\mathbf{g}}_{j\alpha}+d_2\widetilde{\mathbf{g}}'_{j\alpha}. \tag{9.2.67}$$

The values d_1, d_2 choose by special way, i.e., from the condition

$$\widetilde{G}_0^2+\widetilde{g}_{\alpha j}+SW'^2_{\alpha}+TW_{\alpha}^2=i_{\alpha j}c^2+j_{\alpha j}\widetilde{g}_{\alpha j}^2. \tag{9.2.68}$$

This means of Eq. (9.2.68) inscription may easy realize the integration over the variable **c**. No difficulties to find that if

$$d_1=-\sqrt{M_{\alpha}M_j/i_{\alpha j}}\widetilde{T},\ d_2=-\sqrt{M_{\alpha}M_j/i_{\alpha j}}\widetilde{S}, \tag{9.2.69}$$

hen

$$i_{\alpha j}=\frac{1-M_j(s+t)+\left(M_j-M_{\alpha}\right)st}{(1-s)(1-t)}, \tag{9.2.70}$$

$$j_{\alpha j}=\frac{1-st\left(M_{\alpha}^2+M_j^2+2M_{\alpha}M_j\cos\chi\right)}{1-M_j(s+t)+\left(M_j-M_{\alpha}\right)st}, \tag{9.2.71}$$

and in this case

$$\mathbf{c}=\widetilde{\mathbf{G}}_0-\frac{\sqrt{M_{\alpha}M_j}\left[s(1-t)\widetilde{\mathbf{g}}'_{j\alpha}+t(1-s)\widetilde{\mathbf{g}}_{j\alpha}\right]}{1-M_j(s+t)+\left(M_j-M_{\alpha}\right)st}, \tag{9.2.72}$$

or

$$\mathbf{c}=\widetilde{\mathbf{G}}_0-\frac{1}{i_{\alpha j}}\sqrt{M_{\alpha}M_j}\left(\widetilde{S}\widetilde{\mathbf{g}}'_{j\alpha}+\widetilde{T}\widetilde{\mathbf{g}}_{j\alpha}\right). \tag{9.2.73}$$

Notice, that

$$\begin{aligned}\mathbf{W}_{\alpha}&=\sqrt{M_{\alpha}}\widetilde{\mathbf{G}}_0-\sqrt{M_j}\widetilde{\mathbf{g}}_{j\alpha}\\ \mathbf{W}'_{\alpha}&=\sqrt{M_{\alpha}}\widetilde{\mathbf{G}}_0-\sqrt{M_j}\widetilde{\mathbf{g}}'_{j\alpha}\end{aligned}. \tag{9.2.74}$$

Then

$$\begin{aligned}\mathbf{W}_{\alpha}&=\sqrt{M_{\alpha}}\left(\mathbf{c}+\frac{1}{i_{\alpha j}}\sqrt{M_{\alpha}M_j}\left(\widetilde{S}\widetilde{\mathbf{g}}'_{j\alpha}+\widetilde{T}\widetilde{\mathbf{g}}_{j\alpha}\right)\right)-\sqrt{M_j}\widetilde{\mathbf{g}}_{j\alpha}\\ \mathbf{W}'_{\alpha}&=\sqrt{M_{\alpha}}\left(\mathbf{c}+\frac{1}{i_{\alpha j}}\sqrt{M_{\alpha}M_j}\left(\widetilde{S}\widetilde{\mathbf{g}}'_{j\alpha}+\widetilde{T}\widetilde{\mathbf{g}}_{j\alpha}\right)\right)-\sqrt{M_j}\widetilde{\mathbf{g}}'_{j\alpha}\end{aligned}. \tag{9.2.75}$$

$$\mathbf{W}_{\alpha}=\sqrt{M_{\alpha}}\mathbf{c}+\sqrt{M_j}\mathbf{c}_{\alpha}, \tag{9.2.76}$$

$$\mathbf{W}'_{\alpha}=\sqrt{M_{\alpha}}\mathbf{c}+\sqrt{M_j}\mathbf{c}_j, \tag{9.2.77}$$

where in similar manner as in Eqs. (9.2.48) and (9.2.49) the next notation is introduced

$$\mathbf{c}_{\alpha}=\frac{M_{\alpha}}{i_{\alpha j}}\left(\widetilde{T}\widetilde{\mathbf{g}}_{j\alpha}+\widetilde{S}\widetilde{\mathbf{g}}'_{j\alpha}\right)-\widetilde{\mathbf{g}}_{j\alpha}, \tag{9.2.78}$$

$$\mathbf{c}_j = \frac{M_\alpha}{i_{\alpha j}}\left(\tilde{T}\tilde{\mathbf{g}}_{j\alpha} + \tilde{S}\tilde{\mathbf{g}}'_{j\alpha}\right) - \tilde{\mathbf{g}}'_{j\alpha}. \tag{9.2.79}$$

Using also Eqs. (9.2.76) and (9.2.77), we find

$$\begin{aligned}\mathbf{W}_\alpha \cdot \mathbf{W}'_\alpha W_{\alpha i} = & \left(M_\alpha c^2 + \sqrt{M_\alpha M_j}\mathbf{c}\cdot\mathbf{c}_j + \sqrt{M_\alpha M_j}\mathbf{c}\cdot\mathbf{c}_\alpha \right.\\ & \left. + M_j \mathbf{c}_\alpha \cdot \mathbf{c}_j\right)\left(\sqrt{M_\alpha}c_i + \sqrt{M_j}c_{ji}\right) = \cdots + M_\alpha\sqrt{M_j}c_i\mathbf{c}\cdot\mathbf{c}_j \\ & + M_\alpha\sqrt{M_j}c_i\mathbf{c}\cdot\mathbf{c}_\alpha + \sqrt{M_j}M_\alpha c^2 c_{ji} + \sqrt{M_j}M_j c_{ji}\mathbf{c}_\alpha\cdot\mathbf{c}_j.\end{aligned} \tag{9.2.80}$$

In the right-hand-side of relation (9.2.80) the terms are omitted which contain odd powers of the velocity components $\mathbf{c}$. Analogously (9.2.50) we have

$$\begin{aligned}H_{\alpha j,i}(\chi) = & 4\pi\sqrt{\frac{2k_B T}{m_\alpha}}\int_0^\infty e^{-i_{\alpha j}c^2}c^2\left[\frac{1}{3}M_\alpha\sqrt{M_j}c^2\left(c_{ji}+c_{\alpha i}\right)\right.\\ & \left. + M_\alpha\sqrt{M_j}c^2 c_{ji} + M_j\sqrt{M_j}c_{ji}\mathbf{c}_\alpha\cdot\mathbf{c}_j\right] dc\, e^{-j_{\alpha j}\tilde{g}^2_{\alpha j}},\end{aligned} \tag{9.2.81}$$

and after integration

$$\begin{aligned}H_{\alpha j,i}(\chi) = & \frac{\pi^{3/2}}{2}\sqrt{\frac{2k_B T}{m_\alpha}}\left\{M_\alpha\sqrt{M_j}\left(c_{ji}+c_{\alpha i}\right)i_{\alpha j}^{-5/2}\right.\\ & \left. + 3c_{ji}M_\alpha\sqrt{M_j}i_{\alpha j}^{-5/2} + 2M_j\sqrt{M_j}c_{ji}\mathbf{c}_\alpha\cdot\mathbf{c}_j i_{\alpha j}^{-3/2}\right\}e^{-j_{\alpha j}\tilde{g}^2_{\alpha j}},\end{aligned}$$

or

$$\begin{aligned}H_{\alpha j,i}(\chi) = & i_{\alpha j}^{-5/2}\frac{\pi^{3/2}}{2}\sqrt{\frac{2k_B T}{m_0}}\sqrt{\frac{M_j}{M_\alpha}}\left\{M_\alpha\left(c_{\alpha i}+4c_{ji}\right)\right.\\ & \left. + 2M_j c_{ji}\mathbf{c}_\alpha\cdot\mathbf{c}_j i_{\alpha j}\right\}e^{-j_{\alpha j}\tilde{g}^2_{\alpha j}}.\end{aligned} \tag{9.2.82}$$

Further practically word for word all considerations can be repeated which led to expressions like (9.2.55) and (9.2.56) equal to zero as odd integrand functions. Then bracket expressions of the type (9.2.31) turn into zero.

Finally notice, that Eq. (9.2.23) lead to bracket expressions of the kind

$$\begin{aligned}&\left[V_{\alpha i}\mathbf{W}^0_\alpha\mathbf{W}_\alpha S^{(p)}_{5/2}\left(W^2_\alpha\right), \mathbf{W}^0_j\mathbf{W}_j S^{(q)}_{5/2}\left(W^2_j\right)\right]_{\alpha j}\\ &= -\frac{1}{n_\alpha n_j}\int f^{(0)}_\alpha f^{(0)}_j \mathbf{W}^0_j\mathbf{W}_j S^{(q)}_{5/2}\left(W^2_j\right) : \left[V'_{\alpha i}\mathbf{W}'^0_\alpha\mathbf{W}'_\alpha S^{(p)}_{5/2}\left(W'^2_\alpha\right)\right.\\ &\left. - V_{\alpha i}\mathbf{W}^0_\alpha\mathbf{W}_\alpha S^{(p)}_{5/2}\left(W^2_\alpha\right)\right] g_{\alpha j} b\,db\,d\varphi\,d\mathbf{v}_\alpha d\mathbf{v}_j,\end{aligned} \tag{9.2.83}$$

$$\begin{aligned}&\left[V_{\alpha i}\mathbf{W}^0_\alpha\mathbf{W}_\alpha S^{(p)}_{5/2}\left(W^2_\alpha\right), \mathbf{W}^0_\alpha\mathbf{W}_\alpha S^{(q)}_{5/2}\left(W^2_\alpha\right)\right]_{\alpha j}\\ &= -\frac{1}{n_\alpha n_j}\int f^{(0)}_\alpha f^{(0)}_j \mathbf{W}^0_\alpha\mathbf{W}_\alpha S^{(q)}_{5/2}\left(W^2_\alpha\right) : \left[V'_{\alpha i}\mathbf{W}'^0_\alpha\mathbf{W}'_\alpha S^{(p)}_{5/2}\left(W'^2_\alpha\right)\right.\\ &\left. - V_{\alpha i}\mathbf{W}^0_\alpha\mathbf{W}_\alpha S^{(p)}_{5/2}\left(W^2_\alpha\right)\right] g_{\alpha j} b\,db\,d\varphi\,d\mathbf{v}_\alpha d\mathbf{v}_j.\end{aligned} \tag{9.2.84}$$

As result of analogous calculations can be shown that bracket integrals (9.2.83) and (9.2.84) are equal to zero like (9.2.30) and (9.2.31).

9.3 KINETIC COEFFICIENT CALCULATION WITH TAKING INTO ACCOUNT THE STATISTICAL FLUCTUATIONS

In the frame of developed approximate method we need to use the following form of distribution function for calculation of kinetic coefficients and hydrodynamic fluxes:

$$f_\alpha = f^{(0)}_\alpha + f^{(0)}_\alpha\psi_\alpha - \frac{D}{Dt}\left[\tau^{(0)}f^{(0)}_\alpha\psi_\alpha\right], \tag{9.3.1}$$

where

$$\psi_\alpha = -A_\alpha \mathbf{W}_\alpha - B_\alpha \mathbf{W}_\alpha^0 \mathbf{W}_\alpha : \frac{\partial}{\partial \mathbf{r}} \mathbf{v}_0 + n \sum_j C_\alpha^{(j)} \mathbf{W}_\alpha \cdot \mathbf{d}_j. \tag{9.3.2}$$

Let us calculate diffusive fluxes of α-species, ($\alpha = 1, \ldots, \eta$).

$$\mathbf{J}_\alpha = \rho_\alpha \overline{\mathbf{V}}_\alpha = m_\alpha \int \mathbf{V}_\alpha f_\alpha^{(0)} \psi_\alpha \mathrm{d}\mathbf{V}_\alpha - m_\alpha \int \mathbf{V}_\alpha \frac{\mathrm{D}}{\mathrm{D}t} \left[\tau^{(0)} f_\alpha^{(0)} \psi_\alpha \right] \mathrm{d}\mathbf{V}_\alpha, \tag{9.3.3}$$

where

$$\frac{\mathrm{D}}{\mathrm{D}t} = \frac{\partial}{\partial t} + \mathbf{v}_\alpha \cdot \frac{\partial}{\partial \mathbf{r}}. \tag{9.3.4}$$

Let us denote

$$\mathbf{J}_\alpha^\wedge = m_\alpha \int \mathbf{V}_\alpha f_\alpha^{(0)} \psi_\alpha \mathrm{d}\mathbf{V}_\alpha, \tag{9.3.5}$$

$$\mathbf{J}_\alpha^\tau = -m_\alpha \int \mathbf{V}_\alpha \frac{\mathrm{D}}{\mathrm{D}t} \left[\tau^{(0)} f_\alpha^{(0)} \psi_\alpha \right] \mathrm{d}\mathbf{V}_\alpha. \tag{9.3.6}$$

Obviously diffusive flux $\mathbf{J}_\alpha^\wedge$ can be obtained from the Chapman-Enskog expression [36]

$$\mathbf{J}_\alpha = \frac{m_\alpha n^2}{\rho} \sum_{j=1}^{\eta} m_j D_{\alpha j} \mathbf{d}_j - D_\alpha^T \frac{\partial \ln T}{\partial \mathbf{r}}, \tag{9.3.7}$$

where $D_{\alpha j}$ and D_α^T are the coefficients of diffusion and thermo-diffusion correspondingly. For calculation of $\mathbf{J}_\alpha^\tau$ we need to substitute Eq. (9.3.2) into integral (9.3.6) and to use the additional conditions which—as in Enskog theory—have the form

$$\int f_\alpha^{(1)} \mathrm{d}\mathbf{v}_\alpha = 0, \tag{9.3.8}$$

$$\sum_{\alpha=1}^{\eta} m_\alpha \int \mathbf{v}_\alpha f_\alpha^{(1)} \mathrm{d}\mathbf{v}_\alpha = 0, \tag{9.3.9}$$

$$\sum_{\alpha=1}^{\eta} \int V_\alpha^2 f_\alpha^{(1)} \mathrm{d}\mathbf{v}_\alpha = 0, \tag{9.3.10}$$

because $f_\alpha^{(1)}$ corresponds to classical solution of Enskog equation in the first approximation $f_\alpha^{(1)} = f_\alpha^{(0)} \psi_\alpha$.

Then one supposes that numerical density of component, mean mass velocity of gas mixture and temperature define by the Maxwellian distribution function. Mentioned transformations lead to result

$$\begin{aligned} \mathbf{J}_\alpha^\tau = & -\frac{\partial}{\partial t}\left(\tau^{(0)} \mathbf{J}_\alpha^\wedge\right) - \tau^{(0)} \left(\mathbf{J}_\alpha^\wedge \cdot \frac{\partial}{\partial \mathbf{r}} \right) \mathbf{v}_0 - \frac{\partial}{\partial \mathbf{r}} \cdot \left(\mathbf{v}_0 \tau^{(0)} \mathbf{J}_\alpha^\wedge \right) \\ & - \frac{\partial}{\partial \mathbf{r}} \cdot \left[\tau^{(0)} m_\alpha \int \mathbf{V}_\alpha \mathbf{V}_\alpha f_\alpha^{(0)} \mathrm{d}\mathbf{V}_\alpha \right]. \end{aligned} \tag{9.3.11}$$

Calculation of integral in the-right-hand side of Eq. (9.3.11) is realizing by the use of Eq. (9.3.2):

$$\int \mathbf{V}_\alpha \mathbf{V}_\alpha f_\alpha^{(1)} \mathrm{d}\mathbf{V}_\alpha = -\frac{m_\alpha}{15 k_B T} \overleftrightarrow{S} \int B_\alpha(W_\alpha) V_\alpha^4 f_\alpha^{(0)} \mathrm{d}\mathbf{V}_\alpha, \tag{9.3.12}$$

where $\overleftrightarrow{S}$ is tensor, defined by expression ($i,k = 1,2,3$)

$$S_{ik} = \frac{1}{2}\left(\frac{\partial v_{0i}}{\partial r_k} + \frac{\partial v_{0k}}{\partial r_i} \right) - \frac{1}{3} \delta_{ik} \frac{\partial}{\partial \mathbf{r}} \cdot \mathbf{v}_0. \tag{9.3.13}$$

Introduce dynamical viscosity of α-component

$$\mu_\alpha = \frac{1}{15}\frac{m_\alpha^2}{2k_B T}\int B_\alpha(W_\alpha)V_\alpha^4 f_\alpha^{(0)}\,d\mathbf{V}_\alpha. \tag{9.3.14}$$

Then Eq. (9.3.11) can be transformed:

$$\begin{aligned}\mathbf{J}_\alpha^\tau = &-\tau^{(0)}\left(\mathbf{J}_\alpha^\wedge\cdot\frac{\partial}{\partial\mathbf{r}}\right)\mathbf{v}_0 - \frac{\partial}{\partial t}\left(\tau^{(0)}\mathbf{J}_\alpha^\wedge\right)\\ &-\frac{\partial}{\partial\mathbf{r}}\cdot\left(\tau^{(0)}\mathbf{v}_0\mathbf{J}_\alpha^\wedge\right) + \frac{\partial}{\partial\mathbf{r}}\cdot 2\tau^{(0)}\mu_\alpha\overleftrightarrow{S}.\end{aligned} \tag{9.3.15}$$

Using condition

$$\sum_{\alpha=1}^{\eta}\mathbf{J}_\alpha^\wedge = 0, \tag{9.3.16}$$

we see that

$$\sum_\alpha \mathbf{J}_\alpha^\tau = 2\frac{\partial}{\partial\mathbf{r}}\cdot\left(\tau^{(0)}\mu\overleftrightarrow{S}\right), \tag{9.3.17}$$

where μ—viscosity of gas mixture

$$\mu = \sum_\alpha \mu_\alpha. \tag{9.3.18}$$

Therefore inconsistency in Eq. (9.3.17) appeared which is not equal to zero and proportional to squared viscosity or—for the dimensionless form of equation—to the squared Knudsen number. The origin of this inconsistency is connected with the use of the additional conditions in the form (9.3.8)–(9.3.10), obtained for the shortened classical solution $f_\alpha = f_\alpha^{(0)}(1+\psi_\alpha)$. Practically this inconsistency related with the generalized Barnett approximation, but nevertheless the sum of diffusive fluxes is turned into zero for another definition of fluxes $\mathbf{J}_\alpha$ for generalized Navier-Stokes approximation

$$\mathbf{J}_\alpha = \mathbf{J}_\alpha^\wedge - \tau^{(0)}\left(\mathbf{J}_\alpha^\wedge\cdot\frac{\partial}{\partial\mathbf{r}}\right)\mathbf{v}_0 - \frac{\partial}{\partial t}\left(\tau^{(0)}\mathbf{J}_\alpha^\wedge\right) - \frac{\partial}{\partial\mathbf{r}}\cdot\left(\tau^{(0)}\mathbf{v}_0\mathbf{J}_\alpha^\wedge\right), \tag{9.3.19}$$

introducing the corresponding inconsistent terms in fluctuation parts of the generalized equations.

Consider now pressure tensor

$$\begin{aligned}\overleftrightarrow{P} = &\, m_\alpha\int\mathbf{V}_\alpha\mathbf{V}_\alpha f_\alpha^{(0)}\,d\mathbf{V}_\alpha + m_\alpha\int\mathbf{V}_\alpha\mathbf{V}_\alpha f_\alpha^{(0)}\psi_\alpha\,d\mathbf{V}_\alpha\\ &- m_\alpha\int\mathbf{V}_\alpha\mathbf{V}_\alpha\frac{\partial}{\partial t}\left(\tau^{(0)}f_\alpha^{(0)}\psi_\alpha\right)d\mathbf{V}_\alpha\\ &- m_\alpha\int\mathbf{V}_\alpha\mathbf{V}_\alpha\frac{\partial}{\partial\mathbf{r}}\cdot\left(\tau^{(0)}\mathbf{v}_\alpha f_\alpha^{(0)}\psi_\alpha\right)d\mathbf{V}_\alpha.\end{aligned} \tag{9.3.20}$$

Let us now pass over cumbersome calculations and present the result of calculation, (see also [30]). The cross-disposition of brackets in Eq. (9.3.21) underlines the order of tensor calculations.

$$\begin{aligned}\overleftrightarrow{P}_\alpha = &\, p_\alpha\overleftrightarrow{I} - 2\mu\overleftrightarrow{S} - \mathbf{J}_\alpha^\wedge\tau^{(0)}\frac{\partial\mathbf{v}_0}{\partial t} - \tau^{(0)}\frac{\partial\mathbf{v}_0}{\partial t}\mathbf{J}_\alpha^\wedge + 2\frac{\partial}{\partial t}\left(\tau^{(0)}\mu_\alpha\overleftrightarrow{S}\right)\\ &- 2\mathbf{v}_0\frac{\partial}{\partial\mathbf{r}}\cdot\left(\tau^{(0)}\mu_\alpha\overleftrightarrow{S}\right) + 2\tau^{(0)}\mu_\alpha\left[\overleftrightarrow{S}\cdot\frac{\partial}{\partial\mathbf{r}}\right]\mathbf{v}_0 + 2\left[\frac{\partial}{\partial\mathbf{r}}\cdot\left(\tau^{(0)}\mathbf{v}_0\right]\mu_\alpha\overleftrightarrow{S}\right)\\ &- \mathbf{J}_\alpha^\wedge\tau^{(0)}\left(\mathbf{v}_0\cdot\frac{\partial}{\partial\mathbf{r}}\right)\mathbf{v}_0 - \left[\left(\mathbf{v}_0\cdot\frac{\partial}{\partial\mathbf{r}}\right)\mathbf{v}_0\right]\mathbf{J}_\alpha^\wedge\tau^{(0)} + \overleftrightarrow{T}_\alpha - \overleftrightarrow{D}_\alpha - {}^{\mathbf{v}_0}\overleftrightarrow{H}_\alpha,\end{aligned} \tag{9.3.21}$$

where $\overleftrightarrow{T}_\alpha$, $\overleftrightarrow{D}_\alpha$, ${}^{\mathbf{v}_0}\overleftrightarrow{H}_\alpha$ are the tensors which components have the form (k, $\ell = 1,2,3$)

$$
\begin{aligned}
T_{k\ell,\alpha} = & \frac{\partial}{\partial r_k}\left(\tau^{(0)}\mathrm{K}_\alpha \frac{\partial T}{\partial r_\ell}\right) + \frac{\partial}{\partial r_\ell}\left(\tau^{(0)}\mathrm{K}_\alpha \frac{\partial T}{\partial r_k}\right) \\
& + \delta_{k\ell}\sum_{i=1}^{3}\frac{\partial}{\partial r_i}\left(\tau^{(0)}\mathrm{K}_\alpha \frac{\partial T}{\partial r_i}\right), \quad \mathrm{K}_\alpha = \frac{k_\mathrm{B}}{m_\alpha}D_\alpha^T + \frac{2}{5}\lambda'_\alpha,
\end{aligned} \tag{9.3.22}
$$

$$
\begin{aligned}
D_{k\ell,\alpha} = & \frac{\partial}{\partial r_k}\left(\tau^{(0)}\ell_{\alpha\ell}\right) + \frac{\partial}{\partial r_\ell}\left(\tau^{(0)}\ell_{\alpha k}\right) + \delta_{k\ell}\sum_{i=1}^{3}\frac{\partial}{\partial r_i}\left(\tau^{(0)}\ell_{\alpha i}\right), \\
\ell_{\alpha k} = & \frac{p}{\rho}n\sum_\beta m_\beta\left(D_{\alpha\beta} - D^1_{\alpha\beta}\right)d_{\beta k}
\end{aligned} \tag{9.3.23}
$$

$$
{}^{v_0}H_{k\ell,\alpha} = -2\sum_i \frac{\partial}{\partial r_i}\left\{\tau^{(0)} v_{0k}\mu_\alpha\left[\frac{1}{2}\left(\frac{\partial v_{0\ell}}{\partial r_i} + \frac{\partial v_{0i}}{\partial r_\ell}\right) - \frac{1}{3}\delta_{i\ell}\frac{\partial v_{0i}}{\partial r_i}\right]\right\}. \tag{9.3.24}
$$

Equation (9.3.22) contains coefficients of thermal diffusion D_α^T and thermal conduction λ'_α [36]

$$
\lambda'_\alpha = -\frac{5}{4}k_\mathrm{B} n\alpha\sqrt{\frac{2k_\mathrm{B}T}{m_\alpha}}a_{\alpha 1}, \tag{9.3.25}
$$

where $a_{\alpha 1}$ is coefficient in expansion

$$
A_\alpha = \sum_{m=0}^{\xi} a_{\alpha m} S_{3/2}^{(m)}\left(W_\alpha^2\right). \tag{9.3.26}
$$

Relation (9.3.23) contain apart of usual "zero" diffusive coefficient

$$
D_{\alpha\beta} = \frac{\rho n_\alpha}{2nm_\beta}\sqrt{\frac{2k_\mathrm{B}T}{m_\alpha}}C_{\alpha 0}^{(\beta,\alpha)}, \tag{9.3.27}
$$

also "first" diffusive coefficient defined by relation

$$
D^1_{\alpha\beta} = \frac{\rho n_\alpha}{2nm_\beta}\sqrt{\frac{2k_\mathrm{B}T}{m_\alpha}}C_{\alpha 1}^{(\beta,\alpha)}. \tag{9.3.28}
$$

Pressure tensor of gas mixture $\overleftrightarrow{P}$ forms by summation over all components of gas mixture

$$
\begin{aligned}
\overleftrightarrow{P} = & p\overleftrightarrow{I} - 2\eta\overleftrightarrow{S} + 2\frac{\partial}{\partial t}\left(\tau^{(0)}\mu\overleftrightarrow{S}\right) - 2\mathbf{v}_0\cdot\left(\tau^{(0)}\mu\overleftrightarrow{S}\right) \\
& + 2\tau^{(0)}\mu\left[\overleftrightarrow{S}\cdot\frac{\partial}{\partial \mathbf{r}}\right]\mathbf{v}_0 + \overleftrightarrow{T} - \overleftrightarrow{D} - {}^{v_0}\overleftrightarrow{H},
\end{aligned} \tag{9.3.29}
$$

where

$$
\overleftrightarrow{T} = \sum_\alpha \overleftrightarrow{T}_\alpha;\ \overleftrightarrow{D} = \sum_\alpha \overleftrightarrow{D}_\alpha;\ {}^{v_0}\overleftrightarrow{H} = \sum_\alpha {}^{v_0}\overleftrightarrow{H}_\alpha. \tag{9.3.30}
$$

To use hydrodynamic equations in the generalized Navier-Stokes approximation we need calculate the thermal flux

$$
\begin{aligned}
\mathbf{q}_\alpha = & \mathbf{q}_\alpha^\wedge - \frac{m_\alpha}{2}\int \mathbf{V}_\alpha V_\alpha^2 \frac{\partial}{\partial t}\left(\tau^{(0)} f_\alpha^{(0)}\psi_\alpha\right)\mathrm{d}\mathbf{V}_\alpha \\
& - \frac{m_\alpha}{2}\int \mathbf{V}_\alpha V_\alpha^2 \frac{\partial}{\partial \mathbf{r}}\cdot\left(\mathbf{v}_\alpha \tau^{(0)} f_\alpha^{(0)}\psi_\alpha\right)\mathrm{d}\mathbf{V}_\alpha.
\end{aligned} \tag{9.3.31}
$$

Calculations lead to the result:

$$
\begin{aligned}
\mathbf{q}_\alpha = {} & \mathbf{q}_\alpha^\wedge - \frac{\partial}{\partial t}\left(\tau^{(0)}\mathbf{q}_\alpha^\wedge\right) + \frac{1}{2}v_0^2\tau^{(0)}\mathbf{J}_\alpha^\wedge\frac{\partial}{\partial \mathbf{r}}\cdot\mathbf{v}_0 + \frac{1}{2}v_0^2\left(\mathbf{v}_0\cdot\frac{\partial}{\partial \mathbf{r}}\right)\left(\tau^{(0)}\mathbf{J}_\alpha\right) \\
& + \frac{1}{2}\mathbf{J}_\alpha^\wedge\tau^{(0)}\left(\mathbf{v}_0\cdot\frac{\partial}{\partial \mathbf{r}}\right)v_0^2 + 2\tau^{(0)}\mu_\alpha\left(\mathbf{v}_0\cdot\left[\frac{\partial}{\partial \mathbf{r}}\right)\left(\mathbf{v}_0\cdot\right]\overleftrightarrow{S}\right) \\
& + 2\tau^{(0)}\mu_\alpha\left(\mathbf{v}_0\cdot\overleftrightarrow{S}\right)\cdot\frac{\partial}{\partial \mathbf{r}}\mathbf{v}_0 - \frac{\partial}{\partial \mathbf{r}}\cdot\left(\tau^{(0)}\mu_\alpha\overleftrightarrow{S}v_0^2\right) + v_0^2\frac{\partial}{\partial \mathbf{r}}\cdot\left(\tau^{(0)}\mu_\alpha\overleftrightarrow{S}\right) \\
& - \frac{\partial}{\partial \mathbf{r}}\cdot\left[\frac{7p_\alpha}{\rho_\alpha}\tau^{(0)}\left(\mu_\alpha^1-\mu_\alpha\right)\overleftrightarrow{S}\right] + \tau^{(0)}\mathbf{K}_\alpha\left[\frac{\partial}{\partial \mathbf{r}}\mathbf{v}_0\cdot\frac{\partial T}{\partial \mathbf{r}} + \frac{\partial T}{\partial \mathbf{r}}\cdot\frac{\partial}{\partial \mathbf{r}}\mathbf{v}_0\right. \\
& \left.+ \frac{\partial T}{\partial \mathbf{r}}\frac{\partial}{\partial \mathbf{r}}\cdot\mathbf{v}_0\right] - \tau^{(0)}\frac{pn}{\rho}\left[\frac{\partial}{\partial \mathbf{r}}\mathbf{v}_0\cdot\sum_\beta m_\beta\left(D_{\alpha\beta}-D_{\alpha\beta}^1\right)\mathbf{d}_\beta\right. \\
& \left.+ \sum_\beta m_\beta\left(D_{\alpha\beta}-D_{\alpha\beta}^1\right)\mathbf{d}_\beta\cdot\frac{\partial}{\partial \mathbf{r}}\mathbf{v}_0 + \sum_\beta m_\beta\left(D_{\alpha\beta}-D_{\alpha\beta}^1\right)\mathbf{d}_\beta\frac{\partial}{\partial \mathbf{r}}\cdot\mathbf{v}_0\right] \\
& - \frac{5}{2}\frac{p}{\rho}\tau^{(0)}\sum_\beta m_\beta\left(D_{\alpha\beta}-D_{\alpha\beta}^1\right)\left(\mathbf{d}_\beta\cdot\frac{\partial}{\partial \mathbf{r}}\right)\mathbf{v}_0 + \frac{5}{2}\tau^{(0)}\mathbf{K}_\alpha\left(\frac{\partial T}{\partial \mathbf{r}}\cdot\frac{\partial}{\partial \mathbf{r}}\right)\mathbf{v}_0.
\end{aligned}
\tag{9.3.32}
$$

The "first" viscosity coefficient is introduced in Eq. (9.3.33)

$$
\mu_\alpha^1 = \frac{1}{2}k_\mathrm{B}Tn_\alpha b_{\alpha 1}.
\tag{9.3.33}
$$

After summation over all components of the gas mixture we obtain heat flux for mixture

$$
\begin{aligned}
\mathbf{q} = {} & \mathbf{q}^\wedge - \frac{\partial}{\partial t}\left(\tau^{(0)}\mathbf{q}^\wedge\right) + 2\tau^{(0)}\mu\left(\mathbf{v}_0\cdot\left[\frac{\partial}{\partial \mathbf{r}}\right)\left(\mathbf{v}_0\cdot\right]\overleftrightarrow{S}\right) \\
& + 2\tau^{(0)}\mu\left(\mathbf{v}_0\cdot\overleftrightarrow{S}\right)\cdot\frac{\partial}{\partial \mathbf{r}}\mathbf{v}_0 - \frac{\partial}{\partial \mathbf{r}}\cdot\left(\tau^{(0)}\overleftrightarrow{S}v_0^2\right) + v_0^2\frac{\partial}{\partial \mathbf{r}}\cdot\left(\tau^{(0)}\mu\overleftrightarrow{S}\right) \\
& - \frac{\partial}{\partial \mathbf{r}}\cdot\left[7\overleftrightarrow{S}\tau^{(0)}\sum_\alpha\frac{p_\alpha}{\rho_\alpha}\left(\mu_\alpha^1-\mu_\alpha\right)\right] + \tau^{(0)}\mathbf{K}\left[\frac{\partial}{\partial \mathbf{r}}\mathbf{v}_0\cdot\frac{\partial T}{\partial \mathbf{r}} + \frac{\partial T}{\partial \mathbf{r}}\cdot\frac{\partial}{\partial \mathbf{r}}\mathbf{v}_0\right. \\
& \left.+ \frac{\partial T}{\partial \mathbf{r}}\frac{\partial}{\partial \mathbf{r}}\cdot\mathbf{v}_0\right] - \tau^{(0)}\frac{pn}{\rho}\left[\frac{\partial}{\partial \mathbf{r}}\mathbf{v}_0\cdot\sum_{\alpha\beta} m_\beta\left(D_{\alpha\beta}-D_{\alpha\beta}^1\right)\mathbf{d}_\beta\right. \\
& \left.+ \sum_{\alpha\beta} m_\beta\left(D_{\alpha\beta}-D_{\alpha\beta}^1\right)\mathbf{d}_\beta\cdot\frac{\partial}{\partial \mathbf{r}}\mathbf{v}_0 + \sum_{\alpha\beta} m_\beta\left(D_{\alpha\beta}-D_{\alpha\beta}^1\right)\mathbf{d}_\beta\frac{\partial}{\partial \mathbf{r}}\cdot\mathbf{v}_0\right] \\
& - \frac{5}{2}\tau^{(0)}\frac{p}{\rho}\sum_{\alpha\beta} m_\beta\left(D_{\alpha\beta}-D_{\alpha\beta}^1\right)\left(\mathbf{d}_\beta\cdot\frac{\partial}{\partial \mathbf{r}}\right)\mathbf{v}_0 + \frac{5}{2}\mathbf{K}\tau^{(0)}\left(\frac{\partial T}{\partial \mathbf{r}}\cdot\frac{\partial}{\partial \mathbf{r}}\right)\mathbf{v}_0.
\end{aligned}
\tag{9.3.34}
$$

As is evident from the foregoing the heat flux—in its fluctuation part—depends also on coefficients of diffusion and viscosity in the considered physical system. It remains only to calculate integrals, corresponding to average values in the generalized Navier-Stokes approximation. These integrals are given in Appendices 6 and 7. We can state that developed theory of calculation of kinetic coefficients lead to appearance of additional τ-terms which contain time and space fluctuations including cross effects of different transport mechanisms.

9.4 SOUND PROPAGATION STUDIED WITH THE GENERALIZED EQUATIONS OF FLUID DYNAMICS

The propagation of sound is a classical problem in kinetic and hydrodynamic theories. Let an infinite plate oscillate in a gas with the frequency ω in the direction of its normal. Put $a=\omega\tau$, where τ is the mean time between collisions. For a Boltzmann gas of hard spheres one has

$$
p\tau = \Pi\mu.
\tag{9.4.1}
$$

In addition to the static pressure p and dynamic viscosity μ, the hydrodynamic relation (9.4.1) contains the parameter Π, which is equal to 0.786 if the distribution function is expanded in terms of the Sonine polynomials, and 0.8 in the first-order (Maxwell) approximation.

The parameter a may be linked to the Reynolds number analog

$$r = \frac{\Pi}{a} = \frac{p}{\omega\mu}. \tag{9.4.2}$$

For large enough values of r, classical hydrodynamics works quite satisfactorily. In linear acoustics, the attenuation of sound tends to zero as $r \to \infty$, and the velocity of sound is given by

$$c_0^2 = \gamma \frac{p_0}{\rho_0}, \tag{9.4.3}$$

where ρ_0 is the density of the unperturbed gas and

$$\gamma = \chi^{-1} = \frac{C_p}{C_V}, \tag{9.4.4}$$

is the ratio of the heat capacity at constant pressure to that at constant volume.

Complications arise when $r \sim 1$ and especially in the limit as $r \to 0$. The Euler equations do not "feel" that the situation has changed and yield a constant velocity of sound and zero attenuation over the entire range of r. The Navier-Stokes equation leads to an entirely unreasonable prediction that the attenuation tends to zero after having reached a maximum at $r \sim 1$, and that the speed of sound tends to infinity as $r \to 0$. Therefore the problem of sound propagation at small r numbers requires a kinetic theory treatment. Without entering into a detailed discussion of the methods mentioned (see, e.g., Ref. [192]), it should be admitted that the situation as a whole is unsatisfactory in this field.

In particular, the increased number of moments employed when solving the Boltzmann equation by moment methods gives a poorer agreement with experimental data. One commonly speaks of the "critical Reynolds number" r_{cr}, below which it is impossible to obtain a plane-wave solution. For each particular type of model or moment equations there exists a unique number r_{cr}, thus revealing the purely mathematical—rather than physical—nature of the effect observed.

Let us apply the generalized equations of fluid dynamics to the propagation of sound waves in a monatomic gas. In linear acoustics, density, and temperature perturbations are written as

$$\rho = \rho_0(1+s), \tag{9.4.5}$$

$$T = T_0(1+\eta), \tag{9.4.6}$$

respectively, and the solution of the generalized hydrodynamic equations is taken in the form

$$s = \bar{s}\exp(i\omega t - k'x), \tag{9.4.7}$$

$$\eta = \bar{\eta}\exp(i\omega t - k'x), \tag{9.4.8}$$

$$v = \bar{v}\exp(i\omega t - k'x), \tag{9.4.9}$$

with v the hydrodynamic velocity, and k' the complex wave number.

We now write down the system of non-stationary generalized Euler equations in a one dimension case [cf. Eqs. (2.7.52)–(2.7.54)]:

$$\frac{\partial}{\partial t}\left\{\rho - \tau\left[\frac{\partial \rho}{\partial t} + \frac{\partial}{\partial t}(\rho v)\right]\right\} + \frac{\partial}{\partial x}\left\{\rho v - \tau\left[\frac{\partial}{\partial t}(\rho v) + \frac{\partial}{\partial x}(\rho v^2) + \frac{\partial p}{\partial x}\right]\right\} = 0, \tag{9.4.10}$$

$$\begin{gathered}\frac{\partial}{\partial t}\left\{\rho v - \tau\left[\frac{\partial}{\partial t}(\rho v) + \frac{\partial}{\partial x}(\rho v^2) + \frac{\partial p}{\partial x}\right]\right\} \\ + \frac{\partial}{\partial x}\left\{\left\{\rho v^2 + p - \tau\left[\frac{\partial}{\partial t}(\rho v^2 + p) + \frac{\partial}{\partial x}(\rho v^3 + 3pv)\right]\right\}\right\} = 0,\end{gathered} \tag{9.4.11}$$

$$\frac{\partial}{\partial t}\left\{\rho v^2+3p-\tau\left[\frac{\partial}{\partial t}(\rho v^2+3p)+\frac{\partial}{\partial x}(\rho v^3+5pv)\right]\right\}$$
$$+\frac{\partial}{\partial x}\left\{\rho v^3+5pv-\tau\left[\frac{\partial}{\partial t}(\rho v^3+5pv)+\frac{\partial}{\partial x}\left(\rho v^4+8pv^2+5\frac{p^2}{\rho}\right)\right]\right\}=0. \tag{9.4.12}$$

Generally speaking, the non-local parameter is equal to the relaxation time in the hydrodynamic limit. It means that we intend to take the non-local parameter in the simplest possible form as the mean time between collisions.

From the equation of state

$$p=\rho RT, \tag{9.4.13}$$

in which R is the universal gas constant and which is valid for the Maxwellian distribution function, it follows that

$$p=p_0(1+s+\eta). \tag{9.4.14}$$

On carrying out the linearization, Eqs. (9.4.10)–(9.4.12) reduce to

$$\frac{\partial s}{\partial t}+\frac{\partial v}{\partial x}-\tau\left[\frac{\partial^2 s}{\partial t^2}+\frac{p_0}{\rho_0}\frac{\partial^2}{\partial x^2}(s+\eta)+2\frac{\partial^2 v}{\partial t\partial x}\right]=0 \tag{9.4.15}$$

$$\frac{\partial v}{\partial t}+\frac{p_0}{\rho_0}\frac{\partial}{\partial x}(s+\eta)-\tau\left[\frac{\partial^2 v}{\partial t^2}+2\frac{p_0}{\rho_0}\frac{\partial^2}{\partial t\partial x}(s+\eta)+3\frac{p_0}{\rho_0}\frac{\partial^2 v}{\partial x^2}\right]=0 \tag{9.4.16}$$

$$3\frac{\partial}{\partial t}(s+\eta)+5\frac{\partial v}{\partial x}-\tau\left[3\frac{\partial^2}{\partial t^2}(s+\eta)+5\frac{p_0}{\rho_0}\frac{\partial^2}{\partial x^2}(s+2\eta)+10\frac{\partial^2 v}{\partial t\partial x}\right]=0 \tag{9.4.17}$$

To consider in somewhat more detail the linearization process, we use the continuity equation (9.4.10) as an example. Let us rewrite it as

$$\frac{\partial \rho}{\partial t}+\frac{\partial}{\partial x}(\rho v)-\tau\left[\frac{\partial^2 \rho}{\partial t^2}+\frac{\partial^2}{\partial x^2}(p+\rho v^2)+2\frac{\partial^2}{\partial t\partial x}(\rho v)\right.$$
$$\left.-\left(\frac{\partial \rho}{\partial t}+\frac{\partial}{\partial x}(\rho v)\right)\frac{\partial \ln\tau}{\partial x}-\left(\frac{\partial}{\partial t}(\rho v)+\frac{\partial}{\partial x}(p+\rho v^2)\right)\frac{\partial \ln\tau}{\partial x}\right]=0. \tag{9.4.18}$$

Because for the one-component ("simple") gas of hard spheres

$$\tau=\frac{1}{\sqrt{2}\pi\sigma^2}\sqrt{\frac{\pi m}{2kT}}=\frac{\text{const}}{n\sqrt{T}}, \tag{9.4.19}$$

we have

$$\frac{\partial \ln\tau}{\partial t}=-\frac{1}{\rho}\frac{\partial \rho}{\partial t}-\frac{1}{2T}\frac{\partial T}{\partial t}, \tag{9.4.20}$$

$$\frac{\partial \ln\tau}{\partial x}=-\frac{1}{\rho}\frac{\partial \rho}{\partial x}-\frac{1}{2T}\frac{\partial T}{\partial x}. \tag{9.4.21}$$

Therefore, in linear acoustics, terms containing derivatives of $\ln\tau$ contribute nothing to the first-order equations.

Using the representations (9.4.7)–(9.4.9) we now arrive at a system of equations

$$i\omega s-k'v-\tau\left[-\omega^2 s+\frac{p_0}{\rho_0}k'^2(s+\eta)-2i\omega k'v\right]=0, \tag{9.4.22}$$

$$i\omega v-\frac{p_0}{\rho_0}k'(s+\eta)-\tau\left[-\omega^2 v-2i\omega\frac{p_0}{\rho_0}k'(s+\eta)+3\frac{p_0}{\rho_0}k'^2 v\right]=0, \tag{9.4.23}$$

$$3i\omega(s+\eta)-5k'v-\tau\left[-3\omega^2(s+\eta)+5\frac{p_0}{\rho_0}k'^2(s+2\eta)-10i\omega k'v\right]=0. \tag{9.4.24}$$

If the rank of matrix equals the number of equations then the system of homogeneous algebraic equations (9.4.22)–(9.4.24) has only a trivial solution. The requirement that there be a nonzero solution to this system is that its determinant must be zero:

$$\begin{vmatrix} i\omega+\omega^2\tau-\tau\frac{p_0}{\rho_0}k'^2 & -\tau\frac{p_0}{\rho_0}k' & -k+2i\omega\tau k' \\ -\frac{p_0}{\rho_0}k'+2i\omega\tau\frac{p_0}{\rho_0}k' & -\frac{p_0}{\rho_0}k'+2i\omega\tau\frac{p_0}{\rho_0}k' & i\omega+\omega^2\tau-3\tau\frac{p_0}{\rho_0}k'^2 \\ 3i\omega+3\omega^2\tau-5\tau\frac{p_0}{\rho_0}k'^2 & 3i\omega+3\omega^2\tau-10\tau\frac{p_0}{\rho_0}k'^2 & -5k'+10i\omega\tau k' \end{vmatrix}=0 \tag{9.4.25}$$

giving an algebraic equation of the sixth order in the wave number k':

$$\begin{aligned} &5\tau^3\frac{p_0^3}{\rho_0^3}k'^6-\frac{p_0^2}{\rho_0^2}\left[5i\omega\tau^2+5\omega^2\tau^3+\frac{5}{3}\tau\right]k'^4+\frac{p_0}{\rho_0}\left[\frac{5}{3}i\omega+2\omega^2\tau-\frac{2}{3}i\omega^3\tau^2-\frac{1}{3}\omega^4\tau^3\right]k'^2 \\ &+i\omega^3+3\omega^4\tau-3i\omega^5\tau^2-\omega^6\tau^3=0 \end{aligned} \tag{9.4.26}$$

This equation reduces to the dimensionless form

$$\begin{aligned} &3a^3\chi^2\hat{k}^6-[3ia^2+3a^3+a]\chi\hat{k}^4 \\ &+\left[i+\frac{6}{5}a-\frac{2}{5}ia^2-\frac{1}{5}a^3\right]\hat{k}^2+\frac{3}{5\chi}i+\frac{9}{5\chi}a-\frac{9}{5\chi}a^2-\frac{9}{5\chi}a^3=0, \end{aligned} \tag{9.4.27}$$

where we have introduced the dimensionless wave number $\hat{k}=k'c_0/\omega$ with the characteristic velocity

$$c_0=\sqrt{\gamma\frac{p_0}{\rho_0}}. \tag{9.4.28}$$

The separation of the real from imaginary part in Eq. (9.4.28) according to the equality

$$\hat{k}=\alpha+i\beta, \tag{9.4.29}$$

now yields the system of equations for α and β:

$$\begin{aligned} &3a^3\chi^2(\alpha^2-\beta^2)(\alpha^4+\beta^4-14\alpha^2\beta^2)+12\alpha\beta a^2(\alpha^2-\beta^2)\chi \\ &-(3a^2+a)\chi(\alpha^4+\beta^4-6\alpha^2\beta^2)+(\alpha^2-\beta^2)\left(\frac{6}{5}a-\frac{1}{5}a^3\right) \\ &-2\alpha\beta\left(1-\frac{2}{5}a^2\right)+\frac{9}{5\chi}-\frac{3}{5\chi}a^3=0, \end{aligned} \tag{9.4.30}$$

$$\begin{aligned} &6a^3\chi^2\alpha\beta(3\alpha^4+3\beta^4-10\alpha^2\beta^2)-3a^2\chi(\alpha^4+\beta^4-6\alpha^2\beta^2) \\ &-4(3a^2+a)\chi\alpha\beta(\alpha^2-\beta^2)+(\alpha^2-\beta^2)\left(1-\frac{1}{5}a^2\right) \\ &+2\alpha\beta\left(\frac{6}{5}a-\frac{1}{5}a^3\right)+\frac{3}{5\chi}-\frac{9}{5\chi}a^2=0. \end{aligned} \tag{9.4.31}$$

From Eq. (9.4.7) it follows:

$$s=\bar{s}\exp\left(-\omega\frac{\alpha}{c_0}x\right)\exp\left[i\omega\left(t-\frac{\beta}{c_0}x\right)\right], \tag{9.4.32}$$

showing that the factor α characterizes the attenuation of sound and that $\beta=c_0/c$ is the ratio of the classical Eulerian speed of sound to its calculated value.

Let us now consider two asymptotic solutions to Eqs. (9.4.30) and (9.4.31).

(1) If $a=\omega\tau\to 0$, then in the limiting case $a=0$ it follows from Eqs. (9.4.27) that

$$\hat{k}^2=-\frac{3}{5}\chi^{-1}. \tag{9.4.33}$$

Using the value

$$\chi = \frac{5}{3}, \tag{9.4.34}$$

for a monatomic gas [see Eq. (9.4.4)], one is led to the classical Euler limit

$$k' = \pm i\frac{\omega}{c_0}. \tag{9.4.35}$$

The wave number k' proves to be imaginary; for the density perturbation, for example, the solution is written as

$$s = \bar{s}\exp\left[i\omega\left(t \pm \frac{x}{c_0}\right)\right]. \tag{9.4.36}$$

Thus, in the classical Euler description sound is undamped and its velocity remains constant and equal to c_0 [cf. Eq. (9.4.28)]. In other words, for oscillations traveling in the positive x direction one obtains

$$\alpha = 0,\ \beta = 1, \tag{9.4.37}$$

(2) If $a \to \infty$, then it follows from Eq. (9.4.27) that

$$\hat{k}^6 - \frac{5}{3}\hat{k}^4 - \frac{5}{27}\hat{k}^2 = \frac{25}{27}, \tag{9.4.38}$$

or

$$-x\left(x^2 - 12y\right) - \frac{5}{3}x^2 + \frac{20}{3}y + \frac{5}{27}x = \frac{25}{27} \tag{9.4.39}$$

$$3x^2 - 4y + \frac{10}{3}x = \frac{5}{27}, \tag{9.4.40}$$

where

$$\beta^2 - \alpha^2 = x,\ \alpha^2\beta^2 = y, \tag{9.4.41}$$

From Eqs. (9.4.39) and (9.4.40) one finds

$$x^3 + \frac{5}{3}x^2 + \frac{35}{54}x - \frac{25}{162} = 0, \tag{9.4.42}$$

and the asymptotic solution then follows readily as

$$\alpha = 0.509,\ \beta = 0.650. \tag{9.4.43}$$

Figures 9.1 and 9.2 plot the dimensionless velocity of sound β and the dimensionless attenuation rate α, as calculated by the generalized Euler equations [30, 193], the Navier-Stokes equations, and the generalized Navier-Stokes equations [194, 195], and compares the results with the experimental data of Greenspan [196], and Meyer and Sessler [197].

Detailed generalized Navier-Stokes calculations are given elsewhere [194, 195], they are rather cumbersome and we present here main steps of transformations. Linearization of the generalized hydrodynamic equations (GHE) with taking into account (9.4.7)–(9.4.9) lead to following equations:

continuity equation

$$\begin{aligned}&\left[i + a - \chi a\hat{k}^2\right]s + \left[-\chi a\hat{k}^2 - \frac{9\chi^2}{2\Pi}a^3\hat{k}^4\right]\eta + \left[-\frac{\hat{k}}{c_0} + 2ia\frac{\hat{k}}{c_0}\right. \\ &\left. -\frac{8}{3}\chi\frac{a^2}{c_0\Pi}\hat{k}^3 + 4i\frac{\chi a^3}{c_0\Pi}\hat{k}^3\right]v = 0,\end{aligned} \tag{9.4.44}$$

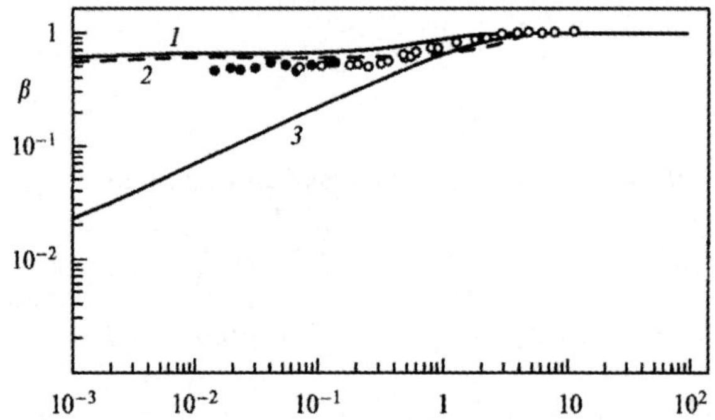

FIGURE 9.1 Comparision of observed (symbols) and calculated (lines) dimensionless velocity of sound $\beta = c_0/c$ as a function of the Reynolds number analog *R*. *1*, generalized Euler equation; *2*, generalized Navier-Stokes equation; *3*, Navier-Stokes equation. Open circles, data by Greenspan; filled circles, data by Meyer and Sessler.

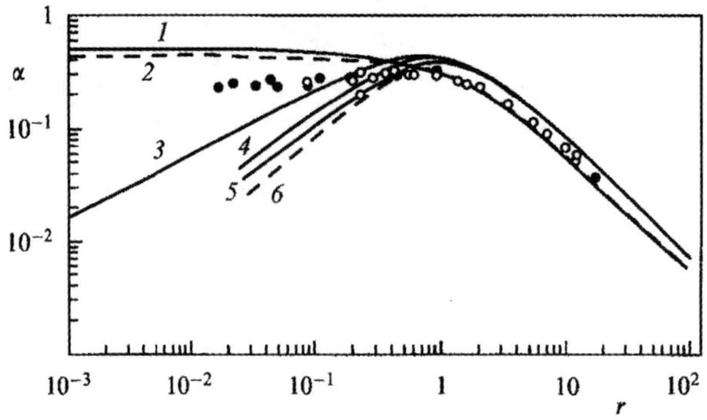

FIGURE 9.2 Comparison of observed (symbols) and calculated (lines) dimensionless attenuation rate α as a function of the Reynolds number analog *r*: *1*, generalized Euler equation; *2*, generalized Navier-Stokes equation; *3*, Navier-Stokes equation; *4*, Burnett equation; *5*, super-Burnett equation; *6*, moment equations (the number of moments $N = 105$). Open circles, data by Greenspan; filled circles, data by Meyer and Sessler.

equation of motion

$$\begin{aligned}&\left[-\chi\hat{k}+2ia\chi\hat{k}\right]s\\&+\left[-\chi\hat{k}-9\frac{\chi^2}{\Pi}a^2\hat{k}^3+\frac{27}{2}i\frac{\chi^2}{\Pi}a^3\hat{k}^3+2i\chi a\hat{k}\right]\eta\\&+\left[\frac{i}{c_0}-\frac{4}{3}\frac{\chi a}{\Pi c_0}\hat{k}^2+4i\frac{\chi a^2}{\Pi c_0}\hat{k}^2+\frac{a}{c_0}-3\chi\frac{a}{c_0}\hat{k}^2-8\chi^2\frac{a^3}{\Pi c_0}A\hat{k}^4+\frac{8}{3}\frac{\chi a^3}{\Pi c_0}\hat{k}^2\right]v=0,\end{aligned} \tag{9.4.45}$$

energy equation

$$\begin{aligned}&\left[-5\chi a\hat{k}^2+3i+3a\right]s+\left[3i-\frac{15}{2}\frac{\chi}{\Pi}a\hat{k}^2+\frac{45}{2}i\frac{\chi}{\Pi}a^2\hat{k}^2+3a\right.\\&\left.-10\chi a\hat{k}^2-126\frac{\chi^2}{\Pi}a^3B\hat{k}^4+15\frac{\chi}{\Pi}a^3\hat{k}^2\right]\eta\\&+\left[-5\frac{\hat{k}}{c_0}-28\frac{\chi a^2}{\Pi c_0}A\hat{k}^3+\frac{112}{3}\frac{\chi a^2}{\Pi c_0}ia^3A\hat{k}^3+10i\frac{a}{c_0}\hat{k}\right]v=0.\end{aligned} \tag{9.4.46}$$

Here

$$A = 1 - \frac{\mu^1}{\mu},\ B = 1 - \frac{\lambda^1}{2\lambda}, \tag{9.4.47}$$

μ^1, λ^1—additional coefficients of viscosity and thermal conductivity (9.3.25) and (9.3.33), which are calculated—in comparison of usual coefficients viscosity μ and thermal conductivity λ, using the first and second coefficients of Sonine polynomial expansions, for example $\mu^1 = pb_1/2$.

As result GHE lead to the dispersion equation of the tenth power:

$$\begin{aligned}
&\{1008AB-180A\}\frac{\chi^5}{\Pi^2}a^7\hat{k}^{10}+\left\{-\frac{1008}{\Pi^2}a^7AB+\frac{156}{\Pi^2}a^7A\right.\\
&+\frac{672}{\Pi^2}a^7B-\frac{210}{\Pi^2}a^7-\frac{66}{\Pi^2}a^5A-\frac{168}{\Pi^2}a^5B+\frac{40}{\Pi}a^5A+\frac{378}{\Pi}a^5B\\
&\left.+\frac{90}{\Pi^2}a^5-\frac{135}{2\Pi}a^5+i\frac{a^6}{\Pi^2}[-1008AB+222A+672B-270]\right\}\chi^4\hat{k}^8\\
&+\left\{-\frac{384}{\Pi^2}a^7A+\frac{336}{\Pi^2}a^7B+\frac{46}{\Pi^2}a^7+\frac{726}{\Pi^2}a^5A-\frac{672}{\Pi^2}a^5B-\frac{80}{\Pi}a^5A\right.\\
&-\frac{36}{\Pi^2}a^5-\frac{137}{6\Pi}a^5-\frac{10}{\Pi^2}a^3+\frac{95}{6\Pi}a^3+15a^3-\frac{126}{\Pi}a^3B\\
&+i\left[-\frac{918}{\Pi^2}a^6A+\frac{840}{\Pi^2}a^6B+\frac{88}{\Pi^2}a^6+\frac{192}{\Pi^2}a^4A-\frac{168}{\Pi^2}a^4B,\right.\\
&\left.\left.-\frac{80}{\Pi}a^4A+\frac{36}{\Pi^2}a^4-\frac{527}{6\Pi}a^4\right]\right\}\chi^3\hat{k}^6+\left\{-\frac{40}{\Pi^2}a^7-\frac{152}{3\Pi}a^5A\right.\\
&+\frac{126}{\Pi}a^5B+\frac{250}{\Pi^2}a^5-\frac{245}{6\Pi}a^5+\frac{124}{3\Pi}a^3A-\frac{126}{\Pi}a^3B+\frac{337}{6\Pi}a^3\\
&-\frac{70}{\Pi^2}a^3-15a^3-\frac{15}{2\Pi}a-5a+i\left[-\frac{160}{\Pi^2}a^6-\frac{120}{\Pi}a^4A+\frac{252}{\Pi}a^4B\right.\\
&\left.\left.+\frac{190}{\Pi^2}a^4-\frac{232}{3\Pi}a^4+\frac{103}{6\Pi}a^2-\frac{28}{\Pi}a^2A-15a^2\right]\right\}\chi^2\hat{k}^4\\
&+\left\{-\frac{23}{\Pi}a^5+\frac{207}{2\Pi}a^3-a^3-\frac{23}{2\Pi}a+6a\right.\\
&\left.+i\left[-\frac{161}{2\Pi}a^4+\frac{115}{2\Pi}a^2-2a^2+5\right]\right\}\chi\hat{k}^2\\
&+\left\{-3a^3+9a+i[-9a^2+3]\right\}=0.
\end{aligned}\tag{9.4.48}$$

Dispersion equation (9.4.48) has two obvious asymptotic solutions:

(a) $a\to 0$ ($r\to\infty$, see also Eq. (9.4.2)). Eq. (9.4.48) is transforming to Eq. (9.4.33):

$$5\chi\hat{k}^2+3=0,$$

(b) $a\to\infty$ ($r\to 0$). From Eq. (9.4.48) can be found:

$$\begin{aligned}
&\{336AB-60A\}\chi^3\hat{k}^6+\{-336AB+52A+224B-70\}\chi^2\hat{k}^4\\
&+\left\{-128A+112B+\frac{46}{3}\right\}\chi\hat{k}^2-\frac{40}{3}=0.
\end{aligned}\tag{9.4.49}$$

For the models of hard spheres $A=0.9415$, $B=0.956$ and solution of Eq. (9.4.49) lead to result (compare with (9.4.43))

$$\alpha=0.428,\ \beta=0.564 \tag{9.4.50}$$

Introduction of Lennard-Jones model of the particle interaction instead of the hard spheres model does not lead to significant influence on result of calculations. For example, in the limit $r\to 0$ ($T=300$ K) calculations lead practically to the same attenuation rate and to $\beta=0.55$.

We now proceed to discuss the numerical results obtained from the generalized hydrodynamic equations and to make comparisons with available published data. We will also consider some aspects of the method of moments as applied to the

Boltzmann equations used in the sound propagation problem. We will base our analysis on the results of Sirovich and Thurber [198], presented in Figs. 9.3 through 9.5 with necessary notational changes.

Figure 9.3 compares the numerically calculated dimensionless velocity of sound $\beta = c_0/c$ and dimensionless attenuation rate α as functions of the Reynolds number analog r for the eight- and eleven moment models using the interaction potential of Maxwellian molecules and that of the hard-sphere models. Notice that the two models yield very close results.

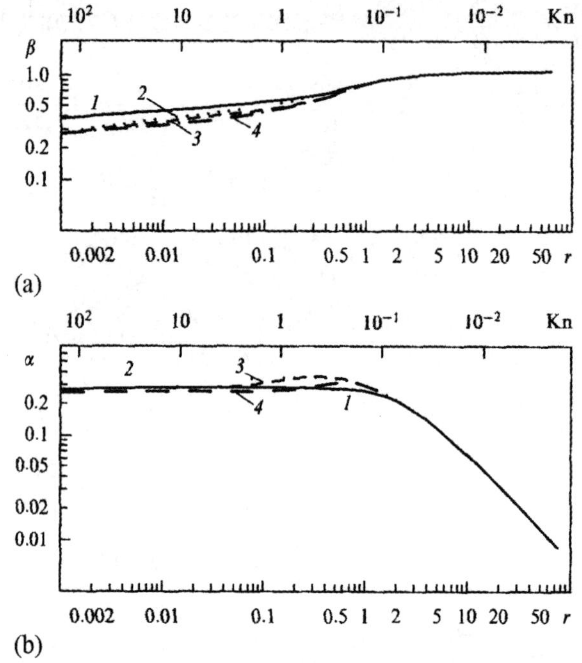

FIGURE 9.3 Comparison of the velocity of sound (a) and the attenuation rate (b) as calculated from the Boltzmann equation for two models: *1*—11 moment hard-sphere model, *2*—11 moment model of Maxwellian molecules, *3*—8 moment hard-sphere model, *4*—8 moment model of Maxwellian molecules.

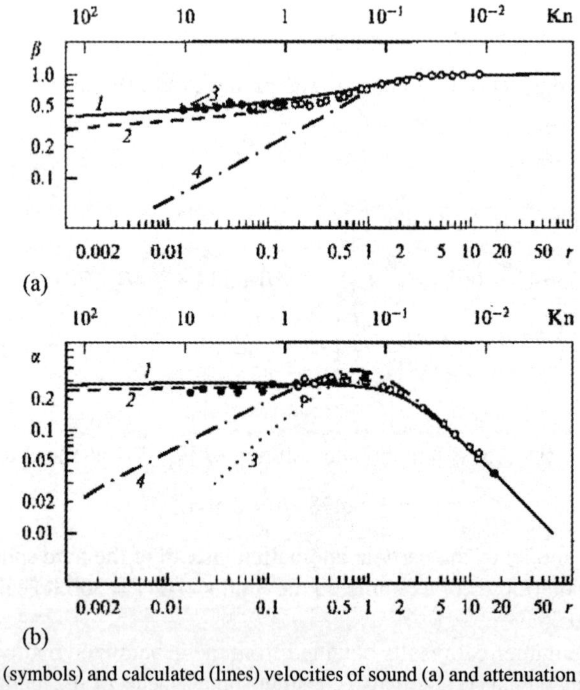

FIGURE 9.4 Comparison of observed (symbols) and calculated (lines) velocities of sound (a) and attenuation rates (b) as calculated from the Boltzmann equation for the hard-sphere model: *1*—11 moment model, *2*—8 moment model, *3*—105 moment model, *4*—Navier-Stokes equation. Open circles—data by Greenspan, filled circles—data by Meyer and Sessler.

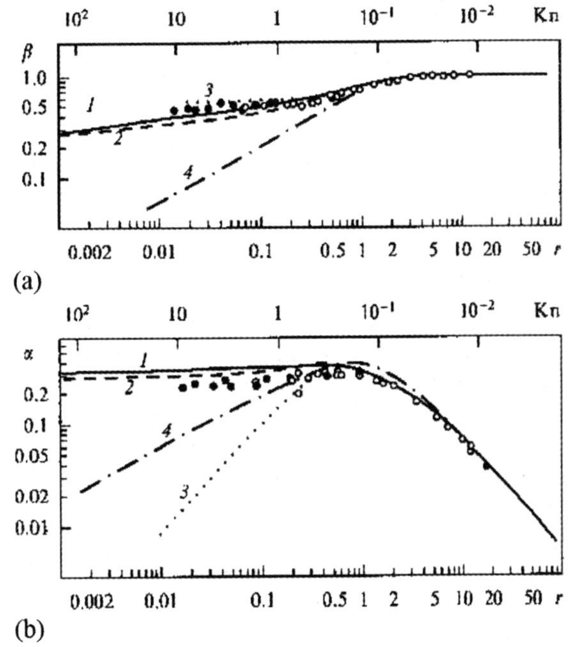

FIGURE 9.5 Comparison of observed (symbols) and calculated (lines) velocities of sound (a) and attenuation rates (b) as calculated from the Boltzmann equation for the model of Maxwelhan molecules: *1*—11 moment model, *2*—8 moment model, *3*—483 moment model, *4*—Navier-Stokes equation. Open circles—data by Greenspan, filled circles—data by Meyer and Sessler.

Figures 9.4 and 9.5 plot similar data for hard spheres and Maxwellian molecules and compare them with the experimental data of Greenspan [196] and Meyer and Sessler [197] mononatomic gases. Also shown are Navier-Stokes results and those obtained by Pekeris et al. [199, 200] in their unprecedently voluminous computations using the 105- and 483 moment models.

A comparison of the experimental data with the theoretical results obtained with the Boltzmann equations suggests what appears at first sight to be a paradoxical conclusion—the more accurate a theoretical model the worse its agreement with experiment. We see indeed that the hard-sphere model of Sirovich and Thurber [198] works better with eight moments than with eleven, and that the 105 moment results of Pekeris et al. [199, 200] are much poorer.

Considering the weak correlation between the molecular interaction model and the velocity of sound and attenuation rate calculations, the results of the 483-moment computations for the Maxwellian molecules within the range $r<1$ should be viewed as simply catastrophic. Note that as the number of the moments used increases, the "critical" number r decreases, apparently raising hopes for a better final result.

A similar situation exists with regard to the hydrodynamic results: while the Navier-Stokes equations are totally invalid for $r<1$, "corrected" models (e.g., the Burnett equation) are even less successful. It can be argued that, paradoxically though it may seem, the best classical theory approach is to employ the Euler equation which, although yielding zero attenuation and a constant velocity of sound, at least does not involve divergences or non-physical "critical" points.

Viewed in the context of the generalized Boltzmann kinetic theory, this effect has a very clear origin. Let us introduce the Knudsen number as the ratio of l/l_ω the mean free path of hard-sphere particles to the wavelength

$$l_\omega = 2\pi\frac{c}{\omega}. \tag{9.4.51}$$

Since in the hard-sphere model

$$l=\frac{1}{\sqrt{2}\pi n\sigma^2},\quad \mu=\frac{5}{16}\frac{\sqrt{mkT}}{\sqrt{\pi}\sigma^2}, \tag{9.4.52}$$

where σ is the particle diameter, it follows from Eqs. (9.4.2) and (9.4.3) that

$$\mathrm{Kn}=\frac{l}{l_\omega}=\frac{8}{5\pi\sqrt{2\pi\gamma}}\frac{1}{r}. \tag{9.4.53}$$

In this case, the ratio of the heat capacities $\gamma = C_p/C_V$ at constant pressure and constant volume is 5/3, and using Eq. (9.4.52) we can recalculate the scale of the Reynolds number analog r to an equivalent scale of the Knudsen number as shown in Figs. 9.3–9.5.

Some conclusions should be done:

1. Discrepancies between the experiment and the "revised" theoretical models based on the Boltzmann equations start to appear for Knudsen number values Kn ~ 1. This is to be expected because the additional terms in the kinetic equation of the generalized Boltzmann theory first become comparable in magnitude and then start to dominate the terms on the left-hand side of the Boltzmann equations as the Knudsen number increases. This means, in particular, that neither the Burnett equations nor, less still, super-Burnett equations hold promise for higher Knudsen number computations.
2. The generalized Boltzmann equation performs much better. The generalized Euler equations and Navier-Stokes equations give quite satisfactory agreement with the experimental data over the entire range of the Knudsen number, including the asymptotic regions. The generalized Navier-Stokes equations fit the experimental points better than the generalized Euler equations. Another important point about this result is that it is obtained from the hydrodynamic equations; this raises hope for a through computation of hydrodynamic flows including shock layers, shock waves, and intermediate Knudsen numbers, thus eliminating the necessity of coupling the hydrodynamic and free-molecular solutions.

   Coupling problems of this kind are discussed widely in the scientific literature (see, e.g., Refs. [201, 202]) In the next section we will see that the generalized hydrodynamic equations make it possible to perform accurate through computations via the shock wave or, in other words, to examine the structure of the shock wave.
3. In the hydrodynamic interpretation of the unified non-local theory (UNLT) the non-local parameter τ plays the role of the kinetic coefficient and should be considered as the external parameter coinciding with the relaxation time τ_{rel}. In the theory developed in this Section, the non-local parameter τ was chosen in the simplest possible form as the mean time between collisions. Even in this approximation we reach the sufficient accuracy. Moreover, practically no reason to use the next generalized Navier-Stokes approximation which effect is not too significant but lead to very complicated relations. If it needs, in the practical applications the best coinciding with the experimental data can be reached using the applicable τ_{rel}.
4. No surprise that generalized Euler equations work even in the regime of the large Knudsen numbers. It was established before that the UNLT equations contain the Madelung quantum hydrodynamics as a deep particular case.

The theory of waves traveling in the atmosphere is of some interest in connection with human health state. Let us consider the solution of the Navier-Stokes equation in the one-dimensional viscid atmosphere. Let ρ is the corresponding value of density, the equilibrium value being denoted by ρ_0. We suppose that $T = \text{const}$, $v_0 = 0$. Then $\rho = \rho_0 + \rho'$, sound speed is $c = \sqrt{RT}$, where R—the gas constant.

The quantities ρ' and v are subject to equation of continuity

$$\frac{\partial \rho'}{\partial t} + \rho_0 \frac{\partial v}{\partial x} = 0, \tag{9.4.54}$$

and dynamical relation (linearized Navier-Stokes equation)

$$\rho_0 \frac{\partial v}{\partial t} + c^2 \frac{\partial \rho'}{\partial x} - b \frac{\partial^2 v}{\partial x^2} = 0. \tag{9.4.55}$$

The viscosity coefficient

$$b = \frac{4}{3}\mu + \xi. \tag{9.4.56}$$

contains also the volume viscosity в ξ. After differentiating Eq. (9.4.55) with respect to time

$$\rho_0 \frac{\partial^2 v}{\partial t^2} + c^2 \frac{\partial}{\partial x} \frac{\partial \rho'}{\partial t} - b \frac{\partial^3 v}{\partial x^2 \partial t} = 0. \tag{9.4.57}$$

and using Eq. (9.4.54) we find:

$$\frac{\partial^2 v}{\partial t^2} - c^2 \frac{\partial^2 v}{\partial x^2} - \frac{b}{\rho_0} \frac{\partial^3 v}{\partial x^2 \partial t} = 0. \tag{9.4.58}$$

The last term of the left-hand-side of Eq. (9.4.58) reflects the attenuation of the acoustic wave. Without this term, Eq. (9.4.58) transforms into classical equation of the hyperbolic type describing the wave motion in the directions $\pm x$ with the phase velocity c. Equation (9.4.59) admits the exact solution

$$v = y(x)w(t). \tag{9.4.59}$$

After the substitution of Eq. (9.4.59) into Eq. (9.4.58) we have

$$\frac{1}{c^2}\frac{\partial^2 w}{\partial t^2} \bigg/ \left(\frac{b}{\rho_0 c^2}\frac{\partial w}{\partial t} + w\right) = \frac{1}{y}\frac{\partial^2 y}{\partial x^2}. \tag{9.4.60}$$

As usual in the Fourier method, we constant that the left side of Eq. (9.4.60) depends on time, whereas the right side depend on x. It means that the both sides separately are equal to the same constant A. As a result:

$$\frac{d^2 y}{dx^2} - Ay = 0, \tag{9.4.61}$$

$$\frac{d^2 w}{dt^2} = Ac^2\left(B\frac{dw}{dt} + w\right), \tag{9.4.62}$$

where $B = \frac{b}{c^2 \rho_0} = \frac{b}{RT\rho_0} = \frac{b}{p_0} \sim \tau^{(0)}$. The result of integration of Eqs. (9.4.61) and (9.4.62) is written as

$$y = C_1 e^{x\sqrt{A}} + C_2 e^{-x\sqrt{A}}, \tag{9.4.63}$$

$$w = C_3 \exp\left[\left(ABc + \sqrt{c^2A^2B^2 + 4A}\right)\frac{ct}{2}\right] + C_4 \exp\left[\left(ABc - \sqrt{c^2A^2B^2 + 4A}\right)\frac{ct}{2}\right]. \tag{9.4.64}$$

As it follows from Eq. (9.4.63) the value A has the dimension in inverse proportion to the square of the distance. The constant B has the time dimension and the order of mean time between collisions $\tau^{(0)}$. Constants A, C_1, C_2, C_3, C_4 are defined by the initial and boundary conditions of the problem. The very essential fact of the structure for the solution (9.4.59), (9.4.63), and (9.4.64). Namely, let the initial velocity perturbation $v(x)$ is given. From relation (9.4.64) follows that for all following time moments the time dependence $w(t)$ has the exponential character.

Let us show that generalized hydrodynamic equations can lead to fundamentally other solutions. We use the same assumptions as it was done for the previous Navier-Stokes model. The system of the generalized hydrodynamic equations

$$\frac{\partial}{\partial t}\left\{\rho - \tau^{(0)}\left[\frac{\partial \rho}{\partial t} + \frac{\partial}{\partial x}(\rho v_0)\right]\right\} + \frac{\partial}{\partial x}\left\{\rho v_0 - \tau^{(0)}\left[\frac{\partial}{\partial t}(\rho v_0) + \frac{\partial}{\partial x}(\rho v_0^2) + \frac{\partial p}{\partial x}\right]\right\} = 0, \tag{9.4.65}$$

$$\frac{\partial}{\partial t}\left\{\rho v_0 - \tau^{(0)}\left[\frac{\partial}{\partial t}(\rho v_0) + \frac{\partial}{\partial x}(\rho v_0^2) + \frac{\partial p}{\partial x}\right]\right\} + \frac{\partial}{\partial x}\left\{\rho v_0^2 + p - \tau^{(0)}\left[\frac{\partial}{\partial t}(\rho v_0^2 + p) + \frac{\partial}{\partial x}(\rho v_0^3 + 3p v_0)\right]\right\} = 0, \tag{9.4.66}$$

takes the form

$$\frac{\partial \rho'}{\partial t} - \frac{\partial}{\partial t}\left\{\tau^{(0)}\left[\frac{\partial \rho'}{\partial t} + \rho_0\frac{\partial v}{\partial x}\right]\right\} + \rho_0\frac{\partial v}{\partial x} - \frac{\partial}{\partial x}\left\{\tau^{(0)}\left[\rho_0\frac{\partial v}{\partial t} + c^2\frac{\partial \rho'}{\partial x}\right]\right\} = 0, \tag{9.4.67}$$

$$\rho_0\frac{\partial v}{\partial t} - \frac{\partial}{\partial t}\left\{\tau^{(0)}\left[\rho_0\frac{\partial v}{\partial t} + c^2\frac{\partial \rho'}{\partial t}\right]\right\} + c^2\frac{\partial \rho'}{\partial t} - \frac{\partial}{\partial x}\left\{\tau^{(0)}\left[c^2\frac{\partial \rho'}{\partial t} + 3p_0\frac{\partial v}{\partial x}\right]\right\} = 0. \tag{9.4.68}$$

The derivatives in Eqs. (9.4.67) and (9.4.68) contain $\tau^{(0)}$ which depend on viscosity and pressure. But in the considered approximation the value $\tau^{(0)}$ is constant. For example,

$$\frac{\partial \tau}{\partial t} = \Pi\mu\frac{\partial}{\partial t}\left(\frac{1}{p}\right) = -\frac{\Pi\mu}{c^2\rho_0^2}\frac{\partial \rho'}{\partial t}. \tag{9.4.69}$$

By the differentiation in Eq. (9.4.69) is taken into account that the dynamical viscosity is the function of temperature which is constant for this problem. As a result the derivatives $\partial\tau/\partial t$, $\partial\tau/\partial x$ are small values of the first-order and after substitution lead to appearance in Eqs. (9.4.67) and (9.4.68) the small values of the second order. These values can be neglected and we reach the system

$$\frac{\partial\rho'}{\partial t}-\tau^{(0)}\frac{\partial^2\rho'}{\partial t^2}-2\rho_0\tau^{(0)}\frac{\partial^2 v}{\partial t\partial x}+\rho_0\frac{\partial v}{\partial x}-\tau^{(0)}c^2\frac{\partial^2\rho'}{\partial x^2}=0, \tag{9.4.70}$$

$$\rho_0\frac{\partial v}{\partial t}-\tau^{(0)}\rho_0\frac{\partial^2 v}{\partial t^2}-2\tau^{(0)}c^2\frac{\partial^2\rho'}{\partial x\partial t}+c^2\frac{\partial\rho'}{\partial x}-3\tau^{(0)}p_0\frac{\partial^2 v}{\partial x^2}=0, \tag{9.4.71}$$

From Eq. (9.4.70) follows:

$$\frac{\partial}{\partial t}\left(\rho'-\tau^{(0)}\frac{\partial\rho'}{\partial t}\right)-\tau^{(0)}c^2\frac{\partial^2\rho'}{\partial x^2}=2\rho_0\tau^{(0)}\frac{\partial^2 v}{\partial t\partial x}-\rho_0\frac{\partial v}{\partial x}, \tag{9.4.72}$$

Leaving the first order values in the left side of Eq. (9.4.72) we find the estimation.

$$\frac{\partial\rho'}{\partial t}\approx 2\rho_0\tau^{(0)}\frac{\partial^2 v}{\partial t\partial x}-\rho_0\frac{\partial v}{\partial x}, \tag{9.4.73}$$

Upon differentiating (9.4.71) on time

$$\rho_0\frac{\partial^2 v}{\partial t^2}-\tau^{(0)}\rho_0\frac{\partial^3 v}{\partial t^3}-2\tau^{(0)}c^2\frac{\partial^2}{\partial x\partial t}\frac{\partial\rho'}{\partial t}+c^2\frac{\partial}{\partial x}\frac{\partial\rho'}{\partial t}-3\tau^{(0)}p_0\frac{\partial^3 v}{\partial x^2\partial t}=0 \tag{9.4.74}$$

and following using of (9.4.73) one obtains

$$\begin{aligned}&\rho_0\frac{\partial^2 v}{\partial t^2}-\tau^{(0)}\rho_0\frac{\partial^3 v}{\partial t^3}-2\tau^{(0)}c^2\frac{\partial^2}{\partial x\partial t}\left(-\rho_0\frac{\partial v}{\partial x}\right)\\&+c^2\frac{\partial}{\partial x}\left(-\rho_0\frac{\partial v}{\partial x}+2\rho_0\tau^{(0)}\frac{\partial^2 v}{\partial t\partial x}\right)-3\tau^{(0)}p_0\frac{\partial^3 v}{\partial x^2\partial t}=0.\end{aligned} \tag{9.4.75}$$

Taking into account that $p_0=\rho_0 RT=\rho_0 c^2$, $\tau^{(0)}=\Pi\mu/p_0 c^2$, where μ is the dynamical viscosity, we have from (9.4.75)

$$\tau^{(0)}\frac{\partial^3 v}{\partial t^3}-\frac{\partial^2 v}{\partial t^2}-\tau^{(0)}c^2\frac{\partial^3 v}{\partial x^2\partial t}+c^2\frac{\partial^2 v}{\partial x^2}=0. \tag{9.4.76}$$

Compare now Eq. (9.4.76) with its Navier-Stokes analog (9.4.58). Equation (9.4.76) is the higher-order equation on time t. If the parameter $\tau^{(0)}$ is not small, the corresponding terms cannot be omitted. But these terms cannot be omitted also in the case of small $\tau^{(0)}$ as the terms with small parameters in front of senior derivatives. Equation (9.4.76) contains also the cross derivative $\partial^3 v/\partial x^2\partial t$ with another sign.

Solution of the hyperbolic equation (9.4.76) is written in the analytical form

$$v=F_1(x-ct)+F_2(x+ct), \tag{9.4.77}$$

where $F_1(x-ct)$ and $F_2(x+ct)$ are the wave functions which depend on $(x-ct)$ and $(x+ct)$ correspondingly. The explicit form of these functions is defined by the initial and boundary conditions of the concrete problem.

Therefore in the adopted suggestions the non-local theory leads to two waves expanding in the viscid media without attenuation in the positive and negative x-directions.

The sound velocity corresponds to the isothermal case typical for the infrasound.

Interesting to notice that the modern versions of Maple (like Maple 10) enable to find the analytical solutions for Eqs. (9.4.70) and (9.4.71), which turn out rather complicated. Solution of these equations confirm the appearance the infrasound waves by the weather change connected with the cyclone and anti-cyclone evolution. Infrasound is sound that is lower in frequency than 20 Hz, the limit of human hearing. The study of such sound waves is referred to sometimes as infrasonics, covering sounds beneath 20 Hz down to 0.001 Hz. Infrasound is characterized by an ability to cover long distances and get around obstacles with little dissipation. Infrasound sometimes results naturally from severe weather, surf, lee waves, avalanches, earthquakes, volcanoes, bolides, waterfalls, calving of icebergs, aurorae, meteors, lightning, and upper-atmospheric lightning. Natural electromagnetism, strong enough to cause weather sensitivity, is present in lightning-induced atmospherics (sferics) and charged particles (ions). The largest infrasound ever recorded by the monitoring system was generated by the 2013 Chelyabinsk meteor.

It is well known the animal and human reactions to infrasound. Rapid and frequent weather changes appear to be the main culprits connected with the meteorological front motion. In Section 11.3 is shown that the cyclone or anticyclone evolution corresponds to moving solitons.

Statistical evidence links increased numbers of many disorders and behavior to certain weather conditions. Biometeorologists subdivided the passage of weather fronts into weather phases and compared the occurrence dates of each phase with hospital records. They found and published some startling relationships between weather and health. Critics could not dismiss the statistical evidence as pure coincidence. Weather-sensitive people become irritated a day or two before the change and are often miserable when a weather front arrives. Cases of suicides, heart attacks, bleeding ulcers, headaches, and migraines all increase. Rheumatics dread the arrival of cold and damp weather, while cold and dry air aggravates asthma symptoms. The weather fronts have something for everybody, it seems.

Our environment includes many factors, and each person on the Earth is permanently influenced by two of them: weather and magnetic field. As we see the evolution of the soliton wave front leads to appearance of infrasound oscillations. But the human being is the resonance system working with the resonance frequency close to the pulse frequaency 1 Hz. Obviously the superposition of the close frequencies leads to the significant disbalance of the physical system.

It is the best approach to solve the puzzle.

9.5 SHOCK WAVE STRUCTURE EXAMINED WITH THE GENERALIZED EQUATIONS OF FLUID DYNAMICS

Let us consider the structure of the shock wave in a monatomic gas based on the solution of the generalized hydrodynamic equations The solution of the usual gas-dynamic equations in this case is given by discontinuous density, velocity, and temperature functions interrelated by the Rankine-Hugoniot equations.

This classical problem of kinetic theory has long become a kind of a testing ground for approximate kinetic theories as well as for methods of solving the Boltzmann equation. Note, that although the solution of this problem has also been obtained with the Navier-Stokes equations, but the conditions for the applicability of the Navier-Stokes equations (small variations of hydrodynamic quantities over the molecular mean free path) are of course not fulfilled—at Mach numbers not too close to unity—and a qualitative description of the transition layer is the most that seems achievable. In this case the viscous terms in the Navier-Stokes equations play the same role as the artificial viscosity terms, which are introduced into the Euler equations in shock wave calculations.

The generalized Euler equations in the one-dimensional steady case are [see Eqs. (2.7.52)–(2.7.54)]

$$\frac{\mathrm{d}}{\mathrm{d}x}\left(\rho v_0 - \tau\frac{\mathrm{d}}{\mathrm{d}x}(\rho v_0^2 + p)\right) = 0, \tag{9.5.1}$$

$$\frac{\mathrm{d}}{\mathrm{d}x}\left(\rho v_0^2 + p - \tau\frac{\mathrm{d}}{\mathrm{d}x}(\rho v_0^3 + 3pv_0)\right) = 0, \tag{9.5.2}$$

$$\frac{\mathrm{d}}{\mathrm{d}x}\left(\rho v_0^3 + 5pv_0 - \tau\frac{\mathrm{d}}{\mathrm{d}x}\left(\rho v_0^4 + 8pv_0^2 + 5\frac{p^2}{\rho}\right)\right) = 0. \tag{9.5.3}$$

Recall that in the simplest case the non-local parameter τ can be estimated using the hard-sphere model $\tau^{(0)}p = 0.8\mu$ (the first-order approximation within the framework of the Enskog method).

Equations (6.3.1) and (6.3.3) in the one-dimensional steady case are readily integrated once to yield

$$\rho v_0 = \tau\frac{\mathrm{d}}{\mathrm{d}x}(p + \rho v_0^2) + C_1, \tag{9.5.4}$$

$$p + \rho v_0^2 = \tau\frac{\mathrm{d}}{\mathrm{d}x}(v_0(3p + \rho v_0^2)) + C_2, \tag{9.5.5}$$

$$v_0(5p + \rho v_0^2) = \tau\frac{\mathrm{d}}{\mathrm{d}x}\left(8pv_0^2 + 5\frac{p^2}{\rho} + \rho v_0^4\right) + C_3. \tag{9.5.6}$$

To integrate Eqs. (9.5.1)–(9.5.3), it is necessary to specify two boundary conditions for the hydrodynamic velocity, density, and pressure These are the so-called Hugoniot conditions The constants C_1, C_2 and C_3 for Eqs. (9.5.4)–(9.5.6) are determined by the conditions before the shock wave

$$(\rho v_0)_{\rm b} = C_1, \tag{9.5.7}$$

$$\left(p + \rho v_0^2\right)_{\rm b} = C_2, \tag{9.5.8}$$

$$\left[v_0\left(5p + \rho v_0^2\right)\right]_{\rm b} = C_3, \tag{9.5.9}$$

where the subscript "b" refers to the flow before the shock wave.

However, the numerical integration of Eqs. (9.5.4)–(9.5.6) is complicated by the necessity of satisfying the boundary conditions at the opposite end of the integration region. A simpler approach in this case is to solve the boundary value problem directly, by applying the three-diagonal method of Gauss elimination technique for the differential second order equation (see Appendix 5).

This is exactly what Polev and the present author did in 1988, [203].

Let us define the width of the shock wave by the relation

$$d = \frac{\rho_{\rm a} - \rho_{\rm b}}{(\mathrm{d}\rho/\mathrm{d}x)_{\max}}, \tag{9.5.10}$$

where the subscript "a" refers to the flow parameters after the shock wave, and $(\mathrm{d}\rho/\mathrm{d}x)_{\max}$ corresponds to the maximum values of the density gradient in the shock wave

Let us next define the dimensionless shock wave width

$$\bar{\delta} = \frac{l_{\rm b}}{d}, \tag{9.5.11}$$

where λ_b is the mean free path in the region before the shock. For the hard-sphere model

$$l_{\rm b} = \frac{m}{\sqrt{2}\pi\rho_{\rm b}\sigma^2}, \tag{9.5.12}$$

It is also useful to define the dimensionless density

$$\bar{\rho} = \frac{\rho - \rho_{\rm b}}{\rho_{\rm a} - \rho_{\rm b}}. \tag{9.5.14}$$

Figure 9.6 plots the dimensionless shock wave width $\bar{\delta}$ as a function of the Mach number Ma. The theoretical curves *1*, *2*, and *3* (computed with the generalized Euler equations, the generalized Navier-Stokes equations, and the ordinary Navier-Stokes equations [203], respectively) are compared with the experimental data of Schmidt [204]. Curves *1* and *2* lie somewhat above the experimental points, the generalized Navier-Stokes calculations giving a better fit. Notice that the Navier-Stokes results (curve *3*) become unsatisfactory for Ma > 1.6.

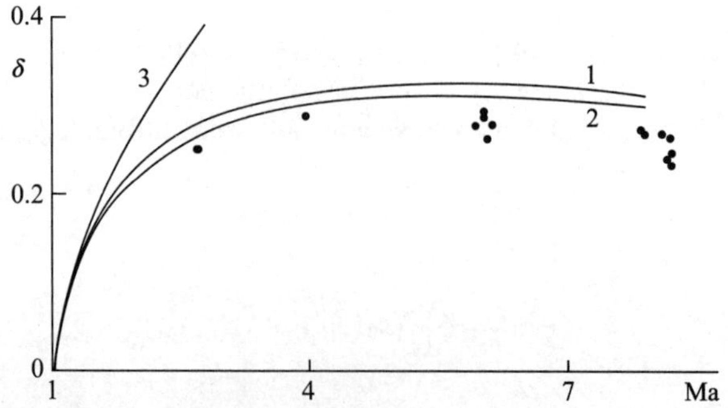

FIGURE 9.6 Comparison of observed (experimental points by Schmidt [204]) and calculated (lines) dimensionless shock wave width $\bar{\delta}$ as a function of the Mach number Ma: 1—generalized Euler equations, 2—generalized Navier-Stokes equations, 3—Navier-Stokes equations.

The use of the Grad method also has proved unsatisfactory. Grad himself used the thirteen-moment approximation to determine the shock wave structure [205]. He found that a solution to this problem does not exist for Ma > 1.65, and Holway later established [206] that the Grad series for the distribution function in the Boltzmann equation diverges when Ma > 1.85.

It is important to note that our results on shock wave structure were obtained in the framework of the generalized hydrodynamic equations, which leads to the expectation that these equations may be used effectively in through calculations for arbitrary Mach and Knudsen numbers.

9.6 BOUNDARY CONDITIONS IN THE THEORY OF THE GENERALIZED HYDRODYNAMIC EQUATIONS

Let us consider now the problem of the additional boundary conditions in the theory of GHE. With this aim write down once more the generalized continuity equation

$$\frac{\partial}{\partial t}\left\{\rho-\tau\left[\frac{\partial \rho}{\partial t}+\frac{\partial}{\partial \mathbf{r}}\cdot(\rho\mathbf{v}_0)\right]\right\}+\frac{\partial}{\partial \mathbf{r}}\cdot\left\{\rho\mathbf{v}_0-\tau\left[\frac{\partial}{\partial t}(\rho\mathbf{v}_0)+\frac{\partial}{\partial \mathbf{r}}\cdot(\rho\overline{\mathbf{v}\mathbf{v}})\right]\right\}=0. \tag{9.6.1}$$

and demonstrate—using this equation as an example—how this problem can be solved.

Let be ρ_∞—the density scale, L—character hydrodynamic length L, v_T is the character mean thermal velocity as the scale for $\mathbf{v}$ and $v_{0\infty}$ is the corresponding scale for the hydrodynamic velocity $\mathbf{v}_0$. The hydrodynamic time scale is $t_\Gamma = L v_{0\infty}^{-1}$, and τ-scale is $\tau l/v_T$. Then from Eq. (9.6.1) we have the dimensionless equation

$$\frac{\partial}{\partial \hat{t}}\left\{\hat{\rho}-A_v\hat{\tau}\frac{\ell}{L}\left[\frac{\partial \hat{\rho}}{\partial \hat{t}}+\frac{\partial}{\partial \hat{\mathbf{r}}}\cdot(\hat{\rho}\hat{\mathbf{v}}_0)\right]\right\}+\frac{\partial}{\partial \hat{\mathbf{r}}}\cdot\left\{\hat{\rho}\hat{\mathbf{v}}_0-A_v\hat{\tau}\frac{\ell}{L}\left[\frac{\partial}{\partial \hat{t}}(\hat{\rho}\hat{\mathbf{v}}_0)+\frac{\partial}{\partial \hat{\mathbf{r}}}\cdot(\hat{\rho}\overline{\hat{\mathbf{v}}\hat{\mathbf{v}}})\right]\right\}=0, \tag{9.6.2}$$

where $A_v = v_{0\infty}/v_T$. Knudsen number is defined as ratio of the mean free path ℓ near the wall to the character hydrodynamic scale L. The length L can be significantly different and should be introduced in correspondence with concrete solving problem. Equation (9.6.2) can be rewritten as

$$\frac{\partial}{\partial \hat{t}}\left\{\hat{\rho}-A_v\hat{\tau}\mathrm{Kn}\left[\frac{\partial \hat{\rho}}{\partial \hat{t}}+\frac{\partial}{\partial \hat{\mathbf{r}}}\cdot(\hat{\rho}\hat{\mathbf{v}}_0)\right]\right\}+\frac{\partial}{\partial \hat{\mathbf{r}}}\cdot\left\{\hat{\rho}\hat{\mathbf{v}}_0-A_v\hat{\tau}\mathrm{Kn}\left[\frac{\partial}{\partial \hat{t}}(\hat{\rho}\hat{\mathbf{v}}_0)+\frac{\partial}{\partial \hat{\mathbf{r}}}\cdot(\hat{\rho}\overline{\hat{\mathbf{v}}\hat{\mathbf{v}}})\right]\right\}=0, \tag{9.6.3}$$

For the boundary condition's construction, we consider an area near the streamed wall taking into account that L is distance to the wall (Fig. 9.7). Tend now L to zero using an arbitrary law of this tendency. It means that in (9.6.3) differential terms—for which Kn is coefficient—in square brackets are far more than algebraic terms from which they are subtracted. Therewith Kn can be considered as value independent on $\hat{t}$ and $\hat{\mathbf{r}}$.

After dividing of Eq. (9.6.3) by Kn, it becomes obvious that this equation can be identically satisfied for arbitrary tendency $L \to 0$ (and therefore $\mathrm{Kn} \to \infty$) if

$$\frac{\partial \rho}{\partial t}+\frac{\partial}{\partial \mathbf{r}}\cdot\rho\mathbf{v}_0=0, \tag{9.6.4}$$

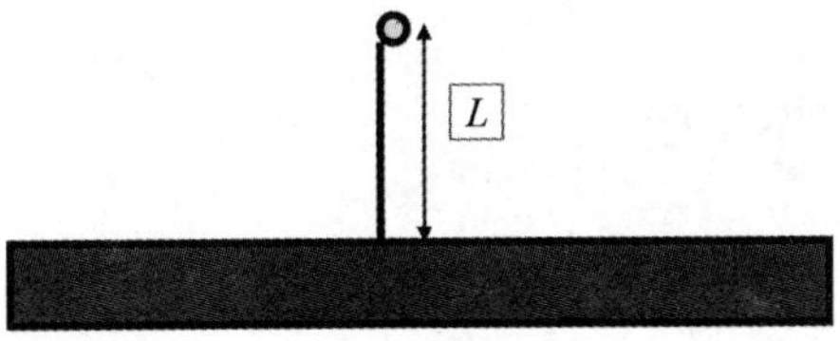

FIGURE 9.7 Domain in the vicinity of the streamed wall.

$$\frac{\partial}{\partial t}(\rho \mathbf{v}_0)+\frac{\partial}{\partial \mathbf{r}}\cdot \rho \overline{\mathbf{v}\mathbf{v}}=0. \tag{9.6.5}$$

But Eqs. (9.6.4) and (9.6.5) are classical continuity and momentum equations written on the wall. This very remarkable fact will be discussed in the next chapter from turbulence positions.

On the whole we state that GHE can be written in the symbolic form (index "i" related to the continuity, momentum, and energy equations)

$$\Phi_{1i}+\mathrm{Kn}\Phi_{2i}=0,\quad (i=1,2,3) \tag{9.6.6}$$

and following considered method for $\mathrm{Kn}\rightarrow\infty$ we reach conditions

$$\Phi_{2i,\mathrm{W}}=0, \tag{9.6.7}$$

where index "W" corresponds to the wall.

Relation (9.6.6) deliver additional boundary conditions we need, and have transparent physical sense—fluctuations disappear on the wall.

In hydrodynamics of weakly rarefied gases the kinetic boundary conditions (in particular velocity slip) are used. These conditions are direct consequence of the Boltzmann equation and allow expanding the area of the formal appliance of the Navier-Stokes description. Further we intend to show that GHE contain known kinetic boundary conditions as asymptotic solutions near the wall. It means from one side that GHE can be applied for description of the rarefied gas without additional kinetic conditions serving for adjusting kinetic and hydrodynamic solutions in boundary kinetic layer. From other side we can state that "jump conditions" can be obtained as solution of *hydrodynamic* equations which can be used for adequate description of kinetic layer near the wall.

With the aim of analytical investigation we introduce reasonable assumptions often used in the theory of kinetic Knudsen layer. We suppose that the flow is one-dimensional (and therefore all gradients along streamed surface can be omitted), gravitation is not significant; static pressure in the layer is constant,

$$p=\mathrm{const}. \tag{9.6.8}$$

The last presumption is not of principal significance, but allows simplifying the problem eliminating from consideration the energy equation and then the temperature jump.

One-dimensional Euler equations in these assumptions are written as:

$$\frac{\partial}{\partial y}\left[\rho v-\Pi\frac{\mu}{p}\left(\frac{\partial}{\partial y}(\rho v^2)+\frac{\partial p}{\partial y}\right)\right]=0, \tag{9.6.9}$$

$$\frac{\partial}{\partial y}\left\{\rho uv-\Pi\frac{\mu}{p}\left(\frac{\partial}{\partial y}(\rho uv^2)+\frac{\partial}{\partial y}(pu)\right)\right\}=0, \tag{9.6.10}$$

$$\frac{\partial}{\partial y}\left\{p+\rho v^2-\Pi\frac{\mu}{p}\left[\frac{\partial}{\partial y}(\rho v^3)+3\frac{\partial}{\partial y}(pv)\right]\right\}=0, \tag{9.6.11}$$

where y is the direction normal to the streamed plane surface and u, v are the components of hydrodynamic velocity. After integration over y one obtains

$$\rho v=\Pi\frac{\mu}{p}\frac{\mathrm{d}}{\mathrm{d}y}(\rho v^2)+C_1, \tag{9.6.12}$$

$$\rho uv=\Pi\frac{\mu}{p}\frac{\mathrm{d}}{\mathrm{d}y}(\rho uv^2)+\Pi\mu\frac{\mathrm{d}u}{\mathrm{d}y}+C_2, \tag{9.6.13}$$

$$\rho v^2=\Pi\frac{\mu}{p}\frac{\mathrm{d}}{\mathrm{d}y}(\rho v^3)+3\Pi\mu\frac{\mathrm{d}v}{\mathrm{d}y}+C_3-p, \tag{9.6.14}$$

where C_1, C_2, and C_3 are constants of integration.

By application of Eq. (9.6.12), Eq. (9.6.14) is written as:

$$\Pi\frac{\mu}{p}(\rho v^2+3p)\frac{\mathrm{d}v}{\mathrm{d}y}=C_1 v-C_3+p. \tag{9.6.15}$$

Write (9.6.12) in the form

$$1=\Pi\frac{\mu}{p}\left[2\frac{\mathrm{d}v}{\mathrm{d}y}+v\frac{\mathrm{d}\ln\rho}{\mathrm{d}y}\right]+\frac{C_1}{\rho v}. \tag{9.6.16}$$

On the bottom of the Knudsen layer particles go into the stream after desorption from the surface with normal velocity component v_w, corresponding to the order of thermal velocity for wall temperature. Particles, directed to the surface, have in average hydrodynamic velocity and known relation

$$(\rho v)_+=(\rho v)_-. \tag{9.6.17}$$

defines impermeability of the surface. In kinetic layer the process of relaxation is being realized by adjusting the dynamic parameters of particles after desorption to the hydrodynamic parameters of flow. For us of interest is the process on the bottom of Knudsen layer and corresponding asymptotic GHE solutions. In this case we omit the second logarithmic in square brackets in Eq. (6.9.16) and eliminate density in Eq. (9.6.15):

$$6\Pi\mu\left(\frac{\mathrm{d}v}{\mathrm{d}y}\right)^2-\left[3C_1v+3p-2C_3'\right]\frac{\mathrm{d}v}{\mathrm{d}y}+\left(C_1v-C_3'\right)\frac{p}{\Pi\mu}=0, \tag{9.6.18}$$

where $C_3'=C_3-p$.

From (9.6.18) follows:

$$\frac{\mathrm{d}v}{\mathrm{d}y}=\frac{3C_1v+3p-2C_3'\pm\sqrt{\left(3C_1v+3p-2C_3'\right)^2-24p\left(C_1v-C_3'\right)}}{12\Pi\mu}. \tag{9.6.19}$$

Expression under square root in (9.6.19) can be transformed:

$$\left(3C_1v+3p-2C_3'\right)^2-24p\left(C_1v-C_3'\right)=\left(3C_1v-3p-2C_3'\right)^2+12C_1pv. \tag{9.6.20}$$

For investigation of approximate solution one supposes that energy of directed hydrodynamic motion is much less the energy of chaotic motion. As result, we have

$$\sqrt{\left(3C_1v-3p-2C_3'\right)^2+12C_1pv}=3C_1v-3p-2C_3'+\frac{6C_1pv}{3C_1v-3p-2C_3'} \tag{9.6.21}$$

or

$$\frac{\mathrm{d}v}{\mathrm{d}y}=\frac{1}{2\tau}\left[1-\frac{C_1v}{3C_1v-3p-2C_3'}\right]. \tag{9.6.22}$$

Further we use the simplest estimation—following from Eq. (9.6.22)—for normal component of velocity on the bottom of Knudsen layer,

$$v=\frac{y}{2\tau}+v_w, \tag{9.6.23}$$

where τ is non-local parameter.

From Eq. (9.6.12) one obtains the equation defining the density change on the bottom of Knudsen layer

$$\left(\frac{y}{2\tau}+v_{\mathrm{w}}\right)^2\frac{\mathrm{d}\rho}{\mathrm{d}y}=-C_1\frac{1}{\tau}, \tag{9.6.24}$$

solution of this equation:

$$\rho=\frac{2C_1}{(y/2\tau)+v_{\mathrm{w}}}+D. \tag{9.6.25}$$

Constant of integration

$$D=\rho_{\mathrm{w}}-\frac{2C_1}{v_{\mathrm{w}}}, \tag{9.6.26}$$

then

$$\rho = \rho_w - \frac{(y/2\tau)}{(y/2\tau) + v_w} 2C_1. \tag{9.6.27}$$

Let us calculate the transverse component of velocity in Knudsen layer. After substitution of (9.6.23) and (9.6.27) in Eq. (9.6.13), one finds

$$\frac{1}{2}\frac{du}{dz}\left(Dz^2 + 2C_1 z + p\right) - uC_1 + C_2 = 0, \tag{9.6.28}$$

where

$$z = \frac{y}{2\tau} + v_w. \tag{9.6.29}$$

General solution of non-homogeneous differential equation (9.6.28) has the form

$$u = Ee^{\int \left(2C_1/\left(Dz^2 + 2C_1 z + p\right)\right)dz} + \frac{C_2}{C_1}, \tag{9.6.30}$$

where E is constant of integration. Evaluate integral

$$I = \int \frac{dz}{Dz^2 + 2C_1 z + p}. \tag{9.6.31}$$

Introduce

$$\delta = 4\left(pD - C_1^2\right). \tag{9.6.32}$$

In the theory of integrals (9.6.31) sign of δ is important

$$\delta \cong 4p[\rho_w - \Delta\rho], \tag{9.6.33}$$

where $\Delta\rho$—density change in Knudsen layer. From Eq. (9.6.33) follows

$$\delta \cong 4p\rho_\infty,\ \delta > 0, \tag{9.6.34}$$

then integral can be evaluated in finite form

$$\begin{aligned} I &= \frac{2}{\sqrt{\delta}} \arctan \frac{2Dz + 2C_1}{\sqrt{\delta}} = \frac{1}{\sqrt{p\rho_\infty}} \arctan\left[\frac{1}{\sqrt{p\rho_\infty}}(Dz + C_1)\right] \\ &= \frac{1}{\sqrt{p\rho_\infty}} \arctan\left[\frac{\rho_\infty z + v_w \Delta\rho/2}{\sqrt{p\rho_\infty}}\right] \end{aligned} \tag{9.6.35}$$

where (see Eqs. (9.6.25) and (9.6.26))

$$C_1 = (\rho_w v_w - \rho_\infty v_w)/2 = \frac{v_w}{2}\Delta\rho,\ D = \rho_\infty. \tag{9.6.36}$$

Index ∞ related to asymptotic hydrodynamic values outside of Knudsen layer. Returning to variable y, we find

$$I = \frac{1}{\sqrt{p\rho_\infty}} \arctan \frac{\rho_\infty \frac{y}{\tau} + v_w(\rho_w + \rho_\infty)}{2\sqrt{p\rho_\infty}} \tag{9.6.37}$$

and corresponding general solution of Eq. (9.6.28)

$$u = Ee^{2IC_1} + \frac{C_2}{C_1}, \tag{9.6.38}$$

where

$$E = -\frac{C_2/C_1}{1+(\Delta\rho/\rho_\infty)v_{\rm w}\left(1/\sqrt{RT_\infty}\right)\arctan\left(v_{\rm w}(1+\rho_{\rm w}/\rho_\infty)/2\sqrt{RT_\infty}\right)}. \quad (9.6.39)$$

Then in indicated suppositions GHE lead to following profile of transversal velocity in Knudsen layer:

$$u = \frac{C_2}{C_1}\left[1-\frac{1+\left(v_{\rm w}\Delta\rho/\sqrt{p\rho_\infty}\right)\arctan\left(\rho_\infty\frac{y}{\tau}+v_{\rm w}(\rho_{\rm w}+\rho_\infty)/2\sqrt{p\rho_\infty}\right)}{1+v_{\rm w}(\Delta\rho/\rho_\infty)\left(1/\sqrt{RT_\infty}\right)\arctan\left(v_{\rm w}(1+\rho_{\rm w}/\rho_\infty)/2\sqrt{RT_\infty}\right)}\right]. \quad (9.6.40)$$

Introduce notation $\overline{C}_2 = -C_2$, $\overline{C}_2 > 0$, and friction coefficient on the wall

$$f_{\rm w} = \mu\left(\frac{{\rm d}u}{{\rm d}y}\right)_{\rm w}, \quad (9.6.41)$$

then

$$\frac{C_1}{\overline{C}_2} \cong \frac{v_{\rm w}\Delta\rho}{2\Pi f_{\rm w}} \quad (9.6.42)$$

Using Eq. (9.6.42) one obtains from Eq. (9.6.40)

$$u = f_w\frac{2\Pi}{\rho_\infty\sqrt{RT_\infty}}\frac{\arctan\left[\frac{y}{\tau}+v_{\rm w}\left(1+\frac{\rho_{\rm w}}{\rho_\infty}\right)/2\sqrt{RT_\infty}\right]-\arctan\left[v_{\rm w}\left(1+\frac{\rho_{\rm w}}{\rho_\infty}\right)/2\sqrt{RT_\infty}\right]}{1+\frac{\Delta\rho}{\rho_\infty}\frac{v_{\rm w}}{\sqrt{RT_\infty}}\arctan\left(v_{\rm w}\left(1+\frac{\rho_{\rm w}}{\rho_\infty}\right)/2\sqrt{RT_\infty}\right)}. \quad (9.6.43)$$

Profile of transversal velocity u—under formulated assumptions—is defined by trigonometric function $\arctan\overline{y}$. To derive the expression (9.6.43) the assumption was introduced that transversal molecular velocity on the wall is equal to zero. In physical chemistry cases are known [207], when absorbed particles move with a velocity along the surface of being bound with mentioned surface. But evaluation of this "chemical" velocity slip is the problem of quantum chemistry and is not here the subject of consideration.

By small y profile is linear

$$u = \frac{C_2}{C_1}\left[1-\frac{1+\frac{v_{\rm w}\Delta\rho}{\sqrt{p\rho_\infty}}\arctan\left(\rho_\infty\frac{y}{\tau}+v_{\rm w}(\rho_{\rm w}+\rho_\infty)/2\sqrt{p\rho_\infty}\right)}{1+v_{\rm w}\frac{\Delta\rho}{\rho_\infty}\frac{1}{\sqrt{RT_\infty}}\arctan\left(v_{\rm w}(1+\rho_{\rm w}/\rho_\infty)/2\sqrt{RT_\infty}\right)}\right]. \quad (9.6.44)$$

and

$$u\frac{1}{f_{\rm w}}\frac{\tau p}{\Pi} = \frac{y}{1+v_{\rm w}\frac{\Delta\rho}{\rho_\infty}\frac{1}{\sqrt{RT_\infty}}\arctan\left(v_{\rm w}(1+\rho_{\rm w}/\rho_\infty)/2\sqrt{RT_\infty}\right)}. \quad (9.6.45)$$

Let us introduce now the kinetic velocity slip $u_{\rm sl}$ of flow as difference between transversal velocity on the external boundary of kinetic layer and transversal velocity on the wall. In classical hydrodynamics $u_{\rm sl}$ cannot be introduced without taking into account of Boltzmann equation. The origin of this fact is well known—classical hydrodynamics is not "working" in kinetic layer.

Formula (9.6.43), which is consequence of GHE, can be rewritten in terminology of slip theory. With this aim introduce asymptotic velocity u_∞ as velocity on the external boundary of Knudsen layer ($u \to u_\infty$ if $y \to \infty$), and slip velocity

$$u_{\rm sl} = u_\infty - u_{\rm w}. \quad (9.6.46)$$

As result, formula (9.6.43) leads to the next slip velocity

$$u_{\rm sl}=f_{\rm w}\frac{2\Pi}{\rho_\infty\sqrt{RT_\infty}}\frac{\dfrac{\pi}{2}-\arctan\left(v_{\rm w}\left(1+\dfrac{\rho_{\rm w}}{\rho_\infty}\right)/2\sqrt{RT_\infty}\right)}{1+\dfrac{\Delta\rho}{\rho_\infty}\dfrac{v_{\rm w}}{\sqrt{RT_\infty}}\arctan\left(v_{\rm w}\left(1+\dfrac{\rho_{\rm w}}{\rho_\infty}\right)/2\sqrt{RT_\infty}\right)}. \tag{9.6.47}$$

Let us compare now the defined slip velocity $u_{\rm sl}$ from (9.6.47) with known expressions of the Boltzmann kinetic theory, particularly with data being given by Cercignani [192].

$$\begin{aligned} u_{\rm sl}&=1.1466\,l\left(\frac{{\rm d}u}{{\rm d}y}\right)_w=1.1466\frac{\mu}{\rho_\infty}\sqrt{\frac{\pi}{2RT_\infty}}\left(\frac{{\rm d}u}{{\rm d}y}\right)_w=1.1466\frac{1}{\rho_\infty}\sqrt{\frac{\pi}{2RT_\infty}}f_{\rm w}\\ &=1.4370\frac{1}{\rho_\infty\sqrt{RT_\infty}}f_{\rm w}. \end{aligned} \tag{9.6.48}$$

If density changing across Knudsen layer is not significant, one obtains from (9.6.47)

$$u_{\rm sl}=f_{\rm w}\frac{2\Pi}{\rho_\infty\sqrt{RT_\infty}}\left[\frac{\pi}{2}-\arctan\left(v_{\rm w}\left(1+\frac{\rho_{\rm w}}{\rho_\infty}\right)/2\sqrt{RT_\infty}\right)\right]-f_{\rm w}\frac{2\Pi}{\rho_\infty\sqrt{RT_\infty}}\frac{\Delta\rho}{\rho_\infty}\frac{v_{\rm w}}{\sqrt{RT_\infty}}\arctan\left(v_{\rm w}\left(1+\frac{\rho_{\rm w}}{\rho_\infty}\right)/2\sqrt{RT_\infty}\right). \tag{9.6.49}$$

Therefore relations for velocity slip (9.6.47) and (9.6.49) contain also density jump. Let us neglect density jump in Knudsen layer. It means

$$\Delta\rho\ll\rho_\infty, \tag{9.6.50}$$

or approximately on the wall

$$\left(\frac{\partial\rho}{\partial y}\right)_{\rm w}=0. \tag{9.6.51}$$

We have from Eq. (9.6.49)

$$u_{\rm sl}=2\Pi\left[\frac{\pi}{2}-\arctan\frac{v_{\rm w}}{\sqrt{RT_\infty}}\right]\frac{1}{\rho_\infty\sqrt{RT_\infty}}f_{\rm w}. \tag{9.6.52}$$

Obviously relations (9.6.48) and (9.6.52) have analogous structure but differ by numerical factors. Multiplier

$$A=2\Pi\left[\frac{\pi}{2}-\arctan\frac{v_{\rm w}}{\sqrt{RT_\infty}}\right]. \tag{9.6.53}$$

in (9.6.52) can vary within wide limits, its value is defined by model of particles interaction—which is used for parameter Π evaluation—and velocity of the particles desorption $v_{\rm w}$ from the surface. For the simplest hard spheres model of particles interactions one obtains $0\le A\le 2.51$. Notice that indicated by Cercignani four ciphers after point in numerical factors in (9.6.48) have sense only as an orientation by comparison of results of different analytical models.

As conclusion we state that GHE incorporate the kinetic effect of slip as an element of adjusting of hydrodynamic and kinetic regimes of flow and allow—avoiding artificial fashions—to describe flows by intermediate flow numbers.

9.7 TO THE KINETIC AND HYDRODYNAMIC THEORY OF LIQUIDS

As is known, the classical Boltzmann equation (BE), describing the processes of transfer in gases is valid only on characteristic scales related to the hydrodynamic time of flow and the mean time between collisions. As it was established, the inclusion of the third possible scale, namely, the particle collision time, leads to the emergence of additional terms in the BE, which, generally speaking, are of the same order as the other terms in the BE.

A fundamental difference between a liquid and a rarefied gas consists in the many-particle interaction of its component particles. Nevertheless, it turns out that the use of fundamental concepts of the generalized Boltzmann kinetic theory (GBKT) and of the many-scale method leads to important results in the kinetic and hydrodynamic theory of liquids.

Let us turn to experimental data in the theory of liquids and theoretical models. In the theory of liquids, some or other microscopic models are usually employed. A liquid is investigated either as a non-ideal gas with many-particle interaction or from the standpoint of the theory of a crystal in which the long-range order is lost. The singularities of these models are described in [208, 209], and we will only refer to two models that are often used in the theory of liquid, namely:

1. The cell model, in which a liquid is treated as a deformed crystal with molecules localized in the vicinity of the "points" of deformed lattice.
2. The hole theory in which it is assumed that the transition of molecules from one state to another is realized owing to the vacancies available in the deformed lattice.

Starting in 1924, Ya. I. Frenkel has been systematically developing the theory of liquid state as a generalization of the theory of real crystals involving the concepts of mobile holes, as well as the "concepts of thermal motion as alternation of small vibrations about some positions of equilibrium with abrupt variation of these positions" [210]. The latter statement is equivalent to introducing the activation energy (or loosening energy using Frenkel's terminology) in the liquid theory. In principle, Frenkel's model reflects the experimental fact of existence of the short-range order in liquid. As a result, the particle motion assumes the behavior of "irregular vibrations with the mean frequency τ_0^{-1} close to that of the vibration frequency of particles in crystals, and with the amplitude defined by the size of the 'free volume' offered to the given particle by its neighbors. The free lifetime of molecule in the temporary position of equilibrium between two activated jumps is related by

$$\tau = \tau_0 \exp(W/k_B T). \tag{9.7.1}$$

where W is the activation energy" [210]. Relation (9.7.1) reflects the physically transparent fact that, with the activation energy (loosening energy, according to the preferable terminology of Frenkel) equal to zero, the particle loses its contact with neighbors during the characteristic vibration time τ_0, while with $W \to \infty$, the molecule does not change its environment at all.

We will discuss Frenkel's model in view of the experimental data of Hildebrand [211, 212]. Hildebrand gives the experimental data on the viscosity of liquid in the form

$$\mu = \mu_a + \mu_b, \tag{9.7.2}$$

where μ_a is the viscosity that includes only the collective effects in the liquid, and μ_b is the viscosity that includes the "individuality" of the particles. This differentiation calls for additional explanations. For this purpose, consider Fig. 9.8 which gives, by way of example, the viscosity μ, and the components μ_a and μ_b for C_3H_8 ($T = 410.9$ K). Also given in Fig. 9.8 is the critical volume V_c, as well as the volume V_t; when this latter volume is reached (on the side of smaller volumes), the individuality of interacting particles starts showing up. If μ_0 is the viscosity of rarefied gas,

$$\mu_0 = \frac{2(mk_B T)^{1/2}}{3\pi^{3/2}\sigma^2}, \tag{9.7.3}$$

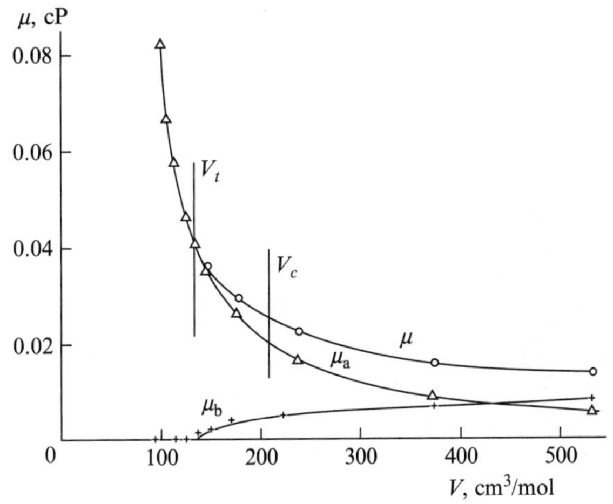

FIGURE 9.8 The viscosity μ and viscosity components μ_a and μ_b, for C_3H_8 vs. volume V.

where σ is the diameter of molecule and m is its mass, then

$$\mu_b = \mu_0 \left(1 - \frac{V_t}{V}\right), \tag{9.7.4}$$

with $\mu_b = 0$ at $V < V_t$.

Therefore, with rarefaction, a term emerges in the total viscosity, which is calculated by the kinetic theory of rarefied gas within the coefficient $1 - V_t/V$. At $V < V_t$, the last "traces" of pair interaction fully disappear, and the viscosity is defined by purely collective effects. The values of volume V_t are found experimentally and tabulated.

As a test of validity of the relation

$$\frac{\mu_b}{\mu_a} = 1 - \frac{V_t}{V}, \tag{9.7.5}$$

Figure 9.9 gives the data of Hildebrand on $(\mu - \mu_a)/\mu_0$ function of V. In so doing, the points in the plot for Ar correspond to the temperature of 323 and 373 K, those for CH_4 to 311 and 341 K, and those for C_3H_8 to 411 and 511 K. The construction of the plot of μ_b/μ_0 as a function of $1 - V_t/V$ leads to a universal dependence for different temperatures. The "collective" viscosity μ_a according to Hildebrand is approximated by the formula

$$\mu_a = \frac{V_0}{B(V - V_0)}, \tag{9.7.6}$$

where V_0 is the eigenvolume of molecules. Table 9.1 contains the data of Hildebrand, according to which it follows that the coefficient B is both universal and little dependent on the sort of matter. The values of B and V_0 are determined by the experimental data from the plot of dependence of the fluidity $1/\mu$ on the volume V in the liquid state region, when the correlations associated with pair interaction are insignificant. The values of V_t, are also determined by the experimental data from Fig. 9.9. Table 9.2 gives the values of these quantities according to Hildebrand's experiments and literature data. The viscosity in liquid and in the transition state may be calculated by the formula

$$\mu = \mu_0 \frac{V - V_t}{V} + \frac{1}{B} \frac{V_0}{V - V_0}. \tag{9.7.7}$$

The viscosity μ_0, calculated by formulas of rarefied gas dynamics, may be expressed in terms of the mean time τ_r between collisions,

$$\tau_r p = \Pi \mu_0, \tag{9.7.8}$$

where Π is the parameter associated with the model of particle interaction; in a first approximation, for the hard-sphere model, $\Pi = 0.8$. From Eq. (9.7.7)—after multiplication of the both sides of this equation by Π/p—and (9.7.8) follows:

$$\Pi \frac{\mu}{p} = \tau_r \frac{V - V_t}{V} + \frac{\Pi}{B} \frac{V_0}{V - V_0} \frac{1}{p}. \tag{9.7.9}$$

where p is the static pressure.

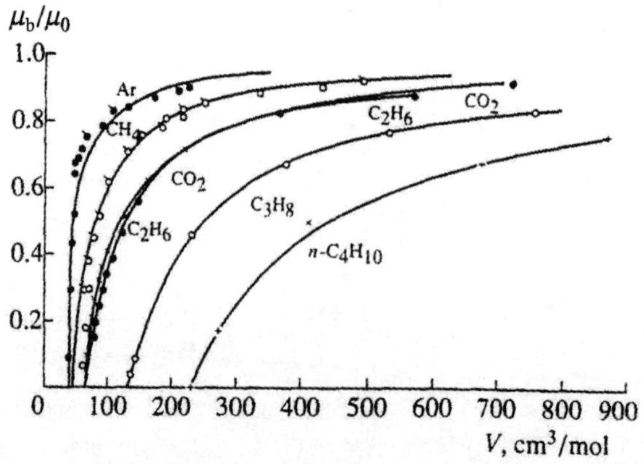

FIGURE 9.9 The data of Hildebrand on $(\mu - \mu_a)/\mu_0$ function of V.

TABLE 9.1 Eigenvolume V_0 and Coefficient B for Different Molecules

Molecule	V_0, cm^3/mol	B, cP^{-1}
C_3H_8	61	18.6
C_5H_{12}	92	18.5
C_6H_{14}	111	18.0
C_7H_{16}	130	18.8
C_8H_{18}	147	17.1
C_9H_{20}	165	16.5
$C_{10}H_{22}$	183	15.2
$C_{12}H_{26}$	217	14.1
$CC1_4$	88.3	17.4
P_4	68.6	16.3
CS_2	50.5	14.0
$CHC1_3$	70	14.0
Hg	14.1	12.6

TABLE 9.2 Parameters of the Liquid State

Molecule	V_0, cm^3/mol	V_t	V_c	B, cP^{-1}
CH_4	32.0	40	99.4	35.6
C_2H_6	44.6	64	148	25.8
C_3H_8	61.5	130	205	22.0
n-C_4H_{10}	78.5	230	255	21.8
CO_2	28.7	60	94	15.7

The time

$$\tau_r^* = \tau_r \frac{V - V_t}{V}, \tag{9.7.10}$$

may be treated as the mean time between particle collisions in the transition mode. With $V < V_t$, the "sample" particle ceases to be predominantly influenced by one of the particles and senses the effect of a self-consistent force alone, in this case, $\tau_r^* = 0$.

We will now introduce the time

$$\tau_{osc} = \frac{\Pi}{B} \frac{V_0}{V - V_0} \frac{1}{p}, \tag{9.7.11}$$

Relation (9.7.9) may be represented in terms of Frenkel's model if one uses the equation of state in the form of Dieterici's equation, which is "the best equation of state, depending on two constants" [36],

$$p(V - b) = RT \exp\left(-\frac{a}{RTV}\right), \tag{9.7.12}$$

The constant a is related to the measure of "cohesive forces" of molecules; and the constant b is proportional to the volume of molecules and taken equal to V_0. The first Dieterici equation describes well the region of the critical state of matter. For example, the critical coefficient s, defined by the thermodynamic parameters of matter p_c, V_c, and T_c in the critical state, $s = RT_c/p_cV_c = 3.695$, which correlates with the experimental data better than $s = 2.67$ according to van der Waals. With high values of V, Dieterici's equation transforms to van der Waals' equation. It follows from Eqs. (9.7.10) and (9.7.11) that

$$\tau_{osc} = \tau_0 e^{W/RT}, \tag{9.7.13}$$

where

$$\tau_0 = \frac{\Pi V_0}{BRT}, \tag{9.7.14}$$

$$W = \frac{a}{V}. \tag{9.7.15}$$

But formula (9.7.13) is Frenkel's formula for the time of residence of molecule in the cell, τ_0 is the particle vibration period, and W is the activation energy (or "loosening energy," according to Frenkel).

As found by Hevesy ([210], p. 104), the activation energy W decreases as the temperature rises. This effect is readily taken into account by the second Dieterici equation,

$$p(V - V_0) = \exp\left(-\frac{a}{RT^{\gamma}V}\right), \tag{9.7.16}$$

which contains the temperature correction T^{γ}, $\gamma = 1.27$. In this case,

$$W = \frac{a}{VT^{0.27}}. \tag{9.7.17}$$

Further detailing of relations (9.7.14) and (9.7.15) is possible within Dieterici's equation. In view of

$$a = 2RT_cV_c, \quad V_0 = V_c/2, \tag{9.7.18}$$

we find that, for the first Dieterici equation,

$$\tau_0 = \frac{\Pi V_c}{2BRT}, \quad W = \frac{2RT_cV_c}{V}. \tag{9.7.19}$$

Needless to say, formulas of the type of (9.7.19) can be expected to coincide with the experimental data only by the order of magnitude. The prospects for rigorous calculation of the above-identified parameters are associated with further development of the trend in statistical physics, based on the calculation of thermodynamic parameters "from the first principles," and with the methods of molecular dynamics. For example, for mercury at $T = 300$ K, Eq. (9.7.19) yields $\tau_0 = 1.4 \times 10^{-13}$ s. Note that the typical value of the particle vibration period in the cell, given by Frenkel ([210], p. 182), is 10^{-13} s.

Therefore, the data of Hildebrand may in fact be treated from the standpoint of Frenkel's theory. Note only that formula (9.7.13) was written by Frenkel for the liquid state (when the pair interaction effects are fully suppressed). In what follows, we will *only* need the Frenkel's model to introduce the *scales* of quantities in the kinetic theory of liquids.

We will now write Eq. (9.7.9) as

$$\Pi\frac{\mu}{p} = \tau_r^* + \tau_0 e^{W/(RT)}, \tag{9.7.20}$$

and introduce the small parameter $\varepsilon = nr_{cell}^3$, where r_{cell} is the characteristic radius of the cell. The parameter ε may be represented as

$$\varepsilon = nr_{cell}^3 = \frac{N}{V}\frac{V - V_0}{N} = \frac{V - V_0}{V}, \tag{9.7.21}$$

where N is the number of particles in the system, and V is the molar volume. We will introduce Hildebrand's parameter $k = (V - V_0)/V_0$ and note that $\varepsilon = k/(k+1)$. Tables 9.3 and 9.4 contain the experimentally obtained values of ε for some liquids, as well as the values of viscosity of liquids and the pressure dependence of the parameter ε.

For liquid He II the eigenvolume of an atom is equal about 9 Å, and in average 45.8 Å^3 of "empty" space corresponds to every atom.

It follows from Tables 9.3 and 9.4 that the quantity ε may be treated as an adequate small parameter in constructing the theory.

TABLE 9.3 Parameter ε and the Viscosity of Some Liquids

Liquid	T, °C	μ, cP	ε
CCl_4	20	0.97	0.0557
	40	0.74	0.0723
C_6H_6	20	0.650	0.0741
	40	0.492	0.1007
CS_2	20	0.366	0.1554
	40	0.349	0.1625
$(C_4F_9)_3N$	20		0.0148
	42.2		0.0393
C_5ONF_{11}	20		0.1007
	42.2		0.1197

TABLE 9.4 The Pressure Dependence of Some Parameters for n-Decanes

p, atm	V, cm^3	$V - V_0$	ε
13.6	232.2	4.821	0.0208
27.2	231.0	4.687	0.0203
54.4	229.9	4.597	0.0200
68.0	227.9	4.396	0.0193
204.0	220.4	3.631	0.0165
340.0	214.8	3.081	0.0143
408.0	212.3	2.829	0.0133

It becomes obvious now that, in constructing the kinetic theory of liquid, one must take into account the following three groups of scales:

(I) the cell scale corresponding to the vibrations of molecule in the "blocked" state, namely, $v_{0\text{cell}}$—the scale of velocity of vibratory motion of particle in the "blocked" state, r_{cell}—the scale of cell size ($r_{\text{cell}} = V_{\text{cell}}^{1/3}$, where V_{cell} is the cell volume), and τ_0—the period of particle vibration in the cell. We will refer to this group of scales as the "τ_0-scale";

(II) the scale of the cell state associated with the characteristic time of particle residence in the cell, or the "τ-scale," namely, $v_{0\tau}$—the scale of particle velocity in the τ-scale (generally speaking, this scale of velocity does not coincide with $v_{0\text{cell}}$, because a high-energy particle is capable of overcoming the energy barrier associated with the activation energy), r_τ—the scale of length in the τ-scale (the characteristic distance covered by a particle during the time of residence in the cell), and τ—the residence time of particle in the cell (for example, under normal conditions for water, $\tau \sim 10^2\tau_0$); and

(III) the hydrodynamic scale that needs no additional explanations, namely, v_L—the hydrodynamic velocity of flow, L—the hydrodynamic length, and τ_L—the hydrodynamic time.

In what follows, these groups of scales will be used to derive the kinetic equation for liquid. We will use the Bogolyubov chain equation for the s-particle distribution function f_s,

$$\frac{\partial f_s}{\partial t} + \sum_{i=1}^{s} \mathbf{v}_i \cdot \frac{\partial f_s}{\partial \mathbf{r}_i} + \sum_{i=1}^{s} \mathbf{F}_i \cdot \frac{\partial f_s}{\partial \mathbf{v}_i} + \sum_{ij=1}^{s} \mathbf{F}_{ij} \cdot \frac{\partial f_s}{\partial \mathbf{v}_i} = -\frac{1}{N} \sum_{i=1}^{s} \sum_{j=s+1}^{N} \int \mathbf{F}_{ij} \cdot \frac{\partial}{\partial \mathbf{v}_i} f_{s+1} \mathrm{d}\Omega_{s+1}, \tag{9.7.22}$$

written in standard notation, where, in particular, $\mathbf{F}_i$, and $\mathbf{F}_{ij}$, are external and internal forces, respectively. We will make Eq. (9.7.22) dimensionless on the τ_0-scale, using the following scales: $n^s v_{0\mathrm{cell}}^{-3s}$ for f_s, and $v_{0\mathrm{cell}}^2/r_{\mathrm{cell}}$, F_0^{ex} for the internal and external forces, respectively. Then, Eq. (9.7.22) assumes the form

$$\frac{\partial \hat{f}_s}{\partial \hat{t}_{\mathrm{cell}}} + \sum_{i=1}^{s} \hat{\mathbf{v}}_{i\mathrm{cell}} \cdot \frac{\partial \hat{f}_s}{\partial \hat{\mathbf{r}}_{i\mathrm{cell}}} + \alpha \sum_{i=1}^{s} \hat{\mathbf{F}}_i \cdot \frac{\partial \hat{f}_s}{\partial \hat{\mathbf{v}}_{i\mathrm{cell}}} + \sum_{ij=1}^{s} \hat{\mathbf{F}}_{ij} \cdot \frac{\partial \hat{f}_s}{\partial \hat{\mathbf{v}}_{i\mathrm{cell}}} = -\varepsilon \frac{1}{N} \sum_{i=1}^{s} \sum_{j=s+1}^{N} \hat{\mathbf{F}}_{ij} \cdot \frac{\partial}{\partial \hat{\mathbf{v}}_{i\mathrm{cell}}} \hat{f}_{s+1} \mathrm{d}\hat{\Omega}_{s+1}, \tag{9.7.23}$$

where $\alpha = F_0^{\mathrm{ex}}/F_{0\mathrm{cell}}$. We will follow the many-scale method and represent the dimensionless s-particle distribution function as a function of three groups of scales and ε,

$$\hat{f}_s = \hat{f}_s(\hat{t}_{\mathrm{cell}}, \hat{\mathbf{r}}_{i\mathrm{cell}}, \hat{\mathbf{v}}_{i\mathrm{cell}}; \hat{t}_\tau, \hat{\mathbf{r}}_{i\tau}, \hat{\mathbf{v}}_{i\tau}; \hat{t}_L, \hat{\mathbf{r}}_{iL}, \hat{\mathbf{v}}_{iL}; \varepsilon), \tag{9.7.24}$$

and introduce the expansion

$$\hat{f}_s = \sum_{\upsilon=0}^{\infty} \hat{f}_s^{\upsilon}(\hat{t}_{\mathrm{cell}}, \hat{\mathbf{r}}_{i\mathrm{cell}}, \hat{\mathbf{v}}_{i\mathrm{cell}}; \hat{t}_\tau, \hat{\mathbf{r}}_{i\tau}, \hat{\mathbf{v}}_{i\tau}; \hat{t}_L, \hat{\mathbf{r}}_{iL}, \hat{\mathbf{v}}_{iL}) \varepsilon^{\upsilon}. \tag{9.7.25}$$

The derivatives in the left-hand part of Eq. (9.7.23) are calculated by the rule of differentiation of composite function,

$$\frac{\mathrm{d}\hat{f}_s}{\mathrm{d}\hat{t}_{\mathrm{cell}}} = \frac{\partial \hat{f}_s}{\partial \hat{t}_{\mathrm{cell}}} + \varepsilon\varepsilon_2 \frac{\partial \hat{f}_s}{\partial \hat{t}_\tau} + \varepsilon\varepsilon_1\varepsilon_2 \frac{\partial \hat{f}_s}{\partial \hat{t}_L}, \tag{9.7.26}$$

$$\frac{\mathrm{d}\hat{f}_s}{\mathrm{d}\hat{\mathbf{r}}_{i\mathrm{cell}}} = \frac{\partial \hat{f}_s}{\partial \hat{\mathbf{r}}_{i\mathrm{cell}}} + \varepsilon\varepsilon_2\varepsilon_3 \frac{\partial \hat{f}_s}{\partial \hat{\mathbf{r}}_{i\tau}} + \varepsilon\varepsilon_1\varepsilon_2\varepsilon_3\varepsilon_4 \frac{\partial \hat{f}_s}{\partial \hat{\mathbf{r}}_{iL}}, \tag{9.7.27}$$

$$\frac{\mathrm{d}\hat{f}_s}{\mathrm{d}\hat{\mathbf{v}}_{i\mathrm{cell}}} = \frac{\partial \hat{f}_s}{\partial \hat{\mathbf{v}}_{i\mathrm{cell}}} + \varepsilon\varepsilon_2 \frac{F_{0\mathrm{cell}}}{F_{0\tau}} \frac{\partial \hat{f}_s}{\partial \hat{\mathbf{v}}_{i\tau}} + \varepsilon\varepsilon_2\varepsilon_4 \frac{F_{0\mathrm{cell}}}{F_{0\tau}} \frac{\partial \hat{f}_s}{\partial \hat{\mathbf{v}}_{iL}}, \tag{9.7.28}$$

using the notation $\varepsilon_1 = \tau/\tau_L$, $\varepsilon_2 = \frac{pV}{RT}$, $\varepsilon_3 = v_{0\mathrm{cell}}/v_{0\tau}$, $\varepsilon_4 = v_{0\tau}/v_{0L}$, $F_{0\tau} = v_{0\tau}/\tau$, $F_{0\mathrm{cell}} = v_{0\mathrm{cell}}^2/r_{\mathrm{cell}}$; the symbol ^ indicates dimensionless quantities. One can introduce the analog of Knudsen number,

$$\mathrm{Kn}_{f\ell} = \frac{\tau v_{0\tau}}{\tau_L v_{0L}} = \varepsilon_1\varepsilon_4, \tag{9.7.29}$$

no restrictions are introduced on the above-identified parameters.

We will now substitute series (9.7.25) in approximation derivatives (9.7.26)–(9.7.28), and the derived expressions—in Eq. (9.7.23). Besides, in the right-hand part of Eq. (9.7.23), we also use the notation $\hat{f}_{s+1}$ in the form of the above-mentioned series. We will equate the expressions for the same values of ε.

For ε^0, we find

$$\frac{\partial \hat{f}_s^0}{\partial \hat{t}_{\mathrm{cell}}} + \sum_{i=1}^{s} \hat{\mathbf{v}}_{i\mathrm{cell}} \cdot \frac{\partial \hat{f}_s^0}{\partial \hat{\mathbf{r}}_{i\mathrm{cell}}} + \sum_{ij=1}^{s} \hat{\mathbf{F}}_{ij} \cdot \frac{\partial \hat{f}_s^0}{\partial \hat{\mathbf{v}}_{i\mathrm{cell}}} + \alpha \sum_{i=1}^{s} \hat{\mathbf{F}}_i \cdot \frac{\partial \hat{f}_s^0}{\partial \hat{\mathbf{v}}_{i\mathrm{cell}}} = 0. \tag{9.7.30}$$

For ε^1, we derive

$$\begin{aligned}
&\frac{\partial \hat{f}_s^1}{\partial \hat{t}_{\text{cell}}}+\sum_{i=1}^{s}\hat{\mathbf{v}}_{i\text{cell}}\cdot\frac{\partial \hat{f}_s^1}{\partial \hat{\mathbf{r}}_{i\text{cell}}}+\sum_{ij=1}^{s}\hat{\mathbf{F}}_{ij}\cdot\frac{\partial \hat{f}_s^1}{\partial \hat{\mathbf{v}}_{i\text{cell}}}+\alpha\sum_{i=1}^{s}\hat{\mathbf{F}}_i\cdot\frac{\partial \hat{f}_s^1}{\partial \hat{\mathbf{v}}_{i\text{cell}}}+\varepsilon_2\frac{\partial \hat{f}_s^0}{\partial \hat{t}_\tau}\\
&+\varepsilon_2\varepsilon_3\sum_{i=1}^{s}\hat{\mathbf{v}}_{i\text{cell}}\cdot\frac{\partial \hat{f}_s^0}{\partial \hat{\mathbf{r}}_{i\tau}}+\varepsilon_2\frac{F_{0\text{cell}}}{F_{0\tau}}\sum_{ij=1}^{s}\hat{\mathbf{F}}_{ij}\cdot\frac{\partial \hat{f}_s^0}{\partial \hat{\mathbf{v}}_{i\tau}}+\alpha\varepsilon_2\frac{F_{0\text{cell}}}{F_{0\tau}}\sum_{i=1}^{s}\hat{\mathbf{F}}_i\cdot\frac{\partial \hat{f}_s^0}{\partial \hat{\mathbf{v}}_{i\tau}}\\
&+\varepsilon_1\varepsilon_2\frac{\partial \hat{f}_s^0}{\partial \hat{t}_L}+\varepsilon_1\varepsilon_2\varepsilon_3\varepsilon_4\sum_{i=1}^{s}\hat{\mathbf{v}}_{i\text{cell}}\cdot\frac{\partial \hat{f}_s^0}{\partial \hat{\mathbf{r}}_{iL}}+\varepsilon_2\varepsilon_4\frac{F_{0\text{cell}}}{F_{0\tau}}\sum_{ij=1}^{s}\hat{\mathbf{F}}_{ij}\cdot\frac{\partial \hat{f}_s^0}{\partial \hat{\mathbf{v}}_{iL}}\\
&+\alpha\varepsilon_2\varepsilon_4\frac{F_{0\text{cell}}}{F_{0\tau}}\sum_{i=1}^{s}\hat{\mathbf{F}}_i\cdot\frac{\partial \hat{f}_s^0}{\partial \hat{\mathbf{v}}_{iL}}=-\frac{1}{N}\sum_{i=1}^{s}\sum_{j=s+1}^{N}\hat{\mathbf{F}}_{ij}\cdot\frac{\partial}{\partial \hat{\mathbf{v}}_{i\text{cell}}}\hat{f}_{s+1}^0\,\mathrm{d}\hat{\Omega}_j.
\end{aligned}\tag{9.7.31}$$

For $s=1$, with due regard for the indistinguishability of particles and for the condition of $s \ll N$, equation (9.7.31) yields

$$\begin{aligned}
&\frac{\partial \hat{f}_1^1}{\partial \hat{t}_{\text{cell}}}+\hat{\mathbf{v}}_{i\text{cell}}\cdot\frac{\partial \hat{f}_1^1}{\partial \hat{\mathbf{r}}_{1\text{cell}}}+\alpha\hat{\mathbf{F}}_1\cdot\frac{\partial \hat{f}_1^1}{\partial \hat{\mathbf{v}}_{1\text{cell}}}+\varepsilon_2\frac{\partial \hat{f}_1^0}{\partial \hat{t}_\tau}+\varepsilon_2\varepsilon_3\hat{\mathbf{v}}_{1\text{cell}}\cdot\frac{\partial \hat{f}_1^0}{\partial \hat{\mathbf{r}}_{1\tau}}\\
&+\alpha\varepsilon_2\frac{F_{0\text{cell}}}{F_{0\tau}}\hat{\mathbf{F}}_1\cdot\frac{\partial \hat{f}_1^0}{\partial \hat{\mathbf{v}}_{1\tau}}+\varepsilon_1\varepsilon_2\frac{\partial \hat{f}_1^0}{\partial \hat{t}_L}+\varepsilon_1\varepsilon_2\varepsilon_3\varepsilon_4\hat{\mathbf{v}}_{1\text{cell}}\cdot\frac{\partial \hat{f}_1^0}{\partial \hat{\mathbf{r}}_{1L}}\\
&+\alpha\varepsilon_2\varepsilon_4\frac{F_{0\text{cell}}}{F_{0\tau}}\hat{\mathbf{F}}_1\cdot\frac{\partial \hat{f}_1^0}{\partial \hat{\mathbf{v}}_{1L}}=-\int\hat{\mathbf{F}}_{12}\cdot\frac{\partial}{\partial \hat{\mathbf{v}}_{1\text{cell}}}\hat{f}_2^0\,\mathrm{d}\hat{\Omega}_2.
\end{aligned}\tag{9.7.32}$$

We will now write Eq. (9.7.23) for $s=2$ and use the derived relation to represent the integral in the right part of Eq. (9.7.32) in the form

$$\begin{aligned}
&-\int\hat{\mathbf{F}}_{12}\cdot\frac{\partial}{\partial \hat{\mathbf{v}}_{1\text{cell}}}\hat{f}_2^0\,\mathrm{d}\hat{\Omega}_2=\int(\hat{\mathbf{v}}_{1\text{cell}}-\hat{\mathbf{v}}_{2\text{cell}})\cdot\frac{\partial \hat{f}_2^0}{\partial \hat{\mathbf{x}}_{12}}\mathrm{d}\hat{\Omega}_2+\int\Bigg[\frac{\partial \hat{f}_2^0}{\partial \hat{t}_{\text{cell}}}\\
&+\hat{\mathbf{v}}_{1\text{cell}}\cdot\frac{\partial \hat{f}_2^0}{\partial \hat{\mathbf{r}}_{1\text{cell}}}+\alpha\hat{\mathbf{F}}_1\cdot\frac{\partial \hat{f}_2^0}{\partial \hat{\mathbf{v}}_{1\text{cell}}}+\alpha\hat{\mathbf{F}}_2\cdot\frac{\partial \hat{f}_2^0}{\partial \hat{\mathbf{v}}_{2\text{cell}}}\Bigg]\mathrm{d}\hat{\Omega}_2+\int\hat{\mathbf{F}}_{21}\cdot\frac{\partial \hat{f}_2}{\partial \hat{\mathbf{v}}_{2\text{cell}}}\mathrm{d}\hat{\Omega}_2,
\end{aligned}\tag{9.7.33}$$

where $\hat{\mathbf{x}}_{12}=\hat{\mathbf{r}}_{1\text{cell}}-\hat{\mathbf{r}}_{2\text{cell}}$ defines distance between some liquid particles designated by the numbers 1 and 2. The last integral terms in Eq. (9.7.33) vanish, because it may be transformed to integrals over the surface that is infinitely removed in the velocity space, where the distribution function goes to zero. Using Eq. (9.7.30) written for $s=1$ (see also Eqs. (7.1.9) and (7.1.12)), as a result the collision integral assumes the form

$$-\int\hat{\mathbf{F}}_{12}\cdot\frac{\partial}{\partial \hat{\mathbf{v}}_{1\text{cell}}}\hat{f}_2^0\,\mathrm{d}\hat{\Omega}_2=A+B+C,\tag{9.7.34}$$

where

$$A=\int\hat{f}_{1,1}^0\frac{\partial \hat{f}_{1,2}^0}{\partial \hat{t}_{\text{cell}}}\mathrm{d}\hat{\Omega}_2,\tag{9.7.35}$$

$$B=\int(\hat{\mathbf{v}}_{1\text{cell}}-\hat{\mathbf{v}}_{2\text{cell}})\cdot\frac{\partial \hat{f}_2^0}{\partial \hat{\mathbf{x}}_{12}}\mathrm{d}\hat{\Omega}_2,\tag{9.7.36}$$

$$C=\int\left[\frac{\mathrm{D}W_2^0}{\mathrm{D}\hat{t}_{\text{cell}}}\right]\mathrm{d}\hat{\Omega}_2,\tag{9.7.37}$$

where the correlation function W_2^0 is introduced

$$f_2^0(t,\mathbf{r}_1,\mathbf{v}_1,\mathbf{r}_2,\mathbf{v}_2)=f_1^0(t,\mathbf{r}_1,\mathbf{v}_1)f_1^0(t,\mathbf{r}_2,\mathbf{v}_2)+W_2^0(t,\mathbf{r}_1,\mathbf{v}_1,\mathbf{r}_2,\mathbf{v}_2).\tag{9.7.38}$$

Integral A lead to appearance of self-consistent forces in the substantive derivatives, integral B may be transformed to the Boltzmann form. Integral C is equal to zero as containing the correlation functions of *zeroth* order in ε.

We return to dimensional variables to derive the kinetic equation that is valid for all three scales in a lower approximation in s

$$\frac{\mathrm{D}f_{1,1}{}^{1}}{\mathrm{D}t}+\frac{\mathrm{D}f_{1,1}{}^{0}}{\mathrm{D}t}=\int\left[f_{1,1}^{0'}f_{1,2}^{0'}-f_{1,1}^{0}f_{1,2}^{0}\right]gb\mathrm{d}b\mathrm{d}\varphi\mathrm{d}\hat{\mathbf{v}}_{2} \tag{9.7.39}$$

Here, the prime indicates the parameters of particles after interaction; integration is performed on the τ_0-scale. Equation (9.7.39) remains linking by the superscript in the first term of the left-hand part of (9.7.39), where the notation

$$\frac{\mathrm{D}}{\mathrm{D}t}\equiv\frac{\partial}{\partial t}+\mathbf{v}_{1}\cdot\frac{\partial}{\partial\mathbf{r}_{1}}+\mathbf{F}_{1}^{\mathrm{sc}}\cdot\frac{\partial}{\partial\mathbf{v}_{1}} \tag{9.7.40}$$

is introduced, $\mathbf{F}_1^{\mathrm{sc}}$ is the effective force acting on the particle on the cell scale

We will treat

$$\frac{\mathrm{D}f_{1,1}^{1}}{\mathrm{D}t}=\frac{\mathrm{D}}{\mathrm{D}t}\left(\frac{\partial f_{1,1}}{\partial\varepsilon}\right)_{\varepsilon=0}, \tag{9.7.41}$$

introduce the assumption of the ε-dependence of the one-particle distribution function on the cell scale in terms of the dynamic variables $\mathbf{r},\mathbf{v},t$ and use the irreversibility of time. Then,

$$\frac{\mathrm{D}f_{1,1}^{1}}{\mathrm{D}t}=-\frac{\mathrm{D}}{\mathrm{D}t}\left[\frac{\varepsilon_{\mathrm{eq}}}{\left(\frac{\partial\varepsilon}{\partial t}\right)_{\varepsilon=0}}\frac{\mathrm{D}f_{1,1}^{0}}{\mathrm{D}t}\right]. \tag{9.7.42}$$

Consider now the relaxation form $(\partial\varepsilon/\partial t)_{\varepsilon=0}$. Physically, this means that it is necessary to find out how many particles will enter the cell per unit time under additional condition of $\varepsilon=0$. Relaxation time

$$\tau_{\mathrm{rel}}^{*}=\varepsilon_{\mathrm{eq}}\left(\frac{\partial\varepsilon}{\partial t}\right)_{\varepsilon=0}^{-1}. \tag{9.7.43}$$

can be introduced in the relaxation equation

$$\frac{\partial\varepsilon}{\partial t}=-\frac{\varepsilon-\varepsilon_{\mathrm{eq}}}{\tau_{\mathrm{rel}}^{*}}. \tag{9.7.44}$$

The non-local parameter $\tau\approx\tau_{\mathrm{rel}}^{*}$ is proportional to the mean time τ^{*} of the particle residence in the cell

$$\tau^{*}=\tau_{0}\mathrm{e}^{W/RT}+\tau_{\mathrm{r}}^{*}. \tag{9.7.45}$$

The second term in the right-hand part of Eq. (9.7.45) includes the correction of the time of particle residence in the cell for pair interaction, when the system in the liquid state approaches the gaseous state. Namely,

$$\tau^{*}=\begin{cases}\tau_{0}\mathrm{e}^{W/RT}+\tau_{\mathrm{r}}^{*}; & (V\geq V_{\mathrm{t}})\\ \tau_{0}\mathrm{e}^{W/RT}; & (V\leq V_{\mathrm{t}}),\end{cases} \tag{9.7.46}$$

$$\tau_{\mathrm{r}}^{*}=\tau_{\mathrm{r}}\frac{V-V_{\mathrm{t}}}{V}.$$

The mean time of the molecule residence τ^{*} in the temporal equilibrium between the activated jumps is varied in the vast limits—from $\tau^{*}\sim10^{-11}$ s for liquids of the low viscosity to many hours for glass [207].

As a result from Eq. (9.7.39) and relations (9.7.42) – (9.7.46), the kinetic equation for liquid takes the form (in what follows, the superscript * is omitted, see also (9.7.39))

$$\frac{\mathrm{D}f_{1,1}}{\mathrm{D}t}-\frac{\mathrm{D}}{\mathrm{D}t}\left(\tau_{\mathrm{rel}}^{*}\frac{\mathrm{D}f_{1,1}}{\mathrm{D}t}\right)=\int\left[f_{1,1}{}'f_{1,2}{}'-f_{1,1}f_{1,2}\right]gb\,\mathrm{d}b\mathrm{d}\varphi\mathrm{d}\mathbf{v}_{2} \tag{9.7.47}$$

The kinetic equation (9.7.47) transforms asymptotically to the generalized Boltzmann equation for rarefied gases. It should be noticed that corrections of local Boltzmann collision integral—related to correlation functions—begin in the following approximation in ε. In this sense situation is analogous to developed plasma kinetic theory where corresponding collision integral incorporates effects of plasma polarization in following approximations.

Further derivation of generalized hydrodynamic Enskog equations may be performed analogously of the previous considerations. In so doing, the parameter τ^* defined by relation (9.7.46) is introduced into the equations (in what follows, we use $\tau \approx \tau^*_{\mathrm{rel}}$).

For a one-component non-reacting medium in the absence of magnetic field, generalized hydrodynamic equations take the following form:

continuity equation,

$$\frac{\partial}{\partial t}\left\{\rho-\tau\left[\frac{\partial\rho}{\partial t}+\frac{\partial}{\partial\mathbf{r}}\cdot(\rho\overline{\mathbf{v}})\right]\right\}+\frac{\partial}{\partial\mathbf{r}}\cdot\left\{\rho\overline{\mathbf{v}}-\tau\left[\frac{\partial}{\partial t}(\rho\overline{\mathbf{v}})+\frac{\partial}{\partial\mathbf{r}}\cdot(\rho\overline{\mathbf{v}\mathbf{v}})-\rho\mathbf{F}^{(1)}\right]\right\}=0, \tag{9.7.48}$$

equation of motion,

$$\frac{\partial}{\partial t}\left\{\rho\overline{\mathbf{v}}-\tau\left[\frac{\partial}{\partial t}(\rho\overline{\mathbf{v}})+\frac{\partial}{\partial\mathbf{r}}\cdot\rho\overline{\mathbf{v}\mathbf{v}}-\rho\mathbf{F}^{(1)}\right]\right\}-\mathbf{F}^{(1)}\left[\rho-\tau\left(\frac{\partial\rho}{\partial t}+\frac{\partial}{\partial\mathbf{r}}\cdot\rho\overline{\mathbf{v}}\right)\right]+\frac{\partial}{\partial\mathbf{r}}\cdot\left\{\rho\overline{\mathbf{v}\mathbf{v}}-\tau\left[\frac{\partial}{\partial t}(\rho\overline{\mathbf{v}\mathbf{v}})+\frac{\partial}{\partial\mathbf{r}}\cdot\rho\overline{(\mathbf{v}\mathbf{v})\mathbf{v}}-\rho\mathbf{F}^{(1)}\overline{\mathbf{v}}-\rho\overline{\mathbf{v}}\mathbf{F}^{(1)}\right]\right\}=0, \tag{9.7.49}$$

energy equation,

$$\begin{aligned}&\frac{\partial}{\partial t}\left\{\frac{\rho\overline{v^2}}{2}-\tau\left[\frac{\partial}{\partial t}\frac{\rho\overline{v^2}}{2}+\frac{\partial}{\partial\mathbf{r}}\cdot\frac{1}{2}\rho\overline{v^2\mathbf{v}}-\mathbf{F}^{(1)}\cdot\rho\overline{\mathbf{v}}\right]\right\}\\&+\frac{\partial}{\partial\mathbf{r}}\cdot\left\{\frac{1}{2}\rho\overline{v^2\mathbf{v}}-\tau\left[\frac{\partial}{\partial t}\frac{1}{2}\rho\overline{v^2\mathbf{v}}+\frac{\partial}{\partial\mathbf{r}}\cdot\frac{1}{2}\rho\overline{v^2\mathbf{v}\mathbf{v}}-\rho\mathbf{F}^{(1)}\cdot\overline{\mathbf{v}\mathbf{v}}-\frac{1}{2}\rho\overline{v^2}\mathbf{F}^{(1)}\right]\right\}\\&-\left\{\rho\mathbf{F}^{(1)}\cdot\overline{\mathbf{v}}-\tau\left[\mathbf{F}^{(1)}\cdot\left(\frac{\partial}{\partial t}(\rho\overline{\mathbf{v}})+\frac{\partial}{\partial\mathbf{r}}\cdot\rho\overline{\mathbf{v}\mathbf{v}}-\rho\mathbf{F}^{(1)}\right)\right]\right\}=0,\end{aligned} \tag{9.7.50}$$

where ρ is the density of liquid, the bar over expressions depending on the own velocity of particles is indicative of averaging over the velocity using the one- particle distribution function.

Note in conclusion that τ transforms to the value of τ_r for rarefied gas, this providing for a "through" description of the liquid-gas system using generalized hydrodynamic equations (9.7.48)–(9.7.50).

Write down now GHE using notations of averaged values, for example

$$\rho^a=\rho-\tau\left[\frac{\partial\rho}{\partial t}+\frac{\partial}{\partial\mathbf{r}}\cdot(\rho\overline{\mathbf{v}})\right]. \tag{9.7.51}$$

Then density ρ^{a} corresponds to mass containing *inside* of control unit volume filled with the particles of *finite* sizes. The other velocity moments have analogous physical sense. As result we have for mentioned case

continuity equation

$$\frac{\partial\rho^{\mathrm{a}}}{\partial t}+\frac{\partial}{\partial\mathbf{r}}\cdot(\rho\overline{\mathbf{v}})^{\mathrm{a}}=0, \tag{9.7.52}$$

momentum equation

$$\frac{\partial}{\partial t}(\rho\overline{\mathbf{v}})^{\mathrm{a}}-\left(\rho\mathbf{F}^{(1)}\right)^{\mathrm{a}}+\frac{\partial}{\partial\mathbf{r}}\cdot(\rho\overline{\mathbf{v}\mathbf{v}})^{\mathrm{a}}=0, \tag{9.7.53}$$

energy equation

$$\frac{\partial}{\partial t}\left(\rho\overline{v^2}\right)^{\mathrm{a}}+\frac{\partial}{\partial\mathbf{r}}\cdot\left(\rho\overline{v^2\mathbf{v}}\right)^{\mathrm{a}}-2\left(\rho\mathbf{F}^{(1)}\cdot\overline{\mathbf{v}}\right)^{\mathrm{a}}=0, \tag{9.7.54}$$

Systems of generalized hydrodynamic equations written in forms (9.7.48)–(9.7.50) and (9.7.52)–(9.7.54) can be considered from positions of theory of super-fluidity developed by Landau [213, 214]. In phenomenological theory of super-fluidity hydrodynamic parameters—in particular density of mass flux and density of quantum liquid—are represented as sum normal and super-fluid components:

$$\mathbf{j}=\mathbf{j}_s+\mathbf{j}_n;\ \rho=\rho_s+\rho_n \tag{9.7.55}$$

Obviously the structure of GHE coincides with Landau equations if correlate now the super-fluid component (s-component) with averaged values in GHE (for example, $\mathbf{j}^{a}$ and ρ^{a}) and normal component (n-component) with fluctuation parameters in GHE (for example $\mathbf{j}^{fl}$ and ρ^{fl}). In this case Eqs. (9.7.52)–(9.7.54) have the character of Euler equations. In general case the separation of normal and super-fluid components of liquid is not possible, the flow of liquid helium is said to consider as non-viscous flow in viscous media.

As it was noticed in Item 9.4 our environment includes many factors, and each person (and by the way electronic devices) on the Earth are permanently influenced by two of them: weather and electromagnetic field. The evolution of the soliton atmospheric wave front leads to appearance of the infrasound oscillations. But let us look now at the structure of GHE (7.1.40)–(7.1.42). The generalized moment and energy equations contain the cross-terms connected with the mass forces (gravitation) and the forces of the electromagnetic origin. It means that the perturbation of the gravitational field in the soliton area (see also Fig. 11.12) leads to the production of the electromagnetic fluctuations (and vice versa) acting on the human being and electronics.

Other important application of GHE (see for example (9.7.48)–(9.7.50)) is connected with the problem of the cold nuclear fusion. In Appendix 8 we revisit the cold fusion phenomenon using the generalized Bolzmann kinetics theory which can represent the non-local physics of this cold fusion phenomenon. This approach can identify the conditions when the cold fusion can take place as the soliton creation under the influence of the intensive sound waves. The vast mathematical modeling leads to affirmation that all parts of soliton move with the same velocity and with the small internal change of the pressure. The zone of the high density is shaped on the soliton's front. It means that the regime of the "acoustic cold fusion" could be realized from the position of the non-local hydrodynamics.

Chapter 10

Strict Theory of Turbulence and Some Applications of the Generalized Hydrodynamic Theory

ABSTRACT

The generalized Boltzmann equation necessarily leads to a new formulation of the hydrodynamic equations, yielding what we call the generalized hydrodynamic equations (GHE). The classical Enskog, Euler, and Navier-Stokes equations of fluid dynamics are special cases of these new equations. The derivation of the GHE is given in the previous chapters. But area of applications of GHE and corresponding principles used for their derivation is much vaster. In this chapter we intend to discuss these new possibilities of GHE in the strict theory of turbulent flows.

Keywords: Strict theory of turbulence, Flow in a cavity, Flow in channels with a step

10.1 ABOUT PRINCIPLES OF CLASSICAL THEORY OF TURBULENT FLOWS

The turbulent fluid motion has been the subject of intense research for over a hundred years because it has numerous applications in aerodynamics, hydraulics, combustion, and explosion processes, and hence is of direct relevance to processes occurring in turbines, engines, compressors, and other modern-day machines. The scientific literature on this subject is enormous, and a detailed analysis of all the existing models is beyond the scope of this book. Here the object is to discuss the currently available turbulence concept in the context of generalized equations of fluid dynamics. In what follows we will discuss "classical" turbulence, usually treated starting from the Navier-Stokes equations, moment methods, and similarity theory. We will also see how this picture corresponds to the generalized Boltzmann kinetics and will try to find out which of the known approaches may be used and which should be abandoned.

It is commonly held that a fully developed turbulence may be characterized by the irregular variation of velocity with time at each point in the flow and that hydrodynamic quantities undergo fluctuations (turbulent ones or pulsations). Their scale varies over the wide range from the external (using the terminology of Ref. [9]) scale comparable to the characteristic flow size, to a small-scale on which the dynamic fluid viscosity begins to dominate.

Because of the major role of the Reynolds criterion in the theory of turbulence, the study of fluid motion on various typical scales crucially depends on the construction of the Reynolds number

$$\mathrm{Re} = \frac{v_l l}{\upsilon_k}.$$

Here, l is the fluctuation scale, v_l is the characteristic velocity, and υ_k is the kinematic viscosity. If $l \sim L$, with L being the typical hydrodynamic size, then the Reynolds number Re is large and the effect of molecular viscosity small—so that one may neglect it altogether and apply the similarity theory (the Kolmogorov-Obukhov law) to get some idea of the fluctuations.

From the large-scale fluctuations, the energy goes (practically undissipated) to the small-scale ones, where viscous dissipation takes place (Richardson model of 1922). And even though the dissipation of mechanical energy ε (falling at the unit mass per unit time) occurs on the least possible scale l_K (referred to below as the Kolmogorov turbulence scale), it is believed that the quantity ε also determines the properties of turbulent motion on larger scales. Between the Kolmogorov (or, using the terminology of Ref. [9], internal) scale l_K and the external scale $l_L \sim L$ there is an inertial interval where the typical size l satisfies the inequality

Unified non-local theory of transport processes. http://dx.doi.org/10.1016/B978-0-444-63478-8.00010-5

$$l_K \ll l \leq l_L.$$

For want of a better model, it is assumed that turbulent motion is described by the same equations of fluid mechanics (Navier-Stokes equations) used for laminar flows, with a consequence that turbulence emerges as a flow instability or, in this particular case, as an instability in the Navier-Stokes flow model. This gives rise to many inconsistencies, however. It is known, for example, that "although no comprehensive theoretical study has thus far been made for flows through a circular pipe, there is compelling evidence that this motion is stable with respect to infinitesimal perturbations (in an absolute as well as a convective sense) for any Reynolds numbers" [9]. This contradicts experimental data.

In 1924, W. Heisenberg published a study on the instability of laminar flows [215]. A year later, E. Noether "published another paper"—we are quoting Heisenberg [216]—"in which she proved with all mathematical rigour that the problem admits of no unstable solutions at all and that a flow must be everywhere stable ... What about the rigorous mathematics then? I think that even now nobody knows what is wrong with Noether's work." It would appear that rather than Noether's mistake, the drawbacks of the Navier-Stokes flow model are to blame.

The notions of averaged and fluctuating motions prompted Reynolds [217] to explicitly isolate the fluctuation terms in the Navier-Stokes equations and to subsequently average them over a certain time interval. But neither this approach nor the later technique of averaging over the masses of liquid volumes (sometimes called Favre averaging [218]) provides close solutions, and indeed neither of them are adequate when it comes to physics because, as we will see below, the Navier-Stokes equations are not written for true physical quantities.

One further approach to the problem involves the evaluation of velocity correlation functions with the aim of establishing the relation between the velocities at two neighboring flow field points within the theory of local turbulence. For example, the simplest correlation function is the second-rank tensor

$$B_{rs} = \langle (v_{2r} - v_{1r})(v_{2s} - v_{1s}) \rangle, \tag{10.1.1}$$

where $\mathbf{v}_1$ and $\mathbf{v}_2$ are the fluid velocities at two neighboring points, and the angle brackets denote time averaging. A question remains, however, what exactly "neighboring points" means and how the time averaging procedure is to be carried out. The theory of correlation functions attracted a great deal of attention after L. Keller and A. Fridman first introduced them into the hydrodynamics of turbulent motions back in 1924.

In 1944, L. Landau gave a comprehensive assessment of this line of research. To quote him from Ref. [9], "One would imagine that in principle it is possible to derive a universal formula, applicable to any turbulent motion, for determining B_{rr}, B_{tt} for all distances r small compared to l_L. In reality, however, such a formula cannot exist at all as the following argument shows. The instantaneous value of the quantity $(v_{2i} - v_{1i})(v_{2k} - v_{1k})$ could in principle be expressed in terms of the energy dissipation ε at the same instant of time t. However, the averaging of these expressions depends significantly on how ε varies in time throughout periods of large-scale (of order l_L) motions. But this variation is different for various specific cases of motion, so the result of such averaging cannot be universal."

One can but agree with this view. To put it another way, if the Kolmogorov scale admits an explicit universal formulation for turbulent fluctuations (as we will show later on), then large-scale fluctuations are determined by solving a specific boundary-value problem.

A Kolmogorov advanced the hypothesis that the statistical regime of the small-scale components is universal and is determined by only two-dimensional parameters, the average rate of energy dissipation ε and the kinematic viscosity υ_K. From dimensional considerations it follows that the Kolmogorov fluctuation scale l_K is of the order of $\upsilon_K{}^{3/4}\varepsilon^{-1/4}$. The following estimations can be introduced:

$$\varepsilon \sim \frac{k_B T}{m\tau}, \quad \left(\frac{p}{\rho}\right)^2 \sim \bar{v}^4, \tag{10.1.2}$$

where $\bar{v}$ is the mean molecular velocity, then

$$l_K \sim \tau \bar{v} \tag{10.1.3}$$

and corresponds to the particle mean free path in a gas. But mean time between collision τ is proportional to viscosity, therefore we can wait that all fluctuations on the Kolmogorov micro-scale of turbulence will be proportional to viscosity.

10.2 THEORY OF TURBULENCE AND GENERALIZED EULER EQUATIONS

We now apply the generalized hydrodynamic equations (GHE) to the theory of turbulence and demonstrate that they enable one to write explicitly the fluctuations of all hydrodynamic quantities on the Kolmogorov turbulence scale l_K. Importantly, these turbulent fluctuations can be tabulated for any type of flow and in this sense can serve as "universal formulas" to use the terminology of monograph [9]. We start by writing down the GHE and, for the sake of simplicity, employ the generalized Euler equations for the special case of a one-component gas flow in a gravitational field. To this end we multiply the generalized Boltzmann equations by the particles' elastic collision invariants ($m, mv, mv^2/2$) and integrate the resulting equations term by term with respect to velocity.

The calculation of the moments using the Maxwellian distribution function yields the system of generalized Euler equations which includes:

the continuity equation

$$\frac{\partial}{\partial t}\left\{\rho-\Pi\frac{\mu}{p}\left[\frac{\partial\rho}{\partial t}+\frac{\partial}{\partial\mathbf{r}}\cdot(\rho\mathbf{v}_0)\right]\right\}+\frac{\partial}{\partial\mathbf{r}}\cdot\left\{\rho\mathbf{v}_0-\Pi\frac{\mu}{p}\left[\frac{\partial}{\partial t}(\rho\mathbf{v}_0)+\frac{\partial}{\partial\mathbf{r}}\cdot\rho\mathbf{v}_0\mathbf{v}_0+\overset{\leftrightarrow}{\mathrm{I}}\cdot\frac{\partial p}{\partial\mathbf{r}}-\rho\mathbf{g}\right]\right\}=0, \tag{10.2.1}$$

the equation of motion

$$\begin{aligned}&\frac{\partial}{\partial t}\left\{\rho\mathbf{v}_0-\Pi\frac{\mu}{p}\left[\frac{\partial}{\partial t}(\rho\mathbf{v}_0)+\frac{\partial}{\partial\mathbf{r}}\cdot\rho\mathbf{v}_0\mathbf{v}_0+\frac{\partial p}{\partial\mathbf{r}}-\rho\mathbf{g}\right]\right\}\\&-\mathbf{g}\left[\rho-\Pi\frac{\mu}{p}\left(\frac{\partial\rho}{\partial t}+\frac{\partial}{\partial\mathbf{r}}\cdot(\rho\mathbf{v}_0)\right)\right]+\frac{\partial}{\partial\mathbf{r}}\cdot\{\rho\mathbf{v}_0\mathbf{v}_0\\&+p\overset{\leftrightarrow}{\mathrm{I}}-\Pi\frac{\mu}{p}\left[\frac{\partial}{\partial t}(\rho\mathbf{v}_0\mathbf{v}_0+p\overset{\leftrightarrow}{\mathrm{I}})+\frac{\partial}{\partial\mathbf{r}}\cdot\rho(\mathbf{v}_0\mathbf{v}_0)\mathbf{v}_0\right.\\&\left.+2\overset{\leftrightarrow}{\mathrm{I}}\left[\frac{\partial}{\partial\mathbf{r}}\cdot(p\mathbf{v}_0)\right]+\frac{\partial}{\partial\mathbf{r}}\cdot\left(\overset{\leftrightarrow}{\mathrm{I}}p\mathbf{v}_0\right)-\mathbf{g}\mathbf{v}_0\rho-\mathbf{v}_0\mathbf{g}\rho\right]\}=0,\end{aligned} \tag{10.2.2}$$

and the equation of energy

$$\begin{aligned}&\frac{\partial}{\partial t}\left\{\frac{\rho v_0^2}{2}+\frac{3}{2}p-\Pi\frac{\mu}{p}\left[\frac{\partial}{\partial t}\left(\frac{\rho v_0^2}{2}+\frac{3}{2}p\right)\right.\right.\\&\left.\left.+\frac{\partial}{\partial\mathbf{r}}\cdot\left(\frac{1}{2}\rho v_0^2\mathbf{v}_0+\frac{5}{2}p\mathbf{v}_0\right)-\mathbf{g}\cdot\rho\mathbf{v}_0\right]\right\}\\&+\frac{\partial}{\partial\mathbf{r}}\cdot\left\{\frac{1}{2}\rho v_0^2\mathbf{v}_0+\frac{5}{2}p\mathbf{v}_0-\Pi\frac{\mu}{p}\left[\frac{\partial}{\partial t}\left(\frac{1}{2}\rho v_0^2\mathbf{v}_0\right.\right.\right.\\&\left.+\frac{5}{2}p\mathbf{v}_0\right)+\frac{\partial}{\partial\mathbf{r}}\cdot\left(\frac{1}{2}\rho v_0^2\mathbf{v}_0\mathbf{v}_0+\frac{7}{2}p\mathbf{v}_0\mathbf{v}_0+\frac{1}{2}pv_0^2\overset{\leftrightarrow}{\mathrm{I}}\right.\\&\left.\left.\left.+\frac{5p^2}{2\rho}\overset{\leftrightarrow}{\mathrm{I}}\right)-\rho\mathbf{g}\cdot\mathbf{v}_0\mathbf{v}_0-p\mathbf{g}\cdot\overset{\leftrightarrow}{\mathrm{I}}-\frac{1}{2}\rho v_0^2\mathbf{g}-\frac{3}{2}\mathbf{g}p\right]\right\}\\&-\left\{\rho\mathbf{g}\cdot\mathbf{v}_0-\Pi\frac{\mu}{p}\left[\mathbf{g}\cdot\left(\frac{\partial}{\partial t}(\rho\mathbf{v}_0)+\frac{\partial}{\partial\mathbf{r}}\cdot\rho\mathbf{v}_0\mathbf{v}_0\right.\right.\right.\\&\left.\left.\left.+\frac{\partial}{\partial\mathbf{r}}\cdot p\overset{\leftrightarrow}{\mathrm{I}}-\rho\mathbf{g}\right)\right]\right\}=0,\end{aligned} \tag{10.2.3}$$

where $\overset{\leftrightarrow}{\mathrm{I}}$ is the unit tensor. In the simplest case in the hydrodynamic approximation $\tau^{(0)}=\Pi\mu/p$; for the hard sphere model, $\Pi=0.8$. The influence of the relaxation time can be taken into account by changing of the Π parameter.

We next introduce ρ_∞, v_∞, p_∞, and μ_∞ as the density, velocity, pressure, and viscosity scales, respectively. We take the characteristic dimension to be L, and the time scale, L/v_∞. Then the dimensionless equation of continuity takes the form

$$\begin{aligned}&\frac{\partial}{\partial\hat{t}}\left\{\hat{\rho}-\Pi\frac{\hat{\mu}\mu_\infty v_\infty}{\hat{p}\,p_\infty L}\left[\frac{\partial\hat{\rho}}{\partial\hat{t}}+\frac{\partial}{\partial\hat{\mathbf{r}}}\cdot(\hat{\rho}\hat{\mathbf{v}}_0)\right]\right\}+\frac{\partial}{\partial\hat{\mathbf{r}}}\cdot\{\hat{\rho}\hat{\mathbf{v}}_0\\&-\Pi\frac{\hat{\mu}\mu_\infty}{\hat{p}p_\infty}\frac{v_\infty}{L}\left[\frac{\partial}{\partial\hat{t}}(\hat{\rho}\hat{\mathbf{v}}_0)+\frac{\partial}{\partial\hat{\mathbf{r}}}\cdot\hat{\rho}\hat{\mathbf{v}}_0\hat{\mathbf{v}}_0+\frac{p_\infty}{\rho_\infty v_\infty^2}\overset{\leftrightarrow}{\mathrm{I}}\cdot\frac{\partial\hat{p}}{\partial\hat{\mathbf{r}}}-\frac{Lg}{v_\infty^2}\hat{\rho}\hat{\mathbf{g}}\right]\}=0.\end{aligned} \tag{10.2.4}$$

Dimensionless combinations of the scale quantities introduced above form the similarity criteria

$$\frac{\mu_\infty v_\infty}{p_\infty L}=\frac{\mu_\infty}{Lv_\infty\rho_\infty}\frac{\rho_\infty v_\infty^2}{p_\infty}=\mathrm{Re}^{-1}\mathrm{Eu}^{-1},\quad\frac{v_\infty^2}{Lg}=\mathrm{Fr}$$

Thus, the continuity equation (10.2.4) contains the Reynolds, Euler, and Froude similarity criteria, and may be rewritten as

$$\begin{aligned}&\frac{\partial}{\partial\hat{t}}\left\{\hat{\rho}-\Pi\frac{\hat{\mu}}{\hat{p}}\frac{1}{\mathrm{EuRe}}\left[\frac{\partial\hat{\rho}}{\partial\hat{t}}+\frac{\partial}{\partial\hat{\mathbf{r}}}\cdot(\hat{\rho}\hat{\mathbf{v}}_0)\right]\right\}+\frac{\partial}{\partial\hat{\mathbf{r}}}\cdot\left\{\hat{\rho}\hat{\mathbf{v}}_0\right.\\&\left.-\Pi\frac{\hat{\mu}\mu_\infty}{\hat{p}p_\infty}\frac{1}{\mathrm{EuRe}}\left[\frac{\partial}{\partial\hat{t}}(\hat{\rho}\hat{\mathbf{v}}_0)+\frac{\partial}{\partial\hat{\mathbf{r}}}\cdot\hat{\rho}\hat{\mathbf{v}}_0\hat{\mathbf{v}}_0+\mathrm{Eu}\overset{\leftrightarrow}{\mathrm{I}}\cdot\frac{\partial\hat{p}}{\partial\hat{\mathbf{r}}}-\frac{1}{\mathrm{Fr}}\hat{\rho}\hat{\mathbf{g}}\right]\right\}=0.\end{aligned} \tag{10.2.5}$$

In a similar fashion we write the dimensionless equation of motion

$$\begin{aligned}&\frac{\partial}{\partial \hat{t}}\left\{\hat{\rho}\hat{\mathbf{v}}_0-\Pi\frac{\hat{\mu}\,\mathrm{Eu}^{-1}}{\hat{p}\ \mathrm{Re}}\left[\frac{\partial}{\partial \hat{t}}(\hat{\rho}\hat{\mathbf{v}}_0)+\frac{\partial}{\partial \hat{\mathbf{r}}}\cdot\hat{\rho}\hat{\mathbf{v}}_0\hat{\mathbf{v}}_0+\mathrm{Eu}\frac{\partial \hat{p}}{\partial \hat{\mathbf{r}}}-\frac{1}{\mathrm{Fr}}\hat{\rho}\hat{\mathbf{g}}\right]\right\}\\&-\frac{1}{\mathrm{Fr}}\hat{\mathbf{g}}\left[\hat{\rho}-\Pi\frac{\hat{\mu}\,\mathrm{Eu}^{-1}}{\hat{p}\ \mathrm{Re}}\left(\frac{\partial \hat{\rho}}{\partial \hat{t}}+\frac{\partial}{\partial \hat{\mathbf{r}}}\cdot(\hat{\rho}\hat{\mathbf{v}}_0)\right)\right]+\frac{\partial}{\partial \hat{\mathbf{r}}}\cdot\left\{\hat{\rho}\hat{\mathbf{v}}_0\hat{\mathbf{v}}_0\cdot\right.\\&+\mathrm{Eu}\hat{p}\overleftrightarrow{\mathrm{I}}-\Pi\frac{\hat{\mu}\,\mathrm{Eu}^{-1}}{\hat{p}\ \mathrm{Re}}\left[\frac{\partial}{\partial \hat{t}}\left(\hat{\rho}\hat{\mathbf{v}}_0\hat{\mathbf{v}}_0+\mathrm{Eu}\hat{p}\overleftrightarrow{\mathrm{I}}\right)+\frac{\partial}{\partial \hat{\mathbf{r}}}\cdot\hat{\rho}(\hat{\mathbf{v}}_0\hat{\mathbf{v}}_0)\hat{\mathbf{v}}_0\right.\\&\left.\left.+2\overleftrightarrow{\mathrm{I}}\left[\frac{\partial}{\partial \hat{\mathbf{r}}}\cdot(\hat{p}\hat{\mathbf{v}}_0)\right]\mathrm{Eu}+\mathrm{Eu}\frac{\partial}{\partial \hat{\mathbf{r}}}\cdot\left(\overleftrightarrow{\mathrm{I}}\hat{p}\hat{\mathbf{v}}_0\right)-\frac{1}{\mathrm{Fr}}\hat{\rho}\hat{\mathbf{g}}\hat{\mathbf{v}}_0-\frac{1}{\mathrm{Fr}}\rho\hat{\mathbf{v}}_0\hat{\mathbf{g}}\right]\right\}=0,\end{aligned}\tag{10.2.6}$$

and the dimensionless equation of energy

$$\begin{aligned}&\frac{\partial}{\partial \hat{t}}\left\{\frac{\hat{\rho}\hat{v}_0^2}{2}+\mathrm{Eu}\frac{3}{2}\hat{p}-\Pi\frac{\hat{\mu}\,\mathrm{Eu}^{-1}}{\hat{p}\ \mathrm{Re}}\left[\frac{\partial}{\partial \hat{t}}\left(\frac{\hat{\rho}\hat{v}_0^2}{2}+\mathrm{Eu}\frac{3}{2}\hat{p}\right)\right.\right.\\&\left.\left.+\frac{\partial}{\partial \hat{\mathbf{r}}}\cdot\left(\frac{1}{2}\hat{\rho}\hat{v}_0^2\hat{\mathbf{v}}_0+\mathrm{Eu}\frac{5}{2}\hat{p}\hat{\mathbf{v}}_0\right)-\frac{1}{\mathrm{Fr}}\hat{\mathbf{g}}\cdot\hat{\rho}\hat{\mathbf{v}}_0\right]\right\}\\&+\frac{\partial}{\partial \hat{\mathbf{r}}}\cdot\left\{\frac{1}{2}\rho\hat{v}_0^2\hat{\mathbf{v}}_0+\mathrm{Eu}\frac{5}{2}\hat{p}\hat{\mathbf{v}}_0-\Pi\frac{\hat{\mu}\,\mathrm{Eu}^{-1}}{\hat{p}\ \mathrm{Re}}\left[\frac{\partial}{\partial t}\left(\frac{1}{2}\hat{\rho}\hat{v}_0^2\hat{\mathbf{v}}_0\right.\right.\right.\\&\left.+\mathrm{Eu}\frac{5}{2}\hat{p}\hat{\mathbf{v}}_0\right)+\frac{\partial}{\partial \hat{\mathbf{r}}}\cdot\left(\frac{1}{2}\hat{\rho}\hat{v}_0^2\hat{\mathbf{v}}_0\hat{\mathbf{v}}_0+\frac{7}{2}\mathrm{Eu}\hat{p}\hat{\mathbf{v}}_0\hat{\mathbf{v}}_0+\frac{1}{2}\mathrm{Eu}\hat{p}\hat{v}_0^2\overleftrightarrow{\mathrm{I}}\right.\\&\left.+5\mathrm{Eu}^2\frac{\hat{p}^2}{2\hat{\rho}}\overleftrightarrow{\mathrm{I}}\right)-\frac{1}{\mathrm{Fr}}\hat{\rho}\hat{\mathbf{g}}\cdot\hat{\mathbf{v}}_0\hat{\mathbf{v}}_0-\frac{\mathrm{Eu}}{\mathrm{Fr}}\hat{p}\hat{\mathbf{g}}\cdot\overleftrightarrow{\mathrm{I}}\\&\left.\left.-\frac{1}{\mathrm{Fr}}\frac{1}{2}\hat{\rho}\hat{v}_0^2\hat{\mathbf{g}}-\frac{\mathrm{Eu}}{\mathrm{Fr}}\frac{3}{2}\hat{\mathbf{g}}\hat{p}\right]\right\}\\&-\left\{\frac{1}{\mathrm{Fr}}\hat{\rho}\hat{\mathbf{g}}\cdot\hat{\mathbf{v}}_0-\Pi\frac{\hat{\mu}\,\mathrm{Eu}^{-1}}{\hat{p}\ \mathrm{Re}}\left[\frac{1}{\mathrm{Fr}}\hat{\mathbf{g}}\cdot\left(\frac{\partial}{\partial \hat{t}}(\hat{\rho}\hat{\mathbf{v}}_0)+\frac{\partial}{\partial \hat{\mathbf{r}}}\cdot\hat{\rho}\hat{\mathbf{v}}_0\hat{\mathbf{v}}_0\right.\right.\right.\\&\left.\left.\left.+\frac{\partial}{\partial \hat{\mathbf{r}}}\cdot\hat{p}\overleftrightarrow{\mathrm{I}}\,\mathrm{Eu}-\hat{\rho}\hat{\mathbf{g}}\frac{1}{\mathrm{Fr}}\right)\right]\right\}=0.\end{aligned}\tag{10.2.7}$$

Equations (10.2.5)–(10.2.7) are notable for their structure. All the generalized equations of Euler fluid dynamics contain the Reynolds, Euler, and Froude numbers (similarity criteria). Naturally, the inclusion of forces of electromagnetic origin would lead to additional similarity criteria. For each hydrodynamic quantity—density, energy, and momentum as well as their fluxes—there is a corresponding temporally and spatially fluctuating term which is proportional to Re^{-1} and, hence, to the viscosity. For small-scale fluctuations (i.e. smaller characteristic dimension l in the Reynolds number), viscosity increases in importance and starts to determine ε, the dissipation of the mechanical energy.

Introduce the Kolmogorov scale length l_{K} as:

$$l_{\mathrm{K}}=\left(\upsilon_{\mathrm{K}}{}^3/\varepsilon\right)^{1/4}.\tag{10.2.8}$$

and find using Eq. (10.1.3) that the length l_{K} is of the mean free path order. The fluctuation terms thus determine turbulent Kolmogorov scale fluctuations (small-scale fluctuations or, using the computational hydrodynamics term, sub-grid turbulence) which are of a universal nature and not problem specific.

To fully understand the situation, however, the following questions remain to be answered:

(1) Are there no contradictions in the system of fluctuations introduced in this way? In other words, is the set of fluctuations self-consistent?

(2) With a system of base (independent) fluctuations on hand, is it possible to derive explicit expressions for other hydrodynamic quantities and their moments?

(3) What do the GHE for averaged quantities look like and how does the procedure for obtaining the averaged equations agree with the Reynolds procedure familiar from the theory of turbulence?

In answering the above questions, the generalized Euler equations for one-component gas will be employed for the sake of clarity. Implicit in the following analysis will be the fact, already noted above, that we are dealing with small-scale fluctuations. The equations to be investigated are:

continuity equation

$$\frac{\partial}{\partial t}\left\{\rho-\tau\left(\frac{\partial \rho}{\partial t}+\frac{\partial}{\partial \mathbf{r}}\cdot(\rho\mathbf{v}_0)\right)\right\}+\frac{\partial}{\partial \mathbf{r}}\cdot\left\{\rho\mathbf{v}_0-\tau\left(\frac{\partial}{\partial t}(\rho\mathbf{v}_0)\right.\right.$$
$$\left.\left.+\frac{\partial}{\partial \mathbf{r}}\cdot(\rho\mathbf{v}_0\mathbf{v}_0)+\frac{\partial p}{\partial \mathbf{r}}-\rho\mathbf{g}\right)\right\}=0, \tag{10.2.9}$$

equation of motion

$$\frac{\partial}{\partial t}\left\{\rho v_{0\beta}-\tau\left[\frac{\partial}{\partial t}(\rho v_{0\beta})+\frac{\partial}{\partial r_\alpha}(p\delta_{\alpha\beta}+\rho v_{0\alpha})-\rho g_\beta\right]\right\}$$
$$-\left\{\rho-\tau\left[\frac{\partial \rho}{\partial t}+\frac{\partial}{\partial r_\alpha}(\rho v_{0\alpha})\right]\right\}g_\beta+\frac{\partial}{\partial r_\alpha}\left\{p\delta_{\alpha\beta}+\rho v_{0\alpha}v_{0\beta}\right.$$
$$-\tau\left[\frac{\partial}{\partial t}(p\delta_{\alpha\beta}+\rho v_{0\alpha}v_{0\beta})+\frac{\partial}{\partial r_\gamma}(p\delta_{\alpha\gamma}v_{0\beta}+p v_{0\alpha}\delta_{\beta\gamma}\right.$$
$$\left.\left.+p v_{0\gamma}\delta_{\alpha\beta}+\rho v_{0\alpha}v_{0\beta}v_{0\gamma})-g_\alpha\rho v_{0\beta}-g_\beta\rho v_{0\alpha}\right]\right\}=0, \tag{10.2.10}$$

where we employ the Einstein summation rule for recurrent subscripts $\alpha,\beta,\gamma=1,2,3$ referring to components of vectors in the Cartesian coordinate system, and equation of energy

$$\frac{\partial}{\partial t}\left\{3p+\rho v_0^2-\tau\left[\frac{\partial}{\partial t}(3p+\rho v_0^2)+\frac{\partial}{\partial \mathbf{r}}\cdot(\mathbf{v}_0(\rho v_0^2+5p))\right.\right.$$
$$\left.\left.-2\mathbf{g}\cdot\rho\mathbf{v}_0\right]\right\}+\frac{\partial}{\partial \mathbf{r}}\cdot\left\{\mathbf{v}_0(\rho v_0^2+5p)-\tau\left[\frac{\partial}{\partial t}(\mathbf{v}_0(\rho v_0^2+5p))\right.\right.$$
$$+\frac{\partial}{\partial \mathbf{r}}\cdot\left[\overset{\leftrightarrow}{\mathrm{I}}p v_0^2+\rho v_0^2\mathbf{v}_0\mathbf{v}_0+7p\mathbf{v}_0\mathbf{v}_0+5\overset{\leftrightarrow}{\mathrm{I}}\frac{p^2}{\rho}\right]$$
$$\left.\left.-2\rho\mathbf{v}_0\mathbf{v}_0\cdot\mathbf{g}-5p\overset{\leftrightarrow}{\mathrm{I}}\cdot\mathbf{g}-\rho v_0^2\overset{\leftrightarrow}{\mathrm{I}}\cdot\mathbf{g}\right]\right\}$$
$$-2\mathbf{g}\cdot\left\{\rho\mathbf{v}_0-\tau\left[\frac{\partial}{\partial t}(\rho\mathbf{v}_0)+\frac{\partial p}{\partial \mathbf{r}}+\frac{\partial}{\partial \mathbf{r}}\cdot(\rho\mathbf{v}_0\mathbf{v}_0)-\rho\mathbf{g}\right]\right\}=0. \tag{10.2.11}$$

To calculate hydrodynamic fluctuations, the Reynolds procedure will be employed. Thus, for example, the product of the true density ρ and the true velocity v_0 can be used to obtain the fluctuation quantity $\mathrm{v}_0^{\mathrm{f}}$. Indeed, we have

$$\rho\mathbf{v}_0=(\rho^{\mathrm{a}}+\rho^{\mathrm{f}})(\mathbf{v}_0^{\mathrm{a}}+\mathbf{v}_0^{\mathrm{f}}), \tag{10.2.12}$$

where the superscript "a" denotes the average hydrodynamic quantities. Ignoring the fluctuation terms squared and keeping only first-order small quantities in relations of type (Eq. (10.2.12)) we find

$$(\rho\mathbf{v}_0)^{\mathrm{f}}=\rho\mathbf{v}_0-\rho^{\mathrm{a}}\mathbf{v}_0^{\mathrm{a}}=\rho^{\mathrm{a}}\mathbf{v}_0^{\mathrm{f}}+\rho^{\mathrm{f}}\mathbf{v}_0^{\mathrm{a}}. \tag{10.2.13}$$

Thus one obtains

$$\mathbf{v}_0^{\mathrm{f}}=\frac{(\rho\mathbf{v}_0)^{\mathrm{f}}-\rho^{\mathrm{f}}\mathbf{v}_0^{\mathrm{a}}}{\rho^{\mathrm{a}}}. \tag{10.2.14}$$

From the continuity equation (10.2.9) we have

$$\rho^{\mathrm{f}}=\tau\left\{\frac{\partial \rho}{\partial t}+\frac{\partial}{\partial \mathbf{r}}\cdot(\rho\mathbf{v}_0)\right\}, \tag{10.2.15}$$

$$(\rho\mathbf{v}_0)^{\mathrm{f}}=\tau\left\{\frac{\partial}{\partial t}(\rho\mathbf{v}_0)+\frac{\partial}{\partial \mathbf{r}}\cdot(\rho\mathbf{v}_0\mathbf{v}_0)+\frac{\partial p}{\partial \mathbf{r}}-\rho\mathbf{g}\right\}, \tag{10.2.16}$$

and therefore from Eq. (10.2.14) it follows:

$$\mathbf{v}_0^{\mathrm{f}}=\tau\left\{\frac{\partial \mathbf{v}_0}{\partial t}+\left(\mathbf{v}_0\cdot\frac{\partial}{\partial \mathbf{r}}\right)\mathbf{v}_0+\frac{1}{\rho}\frac{\partial p}{\partial \mathbf{r}}-\mathbf{g}\right\}. \tag{10.2.17}$$

Using Eq. (10.2.10), we find that the fluctuation of the combined hydrodynamic quantity $p\delta_{\alpha\beta}+\rho v_{0\alpha}v_{0\beta}$ is given by

$$\begin{aligned}(p\delta_{\alpha\beta}+\rho v_{0\alpha}v_{0\beta})^{\mathrm{f}}=\tau\Bigg[&\frac{\partial}{\partial t}(p\delta_{\alpha\beta}+\rho v_{0\alpha}v_{0\beta})\\&+\frac{\partial}{\partial r_\gamma}(p\delta_{\alpha\beta}v_{0\beta}+p v_{0\alpha}\delta_{\beta\gamma}+p v_{0\gamma}\delta_{\alpha\beta}+\rho v_{0\alpha}v_{0\beta}v_{0\gamma})\\&-g_\alpha\rho v_{0\beta}-g_\beta\rho v_{0\alpha}\Bigg].\end{aligned}\tag{10.2.18}$$

With the help of Eq. (10.2.18) we obtain:

$$(3p+\rho v_0^2)^{\mathrm{f}}=\tau\left[\frac{\partial}{\partial t}(3p+\rho v_0^2)+\frac{\partial}{\partial\mathbf{r}}\cdot\left(\mathbf{v}_0\left(\rho v_0^2+5p\right)\right)-2\mathbf{g}\cdot\rho\mathbf{v}_0\right].\tag{10.2.19}$$

We now proceed to calculate $(\rho v_0^2)^{\mathrm{f}}$:

$$\begin{aligned}(\rho v_0^2)^{\mathrm{f}}&=\rho v_0^2-\rho^{\mathrm{a}}v_0^{\mathrm{a}2}=(\rho^{\mathrm{a}}+\rho^{\mathrm{f}})(\mathbf{v}_0^{\mathrm{a}}+\mathbf{v}_0^{\mathrm{f}})^2-\rho^{\mathrm{a}}v_0^{\mathrm{a}2}\\&\cong(\rho^{\mathrm{a}}+\rho^{\mathrm{f}})(v_0^{\mathrm{a}}+2\mathbf{v}_0^{\mathrm{a}}\cdot\mathbf{v}_0^{\mathrm{f}})-\rho^{\mathrm{a}}v_0^{\mathrm{a}}\cong\rho^{\mathrm{f}}v_0^{\mathrm{a}}+2\rho^{\mathrm{a}}\mathbf{v}_0^{\mathrm{a}}\cdot\mathbf{v}_0^{\mathrm{f}},\end{aligned}\tag{10.2.20}$$

which, when combined with Eqs. (10.2.15) and (10.2.17), yields

$$(\rho v_0^2)^{\mathrm{f}}=\tau\left\{v_0^{\mathrm{a}2}\left[\frac{\partial\rho}{\partial t}+\frac{\partial}{\partial\mathbf{r}}\cdot(\rho\mathbf{v}_0)\right]+2\rho^{\mathrm{a}}\mathbf{v}_0^{\mathrm{a}}\cdot\left[\frac{\partial\mathbf{v}_0}{\partial t}+\left(\mathbf{v}_0\cdot\frac{\partial}{\partial\mathbf{r}}\right)\mathbf{v}_0+\frac{1}{\rho}\frac{\partial p}{\partial\mathbf{r}}-\mathbf{g}\right]\right\}.\tag{10.2.21}$$

From Eq. (10.2.19) fluctuation of static pressure is found

$$\left[3p+\rho v_0^2\right]^{\mathrm{f}}=3p^{\mathrm{f}}+(\rho v_0^2)^{\mathrm{f}},\tag{10.2.22}$$

$$\begin{aligned}3p^{\mathrm{f}}=\tau\Bigg\{&\frac{\partial}{\partial t}(3p+\rho v_0^2)+\frac{\partial}{\partial\mathbf{r}}\cdot\left(\mathbf{v}_0\left(\rho v_0^2+5p\right)\right)-2\mathbf{g}\cdot\rho\mathbf{v}_0\\&-v_0^{\mathrm{a}}\left[\frac{\partial\rho}{\partial t}+\frac{\partial}{\partial\mathbf{r}}\cdot(\rho\mathbf{v}_0)\right]-2\rho^{\mathrm{a}}\mathbf{v}_0^{\mathrm{a}}\cdot\left[\frac{\partial\mathbf{v}_0}{\partial t}+\left(\mathbf{v}_0\cdot\frac{\partial}{\partial\mathbf{r}}\right)\mathbf{v}_0\right.\\&\left.\left.+\frac{1}{\rho}\frac{\partial p}{\partial\mathbf{r}}-\mathbf{g}\right]\right\}.\end{aligned}\tag{10.2.23}$$

Neglecting squared fluctuations—this tantamount of omitting terms proportional to τ^2—we find:

$$\begin{aligned}3p^{\mathrm{f}}&=\tau\left\{\frac{\partial}{\partial t}(3p+\rho v_0^2)+\frac{\partial}{\partial\mathbf{r}}\cdot\left(\mathbf{v}_0\left(\rho v_0^2+5p\right)\right)\right.\\&\quad\left.-v_0^2\left[\frac{\partial\rho}{\partial t}+\frac{\partial}{\partial\mathbf{r}}\cdot(\rho\mathbf{v}_0)\right]-2\rho\mathbf{v}_0\cdot\left[\frac{\partial\mathbf{v}_0}{\partial t}+\left(\mathbf{v}_0\cdot\frac{\partial}{\partial\mathbf{r}}\right)\mathbf{v}_0+\frac{1}{\rho}\frac{\partial p}{\partial\mathbf{r}}\right]\right\}\\&=\tau\left\{3\frac{\partial p}{\partial t}+\frac{\partial}{\partial\mathbf{r}}\cdot(\rho v_0^2\mathbf{v}_0)-v_0^2\frac{\partial}{\partial\mathbf{r}}\cdot(\rho\mathbf{v}_0)-2\rho\mathbf{v}_0\cdot\left(\mathbf{v}_0\cdot\frac{\partial}{\partial\mathbf{r}}\right)\mathbf{v}_0\right.\\&\quad\left.+5\frac{\partial}{\partial\mathbf{r}}\cdot(p\mathbf{v}_0)-2\mathbf{v}_0\cdot\frac{\partial p}{\partial\mathbf{r}}\right\}\\&=\tau\left\{3\frac{\partial p}{\partial t}+\left(\rho\mathbf{v}_0\cdot\frac{\partial}{\partial\mathbf{r}}\right)v_0^2-2\rho\mathbf{v}_0\cdot\left(\mathbf{v}_0\cdot\frac{\partial}{\partial\mathbf{r}}\right)\mathbf{v}_0\right.\\&\quad\left.+3\frac{\partial}{\partial\mathbf{r}}\cdot(p\mathbf{v}_0)+2p\left(\frac{\partial}{\partial\mathbf{r}}\cdot\mathbf{v}_0\right)\right\}=\tau\left\{3\frac{\partial p}{\partial t}+3\frac{\partial}{\partial\mathbf{r}}\cdot(p\mathbf{v}_0)+2p\frac{\partial}{\partial\mathbf{r}}\cdot\mathbf{v}_0\right\}\end{aligned}\tag{10.2.24}$$

or

$$p^{\mathrm{f}}=\tau\left\{\frac{\partial p}{\partial t}+\frac{\partial}{\partial\mathbf{r}}\cdot(p\mathbf{v}_0)+\frac{2}{3}p\frac{\partial}{\partial\mathbf{r}}\cdot\mathbf{v}_0\right\}.\tag{10.2.25}$$

Then formulated procedure allows to obtain fluctuations of the zeroth-order velocity's moment (Eq. (10.2.15)), fluctuations of moments of the first-order (Eq. (10.2.17)), and the second-order ones (Eq. (10.2.20)). Dependent fluctuations (for example, p^{f}) should be calculated using independent fluctuations of hydrodynamic values. Senior velocity's moment in the system of Eq. (10.2.9)–(10.2.11) is $\mathbf{v}_0(\rho v_0^2+5p)$. Let us find the fluctuation of this moment.

$$\left[\mathbf{v}_0\left(\rho v_0^2+5p\right)\right]^{\mathrm{f}}=\mathbf{v}_0\left(\rho v_0^2+5p\right)-\mathbf{v}_0^{\mathrm{a}}\left(\rho^{\mathrm{a}}v_0^{\mathrm{a}}+5p^{\mathrm{a}}\right). \tag{10.2.26}$$

The corresponding result is of principal significance and we deliver its derivation in details.

$$\begin{aligned}\left[\mathbf{v}_0\left(\rho v_0^2+5p\right)\right]^{\mathrm{f}}&\cong\left(\mathbf{v}_0^{\mathrm{a}}+\mathbf{v}_0^{\mathrm{f}}\right)\left(\rho^{\mathrm{a}}+\rho^{\mathrm{f}}\right)\left(v_0^{\mathrm{a}2}+2\mathbf{v}_0^{\mathrm{a}}\cdot\mathbf{v}_0^{\mathrm{f}}\right)\\&\quad+5\left(\mathbf{v}_0^{\mathrm{a}}+\mathbf{v}_0^{\mathrm{f}}\right)\left(p^{\mathrm{a}}+p^{\mathrm{f}}\right)-\mathbf{v}_0^{\mathrm{a}}\left(\rho^{\mathrm{a}}v_0^{\mathrm{a}2}+5p^{\mathrm{a}}\right)\\&\cong\left(\mathbf{v}_0^{\mathrm{a}}\rho^{\mathrm{f}}+\mathbf{v}_0^{\mathrm{f}}\rho^{\mathrm{a}}\right)v_0^{\mathrm{a}2}+5\mathbf{v}_0^{\mathrm{f}}p^{\mathrm{a}}+5p^{\mathrm{f}}\mathbf{v}_0^{\mathrm{a}}+2\mathbf{v}_0^{\mathrm{a}}\cdot\mathbf{v}_0^{\mathrm{f}}\mathbf{v}_0^{\mathrm{a}}\rho^{\mathrm{a}}.\end{aligned} \tag{10.2.27}$$

Use now Eqs. (10.2.15), (10.2.17), and (10.2.25):

$$\begin{aligned}\left[\mathbf{v}_0\left(\rho v_0^2+5p\right)\right]^{\mathrm{f}}\cong\tau\Bigg\{&\mathbf{v}_0^{\mathrm{a}}v_0^{\mathrm{a}2}\left(\frac{\partial\rho}{\partial t}+\frac{\partial}{\partial\mathbf{r}}\cdot(\rho\mathbf{v}_0)\right)+\left(5p^{\mathrm{a}}\right.\\&\left.+\rho^{\mathrm{a}}v_0^{\mathrm{a}2}\right)\left[\frac{\partial\mathbf{v}_0}{\partial t}+\left(\mathbf{v}_0\cdot\frac{\partial}{\partial\mathbf{r}}\right)\mathbf{v}_0+\frac{1}{\rho}\frac{\partial p}{\partial\mathbf{r}}-\mathbf{g}\right]\\&+5\mathbf{v}_0^{\mathrm{a}}\left(\frac{\partial p}{\partial t}+\frac{\partial}{\partial\mathbf{r}}\right)\cdot(p\mathbf{v}_0)+\frac{2}{3}p\frac{\partial}{\partial\mathbf{r}}\cdot\mathbf{v}_0\\&+2\mathbf{v}_0^{\mathrm{a}}\rho^{\mathrm{a}}\mathbf{v}_0^{\mathrm{a}}\cdot\left[\frac{\partial\mathbf{v}_0}{\partial t}+\left(\mathbf{v}_0\cdot\frac{\partial}{\partial\mathbf{r}}\right)\mathbf{v}_0+\frac{1}{\rho}\frac{\partial p}{\partial\mathbf{r}}-\mathbf{g}\right]\Bigg\}.\end{aligned} \tag{10.2.28}$$

Consider first of all the temporal part of Eq. (10.2.28):

$$\begin{aligned}&\tau\left\{\mathbf{v}_0^{\mathrm{a}}v_0^{\mathrm{a}2}\frac{\partial\rho}{\partial t}+\left(5p^{\mathrm{a}}+\rho^{\mathrm{a}}v_0^{\mathrm{a}2}\right)\frac{\partial\mathbf{v}_0}{\partial t}+5\mathbf{v}_0^{\mathrm{a}}\frac{\partial p}{\partial t}+2\rho_a\mathbf{v}_0^{\mathrm{a}}\mathbf{v}_0^{\mathrm{a}}\cdot\frac{\partial\mathbf{v}_0}{\partial t}\right\}\\&\cong\tau\frac{\partial}{\partial t}\left\{\mathbf{v}_0\left(5p+\rho v_0^2\right)\right\}.\end{aligned} \tag{10.2.29}$$

The temporal part of fluctuation of the considered moment, calculated with help of fluctuations of the minor moments, coincides with the time derivative of this moment $\mathbf{v}_0(5p+\rho v_0^2)$ in right-hand side of Eq. (10.2.11). Let us go now to the spatial part of fluctuation connected with the derivative $\partial/\partial\mathbf{r}$,

$$\begin{aligned}M_{v^3}^{<}=\tau\Bigg\{&\mathbf{v}_0^{\mathrm{a}}v_0^{\mathrm{a}2}\frac{\partial}{\partial\mathbf{r}}\cdot(\rho\mathbf{v}_0)+\left(5p^{\mathrm{a}}+\rho^{\mathrm{a}}v_0^{\mathrm{a}2}\right)\left[\left(\mathbf{v}_0\cdot\frac{\partial}{\partial\mathbf{r}}\right)\mathbf{v}_0\right.\\&\left.+\frac{1}{\rho}\frac{\partial p}{\partial\mathbf{r}}-\mathbf{g}\right]+5\mathbf{v}_0^{\mathrm{a}}\left[\frac{\partial}{\partial\mathbf{r}}\cdot(p\mathbf{v}_0)+\frac{2}{3}p\frac{\partial}{\partial\mathbf{r}}\cdot\mathbf{v}_0\right]\\&+2\mathbf{v}_0^{\mathrm{a}}\rho^{\mathrm{a}}\mathbf{v}_0^{\mathrm{a}}\cdot\left[\frac{\partial\mathbf{v}_0}{\partial t}+\left(\mathbf{v}_0\cdot\frac{\partial}{\partial\mathbf{r}}\right)\mathbf{v}_0+\frac{1}{\rho}\frac{\partial p}{\partial\mathbf{r}}-\mathbf{g}\right]\Bigg\},\end{aligned} \tag{10.2.30}$$

where $M_{v^3}^{<}$ denotes fluctuation of the third-order moment fond with help of minor moments. The corresponding term M_{v^3} in Eq. (10.2.11) has the form

$$\begin{aligned}M_{v^3}=\tau\Bigg\{&\frac{\partial}{\partial\mathbf{r}}\cdot\left[\overleftrightarrow{\mathrm{I}}pv_0^2+\rho v_0^2\mathbf{v}_0\mathbf{v}_0+7p\mathbf{v}_0\mathbf{v}_0+5\overleftrightarrow{\mathrm{I}}\frac{p^2}{\rho}\right]\\&-2\rho\mathbf{v}_0\mathbf{v}_0\cdot\mathbf{g}-5p\overleftrightarrow{\mathrm{I}}\cdot\mathbf{g}-\rho v_0^2\overleftrightarrow{\mathrm{I}}\cdot\mathbf{g}\Bigg\}.\end{aligned} \tag{10.2.31}$$

Then the spatial fluctuation of the third-order velocity's moments contains the moments of the fourth-order. Are they coincided, $M_{v^3}^{<}$ and M_{v^3}? It turns out that not. Investigate discrepancy, but before transform the derivative.

$$\begin{aligned}\frac{\partial}{\partial\mathbf{r}}\cdot\left(\rho v_0^2\mathbf{v}_0\mathbf{v}_0\right)&=\rho v_0^2\frac{\partial}{\partial\mathbf{r}}\cdot(\mathbf{v}_0\mathbf{v}_0)+\left(\mathbf{v}_0\mathbf{v}_0\cdot\frac{\partial}{\partial\mathbf{r}}\right)\left(\rho v_0^2\right)\\&=\rho v_0^2\left(\mathbf{v}_0\cdot\frac{\partial}{\partial\mathbf{r}}\right)\mathbf{v}_0+\rho v_0^2\mathbf{v}_0\left(\frac{\partial}{\partial\mathbf{r}}\cdot\mathbf{v}_0\right)+\left(\mathbf{v}_0\mathbf{v}_0\cdot\frac{\partial}{\partial\mathbf{r}}\right)\left(\rho v_0^2\right)\\&=\rho v_0^2\left(\mathbf{v}_0\cdot\frac{\partial}{\partial\mathbf{r}}\right)\mathbf{v}_0+v_0^2\mathbf{v}_0\frac{\partial}{\partial\mathbf{r}}\cdot(\rho\mathbf{v}_0)+\left(\mathbf{v}_0\mathbf{v}_0\cdot\frac{\partial}{\partial\mathbf{r}}\right)\left(\rho v_0^2\right)-v_0^2\mathbf{v}_0\left(\mathbf{v}_0\cdot\frac{\partial}{\partial\mathbf{r}}\right)\rho.\end{aligned} \tag{10.2.32}$$

Let us consider the difference of two last terms in Eq. (10.2.32):

$$\begin{aligned}
&\left(\mathbf{v}_0\mathbf{v}_0\cdot\frac{\partial}{\partial\mathbf{r}}\right)(\rho v_0^2)-v_0^2\mathbf{v}_0\left(\mathbf{v}_0\cdot\frac{\partial}{\partial\mathbf{r}}\right)\rho\\
&=\left(\mathbf{v}_0\mathbf{v}_0v_0^2\cdot\frac{\partial}{\partial\mathbf{r}}\right)\rho+\left(\rho\mathbf{v}_0\mathbf{v}_0\cdot\frac{\partial}{\partial\mathbf{r}}\right)v_0^2-v_0^2\mathbf{v}_0\left(\mathbf{v}_0\cdot\frac{\partial}{\partial\mathbf{r}}\right)\rho\\
&=\left(\rho\mathbf{v}_0\mathbf{v}_0\cdot\frac{\partial}{\partial\mathbf{r}}\right)v_0^2\\
&=2\rho\mathbf{v}_0\mathbf{v}_0\cdot\left[\left(\mathbf{v}_0\cdot\frac{\partial}{\partial\mathbf{r}}\right)\mathbf{v}_0\right].
\end{aligned}\tag{10.2.33}$$

Now let us find the discrepancy of values $M_{v^3}-M_{v^3}^{<}$, using Eqs. (10.2.31) and (10.2.32):

$$\begin{aligned}
M_{v^3}-M_{v^3}^{<}=&\tau\left\{\frac{\partial}{\partial\mathbf{r}}\cdot\left[\overleftrightarrow{I}pv_0^2+7p\mathbf{v}_0\mathbf{v}_0+5\overleftrightarrow{I}\frac{p^2}{\rho}\right]\right.\\
&-5p\left[\left(\mathbf{v}_0\cdot\frac{\partial}{\partial\mathbf{r}}\right)\mathbf{v}_0+\frac{1}{\rho}\frac{\partial p}{\partial\mathbf{r}}\right]-v_0^2\frac{\partial p}{\partial\mathbf{r}}-5\mathbf{v}_0\left[\frac{\partial}{\partial\mathbf{r}}\cdot(p\mathbf{v}_0)\right.\\
&\left.\left.+\frac{2}{3}p\frac{\partial}{\partial\mathbf{r}}\cdot\mathbf{v}_0\right]-2\mathbf{v}_0\mathbf{v}_0\cdot\frac{\partial p}{\partial\mathbf{r}}\right\}\\
=&\tau\left\{\frac{\partial}{\partial\mathbf{r}}\cdot\left[\overleftrightarrow{I}pv_0^2+7p\mathbf{v}_0\mathbf{v}_0+5\overleftrightarrow{I}\frac{p^2}{\rho}\right]-5p\left(\mathbf{v}_0\cdot\frac{\partial}{\partial\mathbf{r}}\right)\mathbf{v}_0\right.\\
&\left.-5\frac{p}{\rho}\frac{\partial p}{\partial\mathbf{r}}-v_0^2\frac{\partial p}{\partial\mathbf{r}}-5\mathbf{v}_0\frac{\partial}{\partial\mathbf{r}}\cdot(p\mathbf{v}_0)-\frac{10}{3}\mathbf{v}_0p\frac{\partial}{\partial\mathbf{r}}\cdot\mathbf{v}_0-2\mathbf{v}_0\mathbf{v}_0\cdot\frac{\partial p}{\partial\mathbf{r}}\right\}\\
=&\tau\left\{p\frac{\partial}{\partial\mathbf{r}}v_0^2+7p\frac{\partial}{\partial\mathbf{r}}\cdot(\mathbf{v}_0\mathbf{v}_0)+7\left(\mathbf{v}_0\mathbf{v}_0\cdot\frac{\partial}{\partial\mathbf{r}}\right)p\right.\\
&+5p\frac{\partial}{\partial\mathbf{r}}\left(\frac{p}{\rho}\right)-5p\left(\mathbf{v}_0\cdot\frac{\partial}{\partial\mathbf{r}}\right)\mathbf{v}_0-5\mathbf{v}_0\frac{\partial}{\partial\mathbf{r}}\cdot(p\mathbf{v}_0)\\
&\left.-\frac{10}{3}p\mathbf{v}_0\frac{\partial}{\partial\mathbf{r}}\cdot\mathbf{v}_0-2\mathbf{v}_0\mathbf{v}_0\cdot\frac{\partial p}{\partial\mathbf{r}}\right\}\\
=&\tau\left\{p\frac{\partial}{\partial\mathbf{r}}v_0^2+2p\left(\mathbf{v}_0\cdot\frac{\partial}{\partial\mathbf{r}}\right)\mathbf{v}_0+2p\mathbf{v}_0\left(\frac{\partial}{\partial\mathbf{r}}\cdot\mathbf{v}_0\right)\right.\\
&\left.+5p\frac{\partial}{\partial\mathbf{r}}\left(\frac{p}{\rho}\right)-\frac{10}{3}p\mathbf{v}_0\frac{\partial}{\partial\mathbf{r}}\cdot\mathbf{v}_0\right\}\\
=&\tau p\left\{\frac{\partial}{\partial\mathbf{r}}\left[5\frac{p}{\rho}+v_0^2\right]+2\left(\mathbf{v}_0\cdot\frac{\partial}{\partial\mathbf{r}}\right)\mathbf{v}_0-\frac{4}{3}\mathbf{v}_0\left(\frac{\partial}{\partial\mathbf{r}}\cdot\mathbf{v}_0\right)\right\}.
\end{aligned}\tag{10.2.34}$$

Then the chain of equalities (10.2.34) leads to result

$$M_{v^3}-M_{v^3}^{<}=\tau p\left\{\frac{\partial}{\partial\mathbf{r}}\left[5\frac{p}{\rho}+v_0^2\right]+2\left(\mathbf{v}_0\cdot\frac{\partial}{\partial\mathbf{r}}\right)\mathbf{v}_0-\frac{4}{3}\mathbf{v}_0\left(\frac{\partial}{\partial\mathbf{r}}\cdot\mathbf{v}_0\right)\right\}.\tag{10.2.35}$$

Discrepancy (10.2.35) can be written in another form having transparent physical sense. Introduce velocity distortion tensor $\overleftrightarrow{S}$ with components

$$S_{ij}=\frac{1}{2}\left(\frac{\partial v_{0i}}{\partial r_j}+\frac{\partial v_{0j}}{\partial r_i}\right)-\frac{1}{3}\delta_{ij}\frac{\partial}{\partial\mathbf{r}}\cdot\mathbf{v}_0,\quad(i,j=1,2,3).\tag{10.2.36}$$

Then j- component of vector equal to the scalar product of vector $\mathbf{v}_0$ by tensor $\overleftrightarrow{S}$, has the form

$$\left(\mathbf{v}_0\cdot\overleftrightarrow{S}\right)_j=\frac{1}{4}\sum_i\frac{\partial}{\partial r_j}v_{0i}^2+\frac{1}{2}\sum_i\left(v_{0i}\frac{\partial}{\partial r_i}\right)v_{0j}-\frac{1}{3}v_{0j}\frac{\partial}{\partial\mathbf{r}}\cdot\mathbf{v}_0.\tag{10.2.37}$$

Then,

$$\mathbf{v}_0\cdot\overleftrightarrow{S}=\frac{1}{4}\frac{\partial}{\partial\mathbf{r}}v_0^2+\frac{1}{2}\left(\mathbf{v}_0\cdot\frac{\partial}{\partial\mathbf{r}}\right)\mathbf{v}_0-\frac{1}{3}\mathbf{v}_0\frac{\partial}{\partial\mathbf{r}}\cdot\mathbf{v}_0.\tag{10.2.38}$$

Using Eq. (10.2.38) one obtains the form for discrepancy

$$M_{\nu 3} - M_{\nu 3}^{<} = 2\tau p\left[\frac{5}{2}\frac{\partial}{\partial \mathbf{r}}\left(\frac{p}{\rho}\right) + 2\mathbf{v}_0 \cdot \overleftrightarrow{S}\right]. \tag{10.2.39}$$

Now the possibility appears to write down the generalized Euler equations in the terminology of the averaged values.

$$\frac{\partial \rho^{\mathrm{a}}}{\partial t} + \frac{\partial}{\partial \mathbf{r}} \cdot (\rho \mathbf{v}_0)^{\mathrm{a}} = 0, \tag{10.2.40}$$

$$\frac{\partial}{\partial t}(\rho \mathbf{v}_0)^{\mathrm{a}} + \frac{\partial}{\partial \mathbf{r}} \cdot \left[p^{\mathrm{a}} \overleftrightarrow{\mathrm{I}} + (\rho \mathbf{v}_0 \mathbf{v}_0)^{\mathrm{a}}\right] = \rho^{\mathrm{a}} \mathbf{g}. \tag{10.2.41}$$

$$\begin{aligned} &\frac{\partial}{\partial t}\left(3p + \rho v_0^2\right)^{\mathrm{a}} + \frac{\partial}{\partial \mathbf{r}} \cdot \left[\mathbf{v}_0\left(\rho v_0^2 + 5p\right)\right] - 2\mathbf{g} \cdot (\rho \mathbf{v}_0)^{\mathrm{a}} \\ &= 2\frac{\partial}{\partial \mathbf{r}} \cdot \left\{\tau p\left[\frac{5}{2}\frac{\partial}{\partial \mathbf{r}}\left(\frac{p}{\rho}\right) + 2\mathbf{v}_0 \cdot \overleftrightarrow{S}\right]\right\}. \end{aligned} \tag{10.2.42}$$

From the system of generalized Euler equations (10.2.40)–(10.2.42) we conclude the following:

(1) The formulation of the hydrodynamic equations in terms of the average quantities is the objective of the "classical" theory of turbulence. However, a rigorous approach based on the generalized Euler equations leads to a residual (with respect to true quantities) on the right of the equation of energy (10.2.42).

(2) The residual

$$\Pi^e = 2\frac{\partial}{\partial \mathbf{r}} \cdot \left\{\tau p\left[\frac{5}{2}\frac{\partial}{\partial \mathbf{r}}\frac{p}{\rho} + 2\mathbf{v}_0 \cdot \overleftrightarrow{S}\right]\right\}. \tag{10.2.43}$$

in Eq. (10.2.43) turns out to be proportional to τp and hence to the viscosity. If one puts then the set of equations (10.2.40)–(10.2.42) reduces formally to the Euler equations for averaged quantities. It follows that the residual Π^e, which reflects the variation in space of the thermal-energy and shear-energy dissipation, stimulates the development of turbulence in the physical system under study.

$$\Pi^e = 0 \tag{10.2.44}$$

(3) The so-called "soft" boundary conditions commonly imposed at the output of the computational flow region follow from condition (10.2.43)

$$\frac{\partial}{\partial \mathbf{r}} \cdot \left\{\mu\left[\frac{5}{2}\frac{\partial}{\partial \mathbf{r}}\frac{p}{\rho} + 2\mathbf{v}_0 \cdot \overleftrightarrow{S}\right]\right\} = 0. \tag{10.2.45}$$

(4) From the position of kinetic theory the appearance of mentioned discrepancy is connected with approximation of the distribution function (DF) with the help of local Maxwellian function. This problem does not exist for generalized Enskog equations written for genuine DF. But situation is more complicated. Even if relation (10.2.44) is fulfilled, Eqs. (10.2.40)–(10.2.42) do not reduce exactly to the classical Euler equations even under condition (10.2.44) because the average of the product of hydrodynamic quantities is not equal to the product of their averages.

Consequently, this system of equations contains more unknowns (namely, $\rho^{\mathrm{a}}, (\rho \mathbf{v}_0)^{\mathrm{a}}, p^{\mathrm{a}}, (\rho \mathbf{v}_0 \mathbf{v}_0)^{\mathrm{a}}$, and $[\mathbf{v}_0(\rho v_0^2 + 5p)]^{\mathrm{a}}$) than equations, thus presenting the typical problem of the classical theory of turbulence, which consists in closing the moment equations.

In our theory the solution of this problem consists simply in returning to equations written for real, genuine hydrodynamic values. And only in the situation when micro-scale turbulent fluctuations are absent or (it is the same) the averaged product of hydrodynamic values is equal to the product of averaged values. Really in this case when the system of equations contains more unknown values—we reach classical Euler equations.

The theory presented here overcomes this problem by simply reverting to the formulation of the hydrodynamic equations in terms of the true quantities. And it is only in the case when turbulent fluctuations are completely absent or, equivalently, when the average product of hydrodynamic quantities is equal to the product of their averages, that we arrive at the classical form of the Euler and, of course, Navier-Stokes equations. Thus, the classical Euler and Navier-Stokes equations are not written for true quantities, and it is physically meaningless to employ the formal Reynolds procedure to try to "extract" small-scale fluctuations from these equations.

Let be

$$\frac{\partial^{n_i} u_i}{\partial t^{n_i}} = F_i\left(t, x_1, \ldots, x_n, u_1, \ldots, u_N, \ldots, \frac{\partial^k u_j}{\partial t^{k_0}\partial x_1^{k_1}\ldots\partial x_n^{k_n}}, \ldots\right) \tag{10.2.46}$$

a system of differential equations in N unknown functions $u_1, u_2, \ldots, u_N$ with partial derivatives with respect to independent variables $t, x_1, x_2, \ldots, x_n$ for which the following conditions fulfilled

$$k_0 + k_1 + \ldots + k_n = k \leq n_i, k_0 < n_i, \quad i, j = 1, 2, \ldots, N. \tag{10.2.47}$$

For $t = t_0$ the "initial conditions" are formulated for all unknown functions and their derivatives

$$\frac{\partial^k u_i}{\partial t^k} = \varphi_i^{(k)}(x_1, x_2, \ldots, x_n), \quad k = 0, 1, 2, \ldots, n_i - 1, \tag{10.2.48}$$

all functions $\varphi_i^{(k)}(x_1, x_2, \ldots, x_n)$ are given in the same domain G_0 of the space $(x_1, x_2, \ldots, x_n)$. In this case the following Kovalevskaya theorem can be formulated for Cauchy problem:

If all functions F_i are analytical in a vicinity of point $\left(t_0, x_1^0, \ldots, x_n^0, \ldots, \varphi^0_{j,k_0,k_1,\ldots,k_n}, \ldots\right)$ and all functions $\varphi_i^{(k)}(x_1, x_2, \ldots, x_n)$ are analytical in the vicinity of point $(x_1, x_2, \ldots, x_n)$, then Cauchy problem has analytical unique solution in class of the analytical functions in vicinity of the mentioned point $(x_1, x_2, \ldots, x_n)$.

The system of GHE (cf. the generalized Euler equations) can be written in the form (10.2.46), satisfy the conditions (10.2.48) and—in reasonable assumptions concerning functions F_i—satisfy the conditions of Kovalevskaya theorem.

System of fluctuations of hydrodynamic values can be used for investigation of the flow stability. Basic guidelines of stability can be obtained on the basement of theory of differential equations with small parameter in front of derivative, [219].

Let us consider the differential equation of the form

$$\tau\frac{dA}{dt} = f(A, t), \tag{10.2.49}$$

where τ is a small parameter. If $\tau = 0$ from equation

$$f(A, t) = 0, \tag{10.2.50}$$

follows algebraic solution

$$A_i = \varphi_i(t), \tag{10.2.51}$$

where i is number of corresponding solution.

Construct the curve $\varphi_i(t)$ in coordinate system A, t and suppose that trajectory of $A(t)$ begins in a point placed under the curve $\varphi_i(t)$ where—for the sake of definiteness—$f > 0$. Then because τ is a small parameter, the derivative dA/dt is very great and integral curve going out from the mentioned point practically parallel to A-axis, directs to algebraic solution $\phi_i(t)$. After that this curve turns along t-axis because for $f = 0$ the derivative dA/dt is equal to zero.

If outgoing point of trajectory is placed above the curve $A_i = \varphi_i(t)$ in domain where $f < 0$, the integral curve will come down to the curve $A_i = \varphi_i(t)$.

Then in the considered scenario of behavior of integral curves and function $f(A, t) = 0$, exists obviously the stability of solution of the differential equation (10.2.49). In this case can be introduced criterion of stability of the form

$$\frac{\partial f}{\partial A} < 0, A_i = \varphi_i(t). \tag{10.2.52}$$

If vice versa function f is negative under the curve $A_i = \varphi_i(t)$ and positive above this curve, the solution (10.2.51) is not stable.

The formal possibility exists of applications of these qualitative considerations for investigation of the problems of stability in the frame of GHE. Notice really that in one-dimensional non-stationary case the velocity fluctuation has the form (see Table 10.1):

$$\tau\frac{du}{dt} = u - u^{a} - \tau\left(\frac{1}{\rho}\frac{\partial p}{\partial x} + u\frac{\partial u}{\partial x}\right). \tag{10.2.53}$$

TABLE 10.1 Fluctuations of Hydrodynamic Quantities on the Kolmogorov Scale in the Framework of the Generalized Euler Equations

No.	Hydrodynamic Value A	Fluctuation A^{f}
1	ρ	$\tau\left[\frac{\partial \rho}{\partial t}+\frac{\partial}{\partial \mathbf{r}}\cdot(\rho \mathbf{v}_0)\right]$
2	$\rho \mathbf{v}_0$	$\tau\left[\frac{\partial}{\partial t}(\rho \mathbf{v}_0)+\frac{\partial}{\partial \mathbf{r}}\cdot(\rho \mathbf{v}_0\mathbf{v}_0)+\frac{\partial p}{\partial \mathbf{r}}-\rho \mathbf{g}\right]$
3	$\mathbf{v}_0$	$\tau\left[\frac{\partial \mathbf{v}_0}{\partial t}+\left(\mathbf{v}_0\cdot\frac{\partial}{\partial \mathbf{r}}\right)\mathbf{v}_0+\frac{1}{\rho}\frac{\partial p}{\partial \mathbf{r}}-\mathbf{g}\right]$
4	$p\delta_{\alpha\beta}+\rho v_{0\alpha}v_{0\beta}$	$\tau\left[\frac{\partial}{\partial t}(p\delta_{\alpha\beta}+\rho v_{0\alpha}v_{0\beta})+\frac{\partial}{\partial r_\gamma}(p v_{0\alpha}\delta_{\beta\gamma}+\right.$ $+p v_{0\beta}\delta_{\alpha\gamma}+p v_{0\gamma}\delta_{\alpha\beta}+\rho v_{0\alpha}v_{0\beta}v_{0\gamma})$ $\left.-g_\beta \rho v_{0\alpha}-g_\alpha \rho v_{0\beta}\right]$
5	$3p+\rho v_0^2$	$\tau\left[\frac{\partial}{\partial t}(3p+\rho v_0^2)\right.$ $\left.+\frac{\partial}{\partial \mathbf{r}}\cdot(\mathbf{v}_0(\rho v_0^2+5p))-2\mathbf{g}\cdot\rho \mathbf{v}_0\right]$
6	p	$\tau\left[\frac{\partial p}{\partial t}+\frac{\partial}{\partial \mathbf{r}}\cdot(p\mathbf{v}_0)+\frac{2}{3}p\frac{\partial}{\partial \mathbf{r}}\cdot\mathbf{v}_0\right]$
7	$\mathbf{v}_0(\rho v_0^2+5p)$	$\tau\left\{\frac{\partial}{\partial t}[\mathbf{v}_0(\rho v_0^2+5p)]+\frac{\partial}{\partial \mathbf{r}}\cdot\left[\rho v_0^2\mathbf{v}_0\mathbf{v}_0+.+\overleftrightarrow{\mathrm{I}}p v_0^2+7p\mathbf{v}_0\mathbf{v}_0+5\overleftrightarrow{\mathrm{I}}\frac{p^2}{\rho}\right]\right.$ $-2\rho\mathbf{v}_0\mathbf{v}_0\cdot\mathbf{g}-5p\overleftrightarrow{\mathrm{I}}\cdot\mathbf{g}-\rho v_0^2\overleftrightarrow{\mathrm{I}}\cdot\mathbf{g}-p\left[\frac{\partial}{\partial \mathbf{r}}\left(5\frac{p}{\rho}+v_0^2\right)\right.$ $\left.\left.+2\left(\mathbf{v}_0\cdot\frac{\partial}{\partial \mathbf{r}}\right)\mathbf{v}_0-\frac{4}{3}\mathbf{v}_0\left(\frac{\partial}{\partial \mathbf{r}}\cdot\mathbf{v}_0\right)\right]\right\}$

The corresponding function f is written as

$$f=u-u^{\mathrm{a}}-\tau\left(\frac{1}{\rho^{\mathrm{a}}}\frac{\partial p^{\mathrm{a}}}{\partial x}+u^{\mathrm{a}}\frac{\partial u^{\mathrm{a}}}{\partial x}\right), \tag{10.2.54}$$

taking into account that the genuine and averaged values differ about $O(\tau)$. Obviously the sign of function f in a vicinity of solution

$$u=u^{\mathrm{a}}+\tau\left(\frac{1}{\rho^{\mathrm{a}}}\frac{\partial p^{\mathrm{a}}}{\partial x}+u^{\mathrm{a}}\frac{\partial u^{\mathrm{a}}}{\partial x}\right) \tag{10.2.55}$$

—and then stability or instability of solution of Eq. (10.2.47)—depends on behavior of averaged hydrodynamic values and therefore on solutions of concrete hydrodynamic problems.

But in all cases we can confirm the Heisenberg affirmation—stated in 1924 on the basement of the stability investigation of the Navier-Stokes equations [215]—that small but a finite liquid viscosity leads in the definite sense to the destabilizing influence on the flow in comparison with the case of ideal liquids.

10.3 THEORY OF TURBULENCE AND THE GENERALIZED ENSKOG EQUATIONS

Generalized hydrodynamic Enskog equations (GHEnE) (2.7.1)–(2.7.3) lead to the most general formulations of the micro-scale (sub-grid or Kolmogorov) fluctuations. Let us consider the procedure of calculations of the dependent fluctuations with the help of the independent fluctuations coming from the minor tensor velocity moments (containing in continuity Eq. (2.7.1)) to the senior moments containing in the energy equations. The general character of procedure corresponds to obtaining of the related values for the generalized Euler equations and, as in previous case, all found fluctuations are summarized in Table 10.2 where all independent fluctuations are underlined.

TABLE 10.2 Fluctuations of Hydrodynamic Quantities on the Kolmogorov Scale in the Framework of the Generalized Enskog Equations

No	Hydrodynamic Value A	Fluctuation A^{f}
1	ρ_α	$\tau_\alpha\left\{\frac{\partial \rho_\alpha}{\partial t}+\frac{\partial}{\partial \mathbf{r}}\cdot(\rho_\alpha \overline{\mathbf{v}}_\alpha)\right\}$
2	ρ	$\sum_\alpha \tau_\alpha\left\{\frac{\partial \rho_\alpha}{\partial t}+\frac{\partial}{\partial \mathbf{r}}\cdot(\rho_\alpha \overline{\mathbf{v}}_\alpha)\right\}$
3	$\rho_\alpha \overline{\mathbf{v}}_\alpha$	$\tau_\alpha\left\{\frac{\partial}{\partial t}(\rho_\alpha \overline{\mathbf{v}}_\alpha)+\frac{\partial}{\partial \mathbf{r}}\cdot(\rho_\alpha \overline{\mathbf{v}_\alpha \mathbf{v}_\alpha})-\rho_\alpha \overline{\mathbf{F}}_\alpha\right\}$
4	$\rho \mathbf{v}_0$	$\sum_\alpha \tau_\alpha\left\{\frac{\partial}{\partial t}(\rho_\alpha \overline{\mathbf{v}}_\alpha)+\frac{\partial}{\partial \mathbf{r}}\cdot(\rho_\alpha \overline{\mathbf{v}_\alpha \mathbf{v}_\alpha})-\rho_\alpha \overline{\mathbf{F}}_\alpha\right\}$
5	$\mathbf{v}_0$	$\frac{1}{\rho}\sum_\alpha \tau_\alpha\left\{\frac{\partial}{\partial t}(\rho_\alpha \overline{\mathbf{v}}_\alpha)+\frac{\partial}{\partial \mathbf{r}}\cdot(\rho_\alpha \overline{\mathbf{v}_\alpha \mathbf{v}_\alpha})\right.$ $\left.-\rho_\alpha \overline{\mathbf{F}}_\alpha-\mathbf{v}_0\left[\frac{\partial \rho_\alpha}{\partial t}+\frac{\partial}{\partial \mathbf{r}}\cdot(\rho_\alpha \overline{\mathbf{v}}_\alpha)\right]\right\}$
6	$\overline{\mathbf{v}}_\alpha$	$\frac{1}{\rho_\alpha}\tau_\alpha\left\{\frac{\partial}{\partial t}(\rho_\alpha \overline{\mathbf{v}}_\alpha)+\frac{\partial}{\partial \mathbf{r}}\cdot(\rho_\alpha \overline{\mathbf{v}_\alpha \mathbf{v}_\alpha})\right.$ $\left.-\rho_\alpha \overline{\mathbf{F}}_\alpha-\overline{\mathbf{v}}_\alpha\left[\frac{\partial \rho_\alpha}{\partial t}+\frac{\partial}{\partial \mathbf{r}}\cdot(\rho_\alpha \overline{\mathbf{v}}_\alpha)\right]\right\}$
7	$\rho_\alpha \overline{\mathbf{v}_\alpha \mathbf{v}_\alpha}$	$\tau_\alpha\left\{\frac{\partial}{\partial t}(\rho_\alpha \overline{\mathbf{v}_\alpha \mathbf{v}_\alpha})+\frac{\partial}{\partial \mathbf{r}}\cdot\rho_\alpha \overline{(\mathbf{v}_\alpha \mathbf{v}_\alpha)\mathbf{v}_\alpha}\right.$ $-\mathbf{F}_\alpha^{(1)}\rho_\alpha \overline{\mathbf{v}}_\alpha-\rho_\alpha \overline{\mathbf{v}}_\alpha \mathbf{F}_\alpha^{(1)}$ $\left.-\frac{q_\alpha}{m_\alpha}\rho_\alpha \overline{[\mathbf{v}_\alpha \times \mathbf{B}]\mathbf{v}_\alpha}-\frac{q_\alpha}{m_\alpha}\rho_\alpha \overline{\mathbf{v}_\alpha[\mathbf{v}_\alpha \times \mathbf{B}]}\right\}$
8	$\rho_\alpha \overline{v_\alpha^2}$	$\tau_\alpha\left\{\frac{\partial}{\partial t}\left(\rho_\alpha \overline{v_\alpha^2}\right)+\frac{\partial}{\partial \mathbf{r}}\cdot\rho_\alpha \overline{v_\alpha^2 \mathbf{v}_\alpha}\right.$ $\left.-2\rho_\alpha \overline{\mathbf{F}_\alpha \cdot \mathbf{v}_\alpha}\right\}$
9	$\varepsilon_\alpha n_\alpha$	$\tau_\alpha\left\{\frac{\partial}{\partial t}(\varepsilon_\alpha n_\alpha)+\frac{\partial}{\partial \mathbf{r}}\cdot(\varepsilon_\alpha n_\alpha \overline{\mathbf{v}}_\alpha)\right\}$
10	$\varepsilon_\alpha n_\alpha \overline{\mathbf{v}}_\alpha$	$\tau_\alpha\left\{\frac{\partial}{\partial t}(\varepsilon_\alpha n_\alpha \overline{\mathbf{v}}_\alpha)+\frac{\partial}{\partial \mathbf{r}}\cdot(\varepsilon_\alpha n_\alpha \overline{\mathbf{v}_\alpha \mathbf{v}_\alpha})\right.$ $\left.-\varepsilon_\alpha n_\alpha \overline{\mathbf{F}}_\alpha\right\}$
11	$\rho_\alpha \mathbf{F}_\alpha^{(1)}\cdot \overline{\mathbf{v}}_\alpha$	$\tau_\alpha \mathbf{F}_\alpha^{(1)}\cdot\left\{\frac{\partial}{\partial t}(\rho_\alpha \overline{\mathbf{v}}_\alpha)+\frac{\partial}{\partial \mathbf{r}}\cdot\rho_\alpha \overline{\mathbf{v}_\alpha \mathbf{v}_\alpha}\right.$ $\left.-\rho_\alpha \overline{\mathbf{F}}_\alpha\right\}$
12	$\rho_\alpha \overline{v_\alpha^2 \mathbf{v}_\alpha}$	$\tau_\alpha\left\{\frac{\partial}{\partial t}\left(\rho_\alpha \overline{v_\alpha^2 \mathbf{v}_\alpha}\right)+\frac{\partial}{\partial \mathbf{r}}\cdot\rho_\alpha \overline{v_\alpha^2 \mathbf{v}_\alpha \mathbf{v}_\alpha}\right.$ $\left.-2\rho_\alpha \mathbf{F}_\alpha^{(1)}\cdot \overline{\mathbf{v}_\alpha \mathbf{v}_\alpha}-\rho_\alpha \overline{v_\alpha^2 \mathbf{F}_\alpha}\right\}$
13	H_α	$\tau_\alpha \frac{\partial H_\alpha}{\partial t}$
14	H	$\sum_\alpha \tau_\alpha \frac{\partial H_\alpha}{\partial t}$

From continuity equation follows

$$\rho_\alpha^{\mathrm{f}} = \tau_\alpha \left\{ \frac{\partial \rho_\alpha}{\partial t} + \frac{\partial}{\partial \mathbf{r}} \cdot \left(\rho_\alpha \overline{\mathbf{v}_\alpha} \right) \right\}, \quad (\alpha = 1, \ldots, \eta), \tag{10.3.1}$$

and the fluctuation of the density of mixture

$$\rho^{\mathrm{f}} = \sum_\alpha \tau_\alpha \left\{ \frac{\partial \rho_\alpha}{\partial t} + \frac{\partial}{\partial \mathbf{r}} \cdot \left(\rho_\alpha \overline{\mathbf{v}_\alpha} \right) \right\}. \tag{10.3.2}$$

If the mean times of free path for species are not too different, $\tau_\alpha \cong \tau$, we have

$$\rho^{\mathrm{f}} = \tau \left\{ \frac{\partial \rho}{\partial t} + \frac{\partial}{\partial \mathbf{r}} \cdot \rho \mathbf{v}_0 \right\}. \tag{10.3.3}$$

The first-order velocity tensor also can be found from the continuity equation

$$\left(\rho_\alpha \overline{\mathbf{v}}_\alpha \right)^{\mathrm{f}} = \tau_\alpha \left\{ \frac{\partial}{\partial t} \left(\rho_\alpha \overline{\mathbf{v}}_\alpha \right) + \frac{\partial}{\partial \mathbf{r}} \cdot \left(\rho_\alpha \overline{\mathbf{v}_\alpha \mathbf{v}_\alpha} \right) - \rho_\alpha \overline{\mathbf{F}}_\alpha \right\}, \tag{10.3.4}$$

and

$$\left(\rho \mathbf{v}_0 \right)^{\mathrm{f}} = \sum_\alpha \tau_\alpha \left\{ \frac{\partial}{\partial t} \left(\rho_\alpha \overline{\mathbf{v}}_\alpha \right) + \frac{\partial}{\partial \mathbf{r}} \cdot \left(\rho_\alpha \overline{\mathbf{v}_\alpha \mathbf{v}_\alpha} \right) - \rho_\alpha \overline{\mathbf{F}}_\alpha \right\}, \tag{10.3.5}$$

From Kolmogorov fluctuations No. 1, No. 3 (see Table 10.2) find the fluctuations of hydrodynamic velocity and averaged velocity of species.

Because

$$\mathbf{v}_0^{\mathrm{f}} = \frac{1}{\rho} \left[\left(\rho \mathbf{v}_0 \right)^{\mathrm{f}} - \rho^{\mathrm{f}} \mathbf{v}_0 \right], \tag{10.3.6}$$

one obtains

$$\begin{aligned} \mathbf{v}_0^{\mathrm{f}} = \frac{1}{\rho} \sum_\alpha \tau_\alpha \Bigg\{ & \frac{\partial}{\partial t} \left(\rho_\alpha \overline{\mathbf{v}}_\alpha \right) + \frac{\partial}{\partial \mathbf{r}} \cdot \left(\rho_\alpha \overline{\mathbf{v}_\alpha \mathbf{v}_\alpha} \right) - \rho_\alpha \overline{\mathbf{F}}_\alpha \\ & - \mathbf{v}_0 \left[\frac{\partial \rho_\alpha}{\partial t} + \frac{\partial}{\partial \mathbf{r}} \cdot \left(\rho_\alpha \overline{\mathbf{v}}_\alpha \right) \right] \Bigg\}. \end{aligned} \tag{10.3.7}$$

Analogously

$$\overline{\mathbf{v}}_\alpha^{\mathrm{f}} = \frac{1}{\rho_\alpha} \left[\left(\rho_\alpha \overline{\mathbf{v}}_\alpha \right)^{\mathrm{f}} - \rho_\alpha^{\mathrm{f}} \overline{\mathbf{v}}_\alpha \right], \tag{10.3.8}$$

then from Eqs. (10.3.1) and (10.3.4) follows

$$\begin{aligned} \overline{\mathbf{v}}_\alpha^{\mathrm{f}} = \frac{1}{\rho_\alpha} \tau_\alpha \Bigg\{ & \frac{\partial}{\partial t} \left(\rho_\alpha \overline{\mathbf{v}}_\alpha \right) + \frac{\partial}{\partial \mathbf{r}} \cdot \left(\rho_\alpha \overline{\mathbf{v}_\alpha \mathbf{v}_\alpha} \right) - \rho_\alpha \overline{\mathbf{F}}_\alpha \\ & - \overline{\mathbf{v}}_\alpha \left[\frac{\partial \rho_\alpha}{\partial t} + \frac{\partial}{\partial \mathbf{r}} \cdot \left(\rho_\alpha \overline{\mathbf{v}}_\alpha \right) \right] \Bigg\}, \end{aligned} \tag{10.3.9}$$

or

$$\overline{\mathbf{v}}_\alpha^{\mathrm{f}} = \tau_\alpha \left\{ \frac{\partial \overline{\mathbf{v}}_\alpha}{\partial t} + \frac{1}{\rho_\alpha} \frac{\partial}{\partial \mathbf{r}} \cdot \left(\rho_\alpha \overline{\mathbf{v}_\alpha \mathbf{v}_\alpha} \right) - \overline{\mathbf{F}}_\alpha - \frac{\overline{\mathbf{v}}_\alpha}{\rho_\alpha} \frac{\partial}{\partial \mathbf{r}} \cdot \left(\rho_\alpha \overline{\mathbf{v}}_\alpha \right) \right\}. \tag{10.3.10}$$

Fluctuations of the second-order tensor velocity moments $\rho_\alpha \overline{\mathbf{v}_\alpha \mathbf{v}_\alpha}$ are contained in momentum equation. As this takes place the fluctuations of tensor moments of the zeroth-order and the first-order ones appearing in momentum equation, coincide (in other words do not contradict) with corresponding fluctuations found from continuity equation.

Then the next independent fluctuation has the form

$$
\begin{aligned}
(\rho_\alpha\overline{\mathbf{v}_\alpha\mathbf{v}_\alpha})^{\mathrm{f}} = \tau_\alpha\Bigg\{&\frac{\partial}{\partial t}(\rho_\alpha\overline{\mathbf{v}_\alpha\mathbf{v}_\alpha}) + \frac{\partial}{\partial\mathbf{r}}\cdot\rho_\alpha\overline{(\mathbf{v}_\alpha\mathbf{v}_\alpha)\mathbf{v}_\alpha}\\
&-\mathbf{F}_\alpha^{(1)}\rho_\alpha\overline{\mathbf{v}}_\alpha - \rho_\alpha\overline{\mathbf{v}}_\alpha\mathbf{F}_\alpha^{(1)} - \frac{q_\alpha}{m_\alpha}\rho_\alpha\overline{[\mathbf{v}_\alpha\times\mathbf{B}]\mathbf{v}_\alpha}\\
&-\frac{q_\alpha}{m_\alpha}\rho_\alpha\overline{\mathbf{v}_\alpha[\mathbf{v}_\alpha\times\mathbf{B}]}\Bigg\}.
\end{aligned}
\tag{10.3.11}
$$

The left side of energy equation contains fluctuation $\left(\rho_\alpha\overline{v_\alpha^2}\right)^{\mathrm{f}}$, which should be consequence of Eq. (10.3.11). Such, indeed, is the case:

$$
\begin{aligned}
\overleftrightarrow{\mathrm{I}} : (\rho_\alpha\overline{\mathbf{v}_\alpha\mathbf{v}_\alpha})^{\mathrm{f}} &= \left(\rho_\alpha\overline{v_\alpha^2}\right)^{\mathrm{f}}\\
&= \tau_\alpha\left\{\frac{\partial}{\partial t}\left(\rho_\alpha\overline{v_\alpha^2}\right) + \frac{\partial}{\partial\mathbf{r}}\cdot\rho_\alpha\overline{v_\alpha^2\mathbf{v}_\alpha} - 2\mathbf{F}_\alpha^{(1)}\cdot\rho_\alpha\overline{\mathbf{v}}_\alpha\right\}.
\end{aligned}
\tag{10.3.12}
$$

The last terms of the right side of (10.3.10) have no contribution in $\left(\rho_\alpha\overline{v_\alpha^2}\right)^{\mathrm{f}}$, because for example,

$$
\begin{aligned}
\overleftrightarrow{\mathrm{I}} : \overline{\mathbf{v}_\alpha[\mathbf{v}_\alpha\times\mathbf{B}]} &= \sum_{i=1}^{3}\overline{v_{\alpha i}[\mathbf{v}_\alpha\times\mathbf{B}]_i}\\
&= v_{\alpha1}(v_{\alpha2}B_3 - v_{\alpha3}B_2) - v_{\alpha2}(v_{\alpha1}B_3 - v_{\alpha3}B_1)\\
&\quad + v_{\alpha3}(v_{\alpha1}B_2 - v_{\alpha2}B_1) = 0
\end{aligned}
\tag{10.3.13}
$$

Then

$$
\left(\rho_\alpha\overline{v_\alpha^2}\right)^{\mathrm{f}} = \tau_\alpha\left\{\frac{\partial}{\partial t}\left(\rho_\alpha\overline{v_\alpha^2}\right) + \frac{\partial}{\partial\mathbf{r}}\cdot\rho_\alpha\overline{v_\alpha^2\mathbf{v}_\alpha} - 2\rho_\alpha\overline{\mathbf{F}_\alpha\cdot\mathbf{v}_\alpha}\right\}.
\tag{10.3.14}
$$

Fluctuation of internal energy (per unit of volume) of α-species has the form

$$
(\varepsilon_\alpha n_\alpha)^{\mathrm{f}} = \tau_\alpha\left\{\frac{\partial}{\partial t}(\varepsilon_\alpha n_\alpha) + \frac{\partial}{\partial\mathbf{r}}\cdot(\varepsilon_\alpha n_\alpha\overline{\mathbf{v}}_\alpha)\right\}.
\tag{10.3.15}
$$

From the other side

$$
(\varepsilon_\alpha n_\alpha)^{\mathrm{f}} = \frac{(\varepsilon_\alpha n_\alpha\overline{\mathbf{v}}_\alpha)^{\mathrm{f}} - \varepsilon_\alpha n_\alpha\overline{\mathbf{v}}_\alpha^{\mathrm{f}}}{\overline{\mathbf{v}}_\alpha},
\tag{10.3.16}
$$

then

$$
\begin{aligned}
\overline{\mathbf{v}}_\alpha(\varepsilon_\alpha n_\alpha)^{\mathrm{f}} = \tau_\alpha\Bigg\{&\frac{\partial}{\partial t}(\varepsilon_\alpha n_\alpha\overline{\mathbf{v}}_\alpha) + \frac{\partial}{\partial\mathbf{r}}\cdot\varepsilon_\alpha n_\alpha\overline{\mathbf{v}_\alpha\mathbf{v}_\alpha}\\
&-\varepsilon_\alpha n_\alpha\mathbf{F}_\alpha - \frac{\varepsilon_\alpha n_\alpha}{\rho_\alpha}\Bigg[\frac{\partial}{\partial t}(\rho_\alpha\overline{\mathbf{v}}_\alpha) + \frac{\partial}{\partial\mathbf{r}}\cdot(\rho_\alpha\overline{\mathbf{v}_\alpha\mathbf{v}_\alpha})\\
&-\rho_\alpha\overline{\mathbf{F}}_\alpha - \overline{\mathbf{v}}_\alpha\frac{\partial\rho_\alpha}{\partial t} - \overline{\mathbf{v}}_\alpha\frac{\partial}{\partial\mathbf{r}}\cdot(\rho_\alpha\overline{\mathbf{v}}_\alpha)\Bigg]\Bigg\}\\
= \tau_\alpha\Bigg\{&\overline{\mathbf{v}}_\alpha\frac{\partial}{\partial t}(\varepsilon_\alpha n_\alpha) + \frac{\varepsilon_\alpha n_\alpha}{\rho_\alpha}\overline{\mathbf{v}}_\alpha\frac{\partial}{\partial\mathbf{r}}\cdot(\rho_\alpha\overline{\mathbf{v}}_\alpha)\Bigg\}\\
= \overline{\mathbf{v}}_\alpha\tau_\alpha\Bigg\{&\frac{\partial}{\partial t}(\varepsilon_\alpha n_\alpha) + \varepsilon_\alpha\frac{\partial}{\partial\mathbf{r}}\cdot(n_\alpha\overline{\mathbf{v}}_\alpha)\Bigg\}\\
= \overline{\mathbf{v}}_\alpha\tau_\alpha\Bigg\{&\frac{\partial}{\partial t}(\varepsilon_\alpha n_\alpha) + \frac{\partial}{\partial\mathbf{r}}\cdot(\varepsilon_\alpha n_\alpha\overline{\mathbf{v}}_\alpha)\Bigg\}.
\end{aligned}
\tag{10.3.17}
$$

This result agrees relation (10.3.15). In Eqs. (10.3.16) and (10.3.17) are realized the typical transformations directed on the investigation of non-contradictions of found fluctuations of the tensor moments.

Finally independent fluctuation of the third-order tensor moment contains the tensor moment of the fourth-order:

$$\left(\overline{\rho_\alpha \mathbf{v}_\alpha v_\alpha^2}\right)^{\mathrm{f}} = \tau_\alpha \left\{ \rho_\alpha \overline{\mathbf{v}_\alpha v_\alpha^2} + \frac{\partial}{\partial \mathbf{r}} \cdot \overline{v_\alpha^2 \mathbf{v}_\alpha \mathbf{v}_\alpha} - 2\rho_\alpha \mathbf{F}_\alpha^{(1)} \cdot \overline{\mathbf{v}_\alpha \mathbf{v}_\alpha} - \rho_\alpha \overline{v_\alpha^2 \mathbf{F}_\alpha} \right\}. \quad (10.3.18)$$

In Table 3.2 is introduced the fluctuation of the Boltzmann H-function.

Now we state that the most general boundary conditions for GHE should we written as follows

$$A_w^{\mathrm{f},ind} = 0. \quad (10.3.19)$$

These conditions imply that all independent fluctuations $A^{\mathrm{f},ind}$ indicated in Table 3.2, on the wall are equal to zero.

In the past many attempts were made at phenomenological refining the Boltzmann equation (see for example [220], p. 69) for the purpose of introducing fluctuation terms into kinetic and hydrodynamic equations; however, these attempts were not rigorous to any extent.

The advent of generalized Boltzmann physical kinetics rendered invalid the statement of Belotserkovskii and Oparin [220] to the effect that "the theory of turbulence remains the science of semi-empirical models on the kinetic level of description as well."

In conclusion several remarks are necessary:

(1) By calculation of turbulent fluctuations containing in Table 10.2, the squared fluctuations were omitted as negligibly small, in other words were neglected by terms proportional to τ^2. It means that calculation of terms in curly brackets of Table 3.2 can be realized for averaged values.

(2) The application of the method of moments to the GBE leads to the hydrodynamic equations, which include pulsation terms corresponding to the small-scale or sub-grid turbulence when, for small values of the Knudsen number, the mean time between collisions becomes proportional to viscosity. These fluctuations of hydrodynamic quantities are universal and tabulated. Every hydrodynamic value considered as a tensor velocity moment of the n-dimension, corresponds the fluctuation containing senior tensor moment of the $(n+1)$-dimension.

In "classical" theory of turbulence this situation corresponds to the problem of closure of moment equations. Remind, this problem does not exist in the generalized Boltzmann physical kinetics; one needs only to return to equations written for genuine variables.

10.4 UNSTEADY FLOW OF A COMPRESSIBLE GAS IN A CAVITY

To demonstrate the research potential of numerical vortex-flow simulation using the GHE, we examine the two-dimensional unsteady flow of a compressible gas. We begin our consideration with calculations of flow in a rectangular cavity and after that the mathematical modeling of flows in channels of different forms will be delivered.

We examine the two-dimensional unsteady flow of a compressible gas in a cavity [221, 222]. The problem to be solved is the following. Consider a flow over a flat surface and suppose there suddenly appears—as a result of some mechanical action, for example—a certain cavity of square cross section, whose length is much longer than the side of the square (Fig. 10.1). It is assumed that at some instant of time gas suddenly starts to move along the segment OL of the axis x with the velocity V_s, which is subsequently maintained constant.

The system of the generalized Euler equations for a two-dimensional, unsteady, and non-isothermic flow of compressbile gas is written in the following way:

$$\frac{\partial}{\partial t}\left\{\rho - \tau\left[\frac{\partial \rho}{\partial t} + \frac{\partial}{\partial x}(\rho u) + \frac{\partial}{\partial y}(\rho v)\right]\right\} + \frac{\partial}{\partial x}\left\{\rho u - \tau\left[\frac{\partial}{\partial t}(\rho u) + \frac{\partial}{\partial x}(\rho u^2) + \frac{\partial}{\partial y}(\rho u v) + \frac{\partial p}{\partial x}\right]\right\} + \frac{\partial}{\partial y}\left\{\rho v - \tau\left[\frac{\partial}{\partial t}(\rho v) + \frac{\partial}{\partial x}(\rho u v) + \frac{\partial}{\partial y}(\rho v^2) + \frac{\partial p}{\partial y}\right]\right\} = 0, \quad (10.4.1)$$

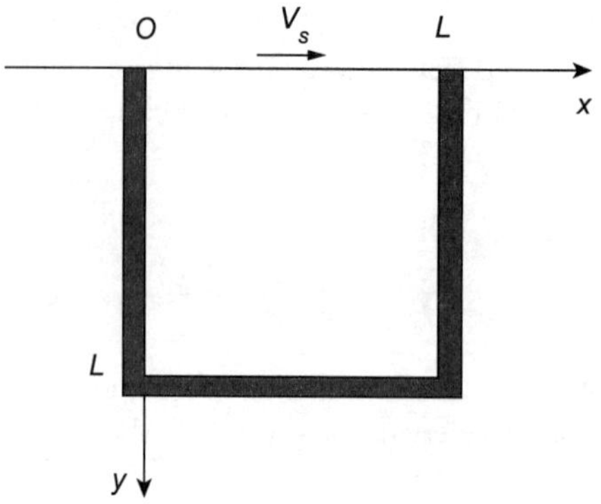

FIGURE 10.1 Unsteady flow of the compressible gas in a cavity.

$$
\begin{aligned}
&\frac{\partial}{\partial t}\left\{\rho u-\tau\left[\frac{\partial}{\partial t}(\rho u)+\frac{\partial}{\partial x}(\rho u^2)+\frac{\partial}{\partial y}(\rho uv)+\frac{\partial p}{\partial x}\right]\right\}\\
&+\frac{\partial}{\partial x}\left\{p+\rho u^2-\tau\left[\frac{\partial}{\partial t}(p+\rho u^2)+3\frac{\partial}{\partial x}(pu)+\frac{\partial}{\partial x}(\rho u^3)\right.\right.\\
&\left.\left.+\frac{\partial}{\partial y}(\rho u^2 v)+2\frac{\partial}{\partial y}(pv)\right]\right\}+\frac{\partial}{\partial y}\left\{\rho uv-\tau\left[\frac{\partial}{\partial t}(\rho uv)+\frac{\partial}{\partial x}(\rho u^2 v)\right.\right.\\
&\left.\left.+\frac{\partial}{\partial y}(pu)+\frac{\partial}{\partial y}(\rho uv^2)\right]\right\}=0,
\end{aligned}
\tag{10.4.2}
$$

$$
\begin{aligned}
&\frac{\partial}{\partial t}\left\{\rho v-\tau\left[\frac{\partial}{\partial t}(\rho v)+\frac{\partial}{\partial x}(\rho uv)+\frac{\partial}{\partial y}(\rho v^2)+\frac{\partial p}{\partial y}\right]\right\}\\
&+\frac{\partial}{\partial x}\left\{\rho uv-\tau\left[\frac{\partial}{\partial t}(\rho uv)+\frac{\partial}{\partial x}(pv)+\frac{\partial}{\partial x}(\rho u^2 v)\right.\right.\\
&\left.\left.+\frac{\partial}{\partial y}(\rho uv^2)\right]\right\}+\frac{\partial}{\partial y}\left\{p+\rho v^2-\tau\left[\frac{\partial}{\partial t}(p+\rho v^2)\right.\right.\\
&\left.\left.+3\frac{\partial}{\partial y}(pv)+\frac{\partial}{\partial y}(\rho v^3)+\frac{\partial}{\partial x}(\rho uv^2)+2\frac{\partial}{\partial x}(pu)\right]\right\}=0,
\end{aligned}
\tag{10.4.3}
$$

$$
\begin{aligned}
&\frac{\partial}{\partial t}\left\{\rho v_0^2+3p-\tau\left[\frac{\partial}{\partial t}(\rho v_0^2+3p)+\frac{\partial}{\partial x}(u(\rho v_0^2+5p))\right.\right.\\
&\left.\left.+\frac{\partial}{\partial y}(v(\rho v_0^2+5p))\right]\right\}+\frac{\partial}{\partial x}\left\{u(\rho v_0^2+5p)-\tau\left[\frac{\partial}{\partial t}(u(\rho v_0^2+5p))\right.\right.\\
&\left.\left.+\frac{\partial}{\partial x}\left(u^2\rho v_0^2+pv_0^2+7pu^2+5\frac{p^2}{\rho}\right)+\frac{\partial}{\partial y}(uv\rho v_0^2+7puv)\right]\right\}\\
&+\frac{\partial}{\partial y}\left\{v(\rho v_0^2+5p)-\tau\left[\frac{\partial}{\partial t}(v(\rho v_0^2+5p))+\frac{\partial}{\partial x}(\rho uvv_0^2+7puv)\right.\right.\\
&\left.\left.+\frac{\partial}{\partial y}\left(v^2\rho v_0^2+pv_0^2+7pv^2+5\frac{p^2}{\rho}\right)\right]\right\}=0.
\end{aligned}
\tag{10.4.4}
$$

Here, mean time between collision is equal

$$\tau=\Pi\mu/p,$$

where the factor Π allows for the model of interaction of gas particles. In the simplest case of so-called “elementary” kinetic theory, $\Pi=1$; $v_0^2=u^2+v^2$, and $\mathbf{v}_0$ is the hydrodynamical flow velocity with components u and v.

The system of generalized equations (10.1.1)–(10.1.4) was made dimensionless by using the dimensionless variables $\hat{p}=p/p_\infty, \hat{\rho}=\rho/\rho_\infty, \hat{u}=u/V_s, \hat{v}=v/V_s$, and $\hat{t}=tV_s/L$.

$$
\begin{aligned}
&\frac{\partial}{\partial \hat{t}}\left\{\hat{\rho}-\Pi \frac{\hat{\mu} \mathrm{Eu}^{-1}}{\hat{p} \mathrm{Re}}\left[\frac{\partial \hat{\rho}}{\partial \hat{t}}+\frac{\partial}{\partial \hat{x}}(\hat{\rho} \hat{u})+\frac{\partial}{\partial \hat{y}}(\hat{\rho} \hat{v})\right]\right\} \\
&\quad \frac{\partial}{\partial \hat{x}}\left\{\hat{\rho} \hat{u}-\Pi \frac{\hat{\mu} \mathrm{Eu}^{-1}}{\hat{p} \mathrm{Re}}\left[\frac{\partial}{\partial t}(\hat{\rho} \hat{u})+\frac{\partial}{\partial \hat{x}}(\hat{\rho} \hat{u}^{2})+\frac{\partial}{\partial \hat{y}}(\hat{\rho} \hat{u} \hat{v})+\mathrm{Eu} \frac{\partial \hat{p}}{\partial \hat{x}}\right]\right\} \\
&\quad \frac{\partial}{\partial \hat{y}}\left\{\hat{\rho} \hat{v}-\Pi \frac{\hat{\mu} \mathrm{Eu}^{-1}}{\hat{p} \mathrm{Re}}\left[\frac{\partial}{\partial \hat{t}}(\hat{\rho} \hat{v})+\frac{\partial}{\partial \hat{x}}(\hat{\rho} \hat{u} \hat{v})+\frac{\partial}{\partial \hat{y}}(\hat{\rho} \hat{v}^{2})+\mathrm{Eu} \frac{\partial \hat{p}}{\partial \hat{y}}\right]\right\}=0,
\end{aligned}
\tag{10.4.5}
$$

$$
\begin{aligned}
&\frac{\partial}{\partial \hat{t}}\left\{\hat{\rho} \hat{u}-\Pi \frac{\hat{\mu} \mathrm{Eu}^{-1}}{\hat{p} \mathrm{Re}}\left[\frac{\partial}{\partial \hat{t}}(\hat{\rho} \hat{u})+\frac{\partial}{\partial \hat{x}}(\hat{\rho} \hat{u}^{2})+\frac{\partial}{\partial \hat{y}}(\hat{\rho} \hat{u} \hat{v})+\mathrm{Eu} \frac{\partial \hat{p}}{\partial \hat{x}}\right]\right\} \\
&\quad+\frac{\partial}{\partial \hat{x}}\left\{\mathrm{Eu} \hat{p}+\hat{\rho} \hat{u}^{2}-\Pi \frac{\hat{\mu} \mathrm{Eu}^{-1}}{\hat{p} \mathrm{Re}}\left[\frac{\partial}{\partial \hat{t}}(\mathrm{Eu} \hat{p}+\hat{\rho} \hat{u}^{2})+3 \mathrm{Eu} \frac{\partial}{\partial \hat{x}}(\hat{p} \hat{u})+\frac{\partial}{\partial \hat{x}}(\hat{\rho} \hat{u}^{3})\right.\right. \\
&\quad\left.\left.+\frac{\partial}{\partial \hat{y}}(\hat{\rho} \hat{u}^{2} \hat{v})+2 \mathrm{Eu} \frac{\partial}{\partial \hat{y}}(\hat{p} \hat{v})\right]\right\} \\
&\quad \times \frac{\partial}{\partial \hat{y}}\left\{\hat{\rho} \hat{u} \hat{v}-\Pi \frac{\hat{\mu} \mathrm{Eu}^{-1}}{\hat{p} \mathrm{Re}}\left[\frac{\partial}{\partial \hat{t}}(\hat{\rho} \hat{u} \hat{v})+\frac{\partial}{\partial \hat{x}}(\hat{\rho} \hat{u}^{2} \hat{v})+\mathrm{Eu} \frac{\partial}{\partial y}(\hat{p} \hat{u})+\frac{\partial}{\partial y}(\hat{\rho} \hat{u} \hat{v}^{2})\right]\right\}=0,
\end{aligned}
\tag{10.4.6}
$$

$$
\begin{aligned}
&\frac{\partial}{\partial \hat{t}}\left\{\hat{\rho} \hat{v}-\Pi \frac{\hat{\mu} \mathrm{Eu}^{-1}}{\hat{p} \mathrm{Re}}\left[\frac{\partial}{\partial \hat{t}}(\hat{\rho} \hat{v})+\frac{\partial}{\partial \hat{x}}(\hat{\rho} \hat{u} \hat{v})+\frac{\partial}{\partial \hat{y}}(\hat{\rho} \hat{v}^{2})+\mathrm{Eu} \frac{\partial \hat{p}}{\partial \hat{y}}\right]\right\} \\
&\quad+\frac{\partial}{\partial \hat{x}}\left\{\hat{\rho} \hat{u} \hat{v}-\Pi \frac{\hat{\mu} \mathrm{Eu}^{-1}}{\hat{p} \mathrm{Re}}\left[\frac{\partial}{\partial \hat{t}}(\hat{\rho} \hat{u} \hat{v})+\mathrm{Eu} \frac{\partial}{\partial \hat{x}}(\hat{p} \hat{v})+\frac{\partial}{\partial \hat{x}}(\hat{\rho} \hat{u}^{2} \hat{v})\right.\right. \\
&\quad\left.\left.+\frac{\partial}{\partial \hat{y}}(\hat{\rho} \hat{u} \hat{v}^{2})\right]\right\}+\frac{\partial}{\partial \hat{y}}\left\{\mathrm{Eu} \hat{p}+\hat{\rho} \hat{v}^{2}-\Pi \frac{\hat{\mu} \mathrm{Eu}^{-1}}{\hat{p} \mathrm{Re}}\left[\frac{\partial}{\partial \hat{t}}(\mathrm{Eu} \hat{p}+\hat{\rho} \hat{v}^{2})\right.\right. \\
&\quad\left.\left.+3 \mathrm{Eu} \frac{\partial}{\partial \hat{y}}(\hat{p} \hat{v})+\frac{\partial}{\partial y}(\hat{\rho} \hat{v}^{3})+\frac{\partial}{\partial \hat{x}}(\hat{\rho} \hat{u} \hat{v}^{2})+2 \mathrm{Eu} \frac{\partial}{\partial \hat{x}}(\hat{p} \hat{u})\right]\right\}=0,
\end{aligned}
\tag{10.4.7}
$$

$$
\begin{aligned}
&\frac{\partial}{\partial \hat{t}}\left\{\hat{\rho} \hat{v}_{0}^{2}+3 \hat{p}-\Pi \frac{\hat{\mu} \mathrm{Eu}^{-1}}{\hat{p} \mathrm{Re}}\left[\frac{\partial}{\partial \hat{t}}(\hat{\rho} \hat{v}_{0}^{2}+3 \mathrm{Eu} \hat{p})+\frac{\partial}{\partial \hat{x}}(\hat{u}(\hat{\rho} \hat{v}_{0}^{2}+5 \mathrm{Eu} \hat{p}))\right.\right. \\
&\quad\left.\left.+\frac{\partial}{\partial \hat{y}}(\hat{v}(\hat{\rho} \hat{v}_{0}^{2}+5 \mathrm{Eu} \hat{p}))\right]\right\}+\frac{\partial}{\partial \hat{x}}\left\{\hat{u}(\hat{\rho} \hat{v}_{0}^{2}+5 \mathrm{Eu} \hat{p})-\Pi \frac{\hat{\mu} \mathrm{Eu}^{-1}}{\hat{p} \mathrm{Re}}\left[\frac{\partial}{\partial \hat{t}}(\hat{u}(\hat{\rho} \hat{v}_{0}^{2}+5 \mathrm{Eu} \hat{p}))\right.\right. \\
&\quad\left.\left.+\frac{\partial}{\partial \hat{x}}\left(\hat{u}^{2} \hat{\rho} \hat{v}_{0}^{2}+\mathrm{Eu} \hat{p} \hat{v}_{0}^{2}+7 \mathrm{Eu} \hat{p} \hat{u}^{2}+5 \mathrm{Eu} \frac{\hat{p}^{2}}{\hat{\rho}}\right)+\frac{\partial}{\partial \hat{y}}(\hat{u} \hat{v} \hat{\rho} \hat{v}_{0}^{2}+7 \mathrm{Eu} \hat{p} \hat{u} \hat{v})\right]\right\} \\
&\quad+\frac{\partial}{\partial \hat{y}}\left\{\hat{v}(\hat{\rho} \hat{v}_{0}^{2}+5 \mathrm{Eu} \hat{p})-\Pi \frac{\hat{\mu} \mathrm{Eu}^{-1}}{\hat{p} \mathrm{Re}}\left[\frac{\partial}{\partial \hat{t}}(\hat{v}(\hat{\rho} \hat{v}_{0}^{2}+5 \mathrm{Eu} \hat{p}))+\frac{\partial}{\partial \hat{x}}(\hat{\rho} \hat{u} \hat{v} \hat{v}_{0}^{2}+7 \mathrm{Eu} \hat{p} \hat{u} \hat{v})\right.\right. \\
&\quad\left.\left.+\frac{\partial}{\partial \hat{y}}\left(\hat{v}^{2} \hat{\rho} \hat{v}_{0}^{2}+\mathrm{Eu} \hat{p} \hat{v}_{0}^{2}+7 \mathrm{Eu} \hat{p} \hat{v}^{2}+5 \mathrm{Eu}^{2} \frac{\hat{p}^{2}}{\hat{\rho}}\right)\right]\right\}=0.
\end{aligned}
\tag{10.4.8}
$$

The effect of the gravity force is neglected, so that there are two similarity criteria in the picture: $\mathrm{Eu}=p_{\infty}/(\rho_{\infty}V_s^2)$ and $\mathrm{Re}=LV_s\rho_{\infty}/\mu_{\infty}$, where V_s is the velocity of the external flow. These systems of dimensional hydrodynamic equations (10.4.1)–(10.4.4) and corresponding dimensionless equations (10.4.5)–(10.4.8) will be applied in following two-dimensional flow calculations in this chapter. The difference consists mainly in geometry of flow boundaries and boundary conditions.

The initial conditions ($t=0$) are as follows:

$$
\begin{aligned}
&\rho=\rho_{\infty}, \quad p=p_{\infty}, \quad v=0 \\
&u=V_s, \quad \text{for}\, y=0, \quad u=0, \quad \text{for}\, y>0 \\
&\frac{\partial u}{\partial t}=0, \quad \frac{\partial v}{\partial t}=0, \quad \frac{\partial \rho}{\partial t}=0, \quad \frac{\partial p}{\partial t}=0
\end{aligned}
.
$$

The boundary conditions to be satisfied are

$$
\begin{gathered}
u(x,0)=V_s, \quad v(x,0)=0, \\
u(x,L)=0, v(x,L)=0, \quad \text{for}\, x\in[0,L], \\
u(0,y)=0, v(0,y)=0, \\
u(L,y)=0, v(L,y)=0, \quad \text{for}\, y\in[0,L],
\end{gathered}
$$

$$\left[\frac{\partial \rho}{\partial x}\right]_{x=0}=0, \left[\frac{\partial \rho}{\partial x}\right]_{x=L}=0, \quad \text{for}\, y \in [0, L],$$

$$\left[\frac{\partial \rho}{\partial y}\right]_{y=0}=0, \left[\frac{\partial \rho}{\partial y}\right]_{y=L}=0, \quad \text{for}\, x \in [0, L],$$

$$\left[\frac{\partial p}{\partial x}\right]_{x=0}=0, \left[\frac{\partial p}{\partial x}\right]_{x=L}=0, \quad \text{for}\, y \in [0, L],$$

$$\left[\frac{\partial p}{\partial y}\right]_{y=0}=0, \left[\frac{\partial p}{\partial y}\right]_{y=L}=0 \quad \text{for}\, x \in [0, L].$$

These boundary conditions imply that there is no-slip, no leakage of compressible gas through the wall, and no heat transfer at the cavity wall—a good enough model to demonstrate the potential of the generalized hydrodynamical equations. The computations performed covered a wide Reynolds number range. Many calculated results, including those for other types of flow (along a heated cylinder and over a step) may be found elsewhere [222, 223], and in what follows only some characteristic results will be given.

Along with the program described above, calculations using the generalized Euler equations and the Navier-Stokes (NS) equations were carried out simultaneously. While the cavity flowfield solutions obtained by the different approaches are qualitatively different for the drastically unsteady regime, they start getting closer for sufficiently large times. Increasing the Reynolds number increases the difference between the flowfield patterns obtained from the generalized Euler equations and the Navier-Stokes equations.

Figure 10.2 illustrates the development of vortex flow with Re$=6.24$, Eu$=2.33$, Kn$=0.13$, and the Mach number M$=0.51$ for the moments of time $\hat{t}=0.20$ and 0.50. The vortex that forms under these conditions corresponding to the intermediate range of the Knudsen numbers values evolves from the left corner of the space towards the center and stays above the middle of the space (see also [57]). The arrows in Figs. 10.2–10.4 and subsequent figures indicate only the direction of the flow, and the arrow length does not depend on the magnitude of velocity vector.

Figures 10.3 and 10.4 correspond to the flow mode with Re$=365$, Eu$=1$, Kn$=0.00343$, and Ma$=0.775$. Figure 10.3 corresponds to calculations by the GHE, and Fig. 10.4, to calculations by the Navier-Stokes equations (NS) for the same moments of time, namely for $\hat{t}=2.0$, 5.0, and 30.0, which enables one to identify the differences in the behavior of flow during transition to the quasi-steady mode. In the essentially unsteady region, the patterns of flow differ considerably. However, for the moments of time $\hat{t}>5.0$ a convergence of the qualitative patterns of flow begins. A "central" eddy and two bottom secondary eddies form, and, at $\hat{t}=30.0$, the flow patterns become very similar from the qualitative

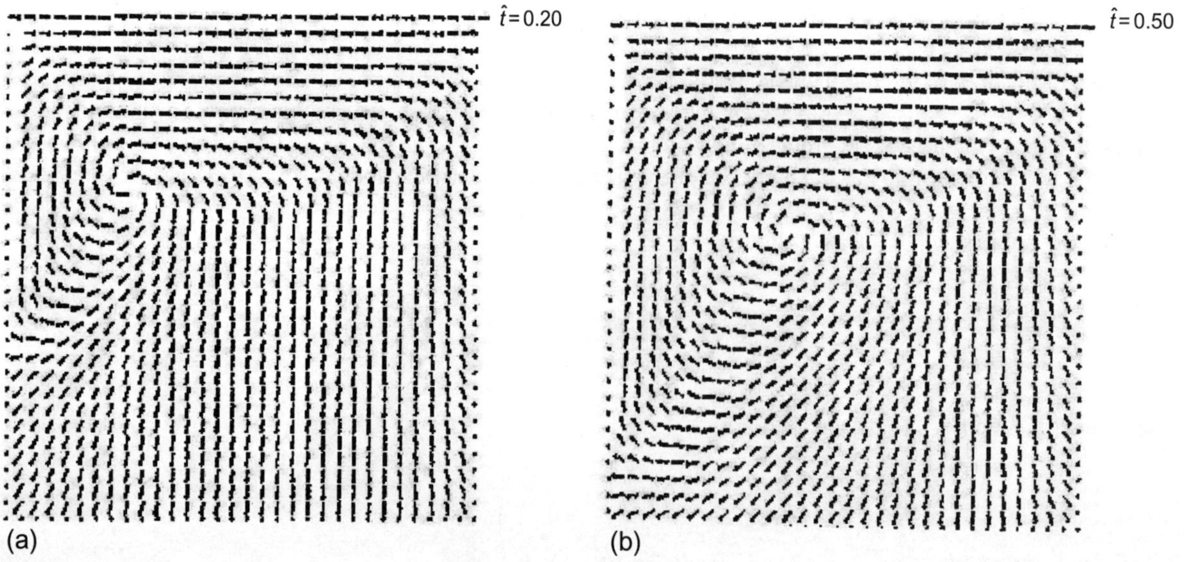

FIGURE 10.2 A flow of gas in the space at Re$=6.24$ and Eu$=2.33$ for the moments of time $\hat{t}$ (a) 0.20, (b) 0.50. The calculations are performed by the GHE.

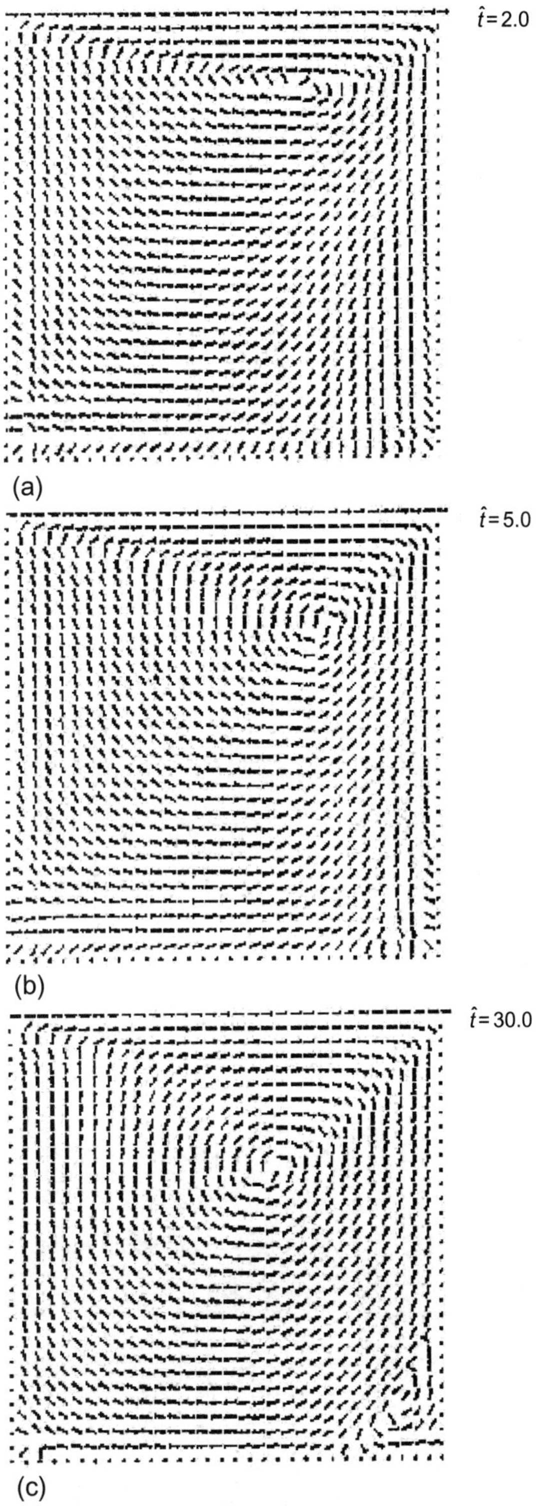

FIGURE 10.3 A flow of gas in the space at Re = 365.0 and Eu = 1.00 for the moments of time $\hat{t}$ = (a) 2.0, (b) 5.0, and (c) 30.0. The calculations are performed by the GHE.

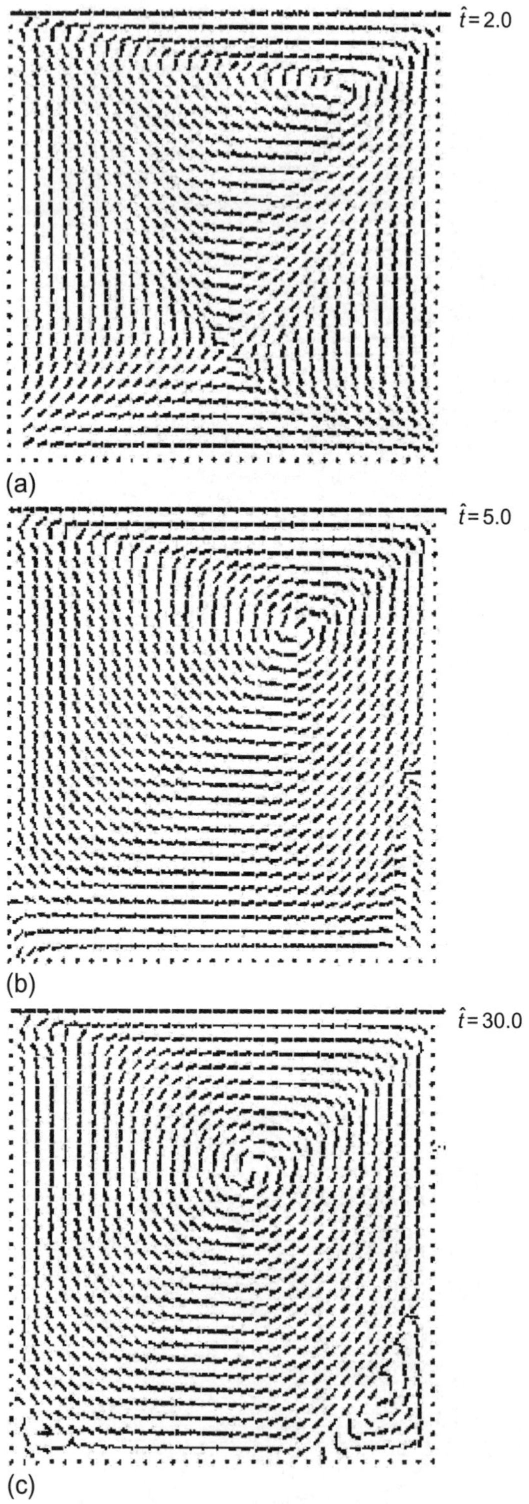

FIGURE 10.4 A flow of gas in the space at Re = 365.0 and Eu = 1.0 for the moments of time $\hat{t}$ = (a) 2.0, (b) 5.0, and (c) 30.0. The calculations are performed by the NS equations.

standpoint. However, in the quasi-steady mode, the profiles of longitudinal and transverse velocities differ as well, as is illustrated by Fig. 10.5, in which the curves *1* are derived by solving the GHE, and the curves 2, by solving NS equations.

As the Reynolds number increases, the flow pattern for the hydrodynamic GHE and NS models in the transition mode become ever more different.

Results for Re = 3200, Eu = 1.0, Kn = 0.0003915, and Ma = 0.775 at (dimensionless) instants of time $\hat{t}=4.0$, 9.5, and 230.0 are shown in Figs 10.6 and 10.7 for the generalized Euler equations and the Navier-Stokes equations, respectively. Notice that the concept of a quasi-stationary regime becomes rather vague in this case.

Figure 10.8 shows the position of the center of the central vortex at large times. Points 1-11 were calculated from the generalized hydrodynamical equations for the dimensionless instants of time $\hat{t}=201.0$, 202.0, 203.5, 204.0, 205.0, 205.8, 207.0, 208.0, 209,0, 210.0, and 211.5. It turns out that the vortex center performs a rotational motion.

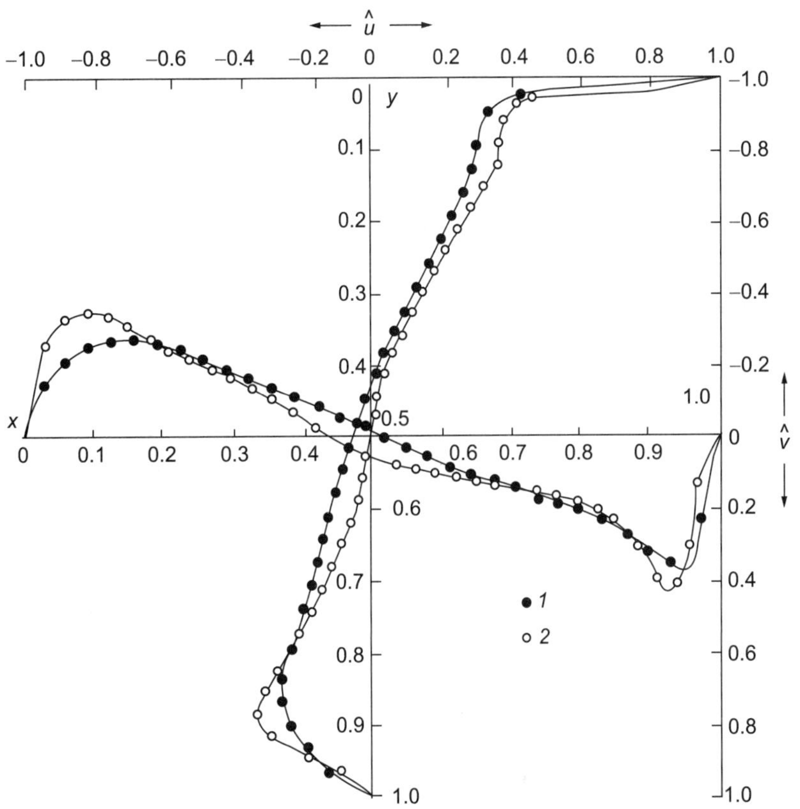

FIGURE 10.5 The profiles of longitudinal and transverse velocity at Re = 365.0 and Eu = 1.0 in the quasi-steady mode. Curves *1* and *2* correspond to the GHE and NS equations, respectively.

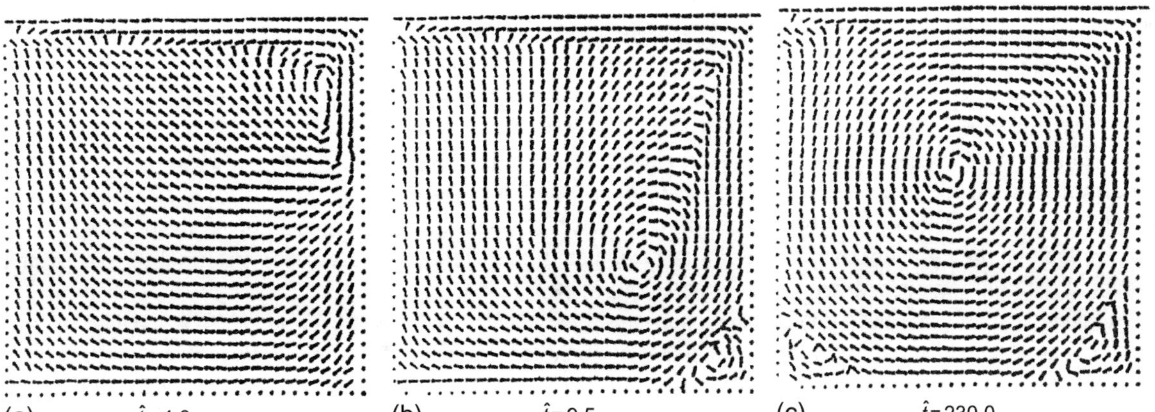

FIGURE 10.6 Gas flow in a cavity at times $\hat{t}=4.9$, 9.5, and 230.0. Calculations are done using the generalized Euler equations for Re = 3200, Eu = 1.

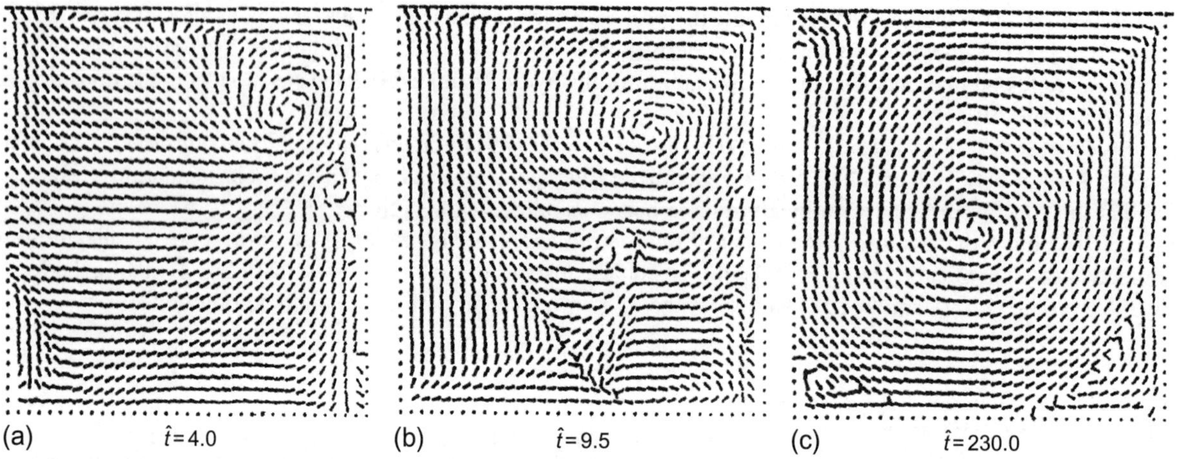

FIGURE 10.7 Gas flow in a cavity at times $\hat{t}$= 4 0, 9 5, and 230.0. Calculations are done using the Navier-Stokes equations for Re = 3200 and Eu = 1.

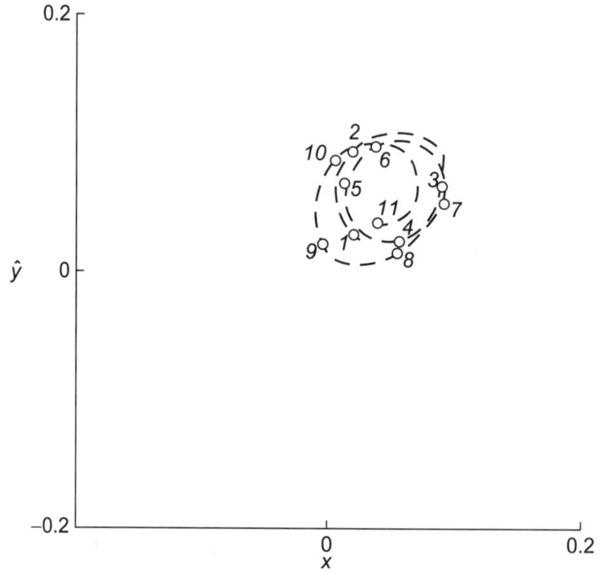

FIGURE 10.8 Position of the center of the central vortex (relative to the center of the cavity) for instants of time t = 201.0, 202.0, 203.5, 204.0, 205.0, 205.8, 207.0, 208.0, 209.0, 210.0, and 211.5 (points 1-11, respectively). Calculations were done using the generalized hydrodynamical equations for Re = 3200 and Eu = 1.

The lid-driven flow in a rectangular cavity has been used extensively as a test bed for computational schemes. From experimental point of view this type of flow belongs to so-called, "benchmark experiments" providing relevant date base for numerical modelers. Moreover an experimental program was beginning as a result of large differences in data obtained by the various schemes.

It is a well-known fact that two-dimensional Navier-Stokes calculations for an incompressible isothermal fluid (water) for the central cross section of a cavity correlate very poorly with experimental data [224–226] even if the length of the cavity is much larger than its width. Nor do three-dimensional Navier-Stokes calculations improve the picture [227, 228]. It has been found that three-dimensional Navier-Stokes calculated results obtained on coarser meshes may agree better with experimental data than do formally more accurate solutions [227, 228].

Detailed experiments had been done by J. Koseff and R. Street with lid-driven cavities [224–226]. The central component of their experimental facility is a rectangular cavity (see Fig. 7.9), which has a width B of 150 mm in the direction of lid motion, a maximum vertical depth D of 925 mm and a span L of 450 mm transverse to the direction of lid motion. The qualitative positions of upper secondary eddy, primary eddy, upstream secondary eddy and downstream secondary eddy are shown in the symmetry plane. As it was discussed above this qualitative picture corresponds to rather high Reynolds numbers. In the J. Koseff and R. Street experiments the visualization method used for water flows in the Reynolds number range of 1000-10,000. In particular, they studied the following aspects of the flow in a square cavity: (1) the presence of turbulence in the flow; (2) the three-dimensionality of the flow and associated flow structures; (3) the size of the downstream secondary eddy as a function of Reynolds number; and (4) the formation of Taylor-type instabilities during the first 30 seconds or so after the lid is started.

Velocity measurements were made for Reynolds numbers of 3200 and 10,000 on two vertical planes. These planes were the symmetry plane and a plane parallel to it and 17.5 mm from the end-wall KTMN (Fig. 10.9, see also [224]).

Unfortunately, we have no experimental results on the situation we are considering—i.e., gas flow in a cavity—and as to the comparison of experimental and theoretical flow data for a gas and a liquid (even at the same Re), this requires great caution. Nevertheless, the calculated Re dependence of the ratio of the size of the bottom near-wall vortex D_3 (downstream secondary eddy), to the cavity width L (see Fig. 10.10) is in general agreement with experiment [221–223]. Interestingly, because of the growing oscillations of hydrodynamic quantities, it is found that even in the quasi-stationary regime fluctuations in the position of the vortex, D_3/L, increase in magnitude. As seen in Fig. 10.10, the region of fluctuations is represented by an expanding band, which shows a transition to fully developed turbulence.

Figure 10.11 demonstrates oscillations in the absolute magnitude of the dimensionless velocity $\hat{v}_0$ at the point $(\hat{x}=0.13, \hat{y}=0.87)$ over a dimensionless time period $\hat{t}=0.6$ for Re = 3200. Note the irregular nature of the oscillations. Thus, already at Re = 3200 the flow starts to exhibit typical features of a turbulent regime.

Figures 10.12 and 10.13 contain comparison of horizontal velocity profiles for numerical (2D and 3D variants for symmetry plane), Re = 3200, Re = 10,000 correspondingly for the GHE and Navier-Stokes models and also for known $k-\varepsilon$ turbulence model. All results obtained by the finite element method for GHE modified for description of liquid media [222, 223].

Details of numerical schemes applied for GHE solution are discussed below. Note that the use of the generalized hydrodynamical equations with viscous terms makes it possible to construct the extremely effective difference schemes, thus making these equations even more attractive.

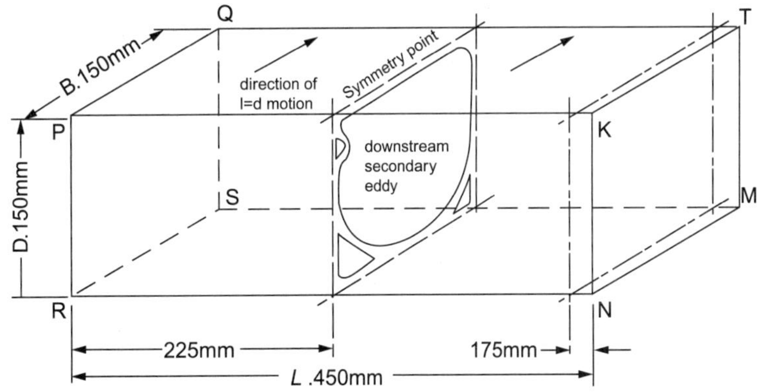

FIGURE 10.9 Experimental definitions for lid-driven cavity with square cross section.

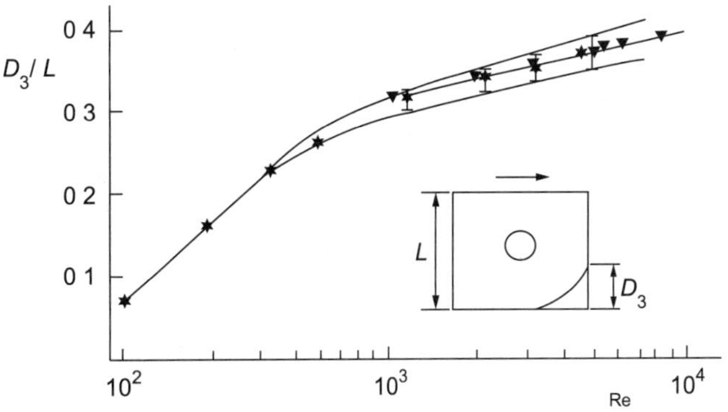

FIGURE 10.10 Relative size D_3/L of the bottom vortex plotted versus Reynolds number Re for Eu = 1. Solid lines represent theoretical results.

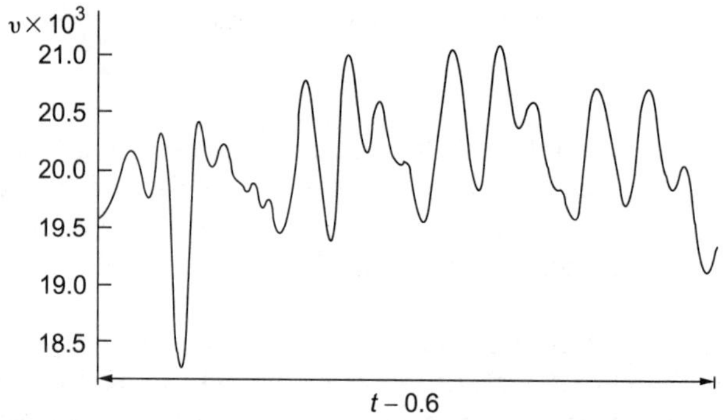

FIGURE 10.11 Oscillations of the absolute value of velocity $\hat{v}_0$ at the point $(\hat{x}=0.13, \hat{y}=0.87)$ over a period of $\hat{t}=0.6$ in the near "quasi-stationary" flow regime for Re = 3200 and Eu = 1. Zero time $\hat{t}_0 = 185.0$.

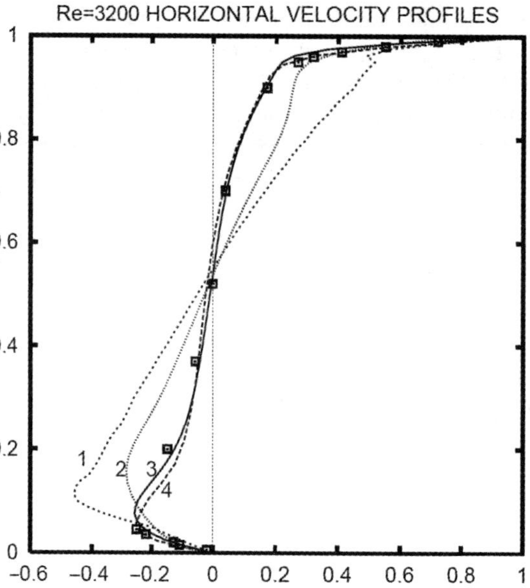

FIGURE 10.12 Comparison of horizontal velocity profiles (1-4) for numerical (solid and dashed lines) and experimental (squares) results, Re = 3200: 1—NS, 2—$k-\varepsilon$ model, 3—GHE (2D), 4—GHE (3D), squares – experimental data by Koseff and Street (1984) in the symmetry vertical plane.

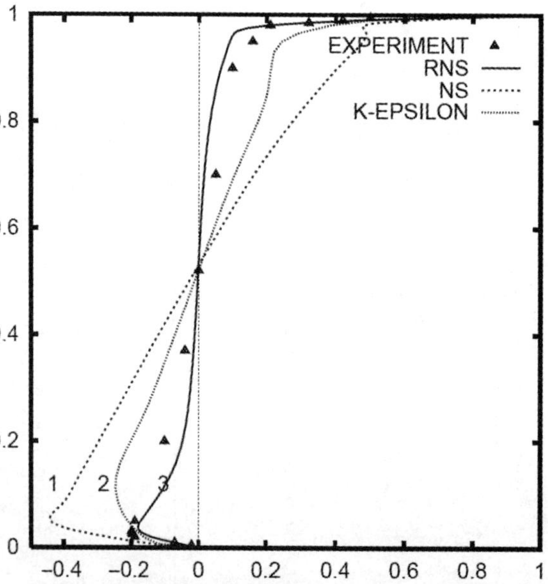

FIGURE 10.13 Comparison of horizontal velocity profiles (1-3) for numerical (solid and dashed lines) and experimental results, Re = 10,000: NS (Navier – Stokes), $k-\varepsilon$ model, RNS – GHE (2D), triangles – experimental data by Koseff and Street (1984) in the symmetry vertical plane.

10.5 APPLICATION OF THE GHE: TO THE INVESTIGATION OF GAS FLOWS IN CHANNELS WITH A STEP

In this section GHE, which include explicitly the Kolmogorov fluctuations of hydrodynamic quantities, are used for mathematical simulation of vortex gas flows. The calculations are performed in the range of Reynolds number values from 64 to 10,000 for two-dimensional unsteady non-isothermal flow of compressible gas in a channel with a step.

Turbulent flows of gases and liquids have been the subject of intensive study for well over a hundred years, which is explained by the large number of possible applications. Considering here backward-facing step flow as investigated above the lid-driven flow in a rectangular cavity, has been used extensively as a test bed for computational schemes. This type of flow also belongs to a set of benchmarks providing relevant date base for numerical modelers.

We will treat the problem in the following formulation. Let a steady flow of gas in a flat channel be initially accompanied by a momentary dip in the bottom that transforms the flow into an unsteady, generally speaking, turbulent, flow in a channel with a step. For the purposes of mathematical simulation, use is made of the geometry of a flat channel with a square step, as shown in Fig. 10.14.

The flow on the left in the narrow channel zone exhibits the behavior of Poiseuille flow and has a parabolic velocity profile. We will demonstrate that GHE, derived using the locally Maxwellian approximation (generalized Euler's equations, GEE), lead to Poiseuille flow in standard, assumptions: (a) the flow is one-dimensional and steady; (b) the density $\rho=\text{const}$; (c) the hydrodynamic velocity depends on y alone, $v_0=v_{0x}(y)$; (d) the static pressure depends on x alone, $p=p(x)$.

We will first turn to the continuity equation, which will be written as

$$\frac{\partial}{\partial \mathbf{r}}\rho\mathbf{v}_0-\tau\left(\Delta p+\frac{\partial^2}{\partial \mathbf{r}\partial \mathbf{r}}:\rho\mathbf{v}_0\mathbf{v}_0\right)=0. \tag{10.5.1}$$

In the foregoing assumptions, Eq. (10.5.1) is satisfied identically. Indeed, the first and third terms of the left-hand part of Eq. (10.5.1) go to zero by virtue of the conditions (a)–(c), and the equation

$$\Delta p=0 \tag{10.5.2}$$

reduces in this case to the condition

$$\frac{\partial^2 p}{\partial x^2}=0, \tag{10.5.3}$$

which leads to the equality

$$\frac{\partial p}{\partial x}=\text{const}, \tag{10.5.4}$$

which is one of the conditions of realization of Poiseuille flow.

We will now treat the generalized equation of motion in a projection onto the x-axis. In view of the conditions (a)–(c) and Eq. (10.5.4), we find from Eq. (10.5.1)

$$\frac{\partial p}{\partial x}=\tau^{(0)}p\frac{\partial^2 u}{\partial y^2}. \tag{10.5.5}$$

However, in view of the relation $\tau=\Pi\mu/p$, we derive the known Poiseuille relation

$$\frac{\partial p}{\partial x}=\Pi\mu\frac{\partial^2 u}{\partial y^2}, \tag{10.5.6}$$

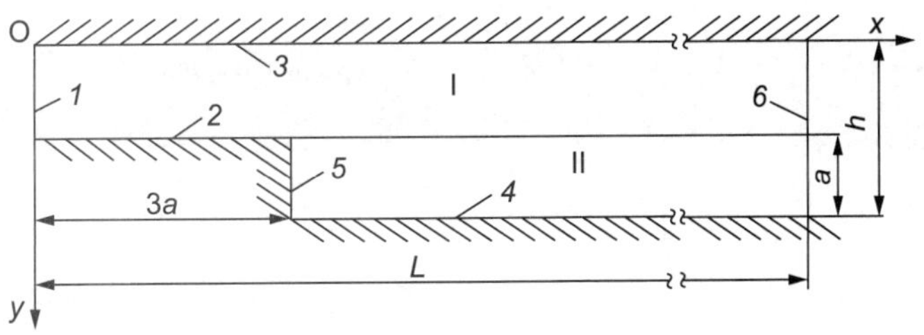

FIGURE 10.14 Schematic of the region being investigated.

in which the factor Π allows for the model of interaction of gas particles. The integration of Eq. (10.5.6) simultaneously with Eq. (10.5.4) yields a parabolic velocity profile. Consequently, the GHE under the above-formulated conditions leads to a parabolic velocity profile which is adopted as the boundary condition for the equation of motion on the left-hand boundary of the calculation region.

We will introduce the characteristic quantities of flow that are used in what follows as scales in reducing the GEE to the dimensionless form: $\rho_\infty, p_\infty, v_\infty$, and μ_∞ denote the density, static pressure, velocity, and dynamic viscosity at the channel inlet; the scale of length h is the channel width at the outlet; and the time scale has the form of h/v_∞. The similarity criteria, i.e., the Reynolds and Euler numbers, are written as $\mathrm{Re}=\rho_\infty v_\infty h/\mu_\infty$, and $\mathrm{Eu}=p_\infty/(\rho_\infty v_\infty^2)$ the force of gravity is neglected, so that there are two similarity criteria in the picture. The dimensionless quantities are further marked by the sign "○" above the respective symbol. We will conventionally divide the flow region into two, "top" and "bottom," parts that are shown in Fig. 10.14 at I and II, respectively; this is necessary solely for convenience of description of the initial conditions and calculation results. We will further describe other geometric characteristics of the calculation region, shown in Fig. 10.14, namely, L, the length of the region being investigated; a, the height of the backward-facing step (in the given concrete conditions, $a=h/2$); *1*, the front section of the region being investigated (flow inlet section); *6*, the rear section of the region being investigated (flow outlet section); *2*, the step plane; *3*, the top plane; *4*, the bottom plane; and *5*, the vertical wall of the step.

It is assumed that a steady laminar flow with the velocity profile in the form of a Poiseuille parabola exists in region I at the initial moment of time, while in region II the flow is absent (in other words, the flow velocity is zero). The results of further calculations readily transform to the model of flow of liquid (water) in a channel; in so doing, it is possible to omit the energy equation and assume the liquid to be incompressible. The systems of dimensional hydrodynamic Eqs. (10.4.1)–(10.4.4) and corresponding dimensionless Eqs. (10.4.5)–(10.4.8) are applied in following two-dimensional flow calculations. The difference with Section 10.4 consists mainly in geometry of flow boundaries and boundary conditions. We begin with initial conditions contained in Tables 10.3 and 10.4.

In Table 10.3, $f(\hat{y})=k_1\hat{y}^2+k_2\hat{y}+k_3$ and the coefficients $k_i(i=1,\ldots,3)$ are found from the conditions:

$$\begin{cases} f(0)=0 \\ f(h-a)/2=1 \\ f(h-a)=0. \end{cases} \tag{10.5.7}$$

The boundary values of flow in the calculation region must be treated in application to each geometric element. Although the flow region is not too complex, we will write out the boundary conditions on each surface for convenience in calculations (Fig. 10.14).

TABLE 10.3 Initial Values of Flow Parameters in Region I

Velocities and Their Derivatives	Pressure and Density and Their Derivatives
$\hat{u}(\hat{t}=0)=f(\hat{y}), \left(\frac{\partial\hat{u}}{\partial\hat{t}}\right)_{\hat{t}=0}=0$	$\hat{p}(\hat{t}=0)=1, \left(\frac{\partial\hat{p}}{\partial\hat{t}}\right)_{\hat{t}=0}=0$
$\hat{v}(\hat{t}=0)=0, \left(\frac{\partial\hat{v}}{\partial\hat{t}}\right)_{\hat{t}=0}=0$	$\hat{\rho}(\hat{t}=0)=1, \left(\frac{\partial\hat{\rho}}{\partial\hat{t}}\right)_{\hat{t}=0}=0$

TABLE 10.4 Initial Values of Flow Parameters in Region II

Velocities and Their Derivatives	Pressure and Density and Their Derivatives
$\hat{u}(\hat{t}=0)=0, \left(\frac{\partial\hat{u}}{\partial\hat{t}}\right)_{\hat{t}=0}=0$	$\hat{p}(\hat{t}=0)=1, \left(\frac{\partial\hat{p}}{\partial\hat{t}}\right)_{\hat{t}=0}=0$
$\hat{v}(\hat{t}=0)=0, \left(\frac{\partial\hat{v}}{\partial\hat{t}}\right)_{\hat{t}=0}=0$	$\hat{\rho}(\hat{t}=0)=1, \left(\frac{\partial\hat{\rho}}{\partial\hat{t}}\right)_{\hat{t}=0}=0$

The boundary conditions specified in Table 10.5 call for further comment. They correspond to the conditions of no-slip on the wall and non-percolation, and the wall is assumed to be absolutely non-heat-conducting. For plane *1*, the conditions are pre-assigned at a distance of $3a$ from the step. This distance may be varied with a view to eliminating the non-physical reactive effect of flow behind the step on the initial flow in the narrow zone of the channel. At sections *1* and *6*, the so-called "soft" boundary conditions are written relative to the averaged (rather than true) hydrodynamic quantities. Such conditions proved to be very effective in the case of numerical realization and eliminated the reactive effect upstream of the flow due to the numerical effect of cutting off the region.

The calculations [229] were performed using an explicit difference scheme of the first order of accuracy with respect to time and of the second order of accuracy on the coordinates with constant dimensionless steps with respect to time and coordinates. The properties of GHE are such that they enable one to realize stable numerical calculation in a wide range of Reynolds number values from units to tens of thousands. In calculations within this problem, the evolution of flow was studied for the values of Re from 64 to 10,000 and Eu = 1.0. Special attention was given to the values of Re of 1000 to 3200 as transition values between the laminar and turbulent modes of flow. Table 10.6 gives the values of steps of the computational mesh and of the geometrical parameters involved in the calculations. Calculations with a smaller spatial step of the mesh failed to reveal substantial changes in the topology of flow in the treated range of Reynolds number values. The similarity criteria, i.e. the Knudsen and Mach numbers, may be represented in terms of Re and Eu using the average velocity of molecules for a Maxwellian distribution,

$$\mathrm{Kn} = \Pi\sqrt{\frac{8}{\pi}\frac{1}{\mathrm{Re}\sqrt{\mathrm{Eu}}}} \tag{10.5.8}$$

$$\mathrm{Kn} = \Pi\sqrt{\frac{8\gamma}{\pi}\frac{\mathrm{Ma}}{\mathrm{Re}}} \tag{10.5.9}$$

where γ is the heat capacity ratio, $\gamma = C_p/C_V$. The respective rescaling of parameters is given in Table 10.7.

Demonstrations of the calculation results using "snapshots" of the flow cannot fully describe its behavior in the calculation region. This is due to the fact that the parameters of a flow of liquid, even if it is laminar, do not remain unvaried in time at every point. We will treat in more detail one of the options of calculation of flow, namely, that with Re = 1000.

It follows from Fig. 10.15 (Re = 1000 and $\hat{t} = 10$) that region *V* is some vortex formation that develops fairly actively with time. Such vortexes arise both during evolution of the region of return flow indicated in Fig. 10.15 by the symbol *P* (a detailed view of return flow is given in Fig. 10.16) and during stagnation of flow in the vicinity of the bottom wall. In viewing successively the entire pattern of flow in time, one can clearly see how independent and fairly intense vortexes

TABLE 10.5 Boundary Conditions

For Section I	For Plane 2	For Plane 3
$x=0$ $\hat{u}(\hat{y})=f(\hat{y})$ $\hat{v}(\hat{y})=0$ $\frac{\partial^2\hat{\bar{p}}}{\partial\hat{x}^2}=0$ $\frac{\partial\hat{\bar{p}}}{\partial x}=0$	$y=h-a$ $\hat{u}(x)=0$ $\hat{v}(x)=0$ $\frac{\partial\hat{p}}{\partial\hat{y}}=0$ $\frac{\partial\hat{p}}{\partial\hat{y}}=0$	$y=0$ $\hat{u}(x)=0$ $\hat{v}(x)=0$ $\frac{\partial\hat{p}}{\partial\hat{y}}=0$ $\frac{\partial\hat{p}}{\partial\hat{y}}=0$
For Plane 4	**For Plane 5**	**For Section 6**
$y=h$ $\hat{u}(x)=0$ $\hat{v}(x)=0$ $\frac{\partial\hat{p}}{\partial\hat{y}}=0$ $\frac{\partial\hat{p}}{\partial\hat{y}}=0$	$x=3a$ $\hat{u}(y)=0$ $\hat{v}(y)=0$ $\frac{\partial\hat{p}}{\partial\hat{x}}=0$ $\frac{\partial\hat{p}}{\partial\hat{x}}=0$	$x=L$ $\frac{\partial\hat{\bar{u}}}{\partial\hat{x}}=0$ $\frac{\partial\hat{\bar{v}}}{\partial\hat{x}}=0$ $\frac{\partial^2\hat{\bar{p}}}{\partial\hat{x}^2}=0$ $\frac{\partial\hat{\bar{p}}}{\partial x}=0$

TABLE 10.6 Values of the Parameters Involved in the Calculations

Re = 64	**Re = 365**
$\Delta\hat{t} = 1.0 \times 10^{-4}$ $\Delta\hat{x} = \Delta\hat{y} = 0.0333$ $\hat{h} = 1.0$ $\hat{L} = 12.0$	$\Delta\hat{t} = 1.0 \times 10^{-3}$ $\Delta\hat{x} = \Delta\hat{y} = 0.0333$ $\hat{h} = 1.0$ $\hat{L} = 12.0$
Re = 1000	**Re = 3200**
$\Delta\hat{t} = 1.0 \times 10^{-3}$ $\Delta\hat{x} = \Delta\hat{y} = 0.0333$ $\hat{h} = 1.0$ $\hat{L} = 12.0$	$\Delta\hat{t} = 1.0 \times 10^{-4}$ $\Delta\hat{x} = \Delta\hat{y} = 0.0333$ $\hat{h} = 1.0$ $\hat{L} = 15.0$
Re = 10,000	
$\Delta\hat{t} = 1.0 \times 10^{-4}$ $\Delta\hat{x} = \Delta\hat{y} = 0.0333$ $\hat{h} = 1.0$ $\hat{L} = 15.0$	

TABLE 10.7 Parameters of Mathematical Simulation

Re = 64, Kn = 1.995×10^{-2} Eu = 1.0, Ma = 0.7747	Re = 365, Kn = 3.498×10^{-3} Eu = 1.0, Ma = 0.7747
Re = 1000, Kn = 1.277×10^{-3} Eu = 1.0, Ma = 0.7747	Re = 3200, Kn = 3.989×10^{-3} Eu = 1.0, Ma = 0.7747
Re = 10,000, Kn = 1.277×10^{-4} Eu = 1.0, Ma = 0.7747	

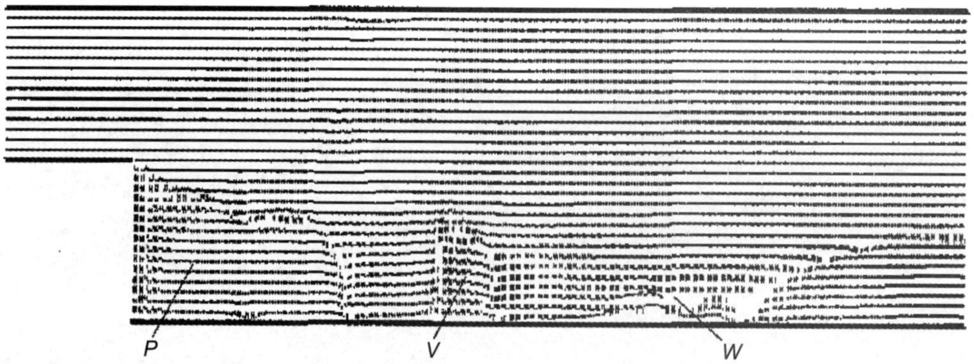

FIGURE 10.15 A general view of the flow at Re = 1000 at the moment of time $\hat{t} = 10$ (for better illustration, the equality of scales on the coordinate axes is not observed).

(a detailed view of vortex V is given in Fig. 10.17), swirling clockwise, originate and start moving along the flow to the outlet behind section 6 (see Fig. 10.14). The transverse dimensions of vortexes V are not constant and amount to 0.20-0.30 of the flow region height h. The extended form of vortex V in Fig. 10.15 does not correspond to the actual pattern; it is defined by the scale difference along the $0x$ and $0y$ axes. The evolution of bottom vortexes strongly resembles the soliton behavior, except for the fact that they are vortexes proper rather than a single wave. In their motion, vortexes V "push out" from the calculation region the vortex formations shown in Fig. 10.15 as W (a detailed view of W is given in Fig. 10.18) to take their place and later move beyond the calculation region. Owing to the special boundary conditions at the flow section

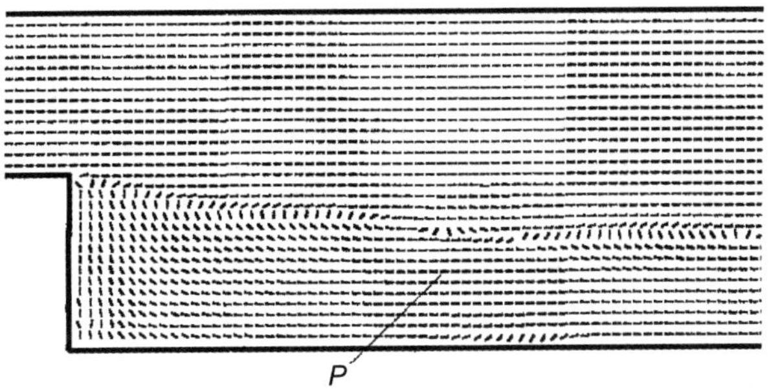

FIGURE 10.16 A detailed view of return flow behind the step (Re = 1000 and $\hat{t} = 10$)

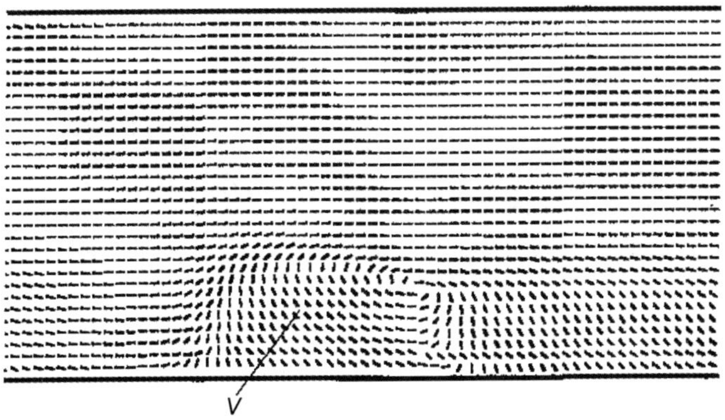

FIGURE 10.17 A detailed view of vortex *V* for Re = 1000 and $\hat{t} = 10$

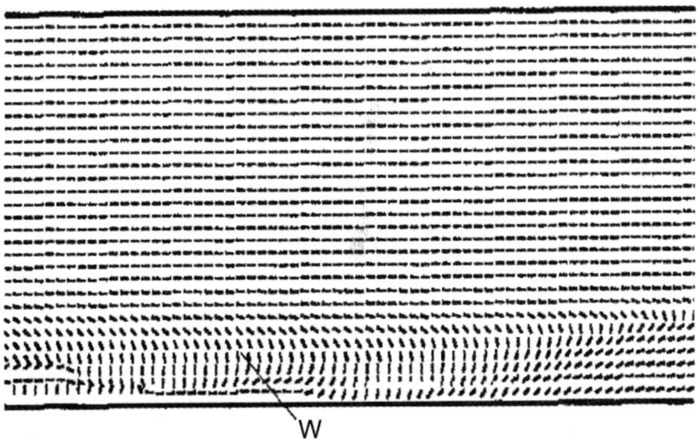

FIGURE 10.18 A detailed view of vortex zones *W* for Re = 1000 and $\hat{t} = 10$.

(Table 10.3, section 6), the vortex formations are not "reflected" but "pass" unobstructed to the non-calculation region behind the section. The motion of bottom vortexes downstream of the flow may lead to additional; erosion of the channel bottom.

However, in passing the calculation region, vortex *V*, in addition to pushing out the vortex formations, introduces some disturbance on its own. This disturbance promotes the formation of wall vortexes on the top plane of the flow region (*3* in Fig. 10.14) that also move along the flow towards section 6 and exist until the moment the flow reaches the quasi-stationary state. It is well seen in Fig. 10.19 (Re = 1000 and $\hat{t} = 110$), namely, region *S* (a more detailed view of region *S* is given in Fig. 10.20), while vortexes of the *V* type degenerate, become smaller, and finally disappear.

At Re = 1000, in contrast to higher values of Re, the flow reaches the quasi-stationary state (Fig. 10.21) in which no further intense origination of vortexes is observed either on the bottom boundary of flow behind the step (*4* in Fig. 10.14) or on the top boundary (*3* in Fig. 10.14). One can clearly see the characteristic point of flow attachment behind the step.

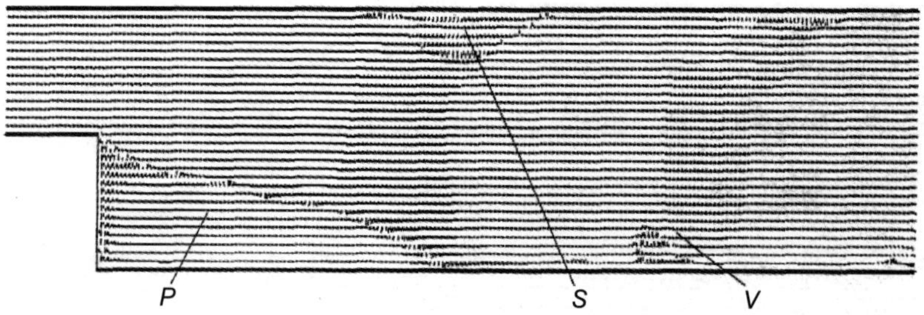

FIGURE 10.19 A general view of flow at Re = 1000 at the moment of time $\hat{t} = 110$ (for better illustration, the equality of scales on the coordinate axes is not observed).

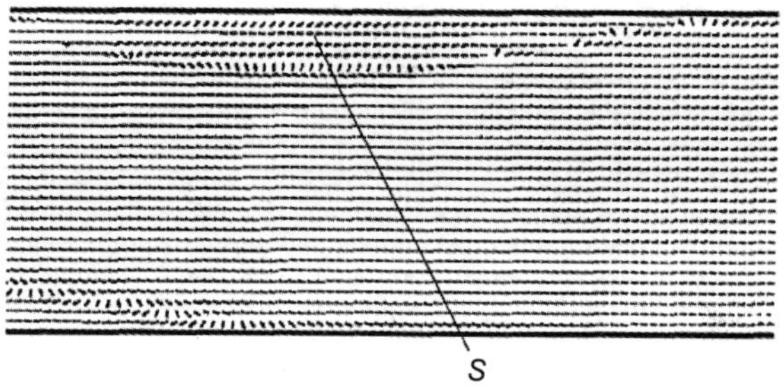

FIGURE 10.20 A detailed view of wall vortex S at Re = 1000 and $\hat{t} = 110$.

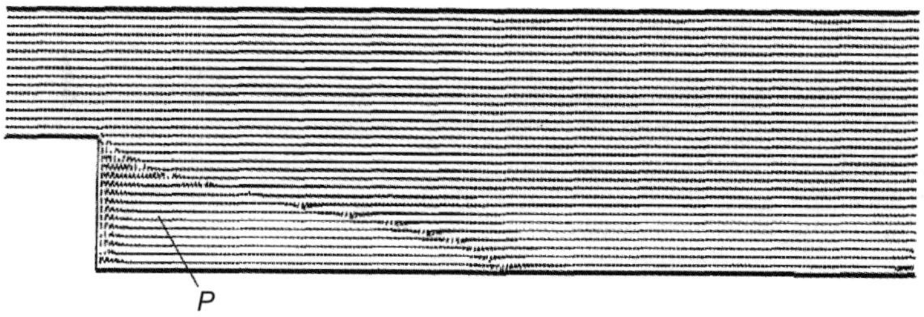

FIGURE 10.21 A general view of flow at Re = 1000 at the moment of time $\hat{t} > 140$; the equality of scales on the coordinate axes is not observed.

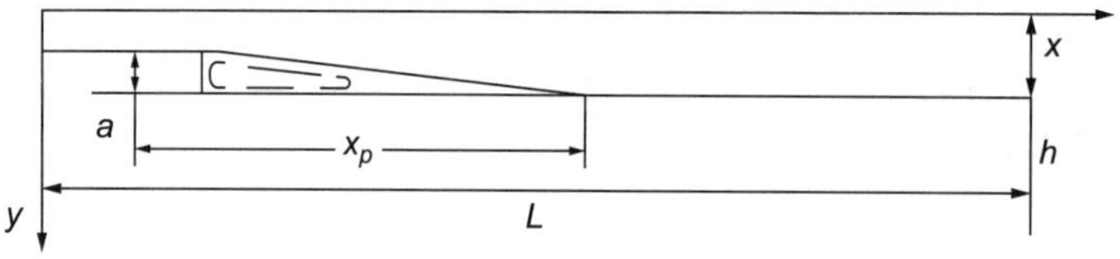

FIGURE 10.22 A diagrammatic view of the point of flow attachment behind the step.

For a more correct representation of the characteristic point of flow attachment to the bottom wall, we will treat a schematic, but properly scaled, Fig. 10.22 in which, according to Fig. 10.14, h is the height of the slot, a is the height for the step, L is the total length of the calculation region, and x_p is the position of the point of flow attachment.

The flow in a channel with a step was investigated by numerous researches (see, e.g., [230]). Moreover, this type of flow is included in the system of benchmark experiments or test experiments aimed at estimating the quality of numerical algorithms. Note that the flow topology and the position of the point of flow attachment are in good agreement with the experimental data. The position of x_p is usually located within five to eight times the step height a, but the position of this point depends strongly on both the flow parameters and the singularities of the experimental setup.

We will point out yet another, not physical but numerical, effect that is observed in the case if insufficient length of calculation region, namely, the existence of a strong dependence of the value x_p on the relative length L/h of the calculation region. This dependence ceased to have an effect when ratio L/h in our calculations exceed 10. Table 10.8 gives values of x_p for different values of Re, obtained in the calculations.

Given for illustration in Figs. 10.23–10.25 are patterns of flows for other values of Re, in addition to the case with Re = 1000 that has already been discussed in detail. Note the absence from Fig. 10.23 of the clearly defined region of return flow behind the step. This region P^* is a turbulized region of return flow with an unstable structure. The region shown at G in Fig. 10.23 is nothing but a turbulized structure moving (similarly to region V in Fig. 10.15) along the flow. Note further region F, which is a "spot" of small-scale turbulence shown in more detail in Fig. 10.24. Also of interest is the representation of the general view of the flow at Re = 10,000. Figure 10.25 illustrates the respective flow. Also marked in Fig. 10.25 are regions analogous with those of the flow shown in Fig. 10.23; however, the absence of region F does not mean that this region does not arise with the given value of Reynolds number: the process of its origination and evolution occurs at subsequent moments of time. We do not illustrate the flows for Re < 1000. The flow topology in this case becomes simpler and less interesting. After a short time of relaxation, such flows repeat in general terms the flow pattern given in Fig. 10.21.

TABLE 10.8 **Values of the Coordinate of the Point of Flow Attachment (for Re < 3200, the Values of x_p are Given for the Flow That Reached the Quasi-Stationary State)**

1	Re = 64	$x_p = 5$
2	Re = 365	$x_p = 7$
3	Re = 1000	$x_p = 10$
4	Re = 3200	$x_p = 5 \div 12$, indeterminacy due to instability of the point of the flow attachment
5	Re = 10,000	x_p indeterminacy due to strong tubulization of the flow

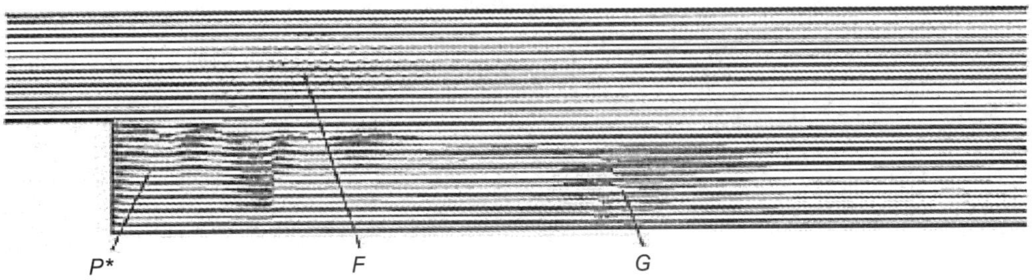

FIGURE 10.23 A general view of flow at Re = 3200 at the moment of time $\hat{t} = 50$; the equality of scales on the coordinate axes is not observed.

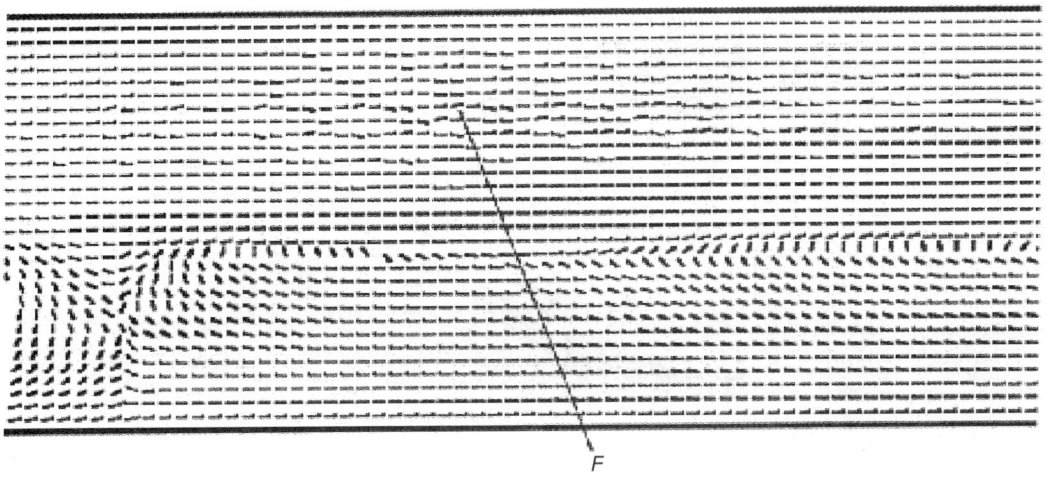

FIGURE 10.24 A detailed view of turbulent "spot" in region F at Re = 3200 and $\hat{t} = 50$.

Mathematical simulation enables one to obtain very extensive information about the hydrodynamic parameters of flow that may hardly be represented graphically to any degree of detail. Figures 10.26 and 10.27 give only the space and time evolution of pressure in a channel. We do not believe it necessary to provide the data of mathematical simulation, based on the Navier-Stokes equations. The appropriate comparison and comments may be found for another flow topology in Ref. [231].

Note that the GHE represent an effective tool for solving problems in hydrodynamics and gas dynamics.

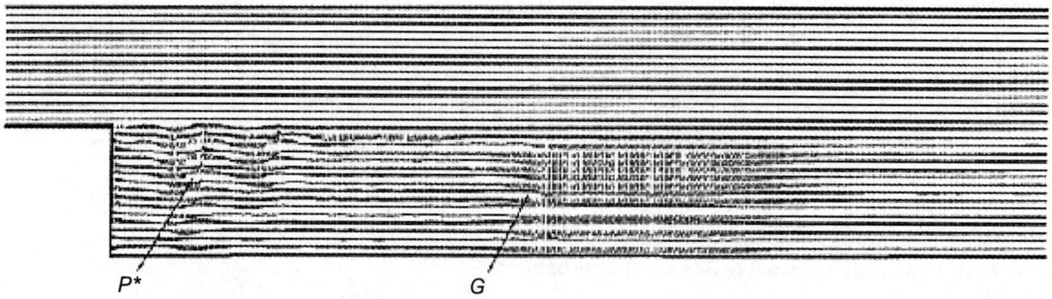

FIGURE 10.25 A general view of flow at Re = 10,000 at the moment of time $\hat{t} = 50$; the equality of scales on the coordinate axes is not observed.

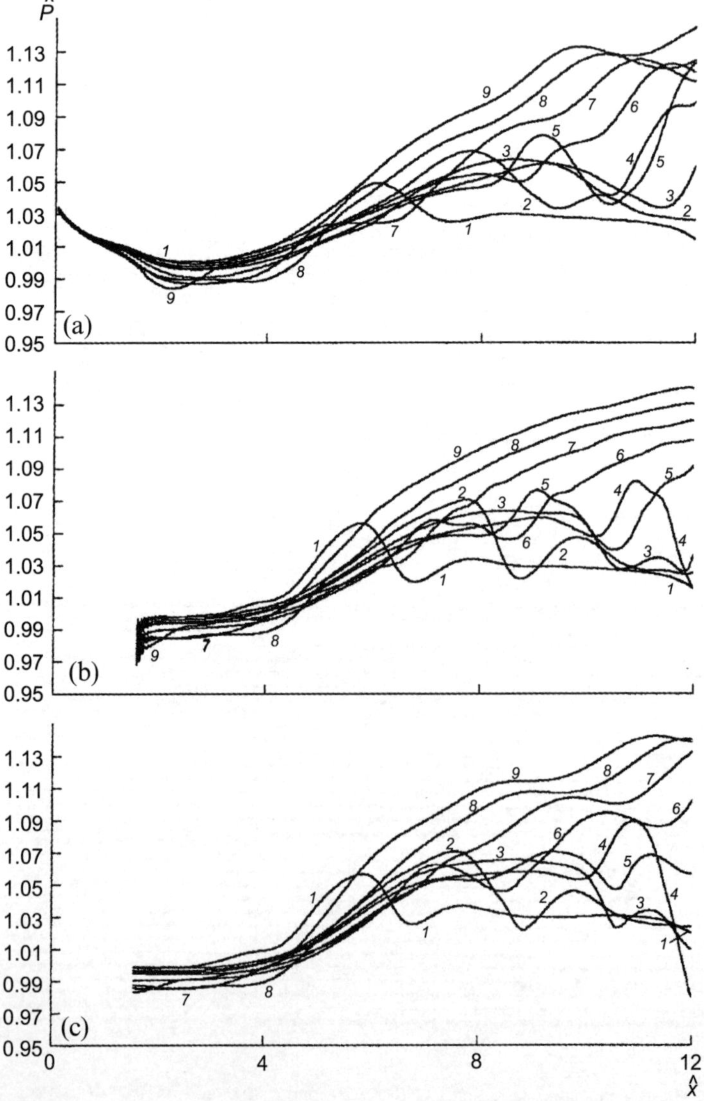

FIGURE 10.26 The distribution of pressure at Re = 1000 along the lines $\hat{y}$ = const for different moments of time: $\hat{y}$ = (a) 0.1, (b) 0.5, and (c) 0.9; $\hat{t}$ = (1) 20, (2) 30, (3) 40, (4) 50, (5) 60, (6) 70, (7) 80, (8) 90, and (9) 100.

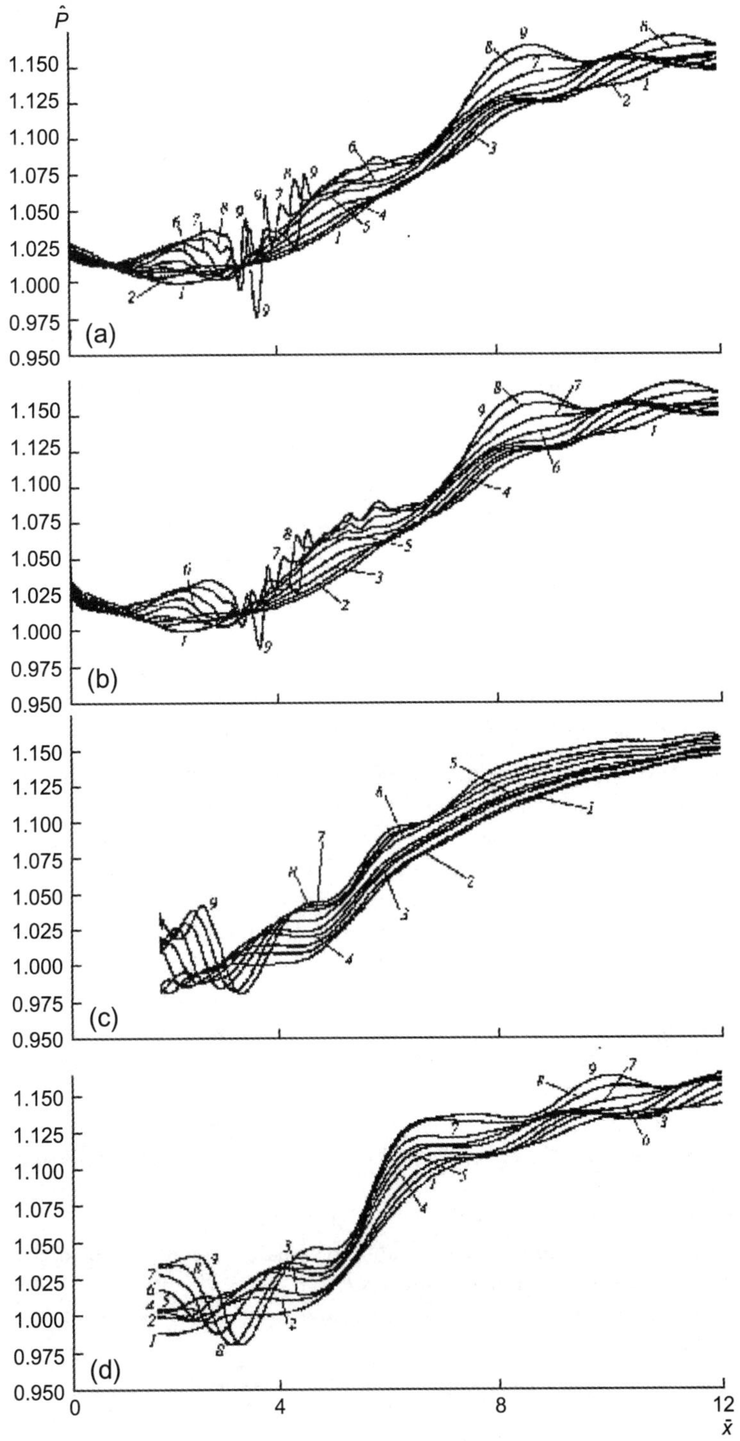

FIGURE 10.27 The distribution of pressure at Re = 3200 along the lines $\hat{y}$ = const for different moments of time: $\hat{y}$ = (a) 0.1, (b) 0.2, (c) 0.5, and (d) 0.9; $\hat{t}$ = (1) 10, (2) 15, (3) 20, (4) 25, (5) 30, (6) 35, (7) 40, (8) 45, and (9) 50.

10.6 VORTEX AND TURBULENT FLOW OF VISCOUS GAS IN CHANNEL WITH FLAT PLATE

We will treat the problem in the following formulation. Let an initially steady-state flow of gas in a flat channel be accompanied in some part of the channel by a momentary separation of flow that transforms the flow into an unsteady-state, generally speaking, turbulent flow. For the purposes of mathematical simulation, we use the geometry of a flat channel with a rectangular plate of finite length, as shown in Fig. 10.28.

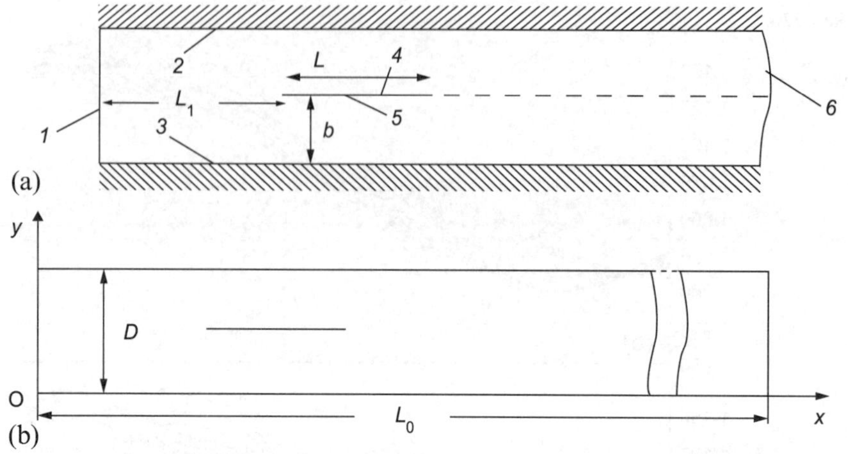

FIGURE 10.28 (a) The flow scheme and (b) the system of coordinates, surfaces are shown on which boundary conditions are pre-assigned (the broken line corresponds to a flow past a semi-infinite plane).

On the left (in a wide zone of the channel before the plate), the flow exhibits a Poiseuille pattern and has a parabolic velocity profile. It was demonstrated in Section 10.4, that the GHE—derived with the help of locally Maxwellian distribution function (generalized Euler equations, GEuE)—lead to Poiseuille flow with standard assumptions:

(a) the flow is one-dimensional and steady-state;
(b) the density $\rho = \text{const}$;
(c) the hydrodynamic velocity depends on y alone, $v_0 = v_{0x}(y)$; and
(d) the static pressure depends on x alone, $p = p(x)$.

Generalized Euler equations (GEuE) under the conditions formulated above lead to a parabolic velocity profile, which is adopted as the boundary condition for the equation of motion on the left-hand boundary of the calculation region.

We will introduce the characteristic quantities of flow that are used in what follows as the scales in reducing the GEuE to a dimensionless form: $\rho_\infty, p_\infty, v_{0\infty} = u_\infty$ and μ_∞ denote the density, static pressure, velocity, and dynamic viscosity at the channel inlet. The channel width at the inlet is used as the scale of length D, and the time scale has the form D/u_∞. The similarity criteria, i.e., the Reynolds and Euler numbers, are written as $\text{Re} = \rho_\infty v_\infty h/\mu_\infty$ and $\text{Eu} = p_\infty/(\rho_\infty v_\infty^2)$. The dimensionless quantities are further marked by the sign ○ above the respective symbol.

The geometric characteristics of the calculation region are given in Fig. 10.28, namely, D, the width of the region (channel) being investigated; L, the length of the plate in the channel; L_1, the distance from the front section to the beginning of the plate; b, the distance from the boundary of the channel to the plate; and L_0, the length of the calculation region.

The boundary and initial conditions are pre-assigned on the following surfaces and sections of the flow region being investigated: *1*, the front section of the region being investigated (flow inlet section); *2*, the top plane; *3*, the bottom plane; *4*, the top surface of the plate; *5*, the bottom surface of the plate; and *6*, the outlet section of the region being investigated (flow outlet section).

Table 10.9 gives the initial conditions for the flow being simulated. These conditions correspond to the existence, at the initial moment of time, of a laminar flow with the velocity profile in the form of Poiseuille parabola. The plate is introduced into the flow at the moment of time zero.

In Table 10.9, $f(\hat{y}) = k_1\hat{y}^2 + k_2\hat{y} + k_3$ and the coefficients $k_i (i = 1, \ldots, 3)$ are found from the conditions $f(0) = 0$, $f(D/2) = 1$, $f(D) = 0$.

TABLE 10.9 Initial Values of Flow Parameters

Velocities and Their Derivatives	Pressure and Density and Their Derivatives
$\hat{u}(\hat{t}=0) = 0, \left(\frac{\partial \hat{u}}{\partial \hat{t}}\right)_{\hat{t}=0} = 0$	$\hat{p}(\hat{t}=0) = 1, \left(\frac{\partial \hat{p}}{\partial \hat{t}}\right)_{\hat{t}=0} = 0$
$\hat{v}(\hat{t}=0) = 0, \left(\frac{\partial \hat{v}}{\partial \hat{t}}\right)_{\hat{t}=0} = 0$	$\hat{\rho}(\hat{t}=0) = 1, \left(\frac{\partial \hat{\rho}}{\partial \hat{t}}\right)_{\hat{t}=0} = 0$

The boundary values in the calculation region of flow must be treated in application to each geometric element. For this purpose, the boundary conditions written out on each surface (see Fig. 10.28) are given in Table 10.10. The boundary conditions given in Table 10.10 call for some comment. They correspond to the conditions of no-slip on the wall and non-percolation, the wall is assumed to be absolutely non-heat-conducting, and the variation of density in the vicinity of a solid surface is taken to be minor. For section *1*, the conditions are pre-assigned at the distance L_1 from the plate tip. This distance may be varied with a view to eliminating the non-physical reactive effect of flow behind the plate on the initial flow. At section *6*, the so-called "soft" boundary conditions are written relative to the *averaged (rather than true)* hydrodynamic quantities.

Such conditions turned out to be very effective in the case of numerical realization and eliminated the upstream reactive effect due to the numerical effect of cutting off of the region. The calculations were performed using an explicit difference scheme of the first order of accuracy with respect to time and of the second order of accuracy with respect to coordinates with constant dimensionless time and coordinate steps.

In calculations [231] within the framework of this problem, the behavior of flow was studied for the values of Re from 1000 to 10,000 and Eu = 1.0. Special attention was given to the range of variation of values of Re from 1000 to 3200 as the region of transition values between the laminar and turbulent modes of flow.

Table 10.11 gives the values of steps of the computational mesh and the geometric parameters of the problem for the employed values of Re. The similarity criteria (Knudsen (Kn) and Mach (Ma) numbers) may be calculated in terms of Re

TABLE 10.10 The Boundary Conditions for the Calculated Region of Flow

For Section "1"	For Surface "2"	For Surface "3"
$\hat{\bar{u}}(0,y)=f(y)$, $\hat{\bar{v}}(0,y)=0$, $\left(\frac{\partial^2 \hat{\bar{p}}}{\partial \hat{x}^2}\right)_{\hat{x}=0}=0$, $\left(\frac{\partial^2 \hat{\bar{p}}}{\partial \hat{x}^2}\right)_{\hat{x}=0}=0$;	$\hat{u}(x,D)=0$, $\hat{v}(x,D)=0$, $\left(\frac{\partial \hat{p}}{\partial \hat{y}}\right)_{\hat{y}=1}=0$, $\left(\frac{\partial \hat{\rho}}{\partial \hat{y}}\right)_{\hat{y}=1}=0$;	$\hat{u}(x,0)=0$, $\hat{v}(x,0)=0$, $\left(\frac{\partial \hat{p}}{\partial \hat{y}}\right)_{\hat{y}=0}=0$, $\left(\frac{\partial \hat{\rho}}{\partial \hat{y}}\right)_{\hat{y}=0}=0$;
For Surface "4"	**For Surface "5"**	**For Section "6"**
$\widehat{u}(x,b)=0$, $\widehat{v}(x,b)=0$, $\left(\frac{\partial \hat{p}}{\partial \hat{y}}\right)_{\hat{y}=\hat{b}+0}=0$, $\left(\frac{\partial \hat{\rho}}{\partial \hat{y}}\right)_{\hat{y}=\hat{b}+0}=0$;	$\widehat{u}(x,b)=0$, $\widehat{v}(x,b)=0$, $\left(\frac{\partial \hat{p}}{\partial \hat{y}}\right)_{\hat{y}=\hat{b}-0}=0$, $\left(\frac{\partial \hat{\rho}}{\partial \hat{y}}\right)_{\hat{y}=\hat{b}-0}=0$;	$\left(\frac{\partial \hat{\bar{u}}}{\partial \hat{x}}\right)_{x=L_0}=0$, $\left(\frac{\partial \hat{\bar{v}}}{\partial \hat{x}}\right)_{x=L_0}=0$, $\left(\frac{\partial^2 \hat{\bar{p}}}{\partial \hat{x}^2}\right)_{x=L_0}=0$, $\left(\frac{\partial^2 \hat{\bar{\rho}}}{\partial \hat{x}^2}\right)_{x=L_0}=0$;

TABLE 10.11 Parameters Involved in the Calculations

Re = 1000	Re = 3200	Re = 10,000
$\Delta\hat{t}=1.0\times10^{-3}$ $\Delta\hat{x}=\Delta\hat{y}=0.0333$ $\hat{D}=1.0, \hat{b}=0.5, \hat{L}=1.0$ $\hat{L}_1=2.5$	$\Delta\hat{t}=1.0\times10^{-4}$ $\Delta\hat{x}=\Delta\hat{y}=0.0333$ $\hat{D}=1.0, \hat{b}=0.5, \hat{L}=1.0$ $\hat{L}_1=2.5$	$\Delta\hat{t}=1.0\cdot10^{-5}$ $\Delta\hat{x}=\Delta\hat{y}=0.0333$ $\hat{D}=1.0, \hat{b}=0.5, \hat{L}=1.0$ $\hat{L}_1=2.5$
1.	2.	3.
Re = 1000, Kn = 1.277×10^{-3} Eu = 1.0, Ma = 0.7747	Re = 3200, Kn = 3.989×10^{-3} Eu = 1.0, Ma = 0.7747	Re = 10,000, Kn = 1.277×10^{-4} Eu = 1.0, Ma = 0.7747

and Eu using the average velocity of molecules for a Maxwellian distribution. The respective rescaling of parameters is given in Table 10.11.

We will now discuss the results of calculation of flow past a plate of finite length. Relevant illustrations are given for the flow past a plate whose length is equal to the channel depth. The plate is located at the channel mid-depth. Also given for some typical cross sections of the channel are flow velocity profiles corresponding to different moments of dimensionless time. The cross sections selected for illustration are located as follows:

(a) before the plate at a distance equal to half its length,
(b) immediately before the plate,
(c) at the middle of the plate,
(d) before the very end of the plate, and
(e) behind the plate at a distance equal to half its length.

The mathematical simulation helps produce a large volume of information including illustrative information about the topology of flow. Therefore, we will first make a description of flow past a plate of finite length at $\mathrm{Re}=1000$ (Fig. 10.29–10.43).

For the initial moments of time, a Poiseuille parabolic profile of flow is retained in section (a), the perturbation due to the plate in this section is still low, and the velocity component v is extremely small. A vertical velocity component emerges due to the deceleration of flow as a result of the effect of the friction forces, and a characteristic asymmetric profile of the velocity component u is formed. In the region where the flow comes off the plate, a marked vorticity of unsteady-state flow arises, which is reflected in the pulsations of the transverse velocity profile. Quite away from the plate downstream of the flow, the longitudinal velocity profile becomes parabolic again, and the pulsations of the transverse velocity are small. By the moment of time $\hat{t}=3.0$, the perturbations of the hydrodynamic characteristics propagate upstream, and a marked profile of transverse velocity arises in the cross section (a) as well.

We will now turn to the results of calculation of the flow past a plate at $\mathrm{Re}=3200$, which corresponds to the mode of transition from the laminar to turbulent flow. By the moment of time $\hat{t}=2.0$, in addition to forming in the vicinity of the plate surface and before and behind the plate, vortexes begin to form at the top and bottom surfaces defining the channel

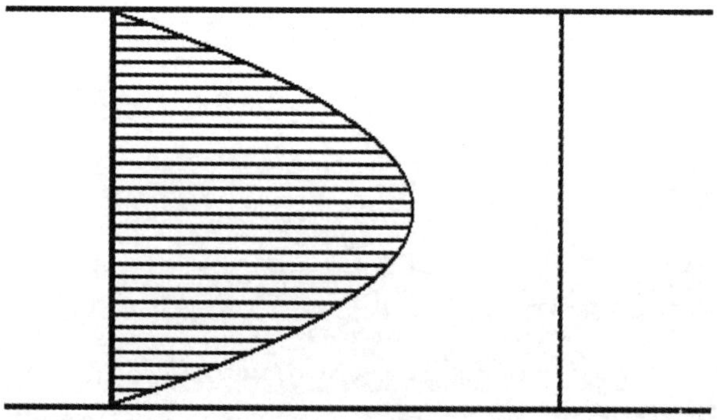

FIGURE 10.29 Velocity profiles on the 0X (left)- and 0Y (right)-axes at a distance $L_x=D/2$ to the left of the beginning of the plate; $\mathrm{Re}=1000$ and $\hat{t}=0.2$.

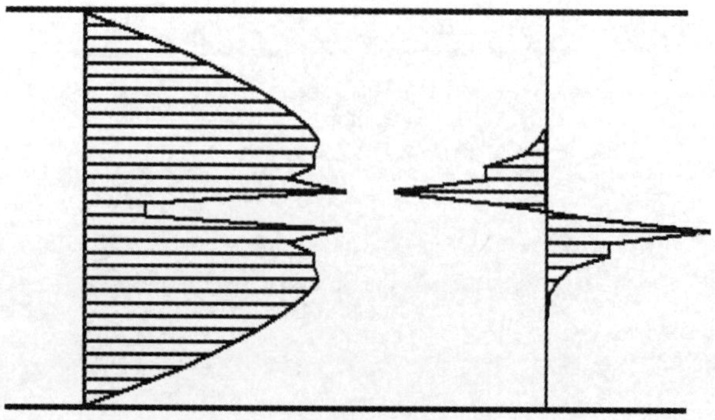

FIGURE 10.30 Velocity profiles on the 0X (left)- and 0Y (right)-axes at a distance $L_x=\Delta x$ to the left of the beginning of the plate; $\mathrm{Re}=1000$ and $\hat{t}=0.2$.

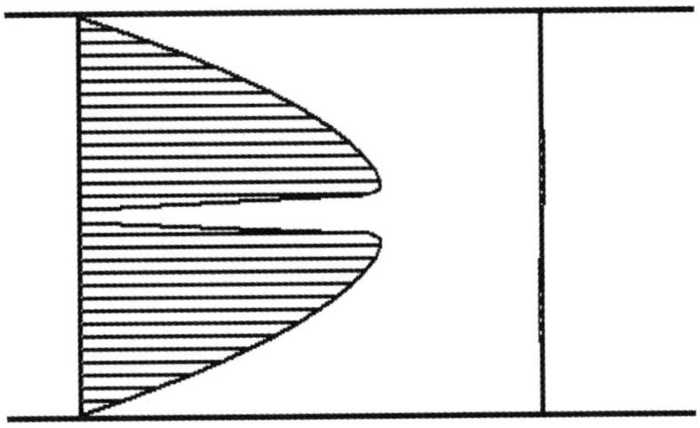

FIGURE 10.31 Velocity profiles on the 0X (left)- and 0Y (right)-axes at a distance $L_x = D/2$ to the right of the beginning of the plate (middle of the plate); Re = 1000 and $\hat{t} = 0.2$.

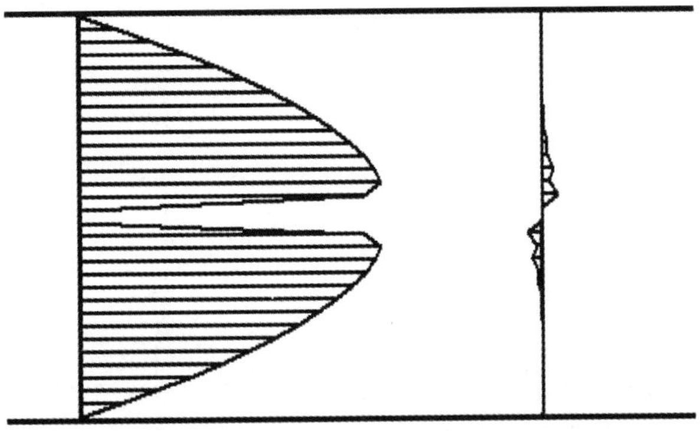

FIGURE 10.32 Velocity profiles on the 0X (left)- and 0Y (right)-axes at a distance $L_x = \Delta x$ to the left of the end of the plate; Re = 1000 and $\hat{t} = 0.2$.

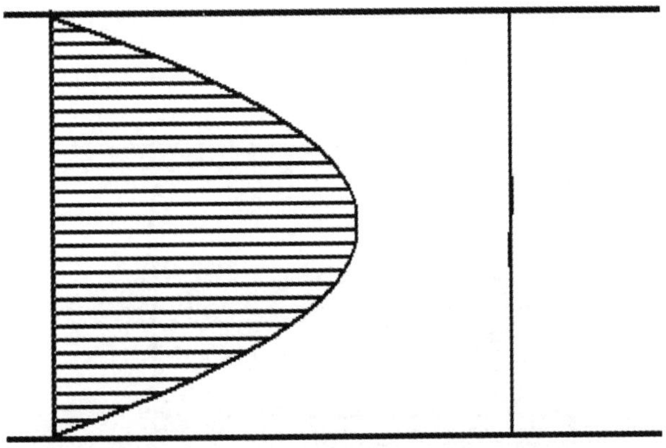

FIGURE 10.33 Velocity profiles on the 0X (left)- and 0Y (right)-axes at a distance $L_x = D/2$ to the right of the end of the plate; Re = 1000 and $\hat{t} = 0.2$.

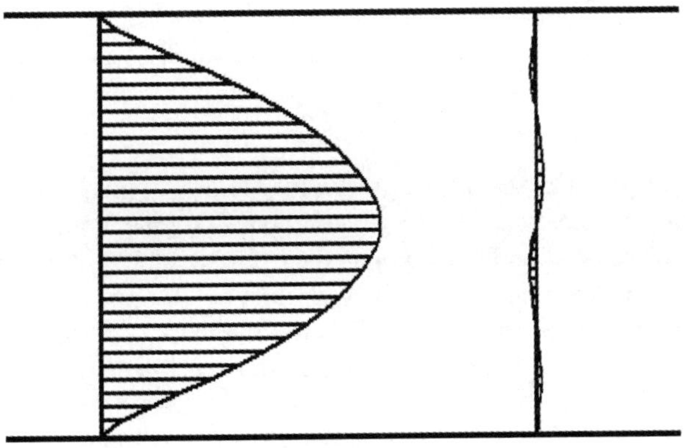

FIGURE 10.34 Velocity profiles on the 0X (left)- and 0Y (right)-axes at a distance $L_x = D/2$ to the left of the beginning of the plate; Re = 1000 and $\hat{t} = 3.0$.

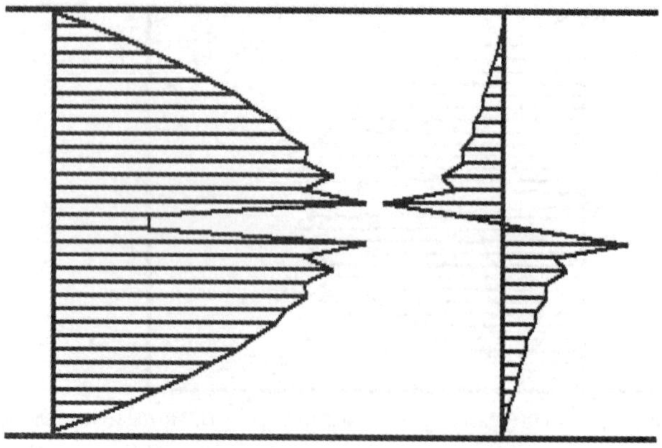

FIGURE 10.35 Velocity profiles on the 0X (left)- and 0Y (right)-axes at a distance $L_x = \Delta x$ to the left of the beginning of the plate; Re = 1000 and $\hat{t} = 3.0$.

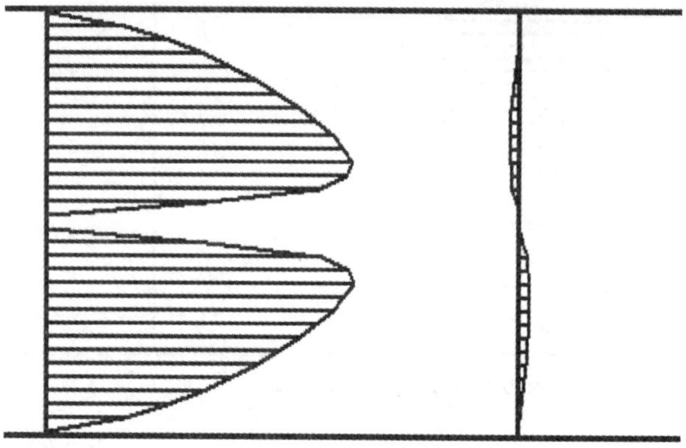

FIGURE 10.36 Velocity profiles on the 0X (left)- and 0Y (right)-axes at a distance $L_x = D/2$ to the right of the beginning of the plate (middle of the plate); Re = 1000 and $\hat{t} = 3.0$.

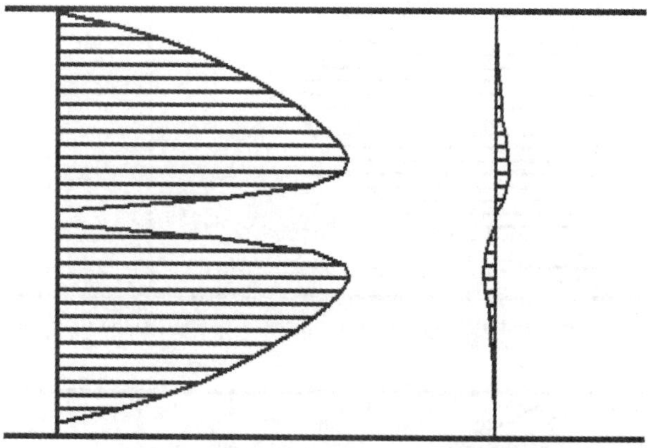

FIGURE 10.37 Velocity profiles on the 0X (left)- and 0Y (right)-axes at a distance $L_x = \Delta x$ to the left of the end of the plate; Re = 1000 and $\hat{t} = 3.0$.

width as well. Figures 10.44–10.49 give the topology of flow in the vicinity of plate, Re = 3200, $\hat{t} = 2.0$, and the profiles of components of flow velocity in the cross sections (a), (b), (c), (d), and (e) for the moment of time $\hat{t} = 5.0$. The dash length is not proportional to the modulus of velocity and corresponds only to the velocity direction.

By this moment of time, the top and bottom wall vortexes weaken but become more extended; with time, an unsteady-state flow changes to a quasi-steady-state mode. In particular, the fluctuations of the longitudinal velocity profile at the middle of the plate decrease somewhat, and in the cross section (e), the fluctuations of the transverse velocity of flow decrease.

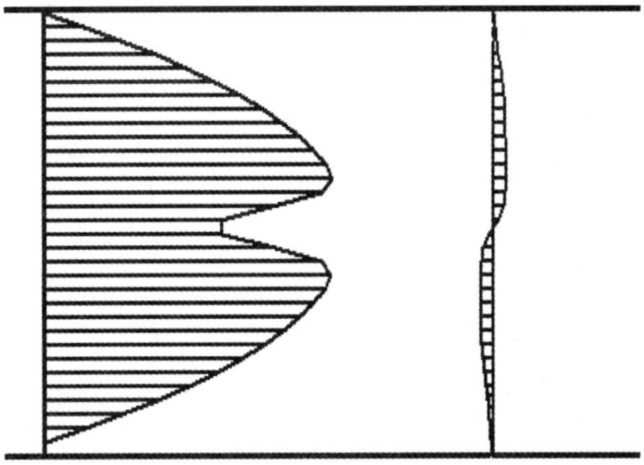

FIGURE 10.38 Velocity profiles on the 0X (left)- and 0Y (right)-axes at a distance $L_x = D/2$ to the right of the end of the plate; Re = 1000 and $\hat{t} = 3.0$.

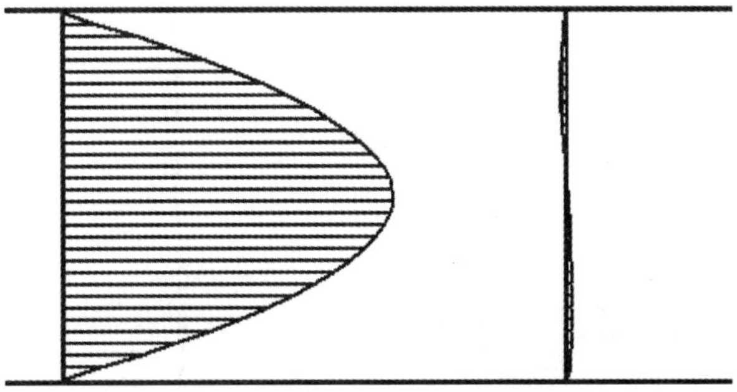

FIGURE 10.39 Velocity profiles on the 0X (left)- and 0Y (right) -axes at a distance $L_x = D/2$ to the left of the beginning of the plate; Re = 1000 and $\hat{t} = 25.0$.

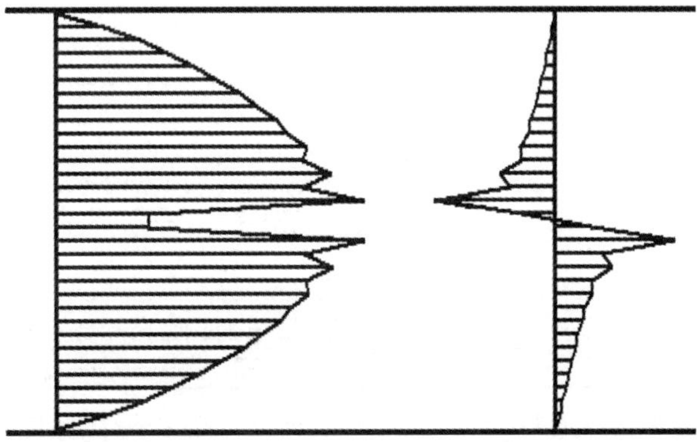

FIGURE 10.40 Velocity profiles on the 0X (left)- ,and 0Y (right)-axes at a distance $L_x = \Delta x$ to the left of the beginning of the plate; Re = 1000 and $\hat{t} = 25.0$.

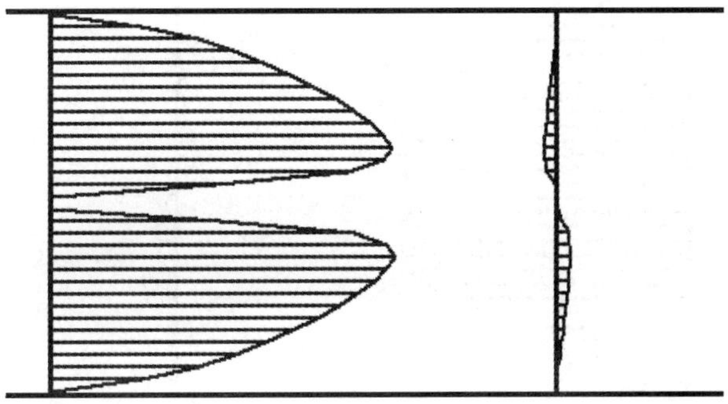

FIGURE 10.41 Velocity profiles on the 0X (left)- and 0Y (right)-axes at a distance $L_x = D/2$ to the right of the beginning of the plate (middle of the plate); Re = 1000 and $\hat{t} = 25.0$.

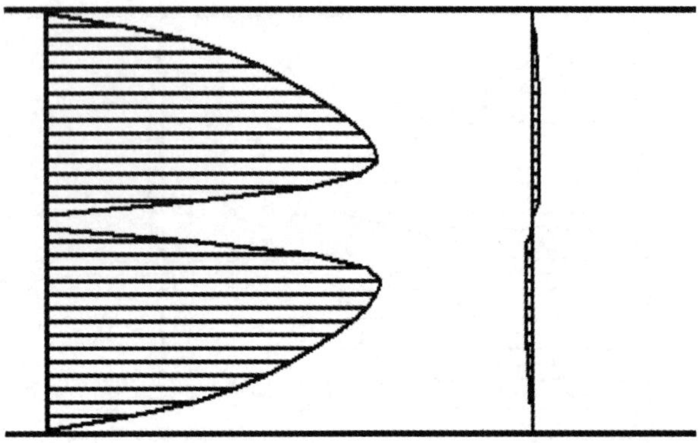

FIGURE 10.42 Velocity profiles on the 0X (left)- and 0Y (right)-axes at a distance $L_x = \Delta x$ to the left of the end of the plate; Re = 1000 and $\hat{t} = 25.0$.

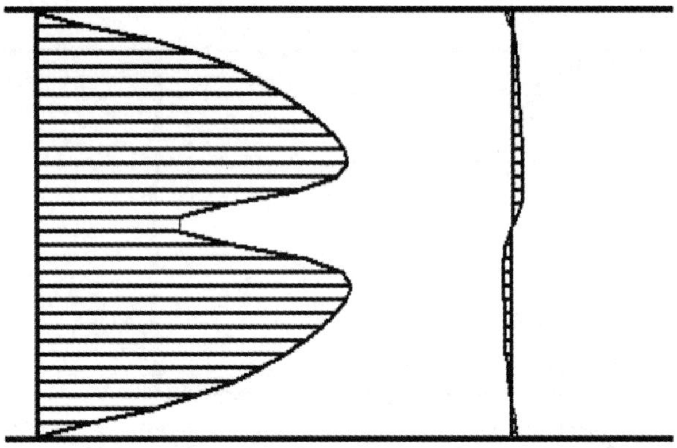

FIGURE 10.43 Velocity profiles on the 0X (left)- and 0Y (right)-axes at a distance $L_x = D/2$ to the right of the end of the plate; Re = 1000 and $\hat{t} = 25.0$.

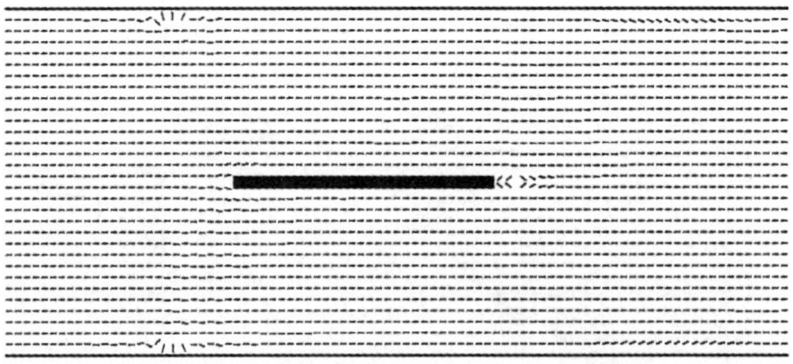

FIGURE 10.44 The topology of flow in the vicinity of plate, Re = 3200, $\hat{t} = 2.0$. The dash length is not proportional to the modulus of velocity.

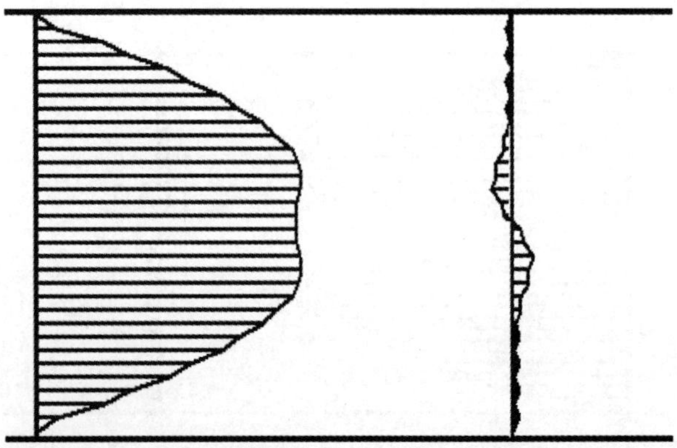

FIGURE 10.45 Velocity profiles on the 0X (left)- and 0Y (right)-axes at a distance $L_x = D/2$ to the left of the beginning of the plate; Re = 3200 and $\hat{t} = 5.0$.

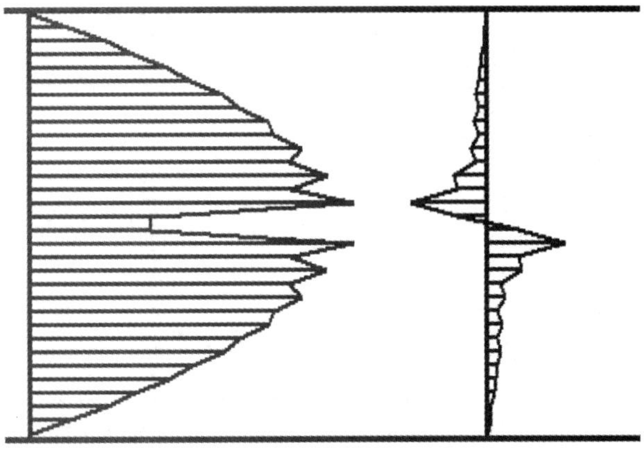

FIGURE 10.46 Velocity profiles on the 0X (left)- and 0Y (right)-axes at a distance $L_x = \Delta x$ to the left of the beginning of the plate; Re = 3200 and $\hat{t} = 5.0$.

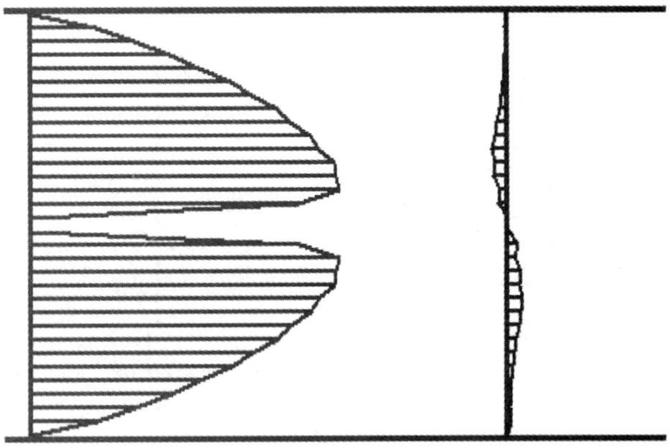

FIGURE 10.47 Velocity profiles on the 0X (left)- and 0Y (right)-axes at a distance $L_x = D/2$ to the right of the beginning of the plate; Re = 3200 and $\hat{t} = 5.0$.

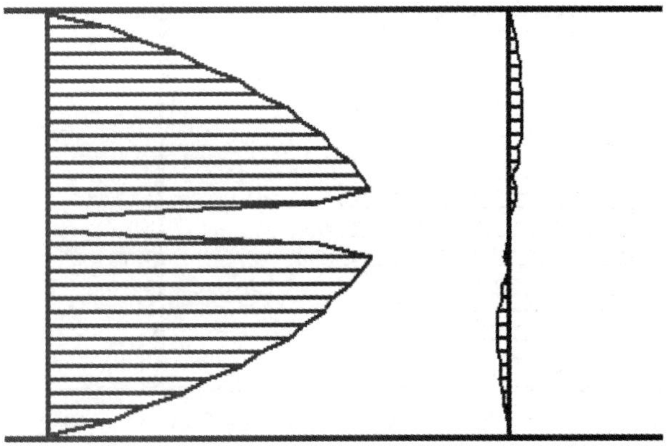

FIGURE 10.48 Velocity profiles on the 0X (left)- and 0Y (right)-axes at a distance $L_x = \Delta x$ to the left of the end of the plate; Re = 3200 and $\hat{t} = 5.0$.

Let, at the initial moment of time, a flow in a channel abruptly separate in two parallel flows. In this case, from the hydrodynamic standpoint, the problem reduces to a flow past a semi-infinite plate of finite thickness that suddenly appeared in the channel. The boundary conditions pre-assigned at the outlet of any of two channels formed are similar to the conditions at section *6* (see Table 10.10).

Figures 10.50–10.53 illustrate, for a flow mode at Re = 1000 and $\hat{t} = 50.0$, the spatial evolution of the profiles of longitudinal and transverse velocities of flow in channels for sections located at distances L_x from the beginning of the

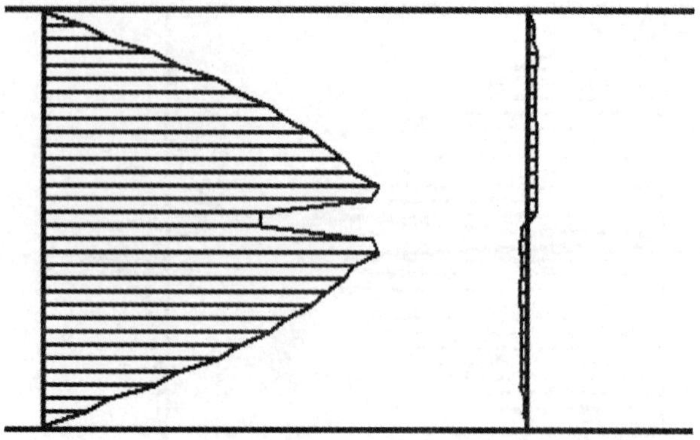

FIGURE 10.49 Velocity profiles on the 0X (left)- and 0Y (right)-axes at a distance $L_x = D/2$ to the right of the end of the plate; Re = 3200 and $\hat{t} = 5.0$.

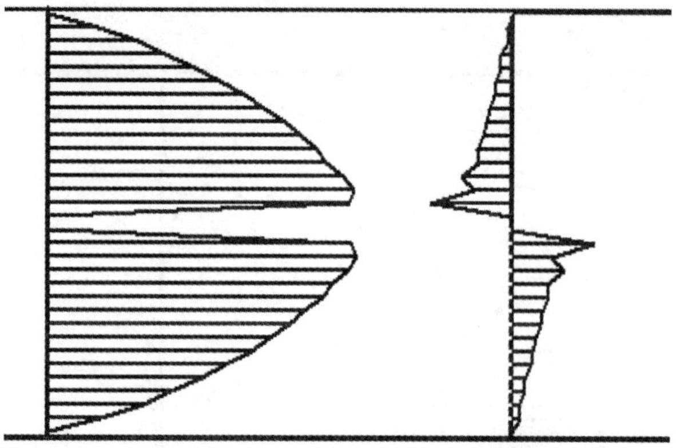

FIGURE 10.50 Velocity profiles on the 0X (left)- and 0Y (right)-axes at a distance $L_x = \Delta x$ to the right of the beginning of the plate; Re = 1000 and $\hat{t} = 50.0$.

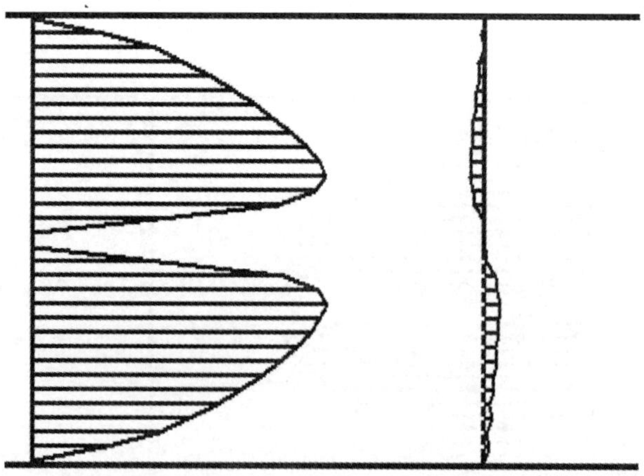

FIGURE 10.51 Velocity profiles on the 0X (left)- and 0Y (right)-axes at a distance $L_x = D/2$ to the right of the beginning of the plate; Re = 1000 and $\hat{t} = 50.0$.

semi-infinite plate equal to Δx (step of the space grid on the OX-axis), $D/2$, D, and $10D$, respectively. The effect of a semi-infinite plate on the upstream rearrangement of flow is by and large similar to the perturbation of flow due to a plate of finite length; therefore, it is not treated in what follows. Note that, at a distance of approximately D, the longitudinal velocity profiles do not become parabolic: the velocity maximum are shifted from the centers of channels to the plate. This shift is caused by the fact that, at the inlet to the wide part of the channel at the initial moment of time, a Poiseuille profile has a maximum of longitudinal velocity located on the level of the plate that suddenly appeared in the channel. By the moment of time $\hat{t} = 50.0$, the longitudinal velocity profiles in the section $10D$ assume a parabolic form, and the fluctuations of longitudinal velocity are very low.

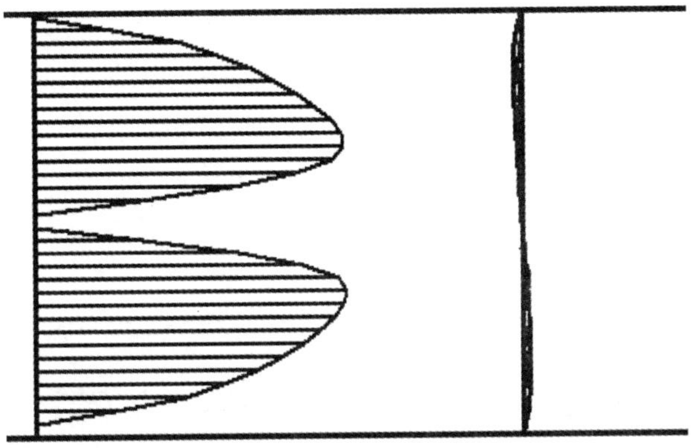

FIGURE 10.52 Velocity profiles on the 0X (left)- and 0Y (right)-axes at a distance $L_x = D$ to the right of the beginning of the plate; Re = 1000 and $\hat{t} = 50.0$.

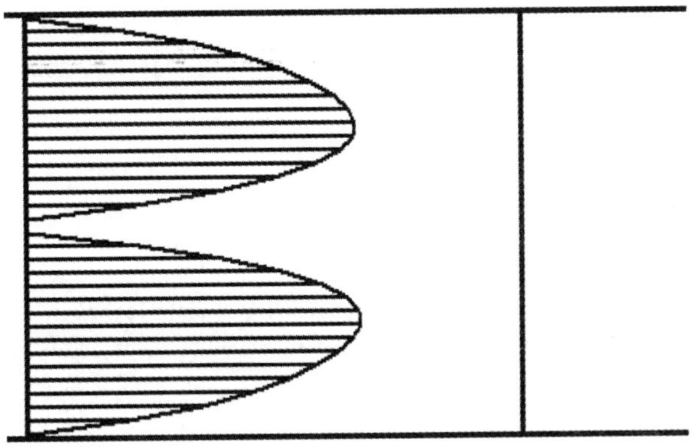

FIGURE 10.53 Velocity profiles on the 0X (left)- and 0Y (right)-axes at a distance $L_x = 10D$ to the right of the beginning of the plate; Re = 1000 and $\hat{t} = 50.0$.

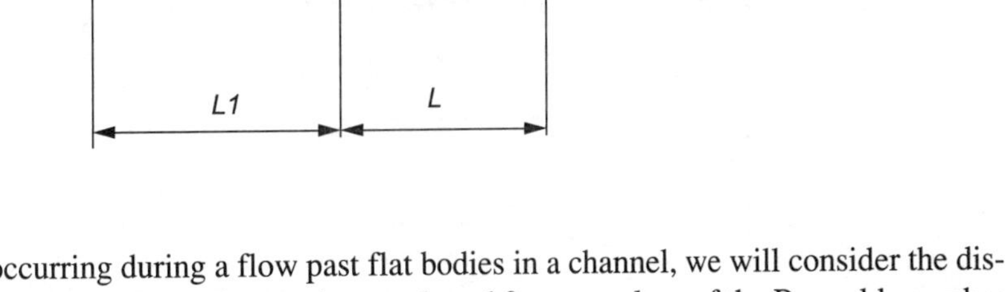

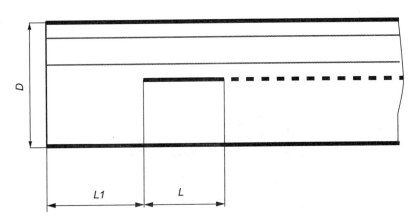

FIGURE 10.54 A schematic representation of pressure chart plotting lines; each line is located at a distance of 0.15 from the nearest surface.

For a better illustration of the processes occurring during a flow past flat bodies in a channel, we will consider the distribution of dimensionless pressure over the channel length. The graphs are plotted for two values of the Reynolds number over two segments parallel to the channel walls (see Figs. 10.54–10.57). It is interesting to investigate the reaction of the GHE to a variation of the boundary conditions. If the "soft" condition at the inlet to the calculation region $d^2\hat{p}/dx^2 = 0$ is replaced by the "rigid" condition $\hat{p} = 1$, the pressure perturbations propagating upstream will not go beyond the limits of the calculation region. As a result, a system of standing pressure waves will be formed in the vicinity of the left-hand boundary of the calculation region; in so doing, the pattern of downstream variation of pressure changes little.

It is known that extensive experimental data exist on the drag coefficients in a flow past a plate, as well as prediction data based, as a rule, on the boundary layer theory. We will compare the results of mathematical simulation using the generalized hydrodynamic equations (GEE) to the Blasius theory.

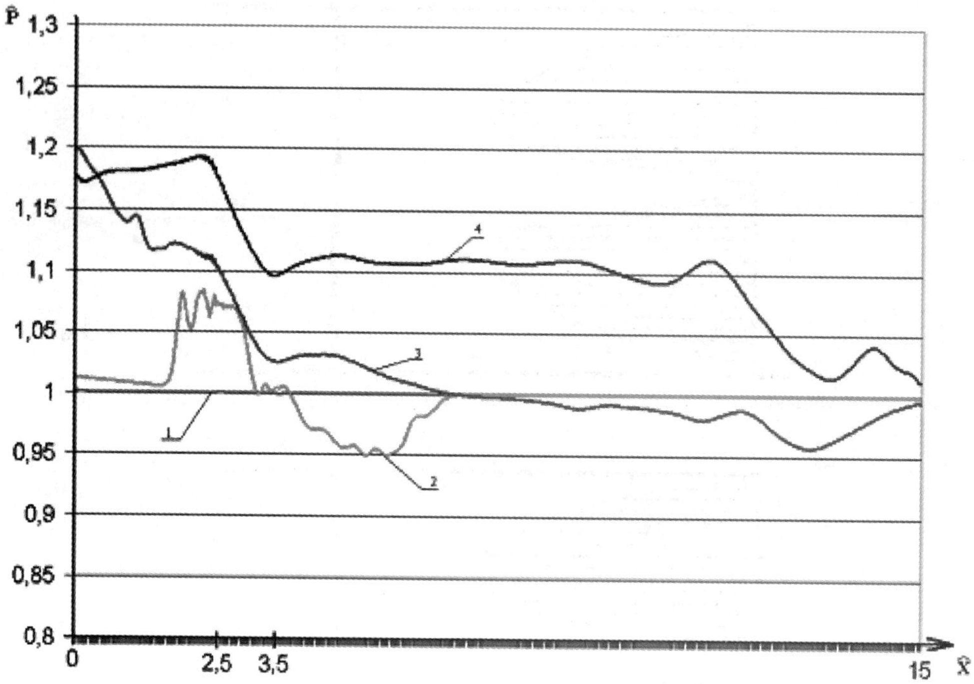

FIGURE 10.55 The distribution of dimensionless pressure over the channel length on a line in the vicinity of the channel wall at Re = 1000 at different moments of time (*1*—$\hat{t}$ = 0.2; *2*—$\hat{t}$ = 3.0; *3*—$\hat{t}$ = 25.0; and *4*—$\hat{t}$ = 100.0). The coordinates of the beginning and end of a plate placed in the channel are indicated.

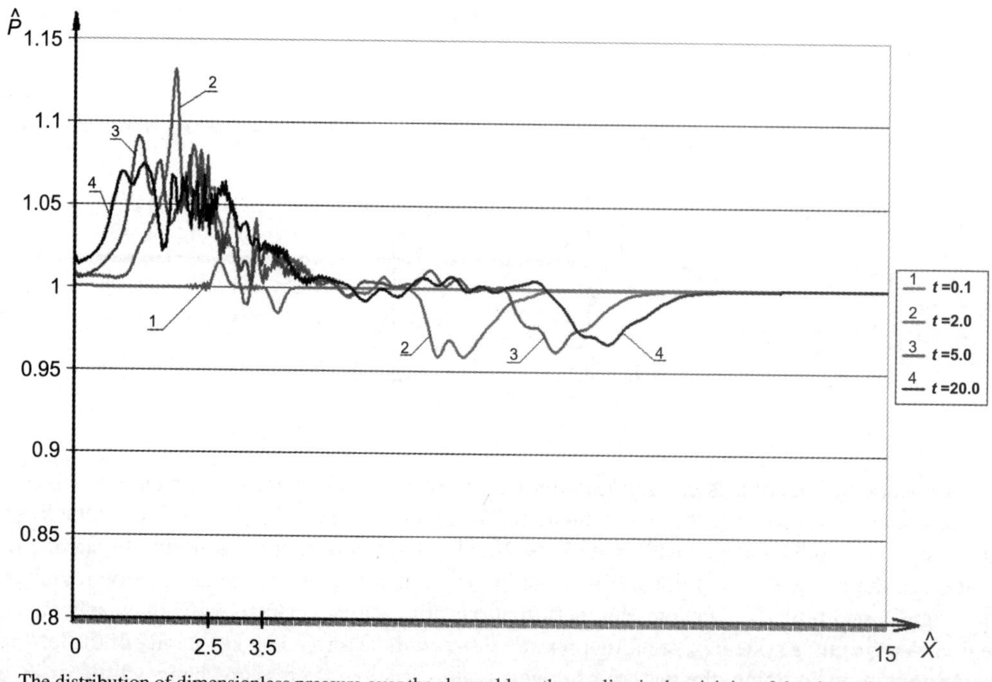

FIGURE 10.56 The distribution of dimensionless pressure over the channel length on a line in the vicinity of the channel wall at Re = 3200 at different moments of time. The coordinates of the beginning and end of a plate placed in the channel are indicated.

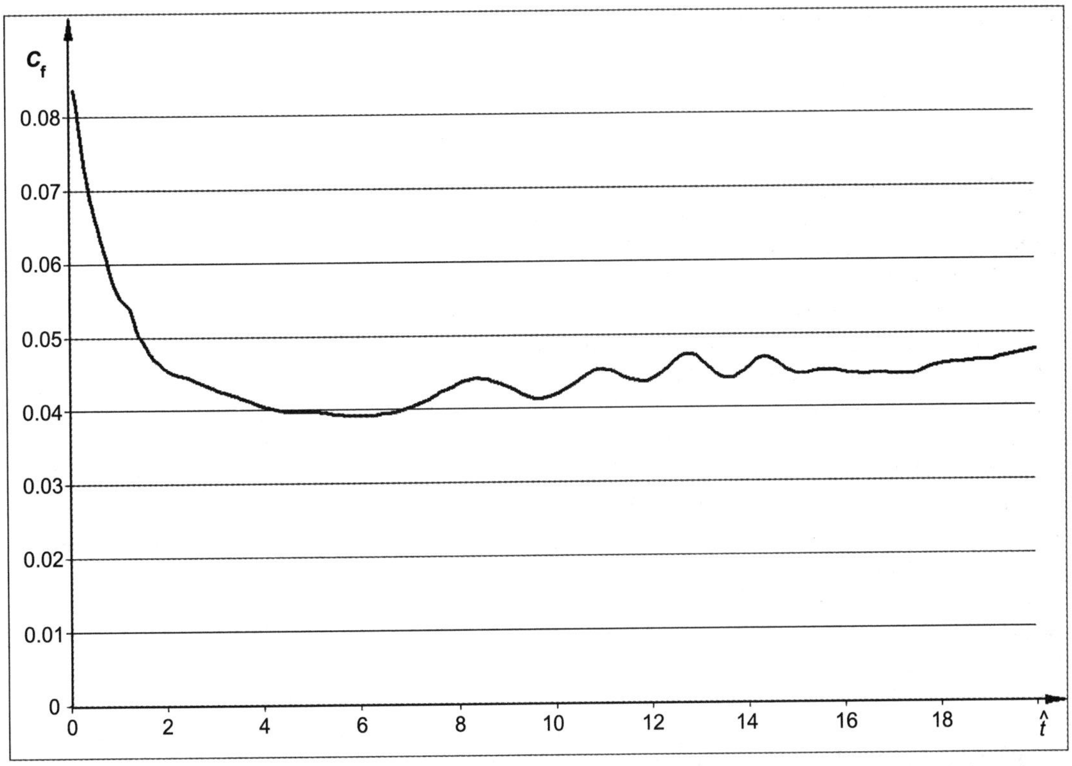

FIGURE 10.57 The time variation of the coefficient of total drag for a plate of finite length at Re = 1000. The number of calculation points over the plate length is 300.

We will write the Blasius formula for the coefficient c_f of total drag of a plate of finite length L:

$$c_f = \frac{1.328}{\sqrt{\mathrm{Re}_L}}, \tag{10.6.1}$$

where Re_L is the Reynolds number calculated using the plate length L and the incident flow velocity. The Blasius formula (10.6.1) was derived within the laminar boundary layer theory [232]. In the general case in a dimensionless form, the total drag coefficient may be represented as

$$c_f = 2W / \left(\frac{1}{2}\rho_\infty u_\infty^2 S\right), \tag{10.6.2}$$

where W is drag of one side of the plate

$$W = h\int_{x=0}^{L} \tau_0 \mathrm{d}x.. \tag{10.6.3}$$

In relations (10.6.2) and (10.6.3) τ_0 is local drag coefficient of the plate corresponding to its surface "w",

$$\tau_0 = \mu\left(\frac{\mathrm{d}u}{\mathrm{d}y}\right)_w, \tag{10.6.4}$$

S is an area of the plated streamed by flow, $S = 2Lh$, h is width of the plate.

As result one obtains the coefficient of the total drag

$$c_f = \frac{\int_{(L)} \mu(\mathrm{d}u/\mathrm{d}y)_w \mathrm{d}x}{(1/2)\rho_\infty u_\infty^2 L}, \tag{10.6.5}$$

(the integral determined in Eq. (10.6.5) is calculated over the plate length) or in the dimensionless form

$$c_f = \frac{2\int_{(\widehat{L})} \widehat{\mu}\left(\mathrm{d}\widehat{u}/\mathrm{d}\widehat{y}\right)_w \mathrm{d}\widehat{x}}{\mathrm{Re}} \tag{10.6.6}$$

The values of c_f calculated by the foregoing formula using the grid functions for Re = 1000 are given in Table 10.12, and for Re = 3200—in Table 10.13; given in the same table for comparison are values; given in the same table for comparison are values found by formula (10.6.1). The Blasius formula corresponds to the steady-state mode of flow; therefore, the values of c_f are duplicated in the right-hand column for all moments of time.

It follows from Tables 10.12 and 10.13 that, at the initial moments of time, the value of total drag coefficient significantly exceeds the value of c_f calculated by the Blasius formula.

Figure 10.57 shows the time evolution of the coefficient of total drag of a plate at Re = 1000. At the initial moments of time, the drag coefficient significantly exceeds the plate resistance in a steady-state laminar flow; then, it decreases with time, experiences oscillation, and approaches the value calculated by the Blasius formula while remaining above that value. Figure 10.58 corresponds to the variation of the local dimensionless friction coefficient over the plate length for different moments of time. The Blasius curve is given along with the results of mathematical simulation using the GHE. Figures 10.59 and 10.60 are similar to Figs. 10.57 and 10.58, but they are plotted for Re = 3200. The results of calculation of local coefficients of friction on a plate using the Blasius formula and the GHEs differ very significantly. However, the otal (integral) drags of the plate (which, in fact, represent the objective of experimental measurements) for these models in the quasi-steady-state mode in the investigated range of Reynolds number values differ little from one another (within 15%, see Tables 10.12 and 9.13). Therefore, the total drag is a "conservative" quantity.

Parameters reflecting the energy of turbulent pulsations are often used in the theory of turbulent flows (see, for example, [233]). In order to demonstrate the results of appropriate mathematical simulation, we will introduce the parameter ε related to the square of pulsating components of velocity,

$$\varepsilon = \sqrt{\frac{1}{3}\left[\left(\hat{u}_{\infty}^{fl}\right)^2 + \left(\hat{v}_{\infty}^{fl}\right)^2\right]}\,100\%, \tag{10.6.7}$$

TABLE 10.12 The Coefficient of Total Drag of a Plate, Found by Numerical Calculation and by Blasius Formula, Re = 1000

	c_f (Numerical Calculation)	c_f (Blasius Formula)
$\widehat{t}=0.2$	0.07908	0.041995
$\widehat{t}=3.0$	0.04730	0.041995
$\widehat{t}=25.0$	0.04826	0.041995
$\widehat{t}=100.0$	0.04931	0.041995

TABLE 10.13 The Coefficient of Total Drag of a Plate, Found by Numerical Calculation and by Blasius Formula, Re = 3200

	c_f (Numerical Calculation)	c_f (Blasius formula)
$\widehat{t}=0.1$	0.03289	0.024246
$\widehat{t}=2.0$	0.02876	0.024246
$\widehat{t}=5.0$	0.02898	0.024246
$\widehat{t}=30.0$	0.02952	0.024246

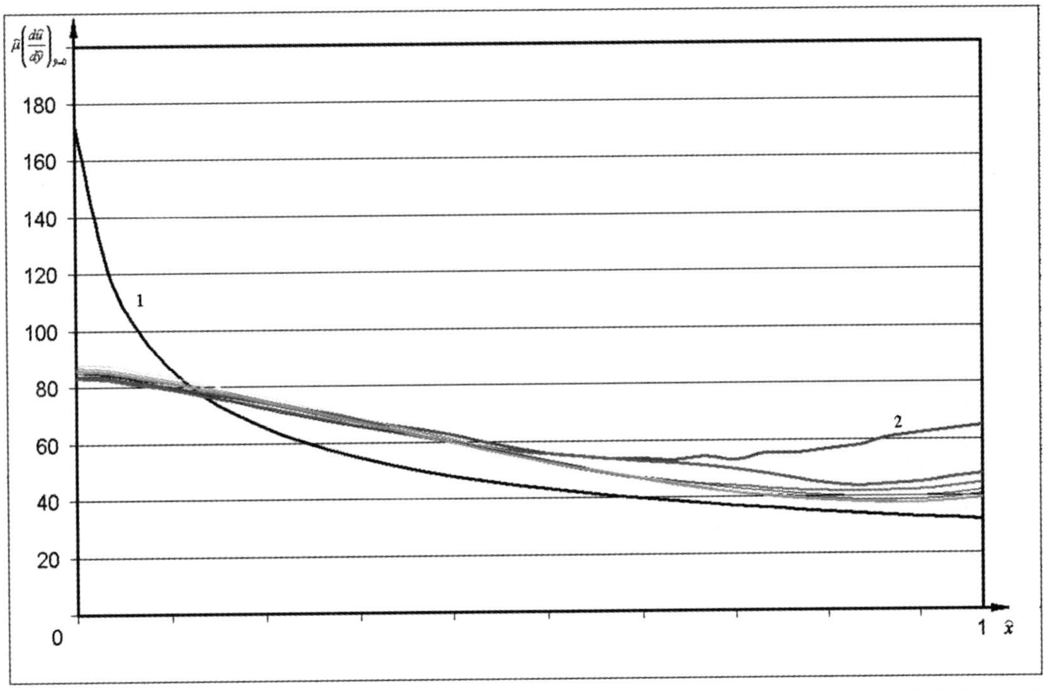

FIGURE 10.58 The variation of the local drag coefficient for a plate of finite length. The dimensionless coordinates 0 and 1 indicate the beginning and end of the plate, Re = 1000: (1) calculation by the Blasius formula and (2) numerical calculation, time $\hat{t} = 1.0$ (the remaining curves for the subsequent moments of time are fairly close and, therefore, are not numbered).

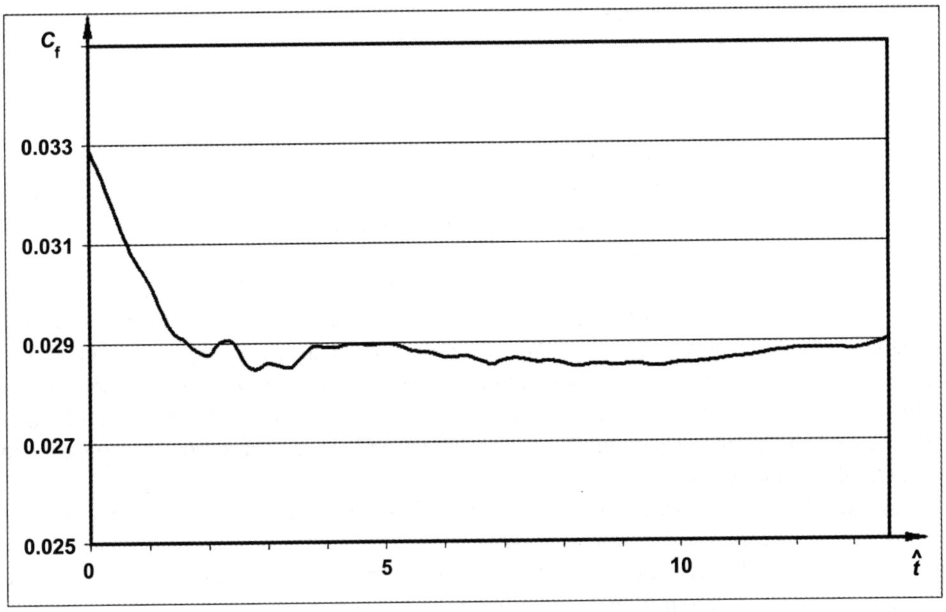

FIGURE 10.59 The time variation of the coefficient of total drag for a plate of finite length at Re = 3200.

where

$$\hat{u}^{fl} = \frac{\Pi\hat{\mu}}{\hat{p}}\frac{1}{\text{ReEu}}\left[\frac{\partial \hat{v}_{0x}}{\partial \hat{t}} + \left(\hat{\mathbf{v}}_0 \cdot \frac{\partial}{\partial \hat{\mathbf{r}}}\right)\hat{v}_{0x} + \frac{\text{Eu}}{\hat{\rho}}\frac{\partial \hat{p}}{\partial \hat{x}}\right], \tag{10.6.8}$$

$$\hat{v}^{fl} = \frac{\Pi\hat{\mu}}{\hat{p}}\frac{1}{\text{ReEu}}\left[\frac{\partial \hat{v}_{0y}}{\partial \hat{t}} + \left(\hat{\mathbf{v}}_0 \cdot \frac{\partial}{\partial \hat{\mathbf{r}}}\right)\hat{v}_{0y} + \frac{\text{Eu}}{\hat{\rho}}\frac{\partial \hat{p}}{\partial \hat{y}}\right]. \tag{10.6.9}$$

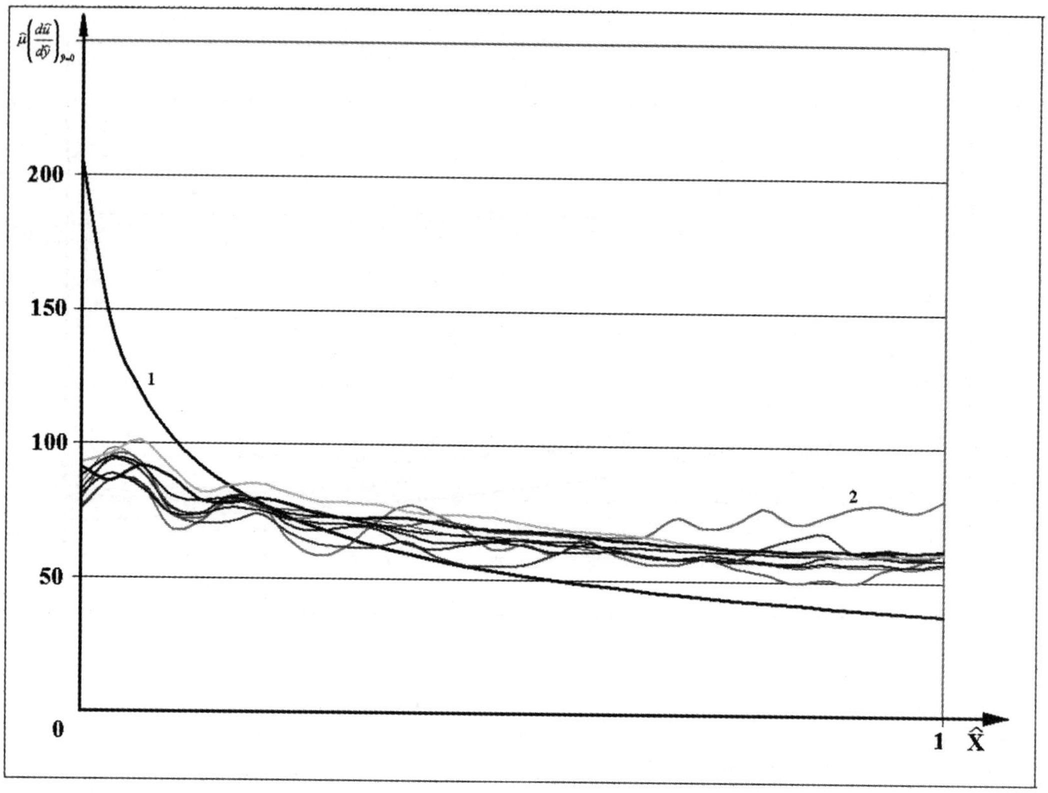

FIGURE 10.60 The variation of the local drag coefficient for a plate of finite length. The dimensionless coordinates 0 and 1 indicate the beginning and end of the plate, Re = 3200: (1) calculation by the Blasius formula and (2) numerical calculation, dimensionless time $\hat{t} = 1.0$ (the remaining curves for the subsequent moments of time are fairly close and, therefore, are not numbered).

These pulsating components of velocity may, of course, be calculated both in the "external" flow (as indicated by the subscript ∞ with pulsating components of velocity) and in the "internal" flow.

Figure 10.61 gives, for different moments of time, the results of calculation of the parameter ε along the top longitudinal line indicated in Fig. 10.54. While, at Re = 1000, a certain stabilization of the predicted parameter ε occurs with time, at Re = 3200 only a certain range of variation of energy parameters is observed.

Note that the "snapshots"" of the behavior of flow, used to demonstrate the calculation results, cannot fully describe the flow behavior in the calculation region. This is due to the fact that the parameters of a flow of liquid, even if this flow is laminar, do not remain unvaried in time at every point. We will point out yet another, not physical but numerical, effect that is observed in the case of insufficient length of the calculation region. The calculation results are strongly dependent on the relative length of the calculation region. However, no such dependence is observed in the case of an adequate length of the calculation region of flow. The choice of the length of the calculation region is based on the comparison of calculation results for different lengths of the region.

It is interesting to notice that in many works (see, for example, [234–237]) have shown that the pressure and temperature profiles in plane Poiseuille flow exhibit a different qualitative behavior from the profiles obtained by Navier-Stokes equations. As comment to these results, F. Uribe and A. Garcia [236] wrote that "Poiseuille flow is the first scenario in which the Navier-Stokes equations have been shown to be susceptible to significant improvement for a flow with relatively small gradients."

In more details the problem formulation looks as follows. Stationary Poiseuille flow is confined between two rigid parallel plates. Two cases are considered. In the acceleration-driven case the boundary conditions are taken as periodic in the flow direction and external force—for example gravity force—is applied in this (x) direction. In the pressure-driven case the boundary conditions are set to create the pressure gradient but without external forces. The results of Navier-Stokes calculations are compared with data of Direct Simulation Monte Carlo (DSMC) [238, 239].

The typical simulated flow is a hard sphere gas for which Kn = 0.1, Ma = 0.5, Eu = 0.5 then Re $\sim$ 5. In boundary conditions for Navier-Stokes (NS) equations the slip and jump corrections for fully accommodating surface were introduced [239].

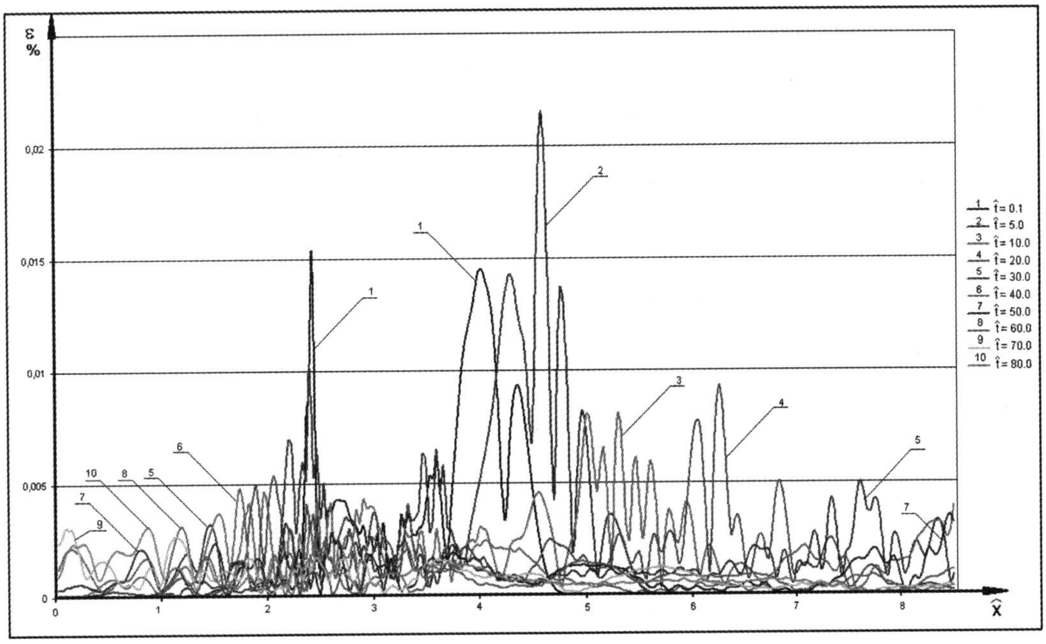

FIGURE 10.61 The values of ε along the top line (Fig. 10.54) in the flow for different time moments; Re=3200.

Two main discrepancy were discovered between NS and DSMC modeling. For acceleration-driven case in DSMC one observes a non-constant pressure profile and a temperature dip in the center of the channel [236]. For NS and DSMC pressure profile in y direction have opposite curvature. Moreover, the DSMC data indicate a reverse temperature jump at the wall—gas temperature near the wall is slightly greater than wall temperature—in spite the global temperature gradient normal to walls is negative.

All these interesting effects need additional investigations. Comparison indicated results with data of modeling based on the generalized Euler equations lead to conclusion that mentioned effects can be observed but as result of non-stationary non-one-dimensional calculations.

As additional explanation let us consider the stationary one-dimensional acceleration-driven case for hard sphere gas, σ is corresponding particle's diameter. Let us suppose that $\rho=\rho(y),\quad u=u(y),\quad p=p(y),\quad T=T(y)$.

In this case the generalized *Euler* Eqs. (2.7.49)–(2.7.51) lead to the system of hydrodynamic equations

$$\frac{\partial p}{\partial y}=0, \tag{10.6.10}$$

$$T^2\frac{\partial^2 u}{\partial y^2}+\frac{T}{2}\left(\frac{\partial T}{\partial y}\right)\left(\frac{\partial u}{\partial y}\right)+\frac{16p\sigma^2}{5k_{\mathrm{B}}\Pi}g\sqrt{\frac{\pi mT}{k_{\mathrm{B}}}}=0, \tag{10.6.11}$$

$$\frac{\partial^2 T}{\partial y^2}+\frac{1}{2T}\left(\frac{\partial T}{\partial y}\right)^2+\frac{2m}{5k_{\mathrm{B}}}\left(\frac{\partial u}{\partial y}\right)^2+\frac{2m^2}{5k_{\mathrm{B}}^2T}g^2=0. \tag{10.6.12}$$

For this case and in this approximation Navier-Stokes equations is written as follows:

$$\frac{\partial p}{\partial y}=0, \tag{10.6.13}$$

$$T^2\frac{\partial^2 u}{\partial y^2}+\frac{T}{2}\left(\frac{\partial T}{\partial y}\right)\left(\frac{\partial u}{\partial y}\right)+\frac{16p\sigma^2}{5k_{\mathrm{B}}}g\sqrt{\frac{\pi mT}{k_{\mathrm{B}}}}=0, \tag{10.6.14}$$

$$\frac{\partial^2 T}{\partial y^2}+\frac{1}{2T}\left(\frac{\partial T}{\partial y}\right)^2+\frac{4m}{15k_{\mathrm{B}}}\left(\frac{\partial u}{\partial y}\right)^2=0. \tag{10.6.15}$$

The main difference between systems of Eqs. (10.6.10)–(10.6.12) and (10.6.13)–(10.6.15) consists in appearance in the generalized Euler energy equation the additional term, which contains—in opposite to NS energy equation—gravitation in explicit form. It can be said more precisely—after transaction to the dimensionless form of energy equation—the ratio of gravitational energy to thermal energy, or in general case the ratio of energy of external field to thermal energy.

Non-local gravitational effects will be discussed in the following chapter in detail.

Note in conclusion that the GHE represent an effective tool for solving problems in gas dynamics.

Chapter 11

Astrophysical Applications

ABSTRACT

The principle of the universal antigravitation is considered from positions of the Newtonian theory of gravitation and non-local kinetic theory. It is found that explanation of the Hubble effect in the universe and peculiar features of the rotational speeds of galaxies need not in introduction of new essence like dark matter and dark energy. The origin of difficulties consists in total Oversimplification following from principles of local physics and reflects the general shortcomings of the local kinetic transport theory.

Keywords: Dark matter and dark energy problems, Disk galaxy rotation, Hubble expansion, Propagation of gravitational waves, Matter movement in black hole, Self-similar non-local solutions

11.1 SOLUTION OF THE DARK MATTER PROBLEM IN THE FRAME OF THE NON-LOCAL PHYSICS

More than 10 years ago, the accelerated cosmological expansion was discovered in direct astronomical observations at distances of a few billion light years, almost at the edge of the observable Universe. This acceleration should be explained because mutual attraction of cosmic bodies is only capable of decelerating their scattering. It means that we reach the revolutionary situation not only in physics but also in the natural philosophy on the whole. Practically we are in front of the new challenge since Newton's *Mathematical Principles of Natural Philosophy* was published. As a result, new idea was introduced in physics about the existence of a force with the opposite sign, which is called universal antigravitation. Its physical source is called as dark energy that manifests itself only because of postulated property of providing antigravitation.

It was postulated that the source of antigravitation is "dark matter" which inferred to exist from gravitational effects on visible matter. However, from the other side dark matter is undetectable by emitted or scattered electromagnetic radiation. It means that new essences—dark matter, dark energy—were introduced in physics only with the aim to account for discrepancies between measurements of the mass of galaxies, clusters of galaxies, and the entire universe made through dynamical and general relativistic means, measurements based on the mass of the visible "luminous" matter. It could be reasonable if we are speaking about small corrections to the system of knowledge achieved by mankind to the time we are living. But mentioned above discrepancies lead to affirmation, that dark matter constitutes 80% of the matter in the universe, while ordinary matter makes up only 20%. There is a variety in the corresponding estimations, but the situation is defined by maybe emotional, but the true exclamation that can be found between thousands Internet cues—"It is humbling, perhaps even humiliating, that we know almost nothing about 96% of what is 'out there'"!!

Dark matter was postulated by Swiss astrophysicist Fritz Zwicky of the California Institute of Technology in 1933 [240, 241]. He applied the virial theorem to the Coma cluster of galaxies and obtained evidence of unseen mass. Zwicky estimated the cluster's total mass based on the motions of galaxies near its edge and compared that estimate to one based on the number of galaxies and total brightness of the cluster. He found that there was about 400 times more estimated mass than was visually observable. The gravity of the visible galaxies in the cluster would be far too small for such fast orbits, so something extra was required. This is known as the "missing mass problem." Based on these conclusions, Zwicky inferred that there must be some non-visible form of matter, which would provide enough of the mass, and gravity to hold the cluster together.

Observations have indicated the presence of dark matter in the universe, including the rotational speeds of galaxies, gravitational lensing of background objects by galaxy clusters such as the Bullet Cluster, and the temperature distribution of hot gas in galaxies and clusters of galaxies.

The work by Vera Rubin (see for example [242, 243]) revealed distant galaxies rotating so fast that they should fly apart. Outer stars rotated at essentially the same rate as inner ones ($\sim$254 km/s). This is in marked contrast to the solar system where planets orbit the sun with velocities that decrease as their distance from the center increases. By the early 1970s, flat rotation curves were routinely detected. It was not until the late 1970s, however, that the community was convinced of the need for dark matter halos around spiral galaxies. The mathematical modeling (based on Newtonian mechanics and local physics) of

Unified non-local theory of transport processes. http://dx.doi.org/10.1016/B978-0-444-63478-8.00011-7

the rotation curves of spiral galaxies was realized for the various visible components of a galaxy (the bulge, thin disk, and thick disk). These models were unable to predict the flatness of the observed rotation curve beyond the stellar disk. The inescapable conclusion, assuming that Newton's law of gravity (and the local physics description) holds on cosmological scales, that the visible galaxy was embedded in a much larger dark matter (DM) halo, which contributes roughly 50-90% of the total mass of a galaxy. As a result another models of gravitation were involved in consideration—from "improved" Newtonian laws (such as modified Newtonian dynamics and tensor-vector-scalar gravity [244]) to the Einstein's theory based on the cosmological constant [245]. Einstein as a mechanism to obtain a stable solution of the gravitational field equation that would lead to a static universe, effectively using dark energy to balance gravity, introduced this term.

Computer simulations with taking into account the hypothetical DM in the local hydrodynamic description include usual moment equations plus Poisson equation with different approximations for the density of DM (ρ_{DM}) containing several free parameters. Computer simulations of cold dark matter (CDM) predict that CDM particles ought to coalesce to peak densities in galactic cores. However, the observational evidence of star dynamics at inner galactic radii of many galaxies, including our own Milky Way, indicates that these galactic cores are entirely devoid of CDM. No valid mechanism has been demonstrated to account for how galactic cores are swept clean of CDM. This is known as the "cuspy halo problem." As a result, the restricted area of CDM influence introduced in the theory. As we see the concept of DM leads to many additional problems.

I do not intend to review the different speculations based on the principles of local physics. I see another problem. It is the problem of Oversimplification—but not "trivial" simplification of the important problem. The situation is much more serious—total Oversimplification based on principles of local physics, and obvious crisis, we see in astrophysics, simply reflects the general shortcomings of the local kinetic transport theory. The antigravitation problem is solved further in the frame of non-local statistical physics and the Newtonian law of gravitation.

In the following sections I intend to answer the questions of principal significance—is it possible using only Newtonian gravitation law and non-local statistical description to forecast the Hubble expansion, and flat gravitational curve of a typical spiral galaxy? Both questions have the positive answers.

11.2 PLASMA-GRAVITATIONAL ANALOGY IN THE GENERALIZED THEORY OF LANDAU DAMPING

The aim of this section consists in the application of plasma-gravitational analogy for the effect of Hubble expansion using the generalized theory of Landau damping and the generalized Boltzmann physical kinetics (see Chapter 7). The collisionless damping of electron plasma waves was predicted by Landau in 1946 [147] and later was confirmed experimentally. In spite of transparent physical sense, the effect of Landau damping has continued to be of great interest to theorist as well. Much of this interest is connected with counterintuitive nature of result itself coupled with the rather abstruse mathematical nature of Landau's original derivation (including so-called Landau's rule of complex integral calculation. In papers [56, 65, 66] the difficulties originated by Landau's derivation were clarified. The mentioned consideration leads to another solution of Vlasov-Landau equation, these ones in agreement with data of experiments. The problem solved in this section consists in consideration of the generalized theory of Landau damping in gravitating systems from viewpoint of Generalized Boltzmann Physical Kinetics and non-local physics. The influence of the particle collisions is taken into account.

Plasma-gravitational analogy is well-known and frequently used effect in physical kinetics. The origin of analogy is simple and is connected with analogy between Coulomb law and Newtonian law of gravitation. From other side electrical charges can have different signs whereas there is just one kind of "gravitational charge" (i.e. masses of particles) corresponding to the force of attraction. This fact leads to the extremely important distinctions in formulation of the generalized theory of Landau damping in gravitational media. In the following, we intend to use the classical non-relativistic Newtonian law of gravitation

$$\mathbf{F}_{21} = \gamma_N \frac{m_1 m_2}{r_{12}^2} \frac{\mathbf{r}_{12}}{r_{12}}, \tag{11.2.1}$$

where $\mathbf{F}_{21}$ is the force acting on the particle "1" from the particle "2", $\mathbf{r}_{12}$ is vector directed from the center-of-mass of the particle "1" to the particle "2", γ_N is gravitational constant $\gamma_N = 6.6 \times 10^{-8}\,\mathrm{cm}^3/\mathrm{gs}^2$; the corresponding force $\mathbf{g}_{21}$ per mass unit is

$$\mathbf{g}_{21} = \mathbf{F}_{21}/m_1. \tag{11.2.2}$$

The flux

$$\Phi = \int_S g_n \mathrm{d}S \tag{11.2.3}$$

for closed surface S can be calculated using (12); one obtains

$$\int_S g_n \mathrm{d}S = -4\pi\gamma_{\mathrm{N}} \int_V \rho^{\mathrm{a}} \mathrm{d}V, \tag{11.2.4}$$

where ρ^{a} is density *inside* of volume V bounded by the surface S. As usual, Eq. (11.2.4) can be rewritten as

$$\int_V \left(\frac{\partial}{\partial \mathbf{r}} \cdot \mathbf{g} + 4\pi\gamma_{\mathrm{N}}\rho^{\mathrm{a}} \right) \mathrm{d}V = 0. \tag{11.2.5}$$

The definite integral (Eq. 11.2.5) is equal to zero for arbitrary volume V, then

$$\frac{\partial}{\partial \mathbf{r}} \cdot \mathbf{g} = -4\pi\gamma_{\mathrm{N}}\rho^{\mathrm{a}}, \tag{11.2.6}$$

and after introduction the gravitational potential ψ

$$\mathbf{g} = -\partial\psi/\partial\mathbf{r} \tag{11.2.7}$$

we reach the known Poisson equation

$$\Delta\psi = 4\pi\gamma_{\mathrm{N}}\rho^{\mathrm{a}}. \tag{11.2.8}$$

Generalized Boltzmann physical kinetics leads to the possibility to calculate the density ρ^{a} using the density ρ (obtained with the help of the one-particle distribution function (DF) f) and the fluctuation term ρ^{f}. All fluctuation terms in the generalized Boltzmann equation (GBE) theory were tabulated (see Chapter 10) and for ρ^{f} we have

$$\rho^{\mathrm{f}} = \tau\left(\frac{\partial\rho}{\partial t} + \frac{\partial}{\partial\mathbf{r}} \cdot \rho\mathbf{v}_0\right), \tag{11.2.9}$$

where $\mathbf{v_0}$ is hydrodynamic velocity. After substitution of ρ^{f} in Eq. (11.2.8) one obtains

$$\Delta\psi = 4\pi\gamma_{\mathrm{N}}\left[\rho - \tau\left(\frac{\partial\rho}{\partial t} + \frac{\partial}{\partial\mathbf{r}} \cdot \rho\mathbf{v}_0\right)\right]. \tag{11.2.10}$$

From Eqs. (11.2.8) to (11.2.10) follow that classical Newtonian field equation

$$\Delta\psi = 4\pi\gamma_{\mathrm{N}}\rho. \tag{11.2.11}$$

valid only for situation when the fluctuations terms can be omitted and then

$$\rho = \rho^{\mathrm{a}}. \tag{11.2.12}$$

This condition can be considered as the simplest closure condition, but in the general case, the other hydrodynamic equations should be involved into consideration because Eq. (11.2.10) contains hydrodynamic velocity $\mathbf{v}_0$. As a result, one obtains the system of moment equations, i.e. gravitation equation

$$\frac{\partial}{\partial \mathbf{r}} \cdot \mathbf{g} = -4\pi\gamma_{\mathrm{N}}\left[\rho - \tau\left(\frac{\partial\rho}{\partial t} + \frac{\partial}{\partial\mathbf{r}} \cdot \rho\mathbf{v}_0\right)\right], \tag{11.2.13}$$

and generalized continuity, motion and energy equations which can be further applied to the theory of the rotation curves of spiral galaxies.

The following mathematical transformations will be obtained on the level of the generalized theory of Landau damping based on the GBE, and need in some preliminary additional explanations from viewpoint of so-called dark energy and dark matter.

As it was mentioned above, the accelerated cosmological expansion was discovered in direct astronomical observations. For explanation of this acceleration, new idea was introduced in physics about existing of a force with the opposite sign, which is called the universal antigravitation. In the simplest interpretation, dark energy is related usually to the Einstein cosmological constant. In review [245] the modified Newton force is written as

$$F(r) = -\frac{\gamma_{\mathrm{N}} M}{r^2} + \frac{8\pi\gamma_{\mathrm{N}}}{3}\rho_{\mathrm{v}} r, \tag{11.2.14}$$

where ρ_{v} is the Einstein-Gliner vacuum density introduced also in [246]. The problem can be solved without the ideology of the Einstein-Gliner vacuum. However, for us it is interesting the interpretation of the modified law (Eq. (11.2.14)) [245]. In the limit of large distances, the influence of central mass M becomes negligibly small and the field of forces is determined only by the second term in the right side of Eq. (11.2.14). It follows from relation (Eq. (11.2.14)) that there is a "equilibrium" distance r_v, at which the sum of the gravitation and antigravitation forces is equal to zero. In other words r_v is "the zero-gravitational radius". For so-called Local Group of galaxies estimation of r_v is about 1 Mpc, [245]. There are no theoretical methods of the density ρ_v calculation. Obviously, the second term in Eq. (11.2.14) should be defined as result of solution of the self-consistent gravitational problem.

Let us return now to the formulation of plasma-gravitational analogy in the frame of generalized theory of Landau damping. I intend to apply the GBE model with the aim to obtain the dispersion relation for one component gas placed in the self-consistent gravitational field and to consider effect of "antigravitation" in the frame of the Newton theory of gravitation.

With this aim let us admit now that there is a gravitational perturbation $\delta\Psi$ in the system of particles as result of the density perturbation $\delta\rho$. These perturbations are connected with the perturbation of DF in the system, which was before in the local equilibrium. In doing so, we will make the additional assumptions (typical in the theory of Landau damping, see Chapter 7) for simplification of the problem, namely:

(a) Consideration of the self-consistent gravitational field corresponds to the area of the large distance r (see Eq. (11.2.14)) from the central mass M where the first term is not significant and in particular, the problem corresponds to the plane case. As mentioned above the second term should be defined as a self-consistent force of the Newtonian origin

$$F = -\frac{\partial \delta\psi}{\partial x}. \tag{11.2.15}$$

(b) The integral local collision term is written in the Bhatnagar-Gross-Krook (BGK) form

$$J = -\frac{f - f_0}{v_r^{-1}}. \tag{11.2.16}$$

into the right-hand side of the GBE. Here, f_0 and $v_r^{-1} = \tau_r$ are respectively a certain equilibrium DF and the relaxation time.

(c) The evolution of particles in a self-consistent gravitational field corresponds to a non-stationary one-dimensional model, u is the velocity component along the x-axis.

(d) The DF f deviate little from its equilibrium counterpart f_0,

$$f = f_0(u) + \delta f(x, u, t). \tag{11.2.17}$$

(e) A wave number k and a complex frequency ω ($\omega = \omega' + i\omega''$) are appropriate to the wave mode considered,

$$\delta f = \langle \delta f \rangle e^{i(kx - \omega t)}, \tag{11.2.18}$$

$$\delta\psi = \langle \delta\psi \rangle e^{i(kx - \omega t)}. \tag{11.2.19}$$

(f) The quadratic GBE terms determining the deviation from the equilibrium DF are neglected.

Under these assumptions listed above, the GBE is written as follows:

$$\begin{aligned} &\frac{\partial f}{\partial t} + u\frac{\partial f}{\partial x} + F\frac{\partial f}{\partial u} - \tau\left\{\frac{\partial^2 f}{\partial t^2} + 2u\frac{\partial^2 f}{\partial t \partial x} + u^2\frac{\partial^2 f}{\partial x^2} + 2F\frac{\partial^2 f}{\partial t \partial u}\right. \\ &\left. + \frac{\partial F}{\partial t}\frac{\partial f}{\partial u} + F\frac{\partial f}{\partial x} + u\frac{\partial F}{\partial x}\frac{\partial f}{\partial u} + F^2\frac{\partial^2 f}{\partial u^2} + 2uF\frac{\partial^2 f}{\partial u \partial x}\right\} = -v_r \delta f, \end{aligned} \tag{11.2.20}$$

where the relations take place for the corresponding terms in Eq. (11.2.20):

$$\begin{aligned} &\frac{\partial f}{\partial t} = -i\omega\,\delta f, u\frac{\partial f}{\partial x} = iku\,\delta f, F\frac{\partial f}{\partial u} = -\frac{\partial \delta\psi}{\partial x}\frac{\partial f_0}{\partial u}, \frac{\partial^2 f}{\partial t^2} = -\omega^2\delta f, 2u\frac{\partial^2 f}{\partial t \partial x} = 2\omega uk\,\delta f, u^2\frac{\partial^2 f}{\partial x^2} = -u^2k^2\,\delta f, \\ &2F\frac{\partial^2 f}{\partial t \partial u} = 0, \frac{\partial F}{\partial t}\frac{\partial f}{\partial u} = -\frac{\partial}{\partial t}\frac{\partial \delta\psi}{\partial x}\frac{\partial f}{\partial u} = -\omega k\,\delta\psi\frac{\partial f_0}{\partial u}, F\frac{\partial f}{\partial x} = 0, \\ &u\frac{\partial f}{\partial u}\frac{\partial F}{\partial x} = -u\frac{\partial f}{\partial u}\frac{\partial}{\partial x}\frac{\partial \delta\psi}{\partial x} = k^2u\,\delta\psi\frac{\partial f_0}{\partial u}, F^2\frac{\partial^2 f}{\partial u^2} = 0, \frac{\partial^2 f}{\partial u \partial x}2uF = 0. \end{aligned} \tag{11.2.21}$$

We are concerned with developing (within the GBE framework) the dispersion relation for gravitational field, and substitution of Eq. (11.2.21) into Eq. (11.2.20) yields

$$\left\{i(ku-\omega)+\nu_r+\tau(ku-\omega)^2\right\}\langle\delta f\rangle-\langle\delta\psi\rangle\frac{\partial f_0}{\partial u}k\{i+\tau(ku-\omega)\}=0. \tag{11.2.22}$$

For the physical system under consideration, the influence of the collision term $\nu_r\langle\delta f\rangle$ is rather small. Using for this case the Poisson equation in the form (Eq. (11.2.11)) and then the relation

$$k^2\langle\delta\psi\rangle=-4\pi\gamma_N\langle\delta n\rangle, \tag{11.2.23}$$

one obtains from Eqs. (11.2.22) and (11.2.23)

$$\langle\delta f\rangle=\frac{4\pi\gamma_N m}{k}\frac{[i-\tau(\omega-ku)]\partial f_0/\partial u}{i(\omega-ku)-\tau(\omega-ku)^2-\nu_r}\langle\delta n\rangle. \tag{11.2.24}$$

After integration over all u we arrive at the dispersion relation

$$1=\frac{4\pi\gamma_N m}{k}\int_{-\infty}^{+\infty}\frac{\partial f_0/\partial u[i-\tau(\omega-ku)]}{i(\omega-ku)-\tau(\omega-ku)^2-\nu_r}du. \tag{11.2.25}$$

Let us suppose that the velocity depending part of DF f_0 corresponds to the Maxwell DF. Then after differentiating in Eq. (11.2.25) under the sign of integral and some transformations, we obtain the integral dispersion equation

$$1-\frac{1}{r_A^2k^2}\left[1-\sqrt{\frac{m}{2\pi k_BT}}\int_{-\infty}^{+\infty}\frac{\{[i-\tau(\omega-ku)]\omega-\nu_r\}e^{-mu^2/2k_BT}}{i(\omega-ku)-\tau(\omega-ku)^2-\nu_r}du\right]=0, \tag{11.2.26}$$

where

$$r_A=\sqrt{\frac{k_BT}{4\pi\gamma_N m^2 n}}. \tag{11.2.27}$$

Poisson equation (11.2.11) has the structure like the Poisson equation for the electrical potential, as result the relation for r_A is analogous to the Debye-Hückel radius $r_D=\sqrt{k_BT/(4\pi e^2 n)}$.

Introducing now the dimensionless variables

$$\widehat{u}=u\sqrt{\frac{m}{2k_BT}},\ \widehat{\omega}=\omega\frac{1}{k}\sqrt{\frac{m}{2k_BT}},\ \widehat{\nu}_r=\nu_r\frac{1}{k}\sqrt{\frac{m}{2k_BT}},\ \widehat{\tau}=\tau k\sqrt{\frac{2k_BT}{m}} \tag{11.2.28}$$

we can rewrite Eq. (11.2.26) in the form

$$1-\frac{1}{r_A^2k^2}\left[1-\frac{1}{\sqrt{\pi}}\int_{-\infty}^{+\infty}\frac{\{[i-\widehat{\tau}(\widehat{\omega}-\widehat{u})]\widehat{\omega}-\widehat{\nu}_r\}e^{-\widehat{u}^2}}{i(\widehat{\omega}-\widehat{u})-\widehat{\tau}(\widehat{\omega}-\widehat{u})^2-\widehat{\nu}_r}d\widehat{u}\right]=0. \tag{11.2.29}$$

Now consider a situation in which the denominator of the complex integrand in Eq. (11.2.29) becomes zero. The quadratic equation

$$\widehat{\tau}y^2-iy+\widehat{\nu}_r=0,\quad y=\widehat{\omega}-\widehat{u}, \tag{11.2.30}$$

has the roots

$$y_1=\frac{i}{2\widehat{\tau}}\left(1+\sqrt{1+4\widehat{\tau}\widehat{\nu}_r}\right),\ y_2=\frac{i}{2\widehat{\tau}}\left(1-\sqrt{1+4\widehat{\tau}\widehat{\nu}_r}\right). \tag{11.2.31}$$

Hence, Eq. (11.2.29) can be rewritten as

$$1-\frac{1}{r_A^2k^2}\left[1+\frac{1}{\widehat{\tau}\sqrt{\pi}}\int_{-\infty}^{+\infty}\frac{\{[i+\widehat{\tau}(\widehat{u}-\widehat{\omega})]\widehat{\omega}-\widehat{\nu}_r\}e^{-\widehat{u}^2}}{(\widehat{u}-\widehat{u}_1)(\widehat{u}-\widehat{u}_2)}d\widehat{u}\right]=0. \tag{11.2.32}$$

where

$$\widehat{u}_1=\widehat{\omega}-y_1,\ \widehat{u}_2=\widehat{\omega}-y_2. \tag{11.2.33}$$

Let us transform Eq. (11.2.29) to the following one:

$$1-\frac{1}{r_A^2k^2}\left\{1+\frac{1}{\sqrt{\pi}}\left[\left(\frac{i\widehat{v}_r+0.5\widehat{\omega}}{\sqrt{1+4\widehat{\tau}\widehat{v}_r}}-0.5\widehat{\omega}\right)\int_{-\infty}^{+\infty}\frac{e^{-\widehat{u}^2}}{(\widehat{u}_1-\widehat{u})}d\widehat{u}\right.\right.$$
$$\left.\left.-\left(\frac{i\widehat{v}_r+0.5\widehat{\omega}}{\sqrt{1+4\widehat{\tau}\widehat{v}_r}}+0.5\widehat{\omega}\right)\int_{-\infty}^{+\infty}\frac{e^{-\widehat{u}^2}}{(\widehat{u}_2-\widehat{u})}d\widehat{u}\right]\right\}=0. \tag{11.2.34}$$

Equation (11.2.34) contains improper Cauchy type integrals. From the theory of complex variables is known Cauchy's integral formula: if the function $f(z)$ is analytic inside and on a simple closed curve C, and z_0 is any point inside C, then

$$f(z_0)=-\frac{1}{2\pi i}\oint_C\frac{f(z)}{z_0-z}dz. \tag{11.2.35}$$

where C is traversed in the positive (counterclockwise) sense.

Let C be the boundary of a simple closed curve placed in lower half plane (for example a semicircle of radius R) with the corresponding element of real axis, z_0 is an interior point. As usual after adding to this semicircle a cross-cut connecting semicircle C with the interior circle (surrounding z_0) of the infinite small radius for analytic $f(z)$, the following formula obtains

$$\oint_C\frac{f(z)}{z_0-z}dz=-\int_{-R}^{R}\frac{f(\widetilde{x})}{z_0-\widetilde{x}}d\widetilde{x}+\int_{C_R}\frac{f(z)}{z_0-z}dz+2\pi i f(z_0), \tag{11.2.36}$$

because the integrals along cross-cut cancel each other, $(z=\widetilde{x}+i\widetilde{y})$.

Analogous for upper half plane

$$\oint_C\frac{f(z)}{z_0-z}dz=\int_{-R}^{R}\frac{f(\widetilde{x})}{z_0-\widetilde{x}}d\widetilde{x}+\int_{C_R}\frac{f(z)}{z_0-z}dz+2\pi i f(z_0). \tag{11.2.37}$$

The formulae (11.2.36) and (11.2.37) could be used for calculation (including the case $R\to\infty$) of the integrals along the real axis with the help of the residual theory *for arbitrary* z_0 if analytical function $f(z)$ satisfies the special conditions of decreasing by $R\to\infty$.

Let us consider now integral $\int_{C_R}\frac{e^{-z^2}}{z_0-z}dz$. Generally speaking, for function $f(z)=e^{-z^2}$ Cauchy's conditions are not satisfied. Really for a point $z=\widetilde{x}+i\widetilde{y}$ this function is $f(z)=e^{\widetilde{y}^2-\widetilde{x}^2}[\cos(2\widetilde{x}\widetilde{y})-i\sin(2\widetilde{x}\widetilde{y})]$ and $f(z)$ is growing by $|\widetilde{y}|>|\widetilde{x}|$ for this part of C_R.

But from physical point of view in ***the linear problem*** *of interaction of individual particles* ***only*** *with waves of potential self-consistent gravitational field the natural assumption can be introduced that solution depends* ***only*** *of concrete* $z_0=\widehat{\omega}'+i\widehat{\omega}''$, *but does not depend of another possible modes of oscillations in physical system.*

It can be realized only if the calculations do not depend of choosing of contour C_R. This fact leads to the additional conditions, for lower half plane

$$\int_{-\infty}^{\infty}\frac{f(\widetilde{x})}{z_0-\widetilde{x}}d\widetilde{x}=2\pi i f(z_0), \tag{11.2.38}$$

and for upper half plane

$$\int_{-\infty}^{\infty}\frac{f(\widetilde{x})}{z_0-\widetilde{x}}d\widetilde{x}=-2\pi i f(z_0). \tag{11.2.39}$$

As it is shown (see Chapter 7) Landau approximation for improper integral contains in implicit form restrictions (valid only for close vicinity of $\widetilde{x}$-axis) for the choice of contour C; these restrictions lead to the continuous spectrum. The question arises, is it possible to find solutions of Eq. (11.2.34) by the restrictions (11.2.38) and (11.2.39)? In the following will be shown that the conditions (11.2.38) and (11.2.39) together with Eq. (11.2.34) lead to the discrete spectrum of $z_0 = \widehat{\omega}' + i\widehat{\omega}''$ and from physical point of view conditions (11.2.38) and (11.2.39) can be considered as condition of quantization.

The relations (11.2.38) and (11.2.39) are the additional conditions which physical sense consists in the extraction of independent oscillations—oscillations which existence does not depend on presence of other oscillations in considering physical system.

Then Eq. (11.2.34) produces the dispersion relation, which admits a damped gravitational wave solution $\left(\widehat{\omega}'' < 0\right)$ (see also Chapter 7):

$$\mp e^{\widehat{u}_2^2} \frac{1 - r_A^2 k^2}{2\sqrt{\pi}} = \frac{\widehat{v}_r}{\sqrt{1 + 4\widehat{\tau}\,\widehat{v}_r}} - \frac{i\widehat{\omega}}{2}\left(1 + \frac{1}{\sqrt{1 + 4\widehat{\tau}\,\widehat{v}_r}}\right), \tag{11.2.40}$$

where

$$\widehat{u}_2^2 = \widehat{\omega}'^2 - \widehat{\omega}''^2 - \widehat{\omega}'' \frac{\sqrt{1 + 4\widehat{\tau}\,\widehat{v}_r} - 1}{\widehat{\tau}} - \frac{1 + 2\widehat{\tau}\,\widehat{v}_r - \sqrt{1 + 4\widehat{\tau}\,\widehat{v}_r}}{2\widehat{\tau}^2} + i\left(2\widehat{\omega}'' + \frac{\sqrt{1 + 4\widehat{\tau}\,\widehat{v}_r} - 1}{\widehat{\tau}}\right)\widehat{\omega}'. \tag{11.2.41}$$

The time of the collision relaxation $\tau_{\rm rel} = v_r^{-1}$ for gravitational physical system can be estimated in terms of the mean time τ between close collisions and the Coulomb logarithm:

$$\tau v_r = \Lambda,\ \widehat{\tau}\,\widehat{v}_r = \Lambda. \tag{11.2.42}$$

We separate the real and imaginary parts in Eq. (11.2.40). One obtains for the real part

$$\begin{aligned} &\mp \frac{1 - r_A^2 k^2}{2\sqrt{\pi}} \exp\left\{\widehat{\omega}'^2 - \widehat{\omega}''^2 - \widehat{\omega}''\widehat{v}_r \frac{\sqrt{1+4\Lambda} - 1}{\Lambda} - \widehat{v}_r^2 \frac{1 + 2\Lambda - \sqrt{1+4\Lambda}}{2\Lambda^2}\right\} \\ &= \left[\frac{\widehat{v}_r}{\sqrt{1+4\Lambda}} + 0.5\widehat{\omega}'' + \frac{0.5\widehat{\omega}''}{\sqrt{1+4\Lambda}}\right] \cos\left[\widehat{\omega}'\left(2\widehat{\omega}'' + \widehat{v}_r \frac{\sqrt{1+4\Lambda} - 1}{\Lambda}\right)\right] \\ &- 0.5\widehat{\omega}'\left[1 + \frac{1}{\sqrt{1+4\Lambda}}\right] \sin\left[\widehat{\omega}'\left(2\widehat{\omega}'' + \widehat{v}_r \frac{\sqrt{1+4\Lambda} - 1}{\Lambda}\right)\right]. \end{aligned} \tag{11.2.43}$$

Similarly, for the imaginary part we find

$$\begin{aligned} &0.5\widehat{\omega}'\left[1 + \frac{1}{\sqrt{1+4\Lambda}}\right] \cos\left[\widehat{\omega}'\left(2\widehat{\omega}'' + \widehat{v}_r \frac{\sqrt{1+4\Lambda} - 1}{\Lambda}\right)\right] \\ &+ \left[\frac{\widehat{v}_r}{\sqrt{1+4\Lambda}} + 0.5\widehat{\omega}'' + \frac{0.5\widehat{\omega}''}{\sqrt{1+4\Lambda}}\right] \sin\left[\widehat{\omega}'\left(2\widehat{\omega}'' + \widehat{v}_r \frac{\sqrt{1+4\Lambda} - 1}{\Lambda}\right)\right] = 0. \end{aligned} \tag{11.2.44}$$

Coulomb logarithm Λ is large for such objects like galaxies, the typical value $\Lambda \sim 200$ and the system of Eqs. (11.2.43) and (11.2.44) for the large Coulomb logarithm Λ simplifies to

$$\mp \frac{1 - r_A^2 k^2}{\sqrt{\pi}} e^{\widehat{\omega}'^2 - \widehat{\omega}''^2} = \widehat{\omega}'' \cos\left(2\widehat{\omega}'\widehat{\omega}''\right) - \widehat{\omega}' \sin\left(2\widehat{\omega}'\widehat{\omega}''\right), \tag{11.2.45}$$

$$\widehat{\omega}' \cos\left(2\widehat{\omega}'\widehat{\omega}''\right) + \widehat{\omega}'' \sin\left(2\widehat{\omega}'\widehat{\omega}''\right) = 0. \tag{11.2.46}$$

Let us introduce the notation

$$\alpha = 2\widehat{\omega}'\widehat{\omega}'',\ \beta = 1 - r_A^2 k^2, \tag{11.2.47}$$

we obtain the universal equation

$$-e^{\sigma \cot \sigma} \sin \sigma = \frac{\pi}{2\beta^2}\sigma, \tag{11.2.48}$$

where $\sigma = -2\alpha = -4\widehat{\omega}'\widehat{\omega}''$. Equation (11.2.48) does not depend on the sign in front of parameter β. The exact solution of Eq. (11.2.48) can be found with the help of the W-function of Lambert

$$\sigma_n = \mathrm{Im}\left[W_n\left(\frac{2\beta^2}{\pi}\right)\right], \tag{11.2.49}$$

frequencies $\widehat{\omega}'_n, \widehat{\omega}''_n$ are (see also Eq. (11.2.28))

$$\omega'_n = k\sqrt{-\frac{k_B T}{2m}\sigma_n \tan\frac{\sigma_n}{2}},\ \omega''_n = -k\sqrt{-\frac{k_B T}{2m}\sigma_n \cot\frac{\sigma_n}{2}} \tag{11.2.50}$$

In asymptotic for large entire positive n (singular point $r_A k = 1$ is considered further in this section)

$$\sigma_n = \left(n+\frac{1}{2}\right)\pi,\ \widehat{\omega}'_n = \frac{\sqrt{\sigma_n}}{2} = \frac{1}{2}\sqrt{\pi\left(n+\frac{1}{2}\right)},\ \widehat{\omega}''_n = -\frac{\sqrt{\sigma_n}}{2} = -\frac{1}{2}\sqrt{\pi\left(n+\frac{1}{2}\right)}. \tag{11.2.51}$$

The exact solution for the nth discrete solution from the spectrum of oscillations follows from Eqs. (11.2.49) and (11.2.50):

$$\begin{aligned}\widehat{\omega}_n = &\frac{1}{2}\sqrt{-\mathrm{Im}\left[W_n\left(\frac{2(1-r_A^2k^2)^2}{\pi}\right)\right]\tan\left[\frac{1}{2}\mathrm{Im}\left[W_n\left(\frac{2(1-r_A^2k^2)^2}{\pi}\right)\right]\right]} \\ &-\frac{i}{2}\sqrt{-\mathrm{Im}\left[W_n\left(\frac{2(1-r_A^2k^2)^2}{\pi}\right)\right]\cot\left[\frac{1}{2}\mathrm{Im}\left[W_n\left(\frac{2(1-r_A^2k^2)^2}{\pi}\right)\right]\right]}.\end{aligned} \tag{11.2.52}$$

The square of the oscillation frequency of the longitudinal gravitational waves $\widehat{\omega}'^2_n$ is proportional to the wave energy. Hence, the energy of waves is quantized, and as n grows one obtains the asymptotic expression analogous to quantum levels of quantum oscillator in one dimension

$$\widehat{\omega}'^2_n = \frac{\pi}{4}\left(n+\frac{1}{2}\right), \tag{11.2.53}$$

the squares of possible dimensionless frequencies become equally spaced:

$$\widehat{\omega}'^2_{n+1} - \widehat{\omega}'^2_n = \frac{\pi}{4}, \tag{11.2.54}$$

or

$$\widehat{\omega}'^2_{n+1} - \omega'^2_n = \frac{\pi}{2}k^2\frac{k_B T}{m}. \tag{11.2.55}$$

This difference can be connected with energy of Newtonian graviton. Figures 11.1 and 11.2 reflect the result of calculations for 200 discrete levels for the case of the large Coulomb logarithm Λ. For high levels, this spectrum contains many very close equidistant curves with partly practically straight lines, which human eyes can perceive as background. Moreover, plotter from the technical point of view has no possibility to reflect the small curvature of lines approximating this curvature as a long step. My suggestion is to turn this shortcoming into merit in explication of topology of high quantum levels in quantum systems.

Really, extremely interesting that this (from the first glance) grave shortcoming of plotters lead to the automatic construction of approximation for derivatives $\mathrm{d}(r_A k)/\mathrm{d}\widehat{\omega}'$ and $\mathrm{d}(r_A k)/\mathrm{d}\widehat{\omega}''$.

You can see this very complicated topology of curves in Figs. 11.1 and 11.2 including the discrete spectrum of the bell-like curves in the mentioned figures. This singularity is connected with the existence of generalized derivatives $\mathrm{d}(r_A k)/\mathrm{d}\widehat{\omega}'$, $\mathrm{d}(r_A k)/\mathrm{d}\widehat{\omega}''$ for discontinuous functions. This effect has no attitude to the mathematical programming and looks in the definite sense like effect of "shroud of Christ"—self-organization of visible information in the human conscience. Enlarging of scaling shows that the complicated curves topology exists also in the black domain. Then Figs. 11.1 and 11.2 can be used for understanding of the future development of events in physical system after the initial linear stage.

For example Fig. 11.1 shows the discrete set of frequencies which vicinity corresponds to passing over from abnormal to normal dispersion (for example, by $\widehat{\omega}' \sim 7$) for discrete systems of $r_A k$. Of course the nonlinear stage needs the special investigation with using of another methods including the method of direct mathematical modeling It seems that the curves

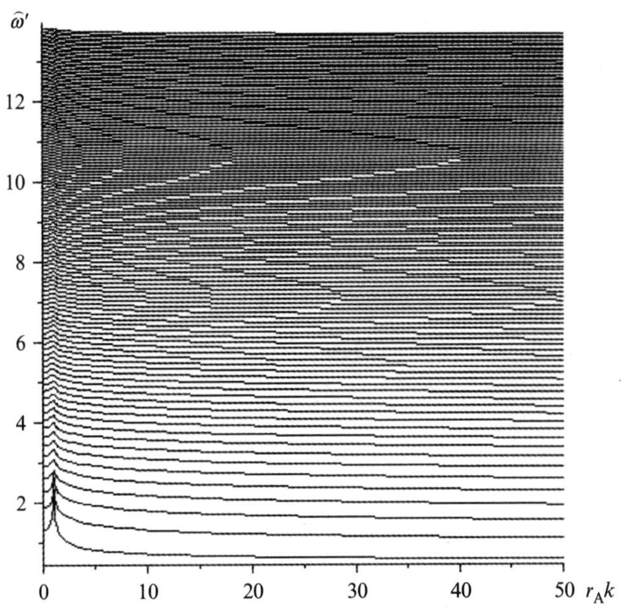

FIGURE 11.1 The dimensionless frequency $\widehat{\omega}'$ versus parameter $r_A k$ (left).

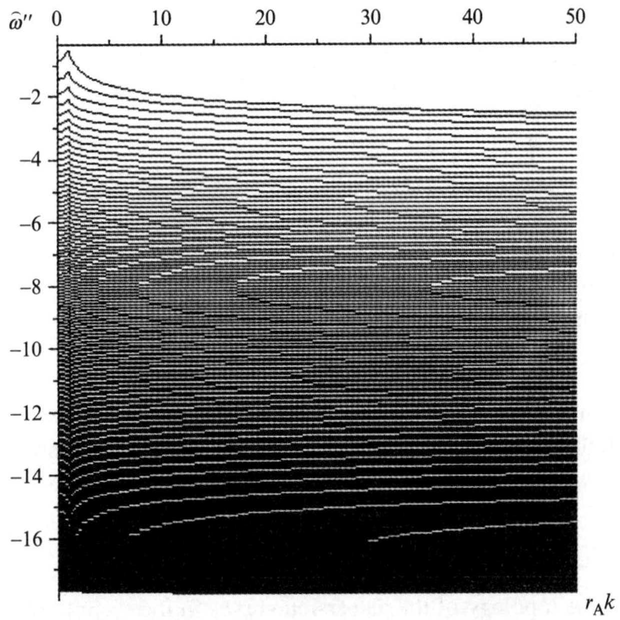

FIGURE 11.2 The dimensionless frequency $\widehat{\omega}''$ versus parameter $r_A k$, (right).

of high levels have different topology in comparison with the low levels. Nevertheless, it is far from reality, the high-level frequencies have the same character features as low frequencies.

Look at Figs. 11.3 and 11.4 and you see for frequencies $\widehat{\omega}'_{200}$ and for $\widehat{\omega}''_{200}$ the same character features as for lower frequencies.

Figure 11.5a and b shows the discrete set of frequencies which vicinity corresponds to passing over from abnormal to normal dispersion for discrete systems of $r_A k$ in more large interval between 0 and 150 $r_A k$. As in the plasma case this discrete systems of $r_A k$ has regular characters with factor of about two.

Let us introduce the parameter $\xi_A = \frac{\pi}{2\beta^2} = \frac{\pi}{2(1 - r_A^2 k^2)^2}$. Then Eq. (11.2.48) can be written as $e^{\sigma \cot \sigma} \sin \sigma = -\xi_A \sigma$.

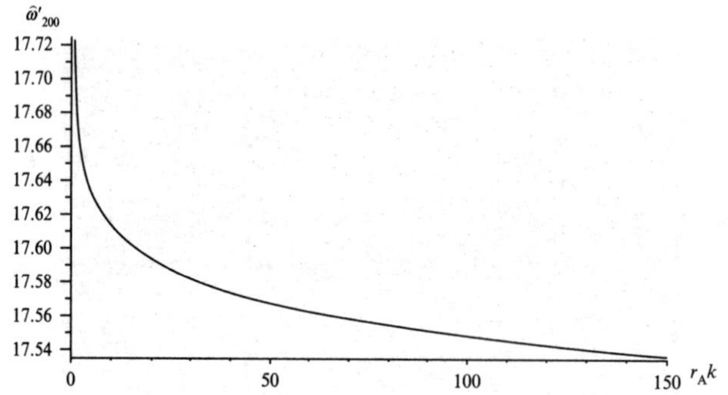

FIGURE 11.3 The dimensionless frequency $\widetilde{\omega}'_{200}$ versus parameter $r_A k$.

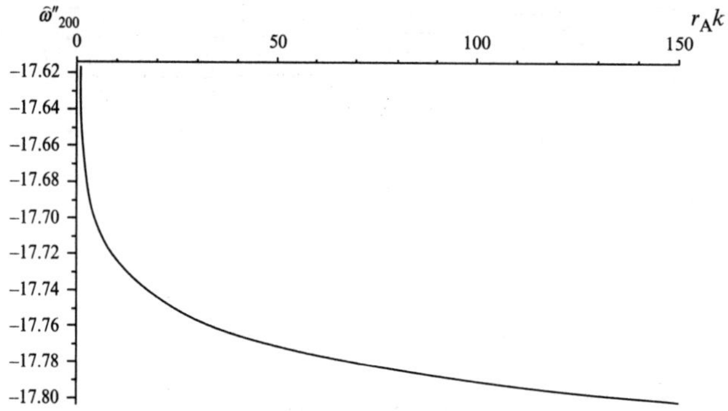

FIGURE 11.4 The dimensionless frequency $\widetilde{\omega}''_{200}$ versus parameter $r_A k$.

Let us consider the concrete examples of calculations. Suppose that $\xi_A = 0.0002$, it corresponds to $r_A k = 9.47$ or $\lambda = 0.663 r_A$. Table 11.1 contains $\gamma_n, \widetilde{\omega}'_n, \widetilde{\omega}''_n$, $n = 1, \ldots, 10$; 300, 1000 for $\xi_A = 0.0002$.

It is of interest to investigate the singular point where

$$r_A k = 1. \tag{11.2.56}$$

Note the solution $\sigma \to \pi + 0$ and therefore $\widetilde{\omega}' \to \infty$ and $\widetilde{\omega}'' \to 0$. But phase velocity of wave $u_\phi = \omega' r_A$ and phase velocity of gravitational wave turns into infinity (in the frame of non-relativistic theory) and damping is equal to zero. In vicinity of $r_A k = 1$ one obtains "gravitational window" with increasing of frequency $\widetilde{\omega}'_n$ and decreasing of damping; the corresponding wave lengths λ_A is

$$\lambda_A = 2\pi r_A. \tag{11.2.57}$$

Figures 11.6 and 11.7 reflect the topology of the dispersion curves in the vicinity of the gravitational window.

Let us make some estimates. The mean density (calculated via luminous matter in stars and galaxies) $\rho \approx (0.01 - 0.02)\rho_c$, where $\rho_c = 0.47 \times 10^{-29}/\text{cm}^3$. Using also for estimates $T = 3\,\text{K}$, $n = 0.3 \times 10^{-7}\,\text{cm}^{-3}$, $m = 1.6 \times 10^{-24}\,\text{g}$, one obtains from relation (11.2.27) $r_A = 0.82 \times 10^{23}\,\text{cm} = 0.027\,\text{Mpc}$ and $\lambda_A = 0.17\,\text{Mpc}$.

Now we can create the physical picture leading to the Hubble flow. *The main origin of Hubble effect (including the matter expansion with acceleration) is self-catching of expanding matter by the self-consistent Newtonian gravitational field in conditions of weak influence of central massive bodies.*

The relation (11.2.57) is the condition of this self-catching as result of explosion with appearance of waves for which the wave lengths is of about λ_A. Gravitational self-catching takes place for Big Bang having given birth to the global expansion of Universe, but also for Little Bang [245] in so-called Local Group of galaxies. The gravitationally bound system of the Local Group can exist only within region $r < r_v$. In this case r_v need not to be connected with the modified Newton force and can be considered as character value where gravitation of the central mass is not significant. Outside the Group at distances $r > r_v$ the Hubble flow of Galaxies starts. This "no reentering radius" was found as result of direct observations of the Local

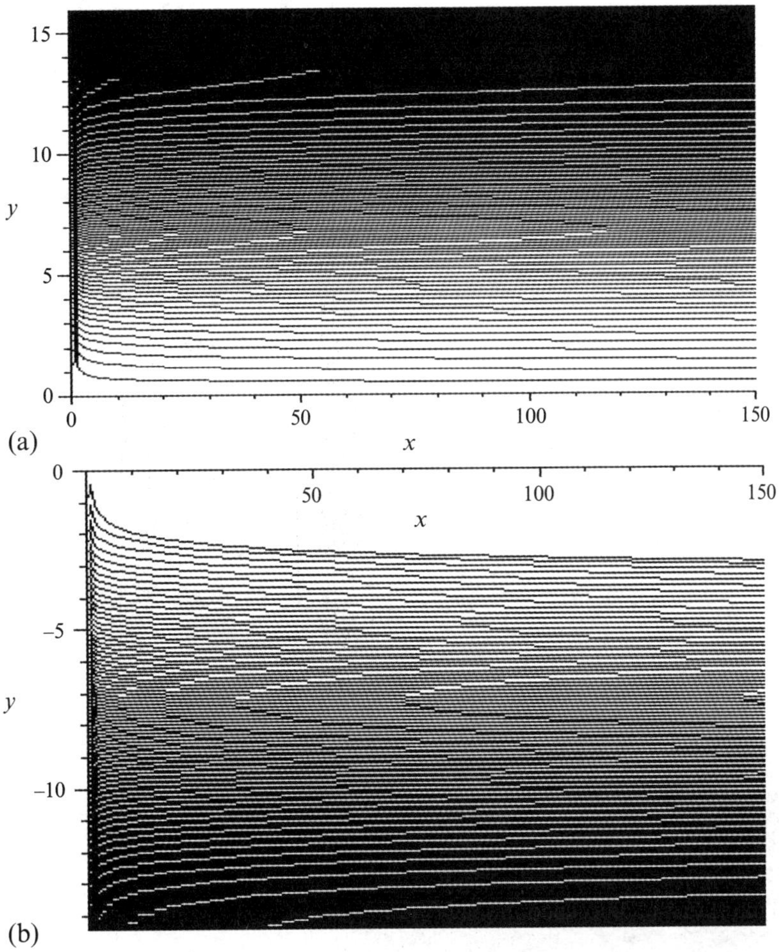

FIGURE 11.5 (a) The dimensionless frequency $\widehat{\omega}'$ (y-axis) versus parameter $r_A k$ (x-axis). (b) The dimensionless frequency $\widehat{\omega}''$ (y-axis) versus parameter $r_A k$ (x-axis).

Group: r_v is of the order (or less) 1 Mpc. This value is in good coincidence with the length $\lambda_A = 0.17$ Mpc. Some important remarks should be done:

1. Effects of gravitational self-catching should be typical for Universe. The existence of "Hubble boxes" discussed in review [243] as typical blocks of the nearby Universe.
2. As it follows from Figs. 11.1, 11.2, 11.6, and 11.7 the effect of the gravitational self-catching exists *in finite region* close to $r_A k = 1$, the phase velocity is defined by discrete spectrum $u_{\phi,n} = \widehat{\omega}'_n \sqrt{2k_B T/m}$.
3. Gravitational self-catching can be significant in the Earth conditions.

The last remark needs to be explained. Gravitational self-catching can be essential in the physical systems, which character lengths correlates with r_A *in conditions of weak influence of the central massive bodies.* The corresponding conditions are realizing in tsunami waves. For water by the earth conditions $T = 300$ K, $\rho = 1$ g/cm^3; $\gamma_N = 6.6 \times 10^{-8}$ cm^3/(g s^2), the lengths $r_A = 407.43$ km and $\lambda_A = 2558.66$ m. For close collisions $r_c \sim 10^{-7}$ cm and Coulomb logarithm $\Lambda_A = \ln(r_A/r_c) \sim 10^2$. The delivered theory can be applied in the Earth conditions if the influence of central mass can be excluded from consideration. This situation realizes in the tsunami motion because the direction to the Earth center supposes perpendicular to the direction of additional self-consistent gravitational force acting in the tsunami wave. In essence, the catching of water mass is realizing by longitudinal self-consistent gravitational wave. The origin of effects of the small attenuation can be qualitatively explained from the position of kinetic theory. Let us calculate the mean velocity $\overline{u}_+$ of particles moving in a chosen direction. If this direction is considering as the positive ones, then $u > 0$ and for the Maxwellian function f_0

$$\overline{u}_+ = \sqrt{\frac{m}{2\pi k_B T}} \int_0^{+\infty} e^{-mu^2/2k_B T} u\,du = \sqrt{\frac{k_B T}{2\pi m}}. \tag{11.2.58}$$

TABLE 11.1 $\xi_A = 0.0002$

n	γ_n	$\widehat{\omega}'_n$	$\widehat{\omega}''_n$
1	1.714π	0.806	−1.671
2	3.602π	1.430	−1.978
3	5.560π	1.900	−2.298
4	$7.541\ \pi$	2.283	−2.594
5	9.530π	2.610	−2.867
6	11.523π	2.901	−3.119
7	13.518π	3.165	−3.354
8	15.515π	3.408	−3.575
9	17.513π	3.635	−3.784
10	19.511π	3.848	−3.982
300	599.500	21.707	−21.691
1000	1999.500	39.636	−39.620

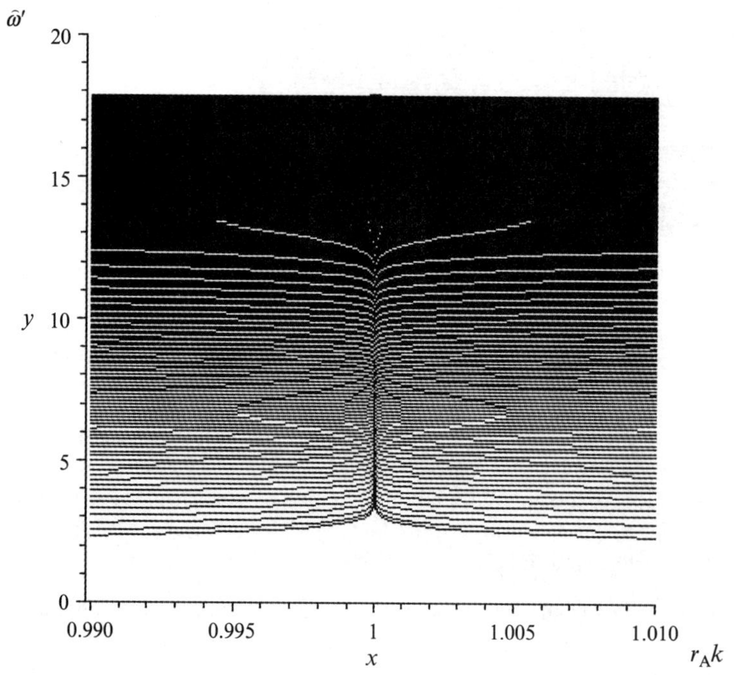

FIGURE 11.6 Topology of the dispersion curves $\widehat{\omega}'$ in the vicinity of the gravitational window.

Kinetic energy, connected with this motion is

$$m\bar{u}_+^2/2 = k_B T/(4\pi). \tag{11.2.59}$$

From relations (11.2.58) and (11.2.59) follow

$$\bar{u}_+ = r_A\sqrt{2\rho\gamma_N}. \tag{11.2.60}$$

Therefore, if the selected direction is opposite to the direction of the wave motion, energy of gravitational field $E_A = \gamma_N m\rho r_A^2$ (per particle) should be applied for exclusion of such kind of particles. For water in considered estimation one obtains $E_A = \gamma_N m\rho r_A^2 = 3.293 \times 10^{-15}$ erg, $\bar{u}_+ = r_A\sqrt{2\rho\gamma_N} = 533$ km/h the typical value of tsunami in ocean.

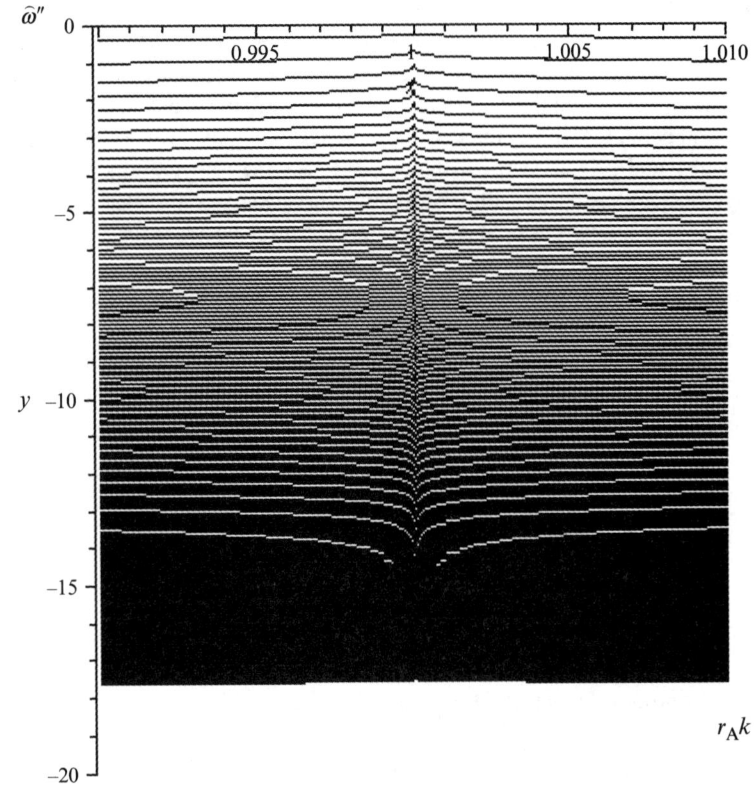

FIGURE 11.7 Topology of the dispersion curves $\widehat{\omega}''$ in the vicinity of the gravitational window.

Otherwise, the wave expansion leads to the energy dissipation of directional motion in the form of the chaotic heat motion. Nevertheless, in the case if the forces of gravitation attraction counteract (or keep to a minimum) these losses, the wave is moving without attenuation.

Other unexpected application of the developed theory in the earth conditions is connected with study of sediment transport in shallow channel flows [247] and the wave formation on a shallow layer of flowing grains [248]. These effects are investigated in [58].

In the following sections I intend to apply the unified generalized non-local theory for nonlinear mathematical modeling of cosmic objects. For the case of galaxies the theory leads to the flat rotation curves known from observations. The transformation of Kepler's regime into the flat rotation curves for different solitons is shown. The Hubble expansion with acceleration is explained as result of mathematical modeling based on the principals of non-local physics.

Therefore the answers for the following questions are formulated:

1. Why the concept of the dark matter is not significant in the Solar system?
2. Why the galaxy rotation curves have the character flat form?
3. Is it possible to obtain the continuous transmission from the Kepler regime to the flat halo curves?
4. Why after Big Bang explosion (or after the explosion in the Hubble boxes) the Hubble expansion exists with acceleration? (Nobel Prize for the observers S. Perlmutter, A. G. Riess, B. Schmidt of the year 2011).

In other words—is it possible using only Newtonian gravitation law and non-local statistical description to forecast the flat gravitational curve of a typical spiral galaxy (Section 11.3) and the Hubble expansion (including the Hubble expansion with acceleration), (Section 11.4)? The last question has the positive answer.

11.3 DISK GALAXY ROTATION AND THE PROBLEM OF DARK MATTER

About 40 years after Zwicky's initial observations Vera Rubin, astronomer at the Department of Terrestrial Magnetism at the Carnegie Institution of Washington presented findings based on a new sensitive spectrograph that could measure the velocity curve of edge-on spiral galaxies to a greater degree of accuracy than had ever before been achieved. Together with

Kent Ford, Rubin announced at a 1975 meeting of the American Astronomical Society the astonishing discovery that most stars in spiral galaxies orbit at roughly the same speed reflected schematically in Fig. 11.8.

For example, the rotation curve of the type B corresponds to the galaxy NGC3198. The following extensive radio observations determined the detailed rotation curve of spiral disk galaxies to be flat (as the curve B), much beyond as seen in the optical band. Obviously the trivial balance between the gravitational and centrifugal forces leads to relation between orbital speed V and galactocentric distance r as $V^2 = \gamma_N M/r$ beyond the physical extent of the galaxy of mass M (the curve A). The obvious contradiction with the velocity curve B having a 'flat' appearance out to a large radius, was explained by introduction of a new physical essence—dark matter because for spherically symmetric case the hypothetical density distribution $\rho(r) \sim 1/r^2$ leads to $V = \text{const}$. The result of this activity is known—undetectable dark matter which does not emit radiation, inferred solely from its gravitational effects. But it means that upwards of 50% of the mass of galaxies was contained in the dark galactic halo.

Strict consideration leads to the following system of the generalized hydrodynamic equations (GHE) written in the generalized Euler form:

(Continuity equation for species α)

$$\frac{\partial}{\partial t}\left\{\rho_\alpha - \tau_\alpha\left[\frac{\partial \rho_\alpha}{\partial t} + \frac{\partial}{\partial \mathbf{r}}\cdot(\rho_\alpha \mathbf{v}_0)\right]\right\}$$
$$+\frac{\partial}{\partial \mathbf{r}}\cdot\left\{\rho_\alpha \mathbf{v}_0 - \tau_\alpha\left[\frac{\partial}{\partial t}(\rho_\alpha \mathbf{v}_0) + \frac{\partial}{\partial \mathbf{r}}\cdot(\rho_\alpha \mathbf{v}_0\mathbf{v}_0) + \overleftrightarrow{\mathbf{I}}\cdot\frac{\partial p_\alpha}{\partial \mathbf{r}} - \rho_\alpha \mathbf{F}_\alpha^{(1)} - \frac{q_\alpha}{m_\alpha}\rho_\alpha \mathbf{v}_0 \times \mathbf{B}\right]\right\} = R_\alpha. \tag{11.3.1}$$

(Continuity equation for mixture)

$$\frac{\partial}{\partial t}\left\{\rho - \sum_\alpha \tau_\alpha\left[\frac{\partial \rho_\alpha}{\partial t} + \frac{\partial}{\partial \mathbf{r}}\cdot(\rho_\alpha \mathbf{v}_0)\right]\right\} + \frac{\partial}{\partial \mathbf{r}}\cdot\left\{\rho \mathbf{v}_0 - \sum_\alpha \tau_\alpha\left[\frac{\partial}{\partial t}(\rho_\alpha \mathbf{v}_0) + \frac{\partial}{\partial \mathbf{r}}\cdot(\rho_\alpha \mathbf{v}_0\mathbf{v}_0)\right.\right.$$
$$\left.\left.+\overleftrightarrow{\mathbf{I}}\cdot\frac{\partial p_\alpha}{\partial \mathbf{r}} - \rho_\alpha \mathbf{F}_\alpha^{(1)} - \frac{q_\alpha}{m_\alpha}\rho_\alpha \mathbf{v}_0 \times \mathbf{B}\right]\right\} = 0. \tag{11.3.2}$$

(Momentum equation for species α)

$$\frac{\partial}{\partial t}\left\{\rho_\alpha \mathbf{v}_0 - \tau_\alpha\left[\frac{\partial}{\partial t}(\rho_\alpha \mathbf{v}_0) + \frac{\partial}{\partial \mathbf{r}}\cdot\rho_\alpha \mathbf{v}_0\mathbf{v}_0 + \frac{\partial p_\alpha}{\partial \mathbf{r}} - \rho_\alpha \mathbf{F}_\alpha^{(1)}\right.\right.$$
$$\left.\left.-\frac{q_\alpha}{m_\alpha}\rho_\alpha \mathbf{v}_0 \times \mathbf{B}\right]\right\} - \mathbf{F}_\alpha^{(1)}\left[\rho_\alpha - \tau_\alpha\left(\frac{\partial \rho_\alpha}{\partial t} + \frac{\partial}{\partial \mathbf{r}}\cdot(\rho_\alpha \mathbf{v}_0)\right)\right]$$
$$-\frac{q_\alpha}{m_\alpha}\left\{\rho_\alpha \mathbf{v}_0 - \tau_\alpha\left[\frac{\partial}{\partial t}(\rho_\alpha \mathbf{v}_0) + \frac{\partial}{\partial \mathbf{r}}\cdot\rho_\alpha \mathbf{v}_0\mathbf{v}_0 + \frac{\partial p_\alpha}{\partial \mathbf{r}} - \rho_\alpha \mathbf{F}_\alpha^{(1)}\right.\right.$$
$$\left.\left.-\frac{q_\alpha}{m_\alpha}\rho_\alpha \mathbf{v}_0 \times \mathbf{B}\right]\right\} \times \mathbf{B} + \frac{\partial}{\partial \mathbf{r}}\cdot\left\{\rho_\alpha \mathbf{v}_0\mathbf{v}_0 + p_\alpha \overleftrightarrow{\mathbf{I}} - \tau_\alpha\left[\frac{\partial}{\partial t}\left(\rho_\alpha \mathbf{v}_0\mathbf{v}_0\right.\right.\right.$$
$$\left.+p_\alpha \overleftrightarrow{\mathbf{I}}\right) + \frac{\partial}{\partial \mathbf{r}}\cdot\rho_\alpha(\mathbf{v}_0\mathbf{v}_0)\mathbf{v}_0 + 2\overleftrightarrow{\mathbf{I}}\left(\frac{\partial}{\partial \mathbf{r}}\cdot(p_\alpha \mathbf{v}_0)\right) + \frac{\partial}{\partial \mathbf{r}}\cdot\left(\overleftrightarrow{\mathbf{I}}p_\alpha \mathbf{v}_0\right)$$
$$\left.\left.-\mathbf{F}_\alpha^{(1)}\rho_\alpha \mathbf{v}_0 - \rho_\alpha \mathbf{v}_0\mathbf{F}_\alpha^{(1)} - \frac{q_\alpha}{m_\alpha}\rho_\alpha[\mathbf{v}_0 \times \mathbf{B}]\mathbf{v}_0 - \frac{q_\alpha}{m_\alpha}\rho_\alpha \mathbf{v}_0[\mathbf{v}_0 \times \mathbf{B}]\right]\right\}$$
$$= \int m_\alpha \mathbf{v}_\alpha J_\alpha^{\text{st,el}} d\mathbf{v}_\alpha + \int m_\alpha \mathbf{v}_\alpha J_\alpha^{\text{st,inel}} d\mathbf{v}_\alpha. \tag{11.3.3}$$

(Momentum equation for mixture)

$$
\begin{aligned}
&\frac{\partial}{\partial t}\left\{\rho \mathbf{v}_0-\sum_\alpha \tau_\alpha\left[\frac{\partial}{\partial t}(\rho_\alpha \mathbf{v}_0)+\frac{\partial}{\partial \mathbf{r}}\cdot\rho_\alpha \mathbf{v}_0\mathbf{v}_0+\frac{\partial p_\alpha}{\partial \mathbf{r}}-\rho_\alpha \mathbf{F}_\alpha^{(1)}\right.\right.\\
&\left.\left.-\frac{q_\alpha}{m_\alpha}\rho_\alpha \mathbf{v}_0\times\mathbf{B}\right]\right\}-\sum_\alpha \mathbf{F}_\alpha^{(1)}\left[\rho_\alpha-\tau_\alpha\left(\frac{\partial \rho_\alpha}{\partial t}+\frac{\partial}{\partial \mathbf{r}}\cdot(\rho_\alpha \mathbf{v}_0)\right)\right]\\
&-\sum_\alpha \frac{q_\alpha}{m_\alpha}\left\{\rho_\alpha \mathbf{v}_0-\tau_\alpha\left[\frac{\partial}{\partial t}(\rho_\alpha \mathbf{v}_0)+\frac{\partial}{\partial \mathbf{r}}\cdot\rho_\alpha \mathbf{v}_0\mathbf{v}_0+\frac{\partial p_\alpha}{\partial \mathbf{r}}-\rho_\alpha \mathbf{F}_\alpha^{(1)}\right.\right.\\
&\left.\left.-\frac{q_\alpha}{m_\alpha}\rho_\alpha \mathbf{v}_0\times\mathbf{B}\right]\right\}\times\mathbf{B}+\frac{\partial}{\partial \mathbf{r}}\cdot\left\{\rho \mathbf{v}_0\mathbf{v}_0+p\overleftrightarrow{\mathbf{I}}-\sum_\alpha \tau_\alpha\left[\frac{\partial}{\partial t}\left(\rho_\alpha \mathbf{v}_0\mathbf{v}_0\right.\right.\right.\\
&\left.+p_\alpha\overleftrightarrow{\mathbf{I}}\right)+\frac{\partial}{\partial \mathbf{r}}\cdot\rho_\alpha(\mathbf{v}_0\mathbf{v}_0)\mathbf{v}_0+2\overleftrightarrow{\mathbf{I}}\left(\frac{\partial}{\partial \mathbf{r}}\cdot(p_\alpha \mathbf{v}_0)\right)+\frac{\partial}{\partial \mathbf{r}}\cdot\left(\overleftrightarrow{\mathbf{I}}p_\alpha \mathbf{v}_0\right)\\
&\left.\left.-\mathbf{F}_\alpha^{(1)}\rho_\alpha \mathbf{v}_0-\rho_\alpha \mathbf{v}_0\mathbf{F}_\alpha^{(1)}-\frac{q_\alpha}{m_\alpha}\rho_\alpha[\mathbf{v}_0\times\mathbf{B}]\mathbf{v}_0-\frac{q_\alpha}{m_\alpha}\rho_\alpha \mathbf{v}_0[\mathbf{v}_0\times\mathbf{B}]\right]\right\}=0
\end{aligned}
\tag{11.3.4}
$$

(Energy equation for α species)

$$
\begin{aligned}
&\frac{\partial}{\partial t}\left\{\frac{\rho_\alpha v_0^2}{2}+\frac{3}{2}p_\alpha+\varepsilon_\alpha n_\alpha-\tau_\alpha\left[\frac{\partial}{\partial t}\left(\frac{\rho_\alpha v_0^2}{2}+\frac{3}{2}p_\alpha+\varepsilon_\alpha n_\alpha\right)\right.\right.\\
&\left.\left.+\frac{\partial}{\partial \mathbf{r}}\cdot\left(\frac{1}{2}\rho_\alpha v_0^2\mathbf{v}_0+\frac{5}{2}p_\alpha \mathbf{v}_0+\varepsilon_\alpha n_\alpha \mathbf{v}_0\right)-\mathbf{F}_\alpha^{(1)}\cdot\rho_\alpha \mathbf{v}_0\right]\right\}\\
&+\frac{\partial}{\partial \mathbf{r}}\cdot\left\{\frac{1}{2}\rho_\alpha v_0^2\mathbf{v}_0+\frac{5}{2}p_\alpha \mathbf{v}_0+\varepsilon_\alpha n_\alpha \mathbf{v}_0-\tau_\alpha\left[\frac{\partial}{\partial t}\left(\frac{1}{2}\rho_\alpha v_0^2\mathbf{v}_0\right.\right.\right.\\
&\left.+\frac{5}{2}p_\alpha \mathbf{v}_0+\varepsilon_\alpha n_\alpha \mathbf{v}_0\right)+\frac{\partial}{\partial \mathbf{r}}\cdot\left(\frac{1}{2}\rho_\alpha v_0^2\mathbf{v}_0\mathbf{v}_0+\frac{7}{2}p_\alpha \mathbf{v}_0\mathbf{v}_0+\frac{1}{2}p_\alpha v_0^2\overleftrightarrow{\mathbf{I}}\right.\\
&\left.+\frac{5}{2}\frac{p_\alpha^2}{\rho_\alpha}\overleftrightarrow{\mathbf{I}}+\varepsilon_\alpha n_\alpha \mathbf{v}_0\mathbf{v}_0+\varepsilon_\alpha\frac{p_\alpha}{m_\alpha}\overleftrightarrow{\mathbf{I}}\right)-\rho_\alpha \mathbf{F}_\alpha^{(1)}\cdot\mathbf{v}_0\mathbf{v}_0-p_\alpha \mathbf{F}_\alpha^{(1)}\cdot\overleftrightarrow{\mathbf{I}}\\
&-\frac{1}{2}\rho_\alpha v_0^2\mathbf{F}_\alpha^{(1)}-\frac{3}{2}\mathbf{F}_\alpha^{(1)}p_\alpha-\frac{\rho_\alpha v_0^2}{2}\frac{q_\alpha}{m_\alpha}[\mathbf{v}_0\times\mathbf{B}]-\frac{5}{2}p_\alpha\frac{q_\alpha}{m_\alpha}[\mathbf{v}_0\times\mathbf{B}]\\
&\left.\left.-\varepsilon_\alpha n_\alpha\frac{q_\alpha}{m_\alpha}[\mathbf{v}_0\times\mathbf{B}]-\varepsilon_\alpha n_\alpha \mathbf{F}_\alpha^{(1)}\right]\right\}-\left\{\rho_\alpha \mathbf{F}_\alpha^{(1)}\cdot\mathbf{v}_0\right.\\
&\left.-\tau_\alpha\left[\mathbf{F}_\alpha^{(1)}\cdot\left(\frac{\partial}{\partial t}(\rho_\alpha \mathbf{v}_0)+\frac{\partial}{\partial \mathbf{r}}\cdot\rho_\alpha \mathbf{v}_0\mathbf{v}_0+\frac{\partial}{\partial \mathbf{r}}\cdot p_\alpha\overleftrightarrow{\mathbf{I}}-\rho_\alpha \mathbf{F}_\alpha^{(1)}-q_\alpha n_\alpha[\mathbf{v}_0\times\mathbf{B}]\right)\right]\right\}\\
&=\int\left(\frac{m_\alpha v_\alpha^2}{2}+\varepsilon_\alpha\right)J_\alpha^{\mathrm{st,el}}\mathrm{d}\mathbf{v}_\alpha+\int\left(\frac{m_\alpha v_\alpha^2}{2}+\varepsilon_\alpha\right)J_\alpha^{\mathrm{st,inel}}\mathrm{d}\mathbf{v}_\alpha.
\end{aligned}
\tag{11.3.5}
$$

(Energy equation for mixture)

$$
\begin{aligned}
&\frac{\partial}{\partial t}\left\{\frac{\rho v_0^2}{2}+\frac{3}{2}p+\sum_\alpha \varepsilon_\alpha n_\alpha-\sum_\alpha \tau_\alpha\left[\frac{\partial}{\partial t}\left(\frac{\rho_\alpha v_0^2}{2}+\frac{3}{2}p_\alpha+\varepsilon_\alpha n_\alpha\right)\right.\right.\\
&\left.\left.+\frac{\partial}{\partial \mathbf{r}}\cdot\left(\frac{1}{2}\rho_\alpha v_0^2\mathbf{v}_0+\frac{5}{2}p_\alpha\mathbf{v}_0+\varepsilon_\alpha n_\alpha\mathbf{v}_0\right)-\mathbf{F}_\alpha^{(1)}\cdot\rho_\alpha\mathbf{v}_0\right]\right\}\\
&+\frac{\partial}{\partial \mathbf{r}}\cdot\left\{\frac{1}{2}\rho v_0^2\mathbf{v}_0+\frac{5}{2}p\mathbf{v}_0+\mathbf{v}_0\sum_\alpha \varepsilon_\alpha n_\alpha-\sum_\alpha \tau_\alpha\left[\frac{\partial}{\partial t}\left(\frac{1}{2}\rho_\alpha v_0^2\mathbf{v}_0\right.\right.\right.\\
&\left.+\frac{5}{2}p_\alpha\mathbf{v}_0+\varepsilon_\alpha n_\alpha\mathbf{v}_0\right)+\frac{\partial}{\partial \mathbf{r}}\cdot\left(\frac{1}{2}\rho_\alpha v_0^2\mathbf{v}_0\mathbf{v}_0+\frac{7}{2}p_\alpha\mathbf{v}_0\mathbf{v}_0+\frac{1}{2}p_\alpha v_0^2\overleftrightarrow{I}\right.\\
&\left.+\frac{5}{2}\frac{p_\alpha^2}{\rho_\alpha}\overleftrightarrow{I}+\varepsilon_\alpha n_\alpha\mathbf{v}_0\mathbf{v}_0+\varepsilon_\alpha\frac{p_\alpha}{m_\alpha}\overleftrightarrow{I}\right)-\rho_\alpha\mathbf{F}_\alpha^{(1)}\cdot\mathbf{v}_0\mathbf{v}_0-p_\alpha\mathbf{F}_\alpha^{(1)}\cdot\overleftrightarrow{I}\\
&-\frac{1}{2}\rho_\alpha v_0^2\mathbf{F}_\alpha^{(1)}-\frac{3}{2}\mathbf{F}_\alpha^{(1)}p_\alpha-\frac{\rho_\alpha v_0^2}{2}\frac{q_\alpha}{m_\alpha}[\mathbf{v}_0\times\mathbf{B}]-\frac{5}{2}p_\alpha\frac{q_\alpha}{m_\alpha}[\mathbf{v}_0\times\mathbf{B}]\\
&\left.\left.-\varepsilon_\alpha n_\alpha\frac{q_\alpha}{m_\alpha}[\mathbf{v}_0\times\mathbf{B}]-\varepsilon_\alpha n_\alpha\mathbf{F}_\alpha^{(1)}\right]\right\}-\left\{\mathbf{v}_0\cdot\sum_\alpha \rho_\alpha\mathbf{F}_\alpha^{(1)}\right.\\
&\left.-\sum_\alpha \tau_\alpha\left[\mathbf{F}_\alpha^{(1)}\cdot\left(\frac{\partial}{\partial t}(\rho_\alpha\mathbf{v}_0)+\frac{\partial}{\partial \mathbf{r}}\cdot\rho_\alpha\mathbf{v}_0\mathbf{v}_0+\frac{\partial}{\partial \mathbf{r}}\cdot p_\alpha\overleftrightarrow{I}-\rho_\alpha\mathbf{F}_\alpha^{(1)}-q_\alpha n_\alpha[\mathbf{v}_0\times\mathbf{B}]\right)\right]\right\}=0.
\end{aligned}
\qquad (11.3.6)
$$

Here $\mathbf{F}_\alpha^{(1)}$ are the forces of the non-magnetic origin, $\mathbf{B}$—magnetic induction, $\overleftrightarrow{I}$—unit tensor, q_α—charge of the α-component particle, p_α—static pressure for α-component, ε_α—internal energy for the particles of α-component, $\mathbf{v}_0$—hydrodynamic velocity for mixture, τ_α—non-local parameter.

GHE can be applied to the physical systems from the Universe to atomic scales. All additional explanations will be done by delivering the results of modeling of corresponding physical systems with the special consideration of non-local parameters τ_α. Generally speaking to GHE should be added the system of generalized Maxwell equations (for example in the form of the generalized Poisson equation for electric potential) and gravitational equations (for example in the form of the generalized Poisson equation for gravitational potential).

In the following I intend to show that the character features reflected on Fig 11.8 can be explained in the frame of Newtonian gravitation law and the non-local kinetic description created by me. With this aim let us consider the formation of the soliton's type of solution of the GHE for gravitational media like galaxy in the self-consistent gravitational field. Our aim consists in calculation of the self-consistent hydrodynamic moments of possible formation like gravitational soliton.

Let us investigate of the gravitational soliton formation in the frame of the non-stationary 1D Cartesian formulation. Then the system of GHE consist from the generalized Poisson equation reflecting the effects of the density and the density

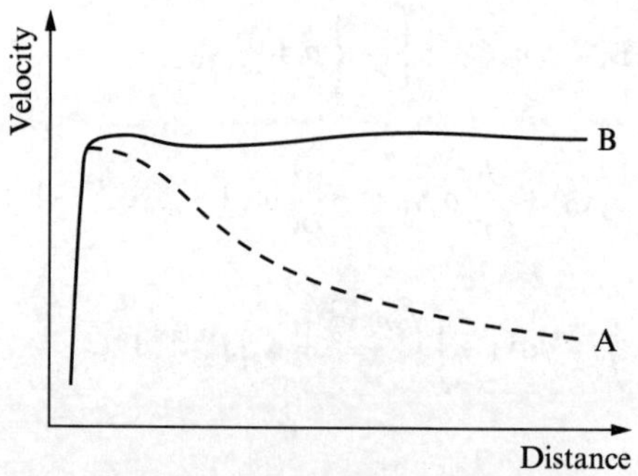

FIGURE 11.8 Rotation curve of a typical spiral galaxy: predicted (A) and observed (B).

flux perturbations, continuity equation, motion, and energy equations. This system of four equations for non-stationary 1D case is written as the deep particular case of Eqs. (11.3.1)–(11.3.6) in the form:

(Poisson equation)

$$\frac{\partial^2 \psi}{\partial x^2} = 4\pi\gamma_{\rm N}\left[\rho - \tau\left(\frac{\partial \rho}{\partial t} + \frac{\partial}{\partial x}(\rho u)\right)\right], \tag{11.3.7}$$

(continuity equation)

$$\frac{\partial}{\partial t}\left\{\rho - \tau\left[\frac{\partial \rho}{\partial t} + \frac{\partial}{\partial x}(\rho u)\right]\right\} + \frac{\partial}{\partial x}\left\{\rho u - \tau\left[\frac{\partial}{\partial t}(\rho u) + \frac{\partial}{\partial x}(\rho u^2) + \frac{\partial p}{\partial x} + \rho\frac{\partial \psi}{\partial x}\right]\right\} = 0, \tag{11.3.8}$$

(motion equation)

$$\frac{\partial}{\partial t}\left\{\rho u - \tau\left[\frac{\partial}{\partial t}(\rho u) + \frac{\partial}{\partial x}(\rho u^2) + \frac{\partial p}{\partial x} + \rho\frac{\partial \psi}{\partial x}\right]\right\} + \frac{\partial \psi}{\partial x}\left[\rho - \tau\left(\frac{\partial \rho}{\partial t} + \frac{\partial}{\partial x}(\rho u)\right)\right] + \frac{\partial}{\partial x}\left\{\rho u^2 + p - \tau\left[\frac{\partial}{\partial t}(\rho u^2 + p) + \frac{\partial}{\partial x}(\rho u^3 + 3pu) + 2\rho u\frac{\partial \psi}{\partial x}\right]\right\} = 0, \tag{11.3.9}$$

(energy equation)

$$\begin{aligned}&\frac{\partial}{\partial t}\left\{\rho u^2 + 3p - \tau\left[\frac{\partial}{\partial t}(\rho u^2 + 3p) + \frac{\partial}{\partial x}(\rho u^3 + 5pu) + 2\rho u\frac{\partial \psi}{\partial x}\right]\right\}\\&+\frac{\partial}{\partial x}\left\{\rho u^3 + 5pu - \tau\left[\frac{\partial}{\partial t}(\rho u^3 + 5pu) + \frac{\partial}{\partial x}\left(\rho u^4 + 8pu^2 + 5\frac{p^2}{\rho}\right) + \frac{\partial \psi}{\partial x}(3\rho u^2 + 5p)\right]\right\}\\&+2\frac{\partial \psi}{\partial x}\left\{\rho u - \tau\left[\frac{\partial}{\partial t}(\rho u) + \frac{\partial}{\partial x}(\rho u^2 + p) + \rho\frac{\partial \psi}{\partial x}\right]\right\} = 0,\end{aligned} \tag{11.3.10}$$

where u is translational velocity of the one species object, ψ—self-consistent gravitational potential ($\mathbf{g} = -\partial\psi/\partial\mathbf{r}$ is acceleration in gravitational field), ρ is density and p is pressure, τ is non-locality parameter, $\gamma_{\rm N}$ is Newtonian gravitation constant.

Let us introduce the coordinate system moving along the positive direction of x-axis in 1D space with velocity $C = u_0$ equal to phase velocity of considering object

$$\xi = x - Ct. \tag{11.3.11}$$

Taking into account the De Broglie relation we should wait that the group velocity $u_{\rm g}$ is equal $2\,u_0$. In moving coordinate system all dependent hydrodynamic values are function of (ξ, t). We investigate the possibility of the object formation of the soliton type. For this solution there is no explicit dependence on time for coordinate system moving with the phase velocity u_0. Write down the system of Eqs. (11.3.7)–(11.3.10) in the dimensionless form, where dimensionless symbols are marked by tildes. For the scales $\rho_0, u_0, x_0 = u_0 t_0, \psi_0 = u_0^2, \gamma_{{\rm N},0} = u_0^2/(\rho_0 x_0^2), p_0 = \rho_0 u_0^2$ and conditions $\widetilde{C} = C/u_0 = 1$, the equations take the form:

(generalized Poisson equation)

$$\frac{\partial^2 \widetilde{\psi}}{\partial \widetilde{\xi}^2} = 4\pi\widetilde{\gamma}_{\rm N}\left[\widetilde{\rho} - \widetilde{\tau}\left(-\frac{\partial \widetilde{\rho}}{\partial \widetilde{\xi}} + \frac{\partial}{\partial \widetilde{\xi}}(\widetilde{\rho}\widetilde{u})\right)\right], \tag{11.3.12}$$

(continuity equation)

$$\frac{\partial\tilde{\rho}}{\partial\tilde{\xi}}-\frac{\partial\widetilde{\rho u}}{\partial\tilde{\xi}}+\frac{\partial}{\partial\tilde{\xi}}\left\{\tilde{\tau}\left[\frac{\partial}{\partial\tilde{\xi}}\left[\tilde{p}+\widetilde{\rho u^2}+\tilde{\rho}-2\widetilde{\rho u}\right]+\tilde{\rho}\frac{\partial\tilde{\psi}}{\partial\tilde{\xi}}\right]\right\}=0, \tag{11.3.13}$$

(motion equation)

$$\begin{aligned}&\frac{\partial}{\partial\tilde{\xi}}\left(\widetilde{\rho u^2}+\tilde{p}-\widetilde{\rho u}\right)+\frac{\partial}{\partial\tilde{\xi}}\left\{\tilde{\tau}\left[\frac{\partial}{\partial\tilde{\xi}}\left(2\widetilde{\rho u^2}-\widetilde{\rho u}+2\tilde{p}-\widetilde{\rho u^3}-3\widetilde{p u}\right)+\tilde{\rho}\frac{\partial\tilde{\psi}}{\partial\tilde{\xi}}\right]\right\}\\&+\frac{\partial\tilde{\psi}}{\partial\tilde{\xi}}\left\{\tilde{\rho}-\tilde{\tau}\left[-\frac{\partial\tilde{\rho}}{\partial\tilde{\xi}}+\frac{\partial}{\partial\tilde{\xi}}(\widetilde{\rho u})\right]\right\}-2\frac{\partial}{\partial\tilde{\xi}}\left\{\widetilde{\tau\rho u}\frac{\partial\tilde{\psi}}{\partial\tilde{\xi}}\right\}=0,\end{aligned} \tag{11.3.14}$$

(energy equation)

$$\begin{aligned}&\frac{\partial}{\partial\tilde{\xi}}\left(\widetilde{\rho u^2}+3\tilde{p}-\widetilde{\rho u^3}-5\widetilde{p u}\right)\\&-\frac{\partial}{\partial\tilde{\xi}}\left\{\tilde{\tau}\frac{\partial}{\partial\tilde{\xi}}\left(2\widetilde{\rho u^3}+10\widetilde{p u}-\widetilde{\rho u^2}-3\tilde{p}-\widetilde{\rho u^4}-8\widetilde{p u^2}-5\frac{\tilde{p}^2}{\tilde{\rho}}\right)\right\}\\&+\frac{\partial}{\partial\tilde{\xi}}\left\{\tilde{\tau}\left(3\widetilde{\rho u^2}+5\tilde{p}\right)\frac{\partial\tilde{\psi}}{\partial\tilde{\xi}}\right\}-2\widetilde{\rho u}\frac{\partial\tilde{\psi}}{\partial\tilde{\xi}}-2\frac{\partial}{\partial\tilde{\xi}}\left\{\widetilde{\tau\rho u}\frac{\partial\tilde{\psi}}{\partial\tilde{\xi}}\right\}\\&+2\tilde{\tau}\frac{\partial\tilde{\psi}}{\partial\tilde{\xi}}\left[-\frac{\partial}{\partial\tilde{\xi}}(\widetilde{\rho u})+\frac{\partial}{\partial\tilde{\xi}}\left(\widetilde{\rho u^2}+\tilde{p}\right)+\tilde{\rho}\frac{\partial\tilde{\psi}}{\partial\tilde{\xi}}\right]=0,\end{aligned} \tag{11.3.15}$$

Some comments to the system of four ordinary nonlinear equations (11.3.12)–(11.3.15):

1. Every equation from the system is of the second order and needs two conditions. The problem belongs to the class of Cauchy problems.
2. In comparison for example, with the Schrödinger theory connected with behavior of the wave function, no special conditions are applied for dependent variables including the domain of the solution existing. This domain is defined automatically in the process of the numerical solution of the concrete variant of calculations.
3. From the introduced scales $\rho_0, u_0, x_0 = u_0 t_0, \psi_0 = u_0^2$, $\gamma_{N0} = u_0^2/(\rho_0 x_0^2)$, $p_0 = \rho_0 u_0^2$, only three parameters are independent, namely, ρ_0, u_0, x_0.
4. Approximation for the dimensionless non-local parameter $\tilde{\tau}$ should be introduced. In the definite sense it is not the problem of the hydrodynamic level of the physical system description (like the calculation of the kinetic coefficients in the classical hydrodynamics). Interesting to notice that quantum GHE were applied with success for calculation of atom structure [63, 64], which is considered as two species charged e, i mixture. The corresponding approximations for non-local parameters τ_i, τ_e, and τ_{ei} are proposed in [63, 64]. In the theory of the atom structure [64] after taking into account the Balmer's relation, non-local parameters τ_e transforms into

$$\tau_e = n\hbar/\left(m_e u^2\right), \tag{11.3.16}$$

where $n = 1, 2, \ldots$ is principal quantum number. As result the length scale relation was written as $x_0 = H/(m_e u_0) = n\hbar/(m_e u_0)$. But the value $\nu^{\mathrm{qu}} = \hbar/m_e$ has the dimension [cm^2/s] and can be titled as *quantum viscosity*, $\nu^{\mathrm{qu}} = 1.1577\,\mathrm{cm}^2/\mathrm{s}$. Then

$$\tau_e = n\nu^{\mathrm{qu}}/u^2. \tag{11.3.17}$$

Introduce now the quantum Reynolds number

$$\mathrm{Re}^{\mathrm{qu}} = u_0 x_0/\nu^{\mathrm{qu}}. \tag{11.3.18}$$

As result from Eqs. (11.3.17) and (11.3.18) follows the condition of quantization for $\mathrm{Re}^{\mathrm{qu}}$. Namely

$$\mathrm{Re}^{\mathrm{qu}} = n, \quad n = 1,2,\ldots \tag{11.3.19}$$

5. Taking into account the previous considerations I introduce the following approximation for the dimensionless non-local parameter

$$\tilde{\tau} = 1/\tilde{u}^2, \tag{11.3.20}$$

$$\tau = u_0 x_0 / u^2 = \nu_0^{\mathrm{k}} / u^2, \tag{11.3.21}$$

where the scale for the kinematical viscosity is introduced $\nu_0^{\mathrm{k}} = u_0 x_0$. Then we have the physically transparent result—non-local parameter is proportional to the kinematical viscosity and in inverse proportion to the square of velocity.

The system of GHE (11.3.12)–(11.3.15) (solved with the help of Maple) have the great possibilities of mathematical modeling as result of changing of eight Cauchy conditions describing the character features of initial perturbations which lead to the soliton formation. The following Maple notations on figures are used: *r*—density $\tilde{\rho}$, *u*—velocity $\tilde{u}$, *p*—pressure $\tilde{p}$, and *v*—self-consistent potential $\tilde{\psi}$.

Explanations placed under all following figures, Maple program contains Maple's notations—for example the expression $D(u)(0)=0$ means in the usual notations $\left(\partial\tilde{u}/\partial\tilde{\xi}\right)(0)=0$, independent variable *t* responds to $\tilde{\xi}$.

We begin with investigation of the problem of principle significance—is it possible after a perturbation (defined by the Cauchy conditions) to obtain the gravitational object of the soliton's kind as result of the self-organization of the matter? With this aim let us consider the initial perturbations (SYSTEM I):

$$\mathbf{u(0)=1, p(0)}=1, \mathbf{r(0)}=1, \mathbf{D(u)(0)}=0, \mathbf{D(p)(0)}=0, \mathbf{D(r)(0)}=0, \mathbf{D(v)(0)}=0, \mathbf{v(0)}=1.$$

Figures 11.9–11.11 reflect the result of solution of Eqs. (11.3.12)–(11.3.15) with the choice of scales leading to $\tilde{\gamma}_{\mathrm{N}} = 1$. Figures 11.9–11.12 correspond to the approximation of the non-local parameter $\tilde{\tau}$ in the form (Eq. (11.3.20)). Figure 11.9 displays the gravitational object placed in bounded region of 1D Cartesian space, all parts of this object are moving with the same velocity. Important to underline that no special boundary conditions were used for this and all following cases. Then this soliton is product of the self-organization of gravitational matter. Figures 11.10 and 11.11 contain the answer for formulated above question about stability of the object. The derivative (see Fig. 11.11) $\partial\tilde{\psi}/\partial\tilde{\xi} = (\partial\psi/\partial\xi)\left(x_0/u_0^2\right) = -g(\xi)/\left(u_0^2/x_0\right) = -\tilde{g}(\xi)$ is proportional to the self-consistent gravitational force acting on the soliton and in its vicinity. Therefore the stability of the object is result of the self-consistent influence of the gravitational potential and pressure.

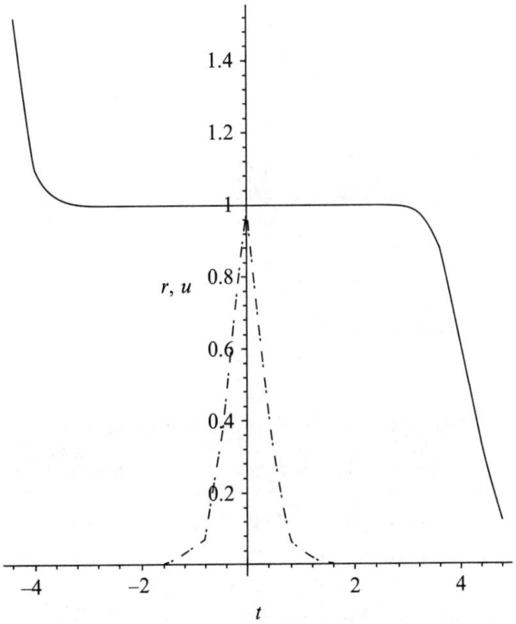

FIGURE 11.9 *r*—density $\tilde{\rho}$ (dash-dotted line), *u*—velocity, $\tilde{u}$ in gravitational soliton.

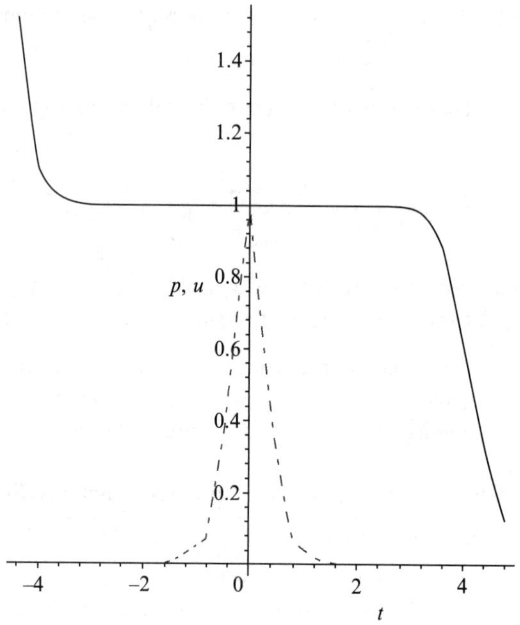

FIGURE 11.10 p—pressure $\widetilde{p}$ (dashed line), u—velocity, $\widetilde{u}$ in gravitational soliton.

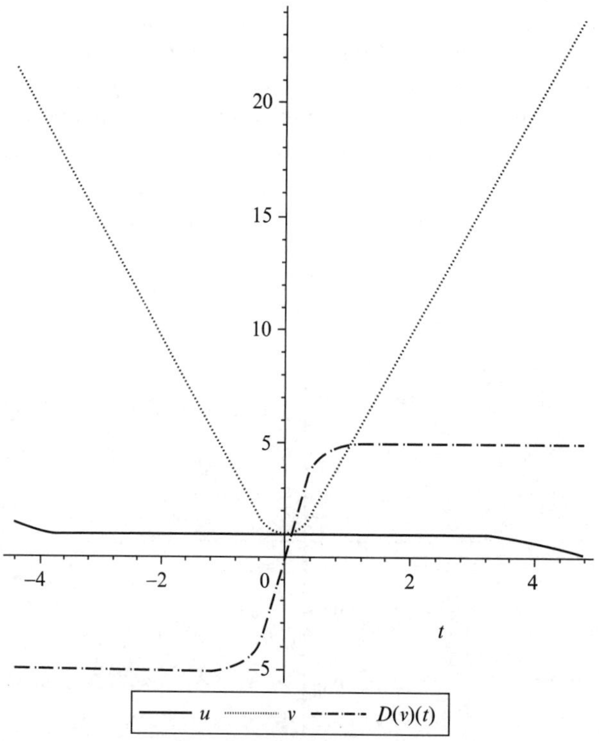

FIGURE 11.11 u—velocity $\widetilde{u}$, v—self-consistent potential $\widetilde{\psi}$, $D(v)(t) = \partial\widetilde{\psi}/\partial\widetilde{\xi}$ in soliton.

Extremely important that the self-consistent gravitational force has the character of the flat area which exists in the vicinity of the object. This solution exists only in the restricted area of space; the corresponding character length is defined automatically as result of the numerical solution of the problem. The non-local parameter $\widetilde{\tau}$, in the definite sense plays the role analogous to kinetic coefficients in the usual Boltzmann kinetic theory. The influence on the results of calculations is

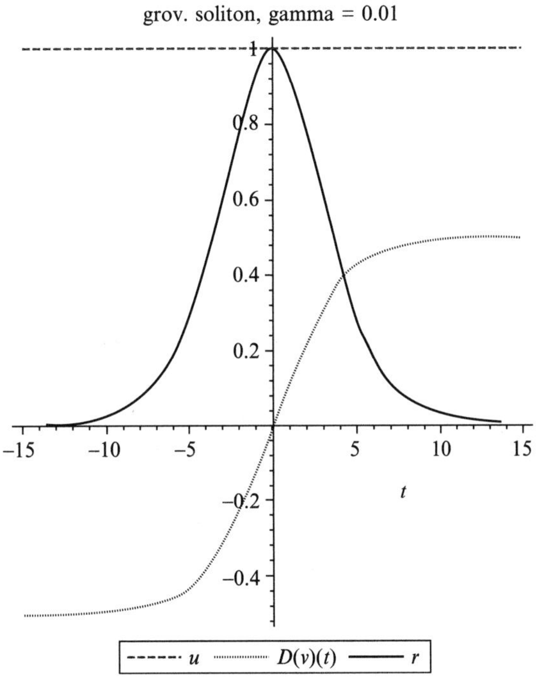

FIGURE 11.12 u—velocity $\widetilde{u}$, r—density $\widetilde{\rho}$, $D(v)(t)=\partial\widetilde{\psi}/\partial\widetilde{\xi}$ in soliton, ($\widetilde{\gamma}_{\mathrm{N}}=0.01$).

not too significant. The same situation exists in the generalized hydrodynamics. Really, let us use the another approximation for $\widetilde{\tau}$ in the simplest possible form, namely

$$\widetilde{\tau}=1. \tag{11.3.22}$$

Figures 11.13 and 11.14 reflect the results of solution of Eqs. (11.3.12)–(11.3.15) with the choice of scales leading to $\widetilde{\gamma}_{\mathrm{N}}=1$, but with the approximation of the non-local parameter $\widetilde{\tau}$ in the form (Eq. (11.3.22)).

Spiral galaxies have rather complicated geometrical forms and 3D calculations should be used. But reasonable to suppose that influence of halo on galaxy kernel is not too significant and use for calculations the spherical coordinate system. The 1D calculations in the Cartesian coordinate system correspond to calculations in the spherical coordinate system by the large radii of curvature, but have also the independent significance in another character scales. Namely for explanations of the meteorological front motion (without taking into account the Earth rotation). In this theory cyclone or anticyclone corresponds to moving solitons. In the Earth scale the scales can be used: $\rho_{\mathrm{air}}=1.29\times10^{-3}\,\mathrm{g/cm^3}$, $u_0=1\,\mathrm{m/s}$, $x_0=10\,\mathrm{km}$, and $\widetilde{\gamma}_{\mathrm{N}}\sim0.01$. Figure 11.12 reflects the results of the corresponding calculation and in particular reflects correctly the wind orientation in front and behind of the soliton.

The full system of 3D non-local hydrodynamic equations in moving (along x-axis) Cartesian coordinate system and the corresponding expression for derivatives in the spherical coordinate system can be found in [58, 59] (see also Appendix 2). The following figures reflect the result of soliton calculations for the case of spherical symmetry for galaxy kernel. The velocity $\widetilde{u}$ corresponds to the direction of the soliton movement for spherical coordinate system on following figures. Self-consistent gravitational force F acting on the unit of mass permits to define the orbital velocity w of objects in halo, $w=\sqrt{Fr}$, or

$$\widetilde{w}=\sqrt{\widetilde{r}\frac{\partial\widetilde{\psi}}{\partial\widetilde{r}}}, \tag{11.3.23}$$

where r is the distance from the center of galaxy. All calculations are realized for the conditions (SYSTEM I) but for different parameter

$$G=\widetilde{\gamma}_{\mathrm{N}}=\gamma_{\mathrm{N}}/\gamma_{\mathrm{N},0}=\gamma_{\mathrm{N}}\rho_0x_0^2/u_0^2. \tag{11.3.24}$$

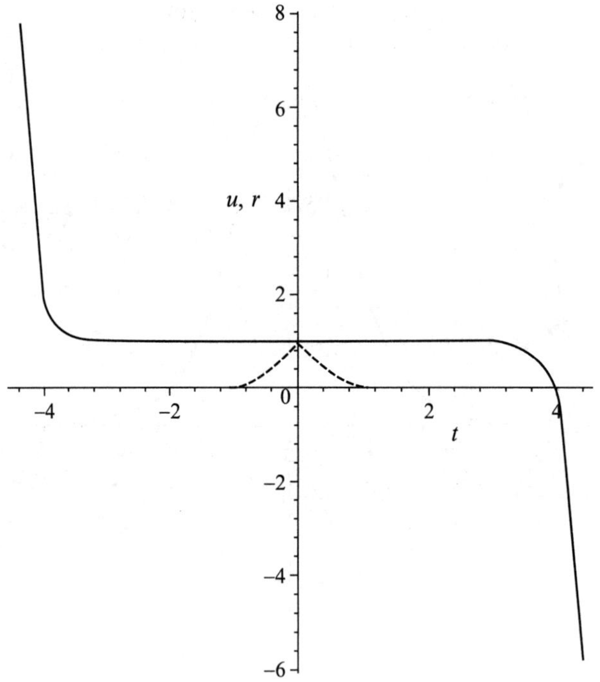

FIGURE 11.13 r—density $\widetilde{\rho}$ (dashed line), u—velocity $\widetilde{u}$ in gravitational soliton.

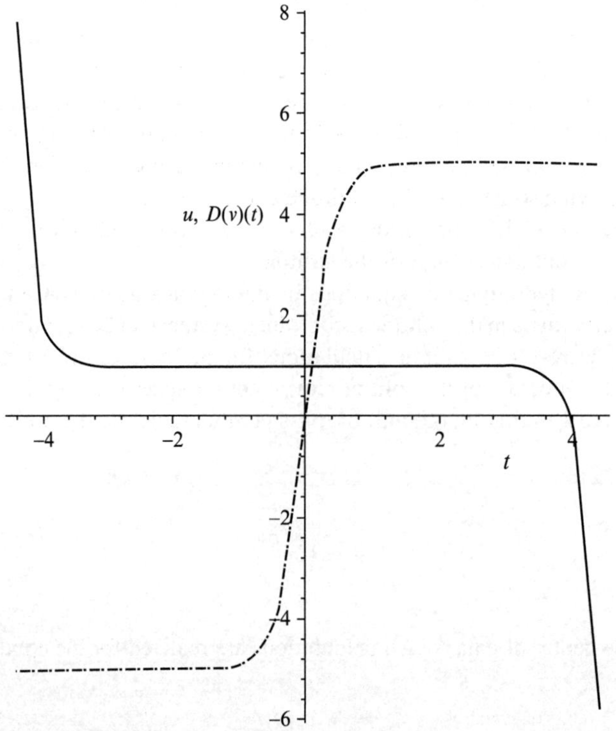

FIGURE 11.14 u—velocity $\widetilde{u}$, $D(v)(t) = \partial\widetilde{\psi}/\partial\widetilde{\xi}$.

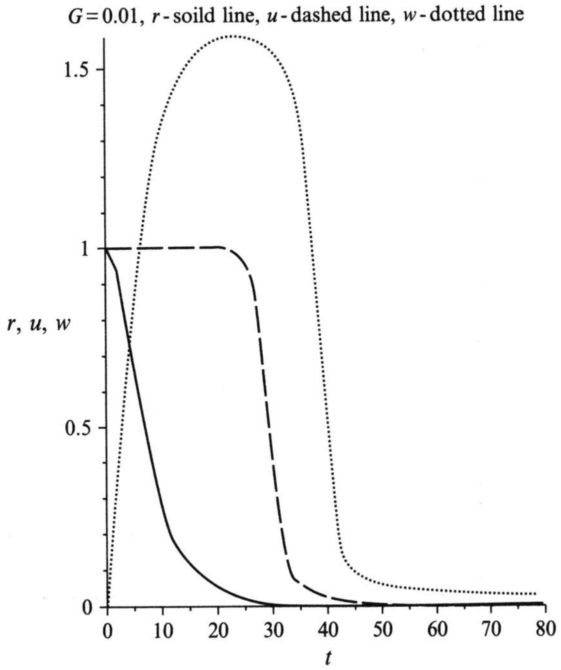

FIGURE 11.15 r—density $\widetilde{\rho}$, u—velocity $\widetilde{u}$, w—orbital velocity $\widetilde{w}$. $G=0.01$.

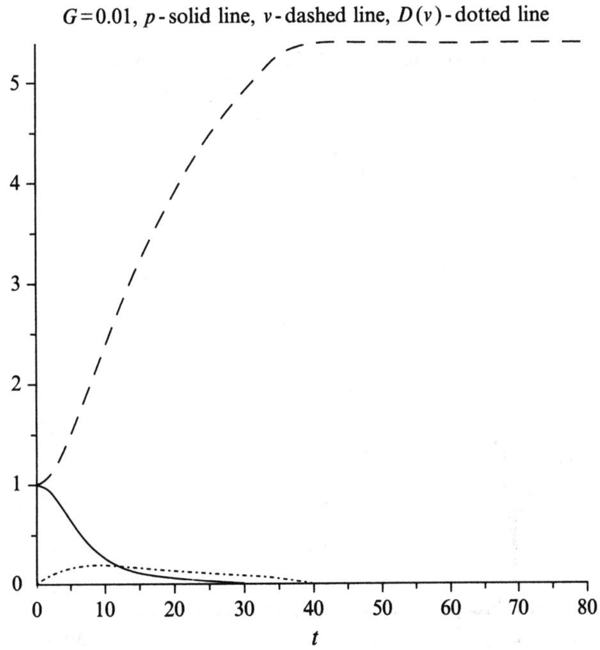

FIGURE 11.16 p—pressure $\widetilde{p}$, v—self consistent potential $\widetilde{\psi}$, $D(v)(t) = \partial\widetilde{\psi}/\partial\widetilde{\xi}$ in gravitational soliton.

Parameter G plays the role of similarity criteria in traditional hydrodynamics. Important conclusions:

1. Figures 11.15–11.22 demonstrate evolution the rotation curves from the Kepler regime (Figs. 11.15 and 11.16; small G, like curve **A** in Fig. 11.8) to observed (Figs. 11.21 and 11.22; large G, like curve **B** in Fig. 11.8) for typical spiral galaxies.
2. The stars with planets (like Sun) correspond to gravitational soliton with small G and therefore originate the Kepler rotation regime.

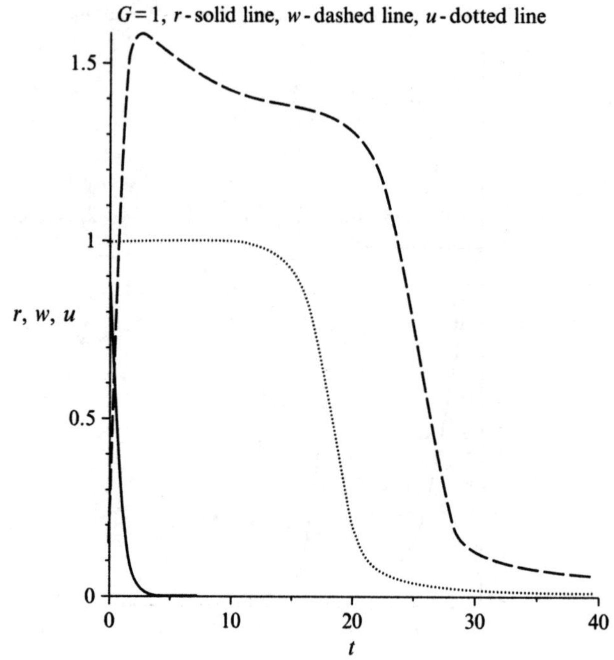

FIGURE 11.17 r—density $\tilde{\rho}$, u—velocity $\tilde{u}$ gravitational soliton, w—orbital velocity $\tilde{w}$. $G=1$.

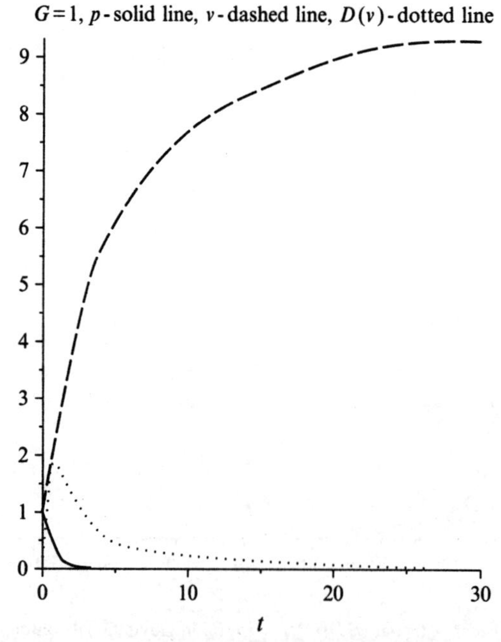

FIGURE 11.18 p—pressure $\tilde{p}$, v—self consistent potential $\tilde{\psi}$, $D(v)(t)=\partial\tilde{\psi}/\partial\tilde{\xi}$ in gravitational soliton.

3. Regime **B** cannot be obtained in the frame of local statistical physics in principal and authors of many papers introduce different approximations for additional "dark matter density" (as usual in Poisson equation) trying to find coincidence with data of observations.
4. From the wrong position of local theories Poisson equation (11.3.7) contains "dark matter density", continuity equation (11.3.8) contains the "flux of dark matter density", motion equation (11.3.9) includes "dark energy", the

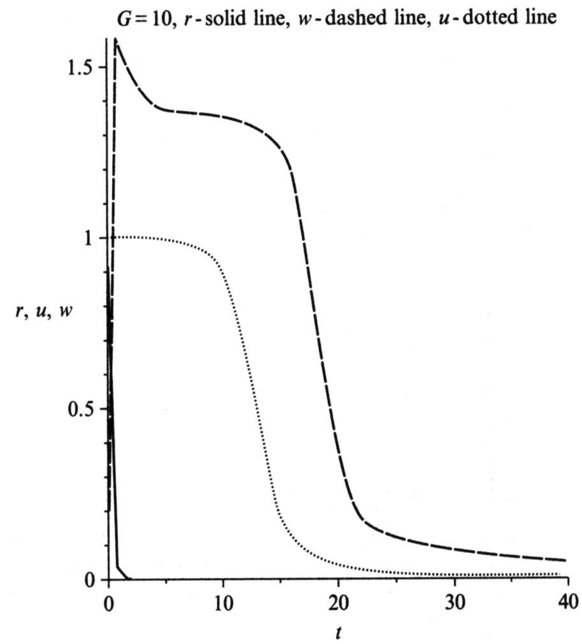

FIGURE 11.19 r—density $\tilde{\rho}$, u—velocity $\tilde{u}$ in gravitational soliton, w—orbital velocity $\tilde{w}$. $G=10$.

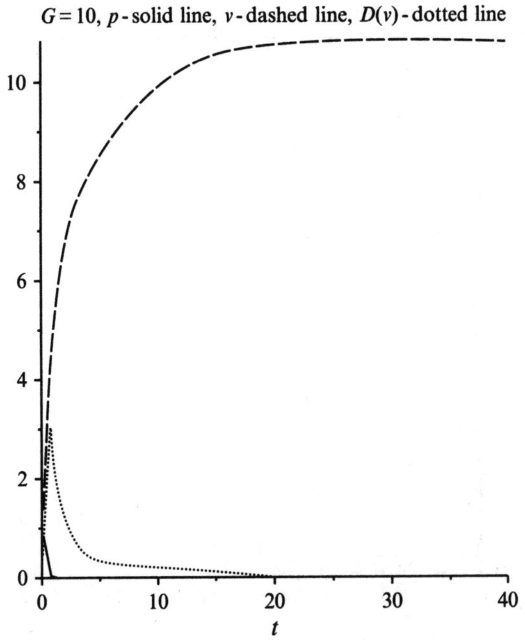

FIGURE 11.20 p—pressure $\tilde{p}$, v—self consistent potential $\tilde{\psi}$, $D(v)(t) = \partial\tilde{\psi}/\partial\tilde{\xi}$ in gravitational soliton.

energy equation (11.3.10) has "the flux of dark energy" and so on to the "senior dark velocity moments". This entire situation is similar to the turbulent theories based on local statistical physics and empirical corrections for velocity moments.

As we see peculiar features of the halo movement can be explained without new concepts like "dark matter" [249]. Important to underline that the shown transformation of the Kepler's regime into the flat rotation curves for different solitons explains the "mysterious" fact of the dark matter absence in the Sun vicinity.

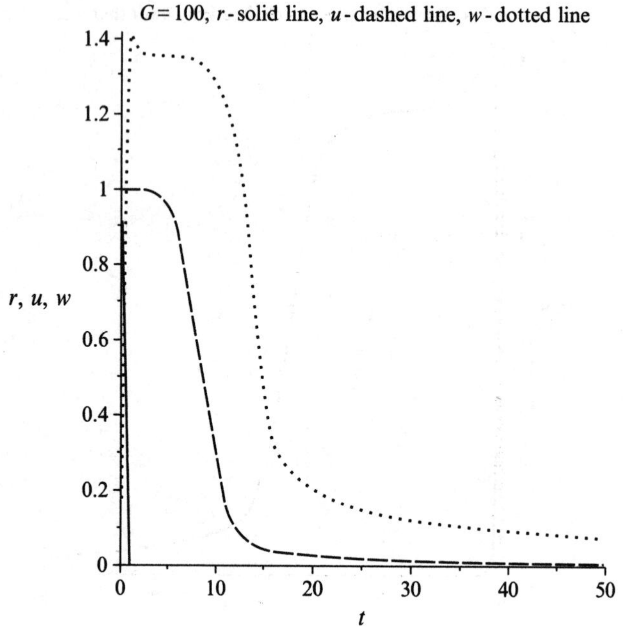

FIGURE 11.21 r—density $\widetilde{\rho}$, u—velocity $\widetilde{u}$ in gravitational soliton, w—orbital velocity $\widetilde{w}$. $G = 100$.

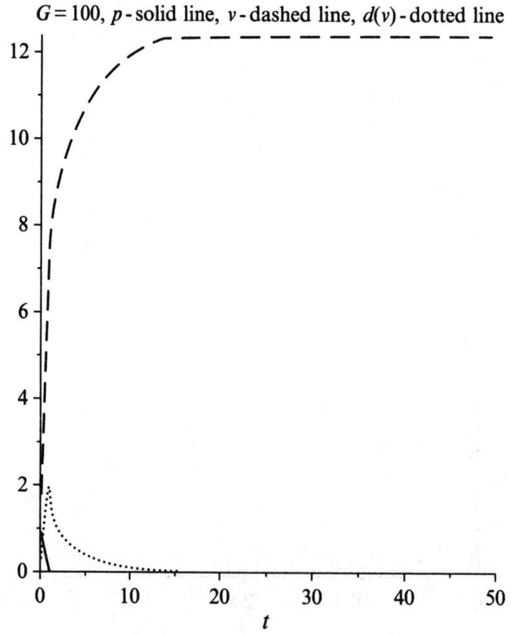

FIGURE 11.22 p—pressure $\widetilde{p}$, v—self consistent potential $\widetilde{\psi}$, $D(v)(t) = \partial\widetilde{\psi}/\partial\widetilde{\xi}$ in gravitational soliton.

11.4 HUBBLE EXPANSION AND THE PROBLEM OF DARK ENERGY

In simplest interpretation of the *local theories* the dark energy is related usually to the Einstein cosmological constant (see Section 11.2).

From the non-local statistical theory the physical picture follows which leading to the Hubble flow without new essence like dark energy and without modification of Newton force like Eq. (11.2.14).

Namely:

The main origin of Hubble effect (including the matter expansion with acceleration) is self-catching of expanding matter by the self-consistent gravitational field in conditions of the weak influence of the central massive bodies.

The formulated result is obtained in Section 11.2 in the frame of the linear theory. Is it possible to obtain the corresponding result on the level of the general nonlinear description? Such an investigation was successfully realized and leads to a direct mathematical model supporting the well-known observations of S. Perlmutter, A. Riess (USA) and B. Schmidt (Australia). These researchers studied Type 1a supernovae and determined that more distant galactic objects seem to move faster. Their observations suggest that not only is the Universe expanding, its expansion is relentlessly speeding up.

Effects of gravitational self-catching should be typical for Universe. The existence of "Hubble boxes" is discussed in review [245] as typical blocks of the nearby Universe. Gravitational self-catching takes place for Big Bang having given birth to the global expansion of Universe, but also for Little Bang in so-called Local Group (using the Hubble's terminology) of galaxies. Then the evolution of the Local Group (the typical Hubble box) is really fruitful field for testing of different theoretical constructions (see Fig. 11.23). The data were obtained by Karachentsev and his collaborators in 2002-2007 in observation with the Hubble Space Telescope [250, 251]]. Each point corresponds to a galaxy with measured values of distance and line-of-site velocity in the reference frame related to the center of the Local Group. The diagram shows two distinct structures, the Local Group and the local flow of galaxies. The galaxies of the Local Group occupy a volume with the radius up to ~1.1-1.2 Mpc, but there are no galaxies in the volume whose radius is less than 0.25 Mpc. These galaxies move both away from the center (positive velocities) and toward the center (negative velocities). These galaxies form a gravitationally bound quasi-stationary system. Their average radial velocity is equal to zero. The galaxies of the local flow are located outside the group and all of them are moving from the center (positive velocities) beginning their motion near $R \approx 1$ Mpc with the velocity $v \sim 50$ km/s. By the way the measured by Karachentsev the average Hubble parameter for the Local Group is $72 \pm 6\,\mathrm{km\,s^{-1}\,Mpc^{-1}}$.

Let us choose these values as scales:

$$x_0 = 1\,\mathrm{Mpc}, \quad u_0 = 50\,\mathrm{kms^{-1}}. \tag{11.4.1}$$

Recession velocities increase as the distance increases in accordance with the Hubble law. The straight line correspond the dependence from observations

$$v = H(r)r. \tag{11.4.2}$$

for the region outside of the Local Group. In the non-dimensional form

$$\tilde{v} = \tilde{H}(\tilde{r})\tilde{r}. \tag{11.4.3}$$

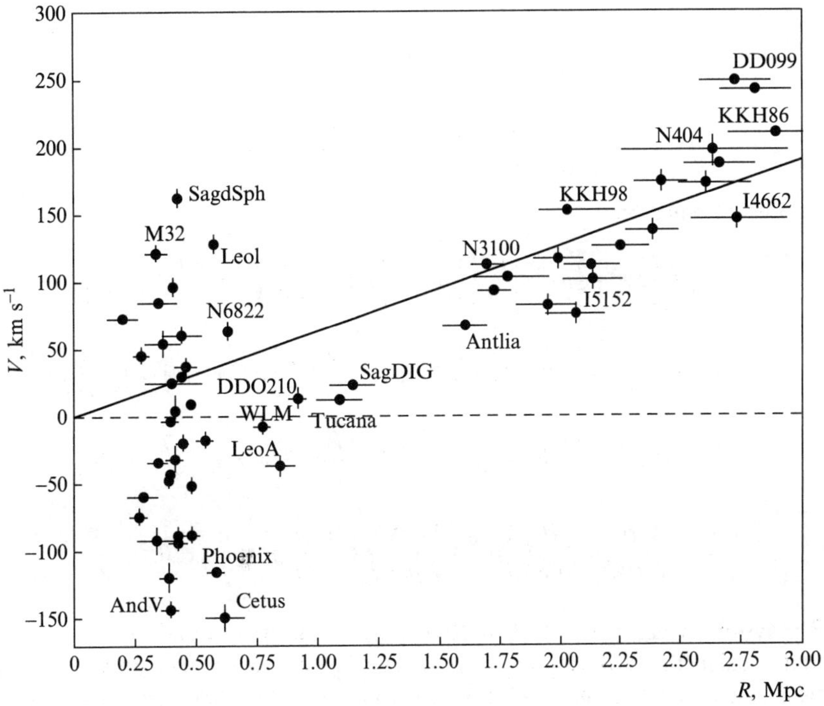

FIGURE 11.23 Velocity-distance diagram for galaxies at distances of up to 3 Mpc for Local Group of galaxies.

where

$$\widetilde{H}=\frac{x_0}{u_0}H(\widetilde{r}). \tag{11.4.4}$$

For the following calculations we should choose the corresponding scales (especially for estimating G) for modeling of the Local Group evolution

$$G\equiv\widetilde{\gamma}_{\mathrm{N}}=\gamma_{\mathrm{N}}/\gamma_{\mathrm{N}0}=\gamma_{\mathrm{N}}\rho_0 x_0^2/u_0^2. \tag{11.4.5}$$

For the density scale estimation the average density of the local flow could be used. But the corresponding data are not accessible and I use the average density of the Local Group which can be taken from references [251, 252] with $\rho_0=4.85\times10^{-29}\,\mathrm{g/cm^3}$. Then from (11.4.5) we have $G\cong1$.

Let us go now to the mathematical modeling. The non-local system of hydrodynamic equations describing the explosion with the spherical symmetry is written as (see [58, 59], Appendix 2)

$$g_r=-\frac{\partial\psi}{\partial r}, \tag{11.4.6}$$

Poisson equation

$$\frac{1}{r^2}\frac{\partial}{\partial r}\left(r^2\frac{\partial\psi}{\partial r}\right)=4\pi\gamma_{\mathrm{N}}\left[\rho-\tau\left(\frac{\partial\rho}{\partial t}+\frac{1}{r^2}\frac{\partial(r^2\rho v_r)}{\partial r}\right)\right], \tag{11.4.7}$$

Continuity equation

$$\frac{\partial}{\partial t}\left\{\rho-\tau\left[\frac{\partial\rho}{\partial t}+\frac{1}{r^2}\frac{\partial(r^2\rho v_r)}{\partial r}\right]\right\}+\frac{1}{r^2}\frac{\partial}{\partial r}\left\{r^2\left\{\rho v_r-\tau\left[\frac{\partial}{\partial t}(\rho v_r)+\frac{1}{r^2}\frac{\partial(r^2\rho v_r^2)}{\partial r}-\rho g_r\right]\right\}\right\}-\frac{1}{r^2}\frac{\partial}{\partial r}\left(\tau r^2\frac{\partial p}{\partial r}\right)=0, \tag{11.4.8}$$

Motion equation

$$\begin{aligned}
&\frac{\partial}{\partial t}\left\{\rho v_r-\tau\left[\frac{\partial}{\partial t}(\rho v_r)+\frac{1}{r^2}\frac{\partial(r^2\rho v_r^2)}{\partial r}+\frac{\partial p}{\partial r}-\rho g_r\right]\right\}-g_r\left[\rho-\tau\left(\frac{\partial\rho}{\partial t}+\frac{1}{r^2}\frac{\partial(r^2\rho v_r)}{\partial r}\right)\right]\\
&+\frac{1}{r^2}\frac{\partial}{\partial r}\left\{r^2\left\{\rho v_r^2-\tau\left[\frac{\partial}{\partial t}(\rho v_r^2)+\frac{1}{r^2}\frac{\partial(r^2\rho v_r^3)}{\partial r}-2g_r\rho v_r\right]\right\}\right\}\\
&+\frac{\partial p}{\partial r}-\frac{\partial}{\partial r}\left(\tau\frac{\partial p}{\partial t}\right)-2\frac{\partial}{\partial r}\left(\frac{\tau}{r^2}\frac{\partial(r^2pv_r)}{\partial r}\right)-\frac{1}{r^2}\frac{\partial}{\partial r}\left(\tau r^2\frac{\partial(pv_r)}{\partial r}\right)=0,
\end{aligned} \tag{11.4.9}$$

Energy equation

$$\begin{aligned}
&\frac{\partial}{\partial t}\left\{\frac{1}{2}\rho v_r^2+\frac{3}{2}p-\tau\left[\frac{\partial}{\partial t}\left(\frac{1}{2}\rho v_r^2+\frac{3}{2}p\right)+\frac{1}{r^2}\frac{\partial}{\partial r}\left(r^2v_r\left(\frac{1}{2}\rho v_r^2+\frac{5}{2}p\right)\right)-\rho g_r v_r\right]\right\}\\
&+\frac{1}{r^2}\frac{\partial}{\partial r}\left\{r^2\left\{\left(\frac{1}{2}\rho v_r^2+\frac{5}{2}p\right)v_r-\tau\left[\frac{\partial}{\partial t}\left(\left(\frac{1}{2}\rho v_r^2+\frac{5}{2}p\right)v_r\right)+\frac{1}{r^2}\frac{\partial}{\partial r}\left(r^2\left(\frac{1}{2}\rho v_r^2+\frac{7}{2}p\right)v_r^2\right)\right.\right.\right.\\
&\left.\left.\left.-\rho g_r v_r^2-\left(\frac{1}{2}\rho v_r^2+\frac{3}{2}p\right)g_r\right]\right\}\right\}-\left\{\rho g_r v_r-\tau\left[g_r\left(\frac{\partial}{\partial t}(\rho v_r)+\frac{1}{r^2}\frac{\partial}{\partial r}(r^2\rho v_r^2)+\frac{\partial p}{\partial r}-\rho g_r\right)\right]\right\}\\
&-\frac{1}{r^2}\frac{\partial}{\partial r}\left(\tau r^2\frac{\partial}{\partial r}\left(\frac{1}{2}pv_r^2+\frac{5}{2}\frac{p^2}{\rho}\right)\right)+\frac{1}{r^2}\frac{\partial}{\partial r}(r^2\tau pg_r)=0.
\end{aligned} \tag{11.4.10}$$

The system of Eqs. (11.4.6)–(11.4.10) belongs to the class of the 1D non-stationary equations and can be solved by known numerical methods. But for the aims of the transparent vast mathematical modeling of self-catching of the expanding matter by the self-consistent gravitational field I introduce the following assumption. Let us allot the quasi-stationary Hubble regime when only the implicit dependence on time for the unknown values exists. It means that for the intermediate (Hubble) regime the substitution

$$\frac{\partial}{\partial t}=\frac{\partial}{\partial r}\frac{\partial r}{\partial t}=v_r\frac{\partial}{\partial r}. \tag{11.4.11}$$

can be introduced. As result we have the following system of the 1D dimensionless equations:

$$\frac{1}{\tilde{r}^2}\frac{\partial}{\partial \tilde{r}}\left(\tilde{r}^2\frac{\partial\tilde{\psi}}{\partial\tilde{r}}\right)=4\pi G\left[\tilde{\rho}-\tilde{\tau}\left(\tilde{v}_r\frac{\partial\tilde{\rho}}{\partial\tilde{r}}+\frac{1}{\tilde{r}^2}\frac{\partial(\tilde{r}^2\tilde{\rho}\tilde{v}_r)}{\partial\tilde{r}}\right)\right], \tag{11.4.12}$$

$$\begin{aligned}&\tilde{v}_r\frac{\partial}{\partial\tilde{r}}\left\{\tilde{\rho}-\tilde{\tau}\left[\tilde{v}_r\frac{\partial\tilde{\rho}}{\partial\tilde{r}}+\frac{1}{\tilde{r}^2}\frac{\partial(\tilde{r}^2\tilde{\rho}\tilde{v}_r)}{\partial\tilde{r}}\right]\right\}+\frac{1}{\tilde{r}^2}\frac{\partial}{\partial\tilde{r}}\left\{\tilde{r}^2\left\{\tilde{\rho}\tilde{v}_r-\tilde{\tau}\left[\tilde{v}_r\frac{\partial}{\partial\tilde{r}}(\tilde{\rho}\tilde{v}_r)\right.\right.\right.\\&\left.\left.\left.+\frac{1}{\tilde{r}^2}\frac{\partial(\tilde{r}^2\tilde{\rho}\tilde{v}_r^2)}{\partial\tilde{r}}+\tilde{\rho}\frac{\partial\tilde{\psi}}{\partial\tilde{r}}\right]\right\}\right\}-\frac{1}{\tilde{r}^2}\frac{\partial}{\partial\tilde{r}}\left(\tilde{\tau}\tilde{r}^2\frac{\partial\tilde{p}}{\partial\tilde{r}}\right)=0\end{aligned} \tag{11.4.13}$$

$$\begin{aligned}&\tilde{v}_r\frac{\partial}{\partial\tilde{r}}\left\{\tilde{\rho}\tilde{v}_r-\tilde{\tau}\left[\tilde{v}_r\frac{\partial}{\partial\tilde{r}}(\tilde{\rho}\tilde{v}_r)+\frac{1}{\tilde{r}^2}\frac{\partial(\tilde{r}^2\tilde{\rho}\tilde{v}_r^2)}{\partial\tilde{r}}+\frac{\partial\tilde{p}}{\partial\tilde{r}}+\tilde{\rho}\frac{\partial\tilde{\psi}}{\partial\tilde{r}}\right]\right\}\\&+\frac{\partial\tilde{\psi}}{\partial\tilde{r}}\left[\tilde{\rho}-\tilde{\tau}\left(\tilde{v}_r\frac{\partial\tilde{\rho}}{\partial\tilde{r}}+\frac{1}{\tilde{r}^2}\frac{\partial(\tilde{r}^2\tilde{\rho}\tilde{v}_r)}{\partial\tilde{r}}\right)\right]\\&+\frac{1}{\tilde{r}^2}\frac{\partial}{\partial\tilde{r}}\left\{\tilde{r}^2\left\{\tilde{\rho}\tilde{v}_r^2-\tilde{\tau}\left[\tilde{v}_r\frac{\partial}{\partial\tilde{r}}(\tilde{\rho}\tilde{v}_r^2)+\frac{1}{\tilde{r}^2}\frac{\partial(\tilde{r}^2\tilde{\rho}\tilde{v}_r^3)}{\partial\tilde{r}}+2\frac{\partial\tilde{\psi}}{\partial\tilde{r}}\tilde{\rho}\tilde{v}_r\right]\right\}\right\}\\&+\frac{\partial\tilde{p}}{\partial\tilde{r}}-\frac{\partial}{\partial\tilde{r}}\left(\tilde{v}_r\tilde{\tau}\frac{\partial\tilde{p}}{\partial\tilde{r}}\right)-2\frac{\partial}{\partial\tilde{r}}\left(\tilde{\tau}\tilde{r}^2\frac{\partial(\tilde{r}^2\tilde{\rho}\tilde{v}_r)}{\partial\tilde{r}}\right)-\frac{1}{\tilde{r}^2}\frac{\partial}{\partial\tilde{r}}\left(\tilde{\tau}\tilde{r}^2\frac{\partial(\tilde{p}\tilde{v}_r)}{\partial\tilde{r}}\right)=0,\end{aligned} \tag{11.4.14}$$

$$\begin{aligned}&\tilde{v}_r\frac{\partial}{\partial\tilde{r}}\left\{\tilde{\rho}\tilde{v}_r^2+3\tilde{p}-\tilde{\tau}\left[\tilde{v}_r\frac{\partial}{\partial\tilde{r}}(\tilde{\rho}\tilde{v}_r^2+3\tilde{p})+\frac{1}{\tilde{r}^2}\frac{\partial}{\partial\tilde{r}}(\tilde{r}^2\tilde{v}_r(\tilde{\rho}\tilde{v}_r^2+5\tilde{p}))+2\tilde{\rho}\frac{\partial\tilde{\psi}}{\partial\tilde{r}}\tilde{v}_r\right]\right\}\\&+\frac{1}{\tilde{r}^2}\frac{\partial}{\partial\tilde{r}}\left\{\tilde{r}^2\left\{(\tilde{\rho}\tilde{v}_r^2+5\tilde{p})\tilde{v}_r-\tilde{\tau}\left[\tilde{v}_r\frac{\partial}{\partial\tilde{r}}((\tilde{\rho}\tilde{v}_r^2+5\tilde{p})\tilde{v}_r)+\frac{1}{\tilde{r}^2}\frac{\partial}{\partial\tilde{r}}(\tilde{r}^2(\tilde{\rho}\tilde{v}_r^2+7\tilde{p})\tilde{v}_r^2)\right.\right.\right.\\&\left.\left.\left.+2\tilde{\rho}\frac{\partial\tilde{\psi}}{\partial\tilde{r}}\tilde{v}_r^2+(\tilde{\rho}\tilde{v}_r^2+3\tilde{p})\frac{\partial\tilde{\psi}}{\partial\tilde{r}}\right]\right\}\right\}+2\left\{\tilde{\rho}\frac{\partial\tilde{\psi}}{\partial\tilde{r}}\tilde{v}_r-\tilde{\tau}\frac{\partial\tilde{\psi}}{\partial\tilde{r}}\left[\tilde{v}_r\frac{\partial}{\partial\tilde{r}}(\tilde{\rho}\tilde{v}_r)+\frac{1}{\tilde{r}^2}\frac{\partial}{\partial\tilde{r}}(\tilde{r}^2\tilde{\rho}\tilde{v}_r^2)+\frac{\partial\tilde{p}}{\partial\tilde{r}}+\tilde{\rho}\frac{\partial\tilde{\psi}}{\partial\tilde{r}}\right]\right\}\\&-\frac{1}{\tilde{r}^2}\frac{\partial}{\partial\tilde{r}}\left(\tilde{\tau}\tilde{r}^2\frac{\partial}{\partial\tilde{r}}\left(\tilde{p}\tilde{v}_r^2+5\frac{\tilde{p}^2}{\tilde{\rho}}\right)\right)-2\frac{1}{\tilde{r}^2}\frac{\partial}{\partial\tilde{r}}\left(\tilde{r}^2\tilde{\tau}\tilde{p}\frac{\partial\tilde{\psi}}{\partial\tilde{r}}\right)=0.\end{aligned} \tag{11.4.15}$$

The system of GHE (11.4.12)–(11.4.15) have the great possibilities of mathematical modeling as result of changing of eight Cauchy conditions describing the character features of the local flow evolution. The following Maple notations on figures are used: *r*—density $\tilde{\rho}$, *u*—velocity $\tilde{v}_r$, *p*—pressure $\tilde{p}$, and *v*—self-consistent potential $\tilde{\psi}$, *h*—$\tilde{H}$, independent variable *t* is $\tilde{r}$. Explanations placed under all following figures, Maple program contains Maple's notations—for example the expression $D(u)(0)=0$ means in the usual notations $(\partial\tilde{u}/\partial\tilde{r})(0)=0$.

As mentioned before, the non-local parameter $\tilde{\tau}$, in the definite sense plays the role analogous to kinetic coefficients in the usual Boltzmann kinetic theory. The influence on the results of calculations is not too significant. The same situation exists in the generalized hydrodynamics. As before I introduce the following approximation for the dimensionless non-local parameter (see Eq. (11.3.20)) $\tilde{\tau}=1/\tilde{u}^2$.

One obtains for the approximation (31) and SYSTEM 2:

$$\mathbf{v(1)=1,u(1)=1,r(1)=1,p(1)=1,D(v)(1)=0,D(u)(1)=0,D(r)(1)=0,D(p)(1)=0.}$$

Figures 11.24 and 11.25 correspond to $G = 1$. From these calculations follow:

1. As was waiting the quasi-stationary regime exists only in the restricted (on the left and on the right sides) area. Out of these boundaries the explicit time dependent regime should be considered. But it is not the Hubble regime.
2. In the Hubble regime one obtains the negative area (low part of the dashed curve). It corresponds to the self-consistent force acting along the expansion of the local flow.

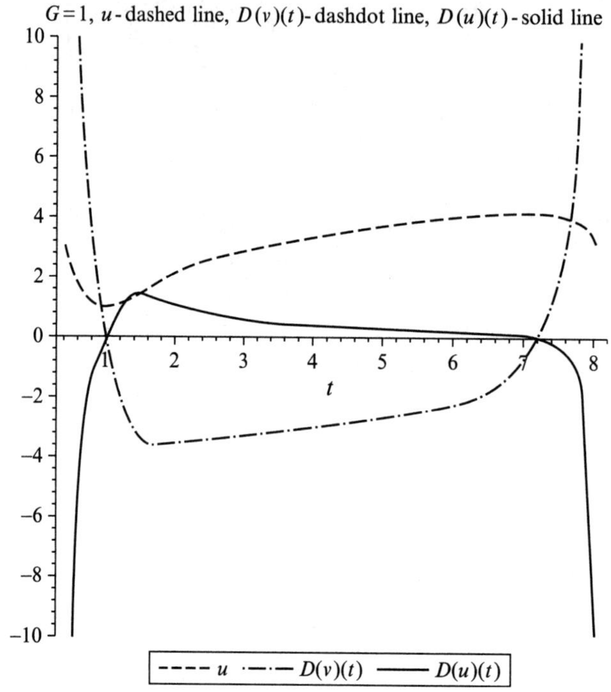

FIGURE 11.24 Dependence of the acceleration-deceleration function (defined as $D(u)(t) = \partial\tilde{u}/\partial\tilde{r}$), derivation of the self-consistent potential $D(v)(t) = \partial\tilde{\psi}/\partial\tilde{\xi}$ and velocity $u = \tilde{u}$ on the radial distance $\tilde{r}$.

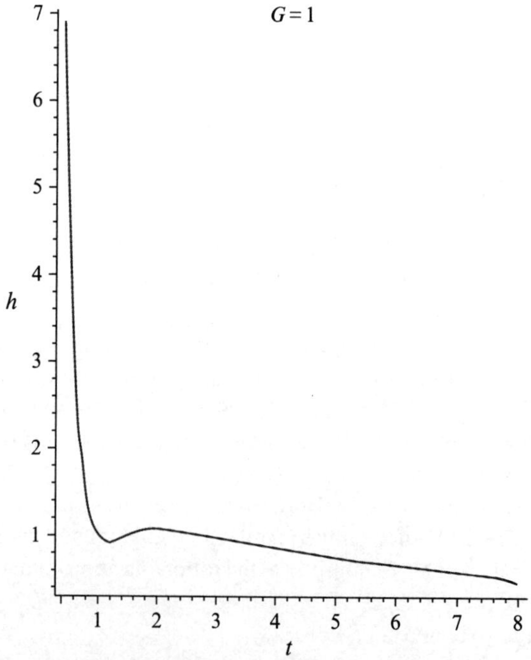

FIGURE 11.25 Dependence of the dimensionless Hubble parameter on the radial distance.

3. The dependence of $\widetilde{H}(\widetilde{r})$ is not linear (see Fig. 11.25), more over the curvature contains maximum. The area of acceleration placed between two areas of the deceleration.

   Let us show now the result of calculations for another $\widetilde{\tau}$ approximation in the simplest possible form, namely (see also Eq. (11.3.22)) $\widetilde{\tau}=1$. One obtains for this $\widetilde{\tau}$-approximation and SYSTEM 2 and $G=1$, see Figs. 11.26–11.29.

   We can add to the previous conclusion:

4. Approximation $\widetilde{\tau}=\text{const}$ conserves all principal characters of the previous dependences, but the area of the Hubble regime becomes larger.
5. Approximation $\widetilde{\tau}=\text{const}$ allows considering of the numerical transmission to the "classical" gas dynamics of explosions. By the $\widetilde{\tau}\to 0$ there no Hubble regime in principal.

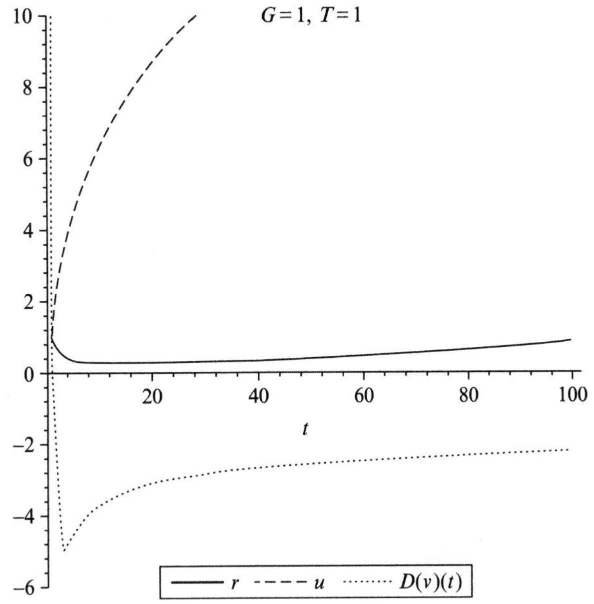

FIGURE 11.26 r—density $\widetilde{\rho}$, u—$\widetilde{u}$, $D(v)(t)=\partial\widetilde{\psi}/\partial\widetilde{r}$, $T=\widetilde{\tau}=1$.

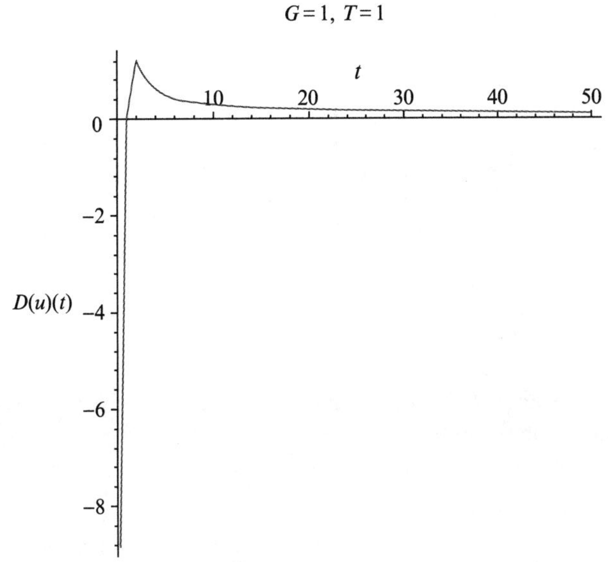

FIGURE 11.27 Dependence of the acceleration–deceleration $D(u)(t)=\partial\widetilde{u}/\partial\widetilde{r}$ on $\widetilde{r}$, $T=\widetilde{\tau}=1$.

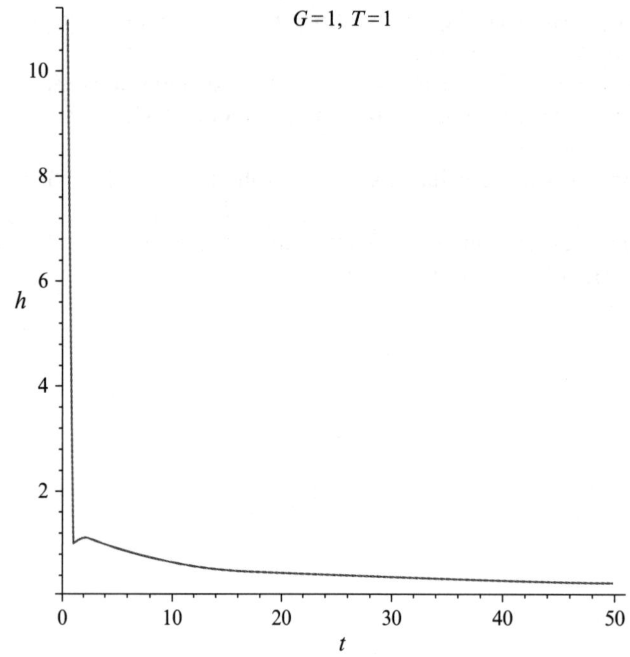

FIGURE 11.28 Dependence of the dimensionless Hubble parameter on the radial distance for $G=1$, $T=\tilde{\tau}=1$.

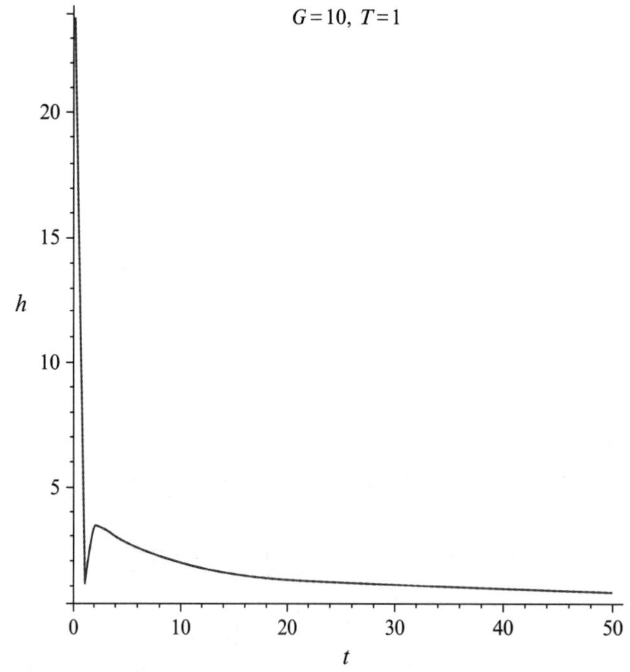

FIGURE 11.29 Dependence of the dimensionless Hubble parameter on the radial distance for $G=10$, $T=\tilde{\tau}=1$.

6. Diminishing of G leads to diminishing of the area of the Hubble regime with the positive acceleration of the matter catched by the self-consistent gravitational field.
7. Dependence of $\widetilde{H}(\widetilde{r})$ does not contain the maximum on the curve for the small value of parameter G.

As we see the Hubble expansion with acceleration is explained as result of mathematical modeling based on the principles of non-local physics. Peculiar features of the rotational speeds of galaxies and the Hubble expansion with acceleration need not in the introduction of new essence like dark matter and dark energy.

11.5 PROPAGATION OF PLANE GRAVITATIONAL WAVES IN VACUUM WITH COSMIC MICROWAVE BACKGROUND

The theory of gravitational waves in the frame of non-local quantum hydrodynamics (NLQH) is considered. From calculations follow that NLQH equations for "empty" space have the traveling wave solutions belonging in particular to the soliton class. The possible influence and reaction of the background microwave radiation is taken into account. These results lead to the principal correction of the inflation theory and serves as the explanation recently discovered the universe's cosmic microwave background anomalies. The simple analytical particular cases and numerical calculations are delivered. Proposal for astronomers—to find in the center domain of the hefty cold spot the smallest hot spot as the origin of the initial burst—Big Bang.

Newtonian gravity propagates with the infinite speed. This conclusion is connected only with the description in the frame of local physics. Usual affirmation—general relativity (GR) reduces to Newtonian gravity in the weak-field, low-velocity limit. In literature you can find criticism of this affirmation because the conservation of angular momentum is implicit in the assumptions on which GR rests. Finite propagation speeds and conservation of angular momentum are incompatible in GR. Therefore, GR was forced to claim that gravity is not a force that propagates in any classical sense, and that aberration does not apply. But here I do not intend to join to this widely discussed topic using only unified non-local model.

Let us apply generalized quantum hydrodynamic equations (11.3.1)–(11.3.6) for investigation of the gravitational wave propagation in vacuum using non-stationary 1D Cartesian description. Call attention to the fact that Eqs. (11.3.1)–(11.3.6) contain two forces of gravitational origin, **F** the force acting on the unit volume of the space and **g**—the force acting on the unit mass. As result we have from Eqs. (11.3.1) to (11.3.6):

(continuity equation)

$$\frac{\partial}{\partial t}\left\{\rho-\tau\left[\frac{\partial\rho}{\partial t}+\frac{\partial}{\partial\mathbf{r}}\cdot(\rho\mathbf{v}_0)\right]\right\}+\frac{\partial}{\partial\mathbf{r}}\cdot\left\{\rho\mathbf{v}_0-\tau\left[\frac{\partial}{\partial t}(\rho\mathbf{v}_0)+\frac{\partial}{\partial\mathbf{r}}\cdot(\rho\mathbf{v}_0\mathbf{v}_0)+\overleftrightarrow{\mathbf{I}}\cdot\frac{\partial p}{\partial\mathbf{r}}-\mathbf{F}\right]\right\}=0, \tag{11.5.1}$$

(continuity equation, 1D case)

$$\frac{\partial}{\partial t}\left\{\rho-\tau\left[\frac{\partial\rho}{\partial t}+\frac{\partial}{\partial x}(\rho v_0)\right]\right\}+\frac{\partial}{\partial x}\left\{\rho v_0-\tau\left[\frac{\partial}{\partial t}(\rho v_0)+\frac{\partial}{\partial x}(\rho v_0^2)+\frac{\partial p}{\partial x}-F\right]\right\}=0, \tag{11.5.2}$$

(momentum equation)

$$\begin{aligned}&\frac{\partial}{\partial t}\left\{\rho\mathbf{v}_0-\tau\left[\frac{\partial}{\partial t}(\rho\mathbf{v}_0)+\frac{\partial}{\partial\mathbf{r}}\cdot\rho\mathbf{v}_0\mathbf{v}_0+\frac{\partial p}{\partial\mathbf{r}}-\mathbf{F}\right]\right\}-\mathbf{g}\left[\rho-\tau\left(\frac{\partial\rho}{\partial t}+\frac{\partial}{\partial\mathbf{r}}\cdot(\rho\mathbf{v}_0)\right)\right]\\&+\frac{\partial}{\partial\mathbf{r}}\cdot\left\{\rho\mathbf{v}_0\mathbf{v}_0+p\overleftrightarrow{\mathbf{I}}-\tau\left[\begin{array}{l}\frac{\partial}{\partial t}\left(\rho\mathbf{v}_0\mathbf{v}_0+p\overleftrightarrow{\mathbf{I}}\right)+\frac{\partial}{\partial\mathbf{r}}\cdot\rho(\mathbf{v}_0\mathbf{v}_0)\mathbf{v}_0\\+2\overleftrightarrow{\mathbf{I}}\left(\frac{\partial}{\partial\mathbf{r}}\cdot(p\mathbf{v}_0)\right)+\frac{\partial}{\partial\mathbf{r}}\cdot\left(\overleftrightarrow{\mathbf{I}}p\mathbf{v}_0\right)-\mathbf{F}\mathbf{v}_0-\mathbf{v}_0\mathbf{F}\end{array}\right]\right\}=0,\end{aligned} \tag{11.5.3}$$

(momentum equation, 1D case)

$$\begin{aligned}&\frac{\partial}{\partial t}\left\{\rho v_0-\tau\left[\frac{\partial}{\partial t}(\rho v_0)+\frac{\partial}{\partial x}(\rho v_0^2)+\frac{\partial p}{\partial x}-F\right]\right\}\\&-g\left[\rho-\tau\left(\frac{\partial\rho}{\partial t}+\frac{\partial}{\partial x}(\rho v_0)\right)\right]\\&+\frac{\partial}{\partial x}\left\{\rho v_0^2+p-\tau\left[\frac{\partial}{\partial t}\left(\rho v_0^2+p\right)+\frac{\partial}{\partial x}(\rho v_0^3+3pv_0)-2Fv_0\right]\right\}=0,\end{aligned} \tag{11.5.4}$$

(energy equation)

$$\frac{\partial}{\partial t}\left\{\frac{\rho v_0^2}{2}+\frac{3}{2}p-\tau\left[\frac{\partial}{\partial t}\left(\frac{\rho v_0^2}{2}+\frac{3}{2}p\right)\right.\right.$$
$$\left.\left.+\frac{\partial}{\partial \mathbf{r}}\cdot\left(\frac{1}{2}\rho v_0^2\mathbf{v}_0+\frac{5}{2}p\mathbf{v}_0\right)-\mathbf{F}\cdot\mathbf{v}_0\right]\right\}$$
$$+\frac{\partial}{\partial \mathbf{r}}\cdot\left\{\frac{1}{2}\rho v_0^2\mathbf{v}_0+\frac{5}{2}p\mathbf{v}_0-\tau\left[\frac{\partial}{\partial t}\left(\frac{1}{2}\rho v_0^2\mathbf{v}_0\right.\right.\right.$$
$$\left.+\frac{5}{2}p\mathbf{v}_0\right)+\frac{\partial}{\partial \mathbf{r}}\cdot\left(\frac{1}{2}\rho v_0^2\mathbf{v}_0\mathbf{v}_0+\frac{7}{2}p\mathbf{v}_0\mathbf{v}_0+\frac{1}{2}pv_0^2\overleftrightarrow{\mathrm{I}}\right. \tag{11.5.5}$$
$$\left.\left.\left.+\frac{5}{2}\frac{p^2}{\rho}\overleftrightarrow{\mathrm{I}}\right)-\mathbf{F}\cdot\mathbf{v}_0\mathbf{v}_0-p\mathbf{g}\cdot\overleftrightarrow{\mathrm{I}}-\frac{1}{2}v_0^2\mathbf{F}-\frac{3}{2}\mathbf{g}p\right]\right\}$$
$$-\left\{\mathbf{F}\cdot\mathbf{v}_0-\tau\left[\mathbf{g}\cdot\left(\frac{\partial}{\partial t}(\rho\mathbf{v}_0)+\frac{\partial}{\partial \mathbf{r}}\cdot\rho\mathbf{v}_0\mathbf{v}_0\right.\right.\right.$$
$$\left.\left.\left.+\frac{\partial}{\partial \mathbf{r}}\cdot p\overleftrightarrow{\mathrm{I}}-\mathbf{F}\right)\right]\right\}=0,$$

(energy equation, 1D case)

$$\frac{\partial}{\partial t}\left\{\rho v_0^2+3p-\tau\left[\frac{\partial}{\partial t}(\rho v_0^2+3p)+\frac{\partial}{\partial x}(\rho v_0^3+5p v_0)-2Fv_0\right]\right\}$$
$$+\frac{\partial}{\partial x}\left\{\rho v_0^3+5pv_0-\tau\left[\frac{\partial}{\partial t}(\rho v_0^3+5pv_0)+\frac{\partial}{\partial x}(\rho v_0^4+8pv_0^2)\right.\right.$$
$$\left.\left.-2Fv_0^2-v_0^2F\right]\right\}+5\frac{\partial}{\partial x}\left\{\tau\left(\frac{p}{\rho}F-\frac{\partial}{\partial x}\frac{p^2}{\rho}\right)\right\} \tag{11.5.6}$$
$$-2\left\{Fv_0-\tau g\left(\frac{\partial}{\partial t}(\rho v_0)+\frac{\partial}{\partial x}(\rho v_0^2)+\frac{\partial p}{\partial x}-F\right)\right\}=0,$$

Nonlinear evolution equations (11.5.1)–(11.5.6) contain forces $\mathbf{F},\mathbf{g}$ acting on space and masses including cross-term (see for example the last line in Eq. (11.5.6)). The relation $\mathbf{F}=\rho\mathbf{g}$ comes into being only after the mass appearance as result of the Big Bang.

The cosmic microwave background (CMB) radiation is an emission of black body thermal energy coming from all parts of the sky. CMB saves the character traces of the initial burst evolution. If the quantum density ρ tends to zero the last term in the third line of the energy equation (11.5.6) can be used for estimation of the initial fluctuation and for the investigation of the influence of this initial fluctuation on the following evolution.

$$5\frac{\partial}{\partial x}\left\{\tau\left(\frac{p}{\rho}F-\frac{\partial}{\partial x}\frac{p^2}{\rho}\right)\right\}\approx\pm \mathrm{L}v_0\frac{\partial F}{\partial x}, \tag{11.5.7}$$

where L is the character length parameter reflecting the fluctuation influence on the initial physical system. Then

$$\frac{\partial}{\partial t}\left\{\rho v_0^2+3p-\tau\left[\frac{\partial}{\partial t}(\rho v_0^2+3p)+\frac{\partial}{\partial x}(\rho v_0^3+5pv_0)-2Fv_0\right]\right\}$$
$$+\frac{\partial}{\partial x}\left\{\rho v_0^3+5pv_0-\tau\left[\frac{\partial}{\partial t}\left(\rho v_0^3+5pv_0\right)+\frac{\partial}{\partial x}(\rho v_0^4+8pv_0^2)-3Fv_0^2\right]\right\} \tag{11.5.8}$$
$$\pm Lv_0\frac{\partial F}{\partial x}-2\left\{Fv_0-\tau g\left(\frac{\partial}{\partial t}(\rho v_0)+\frac{\partial}{\partial x}(\rho v_0^2)+\frac{\partial p}{\partial x}-F\right)\right\}=0.$$

The system of Eqs. (11.5.2), (11.5.4), and (11.5.8) can be transformed as follows (u—velocity in the x-direction):

(continuity equation, 1D case)

$$\frac{\partial}{\partial x}\left\{\tau\left[\frac{\partial p}{\partial x}-F\right]\right\}=0, \tag{11.5.9}$$

(momentum equation, 1D case)

$$\begin{aligned}&\frac{\partial p}{\partial x}-\frac{\partial}{\partial t}\left\{\tau\left[\frac{\partial p}{\partial x}-F\right]\right\}-2\tau\left(\frac{\partial p}{\partial x}-F\right)\frac{\partial u}{\partial x}\\&-\frac{\partial}{\partial x}\left\{\tau\left[\frac{\partial p}{\partial t}+u\frac{\partial p}{\partial x}+3p\frac{\partial u}{\partial x}\right]\right\}=0,\end{aligned} \tag{11.5.10}$$

(energy equation, 1D case)

$$\begin{aligned}&3\frac{\partial p}{\partial t}+3u\frac{\partial p}{\partial x}+5p\frac{\partial u}{\partial x}+2u\left(\frac{\partial p}{\partial x}-F\right)\\&-\frac{\partial}{\partial t}\left\{\tau\left[3\frac{\partial p}{\partial t}+3u\frac{\partial p}{\partial x}+5p\frac{\partial u}{\partial x}+2u\left(\frac{\partial p}{\partial x}-F\right)\right]\right\}\\&-\frac{\partial}{\partial x}\left\{\tau\left[5\frac{\partial}{\partial t}(pu)+5u\frac{\partial}{\partial x}(pu)+11up\frac{\partial u}{\partial x}\right]\right\}\\&-6\tau\left(\frac{\partial p}{\partial x}-F\right)u\frac{\partial u}{\partial x}\pm \mathrm{L}u\frac{\partial F}{\partial x}=0,\end{aligned} \tag{11.5.11}$$

Non-local equations are closed system of three differential equations with three dependent variables. In this case *no needs* to use the additional Poisson equation leading to the Newton gravitational description. If non-locality parameter τ is equal to zero the mentioned system becomes unclosed.

Let us introduce the length scale ξ_0, quantum pressure scale p_0, the force scale F_0, the velocity scale u_0, and approximation for non-local parameter

$$\tau=\frac{\mathrm{H}p_0}{F_0|u|}. \tag{11.5.12}$$

The length scale is taken as $\xi_0=p_0/F_0$, then H is dimensionless parameter. The principles of the τ-approximation discussed before, here I remark only that the approximation (11.5.12) is compatible with the Heisenberg relation.

Let us introduce the coordinate system moving along the positive direction of x-axis in 1D space with velocity $C=u_0$ equal to phase velocity of considering object

$$\xi=x-Ct. \tag{11.5.13}$$

Taking into account the De Broglie relation we should wait that the group velocity u_g is equal to $2u_0$. In moving coordinate system all dependent hydrodynamic values are function of (ξ,t). We investigate the possibility of the object formation of the soliton type. For this solution there is no explicit dependence on time for coordinate system moving with the phase velocity u_0. Write down the system of Eqs. (11.5.9)–(11.5.11) in the moving coordinate system:

(continuity equation, 1D case)

$$\frac{\partial}{\partial \xi}\left\{\tau\left[\frac{\partial p}{\partial \xi}-F\right]\right\}=0, \tag{11.5.14}$$

(momentum equation, 1D case)

$$\frac{\partial p}{\partial \xi}-2\frac{\partial u}{\partial \xi}\tau\left[\frac{\partial p}{\partial \xi}-F\right]-3\frac{\partial}{\partial \xi}\left\{\tau p\frac{\partial u}{\partial \xi}\right\}=0, \tag{11.5.15}$$

(energy equation, 1D case)

$$2u\left(\frac{\partial p}{\partial \xi}-F\right)+5p\frac{\partial u}{\partial \xi}-4\tau\left(\frac{\partial p}{\partial \xi}-F\right)u\frac{\partial u}{\partial \xi}$$
$$-6u\frac{\partial}{\partial \xi}\left\{\tau p\frac{\partial u}{\partial \xi}\right\}-11\tau p\left(\frac{\partial u}{\partial \xi}\right)^2\pm \mathrm{L}u\frac{\partial F}{\partial \xi}=0, \tag{11.5.16}$$

Let us write down these equations in the dimensionless form, where dimensionless symbols are marked by tildes, using the introduced scales and approximation (11.5.12) for the non-local parameter. The mentioned equations take the form

$$\frac{\partial}{\partial \tilde{\xi}}\left\{\frac{1}{\sqrt{\tilde{u}^2}}\left[\frac{\partial \tilde{p}}{\partial \tilde{\xi}}-\tilde{F}\right]\right\}=0, \tag{11.5.17}$$

$$\frac{\partial \tilde{p}}{\partial \tilde{\xi}}-2\frac{\partial \tilde{u}}{\partial \tilde{\xi}}\frac{\mathrm{H}}{\sqrt{\tilde{u}^2}}\left[\frac{\partial \tilde{p}}{\partial \tilde{\xi}}-\tilde{F}\right]-3\frac{\partial}{\partial \tilde{\xi}}\left\{\frac{\mathrm{H}}{\sqrt{\tilde{u}^2}}\tilde{p}\frac{\partial \tilde{u}}{\partial \tilde{\xi}}\right\}=0, \tag{11.5.18}$$

$$2\tilde{u}\left(\frac{\partial \tilde{p}}{\partial \tilde{\xi}}-\tilde{F}\right)+5\tilde{p}\frac{\partial \tilde{u}}{\partial \tilde{\xi}}-4\frac{\mathrm{H}}{\sqrt{\tilde{u}^2}}\tilde{u}\left(\frac{\partial \tilde{p}}{\partial \tilde{\xi}}-\tilde{F}\right)\frac{\partial \tilde{u}}{\partial \tilde{\xi}}$$
$$-6\tilde{u}\frac{\partial}{\partial \tilde{\xi}}\left\{\frac{\mathrm{H}}{\sqrt{\tilde{u}^2}}\tilde{p}\frac{\partial \tilde{u}}{\partial \tilde{\xi}}\right\}-11\frac{\mathrm{H}}{\sqrt{\tilde{u}^2}}\tilde{p}\left(\frac{\partial \tilde{u}}{\partial \tilde{\xi}}\right)^2\pm\frac{\mathrm{L}}{\xi_0}\tilde{u}\frac{\partial \tilde{F}}{\partial \tilde{\xi}}=0, \tag{11.5.19}$$

Now we are ready to display the results of the mathematical modeling [253] realized with the help of Maple (the versions Maple 9 or more can be used).

First of all from Eqs. (11.5.17)–(11.5.19) it is possible to make the analytical estimates using the condition

$$\frac{\partial \tilde{p}}{\partial \tilde{\xi}}-\tilde{F}=0, \tag{11.5.20}$$

In this case Eq. (11.5.17) is satisfied identically and Eqs. (11.5.18) and (11.5.19) can be written as follows:

$$\frac{\partial \tilde{p}}{\partial \tilde{\xi}}-3\frac{\partial}{\partial \tilde{\xi}}\left\{\frac{\mathrm{H}}{\sqrt{\tilde{u}^2}}\tilde{p}\frac{\partial \tilde{u}}{\partial \tilde{\xi}}\right\}=0, \tag{11.5.21}$$

$$5\tilde{p}\frac{\partial \tilde{u}}{\partial \tilde{\xi}}-6\tilde{u}\frac{\partial}{\partial \tilde{\xi}}\left\{\frac{\mathrm{H}}{\sqrt{\tilde{u}^2}}\tilde{p}\frac{\partial \tilde{u}}{\partial \tilde{\xi}}\right\}-11\frac{\mathrm{H}}{\sqrt{\tilde{u}^2}}\tilde{p}\left(\frac{\partial \tilde{u}}{\partial \tilde{\xi}}\right)^2\pm\frac{\mathrm{L}}{\xi_0}\tilde{u}\frac{\partial \tilde{F}}{\partial \tilde{\xi}}=0, \tag{11.5.22}$$

Multiplying Eq. (11.5.21) by $(-2u)$ and adding to (11.5.22) one obtains

$$5\tilde{p}\frac{\partial \tilde{u}}{\partial \tilde{\xi}}-2\tilde{u}\frac{\partial \tilde{p}}{\partial \tilde{\xi}}-11\frac{\mathrm{H}}{\sqrt{\tilde{u}^2}}\tilde{p}\left(\frac{\partial \tilde{u}}{\partial \tilde{\xi}}\right)^2\pm\frac{\mathrm{L}}{\xi_0}\tilde{u}\frac{\partial \tilde{F}}{\partial \tilde{\xi}}=0, \tag{11.5.23}$$

or

$$5\frac{\partial \tilde{u}}{\partial \tilde{\xi}}-2\tilde{u}\frac{\partial \ln \tilde{p}}{\partial \tilde{\xi}}-11\frac{\mathrm{H}}{\sqrt{\tilde{u}^2}}\left(\frac{\partial \tilde{u}}{\partial \tilde{\xi}}\right)^2\pm\frac{\mathrm{L}}{\xi_0}\frac{\tilde{u}}{\tilde{p}}\frac{\partial \tilde{F}}{\partial \tilde{\xi}}=0, \tag{11.5.24}$$

Omitting the derivative of the logarithmic function and the last term in Eq. (11.5.24) one obtains the equation

$$\tilde{u}\frac{\partial \tilde{u}}{\partial \tilde{\xi}}=2.2\mathrm{H}\left(\frac{\partial \tilde{u}}{\partial \tilde{\xi}}\right)^2, \tag{11.5.25}$$

which non-trivial solution is

$$\tilde{u}=\tilde{u}(\tilde{x}=0,\tilde{t}=0)\exp\left(\frac{\tilde{x}-\tilde{u}\tilde{t}}{2.2\mathrm{H}}\right). \tag{11.5.26}$$

for waves propagating in the positive direction of x-axis. It can be shown that for $\tilde{\xi}=\tilde{x}+\tilde{u}\tilde{t}$ exists the solution

$$\tilde{u}=\tilde{u}(\tilde{x}=0,\tilde{t}=0)\exp\left(\frac{\tilde{x}+\tilde{u}\tilde{t}}{2.2\mathrm{H}}\right). \tag{11.5.27}$$

It means that qualitative consideration leads to traveling waves with exponential evolution and no surprise that the solution of full system of Eqs. (11.5.17)–(11.5.19) defines solitons.

I emphasize that:

1. Relations (11.5.26) and (11.5.27) reflecting the exponential law of the perturbation evolution, are the particular case of the generalized H-theorem proved by me (see [31]).
2. If $\tilde{u}=\tilde{u}(\tilde{x}=0,\tilde{t}=0)=0$, then (this follows from Eqs. (11.4.1) and (11.4.2)) $\tilde{p}=\mathrm{const}$, $\tilde{F}=0$.
3. The physical system is at rest until the appearance the external perturbations.

The system of Eqs. (11.5.17)–(11.5.19) have the great possibilities of mathematical modeling as result of changing of two parameters H and $\tilde{L}=L/\xi_0$ and six Cauchy conditions describing the character features of initial perturbations which lead to the soliton formation. Maple program contains Maple's notations—for example the expression $D(u)(0)=0$ means in the usual notations $\left(\partial\tilde{u}/\partial\tilde{\xi}\right)(0)=0$, independent variable t responds to $\tilde{\xi}$.

We begin with investigation of the problem of principle significance—is it possible after a perturbation (defined by the Cauchy conditions) to obtain the object of the soliton's kind as result of the self-organization? With this aim let us consider the initial perturbations:

$$\mathbf{u(0)=1, p(0)=1, f(0)=1, D(u)(0)=0, D(p)(0)=0, D(f)(0)=0.}$$

The following Maple notations on figures are used: u—velocity $\tilde{u}$, p—pressure $\tilde{p}$, and f—the self-consistent force $\tilde{F}$. Explanations placed under all following figures. In the soliton regime the solution exists only in the restricted domain of the 1D space and the obtained object in the moving coordinate system ($\tilde{\xi}=\tilde{x}-\tilde{t}$) has the constant velocity $\tilde{u}=1$ for all parts of the object. In this case the domain of the solution existence defines the character soliton size. The following numerical results reflect two principally different regime of the physical system evolution. The distinctive features of evolution are defined by the sign plus or minus in front of the term D in the energy equation,

$$\mathrm{D}=\pm\frac{\mathrm{L}}{\xi_0}\tilde{u}\frac{\partial\tilde{F}}{\partial\tilde{\xi}}=\pm\tilde{\mathrm{L}}\tilde{u}\frac{\partial\tilde{F}}{\partial\tilde{\xi}}. \tag{11.5.28}$$

The term D reflects the interaction of physical system with the surrounding media and defines the value of perturbation. The Maple inscription L on figures corresponds to the dimensionless value $\tilde{\mathrm{L}}$ including the sign in front of $\tilde{\mathrm{L}}$, independent variable t corresponds to $\tilde{\xi}$.

Therefore, Regime I is characterized by the negative sign in front of $\tilde{\mathrm{L}}$. The mentioned calculations are displayed in Figs. 11.30–11.41.

One can see that the first regime is characterized by the force directed basically (in front of the wave, Figs. 11.31, 11.34, 11.37, and 11.40) against the direction of the wave propagation. This fact leads to the effect of attraction. Domains of the solution existence in regime I:

(a) For $\mathrm{H}=1, \tilde{\mathrm{L}}=-1. \rightarrow (-0.7298;\ 1.1618)$.
(b) For $\mathrm{H}=1, \tilde{\mathrm{L}}=-0.1. \rightarrow (-0.1571;\ 0.6347)$.
(c) For $\mathrm{H}=1, \tilde{\mathrm{L}}=-0.01. \rightarrow (-0.0265;\ 0.5327)$.
(d) For $\mathrm{H}=1, \tilde{\mathrm{L}}=-0.001. \rightarrow (-0.003801;\ 0.5186)$.

Regime II is characterized by the positive sign in front of $\tilde{\mathrm{L}}$. The mentioned calculations are displayed in Figs. 11.42–11.51.

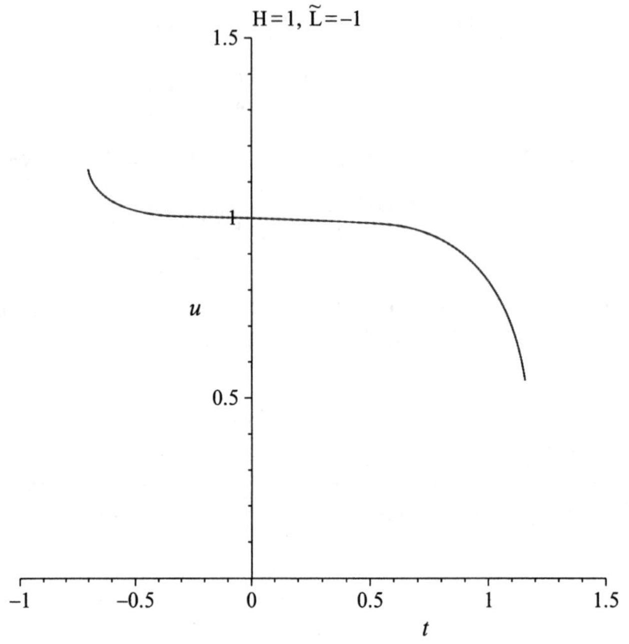

FIGURE 11.30 u—velocity $\tilde{u}$, H = 1, $\tilde{L} = -1$.

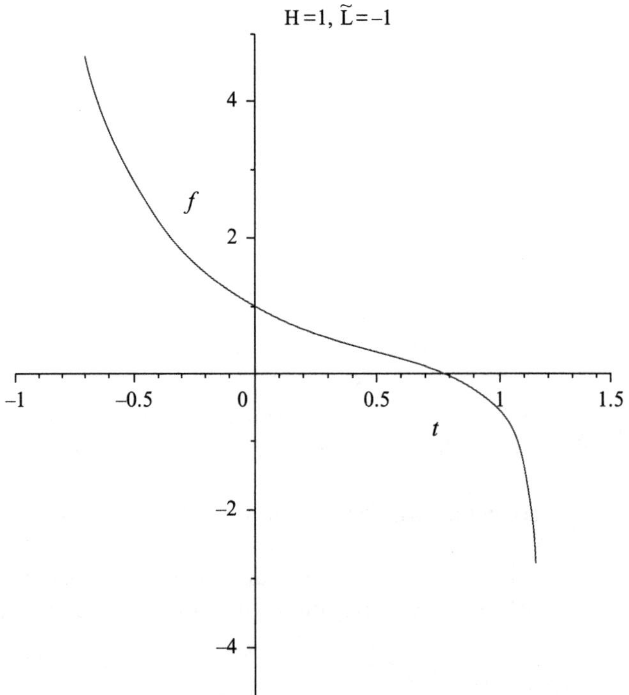

FIGURE 11.31 f—the self-consistent force $\tilde{F}$, H = 1, $\tilde{L} = -1$.

The second regime is characterized by the force directed along the direction of the wave propagation. This fact leads to the effect of anti-attraction during the Big Bang, (Figs. 11.42, 11.46, 11.48, and 11.50). The term "antigravitation" is deeply embedded in the physical literature but this term unlikely applicable for the vacuum explosion. As follow from Figs. 11.44 and 11.45 the fist regime of traveling waves corresponds only the early life of evolution and later gives way to regime of the

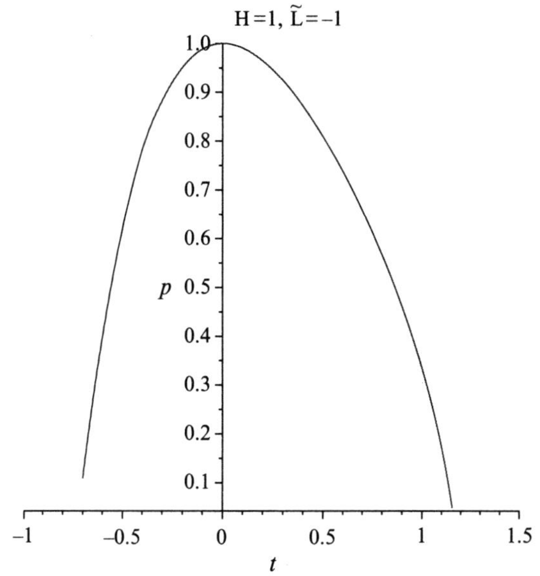

FIGURE 11.32 *p*—pressure $\tilde{p}$, $\text{H}=1$, $\tilde{\text{L}}=-1$.

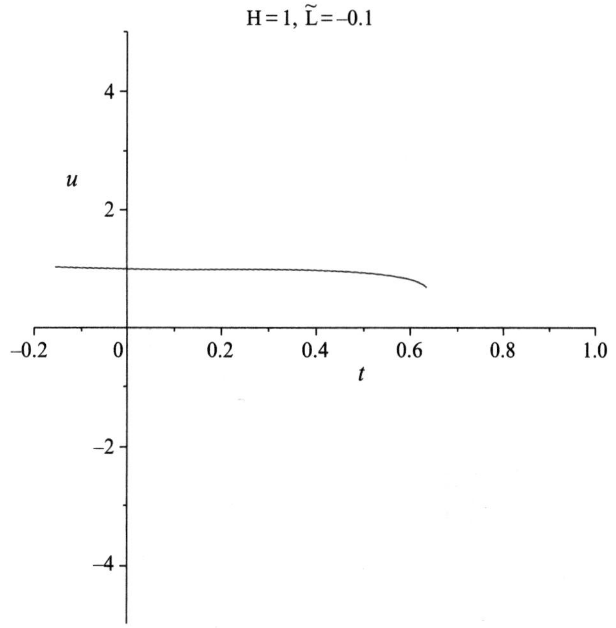

FIGURE 11.33 *u*—velocity $\tilde{u}$, $\text{H}=1$, $\tilde{\text{L}}=-0.1$.

very intensive explosion which details should be investigated using non-stationary 3D models on the basement of Eqs. (11.5.1)–(11.5.6). For example domain of the solution existence for the case shown on Figs. 11.44 and 11.45: $t=\tilde{\xi}_{\text{lim}} \leq 34.582$.

Important to notice that Hubble expansion can be explained as result of the matter self-catching in the frame of the Newtonian law of gravitation. As you see during all investigations we needn't to use the theory Newtonian gravitation

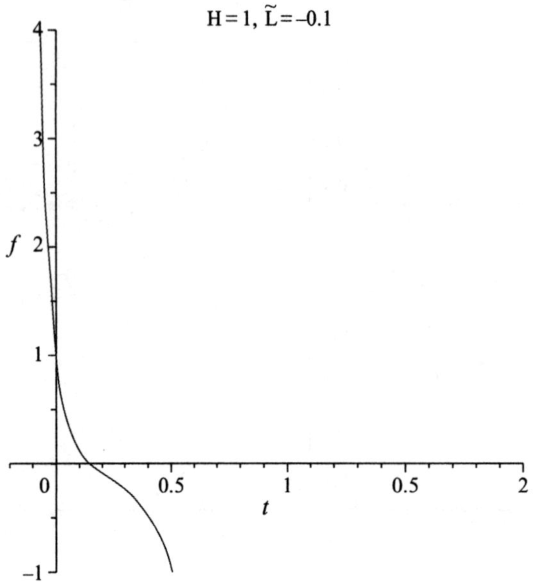

FIGURE 11.34 f—the self-consistent force $\tilde{F}$, H = 1, $\tilde{L} = -0.1$.

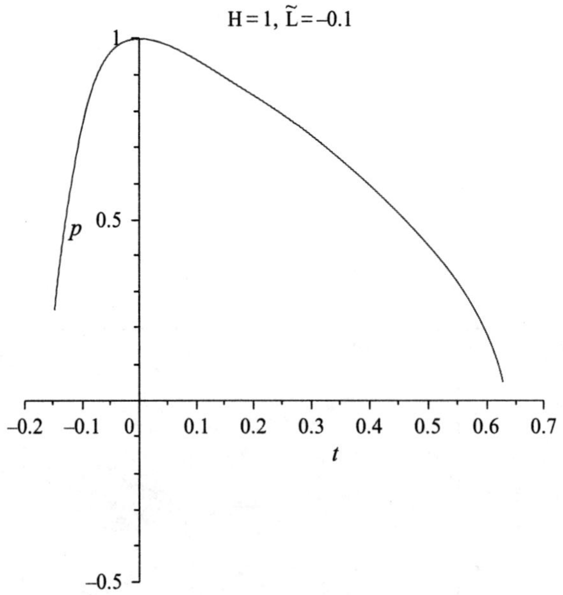

FIGURE 11.35 p—pressure $\tilde{p}$, H = 1, $\tilde{L} = -0.1$.

for solution of nonlinear non-local evolution equations (EE). In contrast with the local physics this approach in the frame of quantum non-local hydrodynamics leads to the closed mathematical description for the physical system under consideration.

If the matter is absent, the gravitational evolution of the system in space and time is containing in EE only so to speak on the "genetic level"; it means the origin of the EE derivation in the macroscopic case for massive system.

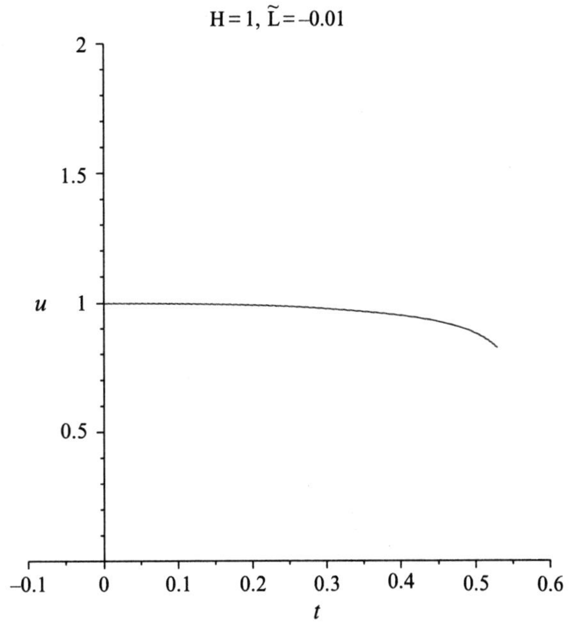

FIGURE 11.36 *u*—velocity $\tilde{u}$, H = 1, $\tilde{L} = -0.1$.

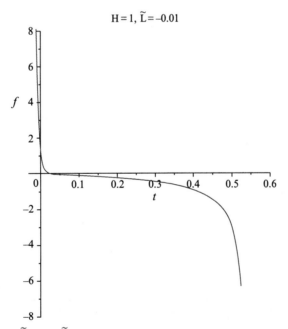

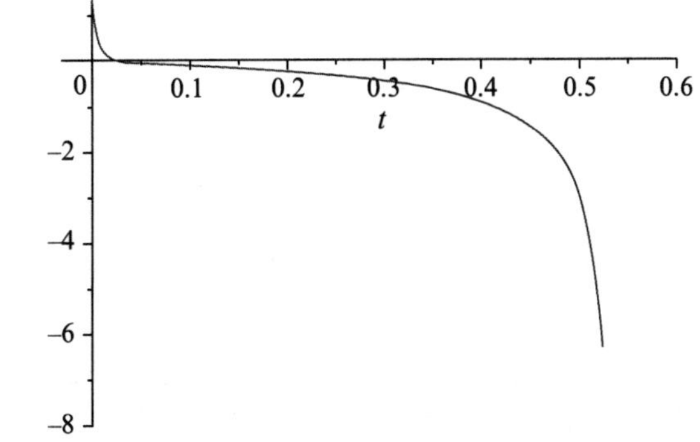

FIGURE 11.37 *f*—the self-consistent force $\tilde{F}$, H = 1, $\tilde{L} = -0.1$.

Better to speak about evolution of "originating vacuum" (OV) which description in time and 3D space on the level of quantum hydrodynamics demands only quantum pressure p, the self-consistent force $\mathbf{F}$ (acting on unit of the space volume) and velocity $\mathbf{v_0}$. The perturbations of OV lead to two different processes—traveling waves including the soliton formation (regime I) and the explosion of the system (regime II, the Big Bang regime). Both regimes can be incorporated in one scenario.

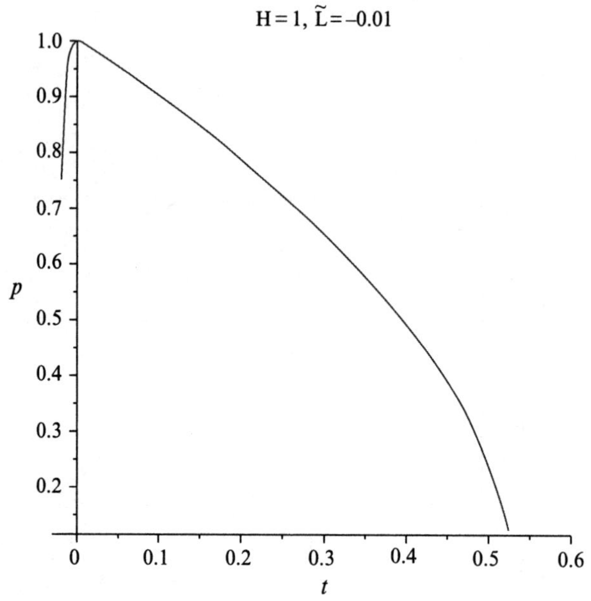

FIGURE 11.38 p—pressure $\tilde{p}$, H = 1, $\tilde{L} = -0.01$.

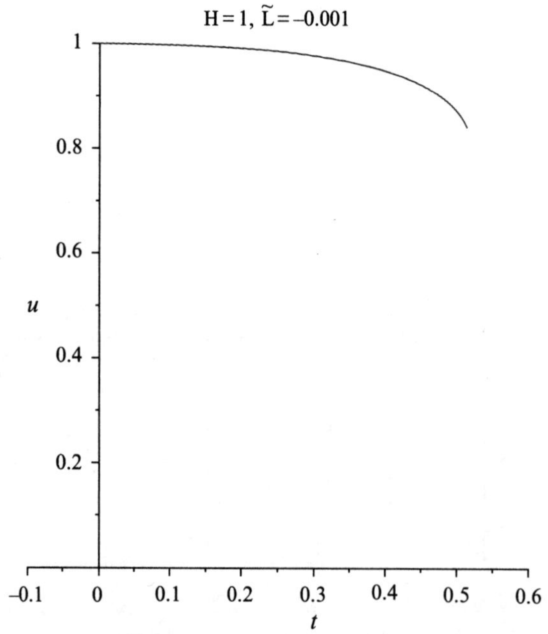

FIGURE 11.39 u—velocity $\tilde{u}$, H = 1, $\tilde{L} = -0.001$.

As follow from calculations (see Figs. 11.42–11.51) the most intensive explosion effect achieves for the smallest perturbations with the positive sign in front of $\tilde{L}$. From the mathematical point of view we have the typical Hadamard instability leading to the Big Bang. Moreover two regimes differ from one another by the directions of forces F.

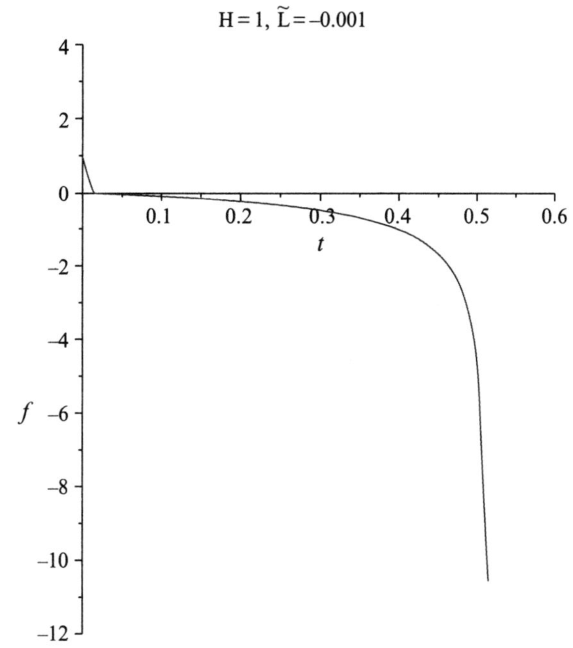

FIGURE 11.40 *f*—the self-consistent force $\widetilde{F}$, H = 1, $\widetilde{\mathrm{L}} = -0.001$.

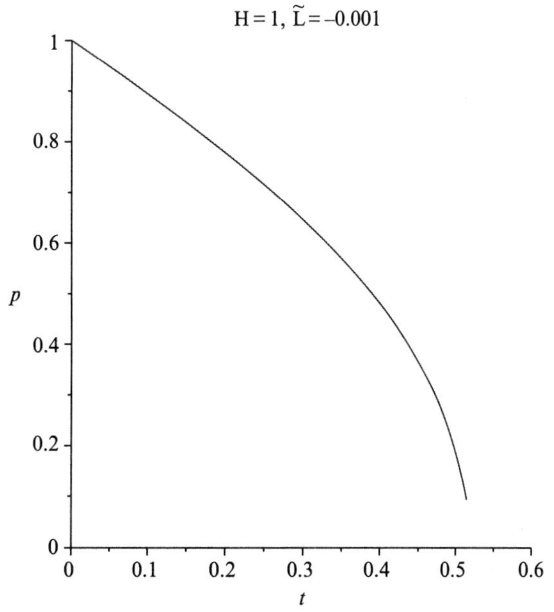

FIGURE 11.41 *p*—pressure $\widetilde{p}$, H = 1, $\widetilde{\mathrm{L}} = -0.001$.

After the Big Bang and interaction of OV with the created matter the microwave background radiation should contain the traces of the traveling waves evolution realized as regime I. Let us look at the last measurements realized in the frame of the Planck program. The temperature variations do not appear to behave the same on large scales as they do on small scales, and there are some particularly large features, such as a hefty cold spot, that were not predicted by basic inflation models.

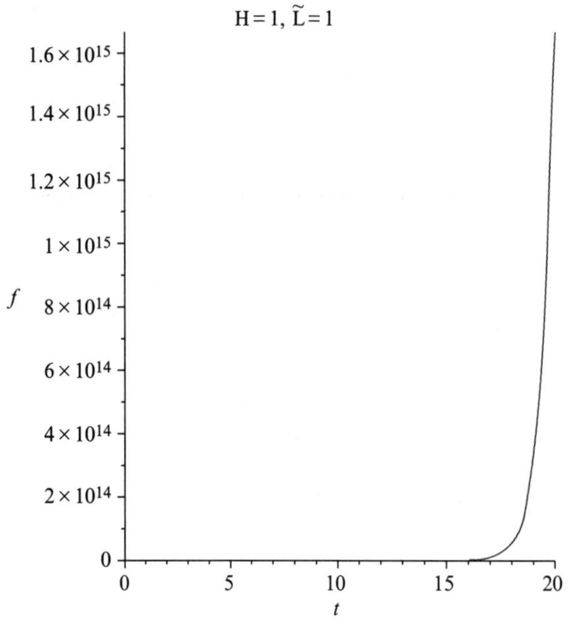

FIGURE 11.42 f—the self-consistent force $\widetilde{F}$, H = 1, $\widetilde{\mathrm{L}}$ = 1.

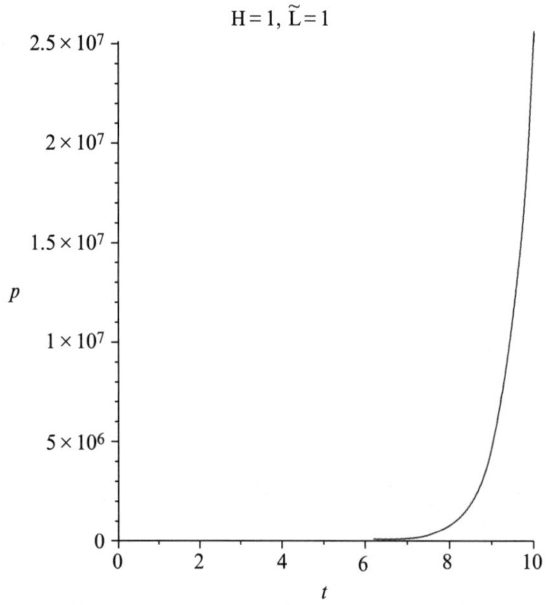

FIGURE 11.43 p—pressure $\widetilde{p}$, H = 1, $\widetilde{\mathrm{L}}$ = 1.

From the position of the developed theory it is no surprise. Really, look at the Planck space observatory's map (Fig. 11.52) of the universe's cosmic microwave background. This map is in open Internet access (see for example SPACE.com Staff. Date: 21 March 2013 Time: 11:15 AM ET).

It was reported that CMB is a snapshot of the oldest light in our Universe, imprinted on the sky when the Universe was just 380,000 years old. It shows tiny temperature fluctuations that correspond to regions of slightly different densities, representing the seeds of all future structure: the stars and galaxies of today.

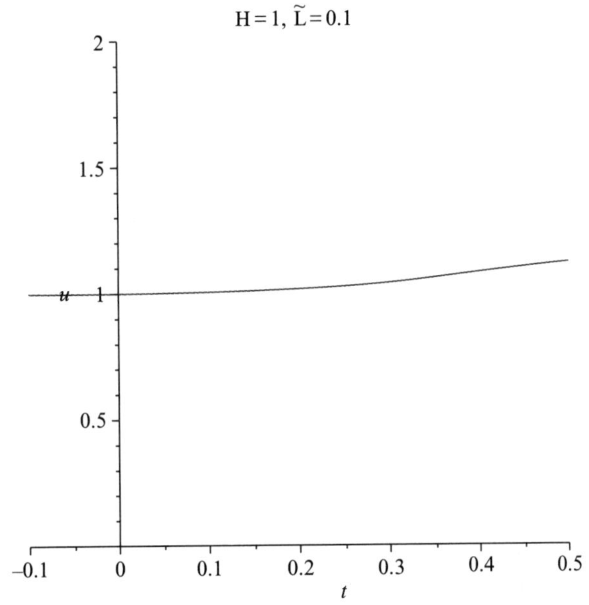

FIGURE 11.44 u—velocity $\tilde{u}$, H = 1, $\tilde{L}$ = 0.1.

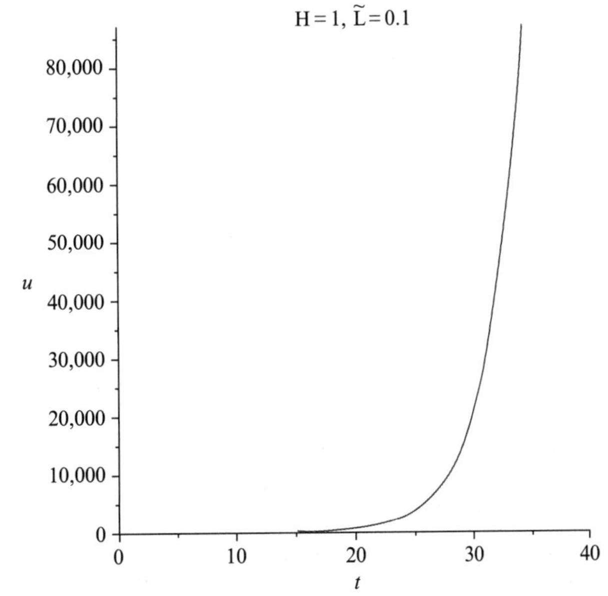

FIGURE 11.45 u—velocity $\tilde{u}$, H = 1, $\tilde{L}$ = 0.1.

From the position of the developed theory Planck's all-sky map contains the regular traces of traveling waves as the alternation of the "hot" (grey) and "cold" (more intensive black color) strips. In Fig. 11.52 the Planck space observatory staff shows the "mysterious" hefty cold spot as the dark black small area bounded by the white circle.

From the position of the developed theory *it is the area reflecting the initial explosion of OV. In this case the center domain of the mentioned hefty cold spot should contain the smallest hot spot as the origin of the initial burst.*

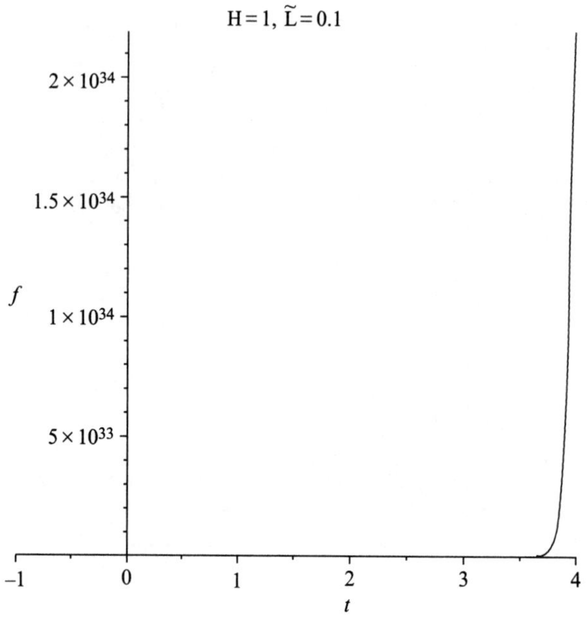

FIGURE 11.46 *f*—the self-consistent force $\widetilde{F}$, H = 1, $\widetilde{\mathrm{L}}$ = 0.1.

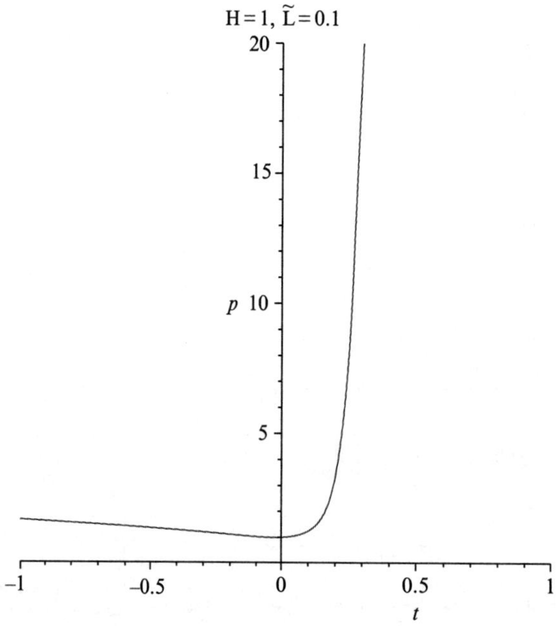

FIGURE 11.47 *p*—pressure $\widetilde{p}$, H = 1, $\widetilde{\mathrm{L}}$ = 0.1.

I hope this fact will be established by astronomers after following more precise observations.

During all investigations we needn't to use the theory Newtonian gravitation for solution of nonlinear non-local evolution equations. In contrast with the local physics this approach in the frame of quantum non-local hydrodynamics leads to the closed mathematical description for the physical system under consideration. If the matter is absent, non-local evolution

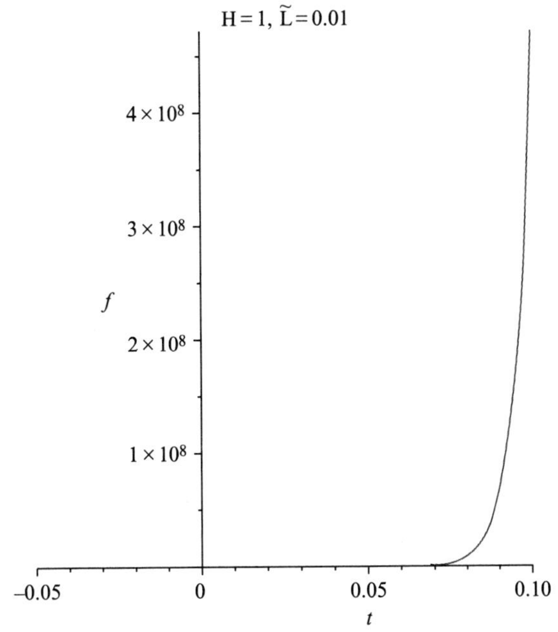

FIGURE 11.48 *f*—the self-consistent force $\widetilde{F}$, H = 1, $\widetilde{\mathrm{L}}$ = 0.01.

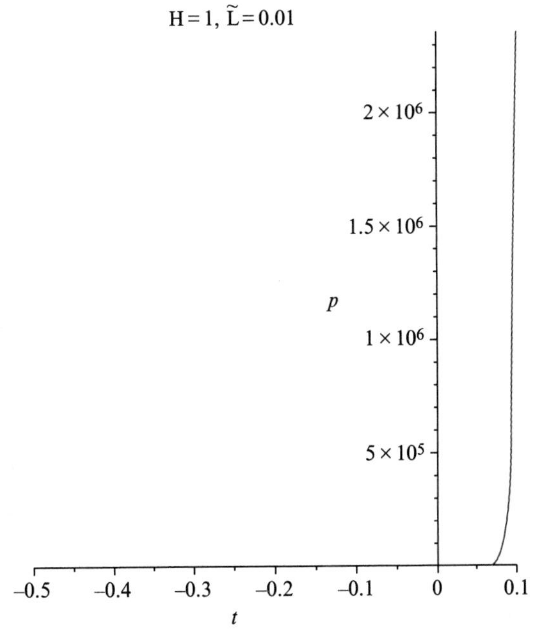

FIGURE 11.49 *p*—pressure $\widetilde{p}$, H = 1, $\widetilde{\mathrm{L}}$ = 0.01.

equations have nevertheless non-trivial solutions corresponding evolution of OV which description in time and 3D space on the level of quantum hydrodynamics demands only quantum pressure p, the self-consistent force $\mathbf{F}$ (acting on unit of the space volume) and velocity $\mathbf{v}_0$.

The perturbations of OV lead to two different processes—traveling waves including the soliton formation (regime I) and the explosion of the system (regime II, the Big Bang regime). Both regimes can be incorporated in one scenario. From the

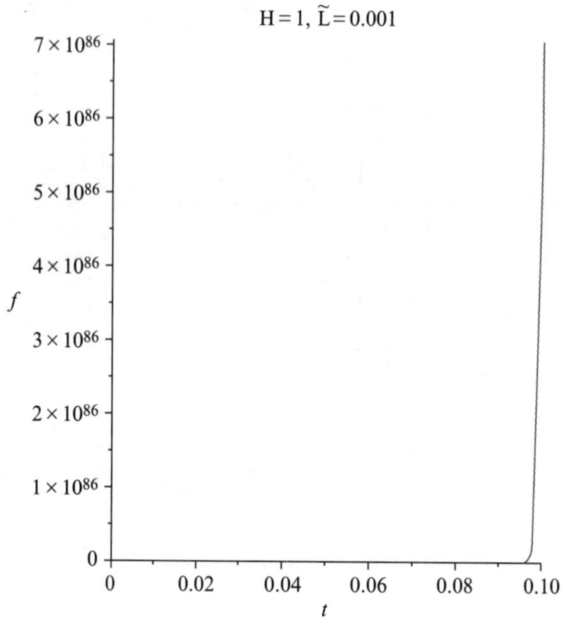

FIGURE 11.50 *f*—the self-consistent force $\widetilde{F}$, H = 1, $\widetilde{\mathrm{L}}$ = 0.001.

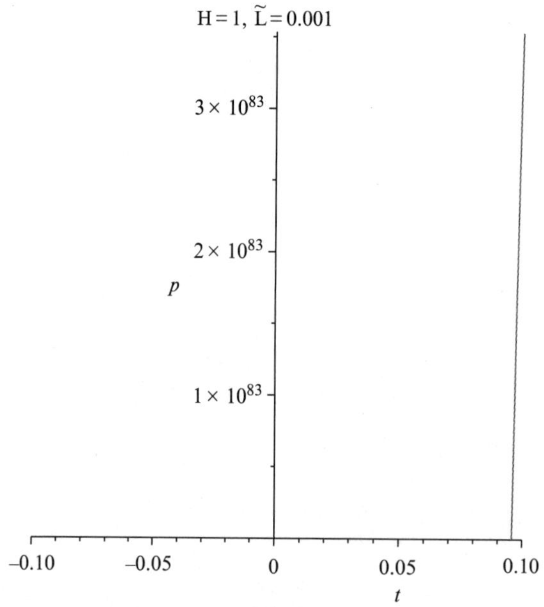

FIGURE 11.51 *p*—pressure $\widetilde{p}$, H = 1, $\widetilde{\mathrm{L}}$ = 0.001.

mathematical point of view we have the typical Hadamard instability (the smaller is an initial perturbation the greater is the burst intensity) leading to the Big Bang. Two regimes differ from one another by the directions of forces **F**.

Finally some words concerning the following investigations. Numerical calculations, realized in the spherical coordinate system for the dependent variables (*r*—radius, *t*—time) cannot change principal results of the shown calculations in the Cartesian coordinate system. Increasing of the character distances between "cold" and "hot" zones (see Fig. 11.52) is obliged to the burst configuration closed to the spherical form. But some other effects obviously need in 3D non-stationary calculations. This remark relates first of all to so-called "dark flow" describing a possible non-random component of the peculiar velocity of galaxy clusters.

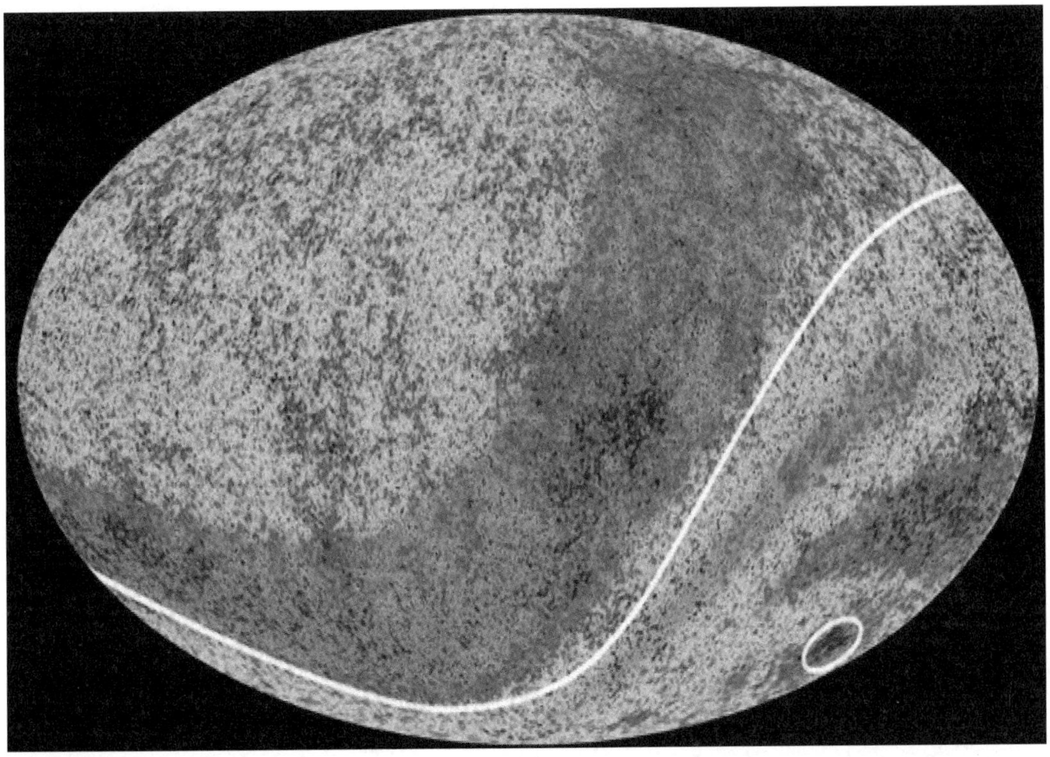

FIGURE 11.52 Planck space observatory's map of the universe's cosmic microwave background.

11.6 APPLICATION OF THE NON-LOCAL PHYSICS IN THE THEORY OF THE MATTER MOVEMENT IN BLACK HOLE

The theory of the matter movement in a black hole in the frame of NLQH is considered. The theory corresponds to the limit case when the matter density tends to infinity. From calculations follow that NLQH equations for the black hole space have the traveling wave solutions. The domain of the solution existence is limited by the event horizon where gravity tends to infinity. The simple analytical particular cases and numerical calculations are delivered.

The first ideas about the existence of cosmic objects which gravitation is be so big that the escape velocity would be faster than the speed of light, were formulated in 1783 by English geologist named John Mitchell. In 1796, Pierre-Simon Laplace promoted the same idea in his book "Exposition du système du Monde". In 1916 Albert Einstein introduced an explanation of gravity called general relativity. According to the general theory of relativity, a black hole is a region of space from which nothing, including light, can escape. It is the result of the denting of spacetime caused by a very compact mass. Around a black hole there is an undetectable surface which marks the point of no return, called an event horizon. It is called "black" because it absorbs all the light that hits it, reflecting nothing, just like a perfect black body in thermodynamics. Black holes possess a temperature (and therefore the internal energy) and emit Hawking radiation through slow dissipation by anti-protons.

In 1930, Subrahmanyan Chandrasekhar predicted that stars heavier than the sun could collapse when they ran out of hydrogen or other nuclear fuels to burn and die. In 1967, John Wheeler gave black holes the name "black hole" for the first time. Astronomers have identified numerous stellar black hole candidates, and have also found evidence of supermassive black holes at the center of every galaxy. In 1970, Stephen Hawking and Roger Penrose proved that black holes must exist.

Let us investigate the possibilities delivering by the unified generalized quantum hydrodynamics for investigation of these problems. From position of NLQH the mentioned theory has two limit cases connected with the density ρ evolution:

1. The density $\rho \to \infty$. From the physical point of view this case corresponds to the matter motion in the Black Hole regime.
2. The density $\rho \to 0$. From the physical point of view this case corresponds to the motion in the Big Bang regime.

Newtonian gravity propagates with the infinite speed. This conclusion is connected only with the description in the frame of local physics. Usual affirmation—GR reduces to Newtonian gravity in the weak-field, low-velocity limit. In literature you can find criticism of this affirmation because the conservation of angular momentum is implicit in the assumptions on which GR rests. Finite propagation speeds and conservation of angular momentum are incompatible in GR. Therefore, GR was forced to claim that gravity is not a force that propagates in any classical sense, and that aberration does not apply. But here I do not intend to join to this widely discussed topic using only unified non-local model.

Strict consideration leads to the system of the GHE (11.3.1)–(11.3.6). Let us apply these generalized quantum hydrodynamic equations for investigation of the traveling wave propagation inside the black hole using non-stationary 1D Cartesian description. It means that consideration corresponds so to speak to "the black channel".

Call attention to the fact that Eqs. (11.3.1)–(11.3.6) contain two forces of gravitational origin, **F** the force acting on the unit volume of the space and **g**—the force acting on the unit mass. As result we have from Eqs. (11.3.1)–(11.3.6):

(continuity equation)

$$\frac{\partial}{\partial t}\left\{\rho-\tau\left[\frac{\partial \rho}{\partial t}+\frac{\partial}{\partial \mathbf{r}}\cdot(\rho\mathbf{v}_0)\right]\right\} + \frac{\partial}{\partial \mathbf{r}}\cdot\left\{\rho\mathbf{v}_0-\tau\left[\frac{\partial}{\partial t}(\rho\mathbf{v}_0)+\frac{\partial}{\partial \mathbf{r}}\cdot(\rho\mathbf{v}_0\mathbf{v}_0)+\overleftrightarrow{\mathbf{I}}\cdot\frac{\partial p}{\partial \mathbf{r}}-\mathbf{F}\right\}=0, \tag{11.6.1}$$

(continuity equation, 1D case)

$$\frac{\partial}{\partial t}\left\{\rho-\tau\left[\frac{\partial \rho}{\partial t}+\frac{\partial}{\partial x}(\rho v_0)\right]\right\}+\frac{\partial}{\partial x}\left\{\rho v_0-\tau\left[\frac{\partial}{\partial t}(\rho v_0)+\frac{\partial}{\partial x}(\rho v_0^2)+\frac{\partial p}{\partial x}-F\right]\right\}=0, \tag{11.6.2}$$

(momentum equation)

$$\begin{aligned}&\frac{\partial}{\partial t}\left\{\rho\mathbf{v}_0-\tau\left[\frac{\partial}{\partial t}(\rho\mathbf{v}_0)+\frac{\partial}{\partial \mathbf{r}}\cdot\rho\mathbf{v}_0\mathbf{v}_0+\frac{\partial p}{\partial \mathbf{r}}-\mathbf{F}\right]\right\}-\mathbf{g}\left[\rho-\tau\left(\frac{\partial \rho}{\partial t}+\frac{\partial}{\partial \mathbf{r}}\cdot(\rho\mathbf{v}_0)\right)\right]\\&+\frac{\partial}{\partial \mathbf{r}}\cdot\left\{\rho\mathbf{v}_0\mathbf{v}_0+p\overleftrightarrow{\mathbf{I}}-\tau\left[\begin{array}{l}\frac{\partial}{\partial t}\left(\rho\mathbf{v}_0\mathbf{v}_0+p\overleftrightarrow{\mathbf{I}}\right)+\frac{\partial}{\partial \mathbf{r}}\cdot\rho(\mathbf{v}_0\mathbf{v}_0)\mathbf{v}_0+2\overleftrightarrow{\mathbf{I}}\left(\frac{\partial}{\partial \mathbf{r}}\cdot(p\mathbf{v}_0)\right)\\+\frac{\partial}{\partial \mathbf{r}}\cdot\left(\overleftrightarrow{\mathbf{I}}p\mathbf{v}_0\right)-\mathbf{F}\mathbf{v}_0-\mathbf{v}_0\mathbf{F}\end{array}\right]\right\}=0,\end{aligned} \tag{11.6.3}$$

(momentum equation, 1D case)

$$\begin{aligned}&\frac{\partial}{\partial t}\left\{\rho v_0-\tau\left[\frac{\partial}{\partial t}(\rho v_0)+\frac{\partial}{\partial x}(\rho v_0^2)+\frac{\partial p}{\partial x}-F\right]\right\}\\&-g\left[\rho-\tau\left(\frac{\partial \rho}{\partial t}+\frac{\partial}{\partial x}(\rho v_0)\right)\right]\\&+\frac{\partial}{\partial x}\left\{\rho v_0^2+p-\tau\left[\frac{\partial}{\partial t}\left(\rho v_0^2+p\right)+\frac{\partial}{\partial x}(\rho v_0^3+3pv_0)-2Fv_0\right]\right\}=0,\end{aligned} \tag{11.6.4}$$

(energy equation)

$$\begin{aligned}&\frac{\partial}{\partial t}\left\{\frac{\rho v_0^2}{2}+\frac{3}{2}p-\tau\left[\frac{\partial}{\partial t}\left(\frac{\rho v_0^2}{2}+\frac{3}{2}p\right)\right.\right.\\&\left.\left.+\frac{\partial}{\partial \mathbf{r}}\cdot\left(\frac{1}{2}\rho v_0^2\mathbf{v}_0+\frac{5}{2}p\mathbf{v}_0\right)-\mathbf{F}\cdot\mathbf{v}_0\right]\right\}\\&+\frac{\partial}{\partial \mathbf{r}}\cdot\left\{\frac{1}{2}\rho v_0^2\mathbf{v}_0+\frac{5}{2}p\mathbf{v}_0-\tau\left[\frac{\partial}{\partial t}\left(\frac{1}{2}\rho v_0^2\mathbf{v}_0\right.\right.\right.\\&\left.+\frac{5}{2}p\mathbf{v}_0\right)+\frac{\partial}{\partial \mathbf{r}}\cdot\left(\frac{1}{2}\rho v_0^2\mathbf{v}_0\mathbf{v}_0+\frac{7}{2}p\mathbf{v}_0\mathbf{v}_0+\frac{1}{2}pv_0^2\overleftrightarrow{\mathbf{I}}\right.\\&\left.\left.\left.+\frac{5p^2}{2\rho}\overleftrightarrow{\mathbf{I}}\right)-\mathbf{F}\cdot\mathbf{v}_0\mathbf{v}_0-p\mathbf{g}\cdot\overleftrightarrow{\mathbf{I}}-\frac{1}{2}v_0^2\mathbf{F}-\frac{3}{2}\mathbf{g}p\right]\right\}\\&-\left\{\mathbf{F}\cdot\mathbf{v}_0-\tau\left[\mathbf{g}\cdot\left(\frac{\partial}{\partial t}(\rho\mathbf{v}_0)+\frac{\partial}{\partial \mathbf{r}}\cdot\rho\mathbf{v}_0\mathbf{v}_0\right.\right.\right.\\&\left.\left.\left.+\frac{\partial}{\partial \mathbf{r}}\cdot p\overleftrightarrow{\mathbf{I}}-\mathbf{F}\right)\right]\right\}=0,\end{aligned} \tag{11.6.5}$$

(energy equation, 1D case)

$$\frac{\partial}{\partial t}\left\{\rho v_0^2+3p-\tau\left[\frac{\partial}{\partial t}(\rho v_0^2+3p)+\frac{\partial}{\partial x}(\rho v_0^3+5pv_0)-2Fv_0\right]\right\}$$
$$+\frac{\partial}{\partial x}\left\{\rho v_0^3+5pv_0-\tau\left[\frac{\partial}{\partial t}(\rho v_0^3+5pv_0)+\frac{\partial}{\partial x}(\rho v_0^4+8pv_0^2)\right.\right.$$
$$\left.\left.-2Fv_0^2-v_0^2F\right]\right\}+5\frac{\partial}{\partial x}\left\{\tau\left(\frac{p}{\rho}F-\frac{\partial}{\partial x}\frac{p^2}{\rho}\right)\right\}$$
$$-2\left\{Fv_0-\tau g\left(\frac{\partial}{\partial t}(\rho v_0)+\frac{\partial}{\partial x}(\rho v_0^2)+\frac{\partial p}{\partial x}-F\right)\right\}=0, \tag{11.6.6}$$

Nonlinear evolution equations (11.6.1)–(11.6.6) contain forces $\mathbf{F},\mathbf{g}$ acting on space and masses including cross-terms (see for example the last line in Eq. (11.6.6)). The relation $\mathbf{F}=\rho\mathbf{g}$ comes into being only after the mass appearance as result of the Big Bang.

Let us introduce now the main mentioned before assumption leading to the theory of motion inside the black holes: the density $\rho\to\infty$. Derivating the basic system of equation, we should take into account two facts:

1. The density can tend to infinity by the arbitrary law.
2. The ratio of pressure to density defines the internal energy of the mass unit $E=p/\rho$ and should be considered as a dependent variable by $\rho\to\infty$.

As result we have the following system of equations:

$$\frac{\partial}{\partial t}\left\{\tau\frac{\partial u}{\partial x}\right\}-\frac{\partial u}{\partial x}+\frac{\partial}{\partial x}\left\{\tau\left[\frac{\partial u}{\partial t}+2u\frac{\partial u}{\partial x}+\frac{\partial E}{\partial x}-g\right]\right\}=0, \tag{11.6.7}$$

$$\frac{\partial}{\partial t}\left\{u-\tau\left[\frac{\partial u}{\partial t}+2u\frac{\partial u}{\partial x}+\frac{\partial E}{\partial x}-g\right]\right\}-g\left[1-\tau\frac{\partial u}{\partial x}\right]$$
$$+\frac{\partial}{\partial x}\left\{u^2+E-\tau\left[\frac{\partial}{\partial t}\left(u^2+E\right)+\frac{\partial}{\partial x}(u^3+3Eu)-2gu\right]\right\}=0, \tag{11.6.8}$$

$$\frac{\partial}{\partial t}\left\{u^2+3E-\tau\left[\frac{\partial}{\partial t}(u^2+3E)+\frac{\partial}{\partial x}(u^3+5Eu)-2gu\right]\right\}$$
$$+\frac{\partial}{\partial x}\left\{u^3+5Eu-\tau\left[\frac{\partial}{\partial t}(u^3+5Eu)+\frac{\partial}{\partial x}(u^4+8Eu^2)-3gu^2\right]\right\}$$
$$+5\frac{\partial}{\partial x}\left\{\tau E\left(g-2\frac{\partial E}{\partial x}\right)\right\}-2gu$$
$$+2\tau g\left(\frac{\partial u}{\partial t}+2u\frac{\partial u}{\partial x}+\frac{\partial E}{\partial x}-g\right)=0, \tag{11.6.9}$$

where u is the velocity component along the x direction. Let us introduce the coordinate system moving along the positive direction of x-axis in 1D space with velocity $C=u_0$ equal to phase velocity of considering object

$$\xi=x-Ct. \tag{11.6.10}$$

Taking into account the De Broglie relation we should wait that the group velocity u_g is equal 2 u_0. In moving coordinate system all dependent hydrodynamic values are function of (ξ,t). We investigate the possibility of the traveling wave formation. For this solution there is no explicit dependence on time for coordinate system moving with the phase velocity u_0. Write down the system of Eqs. (11.6.7)–(11.6.9) in the moving coordinate system using the relation $\xi=x-ut$:

(continuity equation, 1D case)

$$\frac{\partial u}{\partial\xi}-\tau\left(\frac{\partial u}{\partial\xi}\right)^2-\frac{\partial}{\partial\xi}\left\{\tau\left[\frac{\partial E}{\partial\xi}-g\right]\right\}=0, \tag{11.6.11}$$

(momentum equation, 1D case)

$$\left(\frac{\partial E}{\partial \xi}-g\right)+3\tau g\frac{\partial u}{\partial \xi}-5\tau\frac{\partial u}{\partial \xi}\frac{\partial E}{\partial \xi}-3E\frac{\partial}{\partial \xi}\left\{\tau\frac{\partial u}{\partial \xi}\right\}=0, \tag{11.6.12}$$

(energy equation, 1D case)

$$\begin{aligned}&2u\left(\frac{\partial E}{\partial \xi}-g\right)+5E\frac{\partial u}{\partial \xi}-10u\tau\frac{\partial u}{\partial \xi}\frac{\partial E}{\partial \xi}-6uE\frac{\partial}{\partial \xi}\left\{\tau\frac{\partial u}{\partial \xi}\right\}\\&-11\tau E\left(\frac{\partial u}{\partial \xi}\right)^2-10\frac{\partial}{\partial \xi}\left\{\tau E\frac{\partial E}{\partial \xi}\right\}+5\frac{\partial}{\partial \xi}\{\tau E g\}+6\tau g u\frac{\partial u}{\partial \xi}\\&+2\tau g\left(\frac{\partial E}{\partial \xi}-g\right)=0.\end{aligned} \tag{11.6.13}$$

Non-local equations are closed system of three differential equations with three dependent variables u, E, g. In this case *no needs* to use the additional Poisson equation leading to the Newton gravitational description.

If the non-locality parameter τ is equal to zero the mentioned system becomes unclosed.

Let us introduce the length scale ξ_0, the velocity scale u_0, time scale $\tau_0=x_0/u_0$, and scales for the gravitation acceleration $g_0=u_0/\tau_0=u_0^2/x_0$ and for the internal energy of the mass unit $E_0=u_0^2$. Using these scales one obtains

$$\frac{\partial \tilde{u}}{\partial \tilde{\xi}}-\tilde{\tau}\left(\frac{\partial \tilde{u}}{\partial \tilde{\xi}}\right)^2-\frac{\partial}{\partial \tilde{\xi}}\left\{\tilde{\tau}\left[\frac{\partial \tilde{E}}{\partial \tilde{\xi}}-\tilde{g}\right]\right\}=0, \tag{11.6.14}$$

$$\left(\frac{\partial \tilde{E}}{\partial \tilde{\xi}}-\tilde{g}\right)+3\tilde{\tau}\tilde{g}\frac{\partial \tilde{u}}{\partial \tilde{\xi}}-5\tilde{\tau}\frac{\partial \tilde{u}}{\partial \tilde{\xi}}\frac{\partial \tilde{E}}{\partial \tilde{\xi}}-3\tilde{E}\frac{\partial}{\partial \tilde{\xi}}\left\{\tilde{\tau}\frac{\partial \tilde{u}}{\partial \tilde{\xi}}\right\}=0, \tag{11.6.15}$$

$$\begin{aligned}&2\tilde{u}\left(\frac{\partial \tilde{E}}{\partial \tilde{\xi}}-\tilde{g}\right)+5\tilde{E}\frac{\partial \tilde{u}}{\partial \tilde{\xi}}-10\tilde{u}\tilde{\tau}\frac{\partial \tilde{u}}{\partial \tilde{\xi}}\frac{\partial \tilde{E}}{\partial \tilde{\xi}}-6\tilde{u}\tilde{E}\frac{\partial}{\partial \tilde{\xi}}\left(\tilde{\tau}\frac{\partial \tilde{u}}{\partial \tilde{\xi}}\right)\\&-11\tilde{\tau}\tilde{E}\left(\frac{\partial \tilde{u}}{\partial \tilde{\xi}}\right)^2-10\frac{\partial}{\partial \tilde{\xi}}\left(\tilde{\tau}\tilde{E}\frac{\partial \tilde{E}}{\partial \tilde{\xi}}\right)+5\frac{\partial}{\partial \tilde{\xi}}\left(\tilde{\tau}\tilde{E}\tilde{g}\right)+6\tilde{\tau}\tilde{g}\tilde{u}\frac{\partial \tilde{u}}{\partial \tilde{\xi}}\\&+2\tilde{\tau}\tilde{g}\left(\frac{\partial \tilde{E}}{\partial \tilde{\xi}}-\tilde{g}\right)=0,\end{aligned} \tag{11.6.16}$$

We need also an approximation for the non-local parameter $\tilde{\tau}$. Take this approximation in the form

$$\tilde{\tau}=H/\tilde{u}^2, \tag{11.6.17}$$

where H is dimensionless value. In the dimension form

$$\tau=u_0x_0\frac{H}{u^2}. \tag{11.6.18}$$

It means that the non-local parameter is proportional to the kinematic velocity and inversely with square of the velocity. Relation (11.6.18) resembles the Heisenberg relation "time-energy". Remark now that (as follow from the numerical calculations) the choice of the non-local parameter in this case has the small influence on the results of modeling.

Now we are ready to display the results of the mathematical modeling [254] realized with the help of Maple (the versions Maple 9 or higher can be used).

The system of Eqs. (11.6.14)–(11.6.16) have the great possibilities of mathematical modeling as result of changing the parameter H five Cauchy conditions describing the character features of initial perturbations which lead to the traveling wave formation. Maple program contains Maple's notations—for example the expression $D(u)(0)=0$ means in the usual notations $\left(\partial\tilde{u}/\partial\tilde{\xi}\right)(0)=0$, independent variable t responds to $\tilde{\xi}$.

We begin with investigation of the problem of principle significance—is it possible after a perturbation (defined by Cauchy conditions) to obtain the traveling wave as result of the self-organization? With this aim let us consider the initial perturbations:

$$\mathbf{u(0)=1, E(0)=1, g(0)=1, D(u)(0)=0, D(E)(0)=1.} \tag{11.6.19}$$

The following Maple notations on figures are used: u—velocity $\widetilde{u}$, E—energy $\widetilde{E}$, and g—acceleration $\widetilde{g}$. Explanations placed under all following figures. The mentioned calculations are displayed in Figs. 11.53–11.57.

All calculations are realized using the conditions (11.6.19) but by the different value of the H parameter, namely $H=0.001; 1; 1000$. Figure 11.53 reflects the evolution of the dependent values in the area of the event horizon in details.

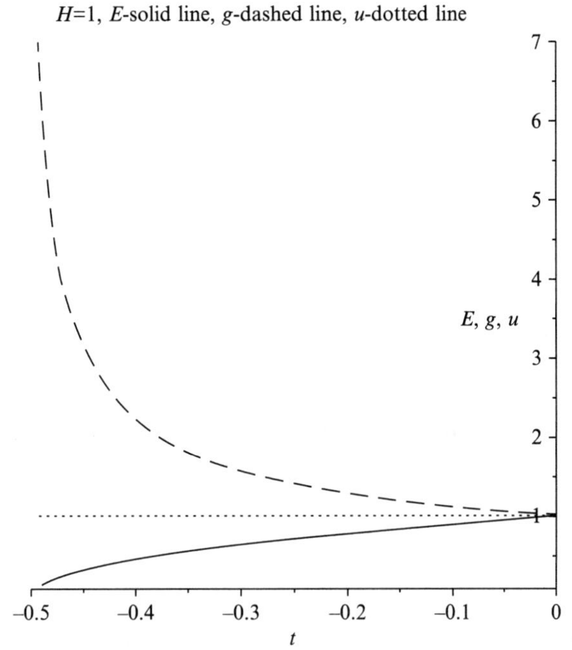

FIGURE 11.53 u—velocity $\widetilde{u}$ (dotted line), $H=1$, E—energy $\widetilde{E}$ (solid line), and g—acceleration $\widetilde{g}$ (dashed line), area of event horizon.

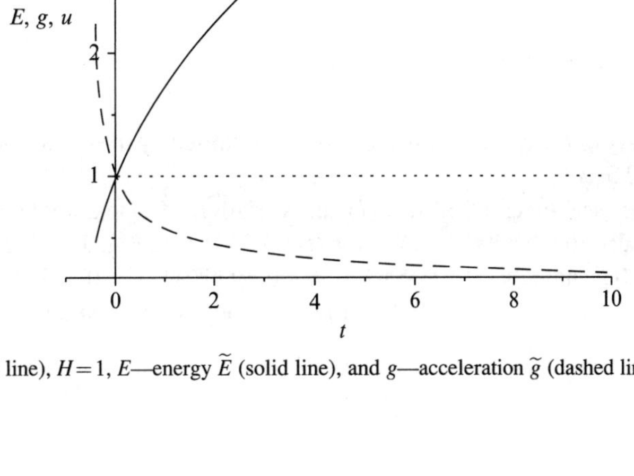

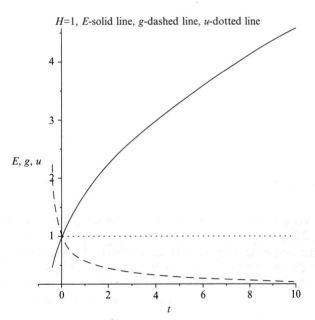

FIGURE 11.54 u—velocity $\widetilde{u}$ (dotted line), $H=1$, E—energy $\widetilde{E}$ (solid line), and g—acceleration $\widetilde{g}$ (dashed line).

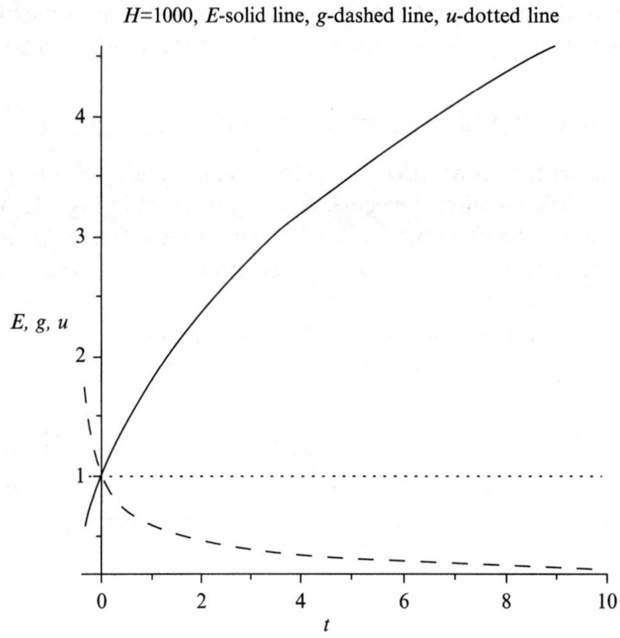

FIGURE 11.55 u—velocity $\widetilde{u}$ (dotted line), $H=1000$, E—energy $\widetilde{E}$ (solid line), and g—acceleration $\widetilde{g}$ (dashed line), area of event horizon.

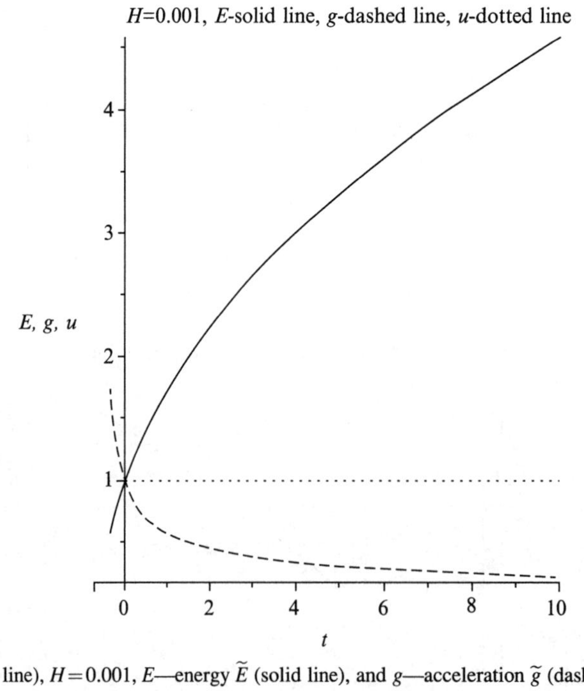

FIGURE 11.56 u—velocity $\widetilde{u}$ (dotted line), $H=0.001$, E—energy $\widetilde{E}$ (solid line), and g—acceleration $\widetilde{g}$ (dashed line).

In all calculations the boundary of the transition area of events is limited by the condition (obtained as the self-consistent result of calculations) $\widetilde{\xi}_{\lim} > -0.5$.

As follow from calculations (see Figs. 11.53–11.57) the variation of H-parameter has the weak influence on the numerical results. Let us show also the results obtained for $H=0.0001$ (see Fig. 11.57) and the corresponding numerical results near singularity $\widetilde{\xi}_{\lim} = -0.5$; namely: $H=0.0001$; $\widetilde{\xi} = -0.4999999$; $\widetilde{E} = 0.382 \times 10^{-3}$; $\widetilde{g} = 2615.014$; $\widetilde{u} = 1$.

As we see the self-consistent solutions lead with the high accuracy to the relation

$$\widetilde{u} = 1. \tag{11.6.20}$$

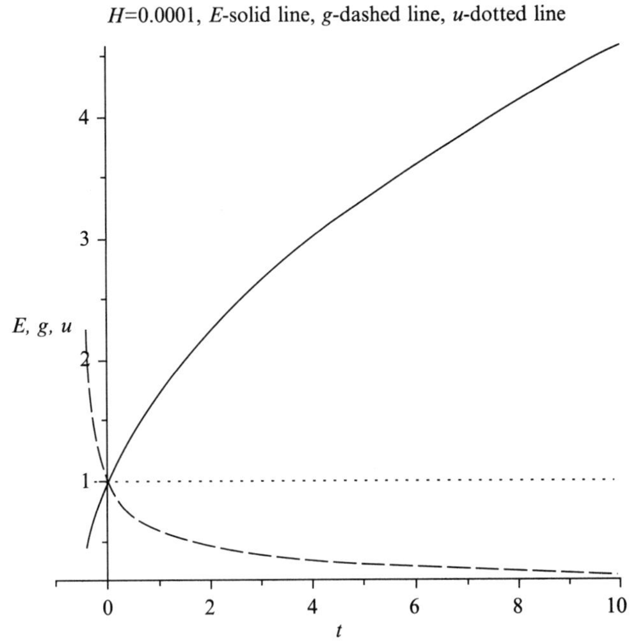

FIGURE 11.57 *u*—velocity $\widetilde{u}$ (dotted line), $H=0.0001$, *E*—energy $\widetilde{E}$ (solid line), and *g*—acceleration $\widetilde{g}$ (dashed line).

Let us use this condition for analytical transformations of the Eqs. (11.6.14)–(11.6.16). We have correspondingly

$$\frac{\partial}{\partial \widetilde{\xi}}\left\{\widetilde{\tau}\left[\frac{\partial \widetilde{E}}{\partial \widetilde{\xi}}-\widetilde{g}\right]\right\}=0, \tag{11.6.21}$$

$$\frac{\partial \widetilde{E}}{\partial \widetilde{\xi}}-\widetilde{g}=0, \tag{11.6.22}$$

$$2\frac{\partial}{\partial \widetilde{\xi}}\left(\widetilde{\tau}\widetilde{E}\frac{\partial \widetilde{E}}{\partial \widetilde{\xi}}\right)-\frac{\partial}{\partial \widetilde{\xi}}\left(\widetilde{\tau}\widetilde{E}\widetilde{g}\right)=0. \tag{11.6.23}$$

From Eqs. (11.6.22) and (11.6.23) follow

$$\frac{\partial}{\partial \widetilde{\xi}}\left(\widetilde{\tau}\widetilde{E}\frac{\partial \widetilde{E}}{\partial \widetilde{\xi}}\right)=0, \tag{11.6.24}$$

$$\widetilde{\tau}\widetilde{E}\frac{\partial \widetilde{E}}{\partial \widetilde{\xi}}=\text{const} \tag{11.6.25}$$

and for chosen $\widetilde{\tau}$ approximation

$$\frac{\partial \widetilde{E}^2}{\partial \widetilde{\xi}}=C, \widetilde{E}^2=C\widetilde{\xi}+C_1, \tag{11.6.26}$$

$$\widetilde{E}^2=\left[\frac{\partial \widetilde{E}^2}{\partial \widetilde{\xi}}\right]_{\widetilde{\xi}=0}\widetilde{\xi}+\widetilde{E}^2(0). \tag{11.6.27}$$

It means that for large $\widetilde{\xi}$

$$\widetilde{E}\cong\sqrt{2\widetilde{E}(0)\left[\frac{\partial \widetilde{E}}{\partial \widetilde{\xi}}\right]_{\widetilde{\xi}=0}}\sqrt{\widetilde{\xi}} \tag{11.6.28}$$

or in the dimensional form

$$E \cong \sqrt{2E(0)\left[\frac{\partial E}{\partial \xi}\right]_{\xi=0}}\sqrt{\xi} \tag{11.6.29}$$

where $\xi = x - ut$. Taking into account Eqs. (11.6.22) and (11.6.25) one obtains

$$\tilde{E}\tilde{g} = \text{const} = \tilde{E}(0)\tilde{g}(0) \tag{11.6.30}$$

$$\tilde{g} \cong \tilde{g}(0)\frac{\tilde{E}(0)}{\sqrt{\left[\frac{\partial \tilde{E}^2}{\partial \tilde{\xi}}\right]_{\tilde{\xi}=0}\tilde{\xi} + \tilde{E}^2(0)}} = \frac{\tilde{g}(0)}{\sqrt{1 + \left[\frac{\partial \ln \tilde{E}^2}{\partial \tilde{\xi}}\right]_{\tilde{\xi}=0}\tilde{\xi}}} \tag{11.6.31}$$

and for large $\tilde{\xi}$

$$\tilde{g} \cong \tilde{g}(0)\sqrt{\frac{\tilde{E}(0)}{2\left[\frac{\partial \tilde{E}}{\partial \tilde{\xi}}\right]_{\tilde{\xi}=0}}}\frac{1}{\sqrt{\tilde{\xi}}} \tag{11.6.32}$$

After the penetration through the frontier barrier the external matter is moving in the black channel in the form of the traveling wave. In this 1D Cartesian model the gravitational acceleration decreases as $\tilde{\xi}^{-0.5}$ with the rise of the $\tilde{\xi}$—distance and, on the contrary, the internal energy of the mass unit increases as $\tilde{\xi}^{0.5}$.

The influence of the tidal force on the object in the black channel can be calculated using Eqs. (11.6.31) and (11.6.32). From Eq. (1.6.31) follows:

$$d\tilde{g} = -\left[\frac{\partial \tilde{E}}{\partial \tilde{\xi}}\right]_{\tilde{\xi}=0}\tilde{g}(0)\frac{\tilde{E}^2(0)}{\left[\left[\frac{\partial \tilde{E}^2}{\partial \tilde{\xi}}\right]_{\tilde{\xi}=0}\tilde{\xi} + \tilde{E}^2(0)\right]^{3/2}}d\tilde{\xi}. \tag{11.6.33}$$

Relation (11.6.33) reflects the change Δg in the tidal force acting at the time moment t across the body element Δx. This change tends to infinity if the point of singularity

$$\tilde{\xi}_s = -\left[\frac{\partial \ln \tilde{E}^2}{\partial \tilde{\xi}}\right]^{-1}_{\tilde{\xi}=0} \tag{11.6.34}$$

which corresponds to the frontier barrier. For example for Cauchy conditions (11.6.19) $\tilde{\xi}_s = -0.5$,

$$\Delta\tilde{g} \cong -\frac{1}{\left[2\tilde{\xi}+1\right]^{3/2}}\Delta\tilde{\xi}. \tag{11.6.35}$$

In this case the change Δg in the tidal force acting at the time moment t across the body element Δx turns into infinity by $\tilde{\xi} = -0.5$. In the following if

$$\left[\frac{\partial \tilde{E}^2}{\partial \tilde{\xi}}\right]_{\tilde{\xi}=0}\tilde{\xi} + \tilde{E}^2(0) \neq 0, \tag{11.6.36}$$

the Δg change of the tidal force acting at the time moment t across the body element Δx has not the catastrophic character.

As one can see during all investigation we needn't to use the theory Newtonian gravitation for solution of nonlinear non-local evolution equations. In contrast with the local physics this approach in the frame of quantum non-local hydrodynamics leads to the closed mathematical description for the physical system under consideration.

If the density tends to infinity the matter evolution inside of "the black channel" (1D Cartesian model) is organizing in the form of the traveling waves.

Numerical modeling leads to appearance of the singularity on the left side of domain where the gravitational acceleration turns into infinity. This singularity corresponds to the event horizon and the whole neighboring area of the strong gravitational variation can be named as the transition area of events, (see Fig. 11.53).

All calculations are realized for the case $\xi = x - ut$, corresponding to the wave traveling along the positive direction of the x-axis. Obviously after the initial perturbations the analogical wave propagates in the opposite direction $\left(-\widetilde{\xi}\right)$ after the sign change $x \to -x$, $u \to -u$. In the theory of Black Hole (BH) with the spherical symmetry it leads near the event horizon to the appearance of black body radiation which was predicted by Stephen Hawking. Hawking radiation reduces the mass and the energy of the black hole and is therefore also known as black hole evaporation. The structure of this radiation significantly depends on the topological features of BH.

Usually the appearance of the analogical picture in the left hand half-plane does not lead to information of the principal significance, but not for the case under consideration.

Really, after rotation the right half-plane picture by 180 ° two domains (see Figs. 11.53 and 11.54) create the joined domain with the width $\widetilde{\xi} = 1$ and minimums for $\widetilde{E}$ and $\widetilde{g}$ in the center of the infinite square well. On the whole the configuration reminds the known quantum mechanical problem of the particle evolution in a box with the infinite potential barriers of the gravitational origin. It is well known that the solution of the analogical problem in the Schrödinger quantum mechanics leads to the discrete energetic levels. Quantum calculations of oscillators in the arbitrary potential fields can be found in [58].

Finally some words concerning the following investigations. Numerical calculations, realized in the spherical coordinate system for the dependent variables (r—radius, t—time) cannot change principal results of the shown calculations in the Cartesian coordinate system. But some other effects (where the real form of the black hole is significant) obviously need in a 3D non-stationary calculation.

11.7 SELF-SIMILAR SOLUTIONS OF THE NON-LOCAL EQUATIONS

11.7.1 Preliminary Remarks

For linear partial differential equations there are various techniques for reducing the partial differential equations (PDE) to the ordinary differential equations (ODE) or at least to equations in a smaller number of independent variables. These include various integral transforms and eigenfunction expansions. Usually such techniques is not applicable are much in dealing with nonlinear partial differential equations. From this point of view much more interesting an approach which identifies equations for which the solution depends on certain groupings of the independent variables rather than depending on each of the independent variables separately. The name of these solutions, *self-similar,* comes from the fact that the spatial distribution of the characteristics of motion remains similar to itself at all times during the motion.

Obviously this fact reflects the intimate relationship between self-similar solutions and uniform-propagation regimes. The uniform-propagation regime in the soliton form is used many times in the previous sections such that the solution at the next instant is obtained by a shift in coordinate from the solution corresponding to the preceding instant.

Roughly speaking the idea is to look whether the solution of a problem $u(x, y)$ can be collapsed in a function $u(x, y) = U(y/f(x))$. The function $f(x)$ may be found by substitution in the PDE, in order to obtain an ODE for U. For example in Navier-Stokes for the flow over a plate $f(x) = \sqrt{x}$; in the heat equation it will be $T(x, t) = \Theta(x/f(t))$ with $f(x) = \sqrt{t}$. First of all we will present a classical example of self-similar solution by the help of the scaling invariance in the simplest possible form and later the scaling applications will be considered for the much more complicated case of the non-local equations of the transport theory.

Consider a rigid stationary obstacle whose surface is flat, and corresponds to the x, y-plane. Let this surface be in contact with a high Reynolds number fluid that occupies the region $x > 0$. Let δ be the typical normal thickness of the boundary layer. The layer thus extends over the region $0 < y \approx \delta$. The fluid that occupies the region $\delta < y < \infty$, is assumed to be both irrotational and inviscid. Viscosity is included in the equation of motion of the fluid within the layer. The fluid both inside and outside the layer is assumed to be incompressible. After the boundary layer approximation (that viscosity acts only within a thin layer and the gradient in the flow direction (x) is much smaller than in the transverse direction (y), and that the pressure is constant in the y direction), the following equations of the boundary layer take place

$$u\frac{\partial u}{\partial x} + v\frac{\partial u}{\partial y} = -\frac{1}{\rho}\frac{\mathrm{d}p}{\mathrm{d}x} + \upsilon\frac{\partial^2 u}{\partial y^2}, \tag{11.7.1}$$

$$\frac{\partial u}{\partial x} + \frac{\partial v}{\partial y} = 0. \tag{11.7.2}$$

Equation (11.7.1) is the x-component of the fluid equation of motion, u—tangential flow velocity. Here, we are assuming, for the sake of simplicity, that the boundary layer is effectively two-dimensional. Likewise, we are also assuming that all flows are steady, so that any time variation can be neglected.

The boundary conditions at the outer edge of the layer, where it interfaces with the irrotational fluid, are

$$u(x,y)\to U(x) \qquad (11.7.3)$$

$$p(x,y)=P(x)$$

as $y/\delta\to\infty$. Here, $P(x)$ is the fluid pressure at the outer edge of the layer, and

$$U\frac{\mathrm{d}U}{\mathrm{d}x}=-\frac{1}{\rho}\frac{\mathrm{d}P}{\mathrm{d}x}, \qquad (11.7.4)$$

since the velocity component $v=0$, and viscosity is negligible, just outside the layer.

Let us introduce the scales

$$x\to L, u\to u_\infty, y\to\sqrt{\upsilon\frac{L}{u_\infty}}, v\to\sqrt{\upsilon\frac{u_\infty}{L}}, p\to\rho u_\infty^2$$

From the set of Eqs. (11.7.1) and (11.7.2) follows the dimensionless equations

$$\tilde{u}\frac{\partial\tilde{u}}{\partial\tilde{x}}+\tilde{v}\frac{\partial\tilde{u}}{\partial\tilde{y}}=-\frac{\mathrm{d}\tilde{p}}{\mathrm{d}\tilde{x}}+\frac{\partial^2\tilde{u}}{\partial\tilde{y}^2} \qquad (11.7.5)$$

$$\frac{\partial\tilde{u}}{\partial\tilde{x}}+\frac{\partial\tilde{v}}{\partial\tilde{y}}=0 \qquad (11.7.6)$$

Since the essence of similarity is that the solution is invariant after certain scaling of the independent and dependent variables, we consider the following stretching transformation, and see if such transformations will leave the PDE and the boundary conditions invariant. Namely

$$\tilde{x}\to a\tilde{x}, \tilde{y}\to b\tilde{y}, \tilde{u}\to c\tilde{u}, \tilde{v}\to d\tilde{v}.$$

We find

$$\begin{aligned}&\tilde{u}(a\tilde{x},b\tilde{y})\frac{\partial\tilde{u}(a\tilde{x},b\tilde{y})}{\partial\tilde{x}}+\frac{ad}{cb}\tilde{v}(a\tilde{x},b\tilde{y})\frac{\partial\tilde{u}(a\tilde{x},b\tilde{y})}{\partial\tilde{y}}\\&=-\frac{1}{c^2}\frac{\mathrm{d}\tilde{p}(a\tilde{x})}{\mathrm{d}\tilde{x}}+\frac{a}{cb^2}\frac{\partial^2\tilde{u}(a\tilde{x},b\tilde{y})}{\partial\tilde{y}^2},\end{aligned} \qquad (11.7.7)$$

$$\frac{\partial\tilde{u}(a\tilde{x},b\tilde{y})}{\partial\tilde{x}}+\frac{ad}{cb}\frac{\partial\tilde{v}(a\tilde{x},b\tilde{y})}{\partial\tilde{y}}=0. \qquad (11.7.8)$$

Solutions of Eqs. (11.7.5), (11.7.6), (11.7.8), and (11.7.9) should be identical, then

$$\frac{ad}{cb}=1, \frac{1}{c^2}=1, \frac{a}{cb^2}=1, \frac{ad}{cb}=1. \qquad (11.7.9)$$

From (11.7.9) follows

$$c=1, \qquad (11.7.10)$$

and

$$ad=b,\ a=b^2,\ d=\frac{1}{b} \qquad (11.7.11)$$

Using the definition

$$\tilde{u}(a\tilde{x},b\tilde{y})=c\tilde{u}(\tilde{x},\tilde{y}), \tilde{v}(a\tilde{x},b\tilde{y})=d\tilde{v}(\tilde{x},\tilde{y}) \qquad (11.7.12)$$

and the relations (11.7.11), one obtains

$$\tilde{u}\left(b^2\tilde{x},b\tilde{y}\right)=\tilde{u}(\tilde{x},\tilde{y}), \tilde{v}\left(b^2\tilde{x},b\tilde{y}\right)=\frac{1}{b}\tilde{v}(\tilde{x},\tilde{y}). \qquad (11.7.13)$$

If the self-similar solutions exist, the relation (11.7.13) should be fulfilled for all $\widetilde{x}$, for example for $\widetilde{x}=1/b_2$. In this case

$$\widetilde{u}\left(1,\frac{\widetilde{y}}{\sqrt{\widetilde{x}}}\right)=\widetilde{u}(\widetilde{x},\widetilde{y}),\widetilde{v}\left(1,\frac{\widetilde{y}}{\sqrt{\widetilde{x}}}\right)=\sqrt{\widetilde{x}}\,\widetilde{v}(\widetilde{x},\widetilde{y}). \tag{11.7.14}$$

Now we can introduce the self-similar variable

$$\eta=\frac{\widetilde{y}}{\sqrt{\widetilde{x}}}, \tag{11.7.15}$$

therefore

$$\widetilde{u}(\widetilde{x},\widetilde{y})=\widetilde{u}\left(1,\frac{\widetilde{y}}{\sqrt{\widetilde{x}}}\right)\equiv f(\eta). \tag{11.7.16}$$

$$\widetilde{v}(\widetilde{x},\widetilde{y})=\frac{1}{\sqrt{\widetilde{x}}}\widetilde{v}\left(1,\frac{\widetilde{y}}{\sqrt{\widetilde{x}}}\right)\equiv\frac{1}{\sqrt{\widetilde{x}}}g(\eta). \tag{11.7.17}$$

Using Eqs. (11.7.16) and (11.7.17), let us transform Eqs. (11.7.5) and (11.7.6) taking into account the derivations

$$\frac{\partial\widetilde{u}}{\partial\widetilde{x}}=\frac{\partial f}{\partial\eta}\frac{\partial\eta}{\partial\widetilde{x}}=f'\left(-\frac{\widetilde{y}}{2\widetilde{x}^{3/2}}\right), \tag{11.7.18}$$

$$\frac{\partial\widetilde{u}}{\partial\widetilde{y}}=\frac{\partial f}{\partial\eta}\frac{\partial\eta}{\partial\widetilde{y}}=f'\frac{1}{\sqrt{\widetilde{x}}}, \tag{11.7.19}$$

$$\frac{d\widetilde{p}}{d\widetilde{x}}=p'\frac{\partial\eta}{\partial\widetilde{x}}=p'\left(-\frac{\widetilde{y}}{2\widetilde{x}^{3/2}}\right), \tag{11.7.20}$$

$$\frac{\partial^2\widetilde{u}}{\partial\widetilde{y}^2}=\frac{\partial}{\partial\widetilde{y}}\left(f'\frac{1}{\sqrt{\widetilde{x}}}\right)=\frac{1}{\sqrt{\widetilde{x}}}\frac{\partial}{\partial\widetilde{y}}(f')=\frac{1}{\sqrt{\widetilde{x}}}\frac{\partial f'}{\partial\eta}\frac{\partial\eta}{\partial\widetilde{y}}=\frac{f''}{\widetilde{x}}. \tag{11.7.21}$$

$$\frac{\partial\widetilde{v}}{\partial\widetilde{y}}=\frac{\partial}{\partial\widetilde{y}}\left(\frac{g(\eta)}{\sqrt{\widetilde{x}}}\right)=\frac{1}{\sqrt{\widetilde{x}}}\frac{\partial g}{\partial\widetilde{y}}=\frac{1}{\sqrt{\widetilde{x}}}\frac{\partial g}{\partial\eta}\frac{\partial\eta}{\partial\widetilde{y}}=\frac{g'}{\widetilde{x}}. \tag{11.7.22}$$

After substitution relations (11.7.18)–(11.7.22) into Eqs. (11.7.5) and (11.7.6), we have

$$ff'\left(-\frac{\widetilde{y}}{2\widetilde{x}^{3/2}}\right)+\frac{1}{\sqrt{\widetilde{x}}}g(\eta)f'\frac{1}{\sqrt{\widetilde{x}}}=p'\left(\frac{\widetilde{y}}{2\widetilde{x}^{3/2}}\right)+\frac{f''}{\widetilde{x}}, \tag{11.7.23}$$

$$f'\left(-\frac{\widetilde{y}}{2\widetilde{x}^{3/2}}\right)+\frac{g'}{\widetilde{x}}=0, \tag{11.7.24}$$

or

$$-\frac{1}{2}ff'\eta+gf'=\frac{1}{2}p'\eta+f'', \tag{11.7.25}$$

$$g'=\frac{1}{2}f'\eta. \tag{11.7.26}$$

The dependent variable $g(\eta)$ can be excluded from Eq. (11.7.25) using (11.7.26). For so-called "Blasius problem" about the flow over a flat semi-plane in the boundary layer $p'=0$, as result

$$f'f'''+\frac{1}{2}ff'^2-f''^2=0. \tag{11.7.27}$$

We have successfully transformed the PDE into an ODE. How about the boundary conditions?

$$\widetilde{u}(\widetilde{x},0)=0,\widetilde{v}(\widetilde{x},0)=0.$$

$$\widetilde{u}(a\widetilde{x},0)=0,\widetilde{v}(a\widetilde{x},0)=0.$$

Obviously after stretching we also have

$$c\tilde{u}(\tilde{x},0)=0, d\tilde{v}(\tilde{x},0)=0. \tag{11.7.28}$$

The tangential component of the fluid velocity at the outer edge of the layer is known

$$\tilde{u}(\tilde{x},\tilde{y})=\tilde{u}_e(\tilde{x}), \quad \text{if } \tilde{y}\to\infty, \tag{11.7.29}$$

then

$$\tilde{u}(a\tilde{x},b\tilde{y})=\tilde{u}_e(a\tilde{x}), \quad \text{if } \tilde{y}\to\infty. \tag{11.7.30}$$

From relation (11.7.12)

$$\tilde{u}(\tilde{x},b\tilde{y})=\tilde{u}_e(\tilde{x}), \quad \text{if } \tilde{y}\to\infty,$$

and we have again Eq. (11.7.29). The compatibility of the boundary conditions is obvious.

11.7.2 Self-similar Solutions of the Non-Local Equations in the Astrophysical Applications

The non-local system of hydrodynamic equations describing the explosion with the spherical symmetry is written as Eqs. (11.4.6)–(11.4.10) (see Appendix 2). This set of equations in the dimensionless form (see Section 11.4):

(Poisson equation)

$$\frac{1}{\tilde{r}^2}\frac{\partial}{\partial\tilde{r}}\left(\tilde{r}^2\frac{\partial\tilde{\psi}}{\partial\tilde{r}}\right)=4\pi G\left[\tilde{\rho}-\tilde{\tau}\left(\frac{\partial\tilde{\rho}}{\partial\tilde{t}}+\frac{1}{\tilde{r}^2}\frac{\partial(\tilde{r}^2\tilde{\rho}\tilde{v}_r)}{\partial\tilde{r}}\right)\right], \tag{11.7.31}$$

(continuity equation)

$$\begin{aligned}&\frac{\partial}{\partial\tilde{t}}\left\{\tilde{\rho}-\tilde{\tau}\left[\frac{\partial\tilde{\rho}}{\partial\tilde{t}}+\frac{1}{\tilde{r}^2}\frac{\partial(\tilde{r}^2\tilde{\rho}\tilde{v}_r)}{\partial\tilde{r}}\right]\right\}+\frac{1}{\tilde{r}^2}\frac{\partial}{\partial\tilde{r}}\left\{\tilde{r}^2\left\{\tilde{\rho}\tilde{v}_r-\tilde{\tau}\left[\frac{\partial}{\partial\tilde{t}}(\tilde{\rho}\tilde{v}_r)\right.\right.\right.\\&\left.\left.\left.+\frac{1}{\tilde{r}^2}\frac{\partial(\tilde{r}^2\tilde{\rho}\tilde{v}_r^2)}{\partial\tilde{r}}+\tilde{\rho}\frac{\partial\tilde{\psi}}{\partial\tilde{r}}\right]\right\}\right\}-\frac{1}{\tilde{r}^2}\frac{\partial}{\partial\tilde{r}}\left(\tilde{\tau}\tilde{r}^2\frac{\partial\tilde{p}}{\partial\tilde{r}}\right)=0,\end{aligned} \tag{11.7.32}$$

(motion equation)

$$\begin{aligned}&\frac{\partial}{\partial\tilde{t}}\left\{\tilde{\rho}\tilde{v}_r-\tilde{\tau}\left[\frac{\partial}{\partial\tilde{t}}(\tilde{\rho}\tilde{v}_r)+\frac{1}{\tilde{r}^2}\frac{\partial(\tilde{r}^2\tilde{\rho}\tilde{v}_r^2)}{\partial\tilde{r}}+\frac{\partial\tilde{p}}{\partial\tilde{r}}+\tilde{\rho}\frac{\partial\tilde{\psi}}{\partial\tilde{r}}\right]\right\}\\&+\frac{\partial\tilde{\psi}}{\partial\tilde{r}}\left[\tilde{\rho}-\tilde{\tau}\left(\frac{\partial\tilde{\rho}}{\partial\tilde{t}}+\frac{1}{\tilde{r}^2}\frac{\partial(\tilde{r}^2\tilde{\rho}\tilde{v}_r)}{\partial\tilde{r}}\right)\right]\\&+\frac{1}{\tilde{r}^2}\frac{\partial}{\partial\tilde{r}}\left\{\tilde{r}^2\left\{\tilde{\rho}\tilde{v}_r^2-\tilde{\tau}\left[\frac{\partial}{\partial\tilde{t}}(\tilde{\rho}\tilde{v}_r^2)+\frac{1}{\tilde{r}^2}\frac{\partial(\tilde{r}^2\tilde{\rho}\tilde{v}_r^3)}{\partial\tilde{r}}+2\frac{\partial\tilde{\psi}}{\partial\tilde{r}}\tilde{\rho}\tilde{v}_r\right]\right\}\right\}\\&+\frac{\partial\tilde{p}}{\partial\tilde{r}}-\frac{\partial}{\partial\tilde{r}}\left(\tilde{\tau}\frac{\partial\tilde{p}}{\partial\tilde{t}}\right)-2\frac{\partial}{\partial\tilde{r}}\left(\frac{\tilde{\tau}}{\tilde{r}^2}\frac{\partial(\tilde{r}^2\tilde{p}\tilde{v}_r)}{\partial\tilde{r}}\right)-\frac{1}{\tilde{r}^2}\frac{\partial}{\partial\tilde{r}}\left(\tilde{\tau}\tilde{r}^2\frac{\partial(\tilde{p}\tilde{v}_r)}{\partial\tilde{r}}\right)=0,\end{aligned} \tag{11.7.33}$$

(energy equation)

$$\frac{\partial}{\partial \tilde{t}}\left\{\tilde{\rho}\tilde{v}_r^2+3\tilde{p}-\tilde{\tau}\left[\frac{\partial}{\partial \tilde{t}}(\tilde{\rho}\tilde{v}_r^2+3\tilde{p})+\frac{1}{\tilde{r}^2}\frac{\partial}{\partial \tilde{r}}(\tilde{r}^2\tilde{v}_r(\tilde{\rho}\tilde{v}_r^2+5\tilde{p}))+2\tilde{\rho}\frac{\partial \tilde{\psi}}{\partial \tilde{r}}\tilde{v}_r\right]\right\}$$
$$+\frac{1}{\tilde{r}^2}\frac{\partial}{\partial \tilde{r}}\left\{\tilde{r}^2\left\{(\tilde{\rho}\tilde{v}_r^2+5\tilde{p})\tilde{v}_r-\tilde{\tau}\left[\frac{\partial}{\partial \tilde{t}}((\tilde{\rho}\tilde{v}_r^2+5\tilde{p})\tilde{v}_r)+\frac{1}{\tilde{r}^2}\frac{\partial}{\partial \tilde{r}}(\tilde{r}^2(\tilde{\rho}\tilde{v}_r^2+7\tilde{p})\tilde{v}_r^2)\right.\right.\right.$$
$$\left.\left.\left.+2\tilde{\rho}\frac{\partial \tilde{\psi}}{\partial \tilde{r}}\tilde{v}_r^2+(\tilde{\rho}\tilde{v}_r^2+3\tilde{p})\frac{\partial \tilde{\psi}}{\partial \tilde{r}}\right]\right\}\right\}+2\left\{\tilde{\rho}\frac{\partial \tilde{\psi}}{\partial \tilde{r}}\tilde{v}_r-\tilde{\tau}\frac{\partial \tilde{\psi}}{\partial \tilde{r}}\left[\frac{\partial}{\partial \tilde{t}}(\tilde{\rho}\tilde{v}_r)+\frac{1}{\tilde{r}^2}\frac{\partial}{\partial \tilde{r}}(\tilde{r}^2\tilde{\rho}\tilde{v}_r^2)+\frac{\partial \tilde{p}}{\partial \tilde{r}}+\tilde{\rho}\frac{\partial \tilde{\psi}}{\partial \tilde{r}}\right]\right\} \tag{11.7.34}$$
$$-\frac{1}{\tilde{r}^2}\frac{\partial}{\partial \tilde{r}}\left(\tilde{\tau}\tilde{r}^2\frac{\partial}{\partial \tilde{r}}\left(\tilde{p}\tilde{v}_r^2+5\frac{\tilde{p}^2}{\tilde{\rho}}\right)\right)-2\frac{1}{\tilde{r}^2}\frac{\partial}{\partial \tilde{r}}\left(\tilde{r}^2\tilde{\tau}\tilde{p}\frac{\partial \tilde{\psi}}{\partial \tilde{r}}\right)=0.$$

The system of Eqs. (11.7.31)–(11.7.34) belongs to the class of the 1D non-stationary equations and can be solved by known numerical methods. But for the aims of the transparent vast mathematical modeling of self-catching of the expanding matter by the self-consistent gravitational field the following assumption was introduced in Section 11.4. Namely, in the quasi-stationary Hubble regime only the implicit dependence on time for the unknown values exists. It means that for the intermediate (Hubble) regime the complicated PDE set can be transformed in the set ODE. This possibility can be realized also in the case if the self-similar solutions exist. Let us investigate this possible situation.

As in Section 11.7.1 we introduce stretching for time $\tilde{t}\to a\tilde{t}$, radial coordinate $\tilde{r}\to b\tilde{r}$ and dependent variables $\tilde{\psi}\to c\tilde{\psi}$, $\tilde{\rho}\to d\tilde{\rho}$, $\tilde{v}_r\to e\tilde{v}_r$, $\tilde{p}\to f\tilde{p}$. The non-local parameter $\tilde{\tau}$ stretches like time $\tilde{t}$. From Eqs. (11.7.31)–(11.7.34) follow

$$\frac{c}{b^2}\frac{1}{\tilde{r}^2}\frac{\partial}{\partial \tilde{r}}\left(\tilde{r}^2\frac{\partial \tilde{\psi}}{\partial \tilde{r}}\right)=4\pi Gd\left[\tilde{\rho}-\tilde{\tau}\left(\frac{\partial \tilde{\rho}}{\partial \tilde{t}}+\frac{ea}{b}\frac{1}{\tilde{r}^2}\frac{\partial(\tilde{r}^2\tilde{\rho}\tilde{v}_r)}{\partial \tilde{r}}\right)\right], \tag{11.7.35}$$

$$\frac{d}{a}\frac{\partial}{\partial \tilde{t}}\left\{\tilde{\rho}-\tilde{\tau}\left[\frac{\partial \tilde{\rho}}{\partial \tilde{t}}+\frac{ea}{b}\frac{1}{\tilde{r}^2}\frac{\partial(\tilde{r}^2\tilde{\rho}\tilde{v}_r)}{\partial \tilde{r}}\right]\right\}+\frac{d}{b}\frac{1}{\tilde{r}^2}\frac{\partial}{\partial \tilde{r}}\left\{\tilde{r}^2\left\{e\tilde{\rho}\tilde{v}_r-\tilde{\tau}\left[e\frac{\partial}{\partial \tilde{t}}(\tilde{\rho}\tilde{v}_r)\right.\right.\right.$$
$$\left.\left.\left.+\frac{e^2a}{b}\frac{1}{\tilde{r}^2}\frac{\partial(\tilde{r}^2\tilde{\rho}\tilde{v}_r^2)}{\partial \tilde{r}}+\frac{ca}{b}\tilde{\rho}\frac{\partial \tilde{\psi}}{\partial \tilde{r}}\right]\right\}\right\}-\frac{af}{b^2}\frac{1}{\tilde{r}^2}\frac{\partial}{\partial \tilde{r}}\left(\tilde{\tau}\tilde{r}^2\frac{\partial \tilde{p}}{\partial \tilde{r}}\right)=0, \tag{11.7.36}$$

$$\frac{1}{a}\frac{\partial}{\partial \tilde{t}}\left\{\tilde{\rho}\tilde{v}_r de-\tilde{\tau}\left[\frac{\partial}{\partial \tilde{t}}(\tilde{\rho}\tilde{v}_r)de+\frac{1}{\tilde{r}^2}\frac{\partial(\tilde{r}^2\tilde{\rho}\tilde{v}_r^2)}{\partial \tilde{r}}\frac{ade^2}{b}+\frac{\partial \tilde{p}}{\partial \tilde{r}}\frac{fa}{b}+\tilde{\rho}\frac{\partial \tilde{\psi}}{\partial \tilde{r}}\frac{cda}{b}\right]\right\}$$
$$+\frac{cd}{b}\frac{\partial \tilde{\psi}}{\partial \tilde{r}}\left[\tilde{\rho}-\tilde{\tau}\left(\frac{\partial \tilde{\rho}}{\partial \tilde{t}}+\frac{ea}{b}\frac{1}{\tilde{r}^2}\frac{\partial(\tilde{r}^2\tilde{\rho}\tilde{v}_r)}{\partial \tilde{r}}\right)\right]$$
$$+\frac{d}{b}\frac{1}{\tilde{r}^2}\frac{\partial}{\partial \tilde{r}}\left\{\tilde{r}^2\left\{\tilde{\rho}\tilde{v}_r^2e^2-\tilde{\tau}\left[\frac{\partial}{\partial \tilde{t}}(\tilde{\rho}\tilde{v}_r^2)e^2+\frac{1}{\tilde{r}^2}\frac{\partial(\tilde{r}^2\tilde{\rho}\tilde{v}_r^3)}{\partial \tilde{r}}\frac{ae^3}{b}+\frac{cea}{b}2\frac{\partial \tilde{\psi}}{\partial \tilde{r}}\tilde{\rho}\tilde{v}_r\right]\right\}\right\} \tag{11.7.37}$$
$$+\frac{f}{b}\frac{\partial \tilde{p}}{\partial \tilde{r}}-\frac{f}{b}\frac{\partial}{\partial \tilde{r}}\left(\tilde{\tau}\frac{\partial \tilde{p}}{\partial \tilde{t}}\right)-\frac{afe}{b^2}2\frac{\partial}{\partial \tilde{r}}\left(\frac{\tilde{\tau}}{\tilde{r}^2}\frac{\partial(\tilde{r}^2\tilde{p}\tilde{v}_r)}{\partial \tilde{r}}\right)-\frac{afe}{b^2}\frac{1}{\tilde{r}^2}\frac{\partial}{\partial \tilde{r}}\left(\tilde{\tau}\tilde{r}^2\frac{\partial(\tilde{p}\tilde{v}_r)}{\partial \tilde{r}}\right)=0,$$

$$\begin{aligned}
&\frac{1}{a}\frac{\partial}{\partial \tilde{t}}\Bigg\{\tilde{\rho}\tilde{v}_r^2 de^2+3\tilde{p}f-\tilde{\tau}\Bigg[\frac{\partial}{\partial \tilde{t}}(\tilde{\rho}\tilde{v}_r^2 de^2+3\tilde{p}f)+\frac{1}{\tilde{r}^2}\frac{ae}{b}\frac{\partial}{\partial \tilde{r}}(\tilde{r}^2\tilde{v}_r(\tilde{\rho}\tilde{v}_r^2 de^2+5\tilde{p}f))+\frac{adec}{b}2\tilde{\rho}\frac{\partial\tilde{\psi}}{\partial \tilde{r}}\tilde{v}_r\Bigg]\Bigg\}\\
&+\frac{1}{\tilde{r}^2}\frac{1}{b}\frac{\partial}{\partial \tilde{r}}\Bigg\{\tilde{r}^2\Bigg\{(\tilde{\rho}\tilde{v}_r^2 de^2+5\tilde{p}f)\tilde{v}_r e-\tilde{\tau}\Bigg[\frac{\partial}{\partial \tilde{t}}((\tilde{\rho}\tilde{v}_r^2 de^2+5\tilde{p}f)\tilde{v}_r e)+\frac{a}{\tilde{r}^2}\frac{1}{b}\frac{\partial}{\partial \tilde{r}}(\tilde{r}^2(\tilde{\rho}\tilde{v}_r^2 de^2+7\tilde{p}f)\tilde{v}_r^2 e^2)\\
&+\frac{acde^2}{b}2\tilde{\rho}\frac{\partial\tilde{\psi}}{\partial \tilde{r}}\tilde{v}_r^2+\frac{ca}{b}(\tilde{\rho}\tilde{v}_r^2 de^2+3\tilde{p}f)\frac{\partial\tilde{\psi}}{\partial \tilde{r}}\Bigg]\Bigg\}\Bigg\}\\
&+2\Bigg\{\frac{dce}{b}\tilde{\rho}\frac{\partial\tilde{\psi}}{\partial \tilde{r}}\tilde{v}_r-\frac{c}{b}\tilde{\tau}\frac{\partial\tilde{\psi}}{\partial \tilde{r}}\Bigg[\frac{\partial}{\partial \tilde{t}}(\tilde{\rho}\tilde{v}_r)de+\frac{ade^2}{b}\frac{1}{\tilde{r}^2}\frac{\partial}{\partial \tilde{r}}(\tilde{r}^2\tilde{\rho}\tilde{v}_r^2)+\frac{af}{b}\frac{\partial\tilde{p}}{\partial \tilde{r}}+\frac{acd}{b}\tilde{\rho}\frac{\partial\tilde{\psi}}{\partial \tilde{r}}\Bigg]\Bigg\}\\
&-\frac{a}{b^2}\frac{1}{\tilde{r}^2}\frac{\partial}{\partial \tilde{r}}\left[\tilde{\tau}\tilde{r}^2\frac{\partial}{\partial \tilde{r}}\left(\tilde{p}\tilde{v}_r^2 fe^2+5\frac{\tilde{p}^2}{\tilde{\rho}}\frac{f^2}{d}\right)\right]-2\frac{acf}{b^2}\frac{1}{\tilde{r}^2}\frac{\partial}{\partial \tilde{r}}\left(\tilde{r}^2\tilde{\tau}\tilde{p}\frac{\partial\tilde{\psi}}{\partial \tilde{r}}\right)=0.
\end{aligned} \tag{11.7.38}$$

From the Poisson equation (11.7.35) follows the relations for the stretching coefficients

$$\frac{db^2}{c}=1, \tag{11.7.39}$$

$$\frac{ea}{b}=1. \tag{11.7.40}$$

Analogical procedure is applied for other Eqs. (11.7.36)–(11.7.38). We have from Eq. (11.7.36)

$$\begin{aligned}
&\frac{\partial}{\partial \tilde{t}}\left\{\tilde{\rho}-\tilde{\tau}\left[\frac{\partial\tilde{\rho}}{\partial \tilde{t}}+\frac{1}{\tilde{r}^2}\frac{\partial(\tilde{r}^2\tilde{\rho}\tilde{v}_r)}{\partial \tilde{r}}\right]\right\}+\frac{ae}{b}\frac{1}{\tilde{r}^2}\frac{\partial}{\partial \tilde{r}}\Bigg\{\tilde{r}^2\Bigg\{\tilde{\rho}\tilde{v}_r-\tilde{\tau}\Bigg[\frac{\partial}{\partial \tilde{t}}(\tilde{\rho}\tilde{v}_r)\frac{ae}{b}\frac{1}{\tilde{r}^2}\frac{\partial}{\partial \tilde{r}}\Bigg\{\tilde{r}^2\Bigg\{\tilde{\rho}\tilde{v}_r-\tilde{\tau}\Bigg[\frac{\partial}{\partial \tilde{t}}(\tilde{\rho}\tilde{v}_r)\\
&+\frac{1}{\tilde{r}^2}\frac{\partial(\tilde{r}^2\tilde{\rho}\tilde{v}_r^2)}{\partial \tilde{r}}+\frac{ca}{be}\tilde{\rho}\frac{\partial\tilde{\psi}}{\partial \tilde{r}}\Bigg]\Bigg\}\Bigg\}-\frac{af}{b^2}\frac{a}{d}\frac{1}{\tilde{r}^2}\frac{\partial}{\partial \tilde{r}}\left(\tilde{\tau}\tilde{r}^2\frac{\partial\tilde{p}}{\partial \tilde{r}}\right)=0,
\end{aligned} \tag{11.7.41}$$

and additional conditions

$$\frac{fa^2}{b^2 d}=1, \tag{11.7.42}$$

$$\frac{ca}{be}=1. \tag{11.7.43}$$

As the consequence of Eqs. (11.7.40), (11.7.42), and (11.7.43) we have the useful dependent relations

$$\frac{fa}{bde}=\frac{fa}{bde}\frac{ab}{ba}=\frac{fa^2}{b^2 d}\frac{b}{ae}=1, \tag{11.7.44}$$

$$\frac{f}{de^2}=\frac{fa^2}{db^2}\frac{b^2}{e^2a^2}=1, \tag{11.7.45}$$

$$\frac{a^2c}{b^2}=\frac{ac}{be}\frac{ae}{b}=1. \tag{11.7.46}$$

Using relations (11.7.42)–(11.7.44), we find the same momentum equation (11.7.33). Call attention now to the energy equation (11.7.38) which after division the both sides of equation by de^2/a, has the form

$$
\begin{aligned}
&\frac{\partial}{\partial\tilde t}\left\{\tilde\rho\tilde v_r^2+3\tilde p\frac{f}{de^2}-\tilde\tau\left[\frac{\partial}{\partial\tilde t}\left(\tilde\rho\tilde v_r^2+3\tilde p\frac{f}{de^2}\right)+\frac{1}{\tilde r^2}\frac{\partial}{\partial\tilde r}\left(\tilde r^2\tilde v_r\left(\tilde\rho\tilde v_r^2+5\tilde p\frac{f}{de^2}\right)\right)+2\frac{ac}{be}\tilde\rho\frac{\partial\tilde\psi}{\partial\tilde r}\tilde v_r\right]\right\}\\
&+\frac{1}{\tilde r^2}\frac{ae}{b}\frac{\partial}{\partial\tilde r}\left\{\tilde r^2\left\{\left(\tilde\rho\tilde v_r^2+5\tilde p\frac{f}{de^2}\right)\tilde v_r-\tilde\tau\left[\frac{\partial}{\partial\tilde t}\left(\left(\tilde\rho\tilde v_r^2+5\tilde p\frac{f}{de^2}\right)\tilde v_r\right)\right.\right.\right.\\
&\left.\left.\left.+\frac{1}{\tilde r^2}\frac{ea}{b}\frac{\partial}{\partial\tilde r}\left(\tilde r^2\left(\tilde\rho\tilde v_r^2+7\tilde p\frac{f}{de^2}\right)\tilde v_r^2\right)+2\frac{ac}{be}\tilde\rho\frac{\partial\psi}{\partial\tilde r}\tilde v_r^2+\frac{ca}{be}\left(\tilde\rho\tilde v_r^2+3\tilde p\frac{f}{de^2}\right)\frac{\partial\tilde\psi}{\partial\tilde r}\right]\right\}\right.\\
&\left.+2\left\{\frac{ac}{be}\tilde\rho\frac{\partial\tilde\psi}{\partial\tilde r}\tilde v_r-\frac{c}{b}\tilde\tau\frac{\partial\tilde\psi}{\partial\tilde r}\left[\frac{\partial}{\partial\tilde t}(\tilde\rho\tilde v_r)\frac{a}{e}+\frac{a^2}{b}\frac{1}{\tilde r^2}\frac{\partial}{\partial\tilde r}(\tilde r^2\tilde\rho\tilde v_r^2)+\frac{af}{b}\frac{a}{de^2}\frac{\partial\tilde p}{\partial\tilde r}+\frac{a^2c}{be^2}\tilde\rho\frac{\partial\tilde\psi}{\partial\tilde r}\right]\right\}\right.\\
&-\frac{a}{b^2}\frac{a}{de^2}\frac{1}{\tilde r^2}\frac{\partial}{\partial\tilde r}\left(\tilde\tau\tilde r^2\frac{\partial}{\partial\tilde r}\left(\tilde p\tilde v_r^2fe^2+5\frac{\tilde p^2}{\tilde\rho}\frac{f^2}{d}\right)\right)-2\frac{acf}{b^2}\frac{a}{de^2}\frac{1}{\tilde r^2}\frac{\partial}{\partial\tilde r}\left(\tilde r^2\tilde\tau\tilde p\frac{\partial\tilde\psi}{\partial\tilde r}\right)=0.
\end{aligned}
\tag{11.7.47}
$$

Taking into account Eqs. (11.7.39), (11.7.40), (11.7.45), (11.7.46), we reach the same energy equation (11.7.34). Finally we have the following independent conditions for the stretching coefficients

$$\frac{db^2}{c}=1,\ \frac{ea}{b}=1,\ \frac{ca}{be}=1,\ \frac{fa^2}{b^2d}=1, \tag{11.7.48}$$

which can be rewritten as

$$e=\frac{b}{a},\ c=\frac{b^2}{a^2},\ d=\frac{1}{a^2},\ f=\frac{b^2}{a^4}. \tag{11.7.49}$$

It is of interest to find the stretching coefficients for analogical local set of equations.

From the local Poisson equation follows:

$$\frac{c}{b^2}\frac{1}{\tilde r^2}\frac{\partial}{\partial\tilde r}\left(\tilde r^2\frac{\partial\tilde\psi}{\partial\tilde r}\right)=4\pi Gd\tilde\rho. \tag{11.7.50}$$

From the local continuity equation one obtains

$$\frac{d}{a}\frac{\partial\tilde\rho}{\partial\tilde t}+\frac{d}{b}\frac{1}{\tilde r^2}\frac{\partial}{\partial\tilde r}(\tilde r^2e\tilde\rho\tilde v_r)=0. \tag{11.7.51}$$

From the local momentum equation we find

$$\frac{1}{a}\frac{\partial}{\partial\tilde t}(\tilde\rho\tilde v_rde)+\frac{cd}{b}\frac{\partial\tilde\psi}{\partial\tilde r}\tilde\rho+\frac{d}{b}\frac{1}{\tilde r^2}\frac{\partial}{\partial\tilde r}(\tilde r^2\tilde\rho\tilde v_r^2e^2)+\frac{f}{b}\frac{\partial\tilde p}{\partial\tilde r}=0, \tag{11.7.52}$$

and from the local energy equation

$$\frac{1}{a}\frac{\partial}{\partial\tilde t}(\tilde\rho\tilde v_r^2de^2+3\tilde pf)+\frac{1}{\tilde r^2}\frac{1}{b}\frac{\partial}{\partial\tilde r}(\tilde r^2(\tilde\rho\tilde v_r^2de^2+5\tilde pf)\tilde v_re)+2\frac{dce}{b}\tilde\rho\frac{\partial\tilde\psi}{\partial\tilde r}\tilde v_r=0. \tag{11.7.53}$$

In a similar manner we obtain from the local Poisson equation

$$\frac{db^2}{c}=1; \tag{11.7.54}$$

from the local continuity equation

$$\frac{ea}{b}=1; \tag{11.7.55}$$

from the local momentum equation

$$\frac{ca}{be}=1; \tag{11.7.56}$$

$$\frac{fa^2}{b^2d}=1. \tag{11.7.57}$$

It means that for the local set of equations the same stretching coefficients (Eq. (11.7.49)) exist as for the non-local case. As we see we have four relations between five independent coefficients. It is convenient to put $a=b$. In this case

$$e=1, c=1. \tag{11.7.59}$$

The function transforms as $\widetilde{\psi}\to c\widetilde{\psi}$ or

$$\widetilde{\psi}\left(b\widetilde{t},b\widetilde{r}\right)=\widetilde{\psi}\left(\widetilde{t},\widetilde{r}\right). \tag{11.7.60}$$

This equation should be fulfilled for all values $\widetilde{r}$, then place $\widetilde{r}=1/b$, or

$$\widetilde{\psi}\left(\widetilde{t},\widetilde{r}\right)=\widetilde{\psi}\left(b\widetilde{t},b\widetilde{r}\right)=\widetilde{\psi}\left(\frac{\widetilde{t}}{\widetilde{r}},1\right)=\breve{\psi}(\eta), \tag{11.7.61}$$

where

$$\eta=\frac{\widetilde{t}}{\widetilde{r}}. \tag{11.7.62}$$

The dependent variables transform by the following way with the help (11.7.59)

$$\widetilde{\psi}\left(\widetilde{t},\widetilde{r}\right)=\breve{\psi}(\eta), \tag{11.7.63}$$

$$\widetilde{v}_r\to e\widetilde{v}_r$$

and

$$\widetilde{v}_r\left(\widetilde{t},\widetilde{r}\right)=\breve{v}_r(\eta). \tag{11.7.64}$$

The density transformation:

$$\widetilde{\rho}\to d\widetilde{\rho}$$

$$\widetilde{\rho}\left(b\widetilde{t},b\widetilde{r}\right)=d\widetilde{\rho}\left(\widetilde{t},\widetilde{r}\right)=\frac{1}{a^2}\widetilde{\rho}\left(\widetilde{t},\widetilde{r}\right)$$

$$\widetilde{\rho}\left(\widetilde{t},\widetilde{r}\right)=a^2\widetilde{\rho}\left(b\widetilde{t},b\widetilde{r}\right)=\frac{1}{\widetilde{r}^2}\widetilde{\rho}\left(\frac{\widetilde{t}}{\widetilde{r}},1\right)=\frac{1}{\widetilde{r}^2}\breve{\rho}(\eta), \tag{11.7.65}$$

The pressure transformation

$$\widetilde{p}\to f\widetilde{p}, f=d=\frac{1}{a^2}$$

and

$$\widetilde{p}\left(\widetilde{t},\widetilde{r}\right)=\frac{1}{\widetilde{r}^2}\breve{p}(\eta). \tag{11.7.66}$$

For the following transformations we need the explicit expressions for the derivatives:

$$\frac{\partial}{\partial\widetilde{r}}=\frac{\partial\eta}{\partial\widetilde{r}}\frac{\partial}{\partial\eta}=\left(-\frac{\widetilde{t}}{\widetilde{r}^2}\right)\frac{\partial}{\partial\eta}=-\frac{\eta}{\widetilde{r}}\frac{\partial}{\partial\eta}, \tag{11.7.67}$$

$$\frac{\partial}{\partial\widetilde{t}}=\frac{\partial\eta}{\partial\widetilde{t}}\frac{\partial}{\partial\eta}=\frac{1}{\widetilde{r}}\frac{\partial}{\partial\eta}. \tag{11.7.68}$$

We proceed to show now the transformations of equations for the independent variable η; transformation of the Poisson equation in detail:

$$\frac{1}{\tilde{r}^2}\frac{\partial}{\partial \tilde{r}}\left(\tilde{r}^2\frac{\partial \breve{\psi}}{\partial \tilde{r}}\right)=4\pi G\left[\breve{\rho}-\tilde{\tau}\left(\frac{\partial \breve{\rho}}{\partial \tilde{t}}+\frac{1}{\tilde{r}^2}\frac{\partial(\tilde{r}^2\breve{\rho}\breve{v}_r)}{\partial \tilde{r}}\right)\right], \tag{11.7.69}$$

$$-\frac{1}{\tilde{r}^2}\frac{\partial}{\partial \tilde{r}}\left(\tilde{r}^2\frac{\eta}{\tilde{r}}\frac{\partial \breve{\psi}}{\partial \eta}\right)=4\pi G\frac{1}{\tilde{r}^2}\left[\breve{\rho}-\tilde{\tau}\left(\frac{\partial \breve{\rho}}{\partial \tilde{t}}+\frac{\partial(\breve{\rho}\breve{v}_r)}{\partial \tilde{r}}\right)\right], \tag{11.7.70}$$

$$-\tilde{r}\frac{\partial}{\partial \tilde{r}}\left(\eta\frac{\partial \breve{\psi}}{\partial \eta}\right)-\eta\frac{\partial \breve{\psi}}{\partial \eta}=4\pi G\left[\breve{\rho}-\tilde{\tau}\left(\frac{1}{\tilde{r}}\frac{\partial \breve{\rho}}{\partial \eta}-\frac{\eta}{\tilde{r}}\frac{\partial(\breve{\rho}\breve{v}_r)}{\partial \eta}\right)\right], \tag{11.7.71}$$

$$\tilde{r}\frac{\eta}{\tilde{r}}\frac{\partial}{\partial \eta}\left(\eta\frac{\partial \breve{\psi}}{\partial \eta}\right)-\eta\frac{\partial \breve{\psi}}{\partial \eta}=4\pi G\left[\breve{\rho}-\frac{\tilde{\tau}}{\tilde{r}}\left(\frac{\partial \breve{\rho}}{\partial \eta}-\eta\frac{\partial(\breve{\rho}\breve{v}_r)}{\partial \eta}\right)\right], \tag{11.7.72}$$

$$\eta^2\frac{\partial^2 \breve{\psi}}{\partial \eta^2}=4\pi G\left[\breve{\rho}-\frac{\tilde{\tau}}{\tilde{r}}\left(\frac{\partial \breve{\rho}}{\partial \eta}-\eta\frac{\partial(\breve{\rho}\breve{v}_r)}{\partial \eta}\right)\right]. \tag{11.7.73}$$

The coefficient $\tilde{\tau}/\tilde{r}$ appears in front of the round brackets in Eq. (11.7.73). We need to introduce the corresponding η-approximation for this non-local coefficient. We have from Eq. (11.7.62) $\tilde{t}=\eta\tilde{r}$. Physically the idea made sense that

$$\tilde{\tau}=T(\eta)\tilde{r}, \tag{11.7.74}$$

or in the simplest case

$$\tilde{\tau}=T\tilde{r}, \tag{11.7.75}$$

where $T=\text{const}$. We have in this case

$$\eta^2\frac{\partial^2 \breve{\psi}}{\partial \eta^2}=4\pi G\left[\breve{\rho}-T\left(\frac{\partial \breve{\rho}}{\partial \eta}-\eta\frac{\partial(\breve{\rho}\breve{v}_r)}{\partial \eta}\right)\right]. \tag{11.7.76}$$

Analogical transformations can be realized for the other Eqs. (11.7.32)–(11.7.34). Namely,

for the continuity equation

$$\begin{aligned}&\frac{\partial}{\partial \eta}\left\{\breve{\rho}-T\left[\frac{\partial \breve{\rho}}{\partial \eta}-\eta\frac{\partial(\breve{\rho}\breve{v}_r)}{\partial \eta}\right]\right\}-\eta\frac{\partial}{\partial \eta}\left\{\breve{\rho}\breve{v}_r-T\left[\frac{\partial}{\partial \eta}(\breve{\rho}\breve{v}_r)\right.\right.\\&\left.\left.-\eta\frac{\partial(\breve{\rho}\breve{v}_r^2)}{\partial \eta}-\breve{\rho}\eta\frac{\partial \breve{\psi}}{\partial \eta}\right]\right\}-\eta\frac{\partial}{\partial \eta}\left(T\left(2\breve{p}+\eta\frac{\partial \breve{p}}{\partial \eta}\right)\right)=0;\end{aligned} \tag{11.7.77}$$

for the momentum equation

$$\begin{aligned}&\frac{\partial}{\partial \eta}\left\{\breve{\rho}\breve{v}_r-T\left[\frac{\partial}{\partial \eta}(\breve{\rho}\breve{v}_r)-\eta\frac{\partial(\breve{\rho}\breve{v}_r^2)}{\partial \eta}-\eta\frac{\partial \breve{p}}{\partial \eta}-2\breve{p}-\eta\breve{\rho}\frac{\partial \breve{\psi}}{\partial \eta}\right]\right\}\\&-\eta\frac{\partial \breve{\psi}}{\partial \eta}\left[\breve{\rho}-T\left(\frac{\partial \breve{\rho}}{\partial \eta}-\eta\frac{\partial(\breve{\rho}\breve{v}_r)}{\partial \eta}\right)\right]\\&-\eta\frac{\partial}{\partial \eta}\left\{\breve{\rho}\breve{v}_r^2-T\left[\frac{\partial}{\partial \eta}(\breve{\rho}\breve{v}_r^2)-\eta\frac{\partial(\breve{\rho}\breve{v}_r^3)}{\partial \eta}-2\eta\frac{\partial \breve{\psi}}{\partial \eta}\breve{\rho}\breve{v}_r\right]\right\}\\&-\eta\frac{\partial \breve{p}}{\partial \eta}-2\breve{p}+\eta\frac{\partial}{\partial \eta}\left(T\frac{\partial \breve{p}}{\partial \eta}\right)+2T\frac{\partial \breve{p}}{\partial \eta}-2\eta\frac{\partial}{\partial \eta}\left(T\eta\frac{\partial(\breve{p}\breve{v}_r)}{\partial \eta}\right)-4T\eta\frac{\partial(\breve{p}\breve{v}_r)}{\partial \eta}\\&-\eta\frac{\partial}{\partial \eta}\left[T\left(\eta\frac{\partial}{\partial \eta}(\breve{p}\breve{v}_r)+2\breve{p}\breve{v}_r\right)\right]=0;\end{aligned} \tag{11.7.78}$$

for the energy equation

$$\begin{aligned}
&\frac{\partial}{\partial\eta}\left\{\breve{\rho}\breve{v}_r^2+3\breve{p}-T\left[\frac{\partial}{\partial\eta}\left(\breve{\rho}\breve{v}_r^2+3\breve{p}\right)-\eta\frac{\partial}{\partial\eta}\left(\breve{v}_r\left(\breve{\rho}\breve{v}_r^2+5\breve{p}\right)\right)-2\breve{\rho}\eta\frac{\partial\breve{\psi}}{\partial\eta}\breve{v}_r\right]\right\}-\eta\frac{\partial}{\partial\eta}\left\{\left(\breve{\rho}\breve{v}_r^2+5\breve{p}\right)\breve{v}_r\right.\\
&\left.-T\left[\frac{\partial}{\partial\eta}\left(\left(\breve{\rho}\breve{v}_r^2+5\breve{p}\right)\breve{v}_r\right)-\eta\frac{\partial}{\partial\eta}\left(\left(\breve{\rho}\breve{v}_r^2+7\breve{p}\right)\breve{v}_r^2\right)-2\breve{\rho}\eta\frac{\partial\breve{\psi}}{\partial\eta}\breve{v}_r^2-\left(\breve{\rho}\breve{v}_r^2+3\breve{p}\right)\eta\frac{\partial\breve{\psi}}{\partial\eta}\right]\right\}+2\left\{-\breve{\rho}\eta\frac{\partial\breve{\psi}}{\partial\eta}\breve{v}_r\right.\\
&\left.+T\eta\frac{\partial\breve{\psi}}{\partial\eta}\left[\frac{\partial}{\partial\eta}\left(\breve{\rho}\breve{v}_r\right)-\eta\frac{\partial}{\partial\eta}\left(\breve{\rho}\breve{v}_r^2\right)-\eta\frac{\partial\breve{p}}{\partial\eta}-2\breve{p}-\breve{\rho}\eta\frac{\partial\breve{\psi}}{\partial\eta}\right]\right\}\\
&+\eta\frac{\partial}{\partial\eta}\left[T\left(-\eta\frac{\partial}{\partial\eta}\left(\breve{p}\breve{v}_r^2+5\frac{\breve{p}^2}{\breve{\rho}}\right)-2\left(\breve{p}\breve{v}_r^2+5\frac{\breve{p}^2}{\breve{\rho}}\right)\right)\right]-2\eta\frac{\partial}{\partial\eta}\left(T\breve{p}\eta\frac{\partial\breve{\psi}}{\partial\eta}\right)=0,
\end{aligned} \tag{11.7.79}$$

In all cases (in particular in Eq. (11.7.79)) the non-local terms are proportional to $\tilde{\tau}/\tilde{r}$, for which the approximations (11.7.74) and (11.7.75) are applicable. Let us accumulate the results in the form of the set of ODE:

$$\eta^2\frac{\partial^2\breve{\psi}}{\partial\eta^2}=4\pi G\left[\breve{\rho}-T\left(\frac{\partial\breve{\rho}}{\partial\eta}-\eta\frac{\partial\left(\breve{\rho}\breve{v}_r\right)}{\partial\eta}\right)\right], \tag{11.7.80}$$

$$\begin{aligned}
&\frac{\partial}{\partial\eta}\left\{\breve{\rho}-T\left[\frac{\partial\breve{\rho}}{\partial\eta}-\eta\frac{\partial\left(\breve{\rho}\breve{v}_r\right)}{\partial\eta}\right]\right\}-\eta\frac{\partial}{\partial\eta}\left\{\breve{\rho}\breve{v}_r-T\left[\frac{\partial}{\partial\eta}\left(\breve{\rho}\breve{v}_r\right)\right.\right.\\
&\left.\left.-\eta\frac{\partial\left(\breve{\rho}\breve{v}_r^2\right)}{\partial\eta}-\breve{\rho}\eta\frac{\partial\breve{\psi}}{\partial\eta}\right]\right\}-\eta\frac{\partial}{\partial\eta}\left[T\left(2\breve{p}+\eta\frac{\partial\breve{p}}{\partial\eta}\right)\right]=0,
\end{aligned} \tag{11.7.81}$$

$$\begin{aligned}
&\frac{\partial}{\partial\eta}\left\{\breve{\rho}\breve{v}_r-T\left[\frac{\partial}{\partial\eta}\left(\breve{\rho}\breve{v}_r\right)-\eta\frac{\partial\left(\breve{\rho}\breve{v}_r^2\right)}{\partial\eta}-\eta\frac{\partial\breve{p}}{\partial\eta}-2\breve{p}-\eta\breve{\rho}\frac{\partial\breve{\psi}}{\partial\eta}\right]\right\}\\
&-\eta\frac{\partial\breve{\psi}}{\partial\eta}\left[\breve{\rho}-T\left(\frac{\partial\breve{\rho}}{\partial\eta}-\eta\frac{\partial\left(\breve{\rho}\breve{v}_r\right)}{\partial\eta}\right)\right]\\
&-\eta\frac{\partial}{\partial\eta}\left\{\breve{\rho}\breve{v}_r^2-T\left[\frac{\partial}{\partial\eta}\left(\breve{\rho}\breve{v}_r^2\right)-\eta\frac{\partial\left(\breve{\rho}\breve{v}_r^3\right)}{\partial\eta}-2\eta\frac{\partial\breve{\psi}}{\partial\eta}\breve{\rho}\breve{v}_r\right]\right\}\\
&-\eta\frac{\partial\breve{p}}{\partial\eta}-2\breve{p}+\eta\frac{\partial}{\partial\eta}\left(T\frac{\partial\breve{p}}{\partial\eta}\right)+2T\frac{\partial\breve{p}}{\partial\eta}-2\eta\frac{\partial}{\partial\eta}\left(T\eta\frac{\partial\left(\breve{p}\breve{v}_r\right)}{\partial\eta}\right)-4T\eta\frac{\partial\left(\breve{p}\breve{v}_r\right)}{\partial\eta}\\
&-\eta\frac{\partial}{\partial\eta}\left[T\left(\eta\frac{\partial}{\partial\eta}\left(\breve{p}\breve{v}_r\right)+2\breve{p}\breve{v}_r\right)\right]=0,
\end{aligned} \tag{11.7.82}$$

$$\begin{aligned}
&\frac{\partial}{\partial\eta}\left\{\breve{\rho}\breve{v}_r^2+3\breve{p}-T\left[\frac{\partial}{\partial\eta}\left(\breve{\rho}\breve{v}_r^2+3\breve{p}\right)-\eta\frac{\partial}{\partial\eta}\left(\breve{v}_r\left(\breve{\rho}\breve{v}_r^2+5\breve{p}\right)\right)-2\breve{\rho}\eta\frac{\partial\breve{\psi}}{\partial\eta}\breve{v}_r\right]\right\}-\eta\frac{\partial}{\partial\eta}\left\{\left(\breve{\rho}\breve{v}_r^2+5\breve{p}\right)\breve{v}_r\right.\\
&\left.-T\left[\frac{\partial}{\partial\eta}\left(\left(\breve{\rho}\breve{v}_r^2+5\breve{p}\right)\breve{v}_r\right)-\eta\frac{\partial}{\partial\eta}\left(\left(\breve{\rho}\breve{v}_r^2+7\breve{p}\right)\breve{v}_r^2\right)-2\breve{\rho}\eta\frac{\partial\breve{\psi}}{\partial\eta}\breve{v}_r^2-\left(\breve{\rho}\breve{v}_r^2+3\breve{p}\right)\eta\frac{\partial\breve{\psi}}{\partial\eta}\right]\right\}+2\left\{-\breve{\rho}\eta\frac{\partial\breve{\psi}}{\partial\eta}\breve{v}_r\right.\\
&\left.+T\eta\frac{\partial\breve{\psi}}{\partial\eta}\left[\frac{\partial}{\partial\eta}\left(\breve{\rho}\breve{v}_r\right)-\eta\frac{\partial}{\partial\eta}\left(\breve{\rho}\breve{v}_r^2\right)-\eta\frac{\partial\breve{p}}{\partial\eta}-2\breve{p}-\breve{\rho}\eta\frac{\partial\breve{\psi}}{\partial\eta}\right]\right\}\\
&-\eta\frac{\partial}{\partial\eta}\left[T\left(\eta\frac{\partial}{\partial\eta}\left(\breve{p}\breve{v}_r^2+5\frac{\breve{p}^2}{\breve{\rho}}\right)+2\left(\breve{p}\breve{v}_r^2+5\frac{\breve{p}^2}{\breve{\rho}}\right)\right)\right]-2\eta\frac{\partial}{\partial\eta}\left(T\breve{p}\eta\frac{\partial\breve{\psi}}{\partial\eta}\right)=0.
\end{aligned} \tag{11.7.83}$$

$\tilde{t}=\eta\tilde{r}, \tilde{\tau}=T\tilde{r}$. Scales $t_0, p_0, u_0, \tau_0=t_0, \tau=\frac{T}{u_0}r$.

The systems of Eqs. (11.7.80)–(11.7.83) have the great possibilities of mathematical modeling as result of changing the parameter T or $\mathsf{T}(\eta)$, eight Cauchy conditions describing the character features of initial perturbations. As in the previous sections Maple program contains Maple's notations—for example the expression $D(u)(0)=0$ means in the usual notations $(\partial\breve{v}_r/\partial\breve{\eta})(0)=0$, independent variable t responds to η. The following Maple notations on figures are used: u—velocity $\breve{v}_r$, r—density $\breve{\rho}$, p—pressure $\breve{p}$, and v—the self-consistent potential $\breve{\psi}$. Explanations placed under all following figures. The mentioned calculations are displayed on Figs. 11.58–11.81. All Cauchy conditions accumulate in Table 11.2.

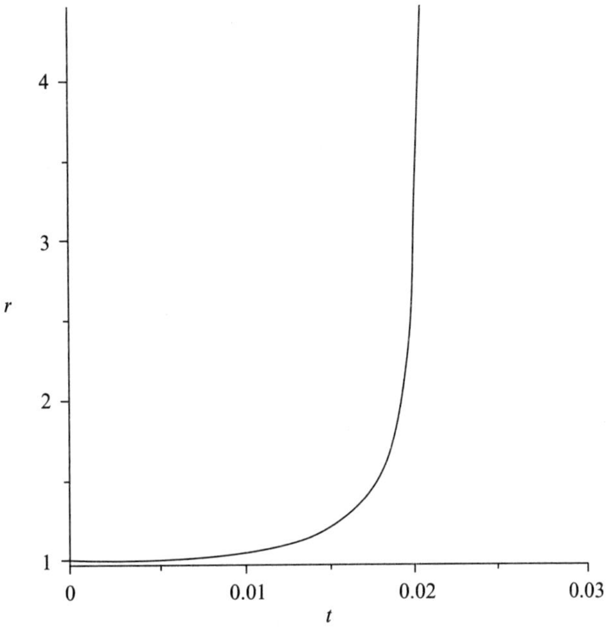

FIGURE 11.58 Dependence $\breve{\rho}(\eta)$, $\eta=\tilde{t}/\tilde{r}$. $G=1$, $T=0$. Cauchy conditions A.

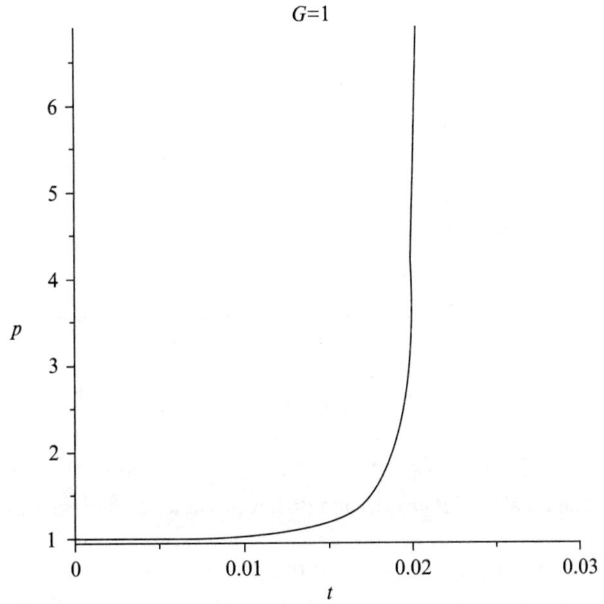

FIGURE 11.59 Dependence $\breve{p}(\eta)$, $\eta=\tilde{t}/\tilde{r}$. $G=1$, $T=0$. Cauchy conditions A.

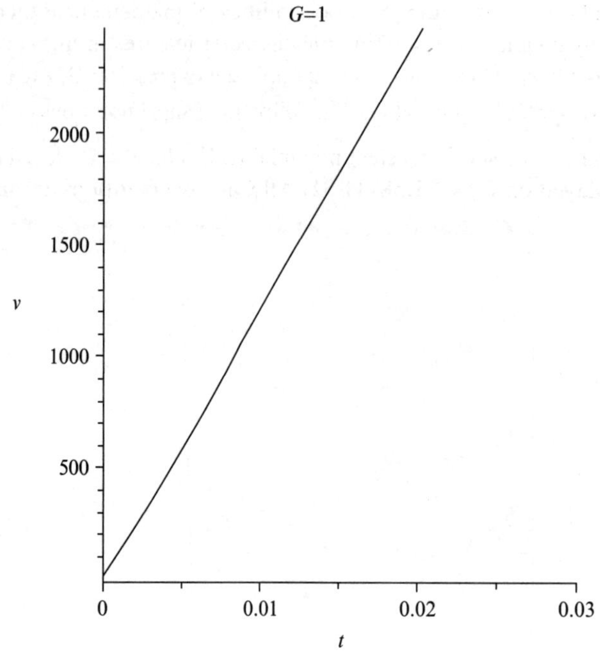

FIGURE 11.60 Dependence $\breve{\psi}(\eta)$, $\eta=\tilde{t}/\tilde{r}$. $G=1$, $T=0$. Cauchy conditions A.

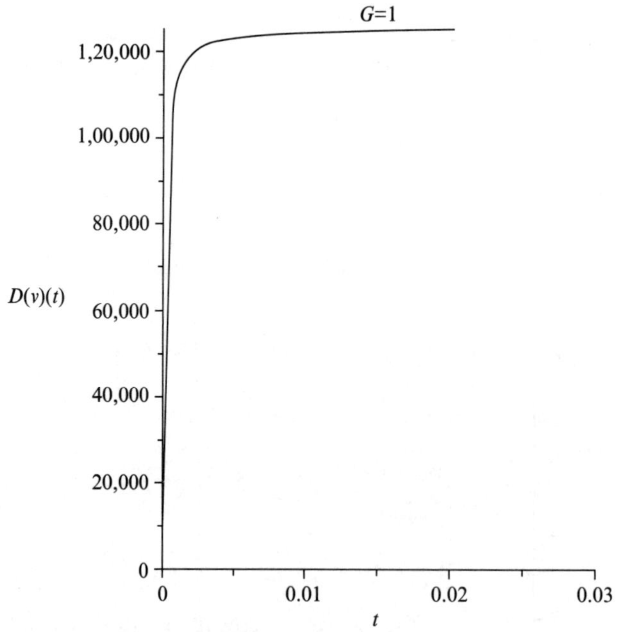

FIGURE 11.61 Dependence $d\breve{\psi}/d\eta$, $\eta=\tilde{t}/\tilde{r}$. $G=1$, $T=0$. Cauchy conditions A.

Figures 11.58–11.65 respond the local model ($T=0$) and different values of dimensionless values G, initial conditions A (see Table 11.2).

From Figs. 11.58 and 11.59 we see that the density and the pressure increase with time in a fixed distance (from the center of the explosion) $\tilde{r}$. Verse versa, the density and the pressure decrease with the distance rising for a given time moment.

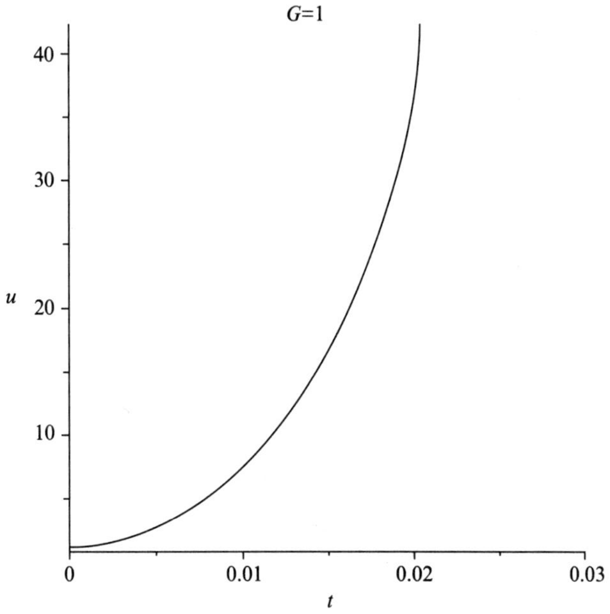

FIGURE 11.62 Dependence $\breve{v}_r(\eta)$, $\eta=\tilde{t}/\tilde{r}$. $G=1$, $T=0$. Cauchy conditions A.

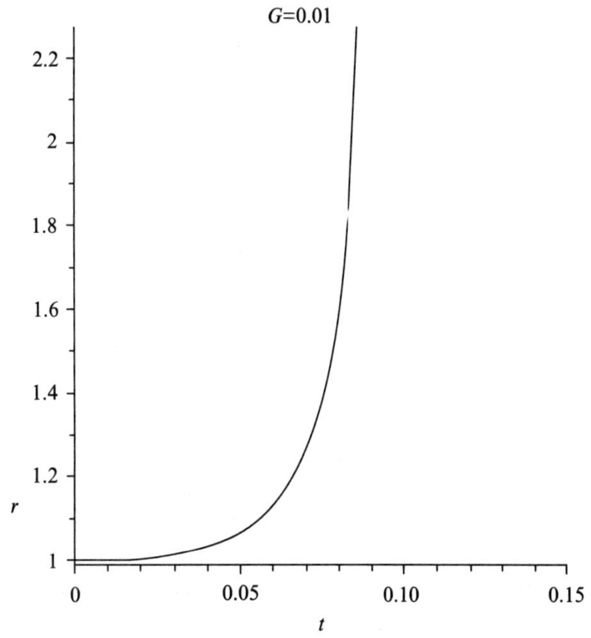

FIGURE 11.63 Dependence $\breve{\rho}(\eta)$, $\eta=\tilde{t}/\tilde{r}$. $G=1$, $T=0$. Cauchy conditions A.

From Figs. 11.60 and 11.61 follow that gravitational potential increases with time at the fixed distance, this dependence becomes linear after $\eta\approx 0,005$. At the time moment the potential decreases with the distance raising.

In this case (Figs. 11.62 and 11.63) the radial velocity of the flow increases (with the approximate dependence $\breve{v}_r \approx \text{const}\cdot\tilde{t}^n$, $n>1$) in the fixed point with time. But for the every fixed time moment the dependence of the flow velocity

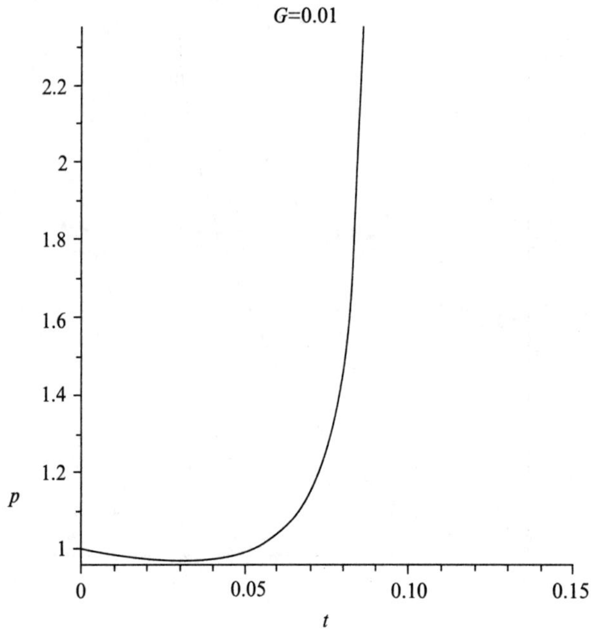

FIGURE 11.64 Dependence $\breve{p}(\eta)$, $\eta=\tilde{t}/\tilde{r}$. $G=0.01$, $T=0$. Cauchy conditions A.

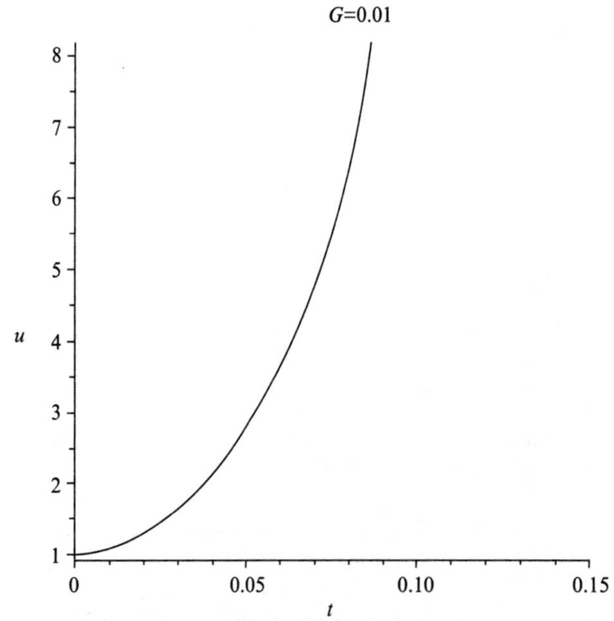

FIGURE 11.65 Dependence $\breve{v}_r(\eta)$, $\eta=\tilde{t}/\tilde{r}$. $G=0.01$, $T=0$. Cauchy conditions A.

with the radial distance can estimated as $\breve{v}_r \approx \text{const}/\tilde{r}^n$, i.e., the velocity decreases with the distance. It means that that the Hubble effect ($\breve{v}_r \approx \text{const}\cdot\tilde{r}$) cannot be discovered in the frame of the local theory of the explosion.

The following Figs. 11.64 and 11.65 reflect the influence of the G decreasing on the calculations result. Decrease of the gravitational constant G leads to the less significant growth of the density, the pressure and the hydrodynamic velocity in

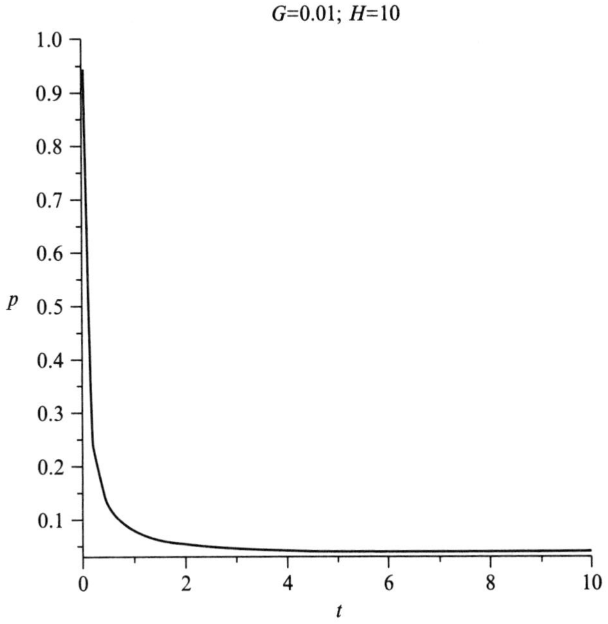

FIGURE 11.66 Dependence $\breve{p}(\eta)$, $\eta=\tilde{t}/\breve{r}$. $G=0.01$, $T=10$. Cauchy conditions A.

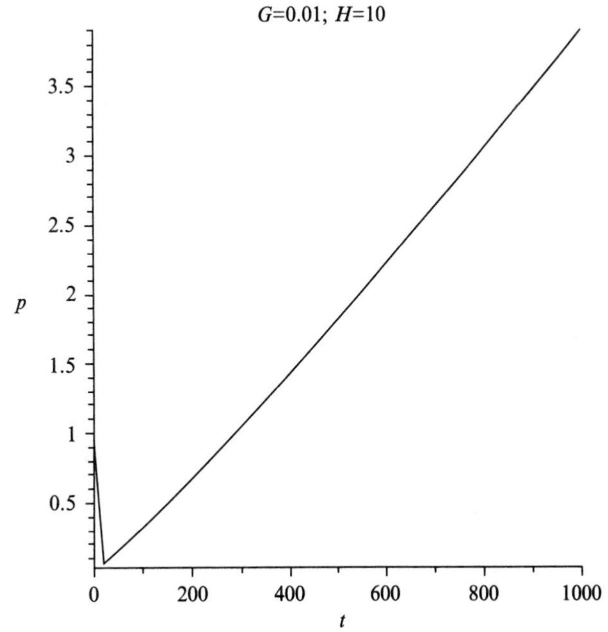

FIGURE 11.67 Dependence $\breve{p}(\eta)$, $\eta=\tilde{t}/\breve{r}$. $G=0.01$, $T=10$. Cauchy conditions A.

time for the every fixed distance $\breve{r}$. It is the awaitng effect—diminishing of the gravitational constant leads in the local theory of the explosion to qualitative the same (but slower) effects.

Let us consider now the results of calculations for the non-local model ($T\neq 0$, Figs. 11.66–11.81). Let be $G=0.01$; $H=10$.

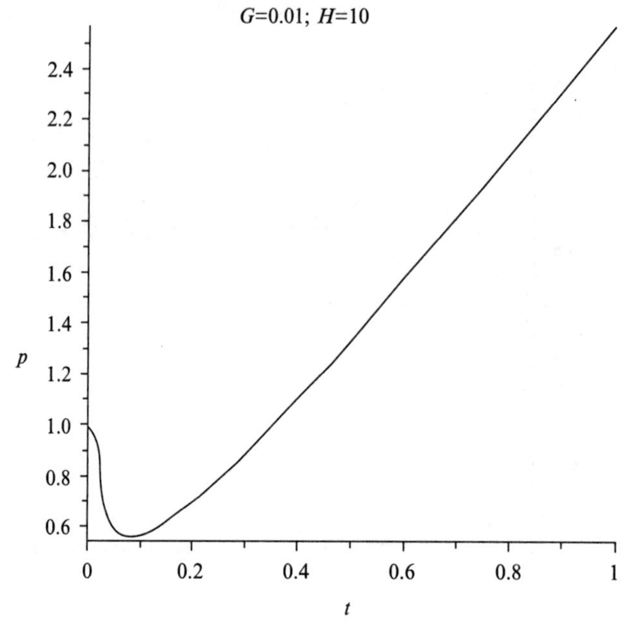

FIGURE 11.68 Dependence $\breve{p}(\eta)$, $\eta=\tilde{t}/\tilde{r}$. $G=0.01$, $T=10$. Cauchy conditions B.

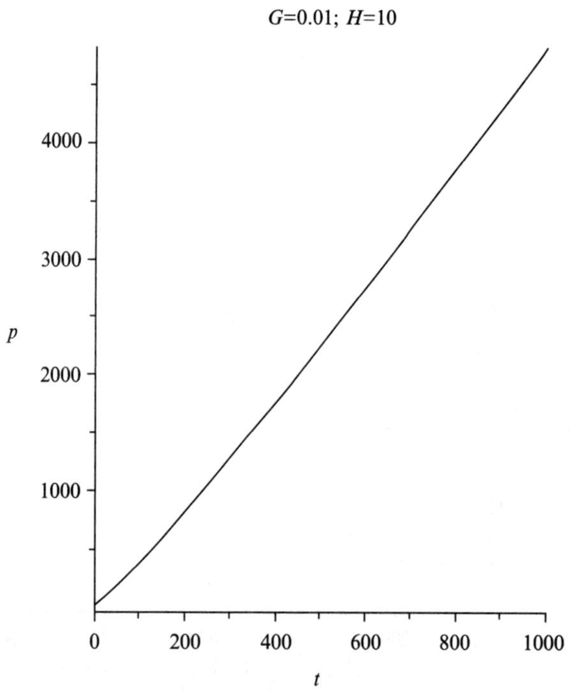

FIGURE 11.69 Dependence $\breve{p}(\eta)$, $\eta=\tilde{t}/\tilde{r}$. $G=0.01$, $T=10$. Cauchy conditions B.

Non-local model attests about the principal other evolution of the physical system—in the every point under consideration the pressure diminishes with time for the beginning and increases with time later. The qualitative behavior of the pressure evolution is the same for the both cases A and B (see Table 11.1). But by going from A to B (increasing of initial derivatives and therefore the explosion energy) the pressure increases to thousand-fold (Figs. 11.67 and 11.69).

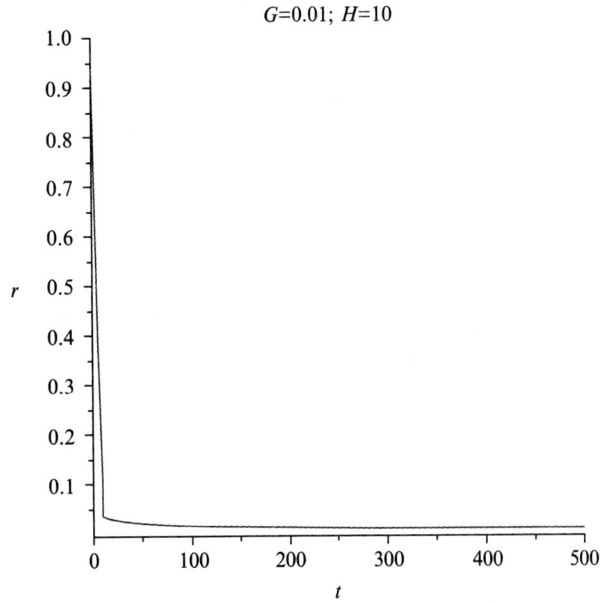

FIGURE 11.70 Dependence $\breve{\rho}(\eta)$, $\eta = \tilde{t}/\tilde{r}$. $G = 0.01$, $T = 10$. Cauchy conditions A.

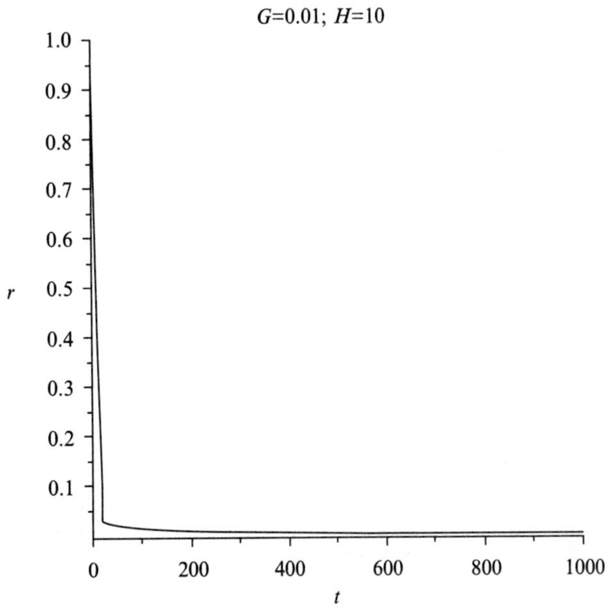

FIGURE 11.71 Dependence $\breve{\rho}(\eta)$, $\eta = \tilde{t}/\tilde{r}$. $G = 0.01$, $T = 10$. Cauchy conditions B.

As one passes from the case A to B, the character run of the density curves (Fig. 11.70 and Fig.11.71) are the same. The density and the gravitational potential decrease with the $\tilde{t}$ increasing if $\tilde{r} = \text{const}$; correspondingly, the density and the gravitational potential increase with the $\tilde{r}$ increasing if $\tilde{t} = \text{const}$. On the general ground of the mentioned behavior of pressure and potential, we observe the hydrodynamic flow directed to the center of the explosion

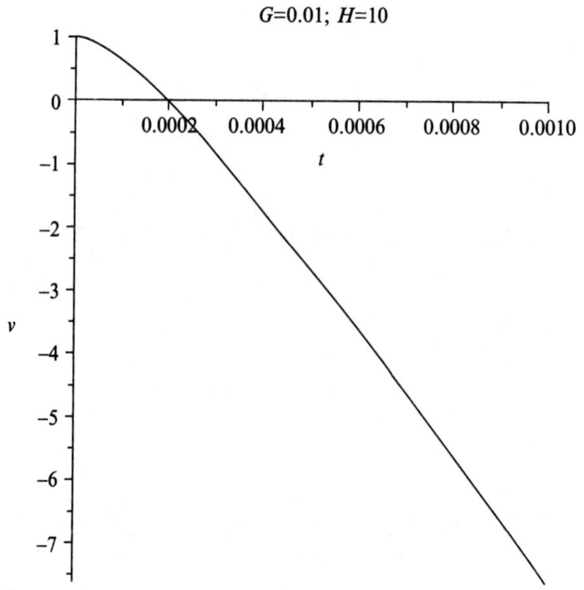

FIGURE 11.72 Dependence $\breve{\psi}(\eta)$, $\eta=\tilde{t}/\tilde{r}$. $G=0.01$, $T=10$. Cauchy conditions A.

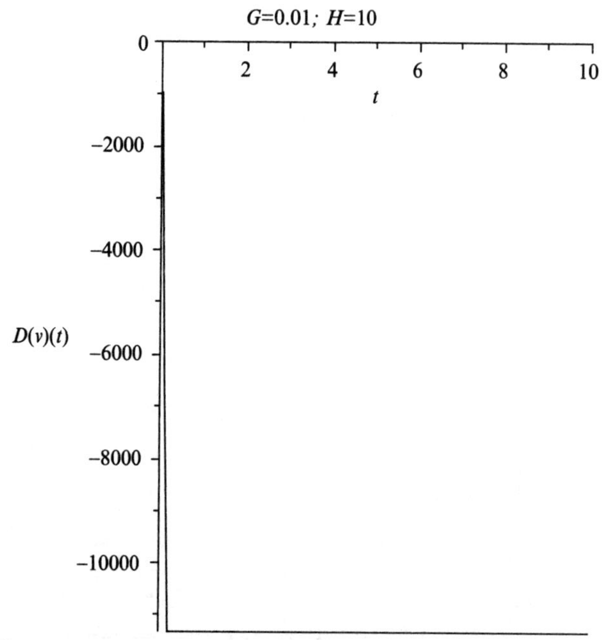

FIGURE 11.73 Dependence $\frac{d\breve{\psi}}{d\eta}$, $\eta=\tilde{t}/\tilde{r}$. $G=0.01$, $T=10$. Cauchy conditions A.

(Figs. 11.74 and 11.75), $\breve{v}_r<0$. But on the same figures you see the area where $\breve{v}_r>0$. For example, Fig. 11.74 demonstrates the solution with $\breve{v}_r>0$ if $\eta<0.01$. Moreover, $\breve{v}_r$ grows with the $\tilde{r}$ increasing for the fixed time moment. It means that application of the non-local physics leads to the Hubble regime. The Hubble regime can span the tremendous area of the $\tilde{r}$-distance because $\tilde{r}\to\infty$ if $\eta\to 0$.

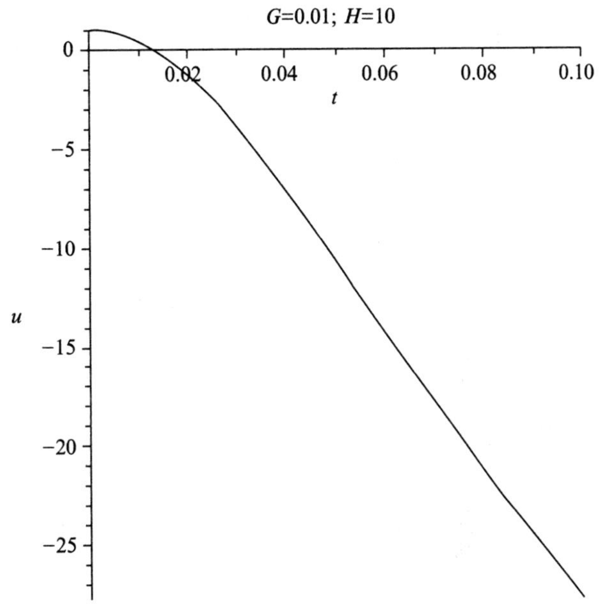

FIGURE 11.74 Dependence $\breve{v}_r(\eta)$, $\eta=\tilde{t}/\tilde{r}$. $G=0.01$, $T=10$. Cauchy conditions A.

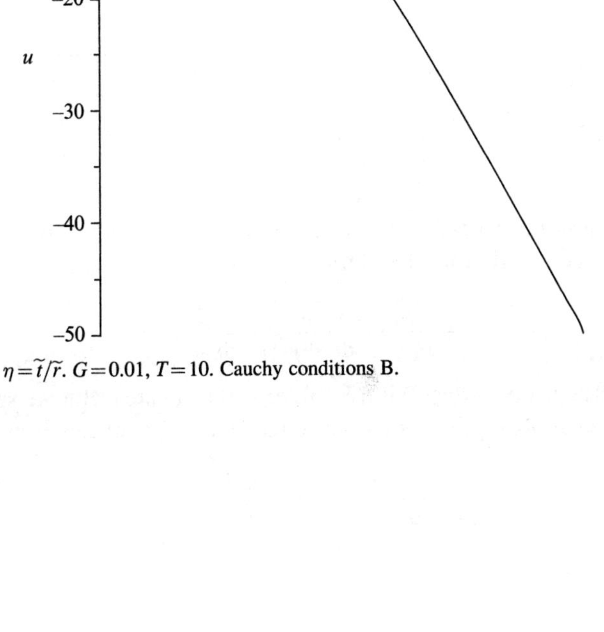

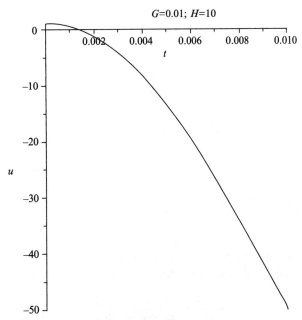

FIGURE 11.75 Dependence $\breve{v}_r(\eta)$, $\eta=\tilde{t}/\tilde{r}$. $G=0.01$, $T=10$. Cauchy conditions B.

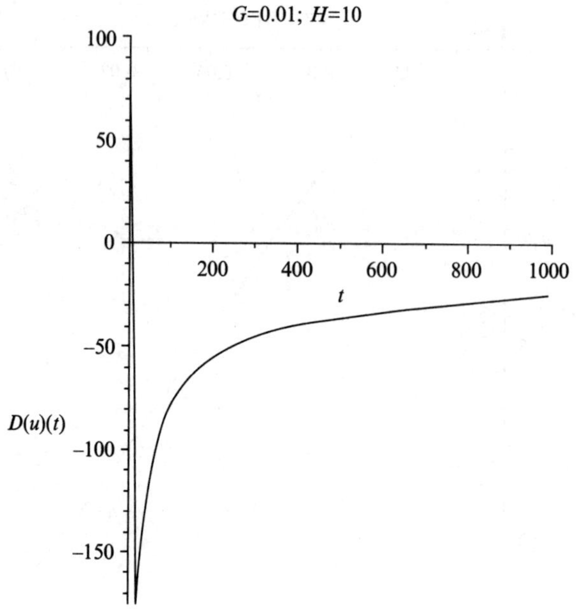

FIGURE 11.76 Dependence $\frac{d\breve{v}_r(\eta)}{d\eta}$, $\eta=\tilde{t}/\tilde{r}$. $G=0.01$, $T=10$. Cauchy conditions B.

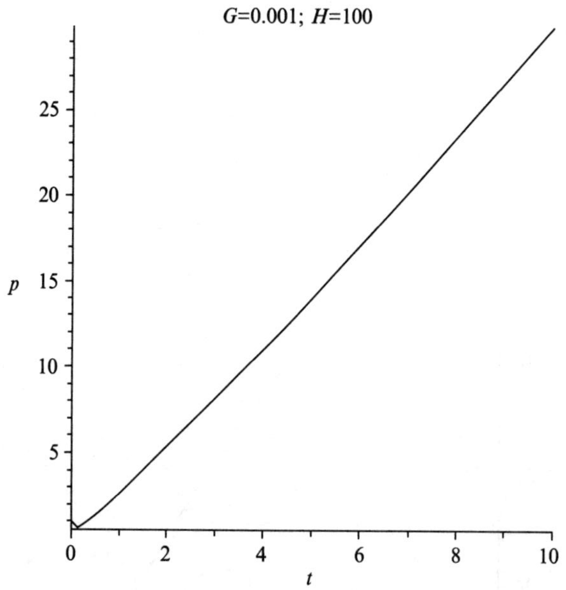

FIGURE 11.77 Dependence $\breve{p}(\eta)$, $\eta=\tilde{t}/\tilde{r}$. $G=0.001$, $T=10$. Cauchy conditions B.

Following Figs. 11.76–11.82 show the dependent values evolution for some other combination of the G, T parameters. Let us consider the derivative $d\breve{v}_r(\eta)/d\eta$ for a fixed time moment,

$$\frac{d\breve{v}_r(\eta)}{d\eta} \sim \frac{d\breve{v}_r}{d\tilde{r}}\frac{d\tilde{r}}{d\eta} \sim -\frac{d\breve{v}_r}{d\tilde{r}}\tilde{r}^2. \tag{11.7.84}$$

It is obvious from relation (11.7.84) that $d\breve{v}_r/d\tilde{r} > 0$ if $d\breve{v}_r(\eta)/d\eta < 0$ in an area. But we see this area on Figs. 11.74 and 11.76. It is the typical Hubble regime. Really, let us introduce the Hubble parameter H as $\breve{v}_r = H\tilde{r}$. In this case

$$\frac{d\breve{v}_r(\eta)}{d\eta} \sim -H\tilde{r}^2 - \frac{dH}{d\tilde{r}}\tilde{r}^3. \tag{11.7.85}$$

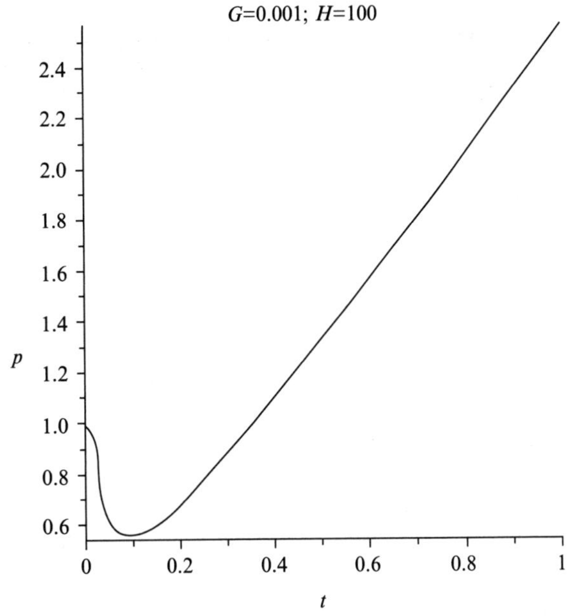

FIGURE 11.78 Dependence $\breve{p}(\eta)$, $\eta=\tilde{t}/\tilde{r}$. $G=0.001$, $T=10$. Cauchy conditions B.

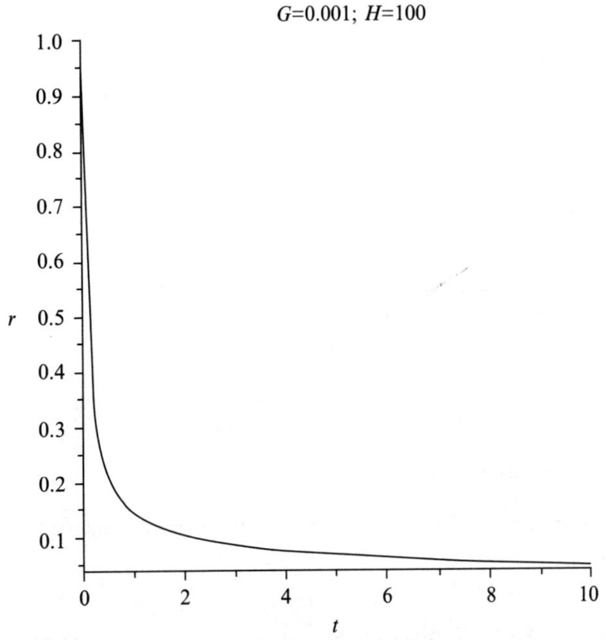

FIGURE 11.79 Dependence $\breve{\rho}(\eta)$, $\eta=\tilde{t}/\tilde{r}$. $G=0.001$, $T=10$. Cauchy conditions B.

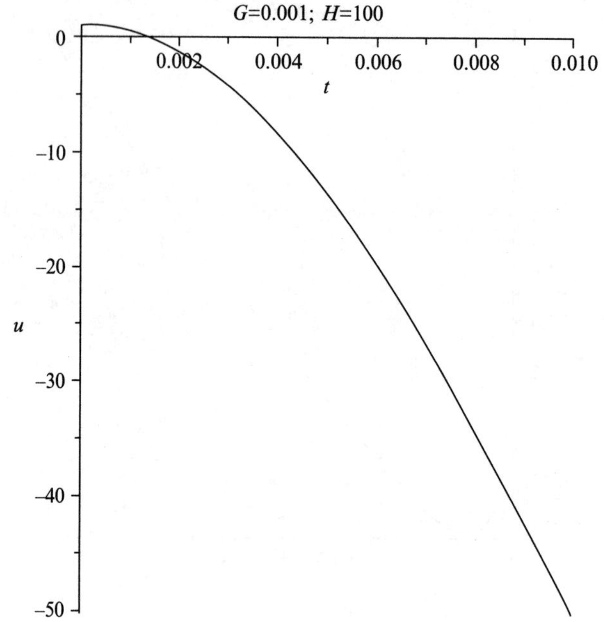

FIGURE 11.80 Dependence $\breve{v}_r(\eta)$, $\eta=\tilde{t}/\tilde{r}$. $G=0.001$, $T=100$. Cauchy conditions B.

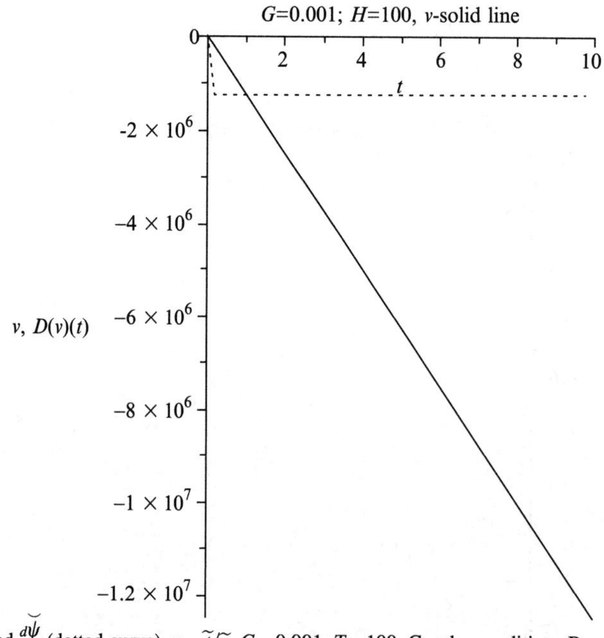

FIGURE 11.81 Dependence $\breve{\psi}(\eta)$ and $\frac{d\breve{\psi}}{d\eta}$ (dotted curve), $\eta=\tilde{t}/\tilde{r}$. $G=0.001$, $T=100$. Cauchy conditions B.

TABLE 11.2 Cauchy Conditions

	$\breve{v}_r(0)$	$\breve{\rho}(0)$	$\breve{p}(0)$	$\breve{\psi}(0)$	$\frac{d\breve{v}_r}{d\eta}(0)$	$\frac{d\breve{\rho}}{d\eta}(0)$	$\frac{d\breve{p}}{d\eta}(0)$	$\frac{d\breve{\psi}}{d\eta}(0)$
A	1	1	1	1	1	1	1	1
B	1	1	1	1	100	100	100	100

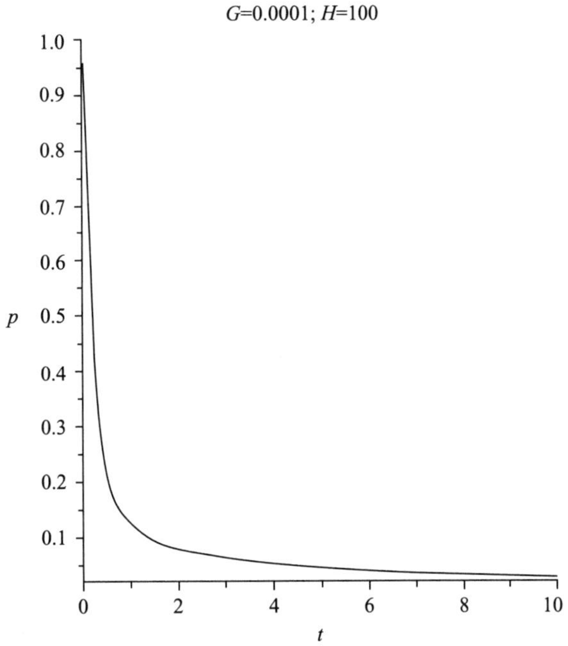

FIGURE 11.82 Dependence $\breve{p}(\eta)$, $\eta = \breve{t}/\breve{r}$. $G = 0.0001$, $T = 100$. Cauchy conditions A.

From relation (11.7.85), Figs. 11.74–11.76 and observations follow (observations of S. Perlmutter, A. Riess (USA) and B. Schmidt (Australia)) that we are living in the η-area (time-space area) of the Universe expansion with acceleration.

Let be now $G = 0001$, $\mathrm{H} = 100$, see Figs. 11.77–11.81.

The final curve (Fig. 11.82) responds to the small G, namely $G = 0.0001$, $H = 100$. All the last Figs. 11.77–11.82 demonstrate the same non-local character features as previous ones.

Two main conclusions should be done:

1. Hubble regimes cannot be described in the frame of local statistic theory of dissipative processes.
2. Unified non-local theory of transport processes adequately reflects the Hubble expansion, including the expansion with the Universe acceleration.

Now we can revisit two fundamental problems:

1. The variable star phenomenon using the generalized Bolzmann kinetics theory. The main goals of investigations consist in answer two questions:
   A) Is it possible to construct the self-consistent non-local theory of explosion after the initial perturbation with taking into account the possible turbulent fluctuations?
   B) Is it possible to take into account the effect of the gravitational self-catching leading to periodic star pulsations?
   Both questions have the positive answers (Appendix 9).
2. The levitation phenomenon using the generalized Bolzmann kinetics theory which can represent the non-local physics of this levitation phenomenon. This approach can identify the conditions when the levitation can take place under the influence of correlated electromagnetic and gravitational fields. The sufficient mathematical conditions of levitation are obtained. It means that the regime of levitation could be realized from the position of the non-local hydrodynamics (Appendix 10).

Chapter 12

The Generalized Relativistic Kinetic Hydrodynamic Theory

ABSTRACT

On the basement of statistical description, the movement of relativistic particles without external forces is considered. The hydrodynamic form of the Dirac equation is obtained. The Lorentz invariant form of the non-local Alexeev's equation (generalized Boltzmann equation) is delivered. The generalized relativistic hydrodynamic equations are derived. These results lead to construction of the relativistic quantum hydrodynamics.

Keywords: Generalized relativistic kinetic equations, Hydrodynamic form of the Dirac equation, Generalized relativistic hydrodynamic equations, Sound propagation in a relativistic gas

12.1 HYDRODYNAMIC FORM OF THE DIRAC QUANTUM RELATIVISTIC EQUATION

In this section the hydrodynamic form of the Dirac quantum equation is obtained and we demonstrate the transmission from this hydrodynamics to the Madelung's ones.

Let us write down the Dirac equation describing the evolution of the particle in the absence of the external forces including the forces of the electromagnetic origin:

$$\left[\left(\frac{i\hbar}{c}\frac{\partial}{\partial t}\right)^2 - (i\hbar\nabla)^2 - m_0^2c^2\right]\psi = 0, \tag{12.1.1}$$

or

$$\frac{1}{\psi}\frac{\partial^2\psi}{\partial t^2} - c^2\frac{1}{\psi}\Delta\psi + \frac{m_0^2c^4}{\hbar^2} = 0. \tag{12.1.2}$$

The usual nomenclature is used: ψ is wave function, which can be written for this case in the form

$$\psi = \alpha e^{i\beta}, \tag{12.1.3}$$

m_0 is the mass of rest of the particle under consideration, c—light velocity in vacuum, Δ is the three dimensional Laplace operator

$$\Delta = \sum_{i=1}^{3}\frac{\partial^2}{\partial x_i^2} = \frac{\partial}{\partial \mathbf{r}}\cdot\frac{\partial}{\partial \mathbf{r}}. \tag{12.1.4}$$

After differentiating Eq. (12.1.3) in time and space we find

$$\frac{1}{\psi}\frac{\partial^2\psi}{\partial t^2} = -\left(\frac{\partial\beta}{\partial t}\right)^2 + \frac{1}{\alpha}\frac{\partial^2\alpha}{\partial t^2} + i\frac{2}{\alpha}\frac{\partial\alpha}{\partial t}\frac{\partial\beta}{\partial t} + i\frac{\partial^2\beta}{\partial t^2}. \tag{12.1.5}$$

and space

$$\frac{\Delta\psi}{\psi} = 2i\frac{1}{\alpha}\frac{\partial\alpha}{\partial \mathbf{r}}\cdot\frac{\partial\beta}{\partial \mathbf{r}} + \frac{1}{\alpha}\Delta\alpha - \left(\frac{\partial\beta}{\partial \mathbf{r}}\right)^2 + i\Delta\beta, \tag{12.1.6}$$

Unified non-local theory of transport processes. http://dx.doi.org/10.1016/B978-0-444-63478-8.00012-9

we find after substitution in Eq. (12.1.2)

$$-\left(\frac{\partial\beta}{\partial t}\right)^2+\frac{1}{\alpha}\frac{\partial^2\alpha}{\partial t^2}+i\frac{2}{\alpha}\frac{\partial\alpha}{\partial t}\frac{\partial\beta}{\partial t}+i\frac{\partial^2\beta}{\partial t^2}$$
$$-c^2\left[\frac{\Delta\alpha}{\alpha}-\left(\frac{\partial\beta}{\partial\mathbf{r}}\right)^2+i\left(\Delta\beta+\frac{2}{\alpha}\frac{\partial\alpha}{\partial\mathbf{r}}\cdot\frac{\partial\beta}{\partial\mathbf{r}}\right)\right]+\frac{m_0^2c^4}{\hbar^2}=0. \tag{12.1.7}$$

Separate the real and imagine parts in Eq. (12.1.7). Real part:

$$-\left(\frac{\partial\beta}{\partial t}\right)^2+\frac{1}{\alpha}\frac{\partial^2\alpha}{\partial t^2}-c^2\left[\frac{\Delta\alpha}{\alpha}-\left(\frac{\partial\beta}{\partial\mathbf{r}}\right)^2\right]+\frac{m_0^2c^4}{\hbar^2}=0 \tag{12.1.8}$$

and the imagine part:

$$\frac{2}{\alpha}\frac{\partial\alpha}{\partial t}\frac{\partial\beta}{\partial t}+\frac{\partial^2\beta}{\partial t^2}-c^2\left[\Delta\beta+\frac{2}{\alpha}\left(\frac{\partial\alpha}{\partial\mathbf{r}}\right)\cdot\left(\frac{\partial\beta}{\partial\mathbf{r}}\right)\right]=0. \tag{12.1.9}$$

or after multiplication of Eq. (12.1.9) by α^2:

$$2\alpha\frac{\partial\alpha}{\partial t}\frac{\partial\beta}{\partial t}+\alpha^2\frac{\partial^2\beta}{\partial t^2}-c^2\left[\alpha^2\Delta\beta+2\alpha\left(\frac{\partial\alpha}{\partial\mathbf{r}}\right)\cdot\left(\frac{\partial\beta}{\partial\mathbf{r}}\right)\right]=0. \tag{12.1.10}$$

The following identity can be proved by the direct differentiation:

$$\frac{\partial}{\partial\mathbf{r}}\cdot\left(\alpha^2\frac{\partial\beta}{\partial\mathbf{r}}\right)=\alpha^2\Delta\beta+2\alpha\frac{\partial\alpha}{\partial\mathbf{r}}\cdot\frac{\partial\beta}{\partial\mathbf{r}} \tag{12.1.11}$$

and for Eq. (12.1.10) one obtains

$$\frac{\partial\alpha^2}{\partial t}\frac{\partial\beta}{\partial t}+\alpha^2\frac{\partial^2\beta}{\partial t^2}-c^2\frac{\partial}{\partial\mathbf{r}}\cdot\left(\alpha^2\frac{\partial\beta}{\partial\mathbf{r}}\right)=0. \tag{12.1.12}$$

Let us introduce the notation:

$$\rho=\alpha^2. \tag{12.1.13}$$

From Eqs. (12.1.12) and (12.1.13) follows

$$\frac{\partial}{\partial t}\left[\rho\frac{\partial\beta}{\partial t}\right]-c^2\frac{\partial}{\partial\mathbf{r}}\cdot\left[\rho\frac{\partial\beta}{\partial\mathbf{r}}\right]=0, \tag{12.1.14}$$

or in the coordinate form

$$\frac{1}{c^2}\frac{\partial}{\partial t}\left[\rho\frac{\partial\beta}{\partial t}\right]-\sum_{i=1}^{3}\frac{\partial}{\partial x_i}\left[\rho\frac{\partial\beta}{\partial x_i}\right]=0. \tag{12.1.15}$$

Introduce the 3-velocity

$$\mathbf{v}=\frac{\partial}{\partial\mathbf{r}}\left(\frac{\hbar}{m_0}\beta\right). \tag{12.1.16}$$

Then

$$-\frac{1}{c}\frac{\partial}{\partial t}\left[\rho\frac{1}{c}\frac{\hbar}{m_0}\frac{\partial\beta}{\partial t}\right]+\frac{\partial}{\partial\mathbf{r}}\cdot[\rho\mathbf{v}]=0. \tag{12.1.17}$$

Define the fourth velocity component in Minkowski space as

$$v_t=-\frac{1}{c}\frac{\hbar}{m_0}\frac{\partial\beta}{\partial t}. \tag{12.1.18}$$

and from Eqs. (12.1.17) and (12.1.18) follow

$$\frac{1}{c}\frac{\partial}{\partial t}[\rho v_t]+\frac{\partial}{\partial\mathbf{r}}\cdot[\rho\mathbf{v}]=0. \tag{12.1.19}$$

Consider now the real part of the relativistic Dirac equation (12.1.8) written in the form

$$-\frac{1}{c^2}\frac{\hbar^2}{m_0^2}\left(\frac{\partial\beta}{\partial t}\right)^2+\frac{1}{c^2}\frac{\hbar^2}{m_0^2}\frac{1}{\alpha}\frac{\partial^2\alpha}{\partial t^2}-\frac{\hbar^2}{m_0^2}\left[\frac{\Delta\alpha}{\alpha}-\left(\frac{\partial\beta}{\partial\mathbf{r}}\right)^2\right]+c^2=0 \tag{12.1.20}$$

or after introduction of the definition (12.1.18)

$$-v_t^2+\frac{1}{c^2}\frac{\hbar^2}{m_0^2}\frac{1}{\alpha}\frac{\partial^2\alpha}{\partial t^2}-\frac{\hbar^2}{m_0^2}\left[\frac{\Delta\alpha}{\alpha}-\left(\frac{\partial\beta}{\partial\mathbf{r}}\right)^2\right]+c^2=0. \tag{12.1.21}$$

Using the 3-velocity (Eq. (12.1.16)) one obtains from Eq. (12.1.21):

$$-v_t^2+\frac{1}{c^2}\frac{\hbar^2}{m_0^2}\frac{1}{\alpha}\frac{\partial^2\alpha}{\partial t^2}-\frac{\hbar^2}{m_0^2}\frac{\Delta\alpha}{\alpha}+v^2+c^2=0. \tag{12.1.22}$$

Apply the gradient operator to the both sides of Eq. (12.1.22):

$$-\frac{\partial}{\partial\mathbf{r}}v_t^2+\frac{1}{c^2}\frac{\hbar^2}{m_0^2}\frac{\partial}{\partial\mathbf{r}}\left[\frac{1}{\alpha}\frac{\partial^2\alpha}{\partial t^2}\right]-\frac{\hbar^2}{m_0^2}\frac{\partial}{\partial\mathbf{r}}\frac{\Delta\alpha}{\alpha}+\frac{\partial}{\partial\mathbf{r}}v^2=0 \tag{12.1.23}$$

and use the known identity of the vector analyses:

$$2\left(\mathbf{v}\cdot\frac{\partial}{\partial\mathbf{r}}\right)\mathbf{v}=\frac{\partial}{\partial\mathbf{r}}v^2-2\mathbf{v}\times\mathbf{rot}\,\mathbf{v}. \tag{12.1.24}$$

Taking into account that

$$\mathbf{rot}\,\mathbf{v}=\frac{\hbar}{m_0}\mathbf{rot}\,\mathbf{grad}\,\beta=0, \tag{12.1.25}$$

we find

$$2\left(\mathbf{v}\cdot\frac{\partial}{\partial\mathbf{r}}\right)\mathbf{v}=\frac{\partial}{\partial\mathbf{r}}v^2. \tag{12.1.26}$$

Now from Eq. (12.1.23) with the help of Eqs. (12.1.13), (12.1.24)–(12.1.26) follows

$$-\frac{\partial}{\partial\mathbf{r}}v_t^2+2\left(\mathbf{v}\cdot\frac{\partial}{\partial\mathbf{r}}\right)\mathbf{v}=-\frac{\hbar^2}{m_0^2}\frac{\partial}{\partial\mathbf{r}}\left[\frac{1}{c^2\sqrt{\rho}}\frac{\partial^2\sqrt{\rho}}{\partial t^2}-\frac{\Delta\sqrt{\rho}}{\sqrt{\rho}}\right]. \tag{12.1.27}$$

Transform the first left hand side term in Eq. (12.1.27)

$$-\frac{\partial}{\partial\mathbf{r}}v_t^2=-\frac{1}{c^2}\left[\frac{\hbar}{m_0}\right]^2\frac{\partial}{\partial\mathbf{r}}\left(\frac{\partial\beta}{\partial t}\right)^2=-2\frac{1}{c^2}\left[\frac{\hbar}{m_0}\right]^2\frac{\partial\beta}{\partial t}\frac{\partial}{\partial t}\left(\frac{\partial\beta}{\partial\mathbf{r}}\right)=2\frac{v_t}{c}\frac{\partial}{\partial t}\mathbf{v}. \tag{12.1.28}$$

and Eq. (12.1.27) takes the form

$$2\frac{v_t}{c}\frac{\partial}{\partial t}\mathbf{v}+2\left(\mathbf{v}\cdot\frac{\partial}{\partial\mathbf{r}}\right)\mathbf{v}=-\frac{\hbar^2}{m_0^2}\frac{\partial}{\partial\mathbf{r}}\left[\frac{1}{c^2\sqrt{\rho}}\frac{\partial^2\sqrt{\rho}}{\partial t^2}-\frac{\Delta\sqrt{\rho}}{\sqrt{\rho}}\right]. \tag{12.1.29}$$

Let us introduce the additional condition

$$v_t=c. \tag{12.1.30}$$

As result, we obtain the relativistic system of equations of the hydrodynamic type from Eqs. (12.1.19) and (12.1.29), namely continuity equation

$$\frac{\partial\rho}{\partial t}+\frac{\partial}{\partial\mathbf{r}}\cdot(\rho\mathbf{v})=0. \tag{12.1.31}$$

and the motion equation

$$\frac{\partial}{\partial t}\mathbf{v}+\left(\mathbf{v}\cdot\frac{\partial}{\partial\mathbf{r}}\right)\mathbf{v}=-\frac{\hbar^2}{2m_0^2}\frac{\partial}{\partial\mathbf{r}}\left[\frac{1}{c^2\sqrt{\rho}}\frac{\partial^2\sqrt{\rho}}{\partial t^2}-\frac{\Delta\sqrt{\rho}}{\sqrt{\rho}}\right]. \tag{12.1.32}$$

Motion equation (12.1.32) can be written in the form

$$\frac{\partial}{\partial t}\mathbf{v}+\left(\mathbf{v}\cdot\frac{\partial}{\partial \mathbf{r}}\right)\mathbf{v}=-\frac{\partial}{\partial \mathbf{r}}U^*, \tag{12.1.33}$$

containing the generalization of the Bohm potential U^*

$$U^*=\frac{\hbar^2}{2m_0^2}\left[\frac{1}{c^2\sqrt{\rho}}\frac{\partial^2\sqrt{\rho}}{\partial t^2}-\frac{\Delta\sqrt{\rho}}{\sqrt{\rho}}\right], \tag{12.1.34}$$

for relativistic case. Significant to notice that relativistic analog of the Bohm potential has non-stationary character. For the non-relativistic case the first tern in the right side of Eq. (12.1.34) can be omitted and we have the usual form of the Bohm potential

$$U^*=-\frac{\hbar^2}{2m^2}\frac{\Delta\sqrt{\rho}}{\sqrt{\rho}}. \tag{12.1.35}$$

Introduce the variable

$$x_0=ct, \tag{12.1.36}$$

which often uses in relativistic theory and operator

$$\Delta_4=\frac{\partial^2}{\partial x_0^2}-\Delta \tag{12.1.37}$$

The relativistic potential U^* takes the form:

$$U^*=\frac{\hbar^2}{2m_0{}^2}\frac{\Delta_4\sqrt{\rho}}{\sqrt{\rho}}. \tag{12.1.38}$$

In the explicit form

$$U^*=-\frac{\hbar^2}{8m_0^2\rho^2}\left[\frac{\partial\rho}{\partial x_0}\right]^2+\frac{\hbar^2}{4m_0^2\rho}\frac{\partial^2\rho}{\partial x_0^2}+\frac{\hbar^2}{8m_0^2\rho^2}\sum_{i=1}^{3}\left[\frac{\partial\rho}{\partial x_i}\right]^2-\frac{\hbar^2}{4m_0^2\rho}\sum_{i=1}^{3}\frac{\partial^2\rho}{\partial x_i^2}. \tag{12.1.39}$$

In particular, for the one-dimensional form we have from Eqs. (12.1.32) and (12.1.39):

$$\frac{\partial v_x}{\partial t}+v_x\frac{\partial v_x}{\partial x}=-\frac{\hbar^2}{4m_0{}^2}\frac{\partial}{\partial x}\left[\frac{1}{\rho}\left(\frac{1}{c^2}\frac{\partial^2\rho}{\partial t^2}-\frac{\partial^2\rho}{\partial x^2}\right)-\frac{1}{2\rho^2}\left[\frac{1}{c^2}\left(\frac{\partial\rho}{\partial t}\right)^2-\left(\frac{\partial\rho}{\partial x}\right)^2\right]\right]. \tag{12.1.40}$$

Introduce the definition of the quantum kinematical viscosity

$$\frac{\hbar}{m_0}=\nu_{\text{qu}}. \tag{12.1.41}$$

As result from Eqs. (12.1.40) and (12.1.41)

$$\frac{\partial v_x}{\partial t}+v_x\frac{\partial v_x}{\partial x}=-\frac{\nu_{\text{qu}}^2}{4}\frac{\partial}{\partial x}\left[\frac{1}{\rho}\left(\frac{1}{c^2}\frac{\partial^2\rho}{\partial t^2}-\frac{\partial^2\rho}{\partial x^2}\right)-\frac{1}{2\rho^2}\left[\frac{1}{c^2}\left(\frac{\partial\rho}{\partial t}\right)^2-\left(\frac{\partial\rho}{\partial x}\right)^2\right]\right]. \tag{12.1.42}$$

For non-relativistic case from Eq. (12.1.42) follows

$$\frac{\partial v_x}{\partial t}+v_x\frac{\partial v_x}{\partial x}=\frac{\hbar^2}{4m_0{}^2}\frac{\partial}{\partial x}\left[\frac{1}{\rho}\left(\frac{\partial^2\rho}{\partial x^2}-\frac{1}{2\rho}\left(\frac{\partial\rho}{\partial x}\right)^2\right)\right]. \tag{12.1.43}$$

Write down the expression

$$\frac{\partial}{\partial t}(\rho v_x)+\frac{\partial}{\partial x}(\rho v_x^2)=\rho\left(\frac{\partial v_x}{\partial t}+v_x\frac{\partial v_x}{\partial x}\right) \tag{12.1.44}$$

using Eq. (12.1.31) written for the 1D case

$$\frac{\partial\rho}{\partial t}+\frac{\partial\rho v_x}{\partial x}=0. \tag{12.1.45}$$

From Eq. (12.1.43) with taking into account Eq. (12.1.44) follows (by $v_x = u$):

$$\frac{\partial}{\partial t}(\rho u) + \frac{\partial}{\partial x}(\rho u^2) = \frac{\hbar^2}{4m_0{}^2}\rho\frac{\partial}{\partial x}\left[\frac{1}{\rho}\left(\frac{\partial^2 \rho}{\partial x^2} - \frac{1}{2\rho}\left(\frac{\partial \rho}{\partial x}\right)^2\right)\right]. \tag{12.1.43a}$$

This equation coincides with equation obtained by me in [64]

$$\frac{\partial}{\partial t}(\rho u) + \frac{\partial}{\partial x}(\rho u^2) = \frac{\hbar^2}{4m^2}\rho\frac{\partial}{\partial x}\left\{\frac{1}{\rho}\left[\gamma\frac{\partial^2 \rho}{\partial x^2} - \delta\frac{1}{\rho}\left(\frac{\partial \rho}{\partial x}\right)^2\right]\right\}, \tag{12.1.43b}$$

for coefficients $\gamma = 1$, $\delta = 0.5$ corresponding the Schrödinger equation.

The expression for velocity component v_t (Eq. (12.1.18)) and additional condition (Eq. (12.1.30)) leads to differential equation:

$$\frac{\partial \beta}{\partial t} = -\frac{m_0 c^2}{\hbar} \tag{12.1.46}$$

and after integration

$$\beta = -\frac{m_0 c^2}{\hbar}t + A(x,y,z) \tag{12.1.47}$$

Therefore, expression (12.1.47) defines the temporal part of the wave function in the hydrodynamic interpretation of the relativistic Dirac theory and

$$\psi = \alpha \exp\left(-\frac{m_0 c^2}{\hbar}t\right). \tag{12.1.48}$$

The value $A(x,y,z)$ is included in the space part α of the wave function ψ. Now we can introduce the relativistic quantum frequency

$$\varpi_{\mathrm{qu}} = \frac{m_0 c^2}{\hbar}. \tag{12.1.49}$$

Let us consider the condition $v_t = c$ from another point of view. The first moment—the particle four-flow—is defined as

$$N^\alpha = c\int p^\alpha f \frac{\mathrm{d}^3 p}{p_0}, \quad \alpha = 0,1,2,3. \tag{12.1.50}$$

Then $N^0 = cn$, that justifies the condition $v_t = c$.

12.2 GENERALIZED RELATIVISTIC KINETIC EQUATION

Let us realize the generalization of the Alexeev equations

$$\frac{\mathrm{D}f}{\mathrm{D}t} = J^{\mathrm{B}} + J^{\mathrm{nl}}, \tag{12.2.1}$$

$$J^{\mathrm{nl}} = \frac{\mathrm{D}}{\mathrm{Dt}}\left(\tau\frac{\mathrm{D}f}{\mathrm{D}t}\right). \tag{12.2.2}$$

for the relativistic case. For better understanding, let us realize the following generalization using the direct Lorentz transformations of the corresponding substantional derivatives. Transform with this aim the non-relativistic substantional derivative

$$D_{\mathrm{nrel}} = \frac{\partial}{\partial t} + v^i\frac{\partial}{\partial x^i}, \quad i = 1,2,3, \tag{12.2.3}$$

in relativistic ones

$$D_{\mathrm{rel}} = p^\alpha\frac{\partial}{\partial x^\alpha}, \quad \alpha = 0,1,2,3, \tag{12.2.4}$$

where the impulse component defined by relations

$$p^0 = m_0\gamma c, p^i = m_0\gamma v^i \tag{12.2.5}$$

m_0—the rest mass of a particle (scalar invariant); $\gamma = \frac{1}{\sqrt{1-(v^2/c^2)}}$, v—velocity module of a particle in the stationary coordinate system K of viewer. For this case

$$p^0 = \sqrt{p^2 + m_0^2 c^2}. \tag{12.2.6}$$

We consider two reference systems K and K' such that K' s moving with the uniform velocity V^1 directed along the x^1 axis. In the reference frame K a material body is moving with the velocity v^1 along the x^1 axis. The following Lorentz relations take place

$$x^1 = \frac{x'^1 + V^1 t'}{\sqrt{1-\left((V^1)^2/c^2\right)}}, x^2 = x'^{,2}, x^3 = x'^{,3}, t = \frac{t' + (V^1/c^2)x'^1}{\sqrt{1-\left((V^1)^2/c^2\right)}} \tag{12.2.7}$$

and, correspondingly

$$\mathrm{d}x^1 = \frac{\mathrm{d}x'^1 + V^1\mathrm{d}t'}{\sqrt{1-\left((V^1)^2/c^2\right)}}, \mathrm{d}x^2 = \mathrm{d}x'^{,2}, \mathrm{d}x^3 = \mathrm{d}x'^{,3}, \mathrm{d}t = \frac{\mathrm{d}t' + (V^1/c^2)\mathrm{d}x'^1}{\sqrt{1-\left((V^1)^2/c^2\right)}}, \tag{12.2.8}$$

From Eq. (12.2.8) follows

$$v^1 = \frac{\left(\mathrm{d}x'^1/\mathrm{d}t'\right) + V^1}{1 + (V^1/c^2)(\mathrm{d}x'^1/\mathrm{d}t')} = \frac{v'^1 + V^1}{1 + (V^1 v'^1/c^2)}, \tag{12.2.9}$$

$$v^{,2} = \frac{v'^{,2}\sqrt{1-\left((V^1)^2/c^2\right)}}{1 + (V^1/c^2)v'^1}, \tag{12.2.10}$$

$$v^{,3} = \frac{v'^{,3}\sqrt{1-\left((V^1)^2/c^2\right)}}{1 + (V^1/c^2)v'^1}. \tag{12.2.11}$$

Relations (12.2.9)–(12.2.11) can be treated by the way what the velocity component the stationary viewer observes in K if in the moving frame K' the corresponding components are v'^1, $v'^{,2}$, $v'^{,3}$? Introduce the distribution function $f[t(t',x'^1), x^1(t',x'^1), x^2, x^3]$.

The following time derivative we need:

$$\begin{aligned}\frac{\partial f}{\partial t'} = &\left[\frac{\partial f}{\partial t}\right]_{x^1,x^{,2},x^{,3}=\mathrm{const}}\left[\frac{\partial t}{\partial t'}\right]_{x'^1=\mathrm{const}} + \left[\frac{\partial f}{\partial x^1}\right]_{t,x^{,2},x^{,3}=\mathrm{const}}\left[\frac{\partial x^1}{\partial t'}\right]_{x'^1=\mathrm{const}}\\ &+ \left[\frac{\partial f}{\partial x^{,2}}\right]_{t,x^1,x^{,3}=\mathrm{const}}\left[\frac{\partial x^{,2}}{\partial t'}\right]_{x'^{,2}=\mathrm{const}} + \left[\frac{\partial f}{\partial x^{,3}}\right]_{t,x^1,x^{,2}=\mathrm{const}}\left[\frac{\partial x^{,3}}{\partial t'}\right]_{x'^{,3}=\mathrm{const}}\end{aligned} \tag{12.2.12}$$

or

$$\frac{\partial f}{\partial t'} = \left[\frac{\partial f}{\partial t}\right]_{x^1,x^{,2},x^{,3}=\mathrm{const}}\frac{1}{\sqrt{1-\left((V^1)^2/c^2\right)}} + \left[\frac{\partial f}{\partial x^1}\right]_{t,x^{,2},x^{,3}=\mathrm{const}}\frac{V^1}{\sqrt{1-\left((V^1)^2/c^2\right)}}. \tag{12.2.13}$$

Analogically for the space partial derivatives

$$\frac{\partial f}{\partial x'^1} = \left[\frac{\partial f}{\partial t}\right]_{x^1,x^{,2},x^{,3}=\mathrm{const}}\frac{V^1/c^2}{\sqrt{1-\left((V^1)^2/c^2\right)}} + \left[\frac{\partial f}{\partial x^1}\right]_{t,x^{,2},x^{,3}=\mathrm{const}}\frac{1}{\sqrt{1-\left((V^1)^2/c^2\right)}}, \tag{12.2.14}$$

$$\frac{\partial f}{\partial x'^{,2}} = \left[\frac{\partial f}{\partial x^{,2}}\right]_{t,x^1,x^{,3}=\text{const}}, \tag{12.2.15}$$

$$\frac{\partial f}{\partial x'^{,3}} = \left[\frac{\partial f}{\partial x^{,3}}\right]_{t,x^1,x^{,2}=\text{const}}. \tag{12.2.16}$$

Consider now the relativistic substantional derivative (12.2.4) in the frame K'. After substitution, Eqs. (12.2.12)–(12.2.16), one obtains:

$$\begin{aligned}
&m_0\gamma'\frac{\partial f}{\partial t'} + m_0\gamma' v'^1\frac{\partial f}{\partial x'^1} + m_0\gamma' v'^{,2}\frac{\partial f}{\partial x'^{,2}} + m_0\gamma' v'^{,3}\frac{\partial f}{\partial x'^{,3}} \\
&= m_0\gamma'\left[\frac{\partial f}{\partial t}\right]\left[\frac{1}{\sqrt{1-\left((V^1)^2/c^2\right)}} + v'^1\frac{V^1/c^2}{\sqrt{1-\left((V^1)^2/c^2\right)}}\right] \\
&+ m_0\gamma'\left[\frac{\partial f}{\partial x^1}\right]\left[\frac{V^1}{\sqrt{1-\left((V^1)^2/c^2\right)}} + v'^1\frac{1}{\sqrt{1-\left((V^1)^2/c^2\right)}}\right] \\
&+ m_0\gamma' v'^{,2}\left[\frac{\partial f}{\partial x^{,2}}\right] + m_0\gamma' v'^{,3}\left[\frac{\partial f}{\partial x^{,3}}\right].
\end{aligned} \tag{12.2.17}$$

Use now the relations (12.2.9)–(12.2.11) for velocities. The values $\left(v'^1 + V^1\right), v'^{,2}, v'^{,3}$ are expressed and substituted in Eq. (12.2.17):

$$\frac{\partial f}{\partial t'} + v'^1\frac{\partial f}{\partial x'^1} + v'^{,2}\frac{\partial f}{\partial x'^{,2}} + v'^{,3}\frac{\partial f}{\partial x'^{,3}} = \frac{1+\left(V^1 v'^1/c^2\right)}{\sqrt{1-\left((V^1)^2/c^2\right)}}\left\{\frac{\partial f}{\partial t} + v^1\frac{\partial f}{\partial x^1} + v^{,2}\frac{\partial f}{\partial x^{,2}} + v^{,3}\frac{\partial f}{\partial x^{,3}}\right\}. \tag{12.2.18}$$

Transform the coefficient in front of the brackets in the right side of relation (12.2.18) using (12.2.9)

$$\frac{(v^1)^2}{c^2} = \frac{1}{c^2}\left[\frac{v'^1 + V^1}{1+\left(v'^1 V^1/c^2\right)}\right]^2. \tag{12.2.19}$$

We have

$$\gamma = \frac{1}{\sqrt{1-\left((V^1)^2/c^2\right)}} = \frac{1+\left(v'^1 V^1/c^2\right)}{\sqrt{1+\left(v'^1 V^1/c^2\right)^2 - \frac{1}{c^2}\left(\left(v'^1\right)^2 + \left(V^1\right)^2\right)}}. \tag{12.2.20}$$

and

$$\sqrt{1+\left((V^1/c^2)v'^1\right)^2 - \frac{1}{c^2}\left((v'^1)^2 + (V^1)^2\right)} = \sqrt{\left(1-\left((V^1)^2/c^2\right)\right)}\sqrt{\left(1-\left((v'^1)^2/c^2\right)\right)}. \tag{12.2.21}$$

From Eqs. (12.2.20) and (12.2.21) follow:

$$\gamma = \frac{1}{\sqrt{1-\left((v^1)^2/c^2\right)}} = \frac{1+\left(v'^1 V^1/c^2\right)}{\sqrt{1-\left((V^1)^2/c^2\right)}\sqrt{1-\left((v'^1)^2/c^2\right)}}, \tag{12.2.22}$$

or

$$\frac{\gamma}{\gamma'} = \frac{1+\left(v'^1 V^1/c^2\right)}{\sqrt{1-\left((V^1)^2/c^2\right)}}. \tag{12.2.23}$$

Using relations (12.2.17) and (12.2.22) we reach the invariant relation obtained by the direct calculations

$$m_0\gamma'\left\{\frac{\partial f}{\partial t'}+v'^1\frac{\partial f}{\partial x'^1}+v'^{,2}\frac{\partial f}{\partial x'^{,2}}+v'^{,3}\frac{\partial f}{\partial x'^{,3}}\right\}=m_0\gamma\left\{\frac{\partial f}{\partial t}+v^1\frac{\partial f}{\partial x^1}+v^{,2}\frac{\partial f}{\partial x^{,2}}+v^{,3}\frac{\partial f}{\partial x^{,3}}\right\}. \tag{12.2.24}$$

or in the equivalent forms

$$D_{\text{rel}}=D'_{\text{rel}} \tag{12.2.25}$$

$$p^\alpha\frac{\partial}{\partial x^\alpha}=p'^\alpha\frac{\partial}{\partial x'^\alpha}\,(\alpha=0,1,2,3). \tag{12.2.26}$$

We are coming now to the creation of the generalized relativistic Boltzmann equation. The procedure of the corresponding transformation of the Alexeev equation consists in the following:

Step 1. Equation (12.2.1) (see also Eq. (12.2.2)) is multiplying term-by term into $m_0\gamma$. As result in the left hand side of equation made its appearance the combination of the Lorentz invariant substantial derivatives because the expression $p^\alpha\frac{\partial}{\partial x^\alpha}=inv$. However,

$$p^\alpha\frac{\partial f}{\partial x^\alpha}-p^\alpha\frac{\partial}{\partial x^\alpha}\left[\frac{\tau}{m_0}\left(p^\beta\frac{\partial}{\partial x^\beta}f\right)\right]=inv, \tag{12.2.27}$$

if the parameter of the non-locality (in this case the value proportional to the mean time between collisions)

$$\tau=\tau'=\tau_0=inv, \tag{12.2.28}$$

does not depend on the choose of the coordinate system and defined only by own time of the particles movement and therefore is also scalar invariant.

Step 2. Write down now the relativistic equation as:

$$p^\alpha\frac{\partial f}{\partial x^\alpha}-p^\alpha\frac{\partial}{\partial x^\alpha}\left[\frac{\tau}{m_0}\left(p^\beta\frac{\partial}{\partial x^\beta}f\right)\right]=m_0\gamma J^{\text{B,rel}},\quad \alpha,\beta=0,1,2,3. \tag{12.2.29}$$

The right side of Eq. (12.2.29) should be the Lorentz invariant local collision integral

$$J^{\text{B,rel}}_{\text{INV}}=m_0\gamma J^{\text{B,rel}}. \tag{12.2.30}$$

Let us prove this affirmation. With this aim, transform the local Boltzmann integral written as (see [255]):

$$J^{\text{B,rel}}\doteq\int\left[f_*^{\text{invert}}f^{\text{invert}}-f_*f\right]v_g\sigma\,\mathrm{d}\Omega\frac{p_\alpha p_*^\alpha\,\mathrm{d}^3p_*}{p^0\quad p_*^0}, \tag{12.2.31}$$

where σ—differential section of collision, $\mathrm{d}\Omega$—elementary solid angle, the sign "invert" is used further for denoting of the characteristic of the backward collisions, and

$$v_g=\frac{p^0p_*^0}{p_\alpha p_*^\alpha}\sqrt{(\mathbf{v}-\mathbf{v}_*)^2-\frac{1}{c^2}(\mathbf{v}\times\mathbf{v}_*)^2}. \tag{12.2.32}$$

Let us show, that the collision integral (12.2.31) converts to the Boltzmann local integral in non-relativistic case. It is convenient to use the relation [255]

$$\frac{p_\alpha p_*^\alpha}{p_*^0p^0}=1-\frac{\mathbf{v}}{c}\cdot\frac{\mathbf{v}_*}{c}, \tag{12.2.33}$$

for the integral transformation in the non-relativistic approach

$$\begin{aligned}J^{\text{B}}&=\int\left[f_*^{\text{invert}}f^{\text{invert}}-f_*f\right]v_g\sigma\,\mathrm{d}\Omega\frac{p_\alpha p_*^\alpha}{p^0p_*^0}\mathrm{d}^3p_*\\&=\int\left[f_*^{\text{invert}}f^{\text{invert}}-f_*f\right]v_g\sigma\,\mathrm{d}\Omega\,\mathrm{d}^3p_*\\&=\int\left[f_*^{\text{invert}}f^{\text{invert}}-f_*f\right]v_g^{\text{nonrel}}\sigma\,\mathrm{d}\Omega\,\mathrm{d}^3p_*\\&=\int\left[f_*^{\text{invert}}f^{\text{invert}}-f_*f\right]|\mathbf{v}-\mathbf{v}_*|\sigma\,\mathrm{d}\Omega\,\mathrm{d}^3p_*.\end{aligned} \tag{12.2.34}$$

Collision integral (12.2.31) is not invariant value relatively the Lorentz transformation because of the presence of p^0 in the denominator. Substitute $p^0 = m_0\gamma c$ in Eq. (12.2.31)

$$J^{\mathrm{B,rel}} = \int \left[f_*^{\mathrm{invert}} f^{\mathrm{invert}} - f_* f \right] v_g \sigma \, \mathrm{d}\Omega \frac{p_\alpha p_*^\alpha}{m_0 \gamma c} \frac{\mathrm{d}^3 p_*}{p_*^0}, \tag{12.2.35}$$

and introduce

$$m_0 \gamma J^{\mathrm{B,rel}} = \int \left[f_*^{\mathrm{invert}} f^{\mathrm{invert}} - f_* f \right] \frac{v_g}{c} \sigma p_\alpha p_*^\alpha \, \mathrm{d}\Omega \frac{\mathrm{d}^3 p_*}{p_*^0}. \tag{12.2.36}$$

For the ratio, v_g/c the relation is valid [255]

$$v_g = c \sqrt{1 - \frac{m_0^2 c^4}{\left(p_\alpha p_*^\alpha\right)^2}}; \tag{12.2.37}$$

using mentioned relation we find from Eq. (12.2.36)

$$J_{\mathrm{INV}}^{\mathrm{B,rel}} = m_0 \gamma J^{\mathrm{B,rel}} = \int \left[f_*^{\mathrm{invert}} f^{\mathrm{invert}} - f_* f \right] \sqrt{\left(p_\alpha p_*^\alpha\right)^2 - m_0^4 c^4} \sigma \, \mathrm{d}\Omega \frac{\mathrm{d}^3 p_*}{p_*^0} = inv. \tag{12.2.38}$$

This is the Lorentz invariant collision integral. Then the right hand side of the generalized relativistic equation is the Lorentz invariant value.

Step 3. Write down the generalized relativistic equation (GRE) as follows:

$$p^\alpha \frac{\partial f}{\partial x^\alpha} - p^\alpha \frac{\partial}{\partial x^\alpha} \left[\frac{\tau}{m_0} \left(p^\beta \frac{\partial}{\partial x^\beta} f \right) \right] = \int \left[f_*^{\mathrm{invert}} f^{\mathrm{invert}} - f_* f \right] F \sigma \, \mathrm{d}\Omega \frac{\mathrm{d}^3 p_*}{p_*^0}, \tag{12.2.39}$$

where $F = \sqrt{\left(p_\alpha p_*^\alpha\right)^2 - m_0^4 c^4}$, or by this way:

$$p^\alpha \frac{\partial f}{\partial x^\alpha} - p^\alpha \frac{\partial}{\partial x^\alpha} \left[\frac{\tau}{m_0} \left(p^\beta \frac{\partial}{\partial x^\beta} f \right) \right] = J_{\mathrm{INV}}^{\mathrm{B,rel}}, \quad \alpha = 0,1,2,3. \tag{12.2.40}$$

It is convenient for practical needs to use the BGK model for the relativistic case. For this aim write down $J_{\mathrm{INV}}^{\mathrm{B,rel}}$ in the form:

$$J_{\mathrm{INV}}^{\mathrm{B,rel}} = -\frac{m_0 \left(f - f^{(0)}\right)}{\breve{\tau}}, \tag{12.2.41}$$

where $\breve{\tau}$ is time of relaxation, which is general case is not equal to the non-local parameter τ.

In the deep particular case when the non-locality is not taking into account, the one-particle distribution function depend only on time, the relation (12.2.40) is written as

$$p^0 \frac{\mathrm{d}f}{\mathrm{d}x^0} = -\frac{m_0 \left(f - f^{(0)}\right)}{\breve{\tau}}, \tag{12.2.42}$$

or

$$\gamma \frac{\mathrm{d}f}{\mathrm{d}t} = -\frac{f - f^{(0)}}{\breve{\tau}}. \tag{12.2.43}$$

For non-relativistic case from Eq. (12.2.43) follows the typical BGK approximation

$$\frac{\mathrm{d}f}{\mathrm{d}t} = -\frac{f - f^{(0)}}{\breve{\tau}}. \tag{12.2.44}$$

Let us consider now the another aspect of the derivation of the generalized relativistic equation. Namely, the both sides of the non-relativistic kinetic equation

$$\frac{\partial f}{\partial t} + v^i \frac{\partial f}{\partial x^i} - \left\{ \frac{\partial}{\partial t} + v^i \frac{\partial}{\partial x^i} \right\} \left[\tau_K \left(\frac{\partial}{\partial t} + v^j \frac{\partial}{\partial x^j} \right) f \right] = J^{\mathrm{B}}, \quad (i = 1,2,3), \tag{12.2.45}$$

multiply by $m_0\gamma$, where m_0—mass of the rest, $\gamma = 1/\sqrt{1-v^2/c^2}$, v—module of the particle velocity in the coordinate system of observer. As result, the left hand side of the obtained equation takes the form

$$p^\alpha \frac{\partial f}{\partial x^\alpha} - p^\alpha \frac{\partial}{\partial x^\alpha}\left[\frac{\tau_K}{m_0\gamma}\left(p^\beta \frac{\partial}{\partial x^\beta} f\right)\right], \tag{12.2.46}$$

whereas before, $p^0 = m_0\gamma c, p^i = m_0\gamma v^i, x^0 = ct, (i = 1,2,3)$. But $p^\alpha \frac{\partial}{\partial x^\alpha}$ is the Lorentz invariant expression, because (see [255]) $\frac{\partial f}{\partial x^\alpha} = \varphi_\alpha$, where φ_α are covariant components of four-vector and therefore $p^\alpha \varphi_\alpha$—is scalar value. As result, the combination of the Lorentz invariant substantive derivations appears in the left side of kinetic equation. Introduce the notation

$$\tau = \tau_K/\gamma. \tag{12.2.47}$$

Then (see Eq. (12.2.46))

$$p^\alpha \frac{\partial f}{\partial x^\alpha} - p^\alpha \frac{\partial}{\partial x^\alpha}\left[\frac{\tau}{m_0}\left(p^\alpha \frac{\partial}{\partial x^\alpha} f\right)\right] = inv,$$

if the non-local parameter $\tau = \tau' = \tau_0 = inv$, does not depend on the choice of the coordinate system and defines by own time of the particle free motion between collisions and also should be considered as invariant.

Really, the own time of the particle motion τ_0 and the corresponding time τ_K in the coordinate system K of observer are tied, as follows from Lorentz transformation, by the relation

$$\tau_0 = \tau_K/\gamma = \tau_K\sqrt{1-v^2/c^2}. \tag{12.2.48}$$

It means, that the time, defined in Eq. (12.2.47) as τ_K, has the sense the value proportional to the mean time between collision in the K-system.

Generalized relativistic equation (12.2.40) is basement for derivation of the relativistic hydrodynamic equations, which are considered in the following sections.

12.3 GENERALIZED ENSKOG RELATIVISTIC HYDRODYNAMIC EQUATIONS

The macroscopic description of the relativistic gas is based on the moments of the DF and defined by tensors

$$T^{\alpha\beta\ldots\gamma\delta} = c\int p^\alpha p^\beta \ldots p^\gamma p^\delta f \frac{d^3 p}{p^0}. \tag{12.3.1}$$

The first moment transforms Eq. (12.3.1) into the particle four-flow, defined by Eq. (12.2.5). The second moment is the energy-momentum tensor

$$T^{\alpha\beta} = c\int p^\alpha p^\beta f \frac{d^3 p}{p^0}, \tag{12.3.2}$$

and so on. Multiply respectively Eq. (12.2.40) by cm_0, cp^l, $cp^0 = mc^2$ and integrate over d^3p/p^0. It is known [255] that the moments of the local collision integral turn into zero

$$\int J_{\rm INV}^{\rm B,rel}\psi \frac{d^3p}{p^0} = \int \psi\left[f_*^{\rm invert} f^{\rm invert} - f_* f\right]\sqrt{\left(p_\alpha p_*^\alpha\right)^2 - m_0^4 c^4}\,\sigma\, d\Omega\left(d^3 p_*/p_*^0\right)\left(d^3 p/p^0\right) = 0 \tag{12.3.3}$$

if

$$\psi(p^\alpha) = A + B_\alpha p^\alpha, \tag{12.3.4}$$

where A—scalar, B_α—arbitrary four-vector, which does not depend on p^α. Then in all mentioned cases, the right side of corresponding equation is equal to zero. The described procedure leads to the generalized relativistic Enskog equations.

12.3.1 Derivation of the Continuity Equation

Multiply (12.2.40) term-by-term into cm_0/p^0

$$\begin{aligned} &\frac{cm_0}{p^0} p^\alpha \frac{\partial f}{\partial x^\alpha} - \frac{c}{p^0} p^\alpha \frac{\partial}{\partial x^\alpha}\left[\tau_0\left(p^\beta \frac{\partial}{\partial x^\beta} f\right)\right] \\ &= cm_0 \int\left[f_*^{\rm invert} f^{\rm invert} - f_* f\right] F\sigma\, d\Omega \frac{1}{p^0}\frac{d^3 p_*}{p_*^0}. \end{aligned} \tag{12.3.5}$$

Integrating (12.3.5) is realizing with respect to the impulses of the collision partners, and then the factor $\frac{1}{p^0}=\frac{1}{m_0\gamma c}$ can be introduced under the integral sign.

Consider further all originated terms, after multiplying Eq. (12.3.5) by d^3p and integrating over all impulses

(a)
$$cm_0\int p^\alpha\frac{\partial f}{\partial x^\alpha}\frac{1}{p^0}\mathrm{d}^3p=cm_0\frac{\partial}{\partial x^\alpha}\int p^\alpha f\frac{1}{p^0}\mathrm{d}^3p=m_0\frac{\partial}{\partial t}\int p^0 f\frac{\mathrm{d}^3p}{p^0}+cm_0\frac{\partial}{\partial x^i}\int p^i f\frac{\mathrm{d}^3p}{p^0}, \tag{12.3.6}$$

or

$$cm_0\int p^\alpha\frac{\partial f}{\partial x^\alpha}\frac{1}{p^0}\mathrm{d}^3p=m_0\frac{\partial}{\partial t}\int f\,\mathrm{d}^3p+\frac{\partial}{\partial x^i}\int\frac{1}{\gamma}mv^i f\,\mathrm{d}^3p=m_0\frac{\partial}{\partial t}\int f\,\mathrm{d}^3p+\frac{\partial}{\partial x^i}\int m_0v^i f\,\mathrm{d}^3p \tag{12.3.7}$$

Here $m=m_0\gamma, p^i=mv^i$. Comments to relations (12.3.6) and (12.3.7):

1. Integration is realizing over all the impulse space. Then the integrations limits do not depend on time and space coordinates and the corresponding derivatives can be written in front of the integral signs.
2. In all transformations is taking into account that DF is function of four independent variables x^1,x^2,x^3,t and impulses. Dependent transformations appear only by transfer to another Lorentz frame.

   From Eq. (12.3.7) follows

$$\int cm_0\frac{p^\alpha}{p^0}\frac{\partial f}{\partial x^\alpha}\mathrm{d}^3p=m_0\frac{\partial n}{\partial t}+m_0\frac{\partial}{\partial x^i}\left[n\overline{v^i}\right], \tag{12.3.8}$$

where

$$\int f\,\mathrm{d}^3p=n,\ \int f v^i\,\mathrm{d}^3p=n\overline{v^i}, \tag{12.3.9}$$

or

$$\int cm_0\frac{p^\alpha}{p^0}\frac{\partial f}{\partial x^\alpha}\mathrm{d}^3p=\frac{\partial\rho}{\partial t}+\frac{\partial}{\partial x^i}\left[\rho\overline{v^i}\right], \tag{12.3.10}$$

because

$$\rho=m_0 n. \tag{12.3.11}$$

(b) Transformation of the next term originated by Eq. (12.3.5):

$$\begin{aligned}-c\int\frac{p^\alpha}{p^0}\frac{\partial}{\partial x^\alpha}\left[\tau_0\left(p^\beta\frac{\partial}{\partial x^\beta}f\right)\right]\mathrm{d}^3p&=-c\frac{\partial}{\partial x^\alpha}\left[\tau_0\frac{\partial}{\partial x^\beta}\int p^\alpha p^\beta f\frac{\mathrm{d}^3p}{p^0}\right]\\&=-\frac{\partial}{\partial t}\left[\tau_0\frac{\partial}{c\partial t}\int p^0p^0 f\frac{\mathrm{d}^3p}{p^0}\right]-\frac{\partial}{\partial x^i}\left[\tau_0\frac{\partial}{\partial t}\int p^ip^0 f\frac{\mathrm{d}^3p}{p^0}\right]\\&\quad-\frac{\partial}{\partial t}\left[\tau_0\frac{\partial}{\partial x^i}\int p^0p^i f\frac{\mathrm{d}^3p}{p^0}\right]-c\frac{\partial}{\partial x^i}\left[\tau_0\frac{\partial}{\partial x^j}\int p^ip^j f\frac{\mathrm{d}^3p}{p^0}\right]\end{aligned} \tag{12.3.12}$$

or

$$\begin{aligned}&-c\int\frac{p^\alpha}{p^0}\frac{\partial}{\partial x^\alpha}\left[\tau_0\left(p^\beta\frac{\partial}{\partial x^\beta}f\right)\right]\mathrm{d}^3p\\&=-\frac{\partial}{\partial t}\left[\tau_0\frac{\partial}{\partial t}\int m_0\gamma f\,\mathrm{d}^3p\right]-\frac{\partial}{\partial t}\left[\tau_0\frac{\partial}{\partial x^i}\int m_0\gamma v^i f\,\mathrm{d}^3p\right]\\&\quad-\frac{\partial}{\partial x^i}\left[\tau_0\frac{\partial}{\partial t}\int m_0\gamma v^i f\,\mathrm{d}^3p\right]-\frac{\partial}{\partial x^i}\left[\tau_0\frac{\partial}{\partial x^j}\int m_0\gamma v^iv^j f\,\mathrm{d}^3p\right]\end{aligned} \tag{12.3.13}$$

Then the left hand side of the generalized continuity equation takes the form:

$$\begin{aligned}&\int cm_0\frac{p^\alpha}{p^0}\frac{\partial f}{\partial x^\alpha}\mathrm{d}^3p-c\int\frac{p^\alpha}{p^0}\frac{\partial}{\partial x^\alpha}\left[\tau_0\left(p^\beta\frac{\partial}{\partial x^\beta}f\right)\right]\mathrm{d}^3p\\&=\frac{\partial\rho}{\partial t}+\frac{\partial}{\partial x^i}\left[\rho\overline{v^i}\right]-\frac{\partial}{\partial t}\left[\tau_0\frac{\partial}{\partial t}\int m_0\gamma f\,\mathrm{d}^3p\right]-\frac{\partial}{\partial t}\left[\tau_0\frac{\partial}{\partial x^i}\int m_0\gamma v^i f\,\mathrm{d}^3p\right]\\&-\frac{\partial}{\partial x^i}\left[\tau_0\frac{\partial}{\partial t}\int m_0\gamma v^i f\,\mathrm{d}^3p\right]-\frac{\partial}{\partial x^i}\left[\tau_0\frac{\partial}{\partial x^j}\int m_0\gamma v^i v^j f\,\mathrm{d}^3p\right],\end{aligned}\tag{12.3.14}$$

or

$$\begin{aligned}&\int cm_0\frac{p^\alpha}{p^0}\frac{\partial f}{\partial x^\alpha}\mathrm{d}^3p-c\int\frac{p^\alpha}{p^0}\frac{\partial}{\partial x^\alpha}\left[\tau_0\left(p^\beta\frac{\partial}{\partial x^\beta}f\right)\right]\mathrm{d}^3p\\&=\frac{\partial}{\partial t}\left\{\rho-\tau_0\left[\frac{\partial}{\partial t}\int m_0\gamma f\,\mathrm{d}^3p+\frac{\partial}{\partial x^i}\int m_0\gamma v^i f\,\mathrm{d}^3p\right]\right\}\\&+\frac{\partial}{\partial x^i}\left\{\rho\overline{v^i}-\tau_0\left[\frac{\partial}{\partial t}\int m_0\gamma v^i f\,\mathrm{d}^3p+\frac{\partial}{\partial x^j}\int m_0\gamma v^i v^j f\,\mathrm{d}^3p\right]\right\}.\end{aligned}\tag{12.3.15}$$

In this case $\psi(p^\alpha)=A$ in relation (12.3.4) and the right side of equation turns into zero. Then generalized continuity equation is

$$\begin{aligned}&m_0\frac{\partial}{\partial t}\int p^0f\frac{\mathrm{d}^3p}{p^0}+cm_0\frac{\partial}{\partial x^i}\int p^if\frac{\mathrm{d}^3p}{p^0}-\frac{\partial}{\partial t}\left[\tau_0\frac{\partial}{c\partial t}\int p^0p^0f\frac{\mathrm{d}^3p}{p^0}\right]-\frac{\partial}{\partial x^i}\left[\tau_0\frac{\partial}{\partial t}\int p^ip^0f\frac{\mathrm{d}^3p}{p^0}\right]\\&-\frac{\partial}{\partial t}\left[\tau_0\frac{\partial}{\partial x^i}\int p^0p^if\frac{\mathrm{d}^3p}{p^0}\right]-c\frac{\partial}{\partial x^i}\left[\tau_0\frac{\partial}{\partial x^j}\int p^ip^jf\frac{\mathrm{d}^3p}{p^0}\right]=0,\end{aligned}\tag{12.3.16}$$

or

$$\frac{\partial}{\partial t}\left\{\rho-\tau_0\left[\frac{\partial}{\partial t}(\rho\overline{\gamma})+\frac{\partial}{\partial x^i}\left(\rho\overline{\gamma v^i}\right)\right]\right\}+\frac{\partial}{\partial x^i}\left\{\rho\overline{v^i}-\tau_0\left[\frac{\partial}{\partial t}\left(\rho\overline{\gamma v^i}\right)+\frac{\partial}{\partial x^j}\left(\rho\overline{\gamma v^iv^j}\right)\right]\right\}=0,\tag{12.3.17}$$

where as usual

$$\overline{\psi}=\frac{1}{n}\int\psi f\,\mathrm{d}^3p.\tag{12.3.18}$$

In the non-relativistic case $\gamma=1$ and from Eq. (12.3.17) follows equation:

$$\frac{\partial}{\partial t}\left\{\rho-\tau_0\left[\frac{\partial\rho}{\partial t}+\frac{\partial}{\partial x^i}\left(\rho\overline{v^i}\right)\right]\right\}+\frac{\partial}{\partial x^i}\left\{\rho\overline{v^i}-\tau_0\left[\frac{\partial}{\partial t}\left(\rho\overline{v^i}\right)+\frac{\partial}{\partial x^j}\left(\rho\overline{v^iv^j}\right)\right]\right\}=0,\tag{12.3.19}$$

this coincides with known continuity equation obtained in [57].

The continuity equation of the second order is obtained by the condition $v\ll c$ after expanding γ into a power series

$$\gamma=\frac{1}{\sqrt{1-(v^2/c^2)}}\approx 1+\frac{v^2}{2c^2}.\tag{12.3.20}$$

We have

$$\begin{aligned}&\frac{\partial}{\partial t}\left\{\rho c^2-\tau_0\left[\frac{\partial}{\partial t}\left(\rho c^2+\rho\frac{\overline{v^2}}{2}\right)+\frac{\partial}{\partial x^i}\left(\rho c^2\overline{v^i}+\rho\frac{\overline{v^2v^i}}{2}\right)\right]\right\}\\&+\frac{\partial}{\partial x^i}\left\{\rho c^2\overline{v^i}-\tau_0\left[\frac{\partial}{\partial t}\left(\rho c^2\overline{v^i}+\frac{1}{2}\rho\overline{v^2v^i}\right)+\frac{\partial}{\partial x^j}\left(\rho c^2\overline{v^iv^j}+\frac{1}{2}\rho\overline{v^2v^iv^j}\right)\right]\right\}=0.\end{aligned}\tag{12.3.21}$$

12.3.2 Derivation of the Motion Equation

Multiply Eq. (12.2.40) term-by-term into $c(p^l/p^0)$ and integrate the obtained equation with respect to the impulses. As before consider further all originated terms and notice that the right side of this equation turn into zero because for this case (see Eq. (12.3.4)) $A=0$, $B_l\neq 0$. Therefore

$$c\int p^{\alpha}p^{l}\frac{\partial f}{\partial x^{\alpha}}\frac{\mathrm{d}^{3}p}{p^{0}}-c\frac{\partial}{\partial x^{\alpha}}\int\left[\frac{\tau_{0}}{m_{0}}\frac{p^{l}}{p^{0}}\left(p^{\alpha}p^{\beta}\frac{\partial}{\partial x^{\beta}}f\right)\right]\mathrm{d}^{3}p=0. \tag{12.3.22}$$

Consider all originated terms in Eq. (12.3.22):

(a)

$$c\int p^{\alpha}p^{l}\frac{\partial f}{\partial x^{\alpha}}\frac{\mathrm{d}^{3}p}{p^{0}}=\frac{\partial}{\partial t}\left(n\overline{p^{l}}\right)+\frac{\partial}{\partial x^{i}}\int v^{i}p^{l}f\,\mathrm{d}^{3}p=\frac{\partial}{\partial t}\left(n\overline{p^{l}}\right)+\frac{\partial}{\partial x^{i}}\left(n\overline{v^{i}p^{l}}\right). \tag{12.3.23}$$

(b)

$$\begin{aligned}
&-c\frac{\partial}{\partial x^{\alpha}}\int\left[\frac{\tau_{0}}{m_{0}}\frac{p^{l}}{p^{0}}\left(p^{\alpha}p^{\beta}\frac{\partial}{\partial x^{\beta}}f\right)\right]\mathrm{d}^{3}p=-\frac{\partial}{\partial t}\left[\frac{\tau_{0}}{m_{0}}\frac{\partial}{c\partial t}\int p^{l}m_{0}\gamma cf\,\mathrm{d}^{3}p\right]-\frac{\partial}{\partial t}\left[\frac{\tau_{0}}{m_{0}}\frac{\partial}{\partial x^{i}}\int p^{l}p^{i}f\,\mathrm{d}^{3}p\right]\\
&-\frac{\partial}{\partial x^{i}}\left[\frac{\tau_{0}}{m_{0}}\frac{\partial}{\partial t}\int p^{l}p^{i}f\,\mathrm{d}^{3}p\right]-c\frac{\partial}{\partial x^{i}}\left[\frac{\tau_{0}}{m_{0}}\frac{\partial}{\partial x^{j}}\int p^{l}m_{0}\gamma v^{i}p^{j}f\frac{\mathrm{d}^{3}p}{m_{0}\gamma c}\right]\\
&=-\frac{\partial}{\partial t}\left\{\tau_{0}\frac{\partial}{\partial t}\left(n\overline{\gamma p^{l}}\right)\right\}-\frac{\partial}{\partial t}\left\{\frac{\tau_{0}}{m_{0}}\frac{\partial}{\partial x^{i}}\left(n\overline{p^{i}p^{l}}\right)\right\}-\frac{\partial}{\partial x^{i}}\left\{\frac{\tau_{0}}{m_{0}}\frac{\partial}{\partial t}\left(n\overline{p^{i}p^{l}}\right)\right\}-\frac{\partial}{\partial x^{i}}\left\{\frac{\tau_{0}}{m_{0}}\frac{\partial}{\partial x^{j}}\left(n\overline{v^{i}p^{j}p^{l}}\right)\right\}.
\end{aligned} \tag{12.3.24}$$

The motion equation takes the following equivalent forms

$$\begin{aligned}
&\frac{\partial}{\partial t}\int p^{l}p^{0}f\frac{\mathrm{d}^{3}p}{p^{0}}+c\frac{\partial}{\partial x^{i}}\int p^{i}p^{l}f\frac{\mathrm{d}^{3}p}{p^{0}}-\frac{\partial}{\partial t}\left[\frac{\tau_{0}}{m_{0}}\frac{\partial}{c\partial t}\int p^{l}p^{0}p^{0}f\frac{\mathrm{d}^{3}p}{p^{0}}\right]\\
&-\frac{\partial}{\partial t}\left[\frac{\tau_{0}}{m_{0}}\frac{\partial}{\partial x^{i}}\int p^{l}p^{0}p^{i}f\frac{\mathrm{d}^{3}p}{p^{0}}\right]-\frac{\partial}{\partial x^{i}}\left[\frac{\tau_{0}}{m_{0}}\frac{\partial}{\partial t}\int p^{l}p^{0}p^{i}f\frac{\mathrm{d}^{3}p}{p^{0}}\right]\\
&-c\frac{\partial}{\partial x^{i}}\left[\frac{\tau_{0}}{m_{0}}\frac{\partial}{\partial x^{j}}\int p^{l}p^{i}p^{j}f\frac{\mathrm{d}^{3}p}{p^{0}}\right]=0,
\end{aligned} \tag{12.3.25}$$

$$\begin{aligned}
&\frac{\partial}{\partial t}\left\{n\overline{p^{l}}-\tau_{0}\left[\frac{\partial}{\partial t}\left(n\overline{\gamma p^{l}}\right)+\frac{1}{m_{0}}\frac{\partial}{\partial x^{i}}\left(n\overline{p^{i}p^{l}}\right)\right]\right\}\\
&+\frac{\partial}{\partial x^{i}}\left\{n\overline{v^{i}p^{l}}-\frac{\tau_{0}}{m_{0}}\left[\frac{\partial}{\partial t}\left(n\overline{p^{i}p^{l}}\right)+\frac{\partial}{\partial x^{j}}\left(n\overline{v^{i}p^{j}p^{l}}\right)\right]\right\}=0,
\end{aligned} \tag{12.3.26}$$

or

$$\begin{aligned}
&\frac{\partial}{\partial t}\left\{\rho\overline{\gamma v^{l}}-\tau_{0}\left[\frac{\partial}{\partial t}\left(\rho\overline{\gamma^{2}v^{l}}\right)+\frac{\partial}{\partial x^{i}}\left(\rho\overline{\gamma^{2}v^{i}v^{l}}\right)\right]\right\}\\
&+\frac{\partial}{\partial x^{i}}\left\{\rho\overline{\gamma v^{i}v^{l}}-\tau_{0}\left[\frac{\partial}{\partial t}\left(\rho\overline{\gamma^{2}v^{i}v^{l}}\right)+\frac{\partial}{\partial x^{j}}\left(\rho\overline{\gamma^{2}v^{i}v^{j}v^{l}}\right)\right]\right\}=0.
\end{aligned} \tag{12.3.27}$$

In the limit case of the non-relativistic case by the condition $\gamma=1$ from Eq. (12.3.27) follow the known generalized non-relativistic motion equation [57]

$$\frac{\partial}{\partial t}\left\{\rho\overline{v^{l}}-\tau_{0}\left[\frac{\partial}{\partial t}\left(\rho\overline{v^{l}}\right)+\frac{\partial}{\partial x^{i}}\left(\rho\overline{v^{i}v^{l}}\right)\right]\right\}+\frac{\partial}{\partial x^{i}}\left\{\rho\overline{v^{i}v^{l}}-\tau_{0}\left[\frac{\partial}{\partial t}\left(\rho\overline{v^{i}v^{l}}\right)+\frac{\partial}{\partial x^{j}}\left(\rho\overline{v^{i}v^{j}v^{l}}\right)\right]\right\}=0. \tag{12.3.28}$$

The motion equation of the second order is obtained by expanding γ into a power series, (see Eq. (12.3.20)):

$$\begin{aligned}
&\frac{\partial}{\partial t}\left\{\rho c^{2}\overline{v^{l}}+\rho\overline{\left(\frac{v^{2}}{2}\right)v^{l}}-\tau_{0}\left[\frac{\partial}{\partial t}\left(\rho c^{2}\overline{v^{l}}+\rho\overline{v^{2}v^{l}}\right)+\frac{\partial}{\partial x^{i}}\left(\rho c^{2}\overline{v^{i}v^{l}}+\rho\overline{v^{2}v^{i}v^{l}}\right)\right]\right\}\\
&+\frac{\partial}{\partial x^{i}}\left\{\rho c^{2}\overline{v^{i}v^{l}}+\rho\overline{\left(\frac{v^{2}}{2}\right)v^{i}v^{l}}-\tau_{0}\left[\frac{\partial}{\partial t}\left(\rho c^{2}\overline{v^{i}v^{l}}+\rho\overline{v^{2}v^{i}v^{l}}\right)+\frac{\partial}{\partial x^{j}}\left(\rho c^{2}\overline{v^{i}v^{j}v^{l}}+\rho\overline{v^{2}v^{i}v^{j}v^{l}}\right)\right]\right\}=0.
\end{aligned} \tag{12.3.29}$$

12.3.3 Derivation of the Energy Equation

Multiply Eq. (12.2.40) term-by-term into $cp^0 = mc^2$ and integrate the obtained equation with respect to the impulses. As before consider further all originated terms and notice that the right side of this equation turn into zero because for this case $A = 0$, $\hat{A}_0 \neq 0$. Therefore:

$$\frac{\partial}{\partial x^\alpha}\int p^\alpha p^0 f\frac{\mathrm{d}^3 p}{p^0} - \frac{\partial}{\partial x^\alpha}\left\{\frac{\tau_0}{m_0}\frac{\partial}{\partial x^\beta}\int \left[p^0 p^\alpha p^\beta f\right]\frac{\mathrm{d}^3 p}{p^0}\right\} = 0. \tag{12.3.30}$$

Consider all originated terms in Eq. (12.3.30), ($\alpha = 0,1,2,3$; $i,j = 1,2,3$):

(a)

$$\begin{aligned}\frac{\partial}{\partial x^\alpha}\int p^\alpha f\,\mathrm{d}^3 p &= m_0\frac{\partial}{\partial t}\int \gamma f\,\mathrm{d}^3 p + m_0\frac{\partial}{\partial x^i}\int \gamma v^i f\,\mathrm{d}^3 p\\ &= \frac{\partial}{\partial t}(\rho\overline{\gamma}) + m_0\frac{\partial}{\partial x^i}\left(n\overline{\gamma v^i}\right) = \frac{\partial}{\partial t}(\rho\overline{\gamma}) + \frac{\partial}{\partial x^i}\left(\rho\overline{\gamma v^i}\right).\end{aligned} \tag{12.3.35}$$

(b)

$$\begin{aligned}&-\frac{\partial}{\partial x^\alpha}\left\{\frac{\tau_0}{m_0}\frac{\partial}{\partial x^\beta}\int p^\alpha p^\beta f\,\mathrm{d}^3 p\right\} = -\frac{\partial}{\partial t}\left\{\tau_0\frac{\partial}{\partial t}\int \gamma m f\,\mathrm{d}^3 p\right\}\\ &-\frac{\partial}{\partial t}\left\{\tau_0\frac{\partial}{\partial x^i}\int \gamma m v^i f\,\mathrm{d}^3 p\right\} - \frac{\partial}{\partial x^i}\left\{\tau_0\frac{\partial}{\partial t}\int \gamma m v^i f\,\mathrm{d}^3 p\right\} - \frac{\partial}{\partial x^i}\left\{\tau_0\frac{\partial}{\partial x^j}\int \gamma v^i m v^j f\,\mathrm{d}^3 p\right\}\\ &= -\frac{\partial}{\partial t}\left\{\tau_0\frac{\partial}{\partial t}\left[\rho\overline{\gamma^2}\right]\right\} - \frac{\partial}{\partial t}\left\{\tau_0\frac{\partial}{\partial x^i}\left[\rho\overline{\gamma^2 v^i}\right]\right\} - \frac{\partial}{\partial x^i}\left\{\tau_0\frac{\partial}{\partial t}\left[\rho\overline{\gamma^2 v^i}\right]\right\} - \frac{\partial}{\partial x^i}\left\{\tau_0\frac{\partial}{\partial x^j}\left[\rho\overline{\gamma^2 v^i v^j}\right]\right\}.\end{aligned} \tag{12.3.36}$$

The energy equation takes the following equivalent forms

$$\begin{aligned}&\frac{\partial}{\partial ct}\int p^0 p^0 f\frac{\mathrm{d}^3 p}{p^0} + \frac{\partial}{\partial x^i}\int p^0 p^i f\frac{\mathrm{d}^3 p}{p^0} - \frac{\partial}{\partial ct}\left\{\frac{\tau_0}{m_0}\frac{\partial}{\partial ct}\int p^0 p^0 p^0 f\frac{\mathrm{d}^3 p}{p^0}\right\}\\ &-\frac{\partial}{\partial ct}\left\{\frac{\tau_0}{m_0}\frac{\partial}{\partial x^i}\int p^0 p^i p^0 f\frac{\mathrm{d}^3 p}{p^0}\right\} - \frac{\partial}{\partial x^i}\left\{\frac{\tau_0}{m_0}\frac{\partial}{\partial ct}\int p^0 p^i p^0 f\frac{\mathrm{d}^3 p}{p^0}\right\}\\ &-\frac{\partial}{\partial x^i}\left\{\frac{\tau_0}{m_0}\frac{\partial}{\partial x^j}\int p^0 p^i p^j f\frac{\mathrm{d}^3 p}{p^0}\right\} = 0,\end{aligned} \tag{12.3.37}$$

or

$$\begin{aligned}&\frac{\partial}{\partial t}(\rho\overline{\gamma}) + \frac{\partial}{\partial x^i}\left(\rho\overline{\gamma v^i}\right) - \frac{\partial}{\partial t}\left\{\tau_0\frac{\partial}{\partial t}\left[\rho\overline{\gamma^2}\right]\right\} - \frac{\partial}{\partial t}\left\{\tau_0\frac{\partial}{\partial x^i}\left[\rho\overline{\gamma^2 v^i}\right]\right\}\\ &-\frac{\partial}{\partial x^i}\left\{\tau_0\frac{\partial}{\partial t}\left[\rho\overline{\gamma^2 v^i}\right]\right\} - \frac{\partial}{\partial x^i}\left\{\tau_0\frac{\partial}{\partial x^j}\left[\rho\overline{\gamma^2 v^i v^j}\right]\right\} = 0,\end{aligned} \tag{12.3.38}$$

or

$$\frac{\partial}{\partial t}\left\{\rho\overline{\gamma} - \tau_0\left[\frac{\partial}{\partial t}\left(\rho\overline{\gamma^2}\right) + \frac{\partial}{\partial x^i}\left(\rho\overline{\gamma^2 v^i}\right)\right]\right\} + \frac{\partial}{\partial x^i}\left\{\rho\overline{\gamma v^i} - \tau_0\left[\frac{\partial}{\partial t}\left(\rho\overline{\gamma^2 v^i}\right) + \frac{\partial}{\partial x^j}\left(\rho\overline{\gamma^2 v^i v^j}\right)\right]\right\} = 0. \tag{12.3.39}$$

Interesting to notice, that the energy equation in the non-relativistic first order approximation ($\gamma = 1$) takes the same form as continuity equation written by $\gamma = 1$ (Eq. (12.3.21)). In other words, continuity equation coincides with the energy equation in the first approximation. Obviously, for the right transition to the non-relativistic energy equation one needs to use the second approximation expanding γ and γ^2 into a power series:

$$\gamma \approx 1 + \frac{v^2}{2c^2},\ \gamma^2 \approx 1 + \frac{v^2}{c^2}, \tag{12.3.40}$$

We have

$$\begin{aligned}&\frac{\partial}{\partial t}\left\{\rho c^2 + \rho\frac{\overline{v^2}}{2} - \tau_0\left[\frac{\partial}{\partial t}\left(\rho c^2 + \rho\overline{v^2}\right) + \frac{\partial}{\partial x^i}\left(\rho c^2\overline{v^i} + \rho\overline{v^2 v^i}\right)\right]\right\}\\ &+\frac{\partial}{\partial x^i}\left\{\rho c^2\overline{v^i} + \rho\frac{\overline{v^2 v^i}}{2} - \tau_0\left[\frac{\partial}{\partial t}\left(\rho c^2\overline{v^i} + \rho\overline{v^2 v^i}\right) + \frac{\partial}{\partial x^j}\left(\rho c^2\overline{v^i v^j} + \rho\overline{v^2 v^i v^j}\right)\right]\right\} = 0.\end{aligned} \tag{12.3.41}$$

Introduce

$$\varepsilon_n = \rho c^2 = m_0 n c^2, \tag{12.3.42}$$

as the internal energy of the unit of volume. As result from Eq. (12.3.41)

$$\begin{aligned} &\frac{\partial}{\partial t}\left\{\varepsilon_n + \rho\frac{\overline{v^2}}{2} - \tau_0\left[\frac{\partial}{\partial t}\left(\varepsilon_n + \rho\overline{v^2}\right) + \frac{\partial}{\partial x^i}\left(\varepsilon_n\overline{v^i} + \rho\overline{v^2 v^i}\right)\right]\right\} \\ &+\frac{\partial}{\partial x^i}\left\{\varepsilon_n\overline{v^i} + \rho\frac{\overline{v^2 v^i}}{2} - \tau_0\left[\frac{\partial}{\partial t}\left(\varepsilon_n\overline{v^i} + \rho\overline{v^2 v^i}\right) + \frac{\partial}{\partial x^j}\left(\varepsilon_n\overline{v^i v^j} + \rho\overline{v^2 v^i v^j}\right)\right]\right\} = 0. \end{aligned} \tag{12.3.43}$$

Nevertheless, Eq. (12.3.43) does not coincide with the generalized non-relativistic equation, obtained in [57]:

$$\begin{aligned} &\frac{\partial}{\partial t}\left\{\varepsilon + \rho\frac{\overline{v^2}}{2} - \tau_K\left[\frac{\partial}{\partial t}\left(\varepsilon + \rho\frac{\overline{v^2}}{2}\right) + \frac{\partial}{\partial x^i}\left(\varepsilon\overline{v^i} + \rho\frac{\overline{v^2 v^i}}{2}\right)\right]\right\}, \\ &+\frac{\partial}{\partial x^i}\left\{\varepsilon\overline{v^i} + \frac{1}{2}\rho\overline{v^2 v^i} - \tau_K\left[\frac{\partial}{\partial t}\left(\varepsilon\overline{v^i} + \frac{1}{2}\rho\overline{v^2 v^i}\right) + \frac{\partial}{\partial x^j}\left(\varepsilon\overline{v^i v^j} + \frac{1}{2}\rho\overline{v^2 v^i v^j}\right)\right]\right\}. \end{aligned} \tag{12.3.44}$$

The transfer to the non-relativistic limit case is not the trivial procedure, as we see, and can be realized by the following way. Subtract term-by-term the continuity equation of the second approximation (12.3.21) from the energy equation of the second approximation (12.3.43). One obtains:

$$\begin{aligned} &\frac{\partial}{\partial t}\left\{\rho\frac{\overline{v^2}}{2} - \tau_0\left[\frac{\partial}{\partial t}\left(\rho\overline{v^2}\right) + \frac{\partial}{\partial x^i}\left(\frac{\rho\overline{v^2 v^i}}{2}\right)\right]\right\} \\ &+\frac{\partial}{\partial x^i}\left\{\rho\frac{\overline{v^2 v^i}}{2} - \tau_0\left[\frac{\partial}{\partial t}\left(\frac{\rho\overline{v^2 v^i}}{2}\right) + \frac{\partial}{\partial x^j}\left(\frac{\rho\overline{v^2 v^i v^j}}{2}\right)\right]\right\} = 0. \end{aligned} \tag{12.3.45}$$

Equation (12.3.45) coincides with the non-relativistic Alexeev equation without taking into account the internal energy, and Eq. (12.3.45) can be named as modified non-relativistic limit of the second approximation for the Alexeev equation. Correspondingly, the difference of Eqs. (12.3.39) and (12.3.17)

$$\begin{aligned} &\frac{\partial}{\partial t}\left\{\rho(\overline{\gamma}-1) - \tau_0\left[\frac{\partial}{\partial t}\left(\rho\overline{(\gamma^2-\gamma)}\right) + \frac{\partial}{\partial x^i}\left(\rho\overline{(\gamma^2-\gamma)v^i}\right)\right]\right\} \\ &+\frac{\partial}{\partial x^i}\left\{\rho\overline{(\gamma-1)v^i} - \tau_0\left[\frac{\partial}{\partial t}\left(\rho\overline{(\gamma^2-\gamma)v^i}\right) + \frac{\partial}{\partial x^j}\left(\rho\overline{(\gamma^2-\gamma)v^i v^j}\right)\right]\right\} = 0, \end{aligned} \tag{12.3.46}$$

which dispatch the direct transfer to the non-relativistic limit, can be named as modified energy equation. Another form of the modified energy equation can be written as difference between Eqs. (12.3.37) and (12.3.16):

$$\begin{aligned} &\frac{\partial}{\partial ct}\int p^0 p^0 f\frac{d^3p}{p^0} - m_0\frac{\partial}{\partial t}\int p^0 f\frac{d^3p}{p^0} + \frac{\partial}{\partial x^i}\int p^0 p^i f\frac{d^3p}{p^0} - cm_0\frac{\partial}{\partial x^i}\int p^i f\frac{d^3p}{p^0} \\ &-\frac{\partial}{\partial ct}\left\{\frac{\tau_0}{m_0}\frac{\partial}{\partial ct}\int p^0 p^0 p^0 f\frac{d^3p}{p^0}\right\} + \frac{\partial}{\partial t}\left\{\tau_0\frac{\partial}{\partial ct}\int p^0 p^0 f\frac{d^3p}{p^0}\right\} \\ &-\frac{\partial}{\partial ct}\left\{\frac{\tau_0}{m_0}\frac{\partial}{\partial x^i}\int p^0 p^i p^0 f\frac{d^3p}{p^0}\right\} + \frac{\partial}{\partial t}\left\{\tau_0\frac{\partial}{\partial x^i}\int p^0 p^i f\frac{d^3p}{p^0}\right\} \\ &-\frac{\partial}{\partial x^i}\left\{\frac{\tau_0}{m_0}\frac{\partial}{\partial ct}\int p^0 p^i p^0 f\frac{d^3p}{p^0}\right\} + \frac{\partial}{\partial x^i}\left\{\tau_0\frac{\partial}{\partial t}\int p^0 p^i f\frac{d^3p}{p^0}\right\} \\ &-\frac{\partial}{\partial x^i}\left\{\frac{\tau_0}{m_0}\frac{\partial}{\partial x^j}\int p^0 p^i p^j f\frac{d^3p}{p^0}\right\} + c\frac{\partial}{\partial x^i}\left\{\tau_0\frac{\partial}{\partial x^j}\int p^i p^j f\frac{d^3p}{p^0}\right\} = 0. \end{aligned} \tag{12.3.47}$$

12.4 GENERALIZED SYSTEM OF THE RELATIVISTIC HYDRODYNAMICS AND TRANSFER TO THE GENERALIZED RELATIVISTIC NON-LOCAL EULER HYDRODYNAMIC EQUATIONS

Additional complications arise for relativistic systems in the frame of the relativistic Boltzmann equation—a first order gradient expansion like Navier and Stokes, Barnett leads to a set of fluid dynamics equations that allow faster-than-light signal propagation, violating causality. It is effect known from the theory of the propagation of plane harmonic waves of small amplitudes through a relativistic gas at rest. A program to determine the precise form for a relativistic viscous fluid was initiated in the last decade; this program was driven by nuclear physics experiments on relativistic heavy-ion collisions and problems of quantum description of transport processes in nano-electronics. The relativistic first-order dissipative theory (created on the basement of the Boltzmann kinetic equation) is not applicable and is discarded in many papers in favor of the second-order one in spite the crash of the first approximation in the local description.

The aim of the following investigation contains in the application of the non-local relativistic theory to the solution of the known test problem of the relativistic theory to the propagation of plane harmonic waves of small amplitudes through a relativistic gas at rest. These results lead to construction of the relativistic quantum hydrodynamics.

Write down now the full system of the generalized hydrodynamic equations (Form 1):

continuity equation

$$\frac{\partial}{\partial t}\left\{\rho-\tau_0\left[\frac{\partial}{\partial t}(\rho\overline{\gamma})+\frac{\partial}{\partial x^i}\left(\rho\overline{\gamma v^i}\right)\right]\right\}+\frac{\partial}{\partial x^i}\left\{\rho\overline{v^i}-\tau_0\left[\frac{\partial}{\partial t}\left(\rho\overline{\gamma v^i}\right)+\frac{\partial}{\partial x^j}\left(\rho\overline{\gamma v^i v^j}\right)\right]\right\}=0, \quad (12.4.1)$$

motion equation

$$\begin{aligned}&\frac{\partial}{\partial t}\left\{\rho\overline{\gamma v^l}-\tau_0\left[\frac{\partial}{\partial t}\left(\rho\overline{\gamma^2 v^l}\right)+\frac{\partial}{\partial x^i}\left(\rho\overline{\gamma^2 v^i v^l}\right)\right]\right\}\\&+\frac{\partial}{\partial x^i}\left\{\rho\overline{\gamma v^i v^l}-\tau_0\left[\frac{\partial}{\partial t}\left(\rho\overline{\gamma^2 v^i v^l}\right)+\frac{\partial}{\partial x^j}\left(\rho\overline{\gamma^2 v^i v^j v^l}\right)\right]\right\}=0,\end{aligned} \quad (12.4.2)$$

energy equation

$$\begin{aligned}&\frac{\partial}{\partial t}\left\{\rho(\overline{\gamma}-1)-\tau_0\left[\frac{\partial}{\partial t}\left(\rho\overline{(\gamma^2-\gamma)}\right)+\frac{\partial}{\partial x^i}\left(\rho\overline{(\gamma^2-\gamma)v^i}\right)\right]\right\}\\&+\frac{\partial}{\partial x^i}\left\{\rho\overline{(\gamma-1)v^i}-\tau_0\left[\frac{\partial}{\partial t}\left(\rho\overline{(\gamma^2-\gamma)v^i}\right)+\frac{\partial}{\partial x^j}\left(\rho\overline{(\gamma^2-\gamma)v^i v^j}\right)\right]\right\}=0.\end{aligned} \quad (12.4.3)$$

The previous notations are used in the hydrodynamic system of Eqs. (12.4.1)–(12.4.3). In particular, $\tau=\tau'=\tau_0=inv$, is the parameter of the non-locality (in this case the mean time between collisions); $\rho=m_0 n$ is density; as before the four-impulse components are $p^0=m_0\gamma c, p^i=m_0\gamma v^i\ (i=1,2,3)$; the relativistic factor $\gamma=\frac{1}{\sqrt{1-(v^2/c^2)}}$, v—velocity module of a particle in the stationary coordinate system K of viewer. For this case $p^0=\sqrt{p^2+m_0^2c^2}$.

Another form of this system has the direct tensor appearance (12.3.2) as the consequence of Eqs. (12.3.16), (12.3.25), (12.3.47). Namely

$$\begin{aligned}&m_0\frac{\partial N^0}{c\partial t}+m_0\frac{\partial N^i}{\partial x^i}-\frac{\partial}{c^2\partial t}\left(\tau_0\frac{\partial T^{00}}{\partial t}\right)\\&-\frac{\partial}{c\partial t}\left(\tau_0\frac{\partial T^{0i}}{\partial x^i}\right)-\frac{\partial}{\partial x^i}\left(\tau_0\frac{\partial T^{0i}}{c\partial t}\right)-\frac{\partial}{\partial x^i}\left(\tau_0\frac{\partial T^{ij}}{\partial x^j}\right)=0,\end{aligned} \quad (12.4.4)$$

$$\begin{aligned}&\frac{\partial T^{l0}}{c\partial t}+\frac{\partial T^{li}}{\partial x^i}-\frac{\partial}{c^2\partial t}\left(\frac{\tau_0}{m_0}\frac{\partial T^{l00}}{\partial t}\right)\\&-\frac{\partial}{c\partial t}\left(\frac{\tau_0}{m_0}\frac{\partial T^{li0}}{\partial x^i}\right)-\frac{\partial}{c\partial x^i}\left(\frac{\tau_0}{m_0}\frac{\partial T^{li0}}{\partial t}\right)-\frac{\partial}{\partial x^i}\left(\frac{\tau_0}{m_0}\frac{\partial T^{lij}}{\partial x^j}\right)=0,\end{aligned} \quad (12.4.5)$$

$$
\begin{aligned}
&\frac{\partial}{c\partial t}\left(T^{00}-m_0cN^0\right)+\frac{\partial}{\partial x^i}\left(T^{0i}-m_0cN^i\right)-\frac{\partial}{c^2\partial t}\left[\frac{\tau_0}{m_0}\frac{\partial}{\partial t}\left(T^{000}-m_0cT^{00}\right)\right]\\
&-\frac{\partial}{c\partial t}\left[\frac{\tau_0}{m_0}\frac{\partial}{\partial x^i}\left(T^{00i}-m_0cT^{0i}\right)\right]-\frac{\partial}{c\partial x^i}\left[\frac{\tau_0}{m_0}\frac{\partial}{\partial t}\left(T^{00i}-m_0cT^{0i}\right)\right]\\
&-\frac{\partial}{\partial x^i}\left[\frac{\tau_0}{m_0}\frac{\partial}{\partial x^j}\left(T^{0ij}-m_0cT^{ij}\right)\right]=0,
\end{aligned}
\tag{12.4.6}
$$

or Form 2:

$$
\frac{\partial}{c\partial t}\left\{m_0N^0-\tau_0\left[\frac{\partial T^{00}}{c\partial t}+\frac{\partial T^{0i}}{\partial x^i}\right]\right\}+\frac{\partial}{\partial x^i}\left\{m_0N^i-\tau_0\left[\frac{\partial T^{0i}}{c\partial t}+\frac{\partial T^{ij}}{\partial x^j}\right]\right\}=0,
\tag{12.4.7}
$$

$$
\frac{\partial}{c\partial t}\left\{T^{l0}-\frac{\tau_0}{m_0}\left[\frac{\partial T^{l00}}{c\partial t}+\frac{\partial T^{li0}}{\partial x^i}\right]\right\}+\frac{\partial}{\partial x^i}\left\{T^{li}-\frac{\tau_0}{m_0}\left[\frac{\partial T^{li0}}{c\partial t}+\frac{\partial T^{lij}}{\partial x^j}\right]\right\}=0,
\tag{12.4.8}
$$

$$
\begin{aligned}
&\frac{\partial}{c\partial t}\left\{T^{00}-m_0cN^0-\frac{\tau_0}{m_0}\left[\frac{\partial}{c\partial t}\left(T^{000}-m_0cT^{00}\right)+\frac{\partial}{\partial x^i}\left(T^{00i}-m_0cT^{0i}\right)\right]\right\}\\
&+\frac{\partial}{\partial x^i}\left\{T^{0i}-m_0cN^i-\frac{\tau_0}{m_0}\left[\frac{\partial}{c\partial t}\left(T^{00i}-m_0cT^{0i}\right)+\frac{\partial}{\partial x^j}\left(T^{0ij}-m_0cT^{ij}\right)\right]\right\}=0.
\end{aligned}
\tag{12.4.9}
$$

The macroscopic description of the relativistic gas is based on the moments of the DF and defined by tensors

$$
T^{\alpha\beta\ldots\gamma\delta}=c\int p^\alpha p^\beta\ldots p^\gamma p^\delta f\frac{\mathrm{d}^3p}{p^0}.
\tag{12.4.10}
$$

For example, the first moment transform (4) into the particle four-flow. The second moment is the energy-momentum tensor

$$
T^{\alpha\beta}=c\int p^\alpha p^\beta f\frac{\mathrm{d}^3p}{p^0},
\tag{12.4.11}
$$

and so on.

Let us calculate the tensor components N^α, $T^{\alpha\beta}$, $T^{\alpha\beta\delta}$ in the local coordinate system of rest R for the equilibrium distribution function; denote as $\mathbf{v}_0$—the hydrodynamic velocity of the gas flow and u^α—the corresponding 4-vector, defined as

$$
u^\alpha=(c\gamma_0,\mathbf{v}_0\gamma_0),
\tag{12.4.12}
$$

where $\gamma_0=\dfrac{1}{\sqrt{1-v_0^2/c^2}}$,

$$
u^\alpha u_\alpha=c^2\gamma_0^2-v_0^2\gamma_0^2=c^2=inv.
\tag{12.4.13}
$$

Consider the local system of rest R, which is moving with the velocity $\mathbf{v}_0$. In this system 4-velocity vector has components

$$
u^\alpha=(c,0,0,0).
\tag{12.4.14}
$$

The following tensor components take place in R

$$
N_R^\alpha=(cn_R;\mathbf{0}),
\tag{12.4.15}
$$

where n_R—particle density in system R;

$$
n_R=\int f\mathrm{d}^3p.
\tag{12.4.16}
$$

$$
T_R^{00}=c\int p^0 f\mathrm{d}^3p=n_R\mathrm{e},
\tag{12.4.17}
$$

where e is the mean energy per one particle in R-system, then T_R^{00} is the particle energy in the unit of volume. In the following it is more informative to use the Cartesian coordinates, where

$$
T_R^{xx}=c\int p^x m_0 v^x\gamma f\frac{\mathrm{d}^3p}{m_0c\gamma}=\int p^x v^x f\mathrm{d}^3p=p_R
\tag{12.4.18}
$$

is the static pressure in R,

$$T_R^{xx} = T_R^{yy} = T_R^{zz}, \tag{12.4.19}$$

$$T^{11} = T^{xx}, T^{12} = T^{xy}, T^{01} = T^{0x}, \tag{12.4.20}$$

and so on.

The equilibrium distribution function (EDF) is written as

$$f^0 = \frac{g_s}{h^3} \exp\left(\frac{\mu}{k_B T} - \frac{u^\alpha p_\alpha}{k_B T}\right), \tag{12.4.21}$$

where μ is chemical potential, g_s—degeneration factor. In the frame R EDF transforms into the form

$$f^0 = \frac{g_s}{h^3} \exp\left(\frac{\mu}{k_B T} - \frac{\zeta}{m_0 c} p^0\right), \tag{12.4.22}$$

where

$$\zeta = \frac{m_0 c^2}{k_B T}. \tag{12.4.23}$$

Only the following tensor components $T_R^{\alpha\beta\delta}$ are not equal to zero for the Maxwellian EDF in the system R: T_R^{000}, $T_R^{xx0} = T_R^{x0x} = T_R^{0xx} = T_R^{yy0} = T_R^{y0y} = T_R^{0yy} = T_R^{zz0} = T_R^{z0z} = T_R^{0zz}$. All mentioned functions n_R, $n_R e$, p_R, T_R^{000}, T_R^{xx0} can be expressed as functions of ζ and the modified Bessel functions

$$K_n(\zeta) = \left(\frac{\zeta}{2}\right)^n \frac{\Gamma(1/2)}{\Gamma\left(n + \frac{1}{2}\right)} \int_1^\infty e^{-\zeta y} \left(y^2 - 1\right)^{n-(1/2)} dy. \tag{12.4.24}$$

After calculations of the tensor values and the substitution in the system of Eqs. (12.4.7)–(12.4.9) we will reach the generalized non-local relativistic Euler equations. Let us realize the mentioned program.

(a) Calculation of the number density n_R.
Substitute (12.4.22) in (12.4.16), one obtains

$$n_R = \frac{g_s}{h^3} e^{\mu/k_B T} \int e^{-\zeta p^0/m_0 c} d^3 p = \frac{g_s}{h^3} e^{\mu/k_B T} \int_0^\infty e^{-\zeta \sqrt{p^2 + m_0^2 c^2}/m_0 c} 4\pi p^2 \, dp. \tag{12.4.25}$$

Introduce the change of the independent variable p

$$y = \sqrt{\frac{p^2}{m_0^2 c^2} + 1}, \tag{12.4.26}$$

$$dp = \frac{m_0 c y}{\sqrt{y^2 - 1}}. \tag{12.4.27}$$

Then

$$n_R = \frac{g_s}{h^3} e^{\mu/k_B T} 4\pi (m_0 c)^3 \int_1^\infty e^{-\zeta y} \left(y^2 - 1\right)^{1/2} y \, dy. \tag{12.4.28}$$

This kind of integrals can be expressed via the modified Bessel function (12.4.24):

$$n_R = \frac{g_s}{h^3} e^{\mu/k_B T} 4\pi (m_0 c)^3 \frac{K_2(\zeta)}{\zeta}. \tag{12.4.29}$$

(b) Calculation of the energy density $n_R e$.
Substitute Eq. (12.4.22) with Eq. (12.4.17),

$$n_R e = \frac{g_s}{h^3} e^{\mu/k_B T} \int e^{-\zeta/m_0 c \, p^0} c p^0 d^3 p = \frac{g_s}{h^3} e^{\mu/k_B T} 4\pi \int_0^\infty e^{-\zeta/m_0 c \sqrt{p^2 + m_0^2 c^2}} c \sqrt{p^2 + m_0^2 c^2} p^2 \, dp \tag{12.4.30}$$

and use Eqs. (12.4.23) and (12.4.24):

$$n_R \mathrm{e} = \frac{g_s}{h^3} \mathrm{e}^{\mu/k_B T} 4\pi c (m_0 c)^4 \int_1^\infty \mathrm{e}^{-\zeta y} y^2 \sqrt{y^2-1}\,\mathrm{d}y. \tag{12.4.31}$$

Introduce the modified Bessel function (12.4.24); we have from Eq. (12.4.31):

$$n_R \mathrm{e} = \frac{g_s}{h^3} \mathrm{e}^{\mu/k_B T} 4\pi c (m_0 c)^4 \left(\frac{K_1(\zeta)}{\zeta} + \frac{3K_2(\zeta)}{\zeta^2} \right). \tag{12.4.32}$$

(c) Calculation of the pressure p_R.
Substitute Eq. (12.4.22) with Eq. (12.4.18),

$$p_R = \frac{g_s}{h^3} \mathrm{e}^{\mu/k_B T} c \int \mathrm{e}^{-\zeta/m_0 c\, p^0} p^{x2} \frac{\mathrm{d}^3 p}{p^0} = \frac{g_s}{h^3} \mathrm{e}^{\mu/k_B T} \frac{4\pi}{3} c \int_0^\infty \mathrm{e}^{-\zeta/m_0 c \sqrt{p^2+m_0^2 c^2}} p^2 p^2 \frac{\mathrm{d}p}{p^0} \tag{12.4.33}$$

and use Eqs. (12.4.26) and (12.4.27)

$$p_R = \frac{4\pi}{3} \frac{g_s}{h^3} c\, \mathrm{e}^{\mu/k_B T} \int_1^\infty \mathrm{e}^{-\zeta y} (m_0 c)^4 \left(y^2-1\right)^{3/2} \mathrm{d}y \tag{12.4.34}$$

we find

$$p_R = 4\pi c \frac{g_s}{h^3} \mathrm{e}^{\mu/k_B T} (m_0 c)^4 \frac{K_2(\zeta)}{\zeta^2} = n_R k_B T. \tag{12.4.35}$$

The values in relations (12.4.29), (12.4.32), and (12.4.35) correspond to the Boltzmann local relativistic part of the kinetic equation. We need in calculation also the two components of the third range tensor $T^{\alpha\beta\delta}$ for the non-local part of the kinetic equation.

(d) Calculation of T_R^{xx0}.
From Eq. (12.4.4) follows

$$T_R^{xx0} = c \int p^x p^x p^0 f \frac{\mathrm{d}^3 p}{p^0} = \frac{g_s}{h^3} \mathrm{e}^{\mu/k_B T} c \int \mathrm{e}^{-\zeta/m_0 c\, p^0} p^{x2} \mathrm{d}^3 p = \frac{g_s}{h^3} \mathrm{e}^{\mu/k_B T} \frac{4\pi}{3} c \int_0^\infty \mathrm{e}^{-\zeta/m_0 c \sqrt{p^2+m_0^2 c^2}} p^2 p^2\, \mathrm{d}p \tag{12.4.36}$$

After substitution (12.4.26), (12.4.27) we have

$$T_R^{xx0} = \frac{g_s}{h^3} \mathrm{e}^{\mu/k_B T} \frac{4\pi}{3} c \int_1^\infty m_0^4 c^4 \left(y^2-1\right)^2 \mathrm{e}^{-\zeta y} \frac{m_0 c y}{\sqrt{y^2-1}} \mathrm{d}y = \frac{g_s}{h^3} \mathrm{e}^{\mu/k_B T} \frac{4\pi}{3} m_0^5 c^6 \int_1^\infty \left(y^2-1\right)^{2-1/2} \mathrm{e}^{-\zeta y} y\, \mathrm{d}y. \tag{12.4.37}$$

Integration (12.4.24) by parts leads to convenient relations

$$K_n(\zeta) = \left(\frac{\zeta}{2}\right)^n \frac{\Gamma(1/2)}{\Gamma(n+1/2)} \frac{1}{\zeta} \int_1^\infty \mathrm{e}^{-\zeta y} 2y \left(y^2-1\right)^{n-(3/2)} (n-1/2)\, \mathrm{d}y, \tag{12.4.38}$$

$$\int_1^\infty y\left(y^2-1\right)^{n-(3/2)} \mathrm{e}^{-\zeta y}\, \mathrm{d}y = \frac{K_n(\zeta)\Gamma(n+1/2)}{\left(\frac{\zeta}{2}\right)^{n-1} \Gamma(1/2)(n-1/2)}. \tag{12.4.39}$$

Using Eq. (12.4.39), we have from Eq. (12.4.37)

$$T_R^{xx0} = \frac{g_s}{h^3} \mathrm{e}^{\mu/k_B T} \frac{4\pi}{3} m_0^5 c^6 \frac{K_3(\zeta)\Gamma(7/2)}{\left(\frac{\zeta}{2}\right)^2 \Gamma(1/2)(3-1/2)} = \frac{g_s}{h^3} \mathrm{e}^{\mu/k_B T} \frac{4\pi}{3} m_0^5 c^6 \frac{K_3(\zeta)(5/2)(3/2)\frac{\sqrt{\pi}}{2}}{\frac{\zeta^2}{4}\sqrt{\pi}5/2}$$
$$= \frac{g_s}{h^3} \mathrm{e}^{\mu/k_B T} \frac{4\pi}{3} m_0^5 c^6 \frac{3K_3(\zeta)}{\zeta^2} \tag{12.4.40}$$

Now using the relations (12.4.29) for n_R and (12.4.23) for ζ, one obtains

$$T_R^{xx0} = m_0 c n_R k_B T \frac{K_3(\zeta)}{K_2(\zeta)}, \tag{12.4.41}$$

or after application of Eq. (12.4.35)

$$T_R^{xx0} = m_0 c p_R \frac{K_3(\zeta)}{K_2(\zeta)}. \tag{12.4.42}$$

(e) Calculation of the tensor component T_R^{000}.
From definition (12.4.10) follows

$$\begin{aligned} T_R^{000} &= c\int p^0 p^0 p^0 \frac{d^3p}{p^0} = c\int_0^\infty (p^0)^2 f 4\pi p^2\,dp = \frac{g_s}{h^3} e^{\mu/k_B T} c4\pi \int_0^\infty e^{-\frac{\zeta}{m_0 c}p^0} (p^0)^2 p^2\,dp \\ &= \frac{g_s}{h^3} e^{\mu/k_B T} c4\pi \int_0^\infty e^{-(\zeta/m_0 c)\sqrt{p^2+m_0^2c^2}} (p^2+m_0^2c^2) p^2\,dp. \end{aligned} \tag{12.4.43}$$

After substituting Eqs. (12.4.26) and (12.4.27) we have

$$\begin{aligned} T_R^{000} &= \frac{g_s}{h^3} e^{\mu/k_B T} c4\pi \int_1^\infty e^{-\zeta y} m_0^2 c^2 y^2 m_0^2 c^2 (y^2-1) \frac{m_0 c y\,dy}{\sqrt{y^2-1}} \\ &= \frac{g_s}{h^3} e^{\mu/k_B T} c4\pi (m_0 c)^5 \int_1^\infty e^{-\zeta y} y^3 \sqrt{y^2-1}\,dy. \end{aligned} \tag{12.4.44}$$

Transform integral

$$\begin{aligned} \int_1^\infty e^{-\zeta y} y^3 \sqrt{y^2-1}\,dy &= \int_1^\infty e^{-\zeta y} y (y^2-1)\sqrt{y^2-1}\,dy + \int_1^\infty e^{-\zeta y} y \sqrt{y^2-1}\,dy \\ &= \int_1^\infty e^{-\zeta y} y (y^2-1)^{3-3/2}\,dy + \int_1^\infty e^{-\zeta y} y (y^2-1)^{2-3/2}\,dy, \end{aligned} \tag{12.4.45}$$

using (12.4.24); one obtains

$$\begin{aligned} \int_1^\infty e^{-\zeta y} y^3 \sqrt{y^2-1}\,dy &= \frac{K_3(\zeta)\Gamma(3+1/2)}{(\zeta/2)^2\Gamma(1/2)(3-1/2)} + \frac{K_2(\zeta)\Gamma(2+1/2)}{(\zeta/2)\Gamma(1/2)(2-1/2)} \\ &= \frac{K_3(\zeta)(5/2)(3/2)(\sqrt{\pi}/2)}{\frac{\zeta^2}{4}\sqrt{\pi}(5/2)} + \frac{K_2(\zeta)(3/2)(\sqrt{\pi}/2)}{\frac{\zeta}{2}\sqrt{\pi}(3/2)} = \frac{3K_3(\zeta)}{\zeta^2} + \frac{K_2(\zeta)}{\zeta} \end{aligned} \tag{12.4.46}$$

Substituting Eq. (12.4.46) into Eq. (12.4.44), we find

$$T_R^{000} = \frac{g_s}{h^3} e^{\mu/k_B T} c4\pi (m_0 c)^5 \left(\frac{3K_3(\zeta)}{\zeta^2} + \frac{K_2(\zeta)}{\zeta}\right). \tag{12.4.47}$$

The expression for the number density n_R from Eq. (12.4.29) can be applied for the following transformation of Eq. (12.4.47).

$$T_R^{000} = m_0^2 c^3 n_R \left(\frac{3K_3(\zeta)}{\zeta K_2(\zeta)} + 1\right). \tag{12.4.48}$$

On this step of investigations, the generalized relativistic non-local kinetic and hydrodynamic theory is constructed without taking into account the external forces. In the following, the developed theory will include the presence of the external forces and, in particular, we intend to present the non-local relativistic Euler equations and show effectiveness of these equations for investigations of propagation of the plane harmonic waves of small amplitudes through a relativistic gas at rest. A program to determine the precise form for a relativistic viscous fluid was initiated in the last decade; this program was driven by nuclear physics experiments in colliders on relativistic heavy-ion collisions (which lead to quark-gluon plasma) and problems of quantum description of transport processes in nano-electronics. We aware, that the developed generalized non-local relativistic theory is adequate to this aims.

12.5 GENERALIZED NON-LOCAL RELATIVISTIC EULER EQUATIONS

Generalized relativistic Euler equations can be obtained after substitution the explicit forms of the tensor components N^α, $T^{\alpha\beta}$, $T^{\alpha\beta\delta}$, calculated above with the help of relativistic EDF, in hydrodynamic equations (12.4.7)–(12.4.9).

For the one dimensional case, which is considered further, more simple way can be applied for derivation of the mentioned equations. Let us admit that the system coordinate R is moving along the velocity $\mathbf{v}_0$ in coordinate system E. Denote as x^α—coordinates in E and x'^α—corresponding coordinates in R. In the general case, the Lorentz transformation (see [6]) has the form

$$x^\alpha = \overline{\Lambda}^\alpha_\beta x'_\beta, \tag{12.5.1}$$

where

$$\overline{\Lambda}^\alpha_\beta = \begin{pmatrix} \gamma_0 & \gamma_0 \frac{v_0'^1}{c} & \gamma_0 \frac{v_0'^2}{c} & \gamma_0 \frac{v_0'^3}{c} \\ \gamma_0 \frac{v_0^1}{c} & 1 + \frac{(\gamma_0 - 1) v_0^1 v_0^1}{\mathbf{v}_0^2} & \frac{(\gamma_0 - 1) v_0^1 v_0'^2}{\mathbf{v}_0^2} & \frac{(\gamma_0 - 1) v_0^1 v_0'^3}{\mathbf{v}_0^2} \\ \gamma_0 \frac{v_0'^2}{c} & \frac{(\gamma_0 - 1) v_0'^2 v_0^1}{\mathbf{v}_0^2} & 1 + \frac{(\gamma_0 - 1) v_0'^2 v_0'^2}{\mathbf{v}_0^2} & \frac{(\gamma_0 - 1) v_0'^2 v_0'^3}{\mathbf{v}_0^2} \\ \gamma_0 \frac{v_0'^3}{c} & \frac{(\gamma_0 - 1) v_0'^3 v_0^1}{\mathbf{v}_0^2} & \frac{(\gamma_0 - 1) v_0'^3 v_0'^2}{\mathbf{v}_0^2} & 1 + \frac{(\gamma_0 - 1) v_0'^3 v_0'^3}{\mathbf{v}_0^2} \end{pmatrix}. \tag{12.5.2}$$

or

$$\overline{\Lambda}^0_0 = \gamma_0, \overline{\Lambda}^i_0 = \gamma_0 \frac{v_0^i}{c}, \overline{\Lambda}^0_i = \gamma_0 \frac{v_0^j}{c}\delta_{ij}, \overline{\Lambda}^i_j = \delta^i_j + (\gamma_0 - 1)\frac{v_0^i v_0^k}{\mathbf{v}_0^2}\delta_{jk} \tag{12.5.3}$$

δ^β_α and $\delta_{\alpha\beta}$—Kronecker symbols.

$$\delta^\beta_\alpha = \begin{cases} 1, & if\, \alpha = \beta \\ 0, & if\, \alpha \neq \beta \end{cases}, \tag{12.5.4}$$

$\delta_{\alpha\beta}$ is defined analogically.

This transformation is applied to all physically small volumes. Write down the tensor components N^α, $T^{\alpha\beta}$, $T^{\alpha\beta\delta}$ in the E-coordinate system using their known form in R-system.

(a) Tensor N^α in the frame E.
After using Eqs. (12.4.15) and (12.5.3), we have

$$N^0 = \overline{\Lambda}^0_0 N^0_R = c\gamma_0 n_R \tag{12.5.5}$$

$$N^i = \overline{\Lambda}^i_0 N^0_R = \gamma_0 \frac{v_0^i}{c} c n_R = \gamma_0 v_0^i n_R. \tag{12.5.6}$$

As result

$$N^\alpha = n_R u^\alpha. \tag{12.5.7}$$

(b) Tensor $T^{\alpha\beta}$ in the frame E.
Using the rule of tensor transformation

$$T^{\alpha\beta} = \overline{\Lambda}^\alpha_\xi \overline{\Lambda}^\beta_\sigma T^{\xi\sigma}_R \tag{12.5.8}$$

one obtains

$$T^{00} = \overline{\Lambda}^0_\xi \overline{\Lambda}^0_\sigma T^{\xi\sigma}_R = \overline{\Lambda}^0_0 \overline{\Lambda}^0_0 T^{00}_R + \overline{\Lambda}^0_i \overline{\Lambda}^0_i T^{ii}_R, \tag{12.5.9}$$

After transformations with the help of Eqs. (12.4.17)–(12.4.19) and (12.5.3), we have from Eqs. (12.5.8) and (12.5.9):

$$\begin{aligned} T^{00} &= \gamma_0^2 n_R \mathrm{e} + \gamma_0^2 \frac{v_0^i v_0^i}{c^2} p_R = \gamma_0^2 n_R \mathrm{e} + \gamma_0^2 \frac{\mathbf{v}_0^2}{c^2} p_R \\ &= \gamma_0^2 n_R \mathrm{e} + \gamma_0^2 \left(1 - \frac{1}{\gamma_0^2}\right) p_R = \gamma_0^2 (n_R \mathrm{e} + p_R) - p_R, \end{aligned} \tag{12.5.10}$$

$$\begin{aligned} T^{i0} &= T^{0i} = \overline{\Lambda}^i_\xi \overline{\Lambda}^0_\sigma T_R^{\xi\sigma} = \overline{\Lambda}^i_0 \overline{\Lambda}^0_0 T_R^{00} + \overline{\Lambda}^i_j \overline{\Lambda}^0_j T_R^{jj} \\ &= \gamma_0 \frac{v_0^i}{c} \gamma_0 n_R \mathrm{e} + \left(\delta_j^i + (\gamma_0 - 1) \frac{v_0^i v_0^j}{\mathbf{v}_0^2} \right) \gamma_0 \frac{v_0^j}{c} p_R \\ &= \gamma_0^2 \frac{v_0^i}{c} n_R \mathrm{e} + \gamma_0 \frac{v_0^i}{c} p_R + (\gamma_0 - 1) \frac{v_0^i}{c} \gamma_0 p_R = (n_R \mathrm{e} + p_R) \gamma_0^2 \frac{v_0^i}{c}, \end{aligned} \tag{12.5.11}$$

$$\begin{aligned} T^{ij} &= \overline{\Lambda}^i_\xi \overline{\Lambda}^j_\sigma T_R^{\xi\sigma} = \overline{\Lambda}^i_0 \overline{\Lambda}^j_0 T_R^{00} + \overline{\Lambda}^i_k \overline{\Lambda}^j_k T_R^{kk} \\ &= \gamma_0 \frac{v_0^i}{c} \gamma_0 \frac{v_0^j}{c} n_R \mathrm{e} + \left(\delta_k^i + (\gamma_0 - 1) \frac{v_0^i v_0^m}{\mathbf{v}_0^2} \delta_{km} \right) \left(\delta_k^j + (\gamma_0 - 1) \frac{v_0^j v_0^l}{\mathbf{v}_0^2} \delta_{lk} \right) p_R \\ &= \gamma_0^2 \frac{v_0^i}{c} \frac{v_0^j}{c} n_R \mathrm{e} + \left(\delta_k^i \delta_k^j p_R + 2(\gamma_0 - 1) \frac{v_0^i v_0^j}{\mathbf{v}_0^2} p_R + (\gamma_0 - 1)^2 \frac{v_0^i v_0^j v_0^k v_0^k}{\mathbf{v}_0^2 \mathbf{v}_0^2} p_R \right) \\ &= \gamma_0^2 \frac{v_0^i}{c} \frac{v_0^j}{c} n_R \mathrm{e} + \left(\delta^{ij} p_R + \frac{v_0^i v_0^j}{\mathbf{v}_0^2} p_R \left(2(\gamma_0 - 1) + (\gamma_0 - 1)^2 \right) \right) \\ &= \gamma_0^2 \frac{v_0^i}{c} \frac{v_0^j}{c} n_R \mathrm{e} + \delta^{ij} p_R + p_R \frac{v_0^i v_0^j}{\mathbf{v}_0^2} (\gamma_0^2 - 1) = \gamma_0^2 \frac{v_0^i}{c} \frac{v_0^j}{c} n_R \mathrm{e} + \delta^{ij} p_R + p_R \frac{v_0^i v_0^j}{\mathbf{v}_0^2} \frac{\mathbf{v}_0^2}{c^2} \gamma_0^2 \\ &= (n_R \mathrm{e} + p_R) \gamma_0^2 \frac{v_0^i v_0^j}{c^2} + p_R \delta^{ij}. \end{aligned} \tag{12.5.12}$$

After introduction of the matrix $\eta^{\alpha\beta}$ as

$$\eta^{\alpha\beta} = \begin{pmatrix} 1 & 0 & 0 & 0 \\ 0 & -1 & 0 & 0 \\ 0 & 0 & -1 & 0 \\ 0 & 0 & 0 & -1 \end{pmatrix}, \tag{12.5.13}$$

the following compact general inscription for the tensor components $T^{\alpha\beta}$ is valid

$$T^{\alpha\beta} = (n_R \mathrm{e} + p_R) \frac{u^\alpha u^\beta}{c^2} - p \eta^{\alpha\beta}. \tag{12.5.14}$$

(c) Tensor $T^{\alpha\beta\delta}$ in the frame E.

Using the rule of tensor transformation

$$T^{\alpha\beta\delta} = \overline{\Lambda}^\alpha_\xi \overline{\Lambda}^\beta_\sigma \overline{\Lambda}^\delta_\chi T_R^{\xi\sigma\chi}. \tag{12.5.15}$$

If the vector of velocity $\mathbf{v}_0$ is directed along x-axis, then $v_0^1 = v_0^x = v_0$, $\overline{\Lambda}^x_0 = \overline{\Lambda}^0_x = \gamma_0 v_0 / c$, $\overline{\Lambda}^x_x = \gamma_0$, $\overline{\Lambda}^0_0 = \gamma_0$, $\overline{\Lambda}^y_y = \overline{\Lambda}^z_z = 1$; the other elements of matrix $\overline{\Lambda}$ turn into zero. Calculate some non-zero tensor components which we need

$$\begin{aligned} T^{xxx} &= \overline{\Lambda}^x_\xi \overline{\Lambda}^x_\sigma \overline{\Lambda}^x_\chi T_R^{\xi\sigma\chi} = \overline{\Lambda}^x_0 \overline{\Lambda}^x_0 \overline{\Lambda}^x_0 T_R^{000} + \overline{\Lambda}^x_x \overline{\Lambda}^x_x \overline{\Lambda}^x_0 T_R^{xx0} + \overline{\Lambda}^x_0 \overline{\Lambda}^x_x \overline{\Lambda}^x_x T_R^{0xx} + \overline{\Lambda}^x_x \overline{\Lambda}^x_0 \overline{\Lambda}^x_x T_R^{x0x} \\ &= \overline{\Lambda}^x_0 \overline{\Lambda}^x_0 \overline{\Lambda}^x_0 T_R^{000} + 3 \overline{\Lambda}^x_x \overline{\Lambda}^x_x \overline{\Lambda}^x_0 T_R^{xx0} = \left(\gamma_0 \tfrac{v_0}{c}\right)^3 T_R^{000} + 3\gamma_0^2 \left(\gamma_0 \frac{v_0}{c}\right) T_R^{xx0}. \end{aligned} \tag{12.5.16}$$

After substitution of Eqs. (12.4.42) and (12.4.48) into Eq. (12.5.16)

$$T^{xxx} = \left(\gamma_0 \frac{v_0}{c}\right)^3 m_0^2 c^3 n_R \left(\frac{3K_3(\zeta)}{\zeta K_2(\zeta)} + 1 \right) + 3\gamma_0^2 \left(\gamma_0 \frac{v_0}{c}\right) m_0 c p_R \frac{K_3(\zeta)}{K_2(\zeta)}. \tag{12.5.17}$$

The next non-zero component in the frame E:

$$
\begin{aligned}
T^{xx0} &= \overline{\Lambda}^{x}_{\xi}\overline{\Lambda}^{x}_{\sigma}\overline{\Lambda}^{0}_{\chi}T_R^{\xi\sigma\chi} = \overline{\Lambda}^{x}_{0}\overline{\Lambda}^{x}_{0}\overline{\Lambda}^{0}_{0}T_R^{000} + \overline{\Lambda}^{x}_{x}\overline{\Lambda}^{x}_{x}\overline{\Lambda}^{0}_{0}T_R^{xx0} + \overline{\Lambda}^{x}_{0}\overline{\Lambda}^{x}_{x}\overline{\Lambda}^{0}_{x}T_R^{0xx} + \overline{\Lambda}^{x}_{x}\overline{\Lambda}^{x}_{0}\overline{\Lambda}^{0}_{x}T_R^{x0x} \\
&= \overline{\Lambda}^{x}_{0}\overline{\Lambda}^{x}_{0}\overline{\Lambda}^{0}_{0}T_R^{000} + \overline{\Lambda}^{x}_{x}\overline{\Lambda}^{x}_{x}\overline{\Lambda}^{0}_{0}T_R^{xx0} + 2\overline{\Lambda}^{x}_{0}\overline{\Lambda}^{x}_{x}\overline{\Lambda}^{0}_{x}T_R^{0xx} \\
&= \left(\gamma_0\frac{v_0}{c}\right)^2\gamma_0 T_R^{000} + \left\{\gamma_0^3 + 2\left(\gamma_0\frac{v_0}{c}\right)^2\gamma_0\right\}T_R^{xx0}.
\end{aligned}
\tag{12.5.18}
$$

After substitution of Eqs. (12.4.42) and (12.4.48) into Eq. (12.5.18)

$$
T^{xx0} = \gamma_0^3\left(\frac{v_0}{c}\right)^2 m_0^2c^3n_R\left(\frac{3K_3(\zeta)}{\zeta K_2(\zeta)}+1\right) + \left\{\gamma_0^3 + 2\gamma_0^3\left(\frac{v_0}{c}\right)^2\right\}m_0cp_R\frac{K_3(\zeta)}{K_2(\zeta)}.
\tag{12.5.19}
$$

The next non-zero component in the frame E

$$
\begin{aligned}
T^{x00} &= \overline{\Lambda}^{x}_{\xi}\overline{\Lambda}^{0}_{\sigma}\overline{\Lambda}^{0}_{\chi}T_R^{\xi\sigma\chi} = \overline{\Lambda}^{x}_{0}\overline{\Lambda}^{0}_{0}\overline{\Lambda}^{0}_{0}T_R^{000} + \overline{\Lambda}^{x}_{x}\overline{\Lambda}^{0}_{x}\overline{\Lambda}^{0}_{0}T_R^{xx0} + \overline{\Lambda}^{x}_{x}\overline{\Lambda}^{0}_{0}\overline{\Lambda}^{0}_{x}T_R^{x0x} + \overline{\Lambda}^{x}_{0}\overline{\Lambda}^{0}_{x}\overline{\Lambda}^{0}_{x}T_R^{0xx} \\
&= \overline{\Lambda}^{x}_{0}\overline{\Lambda}^{0}_{0}\overline{\Lambda}^{0}_{0}T_R^{000} + 2\overline{\Lambda}^{x}_{x}\overline{\Lambda}^{0}_{x}\overline{\Lambda}^{0}_{0}T_R^{xx0} + \overline{\Lambda}^{x}_{0}\overline{\Lambda}^{0}_{x}\overline{\Lambda}^{0}_{x}T_R^{0xx} \\
&= \left(\gamma_0\frac{v_0}{c}\right)\gamma_0^2 T_R^{000} + \left\{2\left(\gamma_0\frac{v_0}{c}\right)\gamma_0^2 + \left(\gamma_0\frac{v_0}{c}\right)^3\right\}T_R^{xx0}.
\end{aligned}
\tag{12.5.20}
$$

After substitution of (12.4.42) and (12.4.48) into (12.5.20)

$$
T^{x00} = \gamma_0^3\frac{v_0}{c}m_0^2c^3n_R\left(\frac{3K_3(\zeta)}{\zeta K_2(\zeta)}+1\right) + \left\{2\left(\frac{v_0}{c}\right)\gamma_0^3 + \left(\gamma_0\frac{v_0}{c}\right)^3\right\}m_0cp_R\frac{K_3(\zeta)}{K_2(\zeta)}.
\tag{12.5.21}
$$

And the last non zero tensor component

$$
\begin{aligned}
T^{000} &= \overline{\Lambda}^{0}_{\xi}\overline{\Lambda}^{0}_{\sigma}\overline{\Lambda}^{0}_{\chi}T_R^{\xi\sigma\chi} = \overline{\Lambda}^{0}_{0}\overline{\Lambda}^{0}_{0}\overline{\Lambda}^{0}_{0}T_R^{000} + \overline{\Lambda}^{0}_{x}\overline{\Lambda}^{0}_{x}\overline{\Lambda}^{0}_{0}T_R^{xx0} + \overline{\Lambda}^{0}_{0}\overline{\Lambda}^{0}_{x}\overline{\Lambda}^{0}_{x}T_R^{0xx} + \overline{\Lambda}^{0}_{x}\overline{\Lambda}^{0}_{0}\overline{\Lambda}^{0}_{x}T_R^{x0x} \\
&= \overline{\Lambda}^{0}_{0}\overline{\Lambda}^{0}_{0}\overline{\Lambda}^{0}_{0}T_R^{000} + 3\overline{\Lambda}^{0}_{x}\overline{\Lambda}^{0}_{x}\overline{\Lambda}^{0}_{0}T_R^{xx0} = \gamma_0^3T_R^{000} + 3\left(\gamma_0\frac{v_0}{c}\right)^2\gamma_0T_R^{xx0}.
\end{aligned}
\tag{12.5.22}
$$

After substitution of Eqs. (12.4.42) and (12.4.48) into Eq. (12.5.22), we have

$$
T^{000} = \gamma_0^3m_0^2c^3n_R\left(\frac{3K_3(\zeta)}{\zeta K_2(\zeta)}+1\right) + 3\gamma_0^3\left(\frac{v_0}{c}\right)^2 m_0cp_R\frac{K_3(\zeta)}{K_2(\zeta)}.
\tag{12.5.23}
$$

Now the generalized non-local relativistic equations (Form 2) for the non-stationary 1D case

$$
\frac{\partial}{c\partial t}\left\{m_0N^0 - \tau_0\left[\frac{\partial T^{00}}{c\partial t} + \frac{\partial T^{0x}}{\partial x}\right]\right\} + \frac{\partial}{\partial x}\left\{m_0N^x - \tau_0\left[\frac{\partial T^{0x}}{c\partial t} + \frac{\partial T^{xx}}{\partial x}\right]\right\} = 0.
\tag{12.5.24}
$$

$$
\frac{\partial}{c\partial t}\left\{T^{x0} - \frac{\tau_0}{m_0}\left[\frac{\partial T^{x00}}{c\partial t} + \frac{\partial T^{xx0}}{\partial x}\right]\right\} + \frac{\partial}{\partial x}\left\{T^{xx} - \frac{\tau_0}{m_0}\left[\frac{\partial T^{xx0}}{c\partial t} + \frac{\partial T^{xxx}}{\partial x}\right]\right\} = 0.
\tag{12.5.25}
$$

$$
\begin{aligned}
&\frac{\partial}{c\partial t}\left\{T^{00} - m_0cN^0 - \frac{\tau_0}{m_0}\left[\frac{\partial}{c\partial t}\left(T^{000} - m_0cT^{00}\right) + \frac{\partial}{\partial x}\left(T^{00x} - m_0cT^{0x}\right)\right]\right\} \\
&+ \frac{\partial}{\partial x}\left\{T^{0x} - m_0cN^x - \frac{\tau_0}{m_0}\left[\frac{\partial}{c\partial t}\left(T^{00x} - m_0cT^{0x}\right) + \frac{\partial}{\partial x}\left(T^{0xx} - m_0cT^{xx}\right)\right]\right\} = 0.
\end{aligned}
\tag{12.5.26}
$$

can be written for the non-stationary 1D case in the explicit form, after taking into account that

$$
N^x = \gamma_0 v_0 n_R,
\tag{12.5.27}
$$

$$
T^{xo} = (n_R\mathrm{e} + p_R)\gamma_0^2\frac{v_0}{c},
\tag{12.5.28}
$$

$$
T^{xx} = (n_R\mathrm{e} + p_R)\gamma_0^2\left(\frac{v_0}{c}\right)^2 + p_R,
\tag{12.5.29}
$$

Using the relations (12.5.5), (12.5.10), (12.5.17), (12.5.19), (12.5.21) and (12.5.23) we yield the non-stationary 1D hydrodynamic equations. Namely,

continuity equation

$$\frac{\partial}{\partial t}\left\{m_0 n_R \gamma_0 - \tau_0\left[\frac{\partial}{c^2\partial t}\left((n_R\mathrm{e}+p_R)\gamma_0^2 - p_R\right) + \frac{\partial}{c^2\partial x}\left((n_R\mathrm{e}+p_R)\gamma_0^2 v_0\right)\right]\right\}$$
$$+\frac{\partial}{\partial x}\left\{m_0 n_R \gamma_0 v_0 - \tau_0\left[\frac{\partial}{c^2\partial t}\left((n_R\mathrm{e}+p_R)\gamma_0^2 v_0\right) + \frac{\partial}{\partial x}\left((n_R\mathrm{e}+p_R)\gamma_0^2\left(\frac{v_0}{c}\right)^2 + p_R\right)\right]\right\} = 0. \quad (12.5.30)$$

Motion equation

$$\frac{\partial}{c\partial t}\left\{(n_R\mathrm{e}+p_R)\gamma_0^2\frac{v_0}{c} - \frac{\tau_0}{m_0}\left[\frac{\partial}{c\partial t}\left(\gamma_0^3 v_0 m_0^2 c^2 n_R\left(\frac{3K_3(\zeta)}{\zeta K_2(\zeta)}+1\right) + \left(\left(\gamma_0\frac{v_0}{c}\right)^3 + 2\gamma_0^3\left(\frac{v_0}{c}\right)\right) m_0 c p_R \frac{K_3(\zeta)}{K_2(\zeta)}\right)\right.\right.$$
$$\left.\left.+\frac{\partial}{\partial x}\left(\gamma_0^3 v_0^2 m_0^2 c n_R\left(\frac{3K_3(\zeta)}{\zeta K_2(\zeta)}+1\right) + \left(\gamma_0^3 + 2\gamma_0^3\left(\frac{v_0}{c}\right)^2\right) m_0 c p_R \frac{K_3(\zeta)}{K_2(\zeta)}\right)\right]\right\} + \frac{\partial}{\partial x}\left\{(n_R\mathrm{e}+p_R)\gamma_0^2\left(\frac{v_0}{c}\right)^2\right.$$
$$+p_R - \frac{\tau_0}{m_0}\left[\frac{\partial}{c\partial t}\left(\gamma_0^3 v_0^2 m_0^2 c n_R\left(\frac{3K_3(\zeta)}{\zeta K_2(\zeta)}+1\right) + \left(\gamma_0^3 + 2\gamma_0^3\left(\frac{v_0}{c}\right)^2\right) m_0 c p_R \frac{K_3(\zeta)}{K_2(\zeta)}\right)\right.$$
$$\left.\left.+\frac{\partial}{\partial x}\left((\gamma_0 v_0)^3 m_0^2 n_R\left(\frac{3K_3(\zeta)}{\zeta K_2(\zeta)}+1\right) + 3\gamma_0^3\left(\frac{v_0}{c}\right) m_0 c p_R \frac{K_3(\zeta)}{K_2(\zeta)}\right)\right]\right\} = 0. \quad (12.5.31)$$

Energy equation

$$\frac{\partial}{c\partial t}\left\{(n_R\mathrm{e}+p_R)\gamma_0^2 - p_R - m_0 c^2 n_R \gamma_0 - \frac{\tau_0}{m_0}\left[\frac{\partial}{c\partial t}\left(\gamma_0^3 m_0^2 c^3 n_R\left(\frac{3K_3(\zeta)}{\zeta K_2(\zeta)}+1\right)\right.\right.\right.$$
$$\left.+3\gamma_0^3\left(\frac{v_0}{c}\right)^2 m_0 c p_R \frac{K_3(\zeta)}{K_2(\zeta)} - m_0 c\left((n_R\mathrm{e}+p_R)\gamma_0^2 - p_R\right)\right) + \frac{\partial}{\partial x}\left(\gamma_0^3 v_0 m_0^2 c^2 n_R\left(\frac{3K_3(\zeta)}{\zeta K_2(\zeta)}+1\right)\right.$$
$$\left.\left.\left.+\left(\left(\gamma_0\frac{v_0}{c}\right)^3 + 2\gamma_0^3\left(\frac{v_0}{c}\right)\right) m_0 c p_R \frac{K_3(\zeta)}{K_2(\zeta)} - (n_R\mathrm{e}+p_R)\gamma_0^2 m_0 v_0\right)\right]\right\} + \frac{\partial}{\partial x}\left\{(n_R\mathrm{e}+p_R)\gamma_0^2\left(\frac{v_0}{c}\right)\right.$$
$$-m_0 c\gamma_0 v_0 n_R - \frac{\tau_0}{m_0}\left[\frac{\partial}{c\partial t}\left(\gamma_0^3 v_0 m_0^2 c^2 n_R\left(\frac{3K_3(\zeta)}{\zeta K_2(\zeta)}+1\right) + \left(\left(\gamma_0\frac{v_0}{c}\right)^3 + 2\gamma_0^3\left(\frac{v_0}{c}\right)\right) m_0 c p_R \frac{K_3(\zeta)}{K_2(\zeta)}\right.\right.$$
$$\left.-(n_R\mathrm{e}+p_R)\gamma_0^2 m_0 v_0\right) + \frac{\partial}{\partial x}\left(\gamma_0^3 v_0^2 m_0^2 c n_R\left(\frac{3K_3(\zeta)}{\zeta K_2(\zeta)}+1\right) + \left(\gamma_0^3 + 2\gamma_0^3\left(\frac{v_0}{c}\right)^2\right) m_0 c p_R \frac{K_3(\zeta)}{K_2(\zeta)}\right.$$
$$\left.\left.\left.-(n_R\mathrm{e}+p_R)\gamma_0^2 v_0\left(\frac{v_0}{c}\right) m_0 - p_R m_0 c\right)\right]\right\} = 0. \quad (12.5.32)$$

12.6 THE LIMIT TRANSFER TO THE NON-RELATIVISTIC GENERALIZED NON-LOCAL EULER EQUATIONS

12.6.1 Some Auxiliary Expressions

Deliver some auxiliary expressions and its asymptotic by the small $\frac{v_0}{c}, \frac{v}{c}$.

(1) Asymptotic of the value $(n_R\mathrm{e}+p_P)\gamma_0^2$.
Asymptotic of the energy density (Eq. (12.4.17))

$$n_R\mathrm{e} = \int \frac{m_0 c^2 f}{\sqrt{1-\left(\frac{v}{c}\right)^2}} \mathrm{d}^3 p \approx \int m_0 c^2\left(1+\frac{v^2}{2c^2}\right) f\,\mathrm{d}^3 p = m_0 n_R c^2 + \frac{m_0 \overline{v^2}}{2} n_R \quad (12.6.1)$$

Taking into account the Lorentz lengths reduction by the one dimensional hydrodynamic motion, we find

$$n_R = n\sqrt{1-\left(\frac{v_0}{c}\right)^2} \tag{12.6.2}$$

and

$$\begin{aligned}(n_R e + p_P)\gamma_0^2 &\approx n\sqrt{1-\left(\frac{v_0}{c}\right)^2}\left(m_0c^2 + \frac{m_0\overline{v^2}}{2}\right)\frac{1}{1-\left(\frac{v_0}{c}\right)^2} + \frac{p_R}{1-\left(\frac{v_0}{c}\right)^2}\\ &\approx \rho c^2 + \frac{\rho v_0^2}{2} + \frac{5}{2}p.\end{aligned} \tag{12.6.3}$$

(2) Asymptotic of $\frac{K_3(\zeta)}{K_2(\zeta)}$.

By $\zeta \gg 1$ $\left(\zeta = \frac{m_0c^2}{k_B T}\right)$ the function $K_n(\zeta)$ can be performed as series expansion:

$$K_n(\zeta) = \sqrt{\frac{\pi}{2\zeta}\frac{1}{e^\zeta}}\left(1 + \frac{4n^2-1}{8\zeta} + \ldots\right). \tag{12.6.4}$$

Then

$$K_3(\zeta) = \sqrt{\frac{\pi}{2\zeta}\frac{1}{e^\zeta}}\left(1 + \frac{35}{8\zeta} + \ldots\right), \tag{12.6.5}$$

$$K_2(\zeta) = \sqrt{\frac{\pi}{2\zeta}\frac{1}{e^\zeta}}\left(1 + \frac{15}{8\zeta} + \ldots\right), \tag{12.6.6}$$

$$\frac{K_3(\zeta)}{K_2(\zeta)} \approx 1 + \frac{35}{8\zeta} - \frac{15}{8\zeta} = 1 + \frac{20}{8\zeta} = 1 + \frac{5}{2\zeta} = 1 + \frac{5k_B T n}{2m_0c^2 n} = 1 + \frac{5}{2}\frac{p}{\rho c^2}. \tag{12.6.7}$$

(3) Asymptotic of $\gamma_0^3 T_R^{000}$ is found using Eqs. (12.4.48), (12.6.2), and (12.6.7):

$$\begin{aligned}\gamma_0^3 T_R^{000} &= \gamma_0^3 m_0^2 c^3 n_R\left(\frac{3K_3(\zeta)}{\zeta K_2(\zeta)} + 1\right) \approx \gamma_0^2 m_0^2 c^3 n\left(\frac{3k_B T}{m_0c^2}\left(1 + \frac{5}{2}\frac{p}{\rho c^2}\right) + 1\right)\\ &\approx \left(1 + \frac{v_0^2}{c^2}\right)m_0^2 c\left(\frac{3nk_B T}{m_0} + nc^2\right) \approx \left(1 + \frac{v_0^2}{c^2}\right)m_0 c(3p + \rho c^2) \approx m_0 c\left(\rho c^2 + \rho v_0^2 + 3p\right).\end{aligned} \tag{12.6.8}$$

12.6.2 Non-Relativistic Generalized Euler Equations as Asymptotic of the Relativistic Equations

(1) Continuity equation.

Consider sequentially all terms in the continuity equation (75) using the first approximation by the decomposition in power series by the small ratios $\frac{v_0}{c}, \frac{v}{c}$:

(a)
$$\frac{\partial}{\partial t}(m_0 n_R \gamma_0) \approx \frac{\partial \rho}{\partial t}, \tag{12.6.9}$$

(b)
$$\frac{\partial}{\partial x}(m_0 n_R \gamma_0 v_0) \approx \frac{\partial}{\partial x}(\rho v_0), \tag{12.6.10}$$

(c)
$$\begin{aligned}&-\frac{\partial}{\partial t}\left\{\tau_0\left[\frac{\partial}{c^2\partial t}\left((n_R e + p_R)\gamma_0^2 - p_R\right)\right]\right\} \approx -\frac{\partial}{\partial t}\left\{\tau_0\left[\frac{\partial}{c^2\partial t}\left(n_R m_0 c^2 + p_R - p_R\right)\right]\right\}\\ &\approx -\frac{\partial}{\partial t}\left\{\tau_0\left[\frac{\partial}{c^2\partial t}(\rho c^2)\right]\right\} = -\frac{\partial}{\partial t}\left\{\tau_0\left[\frac{\partial \rho}{\partial t}\right]\right\},\end{aligned} \tag{12.6.11}$$

(d)
$$-\frac{\partial}{\partial t}\left\{\tau_0\left[\frac{\partial}{c^2\partial x}\left((n_R e+p_R)\gamma_0^2 v_0\right)\right]\right\}\approx-\frac{\partial}{\partial t}\left\{\tau_0\left[\frac{\partial}{c^2\partial x}\left((n_R m_0 c^2+p_R)v_0\right)\right]\right\}$$
$$\approx-\frac{\partial}{\partial t}\left\{\tau_0\left[\frac{\partial}{c^2\partial x}\left((n_R m_0 c^2)v_0\right)\right]\right\}\approx-\frac{\partial}{\partial t}\left\{\tau_0\left[\frac{\partial}{c^2\partial x}\left((\rho c^2)v_0\right)\right]\right\}=-\frac{\partial}{\partial t}\left\{\tau_0\left[\frac{\partial}{\partial x}(\rho v_0)\right]\right\}, \tag{12.6.12}$$

(e)
$$-\frac{\partial}{\partial x}\left\{\tau_0\left[\frac{\partial}{c^2\partial t}\left((n_R e+p_R)\gamma_0^2 v_0\right)\right]\right\}\approx-\frac{\partial}{\partial x}\left\{\tau_0\left[\frac{\partial}{\partial t}(\rho v_0)\right]\right\}, \tag{12.6.13}$$

(f)
$$-\frac{\partial}{\partial x}\left\{\tau_0\left[\frac{\partial}{\partial x}\left((n_R e+p_R)\gamma_0^2\left(\frac{v_0}{c}\right)^2+p_R\right)\right]\right\}\approx-\frac{\partial}{\partial x}\left\{\tau_0\left[\frac{\partial}{\partial x}\left((n_R m_0 c^2)\left(\frac{v_0}{c}\right)^2+p\right)\right]\right\}$$
$$\approx-\frac{\partial}{\partial x}\left\{\tau_0\left[\frac{\partial}{\partial x}(\rho v_0^2+p)\right]\right\}. \tag{12.6.14}$$

Summation all terms in items (a)–(f) lead to the non-relativistic generalized Euler continuity equation obtained in [57].

$$\frac{\partial}{\partial t}\left\{\rho-\tau_0\left[\frac{\partial\rho}{\partial t}+\frac{\partial}{\partial x}(\rho v_0)\right]\right\}+\frac{\partial}{\partial x}\left\{\rho v_0-\tau_0\left[\frac{\partial}{\partial t}(\rho v_0)+\frac{\partial}{\partial x}(\rho v_0^2)+\frac{\partial p}{\partial x}\right]\right\}=0. \tag{12.6.15}$$

(2) Motion equation.
Consider sequentially all terms in the motion equation (12.5.31) using the first approximation by the decomposition in power series by the small ratios $\frac{v_0}{c},\frac{v}{c}$:

(a)
$$\frac{\partial}{c\partial t}\left\{(n_R e+p_R)\gamma_0^2\frac{v_0}{c}\right\}\approx\frac{\partial}{c\partial t}\left\{(n_R m_0 c^2)\frac{v_0}{c}\right\}\approx\frac{\partial}{\partial t}(\rho v_0), \tag{12.6.16}$$

(b)
$$\frac{\partial}{\partial x}\left\{(n_R e+p_R)\gamma_0^2\left(\frac{v_0}{c}\right)^2+p_R\right\}\approx\frac{\partial}{\partial x}\left\{(n_R m_0 c^2)\left(\frac{v_0}{c}\right)^2+p\right\}\approx\frac{\partial}{\partial x}(\rho v_0^2+p), \tag{12.6.17}$$

(c) The relation (12.6.7) is used by the transformations in this item,

$$-\frac{\partial}{c\partial t}\left\{\frac{\tau_0}{m_0}\frac{\partial}{c\partial t}\left(\gamma_0^3 v_0 m_0^2 c^2 n_R\left(\frac{3K_3(\zeta)}{\zeta K_2(\zeta)}+1\right)+\left(\left(\gamma_0\frac{v_0}{c}\right)^3+2\gamma_0^3\left(\frac{v_0}{c}\right)\right)m_0 c p_R\frac{K_3(\zeta)}{K_2(\zeta)}\right)\right\}$$
$$\approx-\frac{\partial}{c\partial t}\left\{\frac{\tau_0}{m_0}\frac{\partial}{c\partial t}\left(v_0 m_0^2 c^2 n_R+\left(\left(\frac{v_0}{c}\right)^3+2\left(\frac{v_0}{c}\right)\right)m_0 c p_R\right)\right\}\approx-\frac{\partial}{c\partial t}\left\{\frac{\tau_0}{m_0}\frac{\partial}{c\partial t}(v_0 m_0^2 c^2 n_R)\right\}$$
$$\approx-\frac{\partial}{\partial t}\left\{\tau_0\frac{\partial}{\partial t}(\rho v_0)\right\}, \tag{12.6.18}$$

(d) The relation (12.6.7) is used by the transformations in this item,

$$-\frac{\partial}{c\partial t}\left\{\frac{\tau_0}{m_0}\frac{\partial}{\partial x}\left(\gamma_0^3 v_0^2 m_0^2 c n_R\left(\frac{3K_3(\zeta)}{\zeta K_2(\zeta)}+1\right)+\left(\gamma_0^3+2\gamma_0^3\left(\frac{v_0}{c}\right)^2\right)m_0 c p_R\frac{K_3(\zeta)}{K_2(\zeta)}\right)\right\}$$
$$\approx-\frac{\partial}{c\partial t}\left\{\frac{\tau_0}{m_0}\frac{\partial}{\partial x}(v_0^2 m_0^2 c n_R+m_0 c p)\right\}\approx-\frac{\partial}{c\partial t}\left\{\tau_0\frac{\partial}{\partial x}(\rho v_0^2+p)\right\} \tag{12.6.19}$$

(e)
$$-\frac{\partial}{\partial x}\left\{\frac{\tau_0}{m_0}\frac{\partial}{c\partial t}\left(\gamma_0^3 v_0^2 m_0^2 c n_R\left(\frac{3K_3(\zeta)}{\zeta K_2(\zeta)}+1\right)+\left(\gamma_0^3+2\gamma_0^3\left(\frac{v_0}{c}\right)^2\right)m_0 c p_R\frac{K_3(\zeta)}{K_2(\zeta)}\right)\right\}$$
$$\approx-\frac{\partial}{\partial x}\left\{\tau_0\frac{\partial}{c\partial t}(\rho v_0^2+p)\right\}, \tag{12.6.20}$$

(f) The relation (12.6.7) is used by the transformations in this item,

$$\begin{aligned}&-\frac{\partial}{\partial x}\left\{\frac{\tau_0}{m_0}\frac{\partial}{\partial x}\left((\gamma_0 v_0)^3 m_0^2 n_R\left(\frac{3K_3(\zeta)}{\zeta K_2(\zeta)}+1\right)+3\gamma_0^3\left(\frac{v_0}{c}\right)m_0 c p_R\frac{K_3(\zeta)}{K_2(\zeta)}\right)\right\}\\&\approx-\frac{\partial}{\partial x}\left\{\frac{\tau_0}{m_0}\frac{\partial}{\partial x}\left(v_0^3 m_0^2 n_R+3\left(\frac{v_0}{c}\right)m_0 c p\right)\right\}\approx-\frac{\partial}{\partial x}\left\{\frac{\tau_0}{m_0}\frac{\partial}{\partial x}\left(\rho v_0^3+3pv_0\right)\right\}.\end{aligned}\tag{12.6.21}$$

Summation all terms in items (a)–(f) lead to the non-relativistic generalized Euler motion equation obtained in [57].

$$\begin{aligned}&\frac{\partial}{\partial t}\left\{\rho v_0-\tau_0\left[\frac{\partial}{\partial t}(\rho v_0)+\frac{\partial}{\partial x}(\rho v_0^2)+\frac{\partial p}{\partial x}\right]\right\}\\&+\frac{\partial}{\partial x}\left\{\rho v_0^2+p-\tau_0\left[\frac{\partial}{\partial t}(\rho v_0^2+p)+\frac{\partial}{\partial x}(\rho v_0^3+3pv_0)\right]\right\}=0.\end{aligned}\tag{12.6.22}$$

(3) Energy equation.

As before consider sequentially all terms in the energy equation (12.5.32) but using the second approximation by the decomposition in power series by the small ratios $\frac{v_0}{c},\frac{v}{c}$:

(a) The relations (12.6.2), (12.6.3) are used by the transformations in the item,

$$\frac{\partial}{c\partial t}\{(n_R e+p_R)\gamma_0^2-p_R-m_0c^2 n_R\gamma_0\}\approx\frac{\partial}{c\partial t}\left\{\rho c^2+\frac{\rho v_0^2}{2}+\frac{3}{2}p-\rho c^2\right\}=\frac{\partial}{c\partial t}\left\{\frac{\rho v_0^2}{2}+\frac{3}{2}p\right\}.\tag{12.6.23}$$

(b) The relations (12.6.2) and (12.6.3) are used by the transformations in the item,

$$\begin{aligned}&\frac{\partial}{\partial x}\left\{(n_R e+p_R)\gamma_0^2\left(\frac{v_0}{c}\right)-m_0 c\gamma_0 v_0 n_R\right\}\approx\frac{\partial}{\partial x}\left\{\left(\rho c^2+\frac{\rho v_0^2}{2}+\frac{5}{2}p\right)\left(\frac{v_0}{c}\right)-m_0 c\gamma_0 v_0\frac{n}{\gamma_0}\right\}\\&=\frac{\partial}{c\partial x}\left\{\left(\rho c^2+\frac{\rho v_0^2}{2}+\frac{5}{2}p\right)v_0-\rho c^2 v_0\right\}=\frac{\partial}{c\partial x}\left\{\frac{\rho v_0^3}{2}+\frac{5}{2}pv_0\right\}\end{aligned}\tag{12.6.24}$$

(c) The relations (12.6.2), (12.6.7), and (12.6.8) are used by the transformations in the item,

$$\begin{aligned}&-\frac{\partial}{c\partial t}\left\{\frac{\tau_0}{m_0}\frac{\partial}{c\partial t}\left(\gamma_0^3 m_0^2 c^3 n_R\left(\frac{3K_3(\zeta)}{\zeta K_2(\zeta)}+1\right)+3\gamma_0^3\left(\frac{v_0}{c}\right)^2 m_0 c p_R\frac{K_3(\zeta)}{K_2(\zeta)}-m_0c\left((n_R e+p_R)\gamma_0^2-p_R\right)\right)\right\}\\&\approx-\frac{\partial}{c\partial t}\left\{\frac{\tau_0}{m_0}\frac{\partial}{c\partial t}\left(m_0c(\rho c^2+\rho v_0^2+3p)+3\left(\frac{v_0}{c}\right)^2 m_0 c p-m_0c\left(\rho c^2+\frac{\rho v_0^2}{2}+\frac{3}{2}p\right)\right)\right\}\\&\approx-\frac{\partial}{c\partial t}\left\{\tau_0\frac{\partial}{\partial t}\left(\frac{\rho v_0^2}{2}+\frac{3}{2}p\right)\right\}\end{aligned}\tag{12.6.25}$$

(d) The relations (12.6.3), (12.6.7), and (12.6.7) are used by the transformations in the item,

$$\begin{aligned}&-\frac{\partial}{c\partial t}\left\{\frac{\tau_0}{m_0}\frac{\partial}{\partial x}\left(\gamma_0^3 v_0 m_0^2 c^2 n_R\left(\frac{3K_3(\zeta)}{\zeta K_2(\zeta)}+1\right)\right.\right.\\&\left.\left.+\left(\left(\gamma_0\frac{v_0}{c}\right)^3+2\gamma_0^3\left(\frac{v_0}{c}\right)\right)m_0 c p_R\frac{K_3(\zeta)}{K_2(\zeta)}-(n_R e+p_R)\gamma_0^2 m_0 v_0\right)\right\}\\&\approx-\frac{\partial}{c\partial t}\left\{\frac{\tau_0}{m_0}\frac{\partial}{\partial x}\left(\left(\frac{v_0}{c}\right)m_0c(\rho c^2+\rho v_0^2+3p)+\left(\left(\gamma_0\frac{v_0}{c}\right)^3+2\gamma_0^3\left(\frac{v_0}{c}\right)\right)m_0 c p_R-\left(\rho c^2+\frac{\rho v_0^2}{2}+\frac{5}{2}p\right)m_0 v_0\right)\right\}\\&\approx-\frac{\partial}{c\partial t}\left\{\tau_0\frac{\partial}{\partial x}\left(\rho c^2 v_0+\rho v_0^3+3pv_0+2pv_0-\rho c^2 v_0-\frac{\rho v_0^3}{2}-\frac{5}{2}pv_0\right)\right\}=-\frac{\partial}{c\partial t}\left\{\tau_0\frac{\partial}{\partial x}\left(\frac{\rho v_0^3}{2}+\frac{5}{2}pv_0\right)\right\}.\end{aligned}\tag{12.6.26}$$

(e)

$$
\begin{aligned}
&-\frac{\partial}{\partial x}\left\{\frac{\tau_0}{m_0}\frac{\partial}{c\partial t}\left(\gamma_0^3 v_0 m_0^2 c^2 n_R\left(\frac{3K_3(\zeta)}{\zeta K_2(\zeta)}+1\right)\right.\right.\\
&\left.\left.+\left(\left(\gamma_0\frac{v_0}{c}\right)^3+2\gamma_0^3\left(\frac{v_0}{c}\right)\right)m_0 c p_R\frac{K_3(\zeta)}{K_2(\zeta)}-(n_R e+p_R)\gamma_0^2 m_0 v_0\right)\right\}\\
&\approx-\frac{\partial}{c\partial t}\left\{\tau_0\frac{\partial}{\partial x}\left(\frac{\rho v_0^3}{2}+\frac{5}{2}p v_0\right)\right\},
\end{aligned}
$$

(f) The relations (12.6.2), (12.6.3), (12.6.7), and (12.6.8) are used by the transformations in the item,

$$
\begin{aligned}
&-\frac{\partial}{\partial x}\left\{\frac{\tau_0}{m_0}\frac{\partial}{\partial x}\left(\gamma_0^3 v_0^2 m_0^2 c n_R\left(\frac{3K_3(\zeta)}{\zeta K_2(\zeta)}+1\right)+\left(\gamma_0^3+2\gamma_0^3\left(\frac{v_0}{c}\right)^2\right)m_0 c p_R\frac{K_3(\zeta)}{K_2(\zeta)}-(n_R e+p_R)\gamma_0^2 v_0\left(\frac{v_0}{c}\right)m_0-p_R m_0 c\right)\right\}\\
&-\frac{\partial}{\partial x}\left\{\frac{\tau_0}{m_0}\frac{\partial}{\partial x}\left(\left(\frac{v_0}{c}\right)^2 m_0 c(\rho c^2+\rho v_0^2+3p)+\left(\gamma_0^3+2\gamma_0^3\left(\frac{v_0}{c}\right)^2\right)m_0 c\frac{nk_{\mathrm{B}}T}{\gamma_0}\left(1+\frac{5}{2}\frac{p}{\rho c^2}\right)\right.\right.\\
&\left.\left.-\left(\rho c^2+\frac{\rho v_0^2}{2}+\frac{5}{2}p\right)v_0\left(\frac{v_0}{c}\right)m_0-\frac{nk_{\mathrm{B}}T}{\gamma_0}m_0 c\right)\right\}\\
&\approx-\frac{\partial}{c\partial x}\left\{\tau_0\frac{\partial}{\partial x}\left(v_0^2(\rho c^2+\rho v_0^2+3p)+\left(\gamma_0^2+2\gamma_0^2\left(\frac{v_0}{c}\right)^2\right)pc^2\left(1+\frac{5}{2}\frac{p}{\rho c^2}\right)\right.\right.\\
&\left.\left.-\left(\rho c^2+\frac{\rho v_0^2}{2}+\frac{5}{2}p\right)v_0^2-pc^2\left(1-\frac{v_0^2}{2c^2}\right)\right)\right\}\\
&\approx-\frac{\partial}{c\partial x}\left\{\tau_0\frac{\partial}{\partial x}\left(\rho c^2 v_0^2+\rho v_0^4+3p v_0^2+\left(1+\left(\frac{v_0}{c}\right)^2+2\left(\frac{v_0}{c}\right)^2\right)pc^2\left(1+\frac{5}{2}\frac{p}{\rho c^2}\right)\right.\right.\\
&\left.\left.-\rho c^2 v_0^2-\frac{\rho v_0^4}{2}-\frac{5}{2}p v_0^2-pc^2+p\frac{v_0^2}{2}\right)\right\}\\
&\approx-\frac{\partial}{c\partial x}\left\{\tau_0\frac{\partial}{\partial x}\left(\frac{\rho v_0^4}{2}+3p v_0^2+pc^2+3p v_0^2+\frac{5p^2}{2\rho}-\frac{5}{2}p v_0^2-pc^2+p\frac{v_0^2}{2}\right)\right\}\\
&\approx-\frac{\partial}{c\partial x}\left\{\tau_0\frac{\partial}{\partial x}\left(\frac{\rho v_0^4}{2}+4p v_0^2+\frac{5p^2}{2\rho}\right)\right\}.
\end{aligned}
\tag{12.6.27}
$$

Summation all terms in items (a)–(f) and multiplication by c lead to the non-relativistic generalized Euler energy equation obtained by Alexeev in [57].

$$
\begin{aligned}
&\frac{\partial}{\partial t}\left\{\frac{\rho v_0^2}{2}+\frac{3}{2}p-\tau_0\left[\frac{\partial}{\partial t}\left(\frac{\rho v_0^2}{2}+\frac{3}{2}p\right)+\frac{\partial}{\partial x}\left(\frac{\rho v_0^3}{2}+\frac{5}{2}p v_0\right)\right]\right\}\\
&+\frac{\partial}{\partial x}\left\{\frac{\rho v_0^3}{2}+\frac{5}{2}p v_0-\tau_0\left[\frac{\partial}{\partial t}\left(\frac{\rho v_0^3}{2}+\frac{5}{2}p v_0\right)+\frac{\partial}{\partial x}\left(\frac{\rho v_0^4}{2}+4p v_0^2+\frac{5p^2}{2\rho}\right)\right]\right\}=0.
\end{aligned}
\tag{12.6.28}
$$

12.7 EXPANSION OF THE FLAT HARMONIC WAVES OF SMALL AMPLITUDES IN ULTRA-RELATIVISTIC MEDIA

Investigation of the expansion of the flat harmonic waves of small amplitudes in the rarefied gas is from the first glance the very particular problem of the transport theory. However, it is far from reality because this problem serves as a test for verification the different kinetic and hydrodynamic models. For example, additional complications arise for relativistic systems in the frame of relativistic Boltzmann equation—a first order gradient expansion like Navier and Stokes, Barnett leads to a set of fluid dynamics equations that allow faster-than-light signal propagation, [256]. This is partly due to the fact that, as indicated in [256], "the simplest" form of viscous hydrodynamics, the relativistic Navier-Stokes equations, are be-riddled by causality problems and instabilities. A program to determine the precise form for a relativistic viscous fluid was initiated in the last decade; this program was driven by nuclear physics experiments on relativistic heavy-ion collisions and problems of quantum description of transport processes in nano-electronics. The relativistic first-order dissipative theory (created on the basement of the Boltzmann kinetic equation) is not applicable and is discarded in many papers in favor of the second-order the Boltzmann based one in spite of the crash of the first approximation in the local description.

The origin of all these difficulties lies in the local origin of the Boltzmann equation. In the Boltzmann local kinetic theory the additional terms of non-local origin are lost. These terms strictly speaking are of the same order as the "classical" ones. From position of non-local physics Boltzmann kinetic theory belongs only to plausible models and do not deliver even minimized model. The GBE, created by Alexeev, introduces a local differential approximation for non-local collision integral.

If Knudsen number tends to unity the local approximation of non-local collision integral needs in taking into account the following terms in series containing substantive orders of highest degrees. Here we are faced in fact with the "price-quality" problem familiar from economics. That is, what price—in terms of increased complexity of kinetic equation—are we ready to pay for improved quality of the theory? The derived kinetic equations of highest orders have mainly theoretical significance because of its complexity in applications. Moreover, the Generalized Boltzmann Equation derived by B.V. Alexeev delivers sufficient approximation for this regime with the correct descriptions of limiting cases. Important, that generalized hydrodynamic equations deliver through approximation in extremely wide diapason of the Knudsen variety sufficient in applications when we enforced to consider much more complicated problems like turbulent flows over blunt bodies or, as the last challenge, the transport processes in colliders in the relativistic variant of the non-local theory. Of course, no reason to pay for unreasonable local approximations and apply to BE, which is not appropriate in the considered physical situation. In literature you can find papers (typical is Ref. [257]) devoted of application of the moment equations of tremendous tensor degrees especially in the theory of sound propagation for the intermediate Knudsen numbers where the local Maxwellian function is not the main term in the local series based on the small Knudsen numbers. As you see this way has no physical sense.

It seems from the first glance that numerical methods connected with the direct simulation of the particle movement (like DSMC) have no mentioned shortages. However, it is not so. Really, these methods unavoidably lead to introduction of physically small volume (PhSV) for the moment calculations. These moments are the same in all points of the cell corresponding of the PhSV. The numerical fluctuations should be damped by the introduction of filter equations, especially in the case by the intermediate Knudsen numbers. It is known from the literature that in many cases the Navies-Stokes equations are taken as the filter equations. This fact returns us to the previous criticism of the Navier-Stokes theory as direct consequence of BE and therefore to inevitable application of the generalized non-local hydrodynamics.

Consider now the expansion of the flat harmonic waves of small amplitudes in the frame of the relativistic generalized physical kinetics. The non-relativistic solution of this problem in the frame of the generalized physical kinetics is considered in this monograph before.

Let us consider the ultra-relativistic media obeyed by condition

$$\zeta = \frac{m_0 c^2}{k_B T} \ll 1, \tag{12.7.1}$$

this corresponds to the high temperature or small rest mass of particles like neutrino. Remark that for modified Bessel function we have

$$\lim_{\zeta \to 0} (K_n(\zeta)\zeta^n) = 2^{n-1}(n-1)! \tag{12.7.2}$$

Eq. (12.7.2) leads to the asymptotic relations for $K_n(\zeta)$ by $\zeta \to 0$:

$$K_1(\zeta) \approx 1/\zeta, K_2(\zeta) \approx 2/\zeta^2, K_3(\zeta) \approx 8/\zeta^3, \tag{12.7.3}$$

and the expression for mean energy e corresponding to one particle (see also Eqs. (12.4.26) and (12.4.29))

$$= m_0 c^2 \left(\frac{K_1(\zeta)}{K_2(\zeta)} + \frac{3}{\zeta} \right). \tag{12.7.4}$$

Using asymptotic relation (12.7.3) one obtains from Eq. (12.7.4) the equation of state known in particular in the theory of radiation

$$e \approx m_0 c^2 \left(\zeta + \frac{3}{\zeta} \right) \approx \frac{3 m_0 c^2}{\zeta} = 3 k_B T \tag{12.7.5}$$

or, using Eq. (12.4.32),

$$n_R \mathrm{e} = 3 p_R. \tag{12.7.6}$$

Consider the wave expansion in the frame of the theory of small perturbations taking into account only the values of the first order

$$n_R = n_{R0}(1+s),$$
$$T = T_0(1+\eta),$$

where $s, \eta \ll 1$. Consequently, we have the following expressions for

density

$$\rho_R = m_0 n_R = m_0 n_{R0}(1+s) = \rho_{R0}(1+s), \tag{12.7.7}$$

pressure

$$p_R = n_R k_B T = n_{R0} k_B T_0 (1+s)(1+\eta) \approx p_{R0}(1+s+\eta), \tag{12.7.8}$$

and the energy density

$$n_R e = 3p_R \approx 3p_{R0}(1+s+\eta). \tag{12.7.9}$$

For perturbation of ζ parameter occurs

$$\zeta = \frac{m_0 c^2}{k_B T} = \frac{m_0 c^2}{k_B T_0 (1+\eta)} \approx \zeta_0 (1-\eta). \tag{12.7.10}$$

Euler equations (12.5.30)–(12.5.32) are applied in the linear formulation; it means that only the first order the velocity perturbations are taking into account, for this case

$$\gamma_0 = \frac{1}{\sqrt{1 - v_0^2/c^2}} \approx 1 + \frac{v_0^2}{2c^2} \approx 1. \tag{12.7.11}$$

As it is shown before, the terms containing the derivations of τ_0, do not play a role in the theory of the first approximation. Then τ_0 does not depend on time and coordinate. Nonlinear case needs in additional consideration.

Taking into account the relations (12.7.5), (12.7.9), and (12.7.11) from the continuity equation (12.5.30)

$$\frac{\partial}{\partial t}\left\{ m_0 n_R \gamma_0 - \tau_0 \left[\frac{\partial}{c^2 \partial t}\left((n_R e + p_R)\gamma_0^2 - p_R \right) + \frac{\partial}{c^2 \partial x}\left((n_R e + p_R)\gamma_0^2 v_0 \right) \right] \right\}$$
$$+ \frac{\partial}{\partial x}\left\{ m_0 n_R \gamma_0 v_0 - \tau_0 \left[\frac{\partial}{c^2 \partial t}\left((n_R e + p_R)\gamma_0^2 v_0 \right) + \frac{\partial}{\partial x}\left((n_R e + p_R)\gamma_0^2 \left(\frac{v_0}{c}\right)^2 + p_R \right) \right] \right\} = 0,$$

one obtains

$$m_0 n_{R0}\frac{\partial s}{\partial t} + m_0 n_{R0}\frac{\partial v_0}{\partial x} - \tau_0 3 p_{R0}\frac{\partial^2}{c^2 \partial t^2}(s+\eta) - \tau_0 8 p_{R0}\frac{\partial^2 v_0}{c^2 \partial t \partial x} - \tau_0 p_{R0}\frac{\partial^2}{\partial x^2}(s+\eta) = 0. \tag{12.7.12}$$

After dividing the both parts of Eq. (12.7.12) by $\rho_{R0} = m_0 n_{R0}$

$$\frac{\partial s}{\partial t} + \frac{\partial v_0}{\partial x} - 3\tau_0 \frac{p_{R0}}{\rho_{R0}}\frac{\partial^2}{c^2 \partial t^2}(s+\eta) - 8\tau_0 \frac{p_{R0}}{\rho_{R0}}\frac{\partial^2 v_0}{c^2 \partial t \partial x} - \tau_0 \frac{p_{R0}}{\rho_{R0}}\frac{\partial^2}{\partial x^2}(s+\eta) = 0. \tag{12.7.13}$$

and using the relation

$$\frac{p_{R0}}{\rho_{R0} c^2} = \frac{n_{R0} k_B T}{n_{R0} m_0 c^2} = \frac{1}{\zeta_0}, \tag{12.7.14}$$

the linearized continuity equation takes the form

$$\frac{\partial s}{\partial t} + \frac{\partial v_0}{\partial x} - \frac{3\tau_0}{\zeta_0}\frac{\partial^2}{\partial t^2}(s+\eta) - \frac{8\tau_0}{\zeta_0}\frac{\partial^2 v_0}{\partial t \partial x} - \frac{\tau_0 c^2}{\zeta_0}\frac{\partial^2}{\partial x^2}(s+\eta) = 0. \tag{12.7.15}$$

Consider the transformation of the motion equation with the help of relations (12.7.16), (12.7.17)

$$\frac{K_3(\zeta)}{K_2(\zeta)} \approx \frac{4}{\zeta} = \frac{4k_B T}{m_0 c^2} = \frac{4}{\zeta_0}(1+\eta), \tag{12.7.16}$$

$$\frac{3K_3(\zeta)}{\zeta K_2(\zeta)} + 1 \approx \frac{12}{\zeta^2} + 1 \approx \frac{12}{\zeta^2} \approx \frac{12}{\zeta_0^2}(1+2\eta). \tag{12.7.17}$$

We find from Eq. (12.5.31)

$$\begin{aligned}
&\frac{\partial}{c\partial t}\left\{(n_R e + p_R)\gamma_0^2 \frac{v_0}{c} - \frac{\tau_0}{m_0}\left[\frac{\partial}{c\partial t}\left(\gamma_0^3 v_0 m_0^2 c^2 n_R\left(\frac{3K_3(\zeta)}{\zeta K_2(\zeta)}+1\right) + \left(\left(\gamma_0 \frac{v_0}{c}\right)^3 + 2\gamma_0^3\left(\frac{v_0}{c}\right)\right) m_0 c p_R \frac{K_3(\zeta)}{K_2(\zeta)}\right)\right.\right.\\
&\left.\left.+\frac{\partial}{\partial x}\left(\gamma_0^3 v_0^2 m_0^2 c n_R\left(\frac{3K_3(\zeta)}{\zeta K_2(\zeta)}+1\right) + \left(\gamma_0^3 + 2\gamma_0^3\left(\frac{v_0}{c}\right)^2\right) m_0 c p_R \frac{K_3(\zeta)}{K_2(\zeta)}\right)\right]\right\} + \frac{\partial}{\partial x}\left\{(n_R e + p_R)\gamma_0^2\left(\frac{v_0}{c}\right)^2\right.\\
&+p_R - \frac{\tau_0}{m_0}\left[\frac{\partial}{c\partial t}\left(\gamma_0^3 v_0^2 m_0^2 c n_R\left(\frac{3K_3(\zeta)}{\zeta K_2(\zeta)}+1\right) + \left(\gamma_0^3 + 2\gamma_0^3\left(\frac{v_0}{c}\right)^2\right) m_0 c p_R \frac{K_3(\zeta)}{K_2(\zeta)}\right)\right.\\
&\left.\left.+\frac{\partial}{\partial x}\left((\gamma_0 v_0)^3 m_0^2 n_R\left(\frac{3K_3(\zeta)}{\zeta K_2(\zeta)}+1\right) + 3\gamma_0^3\left(\frac{v_0}{c}\right) m_0 c p_R \frac{K_3(\zeta)}{K_2(\zeta)}\right)\right]\right\} = 0,
\end{aligned} \tag{12.7.18}$$

or using Eqs. (12.7.8), (12.7.9), and (12.7.11)

$$\begin{aligned}
&\frac{4p_{R0}}{c^2}\frac{\partial v_0}{\partial t} + p_{R0}\frac{\partial}{\partial x}(s+\eta) - \tau_0 \rho_{R0}\frac{12}{\zeta_0^2}\frac{\partial^2 v_0}{\partial t^2} - \frac{\tau_0}{c^2}p_{R0}\frac{8}{\zeta_0}\frac{\partial^2 v_0}{\partial t^2}\\
&-\tau_0 p_{R0}\frac{8}{\zeta_0}\frac{\partial^2}{\partial x\partial t}(s+2\eta) - \tau_0 p_{R0}\frac{12}{\zeta_0}\frac{\partial^2 v_0}{\partial x^2} = 0.
\end{aligned} \tag{12.7.19}$$

After multiplication the both sides of Eq. (12.7.19) by $\frac{c^2}{4p_{R0}}$ the motion equation takes the form

$$\frac{\partial v_0}{\partial t} + \frac{c^2}{4}\frac{\partial}{\partial x}(s+\eta) - \frac{5\tau_0}{\zeta_0}\frac{\partial^2 v_0}{\partial t^2} - \frac{2\tau_0 c^2}{\zeta_0}\frac{\partial^2}{\partial x\partial t}(s+2\eta) - \frac{3\tau_0 c^2}{\zeta_0}\frac{\partial^2 v_0}{\partial x^2} = 0. \tag{12.7.20}$$

Linearize now the energy equation (12.5.32) with the help of relations (12.7.5), (12.7.6), (12.7.7), (12.7.8), (12.7.16), and (12.7.17)

$$\begin{aligned}
&\frac{\partial}{\partial t}(s+\eta) - \frac{\zeta_0}{3}\frac{\partial s}{\partial t} + \frac{4}{3}\frac{\partial v_0}{\partial x} - \frac{\zeta_0}{3}\frac{\partial v_0}{\partial x} - \frac{4\tau_0}{\zeta_0}\frac{\partial^2}{\partial t^2}(s+2\eta) + \tau_0\frac{\partial^2}{\partial t^2}(s+\eta) - \frac{8\tau_0}{\zeta_0}\frac{\partial^2 v_0}{\partial x\partial t}\\
&-\frac{16\tau_0}{3\zeta_0}\frac{\partial^2 v_0}{\partial x\partial t} + \frac{8\tau_0}{3}\frac{\partial^2 v_0}{\partial x\partial t} - \frac{4\tau_0 c^2}{3\zeta_0}\frac{\partial^2}{\partial x^2}(s+2\eta) + \frac{\tau_0 c^2}{3}\frac{\partial^2}{\partial x^2}(s+\eta) = 0.
\end{aligned} \tag{12.7.21}$$

In the following transformations of Eq. (12.7.21) we use the condition $\zeta_0 \ll 1$, omitting the small terms like this one

$$\frac{\partial s}{\partial t} - \frac{\zeta_0}{3}\frac{\partial s}{\partial t} \approx \frac{\partial s}{\partial t} \tag{12.7.22}$$

and so on. Then the energy equation is written as

$$\frac{\partial}{\partial t}(s+\eta) + \frac{4}{3}\frac{\partial v_0}{\partial x} - \frac{4\tau_0}{\zeta_0}\frac{\partial^2}{\partial t^2}(s+2\eta) - \frac{40\tau_0}{3\zeta_0}\frac{\partial^2 v_0}{\partial x\partial t} - \frac{4\tau_0 c^2}{3\zeta_0}\frac{\partial^2}{\partial x^2}(s+2\eta) = 0. \tag{12.7.23}$$

Let us write down the full system of the generalized 1D relativistic equations (12.7.15), (12.7.20), and (12.7.23):

$$\frac{\partial s}{\partial t} + \frac{\partial v_0}{\partial x} - \frac{3\tau_0}{\zeta_0}\frac{\partial^2}{\partial t^2}(s+\eta) - \frac{8\tau_0}{\zeta_0}\frac{\partial^2 v_0}{\partial t\partial x} - \frac{\tau_0 c^2}{\zeta_0}\frac{\partial^2}{\partial x^2}(s+\eta) = 0,$$

$$\frac{\partial v_0}{\partial t} + \frac{c^2}{4}\frac{\partial}{\partial x}(s+\eta) - \frac{5\tau_0}{\zeta_0}\frac{\partial^2 v_0}{\partial t^2} - \frac{2\tau_0 c^2}{\zeta_0}\frac{\partial^2}{\partial x\partial t}(s+2\eta) - \frac{3\tau_0 c^2}{\zeta_0}\frac{\partial^2 v_0}{\partial x^2} = 0,$$

$$\frac{\partial}{\partial t}(s+\eta) + \frac{4}{3}\frac{\partial v_0}{\partial x} - \frac{4\tau_0}{\zeta_0}\frac{\partial^2}{\partial t^2}(s+2\eta) - \frac{40\tau_0}{3\zeta_0}\frac{\partial^2 v_0}{\partial x\partial t} - \frac{4\tau_0 c^2}{3\zeta_0}\frac{\partial^2}{\partial x^2}(s+2\eta) = 0.$$

The solutions of this system are written as

$$s = \bar{s}\exp(i\omega t - k'x), \tag{12.7.24}$$

$$\eta = \bar{\eta}\exp(i\omega t - k'x), \tag{12.7.25}$$

$$v_0 = \bar{v}_0\exp(i\omega t - k'x). \tag{12.7.26}$$

After substitutions (132)–(134) in the system (123), (128), (131) we find

$$\bar{s}\left(i\omega + \frac{3\tau_0}{\zeta_0}\omega^2 - \frac{\tau_0 c^2}{\zeta_0}k'^2\right) + \bar{\eta}\left(\frac{3\tau_0}{\zeta_0}\omega^2 - \frac{\tau_0 c^2}{\zeta_0}k'^2\right) + \bar{v}_0\left(-k' + i\frac{8\tau_0}{\zeta_0}\omega k'\right) = 0, \tag{12.7.27}$$

$$\bar{s}\left(-\frac{c^2}{4}k' + i\frac{2\tau_0 c^2}{\zeta_0}\omega k'\right) + \bar{\eta}\left(-\frac{c^2}{4}k' + i\frac{4\tau_0 c^2}{\zeta_0}\omega k'\right) + \bar{v}_0\left(i\omega + \frac{5\tau_0}{\zeta_0}\omega^2 - \frac{3\tau_0 c^2}{\zeta_0}k'^2\right) = 0, \tag{12.7.28}$$

$$\bar{s}\left(i\omega + \frac{4\tau_0}{\zeta_0}\omega^2 - \frac{4\tau_0 c^2}{3\zeta_0}k'^2\right) + \bar{\eta}\left(i\omega + \frac{8\tau_0}{\zeta_0}\omega^2 - \frac{8\tau_0 c^2}{3\zeta_0}k'^2\right) + \bar{v}_0\left(-\frac{4}{3}k' + i\frac{40\tau_0}{3\zeta_0}\omega k'\right) = 0. \tag{12.7.29}$$

The system of Eqs. (12.7.27)–(12.7.29) of homogeneous linear equations has a non-trivial solution if and only is the determinant of the corresponding coefficients vanishes. Namely,

$$\begin{vmatrix} i\omega + \frac{3\tau_0}{\zeta_0}\omega^2 - \frac{\tau_0 c^2}{\zeta_0}k'^2 & \frac{3\tau_0}{\zeta_0}\omega^2 - \frac{\tau_0 c^2}{\zeta_0}k'^2 & -k' + i\frac{8\tau_0}{\zeta_0}\omega k' \\ -\frac{c^2}{4}k' + i\frac{2\tau_0 c^2}{\zeta_0}\omega k' & -\frac{c^2}{4}k' + i\frac{4\tau_0 c^2}{\zeta_0}\omega k' & i\omega + \frac{5\tau_0}{\zeta_0}\omega^2 - \frac{3\tau_0 c^2}{\zeta_0}k'^2 \\ i\omega + \frac{4\tau_0}{\zeta_0}\omega^2 - \frac{4\tau_0 c^2}{3\zeta_0}k'^2 & i\omega + \frac{8\tau_0}{\zeta_0}\omega^2 - \frac{8\tau_0 c^2}{3\zeta_0}k'^2 & -\frac{4}{3}k' + i\frac{40\tau_0}{3\zeta_0}\omega k' \end{vmatrix} = 0. \tag{12.7.30}$$

The mentioned relation leads to the algebraic equation of the sixth order relatively k':

$$\begin{aligned} &4\frac{\tau_0^3}{\zeta_0^3}c^6 k'^6 + \left(-\frac{1}{3}\frac{\tau_0}{\zeta_0}c^4 - 4i\frac{\tau_0^2}{\zeta_0^2}c^4\omega - 4\frac{\tau_0^3}{\zeta_0^3}c^4\omega^2\right)k'^4 \\ &+ \left(\frac{1}{3}ic^2\omega + 2\frac{\tau_0}{\zeta_0}c^2\omega^2 - 8i\frac{\tau_0^2}{\zeta_0^2}c^2\omega^3 - 4\frac{\tau_0^3}{\zeta_0^3}c^2\omega^4\right)k'^2 \\ &+ \left(i\omega^3 + 13\frac{\tau_0}{\zeta_0}\omega^4 - 52i\frac{\tau_0^2}{\zeta_0^2}\omega^5 - 60\frac{\tau_0^3}{\zeta_0^3}\omega^6\right) = 0. \end{aligned} \tag{12.7.31}$$

After introduction of the dimensionless wave number

$$\widehat{k} = \frac{k'c}{\omega}, \tag{12.7.32}$$

dispersion equation (12.7.33) takes the form

$$\begin{aligned} &4\frac{\tau_0^3\omega^3}{\zeta_0^3}\widehat{k}^6 + \left(-\frac{1}{3}\frac{\tau_0\omega}{\zeta_0} - 4i\frac{\tau_0^2\omega^2}{\zeta_0^2} - 4\frac{\tau_0^3\omega^3}{\zeta_0^3}\right)\widehat{k}^4 \\ &+ \left(\frac{1}{3}i + 2\frac{\tau_0\omega}{\zeta_0} - 8i\frac{\tau_0^2\omega^2}{\zeta_0^2} - 4\frac{\tau_0^3\omega^3}{\zeta_0^3}\right)\widehat{k}^2 + \left(i + 13\frac{\tau_0\omega}{\zeta_0} - 52i\frac{\tau_0^2\omega^2}{\zeta_0^2} - 60\frac{\tau_0^3\omega^3}{\zeta_0^3}\right) = 0. \end{aligned} \tag{12.7.33}$$

Denote as

$$a = \frac{\omega\tau_0}{\zeta_0}, \tag{12.7.34}$$

and rewrite the dispersion equation (12.7.33)

$$\begin{aligned} &4a^3\widehat{k}^6 - \left(\frac{1}{3}a + 4ia^2 + 4a^3\right)\widehat{k}^4 + \left(\frac{1}{3}i + 2a - 8ia^2 - 4a^3\right)\widehat{k}^2 \\ &+ (i + 13a - 52ia^2 - 60a^3) = 0. \end{aligned} \tag{12.7.35}$$

Write down the complex number $\widehat{k}$ as

$$\widehat{k} = \alpha + i\beta \tag{12.7.36}$$

and separate the real and imaginary parts. For real part we have

$$\begin{aligned}&4a^3\left(\alpha^6 - 15\alpha^4\beta^2 + 15\alpha^2\beta^4 - \beta^6\right) - \left(\frac{1}{3}a + 4a^3\right)\left(\alpha^4 - 6\alpha^2\beta^2 + \beta^4\right)\\&+4a^2\left(4\alpha^3\beta - 4\alpha\beta^3\right) + \left(2a - 4a^3\right)\left(\alpha^2 - \beta^2\right) + \left(8a^2 - \frac{1}{3}\right)2\alpha\beta + \left(13a - 60a^3\right) = 0,\end{aligned} \tag{12.7.36}$$

Imaginary part:

$$\begin{aligned}&4a^3\left(6\alpha^5\beta - 20\alpha^3\beta^3 + 6\alpha\beta^5\right) - \left(\frac{1}{3}a + 4a^3\right)\left(4\alpha^3\beta - 4\alpha\beta^3\right)\\&-4a^2\left(\alpha^4 - 6\alpha^2\beta^2 + \beta^4\right) + \left(\frac{1}{3} - 8a^2\right)\left(\alpha^2 - \beta^2\right) + \left(2a - 4a^3\right)2\alpha\beta + \left(1 - 52a^2\right) = 0.\end{aligned} \tag{12.7.37}$$

Inscription of $\widehat{k}$ as $\widehat{k} = \alpha + i\beta$ allows to write down relations (12.7.24)–(12.7.26) in the form of the damped harmonic waves. For example,

$$s = \bar{s}\exp\left(-\frac{\omega}{c}\alpha x\right)\exp\left(i\omega\left(t - \frac{\beta x}{c}\right)\right), \tag{12.7.38}$$

then α characterizes the wave attenuation, and β—is ratio of the light velocity c to the phase velocity of the wave expansion. Investigate the asymptotic solutions of Eq. (12.7.37).

(1) For

$$a = \frac{\omega\tau_0}{\zeta_0} \to 0, \tag{12.7.39}$$

from relation (12.7.37) follow the system of equations

$$\alpha\beta = 0, \tag{12.7.40}$$

$$\alpha^2 - \beta^2 + 3 = 0, \tag{12.7.41}$$

with solutions

$$\alpha = 0, \beta = \sqrt{3}. \tag{12.7.42}$$

For this asymptotic case (147) turns into

$$s = \bar{s}\exp\left(i\omega\left(t - \frac{\sqrt{3}x}{c}\right)\right), \tag{12.7.43}$$

and the expansion of waves realizes without attenuation. Obviously the phase velocity $v_s^{ur} = c/\sqrt{3}$, and coincides with the phase velocity v_s^{ur}, obtained in [255].

(2) For

$$a = \frac{\omega\tau_0}{\zeta_0} \to \infty, \tag{12.7.44}$$

from equations (12.7.37) follow

$$\left(\alpha^6 - 15\alpha^4\beta^2 + 15\alpha^2\beta^4 - \beta^6\right) - \left(\alpha^4 - 6\alpha^2\beta^2 + \beta^4\right) - \left(\alpha^2 - \beta^2\right) - 15 = 0, \tag{12.7.45}$$

$$\left(6\alpha^5\beta - 20\alpha^3\beta^3 + 6\alpha\beta^5\right) - \left(4\alpha^3\beta - 4\alpha\beta^3\right) - 2\alpha\beta = 0. \tag{12.7.46}$$

After introduction of the variables

$$x = \beta^2 - \alpha^2, y = \alpha^2\beta^2 \tag{12.7.47}$$

The system of Eqs. (12.7.45) and (12.7.46) yields

$$x^3 - 12xy + x^2 - x - 4y + 15 = 0, \quad (12.7.48)$$

$$3x^2 - 4y + 2x - 1 = 0. \quad (12.7.49)$$

The single real solution of the system of Eqs. (12.7.48), (12.7.49) is

$$x = 1, y = 1 \quad (12.7.50)$$

and relations (156), (159), corresponding the condition $\omega \to \infty$, yield

$$\alpha_\infty = \frac{1}{\beta_\infty} = \sqrt{\frac{2}{1+\sqrt{5}}} \approx 0.786, \quad (12.7.51)$$

$$\beta_\infty = \sqrt{\frac{1+\sqrt{5}}{2}} \approx 1.272. \quad (12.7.52)$$

Founding of the full solution of the transcendent algebraic Eqs. (12.7.36) and (12.7.37) is not trivial problem. The used method follows [57] and consists in reduction of Eqs. (12.7.36) and (12.7.37) to the system of the ordinary differential equations with differential with respect to parameter a denoting in the following as x. Differential equation, which corresponds to the real part of the dispersion equation:

$$\begin{aligned} &y'\left[2x^3(3y^2 - 12z^2 - 1) + 8x^2 z + y\left(\frac{1}{3}x + 4x^3\right) + x\right] \\ &+ z'\left[8x^2(y-1) - 4z\left(\frac{1}{3}x + 4x^3 + 12x^3 y\right) + \frac{1}{3}\right] + \left(\frac{1}{2}y^2 - 2z^2\right)\left(\frac{1}{3} + 12x^2\right) \\ &+ 6x^2(y^3 - 12yz^2) + 16xyz + y(1 - 6x^2) - 16xz + 90x^2 - \frac{13}{2} = 0. \end{aligned} \quad (12.7.53)$$

Differential equation, which corresponds to the imaginary part of the dispersion equation:

$$\begin{aligned} &y'\left(12yzx^3 - 2x^2 y + 4x^3 z + \frac{1}{3}xz + 2x^2 - \frac{1}{12}\right) + z'\left[x^3(6y^2 - 24z^2 + 4y - 2) + 8x^2 z + x + \frac{1}{3}xy\right] \\ &+ 3x^2(6y^2 z - 8z^3) + 2x(4z^2 - y^2) + 12x^2 yz + \frac{1}{3}yz + 4xy + z(1 - 6x^2) - 26x = 0. \end{aligned} \quad (12.7.54)$$

The following notations are used in Eqs. (12.7.53) and (12.7.54): $\beta^2 - \alpha^2 = y, \alpha\beta = z, a = x$. Other notations and limit values are collected in the following Table 12.1.

The data of the left column are used as the initial conditions for numerical solution of the Cauchy problem. The results of calculations are represented as Fig. 12.1–12.4.

The following results are obtained (see also Figs. 12.5–12.8) in the frame of relativistic Navier–Stokes and Barnett equations based on the local relativistic Boltzmann equation. All numerical results are obtained by the reduction of the corresponding algebraic equations to the differential equations like in the previous method for the solution of the dispersion equation for the non-local case. Namely, for the relativistic local Navier-Stokes one obtains the dispersion equation in the previous notations

TABLE 12.1 Notations and Limit Values

$x = 0$	$x = \infty$
$y = \beta^2 - \alpha^2, y = 3,$ $z = \alpha\beta, z = 0,$ $u = \alpha^2, u = 0,$ $v = \beta^2, v = 3,$ $\alpha = 0,$ $\beta = \sqrt{3} = 1.732.$	$y = \beta^2 - \alpha^2, y = 1,$ $z = \alpha\beta, z = 1,$ $u = \alpha^2, u = \frac{2}{1+\sqrt{5}} = 0.618,$ $v = \beta^2, v = \frac{1+\sqrt{5}}{2} = 1.618,$ $\alpha = 0.786, \beta = 1.272.$

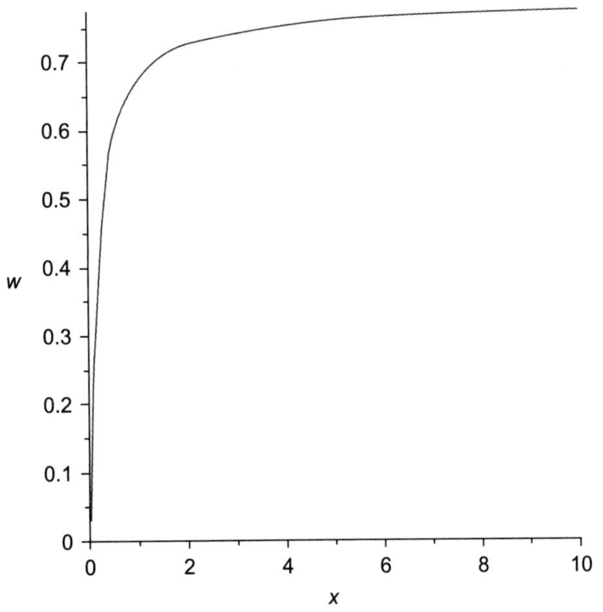

FIGURE 12.1 The attenuation rate, $w(x) = \alpha(a)$, calculated in the non-local relativistic Euler theory.

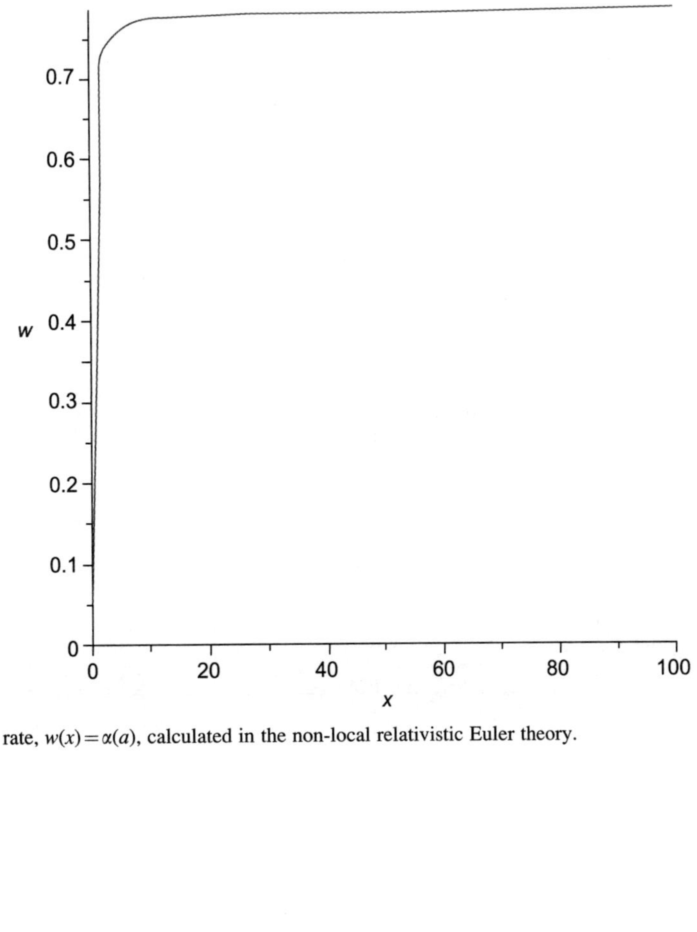

FIGURE 12.2 The attenuation rate, $w(x) = \alpha(a)$, calculated in the non-local relativistic Euler theory.

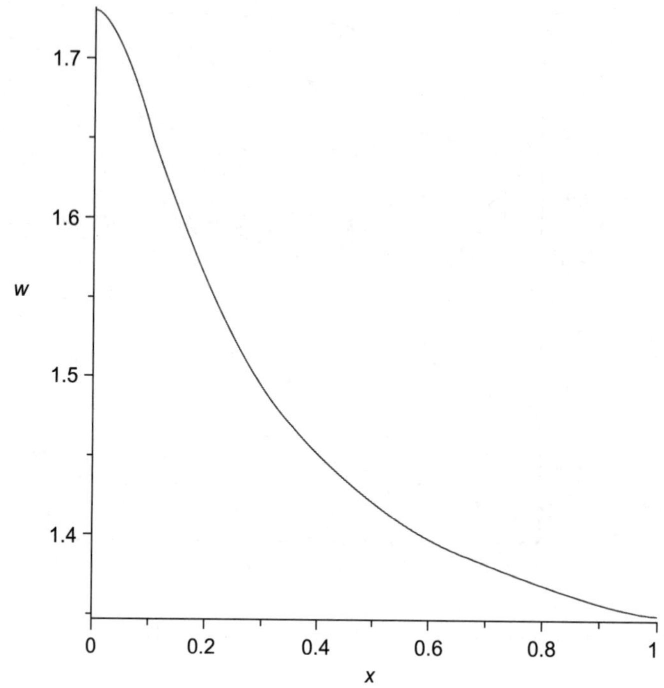

FIGURE 12.3 The velocity rate, here $w(x) = \beta(a)$, calculated in the non-local relativistic Euler theory.

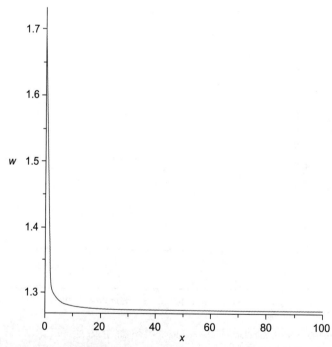

FIGURE 12.4 The velocity rate, here $w(x) = \beta(a)$, calculated in the non-local relativistic Euler theory.

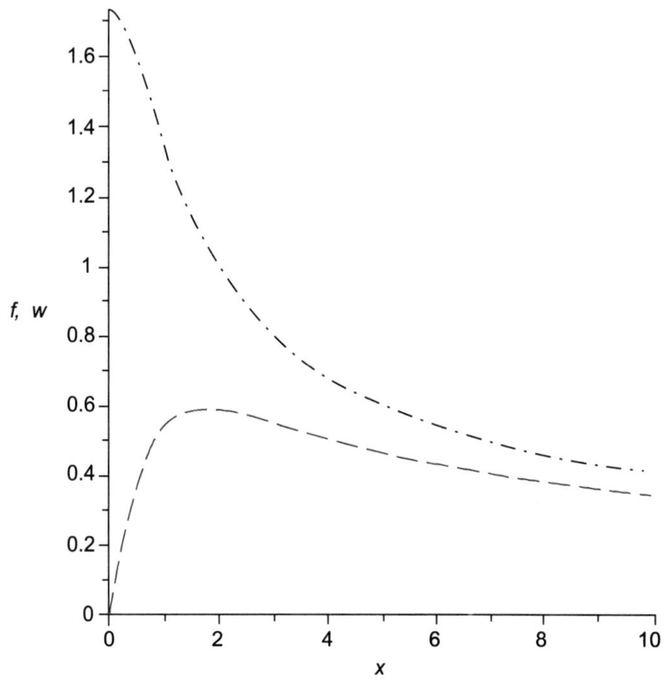

FIGURE 12.5 The attenuation rate $\alpha = f(x)$, (dashed line); the velocity rate, here $\beta = w(x)$, (dash-dot line) calculated in the local relativistic Navier-Stokes theory.

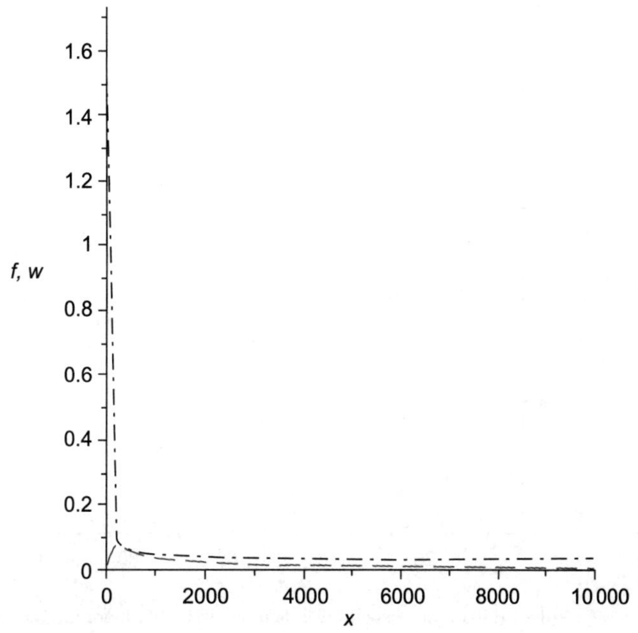

FIGURE 12.6 The attenuation rate $\alpha = f(x)$, (dashed line); the velocity rate, here $\beta = w(x)$, (dash-dot line) calculated in the local relativistic Navier-Stokes theory.

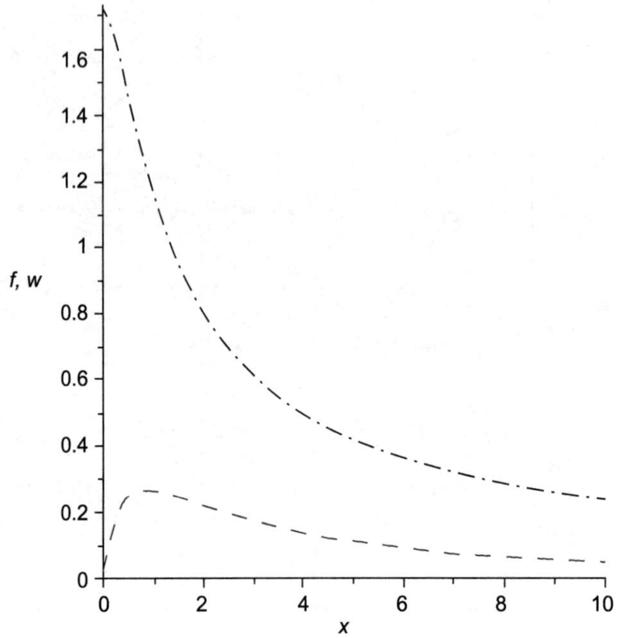

FIGURE 12.7 The attenuation rate $\alpha = f(x)$, (dashed line); the velocity rate, here $\beta = w(x)$, (dash-dot line) calculated in the local relativistic Barnett theory, (left, $0 \le x \le 10$).

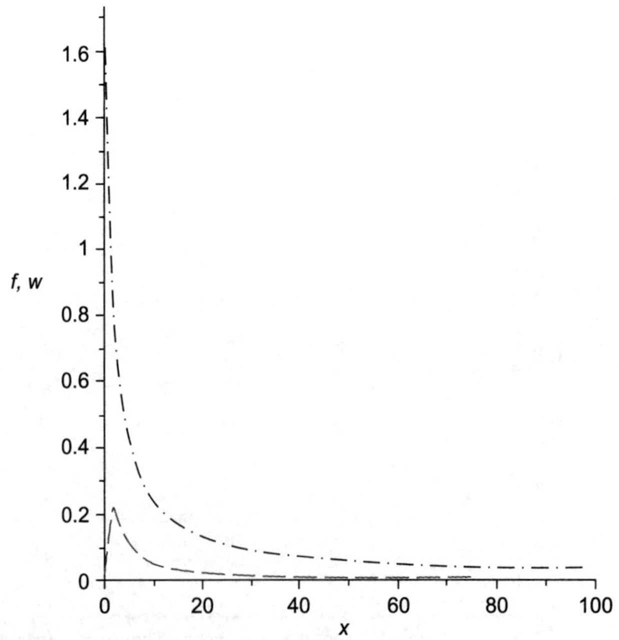

FIGURE 12.8 The attenuation rate $\alpha = f(x)$, (dashed line); the velocity rate, here $\beta = w(x)$, (dash-dot line) calculated in the local relativistic Barnett theory, (right, $0 \le x \le 100$).

$$\frac{\overset{\frown 4}{k}}{9}\left(\frac{5}{4}x^2-\frac{5}{4}ix\right)+\frac{\overset{\frown 2}{k}}{3}\left(-\frac{9}{4}ix-1\right)-1=0, \tag{12.7.55}$$

and corresponding real part of Eq. (12.7.55)

$$(y^2-4z^2)x^2-4zxy+\frac{12}{5}y+\frac{54}{5}zx-\frac{36}{5}=0 \tag{12.7.56}$$

and imaginary part of Eq. (12.7.55)

$$-(y^2-4z^2)x-4zyx^2+\frac{27}{5}xy-\frac{24}{5}z=0. \tag{12.7.57}$$

Asymptotic solutions:

By $x=0$, the solution is $y=3$, $z=0$. By $x\to\infty$, the asymptotic solution is $y=0$, $z=0$.

For the relativistic local Barnett equation one obtains the dispersion equation in the previous notations

$$-\frac{\overset{\frown 6}{k}}{27}\left(\frac{5}{12}x^4-\frac{15}{8}ix^3\right)+\frac{\overset{\frown 4}{k}}{9}\left(\frac{7}{3}x^2-\frac{5}{4}ix\right)-\frac{\overset{\frown 2}{k}}{3}\left(\frac{5}{12}x^2+\frac{9}{4}ix+1\right)-1=0, \tag{12.7.58}$$

and corresponding real part of Eq. (12.7.58):

$$\begin{aligned}&\left(\frac{5}{12}y^3-5z^2y\right)x^4-\frac{15}{4}x^3(3zy^2-4z^3)\\&+7x^2(y^2-4z^2)-15xyz+y\left(\frac{15}{4}x^2+9\right)+\frac{81}{2}xz-27=0.\end{aligned} \tag{12.7.59}$$

and imaginary part of Eq. (12.7.58):

$$\begin{aligned}&\left(\frac{10}{3}z^3-\frac{5}{2}zy^2\right)x^4+\frac{15}{8}(12z^2y-y^3)x^3-\frac{15}{4}x(y^2-4z^2)\\&-28zyx^2+\frac{81}{4}xy-18z\left(\frac{5}{12}x^2+1\right)=0.\end{aligned} \tag{12.7.60}$$

Asymptotic solutions:

By $x=0$, the solution is $y=3$, $z=0$. By $x\to\infty$, the asymptotic solution is $y=0$, $z=0$.

For solution in the frame of the non-local relativistic hydrodynamic Euler equations, is found by $\omega\to\infty$

$$v_{\text{ph}}^{\text{ur}}=c/\beta_\infty\approx 0.786c, \tag{12.7.61}$$

and

$$v_{\text{ph}}^{\text{ur}}/v_{\text{s}}^{\text{ur}}=\sqrt{3}/\beta_\infty=\sqrt{6/\left(1+\sqrt{5}\right)}\approx 1.36; \tag{12.7.62}$$

for attenuation rate from Eq. (12.7.51) follows $\alpha_\infty v_{\text{s}}^{\text{ur}}/c=\alpha_\infty/\sqrt{3}\cong 0.45$.

Results of the theories of "13 fields" and "14 fields" [255] lead to the zero attenuation by $a=\omega\tau_0\to\infty$ and to appearance of two sound velocities. This effect has only the mathematical origin. For example, by $\omega\tau_0=3$ local equations of Navier-Stokes, Barnett and many fields lead for $\alpha_\omega v_{\text{s}}^{\text{ur}}/c$ to the following values 0.33; 0.1; 0.05 correspondingly. Important to notice that for solutions, found with the help of relativistic local Navier-Stokes and Barnett equations, the phase velocity of waves tends to infinity by $\omega\to\infty$; this fact has no physical sense. Therefore, we have typical situation in the theories based on the local Boltzmann equation—divergence of successive approximations in the moment methods.

SOME REMARKS TO THE CONCLUSION OF THE MONOGRAPH

We are in front of the tremendous catastrophe in modern theoretical physics. Moreover we have reached the revolutionary situation not only in physics but in natural philosophy on the whole. Practically we are in front of the new challenge since Newton's Mathematical Principles of Natural Philosophy was first published in 1687. It is impossible to believe that in more than three hundred years after Newton we have the situation when 96% of matter and energy is of unknown origin.

As it is shown in this monograph the origin of difficulties consists in the total Oversimplification inherent in local physics of the dissipative processes.

In the latter part of 20th century two very important results were obtained:

(1) The Irish physicist John Stewart Bell (1928–1990) was to show that all local statistical theories of dissipative processes are wrong in principal.
(2) The Russian physicist Boris V. Alexeev was to show that the derivation of KE_{f_1} from the BBGKY equations (*prior* to introducing any approximation destined to break the Bogolyubov chain) leads to additional terms of the non-local origin, generally of the same order of magnitude, appear in the Boltzmann equation (BE). Then the passage to the BE means the neglect of non-local effects. These additional terms cannot be omitted even in the limit cases of kinetic theory, therefore BE is only a plausible equation.

The scientific community was convinced that the mentioned results could lead only to rather small corrections in the modern theoretical physics. So to speak—4% corrections to 96% of the known results, but not quite the reverse! Many scientists are aware that some way out will be achieved after creation of the unified theory of transport processes working from the structure of so called elementary particles to the Universe evolution. This theory is in front of you.

I had possibility to discuss extremely significant problems challenged modern fundamental physics, which can be titled as "Non-solved problems of the fundamental physics" or more precisely – of local physical kinetics of dissipative processes, like:

(1) Kinetic theory of entropy and the problem of the initial perturbation;
(2) Strict theory of turbulence;
(3) Quantum non-relativistic and relativistic hydrodynamics, theory of charges separation in the atom structure;
(4) Theory of ball lightning;
(5) Theory of dark matter;
(6) Theory of dark energy, Hubble expansion of the Universe; and
(7) The destiny of anti-matter after the Big Bang.

All these problems have the common main feature – these problems cannot be solved in the frame of local physics.

FINALLY, Physics of 21st century is non-local physics.

Appendix 1

Perturbation Method of the Equation Solution Related to $T[f]$

Let us consider the solution of Eq. (3.3.8) by the perturbation method. Equation (3.3.8) is valid only by the small y and moreover the function $T[f]$ receives the physical meaning only by $y=0$. Let us expand $T[f]$ in series:

$$T[f](t,x,y)=\sum_{n=0}^{\infty}T_n[f](t,x)y^n. \tag{A.1.1}$$

After substitution (Eq. A.1.1) in Eq. (3.3.8):

$$\mathrm{i}\alpha\frac{\partial}{\partial t}T[f](x,y,t)=-\frac{\alpha^2}{2m}\frac{\partial^2}{\partial x\partial y}T[f](x,y,t)-2yF(x)T[f](x,y,t),$$

we find after equalizing the terms in front of the same powers of y:
for $y^0_{,}$

$$\mathrm{i}\frac{\partial}{\partial t}T_0[f](x,t)+\frac{\alpha}{2m}\frac{\partial}{\partial x}T_1[f](x,t)=0; \tag{A.1.2}$$

for $y^1_{,}$

$$\mathrm{i}\alpha\frac{\partial}{\partial t}T_1[f](x,t)=-\frac{\alpha^2}{m}\frac{\partial}{\partial x}T_2[f](x,t)-2F(x)T_0[f](x,t); \tag{A.1.3}$$

for $y^2_{,}$

$$\mathrm{i}\alpha\frac{\partial}{\partial t}T_2[f](x,t)=-\frac{3\alpha^2}{2m}\frac{\partial}{\partial x}T_3[f](x,t)-2F(x)T_1[f](x,t) \tag{A.1.4}$$

As we see, the successive approximations lead to the chain of equations and the first links of the chain are Eqs. (A.1.2)–(A.1.4).

The coefficients of the expansion (Eq. A.1.1) are, generally speaking, the complex functions with one exception concerning the coefficient $T_0[f](x,t)$ because of the condition

$$T[f](x,y=0,t)=T_0[f](x,t)=|\Psi(x,t)|^2.$$

Denoting ρ as real probability density ($|\Psi(x,t)|^2=\rho$), one obtains from Eq. (A.1.2):

$$\mathrm{i}\frac{\partial}{\partial t}\rho(x,t)+\frac{\alpha}{2m}\frac{\partial}{\partial x}T_1[f](x,t)=0 \tag{A.1.5}$$

But (see Eq. 3.3.4)

$$T[f](x,y,t)=\Psi^*(t,x-y)\Psi(t,x+y)$$

and

$$\Psi^*(t,x-y)=\sum_{n=0}^{\infty}\Psi^*_n(t,x)(-y)^n=\Psi^*_0(t,x)-\left(\frac{\partial\Psi^*_0}{\partial x}\right)_{y=0}y+\frac{1}{2}\left(\frac{\partial^2\Psi^*_0}{\partial x^2}\right)_{y=0}y^2-\cdots, \tag{A.1.6}$$

$$\Psi(t,x+y)=\sum_{n=0}^{\infty}\Psi_n^*(t,x)y^n=\Psi_0(t,x)+\left(\frac{\partial\Psi_0}{\partial x}\right)_{y=0}y+\frac{1}{2}\left(\frac{\partial^2\Psi_0}{\partial x^2}\right)_{y=0}y^2+\cdots, \tag{A.1.7}$$

then

$$T_1(x,t)=\Psi_0^*(t,x)\left(\frac{\partial\Psi_0}{\partial x}\right)_{y=0}-\Psi_0(t,x)\left(\frac{\partial\Psi_0^*}{\partial x}\right)_{y=0}. \tag{A.1.8}$$

From Eq. (A.1.8) follows that $T_1(x,t)$ is an imagine value. Really,

$$T_1(x,t)=\Psi_0^*(t,x)\left(\frac{\partial\Psi_0}{\partial x}\right)_{y=0}-\Psi_0(t,x)\left(\frac{\partial\Psi_0^*}{\partial x}\right)_{y=0}=\Psi_0^*(t,x)\left(\frac{\partial\Psi_0}{\partial x}\right)_{y=0}-\left[\Psi_0^*(t,x)\left(\frac{\partial\Psi_0}{\partial x}\right)_{y=0}\right]^*. \tag{A.1.9}$$

This result coincides with Eq. (A.1.5), from which follows

$$\frac{\partial}{\partial x}T_1^{\text{real}}[f](x,t)=0, \tag{A.1.10}$$

or

$$T_1^{\text{real}}[f](x,t)=\text{const}. \tag{A.1.11}$$

Then, const = 0 in Eqs. (A.1.11) and (A.1.5) takes the form

$$\frac{\partial}{\partial t}\rho(x,t)+\frac{\alpha}{2m}\frac{\partial}{\partial x}T_1^{\text{imagine}}[f](x,t)=0, \tag{A.1.12}$$

because

$$T_1[f](x,t)=\mathrm{i}T_1^{\text{imagine}}[f](x,t) \tag{A.1.13}$$

From Eqs. (A.1.5) and (A.1.8) follow

$$\mathrm{i}\frac{\partial}{\partial t}\rho(x,t)+\frac{\alpha}{2m}\frac{\partial}{\partial x}\left[\Psi_0^*(t,x)\left(\frac{\partial\Psi_0}{\partial x}\right)_{y=0}-\Psi_0(t,x)\left(\frac{\partial\Psi_0^*}{\partial x}\right)_{y=0}\right]=0 \tag{A.1.14}$$

or

$$\mathrm{i}\Psi_0(t,x)\frac{\partial}{\partial t}\Psi_0^*(t,x)+\mathrm{i}\Psi_0^*(t,x)\frac{\partial}{\partial t}\Psi_0(t,x)+\frac{\alpha}{2m}\left[\Psi_0^*(t,x)\left(\frac{\partial^2\Psi_0}{\partial x^2}\right)_{y=0}-\Psi_0(t,x)\left(\frac{\partial^2\Psi_0^*}{\partial x^2}\right)_{y=0}\right]=0. \tag{A.1.15}$$

Equation (A.1.15) is satisfied identically if

$$\mathrm{i}\frac{\partial}{\partial t}\Psi_0(t,x)+\frac{\alpha}{2m}\left(\frac{\partial^2\Psi_0}{\partial x^2}\right)_{y=0}=0. \tag{A.1.16}$$

Equation (A.1.16) is Schrödinger equation of the first approximation.
For the second approximation (see Eqs. A.1.3 and A.1.8)

$$\mathrm{i}\alpha\frac{\partial}{\partial t}\left[\Psi_0^*\frac{\partial\Psi_0}{\partial x}-\Psi_0\frac{\partial\Psi_0^*}{\partial x}\right]=-\frac{\alpha^2}{m}\frac{\partial}{\partial x}T_2[f](x,t)-2F(x)T_0[f](x,t), \tag{A.1.17}$$

and

$$T_2(x,t)=\frac{1}{2}\Psi_0^*(t,x)\left(\frac{\partial^2\Psi_0}{\partial x^2}\right)_{y=0}+\frac{1}{2}\Psi_0(t,x)\left(\frac{\partial^2\Psi_0^*}{\partial x^2}\right)_{y=0}-\left(\frac{\partial\Psi_0}{\partial x}\right)_{y=0}\left(\frac{\partial\Psi_0^*}{\partial x}\right)_{y=0}. \tag{A.1.18}$$

After substitution (Eq. A.1.18) in Eq. (A.1.17)

$$i\alpha\frac{\partial}{\partial t}\left[\Psi_0^*\frac{\partial\Psi_0}{\partial x}-\Psi_0\frac{\partial\Psi_0^*}{\partial x}\right]=-\frac{\alpha^2}{2m}\frac{\partial}{\partial x}\left[\Psi_0^*\frac{\partial^2\Psi_0}{\partial x^2}+\Psi_0\frac{\partial^2\Psi_0^*}{\partial x^2}-2\frac{\partial\Psi_0}{\partial x}\frac{\partial\Psi_0^*}{\partial x}\right]-2F(x)\Psi_0^*\Psi_0 \tag{A.1.19}$$

Let us consider the equation

$$i\alpha\frac{\partial}{\partial t}\left[\Psi_0^*\frac{\partial\Psi_0}{\partial x}\right]=-\frac{\alpha^2}{2m}\frac{\partial}{\partial x}\left[\Psi_0^*\frac{\partial^2\Psi_0}{\partial x^2}-\frac{\partial\Psi_0}{\partial x}\frac{\partial\Psi_0^*}{\partial x}\right]-F(x)\Psi_0^*\Psi_0. \tag{A.1.20}$$

The conjugate equation can be written as

$$-i\alpha\frac{\partial}{\partial t}\left[\Psi_0\frac{\partial\Psi_0^*}{\partial x}\right]=-\frac{\alpha^2}{2m}\frac{\partial}{\partial x}\left[\Psi_0\frac{\partial^2\Psi_0^*}{\partial x^2}-\frac{\partial\Psi_0^*}{\partial x}\frac{\partial\Psi_0}{\partial x}\right]-F(x)\Psi_0^*\Psi_0. \tag{A.1.21}$$

Summation of Eqs. (A.1.20) and (A.1.21) gives Eq. (A.1.19). From Eq. (A.1.20) follows equation which could be titled as Schrödinger equation of the second approximation

$$i\alpha\frac{\partial\Psi_0^*}{\partial t}\frac{\partial\Psi_0}{\partial x}+i\alpha\Psi_0^*\frac{\partial^2\Psi_0}{\partial t\partial x}=-\frac{\alpha^2}{2m}\left[\Psi_0^*\frac{\partial^3\Psi_0}{\partial x^3}-\frac{\partial\Psi_0}{\partial x}\frac{\partial^2\Psi_0^*}{\partial x^2}\right]-F(x)\Psi_0\Psi_0^*. \tag{A.1.22}$$

Equation (A.1.22) can be transformed as

$$i\alpha\frac{\partial\ln\Psi_0^*}{\partial t}\frac{\partial\Psi_0}{\partial x}+i\alpha\frac{\partial^2\Psi_0}{\partial t\partial x}=-\frac{\alpha^2}{2m}\left[\frac{\partial^3\Psi_0}{\partial x^3}-\frac{1}{2\rho}\frac{\partial\Psi_0^2}{\partial x}\frac{\partial^2\Psi_0^*}{\partial x^2}\right]-F(x)\Psi_0. \tag{A.1.23}$$

As we see, Schrödinger equation of the second approximation is non-linear equation of the third order in space, containing the cross derivative "time-space" and derivative on time of the logarithmic term.

Appendix 2

Using of Curvilinear Coordinates in the Generalized Hydrodynamic Theory

Curvilinear coordinates are often used to define the location or distribution of physical quantities using the non-local description. Let be x, y, z—Cartesian coordinates, the unit vectors of the rectangular coordinate system **i**, **j**, **k** and x^1, x^2,x^3—curvilinear coordinates. A basis whose vectors change their direction and/or magnitude from point to point is called local basis. All bases associated with curvilinear coordinates are necessarily local. Basis vectors that are the same at all points are global bases. A point P in 3d space can be defined using Cartesian coordinates (x, y, z), or in another system x^1,x^2,x^3. The latter is a curvilinear coordinate system, and x^1,x^2,x^3 are the curvilinear coordinates of the point P. The surfaces $x^1=$constant, $x^2=$constant, $x^3=$constant are called the coordinate surfaces; and the space curves formed by their intersection in pairs are called the coordinate curves.

Let us construct three vectors

$$\mathbf{r}_k = \left(\frac{\partial x}{\partial x^k}, \frac{\partial y}{\partial x^k}, \frac{\partial z}{\partial x^k}\right), \quad k = 1, 2, 3. \tag{A.2.1}$$

These may not have unit length, and may also not be orthogonal. In the case that they are orthogonal at all points where the derivatives are well-defined, we define the Lamé coefficients (after Gabriel Lamé) and introduce the Lamé coefficients

$$|\mathbf{r}_1| = H_1, \quad |\mathbf{r}_2| = H_2, \quad |\mathbf{r}_3| = H_3. \tag{A.2.2}$$

and the curvilinear orthonormal basis vectors by

$$\mathbf{e}_1 = \frac{\mathbf{r}_1}{H_1}, \quad \mathbf{e}_2 = \frac{\mathbf{r}_2}{H_2}, \quad \mathbf{e}_3 = \frac{\mathbf{r}_3}{H_3}. \tag{A.2.3}$$

It is important to note that these basis vectors may well depend upon the position of P; it is therefore necessary that they are not assumed to be constant over a region.

Let us introduce the reciprocal basis:

$$\mathbf{r}^1 = \frac{[\mathbf{r}_2 \times \mathbf{r}_3]}{(\mathbf{r}_1 \cdot [\mathbf{r}_2 \times \mathbf{r}_3])}, \quad \mathbf{r}^2 = \frac{[\mathbf{r}_3 \times \mathbf{r}_1]}{(\mathbf{r}_1 \cdot [\mathbf{r}_2 \times \mathbf{r}_3])}, \quad \mathbf{r}^3 = \frac{[\mathbf{r}_1 \times \mathbf{r}_2]}{(\mathbf{r}_1 \cdot [\mathbf{r}_2 \times \mathbf{r}_3])}. \tag{A.2.4}$$

In this

$$\mathbf{r}_i \cdot \mathbf{r}^k = \delta_i^k = \begin{cases} 1, & \text{if } i = k \\ 0, & \text{if } i \neq k \end{cases}. \tag{A.2.5}$$

Any point can be written as a position vector **x** in the standard basis:

$$\mathbf{x} = x^i \mathbf{r}_i, \tag{A.2.6}$$

and in the reciprocal basis

$$\mathbf{x} = x_i \mathbf{r}^i. \tag{A.2.7}$$

After multiplying Eq. (A.2.6) by $\mathbf{r}^k$ and Eq. (A.2.7) by $\mathbf{r}_k$, we have

$$\mathbf{x} \cdot \mathbf{r}^k = x^i \mathbf{r}_i \cdot \mathbf{r}^k = x^k. \tag{A.2.8}$$

$$\mathbf{x} \cdot \mathbf{r}_k = x_i \mathbf{r}^i \cdot \mathbf{r}_k = x_k. \tag{A.2.9}$$

Using Eqs. (A.2.5)–(A.2.9), we obtain relations known as Gibbs formulae:

$$\mathbf{x} = (\mathbf{x} \cdot \mathbf{r}_i)\mathbf{r}^i, \tag{A.2.10}$$

$$\mathbf{x} = \left(\mathbf{x} \cdot \mathbf{r}^i\right)\mathbf{r}_i. \tag{A.2.11}$$

Let us express the gradient of a scalar field $u = u(x, y, z)$. As it usually is,

$$\operatorname{grad} u = \frac{\partial u}{\partial x}\mathbf{i} + \frac{\partial u}{\partial y}\mathbf{j} + \frac{\partial u}{\partial z}\mathbf{k}. \tag{A.2.12}$$

Introducing the curvilinear coordinates $u = u(x^1, x^2, x^3)$, using the differentiation rule

$$\frac{\partial u}{\partial x^i} = \frac{\partial u}{\partial x}\frac{\partial x}{\partial x^i} + \frac{\partial u}{\partial y}\frac{\partial y}{\partial x^i} + \frac{\partial u}{\partial z}\frac{\partial z}{\partial x^i}, \tag{A.2.13}$$

Eq. (A.2.12) and relation (A.2.1) for $\mathbf{r}_i$, we find

$$\frac{\partial u}{\partial x^i} = (\mathbf{r}_i \cdot \operatorname{grad} u). \tag{A.2.14}$$

Using the Gibbs formulae one obtains the imagination for grad$u$

$$\operatorname{grad} u = (\mathbf{r}_i \cdot \operatorname{grad} u)\mathbf{r}^i = \frac{\partial u}{\partial x^i}\mathbf{r}^i. \tag{A.2.15}$$

Divergence of the linear operator $\hat{A}$ in an arbitrary basis is defined by relation

$$\operatorname{div}\hat{A} = \left(\mathbf{r}^i \cdot \hat{A}\mathbf{r}_i\right), \tag{A.2.16}$$

and

$$\hat{A}\mathbf{r}_i = a_i^k \mathbf{r}_k, \tag{A.2.17}$$

where the coefficients (a_i^k) form the matrices of coefficients of this linear operator.

Substitute Eq. (A.2.17) into Eq. (A.2.16)

$$\operatorname{div}\hat{A} = \mathbf{r}^i \cdot \left(a_i^k \mathbf{r}_k\right) = a_i^k \delta_k^i = a_1^1 + a_2^2 + a_3^3. \tag{A.2.18}$$

Let be $\mathbf{p}\ (x,y,z) = P\mathbf{i} + Q\mathbf{j} + R\mathbf{k}$ is a vector field and

$$\operatorname{div}\mathbf{p} = \operatorname{div}\hat{A}, \tag{A.2.19}$$

where

$$\begin{aligned} \hat{A}\mathbf{i} &= \frac{\partial \mathbf{p}}{\partial x} = \frac{\partial P}{\partial x}\mathbf{i} + \frac{\partial Q}{\partial x}\mathbf{j} + \frac{\partial R}{\partial x}\mathbf{k}, \\ \hat{A}\mathbf{j} &= \frac{\partial \mathbf{p}}{\partial y} = \frac{\partial P}{\partial y}\mathbf{i} + \frac{\partial Q}{\partial y}\mathbf{j} + \frac{\partial R}{\partial y}\mathbf{k}, \\ \hat{A}\mathbf{k} &= \frac{\partial \mathbf{p}}{\partial z} = \frac{\partial P}{\partial z}\mathbf{i} + \frac{\partial Q}{\partial z}\mathbf{j} + \frac{\partial R}{\partial z}\mathbf{k}. \end{aligned} \tag{A.2.20}$$

From relations Eq. (A.2.20) follow

$$a_1^1 = \frac{\partial P}{\partial x},\quad a_2^2 = \frac{\partial Q}{\partial y},\quad a_3^3 = \frac{\partial R}{\partial z}. \tag{A.2.21}$$

From Eqs. (A.2.18), (A.2.19), and (A.2.21) we have well-known expression

$$\operatorname{div}\mathbf{p} = \frac{\partial P}{\partial x} + \frac{\partial Q}{\partial y} + \frac{\partial R}{\partial z}. \tag{A.2.22}$$

Divergence of the vector field can be expressed in the curvilinear coordinates $\mathbf{p} = \mathbf{p}(x^1, x^2, x^3)$. Namely

$$\begin{aligned} \frac{\partial \mathbf{p}}{\partial x^i} &= \frac{\partial \mathbf{p}}{\partial x}\frac{\partial x}{\partial x^i} + \frac{\partial \mathbf{p}}{\partial y}\frac{\partial y}{\partial x^i} + \frac{\partial \mathbf{p}}{\partial z}\frac{\partial z}{\partial x^i} \\ &= (\hat{A}\mathbf{i})\frac{\partial x}{\partial x^i} + (\hat{A}\mathbf{j})\frac{\partial y}{\partial x^i} + (\hat{A}\mathbf{k})\frac{\partial z}{\partial x^i} = \hat{A}\left(\frac{\partial x}{\partial x^i}\mathbf{i} + \frac{\partial y}{\partial x^i}\mathbf{j} + \frac{\partial z}{\partial x^i}\mathbf{k}\right) = \hat{A}\mathbf{r}_i. \end{aligned} \tag{A.2.23}$$

Using the definition (A.2.16)

$$\mathrm{div}\mathbf{p} = \mathrm{div}\hat{A} = \left(\mathbf{r}^i \cdot \hat{A}\mathbf{r}_i\right) \tag{A.2.24}$$

and Eq. (A.2.23), we reach the expression for div**p** in the curvilinear coordinate system

$$\mathrm{div}\mathbf{p} = r^i \cdot \frac{\partial \mathbf{p}}{\partial x^i}. \tag{A.2.25}$$

Let us consider the case when vectors $\mathbf{r}_1, \mathbf{r}_2, \mathbf{r}_3$, are orthogonal at all points and define right-handed system. In this case

$$\mathbf{r}_2 \times \mathbf{r}_3 = \frac{H_2 H_3}{H_1}\mathbf{r}_1, \tag{A.2.26}$$

$$\mathbf{r}_1 \cdot [\mathbf{r}_2 \times \mathbf{r}_3] = H_1 H_2 H_3. \tag{A.2.27}$$

We find from Eq. (A.2.3)

$$\mathbf{r}^1 = \frac{1}{H_1^2}\mathbf{r}_1 \tag{A.2.28}$$

and analogically

$$\mathbf{r}^2 = \frac{1}{H_2^2}\mathbf{r}_2, \tag{A.2.29}$$

$$\mathbf{r}^3 = \frac{1}{H_3^2}\mathbf{r}_3. \tag{A.2.30}$$

Expressions for the gradient of the scalar field and the divergence of the vector field follow from Eqs. (A.2.15), (A.2.25) and (A.2.28)–(A.2.30):

$$\mathrm{grad}\,u = \frac{1}{H_1^2}\frac{\partial u}{\partial x^1}\mathbf{r}_1 + \frac{1}{H_2^2}\frac{\partial u}{\partial x^2}\mathbf{r}_2 + \frac{1}{H_3^2}\frac{\partial u}{\partial x^3}\mathbf{r}_3, \tag{A.2.31}$$

$$\mathrm{div}\mathbf{p} = \frac{1}{H_1^2}\frac{\partial \mathbf{p}}{\partial x^1}\cdot\mathbf{r}_1 + \frac{1}{H_2^2}\frac{\partial \mathbf{p}}{\partial x^2}\cdot\mathbf{r}_2 + \frac{1}{H_3^2}\frac{\partial \mathbf{p}}{\partial x^3}\cdot\mathbf{r}_3. \tag{A.2.32}$$

In the curvilinear orthonormal basis created by vectors $\mathbf{e}_i = \frac{\mathbf{r}_i}{H_i}$, relations (A.2.31) and (A.2.32) are written as

$$\mathrm{grad}\,u = \frac{1}{H_1}\frac{\partial u}{\partial x^1}\mathbf{e}_1 + \frac{1}{H_2}\frac{\partial u}{\partial x^2}\mathbf{e}_2 + \frac{1}{H_3}\frac{\partial u}{\partial x^3}\mathbf{e}_3, \tag{A.2.33}$$

$$\mathrm{div}\,\mathbf{p} = \frac{1}{H_1}\frac{\partial \mathbf{p}}{\partial x^1}\cdot\mathbf{e}_1 + \frac{1}{H_2}\frac{\partial \mathbf{p}}{\partial x^2}\cdot\mathbf{e}_2 + \frac{1}{H_3}\frac{\partial \mathbf{p}}{\partial x^3}\cdot\mathbf{e}_3. \tag{A.2.34}$$

Let be

$$\mathbf{p} = P^1\mathbf{e}_1 + P^2\mathbf{e}_2 + P^3\mathbf{e}_3. \tag{A.2.35}$$

and transform the relation (A.2.35) taking into account that

$$\mathbf{e}_1^2 = \mathbf{e}_2^2 = \mathbf{e}_3^2 = 1; \quad \frac{\partial \mathbf{e}_1^2}{\partial x^1} = \frac{\partial \mathbf{e}_2^2}{\partial x^2} = \frac{\partial \mathbf{e}_3^2}{\partial x^3} = 0.$$

$$\begin{aligned} \mathrm{div}\mathbf{p} &= \frac{\mathbf{e}_1}{H_1}\cdot\frac{\partial}{\partial x^1}\left(P^1\mathbf{e}_1 + P^2\mathbf{e}_2 + P^3\mathbf{e}_3\right) \\ &+ \frac{\mathbf{e}_2}{H_2}\cdot\frac{\partial}{\partial x^2}\left(P^1\mathbf{e}_1 + P^2\mathbf{e}_2 + P^3\mathbf{e}_3\right) \\ &+ \frac{\mathbf{e}_3}{H_3}\cdot\frac{\partial}{\partial x^3}\left(P^1\mathbf{e}_1 + P^2\mathbf{e}_2 + P^3\mathbf{e}_3\right), \end{aligned} \tag{A.2.36}$$

$$\operatorname{div}\mathbf{p} = \frac{1}{H_1}\frac{\partial P^1}{\partial x^1} + \frac{1}{H_2}\frac{\partial P^2}{\partial x^2} + \frac{1}{H_3}\frac{\partial P^3}{\partial x^3} + P^1\left(\frac{\mathbf{e}_2}{H_2}\cdot\frac{\partial \mathbf{e}_1}{\partial x^2} + \frac{\mathbf{e}_3}{H_3}\cdot\frac{\partial \mathbf{e}_1}{\partial x^3}\right) + P^2\left(\frac{\mathbf{e}_1}{H_1}\cdot\frac{\partial \mathbf{e}_2}{\partial x^1} + \frac{\mathbf{e}_3}{H_3}\cdot\frac{\partial \mathbf{e}_2}{\partial x^3}\right) + P^3\left(\frac{\mathbf{e}_1}{H_1}\cdot\frac{\partial \mathbf{e}_3}{\partial x^1} + \frac{\mathbf{e}_2}{H_2}\cdot\frac{\partial \mathbf{e}_3}{\partial x^2}\right). \tag{A.2.37}$$

Transform as an example one item from Eq. (A.2.37) using the consequence of Eqs. (A.2.1)–(A.2.3).

$$\mathbf{e}_k = \frac{\mathbf{r}_k}{H_k} = \frac{1}{H_k}\frac{\partial \mathbf{r}}{\partial x^k}. \tag{A.2.38}$$

Then

$$P^1\frac{\mathbf{e}_2}{H_2}\cdot\frac{\partial \mathbf{e}_1}{\partial x^2} = P^1\frac{\mathbf{e}_2}{H_2}\cdot\frac{\partial}{\partial x^2}\left(\frac{1}{H_1}\frac{\partial \mathbf{r}}{\partial x^1}\right) = P^1\frac{\mathbf{e}_2}{H_2}\cdot\left(\frac{\partial \mathbf{r}}{\partial x^1}\frac{\partial}{\partial x^2}\left(\frac{1}{H_1}\right) + \frac{1}{H_1}\frac{\partial^2 \mathbf{r}}{\partial x^2 \partial x^1}\right)$$
$$= P^1\frac{1}{H_2}\left(\mathbf{e}_2\cdot\mathbf{e}_1 H_1\frac{\partial}{\partial x^2}\left(\frac{1}{H_1}\right) + \frac{\mathbf{e}_2}{H_1}\frac{\partial}{\partial x^1}\left(\frac{\partial \mathbf{r}}{\partial x^2}\right)\right) = P^1\frac{1}{H_2}\left(\frac{\mathbf{e}_2}{H_1}\cdot\frac{\partial}{\partial x^1}(H_2\mathbf{e}_2)\right) \tag{A.2.39}$$
$$= P^1\frac{1}{H_2}\left(\frac{\mathbf{e}_2\cdot\mathbf{e}_2}{H_1}\frac{\partial H_2}{\partial x^1} + \frac{H_2\mathbf{e}_2}{H_1}\cdot\frac{\partial \mathbf{e}_2}{\partial x^1}\right) = P^1\frac{1}{H_2}\left(\frac{1}{H_1}\frac{\partial H_2}{\partial x^1} + \frac{H_2}{2H_1}\frac{\partial \mathbf{e}_2^2}{\partial x^1}\right) = \frac{P^1}{H_1 H_2}\frac{\partial H_2}{\partial x^1}$$

Analogically for another term

$$P^1\frac{\mathbf{e}_3}{H_3}\cdot\frac{\partial \mathbf{e}_1}{\partial x^3} = \frac{P^1}{H_1 H_3}\frac{\partial H_3}{\partial x^1}. \tag{A.2.40}$$

and

$$P^1\left(\frac{\mathbf{e}_2}{H_2}\cdot\frac{\partial \mathbf{e}_1}{\partial x^2} + \frac{\mathbf{e}_3}{H_3}\cdot\frac{\partial \mathbf{e}_1}{\partial x^3}\right) = \frac{P^1}{H_1}\left(\frac{1}{H_2}\frac{\partial H_2}{\partial x^1} + \frac{1}{H_3}\frac{\partial H_3}{\partial x^1}\right) = \frac{P^1}{H_1 H_2 H_3}\frac{\partial (H_2 H_3)}{\partial x^1}. \tag{A.2.41}$$

By the similar way we have

$$P^2\left(\frac{\mathbf{e}_1}{H_1}\cdot\frac{\partial \mathbf{e}_2}{\partial x^1} + \frac{\mathbf{e}_3}{H_3}\cdot\frac{\partial \mathbf{e}_2}{\partial x^3}\right) = \frac{P^2}{H_1 H_2 H_3}\frac{\partial (H_1 H_3)}{\partial x^2}, \tag{A.2.42}$$

$$P^3\left(\frac{\mathbf{e}_1}{H_1}\cdot\frac{\partial \mathbf{e}_3}{\partial x^1} + \frac{\mathbf{e}_2}{H_2}\cdot\frac{\partial \mathbf{e}_3}{\partial x^2}\right) = \frac{P^3}{H_1 H_2 H_3}\frac{\partial (H_1 H_2)}{\partial x^3} \tag{A.2.43}$$

and after the substitution of Eqs. (A.2.41)–(A.2.43) into Eq. (A.2.37)

$$\operatorname{div}\mathbf{p} = \frac{1}{H_1}\frac{\partial P^1}{\partial x^1} + \frac{1}{H_2}\frac{\partial P^2}{\partial x^2} + \frac{1}{H_3}\frac{\partial P^3}{\partial x^3} + \frac{1}{H_1 H_2 H_3}\left(P^1\frac{\partial (H_2 H_3)}{\partial x^1} + P^2\frac{\partial (H_3 H_1)}{\partial x^2} + P^3\frac{\partial (H_1 H_2)}{\partial x^3}\right), \tag{A.2.44}$$

or

$$\operatorname{div}\mathbf{p} = \frac{1}{H_1 H_2 H_3}\left(\frac{\partial (P^1 H_2 H_3)}{\partial x^1} + \frac{\partial (P^2 H_3 H_1)}{\partial x^2} + \frac{\partial (P^3 H_1 H_2)}{\partial x^3}\right). \tag{A.2.45}$$

Laplace operator of the scalar function u is defined as

$$\Delta u = \operatorname{div}\operatorname{grad} u. \tag{A.2.46}$$

In the Cartesian coordinate system

$$\Delta u = \frac{\partial^2 u}{\partial x^2} + \frac{\partial^2 u}{\partial y^2} + \frac{\partial^2 u}{\partial z^2}. \tag{A.2.47}$$

and in the curvilinear coordinate system (see Eqs. A.2.33 and A.2.45)

$$\Delta u = \frac{1}{H_1 H_2 H_3}\left(\frac{\partial}{\partial x^1}\left(\frac{H_2 H_3}{H_1}\frac{\partial u}{\partial x^1}\right) + \frac{\partial}{\partial x^2}\left(\frac{H_3 H_1}{H_2}\frac{\partial u}{\partial x^2}\right) + \frac{\partial}{\partial x^3}\left(\frac{H_1 H_2}{H_3}\frac{\partial u}{\partial x^3}\right)\right). \tag{A.2.48}$$

Introduce the spherical coordinate system using relations

$$\begin{aligned} x &= r\sin\theta\cos\varphi, \quad (1)\\ y &= r\sin\theta\sin\varphi, \quad (2)\\ z &= r\cos\theta. \quad (3) \end{aligned} \tag{A.2.49}$$

Here $0 \le r < \infty$, $0 \le \varphi \le 2\pi$, $0 \le \theta \le \pi$ and $x^1 = r$, $x^2 = \varphi$, $x^3 = \theta$. Using relations (1)–(3) in Eq. (A.2.49), we find

$$\begin{aligned} \mathbf{r}_1 &= \left(\frac{\partial x}{\partial r}, \frac{\partial y}{\partial r}, \frac{\partial z}{\partial r}\right) = (\sin\theta\cos\varphi, \sin\theta\sin\varphi, \cos\theta), \quad (1)\\ \mathbf{r}_2 &= \left(\frac{\partial x}{\partial \varphi}, \frac{\partial y}{\partial \varphi}, \frac{\partial z}{\partial \varphi}\right) = (-r\sin\theta\sin\varphi, r\sin\theta\cos\varphi, 0), \quad (2)\\ \mathbf{r}_3 &= \left(\frac{\partial x}{\partial \theta}, \frac{\partial y}{\partial \theta}, \frac{\partial z}{\partial \theta}\right) = (r\cos\theta\cos\varphi, r\cos\theta\sin\varphi, -r\sin\theta), \quad (3) \end{aligned} \tag{A.2.50}$$

The orthogonality of basis $(\mathbf{r}_1, \mathbf{r}_2, \mathbf{r}_3)$ can be directly verified using conditions $(\mathbf{r}_1 \cdot \mathbf{r}_2) = (\mathbf{r}_2 \cdot \mathbf{r}_3) = (\mathbf{r}_1 \cdot \mathbf{r}_3) = 0$. The Lamé coefficients for the spherical coordinate system:

$$\begin{aligned} H_1 &= |\mathbf{r}_1| = \sqrt{\sin^2\theta\cos^2\varphi + \sin^2\theta\sin^2\varphi + \cos^2\theta} = 1, \quad (1)\\ H_2 &= |\mathbf{r}_2| = \sqrt{r^2\sin^2\theta\sin^2\varphi + r^2\sin^2\theta\cos^2\varphi} = r\sin\theta, \quad (2)\\ H_3 &= |\mathbf{r}_3| = \sqrt{r^2\cos^2\theta\cos^2\varphi + r^2\cos^2\theta\sin^2\varphi + r^2\sin^2\theta} = r. \quad (3) \end{aligned} \tag{A.2.51}$$

Relation (A.2.33) in the spherical coordinate system takes the form

$$\operatorname{grad} u = \frac{\partial u}{\partial r}\mathbf{e}_r + \frac{1}{r\sin\theta}\frac{\partial u}{\partial \varphi}\mathbf{e}_\varphi + \frac{1}{r}\frac{\partial u}{\partial \theta}\mathbf{e}_\theta. \tag{A.2.52}$$

Write down the relation for the divergence using Eq. (A.2.45)

$$\operatorname{div}\mathbf{p} = \frac{1}{r^2\sin\theta}\left(\frac{\partial(P_r r^2\sin\theta)}{\partial r} + \frac{\partial(P_\varphi r)}{\partial \varphi} + \frac{\partial(P_\theta r\sin\theta)}{\partial \theta}\right), \tag{A.2.53}$$

or

$$\operatorname{div}\mathbf{p} = \frac{1}{r^2}\frac{\partial(r^2 P_r)}{\partial r} + \frac{1}{r\sin\theta}\frac{\partial P_\varphi}{\partial \varphi} + \frac{1}{r\sin\theta}\frac{\partial(\sin\theta P_\theta)}{\partial \theta}. \tag{A.2.54}$$

The Laplace operator (see Eq. A.2.48) leads to relation

$$\Delta u = \frac{1}{r^2}\frac{\partial}{\partial r}\left(r^2\frac{\partial u}{\partial r}\right) + \frac{1}{r^2\sin\theta}\frac{\partial}{\partial \theta}\left(\sin\theta\frac{\partial u}{\partial \theta}\right) + \frac{1}{r^2\sin^2\theta}\frac{\partial^2 u}{\partial \varphi^2}. \tag{A.2.55}$$

Now we are ready to find the derivatives we need, in the spherical coordinate system using Eq. (A.2.49) and taking into account that

$$r^2 = x^2 + y^2 + z^2, \tag{A.2.56}$$

$$\sin\theta = \frac{\sqrt{x^2 + y^2}}{r}, \tag{A.2.57}$$

$$\operatorname{tg}\varphi = \frac{y}{x}. \tag{A.2.58}$$

From Eq. (A.2.56)

$$\begin{aligned} \frac{\partial r}{\partial x} &= \frac{1}{2}\left(x^2 + y^2 + z^2\right)^{-1/2} 2x = \frac{x}{r} = \sin\theta\cos\varphi,\\ \frac{\partial r}{\partial x} &= \sin\theta\cos\varphi. \end{aligned} \tag{A.2.59}$$

From Eq. (A.2.57)

$$\begin{aligned}\cos\theta\frac{\partial\theta}{\partial x}&=\frac{1}{2r}\left(x^2+y^2\right)^{-1/2}2x-\frac{\left(x^2+y^2\right)^{1/2}}{r^2}\sin\theta\cos\varphi\\&=\frac{1}{r^2\sin\theta}x-\frac{1}{r}\sin^2\theta\cos\varphi=\sin\theta\cos\varphi\frac{1}{r\sin\theta}-\frac{1}{r}\sin^2\theta\cos\varphi\\&=\cos\varphi\frac{1}{r}-\frac{1}{r}\sin^2\theta\cos\varphi=\frac{1}{r}\cos\varphi\left(1-\sin^2\theta\right)=\frac{\cos^2\theta\cos\varphi}{r}.\end{aligned}\tag{A.2.60}$$

Then from Eq. (A.2.60)

$$\frac{\partial\theta}{\partial x}=\frac{\cos\theta\cos\varphi}{r}.\tag{A.2.61}$$

From Eq. (A.2.58) we find

$$\frac{1}{\cos^2\varphi}\frac{\partial\varphi}{\partial x}=-\frac{y}{x^2}=-\frac{r\sin\theta\sin\varphi}{r^2\sin^2\theta\cos^2\varphi}=-\frac{\sin\varphi}{r\sin\theta\cos^2\varphi},\tag{A.2.62}$$

or

$$\frac{\partial\varphi}{\partial x}=-\frac{\sin\varphi}{r\sin\theta}.\tag{A.2.63}$$

Analogically

$$\frac{\partial r}{\partial y}=\frac{1}{2}\left(x^2+y^2+z^2\right)^{-1/2}2y=\frac{y}{r}=\sin\theta\sin\varphi,\tag{A.2.64}$$

$$\frac{\partial r}{\partial z}=\frac{1}{2}\left(x^2+y^2+z^2\right)^{-1/2}2z=\frac{z}{r}=\cos\theta,\tag{A.2.65}$$

$$\begin{aligned}\cos\theta\frac{\partial\theta}{\partial y}&=\frac{1}{2r}\left(x^2+y^2\right)^{-1/2}2y-\frac{\left(x^2+y^2\right)^{1/2}}{r^2}\sin\theta\sin\varphi\\&=\frac{1}{r^2\sin\theta}y-\frac{1}{r}\sin^2\theta\sin\varphi=\sin\theta\sin\varphi\frac{1}{r\sin\theta}-\frac{1}{r}\sin^2\theta\sin\varphi\\&=\sin\varphi\frac{1}{r}-\frac{1}{r}\sin^2\theta\sin\varphi=\frac{1}{r}\sin\varphi\left(1-\sin^2\theta\right)=\frac{\cos^2\theta\sin\varphi}{r}\end{aligned}\tag{A.2.66}$$

or

$$\frac{\partial\theta}{\partial y}=\frac{\cos\theta\sin\varphi}{r}.\tag{A.2.67}$$

$$\cos\theta\frac{\partial\theta}{\partial z}=-\frac{\left(x^2+y^2\right)^{1/2}}{r^2}\cos\theta=-\frac{1}{r}\sin\theta\cos\theta,\tag{A.2.68}$$

or

$$\frac{\partial\theta}{\partial z}=-\frac{1}{r}\sin\theta.\tag{A.2.69}$$

From Eq. (A.2.58)

$$\frac{1}{\cos^2\varphi}\frac{\partial\varphi}{\partial y}=\frac{1}{x}=\frac{1}{r\sin\theta\cos\varphi},\tag{A.2.70}$$

or

$$\frac{\partial\varphi}{\partial y}=\frac{\cos\varphi}{r\sin\theta}.\tag{A.2.71}$$

Finally

$$\frac{\partial\varphi}{\partial z}=0.\tag{A.2.72}$$

Let us calculate the derivative $\frac{\partial}{\partial x}$ in the spherical coordinate system

$$\begin{aligned}\frac{\partial}{\partial x} &= \left(\frac{\partial r}{\partial x}\right)\frac{\partial}{\partial r} + \left(\frac{\partial \theta}{\partial x}\right)\frac{\partial}{\partial \theta} + \left(\frac{\partial \varphi}{\partial x}\right)\frac{\partial}{\partial \varphi} \\ &= \sin\theta\cos\varphi\frac{\partial}{\partial r} + \frac{\cos\theta\cos\varphi}{r}\frac{\partial}{\partial \theta} - \frac{\sin\varphi}{r\sin\theta}\frac{\partial}{\partial \varphi},\end{aligned} \tag{A.2.73}$$

Analogical expressions take place for other operators:

$$\begin{aligned}\frac{\partial}{\partial y} &= \left(\frac{\partial r}{\partial y}\right)\frac{\partial}{\partial r} + \left(\frac{\partial \theta}{\partial y}\right)\frac{\partial}{\partial \theta} + \left(\frac{\partial \varphi}{\partial y}\right)\frac{\partial}{\partial \varphi} \\ &= \sin\theta\sin\varphi\frac{\partial}{\partial r} + \frac{\cos\theta\sin\varphi}{r}\frac{\partial}{\partial \theta} + \frac{\cos\varphi}{r\sin\theta}\frac{\partial}{\partial \varphi},\end{aligned} \tag{A.2.74}$$

$$\frac{\partial}{\partial z} = \left(\frac{\partial r}{\partial z}\right)\frac{\partial}{\partial r} + \left(\frac{\partial \theta}{\partial z}\right)\frac{\partial}{\partial \theta} + \left(\frac{\partial \phi}{\partial z}\right)\frac{\partial}{\partial \phi} = \cos\theta\frac{\partial}{\partial r} - \frac{\sin\theta}{r}\frac{\partial}{\partial \theta}. \tag{A.2.75}$$

A summary set for the first derivatives (Eqs. A.2.73–A.2.75):

$$\begin{aligned}\frac{\partial}{\partial x} &= \sin\theta\cos\varphi\frac{\partial}{\partial r} + \frac{\cos\theta\cos\varphi}{r}\frac{\partial}{\partial \theta} - \frac{\sin\varphi}{r\sin\theta}\frac{\partial}{\partial \varphi}, \\ \frac{\partial}{\partial y} &= \sin\theta\sin\varphi\frac{\partial}{\partial r} + \frac{\cos\theta\sin\varphi}{r}\frac{\partial}{\partial \theta} + \frac{\cos\varphi}{r\sin\theta}\frac{\partial}{\partial \varphi}, \\ \frac{\partial}{\partial z} &= \cos\theta\frac{\partial}{\partial r} - \frac{\sin\theta}{r}\frac{\partial}{\partial \theta}.\end{aligned} \tag{A.2.76}$$

We are coming now to the calculations of the second derivatives:

$$\begin{aligned}\frac{\partial}{\partial x}\frac{\partial}{\partial x} &= \left(\sin\theta\cos\varphi\frac{\partial}{\partial r} + \frac{\cos\theta\cos\varphi}{r}\frac{\partial}{\partial \theta} - \frac{\sin\varphi}{r\sin\theta}\frac{\partial}{\partial \varphi}\right)\left(\sin\theta\cos\varphi\frac{\partial}{\partial r} + \frac{\cos\theta\cos\varphi}{r}\frac{\partial}{\partial \theta} - \frac{\sin\varphi}{r\sin\theta}\frac{\partial}{\partial \varphi}\right) \\
&= \left(\sin\theta\cos\varphi\frac{\partial}{\partial r}\right)\left(\sin\theta\cos\varphi\frac{\partial}{\partial r} + \frac{\cos\theta\cos\varphi}{r}\frac{\partial}{\partial \theta} - \frac{\sin\varphi}{r\sin\theta}\frac{\partial}{\partial \varphi}\right) \\
&+ \left(\frac{\cos\theta\cos\varphi}{r}\frac{\partial}{\partial \theta}\right)\left(\sin\theta\cos\varphi\frac{\partial}{\partial r} + \frac{\cos\theta\cos\varphi}{r}\frac{\partial}{\partial \theta} - \frac{\sin\varphi}{r\sin\theta}\frac{\partial}{\partial \varphi}\right) \\
&- \frac{\sin\varphi}{r\sin\theta}\frac{\partial}{\partial \varphi}\left(\sin\theta\cos\varphi\frac{\partial}{\partial r} + \frac{\cos\theta\cos\varphi}{r}\frac{\partial}{\partial \theta} - \frac{\sin\varphi}{r\sin\theta}\frac{\partial}{\partial \varphi}\right) \\
&= \sin\theta\cos\varphi\left[\frac{\partial}{\partial r}\left(\sin\theta\cos\varphi\frac{\partial}{\partial r} + \frac{\cos\theta\cos\varphi}{r}\frac{\partial}{\partial \theta} - \frac{\sin\varphi}{r\sin\theta}\frac{\partial}{\partial \varphi}\right)\right] \\
&+ \frac{\cos\theta\cos\varphi}{r}\left[\frac{\partial}{\partial \theta}\left(\sin\theta\cos\varphi\frac{\partial}{\partial r} + \frac{\cos\theta\cos\varphi}{r}\frac{\partial}{\partial \theta} - \frac{\sin\varphi}{r\sin\theta}\frac{\partial}{\partial \varphi}\right)\right] \\
&- \frac{\sin\varphi}{r\sin\theta}\left[\frac{\partial}{\partial \varphi}\left(\sin\theta\cos\varphi\frac{\partial}{\partial r} + \frac{\cos\theta\cos\varphi}{r}\frac{\partial}{\partial \theta} - \frac{\sin\varphi}{r\sin\theta}\frac{\partial}{\partial \varphi}\right)\right] \\
&= \sin\theta\cos\varphi\left[\sin\theta\cos\varphi\frac{\partial^2}{\partial r^2} + \cos\theta\cos\varphi\frac{\partial}{\partial r}\left(\frac{1}{r}\frac{\partial}{\partial \theta}\right) - \frac{\sin\varphi}{\sin\theta}\frac{\partial}{\partial r}\left(\frac{1}{r}\frac{\partial}{\partial \varphi}\right)\right] \\
&+ \frac{\cos\theta\cos\varphi}{r}\left[\begin{array}{l}\sin\theta\cos\varphi\dfrac{\partial^2}{\partial r\partial \theta} + \cos\theta\cos\varphi\dfrac{\partial}{\partial r} + \dfrac{\cos\theta\cos\varphi}{r}\dfrac{\partial^2}{\partial \theta^2} - \dfrac{\sin\theta\cos\varphi}{r}\dfrac{\partial}{\partial \theta} \\ - \dfrac{\sin\varphi}{r\sin\theta}\dfrac{\partial^2}{\partial \theta\partial \varphi} + \dfrac{\sin\varphi\cos\theta}{r\sin^2\theta}\dfrac{\partial}{\partial \varphi}\end{array}\right] \\
&- \frac{\sin\varphi}{r\sin\theta}\left[\begin{array}{l}\sin\theta\cos\varphi\dfrac{\partial^2}{\partial r\partial \varphi} - \sin\theta\sin\varphi\dfrac{\partial}{\partial r} + \dfrac{\cos\theta\cos\varphi}{r}\dfrac{\partial^2}{\partial \theta\partial \varphi} \\ - \dfrac{\cos\theta\sin\varphi}{r}\dfrac{\partial}{\partial \theta} - \dfrac{\sin\varphi}{r\sin\theta}\dfrac{\partial^2}{\partial \varphi^2} - \dfrac{\cos\varphi}{r\sin\theta}\dfrac{\partial}{\partial \varphi}\end{array}\right],\end{aligned} \tag{A.2.77}$$

$$
\begin{aligned}
\frac{\partial^2}{\partial x^2} = {} & \sin\theta\cos\varphi\left[\sin\theta\cos\varphi\frac{\partial^2}{\partial r^2}+\frac{\cos\theta\cos\varphi}{r}\frac{\partial^2}{\partial r\partial\theta}-\frac{\cos\theta\cos\varphi}{r^2}\frac{\partial}{\partial\theta}-\frac{\sin\varphi}{r\sin\theta}\frac{\partial^2}{\partial r\partial\varphi}+\frac{\sin\varphi}{r^2\sin\theta}\frac{\partial}{\partial\varphi}\right] \\
& +\frac{\cos\theta\cos\varphi}{r}\left[\begin{aligned}&\sin\theta\cos\varphi\frac{\partial^2}{\partial r\partial\theta}+\cos\theta\cos\varphi\frac{\partial}{\partial r}+\frac{\cos\theta\cos\varphi}{r}\frac{\partial^2}{\partial\theta^2}-\frac{\sin\theta\cos\varphi}{r}\frac{\partial}{\partial\theta}\\&-\frac{\sin\varphi}{r\sin\theta}\frac{\partial^2}{\partial\theta\,\partial\varphi}+\frac{\sin\varphi\cos\theta}{r\sin^2\theta}\frac{\partial}{\partial\varphi}\end{aligned}\right] \\
& -\frac{\sin\varphi}{r\sin\theta}\left[\begin{aligned}&\sin\theta\cos\varphi\frac{\partial^2}{\partial r\partial\varphi}-\sin\theta\sin\varphi\frac{\partial}{\partial r}+\frac{\cos\theta\cos\varphi}{r}\frac{\partial^2}{\partial\theta\,\partial\varphi}\\&-\frac{\cos\theta\sin\varphi}{r}\frac{\partial}{\partial\theta}-\frac{\sin\varphi}{r\sin\theta}\frac{\partial^2}{\partial\varphi^2}-\frac{\cos\varphi}{r\sin\theta}\frac{\partial}{\partial\varphi}\end{aligned}\right].
\end{aligned}
\tag{A.2.78}
$$

Analogically

$$
\begin{aligned}
\frac{\partial}{\partial y}\frac{\partial}{\partial y} = {} & \left(\sin\theta\sin\varphi\frac{\partial}{\partial r}+\frac{\cos\theta\sin\varphi}{r}\frac{\partial}{\partial\theta}+\frac{\cos\varphi}{r\sin\theta}\frac{\partial}{\partial\varphi}\right)\left(\sin\theta\sin\varphi\frac{\partial}{\partial r}+\frac{\cos\theta\sin\varphi}{r}\frac{\partial}{\partial\theta}+\frac{\cos\varphi}{r\sin\theta}\frac{\partial}{\partial\varphi}\right) \\
= {} & \left(\sin\theta\sin\varphi\frac{\partial}{\partial r}\right)\left(\sin\theta\sin\varphi\frac{\partial}{\partial r}+\frac{\cos\theta\sin\varphi}{r}\frac{\partial}{\partial\theta}+\frac{\cos\varphi}{r\sin\theta}\frac{\partial}{\partial\varphi}\right) \\
& +\left(\frac{\cos\theta\sin\varphi}{r}\frac{\partial}{\partial\theta}\right)\left(\sin\theta\sin\varphi\frac{\partial}{\partial r}+\frac{\cos\theta\sin\varphi}{r}\frac{\partial}{\partial\theta}+\frac{\cos\varphi}{r\sin\theta}\frac{\partial}{\partial\varphi}\right) \\
& +\frac{\cos\varphi}{r\sin\theta}\frac{\partial}{\partial\varphi}\left(\sin\theta\sin\varphi\frac{\partial}{\partial r}+\frac{\cos\theta\sin\varphi}{r}\frac{\partial}{\partial\theta}+\frac{\cos\varphi}{r\sin\theta}\frac{\partial}{\partial\varphi}\right) \\
= {} & \sin\theta\sin\varphi\left[\frac{\partial}{\partial r}\left(\sin\theta\sin\varphi\frac{\partial}{\partial r}+\frac{\cos\theta\sin\varphi}{r}\frac{\partial}{\partial\theta}+\frac{\cos\varphi}{r\sin\theta}\frac{\partial}{\partial\varphi}\right)\right] \\
& +\frac{\cos\theta\sin\varphi}{r}\left[\frac{\partial}{\partial\theta}\left(\sin\theta\sin\varphi\frac{\partial}{\partial r}+\frac{\cos\theta\sin\varphi}{r}\frac{\partial}{\partial\theta}+\frac{\cos\varphi}{r\sin\theta}\frac{\partial}{\partial\varphi}\right)\right] \\
& +\frac{\cos\varphi}{r\sin\theta}\left[\frac{\partial}{\partial\varphi}\left(\sin\theta\sin\varphi\frac{\partial}{\partial r}+\frac{\cos\theta\sin\varphi}{r}\frac{\partial}{\partial\theta}+\frac{\cos\varphi}{r\sin\theta}\frac{\partial}{\partial\varphi}\right)\right] \\
= {} & \sin\theta\sin\varphi\left[\sin\theta\sin\varphi\frac{\partial^2}{\partial r^2}+\cos\theta\sin\varphi\frac{\partial}{\partial r}\left(\frac{1}{r}\frac{\partial}{\partial\theta}\right)+\frac{\cos\varphi}{\sin\theta}\frac{\partial}{\partial r}\left(\frac{1}{r}\frac{\partial}{\partial\varphi}\right)\right] \\
& +\frac{\cos\theta\sin\varphi}{r}\left[\begin{aligned}&\sin\theta\sin\varphi\frac{\partial^2}{\partial r\partial\theta}+\cos\theta\sin\varphi\frac{\partial}{\partial r}+\frac{\cos\theta\sin\varphi}{r}\frac{\partial^2}{\partial\theta^2}-\frac{\sin\theta\sin\varphi}{r}\frac{\partial}{\partial\theta}\\&+\frac{\cos\varphi}{r\sin\theta}\frac{\partial^2}{\partial\theta\,\partial\varphi}-\frac{\cos\varphi\cos\theta}{r\sin^2\theta}\frac{\partial}{\partial\varphi}\end{aligned}\right] \\
& +\frac{\cos\varphi}{r\sin\theta}\left[\begin{aligned}&\sin\theta\sin\varphi\frac{\partial^2}{\partial r\partial\varphi}+\sin\theta\cos\varphi\frac{\partial}{\partial r}+\frac{\cos\theta\sin\varphi}{r}\frac{\partial^2}{\partial\theta\,\partial\varphi}\\&+\frac{\cos\theta\cos\varphi}{r}\frac{\partial}{\partial\theta}+\frac{\cos\varphi}{r\sin\theta}\frac{\partial^2}{\partial\varphi^2}-\frac{\sin\varphi}{r\sin\theta}\frac{\partial}{\partial\varphi}\end{aligned}\right],
\end{aligned}
\tag{A.2.79}
$$

$$\begin{aligned}
\frac{\partial^2}{\partial y^2} = \sin\theta\sin\varphi\left[\sin\theta\sin\varphi\frac{\partial^2}{\partial r^2} + \frac{\cos\theta\sin\varphi}{r}\frac{\partial^2}{\partial r\partial\theta} - \frac{\cos\theta\sin\varphi}{r^2}\frac{\partial}{\partial\theta} + \frac{\cos\varphi}{r\sin\theta}\frac{\partial^2}{\partial r\partial\varphi} - \frac{\cos\varphi}{r^2\sin\theta}\frac{\partial}{\partial\varphi}\right] \\
+ \frac{\cos\theta\sin\varphi}{r}\left[\begin{array}{l} \sin\theta\sin\varphi\dfrac{\partial^2}{\partial r\partial\theta} + \cos\theta\sin\varphi\dfrac{\partial}{\partial r} + \dfrac{\cos\theta\sin\varphi}{r}\dfrac{\partial^2}{\partial\theta^2} - \dfrac{\sin\theta\sin\varphi}{r}\dfrac{\partial}{\partial\theta} \\ + \dfrac{\cos\varphi}{r\sin\theta}\dfrac{\partial^2}{\partial\theta\,\partial\varphi} - \dfrac{\cos\varphi\cos\theta}{r\sin^2\theta}\dfrac{\partial}{\partial\varphi} \end{array}\right] \\
+ \frac{\cos\varphi}{r\sin\theta}\left[\begin{array}{l} \sin\theta\sin\varphi\dfrac{\partial^2}{\partial r\partial\varphi} + \sin\theta\cos\varphi\dfrac{\partial}{\partial r} + \dfrac{\cos\theta\sin\varphi}{r}\dfrac{\partial^2}{\partial\theta\,\partial\varphi} \\ + \dfrac{\cos\theta\cos\varphi}{r}\dfrac{\partial}{\partial\theta} + \dfrac{\cos\varphi}{r\sin\theta}\dfrac{\partial^2}{\partial\varphi^2} - \dfrac{\sin\varphi}{r\sin\theta}\dfrac{\partial}{\partial\varphi} \end{array}\right].
\end{aligned} \tag{A.2.80}$$

Finally,

$$\begin{aligned}
\frac{\partial}{\partial z}\frac{\partial}{\partial z} &= \left(\cos\theta\frac{\partial}{\partial r} - \frac{\sin\theta}{r}\frac{\partial}{\partial\theta}\right)\left(\cos\theta\frac{\partial}{\partial r} - \frac{\sin\theta}{r}\frac{\partial}{\partial\theta}\right) \\
&= \left(\cos\theta\frac{\partial}{\partial r}\right)\left(\cos\theta\frac{\partial}{\partial r} - \frac{\sin\theta}{r}\frac{\partial}{\partial\theta}\right) \\
&\quad - \frac{\sin\theta}{r}\frac{\partial}{\partial\theta}\left(\cos\theta\frac{\partial}{\partial r} - \frac{\sin\theta}{r}\frac{\partial}{\partial\theta}\right) \\
&= \cos^2\theta\frac{\partial^2}{\partial r^2} - \cos\theta\sin\theta\frac{\partial}{\partial r}\left(\frac{1}{r}\frac{\partial}{\partial\theta}\right) \\
&\quad - \frac{\sin\theta}{r}\frac{\partial}{\partial\theta}\left(\cos\theta\frac{\partial}{\partial r}\right) + \frac{\sin\theta}{r}\frac{\partial}{\partial\theta}\left(\frac{\sin\theta}{r}\frac{\partial}{\partial\theta}\right) \\
&= \cos^2\theta\frac{\partial^2}{\partial r^2} + \frac{\cos\theta\sin\theta}{r^2}\frac{\partial}{\partial\theta} - \frac{\cos\theta\sin\theta}{r}\frac{\partial^2}{\partial\theta\,\partial r} \\
&\quad + \frac{\sin^2\theta}{r}\frac{\partial}{\partial r} - \frac{\cos\theta\sin\theta}{r}\frac{\partial}{\partial\theta\,\partial r} + \frac{\sin\theta}{r^2}\frac{\partial}{\partial\theta}\left(\sin\theta\frac{\partial}{\partial\theta}\right) \\
&= \cos^2\theta\frac{\partial^2}{\partial r^2} + \frac{\cos\theta\sin\theta}{r^2}\frac{\partial}{\partial\theta} - \frac{2\cos\theta\sin\theta}{r}\frac{\partial^2}{\partial\theta\,\partial r} \\
&\quad + \frac{\sin^2\theta}{r}\frac{\partial}{\partial r} + \frac{\sin^2\theta}{r^2}\frac{\partial^2}{\partial\theta^2} + \frac{\sin\theta\cos\theta}{r^2}\frac{\partial}{\partial\theta},
\end{aligned} \tag{A.2.81}$$

$$\begin{aligned}
\frac{\partial^2}{\partial z^2} = \cos^2\theta\frac{\partial^2}{\partial r^2} - \frac{2\cos\theta\sin\theta}{r}\frac{\partial^2}{\partial\theta\,\partial r} + \frac{\sin^2\theta}{r}\frac{\partial}{\partial r} \\
+ \frac{\sin^2\theta}{r^2}\frac{\partial^2}{\partial\theta^2} + \frac{2\sin\theta\cos\theta}{r^2}\frac{\partial}{\partial\theta}.
\end{aligned} \tag{A.2.82}$$

A summary set for the second derivatives (Eqs. A.2.78), (A.2.80, and A.2.82):

$$\begin{aligned}
\frac{\partial^2}{\partial x^2} = \left[\sin^2\theta\cos^2\varphi\frac{\partial^2}{\partial r^2} + \frac{2\sin\theta\cos\theta\cos^2\varphi}{r}\frac{\partial^2}{\partial r\partial\theta} - \frac{2\sin\theta\cos\theta\cos^2\varphi}{r^2}\frac{\partial}{\partial\theta} - \frac{2\cos\varphi\sin\varphi}{r}\frac{\partial^2}{\partial r\partial\varphi}\right] \\
+ \left[\frac{\cos^2\theta\cos^2\varphi}{r}\frac{\partial}{\partial r} + \frac{\cos^2\theta\cos^2\varphi}{r^2}\frac{\partial^2}{\partial\theta^2} - \frac{2\sin\varphi\cos\theta\cos\varphi}{\sin\theta}\frac{\partial^2}{r^2\,\partial\theta\,\partial\varphi} + \frac{2\sin\varphi\cos\varphi}{r^2\sin^2\theta}\frac{\partial}{\partial\varphi}\right] \\
+ \left[\frac{\sin^2\varphi}{r}\frac{\partial}{\partial r} + \frac{\cos\theta\sin^2\varphi}{r^2\sin\theta}\frac{\partial}{\partial\theta} + \frac{\sin^2\varphi}{r^2\sin^2\theta}\frac{\partial^2}{\partial\varphi^2}\right],
\end{aligned}$$

$$\frac{\partial^2}{\partial y^2}=\left[\sin^2\theta\sin^2\varphi\frac{\partial^2}{\partial r^2}-\frac{2\sin\theta\cos\theta\sin^2\varphi}{r^2}\frac{\partial}{\partial\theta}+\frac{2\sin\varphi\cos\varphi}{r}\frac{\partial^2}{\partial r\partial\varphi}-\frac{2\sin\varphi\cos\varphi}{r^2\sin^2\theta}\frac{\partial}{\partial\varphi}\right]$$
$$+\left[2\sin\theta\frac{\cos\theta\sin^2\varphi}{r}\frac{\partial^2}{\partial r\partial\theta}+\frac{\cos^2\theta\sin^2\varphi}{r}\frac{\partial}{\partial r}+\frac{\cos^2\theta\sin^2\varphi}{r^2}\frac{\partial^2}{\partial\theta^2}+\frac{2\cos\varphi\cos\theta\sin\varphi}{\sin\theta}\frac{\partial^2}{r^2}\frac{\partial^2}{\partial\theta\,\partial\varphi}\right]\quad\text{(A.2.83)}$$
$$+\left[\frac{\cos^2\varphi}{r}\frac{\partial}{\partial r}+\frac{\cos\theta\cos^2\varphi}{r^2\sin\theta}\frac{\partial}{\partial\theta}+\frac{\cos^2\varphi}{r^2\sin^2\theta}\frac{\partial^2}{\partial\varphi^2}\right],$$
$$\frac{\partial^2}{\partial z^2}=\cos^2\theta\frac{\partial^2}{\partial r^2}-\frac{2\cos\theta\sin\theta}{r}\frac{\partial^2}{\partial\theta\,\partial r}+\frac{\sin^2\theta}{r}\frac{\partial}{\partial r}+\frac{\sin^2\theta}{r^2}\frac{\partial^2}{\partial\theta^2}+\frac{2\sin\theta\cos\theta}{r^2}\frac{\partial}{\partial\theta}.$$

After summation of all three relations in Eq. (A.2.83), we find

$$\frac{\partial^2}{\partial x^2}+\frac{\partial^2}{\partial y^2}+\frac{\partial^2}{\partial z^2}=\frac{1}{r^2}\left[\frac{\partial}{\partial r}\left(r^2\frac{\partial}{\partial r}\right)\right]+\frac{1}{r^2\sin\theta}\frac{\partial}{\partial\theta}\left(\sin\theta\frac{\partial}{\partial\theta}\right)+\frac{1}{r^2\sin^2\theta}\frac{\partial^2}{\partial\varphi^2},\quad\text{(A.2.84)}$$

which corresponds to the standard formula for the Laplace operator in the spherical coordinate system.

Deduce the expression for gradient of a scalar function u in the spherical coordinate system. From relations Eqs. (A.2.50) and (A.2.51) follow

$$\begin{aligned}&\mathbf{e}_r=\frac{\mathbf{r}_1}{H_1}=\sin\theta\cos\varphi\mathbf{i}+\sin\theta\sin\varphi\mathbf{j}+\cos\theta\mathbf{k}\quad(1)\\&\mathbf{e}_\varphi=\frac{\mathbf{r}_2}{H_2}=-\sin\varphi\mathbf{i}+\cos\varphi\mathbf{j}\quad(2)\\&\mathbf{e}_\theta=\frac{\mathbf{r}_3}{H_3}=\cos\theta\cos\varphi\mathbf{i}+\cos\theta\sin\varphi\mathbf{j}-\sin\theta\mathbf{k}\quad(3)\end{aligned}\quad\text{(A.2.85)}$$

Therefore

$$\begin{aligned}(\mathbf{e}_r\cdot\mathbf{i})&=\sin\theta\cos\varphi,\\(\mathbf{e}_\phi\cdot\mathbf{i})&=-\sin\varphi,\\(\mathbf{e}_\theta\cdot\mathbf{i})&=\cos\theta\cos\varphi,\end{aligned}\quad\text{(A.2.86)}$$

and

$$\mathbf{i}=\sin\theta\cos\varphi\mathbf{e}_r-\sin\varphi\mathbf{e}_\phi+\cos\theta\cos\varphi\mathbf{e}_\theta.\quad\text{(A.2.87)}$$

Analogically

$$\mathbf{j}=\sin\theta\sin\varphi\mathbf{e}_r+\cos\varphi\mathbf{e}_\phi+\cos\theta\sin\varphi\mathbf{e}_\theta,\quad\text{(A.2.88)}$$

$$\mathbf{k}=\cos\theta\mathbf{e}_r-\sin\theta\mathbf{e}_\theta.\quad\text{(A.2.89)}$$

Using Eqs. (A.2.12), (A.2.76), and (A.2.87)–(A.2.89), we find the formula

$$\operatorname{grad}u=\frac{\partial u}{\partial x}\mathbf{i}+\frac{\partial u}{\partial y}\mathbf{j}+\frac{\partial u}{\partial z}\mathbf{k}=\left(\sin\theta\cos\varphi\frac{\partial u}{\partial r}+\frac{\cos\theta\cos\varphi}{r}\frac{\partial u}{\partial\theta}-\frac{\sin\varphi}{r\sin\theta}\frac{\partial u}{\partial\varphi}\right)(\sin\theta\cos\varphi\mathbf{e}_r-\sin\varphi\mathbf{e}_\phi+\cos\theta\cos\varphi\mathbf{e}_\theta)$$
$$+\left(\sin\theta\sin\varphi\frac{\partial u}{\partial r}+\frac{\cos\theta\sin\varphi}{r}\frac{\partial u}{\partial\theta}+\frac{\cos\varphi}{r\sin\theta}\frac{\partial u}{\partial\varphi}\right)(\sin\theta\sin\varphi\mathbf{e}_r+\cos\varphi\mathbf{e}_\phi+\cos\theta\sin\varphi\mathbf{e}_\theta)+\left(\cos\theta\frac{\partial u}{\partial r}-\frac{\sin\theta}{r}\frac{\partial u}{\partial\theta}\right)(\cos\theta\mathbf{e}_r-\sin\theta\mathbf{e}_\theta)$$
$$=\left(\sin^2\theta\cos^2\varphi\frac{\partial u}{\partial r}+\frac{\cos\theta\sin\theta\cos^2\varphi}{r}\frac{\partial u}{\partial\theta}-\frac{\sin\varphi\cos\varphi}{r}\frac{\partial u}{\partial\varphi}+\sin^2\theta\sin^2\varphi\frac{\partial u}{\partial r}+\frac{\cos\theta\sin\theta\sin^2\varphi}{r}\frac{\partial u}{\partial\theta}\right.$$
$$\left.+\frac{\cos\varphi\sin\varphi}{r}\frac{\partial u}{\partial\varphi}+\cos^2\theta\frac{\partial u}{\partial r}-\frac{\sin\theta\cos\theta}{r}\frac{\partial u}{\partial\theta}\right)\mathbf{e}_r+\left(-\sin\theta\cos\varphi\sin\varphi\frac{\partial u}{\partial r}-\frac{\cos\theta\cos\varphi\sin\varphi}{r}\frac{\partial u}{\partial\theta}+\frac{\sin^2\varphi}{r\sin\theta}\frac{\partial u}{\partial\varphi}\right.$$
$$\left.+\sin\theta\sin\varphi\cos\varphi\frac{\partial u}{\partial r}+\frac{\cos\theta\sin\varphi\cos\varphi}{r}\frac{\partial u}{\partial\theta}+\frac{\cos^2\varphi}{r\sin\theta}\frac{\partial u}{\partial\varphi}\right)\mathbf{e}_\varphi+\left(\sin\theta\cos\theta\cos^2\varphi\frac{\partial u}{\partial r}+\frac{\cos^2\theta\cos^2\varphi}{r}\frac{\partial u}{\partial\theta}-\frac{\cos\theta\cos\varphi\sin\varphi}{r\sin\theta}\frac{\partial u}{\partial\varphi}\right.$$
$$\left.+\sin\theta\cos\theta\sin^2\varphi\frac{\partial u}{\partial r}+\frac{\cos^2\theta\sin^2\varphi}{r}\frac{\partial u}{\partial\theta}+\frac{\cos\theta\cos\varphi\sin\varphi}{r\sin\theta}\frac{\partial u}{\partial\varphi}-\cos\theta\sin\theta\frac{\partial u}{\partial r}+\frac{\sin^2\theta}{r}\frac{\partial u}{\partial\theta}\right)\mathbf{e}_\theta=\frac{\partial u}{\partial r}\mathbf{e}_r+\frac{1}{r\sin\theta}\frac{\partial u}{\partial\varphi}\mathbf{e}_\varphi+\frac{1}{r}\frac{\partial u}{\partial\theta}\mathbf{e}_\theta,$$
$$\text{(A.2.90)}$$

which coincides with Eq. (A.2.52).

Let us express the divergence of the vector field **p**

$$\mathbf{p} = P_r \mathbf{e}_r + P_\varphi \mathbf{e}_\varphi + P_\theta \mathbf{e}_\theta = P\mathbf{i} + Q\mathbf{j} + R\mathbf{k} \tag{A.2.91}$$

in the spherical coordinate system. After multiplication of the both parts of Eq. (A.2.91) by **i** we have

$$P = P_r(\mathbf{e}_r \cdot \mathbf{i}) + P_\varphi(\mathbf{e}_\varphi \cdot \mathbf{i}) + P_\theta(\mathbf{e}_\theta \cdot \mathbf{i}), \tag{A.2.92}$$

or using Eq. (A.2.86),

$$P = P_r \sin\theta\cos\varphi - P_\varphi \sin\varphi + P_\theta \cos\theta\cos\varphi. \tag{A.2.93}$$

Analogically,

$$Q = P_r \sin\theta\sin\varphi + P_\varphi \cos\varphi + P_\theta \cos\theta\sin\varphi, \tag{A.2.94}$$

$$R = P_r \cos\theta - P_\theta \sin\theta. \tag{A.2.95}$$

From relations Eqs. (A.2.22), (A.2.76), and (A.2.93)–(A.2.95) one obtains the formula

$$\begin{aligned}
\operatorname{div}\mathbf{p} &= \frac{\partial P}{\partial x} + \frac{\partial Q}{\partial y} + \frac{\partial R}{\partial z} = \left(\sin\theta\cos\varphi\frac{\partial}{\partial r} + \frac{\cos\theta\cos\varphi}{r}\frac{\partial}{\partial\theta} - \frac{\sin\varphi}{r\sin\theta}\frac{\partial}{\partial\varphi}\right)(P_r\sin\theta\cos\varphi - P_\varphi\sin\varphi + P_\theta\cos\theta\cos\varphi) \\
&+ \left(\sin\theta\sin\varphi\frac{\partial}{\partial r} + \frac{\cos\theta\sin\varphi}{r}\frac{\partial}{\partial\theta} + \frac{\cos\varphi}{r\sin\theta}\frac{\partial}{\partial\varphi}\right)(P_r\sin\theta\sin\varphi + P_\varphi\cos\varphi + P_\theta\cos\theta\sin\varphi) \\
&+ \left(\cos\theta\frac{\partial}{\partial r} - \frac{\sin\theta}{r}\frac{\partial}{\partial\theta}\right)(P_r\cos\theta - P_\theta\sin\theta) \\
&= \left(\sin^2\theta\cos^2\varphi\frac{\partial P_r}{\partial r} - \sin\theta\cos\varphi\sin\varphi\frac{\partial P_\varphi}{\partial r} + \sin\theta\cos\theta\cos^2\varphi\frac{\partial P_\theta}{\partial r} + \sin^2\theta\sin^2\varphi\frac{\partial P_r}{\partial r} + \sin\theta\cos\varphi\sin\varphi\frac{\partial P\varphi}{\partial r}\right. \\
&\left. + \sin\theta\cos\theta\sin^2\varphi\frac{\partial P_\theta}{\partial r} + \cos^2\theta\frac{\partial P_r}{\partial r} - \sin\theta\cos\theta\frac{\partial P_\theta}{\partial r}\right) + \left(\frac{\cos\theta\sin\theta\cos^2\varphi}{r}\frac{\partial P_r}{\partial\theta} - \frac{\cos\theta\cos\phi\sin\varphi}{r}\frac{\partial P_\varphi}{\partial\theta}\right. \\
&+ \frac{\cos^2\theta\cos^2\varphi}{r}\frac{\partial P_\theta}{\partial\theta} + \frac{\cos\theta\sin\theta\sin^2\phi}{r}\frac{\partial P_r}{\partial\theta} + \frac{\cos\theta\cos\varphi\sin\varphi}{r}\frac{\partial P_\varphi}{\partial\theta} + \frac{\cos^2\theta\sin^2\varphi}{r}\frac{\partial P_\theta}{\partial\theta} - \frac{\sin\theta\cos\theta}{r}\frac{\partial P_r}{\partial\theta} \\
&\left. + \frac{\sin^2\theta}{r}\frac{\partial P_\theta}{\partial\theta}\right) + \left(-\frac{\sin\varphi\cos\varphi}{r}\frac{\partial P_r}{\partial\varphi} + \frac{\sin^2\varphi}{r\sin\theta}\frac{\partial P_\varphi}{\partial\varphi} - \frac{\sin\varphi\cos\varphi\cos\theta}{r\sin\theta}\frac{\partial P_\theta}{\partial\varphi} + \frac{\cos\varphi\sin\varphi}{r}\frac{\partial P_r}{\partial\varphi}\right. \\
&\left. + \frac{\cos^2\varphi}{r\sin\theta}\frac{\partial P_\varphi}{\partial\varphi} + \frac{\cos\varphi\sin\varphi\cos\theta}{r\sin\theta}\frac{\partial P_\theta}{\partial\varphi}\right) + P_r\frac{\cos^2\theta\cos^2\varphi}{r} - P_\theta\frac{\cos\theta\sin\theta\cos^2\varphi}{r} + P_r\frac{\sin^2\varphi}{r} \\
&+ P_\varphi\frac{\sin\varphi\cos\varphi}{r\sin\theta} + P_\theta\frac{\sin^2\varphi\cos\theta}{r\sin\theta} + P_r\frac{\cos^2\theta\sin^2\varphi}{r} - P_\theta\frac{\cos\theta\sin\theta\sin^2\varphi}{r} \\
&+ P_r\frac{\cos^2\varphi}{r} - P_\varphi\frac{\cos\varphi\sin\varphi}{r\sin\theta} + P_\theta\frac{\cos\theta\cos^2\varphi}{r\sin\theta} + P_r\frac{\sin^2\theta}{r} + P_\theta\frac{\sin\theta\cos\theta}{r} \\
&= \frac{\partial P_r}{\partial r} + \frac{1}{r\sin\theta}\frac{\partial P_\varphi}{\partial\varphi} + \frac{1}{r}\frac{\partial P_\theta}{\partial\theta} + \frac{2P_r}{r} + \frac{\cos\theta P_\theta}{r\sin\theta} = \frac{1}{r^2}\frac{\partial(r^2P_r)}{\partial r} + \frac{1}{r\sin\theta}\frac{\partial P_\varphi}{\partial\varphi} + \frac{1}{r\sin\theta}\frac{\partial(\sin\theta P_\theta)}{\partial\theta},
\end{aligned} \tag{A.2.96}$$

which coincides with Eq. (A.2.54).

It should be noted that the developed method allows to obtain any one from the complicated differential expressions in the spherical coordinate system using Eqs. (A.2.76) and (A.2.93)–(A.2.95). For example

$$\begin{aligned}
\frac{\partial^2 Q}{\partial x\partial z} &= \left(\sin\theta\cos\varphi\frac{\partial}{\partial r} + \frac{\cos\theta\cos\varphi}{r}\frac{\partial}{\partial\theta} - \frac{\sin\varphi}{r\sin\theta}\frac{\partial}{\partial\varphi}\right) \\
&\times\left(\cos\theta\frac{\partial}{\partial r} - \frac{\sin\theta}{r}\frac{\partial}{\partial\theta}\right)(P_r\sin\theta\sin\varphi + P_\phi\cos\varphi + P_\theta\cos\theta\sin\varphi)
\end{aligned} \tag{A.2.97}$$

Write down now the basic non-local hydrodynamic equations in the gravitational field for the spherical coordinate system. Continuity equation

$$\begin{aligned}
&\frac{\partial}{\partial t}\left\{\rho - \tau\left[\frac{\partial\rho}{\partial t} + \frac{\partial}{\partial\mathbf{r}}\cdot(\rho\mathbf{v}_0)\right]\right\} + \frac{\partial}{\partial\mathbf{r}}\cdot\left\{\rho\mathbf{v}_0 - \tau\left[\frac{\partial}{\partial t}(\rho\mathbf{v}_0)\right.\right. \\
&\left.\left. + \frac{\partial}{\partial\mathbf{r}}\cdot\rho\mathbf{v}_0\mathbf{v}_0 + \overleftrightarrow{\mathrm{I}}\cdot\frac{\partial p}{\partial\mathbf{r}} - \rho\mathbf{g}\right]\right\} = 0,
\end{aligned} \tag{A.2.98}$$

or

$$\frac{\partial}{\partial t}\left\{\rho-\tau\left[\frac{\partial \rho}{\partial t}+\operatorname{div}(\rho \mathbf{v}_0)\right]\right\}+\operatorname{div}\left\{\rho \mathbf{v}_0-\tau\left[\frac{\partial}{\partial t}(\rho \mathbf{v}_0)+\frac{\partial}{\partial \mathbf{r}}\cdot \rho \mathbf{v}_0\mathbf{v}_0-\rho \mathbf{g}\right]\right\}-\operatorname{div}(\tau \operatorname{grad} p)=0. \tag{A.2.99}$$

Using the expressions (A.2.52) and (A.2.54) for gradient and divergence in the spherical coordinate system we have

$$\begin{aligned}
&\frac{\partial}{\partial t}\left\{\rho-\tau\left[\frac{\partial \rho}{\partial t}+\frac{1}{r^2}\frac{\partial(r^2\rho v_{0r})}{\partial r}+\frac{1}{r\sin\theta}\frac{\partial(\rho v_{0\varphi})}{\partial \varphi}+\frac{1}{r\sin\theta}\frac{\partial(\rho v_{0\theta}\sin\theta)}{\partial \theta}\right]\right\}\\
&+\frac{1}{r^2}\frac{\partial}{\partial r}\left\{r^2\left\{\rho v_{0r}-\tau\left[\frac{\partial}{\partial t}(\rho v_{0r})+\operatorname{div}(\rho \mathbf{v}_0 v_{0r})-\rho g_r\right]\right\}\right\}\\
&+\frac{1}{r\sin\theta}\frac{\partial}{\partial \varphi}\left\{\rho v_{0\varphi}-\tau\left[\frac{\partial}{\partial t}(\rho v_{0\varphi})+\operatorname{div}(\rho \mathbf{v}_0 v_{0\varphi})-\rho g_\varphi\right]\right\}\\
&+\frac{1}{r\sin\theta}\frac{\partial}{\partial \theta}\left\{\sin\theta\left\{\rho v_{0\theta}-\tau\left[\frac{\partial}{\partial t}(\rho v_{0\theta})+\operatorname{div}(\rho \mathbf{v}_0 v_{0\theta})-\rho g_\theta\right]\right\}\right\}\\
&-\frac{1}{r^2}\frac{\partial}{\partial r}\left(\tau r^2\frac{\partial p}{\partial r}\right)-\frac{1}{r^2\sin\theta}\frac{\partial}{\partial \theta}\left(\tau\sin\theta\frac{\partial p}{\partial \theta}\right)-\frac{1}{r^2\sin^2\theta}\frac{\partial}{\partial \varphi}\left(\tau\frac{\partial p}{\partial \varphi}\right)=0
\end{aligned} \tag{A.2.100}$$

$$\begin{aligned}
&\frac{\partial}{\partial t}\left\{\rho-\tau\left[\frac{\partial \rho}{\partial t}+\frac{1}{r^2}\frac{\partial(r^2\rho v_{0r})}{\partial r}+\frac{1}{r\sin\theta}\frac{\partial(\rho v_{0\varphi})}{\partial \varphi}+\frac{1}{r\sin\theta}\frac{\partial(\rho v_{0\theta}\sin\theta)}{\partial \theta}\right]\right\}\\
&+\frac{1}{r^2}\frac{\partial}{\partial r}\left\{r^2\left\{\rho v_{0r}-\tau\left[\frac{\partial}{\partial t}(\rho v_{0r})+\frac{1}{r^2}\frac{\partial(r^2\rho v_{0r}^2)}{\partial r}+\frac{1}{r\sin\theta}\frac{\partial(\rho v_{0\varphi}v_{0r})}{\partial \varphi}\right.\right.\right.\\
&\left.\left.\left.+\frac{1}{r\sin\theta}\frac{\partial(\rho v_{0\theta}v_{0r}\sin\theta)}{\partial \theta}-\rho g_r\right]\right\}\right\}+\frac{1}{r\sin\theta}\frac{\partial}{\partial \varphi}\left\{\rho v_{0\varphi}-\tau\left[\frac{\partial}{\partial t}(\rho v_{0\varphi})+\frac{1}{r^2}\frac{\partial(r^2\rho v_{0r}v_{0\varphi})}{\partial r}\right.\right.\\
&\left.\left.+\frac{1}{r\sin\theta}\frac{\partial\left(\rho v_{0\varphi}^2\right)}{\partial \varphi}+\frac{1}{r\sin\theta}\frac{\partial\left(\rho v_{0\theta}v_{0\varphi}\sin\theta\right)}{\partial \theta}-\rho g_\varphi\right]\right\}+\frac{1}{r\sin\theta}\frac{\partial}{\partial \theta}\left\{\sin\theta\left\{\rho v_{0\theta}-\tau\left[\frac{\partial}{\partial t}(\rho v_{0\theta})\right.\right.\right.\\
&\left.\left.\left.+\frac{1}{r^2}\frac{\partial(r^2\rho v_{0r}v_{0\theta})}{\partial r}+\frac{1}{r\sin\theta}\frac{\partial(\rho v_{0\varphi}v_{0\theta})}{\partial \varphi}+\frac{1}{r\sin\theta}\frac{\partial(\rho v_{0\theta}^2\sin\theta)}{\partial \theta}-\rho g_\theta\right]\right\}\right\}\\
&-\frac{1}{r^2}\frac{\partial}{\partial r}\left(\tau r^2\frac{\partial p}{\partial r}\right)-\frac{1}{r^2\sin\theta}\frac{\partial}{\partial \theta}\left(\tau\sin\theta\frac{\partial p}{\partial \theta}\right)-\frac{1}{r^2\sin^2\theta}\frac{\partial}{\partial \varphi}\left(\tau\frac{\partial p}{\partial \varphi}\right)=0.
\end{aligned} \tag{A.2.101}$$

The continuity equation can be simplified in the case of the spherical symmetry:

$$\begin{aligned}
&\frac{\partial}{\partial t}\left\{\rho-\tau\left[\frac{\partial \rho}{\partial t}+\frac{1}{r^2}\frac{\partial(r^2\rho v_{0r})}{\partial r}\right]\right\}+\frac{1}{r^2}\frac{\partial}{\partial r}\left\{r^2\left\{\rho v_{0r}-\tau\left[\frac{\partial}{\partial t}(\rho v_{0r})+\frac{1}{r^2}\frac{\partial(r^2\rho v_{0r}^2)}{\partial r}-\rho g_r\right]\right\}\right\}\\
&-\frac{1}{r^2}\frac{\partial}{\partial r}\left(\tau r^2\frac{\partial p}{\partial r}\right)=0.
\end{aligned} \tag{A.2.102}$$

Transform now the momentum equation:

$$\begin{aligned}
&\frac{\partial}{\partial t}\left\{\rho \mathbf{v}_0-\tau\left[\frac{\partial}{\partial t}(\rho \mathbf{v}_0)+\frac{\partial}{\partial \mathbf{r}}\cdot \rho \mathbf{v}_0\mathbf{v}_0+\frac{\partial p}{\partial \mathbf{r}}-\rho \mathbf{g}\right]\right\}\\
&-\mathbf{g}\left[\rho-\tau\left(\frac{\partial \rho}{\partial t}+\frac{\partial}{\partial \mathbf{r}}\cdot(\rho \mathbf{v}_0)\right)\right]+\frac{\partial}{\partial \mathbf{r}}\cdot\left\{\rho \mathbf{v}_0\mathbf{v}_0\right.\\
&+p\overset{\leftrightarrow}{\mathrm{I}}-\tau\left[\frac{\partial}{\partial t}\left(\rho \mathbf{v}_0\mathbf{v}_0+p\overset{\leftrightarrow}{\mathrm{I}}\right)+\frac{\partial}{\partial \mathbf{r}}\cdot \rho(\mathbf{v}_0\mathbf{v}_0)\mathbf{v}_0\right.\\
&\left.\left.+2\overset{\leftrightarrow}{\mathrm{I}}\left[\frac{\partial}{\partial \mathbf{r}}\cdot(p\mathbf{v}_0)\right]+\frac{\partial}{\partial \mathbf{r}}\cdot\left(\overset{\leftrightarrow}{\mathrm{I}}p\mathbf{v}_0\right)-\mathbf{g}\rho \mathbf{v}_0-\mathbf{v}_0\mathbf{g}\rho\right]\right\}=0.
\end{aligned} \tag{A.2.103}$$

Projection of the momentum equation onto $\mathbf{e}_r$-direction can be obtained with the help of relation (A.2.52) for gradient.

$$
\begin{aligned}
&\frac{\partial}{\partial t}\Big\{\rho v_{0r}-\tau\Big[\frac{\partial}{\partial t}(\rho v_{0r})+\frac{\partial}{\partial \mathbf{r}}\cdot\rho\mathbf{v}_0 v_{0r}+\frac{\partial p}{\partial r}-\rho g_r\Big]\Big\}-g_r\Big[\rho-\tau\Big(\frac{\partial \rho}{\partial t}+\frac{\partial}{\partial \mathbf{r}}\cdot(\rho\mathbf{v}_0)\Big)\Big]\\
&+\frac{\partial}{\partial \mathbf{r}}\cdot\Big\{\rho\mathbf{v}_0 v_{0r}-\tau\Big[\frac{\partial}{\partial t}(\rho\mathbf{v}_0 v_{0r})+\frac{\partial}{\partial \mathbf{r}}\cdot\rho(\mathbf{v}_0\mathbf{v}_0)v_{0r}-\mathbf{g}\rho v_{0r}-\mathbf{v}_0 g_r\rho\Big]\Big\}\\
&+\frac{\partial p}{\partial r}-\frac{\partial}{\partial r}\Big(\tau\frac{\partial p}{\partial t}\Big)-2\frac{\partial}{\partial r}\Big(\tau\frac{\partial}{\partial \mathbf{r}}\cdot(p\mathbf{v}_0)\Big)-\mathrm{div}\,(\tau\,\mathrm{grad}\,(p v_{0r}))=0.
\end{aligned}
\tag{A.2.104}
$$

Transform Eq. (A.2.104) using relation (A.2.54) for divergence in the spherical coordinates:

$$
\begin{aligned}
&\frac{\partial}{\partial t}\Big\{\rho v_{0r}-\tau\Big[\frac{\partial}{\partial t}(\rho v_{0r})+\frac{1}{r^2}\frac{\partial(r^2\rho v_{0r}^2)}{\partial r}+\frac{1}{r\sin\theta}\frac{\partial(\rho v_{0\varphi}v_{0r})}{\partial\varphi}+\frac{1}{r\sin\theta}\frac{\partial(\rho v_{0\theta}v_{0r}\sin\theta)}{\partial\theta}\\
&+\frac{\partial p}{\partial r}-\rho g_r\Big]\Big\}-g_r\Big[\rho-\tau\Big(\frac{\partial\rho}{\partial t}+\frac{1}{r^2}\frac{\partial(r^2\rho v_{0r})}{\partial r}+\frac{1}{r\sin\theta}\frac{\partial(\rho v_{0\varphi})}{\partial\varphi}+\frac{1}{r\sin\theta}\frac{\partial(\rho v_{0\theta}\sin\theta)}{\partial\theta}\Big)\Big]\\
&+\frac{1}{r^2}\frac{\partial}{\partial r}\Big\{r^2\Big\{\rho v_{0r}^2-\tau\Big[\frac{\partial}{\partial t}(\rho v_{0r}^2)+\frac{\partial}{\partial \mathbf{r}}\cdot(\rho\mathbf{v}_0 v_{0r}^2)-g_r\rho v_{0r}-v_{0r}\rho g_r\Big]\Big\}\Big\}\\
&+\frac{1}{r\sin\theta}\frac{\partial}{\partial\varphi}\Big\{\rho v_{0\varphi}v_{0r}-\tau\Big[\frac{\partial}{\partial t}(\rho v_{0\varphi}v_{0r})+\frac{\partial}{\partial \mathbf{r}}\cdot(\rho\mathbf{v}_0 v_{0\varphi}v_{0r})-g_\varphi\rho v_{0r}-v_{0\varphi}\rho g_r\Big]\Big\}\\
&+\frac{1}{r\sin\theta}\frac{\partial}{\partial\theta}\Big\{\sin\theta\Big\{\rho v_{0\theta}v_{0r}-\tau\Big[\frac{\partial}{\partial t}(\rho v_{0\theta}v_{0r})+\frac{\partial}{\partial \mathbf{r}}\cdot(\rho\mathbf{v}_0 v_{0\theta}v_{0r})-g_\theta\rho v_{0r}-v_{0\theta}\rho g_r\Big]\Big\}\Big\}\\
&+\frac{\partial p}{\partial r}-\frac{\partial}{\partial r}\Big(\tau\frac{\partial p}{\partial t}\Big)-2\frac{\partial}{\partial r}\Big(\tau\Big(\frac{1}{r^2}\frac{\partial(r^2 p v_{0r})}{\partial r}+\frac{1}{r\sin\theta}\frac{\partial(p v_{0\varphi})}{\partial\varphi}+\frac{1}{r\sin\theta}\frac{\partial(p v_{0\theta}\sin\theta)}{\partial\theta}\Big)\Big)\\
&-\frac{1}{r^2}\frac{\partial}{\partial r}\Big(\tau r^2\frac{\partial(p v_{0r})}{\partial r}\Big)-\frac{1}{r^2\sin\theta}\frac{\partial}{\partial\theta}\Big(\tau\sin\theta\frac{\partial(p v_{0r})}{\partial\theta}\Big)-\frac{1}{r^2\sin^2\theta}\frac{\partial}{\partial\varphi}\Big(\tau\frac{\partial(p v_{0r})}{\partial\varphi}\Big)=0,
\end{aligned}
\tag{A.2.105}
$$

$$
\begin{aligned}
&\frac{\partial}{\partial t}\Big\{\rho v_{0r}-\tau\Big[\frac{\partial}{\partial t}(\rho v_{0r})+\frac{1}{r^2}\frac{\partial(r^2\rho v_{0r}^2)}{\partial r}+\frac{1}{r\sin\theta}\frac{\partial(\rho v_{0\varphi}v_{0r})}{\partial\varphi}+\frac{1}{r\sin\theta}\frac{\partial(\rho v_{0\theta}v_{0r}\sin\theta)}{\partial\theta}\\
&+\frac{\partial p}{\partial r}-\rho g_r\Big]\Big\}-g_r\Big[\rho-\tau\Big(\frac{\partial\rho}{\partial t}+\frac{1}{r^2}\frac{\partial(r^2\rho v_{0r})}{\partial r}+\frac{1}{r\sin\theta}\frac{\partial(\rho v_{0\varphi})}{\partial\varphi}+\frac{1}{r\sin\theta}\frac{\partial(\rho v_{0\theta}\sin\theta)}{\partial\theta}\Big)\Big]\\
&+\frac{1}{r^2}\frac{\partial}{\partial r}\Big\{r^2\Big\{\rho v_{0r}^2-\tau\Big[\frac{\partial}{\partial t}(\rho v_{0r}^2)+\frac{1}{r^2}\frac{\partial(r^2\rho v_{0r}^3)}{\partial r}+\frac{1}{r\sin\theta}\frac{\partial(\rho v_{0\varphi}v_{0r}^2)}{\partial\varphi}\\
&+\frac{1}{r\sin\theta}\frac{\partial(\rho v_{0\theta}v_{0r}^2\sin\theta)}{\partial\theta}-2g_r\rho v_{0r}\Big]\Big\}\Big\}+\frac{1}{r\sin\theta}\frac{\partial}{\partial\varphi}\Big\{\rho v_{0\varphi}v_{0r}-\tau\Big[\frac{\partial}{\partial t}(\rho v_{0\varphi}v_{0r})\\
&+\frac{1}{r^2}\frac{\partial(r^2\rho v_{0\varphi}v_{0r}^2)}{\partial r}+\frac{1}{r\sin\theta}\frac{\partial(\rho v_{0\varphi}^2 v_{0r})}{\partial\varphi}+\frac{1}{r\sin\theta}\frac{\partial(\rho v_{0\theta}v_{0\varphi}v_{0r}\sin\theta)}{\partial\theta}-g_\varphi\rho v_{0r}-v_{0\varphi}\rho g_r\Big]\Big\}\\
&+\frac{1}{r\sin\theta}\frac{\partial}{\partial\theta}\Big\{\sin\theta\Big\{\rho v_{0\theta}v_{0r}-\tau\Big[\frac{\partial}{\partial t}(\rho v_{0\theta}v_{0r})+\frac{1}{r^2}\frac{\partial(r^2\rho v_{0\theta}v_{0r}^2)}{\partial r}+\frac{1}{r\sin\theta}\frac{\partial(\rho v_{0\varphi}v_{0\theta}v_{0r})}{\partial\varphi}\\
&+\frac{1}{r\sin\theta}\frac{\partial(\rho v_{0\theta}^2 v_{0r}\sin\theta)}{\partial\theta}-g_\theta\rho v_{0r}-v_{0\theta}\rho g_r\Big]\Big\}\Big\}\\
&+\frac{\partial p}{\partial r}-\frac{\partial}{\partial r}\Big(\tau\frac{\partial p}{\partial t}\Big)-2\frac{\partial}{\partial r}\Big(\tau\Big(\frac{1}{r^2}\frac{\partial(r^2 p v_{0r})}{\partial r}+\frac{1}{r\sin\theta}\frac{\partial(p v_{0\varphi})}{\partial\varphi}+\frac{1}{r\sin\theta}\frac{\partial(p v_{0\theta}\sin\theta)}{\partial\theta}\Big)\Big)\\
&-\frac{1}{r^2}\frac{\partial}{\partial r}\Big(\tau r^2\frac{\partial(p v_{0r})}{\partial r}\Big)-\frac{1}{r^2\sin\theta}\frac{\partial}{\partial\theta}\Big(\tau\sin\theta\frac{\partial(p v_{0r})}{\partial\theta}\Big)-\frac{1}{r^2\sin^2\theta}\frac{\partial}{\partial\varphi}\Big(\tau\frac{\partial(p v_{0r})}{\partial\varphi}\Big)=0
\end{aligned}
\tag{A.2.106}
$$

For the case of the radial symmetry momentum equation takes the form

$$
\begin{aligned}
&\frac{\partial}{\partial t}\left\{\rho v_{0r}-\tau\left[\frac{\partial}{\partial t}(\rho v_{0r})+\frac{1}{r^2}\frac{\partial(r^2\rho v_{0r}^2)}{\partial r}+\frac{\partial p}{\partial r}-\rho g_r\right]\right\}-g_r\left[\rho-\tau\left(\frac{\partial\rho}{\partial t}+\frac{1}{r^2}\frac{\partial(r^2\rho v_{0r})}{\partial r}\right)\right]\\
&+\frac{1}{r^2}\frac{\partial}{\partial r}\left\{r^2\left\{\rho v_{0r}^2-\tau\left[\frac{\partial}{\partial t}(\rho v_{0r}^2)+\frac{1}{r^2}\frac{\partial(r^2\rho v_{0r}^3)}{\partial r}-2g_r\rho v_{0r}\right]\right\}\right\}\\
&+\frac{\partial p}{\partial r}-\frac{\partial}{\partial r}\left(\tau\frac{\partial p}{\partial t}\right)-2\frac{\partial}{\partial r}\left(\frac{\tau}{r^2}\frac{\partial(r^2 p v_{0r})}{\partial r}\right)-\frac{1}{r^2}\frac{\partial}{\partial r}\left(\tau r^2\frac{\partial(p v_{0r})}{\partial r}\right)=0.
\end{aligned}
\tag{A.2.107}
$$

Projection of the momentum equation Eq. (A.2.103) onto $\mathbf{e}_\varphi$-direction:

$$
\begin{aligned}
&\frac{\partial}{\partial t}\left\{\rho v_{0\varphi}-\tau\left[\frac{\partial}{\partial t}(\rho v_{0\varphi})+\frac{\partial}{\partial \mathbf{r}}\cdot\rho\mathbf{v}_0 v_{0\varphi}+\frac{1}{r\sin\theta}\frac{\partial p}{\partial\varphi}-\rho g_\varphi\right]\right\}-g_\varphi\left[\rho-\tau\left(\frac{\partial\rho}{\partial t}+\frac{\partial}{\partial \mathbf{r}}\cdot(\rho\mathbf{v}_0)\right)\right]\\
&+\frac{\partial}{\partial \mathbf{r}}\cdot\left\{\rho\mathbf{v}_0 v_{0\varphi}-\tau\left[\frac{\partial}{\partial t}\left(\rho\mathbf{v}_0 v_{0\varphi}+\frac{\partial}{\partial \mathbf{r}}\cdot\rho(\mathbf{v}_0\mathbf{v}_0)v_{0\varphi}-\mathbf{g}\rho v_{0\varphi}-\mathbf{v}_0 g_\varphi\rho\right)\right]\right\}\\
&+\frac{1}{r\sin\theta}\frac{\partial p}{\partial\varphi}-\frac{1}{r\sin\theta}\frac{\partial}{\partial\varphi}\left(\tau\frac{\partial p}{\partial t}\right)-2\frac{1}{r\sin\theta}\frac{\partial}{\partial\varphi}\left(\tau\frac{\partial}{\partial \mathbf{r}}\cdot(p\mathbf{v}_0)\right)-\operatorname{div}(\tau\operatorname{grad}(p v_{0\varphi}))=0,
\end{aligned}
\tag{A.2.108}
$$

$$
\begin{aligned}
&\frac{\partial}{\partial t}\left\{\rho v_{0\varphi}-\tau\left[\frac{\partial}{\partial t}(\rho v_{0\varphi})+\frac{1}{r^2}\frac{\partial(r^2\rho v_{0r}v_{0\varphi})}{\partial r}+\frac{1}{r\sin\theta}\frac{\partial\left(\rho v_{0\varphi}^2\right)}{\partial\varphi}+\frac{1}{r\sin\theta}\frac{\partial(\rho v_{0\theta}v_{0\varphi}\sin\theta)}{\partial\theta}\right.\right.\\
&\left.\left.+\frac{1}{r\sin\theta}\frac{\partial p}{\partial\varphi}-\rho g_\varphi\right]\right\}-g_\varphi\left[\rho-\tau\left(\frac{\partial\rho}{\partial t}+\frac{1}{r^2}\frac{\partial(r^2\rho v_{0r})}{\partial r}+\frac{1}{r\sin\theta}\frac{\partial(\rho v_{0\varphi})}{\partial\varphi}+\frac{1}{r\sin\theta}\frac{\partial(\rho v_{0\theta}\sin\theta)}{\partial\theta}\right)\right]\\
&+\frac{1}{r^2}\frac{\partial}{\partial r}\left\{r^2\left\{\rho v_{0r}v_{0\varphi}-\tau\left[\frac{\partial}{\partial t}(\rho v_{0r}v_{0\varphi})+\frac{\partial}{\partial \mathbf{r}}\cdot(\rho\mathbf{v}_0 v_{0r}v_{0\varphi})-g_r\rho v_{0\varphi}-v_{0r}\rho g_\varphi\right]\right\}\right\}\\
&+\frac{1}{r\sin\theta}\frac{\partial}{\partial\varphi}\left\{\rho v_{0\varphi}^2-\tau\left[\frac{\partial}{\partial t}\left(\rho v_{0\varphi}^2\right)+\frac{\partial}{\partial \mathbf{r}}\cdot\left(\rho\mathbf{v}_0 v_{0\varphi}^2\right)-2g_\varphi\rho v_{0\varphi}\right]\right\}+\frac{1}{r\sin\theta}\frac{\partial}{\partial\theta}\left\{\sin\theta\left\{\rho v_{0\theta}v_{0\varphi}\right.\right.\\
&\left.\left.-\tau\left[\frac{\partial}{\partial t}(\rho v_{0\theta}v_{0\varphi})+\frac{\partial}{\partial \mathbf{r}}\cdot(\rho\mathbf{v}_0 v_{0\theta}v_{0\varphi})-g_\theta\rho v_{0\varphi}-v_{0\theta}\rho g_\varphi\right]\right\}\right\}+\frac{1}{r\sin\theta}\frac{\partial p}{\partial\varphi}-\frac{1}{r\sin\theta}\frac{\partial}{\partial\varphi}\left(\tau\frac{\partial p}{\partial t}\right)\\
&-\frac{2}{r\sin\theta}\frac{\partial}{\partial\varphi}\left(\tau\left(\frac{1}{r^2}\frac{\partial(r^2 p v_{0r})}{\partial r}+\frac{1}{r\sin\theta}\frac{\partial(p v_{0\varphi})}{\partial\varphi}+\frac{1}{r\sin\theta}\frac{\partial(p v_{0\theta}\sin\theta)}{\partial\theta}\right)\right)\\
&-\frac{1}{r^2}\frac{\partial}{\partial r}\left(\tau r^2\frac{\partial(p v_{0\varphi})}{\partial r}\right)-\frac{1}{r^2\sin\theta}\frac{\partial}{\partial\theta}\left(\tau\sin\theta\frac{\partial(p v_{0\varphi})}{\partial\theta}\right)-\frac{1}{r^2\sin^2\theta}\frac{\partial}{\partial\varphi}\left(\tau\frac{\partial(p v_{0\varphi})}{\partial\varphi}\right)=0
\end{aligned}
\tag{A.2.109}
$$

$$
\begin{aligned}
&\frac{\partial}{\partial t}\left\{\rho v_{0\varphi}-\tau\left[\frac{\partial}{\partial t}(\rho v_{0\varphi})+\frac{1}{r^2}\frac{\partial(r^2\rho v_{0r}v_{0\varphi})}{\partial r}+\frac{1}{r\sin\theta}\frac{\partial\left(\rho v_{0\varphi}^2\right)}{\partial\varphi}+\frac{1}{r\sin\theta}\frac{\partial(\rho v_{0\theta}v_{0\varphi}\sin\theta)}{\partial\theta}\right.\right.\\
&\left.\left.+\frac{1}{r\sin\theta}\frac{\partial p}{\partial\varphi}-\rho g_\varphi\right]\right\}-g_\varphi\left[\rho-\tau\left(\frac{\partial\rho}{\partial t}+\frac{1}{r^2}\frac{\partial(r^2\rho v_{0r})}{\partial r}+\frac{1}{r\sin\theta}\frac{\partial(\rho v_{0\varphi})}{\partial\varphi}+\frac{1}{r\sin\theta}\frac{\partial(\rho v_{0\theta}\sin\theta)}{\partial\theta}\right)\right]\\
&+\frac{1}{r^2}\frac{\partial}{\partial r}\left\{r^2\left\{\rho v_{0r}v_{0\varphi}-\tau\left[\frac{\partial}{\partial t}(\rho v_{0r}v_{0\varphi})+\frac{1}{r^2}\frac{\partial(r^2\rho v_{0r}^2v_{0\varphi})}{\partial r}+\frac{1}{r\sin\theta}\frac{\partial\left(\rho v_{0\varphi}^2v_{0r}\right)}{\partial\varphi}\right.\right.\right.\\
&\left.\left.\left.+\frac{1}{r\sin\theta}\frac{\partial(\rho v_{0\theta}v_{0r}v_{0\varphi}\sin\theta)}{\partial\theta}-g_r\rho v_{0\varphi}-v_{0r}\rho g_\varphi\right]\right\}\right\}+\frac{1}{r\sin\theta}\frac{\partial}{\partial\varphi}\left\{\rho v_{0\varphi}^2-\tau\left[\frac{\partial}{\partial t}\left(\rho v_{0\varphi}^2\right)\right.\right.\\
&\left.\left.+\frac{1}{r^2}\frac{\partial\left(r^2\rho v_{0r}v_{0\varphi}^2\right)}{\partial r}+\frac{1}{r\sin\theta}\frac{\partial\left(\rho v_{0\varphi}^3\right)}{\partial\varphi}+\frac{1}{r\sin\theta}\frac{\partial\left(\rho v_{0\theta}v_{0\varphi}^2\sin\theta\right)}{\partial\theta}-2g_\varphi\rho v_{0\varphi}\right]\right\}\\
&+\frac{1}{r\sin\theta}\frac{\partial}{\partial\theta}\left\{\sin\theta\left\{\rho v_{0\theta}v_{0\varphi}-\tau\left[\frac{\partial}{\partial t}(\rho v_{0\theta}v_{0\varphi})+\frac{1}{r^2}\frac{\partial(r^2\rho v_{0r}v_{0\theta}v_{0\varphi})}{\partial r}+\frac{1}{r\sin\theta}\frac{\partial\left(\rho v_{0\varphi}^2v_{0\theta}\right)}{\partial\varphi}\right.\right.\right.\\
&\left.\left.\left.+\frac{1}{r\sin\theta}\frac{\partial(\rho v_{0\theta}^2v_{0\varphi}\sin\theta)}{\partial\theta}-g_\theta\rho v_{0\varphi}-v_{0\theta}\rho g_\varphi\right]\right\}\right\}+\frac{1}{r\sin\theta}\frac{\partial p}{\partial\varphi}-\frac{1}{r\sin\theta}\frac{\partial}{\partial\varphi}\left(\tau\frac{\partial p}{\partial t}\right)\\
&-\frac{2}{r\sin\theta}\frac{\partial}{\partial\varphi}\left(\tau\left(\frac{1}{r^2}\frac{\partial(r^2pv_{0r})}{\partial r}+\frac{1}{r\sin\theta}\frac{\partial(pv_{0\varphi})}{\partial\varphi}+\frac{1}{r\sin\theta}\frac{\partial(pv_{0\theta}\sin\theta)}{\partial\theta}\right)\right)\\
&-\frac{1}{r^2}\frac{\partial}{\partial r}\left(\tau r^2\frac{\partial(pv_{0\varphi})}{\partial r}\right)-\frac{1}{r^2\sin\theta}\frac{\partial}{\partial\theta}\left(\tau\sin\theta\frac{\partial(pv_{0\varphi})}{\partial\theta}\right)-\frac{1}{r^2\sin^2\theta}\frac{\partial}{\partial\varphi}\left(\tau\frac{\partial(pv_{0\varphi})}{\partial\varphi}\right)=0.
\end{aligned}
\tag{A.2.110}
$$

Projection of the momentum equation (A.2.103) onto $\mathbf{e}_\theta$—direction:

$$
\begin{aligned}
&\frac{\partial}{\partial t}\left\{\rho v_{0\theta}-\tau\left[\frac{\partial}{\partial t}(\rho v_{0\theta})+\frac{\partial}{\partial\mathbf{r}}\cdot\rho\mathbf{v}_0v_{0\theta}+\frac{1}{r}\frac{\partial p}{\partial\theta}-\rho g_\theta\right]\right\}-g_\theta\left[\rho-\tau\left(\frac{\partial\rho}{\partial t}+\frac{\partial}{\partial\mathbf{r}}\cdot(\rho\mathbf{v}_0)\right)\right]\\
&+\frac{\partial}{\partial\mathbf{r}}\cdot\left\{\rho\mathbf{v}_0v_{0\theta}-\tau\left[\frac{\partial}{\partial t}(\rho\mathbf{v}_0v_{0\theta})+\frac{\partial}{\partial\mathbf{r}}\cdot\rho(\mathbf{v}_0\mathbf{v}_0)v_{0\theta}-\mathbf{g}\rho v_{0\theta}-\mathbf{v}_0g_\theta\rho\right]\right\}\\
&+\frac{1}{r}\frac{\partial p}{\partial\theta}-\frac{1}{r}\frac{\partial}{\partial\theta}\left(\tau\frac{\partial p}{\partial t}\right)-2\frac{1}{r}\frac{\partial}{\partial\theta}\left(\tau\frac{\partial}{\partial\mathbf{r}}\cdot(p\mathbf{v}_0)\right)-\mathrm{div}(\tau\,\mathrm{grad}\,(pv_{0\theta}))=0,
\end{aligned}
\tag{A.2.111}
$$

$$
\begin{aligned}
&\frac{\partial}{\partial t}\left\{\rho v_{0\theta}-\tau\left[\frac{\partial}{\partial t}(\rho v_{0\theta})+\frac{1}{r^2}\frac{\partial(r^2\rho v_{0r}v_{0\theta})}{\partial r}+\frac{1}{r\sin\theta}\frac{\partial(\rho v_{0\varphi}v_{0\theta})}{\partial\varphi}+\frac{1}{r\sin\theta}\frac{\partial(\rho v_{0\theta}^2\sin\theta)}{\partial\theta}\right.\right.\\
&\left.\left.+\frac{1}{r}\frac{\partial p}{\partial\theta}-\rho g_\theta\right]\right\}-g_\theta\left[\rho-\tau\left(\frac{\partial\rho}{\partial t}+\frac{1}{r^2}\frac{\partial(r^2\rho v_{0r})}{\partial r}+\frac{1}{r\sin\theta}\frac{\partial(\rho v_{0\varphi})}{\partial\varphi}+\frac{1}{r\sin\theta}\frac{\partial(\rho v_{0\theta}\sin\theta)}{\partial\theta}\right)\right]\\
&+\frac{1}{r^2}\frac{\partial}{\partial r}\left\{r^2\left\{\rho v_{0r}v_{0\theta}-\tau\left[\frac{\partial}{\partial t}(\rho v_{0r}v_{0\theta})+\frac{\partial}{\partial\mathbf{r}}\cdot(\rho\mathbf{v}_0v_{0r}v_{0\theta})-g_r\rho v_{0\theta}-v_{0r}\rho g_\theta\right]\right\}\right\}\\
&+\frac{1}{r\sin\theta}\frac{\partial}{\partial\varphi}\left\{\rho v_{0\varphi}v_{0\theta}-\tau\left[\frac{\partial}{\partial t}(\rho v_{0\varphi}v_{0\theta})+\frac{\partial}{\partial\mathbf{r}}\cdot(\rho\mathbf{v}_0v_{0\varphi}v_{0\theta})-g_\varphi\rho v_{0\theta}-v_{0\varphi}\rho g_\theta\right]\right\}\\
&+\frac{1}{r\sin\theta}\frac{\partial}{\partial\theta}\left\{\sin\theta\left\{\rho v_{0\theta}^2-\tau\left[\frac{\partial}{\partial t}(\rho v_{0\theta}^2)+\frac{\partial}{\partial\mathbf{r}}\cdot(\rho\mathbf{v}_0v_{0\theta}^2)-2g_\theta\rho v_{0\theta}\right]\right\}\right\}+\frac{1}{r}\frac{\partial p}{\partial\theta}-\frac{1}{r}\frac{\partial}{\partial\theta}\left(\tau\frac{\partial p}{\partial t}\right)\\
&-\frac{2}{r}\frac{\partial}{\partial\theta}\left(\tau\left(\frac{1}{r^2}\frac{\partial(r^2pv_{0r})}{\partial r}+\frac{1}{r\sin\theta}\frac{\partial(pv_{0\varphi})}{\partial\varphi}+\frac{1}{r\sin\theta}\frac{\partial(pv_{0\theta}\sin\theta)}{\partial\theta}\right)\right)\\
&-\frac{1}{r^2}\frac{\partial}{\partial r}\left(\tau r^2\frac{\partial(pv_{0\theta})}{\partial r}\right)-\frac{1}{r^2\sin\theta}\frac{\partial}{\partial\theta}\left(\tau\sin\theta\frac{\partial(pv_{0\theta})}{\partial\theta}\right)-\frac{1}{r^2\sin^2\theta}\frac{\partial}{\partial\varphi}\left(\tau\frac{\partial(pv_{0\theta})}{\partial\varphi}\right)=0,
\end{aligned}
\tag{A.2.112}
$$

Appendix | 2 Using of Curvilinear Coordinates in the Generalized Hydrodynamic Theory

$$\frac{\partial}{\partial t}\left\{\rho v_{0\theta} - \tau\left[\frac{\partial}{\partial t}(\rho v_{0\theta}) + \frac{1}{r^2}\frac{\partial(r^2\rho v_{0r}v_{0\theta})}{\partial r} + \frac{1}{r\sin\theta}\frac{\partial(\rho v_{0\varphi}v_{0\theta})}{\partial \varphi} + \frac{1}{r\sin\theta}\frac{\partial(\rho v_{0\theta}^2\sin\theta)}{\partial \theta}\right.\right.$$
$$\left.\left.+\frac{1}{r}\frac{\partial p}{\partial \theta} - \rho g_\theta\right]\right\} - g_\theta\left[\rho - \tau\left(\frac{\partial \rho}{\partial t} + \frac{1}{r^2}\frac{\partial(r^2\rho v_{0r})}{\partial r} + \frac{1}{r\sin\theta}\frac{\partial(\rho v_{0\varphi})}{\partial \varphi} + \frac{1}{r\sin\theta}\frac{\partial(\rho v_{0\theta}\sin\theta)}{\partial \theta}\right)\right]$$
$$+\frac{1}{r^2}\frac{\partial}{\partial r}\left\{r^2\left\{\rho v_{0r}v_{0\theta} - \tau\left[\frac{\partial}{\partial t}(\rho v_{0r}v_{0\theta}) + \frac{1}{r^2}\frac{\partial(r^2\rho v_{0r}^2 v_{0\theta})}{\partial r} + \frac{1}{r\sin\theta}\frac{\partial(\rho v_{0\varphi}v_{0r}v_{0\theta})}{\partial \varphi}\right.\right.\right.$$
$$\left.\left.\left.+\frac{1}{r\sin\theta}\frac{\partial(\rho v_{0\theta}^2 v_{0r}\sin\theta)}{\partial \theta} - g_r\rho v_{0\theta} - v_{0r}\rho g_\theta\right]\right\}\right\} + \frac{1}{r\sin\theta}\frac{\partial}{\partial \varphi}\left\{\rho v_{0\varphi}v_{0\theta} - \tau\left[\frac{\partial}{\partial t}(\rho v_{0\varphi}v_{0\theta})\right.\right.$$
$$+\frac{1}{r^2}\frac{\partial(r^2\rho v_{0r}v_{0\varphi}v_{0\theta})}{\partial r} + \frac{1}{r\sin\theta}\frac{\partial(\rho v_{0\varphi}^2 v_{0\theta})}{\partial \varphi} + \frac{1}{r\sin\theta}\frac{\partial(\rho v_{0\theta}^2 v_{0\varphi}\sin\theta)}{\partial \theta} - g_\varphi\rho v_{0\theta} - v_{0\varphi}\rho g_\theta\Bigg]\Bigg\} \quad (A.2.113)$$
$$+\frac{1}{r\sin\theta}\frac{\partial}{\partial \theta}\left\{\sin\theta\left\{\rho v_{0\theta}^2 - \tau\left[\frac{\partial}{\partial t}(\rho v_{0\theta}^2) + \frac{1}{r^2}\frac{\partial(r^2\rho v_{0r}v_{0\theta}^2)}{\partial r} + \frac{1}{r\sin\theta}\frac{\partial(\rho v_{0\varphi}v_{0\theta}^2)}{\partial \varphi}\right.\right.\right.$$
$$\left.\left.\left.+\frac{1}{r\sin\theta}\frac{\partial(\rho v_{0\theta}^3\sin\theta)}{\partial \theta} - 2g_\theta\rho v_{0\theta}\right]\right\}\right\} + \frac{1}{r}\frac{\partial p}{\partial \theta} - \frac{1}{r}\frac{\partial}{\partial \theta}\left(\tau\frac{\partial p}{\partial t}\right)$$
$$-\frac{2}{r}\frac{\partial}{\partial \theta}\left(\tau\left(\frac{1}{r^2}\frac{\partial(r^2 p v_{0r})}{\partial r} + \frac{1}{r\sin\theta}\frac{\partial(p v_{0\varphi})}{\partial \varphi} + \frac{1}{r\sin\theta}\frac{\partial(p v_{0\theta}\sin\theta)}{\partial \theta}\right)\right)$$
$$-\frac{1}{r^2}\frac{\partial}{\partial r}\left(\tau r^2\frac{\partial(p v_{0\theta})}{\partial r}\right) - \frac{1}{r^2\sin\theta}\frac{\partial}{\partial \theta}\left(\tau\sin\theta\frac{\partial(p v_{0\theta})}{\partial \theta}\right) - \frac{1}{r^2\sin^2\theta}\frac{\partial}{\partial \varphi}\left(\tau\frac{\partial(p v_{0\theta})}{\partial \varphi}\right) = 0.$$

The left hand sides of equations Eq. (A.2.110) and Eq. (A.2.113) are equal zero identically if the problem has the spherical symmetry.

Let us transform now the energy equation:

$$\frac{\partial}{\partial t}\left\{\frac{\rho v_0^2}{2} + \frac{3}{2}p - \tau\left[\frac{\partial}{\partial t}\left(\frac{\rho v_0^2}{2} + \frac{3}{2}p\right)\right.\right.$$
$$\left.\left.+\frac{\partial}{\partial \mathbf{r}}\cdot\left(\frac{1}{2}\rho v_0^2\mathbf{v}_0 + \frac{5}{2}p\mathbf{v}_0\right) - \mathbf{g}\cdot\rho\mathbf{v}_0\right]\right\}$$
$$+\frac{\partial}{\partial \mathbf{r}}\cdot\left\{\frac{1}{2}\rho v_0^2\mathbf{v}_0 + \frac{5}{2}p\mathbf{v}_0 - \tau\left[\frac{\partial}{\partial t}\left(\frac{1}{2}\rho v_0^2\mathbf{v}_0\right.\right.\right. \quad (A.2.114)$$
$$\left.\left.\left.+\frac{5}{2}p\mathbf{v}_0\right) + \frac{\partial}{\partial \mathbf{r}}\cdot\left(\frac{1}{2}\rho v_0^2\mathbf{v}_0\mathbf{v}_0 + \frac{7}{2}p\mathbf{v}_0\mathbf{v}_0 + \frac{1}{2}pv_0^2\overleftrightarrow{\mathbf{I}} + \frac{5p^2}{2\rho}\overleftrightarrow{\mathbf{I}}\right) - \rho\mathbf{g}\cdot\mathbf{v}_0\mathbf{v}_0 - p\mathbf{g}\cdot\overleftrightarrow{\mathbf{I}} - \frac{1}{2}\rho v_0^2\mathbf{g} - \frac{3}{2}\mathbf{g}p\right]\right\}$$
$$-\left\{\rho\mathbf{g}\cdot\mathbf{v}_0 - \tau\left[\mathbf{g}\cdot\left(\frac{\partial}{\partial t}(\rho\mathbf{v}_0) + \frac{\partial}{\partial \mathbf{r}}\cdot\rho\mathbf{v}_0\mathbf{v}_0 + \frac{\partial}{\partial \mathbf{r}}\cdot p\overleftrightarrow{\mathbf{I}} - \rho\mathbf{g}\right)\right]\right\} = 0,$$

$$\frac{\partial}{\partial t}\left\{\frac{1}{2}\rho v_0^2+\frac{3}{2}p-\tau\left[\frac{\partial}{\partial t}\left(\frac{1}{2}\rho v_0^2+\frac{3}{2}p\right)+\frac{1}{r^2}\frac{\partial}{\partial r}\left(r^2 v_{0r}\left(\frac{1}{2}\rho v_0^2+\frac{5}{2}p\right)\right)\right.\right.$$
$$+\frac{1}{r\sin\theta}\frac{\partial}{\partial\varphi}\left(v_{0\varphi}\left(\frac{1}{2}\rho v_0^2+\frac{5}{2}p\right)\right)+\frac{1}{r\sin\theta}\frac{\partial}{\partial\theta}\left(\sin\theta v_{0\theta}\left(\frac{1}{2}\rho v_0^2+\frac{5}{2}p\right)\right)$$
$$\left.\left.-\rho\left(g_r v_{0r}+g_\varphi v_{0\varphi}+g_\theta v_{0\theta}\right)\right]\right\}+\frac{1}{r^2}\frac{\partial}{\partial r}\left\{r^2\left\{\left(\frac{1}{2}\rho v_0^2+\frac{5}{2}p\right)v_{0r}-\tau\left[\frac{\partial}{\partial t}\left(\left(\frac{1}{2}\rho v_0^2+\frac{5}{2}p\right)v_{0r}\right)\right.\right.\right.$$
$$\left.\left.\left.+\frac{\partial}{\partial \mathbf{r}}\cdot\left(\left(\frac{1}{2}\rho v_0^2+\frac{7}{2}p\right)\mathbf{v}_0 v_{0r}\right)-\rho\mathbf{g}\cdot\mathbf{v}_0 v_{0r}-\left(\frac{1}{2}\rho v_0^2+\frac{3}{2}p\right)g_r\right]\right\}\right\}$$
$$+\frac{1}{r\sin\theta}\frac{\partial}{\partial\varphi}\left\{\left(\frac{1}{2}\rho v_0^2+\frac{5}{2}p\right)v_{0\varphi}-\tau\left[\frac{\partial}{\partial t}\left(\left(\frac{1}{2}\rho v_0^2+\frac{5}{2}p\right)v_{0\varphi}\right)+\frac{\partial}{\partial \mathbf{r}}\cdot\left(\left(\frac{1}{2}\rho v_0^2+\frac{7}{2}p\right)\mathbf{v}_0 v_{0\varphi}\right)\right.\right. \quad (A.2.115)$$
$$\left.\left.-\rho\mathbf{g}\cdot\mathbf{v}_0 v_{0\varphi}-\left(\frac{1}{2}\rho v_0^2+\frac{3}{2}p\right)g_\varphi\right]\right\}+\frac{1}{r\sin\theta}\frac{\partial}{\partial\theta}\left\{\sin\theta\left\{\left(\frac{1}{2}\rho v_0^2+\frac{5}{2}p\right)v_{0\theta}\right.\right.$$
$$\left.\left.-\tau\left[\frac{\partial}{\partial t}\left(\left(\frac{1}{2}\rho v_0^2+\frac{5}{2}p\right)v_{0\theta}\right)+\frac{\partial}{\partial \mathbf{r}}\cdot\left(\left(\frac{1}{2}\rho v_0^2+\frac{7}{2}p\right)\mathbf{v}_0 v_{0\theta}\right)-\rho\mathbf{g}\cdot\mathbf{v}_0 v_{0\theta}-\left(\frac{1}{2}\rho v_0^2+\frac{3}{2}p\right)g_\theta\right]\right\}\right\}$$
$$-\left\{\rho\left(g_r v_{0r}+g_\varphi v_{0\varphi}+g_\theta v_{0\theta}\right)-\tau\left[g_r\left(\frac{\partial}{\partial t}(\rho v_{or})+\frac{\partial}{\partial \mathbf{r}}\cdot(\rho\mathbf{v}_0 v_{or})+\frac{\partial p}{\partial r}-\rho g_r\right)\right.\right.$$
$$\left.\left.+g_\varphi\left(\frac{\partial}{\partial t}(\rho v_{o\varphi})+\frac{\partial}{\partial \mathbf{r}}\cdot(\rho\mathbf{v}_0 v_{o\varphi})+\frac{1}{r\sin\theta}\frac{\partial p}{\partial\varphi}-\rho g_\varphi\right)+g_\theta\left(\frac{\partial}{\partial t}(\rho v_{o\theta})+\frac{\partial}{\partial \mathbf{r}}\cdot(\rho\mathbf{v}_0 v_{o\theta})+\frac{1}{r}\frac{\partial p}{\partial\theta}-\rho g_\theta\right)\right]\right\}$$
$$-\operatorname{div}\left(\tau\operatorname{grad}\left(\frac{1}{2}pv_0^2+\frac{5}{2}\frac{p^2}{\rho}\right)\right)+\operatorname{div}(\tau p\mathbf{g})=0,$$

$$\frac{\partial}{\partial t}\left\{\frac{1}{2}\rho v_0^2+\frac{3}{2}p-\tau\left[\frac{\partial}{\partial t}\left(\frac{1}{2}\rho v_0^2+\frac{3}{2}p\right)+\frac{1}{r^2}\frac{\partial}{\partial r}\left(r^2 v_{0r}\left(\frac{1}{2}\rho v_0^2+\frac{5}{2}p\right)\right)\right.\right.$$
$$+\frac{1}{r\sin\theta}\frac{\partial}{\partial\varphi}\left(v_{0\varphi}\left(\frac{1}{2}\rho v_0^2+\frac{5}{2}p\right)\right)+\frac{1}{r\sin\theta}\frac{\partial}{\partial\theta}\left(\sin\theta v_{0\theta}\left(\frac{1}{2}\rho v_0^2+\frac{5}{2}p\right)\right)$$
$$\left.\left.-\rho\left(g_r v_{0r}+g_\varphi v_{0\varphi}+g_\theta v_{0\theta}\right)\right]\right\}+\frac{1}{r^2}\frac{\partial}{\partial r}\left\{r^2\left\{\left(\frac{1}{2}\rho v_0^2+\frac{5}{2}p\right)v_{0r}-\tau\left[\frac{\partial}{\partial t}\left(\left(\frac{1}{2}\rho v_0^2+\frac{5}{2}p\right)v_{0r}\right)\right.\right.\right.$$
$$+\frac{1}{r^2}\frac{\partial}{\partial r}\left(r^2\left(\frac{1}{2}\rho v_0^2+\frac{7}{2}p\right)v_{0r}^2\right)+\frac{1}{r\sin\theta}\frac{\partial}{\partial\varphi}\left(\left(\frac{1}{2}\rho v_0^2+\frac{7}{2}p\right)v_{0\varphi}v_{0r}\right)$$
$$\left.\left.\left.+\frac{1}{r\sin\theta}\frac{\partial}{\partial\theta}\left(\sin\theta\left(\frac{1}{2}\rho v_0^2+\frac{7}{2}p\right)v_{0\theta}v_{0r}\right)-\rho\left(g_r v_{0r}+g_\varphi v_{0\varphi}+g_\theta v_{0\theta}\right)v_{0r}-\left(\frac{1}{2}\rho v_0^2+\frac{3}{2}p\right)g_r\right]\right\}\right\}$$
$$+\frac{1}{r\sin\theta}\frac{\partial}{\partial\varphi}\left\{\left(\frac{1}{2}\rho v_0^2+\frac{5}{2}p\right)v_{0\varphi}-\tau\left[\frac{\partial}{\partial t}\left(\left(\frac{1}{2}\rho v_0^2+\frac{5}{2}p\right)v_{0\varphi}\right)+\frac{1}{r^2}\frac{\partial}{\partial r}\left(r^2\left(\frac{1}{2}\rho v_0^2+\frac{7}{2}p\right)v_{0r}v_{0\varphi}\right)\right.\right.$$
$$+\frac{1}{r\sin\theta}\frac{\partial}{\partial\varphi}\left(\left(\frac{1}{2}\rho v_0^2+\frac{7}{2}p\right)v_{0\varphi}^2\right)+\frac{1}{r\sin\theta}\frac{\partial}{\partial\theta}\left(\sin\theta\left(\frac{1}{2}\rho v_0^2+\frac{7}{2}p\right)v_{0\theta}v_{0\varphi}\right)$$
$$\left.\left.-\rho\left(g_r v_{0r}+g_\varphi v_{0\varphi}+g_\theta v_{0\theta}\right)v_{0\varphi}-\left(\frac{1}{2}\rho v_0^2+\frac{3}{2}p\right)g_\varphi\right]\right\}+\frac{1}{r\sin\theta}\frac{\partial}{\partial\theta}\left\{\sin\theta\left\{\left(\frac{1}{2}\rho v_0^2+\frac{5}{2}p\right)v_{0\theta}\right.\right.$$

$$
\begin{aligned}
&-\tau\left[\frac{\partial}{\partial t}\left(\left(\frac{1}{2}\rho v_0^2+\frac{5}{2}p\right)v_{0\theta}\right)+\frac{1}{r^2}\frac{\partial}{\partial r}\left(r^2\left(\frac{1}{2}\rho v_0^2+\frac{7}{2}p\right)v_{0r}v_{0\theta}\right)+\frac{1}{r\sin\theta}\frac{\partial}{\partial\varphi}\left(r^2\left(\frac{1}{2}\rho v_0^2+\frac{7}{2}p\right)v_{0\varphi}v_{0\theta}\right)\right.\\
&\left.+\frac{1}{r\sin\theta}\frac{\partial}{\partial\theta}\left(\sin\theta\left(\frac{1}{2}\rho v_0^2+\frac{7}{2}p\right)v_{0\theta}^2\right)-\rho\left(g_r v_{0r}+g_\varphi v_{0\varphi}+g_\theta v_{0\theta}\right)v_{0\theta}-\left(\frac{1}{2}\rho v_0^2+\frac{3}{2}p\right)g_\theta\right]\Bigg\}\Bigg\}\\
&-\Bigg\{\rho\left(g_r v_{0r}+g_\varphi v_{0\varphi}+g_\theta v_{0\theta}\right)-\\
&-\tau\left[g_r\left(\frac{\partial}{\partial t}(\rho v_{0r})+\frac{1}{r^2}\frac{\partial}{\partial r}(r^2\rho v_{0r}^2)+\frac{1}{r\sin\theta}\frac{\partial}{\partial\varphi}(\rho v_{0\varphi}v_{0r})+\frac{1}{r\sin\theta}\frac{\partial}{\partial\theta}(\rho v_{0\theta}v_{0r}\sin\theta)+\frac{\partial p}{\partial r}-\rho g_r\right)\right.\\
&+g_\varphi\left(\frac{\partial}{\partial t}(\rho v_{0\varphi})+\frac{1}{r^2}\frac{\partial}{\partial r}(r^2\rho v_{0r}v_{0\varphi})+\frac{1}{r\sin\theta}\frac{\partial}{\partial\varphi}\left(\rho v_{0\varphi}^2\right)+\frac{1}{r\sin\theta}\frac{\partial}{\partial\theta}(\rho v_{0\theta}v_{0\varphi}\sin\theta)\right.\\
&\left.+\frac{1}{r\sin\theta}\frac{\partial p}{\partial\varphi}-\rho g_\varphi\right)+g_\theta\left(\frac{\partial}{\partial t}(\rho v_{0\theta})+\frac{1}{r^2}\frac{\partial}{\partial r}(r^2\rho v_{0r}v_{0\theta})+\frac{1}{r\sin\theta}\frac{\partial}{\partial\varphi}(\rho v_{0\varphi}v_{0\theta})\right.\\
&\left.\left.+\frac{1}{r\sin\theta}\frac{\partial}{\partial\theta}(\rho v_{0\theta}^2\sin\theta)+\frac{1}{r}\frac{\partial p}{\partial\theta}-\rho g_\theta\right)\right]\Bigg\}-\frac{1}{r^2}\frac{\partial}{\partial r}\left(\tau r^2\frac{\partial}{\partial r}\left(\frac{1}{2}pv_0^2+\frac{5}{2}\frac{p^2}{\rho}\right)\right)\\
&-\frac{1}{r^2\sin\theta}\frac{\partial}{\partial\theta}\left(\tau\sin\theta\frac{\partial}{\partial\theta}\left(\frac{1}{2}pv_0^2+\frac{5}{2}\frac{p^2}{\rho}\right)\right)-\frac{1}{r^2\sin^2\theta}\frac{\partial}{\partial\varphi}\left(\tau\frac{\partial}{\partial\varphi}\left(\frac{1}{2}pv_0^2+\frac{5}{2}\frac{p^2}{\rho}\right)\right)\\
&+\frac{1}{r^2}\frac{\partial}{\partial r}(r^2\tau p g_r)+\frac{1}{r\sin\theta}\frac{\partial}{\partial\varphi}\left(\tau p g_\varphi\right)+\frac{1}{r\sin\theta}\frac{\partial}{\partial\theta}(\tau p g_\theta\sin\theta)=0
\end{aligned}
\tag{A.2.116}
$$

Energy equation for the case of the spherical symmetry:

$$
\begin{aligned}
&\frac{\partial}{\partial t}\left\{\frac{1}{2}\rho v_{0r}^2+\frac{3}{2}p-\tau\left[\frac{\partial}{\partial t}\left(\frac{1}{2}\rho v_{0r}^2+\frac{3}{2}p\right)+\frac{1}{r^2}\frac{\partial}{\partial r}\left(r^2 v_{0r}\left(\frac{1}{2}\rho v_{0r}^2+\frac{5}{2}p\right)\right)-\rho g_r v_{0r}\right]\right\}\\
&+\frac{1}{r^2}\frac{\partial}{\partial r}\left\{r^2\left\{\left(\frac{1}{2}\rho v_{0r}^2+\frac{5}{2}p\right)v_{0r}-\tau\left[\frac{\partial}{\partial t}\left(\left(\frac{1}{2}\rho v_{0r}^2+\frac{5}{2}p\right)v_{0r}\right)+\frac{1}{r^2}\frac{\partial}{\partial r}\left(r^2\left(\frac{1}{2}\rho v_{0r}^2+\frac{7}{2}p\right)v_{0r}^2\right)\right.\right.\right.\\
&\left.\left.\left.-\rho g_r v_{0r}^2-\left(\frac{1}{2}\rho v_{0r}^2+\frac{3}{2}p\right)g_r\right]\right\}\right\}-\left\{\rho g_r v_{0r}-\tau\left[g_r\left(\frac{\partial}{\partial t}(\rho v_{0r})+\frac{1}{r^2}\frac{\partial}{\partial r}(r^2\rho v_{0r}^2)+\frac{\partial p}{\partial r}-\rho g_r\right)\right]\right\}\\
&-\frac{1}{r^2}\frac{\partial}{\partial r}\left(\tau r^2\frac{\partial}{\partial r}\left(\frac{1}{2}pv_{0r}^2+\frac{5}{2}\frac{p^2}{\rho}\right)\right)+\frac{1}{r^2}\frac{\partial}{\partial r}(r^2\tau p g_r)=0.
\end{aligned}
\tag{A.2.117}
$$

By the solution of the self-consistent problem in the potential field we use the transformation

$$
\mathbf{g}=-\frac{\partial\psi}{\partial\mathbf{r}},
\tag{A.2.118}
$$

where ψ defined by the Poisson equation

$$
\Delta\psi=4\pi\gamma_N\left[\rho-\tau\left(\frac{\partial\rho}{\partial t}+\frac{\partial}{\partial\mathbf{r}}\cdot\rho\mathbf{v}_0\right)\right].
\tag{A.2.119}
$$

Write down equations Eq. (A.4.118) and Eq. (A.2.119) in the spherical coordinate system:

$$
\mathbf{g}=-\frac{\partial\psi}{\partial r}\mathbf{e}_r-\frac{1}{r\sin\theta}\frac{\partial\psi}{\partial\varphi}\mathbf{e}_\varphi-\frac{1}{r}\frac{\partial\psi}{\partial\theta}\mathbf{e}_\theta,
\tag{A.2.120}
$$

$$\frac{1}{r^2}\frac{\partial}{\partial r}\left(r^2\frac{\partial\psi}{\partial r}\right)+\frac{1}{r^2\sin\theta}\frac{\partial}{\partial\theta}\left(\sin\theta\frac{\partial\psi}{\partial\theta}\right)+\frac{1}{r^2\sin^2\theta}\frac{\partial^2\psi}{\partial\varphi^2}$$
$$=4\pi\gamma_N\left[\rho-\tau\left(\frac{\partial\rho}{\partial t}+\frac{1}{r^2}\frac{\partial(r^2\rho v_{0r})}{\partial r}+\frac{1}{r\sin\theta}\frac{\partial(\rho v_{0\varphi})}{\partial\varphi}+\frac{1}{r\sin\theta}\frac{\partial(\rho v_{0\theta}\sin\theta)}{\partial\theta}\right)\right] \tag{A.2.121}$$

For the problem with the spherical symmetry from Eqs. (A.2.120) and (A.2.121) follow

$$g_r=-\frac{\partial\psi}{\partial r}, \tag{A.2.122}$$

$$\frac{1}{r^2}\frac{\partial}{\partial r}\left(r^2\frac{\partial\psi}{\partial r}\right)=4\pi\gamma_N\left[\rho-\tau\left(\frac{\partial\rho}{\partial t}+\frac{1}{r^2}\frac{\partial(r^2\rho v_{0r})}{\partial r}\right)\right]. \tag{A.2.123}$$

Finally the set Eq. (A.2.124) of the non-local equations defining the evolution of the physical system in the gravitational field with the spherical symmetry takes the form:

$$g_r=-\frac{\partial\psi}{\partial r}.$$

Poisson equation

$$\frac{1}{r^2}\frac{\partial}{\partial r}\left(r^2\frac{\partial\psi}{\partial r}\right)=4\pi\gamma_N\left[\rho-\tau\left(\frac{\partial\rho}{\partial t}+\frac{1}{r^2}\frac{\partial(r^2\rho v_{0r})}{\partial r}\right)\right].$$

Continuity equation

$$\frac{\partial}{\partial t}\left\{\rho-\tau\left[\frac{\partial\rho}{\partial t}+\frac{1}{r^2}\frac{\partial(r^2\rho v_{0r})}{\partial r}\right]\right\}+\frac{1}{r^2}\frac{\partial}{\partial r}\left\{r^2\left\{\rho v_{0r}-\tau\left[\frac{\partial}{\partial t}(\rho v_{0r})+\frac{1}{r^2}\frac{\partial(r^2\rho v_{0r}^2)}{\partial r}-\rho g_r\right]\right\}\right\}-\frac{1}{r^2}\frac{\partial}{\partial r}\left(\tau r^2\frac{\partial p}{\partial r}\right)=0.$$

Momentum equation

$$\frac{\partial}{\partial t}\left\{\rho v_{0r}-\tau\left[\frac{\partial}{\partial t}(\rho v_{0r})+\frac{1}{r^2}\frac{\partial(r^2\rho v_{0r}^2)}{\partial r}+\frac{\partial p}{\partial r}-\rho g_r\right]\right\}-g_r\left[\rho-\tau\left(\frac{\partial\rho}{\partial t}+\frac{1}{r^2}\frac{\partial(r^2\rho v_{0r})}{\partial r}\right)\right]$$
$$+\frac{1}{r^2}\frac{\partial}{\partial r}\left\{r^2\left\{\rho v_{0r}^2-\tau\left[\frac{\partial}{\partial t}(\rho v_{0r}^2)+\frac{1}{r^2}\frac{\partial(r^2\rho v_{0r}^3)}{\partial r}-2g_r\rho v_{0r}\right]\right\}\right\}$$
$$+\frac{\partial p}{\partial r}-\frac{\partial}{\partial r}\left(\tau\frac{\partial p}{\partial t}\right)-2\frac{\partial}{\partial r}\left(\frac{\tau}{r^2}\frac{\partial(r^2 p v_{0r})}{\partial r}\right)-\frac{1}{r^2}\frac{\partial}{\partial r}\left(\tau r^2\frac{\partial(p v_{0r})}{\partial r}\right)=0.$$

Energy equation

$$\frac{\partial}{\partial t}\left\{\frac{1}{2}\rho v_{0r}^2+\frac{3}{2}p-\tau\left[\frac{\partial}{\partial t}\left(\frac{1}{2}\rho v_{0r}^2+\frac{3}{2}p\right)+\frac{1}{r^2}\frac{\partial}{\partial r}\left(r^2 v_{0r}\left(\frac{1}{2}\rho v_{0r}^2+\frac{5}{2}p\right)\right)-\rho g_r v_{0r}\right]\right\}$$
$$+\frac{1}{r^2}\frac{\partial}{\partial r}\left\{r^2\left\{\left(\frac{1}{2}\rho v_{0r}^2+\frac{5}{2}p\right)v_{0r}-\tau\left[\frac{\partial}{\partial t}\left(\left(\frac{1}{2}\rho v_{0r}^2+\frac{5}{2}p\right)v_{0r}\right)+\frac{1}{r^2}\frac{\partial}{\partial r}\left(r^2\left(\frac{1}{2}\rho v_{0r}^2+\frac{7}{2}p\right)v_{0r}^2\right)\right.\right.\right.$$
$$\left.\left.\left.-\rho g_r v_{0r}^2-\left(\frac{1}{2}\rho v_{0r}^2+\frac{3}{2}p\right)g_r\right]\right\}\right\}-\left\{\rho g_r v_{0r}-\tau\left[g_r\left(\frac{\partial}{\partial t}(\rho v_{0r})+\frac{1}{r^2}\frac{\partial}{\partial r}(r^2\rho v_{0r}^2)+\frac{\partial p}{\partial r}-\rho g_r\right)\right]\right\}$$
$$-\frac{1}{r^2}\frac{\partial}{\partial r}\left(\tau r^2\frac{\partial}{\partial r}\left(\frac{1}{2}p v_{0r}^2+\frac{5}{2}\frac{p^2}{\rho}\right)\right)+\frac{1}{r^2}\frac{\partial}{\partial r}(r^2\tau p g_r)=0 \tag{A.2.124}$$

Appendix 3

Characteristic Scales in Plasma Physics

The fundamental feature of plasma physics is the existence of a multiparticle interaction in the system under study. Consequently, care must be exercised when choosing typical scales for describing the evolution of a plasma volume. The Landau length l over which the characteristic kinetic energy $k_B T$ of thermal motion equals the potential energy of interaction between charges e is determined by the relation

$$l = \frac{e^2}{4\pi\varepsilon_0 k_B T} = \frac{1.67 \cdot 10^{-5}}{T}\,\mathrm{m}, \tag{A.3.1}$$

where ε_0 is the electric constant and k_B is the Boltzmann constant.

Binary collisions for which impact parameters are less than or equal to the Landau length are said to be "close." It is useful to introduce the ratio of the Landau length to the average distance $n^{-1/3}$ between plasma particles:

$$\beta = l n^{1/3} = 1.67 \cdot 10^{-5} n^{1/3} T^{-1}, \tag{A.3.2}$$

where n is the number density of particles, in m^{-3}. While the interaction parameter β is usually small in laboratory plasma, the solar corona, and the solar atmosphere, in the ionosphere and interstellar gas, but for free electrons in a metal it can reach a value of $\approx 10^2$. The cross section σ_b for close collisions is determined by the relation

$$\sigma_b = \pi l^2, \tag{A.3.3}$$

and the mean free path of a probe particle between binary close collisions is

$$\lambda = \frac{1}{\pi n l^2} = \frac{1}{\pi n^{1/3} \beta^2} = 1.1 \cdot 10^9 \frac{T^2}{n}\,\mathrm{m}. \tag{A.3.4}$$

The pair interaction between particles in plasma effectively extends to the distance determined by the Debye–Hückel radius r_D defined as follows

$$r_D = \sqrt{\frac{\varepsilon_0 k_B T}{n e^2}} = \frac{1}{n^{1/3}\sqrt{4\pi\beta}} = 0.69 \cdot 10^2 \sqrt{\frac{T}{n}}\,\mathrm{m}. \tag{A.3.5}$$

It can be argued that collective plasma properties disappear in systems less than r_D in size. The following relationship between the characteristic plasma lengths should be noted:

$$l : n^{-1/3} : r_D : \lambda = \beta : 1 : \frac{1}{2\sqrt{\pi\beta}} : \frac{1}{\pi\beta^2}. \tag{A.3.6}$$

Equation (A.3.6) should be complemented by the hydrodynamic scale L being the characteristic size of the system; L is usually much larger than λ.

The above list of characteristic plasma scales is not exhaustive, though for processes in rapidly alternating fields, when the distance a particle travels over a period of oscillation of the field is less than the range of the forces involved. Additional scales may appear in the problem.

Appendix 4

Dispersion Relations in the Generalized Boltzmann Kinetic Theory Neglecting the Integral Collision Term

We are concerned with developing (within the GBE framework) the dispersion relation for plasma in the absence of a magnetic field. We make the same assumptions used in developing this relation within the BE model, namely:

(a) the integral collision term is neglected;
(b) the evolution of electrons and ions in a self-consistent electric field corresponds to a one-dimensional, unsteady model;
(c) distribution functions for ions f_i and electrons f_e deviate only slightly from their respective equilibrium values f_{0i} and f_{0e}:

$$f_i = f_{0i}(u) + \delta f_i(x, u, t), \tag{A.4.1}$$

$$f_e = f_{0e}(u) + \delta f_e(x, u, t); \tag{A.4.2}$$

(d) we consider a wave mode corresponding to a certain wave number k and a complex frequency ω, so that the solution of the GBE can be written in the form:

$$\delta f_i = \langle \delta f_i \rangle e^{i(kx-\omega t)}, \tag{A.4.3}$$

$$\delta f_e = \langle \delta f_e \rangle e^{i(kx-\omega t)}; \tag{A.4.4}$$

(e) the quadratic terms in the GBE, determining the deviation from the equilibrium DFs, are neglected, and
(f) the self-consistent forces F_i and F_e are small:

$$F_i = -\frac{e}{m_i}\frac{\partial \psi}{\partial x}, \tag{A.4.5}$$

$$F_e = \frac{e}{m_e}\frac{\partial \psi}{\partial x}, \tag{A.4.6}$$

where e is the absolute electron charge, m_i are the ion masses, m_e the electron mass, and finally

$$\psi = \langle \psi \rangle e^{i(kx-\omega t)}. \tag{A.4.7}$$

Under these assumptions, the GBE is written as follows (we seek the solution for the ion plasma component, to be specific):

$$\begin{aligned} &\frac{\partial f_i}{\partial t} + u\frac{\partial f_i}{\partial x} + F_i\frac{\partial f_i}{\partial u} - \tau_i\left\{\frac{\partial^2 f_i}{\partial t^2} + 2u\frac{\partial^2 f_i}{\partial t \partial x} + u^2\frac{\partial^2 f_i}{\partial x^2} + 2F_i\frac{\partial^2 f_i}{\partial t \partial u}\right. \\ &\left. + \frac{\partial F_i}{\partial t}\frac{\partial f_i}{\partial u} + F_i\frac{\partial f_i}{\partial x} + u\frac{\partial F_i}{\partial x}\frac{\partial f_i}{\partial u} + F_i^2\frac{\partial^2 f_i}{\partial u^2} + 2uF_i\frac{\partial^2 f_i}{\partial u \partial x}\right\} = 0. \end{aligned} \tag{A.4.8}$$

Using the assumptions listed above, we find the relations

$$\begin{aligned}
&\frac{\partial f_{\mathrm{i}}}{\partial t}=-\mathrm{i}\omega\delta f_{\mathrm{i}},\quad u\frac{\partial f_{\mathrm{i}}}{\partial x}=\mathrm{i}ku\delta f_{\mathrm{i}},\quad F_{\mathrm{i}}\frac{\partial f_{\mathrm{i}}}{\partial u}=-\frac{e}{m_{\mathrm{i}}}\frac{\partial\psi}{\partial x}\frac{\partial f_{0\mathrm{i}}}{\partial u},\\
&\frac{\partial^2 f_{\mathrm{i}}}{\partial t^2}=-\omega^2\delta f_{\mathrm{i}},\quad 2u\frac{\partial^2 f_{\mathrm{i}}}{\partial t\partial x}=2\omega uk\delta f_{\mathrm{i}},\quad u^2\frac{\partial^2 f_{\mathrm{i}}}{\partial x^2}=-u^2k^2\delta f_{\mathrm{i}},\\
&2F_{\mathrm{i}}\frac{\partial^2 f_{\mathrm{i}}}{\partial u\partial t}=0,\quad \frac{\partial F_{\mathrm{i}}}{\partial t}\frac{\partial f_{\mathrm{i}}}{\partial u}=-\frac{e}{m_{\mathrm{i}}}\omega k\psi\frac{\partial f_{0\mathrm{i}}}{\partial u},\quad F_{\mathrm{i}}\frac{\partial f_{\mathrm{i}}}{\partial x}=0,\\
&u\frac{\partial f_{\mathrm{i}}}{\partial u}\frac{\partial F_{\mathrm{i}}}{\partial x}=\frac{e}{m_{\mathrm{i}}}k^2u\psi\frac{\partial f_{0\mathrm{i}}}{\partial u},\quad F_{\mathrm{i}}^2\frac{\partial^2 f_{\mathrm{i}}}{\partial u^2}=0,\quad \frac{\partial^2 f_{\mathrm{i}}}{\partial u\partial x}2uF_{\mathrm{i}}=0,
\end{aligned}\tag{A.4.9}$$

which when substituted into Eq. (A.4.8) yield

$$\mathrm{i}(ku-\omega)\langle\delta f_{\mathrm{i}}\rangle-\mathrm{i}\frac{e}{m_{\mathrm{i}}}k\langle\psi\rangle\frac{\partial f_{0\mathrm{i}}}{\partial u}-(ku-\omega)\tau_{\mathrm{i}}\left\{-(ku-\omega)\langle\delta f_{\mathrm{i}}\rangle+\langle\psi\rangle\frac{ek}{m_{\mathrm{i}}}\frac{\partial f_{0\mathrm{i}}}{\partial u}\right\}=0,\tag{A.4.10}$$

giving the ion density fluctuation

$$\langle\delta n_{\mathrm{i}}\rangle=-\frac{e}{m_{\mathrm{i}}}\langle\psi\rangle k\int\frac{\partial f_{0\mathrm{i}}/\partial u}{\omega-ku}du\tag{A.4.11}$$

and the electron density fluctuation

$$\langle\delta n_{\mathrm{e}}\rangle=\frac{e}{m_{\mathrm{e}}}\langle\psi\rangle k\int\frac{\partial f_{0\mathrm{e}}/\partial u}{\omega-ku}du.\tag{A.4.12}$$

Equations (A.4.11) and (A.4.12) are identical to their BE analogues. Substituting Eqs. (A.4.11) and (A.4.12) into the Poisson equation

$$\varepsilon_0 k^2\psi=e(\delta n_{\mathrm{i}}-\delta n_{\mathrm{e}})\tag{A.4.13}$$

we arrive at the classical dispersion relation (see, for instance, Ref. [144])

$$1=-\frac{e^2}{\varepsilon_0 k}\left\{\frac{1}{m_{\mathrm{e}}}\int_{-\infty}^{+\infty}\frac{\partial f_{0\mathrm{e}}/\partial u}{\omega-ku}du+\frac{1}{m_{\mathrm{i}}}\int_{-\infty}^{+\infty}\frac{\partial f_{0\mathrm{i}}/\partial u}{\omega-ku}du\right\}.\tag{A.4.14}$$

Although Eqs. (A.4.11) and (A.4.12) are a consequence of the general statement that in the absence of the integral collision term the relation

$$\frac{\mathrm{D}f_\alpha}{\mathrm{D}t}=0\tag{A.4.15}$$

(the Vlasov equation) is the solution of the equation

$$\frac{\mathrm{D}f_\alpha}{\mathrm{D}t}-\frac{\mathrm{D}}{\mathrm{D}t}\left(\tau_\alpha\frac{\mathrm{D}f_\alpha}{\mathrm{D}t}\right)=0.\tag{A.4.16}$$

The above argument shows that the GBE can produce correct and expected results, when treated perturbatively.

Appendix 5

Three-Diagonal Method of Gauss Elimination Techniques for the Differential Third- and Second-Order Equations

Let us consider a highly effective numerical method for the solution of boundary problems described by ordinary differential equation of the third order. This method is known in the West as three-diagonal method of Gauss elimination techniques, and in the East as "progonka" ("sweep" method). "Progonka" was applied in the theory of the gas dynamic boundary layer with chemical reactions as early as in the beginning of 1960s (see for example [258]) and introduced by the author as the iterative procedure of suppression of arising oscillations and is now a universally adopted element of numerical solution of differential equations.

The aim of this Appendix is to show how to construct the "progonka" method for the differential equations. We begin with the third-order equation

$$b_1 y''' + b_2 y'' + b_3 y' + b_4 y + b_5 = 0 \tag{A.5.1}$$

It is assumed that coefficients $b_i (i=1,\ldots,5)$ are functions only of the independent variable x; in the following, we intend to discuss how to avoid this restriction.

As an example, the boundary conditions are chosen in the form

$$\begin{aligned} x=a: y=\alpha, \quad y' = \beta \\ x=b: y=\gamma \end{aligned} \tag{A.5.2}$$

The second-order scheme is introduced using a uniform partition of the interval $[a,b]$ with the mesh width h ($x_0=a,\ldots, x_n=b$), $x_k=x_0+kh$, $k=0,\ldots,n$, and

$$y_k' = \frac{y_{k+1}-y_{k-1}}{2h}, \tag{A.5.3}$$

$$y_k'' = \frac{y_{k+1}-2y_k+y_{k-1}}{h^2}, \tag{A.5.4}$$

$$y_k''' = \frac{y_{k+2}-2y_{k+1}+2y_{k-1}-y_{k-2}}{2h^3}. \tag{A.5.5}$$

Using Eqs. (A.5.1)–(A.5.5), we find

$$b_{1k}\frac{y_{k+2}-2y_{k+1}+2y_{k-1}-y_{k-2}}{2h^3} + b_{2k}\frac{y_{k+1}-2y_k+y_{k-1}}{h^2} + b_{3k}\frac{y_{k+1}-y_{k-1}}{2h} + b_{4k}y_k + b_{5k} = 0, \tag{A.5.6}$$

and for $k=2,\ldots,n-2$,

$$y_k = y_{k-2}\frac{-b_{1k}}{4hb_{2k}-2h^3b_{4k}} + y_{k-1}\frac{2b_{1k}+2hb_{2k}-b_{3k}h^2}{4hb_{2k}-2h^3b_{4k}} + y_{k+1}\frac{-2b_{1k}+2hb_{2k}-b_{3k}h^2}{4hb_{2k}-2h^3b_{4k}}$$
$$+ y_{k+2}\frac{b_{1k}}{4hb_{2k}-2h^3b_{4k}} + \frac{2h^3b_{5k}}{4hb_{2k}-2h^3b_{4k}}. \tag{A.5.7}$$

Newton's formula leads to the following approximation for function

$$y(x_m - th) = y(x_m) - \frac{t}{1!}\Delta y_m + \frac{t(t-1)}{2!}\Delta^2 y_m$$
$$-\frac{t(t-1)(t-2)}{3!}\Delta^3 y_m + \frac{t(t-1)(t-2)(t-3)}{4!}\Delta^4 y_m + R_5; \tag{A.5.8}$$

in the point $x_m - th$, where t is an integer or fractional number, finite differences can be calculated as

$$\begin{aligned}
\Delta y_m &= y_m - y_{m-1},\\
\Delta y_{m-1} &= y_{m-1} - y_{m-2},\\
\Delta y_{m-2} &= y_{m-2} - y_{m-3},\\
\Delta y_{m-4} &= y_{m-4} - y_{m-5},\\
\Delta^2 y_m &= \Delta y_m - \Delta y_{m-1} = y_m - 2y_{m-1} + y_{m-2},\\
\Delta^2 y_{m-1} &= \Delta y_{m-1} - \Delta y_{m-2} = y_{m-1} - 2y_{m-2} + y_{m-3},\\
\Delta^3 y_m &= \Delta^2 y_m - \Delta^2 y_{m-1} = y_m - 3y_{m-1} + 3y_{m-2} - y_{m-3},\\
\Delta^3 y_{m-1} &= \Delta^2 y_{m-1} - \Delta^2 y_{m-2} = y_{m-1} - 3y_{m-2} + 3y_{m-3} - y_{m-4},\\
\Delta^4 y_m &= \Delta^3 y_m - \Delta^3 y_{m-1} = y_m - 4y_{m-1} + 6y_{m-2} - 4y_{m-3} + y_{m-4}.
\end{aligned} \tag{A.5.9}$$

Let us introduce the variable $x = x_m - th$, $t = \frac{(x_m - x)}{h}$, and $t_x' = -h^{-1}$. After differentiating Eq. (A.5.8) with respect to x, we find

$$\begin{aligned}
y_x' &= \Delta y t_x' + \frac{2t-1}{2!}\Delta^2 y t_x',\\
y_x'' &= \Delta^2 y (t_x')^2 - \frac{6t-6}{3!}\Delta^3 y (t_x')^2 + \frac{12t^2 - 36t + 22}{4!}\Delta^4 y (t_x')^2,\\
y_x''' &= -\Delta^3 y (t_x')^3 + \frac{24t-36}{4!}\Delta^4 y (t_x')^3,
\end{aligned} \tag{A.5.10}$$

or for the point x_{m-1}:

$$y_x'(x_{m-1}) = \frac{y_m - y_{m-2}}{2h}, \tag{A.5.11}$$

$$y_x''(x_{m-1}) = \frac{1}{h^2}(\Delta^2 y)_m - \frac{1}{12h^2}(\Delta^4 y)_m = \frac{1}{h^2}\left[y_m - 2y_{m-1} + y_{m-2} - \frac{1}{12}y_m\right.$$
$$\left.+\frac{1}{3}y_{m-1} - \frac{1}{2}y_{m-2} + \frac{1}{3}y_{m-3} - \frac{1}{12}y_{m-4}\right] = \frac{1}{h^2}\left[\frac{11}{12}y_m - \frac{5}{3}y_{m-1} + \frac{1}{2}y_{m-2} + \frac{1}{3}y_{m-3} - \frac{1}{12}y_{m-4}\right], \tag{A.5.12}$$

$$y_x'''(x_{m-1}) = \frac{1}{h^3}\left[\frac{3}{2}y_m - 5y_{m-1} + 6y_{m-2} - 3y_{m-3} + \frac{1}{2}y_{m-4}\right] \tag{A.5.13}$$

After substitution (Eqs. A.5.11–A.5.13) in (A.5.1) we have:

$$\begin{aligned}
&y_m\left[18b_{1,m-1} + 11hb_{2,m-1} + 6h^2b_{3,m-1}\right] + y_{m-1}\left[-60b_{1,m-1} - 20hb_{2,m-1} + 12h^3b_{4,m-1}\right]\\
&+ y_{m-2}\left[72b_{1,m-1} + 6hb_{2,m-1} - 6h^2b_{3,m-1}\right] + y_{m-3}\left[-36b_{1,m-1} + 4hb_{2,m-1}\right]\\
&+ y_{m-4}\left[6b_{1,m-1} - hb_{2,m-1}\right] + 12h^3b_{5,m-1} = 0.
\end{aligned} \tag{A.5.14}$$

Three-point difference equation is written as

$$y_k = a_1^{(k)} y_{k+1} + a_2^{(k)} y_{k+2} + a_3^{(k)}, \tag{A.5.15}$$

where $a_1^{(k)}$, $a_2^{(k)}$, $a_3^{(k)}$ are coefficients ("progonka" coefficients), which should be calculated in every point k. On the left boundary ($k=0$), we have $y_0=\alpha$, then

$$\begin{gathered} y_0 = a_1^{(0)} y_1 + a_2^{(0)} y_2 + a_3^{(0)}, \\ a_1^{(0)} = 0, \ \ a_2^{(0)} = 0, \ \ a_3^{(0)} = \alpha \end{gathered} \tag{A.5.16}$$

For $k=1$ ($m=2$) from Eqs. (A.5.9), (A.5.10) is found:

$$y_0' = \frac{-y_2 + 4y_1 - 3y_0}{2h}, \tag{A.5.17}$$

or

$$y_1 = \frac{1}{4} y_2 + \frac{1}{4}(2h\beta + 3\alpha).$$

Then,

$$a_1^{(1)} = \frac{1}{4};\ a_2^{(1)} = 0;\ a_3^{(1)} = \frac{1}{4}(3\alpha + 2h\beta). \tag{A.5.18}$$

Let us find the values y_{k+1} and y_{k-2} using the corresponding "progonka" coefficients and y_{k+1}, y_{k+2}. After substituting the expressions y_{k-1} and y_{k-2}, we find using Eqs. (A.5.7) and (A.5.15)

$$y_{k-1} = a_1^{(k-1)} a_1^{(k)} y_{k+1} + a_1^{(k-1)} a_2^{(k)} y_{k+2} + a_1^{(k-1)} a_3^{(k)} + a_2^{(k-1)} y_{k+1} + a_3^{(k-1)}, \tag{A.5.19}$$

$$\begin{aligned} y_{k-2} = {} & y_{k+1}\left(a_1^{(k-2)} a_1^{(k-1)} a_1^{(k)} + a_1^{(k-2)} a_2^{(k-1)} + a_2^{(k-2)} a_1^{(k)}\right) + y_{k+2}\left(a_1^{(k-2)} a_1^{(k-1)} a_2^{(k)} + a_2^{(k-2)} a_2^{(k)}\right) \\ & + a_1^{(k-2)} a_3^{(k-1)} + a_2^{(k-2)} a_3^{(k)} + a_3^{(k-2)} + a_1^{(k-2)} a_1^{(k-1)} a_3^{(k)}, \end{aligned} \tag{A.5.20}$$

$$\begin{aligned} y_k = {} & y_{k+1}\left[\frac{-b_{1k}\left(a_1^{(k-2)} a_1^{(k-1)} a_1^{(k)} + a_1^{(k-2)} a_2^{(k-1)} + a_2^{(k-2)} a_1^{(k)}\right)}{4hb_{2k} - 2h^3 b_{4k}} + \frac{\left(a_1^{(k-1)} a_1^{(k)} + a_2^{(k-1)}\right)\left(2b_{1k} + 2hb_{2k} - b_{3k}h^2\right) + \left(-2b_{1k} + 2hb_{2k} + b_{3k}h^2\right)}{4hb_{2k} - 2h^3 b_{4k}}\right] \\ & + y_{k+2}\frac{-b_{1k}\left(a_1^{(k-2)} a_1^{(k-1)} a_2^{(k)} + a_2^{(k-2)} a_2^{(k)}\right) + a_1^{(k-1)} a_2^{(k)}\left(2b_{1k} + 2hb_{2k} - b_{3k}h^2\right) + b_{1k}}{4hb_{2k} - 2h^3 b_{4k}} \\ & + \frac{2h^3 b_{5k} - b_{1k}\left(a_1^{(k-2)} a_1^{(k-1)} a_3^{(k)} + a_1^{(k-2)} a_3^{(k-1)} + a_2^{(k-2)} a_3^{(k)} + a_3^{(k-2)}\right)}{4hb_{2k} - 2h^3 b_{4k}} + \frac{\left(a_1^{(k-1)} a_3^{(k)} + a_3^{(k-1)}\right)\left(2b_{1k} + 2hb_{2k} - b_{3k}h^2\right)}{4hb_{2k} - 2h^3 b_{4k}}. \end{aligned} \tag{A.5.21}$$

But relation (A.5.21) is identity and as a result the following formulae are valid

$$a_1^{(k)} = \frac{-M_k a_2^{(k-1)} + \left(-2b_{1k} + 2hb_{2k} + b_{3k}h^2\right)}{M_k a_1^{(k-1)} + b_{1k} a_2^{(k-2)} + 4hb_{2k} - 2h^3 b_{4k}}, \tag{A.5.22}$$

$$a_2^{(k)} = \frac{b_{1k}}{M_k a_1^{(k-1)} + b_{1k} a_2^{(k-2)} + 4hb_{2k} - 2h^3 b_{4k}}, \tag{A.5.23}$$

$$a_3^{(k)} = \frac{-M_k a_3^{(k-1)} - b_{1k} a_3^{(k-2)} + 2h^3 b_{5k}}{M_k a_1^{(k-1)} + b_{1k} a_2^{(k-2)} + 4hb_{2k} - 2h^3 b_{4k}}, \tag{A.5.24}$$

where

$$M_k = -2b_{1k} - 2hb_{2k} + h^2 b_{3k} + b_{1k} a_1^{(k-2)}.$$

On the right side of the interval can be written the system of two equations for definition of y_{n-1}, y_{n-2}. Really, the first equation of this system is

$$y_{n-2} = a_1^{(n-2)} y_{n-1} + a_2^{(n-2)} y_n + a_3^{(n-2)}, \tag{A.5.25}$$

and $y_n = \gamma$.

But from Eq. (A.5.14) follows

$$\begin{aligned} &y_n\left[18b_{1,n-1} + 11hb_{2,n-1} + 6h^2 b_{3,n-1}\right] + y_{n-1}\left[-60b_{1,n-1} - 20hb_{2,n-1} + 12h^3 b_{4,n-1}\right] \\ &+ y_{n-2}\left[72b_{1,n-1} + 6hb_{2,n-1} - 6h^2 b_{3,n-1}\right] + y_{n-3}\left[-36b_{1,n-1} + 4hb_{2,n-1}\right] \\ &+ y_{n-4}\left[6b_{1,n-1} - hb_{2,n-1}\right] + 12h^3 b_{5,n-1} = 0, \end{aligned} \tag{A.5.26}$$

and

$$y_{n-3} = a_1^{(n-3)} y_{n-3} + a_2^{(n-3)} y_{n-2} + a_3^{(n-3)} \tag{A.5.27}$$

$$y_{n-4} = a_1^{(n-4)} y_{n-3} + a_2^{(n-4)} y_{n-2} + a_3^{(n-4)}. \tag{A.5.28}$$

Substitution (Eqs. A.5.27 and A.5.28) in Eq. (A.5.26) leads to the second equation we need:

$$Ay_{n-2} = By_{n-1} + C, \tag{A.5.29}$$

$$\begin{aligned} A = {} & 72b_{1,n-1} + 6hb_{2,n-1} - 6h^2 b_{3,n-1} - 36b_{1,n-1} a_1^{(n-3)} + \\ & + 4hb_{2,n-1} a_1^{(n-3)} + \left(a_1^{(n-4)} a_1^{(n-3)} + a_2^{(n-4)}\right)(6b_{1,n-1} - hb_{2,n-1}), \end{aligned} \tag{A.5.30}$$

$$\begin{aligned} B = {} & 60b_{1,n-1} + 20hb_{2,n-1} - 12h^3 b_{4,n-1} + a_2^{(n-3)}(36b_{1,n-1} - 4hb_{2,n-1}) \\ & - a_1^{(n-4)} a_2^{(n-3)}(6b_{1,n-1} - hb_{2,n-1}), \end{aligned} \tag{A.5.31}$$

$$\begin{aligned} C = {} & -\gamma\left(18b_{1,n-1} + 11hb_{2,n-1} + 6h^2 b_{3,n-1}\right) + a_3^{(n-3)}(36b_{1,n-1} - 4hb_{2,n-1}) \\ & - \left(a_1^{(n-4)} a_3^{(n-3)} + a_3^{(n-4)}\right)(6b_{1,n-1} - hb_{2,n-1}) - 12h^3 b_{5,n-1}. \end{aligned} \tag{A.5.32}$$

Practical application of the "progonka" method can be realized as follows:

(a) From relations (A.5.16), (A.5.18), and (A.5.22)–(A.5.24) the Gauss ("progonka") coefficients should be found for $0 \le k \le n$.
(b) From the system of two Eqs. (A.5.25), (A.5.29) we find y_{n-1}, y_{n-2}
(c) From the equation

$$y_{k-2} = a_1^{(k-2)} + a_2^{(k-2)} y_k + a_3^{(k-2)}, \tag{A.5.33}$$

(backward "progonka") we find the rest discrete functions $3 \le k \le n-1$.

Let us now formulate the three-diagonal method of Gauss elimination techniques for the differential second-order equation.

We consider the solution of the ordinary differential equation of the second order of the three-diagonal method of Gauss elimination techniques,

$$c_1 y'' + c_2 y' + c_3 y + c_4 = 0. \tag{A.5.34}$$

The boundary conditions are chosen in the form

$$x = a; \quad y = \alpha; \tag{A.5.35}$$

$$x = b; \quad y = \gamma. \tag{A.5.36}$$

The second-order scheme $O(h^2)$ is introduced using the following approximation in the node k

$$y'_k = \frac{y_{k+1} - y_{k-1}}{2h} + \frac{h^2}{6} f'''(\xi); \tag{A.5.37}$$

$$y_k'' = \frac{y_{k-1} - 2y_k + y_{k-1}}{h^2} + \frac{h^2}{12} f^{(IV)}(\xi). \tag{A.5.38}$$

for the uniform partition of the interval $[a, b]$

$$a = x_0 < x_1 < \ldots < x_{n-1} < x_n = b.$$

Therefore

$$x_k - x_{k-1} = (b - a)/n. \tag{A.5.39}$$

Using Eqs. (A.5.37) and (A.5.38) we find the finite-difference approximation for Eq. (A.5.34)

$$c_{1k} \frac{y_{k+1} - 2y_k + y_{k-1}}{h^3} + c_{2k} \frac{y_{k+1} - y_{k-1}}{2h} + c_{3k} y + c_{4k} = 0. \tag{A.5.40}$$

Eq. (A.5.40)—together with boundary conditions $y_0 = \alpha$, $y_n = \gamma$—constitute a coupled system of linear algebraic equations of $(n+1)$-order for obtaining of unknown values y_k.

Eq. (A.5.40) in the point x_1 is written as:

$$c_{11} \frac{y_2 - 2y_1 + y_0}{h^2} + c_{21} \frac{y_2 - y_0}{2h} + c_{31} y_1 + c_{41} = 0. \tag{A.5.41}$$

Because y_0 is known value, then Eq. (A.5.41) can be transformed as:

$$y_1 = a_1^{(1)} y_2 + a_2^{(1)}, \tag{A.5.42}$$

where a_1^1, a_2^1 are known coefficients. The analogs relation can be written for every step of calculations. Then:

$$y_k = a_1^{(k)} y_{k+1} + a_2^{(k)}. \tag{A.5.43}$$

Coefficients $a_1^{(k)}$, $a_2^{(k)}$ are Gauss coefficients in the three-diagonal method of elimination technique for the differential second-order equation (“progonka” coefficients in the east scientific literature).

It is convenient to formulate recurrence formulae for their definitions. With this aim first of all let us find y_{k-1} as function of y_{k+1}

$$\begin{aligned} y_{k-1} &= a_1^{(k-1)} y_k + a_2^{(k-1)}; \\ y_{k-1} &= a_1^{(k-1)} y_{k+1} a_1^{(k)} + a_1^{(k-1)} a_2^{(k)} + a_2^{(k-1)} \end{aligned} \tag{A.5.44}$$

After substitution y_{k-1}, y_k from Eqs. (A.5.43), (A.5.44) in Eq. (A.5.40) we have

$$\begin{aligned} & c_{1k} y_{k+1} - 2c_{1k} a_1^{(k)} y_{k+1} - 2c_{1k} a_2^{(k)} + c_{1k} a_1^{(k-1)} y_{k+1} a_1^{(k)} + c_{1k} a_1^{(k-1)} a_2^{(k)} + c_{1k} a_2^{(k-1)} \\ & + c_{2k} \frac{h}{2} y_{k+1} - \frac{c_{2k}}{2} h a_1^{(k-1)} y_{k+1} a_1^{(k)} - c_{2k} \frac{h}{2} a_1^{(k-1)} a_2^{(k)} - c_{2k} \frac{h}{2} a_2^{(k-1)} + c_{3k} h^2 a_1^{(k)} y_{k+1} + c_{3k} h^2 a_2^{(k)} + h^2 c_{4k} = 0. \end{aligned} \tag{A.5.45}$$

or

$$\begin{aligned} & y_{k+1} \left(c_{1k} - 2c_{1k} a_1^{(k)} + c_{1k} a_1^{(k-1)} a_1^{(k)} + c_{2k} \frac{h}{2} - c_{2k} \frac{h}{2} a_1^{(k-1)} a_1^{(k)} + c_{3k} h^2 a_1^{(k)} \right) \\ & + \left(-2c_{1k} a_2^{(k)} + c_{1k} a_1^{(k-1)} a_2^{(k)} + c_{1k} a_2^{(k-1)} - c_{2k} \frac{h}{2} a_1^{(k-1)} a_2^{(k)} - c_{2k} \frac{h}{2} a_2^{(k-1)} + c_{3k} h^2 a_2^{(k)} + h^2 b_{4k} \right) = 0. \end{aligned} \tag{A.5.46}$$

But relation (A.5.46) is identity which is valid for every y_{k+1}. Then:

$$a_1^{(k)} = \frac{c_{1k} + c_{2k} \frac{h}{2}}{a_1^{(k-1)} \left[-c_{1k} + c_{2k} \frac{h}{2} \right] + 2c_{1k} - c_{3k} h^2}; \tag{A.5.47}$$

$$a_2^{(k)} = \frac{-a_2^{(k-1)} \left(-c_{1k} + c_{2k} \frac{h}{2} \right) + c_{4k} h^2}{a_1^{(k-1)} \left(-c_{1k} + c_{2k} \frac{h}{2} \right) + 2c_{1k} - c_{3k} h^2}; \tag{A.5.48}$$

No difficulties to find the first pair of coefficients $a_1^{(0)}$, $a_2^{(0)}$. Really, if boundary condition has the form ($k=0$) has the form $y_0=\alpha$, then from

$$y_0=a_1^{(0)}y_1+a_2^{(0)}, \tag{A.5.49}$$

we have

$$a_1^{(0)}=0,\ \ a_2^{(0)}=\alpha. \tag{A.5.50}$$

For non-linear equations, the formulated procedure can be realized after linearization of corresponding differential equation and introduction of iterative calculation. But as it was written in [258]: "The formulated procedure led to increasing oscillations of numerical solutions. The way to avoid this non-stability consists in very simple but effective consideration, i.e. introduction of suppressive coefficients δ (in the east scientific literature 'damper coefficients')." For construction of $(n+1)$, iteration the outlet numerical massive of nth iteration $\phi_i^{(n)}$ was used in corrected form $\phi_i'^{(n)}$, which were connected by the following way

$$\phi_i'^{(n)}=\phi_i'^{(n-1)}+\delta\left(\phi_i^{(n)}-\phi_i'^{(n-1)}\right). \tag{A.5.51}$$

The coefficient δ was selected by experimental way during the process of numerical calculations, in many cases is sufficient to use $\delta\approx 0.1$.

Appendix 6

Some Integral Calculations in the Generalized Navier-Stokes Approximation

Some integrals should be calculated in the generalized Navier-Stokes approximation connected with the following distribution function

$$f_\alpha = f_\alpha^{(0)} + f_\alpha^{(1)},$$

where

$$f_\alpha^{(1)} = f_\alpha^{(0)} \left\{ -\mathbf{A}_\alpha \cdot \frac{\partial \ln T}{\partial \mathbf{r}} - \overleftrightarrow{B}_\alpha : \frac{\partial}{\partial \mathbf{r}} \mathbf{v}_0 + n \sum_j \mathbf{C}_\alpha^{(j)} \cdot \mathbf{d}_j \right\}.$$

Here the results of the corresponding tensor transformations are presented:

1.
$$\frac{\partial^2}{\partial \mathbf{r} \partial \mathbf{r}} : \sum_\alpha \rho_\alpha \overline{(\mathbf{V}_\alpha \mathbf{v}_0)\mathbf{V}_\alpha} = \frac{\partial}{\partial \mathbf{r}} \left[\frac{\partial}{\partial \mathbf{r}} \cdot (p \mathbf{v}_0) \right] - 2 \frac{\partial^2}{\partial \mathbf{r} \partial \mathbf{r}} : \mathbf{v}_0 \mu \overleftrightarrow{S}, \tag{A.6.1}$$

where in accordance with the definition, $\frac{\partial^2}{\partial \mathbf{r} \partial \mathbf{r}} : \mathbf{v}_0 \mu \overleftrightarrow{S}$ is vector with components, $k(1, 2, 3)$

$$\left[\frac{\partial^2}{\partial \mathbf{r} \partial \mathbf{r}} : \mathbf{v}_0 \mu \overleftrightarrow{S} \right]_k = \sum_{i,j=1}^{3} \frac{\partial^2}{\partial r_i \partial r_j} \left\{ \mu v_{0j} \left[\frac{1}{2} \left(\frac{\partial v_{0k}}{\partial r_i} + \frac{\partial v_{0i}}{\partial r_k} \right) - \frac{1}{3} \delta_{ki} \frac{\partial}{\partial \mathbf{r}} \cdot \mathbf{v}_0 \right] \right\}. \tag{A.6.2}$$

2.
$$\begin{aligned} \frac{\partial^2}{\partial \mathbf{r} \partial \mathbf{r}} : \sum_\alpha \rho_\alpha \overline{(\mathbf{V}_\alpha \mathbf{V}_\alpha)\mathbf{V}_\alpha} = &-2 \frac{\partial^2}{\partial \mathbf{r} \partial \mathbf{r}} \cdot \left(K \frac{\partial T}{\partial \mathbf{r}} \right) - \Delta \left(K \frac{\partial T}{\partial \mathbf{r}} \right) + 2 \frac{\partial^2}{\partial \mathbf{r} \partial \mathbf{r}} \cdot \left[\frac{pn}{\rho} \sum_{\alpha,\beta} m_\beta \left(D_{\alpha\beta} - D_{\alpha\beta}^1 \right) \mathbf{d}_\beta \right] \\ &+ \Delta \left(\frac{pn}{\rho} \sum_{\alpha,\beta} m_\beta \left(D_{\alpha\beta} - D_{\alpha\beta}^1 \right) \mathbf{d}_\beta \right), \end{aligned} \tag{A.6.3}$$

where Δ is Laplacian, $K = \sum_\alpha K_\alpha$, and $K_\alpha = \frac{k_B}{m_\alpha} D_\alpha^T + \frac{2}{5} \lambda'_\alpha$.

3.
$$\begin{aligned} \frac{\partial^2}{\partial \mathbf{r} \partial \mathbf{r}} : \sum_\alpha \rho_\alpha \overline{(\mathbf{V}_\alpha \mathbf{V}_\alpha) V_\alpha^2} = 5 k_B \Delta \left(T \sum_\alpha \frac{p_\alpha}{m_\alpha} \right) &- \frac{112}{9} \frac{\partial^2}{\partial \mathbf{r} \partial \mathbf{r}} : \left[\overleftrightarrow{S} k_B T \sum_\alpha \frac{1}{m_\alpha} (\mu_\alpha - \mu_\alpha^1) \right] \\ &- \frac{280}{27} \frac{\partial^2}{\partial \mathbf{r} \partial \mathbf{r}} : \left[\overleftrightarrow{D} k_B T \sum_\alpha \frac{1}{m_\alpha} (\mu_\alpha - \mu_\alpha^1) \right], \end{aligned} \tag{A.6.4}$$

where $\overleftrightarrow{D}$ is tensor with components

$$D_{ij} = \delta_{ij} \frac{\partial v_{0i}}{\partial r_j},$$

and viscosity and "the first" viscosity are defined by the relations

$$\mu_\alpha = \frac{1}{2} k_B n_\alpha T b_{\alpha 0}, \tag{A.6.5}$$

$$\mu_\alpha^1 = \frac{1}{2} k_B n_\alpha T b_{\alpha 1}. \tag{A.6.6}$$

4.

$$\frac{\partial^2}{\partial \mathbf{r} \partial \mathbf{r}} : \sum_\alpha \rho_\alpha \overline{(\mathbf{V}_\alpha \mathbf{v}_0) \mathbf{V}_\alpha \cdot \mathbf{v}_0} = \frac{\partial^2}{\partial \mathbf{r} \partial \mathbf{r}} : p \mathbf{v}_0 \mathbf{v}_0 - 2 \frac{\partial^2}{\partial \mathbf{r} \partial \mathbf{r}} : \mu \left(\mathbf{v}_0 \mathbf{v}_0 \cdot \overleftrightarrow{S} \right),$$

where,

$$S_{ij} = \frac{1}{2} \left(\frac{\partial v_{0i}}{\partial r_j} + \frac{\partial v_{0j}}{\partial r_i} \right) - \frac{1}{3} \delta_{ij} \frac{\partial}{\partial \mathbf{r}} \cdot \mathbf{v}_0. \tag{A.6.7}$$

5.

$$\frac{\partial^2}{\partial \mathbf{r} \partial \mathbf{r}} : \sum_\alpha \rho_\alpha \overline{(\mathbf{V}_\alpha \mathbf{v}_0) V_\alpha^2} = -5 \frac{\partial^2}{\partial \mathbf{r} \partial \mathbf{r}} : K \frac{\partial T}{\partial \mathbf{r}} \mathbf{v}_0 + 5 \frac{\partial^2}{\partial \mathbf{r} \partial \mathbf{r}} : \frac{pn}{\rho} \sum_{\alpha,\beta} m_\beta \left(D_{\alpha\beta} - D_{\alpha\beta}^1 \right) \mathbf{d}_\beta \mathbf{v}_0. \tag{A.6.8}$$

6.

$$\begin{aligned} \frac{\partial^2}{\partial \mathbf{r} \partial \mathbf{r}} : \sum_\alpha \rho_\alpha \overline{(\mathbf{V}_\alpha \mathbf{V}_\alpha)(\mathbf{V}_\alpha \cdot \mathbf{v}_0)} = & -2 \frac{\partial^2}{\partial \mathbf{r} \partial \mathbf{r}} : K \mathbf{v}_0 \frac{\partial T}{\partial \mathbf{r}} - \Delta \left(K \mathbf{v}_0 \cdot \frac{\partial T}{\partial \mathbf{r}} \right) + 2 \frac{\partial^2}{\partial \mathbf{r} \partial \mathbf{r}} : \frac{pn}{\rho} \mathbf{v}_0 \sum_{\alpha,\beta} m_\beta \left(D_{\alpha\beta} - D_{\alpha\beta}^1 \right) \mathbf{d}_\beta \\ & + \Delta \left[\frac{pn}{\rho} \mathbf{v}_0 \cdot \sum_{\alpha,\beta} m_\beta \left(D_{\alpha\beta} - D_{\alpha\beta}^1 \right) \mathbf{d}_\beta \right]. \end{aligned} \tag{A.6.9}$$

Appendix 7

Derivation of Energy Equation for Invariant $E_\alpha = \frac{m_\alpha V_\alpha^2}{2} + \varepsilon_\alpha$

Generalized hydrodynamic Enskog energy equation for invariant $E_\alpha = \frac{m_\alpha V_\alpha^2}{2} + \varepsilon_\alpha$ is written as

$$\int E_\alpha \left\{ \frac{\partial f_\alpha}{\partial t}\left[1 - \frac{\partial \tau_\alpha}{\partial t} - \mathbf{v}_\alpha \cdot \frac{\partial \tau_\alpha}{\partial \mathbf{r}}\right] + \mathbf{v}_\alpha \cdot \frac{\partial f_\alpha}{\partial \mathbf{r}}\left[1 - \frac{\partial \tau_\alpha}{\partial t} - \mathbf{v}_\alpha \cdot \frac{\partial \tau_\alpha}{\partial \mathbf{r}}\right] + \mathbf{F}_\alpha \cdot \frac{\partial f_\alpha}{\partial \mathbf{v}_\alpha}\left[1 - \frac{\partial \tau_\alpha}{\partial t} - \mathbf{v}_\alpha \cdot \frac{\partial \tau_\alpha}{\partial \mathbf{r}}\right] \right.$$
$$- \tau_\alpha \left[\frac{\partial^2 f_\alpha}{\partial t^2} + 2\frac{\partial^2 f_\alpha}{\partial \mathbf{r}\partial t} \cdot \mathbf{v}_\alpha + \frac{\partial^2 f_\alpha}{\partial \mathbf{r}\partial \mathbf{r}} : \mathbf{v}_\alpha \mathbf{v}_\alpha + 2\frac{\partial^2 f_\alpha}{\partial \mathbf{v}_\alpha \partial t} \cdot \mathbf{F}_\alpha + \frac{\partial \mathbf{F}_\alpha}{\partial t} \cdot \frac{\partial f_\alpha}{\partial \mathbf{v}_\alpha} + \mathbf{F}_\alpha \cdot \frac{\partial f_\alpha}{\partial \mathbf{r}} + \frac{q_\alpha}{m_\alpha}\frac{\partial f_\alpha}{\partial \mathbf{v}_\alpha}\left(\mathbf{F}_\alpha^{(1)} \times \mathbf{B}\right) \right. \tag{A.7.1}$$
$$\left.\left. + \frac{\partial f_\alpha}{\partial \mathbf{v}_\alpha} \cdot \left[\mathbf{B}(\mathbf{v}_\alpha \cdot \mathbf{B}) - B^2 \mathbf{v}_\alpha\right]\left(\frac{q_\alpha}{m_\alpha}\right)^2 + \frac{\partial f_\alpha}{\partial \mathbf{v}_\alpha}\mathbf{v}_\alpha : \frac{\partial}{\partial \mathbf{r}}\mathbf{F}_\alpha + \frac{\partial^2 f_\alpha}{\partial \mathbf{v}_\alpha \partial \mathbf{v}_\alpha} : \mathbf{F}_\alpha \mathbf{F}_\alpha + 2\frac{\partial^2 f_\alpha}{\partial \mathbf{v}_\alpha \partial \mathbf{r}} : \mathbf{v}_\alpha \mathbf{F}_\alpha \right]\right\} d\mathbf{v}_\alpha = J_\alpha^{en}.$$

Transformation of the terms in the left-hand side of Eq. (A.7.1) is realizing in a similar manner as it was indicated in Appendix 6. Then in the transformations that follow, we present as reference source the results of the integral calculations, which deserve further comments.

1.

$$\int E_\alpha \frac{\partial f_\alpha}{\partial t} d\mathbf{v}_\alpha = \frac{\partial}{\partial t}\left(n_\alpha \overline{E}_\alpha\right) + \mathbf{j}_\alpha \cdot \frac{\partial \mathbf{v}_0}{\partial t}. \tag{A.7.2}$$

Internal energy $\widetilde{U}_\alpha$ of particles of species α (per mass unit) defines as

$$n_\alpha \overline{E}_\alpha = \rho_\alpha \widetilde{U}_\alpha. \tag{A.7.3}$$

Then

$$n_\alpha \overline{E}_\alpha = \frac{3}{2} n_\alpha k_B T_\alpha + \varepsilon_\alpha n_\alpha = \frac{3}{2} p_\alpha + \varepsilon_\alpha n_\alpha = \rho_\alpha \widetilde{U}_\alpha. \tag{A.7.4}$$

For gas mixture we have

$$\rho \widetilde{U} = \frac{3}{2} p + \sum_\alpha \varepsilon_\alpha n_\alpha. \tag{A.7.5}$$

2.

$$\int E_\alpha \mathbf{v}_\alpha \cdot \frac{\partial f_\alpha}{\partial \mathbf{r}} d\mathbf{v}_\alpha = \frac{\partial}{\partial \mathbf{r}} \cdot \left(\mathbf{v}_0 \rho_\alpha \widetilde{U}_\alpha\right) + \frac{\partial}{\partial \mathbf{r}} \cdot \mathbf{q}_\alpha^{\mathrm{T}} + \mathbf{j}_\alpha \cdot \left(\mathbf{v}_0 \cdot \frac{\partial}{\partial \mathbf{r}}\right)\mathbf{v}_0 - \rho_\alpha \overline{\mathbf{V}_\alpha \cdot \left(\mathbf{V}_\alpha \cdot \frac{\partial}{\partial \mathbf{r}}\right)\mathbf{V}_\alpha}. \tag{A.7.6}$$

Let us remind that flux densities of heat and mass can be written as

$$\mathbf{q}_\alpha^{\mathrm{T}} = \frac{1}{2}\rho_\alpha \overline{V_\alpha^2 \mathbf{V}_\alpha} + \varepsilon_\alpha n_\alpha \overline{\mathbf{V}}_\alpha$$
$$\mathbf{j}_\alpha = m_\alpha n_\alpha \overline{\mathbf{V}}_\alpha.$$

3.

$$\int E_\alpha \mathbf{F}_\alpha \cdot \frac{\partial f_\alpha}{\partial \mathbf{v}_\alpha} \mathrm{d}\mathbf{v}_\alpha = -\rho_\alpha \overline{\mathbf{F}_\alpha \cdot \mathbf{V}_\alpha}. \tag{A.7.7}$$

4.

$$\int E_\alpha \frac{\partial f_\alpha}{\partial t} \mathbf{v}_\alpha \cdot \frac{\partial \tau_\alpha}{\partial \mathbf{r}} \mathrm{d}\mathbf{v}_\alpha = \frac{\partial \mathbf{q}_\alpha^{\mathrm{T}}}{\partial t} \cdot \frac{\partial \tau_\alpha}{\partial \mathbf{r}} + \frac{\partial \tau_\alpha}{\partial \mathbf{r}} \cdot \frac{\partial}{\partial t}\left(\mathbf{v}_0 \rho_\alpha \widetilde{U}_\alpha\right) + \rho_\alpha \overline{\mathbf{V}_\alpha \mathbf{V}_\alpha} : \frac{\partial \mathbf{v}_0}{\partial t} \frac{\partial \tau_\alpha}{\partial \mathbf{r}} \tag{A.7.8}$$

or

$$\int E_\alpha \frac{\partial f_\alpha}{\partial t} \mathbf{v}_\alpha \cdot \frac{\partial \tau_\alpha}{\partial \mathbf{r}} \mathrm{d}\mathbf{v}_\alpha = \frac{\partial \mathbf{q}_\alpha^{\mathrm{T}}}{\partial t} \cdot \frac{\partial \tau_\alpha}{\partial \mathbf{r}} + \frac{\partial \tau_\alpha}{\partial \mathbf{r}} \cdot \frac{\partial}{\partial t}\left(\mathbf{v}_0 \rho_\alpha \widetilde{U}_\alpha\right) + \frac{\partial \tau_\alpha}{\partial \mathbf{r}} \cdot \mathbf{v}_0 \left(\mathbf{j}_\alpha \cdot \frac{\partial \mathbf{v}_0}{\partial t}\right) + \frac{\partial \tau_\alpha}{\partial \mathbf{r}} \cdot \left(\overleftrightarrow{P}_\alpha \cdot \frac{\partial \mathbf{v}_0}{\partial t}\right), \tag{A.7.9}$$

where pressure tensor $\overleftrightarrow{P}_\alpha$ is

$$\overleftrightarrow{P}_\alpha = \rho_\alpha \overline{\mathbf{V}_\alpha \mathbf{V}_\alpha}.$$

5.

$$\begin{aligned} &\int E_\alpha \mathbf{v}_\alpha \cdot \frac{\partial f_\alpha}{\partial \mathbf{r}} \left(\mathbf{v}_\alpha \cdot \frac{\partial \tau_\alpha}{\partial \mathbf{r}}\right) d\mathbf{v}_\alpha \\ &= \frac{\partial \tau_\alpha}{\partial \mathbf{r}} \frac{\partial}{\partial \mathbf{r}} : n_\alpha \overline{E_\alpha \mathbf{v}_\alpha \mathbf{v}_\alpha} + \rho_\alpha \overline{(\frac{\partial \tau_\alpha}{\partial \mathbf{r}} \cdot \mathbf{v}_\alpha) \mathbf{V}_\alpha \cdot (\mathbf{v}_\alpha \cdot \frac{\partial}{\partial \mathbf{r}}})\mathbf{v}_0. \end{aligned} \tag{A.7.10}$$

6.

$$\begin{aligned} &\int E_\alpha \left(\mathbf{F}_\alpha \cdot \frac{\partial f_\alpha}{\partial \mathbf{v}_\alpha}\right) \mathbf{v}_\alpha \cdot \frac{\partial \tau_\alpha}{\partial \mathbf{r}} d\mathbf{v}_\alpha = -\frac{\partial \tau_\alpha}{\partial \mathbf{r}} \cdot \left(n_\alpha \overline{E_\alpha \mathbf{F}_\alpha}\right) \\ &- \left(\frac{\partial \tau_\alpha}{\partial \mathbf{r}} \cdot \mathbf{v}_0\right) \rho_\alpha \overline{\mathbf{V}_\alpha \cdot \mathbf{F}_\alpha} - \rho_\alpha \overline{(\frac{\partial \tau_\alpha}{\partial \mathbf{r}} \cdot \mathbf{V}_\alpha) \mathbf{V}_\alpha \cdot \mathbf{F}_\alpha}. \end{aligned} \tag{A.7.11}$$

7.

$$\int E_\alpha \frac{\partial^2 f_\alpha}{\partial t^2} \mathrm{d}\mathbf{v}_\alpha = \frac{\partial^2}{\partial t^2}\left(\rho_\alpha \widetilde{U}_\alpha\right) + \rho_\alpha \left(\frac{\partial \mathbf{v}_0}{\partial t}\right)^2 - \mathbf{j}_\alpha \cdot \frac{\partial^2 \mathbf{v}_0}{\partial t^2} + 2 \frac{\partial \mathbf{j}_\alpha}{\partial t} \cdot \frac{\partial \mathbf{v}_0}{\partial t}. \tag{A.7.12}$$

8.

$$\int E_\alpha \frac{\partial^2 f_\alpha}{\partial \mathbf{r} \partial t} \cdot \mathbf{v}_\alpha \mathrm{d}\mathbf{v}_\alpha = \frac{1}{2} \frac{\partial^2}{\partial \mathbf{r} \partial t} \cdot \rho_\alpha \overline{\mathbf{v}_\alpha v_\alpha^2} - \mathbf{v}_0 \cdot \left(\frac{\partial^2}{\partial \mathbf{r} \partial t} \cdot \rho_\alpha \overline{\mathbf{V}_\alpha \mathbf{V}_\alpha}\right) + \frac{1}{2} v_0^2 \frac{\partial^2}{\partial \mathbf{r} \partial t} \cdot \rho_\alpha \overline{\mathbf{v}_\alpha} + \frac{\partial^2}{\partial \mathbf{r} \partial t} \cdot \varepsilon_\alpha n_\alpha \overline{\mathbf{v}_\alpha} \tag{A.7.13}$$

9.

$$\begin{aligned} &\int E_\alpha \frac{\partial^2 f_\alpha}{\partial \mathbf{r} \partial \mathbf{r}} : \mathbf{v}_\alpha \mathbf{v}_\alpha d\mathbf{v}_\alpha = \frac{1}{2} \frac{\partial^2}{\partial \mathbf{r} \partial \mathbf{r}} : \rho_\alpha \overline{v_\alpha^2 \mathbf{v}_\alpha \mathbf{v}_\alpha} - \left[(\frac{\partial^2}{\partial \mathbf{r} \partial \mathbf{r}} : \rho_\alpha \overline{\mathbf{v}_\alpha \mathbf{v}_\alpha}) \mathbf{v}_\alpha\right] \cdot \mathbf{v}_0 \\ &+ \frac{1}{2} v_0^2 \frac{\partial^2}{\partial \mathbf{r} \partial \mathbf{r}} : \rho_\alpha \overline{\mathbf{v}_\alpha \mathbf{v}_\alpha} + \frac{\varepsilon_\alpha}{m_\alpha} \frac{\partial^2}{\partial \mathbf{r} \partial \mathbf{r}} : \rho_\alpha \overline{\mathbf{v}_\alpha \mathbf{v}_\alpha} \end{aligned} \tag{A.7.14}$$

10.

$$\int E_\alpha \frac{\partial^2 f_\alpha}{\partial \mathbf{v}_\alpha \partial t} \cdot \mathbf{F}_\alpha \mathrm{d}\mathbf{v}_\alpha = -\frac{\partial}{\partial t}\left(\mathbf{F}_\alpha^{(1)} \cdot \mathbf{j}_\alpha\right) - q_\alpha \frac{\partial}{\partial t}\left[n_\alpha \overline{(\mathbf{v}_\alpha \times \mathbf{B}) \cdot \mathbf{V}_\alpha}\right] + \rho_\alpha \overline{\frac{\partial}{\partial t}\left(\mathbf{F}_\alpha^{(1)} \cdot \mathbf{V}_\alpha\right)} + q_\alpha n_\alpha \overline{\frac{\partial}{\partial t}[(\mathbf{v}_\alpha \times \mathbf{B}) \cdot \mathbf{V}_\alpha]}. \tag{A.7.15}$$

11.

$$\int E_\alpha \frac{\partial \mathbf{F}_\alpha}{\partial t} \cdot \frac{\partial f_\alpha}{\partial \mathbf{v}_\alpha} \mathrm{d}\mathbf{v}_\alpha = -\left[\frac{q_\alpha}{m_\alpha}\left(\mathbf{v}_0 \times \frac{\partial \mathbf{B}}{\partial t}\right) + \frac{\partial \mathbf{F}_\alpha^{(1)}}{\partial t}\right] \cdot \mathbf{j}_\alpha. \tag{A.7.16}$$

12.

$$\begin{aligned} &\int E_\alpha \mathbf{F}_\alpha \cdot \frac{\partial f_\alpha}{\partial \mathbf{r}} \mathrm{d}\mathbf{v}_\alpha = \left(\mathbf{F}_\alpha^{(1)} \cdot \frac{\partial}{\partial \mathbf{r}}\right)\left(\rho_\alpha \widetilde{U}_\alpha\right) + \rho_\alpha \overline{\mathbf{v}}_\alpha \mathbf{F}_\alpha^{(1)} : \frac{\partial}{\partial \mathbf{r}} \mathbf{v}_0 - \frac{1}{2} \rho_\alpha \left(\mathbf{F}_\alpha^{(1)} \cdot \frac{\partial}{\partial \mathbf{r}}\right) v_0^2 \\ &+ \frac{q_\alpha}{m_\alpha} \frac{\partial}{\partial \mathbf{r}} \cdot \left[\rho_\alpha \widetilde{U}_\alpha \mathbf{v}_0 \times \mathbf{B}\right] + \frac{q_\alpha}{m_\alpha} \frac{\partial}{\partial \mathbf{r}} \cdot \left[\mathbf{q}_\alpha^{\mathrm{T}} \times \mathbf{B}\right] \\ &+ \frac{1}{2} q_\alpha n_\alpha \overline{v_\alpha^2 \mathbf{v}_\alpha} \cdot \mathrm{rot}\mathbf{B} - q_\alpha n_\alpha \overline{\mathbf{v}_\alpha \cdot \mathrm{rot}[(\mathbf{v}_\alpha \cdot \mathbf{v}_0)} \mathbf{B} + \frac{1}{2} q_\alpha n_\alpha \overline{\mathbf{v}}_\alpha \cdot \mathbf{rot}\left(v_0^2 \mathbf{B}\right) + \frac{q_\alpha}{m_\alpha} \varepsilon_\alpha n_\alpha \overline{\mathbf{v}}_\alpha \cdot \mathrm{rot}\mathbf{B}. \end{aligned} \tag{A.7.17}$$

13.

$$\int E_\alpha \frac{\partial f_\alpha}{\partial \mathbf{v}_\alpha}\cdot\left(\mathbf{F}_\alpha^{(1)}\times\mathbf{B}\right)\mathrm{d}\mathbf{v}_\alpha = -\left(\mathbf{F}_\alpha^{(1)}\times\mathbf{B}\right)\cdot\mathbf{j}_\alpha. \tag{A.7.18}$$

14.

$$\int E_\alpha \frac{\partial f_\alpha}{\partial \mathbf{v}_\alpha}\cdot\mathbf{v}_\alpha \mathrm{d}\mathbf{v}_\alpha = -\left[3\rho_\alpha \widetilde{U}_\alpha + \rho_\alpha \overline{V_\alpha^2} + \mathbf{v}_0\cdot\mathbf{j}_\alpha\right]. \tag{A.7.19}$$

15.

$$\int E_\alpha \frac{\partial f_\alpha}{\partial \mathbf{v}_\alpha}\mathbf{v}_\alpha : \frac{\partial}{\partial \mathbf{r}}\mathbf{F}_\alpha \mathrm{d}\mathbf{v}_\alpha = -\rho_\alpha \overline{\mathbf{V}_\alpha\mathbf{V}_\alpha} : \frac{\partial}{\partial \mathbf{r}}\mathbf{F}_\alpha - \rho_\alpha \widetilde{U}_\alpha \frac{\partial}{\partial \mathbf{r}}\cdot\mathbf{F}_\alpha^{(1)} + n_\alpha \frac{q_\alpha}{m_\alpha}\overline{E_\alpha \mathbf{V}_\alpha}\cdot\mathrm{rot}\mathbf{B}. \tag{A.7.20}$$

16.

$$\int E_\alpha \frac{\partial^2 f_\alpha}{\partial \mathbf{v}_\alpha \partial \mathbf{v}_\alpha} : \mathbf{F}_\alpha^L \mathbf{F}_\alpha^L \mathrm{d}\mathbf{v}_\alpha = \left(\frac{q_\alpha}{m_\alpha}\right)^2 \rho_\alpha \left\{\left[B^2\left(v_0^2 - 2\widetilde{U}_\alpha - \overline{V_\alpha^2}\right)\right] + \overline{(\mathbf{V}_\alpha\cdot\mathbf{B})^2} - (\mathbf{v}_0\cdot\mathbf{B})^2\right\}, \tag{A.7.21}$$

where $\mathbf{F} = \mathbf{F}^{(1)} + \mathbf{F}^L$, $\mathbf{F}^L$ is the Lorentz force $\mathbf{F}^L = \frac{q_\alpha}{m_\alpha}\overline{\mathbf{v}}_\alpha \times \mathbf{B}$.

17.

$$\int E_\alpha \frac{\partial^2 f_\alpha}{\partial \mathbf{v}_\alpha \partial \mathbf{v}_\alpha} : \mathbf{F}_\alpha^L \mathbf{F}_\alpha^{(1)} \mathrm{d}\mathbf{v}_\alpha = \frac{q_\alpha}{m_\alpha} n_\alpha \mathbf{F}_\alpha^{(1)}\cdot[\mathbf{v}_0 \times \mathbf{B}] \tag{A.7.22}$$

18.

$$\int E_\alpha \frac{\partial^2 f_\alpha}{\partial \mathbf{v}_\alpha \partial \mathbf{v}_\alpha} : \mathbf{F}_\alpha^{(1)} \mathbf{F}_\alpha^{(1)} \mathrm{d}\mathbf{v}_\alpha = \rho_\alpha F_\alpha^{(1)^2} \tag{A.7.23}$$

19.

$$\begin{aligned}
\int E_\alpha \frac{\partial^2 f_\alpha}{\partial \mathbf{v}_\alpha \partial \mathbf{r}} : \mathbf{v}_\alpha \mathbf{F}_\alpha d\mathbf{v}_\alpha = & -\mathbf{F}_\alpha^{(1)}\cdot\left[\frac{\partial}{\partial \mathbf{r}}\cdot\overleftrightarrow{P}_\alpha + \frac{\partial}{\partial \mathbf{r}}\cdot\mathbf{v}_0\mathbf{j}_\alpha + \left(\mathbf{j}_\alpha\cdot\frac{\partial}{\partial \mathbf{r}}\right)\mathbf{v}_0\right.\\
& \left. + \rho_\alpha\left(\mathbf{v}_0\cdot\frac{\partial}{\partial \mathbf{r}}\right)\mathbf{v}_0\right] - \left(\mathbf{F}_\alpha^{(1)}\cdot\frac{\partial}{\partial \mathbf{r}}\right)\rho_\alpha\widetilde{U}_\alpha - \left(\mathbf{F}_\alpha^{(1)}\cdot\frac{\partial}{\partial \mathbf{r}}\mathbf{v}_0\right)\cdot\mathbf{j}_\alpha\\
& + \frac{\partial}{\partial \mathbf{r}}\cdot[\mathbf{v}_0\mathbf{v}_0\cdot(q_\alpha n_\alpha \overline{\mathbf{v}}_\alpha \times \mathbf{B})] + q_\alpha\frac{\partial}{\partial \mathbf{r}}\cdot\left[n_\alpha \mathbf{v}_0\cdot\overline{(\mathbf{v}_\alpha \times \mathbf{B})\mathbf{V}_\alpha}\right]\\
& + \overline{(\mathbf{v}_0\cdot q_\alpha n_\alpha \frac{\partial}{\partial \mathbf{r}})[(\mathbf{v}_0 \times \mathbf{B})\cdot\mathbf{V}_\alpha]} + \frac{q_\alpha}{m_\alpha}\overleftrightarrow{P}_\alpha : \frac{\partial}{\partial \mathbf{r}}(\mathbf{v}_0 \times \mathbf{B})\\
& + (\mathbf{v}_0 \times \mathbf{B})\cdot q_\alpha n_\alpha \overline{\left(\mathbf{V}_\alpha\cdot\frac{\partial}{\partial \mathbf{r}}\right)\mathbf{V}_\alpha} - \frac{\partial}{\partial \mathbf{r}}\cdot\left[(\mathbf{v}_0 \times \mathbf{B})\frac{q_\alpha}{m_\alpha}\rho_\alpha\widetilde{U}_\alpha\right]\\
& - \frac{q_\alpha}{m_\alpha}\frac{\partial}{\partial \mathbf{r}}\cdot\left(\mathbf{q}_\alpha^{\mathrm{T}} \times \mathbf{B}\right) + \frac{3}{2}\frac{q_\alpha}{m_\alpha}\left[(\mathbf{v}_0 \times \mathbf{B})\cdot\frac{\partial}{\partial \mathbf{r}}\right]p_\alpha + \frac{1}{2}q_\alpha n_\alpha \overline{\left[(\mathbf{V}_\alpha \times \mathbf{B})\cdot\frac{\partial}{\partial \mathbf{r}}\right]V_\alpha^2}\\
& - \frac{q_\alpha}{m_\alpha}\rho_\alpha\widetilde{U}_\alpha \mathbf{v}_0\cdot\mathrm{rot}\mathbf{B} - \frac{q_\alpha}{m_\alpha}\mathbf{q}_\alpha^{\mathrm{T}}\cdot\mathrm{rot}\mathbf{B},
\end{aligned} \tag{A.7.24}$$

where $p_\alpha = n_\alpha k_B T_\alpha$

Using relations (A.7.2)–(A.7.24) for transformation of energy Eq. (A.7.1), the following form of energy equation can be found for invariant E_α:

$$\begin{aligned}
& \left[\frac{\partial}{\partial t}\left(\rho_\alpha\widetilde{U}_\alpha\right) + \mathbf{j}_\alpha\cdot\frac{\partial \mathbf{v}_0}{\partial t} + \frac{\partial}{\partial \mathbf{r}}\cdot\left(\mathbf{v}_0\rho_\alpha\widetilde{U}_\alpha\right) + \frac{\partial}{\partial \mathbf{r}}\cdot\mathbf{q}_\alpha^{\mathrm{T}} + \mathbf{j}_\alpha\cdot\left(\mathbf{v}_0\cdot\frac{\partial}{\partial \mathbf{r}}\right)\mathbf{v}_0\right.\\
& \left. - \rho_\alpha\overline{\mathbf{F}_\alpha\cdot\mathbf{V}_\alpha} - \rho_\alpha\overline{\mathbf{V}_\alpha\cdot\left(\mathbf{V}_\alpha\cdot\frac{\partial}{\partial \mathbf{r}}\right)\mathbf{V}_\alpha}\right]\left(1 - \frac{\partial \tau_\alpha}{\partial t}\right) - \frac{\partial \mathbf{q}_\alpha^{\mathrm{T}}}{\partial t}\cdot\frac{\partial \tau_\alpha}{\partial \mathbf{r}}\\
& - \frac{\partial \tau_\alpha}{\partial \mathbf{r}}\cdot\frac{\partial}{\partial t}\left(\mathbf{v}_0\rho_\alpha\widetilde{U}_\alpha\right) - \rho_\alpha\overline{\mathbf{v}_\alpha\mathbf{V}_\alpha} : \frac{\partial \mathbf{v}_0}{\partial t}\frac{\partial \tau_\alpha}{\partial \mathbf{r}} - \frac{\partial \tau_\alpha}{\partial \mathbf{r}}\frac{\partial}{\partial \mathbf{r}} : n_\alpha\overline{E_\alpha\mathbf{v}_\alpha\mathbf{v}_\alpha}\\
& - \rho_\alpha\overline{(\frac{\partial \tau_\alpha}{\partial \mathbf{r}}\cdot\mathbf{v}_\alpha)\mathbf{V}_\alpha\cdot(\mathbf{v}_\alpha\cdot\frac{\partial}{\partial \mathbf{r}})}\mathbf{v}_0 + \frac{\partial \tau_\alpha}{\partial \mathbf{r}}\cdot\left(n_\alpha\overline{E_\alpha\mathbf{F}_\alpha}\right) + \left(\frac{\partial \tau_\alpha}{\partial \mathbf{r}}\cdot\mathbf{v}_0\right)\rho_\alpha\overline{\mathbf{F}_\alpha\cdot\mathbf{V}_\alpha}
\end{aligned}$$

$$
\begin{aligned}
&+\rho_\alpha\Big(\frac{\partial \tau_\alpha}{\partial \mathbf{r}}\cdot\overline{\mathbf{V}_\alpha\Big)\mathbf{V}_\alpha\cdot\mathbf{F}_\alpha} - \tau_\alpha\Big\{\rho_\alpha\overline{\mathbf{v}_\alpha}\mathbf{F}_\alpha^{(1)} : \frac{\partial}{\partial\mathbf{r}}\mathbf{v}_0 - \frac{1}{2}\rho_\alpha\left(\mathbf{F}_\alpha^{(1)}\cdot\frac{\partial}{\partial\mathbf{r}}\right)v_0^2 \\
&+\frac{1}{2}q_\alpha n_\alpha\overline{v_\alpha^2\mathbf{v}_\alpha}\cdot\mathrm{rot}\mathbf{B} - q_\alpha n_\alpha\overline{\mathbf{v}_\alpha\cdot\mathrm{rot}[(\mathbf{v}_\alpha\cdot\mathbf{v}_0)\mathbf{B}]} + \frac{1}{2}q_\alpha n_\alpha\overline{\mathbf{v}_\alpha}\cdot\mathrm{rot}\left(v_0^2\mathbf{B}\right) \\
&+\frac{q_\alpha}{m_\alpha}\varepsilon_\alpha n_\alpha\overline{\mathbf{v}_\alpha}\cdot\mathrm{rot}\mathbf{B} - \left[\frac{q_\alpha}{m_\alpha}\left(\mathbf{v}_0\times\frac{\partial\mathbf{B}}{\partial t}\right)+\frac{\partial\mathbf{F}_\alpha^{(1)}}{\partial t}\right]\cdot\mathbf{j}_\alpha - \frac{q_\alpha}{m_\alpha}\left(\mathbf{F}_\alpha^{(1)}\times\mathbf{B}\right)\cdot\mathbf{j}_\alpha + 2\frac{q_\alpha}{m_\alpha}n_\alpha\mathbf{F}_\alpha^{(1)}\cdot[\mathbf{v}_0\times\mathbf{B}] \\
&+\left(\frac{q_\alpha}{m_\alpha}\right)^2 B^2\mathbf{v}_0\cdot\mathbf{j}_\alpha - \rho_\alpha\overline{\mathbf{V}_\alpha\mathbf{v}_\alpha} : \frac{\partial}{\partial\mathbf{r}}\mathbf{F}_\alpha - \rho_\alpha\widetilde{U}_\alpha\frac{\partial}{\partial\mathbf{r}}\cdot\mathbf{F}_\alpha^{(1)} + \frac{\partial^2}{\partial t^2}\left(\rho_\alpha\widetilde{U}_\alpha\right) \\
&+\rho_\alpha\left(\frac{\partial\mathbf{v}_0}{\partial t}\right)^2 - \mathbf{j}_\alpha\cdot\frac{\partial^2\mathbf{v}_0}{\partial t^2} + 2\frac{\partial\mathbf{j}_\alpha}{\partial t}\cdot\frac{\partial\mathbf{v}_0}{\partial t} + \frac{1}{2}\frac{\partial^2}{\partial\mathbf{r}\partial\mathbf{r}} : \rho\overline{v_\alpha^2\mathbf{v}_\alpha\mathbf{v}_\alpha} \\
&-\left[\Big(\frac{\partial^2}{\partial\mathbf{r}\partial\mathbf{r}} : \rho_\alpha\overline{\mathbf{v}_\alpha\mathbf{v}_\alpha\Big)\mathbf{v}_\alpha}\right]\cdot\mathbf{v}_0 + \frac{1}{2}v_0^2\frac{\partial^2}{\partial\mathbf{r}\partial\mathbf{r}} : \rho_\alpha\overline{\mathbf{v}_\alpha\mathbf{v}_\alpha} \\
&+\frac{\varepsilon_\alpha}{m_\alpha}\frac{\partial^2}{\partial\mathbf{r}\partial\mathbf{r}} : \rho_\alpha\overline{\mathbf{v}_\alpha\mathbf{v}_\alpha} + \rho_\alpha F_\alpha^{(1)2} + \rho_\alpha\left(\frac{q_\alpha}{m_\alpha}\right)^2\Big[B^2v_0^2 - (\mathbf{v}_0\cdot\mathbf{B})^2 \\
&-(\mathbf{v}_0\cdot\mathbf{B})\left(\overline{\mathbf{V}}_\alpha\cdot\mathbf{B}\right)\Big] - 2\mathbf{F}_\alpha^{(1)}\cdot\Big[\frac{\partial}{\partial\mathbf{r}}\cdot\overleftrightarrow{\mathbf{P}}_\alpha + \frac{\partial}{\partial\mathbf{r}}\cdot\mathbf{v}_0\mathbf{j}_\alpha + \left(\mathbf{j}_\alpha\cdot\frac{\partial}{\partial\mathbf{r}}\right)\mathbf{v}_0 \\
&+\rho_\alpha\left(\mathbf{v}_0\cdot\frac{\partial}{\partial\mathbf{r}}\right)\mathbf{v}_0\Big] - \left(\mathbf{F}_\alpha^{(1)}\cdot\frac{\partial}{\partial\mathbf{r}}\right)\rho_\alpha\widetilde{U}_\alpha - 2\left(\mathbf{F}_\alpha^{(1)}\cdot\frac{\partial}{\partial\mathbf{r}}\mathbf{v}_0\right)\cdot\mathbf{j}_\alpha \\
&+2\frac{\partial}{\partial\mathbf{r}}\cdot[\mathbf{v}_0\mathbf{v}_0\cdot(q_\alpha n_\alpha\overline{\mathbf{v}}_\alpha\times\mathbf{B})] + 2q_\alpha\frac{\partial}{\partial\mathbf{r}}\cdot\left[n_\alpha\mathbf{v}_0\overline{(\mathbf{v}_\alpha\times\mathbf{B})\mathbf{V}_\alpha}\right] \\
&+2\Big(\mathbf{v}_0\cdot q_\alpha n_\alpha\overline{\frac{\partial}{\partial\mathbf{r}}\Big)[(\mathbf{v}_0\times\mathbf{B})\cdot\mathbf{V}_\alpha]} + 2\frac{q_\alpha}{m_\alpha}\overleftrightarrow{P}_\alpha : \frac{\partial}{\partial\mathbf{r}}(\mathbf{v}_0\times\mathbf{B}) \\
&+2(\mathbf{v}_0\times\mathbf{B})\cdot q_\alpha n_\alpha\overline{\left(\mathbf{V}_\alpha\cdot\frac{\partial}{\partial\mathbf{r}}\right)\mathbf{V}_\alpha} - \frac{q_\alpha}{m_\alpha}\frac{\partial}{\partial\mathbf{r}}\cdot\left[(\mathbf{v}_0\times\mathbf{B})\rho_\alpha\widetilde{U}_\alpha\right] \\
&-\frac{q_\alpha}{m_\alpha}\frac{\partial}{\partial\mathbf{r}}\cdot\left(\mathbf{q}_\alpha^{\mathrm{T}}\times\mathbf{B}\right) + \frac{q_\alpha}{m_\alpha}\left[(\mathbf{v}_0\times\mathbf{B})\cdot\frac{\partial}{\partial\mathbf{r}}\right]\left(\rho_\alpha\overline{V_\alpha^2}\right) \\
&+\frac{q_\alpha}{m_\alpha}\rho_\alpha\overline{\left[(\mathbf{V}_\alpha\times\mathbf{B})\cdot\frac{\partial}{\partial\mathbf{r}}\right]V_\alpha^2} - \mathbf{v}_0\cdot\mathrm{rot}\mathbf{B}\frac{q_\alpha}{m_\alpha}\rho_\alpha\widetilde{U}_\alpha - \frac{q_\alpha}{m_\alpha}\mathbf{q}_\alpha^{\mathrm{T}}\cdot\mathrm{rot}\mathbf{B} \\
&-2\frac{q_\alpha}{m_\alpha}\frac{\partial}{\partial t}\left[\rho_\alpha\overline{(\mathbf{v}_\alpha\times\mathbf{B})\cdot\mathbf{V}_\alpha}\right] + 2\frac{q_\alpha}{m_\alpha}\rho_\alpha\overline{\frac{\partial}{\partial t}[(\mathbf{v}_\alpha\times\mathbf{B})\cdot\mathbf{V}_\alpha]} \\
&-2\frac{\partial}{\partial t}\left(\mathbf{F}_\alpha^{(1)}\cdot\mathbf{j}_\alpha\right) + 2\rho_\alpha\overline{\frac{\partial}{\partial t}\left(\mathbf{F}_\alpha^{(1)}\cdot\mathbf{V}_\alpha\right)} + \frac{\partial^2}{\partial\mathbf{r}\partial t}\cdot\rho_\alpha\overline{\mathbf{v}_\alpha v_\alpha^2} \\
&-2\mathbf{v}_0\cdot\left[\frac{\partial^2}{\partial\mathbf{r}\partial t}\cdot\rho_\alpha\overline{\mathbf{v}_\alpha\mathbf{v}_\alpha}\right] + v_0^2\frac{\partial^2}{\partial\mathbf{r}\partial t}\cdot\rho_\alpha\overline{\mathbf{v}}_\alpha + 2\frac{\partial^2}{\partial\mathbf{r}\partial t}\cdot\left(\varepsilon_\alpha n_\alpha\overline{\mathbf{v}}_\alpha\right)\Big\} = J_\alpha^{en,V}.
\end{aligned}
\tag{A.7.25}
$$

In the generalized Euler approximation (for local Maxwellian distribution function), the following relations are valid:

20.

$$\rho_\alpha\overline{\mathbf{V}_\alpha\cdot\left(\mathbf{V}_\alpha\cdot\frac{\partial}{\partial\mathbf{r}}\right)\mathbf{V}_\alpha} = -\frac{\rho_\alpha\overline{V_\alpha^2}}{3}\frac{\partial}{\partial\mathbf{r}}\cdot\mathbf{v}_0. \tag{A.7.26}$$

21. $$\frac{\partial \tau_\alpha^{(0)}}{\partial \mathbf{r}}\frac{\partial}{\partial \mathbf{r}}:n_\alpha\overline{E_\alpha \mathbf{v}_\alpha \mathbf{v}_\alpha}=\frac{\partial \tau_\alpha^{(0)}}{\partial \mathbf{r}}\frac{\partial}{\partial \mathbf{r}}:\left(\rho_\alpha \widetilde{U}_\alpha \mathbf{v}_0\mathbf{v}_0\right)+\frac{\partial \tau_\alpha^{(0)}}{\partial \mathbf{r}}\frac{\partial}{\partial \mathbf{r}}:\left(\frac{5}{2}k_BT_\alpha\frac{p_\alpha}{m_\alpha}\overleftrightarrow{\mathrm{I}}\right)+\frac{\partial \tau_\alpha^{(0)}}{\partial \mathbf{r}}\frac{\partial}{\partial \mathbf{r}}:\frac{\varepsilon_\alpha}{m_\alpha}p_\alpha\overleftrightarrow{\mathrm{I}}, \tag{A.7.27}$$

where

$$p_\alpha=n_\alpha k_BT_\alpha$$

22. $$\rho_\alpha\overline{\left(\frac{\partial \tau_\alpha^{(0)}}{\partial \mathbf{r}}\cdot\mathbf{v}_\alpha\right)\mathbf{V}_\alpha\cdot\left(\mathbf{v}_\alpha\cdot\frac{\partial}{\partial \mathbf{r}}\right)}\mathbf{v}_0=\frac{\partial \tau_\alpha^{(0)}}{\partial \mathbf{r}}\cdot\mathbf{v}_0\left(\frac{\partial}{\partial \mathbf{r}}\cdot\mathbf{v}_0\right)p_\alpha+p_\alpha\frac{\partial \tau_\alpha^{(0)}}{\partial \mathbf{r}}\cdot\mathbf{v}_0:\frac{\partial}{\partial \mathbf{r}}\mathbf{v}_0 \tag{A.7.28}$$

23. $$\frac{\partial^2}{\partial\mathbf{r}\partial\mathbf{r}}:\rho_\alpha\overline{V_\alpha^2\mathbf{V}_\alpha\mathbf{V}_\alpha}=5k_B\Delta\left(T_\alpha\frac{p_\alpha}{m_\alpha}\right), \tag{A.7.29}$$

where operator $\Delta\equiv\frac{\partial^2}{\partial x^2}+\frac{\partial^2}{\partial y^2}+\frac{\partial^2}{\partial z^2}$ is Laplacian.

24. $$\frac{\partial^2}{\partial\mathbf{r}\partial\mathbf{r}}:\rho_\alpha\overline{(\mathbf{V}_\alpha\mathbf{v}_0)\mathbf{V}_\alpha\cdot\mathbf{v}_0}=\frac{\partial^2}{\partial\mathbf{r}\partial\mathbf{r}}:p_\alpha\mathbf{v}_0\mathbf{v}_0. \tag{A.7.30}$$

25. $$\frac{\partial^2}{\partial\mathbf{r}\partial\mathbf{r}}:\rho_\alpha\overline{\mathbf{v}_\alpha\mathbf{v}_\alpha}=\frac{\partial^2}{\partial\mathbf{r}\partial\mathbf{r}}:\rho_\alpha\overline{\mathbf{v}_0\mathbf{v}_0}+\frac{\partial^2}{\partial\mathbf{r}\partial\mathbf{r}}:p_\alpha\overleftrightarrow{\mathrm{I}}. \tag{A.7.31}$$

26. $$\left[\left(\frac{\partial^2}{\partial\mathbf{r}\partial\mathbf{r}}:\rho_\alpha\overline{\mathbf{v}_0\mathbf{V}_\alpha)\mathbf{V}_\alpha}\right]\cdot\mathbf{v}_0=\frac{1}{3}\left(\frac{\partial^2}{\partial\mathbf{r}\partial\mathbf{r}}\cdot\rho_\alpha\overline{V_\alpha^2}\mathbf{v}_0\right)\cdot\mathbf{v}_0. \tag{A.7.32}$$

27. $$\frac{\partial^2}{\partial\mathbf{r}\partial t}\cdot\left(\rho_\alpha\overline{\mathbf{v}_\alpha v_\alpha^2}\right)=\frac{\partial^2}{\partial\mathbf{r}\partial t}\cdot\rho_\alpha\mathbf{v}_0v_0^2+\frac{\partial^2}{\partial\mathbf{r}\partial t}\cdot\rho_\alpha\mathbf{v}_0\overline{V_\alpha^2}+2\frac{\partial^2}{\partial\mathbf{r}\partial t}\cdot\left(\rho_\alpha\overline{\mathbf{V}_\alpha(\mathbf{v}_0\cdot\mathbf{V}_\alpha)}\right). \tag{A.7.33}$$

As a result, the generalized Euler energy equation (for non-reacting mixture of gases ($\varepsilon_\alpha=0$) in the absence of external forces) is written as

$$\begin{aligned}
&\left[\frac{\partial}{\partial t}\left(\rho_\alpha\widetilde{U}_\alpha\right)+\frac{1}{3}\rho_\alpha\overline{V_\alpha^2}\frac{\partial}{\partial\mathbf{r}}\cdot\mathbf{v}_0+\frac{\partial}{\partial\mathbf{r}}\cdot\left(\mathbf{v}_0\rho_\alpha\widetilde{U}_\alpha\right)\right]\left(1-\frac{\partial\tau_\alpha^{(0)}}{\partial t}\right)\\
&-\frac{\partial\tau_\alpha^{(0)}}{\partial\mathbf{r}}\cdot\frac{\partial}{\partial t}\left(\mathbf{v}_0\rho_\alpha\widetilde{U}_\alpha\right)-p_\alpha\overleftrightarrow{\mathrm{I}}:\frac{\partial\mathbf{v}_0}{\partial t}\frac{\partial\tau_\alpha^{(0)}}{\partial\mathbf{r}}\\
&-\frac{\partial\tau_\alpha^{(0)}}{\partial\mathbf{r}}\frac{\partial}{\partial\mathbf{r}}:\left(\rho_\alpha\widetilde{U}_\alpha\mathbf{v}_0\mathbf{v}_0\right)-\frac{\partial\tau_\alpha^{(0)}}{\partial\mathbf{r}}\frac{\partial}{\partial\mathbf{r}}:\left(\frac{5}{2}k_BT_\alpha\frac{p_\alpha}{m_\alpha}\overleftrightarrow{\mathrm{I}}\right)-\frac{\partial\tau_\alpha^{(0)}}{\partial\mathbf{r}}\cdot\mathbf{v}_0p_\alpha\left(\frac{\partial}{\partial\mathbf{r}}\cdot\mathbf{v}_0\right)\\
&-p_\alpha\frac{\partial\tau_\alpha^{(0)}}{\partial\mathbf{r}}\mathbf{v}_0:\frac{\partial}{\partial\mathbf{r}}\mathbf{v}_0-\tau_\alpha^{(0)}\left\{\frac{\partial^2}{\partial t^2}\left(\rho_\alpha\widetilde{U}_\alpha\right)+\rho_\alpha\left(\frac{\partial\mathbf{v}_0}{\partial t}\right)^2\right.\\
&+\frac{5}{2}k_B\Delta\left(T_\alpha\frac{p_\alpha}{m_\alpha}\right)+\frac{1}{2}\frac{\partial^2}{\partial\mathbf{r}\partial\mathbf{r}}:\rho_\alpha v_0^2\mathbf{v}_0\mathbf{v}_0+\frac{1}{2}\frac{\partial^2}{\partial\mathbf{r}\partial\mathbf{r}}:\left(\rho_\alpha\overline{V_\alpha^2\mathbf{v}_0\mathbf{v}_0}\right)\\
&+\frac{1}{2}\frac{\partial^2}{\partial\mathbf{r}\partial\mathbf{r}}:\left(v_0^2p_\alpha\overleftrightarrow{\mathrm{I}}\right)+2\frac{\partial^2}{\partial\mathbf{r}\partial\mathbf{r}}:p_\alpha\mathbf{v}_0\mathbf{v}_0-\left[\left(\frac{\partial^2}{\partial\mathbf{r}\partial\mathbf{r}}:\rho_\alpha\mathbf{v}_0\mathbf{v}_0\right)\mathbf{v}_0\right]\cdot\mathbf{v}_0\\
&-\left[\left(\frac{\partial^2}{\partial\mathbf{r}\partial\mathbf{r}}:p_\alpha\overleftrightarrow{\mathrm{I}}\right)\mathbf{v}_0\right]\cdot\mathbf{v}_0-\frac{2}{3}\left[\frac{\partial^2}{\partial\mathbf{r}\partial\mathbf{r}}\cdot\rho_\alpha\overline{V_\alpha^2}\mathbf{v}_0\right]\cdot\mathbf{v}_0+\frac{1}{2}\frac{\partial^2}{\partial\mathbf{r}\partial\mathbf{r}}:\rho_\alpha\mathbf{v}_0\mathbf{v}_0\\
&+\frac{1}{2}v_0^2\frac{\partial^2}{\partial\mathbf{r}\partial\mathbf{r}}:p_\alpha\overleftrightarrow{\mathrm{I}}+\frac{\partial^2}{\partial\mathbf{r}\partial t}\cdot\rho_\alpha\mathbf{v}_0v_0^2+\frac{5}{3}\frac{\partial^2}{\partial\mathbf{r}\partial t}\cdot\rho_\alpha\mathbf{v}_0\overline{V_\alpha^2}\\
&\left.-2\mathbf{v}_0\cdot\left[\frac{\partial^2}{\partial\mathbf{r}\partial t}\cdot\rho_\alpha\mathbf{v}_0\mathbf{v}_0\right]-2\mathbf{v}_0\left(\frac{\partial^2}{\partial\mathbf{r}\partial t}\cdot p_\alpha\overleftrightarrow{\mathrm{I}}\right)+v_0^2\frac{\partial^2}{\partial\mathbf{r}\partial t}\cdot\rho_\alpha\mathbf{v}_0\right\}=J_\alpha^{en,V}.
\end{aligned} \tag{A.7.34}$$

Equation (A.7.34) can be simplified for one-species one-dimensional case as

$$\left[\frac{\partial}{\partial t}\left(\rho\widetilde{U}\right)+p\frac{\partial v_0}{\partial x}+\frac{\partial}{\partial x}\left(\rho v_0\widetilde{U}\right)\right]\left(1-\frac{\partial\tau^{(0)}}{\partial t}\right)-\left[\frac{\partial}{\partial t}\left(v_0\rho\widetilde{U}\right)+p\frac{\partial v_0}{\partial t}+\frac{\partial}{\partial x}\left(v_0^2\rho\widetilde{U}\right)+\frac{5}{2}\frac{\partial}{\partial x}\left(k_BT\frac{p}{m}\right)\right.$$
$$\left.+2pv_0\frac{\partial v_0}{\partial x}\right]\frac{\partial\tau^{(0)}}{\partial x}-\tau^{(0)}\left\{\frac{\partial^2}{\partial t^2}\left(\rho\widetilde{U}\right)+\rho\left(\frac{\partial v_0}{\partial t}\right)^2+\frac{5}{2}k_B\frac{\partial^2}{\partial x^2}\left(\frac{pT}{m}\right)+\frac{1}{2}\frac{\partial^2}{\partial x^2}\left(\rho v_0^4\right)\right.$$
$$+4\frac{\partial^2}{\partial x^2}\left(pv_0^2\right)-v_0\frac{\partial^2}{\partial x^2}\left(\rho v_0^3\right)-3v_0\frac{\partial^2}{\partial x^2}(pv_0)+\frac{1}{2}v_0^2\frac{\partial^2}{\partial x^2}\left(\rho v_0^2\right)+\frac{1}{2}v_0^2\frac{\partial^2 p}{\partial x^2}+\frac{\partial^2}{\partial x\partial t}\left(\rho v_0^3\right)$$
$$\left.+5\frac{\partial^2}{\partial x\partial t}(pv_0)-2v_0\frac{\partial^2}{\partial x\partial t}\left(\rho v_0^2\right)-2v_0\frac{\partial^2 p}{\partial x\partial t}+v_0^2\frac{\partial^2}{\partial x\partial t}(\rho v_0)\right\}=0, \tag{A.7.35}$$

In this case, the $\rho\widetilde{U}=\frac{3}{2}p$. Energy equation for invariant $\frac{m_\alpha v_\alpha^2}{2}+\varepsilon_\alpha$ can be obtained from the energy equation for invariant E_α as linear combination of hydrodynamic transport equations. Particularly, we obtain energy Eq. (2.7.54) after summation of Eq. (A.7.35) (multiplied by two), one-dimensional motion Eq. (2.7.53) (multiplied term-by-term on hydrodynamic velocity v_0), and continuity Eq. (2.7.52), multiplied on v_0^2. Then the question about the application of one or other forms of equations is the question of easy-to-use form by the concrete calculation.

Appendix 8

To the Non-Local Theory of Cold Nuclear Fusion

In 1989 two electro-chemists, Martin Fleischmann and Stanley Pons announced [259] about nuclear fusion reactions between deuterium nuclei in a table-top experiment, under ordinary conditions of temperature and pressure, by using electrochemistry. The experimental evidence consisted of the production of large amounts of heat, which could not be attributed to chemical reactions. The reactions were termed "cold fusion" (CF), by comparison with the high temperature of thermonuclear fusion. The typical chain of nuclear reactions can be written as follows

$$
\begin{aligned}
&d + d \rightarrow 3He + n + 4.0\,MeV, \\
&d + d \rightarrow t + p + 3.25\,MeV, \\
&t + d \rightarrow 4He + n + 17.6\,MeV, \\
&3He + d \rightarrow 4He + p + 18.3\,MeV.
\end{aligned}
$$

Obviously the following criteria need to be met in order to establish conventional thermonuclear deuterium fusion unquestionably:

1. The experiment has been repeatable by other investigators;
2. There has to be a significant neutron emission statistically well above background level;
3. The energy spectrum of the detected neutrons must match the energy spectrum of neutrons produced in deuterium fusion;
4. There should be no neutron sources in the laboratory that can confuse the fusion neutron measurements.

Many scientists cannot reproduce the mentioned experimental results. Scientific community concluded that there were no nuclear reactions and that the reported experiments were in error. CF was considered as an example of wrong science. This produced a partition between the traditional scientific world and the community which continued the CF research. In the 20 years elapsed since the announcement by Fleischmann and Pons that the excess enthalpy generated in the negatively polarized Pd-D-D_2O system was attributable to nuclear reactions occurring inside the Pd lattice, there have been reports of other manifestations of nuclear activities in this system. In particular, there have been reports of tritium and helium-4 production; emission of energetic particles, gamma or X-rays, and neutrons; as well as the transmutation of elements. Reproducibility was improved and Mosier-Boss et al. [260] declared as result of accurate measurements about the real existence of "Fleischmann-Pons Effect" (FPE).

Many scientists convinced that the unlikelihood of a chemical reaction being strong enough to overcome the Coulomb barrier, the lack of gamma rays, the lack of explanation for the origin of the extra energy, the lack of the expected radioactivity and so on.

In the following investigation I intend to construct the non-local theory of CF for the system which can be analyzed by the methods of theoretical physics. Then we have no reason to discuss the physical systems with unknown influence of unknown catalysts. Much more interesting for the theoretical investigation the situation when light nuclei are forced together using the external forces like effects of cavitations, shock waves, or combination of the possible force effects. In this case particles will fuse with a yield of energy because the mass of the combination will be less than the sum of the masses of the individual nuclei.

Many successful the cavitation-induced fusion experiments have been performed and reported in peer-reviewed literature (see for example, [261–271]). Subsequent work was carried out by Bityurin and coauthors [261] at the Joint Institute for High Temperatures of the Russian Academy of Sciences. The group studied the effect of shockwaves on deuterated liquid (D_2O) with high (20-95%) bubble content. Their experimental setup is includes admission of deuterium bubbles into deuterated liquid and crushing them with a shockwave generated via explosion of a semicircular wire due to high current pulse. The resulting shockwave propagates in the bubble-liquid phase and focuses much stronger than in the pure liquid due to shockwave amplification effects in the gaseous phase. As is well-known, the fusion of D atoms results in the emission of a proton, helium-3, a neutron (of 2.45 MeV energy) and tritium. Protons (in the MeV range) are charged particles which cannot traverse more than 1 mm in the liquid before getting absorbed, and therefore, cannot be measured with detectors outside of the apparatus. The same problem holds true for helium-3 atoms which are nonradioactive and difficult to detect in small quantities. Neutrons are uncharged particles which can leak out of the test chamber and can be detected with suitable instrumentation. Also, tritium being a radioactive gas which remains in the test liquid can be counted for beta-decay activity (if a suitable state-of-the-art beta spectrometer is available). Therefore, testing was initiated systematically for monitoring the key signatures consisting of tritium and neutron emissions. The group used Indium (beta-decay) detectors to measure neutron flux and estimate total neutron yield at 10^8-10^{10} per explosion.

Naturally, the possibility of attaining nanoscale nuclear fusion in the cores of collapsing bubbles in liquids leads to tremendous difficulties for the theoretical description. Really we should apply the unified theory which realizes the "through" description from macroscopic level to the nuclei scale. This description cannot be realized not only on the "classical" hydrodynamic level but on the level of Schrödinger quantum mechanics (SQM) because SQM is not applicable to the nuclei problems. From this point of view no reason to estimate of fusion efficiency by solving Rayleigh-Plesset-Keller (RPK) differential equation for bubble collapse. By the way the RPK equation must be solved numerically together with the equation of state for the bubble gas. No surprise that the resulting solution is quite sensitive to the choice of the equation of state during the last stage of collapse. But the equation of state for ideal gas cannot be used for this stage which is the most interesting stage from the standpoint of nuclear fusion.

The adversaries of the cavitation-induced fusion affirm that theoretical physics does not lead to the acoustic regimes providing CF and all effect is within the experimental error.

The main objective is to investigate the dynamics of matter under the influence of the sound wave by the methods of non-local physics.

Let be the plane sound wave interacts with matter. In this case we say about the sound pressure P which can be observed for example in the stiff tube. This pressure P was calculated by Rayleigh [J.W. Strutt (Rayleigh), 1902] [272,273] and can be written as

$$P=\frac{\gamma+1}{8}\rho_{\mathrm{m}}v^2=(\gamma+1)E_{\mathrm{k}}, \tag{A.8.1}$$

where ρ_{m} is the density a surrounding medium without perturbations, v is amplitude of the sound particle velocity in the wave antinodes, E_{k} is the time- and space-average of the kinetic energy density of the sound wave, $\gamma=c_{\mathrm{p}}/c_{\mathrm{v}}$ (the ratio of the specific heat at constant pressure to the specific heat at constant volume).

We consider the 1D nonstationary matter movement under acting of the wave front. In this case the GHE equations take the form:
(continuity equation)

$$\frac{\partial}{\partial t}\left\{\rho-\tau\left[\frac{\partial\rho}{\partial t}+\frac{\partial}{\partial x}(\rho u)\right]\right\}+\frac{\partial}{\partial x}\left\{\rho u-\tau\left[\frac{\partial}{\partial t}(\rho u)+\frac{\partial}{\partial x}(\rho u^2)+\frac{\partial p}{\partial x}-F\right]\right\}=0, \tag{A.8.2}$$

(motion equation)

$$\begin{aligned}&\frac{\partial}{\partial t}\left\{\rho u-\tau\left[\frac{\partial}{\partial t}(\rho u)+\frac{\partial}{\partial x}(p+\rho u^2)-F\right]\right\}-F+\frac{F}{\rho}\tau\left(\frac{\partial\rho}{\partial t}+\frac{\partial}{\partial x}(\rho u)\right)\\&+\frac{\partial}{\partial x}\left\{\rho u^2+p-\tau\left[\frac{\partial}{\partial t}(\rho u^2+p)+\frac{\partial}{\partial x}(\rho u^3+3pu)-2uF\right]\right\}=0,\end{aligned} \tag{A.8.3}$$

(energy equation)

$$\begin{aligned}&\frac{\partial}{\partial t}\left\{\rho u^2+3p-\tau\left[\frac{\partial}{\partial t}(\rho u^2+3p)+\frac{\partial}{\partial x}(\rho u^3+5pu)-2Fu\right]\right\}\\&+\frac{\partial}{\partial x}\left\{\rho u^3+5pu-\tau\left[\frac{\partial}{\partial t}(\rho u^3+5pu)+\frac{\partial}{\partial x}\left(\rho u^4+8pu^2+5\frac{p^2}{\rho}\right)-F\left(3u^2+5\frac{p}{\rho}\right)\right]\right\}\\&-2uF+2\tau\frac{F}{\rho}\left[\frac{\partial}{\partial t}(\rho u)+\frac{\partial}{\partial x}(\rho u^2+p)-\rho F\right]=0.\end{aligned} \tag{A.8.4}$$

Introduce the coordinate system moving along the positive x direction of the 1D space with the velocity $C = u_0$ which is equal to phase velocity of the investigated quantum object

$$\xi = x - Ct. \tag{A.8.5}$$

In the coordinate system moving with the phase velocity, indestructible soliton has the velocity which is equal to the phase velocity.

We extend the usual definition of the soliton object, which should satisfy now two important conditions:

1. In moving coordinate system this object is located in the same restricted area for all time moments including the movement under the influence of external forces.
2. In the coordinate system moving with the phase velocity, indestructible soliton has the velocity which is equal to the phase velocity for all parts of moving object.

Therefore we use the moving coordinate system $\xi = x - ut$. In this system all dependent hydrodynamic values are functions of (ξ, t). We investigate the possibility of the soliton creation. For this type of solution the explicit time dependence does not exist in the considered coordinate system. Write down the system of Eqs. (A.8.2)–(A.8.4) in the dimensionless form. All dimensionless values are marked by tilde. Introduce the scales $\rho \to \rho_0$, $u \to u_0$, $p \to \rho_0 u_0^2$, $\xi \to \xi_0$. We use the approximation of the nonlocality parameter τ in the form

$$\tau = \frac{\hbar}{mu^2}, \tag{A.8.6}$$

where $\hbar$ is the Plank constant. Dimensionless continuity equation is rewritten as

$$\tilde{\rho}\frac{\partial \tilde{u}}{\partial \tilde{\xi}} - \frac{\hbar}{m\tilde{u}^2}\frac{1}{u_0 \xi_0}\tilde{\rho}\left(\frac{\partial \tilde{u}}{\partial \tilde{\xi}}\right)^2 - \frac{\hbar}{mu_0}\frac{1}{\xi_0}\frac{\partial}{\partial \tilde{\xi}}\left\{\frac{1}{\tilde{u}^2}\left[\frac{\partial \tilde{p}}{\partial \tilde{\xi}} - \frac{\xi_0}{\rho_0 u_0^2}F\right]\right\} = 0. \tag{A.8.7}$$

Introduce the parameter

$$H = \frac{\hbar}{mu_0 \xi_0} \tag{A.8.8}$$

and (A.8.7), (A.8.8) yield

$$\tilde{\rho}\frac{\partial \tilde{u}}{\partial \tilde{\xi}} - H\frac{1}{\tilde{u}^2}\tilde{\rho}\left(\frac{\partial \tilde{u}}{\partial \tilde{\xi}}\right)^2 - H\frac{\partial}{\partial \tilde{\xi}}\left\{\frac{1}{\tilde{u}^2}\left[\frac{\partial \tilde{p}}{\partial \tilde{\xi}} - \tilde{F}\right]\right\} = 0. \tag{A.8.9}$$

The force scale F_0 is used in (A.8.9)

$$F_0 = \frac{\xi_0}{\rho_0 u_0^2}. \tag{A.8.10}$$

The motion equation

$$\begin{aligned} &H\tilde{u}\frac{\partial}{\partial \tilde{\xi}}\left\{\frac{1}{\tilde{u}^2}\left[\frac{\partial \tilde{p}}{\partial \tilde{\xi}} - \tilde{F}\right]\right\} - \tilde{F}\left[1 - H\frac{1}{\tilde{u}^2}\frac{\partial \tilde{u}}{\partial \tilde{\xi}}\right] + \tilde{\rho}\tilde{u}\frac{\partial \tilde{u}}{\partial \tilde{\xi}} + \frac{\partial \tilde{p}}{\partial \tilde{\xi}} - H\frac{1}{\tilde{u}}\tilde{\rho}\left(\frac{\partial \tilde{u}}{\partial \tilde{\xi}}\right)^2 \\ &-H\frac{\partial}{\partial \tilde{\xi}}\left\{\frac{1}{\tilde{u}^2}\left[3\tilde{p}\frac{\partial \tilde{u}}{\partial \tilde{\xi}} + 2\tilde{u}\left(\frac{\partial \tilde{p}}{\partial \tilde{\xi}} - \tilde{F}\right)\right]\right\} = 0 \end{aligned} \tag{A.8.11}$$

and the energy equation

$$\begin{aligned} &\tilde{\rho}\tilde{u}^2\frac{\partial \tilde{u}}{\partial \tilde{\xi}} + 2\tilde{u}\frac{\partial \tilde{p}}{\partial \tilde{\xi}} + 5\tilde{p}\frac{\partial \tilde{u}}{\partial \tilde{\xi}} - 2\tilde{F}\tilde{u} - H\tilde{u}\frac{\partial}{\partial \tilde{\xi}}\left\{\frac{1}{\tilde{u}}\frac{\partial \tilde{p}}{\partial \tilde{\xi}}\right\} - 6H\tilde{u}\frac{\partial}{\partial \tilde{\xi}}\left\{\frac{1}{\tilde{u}^2}\tilde{p}\frac{\partial \tilde{u}}{\partial \tilde{\xi}}\right\} \\ &-3H\frac{1}{\tilde{u}}\frac{\partial \tilde{p}}{\partial \tilde{\xi}}\frac{\partial \tilde{u}}{\partial \tilde{\xi}} - 11H\frac{1}{\tilde{u}^2}\tilde{p}\left(\frac{\partial \tilde{u}}{\partial \tilde{\xi}}\right)^2 - H\tilde{\rho}\left(\frac{\partial \tilde{u}}{\partial \tilde{\xi}}\right)^2 - 5H\frac{\partial}{\partial \tilde{\xi}}\left\{\frac{1}{\tilde{u}^2}\frac{\partial}{\partial \tilde{\xi}}\left(\frac{\tilde{p}^2}{\tilde{\rho}}\right)\right\} + H\tilde{u}\frac{\partial}{\partial \tilde{\xi}}\left\{\frac{1}{\tilde{u}}\tilde{F}\right\} \\ &+5H\frac{1}{\tilde{u}}\tilde{F}\frac{\partial \tilde{u}}{\partial \tilde{\xi}} + 5H\frac{\partial}{\partial \tilde{\xi}}\left\{\frac{1}{\tilde{u}^2}\tilde{F}\frac{\tilde{p}}{\tilde{\rho}}\right\} + 2H\frac{1}{\tilde{u}^2}\frac{\tilde{F}}{\tilde{\rho}}\left(\frac{\partial \tilde{p}}{\partial \tilde{\xi}} - \tilde{F}\right) = 0 \end{aligned} \tag{A.8.12}$$

are subjected to the same transformations.

In the Rayleigh theory the value $\widetilde{F}$ is the constant dimensionless parameter. The solution of Eqs. (A.8.9), (A.8.11), and (A.8.12) can be simplified after transformation of the mentioned equations into the one parametric system using the special choice of the length scale ξ_0. Namely

$$\xi_0 = \frac{\hbar}{mu_0}, \tag{A.8.13}$$

then $H = 1$, and the force scale is

$$F_0 = \rho_0 u_0^3 \frac{m}{\hbar}\left[\frac{\text{dyne}}{\text{cm}^3}\right]. \tag{A.8.14}$$

The introduced scales have the transparent physical sense. Really, let us introduce the quantum Reynolds number and transform this number using the introduced scales:

$$\mathrm{Re} = \frac{\rho_0 u_0 \xi_0}{\mu_0} = \frac{\rho_0 u_0 \xi_0}{\upsilon_0 \rho_0} = \frac{u_0 \xi_0}{\upsilon_0} = \frac{u_0}{\upsilon_0}\frac{\hbar}{mu_0} = \frac{mu_0}{\hbar}\frac{\hbar}{mu_0} = 1. \tag{A.8.15}$$

Then we are dealing with the matter flow for $\mathrm{Re} = 1$. We need to solve the following equations

$$\widetilde{\rho}\frac{\partial \widetilde{u}}{\partial \widetilde{\xi}} - \frac{1}{\widetilde{u}^2}\widetilde{\rho}\left(\frac{\partial \widetilde{u}}{\partial \widetilde{\xi}}\right)^2 - \frac{\partial}{\partial \widetilde{\xi}}\left\{\frac{1}{\widetilde{u}^2}\left[\frac{\partial \widetilde{p}}{\partial \widetilde{\xi}} - \widetilde{F}\right]\right\} = 0, \tag{A.8.16}$$

$$\begin{gathered}\widetilde{u}\frac{\partial}{\partial \widetilde{\xi}}\left\{\frac{1}{\widetilde{u}^2}\left[\frac{\partial \widetilde{p}}{\partial \widetilde{\xi}} - \widetilde{F}\right]\right\} - \widetilde{F}\left[1 - \frac{1}{\widetilde{u}^2}\frac{\partial \widetilde{u}}{\partial \widetilde{\xi}}\right] + \widetilde{\rho}\widetilde{u}\frac{\partial \widetilde{u}}{\partial \widetilde{\xi}} + \frac{\partial \widetilde{p}}{\partial \widetilde{\xi}} \\ - \frac{1}{\widetilde{u}}\widetilde{\rho}\left(\frac{\partial \widetilde{u}}{\partial \widetilde{\xi}}\right)^2 - \frac{\partial}{\partial \widetilde{\xi}}\left\{\frac{1}{\widetilde{u}^2}\left[3\widetilde{p}\frac{\partial \widetilde{u}}{\partial \widetilde{\xi}} + 2\widetilde{u}\left(\frac{\partial \widetilde{p}}{\partial \widetilde{\xi}} - \widetilde{F}\right)\right]\right\} = 0,\end{gathered} \tag{A.8.17}$$

$$\begin{gathered}\widetilde{\rho}\widetilde{u}^2\frac{\partial \widetilde{u}}{\partial \widetilde{\xi}} + 2\widetilde{u}\frac{\partial \widetilde{p}}{\partial \widetilde{\xi}} + 5\widetilde{p}\frac{\partial \widetilde{u}}{\partial \widetilde{\xi}} - 2\widetilde{F}\widetilde{u} - \widetilde{u}\frac{\partial}{\partial \widetilde{\xi}}\left\{\frac{1}{\widetilde{u}}\frac{\partial \widetilde{p}}{\partial \widetilde{\xi}}\right\} - 6\widetilde{u}\frac{\partial}{\partial \widetilde{\xi}}\left\{\frac{1}{\widetilde{u}^2}\widetilde{p}\frac{\partial \widetilde{u}}{\partial \widetilde{\xi}}\right\} \\ -3\frac{1}{\widetilde{u}}\frac{\partial \widetilde{p}}{\partial \widetilde{\xi}}\frac{\partial \widetilde{u}}{\partial \widetilde{\xi}} - 11\frac{1}{\widetilde{u}^2}\widetilde{p}\left(\frac{\partial \widetilde{u}}{\partial \widetilde{\xi}}\right)^2 - \widetilde{\rho}\left(\frac{\partial \widetilde{u}}{\partial \widetilde{\xi}}\right)^2 - 5\frac{\partial}{\partial \widetilde{\xi}}\left\{\frac{1}{\widetilde{u}^2}\frac{\partial}{\partial \widetilde{\xi}}\left(\frac{\widetilde{p}^2}{\widetilde{\rho}}\right)\right\} + \widetilde{u}\frac{\partial}{\partial \widetilde{\xi}}\left\{\frac{1}{\widetilde{u}}\widetilde{F}\right\} \\ +5\frac{1}{\widetilde{u}}\widetilde{F}\frac{\partial \widetilde{u}}{\partial \widetilde{\xi}} + 5\frac{\partial}{\partial \widetilde{\xi}}\left\{\frac{1}{\widetilde{u}^2}\widetilde{F}\frac{\widetilde{p}}{\widetilde{\rho}}\right\} + 2\frac{1}{\widetilde{u}^2}\frac{\widetilde{F}}{\widetilde{\rho}}\left(\frac{\partial \widetilde{p}}{\partial \widetilde{\xi}} - \widetilde{F}\right) = 0.\end{gathered} \tag{A.8.18}$$

Equations (A.8.16)–(A.8.18) constitute the one parametric Cauchy problem as the system of the three ordinary differential equations of the second order. Technical computing software Maple allows realizing the vast mathematical modeling using the variation of the six Cauchy conditions and the $\widetilde{F}$ parameter. Let us show the results of some calculations using the Maple notations: $\widetilde{\xi} \to t$, $\widetilde{u}\left(\widetilde{\xi}\right) \to u(t)$, $\widetilde{\rho}\left(\widetilde{\xi}\right) \to r(t)$, $\widetilde{p}\left(\widetilde{\xi}\right) \to p(t)$, $\widetilde{F} \to F$. For the chosen Cauchy conditions we find

$$r(0) = 1,\quad u(0) = 1,\quad p(0) = 1,\quad D(r)(0) = 0,\quad D(u)(0) = 0,\quad D(p)(0) = 0. \tag{A.8.19}$$

Figures A.8.1–A.8.6 contain the result of calculations for the following set of $\widetilde{F}$ data, namely $\widetilde{F}$ is equal to: 0.01; 0.1; 1; 10; 100; 1000.

Parameter $\widetilde{F}$ defines the force of the sound action on matter; varying over 8 orders of this parameter $\widetilde{F}$ leads to the radical reconstruction of the flow. Namely:

1. At the condition $\widetilde{F} < 1$ there are no solitons (see Figs. A.8.1 and A.8.2).
2. The condition $\widetilde{F} = 0$ leads to the trivial solution $\widetilde{u} = \widetilde{\rho} = \widetilde{p} = 1$.
3. The condition $\widetilde{F} > 1$ leads to the soliton creation (Figs. A.8.3–A.8.6).

The structure of the creating solitons has the very remarkable features:

1. As it can be expected in the soliton theory, all soliton parts move with the same velocity—the condition $\widetilde{u} = 1$ fulfills with the high accuracy. The soliton is placed in the bounded region of space. Important to underline, that we deal with the Cauchy problem. It means that the mentioned effect is a product of the matter self-organization.

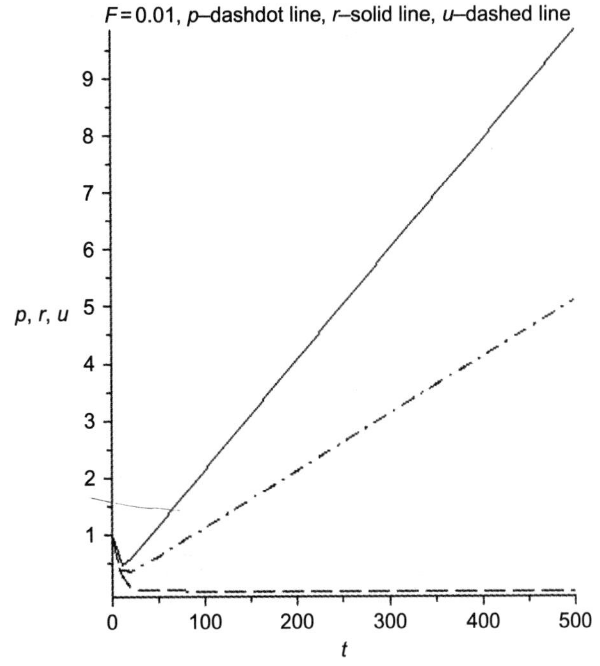

FIGURE A.8.1 The change of the density (solid line), the pressure (dash dot line), the velocity (dashed line) in the moving coordinate system at $\widetilde{F}=0.01$.

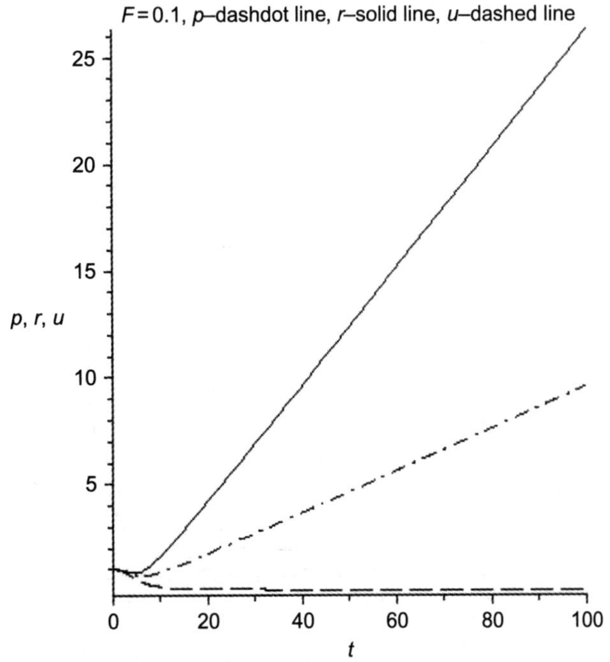

FIGURE A.8.2 The change of the density (solid line), the pressure (dash dot line), the velocity (dashed line) in the moving coordinate system at $\widetilde{F}=0.1$.

2. The linear soliton size diminishes with the $\widetilde{F}$ increase. Let us give the concrete data: if $\widetilde{F}=1 \rightarrow \widetilde{\xi}_{\lim}=2.797$; if $\widetilde{F}=10 \rightarrow \widetilde{\xi}_{\lim}=0.182$; if $\widetilde{F}=100 \rightarrow \widetilde{\xi}_{\lim}=0.017131754$; if $\widetilde{F}=1000 \rightarrow \widetilde{\xi}_{\lim}=0.0017028$; if $\widetilde{F}=10000 \rightarrow \widetilde{\xi}_{\lim}=0.00017017951$; if $\widetilde{F}=10^5 \rightarrow \widetilde{\xi}_{\lim}=0.000017016920$.

3. If $\widetilde{F}>100$, then $\widetilde{F}$ increasing tenfold more leads to diminishing tenfold less of the soliton size. It means that for the large $\widetilde{F}$ the value $\widetilde{x}$ is close to $\widetilde{t}$.

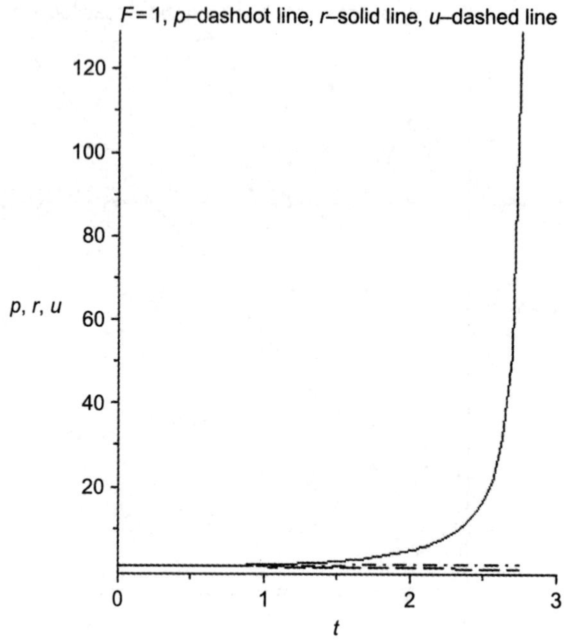

FIGURE A.8.3 The change of the density (solid line), the pressure (dash dot line), the velocity (dashed line) in the moving coordinate system at $\widetilde{F}=1$.

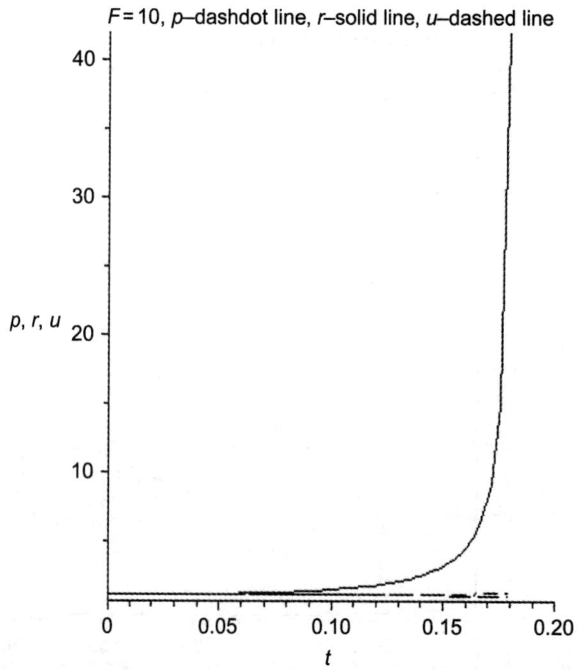

FIGURE A.8.4 The change of the density (solid line), the pressure (dash dot line), the velocity (dashed line) in the moving coordinate system at $\widetilde{F}=10$.

4. We can watch the grossly density change on the soliton front without significant pressure changing. It is desired effect which the CF adherents try to prove. Let us deliver the concrete numerical results because the graphic illustrations do not reflect the scale of dramatic changing.

For $\widetilde{F}=100$ and $\widetilde{\xi}=0.00171317539$ we find $\widetilde{p}=1.000884428$; $\widetilde{u}=0.990726$; $\widetilde{\rho}=2.7810799\cdot 10^7$. For the point placed closer to $\widetilde{\xi}_{\lim}=0.017131754$ we have: if $\widetilde{\xi}=0.0017131753995$ then $\widetilde{p}=1.000884429$, $\widetilde{u}=0.990726$, $\widetilde{\rho}=4.4694082\cdot 10^7$.

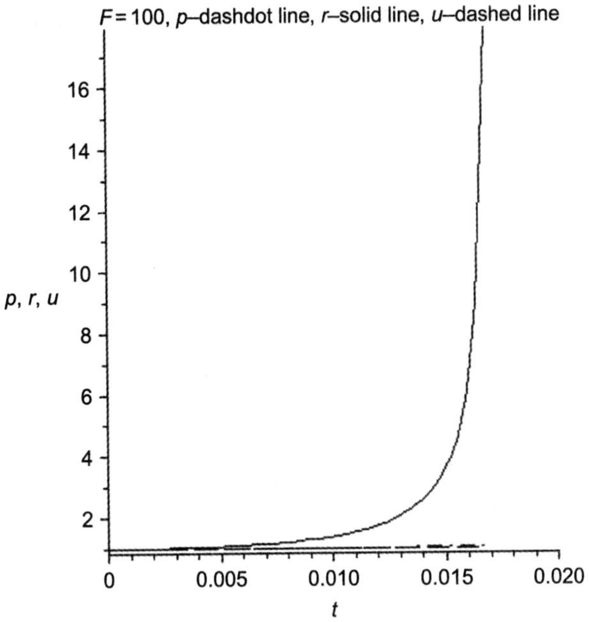

FIGURE A.8.5 The change of the density (solid line), the pressure (dash dot line), the velocity (dashed line) in the moving coordinate system at $\widetilde{F} = 100$.

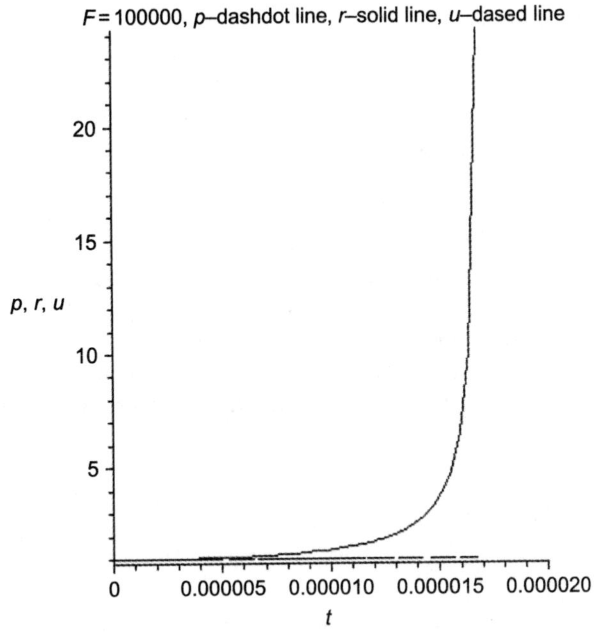

FIGURE A.8.6 The change of the density (solid line), the pressure (dash dot line), the velocity (dashed line) in the moving coordinate system at $\widetilde{F} = 10^5$.

For $\widetilde{F} = 10^5$ ($\widetilde{\xi}_{\lim} = 0.17016920 \cdot 10^{-4}$) and $\widetilde{\xi} = 0.17016919 \cdot 10^{-4}$ we find $\widetilde{p} = 1.0000088907$; $\widetilde{u} = 0.9999907284$; $\widetilde{\rho} = 1.2247653 \cdot 10^7$. If $\widetilde{\xi} = 0.1701691955 \cdot 10^{-4}$ then $\widetilde{p} = 1.0000088907$, $\widetilde{u} = 0.9999907284$, $\widetilde{\rho} = 5.05040680 \cdot 10^8$.

5. The density increase in the milliard times on the soliton front can lead to the effect of CF. If required, the density can be normalized to the total mass $\widetilde{M}$ (known from an experiment) which falls on the unit of the wave front area

$$\widetilde{M} = \int_0^{\xi_{\exp}} \widetilde{\rho} \, d\widetilde{\xi}. \quad \text{(A.8.20)}$$

Let us go now to the formal solution of so called "classical" hydrodynamic equations in the frame of choosing of the plane travelling longitudinal wave. It seems that for this case we should solve numerically the system of Eqs. (A.8.9), (A.8.11), and (A.8.12) for $H=0$. But we need not to do it, because the analytical solution can be found. Really the system of equation has the form for the mentioned case:
(continuity equation)

$$-u\frac{\partial \rho}{\partial \xi}+\frac{\partial}{\partial \xi}(\rho u)=0, \tag{A.8.21}$$

(motion equation)

$$-u\frac{\partial}{\partial \xi}(\rho u)-F+\frac{\partial}{\partial \xi}(\rho u^2+p)=0, \tag{A.8.22}$$

(energy equation)

$$-u\frac{\partial}{\partial \xi}(\rho u^2+3p)+\frac{\partial}{\partial \xi}(\rho u^3+5pu)-2Fu=0. \tag{A.8.23}$$

Equation (A.8.21) yields

$$\rho\frac{\partial u}{\partial \xi}=0, \tag{A.8.24}$$

or

$$u=\text{const.} \tag{A.8.25}$$

From the motion equation (A.8.22) follows

$$\partial p/\partial \xi=F. \tag{A.8.26}$$

Energy equation (A.8.23) does not deliver a new independent relation and returns us to the relation (A.8.26). Then for the constant F Eq. (A.8.26) leads to the trivial (and as we see, wrong) dependence, which does not contain the matter density

$$p=F\xi+\text{const.} \tag{A.8.27}$$

In spite of all experimental problems and difficulties, the cavitation-induced fusion or generally speaking, acoustic cold fusion (ACF) has the serious experimental confirmation. There is the obvious contradiction between the mentioned experimental results and conclusions of the classical local hydrodynamics. As we see local hydrodynamics is not applicable to the description of the ACF in principal. The realized mathematical modeling leads to the grossly density change on the soliton front without significant pressure changing. It is desired effect which the CF adherents try to prove. Then the following step of the investigation consists in the numerical solution of the 3D nonstationary boundary problem on the basement of unified non-local transport theory.

Appendix 9

To the Non-Local Theory of Variable Stars

A star is classified as variable if its apparent brightness as seen from Earth changes over time. Most stars have at least some variation in luminosity: the energy output of our Sun, for example, varies by about 0.1% over an 11 year solar cycle. Moreover, non-local theory of the global Sun oscillations can lead to the solution of the neutrino deficit problem without application the theory of neutrino transformations [274]. Edition of the General Catalogue of Variable Stars (2003) [http://www.sai.msu.su/gcvs/gcvs] lists nearly 40,000 variable stars in our own galaxy, as well as 10,000 in other galaxies, and over 10,000 "suspected" variables.

Variable stars can be classified as intrinsic variable stars (IVS) or extrinsic variable stars (EVS). The variability IVS is being caused by changes in the physical properties of the stars themselves. This category can be divided into three subgroups:

(a) Pulsating variables, stars whose radius alternately expands and contracts as part of their natural evolutionary ageing processes.
(b) Eruptive variables, stars that experience eruptions on their surfaces like flares or mass ejections.
(c) Cataclysmic or explosive variables, stars that undergo a cataclysmic change in their properties like novae and supernovae.

EVS change the brightness due to changes in the amount of their light that can reach Earth; for example, because the star has an orbiting companion that sometimes eclipses it.

In following we intend to apply methods of non-local physics to the stellar pulsation modeling for the case IVSa.

The stellar pulsation modeling has a long history. In 1918 Eddington linearized the equations of the adiabatic model and obtained the second-order ODE describing small harmonic oscillations near the adiabatic equilibrium of a star [275], see also [276,277]. This model was called the standard one. But real pulsations (for example of the most of Cepheids) are distinctly nonlinear [278]. A considerable amount of time has passed before the next step in the stellar pulsation modeling was done. Zhevakin considered linear nonadiabatic oscillations of stars involving multiple "discrete" concentric spherical gas layers [279]. It was supposed that each layer is infinitely thin but encloses a certain part of the complete mass of the star. An analogous "discrete" multilayer model was used by Aleshin for investigation of nonlinear adiabatic stellar pulsations [280]. In 1964 Aleshin applied this technique for integration of the problem of nonlinear nonadiabatic stellar pulsations [281]. His "discrete" model included 10 concentric layers in a stellar envelope and he chases an initial motion to be a small-amplitude solution of the linearized system. The pulsations found were near-sinusoidal in deeper layers but greatly diverged from the sinusoidal form in upper layers after few pulsation periods.

Emergence of high-speed computers has allowed making more detailed calculations. No reason to discuss here the results of mathematical modeling which are based on the local hydrodynamic models. The nonstationary non-local 3D system of hydrodynamic equations (see Appendix 2) describing the star explosion leads to the significant computational difficulties. But the nonstationary 3D challenge can be reduced to the nonstationary 1D problem if we consider the regular oscillations (travelling waves) for the case of the spherical symmetry. Let us consider the transformation of the Poisson equation following this idea. Poisson equation is written as

$$\frac{1}{r^2}\frac{\partial}{\partial r}\left(r^2\frac{\partial\psi}{\partial r}\right)=4\pi\gamma_{\mathrm{N}}\left[\rho-\tau\left(\frac{\partial\rho}{\partial t}+\frac{\partial(\rho v_{\mathrm{r}})}{\partial r}+2\rho\frac{v_{\mathrm{r}}}{r}\right)\right]. \tag{A.9.1}$$

Relation

$$v_{\mathrm{r}}/r=\upsilon_{\mathrm{p}} \tag{A.9.2}$$

defines the local value of the pulsation frequency which is a parameter of calculations. Using (A.9.2) one obtains

$$\frac{\partial^2\psi}{\partial r^2}+2\upsilon_{\mathrm{p}}\frac{1}{v_{\mathrm{r}}}\frac{\partial\psi}{\partial r}=4\pi\gamma_{\mathrm{N}}\left[\rho-\tau\left(\frac{\partial\rho}{\partial t}+\frac{\partial(\rho v_{\mathrm{r}})}{\partial r}+2\upsilon_{\mathrm{p}}\rho\right)\right]. \tag{A.9.3}$$

Analogically we find continuity equation

$$\begin{aligned}&\frac{\partial}{\partial t}\left\{\rho-\tau\left[\frac{\partial \rho}{\partial t}+\frac{\partial(\rho v_{\mathrm{r}})}{\partial r}+2\upsilon_{\mathrm{p}}\rho\right]\right\}+\frac{\partial}{\partial r}\left\{\rho v_{\mathrm{r}}-\tau\left[\frac{\partial}{\partial t}(\rho v_{\mathrm{r}})+\frac{\partial(\rho v_{\mathrm{r}}^2)}{\partial r}+2\upsilon_{\mathrm{p}}\rho v_{\mathrm{r}}-\rho g_{\mathrm{r}}\right]\right\}\\&+2\upsilon_{\mathrm{p}}\left\{\rho-\tau\left[\frac{1}{v_{\mathrm{r}}}\frac{\partial}{\partial t}(\rho v_{\mathrm{r}})+\frac{\partial(\rho v_{\mathrm{r}})}{\partial r}+\rho\frac{\partial v_{\mathrm{r}}}{\partial r}+2\upsilon_{\mathrm{p}}\rho-\rho\frac{g_{\mathrm{r}}}{v_{\mathrm{r}}}\right]\right\}\\&-\frac{\partial}{\partial r}\left(\tau\frac{\partial p}{\partial r}\right)-2\tau\upsilon_{\mathrm{p}}\frac{1}{v_{\mathrm{r}}}\frac{\partial p}{\partial r}=0,\end{aligned}\tag{A.9.4}$$

motion equation

$$\begin{aligned}&\frac{\partial}{\partial t}\left\{\rho v_{\mathrm{r}}-\tau\left[\frac{\partial}{\partial t}(\rho v_{\mathrm{r}})+\frac{\partial(\rho v_{\mathrm{r}}^2)}{\partial r}+2\upsilon_{\mathrm{p}}\rho v_{\mathrm{r}}+\frac{\partial p}{\partial r}-\rho g_{\mathrm{r}}\right]\right\}-g_{\mathrm{r}}\left[\rho-\tau\left(\frac{\partial \rho}{\partial t}+\frac{\partial(\rho v_{\mathrm{r}})}{\partial r}+2\upsilon_{\mathrm{p}}\rho\right)\right]\\&+\frac{\partial}{\partial r}\left\{\rho v_{\mathrm{r}}^2-\tau\left[\frac{\partial}{\partial t}(\rho v_{\mathrm{r}}^2)+\frac{\partial(\rho v_{\mathrm{r}}^3)}{\partial r}+2\upsilon_{\mathrm{p}}\rho v_{\mathrm{r}}^2-2g_{\mathrm{r}}\rho v_{\mathrm{r}}\right]\right\}\\&+2\upsilon_{\mathrm{p}}\left\{\rho v_{\mathrm{r}}-\tau\left[\frac{1}{v_{\mathrm{r}}}\frac{\partial}{\partial t}(\rho v_{\mathrm{r}}^2)+\frac{\partial(\rho v_{\mathrm{r}}^2)}{\partial r}+\rho v_{\mathrm{r}}\frac{\partial v_{\mathrm{r}}}{\partial r}+2\upsilon_{\mathrm{p}}\rho v_{\mathrm{r}}-2g_{\mathrm{r}}\rho\right]\right\}+\frac{\partial p}{\partial r}-\frac{\partial}{\partial r}\left(\tau\frac{\partial p}{\partial t}\right)\\&-3\frac{\partial}{\partial r}\left(\tau\frac{\partial(p v_{\mathrm{r}})}{\partial r}\right)-6\upsilon_{\mathrm{p}}\tau\frac{\partial p}{\partial r}-4\upsilon_{\mathrm{p}}p\frac{\partial \tau}{\partial r}-2\upsilon_{\mathrm{p}}\tau p\frac{1}{v_{\mathrm{r}}}\frac{\partial v_{\mathrm{r}}}{\partial r}=0,\end{aligned}\tag{A.9.5}$$

and energy equation

$$\begin{aligned}&\frac{\partial}{\partial t}\left\{\rho v_{\mathrm{r}}^2+3p+2\varepsilon n-\tau\left[\frac{\partial}{\partial t}(\rho v_{\mathrm{r}}^2+3p+2\varepsilon n)+\frac{\partial}{\partial r}(v_{\mathrm{r}}(\rho v_{\mathrm{r}}^2+5p+2\varepsilon n))+2\upsilon_{\mathrm{p}}(\rho v_{\mathrm{r}}^2+5p+2\varepsilon n)-2\rho g_{\mathrm{r}}v_{\mathrm{r}}\right]\right\}\\&+\frac{\partial}{\partial r}\left\{(\rho v_{\mathrm{r}}^2+5p+2\varepsilon n)v_{\mathrm{r}}-\tau\left[\frac{\partial}{\partial t}((\rho v_{\mathrm{r}}^2+5p+2\varepsilon n)v_{\mathrm{r}})+\frac{\partial}{\partial r}((\rho v_{\mathrm{r}}^2+7p+2\varepsilon n)v_{\mathrm{r}}^2)\right.\right.\\&\left.\left.+2\upsilon_{\mathrm{p}}v_{\mathrm{r}}(\rho v_{\mathrm{r}}^2+7p+2\varepsilon n)-(3\rho v_{\mathrm{r}}^2+3p+2\varepsilon n)g_{\mathrm{r}}\right]\right\}\\&+2\upsilon_{\mathrm{p}}\left\{\rho v_{\mathrm{r}}^2+5p+2\varepsilon n-\frac{\tau}{v_{\mathrm{r}}}\left[\frac{\partial}{\partial t}((\rho v_{\mathrm{r}}^2+5p+2\varepsilon n)v_{\mathrm{r}})+\frac{\partial}{\partial r}((\rho v_{\mathrm{r}}^2+7p+2\varepsilon n)v_{\mathrm{r}}^2)+2\upsilon_{\mathrm{p}}v_{\mathrm{r}}(\rho v_{\mathrm{r}}^2+7p+2\varepsilon n)\right.\right.\\&\left.\left.-(3\rho v_{\mathrm{r}}^2+3p+2\varepsilon n)g_{\mathrm{r}}\right]\right\}\\&-2\left\{\rho g_{\mathrm{r}}v_{\mathrm{r}}-\tau\left[g_{\mathrm{r}}\left(\frac{\partial}{\partial t}(\rho v_{\mathrm{r}})+\frac{\partial}{\partial r}(\rho v_{\mathrm{r}}^2)+2\upsilon_{\mathrm{p}}\rho v_{\mathrm{r}}+\frac{\partial p}{\partial r}-\rho g_{\mathrm{r}}\right)\right]\right\}\\&-\frac{\partial}{\partial r}\left(\tau\frac{\partial}{\partial r}\left(pv_{\mathrm{r}}^2+5\frac{p^2}{\rho}+2\varepsilon\frac{p}{m}\right)\right)-2\upsilon_{\mathrm{p}}\tau\frac{1}{v_{\mathrm{r}}}\frac{\partial}{\partial r}\left(pv_{\mathrm{r}}^2+5\frac{p^2}{\rho}+2\varepsilon\frac{p}{m}\right)+2\frac{\partial}{\partial r}(\tau p g_{\mathrm{r}})+4\upsilon_{\mathrm{p}}\tau\frac{g_{\mathrm{r}}}{v_{\mathrm{r}}}p=0.\end{aligned}\tag{A.9.6}$$

Here ε—internal energy for the particles, v_{r}—radial hydrodynamic velocity, τ—non-local parameter and

$$g_{\mathrm{r}}=-\partial\psi/\partial r,\tag{A.9.7}$$

where ψ is gravitational potential. The system of Eqs. (A.9.3)–(A.9.7) belongs to the class of the 1D nonstationary equations and can be solved by known numerical methods. Let us transform this system of equations.

In following we intend to investigate the possibility of the travelling wave's appearance in variable stars. Two problems of the principal significance are investigated:

(A) Is it possible to find the solution of non-local equations after the initial perturbations in the form of travelling waves?
(B) Is it possible to take into account the effect of the gravitational self-catching leading to periodic star pulsations? Both questions have the positive answers.

For solution of these problems we suppose:

1. One species object has spherical symmetry,
2. For the periodic star's pulsation we use the condition

$$\tilde{\upsilon}_{\mathrm{p}} = \mathrm{const.} \tag{A.9.8}$$

Let us introduce now new variable

$$\xi = r - Ct, \tag{A.9.9}$$

where C is phase velocity of the travelling wave. We apply the dimensionless form of non-local equations (A.9.3)–(A.9.7) using the scales ρ_0, u_0, $r_0 = u_0 t_0$, $\psi_0 = u_0^2$, $\gamma_{\mathrm{N0}} = u_0^2/(\rho_0 r_0^2)$, $p_0 = \rho_0 u_0^2$ and condition $\widetilde{C} = C/u_0 = 1$. It means in following $G = \widetilde{\gamma}_{\mathrm{N}} = \gamma_{\mathrm{N}} \frac{\rho_0 r_0^2}{u_0^2} = \gamma_{\mathrm{N}} \rho_0 t_0^2$, $\gamma_{\mathrm{N}} = 6.6 \times 10^{-8}\,\mathrm{cm}^3/(\mathrm{g s}^2)$. White dwarf stars have the density, which come in range of about 10^4-10^7 g/cm^3. Our sun has a density of about 1 g/cm^3. Neutron stars are even more dense than white dwarfs, estimated at around 10^{13} g/cm^3. Pulsars have different periods of pulsations in range from 640 impulses per second up to one impulse for every 5 s. The Sun particularly has 11 year activity cycle, which leads to the timescale $t_0 = 3.46896 \times 10^8$ s. The equations take the form:
generalized Poisson equation

$$\frac{\partial^2 \tilde{\psi}}{\partial \tilde{\xi}^2} + 2\tilde{\upsilon}_{\mathrm{p}} \frac{1}{\tilde{v}_{\mathrm{r}}} \frac{\partial \tilde{\psi}}{\partial \tilde{\xi}} = 4\pi \tilde{\gamma}_{\mathrm{N}} \left[\tilde{\rho} - \tilde{\tau} \left(-\frac{\partial \tilde{\rho}}{\partial \tilde{\xi}} + \frac{\partial(\tilde{\rho}\tilde{v}_{\mathrm{r}})}{\partial \tilde{\xi}} + 2\tilde{\upsilon}_{\mathrm{p}} \tilde{\rho} \right) \right], \tag{A.9.10}$$

continuity equation

$$\begin{aligned}
&-\frac{\partial}{\partial \tilde{\xi}} \left\{ \tilde{\rho} - \tilde{\tau} \left[-\frac{\partial \tilde{\rho}}{\partial \tilde{\xi}} + \frac{\partial(\tilde{\rho}\tilde{v}_{\mathrm{r}})}{\partial \tilde{\xi}} + 2\tilde{\upsilon}_{\mathrm{p}} \tilde{\rho} \right] \right\} \\
&+\frac{\partial}{\partial \tilde{\xi}} \left\{ \tilde{\rho}\tilde{v}_{\mathrm{r}} - \tilde{\tau} \left[-\frac{\partial}{\partial \tilde{\xi}} (\tilde{\rho}\tilde{v}_{\mathrm{r}}) + \frac{\partial(\tilde{\rho}\tilde{v}_{\mathrm{r}}^2)}{\partial \tilde{\xi}} + 2\tilde{\upsilon}_{\mathrm{p}} \tilde{\rho}\tilde{v}_{\mathrm{r}} + \tilde{\rho} \frac{\partial \tilde{\psi}}{\partial \tilde{\xi}} \right] \right\} \\
&+2\tilde{\upsilon}_{\mathrm{p}} \left\{ \tilde{\rho} - \tilde{\tau} \left[-\frac{1}{\tilde{v}_{\mathrm{r}}} \frac{\partial}{\partial \tilde{\xi}} (\tilde{\rho}\tilde{v}_{\mathrm{r}}) + \frac{\partial(\tilde{\rho}\tilde{v}_{\mathrm{r}})}{\partial \tilde{\xi}} + \tilde{\rho} \frac{\partial \tilde{v}_{\mathrm{r}}}{\partial \tilde{\xi}} + 2\tilde{\upsilon}_{\mathrm{p}} \tilde{\rho} + \tilde{\rho} \frac{1}{\tilde{v}_{\mathrm{r}}} \frac{\partial \tilde{\psi}}{\partial \tilde{\xi}} \right] \right\} \\
&-\frac{\partial}{\partial \tilde{\xi}} \left(\tilde{\tau} \frac{\partial \tilde{p}}{\partial \tilde{\xi}} \right) - 2\tilde{\tau}\tilde{\upsilon}_{\mathrm{p}} \frac{1}{\tilde{v}_{\mathrm{r}}} \frac{\partial \tilde{p}}{\partial \tilde{\xi}} = 0,
\end{aligned} \tag{A.9.11}$$

motion equation

$$\begin{aligned}
&-\frac{\partial}{\partial \tilde{\xi}} \left\{ \tilde{\rho}\tilde{v}_{\mathrm{r}} - \tilde{\tau} \left[-\frac{\partial}{\partial \tilde{\xi}} (\tilde{\rho}\tilde{v}_{\mathrm{r}}) + \frac{\partial(\tilde{\rho}\tilde{v}_{\mathrm{r}}^2)}{\partial \tilde{\xi}} + 2\tilde{\upsilon}_{\mathrm{p}} \tilde{\rho}\tilde{v}_{\mathrm{r}} + 2\frac{\partial \tilde{p}}{\partial \tilde{\xi}} + \tilde{\rho} \frac{\partial \tilde{\psi}}{\partial \tilde{\xi}} \right] \right\} \\
&+\frac{\partial \tilde{\psi}}{\partial \tilde{\xi}} \left[\tilde{\rho} - \tilde{\tau} \left(-\frac{\partial \tilde{\rho}}{\partial \tilde{\xi}} + \frac{\partial(\tilde{\rho}\tilde{v}_{\mathrm{r}})}{\partial \tilde{\xi}} + 2\tilde{\upsilon}_{\mathrm{p}} \tilde{\rho} \right) \right] \\
&+\frac{\partial}{\partial \tilde{\xi}} \left\{ \tilde{\rho}\tilde{v}_{\mathrm{r}}^2 - \tilde{\tau} \left[-\frac{\partial}{\partial \tilde{\xi}} (\tilde{\rho}\tilde{v}_{\mathrm{r}}^2) + \frac{\partial(\tilde{\rho}\tilde{v}_{\mathrm{r}}^3)}{\partial \tilde{\xi}} + 2\tilde{\upsilon}_{\mathrm{p}} \tilde{\rho}\tilde{v}_{\mathrm{r}}^2 - 2\tilde{g}_{\mathrm{r}} \tilde{\rho}\tilde{v}_{\mathrm{r}} \right] \right\} \\
&+2\tilde{\upsilon}_{\mathrm{p}} \left\{ \tilde{\rho}\tilde{v}_{\mathrm{r}} - \tilde{\tau} \left[-\frac{1}{\tilde{v}_{\mathrm{r}}} \frac{\partial}{\partial \tilde{\xi}} (\tilde{\rho}\tilde{v}_{\mathrm{r}}^2) + \frac{\partial(\tilde{\rho}\tilde{v}_{\mathrm{r}}^2)}{\partial \tilde{\xi}} + \tilde{\rho}\tilde{v}_{\mathrm{r}} \frac{\partial \tilde{v}_{\mathrm{r}}}{\partial \tilde{\xi}} + 2\tilde{\upsilon}_{\mathrm{p}} \tilde{\rho}\tilde{v}_{\mathrm{r}} + 2\frac{\partial \tilde{\psi}}{\partial \tilde{\xi}} \tilde{\rho} \right] \right\} \\
&+\frac{\partial \tilde{p}}{\partial \tilde{\xi}} - 3\frac{\partial}{\partial \tilde{\xi}} \left(\tilde{\tau} \frac{\partial(\tilde{p}\tilde{v}_{\mathrm{r}})}{\partial \tilde{\xi}} \right) - 6\tilde{\tau}\tilde{\upsilon}_{\mathrm{p}} \frac{\partial \tilde{p}}{\partial \tilde{\xi}} - 4\tilde{p}\tilde{\upsilon}_{\mathrm{p}} \frac{\partial \tilde{\tau}}{\partial \tilde{\xi}} - 2\tilde{\tau}\tilde{p}\tilde{\upsilon}_{\mathrm{p}} \frac{1}{\tilde{v}_{\mathrm{r}}} \frac{\partial \tilde{v}_{\mathrm{r}}}{\partial \tilde{\xi}} = 0,
\end{aligned} \tag{A.9.12}$$

energy equation

$$
\begin{aligned}
&-\frac{\partial}{\partial\tilde{\xi}}\left\{\tilde{\rho}\tilde{v}_{\mathrm{r}}^{2}+3\tilde{p}+2\widetilde{\varepsilon n}-\tilde{\tau}\left[-\frac{\partial}{\partial\tilde{\xi}}\left(\tilde{\rho}\tilde{v}_{\mathrm{r}}^{2}+3\tilde{p}+2\widetilde{\varepsilon n}\right)+\frac{\partial}{\partial\tilde{\xi}}\left(\tilde{v}_{\mathrm{r}}\left(2\tilde{\rho}\tilde{v}_{\mathrm{r}}^{2}+10\tilde{p}+4\widetilde{\varepsilon n}\right)\right)\right.\right.\\
&\left.\left.+2\tilde{\upsilon}_{\mathrm{p}}\left(\tilde{\rho}\tilde{v}_{\mathrm{r}}^{2}+5\tilde{p}+2\widetilde{\varepsilon n}\right)+2\tilde{\rho}\frac{\partial\tilde{\psi}}{\partial\tilde{\xi}}\tilde{v}_{\mathrm{r}}\right]\right\}\\
&+\frac{\partial}{\partial\tilde{\xi}}\left\{\left(\tilde{\rho}\tilde{v}_{\mathrm{r}}^{2}+5\tilde{p}+2\widetilde{\varepsilon n}\right)\tilde{v}_{\mathrm{r}}-\tilde{\tau}\left[\frac{\partial}{\partial\tilde{\xi}}\left(\left(\rho\tilde{v}_{\mathrm{r}}^{2}+7\tilde{p}+2\widetilde{\varepsilon n}\right)\tilde{v}_{\mathrm{r}}^{2}\right)\right.\right.\\
&\left.\left.+2\tilde{\upsilon}_{\mathrm{p}}\tilde{v}_{\mathrm{r}}\left(\tilde{\rho}\tilde{v}_{\mathrm{r}}^{2}+7\tilde{p}+2\widetilde{\varepsilon n}\right)+\left(3\tilde{\rho}\tilde{v}_{\mathrm{r}}^{2}+2\widetilde{\varepsilon n}\right)\frac{\partial\tilde{\psi}}{\partial\tilde{\xi}}\right]\right\}\\
&+2\tilde{\upsilon}_{\mathrm{p}}\left\{\tilde{\rho}\tilde{v}_{\mathrm{r}}^{2}+5\tilde{p}+2\widetilde{\varepsilon n}-\frac{\tilde{\tau}}{\tilde{v}_{\mathrm{r}}}\left[-\frac{\partial}{\partial\tilde{\xi}}\left(\left(\tilde{\rho}\tilde{v}_{\mathrm{r}}^{2}+5\tilde{p}+2\widetilde{\varepsilon n}\right)\tilde{v}_{\mathrm{r}}\right)\right.\right.\\
&\left.\left.+\frac{\partial}{\partial\tilde{\xi}}\left(\left(\tilde{\rho}\tilde{v}_{\mathrm{r}}^{2}+7\tilde{p}+2\widetilde{\varepsilon n}\right)\tilde{v}_{\mathrm{r}}^{2}\right)+2\tilde{\upsilon}_{\mathrm{p}}\tilde{v}_{\mathrm{r}}\left(\tilde{\rho}\tilde{v}_{\mathrm{r}}^{2}+7\tilde{p}+2\widetilde{\varepsilon n}\right)+2\widetilde{\varepsilon n}\frac{\partial\tilde{\psi}}{\partial\tilde{\xi}}\right]\right\}\\
&+2\frac{\partial\tilde{\psi}}{\partial\tilde{\xi}}\left\{\tilde{\rho}\tilde{v}_{\mathrm{r}}-\tilde{\tau}\left[-\frac{\partial}{\partial\tilde{\xi}}\left(\tilde{\rho}\tilde{v}_{\mathrm{r}}\right)+\frac{\partial}{\partial\tilde{\xi}}\left(\tilde{\rho}\tilde{v}_{\mathrm{r}}^{2}\right)+\frac{\partial\tilde{p}}{\partial\tilde{\xi}}+\tilde{\rho}\frac{\partial\tilde{\psi}}{\partial\tilde{\xi}}\right]\right\}-10\tilde{\upsilon}_{\mathrm{p}}\tilde{\rho}\tilde{v}_{\mathrm{r}}\tilde{\tau}\frac{\partial\tilde{\psi}}{\partial\tilde{\xi}}\\
&-\frac{\partial}{\partial\tilde{\xi}}\left(\tilde{\tau}\frac{\partial}{\partial\tilde{\xi}}\left(\tilde{p}\tilde{v}_{\mathrm{r}}^{2}+5\frac{\tilde{p}^{2}}{\tilde{\rho}}+2\widetilde{\varepsilon n}\frac{\tilde{p}}{\tilde{\rho}}\right)\right)-2\tilde{\upsilon}_{\mathrm{p}}\tilde{\tau}\frac{1}{\tilde{v}_{\mathrm{r}}}\frac{\partial}{\partial\tilde{\xi}}\left(\tilde{p}\tilde{v}_{\mathrm{r}}^{2}+5\frac{p^{2}}{\rho}+2\widetilde{\varepsilon n}\frac{p}{\rho}\right)\\
&-5\frac{\partial}{\partial\tilde{\xi}}\left(\tilde{\tau}\tilde{p}\frac{\partial\tilde{\psi}}{\partial\tilde{\xi}}\right)-10\tilde{\upsilon}_{\mathrm{p}}\tilde{\tau}\frac{1}{\tilde{v}_{\mathrm{r}}}\frac{\partial\tilde{\psi}}{\partial\tilde{\xi}}\tilde{p}=0.
\end{aligned}
\tag{A.9.13}
$$

Some comments to the system of four ordinary nonlinear equations (A.9.10)–(A.9.13):

1. Every equation from the system is of the second order and needs two conditions. The problem belongs to the class of Cauchy problems.
2. In comparison for example, with the Schrödinger theory connected with behavior of the wave function, no special conditions are applied for dependent variables including the domain of the solution existing. This domain is defined automatically in the process of the numerical solution of the concrete variant of calculations.
3. From the introduced scales ρ_0, u_0, $r_0=u_0t_0$, $\psi_0=u_0^2$, $\gamma_{\mathrm{N}0}=u_0^2/(\rho_0r_0^2)$, $p_0=\rho_0u_0^2$, only three parameters are independent, namely, ρ_0, u_0, r_0.
4. Taking into account the previous considerations I introduce the following approximation for the dimensionless non-local parameter

$$\tilde{\tau}=1/\tilde{u}^2, \tag{A.9.14}$$

$$\tau=u_0r_0/u^2=v_0^k/u^2, \tag{A.9.15}$$

where the scale for the kinematical viscosity is introduced $v_0^k=u_0r_0$. Then we have the physically transparent result—non-local parameter is proportional to the kinematical viscosity and in inverse proportion to the square of velocity.

The system of generalized hydrodynamic equations (A.9.10)–(A.9.13) (solved with the help of Maple) have the great possibilities of mathematical modeling as result of changing of eight Cauchy conditions describing the character features of initial perturbations which lead to the soliton formation. The following Maple notations on figures are used: *r*—density $\tilde{\rho}$, *u*—velocity $\tilde{u}\equiv\tilde{v}_r$, *p*—pressure $\tilde{p}$ and *v*—self consistent potential $\tilde{\psi}$, *G*—$\tilde{\gamma}_{\mathrm{N}}$, independent variable *t* is $\tilde{\xi}$ and $A=\tilde{\upsilon}_{\mathrm{p}}$, *D*(*v*)—$\partial\tilde{\psi}/\partial\tilde{\xi}$, *D*(*p*)—$\partial\tilde{p}/\partial\tilde{\xi}$. We need also an approximation for the dimensionless value $\widetilde{\varepsilon n}$ having the sense as the internal energy of the unit volume. This energy $\widetilde{\varepsilon n}$ should be approximated taking into account the concrete thermonuclear processes developing in a star. For example we can use the following approximation

$$\widetilde{\varepsilon n}=e^{B\tilde{\rho}}-1, \tag{A.9.16}$$

where *B* is dimensionless parameter in calculations. Obviously the form of the mentioned approximation (A.9.16) is not of crucial significance.

Figures A.9.1–A.9.10 reflect the results numerical calculations for different parameters A, B, G, and the Cauchy conditions:

$$\begin{gathered} p(0)=1,\ u(0)=1,\ v(0)=1,\ r(0)=1, \\ D(p)(0)=0,\ D(u)(0)=0,\ D(v)(0)=0,\ D(r)(0)=0. \end{gathered} \tag{A.9.17}$$

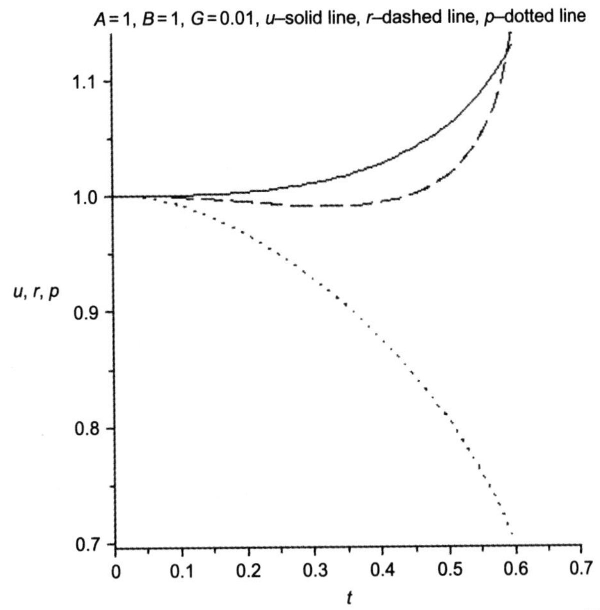

FIGURE A.9.1 The change of the velocity $\tilde{v}_r(\tilde{\xi})$ (solid line), the density $\tilde{\rho}(\tilde{\xi})$ (dashed line) and pressure $\tilde{p}(\tilde{\xi})$ (dotted line) at $A=1, B=1, G=10^{-2}$.

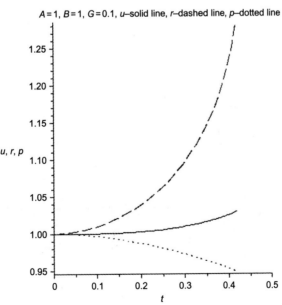

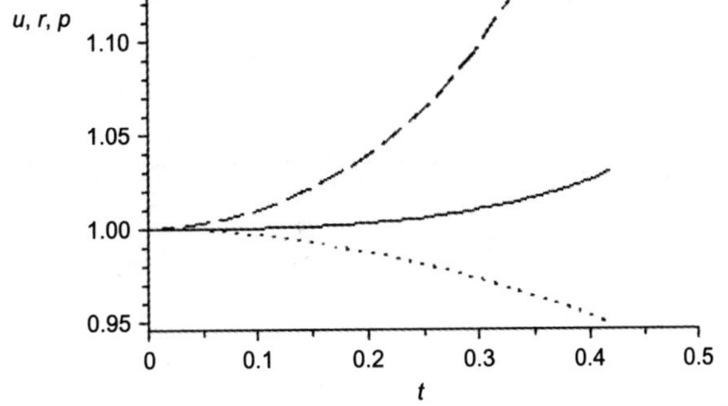

FIGURE A.9.2 The change of the velocity $\tilde{v}_r(\tilde{\xi})$ (solid line), the density $\tilde{\rho}(\tilde{\xi})$ (dashed line) and pressure $\tilde{p}(\tilde{\xi})$ (dotted line) at $A=1, B=1, G=10^{-1}$.

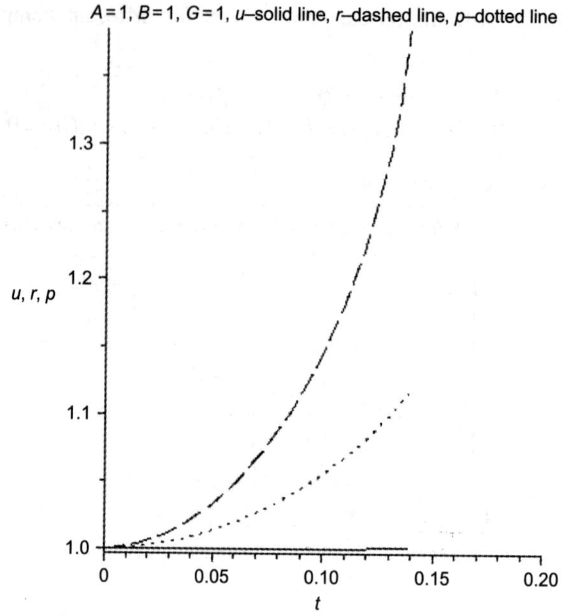

FIGURE A.9.3 The change of the velocity $\tilde{v}_r(\tilde{\xi})$ (solid line), the density $\tilde{\rho}(\tilde{\xi})$ (dashed line) and pressure $\tilde{p}(\tilde{\xi})$ (dotted line) at $A=1$, $B=1$, $G=1$.

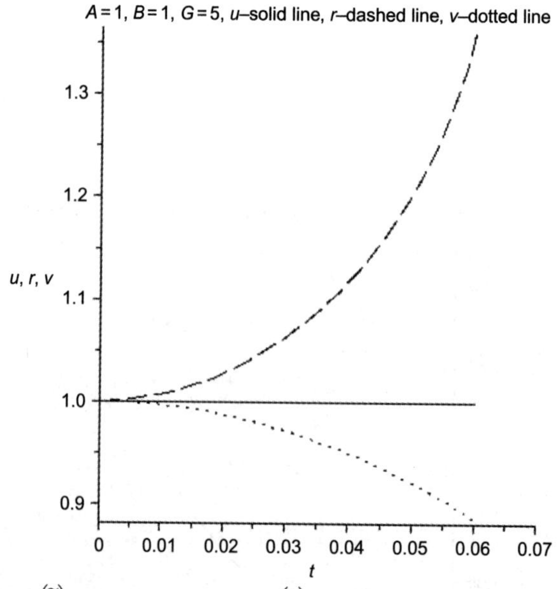

FIGURE A.9.4 The change of the velocity $\tilde{v}_r(\tilde{\xi})$ (solid line), the density $\tilde{\rho}(\tilde{\xi})$ (dashed line) and gravitational potential v—$\tilde{\psi}(\tilde{\xi})$ (dotted line) at $A=1$, $B=1$, $G=5$.

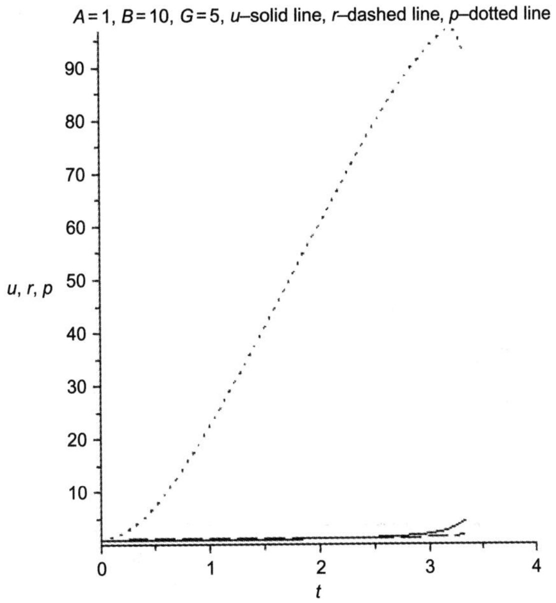

FIGURE A.9.5 The change of the velocity $\tilde{v}_r(\tilde{\xi})$ (solid line), the density $\tilde{\rho}(\tilde{\xi})$ (dashed line) and gravitational potential v—$\tilde{\psi}(\tilde{\xi})$ (dotted line) at $A=1$, $B=10$, $G=5$.

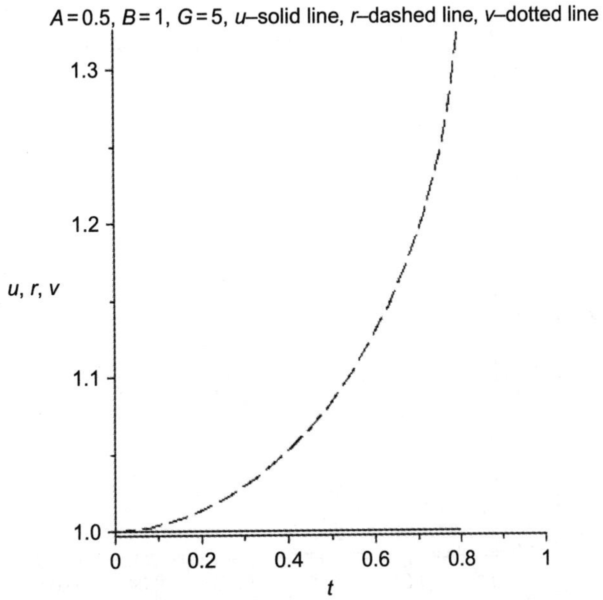

FIGURE A.9.6 The change of the velocity $\tilde{v}_r(\tilde{\xi})$ (solid line), the density $\tilde{\rho}(\tilde{\xi})$ (dashed line) and gravitational potential v—$\tilde{\psi}(\tilde{\xi})$ (dotted line) at $A=0.5$, $B=1$, $G=5$.

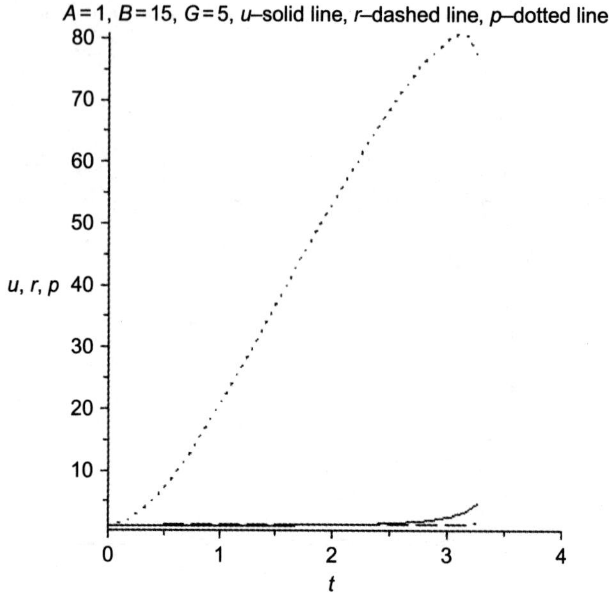

FIGURE A.9.7 The change of the velocity $\tilde{v}_r(\tilde{\xi})$ (solid line), the density $\tilde{\rho}(\tilde{\xi})$ (dashed line) and pressure $\tilde{p}(\tilde{\xi})$ (dotted line) at $A=1$, $B=15$, $G=5$.

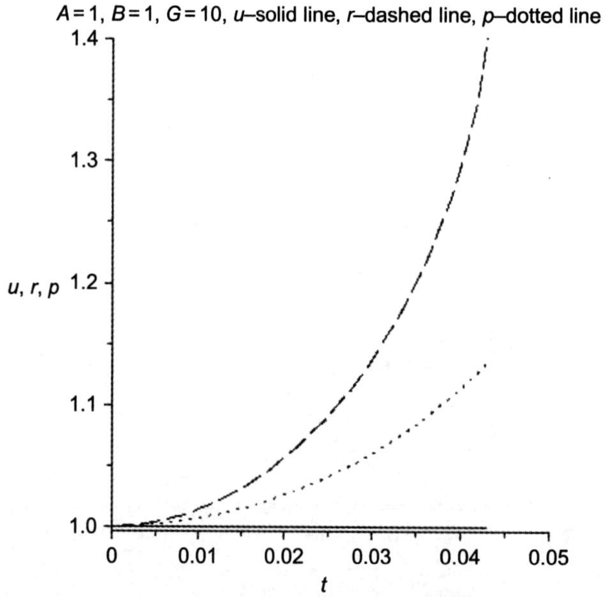

FIGURE A.9.8 The change of the velocity $\tilde{v}_r(\tilde{\xi})$ (solid line), the density $\tilde{\rho}(\tilde{\xi})$ (dashed line) and pressure $\tilde{p}(\tilde{\xi})$ (dotted line) at $A=1$, $B=1$, $G=10$.

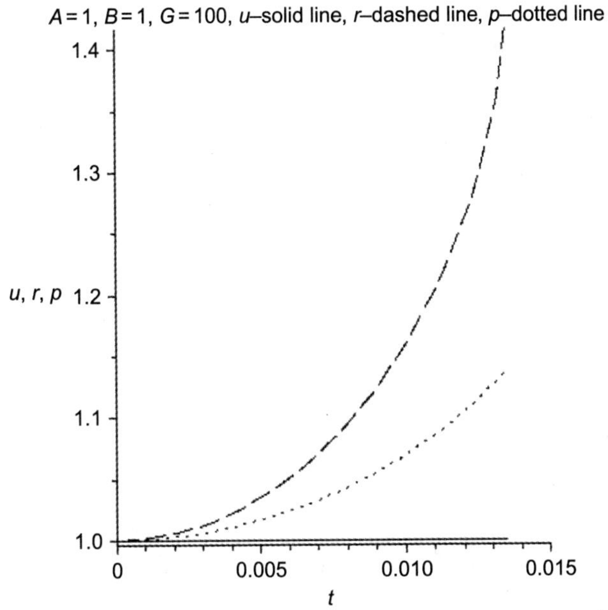

FIGURE A.9.9 The change of the velocity $\tilde{v}_r\left(\tilde{\xi}\right)$ (solid line), the density $\tilde{\rho}\left(\tilde{\xi}\right)$ (dashed line) and pressure $\tilde{p}\left(\tilde{\xi}\right)$ (dotted line) at $A=1$, $B=1$, $G=100$.

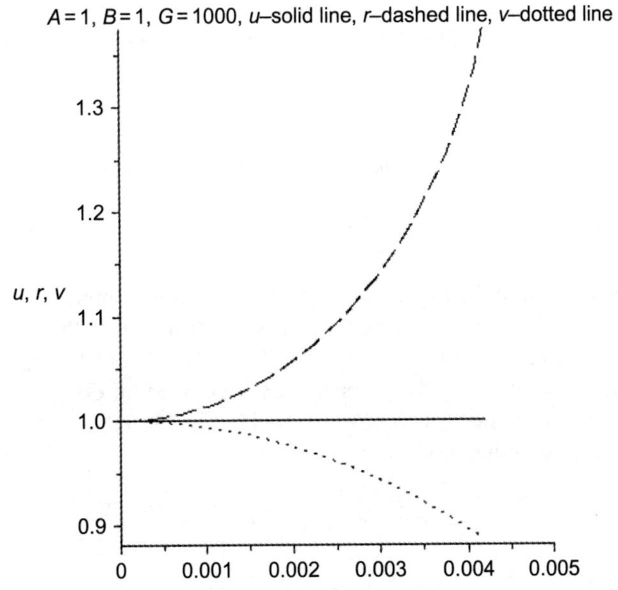

FIGURE A.9.10 The change of the velocity $\tilde{v}_r\left(\tilde{\xi}\right)$ (solid line), the density $\tilde{\rho}\left(\tilde{\xi}\right)$ (dashed line) and gravitational potential v—$\tilde{\psi}\left(\tilde{\xi}\right)$ (dotted line) at $A=1$, $B=1$, $G=1000$.

TABLE A.9.1 Area Occupied by Travelling Wave

A	B	G	$\tilde{\xi}_{\lim}$
1	1	0.01	0.6098
1	1	0.1	0.4266
1	1	1	0.1400
1	1	5	0.06134
1	10	5	3.4006
0.5	1	5	0.8030
1	15	5	3.3244
1	1	10	0.04312
1	1	100	0.01350
1	1	1000	0.004254

From the results of mathematical modeling follow:

1. All self-consistent solutions of non-local equations exist only in the restricted area $\tilde{\xi}_{\lim}$. Table A.9.1 contains the mentioned $\tilde{\xi}_{\lim}$ for the realized calculations.
2. The area occupied by travelling wave can be defined as the area of the solution existence. This size practically does not depend on choice of reasonable numerical schemes.
3. The self-consistent solution of non-local equations admits the existence of travelling indestructible waves (see Figs. A.9.3–A.9.10) if $G>1$.
4. For small gravitational parameter G ($G=\tilde{\gamma}_N$) indestructible waves do not exist (Figs. A.9.1 and A.9.2).
5. With increase in parameter G, the oscillations have acquired more regular character (Figs. A.9.1–A.9.10). It means that the usual pulsar theory (as a rotating star with the large G) can be reconsidered using the oscillation model.
6. For $G<1$ the character wave size is of the star radii order (Figs. A.9.1 and A.9.2); in this case the condition (A.9.8) looses the reasonable exactness.
7. Parameter G increasing leads to diminishing the area occupied by travelling wave (Figs. A.9.3, A.9.4, A.9.8, and A.9.10).
8. The area occupied by travelling wave tends to increasing with increasing of energy $\widetilde{\varepsilon n}$ (Figs. A.9.3, A.9.5, and A.9.7). It can lead to the star explosion.
9. The developed theory can be applied to the matter movement in the Hubble boxes, for example in the Local Group of galaxies. The galaxies of the local flow are located outside the group and all of them are moving from the center. The measured by Karachentsev the average Hubble parameter H for the Local Group is $72 \pm 6\,\mathrm{km \cdot s^{-1} Mpc^{-1}}$. Using the scale relation $u_0/r_0 = \mathrm{H}$ and $\mathrm{G} = \tilde{\gamma}_N = \gamma_N \rho_0 \mathrm{H}^{-2}$, for $r_0 = 0.5\,\mathrm{Mpc}$, we find $\mathrm{G} \sim 2$. It means that the explosion in the Hubble box of Local Group has the character features reflected in Figs. A.9.3–A.9.5. The Hubble movement is a deep particular case of the non-local hydrodynamics.

Appendix 10

To the Non-Local Theory of Levitation

The phenomenon of levitation has attracted attention from philosophers and scientists in the past and now. In spite of the tremendous recent advances, notably in power electronics, magnetic materials, on the application of electromagnetic suspension and levitation techniques to advanced ground transportation, physics of levitations needs in following significant investigations. Here we revisit the levitation phenomenon using the generalized Bolzmann kinetics theory which can represent the non-local physics of this levitation phenomenon.

The investigations of the levitation stability have a long history and are considered in details in [282–285]. As usual the problem review begins with the citation of the Earnshaw paper [286]. Earnshaw's theorem depends on a mathematical property of the $1/r$ type energy potential valid for magnetostatic and electrostatic events and gravitation. At any point where there is force balance is equal to zero, the equilibrium is unstable because there can be no local minimum in the potential energy. There must be some loopholes though, because magnets above superconductors and the magnet configuration do stably levitate including frogs [285] and toys like levitron (spinning magnet tops), flying globe and so on [287,288]. It means that diamagnetic material can stabilize the levitation of permanent magnets. It is well known that the potential energy density of the magnetic field can be written as

$$w_{\mathrm{m}} = -\mathbf{M}\cdot\mathbf{B}, \tag{A.10.1}$$

where $\mathbf{B}$ is magnetic induction, $\mathbf{M}$ is magnetization. Using the phenomenological relation

$$\mathbf{M} = \chi\mathbf{H}, \tag{A.10.2}$$

where χ is magnetic susceptibility, we have for the unit volume of a magnetic material

$$w_{\mathrm{m}} = -\frac{\chi}{\mu\mu_0}B^2. \tag{A.10.3}$$

The force acting on the unit volume of a levitating object is

$$\mathbf{F} = \frac{\chi}{\mu_0\mu}\operatorname{grad} B^2, \tag{A.10.4}$$

if the phenomenological parameters are constant. Diamagnets (for which $\chi < 0$) are repelled by magnetic fields and attracted to field minima. As a result, diamagnets can satisfy the stability conditions [282–285] and the following conditions are exceptions to Earnshaw's theorem:

(a) Diamagnetism occurs in materials which have a relative permeability less than one. The result is that eddy currents are induced in a diamagnetic material, it will repel magnetic flux.
(b) The Meissner effect which occurs in superconductors. Superconductors have zero internal resistance. As such induced currents tend to persist, and as a result the magnetic field they cause will persist as well.
(c) As result of oscillations, when an alternating current is passed through an electromagnet, it behaves like a diamagnetic material.
(d) Rotation: employed by the Levitron, it uses gyroscopic motion to overcome levitation instability.
(e) Feedback can be used in conjunction with electromagnets to dynamically adjust magnetic flux in order to maintain levitation.

The main shortcoming of the Earnshaw theory consists in application of principles of local physics to the nonequilibrium non-local statistical systems. Non-local hydrodynamic equations have the form:

(continuity equation for a mixture)

$$\begin{aligned}&\frac{\partial}{\partial t}\left\{\rho-\sum_{\alpha}\tau_{\alpha}^{(0)}\left[\frac{\partial\rho_{\alpha}}{\partial t}+\frac{\partial}{\partial\mathbf{r}}\cdot(\rho_{\alpha}\mathbf{v}_0)\right]\right\}\\&+\frac{\partial}{\partial\mathbf{r}}\cdot\left\{\rho\mathbf{v}_0-\sum_{\alpha}\tau_{\alpha}^{(0)}\left[\frac{\partial}{\partial t}(\rho_{\alpha}\mathbf{v}_0)+\frac{\partial}{\partial\mathbf{r}}\cdot(\rho_{\alpha}\mathbf{v}_0\mathbf{v}_0)+\overleftrightarrow{\mathbf{I}}\cdot\frac{\partial p_{\alpha}}{\partial\mathbf{r}}\right.\right.\\&\left.\left.-\rho_{\alpha}\mathbf{F}_{\alpha}^{(1)}-\frac{q_{\alpha}}{m_{\alpha}}\rho_{\alpha}\mathbf{v}_0\times\mathbf{B}\right]\right\}=0,\end{aligned}\tag{A.10.5}$$

(motion equation)

$$\begin{aligned}&\frac{\partial}{\partial t}\left\{\rho\mathbf{v}_0-\sum_{\alpha}\tau_{\alpha}^{(0)}\left[\frac{\partial}{\partial t}(\rho_{\alpha}\mathbf{v}_0)+\frac{\partial}{\partial\mathbf{r}}\cdot\rho_{\alpha}\mathbf{v}_0\mathbf{v}_0+\frac{\partial p_{\alpha}}{\partial\mathbf{r}}-\rho_{\alpha}\mathbf{F}_{\alpha}^{(1)}\right.\right.\\&\left.\left.-\frac{q_{\alpha}}{m_{\alpha}}\rho_{\alpha}\mathbf{v}_0\times\mathbf{B}\right]\right\}-\sum_{\alpha}\mathbf{F}_{\alpha}^{(1)}\left[\rho_{\alpha}-\tau_{\alpha}^{(0)}\left(\frac{\partial\rho_{\alpha}}{\partial t}+\frac{\partial}{\partial\mathbf{r}}\cdot(\rho_{\alpha}\mathbf{v}_0)\right)\right]\\&-\sum_{\alpha}\frac{q_{\alpha}}{m_{\alpha}}\left\{\rho_{\alpha}\mathbf{v}_0-\tau_{\alpha}^{(0)}\left[\frac{\partial}{\partial t}(\rho_{\alpha}\mathbf{v}_0)+\frac{\partial}{\partial\mathbf{r}}\cdot\rho_{\alpha}\mathbf{v}_0\mathbf{v}_0+\frac{\partial p_{\alpha}}{\partial\mathbf{r}}-\rho_{\alpha}\mathbf{F}_{\alpha}^{(1)}\right.\right.\\&\left.\left.-\frac{q_{\alpha}}{m_{\alpha}}\rho_{\alpha}\mathbf{v}_0\times\mathbf{B}\right]\right\}\times\mathbf{B}+\frac{\partial}{\partial\mathbf{r}}\cdot\left\{\rho\mathbf{v}_0\mathbf{v}_0+p\overleftrightarrow{\mathbf{I}}-\sum_{\alpha}\tau_{\alpha}^{(0)}\left[\frac{\partial}{\partial t}(\rho_{\alpha}\mathbf{v}_0\mathbf{v}_0\right.\right.\\&+p_{\alpha}\overleftrightarrow{\mathbf{I}})+\frac{\partial}{\partial\mathbf{r}}\cdot\rho_{\alpha}(\mathbf{v}_0\mathbf{v}_0)\mathbf{v}_0+2\overleftrightarrow{\mathbf{I}}\left(\frac{\partial}{\partial\mathbf{r}}\cdot(p_{\alpha}\mathbf{v}_0)\right)+\frac{\partial}{\partial\mathbf{r}}\cdot\left(\overleftrightarrow{\mathbf{I}}p_{\alpha}\mathbf{v}_0\right)\\&\left.\left.-\mathbf{F}_{\alpha}^{(1)}\rho_{\alpha}\mathbf{v}_0-\rho_{\alpha}\mathbf{v}_0\mathbf{F}_{\alpha}^{(1)}-\frac{q_{\alpha}}{m_{\alpha}}\rho_{\alpha}[\mathbf{v}_0\times\mathbf{B}]\mathbf{v}_0-\frac{q_{\alpha}}{m_{\alpha}}\rho_{\alpha}\mathbf{v}_0[\mathbf{v}_0\times\mathbf{B}]\right]\right\}=0,\end{aligned}\tag{A.10.6}$$

(energy equation)

$$\begin{aligned}&\frac{\partial}{\partial t}\left\{\frac{\rho v_0^2}{2}+\frac{3}{2}p+\sum_{\alpha}\varepsilon_{\alpha}n_{\alpha}-\sum_{\alpha}\tau_{\alpha}^{(0)}\left[\frac{\partial}{\partial t}\left(\frac{\rho_{\alpha}v_0^2}{2}+\frac{3}{2}p_{\alpha}+\varepsilon_{\alpha}n_{\alpha}\right)\right.\right.\\&\left.\left.+\frac{\partial}{\partial\mathbf{r}}\cdot\left(\frac{1}{2}\rho_{\alpha}v_0^2\mathbf{v}_0+\frac{5}{2}p_{\alpha}\mathbf{v}_0+\varepsilon_{\alpha}n_{\alpha}\mathbf{v}_0\right)-\mathbf{F}_{\alpha}^{(1)}\cdot\rho_{\alpha}\mathbf{v}_0\right]\right\}\\&+\frac{\partial}{\partial\mathbf{r}}\cdot\left\{\frac{1}{2}\rho v_0^2\mathbf{v}_0+\frac{5}{2}p\mathbf{v}_0+\mathbf{v}_0\sum_{\alpha}\varepsilon_{\alpha}n_{\alpha}-\sum_{\alpha}\tau_{\alpha}^{(0)}\left[\frac{\partial}{\partial t}\left(\frac{1}{2}\rho_{\alpha}v_0^2\mathbf{v}_0\right.\right.\right.\\&\left.+\frac{5}{2}p_{\alpha}\mathbf{v}_0+\varepsilon_{\alpha}n_{\alpha}\mathbf{v}_0\right)+\frac{\partial}{\partial\mathbf{r}}\cdot\left(\frac{1}{2}\rho_{\alpha}v_0^2\mathbf{v}_0\mathbf{v}_0+\frac{7}{2}p_{\alpha}\mathbf{v}_0\mathbf{v}_0+\frac{1}{2}p_{\alpha}v_0^2\overleftrightarrow{\mathbf{I}}\right.\\&\left.+\frac{5p_{\alpha}^2}{2\rho_{\alpha}}\overleftrightarrow{\mathbf{I}}+\varepsilon_{\alpha}n_{\alpha}\mathbf{v}_0\mathbf{v}_0+\varepsilon_{\alpha}\frac{p_{\alpha}}{m_{\alpha}}\overleftrightarrow{\mathbf{I}}\right)-\rho_{\alpha}\mathbf{F}_{\alpha}^{(1)}\cdot\mathbf{v}_0\mathbf{v}_0-p_{\alpha}\mathbf{F}_{\alpha}^{(1)}\cdot\overleftrightarrow{\mathbf{I}}\\&-\frac{1}{2}\rho_{\alpha}v_0^2\mathbf{F}_{\alpha}^{(1)}-\frac{3}{2}\mathbf{F}_{\alpha}^{(1)}p_{\alpha}-\frac{\rho_{\alpha}v_0^2}{2}\frac{q_{\alpha}}{m_{\alpha}}[\mathbf{v}_0\times\mathbf{B}]-\frac{5}{2}p_{\alpha}\frac{q_{\alpha}}{m_{\alpha}}[\mathbf{v}_0\times\mathbf{B}]\\&\left.\left.-\varepsilon_{\alpha}n_{\alpha}\frac{q_{\alpha}}{m_{\alpha}}[\mathbf{v}_0\times\mathbf{B}]-\varepsilon_{\alpha}n_{\alpha}\mathbf{F}_{\alpha}^{(1)}\right]\right\}-\mathbf{v}_0\cdot\sum_{\alpha}\rho_{\alpha}\mathbf{F}_{\alpha}^{(1)}\\&+\sum_{\alpha}\tau_{\alpha}^{(0)}\mathbf{F}_{\alpha}^{(1)}\cdot\left[\frac{\partial}{\partial t}(\rho_{\alpha}\mathbf{v}_0)+\frac{\partial}{\partial\mathbf{r}}\cdot\rho_{\alpha}\mathbf{v}_0\mathbf{v}_0+\frac{\partial}{\partial\mathbf{r}}\cdot p_{\alpha}\overleftrightarrow{\mathbf{I}}-\rho_{\alpha}\mathbf{F}_{\alpha}^{(1)}-q_{\alpha}n_{\alpha}[\mathbf{v}_0\times\mathbf{B}]\right]=0,\end{aligned}\tag{A.10.7}$$

where $\mathbf{v}_0$ is the hydrodynamic velocity in the coordinate system at rest, ρ_α is the density of α-species, p is the pressure, $\overleftrightarrow{\mathbf{I}}$ is unit tensore, $\mathbf{F}_\alpha^{(1)}$ is the force of the nonmagnetic origin acting on the unit of volume, ε_α is the internal energy of a particle of the α-species, τ is non-local parameter. Important remarks:

1. Equations (A.10.5)–(A.10.7) should be considered as local approximation of non-local equations (NLEs) written in the hydrodynamic form. NLE include quantum hydrodynamics of Schrödinger-Madelung as a deep particular case and can be applied in the frame of the unified theory from the atom scale to the Universe evolution.

2. The basic system contains the cross terms for the forces of the mass and electro-magneto-dynamic origin. It means that the fluctuation of the gravitational field leads to the electro-magneto dynamical fluctuations and verse versa.

Sufficient conditions of levitation can be obtained from Eqs. (A.10.5)–(A.10.7) after equalizing all terms containing forces to zero. Namely, from the continuity equation

$$\frac{\partial}{\partial \mathbf{r}}\cdot\left\{\sum_\alpha \tau_\alpha^{(0)}\left[\rho_\alpha \mathbf{F}_\alpha^{(1)}+\frac{q_\alpha}{m_\alpha}\rho_\alpha \mathbf{v}_0\times\mathbf{B}\right]=0,\right. \tag{A.10.8}$$

from the motion equation follows

$$\begin{aligned}&\frac{\partial}{\partial t}\left\{\sum_\alpha \tau_\alpha^{(0)}\left[\rho_\alpha \mathbf{F}_\alpha^{(1)}+\frac{q_\alpha}{m_\alpha}\rho_\alpha \mathbf{v}_0\times\mathbf{B}\right]\right\}-\sum_\alpha \mathbf{F}_\alpha^{(1)}\left[\rho_\alpha-\tau_\alpha^{(0)}\left(\frac{\partial\rho_\alpha}{\partial t}+\frac{\partial}{\partial \mathbf{r}}\cdot(\rho_\alpha \mathbf{v}_0)\right)\right]\\&-\sum_\alpha\frac{q_\alpha}{m_\alpha}\left\{\tau_\alpha^{(0)}\left[\rho_\alpha \mathbf{F}_\alpha^{(1)}+\frac{q_\alpha}{m_\alpha}\rho_\alpha \mathbf{v}_0\times\mathbf{B}\right]\right\}\times\mathbf{B}\\&+\frac{\partial}{\partial \mathbf{r}}\cdot\left\{\sum_\alpha \tau_\alpha^{(0)}\left[\mathbf{F}_\alpha^{(1)}\rho_\alpha \mathbf{v}_0+\rho_\alpha \mathbf{v}_0\mathbf{F}_\alpha^{(1)}+\frac{q_\alpha}{m_\alpha}\rho_\alpha[\mathbf{v}_0\times\mathbf{B}]\mathbf{v}_0+\frac{q_\alpha}{m_\alpha}\rho_\alpha \mathbf{v}_0[\mathbf{v}_0\times\mathbf{B}]\right]\right\}=0,\end{aligned} \tag{A.10.9}$$

and from the energy equation we find

$$\begin{aligned}&\frac{\partial}{\partial t}\sum_\alpha \tau_\alpha^{(0)}\left(\rho_\alpha \mathbf{F}_\alpha^{(1)}\cdot\mathbf{v}_0\right)\\&+\frac{\partial}{\partial \mathbf{r}}\cdot\left\{\sum_\alpha \tau_\alpha^{(0)}\left[\rho_\alpha \mathbf{F}_\alpha^{(1)}\cdot\mathbf{v}_0\mathbf{v}_0+p_\alpha \mathbf{F}_\alpha^{(1)}\cdot\overleftrightarrow{\mathrm{I}}+\frac{1}{2}\rho_\alpha v_0^2\mathbf{F}_\alpha^{(1)}+\frac{3}{2}\mathbf{F}_\alpha^{(1)}p_\alpha+\frac{\rho_\alpha v_0^2}{2}\frac{q_\alpha}{m_\alpha}[\mathbf{v}_0\times\mathbf{B}]\right.\right.\\&\left.\left.+\frac{5}{2}p_\alpha\frac{q_\alpha}{m_\alpha}[\mathbf{v}_0\times\mathbf{B}]+\varepsilon_\alpha n_\alpha\frac{q_\alpha}{m_\alpha}[\mathbf{v}_0\times\mathbf{B}]+\varepsilon_\alpha n_\alpha \mathbf{F}_\alpha^{(1)}\right]\right\}\\&-\mathbf{v}_0\cdot\sum_\alpha\rho_\alpha \mathbf{F}_\alpha^{(1)}+\sum_\alpha\tau_\alpha^{(0)}\mathbf{F}_\alpha^{(1)}\cdot\left[\frac{\partial}{\partial t}(\rho_\alpha \mathbf{v}_0)+\frac{\partial}{\partial \mathbf{r}}\cdot\rho_\alpha \mathbf{v}_0\mathbf{v}_0+\frac{\partial}{\partial \mathbf{r}}\cdot p_\alpha\overleftrightarrow{\mathrm{I}}-\rho_\alpha \mathbf{F}_\alpha^{(1)}-q_\alpha n_\alpha[\mathbf{v}_0\times\mathbf{B}]\right]=0.\end{aligned} \tag{A.10.10}$$

From Eq. (A.10.8) we have

$$\sum_\alpha \tau_\alpha^{(0)}\left[\rho_\alpha \mathbf{F}_\alpha^{(1)}+\frac{q_\alpha}{m_\alpha}\rho_\alpha \mathbf{v}_0\times\mathbf{B}\right]=\mathbf{L}(t) \tag{A.10.11}$$

or

$$\sum_\alpha \tau_\alpha^{(0)}\left[\rho_\alpha \mathbf{F}_\alpha^{(1)}+\frac{q_\alpha}{m_\alpha}\rho_\alpha \mathbf{v}_0\times\mathbf{B}\right]=\mathbf{L}, \tag{A.10.12}$$

where $\mathbf{L}$ is constant vector. Let us introduce vector $\mathbf{L}_\alpha(t)$

$$\mathbf{L}_\alpha(t)=\tau_\alpha^{(0)}\left[\rho_\alpha \mathbf{F}_\alpha^{(1)}+\frac{q_\alpha}{m_\alpha}\rho_\alpha \mathbf{v}_0\times\mathbf{B}\right]. \tag{A.10.13}$$

Rewrite now Eq. (A.10.9), which contains the density fluctuation

$$\rho_\alpha^{\mathrm{fl}}=\tau_\alpha^{(0)}\left(\frac{\partial\rho_\alpha}{\partial t}+\frac{\partial}{\partial \mathbf{r}}\cdot(\rho_\alpha \mathbf{v}_0)\right). \tag{A.10.14}$$

We have

$$\begin{aligned}&\frac{\partial}{\partial t}\mathbf{L}(t)-\sum_\alpha \mathbf{F}_\alpha^{(1)}\rho_\alpha^{\mathrm{a}}-\sum_\alpha\frac{q_\alpha}{m_\alpha}\mathbf{L}_\alpha(t)\times\mathbf{B}\\&+\mathbf{v}_0\left\{\frac{\partial}{\partial \mathbf{r}}\cdot\sum_\alpha \tau_\alpha^{(0)}\left[\mathbf{F}_\alpha^{(1)}\rho_\alpha+\frac{q_\alpha}{m_\alpha}\rho_\alpha[\mathbf{v}_0\times\mathbf{B}]\right]\right\}+\left(\sum_\alpha \tau_\alpha^{(0)}\left[\mathbf{F}_\alpha^{(1)}\rho_\alpha+\frac{q_\alpha}{m_\alpha}\rho_\alpha[\mathbf{v}_0\times\mathbf{B}]\right]\cdot\frac{\partial}{\partial \mathbf{r}}\right)\mathbf{v}_0\\&+\sum_\alpha \tau_\alpha^{(0)}\left[\rho_\alpha \mathbf{F}_\alpha^{(1)}+\frac{q_\alpha}{m_\alpha}\rho_\alpha[\mathbf{v}_0\times\mathbf{B}]\right]\frac{\partial}{\partial \mathbf{r}}\cdot\mathbf{v}_0+\left(\mathbf{v}_0\cdot\frac{\partial}{\partial \mathbf{r}}\right)\left\{\sum_\alpha \tau_\alpha^{(0)}\left[\rho_\alpha \mathbf{F}_\alpha^{(1)}+\frac{q_\alpha}{m_\alpha}\rho_\alpha[\mathbf{v}_0\times\mathbf{B}]\right]\right\}=0,\end{aligned} \tag{A.10.15}$$

where

$$\rho_\alpha^{\mathrm{a}} = \rho_\alpha - \rho_\alpha^{\mathrm{fl}}. \tag{A.10.16}$$

Using also (A.10.11), we find

$$\sum_\alpha \mathbf{F}_\alpha^{(1)} \rho_\alpha^{\mathrm{a}} = \frac{\partial}{\partial t}\mathbf{L}(t) - \sum_\alpha \frac{q_\alpha}{m_\alpha}\mathbf{L}_\alpha(t)\times\mathbf{B} + \left(\mathbf{L}(t)\cdot\frac{\partial}{\partial \mathbf{r}}\right)\mathbf{v}_0 + \mathbf{L}(t)\left(\frac{\partial}{\partial \mathbf{r}}\cdot\mathbf{v}_0\right) + \left(\mathbf{v}_0\cdot\frac{\partial}{\partial \mathbf{r}}\right)\mathbf{L}(t). \tag{A.10.17}$$

The vector product in Eq. (A.10.17) can be transformed as

$$\mathbf{L}_\alpha(t)\times\mathbf{B} = \tau_\alpha^{(0)}\rho_\alpha \mathbf{F}_\alpha^{(1)}\times\mathbf{B} - \tau_\alpha^{(0)} q_\alpha n_\alpha\left[\mathbf{v}_0 B^2 - \mathbf{B}(\mathbf{v}_0\cdot\mathbf{B})\right], \tag{A.10.18}$$

where $q_\alpha n_\alpha$ is the charge of α-species in the unit volume.

Taking into account the relations (A.10.13)–(A.10.15), we can realize the analogical transformation of the energy condition (A.10.10):

$$\begin{aligned}
\mathbf{v}_0\cdot\sum_\alpha \mathbf{F}_\alpha^{(1)}\rho_\alpha^{\mathrm{a}} &= \frac{\partial}{\partial t}\sum_\alpha \tau_\alpha^{(0)}\left(\rho_\alpha \mathbf{F}_\alpha^{(1)}\cdot\mathbf{v}_0\right) + \sum_\alpha \tau_\alpha^{(0)}\mathbf{F}_\alpha^{(1)}\cdot\left[\rho_\alpha\frac{\partial \mathbf{v}_0}{\partial t} + \rho_\alpha\left(\mathbf{v}_0\cdot\frac{\partial}{\partial \mathbf{r}}\right)\mathbf{v}_0 + \frac{\partial}{\partial \mathbf{r}}\cdot p_\alpha \overleftrightarrow{\mathrm{I}}\right] \\
&+ \frac{\partial}{\partial \mathbf{r}}\cdot\left\{\sum_\alpha \tau_\alpha^{(0)}\left[\begin{array}{l}\rho_\alpha \mathbf{F}_\alpha^{(1)}\cdot\mathbf{v}_0\mathbf{v}_0 + p_\alpha \mathbf{F}_\alpha^{(1)}\cdot\overleftrightarrow{\mathrm{I}} + \frac{1}{2}\rho_\alpha v_0^2 \mathbf{F}_\alpha^{(1)} + \frac{3}{2}\mathbf{F}_\alpha^{(1)} p_\alpha \\ + \frac{\rho_\alpha v_0^2}{2}\frac{q_\alpha}{m_\alpha}[\mathbf{v}_0\times\mathbf{B}] + \frac{5}{2}p_\alpha\frac{q_\alpha}{m_\alpha}[\mathbf{v}_0\times\mathbf{B}]\end{array}\right]\right\} \\
&+ \frac{\partial}{\partial \mathbf{r}}\cdot\sum_\alpha \varepsilon_\alpha n_\alpha \mathbf{L}_\alpha - \sum_\alpha \tau_\alpha^{(0)}\mathbf{F}_\alpha^{(1)}\cdot\mathbf{L}_\alpha.
\end{aligned} \tag{A.10.19}$$

Equations (A.10.11), (A.10.17), and (A.10.19) define the system of the sufficient conditions for levitation. Write down the system of the sufficient levitation conditions for the quasi-stationary case neglecting dissipation and the space derivatives in Eq. (A.10.17). We find

$$\sum_\alpha \mathbf{F}_\alpha^{(1)}\rho_\alpha^{\mathrm{a}} = -\sum_\alpha \frac{q_\alpha}{m_\alpha}\mathbf{L}_\alpha(t)\times\mathbf{B}, \tag{A.10.20}$$

$$\sum_\alpha \mathbf{F}_\alpha^{(1)}\rho_\alpha^{\mathrm{a}} = \mathbf{B}\times\sum_\alpha \tau_\alpha^{(0)}\frac{q_\alpha}{m_\alpha}\left[\rho_\alpha \mathbf{F}_\alpha^{(1)} + \frac{q_\alpha}{m_\alpha}\rho_\alpha \mathbf{v}_0\times\mathbf{B}\right]. \tag{A.10.21}$$

Introducing the current density

$$\mathbf{j}_\alpha = \tau_\alpha^{(0)} q_\alpha n_\alpha \mathbf{v}_0, \tag{A.10.22}$$

one obtains

$$\sum_\alpha \mathbf{F}_\alpha^{(1)}\rho_\alpha^{\mathrm{a}} = \mathbf{B}\times\sum_\alpha \tau_\alpha^{(0)} q_\alpha n_\alpha \mathbf{F}_\alpha^{(1)} + \mathbf{B}\times\sum_\alpha \frac{q_\alpha}{m_\alpha}[\mathbf{j}_\alpha\times\mathbf{B}]. \tag{A.10.23}$$

The right-hand-side of Eq. (A.10.23) contains the cross terms for the forces of the mass and electro-magneto-dynamic origin. The last term in Eq. (A.10.23) can be written also in the form

$$\mathbf{B}\times\sum_\alpha \frac{q_\alpha}{m_\alpha}[\mathbf{j}_\alpha\times\mathbf{B}] = \sum_\alpha \rho_\alpha \tau_\alpha^{(0)}\left[\frac{q_\alpha}{m_\alpha}\right]^2\left\{\mathbf{v}_0 B^2 - \mathbf{B}(\mathbf{v}_0\cdot\mathbf{B})\right\}. \tag{A.10.24}$$

It is naturally to suppose that

$$k_{\mathrm{B}} T \tau_\alpha^{(0)} \geq \hbar. \tag{A.10.25}$$

Introduce now

$$\tau_\alpha^{(0)} = A\frac{\hbar}{k_{\mathrm{B}} T}, \tag{A.10.26}$$

where A is a parameter which leads to appearance the effective temperature T_{eff}. Other approximations can be used, for example

$$\tau_\alpha^{(0)} = \frac{\hbar}{k_{\text{B}} T_{\alpha,\,\text{eff}}}. \tag{A.10.27}$$

Let us consider now other particular case when $\mathbf{v}_0 = 0$. In Eqs. (A.10.11), (A.10.17), and (A.10.19) we conserve the terms up to the $\tau_\alpha^{(0)}$ order. From Eq. (A.10.17) follows

$$\sum_\alpha \mathbf{F}_\alpha^{(1)} \rho_\alpha^{\text{a}} = \frac{\partial}{\partial t} \mathbf{L}(t) - \sum_\alpha \frac{q_\alpha}{m_\alpha} \mathbf{L}_\alpha(t) \times \mathbf{B}, \tag{A.10.28}$$

where now

$$\mathbf{L}_\alpha(t) = \tau_\alpha^{(0)} \rho_\alpha \mathbf{F}_\alpha^{(1)}, \tag{A.10.29}$$

$$\mathbf{L}(t) = \sum_\alpha \tau_\alpha^{(0)} \rho_\alpha \mathbf{F}_\alpha^{(1)}. \tag{A.10.30}$$

Then

$$\sum_\alpha \mathbf{F}_\alpha^{(1)} \rho_\alpha^{\text{a}} = \frac{\partial}{\partial t} \sum_\alpha \tau_\alpha^{(0)} \rho_\alpha^{\text{a}} \mathbf{F}_\alpha^{(1)} - \sum_\alpha q_\alpha n_\alpha^{\text{a}} \tau_\alpha^{(0)} \mathbf{F}_\alpha^{(1)} \times \mathbf{B}. \tag{A.10.31}$$

Introduce the explicit expression for the mass force

$$\mathbf{F}_\alpha^{(1)} = \mathbf{g} + \frac{q_\alpha}{m_\alpha} \mathbf{E} \tag{A.10.32}$$

in Eq. (A.10.28)

$$\sum_\alpha \mathbf{F}_\alpha^{(1)} \rho_\alpha^{\text{a}} = \frac{\partial}{\partial t} \sum_\alpha \tau_\alpha^{(0)} \rho_\alpha^{\text{a}} \left(\mathbf{g} + \frac{q_\alpha}{m_\alpha} \mathbf{E} \right) - \sum_\alpha q_\alpha n_\alpha^{\text{a}} \tau_\alpha^{(0)} \left(\mathbf{g} + \frac{q_\alpha}{m_\alpha} \mathbf{E} \right) \times \mathbf{B}, \tag{A.10.33}$$

where $n_\alpha^{\text{a}} = \rho_\alpha^{\text{a}} / m_\alpha$. From Eq. (A.10.33) follows

$$\sum_\alpha \mathbf{F}_\alpha^{(1)} \rho_\alpha^{\text{a}} = \frac{\partial}{\partial t} \sum_\alpha \tau_\alpha^{(0)} \rho_\alpha^{\text{a}} \mathbf{g} + \frac{\partial}{\partial t} \left[\mathbf{E} \sum_\alpha \tau_\alpha^{(0)} n_\alpha^{\text{a}} q_\alpha \right] - \mathbf{g} \times \mathbf{B} \sum_\alpha q_\alpha n_\alpha^{\text{a}} \tau_\alpha^{(0)} - \mathbf{E} \times \mathbf{B} \sum_\alpha \left(\frac{q_\alpha}{m_\alpha} \right)^2 \rho_\alpha^{\text{a}} \tau_\alpha^{(0)}. \tag{A.10.34}$$

Let us introduce in Eq. (A.10.34) the Umov-Pointing vector $\mathbf{S}$ and Alexeev vector $\mathbf{S}_{\text{A}}$ in the forms

$$\mathbf{S} = \mathbf{E} \times \mathbf{B}, \tag{A.10.35}$$

$$\mathbf{S}_{\text{A}} = \mathbf{g} \times \mathbf{B}. \tag{A.10.36}$$

In this case

$$\sum_\alpha \mathbf{F}_\alpha^{(1)} \rho_\alpha^{\text{a}} = \frac{\partial}{\partial t} \left(\sum_\alpha \tau_\alpha^{(0)} \rho_\alpha^{\text{a}} \mathbf{g} \right) + \frac{\partial}{\partial t} \left[\mathbf{E} \sum_\alpha \tau_\alpha^{(0)} n_\alpha^{\text{a}} q_\alpha \right] - \mathbf{S}_{\text{A}} \sum_\alpha q_\alpha n_\alpha^{\text{a}} \tau_\alpha^{(0)} - \mathbf{S} \sum_\alpha \left(\frac{q_\alpha}{m_\alpha} \right)^2 \rho_\alpha^{\text{a}} \tau_\alpha^{(0)}. \tag{A.10.37}$$

For the approximation (A.10.26) one obtains

$$\sum_\alpha \mathbf{F}_\alpha^{(1)} \rho_\alpha^{\text{a}} = A \frac{\hbar}{k_{\text{B}}} \left[\frac{\partial}{\partial t} \left(\frac{\rho^{\text{a}}}{T} \mathbf{g} \right) + \frac{\partial}{\partial t} \left(\mathbf{E} \frac{Q^{\text{a}}}{T} \right) - \mathbf{S}_{\text{A}} \frac{Q^{\text{a}}}{T} - \mathbf{S} \frac{1}{T} \sum_\alpha \left(\frac{q_\alpha}{m_\alpha} \right)^2 \rho_\alpha^{\text{a}} \right], \tag{A.10.38}$$

where the average charge density is introduced

$$Q^{\text{a}} = \sum_\alpha q_\alpha n_\alpha^{\text{a}}. \tag{A.10.39}$$

The analogical transformations of the energy condition (A.10.10) can be realized for this particular case when $\mathbf{v}_0^{\text{a}} = 0$. Namely

$$-\sum_\alpha \tau_\alpha^{(0)}\mathbf{F}_\alpha^{(1)}\cdot\sum_\alpha \mathbf{F}_\alpha^{(1)}\rho_\alpha^a=\sum_\alpha \tau_\alpha^{(0)}\mathbf{F}_\alpha^{(1)}\cdot\left[\frac{\partial}{\partial\mathbf{r}}\cdot p_\alpha^a\overleftrightarrow{\mathrm{I}}\right]+\frac{\partial}{\partial\mathbf{r}}\cdot\left\{\sum_\alpha \tau_\alpha^{(0)}\left[p_\alpha^a\mathbf{F}_\alpha^{(1)}\cdot\overleftrightarrow{\mathrm{I}}+\frac{3}{2}\mathbf{F}_\alpha^{(1)}p_\alpha^a\right]\right\}$$
$$+\frac{\partial}{\partial\mathbf{r}}\cdot\sum_\alpha \varepsilon_\alpha n_\alpha^a\tau_\alpha^{(0)}\mathbf{F}_\alpha^{(1)}. \tag{A.10.40}$$

For the approximation (A.10.26) we find from Eq. (A.10.40)

$$\frac{\partial}{\partial\mathbf{r}}\cdot\sum_\alpha\left[\frac{5}{2}p_\alpha^a+\varepsilon_\alpha n_\alpha^a\right]\mathbf{F}_\alpha^{(1)}-\sum_\alpha\left(\frac{5}{2}p_\alpha^a+\varepsilon_\alpha n_\alpha^a\right)\mathbf{F}_\alpha^{(1)}\cdot\frac{\partial\ln T}{\partial\mathbf{r}}$$
$$=-\mathbf{F}^{(1)}\cdot\sum_\alpha\rho_\alpha^a\mathbf{F}_\alpha^{(1)}-\sum_\alpha\mathbf{F}_\alpha^{(1)}\cdot\frac{\partial p_\alpha^a}{\partial\mathbf{r}}. \tag{A.10.41}$$

where $\mathbf{F}^{(1)}=\sum_\alpha\mathbf{F}_\alpha^{(1)}$.

Eq. (A.10.41) is a relation defining the energy consumption needed for the levitation. From (A.10.34) follows a relation

$$\sum_\alpha\mathbf{F}_\alpha^{(1)}\rho_\alpha^a=\mathbf{g}\rho^a+\mathbf{E}\sum\frac{q_\alpha}{m_\alpha}\rho_\alpha^a=\mathbf{g}\rho^a+\mathbf{E}\sum_\alpha q_\alpha n_\alpha^a=\mathbf{g}\rho^a+\mathbf{E}Q^a, \tag{A.10.42}$$

which can be used for the transformation of Eq. (A.10.41). For a tentative estimate we can omit the derivatives of the logarithmic terms and the time derivatives for a quasi-neutral media. As a result from (A.10.41)

$$\frac{\partial}{\partial\mathbf{r}}\cdot\left\{\mathbf{g}\sum_\alpha\left[\frac{5}{2}p_\alpha^a+\varepsilon_\alpha n_\alpha^a\right]\right\}+\frac{\partial}{\partial\mathbf{r}}\cdot\left\{\mathbf{E}\sum_\alpha\frac{q_\alpha}{m_\alpha}\left[\frac{5}{2}p_\alpha^a+\varepsilon_\alpha n_\alpha^a\right]\right\}$$
$$-\mathbf{F}^{(1)}\cdot[\rho^a\mathbf{g}+\mathbf{E}Q^a]-\mathbf{g}\cdot\frac{\partial p^a}{\partial\mathbf{r}}-\mathbf{E}\cdot\sum_\alpha\frac{q_\alpha}{m_\alpha}\frac{\partial p_\alpha^a}{\partial\mathbf{r}}. \tag{A.10.43}$$

For a quasi-neutral media $Q^a=0$, then

$$\frac{\partial}{\partial\mathbf{r}}\cdot\left[\mathbf{g}\left(\frac{5}{2}p^a+\Xi\right)\right]+\frac{\partial}{\partial\mathbf{r}}\cdot\left\{\mathbf{E}\sum_\alpha\frac{q_\alpha}{m_\alpha}\left[\frac{5}{2}p_\alpha^a+\Xi_\alpha\right]\right\}$$
$$=-\rho^a\mathbf{F}^{(1)}\cdot\mathbf{g}-\mathbf{E}\cdot\sum_\alpha\frac{q_\alpha}{m_\alpha}\frac{\partial p_\alpha^a}{\partial\mathbf{r}}-\mathbf{g}\cdot\frac{\partial p^a}{\partial\mathbf{r}}, \tag{A.10.44}$$

where

$$\Xi=\sum_\alpha\varepsilon_\alpha n_\alpha^a,\quad \Xi_\alpha=\varepsilon_\alpha n_\alpha^a. \tag{A.10.45}$$

Let us obtain a tentative estimate from (A.10.34) for the quasi-stationary case in a quasi-neutral media. From (A.10.34) for the case under consideration we have

$$\sum_\alpha\mathbf{F}_\alpha^{(1)}\rho_\alpha^a=-\mathbf{S}_\mathrm{A}\sum_\alpha q_\alpha n_\alpha^a\tau_\alpha^{(0)}-\mathbf{S}\sum_\alpha\left(\frac{q_\alpha}{m_\alpha}\right)^2\rho_\alpha^a\tau_\alpha^{(0)}. \tag{A.10.46}$$

From (A.10.42) and (A.10.46) we find

$$\rho^a\mathbf{g}=-\mathbf{S}_\mathrm{A}\sum_\alpha q_\alpha n_\alpha^a\tau_\alpha^{(0)}-\mathbf{S}\sum_\alpha\left(\frac{q_\alpha}{m_\alpha}\right)^2\rho_\alpha^a\tau_\alpha^{(0)} \tag{A.10.47}$$

or

$$\rho^a\mathbf{g}=-A\frac{\hbar}{k_\mathrm{B}}\mathbf{S}\frac{1}{T}\sum_\alpha\left(\frac{q_\alpha}{m_\alpha}\right)^2\rho_\alpha^a \tag{A.10.48}$$

in the case of (A.10.26) approximation. Relation leads in SI to the estimate

$$\rho^a\mathbf{g}\cong-A\frac{\mathbf{S}}{T}\cdot 2.138\times10^{-19}n_\mathrm{e}^a \tag{A.10.49}$$

or

$$\rho^{\mathrm{a}}\mathbf{g} \cong 2.138\times10^{-19}\frac{A}{T}[\mathbf{B}\times\mathbf{E}]n_{\mathrm{e}}^{\mathrm{a}}. \tag{A.10.50}$$

For example for dry air at 1 atm and 20 °C we have number density $n=0.2504\times10^{26}\,\mathrm{m}^{-3}$, density $\rho=1.2041\,\mathrm{kg/m^3}$. Using (A.10.4) and the phenomenological condition of the force balance (see also [285]) we have

$$\mathbf{F}=\frac{\chi}{\mu_0\mu}\mathrm{grad}\,B^2=\rho g\hat{\mathbf{e}}_z, \tag{A.10.51}$$

where ρ is the mass density of the material to be levitated and $\hat{\mathbf{e}}_z$ is the unit vector in the vertical direction, magnetic susceptibility χ is negative for diamagnetic materials. In the frame of the phenomenological description of the magnetic and gravitational field we have

$$w=w_{\mathrm{m}}+w_{\mathrm{g}}=-\frac{\chi}{\mu\mu_0}B^2+\rho gz. \tag{A.10.52}$$

A necessary condition for stability is

$$\int_S \mathbf{F}\cdot\mathbf{ds}<0, \tag{A.10.53}$$

where S is any small closed surface surrounding the equilibrium point. It leads to the condition

$$\mathrm{div}\,\mathbf{F}<0. \tag{A.10.54}$$

This relation leads to the stability condition

$$\Delta w=\mathrm{div}\,\mathrm{grad}\,w=\mathrm{div}\,\mathrm{grad}\,w_{\mathrm{m}}=-\frac{\chi}{\mu\mu_0}\Delta B^2=-\mathrm{div}\,\mathbf{F}>0, \tag{A.10.55}$$

if $\chi<0$ (diamagnetic materials) and $\Delta B^2>0$. The corresponding stability investigation from the phenomenological point of view was realized in [285].

From the relation (A.10.51) follows ($\mu\sim1$)

$$\rho g=2\frac{\chi}{\mu_0}B\frac{\partial B}{\partial z} \tag{A.10.56}$$

and from (A.10.47)

$$\rho^{\mathrm{a}}\mathbf{g}=-\mathbf{S}\tau^{(0)}\sum_\alpha\left(\frac{q_\alpha}{m_\alpha}\right)^2\rho_\alpha^{\mathrm{a}}, \tag{A.10.57}$$

if the non-local parameter does not depend on the sort of species α. After equalizing the right-hand-sides of relations (A.10.56) and (A.10.57) one obtains

$$\tau^{(0)}\mu_0 E\frac{q_{\mathrm{e}}}{2m_e}q_{\mathrm{e}}n_{\mathrm{e}}^{\mathrm{a}}=\chi\frac{\partial B}{\partial z}, \tag{A.10.58}$$

because

$$\sum_\alpha\left(\frac{q_\alpha}{m_\alpha}\right)^2\rho_\alpha^{\mathrm{a}}\cong\left(\frac{q_{\mathrm{i}}}{m_{\mathrm{i}}}\right)^2\rho^{\mathrm{a}}+\left(\frac{q_{\mathrm{e}}}{m_{\mathrm{e}}}\right)^2\rho_{\mathrm{e}}^{\mathrm{a}}=\frac{q_{\mathrm{i}}^2}{m_{\mathrm{i}}}n_{\mathrm{i}}^{\mathrm{a}}+\frac{q_{\mathrm{e}}^2}{m_{\mathrm{e}}}n_{\mathrm{e}}^{\mathrm{a}}\cong\left(\frac{q_{\mathrm{e}}}{m_{\mathrm{e}}}\right)^2\rho_{\mathrm{e}}^{\mathrm{a}}. \tag{A.10.59}$$

Let us introduce the character length l_{m}

$$l_{\mathrm{m}}=\mu_0\frac{q_{\mathrm{e}}^2}{2m_{\mathrm{e}}}, \tag{A.10.60}$$

hence from (A.10.58) and (A.10.60)

$$l_{\mathrm{m}}En_{\mathrm{e}}^{\mathrm{a}}=-|\chi|\frac{1}{\tau^{(0)}}\frac{\partial B}{\partial z}. \tag{A.10.61}$$

Introduce the electromotive force (EMF) for a particle

$$E_{\text{ind}} = l_{\text{m}} E, \tag{A.10.62}$$

and for n_{e}^{a} particles

$$E_{\text{ind},n} = l_{\text{m}} E n_{\text{e}}^{\text{a}}. \tag{A.10.63}$$

Hence from (A.10.61) and (A.10.63) we find

$$E_{\text{ind},n} = -\frac{|\chi|}{\tau^{(0)}} \frac{\partial B}{\partial z}. \tag{A.10.64}$$

Formally Eq. (A.10.64) can be written in the form of Faraday's law of induction, the most widespread version of this law states that the induced EMF in any closed circuit is equal to the rate of change of the magnetic flux through the circuit:

$$E_{\text{ind}} = -\partial \Phi_{\text{B}} / \partial t, \tag{A.10.65}$$

where Φ_{B} is the magnetic flux. This version of Faraday's law strictly holds only when the closed circuit is a loop of infinitely thin wire and is invalid in some other circumstances. Nevertheless formally

$$E_{\text{ind},n} = -|\chi| \frac{1}{\tau^{(0)}} \frac{\partial B}{\partial t} \frac{\partial t}{\partial z} \tag{A.10.66}$$

or

$$E_{\text{ind},n} = -|\chi| \frac{1}{\tau^{(0)}} \frac{\partial B}{\partial t} \frac{1}{v_{\text{m}}}. \tag{A.10.67}$$

After introduction of the character counter square

$$S_{\text{m}} = \frac{|\chi|}{\tau^{(0)} v_{\text{m}} n_{\text{e}}^{\text{a}}}, \tag{A.10.68}$$

we reach the relation in the form of Faraday's law (A.10.65).

The following conclusions of the principal significance can be done:

1. The levitation effects are the direct consequence of the NLEs (A.10.5)–(A.10.7).
2. The sufficient conditions of levitation are the particular case of Eqs. (A.10.5)–(A.10.7).
3. The strict theory of levitation can be constructed only in the frame of non-local physics.
4. Fluctuations of the gravitational field lead to the electro-magneto dynamical fluctuations and verse versa. This fact can effect on the work of electronic devices during the evolution of the wave atmospheric fronts.
5. Levitation effects are connected not only with the electro-magnetic energy flux **S**, but also with the cross flux $\mathbf{S}_{\text{A}}$.
6. Usual local conditions of levitation are the deep particular cases of the non-local theory.

References

[1] L. Boltzmann, Sitz. Ber. Kaiserl. Akad. Wiss. 66 (2) (1872) 275.
[2] S. Chapmsan, Philos. Trans. R. Soc. A216 (1916) 279.
[3] S. Chapman, Philos. Trans. R. Soc. A217 (1917) 115.
[4] D. Enskog, The Kinetic Theory of Phenomena in Fairly Rare Gases, University of Upsala, Upsala, 1917.
[5] D. Enskog, Svensk. Vet. Akad. Arkiv. f. Math. Ast. och Fys. 16 (1921) 1.
[6] L. Boltzmann, Vorlesungen über Gastheorie, Verlag von Johann Barth, Leipzig, 1912, Zweiter unveränderten Abdruck.2 Teile.
[7] P. Ehrenfest, *Otnositelnost', kvanty, statistika.* (Relativity, quantum, statistics.), Nauka, Moscow, 1972.
[8] P. Ehrenfest, in: M. Klein (Ed.), Collected Scientific Papers, North-Holland Publ. Co, Amsterdam, 1979.
[9] L.D. Landau, E.M. Lifshitz, *Gidrodinamika* (Kurs Teoreticheskoi Fiziki T 6) (Hydrodynamics Course of Theoretical Physics Vol 6), Nauka, Moscow, 1988 [Translated into English (Oxford: Pergamon Press 1990)].
[10] L.C. Woods, An Introduction to the Kinetic Theory of Gases and Magnetoplasmas, Oxford Univ. Press, Oxford, 1993.
[11] B.I. Davydov, Dokl. Akad. Nauk SSSR 2 (7) (1935) 474.
[12] P. Vernotte, C. R. Acad. Sci. Paris 246 (1958) 3154–3155.
[13] C. Cattaneo, C. R. Acad. Sci. Paris. 247 (1958) 431–433.
[14] C. Cattaneo, Atti del Seminar. Mat. Fis. Univ. Modena 3 (1948) 3–21.
[15] P. Vernotte, C. R. Acad. Sci. Paris 251 (1958) 2103–2105.
[16] N.N. Bogolyubov, Problemy Dinamicheskoi Teorii v Statisticheskoi Fizike (Dynamic Theory Problems in Statistical Physics), Gostekhizdat, Moscow, Leningrad, 1946 [Translated into English The Dynamical Theory in Statistical Physics (Delhi Hindustan Publ. Corp. 1965)].
[17] M. Born, H.S. Green, Proc. R. Soc. 188 (1012) (1946) 10.
[18] H.S. Green, The Molecular Theory of Fluids, North-Holland Pub. Co., Amsterdam, 1952.
[19] J.G. Kirkwood, J. Chem. Phys. 15 (1) (1947) 72.
[20] J. Yvon, La Theorie Statistique des Fluide et l'Equation d'etat, Hermann, Paris, 1935.
[21] B.V. Alekseev, Matematicheskaya Kinetika Reagiruyushchikh Gazov (Mathematical Theory of Reacting Gases), Nauka, Moscow, 1982.
[22] N.A. Slezkin, Dokl. Akad. Nauk SSSR 79 (1951) 33.
[23] N.A. Slezkin, Dokl. Akad. Nauk SSSR 77 (1951) 205.
[24] I.V. Meshcherskii, Dinamika Tochki Peremennoi Massy (Dynamics of a Variable Mass Point), Petersburg Publishing, Petersburg, 1897.
[25] S.V. Vallander, Dokl. Akad. Nauk SSSR 78 (1951) 25.
[26] I.G. Shaposhnikov, Zh. Ekip. Teor. Fiz. 21 (1951) 1310.
[27] Ya.P. Terletski, Zh. Eksp. Teor. Fiz. 22 (1952) 506.
[28] Ya.P. Terletski, Statisticheskaya Fizika (Statistical Physics), Vyssh. Shkola, Moscow, 1994 [Translated into English (Amsterdam: North-Holland, 1971)].
[29] J.W. Gibbs, The Collected Works of J. Willard Gibbs, Vol. 2, Longmans, Green & Co., New York, 1934 [Translated into Russian (Moscow: Gostekhizdat, 1946)].
[30] B.V. Alexeev, Philos. Trans. R. Soc. Lond. 349 (1994) 417–443.
[31] B.V. Alexeev, Physica A 216 (1995) 459.
[32] B.V. Alexeev, Phys.-Usp. 170 (6) (2000) 649.
[33] L.C. Woods, 17-th Internat. Symp.RGD. Book of Abstracts Aachen. 2 (1990) 442–444.
[34] J.C. Maxwell, Philos. Mag. 19 (1860) 19.
[35] S. Chapman, T.G. Cowling, The Mathematical Theory of Non-uniform Gases, Cambridge University Press, Cambridge, 1952.
[36] I.O. Hirschfelder, Ch.F. Curtiss, R.B. Bird, Molecular Theory of Gases and Liquids, John Wiley and sons, inc., New York, 1954, Chamman and Hall, lim., London.
[37] S.L. Sobolev, J. Phys. France 3 (1993) 2261–2269.
[38] V.M. Zhdanov, V.I. Roldugin, Zh. Eksp. Teor. Fiz. 122 (2002) 789 (JETP 95 (2002) 682).
[39] A.B. Jarzebski, J.W. Thulli, J. Comput. Phys. 63 (1986) 236–239.
[40] Ph. Rosenau, Am. Phys. Soc. 48 (7) (1993) 655–657.
[41] Yu.L. Klimontovich, Statisticheskaya Teoriya Otkrytykh Sistem (Statistical Theory of Open Systems), Yanus-K, Moscow, 1995 [Translated into English (Dordrecht: Kluwer Acad. Publ., 1995)].
[42] M.N. Saha, Philos. Mag. 40 (1920) 472.
[43] J. Eggert, J. Phys. Z. 20 (1919) 570.
[44] R.H. Fowler, E.A. Guggenheim, Statistical Thermodynamics, Cambridge, Hamburg, 1939.
[45] E.M. Dewan, Phys. Fluids 4 (6) (1961).

[46] M. Gryzinski, J. Kunc, M. Zgorzelski, J. Phys. B 6 (1973) 2292.
[47] D. Enskog, Kungl. Svenska. Vetenskaps. Akad. Handl. 63 (4) (1921) 3.
[48] J.M.J. Koremans, J. Beenaker, Physica 26 (1960) 653.
[49] B.N. Chetverushkin, Kineticheski-Soglasovannye Skhemy v Gazovoi Dinamike: Novaya Model' Vyazkogo Gaza, Algoritmy, Parallel'naya Realizatsiya, Prilozheniya (Kinetically Consistent Schemes in Gas Dynamics: New Viscid Gas Model, Algorithms, Parallel Realization, and Applications), Izd-vo MGU, Moscow, 1999.
[50] Yu.L. Klimontovich, Phys. Usp. 167 (1) (1997) 23.
[51] A.S. Bakai, Yu.S Sigov, Mnogolikaya turbulentnost (Multifacial turbulence), in: I.M. Makarov (Ed.), Novoe v Sinergetike. Zagadki Mira Neravnovesnykh Struktur, Nauka, Moscow, 1996, p. 10.
[52] S.A. Reshetnyak, L.A. Shelepin, Kvazistatsionarnye Raspredeleniya v Kinetike (Quasi-stationary Distributions in Kinetics), IPO Avtor, Moscow, 1996.
[53] A.A. Vlasov, Nelokal'naya Statisticheskaya Mekhanika (Nonlocal Statistical Mechanics), Nauka, Moscow, 1978.
[54] B.V. Alexeev, High Temp. 35 (1) (1997) 125.
[55] J.S. Bell, Physics 1 (1964) 195.
[56] B.V. Alekseev, Phys.-Usp. 46 (2) (2003) 139.
[57] B.V. Alexeev, Generalized Boltzmann Physical Kinetics, Elsevier, Amsterdam, The Netherlands. 2004.
[58] Б.В. Алексеев, Нелокальная физика. Нерелятивистская теория, (Non-local Physics. Non-relativistic Theory, in Russian), Lambert, Saarbrücken, 2011.
[59] Б.В. Алексеев, И.В. Овчинникова, Нелокальная физика. Релятивистская теория, (Non-local Physics. Relativistic Theory, in Russian), Lambert, Saarbrücken, 2011.
[60] G. Uhlenbeck, Phys.-Usp. 103 (2) (1971) 275.
[61] H.A. Kramers, Physica 7 (4) (1940) 284.
[62] A.N. Morozov, Zh. Eksp. Teor. Fiz. 109 (4) (1996) 1304 (JETP 82 (1996) 703).
[63] B.V. Alexeev, J. Nanoelectron. Optoelectron. 3 (2008) 143–158.
[64] B.V. Alexeev, J. Nanoelectron. Optoelectron. 3 (2008) 316–328.
[65] B.V. Alexeev, J. Nanoelectron. Optoelectron. 4 (2009) 186–199.
[66] B.V. Alexeev, J. Nanoelectron. Optoelectron. 4 (2009) 379–393.
[67] E. Madelung, Z. Phys. 40 (1927) 322–326.
[68] A.H. Nayfeh, Perturbation Methods, Wiley, N.Y., London, Sydney, Toronto, 1972.
[69] B.V. Alexeev, Book of abstacts. Conference on Mechanics of Media with Chemical Reactions. Krasnoyarsk, 1988, pp. 2–4.
[70] T.G. Elizarova, B.N. Chetverushkin, Intern. Symposium Computational Fluid Dynamics, Japan Society of Computational Fluid Dynamics, Nagoya, Japan, 1989.
[71] P.L. Bhatnagar, E.P. Gross, M.A. Krook, Phys. Rev. 511 (1954) 94.
[72] I. Prigogine, Introduction to Thermodynamics of Irreversible Processes, second ed. Wiley, London, 1962.
[73] G. Nikolis, I. Prigogine, Self-organization in Nonequilibrium Systems, Wiley, N.Y., 1977.
[74] B.V. Alekseev, I.T. Grushin, Protsessy perenosa v reagiruyushchikh gazakh i plazme, (Transport Processesin Reacting Gases and Plasma), Energoatomizdat, Moscow, 1994.
[75] B.V. Alekseev, J. Comp. Math. Math. Phys. 27 (1987) 730–740.
[76] B.V. Alekseev, High Temp. 31 (4) (1993) 624–633.
[77] R.D. Reitz, J. Comput. Phys. 42 (1) (1981) 103–123.
[78] T.G. Elizarova, Aerothermochemistry of spacecraft and associated hypersonic flows, in: IUTAM Symposium, Marseille, France, 1992, pp. 211–213.
[79] Yu.L. Klimontovich, Theor. Math. Phys. 92 (2) (1992) 312.
[80] T.G. Elizarova, B.N. Chetverushkin, Mathematical Modeling, Nauka, Moscow, 1986, pp. 261–278.
[81] V.P. Silin, Introduction in Kinetic Theory of Gases (in Russian), Nauka, Moscow, 1971.
[82] B.N. Rodimov, Auto-Oscillating Quantum Mechanics, Tomsk State University, Tomsk, 1976.
[83] R.D. Richtmyer, K.W. Morton, Difference Methods for Initial-Value Problems, Interscience Publishers, New York, 1967.
[84] Ch.R. Doering, J.D. Gibbon, Applied Analysis of the Navier-Stokes Equations, Cambridge University Press, Cambridge, 1995.
[85] C.A. Fletcher, Computational Techniques for Fluid Dynamics I, Springer Verlag, Sydney, 1987.
[86] B.V. Alexeev, A.I. Abakumov, Dokl. AN SSSR 262 (1982) 5.
[87] A.I. Abakumov, V.S. Vinogradov, Izvestiya vuzov Phys. 1 (1981) 8–12.
[88] G. Moliere, Z. Naturforsch. 2A (3) (1947) 133–145.
[89] L.V. Spencer, Phys. Rev. 98 (6) (1955) 1597–1615.
[90] C.A. Fletcher, Computational Techniques for Fluid Dynamics II, Springer Verlag, Sydney, 1987.
[91] E. Carnovalli Jr., H.M. Franca. Arxiv, quant-ph/0512049 vol. 2, 17, 2006.
[92] T. Scheidla, R. Ursina, J. Koflera, S. Ramelowa, Xiao-Song Maa, T. Herbst, et al., Violation of local realism with freedom of choice, PNAS 107 (46) (2010) 119708–119713, November 16.
[93] S.E. Shnoll, V.A. Kolombet, E.V. Pozharskii, T.A. Zenchenko, I.M. Zvereva, A.A. Konradov, Phys.-Usp. 41 (1998) 1025, http://dx.doi.org/10.1070/PU1998v041n10ABEH000463.
[94] S.E. Shnoll, T.A. Zenchenko, K.I. Zenchenko, E.V. Pozharski, V.A. Kolombet, A.A. Konradov, Phys.-Usp. 43 (2) (2000) 205–209.
[95] T.V. Perevertun, N.V. Udaltzova, V.A. Kolombet, N.P. Ivanova, T.Ya. Britsina, S.E. Shnol', Biophysics 26 (4) (1981) 613–624.

[96] S.E. Shnol', V.A. Namiot, V.E. Zhvirblis, V.N. Morozov, A.V. Temnov, T.Ya. Morozova, Biophysics 28 (1) (1983) 164–168.

[97] S.E. Shnoll, Cosmo-physical factors in the probability processes (in Russian), Svenska fysikarkived, Nasbydalsvagen, 2009. ISBN: 978-91-85917-06-8.

[98] B.V. Alexeev, J. Mod. Phys. 3 (2012) 1895–1906, http://dx.doi.org/10.4236/jmp.2012.312239, Published, Online December 2012 (http://www.SciRP.org/journal/jmp).

[99] H. Dehmelt, Phys. Scr. 22 (1988) 102–110, http://dx.doi.org/10.1088/0031-8949/1988/T22/016, Bibcode 1988 PhST.22.102D.

[100] Ph. Ekstrom, D. Wineland, The isolated electron, Sci. Am. 243 (2) (1980) 90–98100–101.

[101] M. Breidebach, J.I. Friedman, H.W. Kendall, E.D. Bloom, D.H. Coward, H. DeStaebler, et al., Phys. Rev. Lett. 23 (16) (1969) 935–939.

[102] A.Yu. Popkov, I.K. Kuzmichev, Radio Phys. Radio Astronomy 14 (4) (2009) 425–432 (in Russian).

[103] M. Popescu, Rom. Rep. Phys. 57 (2005) 795–799.

[104] D.I. Blokhintsev, Phys.-Usp. 3 (1957) 381–383, Letter to Editor, LXII.

[105] M. Stenhoff, Ball Lightning. An Unsolved Problem in Atmospheric Physics, Kluwer/Plenum, New York, 1999.

[106] S. Singer, The Nature of Ball Lightning, Plenum, New York, 1971.

[107] J.D. Barry, Ball Lightning and Bead Lightning, Plenum, New York, 1980, pp. 25–27, 59–60, 180–182.

[108] P.L. Kapitza, Phys. Blatter 14 (1958) 11.

[109] P.L. Kapitza, Sov. Phys. Tech. Phys. 13 (1969) 1475 (Engl. Transl.).

[110] N. Tesla, Colorado Springs Notes 1899–1900, in: A. Marincic (Ed.), Nolit, Beograd, Yugoslavia, 1978, pp. 368–370.

[111] A.I. Egorov, S.I. Stepanov, J. Tech. Phys. 72 (12) (2002) 104.

[112] A.G. Oreshko, Int. J. Unconventional Electromagn. Plasmas (UEP) 3 (1–2) (2011) 77.

[113] B.V. Alexeev, *To the Non local Theory of Plasmoids. Gagarin Catastrophe Version.* Vestnik MITHT, Vol. 8, No. 2, 2013 (in Russian).

[114] A. Leonov, D. Scott, Two Sides of the Moon, Thomas Dunne Books, New York, 2004. ISBN: 0-312-30865-5, OCLC 56587777, p. 218.

[115] http://ria.ru/gagarin_news/20110408/362309350.html#ixzz2NwO1pBuJ

[116] С.А. Микоян, А.И. Пушкин, С.В. Петров, Г.С. Титов, А.А. Леонов, С.М. Белоцерковский, et al., Заключение специалистов о причинах гибели Гагарина и Серёгина, //Гражданская авиация, № 7, 1989.

[117] Б.Е. Черток. Ракеты и люди. Горячие дни Холодной Войны. 2-е издание. Москва. Машиностроение. 1999 (B.E. Chertok. Rockets and People. Hot days of Cold War. Chapter 6).

[118] http://diary-news.com/intresting/33042-taynu-gibeli-gagarina-ne-uznayut-nikogda.html

[119] http://militera.lib.ru/memo/russian/mikoyan_sa/23.html

[120] http://sobesednik.ru/cosmos/taina-gibeli-gagarina-byla-rassekrechena-tri-goda-nazad.

[121] http://ru.wikipedia.org/wiki.

[122] S. Kawano, Int. J. Unconventional Electromagn. Plasmas (UEP) 3 (1–2) (2011) 41.

[123] B.V. Alexeev, Vestnik MITHT (Вестник МИТХТ). 3, 2014, p. 2.

[124] T. Dürkop, S.A. Getty, E. Cobas, M.S. Fuhrer, Nano Lett. 4 (2004) 35.

[125] Y. Kohsaka, T. Hanaguri, M. Azuma, M. Takano, J.C. Davis, H. Takagi, Nat. Phys. 8 (2012) 534–538, http://dx.doi.org/10.1038/nphys2321.

[126] J. Chang, E. Blackburn, A.T. Holmes, N.B. Christensen, J. Larsen, J. Mesot, et al., ArXiv:1206.4333[v2], Condensed Matter. Superconductivity, 3 Jul 2012.

[127] Ю.Е. Лозовик, С.П. Меркулова, А.А. Соколик А.А. *Успехи физических наук*. 178 №7, 2008, pp. 757–776.

[128] Y. Barlas, T. Pereg-Barnea, M. Polini, R. Asgari, A.H. MacDonald, Phys. Rev. Lett. 98 (2007) 236601.

[129] A.H. Castro Neto, F. Guinea, N.M.R. Peres, K.S. Novoselov, A.K. Geim, Rev. Mod. Phys. 81 (2009) 109–162.

[130] F.T. Vasko, V. Ryzhii V. ArXiv,0801/3476v2 [cond-mat.mtrl-sci], 27 May 2008.

[131] S. Pisana, M. Lazzeri, C. Casiraghi, K.S. Novoselov, A.K. Geim, A.C. Ferrari, et al., Nat. Mater. 6 (2007) 198–201.

[132] Д.В. Завьялов, С.В. Крючков, Т.А. Тюлькина Т.А. Физика и техника полупроводников 44, вып.7, 2010, pp. 910–914 (Zavyalov D.V., Krychkov S.V., Tyul'kina T.A. Fisika and Tehnika Poluprovodnikov, in Russian).

[133] Д.В. Завьялов, С.В. Крючков, Н.Е. Мещерякова Н.Е. Физика и техника полупроводников 39, вып.2, 2010, pp. 214–217 (Zavjalov D.V., Kruchkov S.V., Mestcheryakova N.E. Fisika and Tehnika Poluprovodnikov, in Russian).

[134] E.H. Hwang, S. Adam, S.D. Sarma, ArXiv, 0610157v2 [cond-mat.mes-hall], 4 May 2007.

[135] М.Б. Белоненко, Н.Г. Лебедев, А.В. Пак, Н.Н. Янюшкина, Журнал технической физики (Journal of Technical Physics, in Russian). Т.81. Вып. 8, 2011, pp. 64–69.

[136] C.A. Reynolds, B. Serin, W.H. Wright, L.B. Nesbitt, Phys. Rev. 78 (1950) 487.

[137] E. Maxwell, Phys. Rev. 78 (1950) 477.

[138] H. Fröhlich, Phys. Rev. 79 (1950) 845–856.

[139] J.E. Hirsch, ArXiv:1108.3835v2 [cond-mat.supr-con], 18 Oct 2011.

[140] J. Bardeen, L.N. Cooper, J.R. Schrieffer, Phys. Rev. 106 (1957) 162–164.

[141] A. Lenard, Ann. Phys. (New York) 10 (1960) 390.

[142] R. Balescu, Phys. Fluids 3 (1960) 52.

[143] R. Balescu, Equilibrium and Nonequilibrium Statistical Mechanics, Wiley, New York, 1975 [Translated into Russian: Vols 1, 2 (Moscow: Mir, 1978)].

[144] Yu.L. Klimontovich, Kineticheskaya Teoriya Neideal'nogo Gaza i Neideal'noy Plazmy (Kinetic Theory of Nonideal Gases and Non-ideal Plasmas), Nauka, Moscow, 1975 [Translated into English (Oxford: Pergamon Press, 1982)].

[145] B.A. Trubnikov, M.A. Leontovich (Ed.), Voprosy Teorii Plazmy, Gosatomizdat, Moscow, 1963, p. 98 [Translated into English: in Reviews of Plasma Physics Vol. 1 (Ed. M A Leontovich) (New York: Consultants Bureau, 1965) p. 105].
[146] L.A. Artsimovich, R.Z. Sagdeev, Fizika Plazmy dlya Fizikov (Plasma Physics for Physicists), Moscow, Atomizdat, 1979.
[147] L.D. Landau, Zh. Eksp. Teor. Fiz. 16 (1946) 574.
[148] L.D. Landau, J. Phys. USSR 5 (1941) 71, Physics-Uspekhi 93 (1967) 527.
[149] L.D. Landau, J. Phys. USSR 10 (1946) 25.
[150] J. Dawson, Phys. Fluids 4 (1961) 869.
[151] V.N. Soshnikov, Some paradoxes of the Landau damping. Supplement to the Russian translation of the book [152].
[152] P.C. Clemmow, J.P. Dougherty, Electrodynamics of Particles and Plasmas, Addison-Wesley Publ. Co, Reading, Mass, 1990 [Translated into Russian (Moscow: Mir, 1996)].
[153] B.V. Alekseev, Teplofiz. Vys. Temp. 38 (2000) 374, [High Temp. 38 (2000) 351].
[154] B.V. Alekseev, Teplofiz. Vys. Temp. 39 (2001) 693, [High Temp. 39 (2001) 641].
[155] N.M. Faddeva, N.M. Terentyev, Tables of Values of Function of complex argument, Gostechizdat, Moscow, 1954.
[156] M. Abramowitz, I.A. Stegun, Handbook of Mathematical Functions, National Bureau of Standards, Gaithersburg, 1964.
[157] B.V. Alexeev, Mathematical modeling of the electric wave expansion in plasma. Part 1. Landau Damping. Part 2; Tables of the Integral Function. Part 3; Tables of the Integral Function. MITHT Publ. Co., Moscow, 2006.
[158] N.S. Buchel'nikova, E.P. Matochkin, Preprint No. 77-15, Novosibirsk: The Budker Institute of Nuclear Physics of the Siberian Branch of the USSR Academy of Science, 1977.
[159] N.S. Buchel'nikova, E.P. Matochkin, Preprint No. 77-39, Novosibirsk: The Budker Institute of Nuclear Physics of the Siberian Branch of the USSR Academy of Sciences, 1977.
[160] N.S. Buchel'nikova, E.P. Matochkin, Preprint No. 79-112, Novosibirsk: The Budker Institute of Nuclear Physics of the Siberian Branch of the USSR Academy of Sciences, 1979.
[161] N.S. Buchel'nikova, E.P. Matochkin, Fiz. Plazmy. 6 (1980) 1097, [Sov. J. Plasma Phys. 6 (1980) 603].
[162] E.M. Kudriavtsev, S.D. Zotov, V.V. Krivov, M. Autric, Physica C 1439 (1994) 234–240.
[163] E.M. Kudriavtsev, S.D. Zotov, V.V. Krivov, M. Autric, Physica C 282–287 (1997) 1145–1146.
[164] E.M. Shakhov, Method of Rarefied Gas Investigation, Nauka, Moscow, 1974.
[165] T.F. Morse, Phys. Fluids 6 (10) (1963) 1420–1427.
[166] F.B. Pidduck, Proc. R. Soc. Lond. Ser. A. 88 (603) (1913) 296.
[167] K.T. Compton, Phys. Rev. 7 (4) (1916) 489.
[168] K.T. Compton, Phys. Rev. 7 (5) (1916) 509.
[169] M.J. Druyvesteyn, Physica 10 (1930) 61.
[170] M.J. Druyvesteyn, Z. Phys. 64 (11–12) (1930) 781.
[171] B.I. Davydov, JETP 6 (5) (1936) 463–480.
[172] Yu.A. Ivanov, Yu.A. Lebedev, L.S. Polak, Methods of Contact Diagnostics in Non-Equilibrium Plasma Chemistry (in Russian), Nauka, Moscow, 1981.
[173] V.L. Ginzburg, A.V. Gurevich, Usp. Fiz. Nauk. 70 (2) (1960) 202.
[174] V.L. Ginzburg, A.V. Gurevich, Phys.-Usp. 70 (3) (1960) 393.
[175] B.M. Smirnov, Physics of Weakly Ionized Gas, Nauka, Moscow, 1985.
[176] А.А. Кокин, В.И. Толстихин В.И., Микроэлектроника, т.13, вып.1, 1984.
[177] K. Seeger, Semiconductors Physics, Springer Verlag (Berlin), 1973. ISBN: 014392674.
[178] D.P. Bhattacharya, T.K. Pramanic, J. Phys. Chem. Solid 52 (1991) 2803.
[179] E.M. Convell, Phys. Rev. 90 (1953), http://dx.doi.org/10.1103/PhysRev.90.769.
[180] E.G. Ryder, Phys. Rev. 90 (5) (1953) 759–763.
[181] C. Canali, C. Jacobini, F. Nava, G. Ottaviani, A. Alberige, Phys. Rev. B 12 (6) (1975) 2265–2284.
[182] B.V. Alekseev, High Temp. 33 (6) (1995) 834–845.
[183] B.V. Alekseev, Yu.A. Lebedev, V.V. Michailov, High Temp. 35 (2) (1997) 207–213.
[184] I. Shkarovsky, T. Johnstone, M. Bachinsky, Kinetika chastits plazmy (Kinetics of Plasma Particles), Atomizdat, Moscow, 1969.
[185] V.E. Golant, Izv. Akad. Nauk. SSSR, Ser. Fis. 23 (1959) 958–961.
[186] V.E. Gal'tsev, A.V. Dem'yanov, I.V. Kochetov, V.G. Pevgov, V.F. Sharkov, Mechanism of Excitation of Gas Discharge Lasers, Preprint of the Institute of Atomic Energy, Moscow, 1979, no. 3156.
[187] Ю.~А.~ЛебеДев, Параметры эЛектронной компоненты метансодержащей волоролноей СВЧ-плазмы в~резонаторе. ТВТ 33 (6) (1995) 850–854. http://mi.mathnet.ru/tvt3016; J. High Temp. 33 (6) (1995) 846–850.
[188] E.B. Karoulina, Yu.A. Lebedev, J. Phys. D: Appl. Phys. 25 (1992) 401.
[189] E.B. Karoulina, Yu.A. Lebedev, J. Phys. D: Appl. Phys. 21 (1988) 411.
[190] A.I. Lukovnikov, I.V. Fetisov, The electron distribution function in molecular plasma, Fizika Gazorazryadnoi Plazmy, (Physics of Gas Discharge Plasma), Atomizdat, Moscow, 1969.
[191] B.V. Alexeev, In Proceedings: IUTAM Symposium Aerothermochemistiry of Spacecraft and Associated Hypersonic Flows, Universite de Provence (IUSTI/MHEQ, 1992), pp. 85–91.

[192] C. Cercignani, Theory and Application of the Boltzmann Equation, Scottish Academic Press, Edinburgh and London, 1975.
[193] B.V. Alekseev, A.B. Poddoskin, V.V. Mikhailov, Investigation of sound propagation in the frame of generalized hydrodynamic equations, in: B.V. Alekseev (Ed.), Mekhanika i Elektrodinamika Sploshnykh Sred, Izd. Mosk. Univ., Moscow, 1990, p. 25.
[194] B.V. Alekseev, Investigation of Sound Propagation in the Frame of Generalized Navier-Stokes Equations, Dokl. Akad. Nauk SSSR 313 (1990) 1078, [Sov. Phys. Dokl. 35 (1990).
[195] B.V. Alekseev, Investigation of sound propagation in the frame of generalized navier-stokes equations, in: B.V. Alekseev (Ed.), Mekhanika i Elektrodinamika Sploshnykh Sred, Izd. Mosk. Univ., Moscow, 1990, pp. 20–24.
[196] M. Greenspan, J. Acoust. Soc. Am. 28 (4) (1956) 644–648.
[197] E. Meyer, G. Sessler, Z. Phys. 149 (15) (1957) 15–39.
[198] L. Sirovich, J.K. Thurber, J. Acoust. Soc. Am. 37 (2) (1965) 329–339.
[199] C.L. Pekeris, Z. Alterman, L. Finkelstein, Sumposium on the Numerical Treatment of Ordinary Differential Equations, Integral and Integro-differential Equation of P.I.C.C, Birkhauser Verlag, Basel, 1960, pp. 338–398.
[200] C.L. Pekeris, Z. Alterman, L. Finkelstein, K. Frankowski, Phys. Fluids 5 (1962) 1608–1616.
[201] E. Longo, L. Preziosi, N. Bellomo, in: Aerothemochemistry of Spacecraft and Associated Hypersonic Flows, Proceedings of IUTAM Symposium, 1992, , pp. 41–46, Marseille, France.
[202] J.F. Bourgat, P. Le Tallec, F. Malinger, D. Tidriri, Y. Qiu, Proceedings of JUTAM Symposium, 1992, Marseille, France; pp. 60–66.
[203] B.V. Alekseev, V.V. Polev, Teplofiz. Vys. Temp. 28 (6) (1990) 614–616, [High Temp. (USSR) 28 (1990) 896].
[204] B. Schmidt, J. Fluids Mech. 39 (2) (1969) 361–373.
[205] H. Grad, Commun. Pure Appl. Math. 2 (4) (1949) 38.
[206] L.H. Holway, Phys. Fluids 7 (6) (1964) 141.
[207] S. Glasstone, K.J. Laidler, H. Eyring, The Theory of Rate Processes. The Kinetics of Chemical Reactions, Viscosity, Diffusion and Electrochemical Phenomena, McGraw-Hill Book Company, inc, NY and London, 1941.
[208] P. Rezibois, M. De Lener, Classical Kinetic Theory of Fluids, Mir, Moscow, 1980 (Russ. transl.).
[209] Fizicheskaya entsiklopediya. T. 2. Zhidkost' (Physical Encyclopedia. Vol. 2. Liquid), Moscow: Sovetskaya entsiklopediya, 1990.
[210] Ya.I. Frenkel, Kineticheskaya teoriya zhidkostei, (Kinetic Theory of Liquids), Izd. AN SSSR, Moscow-Leningrad, 1945.
[211] B.V. Alekseev, Teplofiz. Vys. Temp. 35 (1) (1997) 129, (High Temp. (Engl. transl.) 35 (1) (1997) 125).
[212] J.H. Hildebrand, Viscosity and Diffusivity, Wiley, New York, 1977.
[213] L.D. Landau, JETP 11 (1941) 592.
[214] L.D. Landau, J. Phys. 11 (1947) 91.
[215] W. Heisenberg, Ann. Phys. (Leipzig) 74 (5) (1924) 577.
[216] W. Heisenberg, Life of Physics; Evening Lectures at the International Centre of Theoretical Physics in Triest, IAIA, Wiena, 1969.
[217] O. Reynolds, Philos. Trans. R. Soc. A CLXXVI (1894) 123.
[218] A.J.A. Favre, Formulation of the statistical equations of turbulent flows with variable density, in: Th.B. Gatski, S. Sarkar, Ch.G. Speziale (Eds.), Studies in Turbulence, Springer Verlag, New York, 1992, pp. 324–341.
[219] L.E. Elsholz, Qualitative Methods in Mathematical Analysis (in Russian), Gos. Izd. Tekhnico-Teoret. Literat, Moscow, 1955.
[220] O.M. Belotserkovski, A.M. Oparin, Numerical Experiment in Turbulence (in Russian), Nauka, Moscow, 2000.
[221] B.V. Alexeev, V.V. Mikhailov, High Temp. 37 (2) (1999) 250–259.
[222] A.I. Fedoseyev, B.V. Alexeev, Mathematical Model for Viscous Incompressible Fluid Flow Using Alexeev Equations and Comparison with Experimental Data, in: S.K. Dey, Ziebarth G. Ferrandiz (Eds.), Proceedings of Advances in Scientific Computing Modelling (Special Proceedings of IMACS' 98), Alicante, Spain, 1998, pp. 158–163.
[223] A.I. Fedoseyev, B.V. Alexeev, Higher order continuum model for incompressible viscous flow and application to the modelling of thermal vibrational convection, in: E. Inan, K.Z. Markov (Eds.), Proceedings of the 9th International Symposium, Continuum Models and Discrete Systems (CMDS9), Istanbul, Turkey, 1998, pp. 130–137.
[224] J.R. Kozeff, R.L. Street, J. Fluids Eng. 106 (1984) 21–29.
[225] J.R. Kozeff, R.L. Street, J Trans. ASME 106 (1984) 390–398.
[226] J.R. Kozeff, R.L. Street, J. Fluids Eng. 106 (1984) 385–389.
[227] M. Arnal, O. Laurel, Z. Liek, M. Reric, A GAMM Workshop (Notes on numerical Fluid Mechanics, 36, Eds M Devill, T-H Le, Yu Morchois), F. Vieweg, Baunsch, 1992, p. 13.
[228] R.W. Verstappen, A.E. Veldman, in Proc. of the Third ECCOMAS Computational Fluid Dynamics Conf. (Paris, France, 9–13 Sept. 1996) (Eds J A Dosidori et al), New York, Wiley, 1996, p. 1073.
[229] B.V. Alexeev, V.V. Mikhailov, High Temp. 39 (4) (2001) 567–577.
[230] J.K. Eaton, A. Johnston, AIAA J. 19 (9) (1981) 1093–1100.
[231] B.V. Alexeev, V.V. Mikhailov, High Temp. 41 (2) (2003) 401–414.
[232] H. Schlichting, Grenzschicht—Theory, Verlag G, Braun, Karlsruhe, 1964.
[233] A.N. Labusov, Yu.V. Lapin, High Temp. 34 (6) (1996) 942.
[234] M. Tij, A.J. Santos, Stat. Phys. 76 (1994) 1399.
[235] M.M. Malek, F. Baras, A.L. Garcia, Physica A 240 (1997) 255.
[236] F.J. Uribe, A.L. Garcia, Phys. Rev. E 60 (4) (1999) 4063–4078.

[237] Kazuo Aoki, Shigeru Takata, Toshiyuki Nakanishi, Phys. Rev. E 65 (026315) (2002) 1–22.
[238] G.A. Bird, Molecular Gas Dynamics and Direct Simulation of Gas Flows, Clarendon, Oxford, 1994.
[239] F.J. Alexander, A.L. Garcia, B.J. Alder, Phys. Rev. Lett. 74 (26) (1995) 5212–5215.
[240] F. Zwicky, Helvetica. Phys. Acta 6 (1933) 110–127, Bibcode: 1933AcHPh...6.110Z.
[241] F. Zwicky, Astrophys. J. 86 (1937) 217, http://dx.doi.org/10.1086/143864, Bibcode:1937ApJ....86.217Z.
[242] V. Rubin, W.K. Ford Jr., Astrophys. J. 159 (1970) 379.
[243] V. Rubin, N. Thonnard, W.K. Ford Jr., Astrophys. J. 238 (1980) 471.
[244] M. Milgrom, ArXiv, preprint. http://arxiv.org/abs/0801.3133v2, (2007).
[245] A.D. Chernin, Phys.-Usp. 51 (3) (2008) 267–300.
[246] E.B. Gliner, Sov. Phys. Dokl. 15 (1970) 559.
[247] D. Pal, S.N. Prasad, M.J.M. Roemkens, Study of sediment transport in shallow channel flows, in: D.E. Statt, R.H. Mohtar, G.C. Steihardt (Eds.), Selected Papers of the 10th International Soil Conservation Organization Meeting, Purdue University, USA, 2001, pp. 717–724.
[248] S.N. Prasad, D. Pal, M.J.M. Roemkens, J. Fluid Mech. 413 (2000) 89–110.
[249] B.V. Alexeev, J. Mod. Phys. 3 (29A) (2012) 1103–1122.
[250] I.D. Karachentsev, O.G. Kashibadze, Astrophysics 49 (2006) 3.
[251] I.D. Karachentsev, O.G. Kashibadze, Astrophysics 49 (2006) 3–18.
[252] L.S. Sparke, J.S. Gallagher, Galaxies in the Universe: Introduction, Cambridge University Press, 2000.
[253] B.V. Alexeev, J. Mod. Phys. 4 (2013) 42–49, http://dx.doi.org/10.4236/jmp.2013.47A1005, Published, Online July 2013 (http://www.scirp.org/journal/jmp).
[254] B.V. Alexeev, J. Mod. Phys. 4 (2013) 26–41, http://dx.doi.org/10.4236/jmp.2013.47A1004, Published, Online July 2013 (http://www.scirp.org/journal/jmp).
[255] C. Cercignani, G.M. Kremer, The Relativistic Boltzmann Equation: Theory and Applications, Birkhäuser Verlag, Basel, Boston, Berlin, 2002.
[256] P. Romatschke, A. Rebhan, Phys. Rev. Lett. 97 (2006) 252301.
[257] G. Napier, B.D. Shizgal, Physica A 387 (2008) 4099.
[258] B.V. Alexeev, Boundary Layer with Chemical Reactions, Computer Center of USSR Academy of Sciences, Moscow, 1967 (in Russian, see also translation realized by Lockheed missiles and space company, 1968).
[259] M. Fleischmann, S. Pons, M. Hawkins, Electrochemically induced nuclear fusion of deuterium, J. Electroanal. Chem. 261 (2) (1989) 301, http://dx.doi.org/10.1016/002-0728(89)80006-3.
[260] P.A. Mosier-Boss, S. Szpak, F.E. Gordon, L.P.G. Forsley, Triple tracks in CR-39 as the result of Pd–D Co-deposition: evidence of energetic neutrons, Naturwissenschaften 96 (1) (2009) 135–142.
[261] V. Bityurin, A. Bykov, V. Velikodny, A. Dyrenkov, B. Tolkunov, Theoretical and experimental investigation of the effect of shockwave on porous deuterated liquid, Fiziko-Khimicheskaja Kinetika v Gazovoy Dinamike 6 (2008).
[262] E. Forringer, D. Robbins, J. Martin, Confirmation of neutron production during self-nucleated acoustic cavitation, Trans. Am. Nucl. Soc. 95 (2006) 736–737.
[263] A. Lipson, V. Klhyuev, B. Deryagin, Y. Toporov, M. Sirotyuk, O. Khavroshkin, et al., Observation of neutrons accompanying cavitation in detuterium-containing media, Pis'ma v Zhurnal Tekhnicheskoj Fiziki 16 (1990) 89–93.
[264] S. Putterman, L. Crum, K. Suslick, Comments on "Evidence for Nuclear Emissions During Acoustic Cavitation"; by R.P. Taleyarkhan et al., Science 295 (2002) 1868.
[265] D. Shapira, M. Saltmarsh, Nuclear fusion in collapsing bubbles—is it there? An attempt to repeat the observation of nuclear emissions from sonoluminescence, Phys. Rev. Lett. 89 (10) (2002) 104302.
[266] E. Smorodov, R. Galiakhmetov, M. Il'gamov, Physics and Chemistry of Cavitation, Nauka, Moscow, 2008.
[267] R. Taleyarkhan, J. Cho, J. West, R. Lahey, R. Nigmatulin, R. Block, Additional evidence of nuclear emissions during acoustic cavitation, Phys. Rev. E 69 (3) (2004) 036109.
[268] R. Taleyarkhan, J. Lapinskas, Y. Xu, J. Cho, R. Block, R. Lahey, et al., Modeling, analysis and prediction of neutron emission spectra from acoustic cavitation bubble fusion experiments, Nucl. Eng. Des. 238 (2008) 2779–2791.
[269] R. Taleyarkhan, C. West, J. Cho, R. Lahey, R. Nigmatulin, R. Block, Evidence for nuclear emissions during acoustic cavitation, Science 295 (5561) (2002) 1868–1873.
[270] R. Taleyarkhan, C. West, R. Lahey, R. Nigmatulin, R. Block, Y. Xu, Nuclear emissions during self-nucleated acoustic cavitation, Phys. Rev. Lett. 96 (2006) 034301.
[271] Y. Xu, A. Butt, Confirmatory experiments for nuclear emissions during acoustic cavitation, Nucl. Eng. Des. 235 (2005) 1317–1324.
[272] J.W. Strutt, Rayleigh, The Theory of Sound, vol. 1, Macmillan, London, 1877, 1894 (Cambridge: University Press, reissued 2011, ISBN 978-1-108-03220-9).
[273] J.W. Strutt, Rayleigh, The Theory of Sound, vol. 2, Macmillan, London, 1878, 1896 (Cambridge: University Press, reissued 2011, ISBN 978-1-108-03221-6).
[274] B.V. Alexeev, V.V. Polev, Dynamic model of solar activity, Dokl. Akad. Nauk SSSR 295 (1987) 1074.
[275] A.S. Eddington, On the pulsation of a gaseous star and the problem of Cepheid variables, MNRAS 79 (1918) 2–22.
[276] A.S. Eddington, The Internal Constitution of the Stars, Cambridge University Press, Cambridge, 1926.
[277] A.S. Eddington, On the cause of Cepheid pulsation, MNRAS 101 (1941) 182–194.

[278] C. Hoffmeister, G. Richter, W. Wenzel, Variable Stars, Springer Verlag, NY, 1985.
[279] S.A. Zhevakin, On the pulsational theory of stellar variability, part V. Multilayer spherical discrete model, Sov. Astron. 3 (1959) 389–403.
[280] V.I. Aleshin, On the asymmetry of the radial velocity curve of Cepheids, Sov. Astron. 3 (1959) 458–465.
[281] V.I. Aleshin, Auto-oscillations of variable stars, Sov. Astron. 8 (1964) 154–162.
[282] M.V. Berry, A.K. Geim, Of flying frogs and levitrons, Eur. J. Phys. 18 (1997) 307–313, http://wwwhfml.sci.kun.nl/hfml/levitate.html.
[283] A.K. Geim, Everyone's magnetism, Phys. Today 51 (September) (1998) 36–39.
[284] A.K. Geim, M.D. Simon, M.I. Boamfa, O.L. Heflinger, Magnet levitation at your fingertips, Nature 400 (1999) 323–324.
[285] M.D. Simon, A.K. Geim, Diamagnetic levitation: flying frogs and floating magnets, J. Appl. Phys. 87 (2000) 6200–6204.
[286] S. Earnshaw, On the nature of the molecular forces which regulate the constitution of the luminiferous ether, Trans. Cambridge Phil. Soc. 7 (1842) 97–112.
[287] M.V. Berry, The LEVITRON and adiabatic trap for spins, Proc. R. Soc. Lond. A 452 (1996) 1207–1220.
[288] R.M. Harrigan, Levitation device, U.S. Patent $,382245, May 3, 1983.

Index

Note: Page numbers followed by *f* indicate figures and *t* indicate tables.

哈尔滨工业大学出版社刘培杰数学工作室已出版(即将出版)图书目录

书　　名	出版时间	定　价	编号
新编中学数学解题方法全书(高中版)上卷	2007－09	38.00	7
新编中学数学解题方法全书(高中版)中卷	2007－09	48.00	8
新编中学数学解题方法全书(高中版)下卷(一)	2007－09	42.00	17
新编中学数学解题方法全书(高中版)下卷(二)	2007－09	38.00	18
新编中学数学解题方法全书(高中版)下卷(三)	2010－06	58.00	73
新编中学数学解题方法全书(初中版)上卷	2008－01	28.00	29
新编中学数学解题方法全书(初中版)中卷	2010－07	38.00	75
新编中学数学解题方法全书(高考复习卷)	2010－01	48.00	67
新编中学数学解题方法全书(高考真题卷)	2010－01	38.00	62
新编中学数学解题方法全书(高考精华卷)	2011－03	68.00	118
新编平面解析几何解题方法全书(专题讲座卷)	2010－01	18.00	61
新编中学数学解题方法全书(自主招生卷)	2013－08	88.00	261
数学眼光透视	2008－01	38.00	24
数学思想领悟	2008－01	38.00	25
数学应用展观	2008－01	38.00	26
数学建模导引	2008－01	28.00	23
数学方法溯源	2008－01	38.00	27
数学史话览胜	2008－01	28.00	28
数学思维技术	2013－09	38.00	260
从毕达哥拉斯到怀尔斯	2007－10	48.00	9
从迪利克雷到维斯卡尔迪	2008－01	48.00	21
从哥德巴赫到陈景润	2008－05	98.00	35
从庞加莱到佩雷尔曼	2011－08	138.00	136
数学奥林匹克与数学文化(第一辑)	2006－05	48.00	4
数学奥林匹克与数学文化(第二辑)(竞赛卷)	2008－01	48.00	19
数学奥林匹克与数学文化(第二辑)(文化卷)	2008－07	58.00	36′
数学奥林匹克与数学文化(第三辑)(竞赛卷)	2010－01	48.00	59
数学奥林匹克与数学文化(第四辑)(竞赛卷)	2011－08	58.00	87
数学奥林匹克与数学文化(第五辑)	2015－06	98.00	370

哈尔滨工业大学出版社刘培杰数学工作室已出版(即将出版)图书目录

书　　名	出版时间	定　价	编号
世界著名平面几何经典著作钩沉——几何作图专题卷(上)	2009－06	48.00	49
世界著名平面几何经典著作钩沉——几何作图专题卷(下)	2011－01	88.00	80
世界著名平面几何经典著作钩沉(民国平面几何老课本)	2011－03	38.00	113
世界著名平面几何经典著作钩沉(建国初期平面三角老课本)	2015－08	38.00	507
世界著名解析几何经典著作钩沉——平面解析几何卷	2014－01	38.00	264
世界著名数论经典著作钩沉(算术卷)	2012－01	28.00	125
世界著名数学经典著作钩沉——立体几何卷	2011－02	28.00	88
世界著名三角学经典著作钩沉(平面三角卷Ⅰ)	2010－06	28.00	69
世界著名三角学经典著作钩沉(平面三角卷Ⅱ)	2011－01	38.00	78
世界著名初等数论经典著作钩沉(理论和实用算术卷)	2011－07	38.00	126

发展空间想象力	2010－01	38.00	57
走向国际数学奥林匹克的平面几何试题诠释(上、下)(第1版)	2007－01	68.00	11,12
走向国际数学奥林匹克的平面几何试题诠释(上、下)(第2版)	2010－02	98.00	63,64
平面几何证明方法全书	2007－08	35.00	1
平面几何证明方法全书习题解答(第1版)	2005－10	18.00	2
平面几何证明方法全书习题解答(第2版)	2006－12	18.00	10
平面几何天天练上卷·基础篇(直线型)	2013－01	58.00	208
平面几何天天练中卷·基础篇(涉及圆)	2013－01	28.00	234
平面几何天天练下卷·提高篇	2013－01	58.00	237
平面几何专题研究	2013－07	98.00	258
最新世界各国数学奥林匹克中的平面几何试题	2007－09	38.00	14
数学竞赛平面几何典型题及新颖解	2010－07	48.00	74
初等数学复习及研究(平面几何)	2008－09	58.00	38
初等数学复习及研究(立体几何)	2010－06	38.00	71
初等数学复习及研究(平面几何)习题解答	2009－01	48.00	42
几何学教程(平面几何卷)	2011－03	68.00	90
几何学教程(立体几何卷)	2011－07	68.00	130
几何变换与几何证题	2010－06	88.00	70
计算方法与几何证题	2011－06	28.00	129
立体几何技巧与方法	2014－04	88.00	293
几何瑰宝——平面几何500名题暨1000条定理(上、下)	2010－07	138.00	76,77
三角形的解法与应用	2012－07	18.00	183
近代的三角形几何学	2012－07	48.00	184
一般折线几何学	2015－08	48.00	203
三角形的五心	2009－06	28.00	51
三角形的六心及其应用	2015－10	68.00	542
三角形趣谈	2012－08	28.00	212
解三角形	2014－01	28.00	265
三角学专门教程	2014－09	28.00	387

哈尔滨工业大学出版社刘培杰数学工作室 已出版(即将出版)图书目录

书　　名	出版时间	定　价	编号
距离几何分析导引	2015—02	68.00	446
圆锥曲线习题集(上册)	2013—06	68.00	255
圆锥曲线习题集(中册)	2015—01	78.00	434
圆锥曲线习题集(下册)	即将出版		
近代欧氏几何学	2012—03	48.00	162
罗巴切夫斯基几何学及几何基础概要	2012—07	28.00	188
罗巴切夫斯基几何学初步	2015—06	28.00	474
用三角、解析几何、复数、向量计算解数学竞赛几何题	2015—03	48.00	455
美国中学几何教程	2015—04	88.00	458
三线坐标与三角形特征点	2015—04	98.00	460
平面解析几何方法与研究(第 1 卷)	2015—05	18.00	471
平面解析几何方法与研究(第 2 卷)	2015—06	18.00	472
平面解析几何方法与研究(第 3 卷)	2015—07	18.00	473
解析几何研究	2015—01	38.00	425
解析几何学教程.上	2016—01	38.00	574
解析几何学教程.下	2016—01	38.00	575
几何学基础	2016—01	58.00	581
初等几何研究	2015—02	58.00	444
俄罗斯平面几何问题集	2009—08	88.00	55
俄罗斯立体几何问题集	2014—03	58.00	283
俄罗斯几何大师——沙雷金论数学及其他	2014—01	48.00	271
来自俄罗斯的 5000 道几何习题及解答	2011—03	58.00	89
俄罗斯初等数学问题集	2012—05	38.00	177
俄罗斯函数问题集	2011—03	38.00	103
俄罗斯组合分析问题集	2011—01	48.00	79
俄罗斯初等数学万题选——三角卷	2012—11	38.00	222
俄罗斯初等数学万题选——代数卷	2013—08	68.00	225
俄罗斯初等数学万题选——几何卷	2014—01	68.00	226
463 个俄罗斯几何老问题	2012—01	28.00	152
超越吉米多维奇.数列的极限	2009—11	48.00	58
超越普里瓦洛夫.留数卷	2015—01	28.00	437
超越普里瓦洛夫.无穷乘积与它对解析函数的应用卷	2015—05	28.00	477
超越普里瓦洛夫.积分卷	2015—06	18.00	481
超越普里瓦洛夫.基础知识卷	2015—06	28.00	482
超越普里瓦洛夫.数项级数卷	2015—07	38.00	489
初等数论难题集(第一卷)	2009—05	68.00	44
初等数论难题集(第二卷)(上、下)	2011—02	128.00	82,83
数论概貌	2011—03	18.00	93
代数数论(第二版)	2013—08	58.00	94
代数多项式	2014—06	38.00	289
初等数论的知识与问题	2011—02	28.00	95
超越数论基础	2011—03	28.00	96
数论初等教程	2011—03	28.00	97
数论基础	2011—03	18.00	98
数论基础与维诺格拉多夫	2014—03	18.00	292

哈尔滨工业大学出版社刘培杰数学工作室已出版(即将出版)图书目录

书　　名	出版时间	定　价	编号
解析数论基础	2012－08	28.00	216
解析数论基础(第二版)	2014－01	48.00	287
解析数论问题集(第二版)	2014－05	88.00	343
数论入门	2011－03	38.00	99
代数数论入门	2015－03	38.00	448
数论开篇	2012－07	28.00	194
解析数论引论	2011－03	48.00	100
Barban Davenport Halberstam 均值和	2009－01	40.00	33
基础数论	2011－03	28.00	101
初等数论 100 例	2011－05	18.00	122
初等数论经典例题	2012－07	18.00	204
最新世界各国数学奥林匹克中的初等数论试题(上、下)	2012－01	138.00	144,145
初等数论(Ⅰ)	2012－01	18.00	156
初等数论(Ⅱ)	2012－01	18.00	157
初等数论(Ⅲ)	2012－01	28.00	158
平面几何与数论中未解决的新老问题	2013－01	68.00	229
代数数论简史	2014－11	28.00	408
代数数论	2015－09	88.00	532
数论导引提要及习题解答	2016－01	48.00	559

谈谈素数	2011－03	18.00	91
平方和	2011－03	18.00	92
复变函数引论	2013－10	68.00	269
伸缩变换与抛物旋转	2015－01	38.00	449
无穷分析引论(上)	2013－04	88.00	247
无穷分析引论(下)	2013－04	98.00	245
数学分析	2014－04	28.00	338
数学分析中的一个新方法及其应用	2013－01	38.00	231
数学分析例选:通过范例学技巧	2013－01	88.00	243
高等代数例选:通过范例学技巧	2015－06	88.00	475
三角级数论(上册)(陈建功)	2013－01	38.00	232
三角级数论(下册)(陈建功)	2013－01	48.00	233
三角级数论(哈代)	2013－06	48.00	254
三角级数	2015－07	28.00	263
超越数	2011－03	18.00	109
三角和方法	2011－03	18.00	112
整数论	2011－05	38.00	120
从整数谈起	2015－10	18.00	538
随机过程(Ⅰ)	2014－01	78.00	224
随机过程(Ⅱ)	2014－01	68.00	235
算术探索	2011－12	158.00	148
组合数学	2012－04	28.00	178
组合数学浅谈	2012－03	28.00	159
丢番图方程引论	2012－03	48.00	172
拉普拉斯变换及其应用	2015－02	38.00	447
高等代数.上	2016－01	38.00	548
高等代数.下	2016－01	38.00	549

哈尔滨工业大学出版社刘培杰数学工作室已出版(即将出版)图书目录

书　　名	出版时间	定　价	编号
高等代数教程	2016－01	58.00	579
数学解析教程.上卷.1	2016－01	58.00	546
数学解析教程.上卷.2	2016－01	38.00	553
函数构造论.上	2016－01	38.00	554
函数构造论.下	即将出版		555
数与多项式	2016－01	38.00	558
概周期函数	2016－01	48.00	572
变叙的项的极限分布律	2016－01	18.00	573
多项式理论	2015－10	88.00	541
整函数	2012－08	18.00	161
近代拓扑学研究	2013－04	38.00	239
多项式和无理数	2008－01	68.00	22
模糊数据统计学	2008－03	48.00	31
模糊分析学与特殊泛函空间	2013－01	68.00	241
谈谈不定方程	2011－05	28.00	119

受控理论与解析不等式	2012－05	78.00	165
解析不等式新论	2009－06	68.00	48
建立不等式的方法	2011－03	98.00	104
数学奥林匹克不等式研究	2009－08	68.00	56
不等式研究(第二辑)	2012－02	68.00	153
不等式的秘密(第一卷)	2012－02	28.00	154
不等式的秘密(第一卷)(第2版)	2014－02	38.00	286
不等式的秘密(第二卷)	2014－01	38.00	268
初等不等式的证明方法	2010－06	38.00	123
初等不等式的证明方法(第二版)	2014－11	38.00	407
不等式・理论・方法(基础卷)	2015－07	38.00	496
不等式・理论・方法(经典不等式卷)	2015－07	38.00	497
不等式・理论・方法(特殊类型不等式卷)	2015－07	48.00	498

同余理论	2012－05	38.00	163
[x]与{x}	2015－04	48.00	476
极值与最值.上卷	2015－06	38.00	486
极值与最值.中卷	2015－06	38.00	487
极值与最值.下卷	2015－06	28.00	488
整数的性质	2012－11	38.00	192

历届美国中学生数学竞赛试题及解答(第一卷)1950－1954	2014－07	18.00	277
历届美国中学生数学竞赛试题及解答(第二卷)1955－1959	2014－04	18.00	278
历届美国中学生数学竞赛试题及解答(第三卷)1960－1964	2014－06	18.00	279
历届美国中学生数学竞赛试题及解答(第四卷)1965－1969	2014－04	28.00	280
历届美国中学生数学竞赛试题及解答(第五卷)1970－1972	2014－06	18.00	281
历届美国中学生数学竞赛试题及解答(第七卷)1981－1986	2015－01	18.00	424

历届 IMO 试题集(1959—2005)	2006－05	58.00	5
历届 CMO 试题集	2008－09	28.00	40
历届中国数学奥林匹克试题集	2014－10	38.00	394
历届加拿大数学奥林匹克试题集	2012－08	38.00	215
历届美国数学奥林匹克试题集:多解推广加强	2012－08	38.00	209

哈尔滨工业大学出版社刘培杰数学工作室
已出版(即将出版)图书目录

书　　名	出版时间	定　价	编号
历届波兰数学竞赛试题集.第1卷,1949～1963	2015－03	18.00	453
历届波兰数学竞赛试题集.第2卷,1964～1976	2015－03	18.00	454
保加利亚数学奥林匹克	2014－10	38.00	393
圣彼得堡数学奥林匹克试题集	2015－01	48.00	429
历届国际大学生数学竞赛试题集(1994－2010)	2012－01	28.00	143
全国大学生数学夏令营数学竞赛试题及解答	2007－03	28.00	15
全国大学生数学竞赛辅导教程	2012－07	28.00	189
全国大学生数学竞赛复习全书	2014－04	48.00	340
历届美国大学生数学竞赛试题集	2009－03	88.00	43
前苏联大学生数学奥林匹克竞赛题解(上编)	2012－04	28.00	169
前苏联大学生数学奥林匹克竞赛题解(下编)	2012－04	38.00	170
历届美国数学邀请赛试题集	2014－01	48.00	270
全国高中数学竞赛试题及解答.第1卷	2014－07	38.00	331
大学生数学竞赛讲义	2014－09	28.00	371
亚太地区数学奥林匹克竞赛题	2015－07	18.00	492
高考数学临门一脚(含密押三套卷)(理科版)	2015－01	24.80	421
高考数学临门一脚(含密押三套卷)(文科版)	2015－01	24.80	422
新课标高考数学题型全归纳(文科版)	2015－05	72.00	467
新课标高考数学题型全归纳(理科版)	2015－05	82.00	468
王连笑教你怎样学数学:高考选择题解题策略与客观题实用训练	2014－01	48.00	262
王连笑教你怎样学数学:高考数学高层次讲座	2015－02	48.00	432
高考数学的理论与实践	2009－08	38.00	53
高考数学核心题型解题方法与技巧	2010－01	28.00	86
高考思维新平台	2014－03	38.00	259
30分钟拿下高考数学选择题、填空题(第二版)	2012－01	28.00	146
高考数学压轴题解题诀窍(上)	2012－02	78.00	166
高考数学压轴题解题诀窍(下)	2012－03	28.00	167
北京市五区文科数学三年高考模拟题详解:2013～2015	2015－08	48.00	500
北京市五区理科数学三年高考模拟题详解:2013～2015	2015－09	68.00	505
向量法巧解数学高考题	2009－08	28.00	54
高考数学万能解题法	2015－09	28.00	534
高考物理万能解题法	2015－09	28.00	537
高考化学万能解题法	2015－11	25.00	557
我一定要赚分:高中物理	2016－01	38.00	580
2011～2015年全国及各省市高考数学文科精品试题审题要津与解法研究	2015－10	68.00	539
2011～2015年全国及各省市高考数学理科精品试题审题要津与解法研究	2015－10	88.00	540
最新全国及各省市高考数学试卷解法研究及点拨评析	2009－02	38.00	41
2011年全国及各省市高考数学试题审题要津与解法研究	2011－10	48.00	139
2013年全国及各省市高考数学试题解析与点评	2014－01	48.00	282
全国及各省市高考数学试题审题要津与解法研究	2015－02	48.00	450
新课标高考数学——五年试题分章详解(2007～2011)(上、下)	2011－10	78.00	140,141
全国中考数学压轴题审题要津与解法研究	2013－04	78.00	248
新编全国及各省市中考数学压轴题审题要津与解法研究	2014－05	58.00	342
全国及各省市5年中考数学压轴题审题要津与解法研究	2015－04	58.00	462
中考数学专题总复习	2007－04	28.00	6

哈尔滨工业大学出版社刘培杰数学工作室
已出版(即将出版)图书目录

书　名	出版时间	定　价	编号
数学奥林匹克在中国	2014—06	98.00	344
数学奥林匹克问题集	2014—01	38.00	267
数学奥林匹克不等式散论	2010—06	38.00	124
数学奥林匹克不等式欣赏	2011—09	38.00	138
数学奥林匹克超级题库(初中卷上)	2010—01	58.00	66
数学奥林匹克不等式证明方法和技巧(上、下)	2011—08	158.00	134,135
新编 640 个世界著名数学智力趣题	2014—01	88.00	242
500 个最新世界著名数学智力趣题	2008—06	48.00	3
400 个最新世界著名数学最值问题	2008—09	48.00	36
500 个世界著名数学征解问题	2009—06	48.00	52
400 个中国最佳初等数学征解老问题	2010—01	48.00	60
500 个俄罗斯数学经典老题	2011—01	28.00	81
1000 个国外中学物理好题	2012—04	48.00	174
300 个日本高考数学题	2012—05	38.00	142
500 个前苏联早期高考数学试题及解答	2012—05	28.00	185
546 个早期俄罗斯大学生数学竞赛题	2014—03	38.00	285
548 个来自美苏的数学好问题	2014—11	28.00	396
20 所苏联著名大学早期入学试题	2015—02	18.00	452
161 道德国工科大学生必做的微分方程习题	2015—05	28.00	469
500 个德国工科大学生必做的高数习题	2015—06	28.00	478
德国讲义日本考题.微积分卷	2015—04	48.00	456
德国讲义日本考题.微分方程卷	2015—04	38.00	457
几何变换(Ⅰ)	2014—07	28.00	353
几何变换(Ⅱ)	2015—06	28.00	354
几何变换(Ⅲ)	2015—01	38.00	355
几何变换(Ⅳ)	2015—12	38.00	356
中国初等数学研究　2009 卷(第 1 辑)	2009—05	20.00	45
中国初等数学研究　2010 卷(第 2 辑)	2010—05	30.00	68
中国初等数学研究　2011 卷(第 3 辑)	2011—07	60.00	127
中国初等数学研究　2012 卷(第 4 辑)	2012—07	48.00	190
中国初等数学研究　2014 卷(第 5 辑)	2014—02	48.00	288
中国初等数学研究　2015 卷(第 6 辑)	2015—06	68.00	493
博弈论精粹	2008—03	58.00	30
博弈论精粹.第二版(精装)	2015—01	88.00	461
数学 我爱你	2008—01	28.00	20
精神的圣徒　别样的人生——60 位中国数学家成长的历程	2008—09	48.00	39
数学史概论	2009—06	78.00	50
数学史概论(精装)	2013—03	158.00	272
数学史选讲	2016—01	48.00	544
斐波那契数列	2010—02	28.00	65
数学拼盘和斐波那契魔方	2010—07	38.00	72
斐波那契数列欣赏	2011—01	28.00	160
数学的创造	2011—02	48.00	85

哈尔滨工业大学出版社刘培杰数学工作室已出版(即将出版)图书目录

书　　名	出版时间	定　价	编号
数学中的美	2011—02	38.00	84
数论中的美学	2014—12	38.00	351
数学王者　科学巨人——高斯	2015—01	28.00	428
振兴祖国数学的圆梦之旅:中国初等数学研究史话	2015—06	78.00	490
二十世纪中国数学史料研究	2015—10	48.00	536
数字谜、数阵图与棋盘覆盖	2016—01	58.00	298
时间的形状	2016—01	38.00	556
数学解题——靠数学思想给力(上)	2011—07	38.00	131
数学解题——靠数学思想给力(中)	2011—07	48.00	132
数学解题——靠数学思想给力(下)	2011—07	38.00	133
我怎样解题	2013—01	48.00	227
数学解题中的物理方法	2011—06	28.00	114
数学解题的特殊方法	2011—06	48.00	115
中学数学计算技巧	2012—01	48.00	116
中学数学证明方法	2012—01	58.00	117
数学趣题巧解	2012—03	28.00	128
高中数学教学通鉴	2015—05	58.00	479
和高中生漫谈:数学与哲学的故事	2014—08	28.00	369
自主招生考试中的参数方程问题	2015—01	28.00	435
自主招生考试中的极坐标问题	2015—04	28.00	463
近年全国重点大学自主招生数学试题全解及研究.华约卷	2015—02	38.00	441
近年全国重点大学自主招生数学试题全解及研究.北约卷	即将出版		
自主招生数学解证宝典	2015—09	48.00	535
格点和面积	2012—07	18.00	191
射影几何趣谈	2012—04	28.00	175
斯潘纳尔引理——从一道加拿大数学奥林匹克试题谈起	2014—01	28.00	228
李普希兹条件——从几道近年高考数学试题谈起	2012—10	18.00	221
拉格朗日中值定理——从一道北京高考试题的解法谈起	2015—10	18.00	197
闵科夫斯基定理——从一道清华大学自主招生试题谈起	2014—01	28.00	198
哈尔测度——从一道冬令营试题的背景谈起	2012—08	28.00	202
切比雪夫逼近问题——从一道中国台北数学奥林匹克试题谈起	2013—04	38.00	238
伯恩斯坦多项式与贝齐尔曲面——从一道全国高中数学联赛试题谈起	2013—03	38.00	236
卡塔兰猜想——从一道普特南竞赛试题谈起	2013—06	18.00	256
麦卡锡函数和阿克曼函数——从一道前南斯拉夫数学奥林匹克试题谈起	2012—08	18.00	201
贝蒂定理与拉姆贝克莫斯尔定理——从一个拣石子游戏谈起	2012—08	18.00	217
皮亚诺曲线和豪斯道夫分球定理——从无限集谈起	2012—08	18.00	211
平面凸图形与凸多面体	2012—10	28.00	218
斯坦因豪斯问题——从一道二十五省市自治区中学数学竞赛试题谈起	2012—07	18.00	196
纽结理论中的亚历山大多项式与琼斯多项式——从一道北京市高一数学竞赛试题谈起	2012—07	28.00	195
原则与策略——从波利亚"解题表"谈起	2013—04	38.00	244

哈尔滨工业大学出版社刘培杰数学工作室已出版(即将出版)图书目录

书　　名	出版时间	定　价	编号
转化与化归——从三大尺规作图不能问题谈起	2012—08	28.00	214
代数几何中的贝祖定理(第一版)——从一道 IMO 试题的解法谈起	2013—08	18.00	193
成功连贯理论与约当块理论——从一道比利时数学竞赛试题谈起	2012—04	18.00	180
磨光变换与范·德·瓦尔登猜想——从一道环球城市竞赛试题谈起	即将出版		
素数判定与大数分解	2014—08	18.00	199
置换多项式及其应用	2012—10	18.00	220
椭圆函数与模函数——从一道美国加州大学洛杉矶分校(UCLA)博士资格考题谈起	2012—10	28.00	219
差分方程的拉格朗日方法——从一道 2011 年全国高考理科试题的解法谈起	2012—08	28.00	200
力学在几何中的一些应用	2013—01	38.00	240
高斯散度定理、斯托克斯定理和平面格林定理——从一道国际大学生数学竞赛试题谈起	即将出版		
康托洛维奇不等式——从一道全国高中联赛试题谈起	2013—03	28.00	337
西格尔引理——从一道第 18 届 IMO 试题的解法谈起	即将出版		
罗斯定理——从一道前苏联数学竞赛试题谈起	即将出版		
拉克斯定理和阿廷定理——从一道 IMO 试题的解法谈起	2014—01	58.00	246
毕卡大定理——从一道美国大学数学竞赛试题谈起	2014—07	18.00	350
贝齐尔曲线——从一道全国高中联赛试题谈起	即将出版		
拉格朗日乘子定理——从一道 2005 年全国高中联赛试题的高等数学解法谈起	2015—05	28.00	480
雅可比定理——从一道日本数学奥林匹克试题谈起	2013—04	48.00	249
李天岩—约克定理——从一道波兰数学竞赛试题谈起	2014—06	28.00	349
整系数多项式因式分解的一般方法——从克朗耐克算法谈起	即将出版		
布劳维不动点定理——从一道前苏联数学奥林匹克试题谈起	2014—01	38.00	273
压缩不动点定理——从一道高考数学试题的解法谈起	即将出版		
伯恩赛德定理——从一道英国数学奥林匹克试题谈起	即将出版		
布查特—莫斯特定理——从一道上海市初中竞赛试题谈起	即将出版		
数论中的同余数问题——从一道普特南竞赛试题谈起	即将出版		
范·德蒙行列式——从一道美国数学奥林匹克试题谈起	即将出版		
中国剩余定理:总数法构建中国历史年表	2015—01	28.00	430
牛顿程序与方程求根——从一道全国高考试题解法谈起	即将出版		
库默尔定理——从一道 IMO 预选试题谈起	即将出版		
卢丁定理——从一道冬令营试题的解法谈起	即将出版		
沃斯滕霍姆定理——从一道 IMO 预选试题谈起	即将出版		
卡尔松不等式——从一道莫斯科数学奥林匹克试题谈起	即将出版		
信息论中的香农熵——从一道近年高考压轴题谈起	即将出版		
约当不等式——从一道希望杯竞赛试题谈起	即将出版		
拉比诺维奇定理	即将出版		
刘维尔定理——从一道《美国数学月刊》征解问题的解法谈起	即将出版		
卡塔兰恒等式与级数求和——从一道 IMO 试题的解法谈起	即将出版		
勒让德猜想与素数分布——从一道爱尔兰竞赛试题谈起	即将出版		

哈尔滨工业大学出版社刘培杰数学工作室
已出版(即将出版)图书目录

书　　名	出版时间	定　价	编号
天平称重与信息论——从一道基辅市数学奥林匹克试题谈起	即将出版		
哈密尔顿一凯莱定理:从一道高中数学联赛试题的解法谈起	2014—09	18.00	376
艾思特曼定理——从一道 CMO 试题的解法谈起	即将出版		
一个爱尔特希问题——从一道西德数学奥林匹克试题谈起	即将出版		
有限群中的爱丁格尔问题——从一道北京市初中二年级数学竞赛试题谈起	即将出版		
贝克码与编码理论——从一道全国高中联赛试题谈起	即将出版		
帕斯卡三角形	2014—03	18.00	294
蒲丰投针问题——从 2009 年清华大学的一道自主招生试题谈起	2014—01	38.00	295
斯图姆定理——从一道"华约"自主招生试题的解法谈起	2014—01	18.00	296
许瓦兹引理——从一道加利福尼亚大学伯克利分校数学系博士生试题谈起	2014—08	18.00	297
拉格朗日中值定理——从一道北京高考试题的解法谈起	2014—01		298
拉姆塞定理——从王诗宬院士的一个问题谈起	2014—01		299
坐标法	2013—12	28.00	332
数论三角形	2014—04	38.00	341
毕克定理	2014—07	18.00	352
数林掠影	2014—09	48.00	389
我们周围的概率	2014—10	38.00	390
凸函数最值定理:从一道华约自主招生题的解法谈起	2014—10	28.00	391
易学与数学奥林匹克	2014—10	38.00	392
生物数学趣谈	2015—01	18.00	409
反演	2015—01		420
因式分解与圆锥曲线	2015—01	18.00	426
轨迹	2015—01	28.00	427
面积原理:从常庚哲命的一道 CMO 试题的积分解法谈起	2015—01	48.00	431
形形色色的不动点定理:从一道 28 届 IMO 试题谈起	2015—01	38.00	439
柯西函数方程:从一道上海交大自主招生的试题谈起	2015—02	28.00	440
三角恒等式	2015—02	28.00	442
无理性判定:从一道 2014 年"北约"自主招生试题谈起	2015—01	38.00	443
数学归纳法	2015—03	18.00	451
极端原理与解题	2015—04	28.00	464
法雷级数	2014—08	18.00	367
摆线族	2015—01	38.00	438
函数方程及其解法	2015—05	38.00	470
含参数的方程和不等式	2012—09	28.00	213
希尔伯特第十问题	2016—01	38.00	543
无穷小量的求和	2016—01	28.00	545

中等数学英语阅读文选	2006—12	38.00	13
统计学专业英语	2007—03	28.00	16
统计学专业英语(第二版)	2012—07	48.00	176
统计学专业英语(第三版)	2015—04	68.00	465
幻方和魔方(第一卷)	2012—05	68.00	173
尘封的经典——初等数学经典文献选读(第一卷)	2012—07	48.00	205
尘封的经典——初等数学经典文献选读(第二卷)	2012—07	38.00	206
代换分析:英文	2015—07	38.00	499

哈尔滨工业大学出版社刘培杰数学工作室已出版(即将出版)图书目录

书　　名	出版时间	定　价	编号
实变函数论	2012－06	78.00	181
复变函数论	2015－08	38.00	504
非光滑优化及其变分分析	2014－01	48.00	230
疏散的马尔科夫链	2014－01	58.00	266
马尔科夫过程论基础	2015－01	28.00	433
初等微分拓扑学	2012－07	18.00	182
方程式论	2011－03	38.00	105
初级方程式论	2011－03	28.00	106
Galois 理论	2011－03	18.00	107
古典数学难题与伽罗瓦理论	2012－11	58.00	223
伽罗华与群论	2014－01	28.00	290
代数方程的根式解及伽罗瓦理论	2011－03	28.00	108
代数方程的根式解及伽罗瓦理论(第二版)	2015－01	28.00	423
线性偏微分方程讲义	2011－03	18.00	110
几类微分方程数值方法的研究	2015－05	38.00	485
N 体问题的周期解	2011－03	28.00	111
代数方程式论	2011－05	18.00	121
动力系统的不变量与函数方程	2011－07	48.00	137
基于短语评价的翻译知识获取	2012－02	48.00	168
应用随机过程	2012－04	48.00	187
概率论导引	2012－04	18.00	179
矩阵论(上)	2013－06	58.00	250
矩阵论(下)	2013－06	48.00	251
对称锥互补问题的内点法:理论分析与算法实现	2014－08	68.00	368
抽象代数:方法导引	2013－06	38.00	257
集论	2016－01	48.00	576
多项式理论研究综述	2016－01	38.00	577
函数论	2014－11	78.00	395
反问题的计算方法及应用	2011－11	28.00	147
初等数学研究(Ⅰ)	2008－09	68.00	37
初等数学研究(Ⅱ)(上、下)	2009－05	118.00	46,47
数阵及其应用	2012－02	28.00	164
绝对值方程—折边与组合图形的解析研究	2012－07	48.00	186
代数函数论(上)	2015－07	38.00	494
代数函数论(下)	2015－07	38.00	495
偏微分方程论:法文	2015－10	48.00	533

闵嗣鹤文集	2011－03	98.00	102
吴从炘数学活动三十年(1951～1980)	2010－07	99.00	32
吴从炘数学活动又三十年(1981～2010)	2015－07	98.00	491

趣味初等方程妙题集锦	2014－09	48.00	388
趣味初等数论选美与欣赏	2015－02	48.00	445
耕读笔记(上卷):一位农民数学爱好者的初数探索	2015－04	28.00	459
耕读笔记(中卷):一位农民数学爱好者的初数探索	2015－05	28.00	483
耕读笔记(下卷):一位农民数学爱好者的初数探索	2015－05	28.00	484
几何不等式研究与欣赏.上卷	2016－01	88.00	547
几何不等式研究与欣赏.下卷	2016－01	48.00	552
初等数列研究与欣赏・上	2016－01	48.00	570
初等数列研究与欣赏・下	2016－01	48.00	571

哈尔滨工业大学出版社刘培杰数学工作室
已出版(即将出版)图书目录

书　　名	出版时间	定　价	编号
数贝偶拾——高考数学题研究	2014－04	28.00	274
数贝偶拾——初等数学研究	2014－04	38.00	275
数贝偶拾——奥数题研究	2014－04	48.00	276
集合、函数与方程	2014－01	28.00	300
数列与不等式	2014－01	38.00	301
三角与平面向量	2014－01	28.00	302
平面解析几何	2014－01	38.00	303
立体几何与组合	2014－01	28.00	304
极限与导数、数学归纳法	2014－01	38.00	305
趣味数学	2014－03	28.00	306
教材教法	2014－04	68.00	307
自主招生	2014－05	58.00	308
高考压轴题(上)	2015－01	48.00	309
高考压轴题(下)	2014－10	68.00	310
从费马到怀尔斯——费马大定理的历史	2013－10	198.00	Ⅰ
从庞加莱到佩雷尔曼——庞加莱猜想的历史	2013－10	298.00	Ⅱ
从切比雪夫到爱尔特希(上)——素数定理的初等证明	2013－07	48.00	Ⅲ
从切比雪夫到爱尔特希(下)——素数定理100年	2012－12	98.00	Ⅲ
从高斯到盖尔方特——二次域的高斯猜想	2013－10	198.00	Ⅳ
从库默尔到朗兰兹——朗兰兹猜想的历史	2014－01	98.00	Ⅴ
从比勃巴赫到德布朗斯——比勃巴赫猜想的历史	2014－02	298.00	Ⅵ
从麦比乌斯到陈省身——麦比乌斯变换与麦比乌斯带	2014－02	298.00	Ⅶ
从布尔到豪斯道夫——布尔方程与格论漫谈	2013－10	198.00	Ⅷ
从开普勒到阿诺德——三体问题的历史	2014－05	298.00	Ⅸ
从华林到华罗庚——华林问题的历史	2013－10	298.00	Ⅹ
吴振奎高等数学解题真经(概率统计卷)	2012－01	38.00	149
吴振奎高等数学解题真经(微积分卷)	2012－01	68.00	150
吴振奎高等数学解题真经(线性代数卷)	2012－01	58.00	151
钱昌本教你快乐学数学(上)	2011－12	48.00	155
钱昌本教你快乐学数学(下)	2012－03	58.00	171
第19～23届"希望杯"全国数学邀请赛试题审题要津详细评注(初一版)	2014－03	28.00	333
第19～23届"希望杯"全国数学邀请赛试题审题要津详细评注(初二、初三版)	2014－03	38.00	334
第19～23届"希望杯"全国数学邀请赛试题审题要津详细评注(高一版)	2014－03	28.00	335
第19～23届"希望杯"全国数学邀请赛试题审题要津详细评注(高二版)	2014－03	38.00	336
第19～25届"希望杯"全国数学邀请赛试题审题要津详细评注(初一版)	2015－01	38.00	416
第19～25届"希望杯"全国数学邀请赛试题审题要津详细评注(初二、初三版)	2015－01	58.00	417
第19～25届"希望杯"全国数学邀请赛试题审题要津详细评注(高一版)	2015－01	48.00	418
第19～25届"希望杯"全国数学邀请赛试题审题要津详细评注(高二版)	2015－01	48.00	419
高等数学解题全攻略(上卷)	2013－06	58.00	252
高等数学解题全攻略(下卷)	2013－06	58.00	253
高等数学复习纲要	2014－01	18.00	384

哈尔滨工业大学出版社刘培杰数学工作室已出版(即将出版)图书目录

书名	出版时间	定价	编号
三角函数	2014－01	38.00	311
不等式	2014－01	38.00	312
数列	2014－01	38.00	313
方程	2014－01	28.00	314
排列和组合	2014－01	28.00	315
极限与导数	2014－01	28.00	316
向量	2014－09	38.00	317
复数及其应用	2014－08	28.00	318
函数	2014－01	38.00	319
集合	即将出版		320
直线与平面	2014－01	28.00	321
立体几何	2014－04	28.00	322
解三角形	即将出版		323
直线与圆	2014－01	28.00	324
圆锥曲线	2014－01	38.00	325
解题通法(一)	2014－07	38.00	326
解题通法(二)	2014－07	38.00	327
解题通法(三)	2014－05	38.00	328
概率与统计	2014－01	28.00	329
信息迁移与算法	即将出版		330
物理奥林匹克竞赛大题典——力学卷	2014－11	48.00	405
物理奥林匹克竞赛大题典——热学卷	2014－04	28.00	339
物理奥林匹克竞赛大题典——电磁学卷	2015－07	48.00	406
物理奥林匹克竞赛大题典——光学与近代物理卷	2014－06	28.00	345
历届中国东南地区数学奥林匹克试题集(2004～2012)	2014－06	18.00	346
历届中国西部地区数学奥林匹克试题集(2001～2012)	2014－07	18.00	347
历届中国女子数学奥林匹克试题集(2002～2012)	2014－08	18.00	348
美国高中数学竞赛五十讲.第1卷(英文)	2014－08	28.00	357
美国高中数学竞赛五十讲.第2卷(英文)	2014－08	28.00	358
美国高中数学竞赛五十讲.第3卷(英文)	2014－09	28.00	359
美国高中数学竞赛五十讲.第4卷(英文)	2014－09	28.00	360
美国高中数学竞赛五十讲.第5卷(英文)	2014－10	28.00	361
美国高中数学竞赛五十讲.第6卷(英文)	2014－11	28.00	362
美国高中数学竞赛五十讲.第7卷(英文)	2014－12	28.00	363
美国高中数学竞赛五十讲.第8卷(英文)	2015－01	28.00	364
美国高中数学竞赛五十讲.第9卷(英文)	2015－01	28.00	365
美国高中数学竞赛五十讲.第10卷(英文)	2015－02	38.00	366
IMO 50年.第1卷(1959－1963)	2014－11	28.00	377
IMO 50年.第2卷(1964－1968)	2014－11	28.00	378
IMO 50年.第3卷(1969－1973)	2014－09	28.00	379
IMO 50年.第4卷(1974－1978)	即将出版		380
IMO 50年.第5卷(1979－1984)	2015－04	38.00	381
IMO 50年.第6卷(1985－1989)	2015－04	58.00	382
IMO 50年.第7卷(1990－1994)	2016－01	48.00	383
IMO 50年.第8卷(1995－1999)	即将出版		384
IMO 50年.第9卷(2000－2004)	2015－04	58.00	385
IMO 50年.第10卷(2005－2008)	即将出版		386

哈尔滨工业大学出版社刘培杰数学工作室已出版(即将出版)图书目录

书　　名	出版时间	定　价	编号
历届美国大学生数学竞赛试题集.第一卷(1938—1949)	2015－01	28.00	397
历届美国大学生数学竞赛试题集.第二卷(1950—1959)	2015－01	28.00	398
历届美国大学生数学竞赛试题集.第三卷(1960—1969)	2015－01	28.00	399
历届美国大学生数学竞赛试题集.第四卷(1970—1979)	2015－01	18.00	400
历届美国大学生数学竞赛试题集.第五卷(1980—1989)	2015－01	28.00	401
历届美国大学生数学竞赛试题集.第六卷(1990—1999)	2015－01	28.00	402
历届美国大学生数学竞赛试题集.第七卷(2000—2009)	2015－08	18.00	403
历届美国大学生数学竞赛试题集.第八卷(2010—2012)	2015－01	18.00	404
新课标高考数学创新题解题诀窍:总论	2014－09	28.00	372
新课标高考数学创新题解题诀窍:必修 1～5 分册	2014－08	38.00	373
新课标高考数学创新题解题诀窍:选修 2－1,2－2,1－1,1－2分册	2014－09	38.00	374
新课标高考数学创新题解题诀窍:选修 2－3,4－4,4－5 分册	2014－09	18.00	375
全国重点大学自主招生英文数学试题全攻略:词汇卷	2015－07	48.00	410
全国重点大学自主招生英文数学试题全攻略:概念卷	2015－01	28.00	411
全国重点大学自主招生英文数学试题全攻略:文章选读卷(上)	即将出版		412
全国重点大学自主招生英文数学试题全攻略:文章选读卷(下)	即将出版		413
全国重点大学自主招生英文数学试题全攻略:试题卷	2015－07	38.00	414
全国重点大学自主招生英文数学试题全攻略:名著欣赏卷	即将出版		415
数学物理大百科全书.第 1 卷	2016－01	418.00	508
数学物理大百科全书.第 2 卷	2016－01	408.00	509
数学物理大百科全书.第 3 卷	2016－01	396.00	510
数学物理大百科全书.第 4 卷	2016－01	408.00	511
数学物理大百科全书.第 5 卷	2016－01	368.00	512
劳埃德数学趣题大全.题目卷.1:英文	2016－01	18.00	516
劳埃德数学趣题大全.题目卷.2:英文	2016－01	18.00	517
劳埃德数学趣题大全.题目卷.3:英文	2016－01	18.00	518
劳埃德数学趣题大全.题目卷.4:英文	2016－01	18.00	519
劳埃德数学趣题大全.题目卷.5:英文	2016－01	18.00	520
劳埃德数学趣题大全.答案卷:英文	2016－01	18.00	521
李成章教练奥数笔记.第 1 卷	2016－01	48.00	522
李成章教练奥数笔记.第 2 卷	2016－01	48.00	523
李成章教练奥数笔记.第 3 卷	2016－01	38.00	524
李成章教练奥数笔记.第 4 卷	2016－01	38.00	525
李成章教练奥数笔记.第 5 卷	2016－01	38.00	526
李成章教练奥数笔记.第 6 卷	2016－01	38.00	527
李成章教练奥数笔记.第 7 卷	2016－01	38.00	528
李成章教练奥数笔记.第 8 卷	2016－01	48.00	529
李成章教练奥数笔记.第 9 卷	2016－01	28.00	530

哈尔滨工业大学出版社刘培杰数学工作室
已出版(即将出版)图书目录

书　　名	出版时间	定　价	编号
zeta 函数,q-zeta 函数,相伴级数与积分	2015－08	88.00	513
微分形式:理论与练习	2015－08	58.00	514
离散与微分包含的逼近和优化	2015－08	58.00	515
艾伦·图灵:他的工作与影响	2016－01	98.00	560
测度理论概率导论,第 2 版	2016－01	88.00	561
带有潜在故障恢复系统的半马尔柯夫模型控制	2016－01	98.00	562
数学分析原理	2016－01	88.00	563
随机偏微分方程的有效动力学	2016－01	88.00	564
图的谱半径	2016－01	58.00	565
量子机器学习中数据挖掘的量子计算方法	2016－01	98.00	566
运输过程的统一非局部理论:广义波尔兹曼物理动力学,第 2 版	2016－01	198.00	567
量子物理的非常规方法	2016－01	118.00	568
量子力学与经典力学之间的联系在原子、分子及电动力学系统建模中的应用	2016－01	58.00	569

联系地址:哈尔滨市南岗区复华四道街 10 号　哈尔滨工业大学出版社刘培杰数学工作室
网　　址:http://lpj.hit.edu.cn/
邮　　编:150006
联系电话:0451－86281378　　13904613167
E-mail:lpj1378@163.com